Signal Processing and Machine Learning for Biomedical Big Data

Signal Processing and Machine Learning for Biomedical Big Data

Edited by
Ervin Sejdić
Tiago H. Falk

CRC Press is an imprint of the
Taylor & Francis Group, an **informa** business

MATLAB® is a trademark of The MathWorks, Inc. and is used with permission. The MathWorks does not warrant the accuracy of the text or exercises in this book. This book's use or discussion of MATLAB® software or related products does not constitute endorsement or sponsorship by The MathWorks of a particular pedagogical approach or particular use of the MATLAB® software.

CRC Press
Taylor & Francis Group
6000 Broken Sound Parkway NW, Suite 300
Boca Raton, FL 33487-2742

© 2018 by Taylor & Francis Group, LLC
CRC Press is an imprint of Taylor & Francis Group, an Informa business

No claim to original U.S. Government works

Printed on acid-free paper

International Standard Book Number-13: 978-1-4987-7345-4 (Hardback)

This book contains information obtained from authentic and highly regarded sources. Reasonable efforts have been made to publish reliable data and information, but the author and publisher cannot assume responsibility for the validity of all materials or the consequences of their use. The authors and publishers have attempted to trace the copyright holders of all material reproduced in this publication and apologize to copyright holders if permission to publish in this form has not been obtained. If any copyright material has not been acknowledged please write and let us know so we may rectify in any future reprint.

Except as permitted under U.S. Copyright Law, no part of this book may be reprinted, reproduced, transmitted, or utilized in any form by any electronic, mechanical, or other means, now known or hereafter invented, including photocopying, microfilming, and recording, or in any information storage or retrieval system, without written permission from the publishers.

For permission to photocopy or use material electronically from this work, please access www.copyright.com (http://www.copyright.com/) or contact the Copyright Clearance Center, Inc. (CCC), 222 Rosewood Drive, Danvers, MA 01923, 978-750-8400. CCC is a not-for-profit organization that provides licenses and registration for a variety of users. For organizations that have been granted a photocopy license by the CCC, a separate system of payment has been arranged.

Trademark Notice: Product or corporate names may be trademarks or registered trademarks, and are used only for identification and explanation without intent to infringe.

Library of Congress Cataloging-in-Publication Data

Names: Sejdic, Ervin, author. | Falk, Tiago H., author.
Title: Signal processing and machine learning for biomedical big data / Ervin Sejdic and Tiago H. Falk.
Description: Boca Raton : Taylor & Francis, 2018. | Includes bibliographical references.
Identifiers: LCCN 2017053075| ISBN 9781498773454 (hardback : alk. paper) | ISBN 9781498773461 (ebook)
Subjects: | MESH: Medical Informatics | Data Collection | Signal Processing, Computer-Assisted | Machine Learning
Classification: LCC R855.3 | NLM W 26.5 | DDC 610.285--dc23
LC record available at https://lccn.loc.gov/2017053075

Visit the Taylor & Francis Web site at
http://www.taylorandfrancis.com

and the CRC Press Web site at
http://www.crcpress.com

To my family! For your unconditional love and support.

Ervin Sejdić

To Eric and Samuel. For your unconditional love and endless supply of giggles and laughter. I love you, forever!

Tiago H. Falk

Contents

Preface..ix
Editors...xi
Contributors...xiii

Section I Introduction

1. **Signal Processing in the Era of Biomedical Big Data**..3
   Tiago H. Falk and Ervin Sejdić

2. **Collecting and Making Sense of Big Data for Improved Health Care**.......................................9
   Thomas R. Clancy

3. **Big Data Era in Magnetic Resonance Imaging of the Human Brain**..21
   Xiaoyu Ding, Elisabeth de Castro Caparelli, and Thomas J. Ross

Section II Signal Processing for Big Data

4. **Data-Driven Approaches for Detecting and Identifying Anomalous Data Streams**...............57
   Shaofeng Zou, Yingbin Liang, H. Vincent Poor, and Xinghua Shi

5. **Time–Frequency Analysis for EEG Quality Measurement and Enhancement with Applications in Newborn EEG Abnormality Detection Multichannel EEG Enhancement and Classification for Newborn Health Outcome Prediction**..73
   Boualem Boashash, Samir Ouelha, Mohammad Al-Sa'd, Ayat Salim, and Paul Colditz

6. **Active Recursive Bayesian State Estimation for Big Biological Data**.....................................115
   Mohammad Moghadamfalahi, Murat Akcakaya, and Deniz Erdogmus

7. **Compressive Sensing Methods for Reconstruction of Big Sparse Signals**............................133
   Ljubiša Stanković, Miloš Daković, and Isidora Stanković

8. **Low-Complexity DCT Approximations for Biomedical Signal Processing in Big Data**.......151
   Renato J. Cintra, Fábio M. Bayer, Yves Pauchard, and Arjuna Madanayake

9. **Dynamic Processes on Complex Networks**..177
   June Zhang and José M.F. Moura

10. **Modeling Functional Networks via Piecewise-Stationary Graphical Models**......................193
    Hang Yu and Justin Dauwels

11. **Topological Data Analysis of Biomedical Big Data**...209
    Angkoon Phinyomark, Esther Ibañez-Marcelo, and Giovanni Petri

12. **Targeted Learning with Application to Health Care Research**..235
    Susan Gruber

Section III   Applications of Signal Processing and Machine Learning for Big Biomedical Data

13. **Scalable Signal Data Processing for Measuring Functional Connectivity in Epilepsy Neurological Disorder**.....259
    Arthur Gershon, Samden D. Lhatoo, Curtis Tatsuoka, Kaushik Ghosh, Kenneth Loparo, and Satya S. Sahoo

14. **Machine Learning Approaches to Automatic Interpretation of EEGs**..............271
    Iyad Obeid and Joseph Picone

15. **Information Fusion in Deep Convolutional Neural Networks for Biomedical Image Segmentation**..........301
    Mohammad Havaei, Nicolas Guizard, Nicolas Chapados, and Yoshua Bengio

16. **Automated Biventricular Cardiovascular Modelling from MRI for Big Heart Data Analysis**.....................313
    Kathleen Gilbert, Xingyu Zhang, Beau Pontré, Avan Suinesiaputra, Pau Medrano-Gracia, and Alistair Young

17. **Deep Learning for Retinal Analysis**...........................329
    Henry A. Leopold, John S. Zelek, and Vasudevan Lakshminarayanan

18. **Dictionary Learning Applications for HEp-2 Cell Classification**................369
    Sadaf Monajemi, Shahab Ensafi, Shijian Lu, Ashraf A. Kassim, Chew Lim Tan, Saeid Sanei, and Sim-Heng Ong

19. **Computational Sequence- and NGS-Based MicroRNA Prediction**................381
    R.J. Peace and James R. Green

20. **Bayesian Classification of Genomic Big Data**................411
    Ulisses M. Braga-Neto, Emre Arslan, Upamanyu Banerjee, and Arghavan Bahadorinejad

21. **Neuroelectrophysiology of Sleep and Insomnia**................429
    Ramiro Chaparro-Vargas, Beena Ahmed, Thomas Penzel, and Dean Cvetkovic

22. **Automated Processing of Big Data in Sleep Medicine**................443
    Sara Mariani, Shaun M. Purcell, and Susan Redline

23. **Integrating Clinical Physiological Knowledge at the Feature and Classifier Levels in Design of a Clinical Decision Support System for Improved Prediction of Intensive Care Unit Outcome**............465
    Ali Jalali, Vinay M. Nadkarni, Robert A. Berg, Mohamed Rehman, and C. Nataraj

24. **Trauma Outcome Prediction in the Era of Big Data: From Data Collection to Analytics**.............477
    Shiming Yang, Peter F. Hu, and Colin F. Mackenzie

25. **Enchancing Medical Problem Solving through the Integration of Temporal Abstractions with Bayesian Networks in Time-Oriented Clinical Domains**................493
    Kalia Orphanou, Athena Stassopoulou, and Elpida Keravnou

26. **Big Data in Critical Care Using Artemis**................519
    Carolyn McGregor

27. **Improving Neurorehabilitation of the Upper Limb through Big Data**................533
    José Zariffa

28. **Multimodal Ambulatory Fall Risk Assessment in the Era of Big Data**................551
    Mina Nouredanesh and James Tung

Index................581

Preface

Big data has been loosely defined as data that is too large or complex for traditional data processing techniques to be applied effectively. The challenges that arise with such massive data sets include capture, storage, analysis, search, and visualization, to name a few. Within the healthcare domain, the definition of biomedical big data has been expanded to encompass "high volume, high diversity biological, clinical, environmental, and lifestyle information collected from single individuals to large cohorts, in relation to their health and wellness status, at one or several time points."[1] As pointed out by the Ponemon Institute, in 2012, 30% of all electronic data stored in the world was occupied by the healthcare industry. It is clear that buried deep within these vast amounts of data lies invaluable hidden knowledge that not only could change a patient's life, but could open doors to new therapies, drugs, diagnostic tools, and gene discoveries, thus ultimately improving our population's quality of life.

While important steps have already been taken to leverage such insights from, e.g., neuroimaging big data or fitness data collected by wearables giants such as Fitbit, we have only scratched the surface of this data iceberg. As more data has become available over the years, it has enabled important advances in machine learning, specifically in deep learning and deep neural network architectures, thus redefining the performance envelope of existing technologies, such as object recognition from images and automatic speech recognition. Black-box machine learning approaches, however, while very useful indeed, are only as good as the data they are trained on. Ultimately, to make significant progress, expert domain knowledge and signal processing techniques are still needed. This is particularly true for biomedical data.

Big biomedical data is often captured via a multitude of sensors and modalities, each of varying sample rate and dimensionality. Patient-generated data is often of varying quality. The majority of existing data is unstructured and unlabeled. Video resolutions are doubling almost every few years, with resolutions now at 8K ultra-high definition. Capturing, analyzing, storing, searching, and visualizing such massive data have required new shifts in signal processing paradigms and new ways of combining signal processing with machine learning tools. This book aims at covering several of these aspects. While it does not touch upon important topics, such as data sharing and privacy, it does touch upon several other factors, such as data quality, data compression, and new statistical and graph signal processing techniques. It also provides a comprehensive overview of existing state-of-the-art signal processing and machine learning techniques applied to big biomedical data within neuroimaging, cardiac, retinal, genomic, sleep, patient outcome prediction, critical care, and rehabilitation domains.

The book was conceptualized with the goal of bringing together more theoretical signal processing chapters describing tools aimed at big data (be it biomedical or not) with more application-driven chapters focusing on existing applications of signal processing and machine learning for big biomedical data. As such, the book is aimed at both researchers already working in the field as well as undergraduate and graduate students eager to learn how signal processing can help with big data analysis. It is hoped that the book will bring together signal processing and machine learning researchers to unlock existing bottlenecks within the healthcare field and, ultimately, improve patient quality of life. We would like to thank all contributing authors for their excellent chapters.

Tiago H. Falk
Ervin Sejdić

Reference

1. Auffray C, Balling R, Barroso I, Bencze L, Benson M, Bergeron J et al. Making sense of big data in health research: Towards an EU action plan. *Genome Med.* 2016;8(1):71.

MATLAB® is a registered trademark of The MathWorks, Inc. For product information, please contact:

The MathWorks, Inc.
3 Apple Hill Drive
Natick, MA 01760-2098 USA
Tel: 508-647-7000
Fax: 508-647-7001
Email: info@mathworks.com
Web: www.mathworks.com

Editors

Prof. Ervin Sejdić received BESc and PhD degrees in electrical engineering from the University of Western Ontario, London, Ontario, Canada, in 2002 and 2008, respectively. From 2008 to 2010, he was a postdoctoral fellow at the University of Toronto with a cross-appointment at Bloorview Kids Rehab, Canada's largest children's rehabilitation teaching hospital. From 2010 until 2011, he was a research fellow at Harvard Medical School with a cross-appointment at Beth Israel Deaconess Medical Center. In 2011, Prof. Sejdić joined the Department of Electrical and Computer Engineering at the University of Pittsburgh, Philadelphia, USA, as a tenure-track assistant professor. In 2017, he was promoted to a tenured associate professor. He also holds secondary appointments in the Department of Bioengineering (Swanson School of Engineering), the Department of Biomedical Informatics (School of Medicine), and the Intelligent Systems Program (Kenneth P. Dietrich School of Arts and Sciences) at the University of Pittsburgh.

Prof. Sejdić is a senior member of the Institute of Electrical and Electronics Engineers (IEEE) and a recipient of many awards. As a graduate student, he was awarded two prestigious awards from the Natural Sciences and Engineering Research Council of Canada. In 2010, he received the Melvin First Young Investigators Award from the Institute for Aging Research at Hebrew Senior Life in Boston, Massachusetts, USA. In February 2016, President Obama named Prof. Sejdić as a recipient of the Presidential Early Career Award for Scientists and Engineers, the highest honor bestowed by the United States Government on science and engineering professionals in the early stages of their independent research careers. In 2017, Prof. Sejdić was awarded the National Science Foundation CAREER Award, which is the National Science Foundation's most prestigious award in support of career-development activities of those scholars who most effectively integrate research and education within the context of the mission of their organization.

From his earliest exposure to research, he has been eager to contribute to the advancement of scientific knowledge through carefully executed experiments and groundbreaking published work. He is an area editor of the *IEEE Signal Processing Magazine* and an associate editor of *Biomedical Engineering Online*. Prof. Sejdić's passion for discovery and innovation drives his constant endeavors to connect advances in engineering to society's most challenging problems. Hence, his research interests include biomedical signal processing, gait analysis, swallowing difficulties, advanced information systems in medicine, rehabilitation engineering, assistive technologies, and anticipatory medical devices.

Prof. Tiago H. Falk received a BSc degree from the Federal University of Pernambuco, Brazil, in 2002, and MSc and PhD degrees from Queens University, Canada, in 2005 and 2008, respectively, all in electrical engineering. In 2007, he was a visiting research fellow at the Sound and Image Processing Lab, Royal Institute of Technology (KTH), Sweden, and in 2008 at the Quality and Usability Lab, Deutsche Telekom/TU Berlin, Germany. From 2009 to 2010, he was a Natural Sciences and Engineering Research Council of Canada (NSERC) postdoctoral fellow at the Holland Bloorview Kids Rehabilitation Hospital, affiliated with the University of Toronto. He joined the Institut National de la Recherche Scientifique (INRS) in Montreal, Canada, in December 2010 as a tenure-track assistant professor. In 2015, he was promoted to tenured associate professor. He is also an adjunct professor at McGill University (ECE Department) and an affiliate researcher at Concordia University (PERFORM Centre). At INRS, he heads the Multimedia/Multimodal Signal Analysis and Enhancement (MuSAE) Laboratory.

Prof. Falk is a senior member of the Institute of Electrical and Electronics Engineers (IEEE), an elected member of the Global Young Academy (GYA), a member of the Sigma Xi Research Society, and academic chair of the Canadian Biomedical and Biological Engineering Society. His research interests lie at the crossroads of multimedia and biomedical signal processing with particular focus on leveraging big data for the development of affective human–machine interfaces and anthropomorphic, human-inspired technologies. His work has resulted in numerous awards, including the EURASIP Best Paper Award (2017), the Sigma Xi Young Investigator Award (2016), the NSERC Discovery Accelerator Supplement Award (2016), the Bell Outstanding Achievement Award (2015), the CMBES Early Career Achievement Award (2015), the IEEE Kingston Section PhD Research Excellence Award (2008), the Prof. Newton Maia Young

Scientist Award (2002), and several best paper awards at international conferences and workshops.

Prof. Falk is a member of the editorial board of the Quality and User Experience (Springer) journal, the Journal of the Canadian Acoustical Association, IEEE Canada Journal of Electrical and Computer Engineering, and the International Journal of Healthcare Engineering. He has served as guest editor to special issues in the Journal of Medical and Biological Engineering, Frontiers in Neurology, the EURASIP Journal on Audio, Speech & Music Processing, and the Journal of Electrical and Computer Engineering.

Contributors

Beena Ahmed
School of Electrical Engineering and Telecommunications
University New South Wales
Sydney, Australia

Murat Akcakaya
Electrical and Computer Engineering Department
University of Pittsburgh
Pittsburgh, Pennsylvania

Mohammad Al-Sa'd
Qatar University College of Engineering
Doha, Qatar

Emre Arslan
Department of Electrical and Computer Engineering
Center for Bioinformatics and Genomics Systems Engineering
Texas A&M University
College Station, Texas

Arghavan Bahadorinejad
Department of Electrical and Computer Engineering
Center for Bioinformatics and Genomics Systems Engineering
Texas A&M University
College Station, Texas

Upamanyu Banerjee
Department of Electrical and Computer Engineering
Center for Bioinformatics and Genomics Systems Engineering
Texas A&M University
College Station, Texas

Fábio M. Bayer
Departamento de Estatística and Laboratório de Ciências Espaciais de Santa Maria (LACESM)
Universidade Federal de Santa Maria
Santa Maria, Brazil

Yoshua Bengio
Montreal Institute for Learning Algorithms (MILA)
and
Department of Computer Science and Operations Research
University of Montreal
Montreal, Canada

Robert A. Berg
Department of Anesthesiology and Critical Care Medicine
Children's Hospital of Philadelphia
Philadelphia, Pennsylvania

Boualem Boashash
The University of Queensland Center for Clinical Research (UQCCR)
The University of Queensland
Brisbane, Australia

Ulisses M. Braga-Neto
Department of Electrical and Computer Engineering
Center for Bioinformatics and Genomics Systems Engineering
Texas A&M University
College Station, Texas

Nicolas Chapados
Imagia Inc.
Montreal, Canada

Ramiro Chaparro-Vargas
eys IT SAS
Bogota, Colombia

Renato J. Cintra
Signal Processing Group
Departamento de Estatística
Universidade Federal de Pernambuco
Recife, Brazil

and

ECE
University of Calgary
Calgary, Canada

Thomas R. Clancy
School of Nursing
The University of Minnesota
Minneapolis, Minnesota

Paul Colditz
The University of Queensland Center for Clinical Research (UQCCR)
The University of Queensland
Brisbane, Australia

Dean Cvetkovic
School of Engineering Cluster
RMIT University
Melbourne, Australia

Miloš Daković
Faculty of Electrical Engineering
University of Montenegro
Podgorica, Montenegro

Justin Dauwels
School of Electrical and Electronic Engineering
Nanyang Technological University
Singapore

Elisabeth de Castro Caparelli
Neuroimaging Research Branch
National Institute on Drug Abuse, Intramural Research Program (NIDA-IRP)
National Institutes of Health (NIH)
Baltimore, Maryland

Xiaoyu Ding
Neuroimaging Research Branch
National Institute on Drug Abuse, Intramural Research Program (NIDA-IRP)
National Institutes of Health (NIH)
Baltimore, Maryland

Shahab Ensafi
Electrical and Computer Engineering
Faculty of Engineering
National University of Singapore
Singapore

Deniz Erdogmus
Cognitive Systems Lab
Electrical and Computer Engineering Department
Northeastern University
Boston, Massachusetts

Tiago H. Falk
Institut National de la Recherche Scientifique (INRS-EMT)
University of Quebec
Montreal, Canada

Arthur Gershon
Department of Population and Quantitative Health Sciences
School of Medicine
Case Western Reserve University
Cleveland, Ohio

Kaushik Ghosh
Department of Mathematical Sciences
College of Sciences
University of Nevada Las Vegas
Las Vegas, Nevada

Kathleen Gilbert
Department of Anatomy and Medical Imaging
School of Medical Sciences
Faculty of Medical and Health Sciences
University of Auckland
Aukland, New Zealand

James R. Green
Systems and Computer Engineering
Carleton University
Ottawa, Canada

Susan Gruber
Department of Population Medicine
Harvard Pilgrim Health Care Institute
and
Harvard Medical School
Boston, Massachusetts

Nicolas Guizard
Imagia Inc.
Montreal, Canada

Mohammad Havaei
Imagia Inc.
Montreal, Canada

Peter F. Hu

Esther Ibáñez-Marcelo
ISI Foundation
Turin, Italy

Ali Jalali
Health Informatics Core
Johns Hopkins All Children's Hospital
St Petersburg, Florida

Ashraf A. Kassim
Electrical and Computer Engineering
Faculty of Engineering
National University of Singapore
Singapore

Elpida Keravnou
Department of Computer Science
University of Cyprus
Nicosia, Cyprus

Vasudevan Lakshminarayanan
School of Optometry and Vision Science
Department of Physics
Department of Systems Design Engineering
Department of Electrical and Computer Engineering
University of Waterloo
Waterloo, Canada

Henry A. Leopold
Department of Systems Design Engineering
School of Optometry and Vision Science
University of Waterloo
Waterloo, Canada

Samden D. Lhatoo
Department of Neurology
School of Medicine
Case Western Reserve University
Cleveland, Ohio

Yingbin Liang
Department of Electrical and Computer Engineering
The Ohio State University
Columbus, Ohio

Kenneth Loparo
Department of Electrical Engineering and Computer Science
Case School of Engineering
Case Western Reserve University
Cleveland, Ohio

Shijian Lu
Institute for Infocomm Research
Agency for Science Technology and Research (A*STAR)
Singapore

Colin F. Mackenzie

Arjuna Madanayake
Department of Electrical and Computer Engineering
University of Akron
Akron, Ohio

Sara Mariani
Division of Sleep and Circadian Disorders
Department of Medicine
Brigham and Women's Hospital
and
Harvard Medical School
Boston, Massachusetts

Carolyn McGregor
University of Ontario Institute of Technology
Oshawa, Canada

and

University of Technology
Sydney, Australia

Pau Medrano-Gracia
Department of Anatomy and Medical Imaging
School of Medical Sciences
Faculty of Medical and Health Sciences
University of Auckland
Aukland, New Zealand

Mohammad Moghadamfalahi
Cognitive Systems Lab
Electrical and Computer Engineering Department
 Northeastern University
Boston, Massachusetts

Sadaf Monajemi
NUS Graduate School for Integrative Sciences
 and Engineering
National University of Singapore
Singapore

José M.F. Moura
Electrical and Computer Engineering Department
Carnegie Mellon University
Pittsburgh, Pennsylvania

Vinay M. Nadkarni
Department of Anesthesiology and Critical Care
 Medicine
Center for Simulation, Advanced Education,
 and Innovation
Children's Hospital of Philadelphia
Philadelphia, Pennsylvania

C. Nataraj
Department of Mechanical Engineering
Villanova Center for Analytics of Dynamic Systems
Villanova University
Villanova, Pennsylvania

Mina Nouredanesh
Department of Mechanical and Mechatronics
 Engineering
University of Waterloo
Ontario, Canada

Iyad Obeid
Department of Electrical and Computer Engineering
College of Engineering
Temple University
Philadelphia, Pennsylvania

Sim-Heng Ong
Electrical and Computer Engineering
Faculty of Engineering
National University of Singapore
Singapore

Kalia Orphanou
Department of Computer Science
University of Cyprus
Nicosia, Cyprus

Samir Ouelha
Qatar University College of Engineering
Doha, Qatar

Yves Pauchard
Department of Radiology
Cumming School of Medicine
University of Calgary
Calgary, Canada

R.J. Peace
Systems and Computer Engineering
Carleton University
Ontario, Canada

Thomas Penzel
Interdisciplinary Center of Sleep Medicine
Charité Universitätsmedizin Berlin
Berlin, Germany

Giovanni Petri
ISI Foundation
Turin, Italy

Angkoon Phinyomark
ISI Foundation
Turin, Italy

and

Institute of Biomedical Engineering
University of New Brunswick
Fredericton, Canada

Joseph Picone
Department of Electrical and Computer Engineering
College of Engineering
Temple University
Philadelphia, Pennsylvania

Beau Pontré
Department of Anatomy and Medical Imaging
School of Medical Sciences
Faculty of Medical and Health Sciences
University of Auckland
Aukland, New Zealand

H. Vincent Poor
Department of Electrical Engineering
Princeton University
Princeton, New Jersey

Shaun M. Purcell
Department of Psychiatry
Brigham & Women's Hospital
and
Department of Psychiatry
Harvard Medical School
Boston, Massachusetts

and

Division of Psychiatric Genomics
Departments of Psychiatry and Genetics and Genomic Sciences
Icahn School of Medicine at Mount Sinai
New York City, New York

Susan Redline
Division of Sleep and Circadian Disorders
Department of Medicine
Brigham and Women's Hospital
and
Harvard Medical School
Boston, Massachusetts

Mohamed Rehman
Department of Anesthesia
Johns Hopkins All Children's Hospital
St Petersburg, Florida

Contributors

Thomas J. Ross
Neuroimaging Research Branch
National Institute on Drug Abuse, Intramural Research Program (NIDA-IRP)
National Institutes of Health (NIH)
Baltimore, Maryland

Satya S. Sahoo
Department of Population and Quantitative Health Sciences
School of Medicine
and
Department of Neurology
and
Department of Electrical Engineering and Computer Science
Case School of Engineering
Case Western Reserve University
Cleveland, Ohio

Ayat Salim
Qatar University College of Engineering
Doha, Qatar

Saeid Sanei
Faculty of Engineering and Physical Sciences
University of Surrey
Guildford, United Kingdom

Ervin Sejdić
Department of Electrical and Computer Engineering
Swanson School of Engineering
University of Pittsburgh
Philadelphia
Pittsburgh, Pennsylvania

Xinghua Shi
Department of Bioinformatics and Genomics
University of North Carolina at Charlotte
Charlotte, North Carolina

Isidora Stanković
Faculty of Electrical Engineering
University of Montenegro
Podgorica, Montenegro

and

GIPSA-lab, Grenoble INP, CNRS
University of Grenoble Alpes
Grenoble, France

Ljubiša Stanković
Faculty of Electrical Engineering
University of Montenegro
Podgorica, Montenegro

Athena Stassopoulou
Department of Computer Science
University of Nicosia
Nicosia, Cyprus

Avan Suinesiaputra
Department of Anatomy and Medical Imaging
School of Medical Sciences
Faculty of Medical and Health Sciences
University of Auckland
Aukland, New Zealand

Chew Lim Tan
Department of Computer Science
School of Computing
National University of Singapore
Singapore

Curtis Tatsuoka
Department of Neurology
School of Medicine
Case Western Reserve University
Cleveland, Ohio

James Tung
Department of Mechanical and Mechatronics Engineering
University of Waterloo
Waterloo, Canada

Shiming Yang

Alistair Young
Department of Anatomy and Medical Imaging
School of Medical Sciences
Faculty of Medical and Health Sciences
University of Auckland
Aukland, New Zealand

Hang Yu
School of Electrical and Electronic Engineering
Nanyang Technological University
Singapore

José Zariffa
Toronto Rehabilitation Institute–University
    Health Network
and
Institute of Biomaterials and Biomedical Engineering
University of Toronto
Toronto, Canada

John S. Zelek
Department of Systems Design Engineering
School of Optometry and Vision Science
University of Waterloo
Waterloo, Canada

June Zhang
Electrical Engineering Department
University of Hawai'i at Mānoa
Honolulu, Hawaii

Xingyu Zhang
Department of Anatomy and Medical Imaging
School of Medical Sciences
Faculty of Medical and Health Sciences
University of Auckland
Aukland, New Zealand

Shaofeng Zou
Coordinated Science Lab
University of Illinois at Urbana-Champaign
Urbana, Illinois

Section I

Introduction

1

Signal Processing in the Era of Biomedical Big Data

Tiago H. Falk and Ervin Sejdić

CONTENTS

1.1 Introduction 3
1.2 Technological Challenges 3
    1.2.1 Data Quality 4
    1.2.2 Data Variety 4
    1.2.3 Parallel/Distributed Data Processing 4
    1.2.4 Data Storage/Compression/Sampling 5
    1.2.5 Analytics 5
1.3 Book Overview 5
Bibliography 6

1.1 Introduction

We live in a data-rich era with the Internet, social media, and sensors collecting massive amounts of data ceaselessly. It has been predicted that by 2020, each human being in the planet will create 1.7 megabytes of data per second. With electronic health records and daily advances in, e.g., health care Internet of Things (IoT), neuroimaging, genomics, and wearables, a large part of this data will come from the health care industry. Such massive data sets are termed *big data*, and here we refer to them as *health/biomedical big data*.

While there is no universally agreed-upon set of properties that formally define big data, different organizations mention (1) volume (amount of generated or stored data), (2) velocity (pace at which data is created and made available), (3) variety (source and type of structured and unstructured data), (4) veracity (varying quality data), (5) validity (correct/incorrect data for the task at hand), (6) volatility (length/duration of data that is valid and should be stored), (7) variability (data constantly changing), (8) visualization (how to visualize the data to gain insights), and (9) value (how to uncover insights for better decision making, outcomes, and savings). The first three are often called the three *V*'s of big data, which were later expanded to the six *V*'s and more recently the nine *V*'s [1,2].

Within the health care domain, knowledge and insights from big health data can lead to innovations in prevention, diagnosis, treatment, and follow-up care, thus ultimately improving our quality of life. In fact, a report by McKinsey Global Institute in 2011 suggested that efficient use of big health care data could save the United States government more than $300 billion in value every year, two-thirds of that in the form of reduced health care expenditures [3]. Gathering such useful insights, however, is not a trivial task, and the challenges with data capture, analysis, storage, search, and visualization are grand. As an example, of the available digital data in existence in 2012, only 3% was tagged, and a mere 0.5% had been analyzed. Notwithstanding, 23% of this data could have generated useful insights had it been labelled and tagged. By 2020, that number is expected to rise to 35% [4].

1.2 Technological Challenges

Many sectors today have already exceeded hundreds of exabytes (10^{18}) of stored data (the US health care sector is an example), and it will not be long before we deal with zettabyte (10^{21}) and yottabyte (10^{24}) data systems [1].

Such data volumes limit the applicability of many well-known technologies used with conventional data sets, including file systems, processing paradigms, as well as sharing and transmission techniques. Social media, mobile technologies, IoT, and high-throughput systems within the health care sector (e.g., genomics, neuroimaging, etc.) are generating data at unprecedented rates, thus creating many technological challenges [5,6]. Below we list a few challenges that are aligned with some of the chapters within this book, namely, data quality, variety, parallel/distributed processing, storage/compression/sampling, and analytics.

1.2.1 Data Quality

Data quality is an often overlooked issue but extremely important in big data analysis, as "bigger data" does not always mean "better data." As the old saying goes: "garbage in, garbage out" when it comes to inferring knowledge from noisy data. Data quality issues can arise in many flavours. With the burgeoning of wearable devices and the proliferation of patient-generated data (e.g., long-term monitoring of daily activities using wearables), issues such as low-cost/low-quality devices can arise, as well as issues with noisy data due to, e.g., excessive movement artifacts [7] or due to data collected erroneously (e.g., sensors not placed correctly). Data provenance is also a very important issue within big health care data, as reproducibility is important, particularly in clinical and pharmacological realms. Documentation of the data collection protocol, signal preprocessing steps, and what (if any) artifacts could have been introduced due to lossy compression (more on this next) during storage are important factors relating to data quality that need special attention.

Medical IoT devices are also prone to failure, and anomalous data streams may generate false alarm rates, which can have detrimental effects within automated systems within a hospital setting [8]. Corrupted measurements or communication errors are likely, as acquisition and transmission techniques are mostly operated to reduce cost, thus resulting in unstable systems. In order to handle such missing data issues, imputation techniques are needed [9].

1.2.2 Data Variety

The health sector is a domain in which data variety plays a crucial challenge for big data analytics [10]. Data can be clinical, genetic, behavioural, environmental, financial, and/or operational. It can come from electronic health care records, patient summaries, genomic readings, pharmaceutical and clinical trial results, imaging, insurance claims, and/or real-time vital signs monitors. They can come from medical-grade devices or from mobile apps installed on patient smartphones. They can be collected at home with minimal surveillance or via carefully monitored clinical trials. While available data can be both structured and unstructured, recent projections suggest that upward of 90% of available data is unstructured. Extracting hidden knowledge from huge amounts of structured and unstructured data is key to big data analysis but an extremely challenging task that requires clever data fusion, context awareness, and resampling strategies, to name a few [11].

1.2.3 Parallel/Distributed Data Processing

In order to handle such amounts of data, new programming frameworks have been designed to distribute and parallelize data analysis to computing clusters and/or grids. Moreover, a new file system called distributed file system (DFS) has been created to distribute huge files over much larger units than the disk blocks in conventional operating systems [2]. DFS also provides data replication and redundancy to avoid data loss in case of media failure that may occur when distributing data over hundreds or thousands of computing nodes. More secure and safe distributed data streaming technologies have also been invented to match storage/computing speeds and the high throughputs of data production. DFS not only provides file storage management but also acts as the basis whereby a new generation of programming frameworks can take advantage of data distribution and redundancy to parallelize computations over multiple independent computing nodes instead of a single special-purpose computer. Working on top of DFS, these frameworks introduce fault-tolerant architectures that enable systems to rely on much cheaper commodity hardware instead of high-cost specialized machines [12].

Proposed by Google, MapReduce is a programming framework for data-intensive applications. It is more like functional programming, where map and reduce functions are defined by the programmer to process chunks of distributed data [2]. MapReduce can efficiently perform most computations on distributed large-scale data; however, it is not suitable for online processing. It boosts computation performance by dividing the larger main task into smaller jobs to be executed across hundreds of nodes. Hadoop is the open-source implementation of MapReduce and mainly provided by Apache. It consists of a master–slave architecture that provides both the DFS (Hadoop DFS) and the MapReduce programming frameworks [2]. Hadoop also provides higher-level languages that generate MapReduce programs such as Hive and Pig and some technologies that handle distributed databases such as HBase.

Considering the speed, ease of use, and the sophisticated analytics, MapReduce has been recently replaced by Apache Spark. Contrary to MapReduce, Spark makes use of memory not only for computations but also for storage to achieve low latency on big data workloads. In addition, Spark provides a unified runtime that deals with multiple big data storage sources like Hadoop DFS, HBase, and Cassandra. Spark also provides ready-to-use high-level libraries for machine learning, graph processing, and real-time streaming. Such parallel/distributed processing tools have enabled, for example, novel visualization of neural connections [13–15], allowed for faster image retrieval and lung texture classification [16], enabled large-scale biometrics [17,18], and took genomics to the next level [19,20].

1.2.4 Data Storage/Compression/Sampling

Storage of zettabytes of data is not trivial and requires novel compression and dimensionality reduction methods. Dimensionality reduction, as the name suggests, is an effective solution to find a meaningful low-dimensional structure from high-dimensional data. Classical algorithms include principal component analysis, linear discriminant analysis, locally linear embedding, backward feature elimination, and Laplacian eigenmaps, to name a few [21]. Other methods have been proposed for nonlinear cases and categorical data. Hochbaum *et al.*, for example, introduced sparse computation [22] for large-scale data, which avoids computing all pairwise similarities and retains only relevant ones computed by mapping the data into low-dimensional spaces in which the groups of objects share the same grid neighbourhood.

Data compression, in turn, is the process of reducing the size of a data file and can be classified as lossless or lossy. Lossless compression reduces data file size by identifying and eliminating statistical redundancy, and hence does not lose any information. Lossy compression, on the other hand, reduces file size by removing unnecessary or less important information, and thus allows for a trade-off between distortion and compression ratio. Within the health care domain, data compression has played a crucial role in medical imaging [23] and in the transmission of biometric data collected via wearables and smart devices [24].

In order to store digital data, analog signals are typically sampled at the so-called Nyquist rate, which assures that the original signal can be reconstructed from a series of sampling measurements. The Nyquist theorem states that if the signal's highest frequency is less than half of the sampling rate, then the signal can be reconstructed perfectly. More recently, however, a signal processing technique termed compressed sensing (or sparse/compressed sampling) has exploited the sparsity of a signal and, through optimization, shown that signals can be recovered with far fewer samples than those required by the Nyquist sampling theorem [25]. Compressed sensing has been used across numerous health domains, including medical imaging [26] and wireless body area networks [27].

1.2.5 Analytics

Big data analytics uncover hidden insights from data to enable new prevention, diagnosis, treatment, and follow-up care procedures. It can also improve health care quality and outcomes, reduce health care costs, and provide support for reformed payment structures. Analysis of such volumes of high-dimensional, unstructured, unlabelled data, however, is not trivial and requires innovations in signal processing and machine learning, amongst other domains. Given the complex nature of the data, domain knowledge should be used to fine tune algorithms [28].

Access to large amounts of data has enabled significant advances in machine learning, particularly in deep learning and reinforcement learning [29]. Applications of deep learning in the health care domain are growing exponentially [30,31]. Notwithstanding, recent work on adversarial samples has shown that existing deep neural networks may be vulnerable to certain input variations, including noise and artifacts [32,33]. As such, it is expected that big data analytics will require a clever combination of machine learning and signal processing advances, including graph-based techniques [34–37], statistical modeling [38], tensors [39], topological data analysis [40], and targeted learning [41], to name a few.

The goal of this book is to overview theoretical advances of such signal processing techniques and showcase applications of both signal processing and machine learning tools for biomedical big data analytics across different domains, including neuroimaging, cardiac, retinal, genomic, sleep, patient outcome prediction, critical care, and rehabilitation.

1.3 Book Overview

This book is divided into three sections, loosely covering the goals of the book: (1) introduce the reader to big biomedical/health data concepts, (2) showcase (theoretical) signal processing tools aimed at big data analysis, in general, and (3) provide the reader with targeted signal processing and machine learning applications within biomedical/health big data.

The first introductory section comprises three chapters. Chapter 1 (this chapter) has introduced the reader to the concept of biomedical big data, as well as some of the existing challenges with which signal processing and machine learning tools are being explored as possible solutions. In Chapter 2, a more in-depth discussion is presented on the sources of big data in health care and how to leverage such data to improve outcomes, as well as develop smart medical devices. Chapter 3, in turn, provides a comprehensive overview of Big Neural Data and ongoing initiatives/projects leveraging neuroimaging data to learn insights about brain functioning and disease. State-of-the-art signal processing and machine learning tools are described, as are issues with translating obtained findings to real clinical utility.

The second section focuses on signal processing contributions aimed at solving some of the big data challenges described in Section 1.2. While the techniques are applicable to big data in general, each chapter presents examples of how they can be applied to biomedical big data. Chapters 4 and 5, for example, focus on the aspect of data quality (Section 1.2.1). While the former presents data-driven methods to detect and identify anomalous data streams, the former presents time-frequency signal processing strategies for quality assessment and enhancement of electroencephalography (EEG) signals, with particular emphasis on newborn health outcome prediction. Chapter 6 proposes a framework for system state inference of multimodal big data that can be processed in parallel, thus addressing the data variety (Section 1.2.2) and parallel data processing (Section 1.2.3) issues. Next, Chapters 7 and 8 touch on the storage/compression/sampling issues discussed in Section 1.2.4. Chapter 7 addresses the issue of compressive sensing for big sparse signals, while Chapter 8 addresses low-complexity discrete cosine transform approximations with special applications in medical image compression and registration. Lastly, Chapters 9–12 touch on the issue of signal processing for big data analytics (Section 1.2.5) and modeling. More specifically, Chapter 9 focuses on complex network-based epidemics modeling and Chapter 10 on functional network modeling via piecewise-stationary graphical models. Chapters 11 and 12, in turn, propose the use of topological data analysis and targeted learning, respectively, as tools to extract knowledge from big data.

The third section presents 16 chapters describing applications of signal processing and machine learning specifically for biomedical big data. Chapters 13–18, for example, address signal processing and machine learning tools for medical imaging analysis, with Chapters 13–15 addressing neuroimaging data, Chapter 16 cardiovascular magnetic resonance imaging, Chapter 17 retinal images, and Chapter 18 human epithelial type-2 cell images. Chapters 19 and 20, in turn, show methods applied to RNA prediction and classification of genomic big data, respectively. Next, Chapters 21 and 22 cover the topic of multimodal signal processing for automated sleep and insomnia analysis. Chapter 23 addresses the important issue of integrating clinical knowledge into the analytics process, while Chapter 24 looks at brain trauma outcome prediction. Chapter 25 covers the application of combining temporal abstractions with Bayesian networks for the clinical domain of coronary heart disease. Chapter 26, in turn, overviews the Artemis big data analytics platform in use at critical care units around the world. Lastly, Chapters 27 and 28 cover the topics of neurorehabilitation of the upper limb and ambulatory fall risk assessment, respectively, using multimodal sensor fusion strategies.

Bibliography

1. J. Andreu-Perez, C. Poon, R. Merrifield, S. Wong, and G.-Z. Yang, "Big data for health," *IEEE Journal of Biomedical and Health Informatics*, vol. 19, pp. 1193–1208, 2015.
2. E. A. Mohammed, B. H. Far, and C. Naugler, "Applications of the MapReduce programming framework to clinical big data analysis: Current landscape and future trends," *BioData Mining*, vol. 7, no. 1, p. 22, 2014.
3. X. Yi, F. Liu, J. Liu, and H. Jin, "Building a network highway for big data: Architecture and challenges," *IEEE Network*, vol. 28, no. 4, pp. 5–13, 2014.
4. IDC, "The digital universe in 2020: Big data, bigger digital shadows and biggest growth in the Far East," International Data Corporation, Tech. Rep., 2012.
5. A. Gandomi and M. Haider, "Beyond the hype: Big data concepts, methods, and analytics," *International Journal of Information Management*, vol. 35, no. 2, pp. 137–144, 2015.
6. C. Yang, Q. Huang, Z. Li, K. Liu, and F. Hu, "Big data and cloud computing: Innovation opportunities and challenges," *International Journal of Digital Earth*, vol. 10, no. 1, pp. 13–53, 2017.
7. D. Vallejo, T. H. Falk, and M. Maier, "MS-QI: A modulation spectrum-based ECG quality index for telehealth applications," *IEEE Transactions on Biomedical Engineering*, vol. 63, no. 8, pp. 1613–1622, 2016.
8. C. Tsimenidis and A. Murray, "False alarms during patient monitoring in clinical intensive care units are highly related to poor quality of the monitored electrocardiogram signals," *Physiological Measurement*, vol. 37, no. 8, p. 1383, 2016.
9. K. Slavakis, G. Giannakis, and G. Mateos, "Modeling and optimization for big data analytics: (statistical) learning tools for our era of data deluge," *IEEE Signal Processing Magazine*, vol. 31, no. 5, pp. 18–31, 2014.
10. J. Chen, Y. Chen, X. Du, C. Li, J. Lu, S. Zhao, and X. Zhou, "Big data challenge: A data management perspective," *Frontiers of Computer Science*, vol. 7, no. 2, pp. 157–164, 2013.
11. D. P. Tobon, T. H. Falk, and M. Maier, "Context awareness in WBANs: A survey on medical and non-medical applications," *IEEE Wireless Communications*, vol. 20, no. 4, pp. 30–37, 2013.

12. L. Wang, D. Chen, R. Ranjan, S. U. Khan, J. Kolodziej, and J. Wang, "Parallel processing of massive EEG data with MapReduce," in *Proceedings of the 18th IEEE International Conference on Parallel and Distributed Systems*, ser. ICPADS, '12, 2012, pp. 164–171.
13. Z. Wu and N. E. Huang, "Ensemble empirical mode decomposition: A noise-assisted data analysis method," *Advances in Adaptive Data Analysis*, vol. 01, no. 01, pp. 1–41, 2009.
14. L. Wang, D. Chen, R. Ranjan, S. U. Khan, J. Kolodziej, and J. Wang, "Parallel processing of massive EEG data with MapReduce," in *2012 IEEE 18th International Conference on Parallel and Distributed Systems*, December 2012, pp. 164–171.
15. A. V. Nguyen, R. Wynden, and Y. Sun, "Hbase, MapReduce, and integrated data visualization for processing clinical signal data." in *AAAI Spring Symposium: Computational Physiology*. AAAI, 2011.
16. D. Markonis, R. Schaer, I. Eggel, H. Müller, and A. Depeursinge, "Using MapReduce for large-scale medical image analysis," *CoRR*, vol. abs/1510.06937, 2015.
17. S. Mangla and N. S. Raghava, "Iris recognition on Hadoop: A biometrics system implementation on cloud computing," in *2011 IEEE International Conference on Cloud Computing and Intelligence Systems*, September 2011, pp. 482–485.
18. F. Omri, R. Hamila, S. Foufou, and M. Jarraya, *Cloud-Ready Biometric System for Mobile Security Access*. Berlin: Springer, 2012, pp. 192–200.
19. W. P. Chen, C. L. Hung, S. J. Tsai, and Y. L. Lin, "Novel and efficient tag SNPs selection algorithms," *Biomedical Materials and Engineering*, vol. 24, no. 1, pp. 1383–1389, 2014.
20. K. Zhang, F. Sun, M. S. Waterman, and T. Chen, "Dynamic programming algorithms for haplotype block partitioning: Applications to human chromosome 21 haplotype data," in *Proceedings of the Seventh Annual International Conference on Research in Computational Molecular Biology*, ser. RECOMB '03. New York, NY, USA: ACM, 2003, pp. 332–340.
21. L. Van Der Maaten, E. Postma, and J. Van den Herik, "Dimensionality reduction: A comparative," *Journal of Machine Learning Research*, vol. 10, pp. 66–71, 2009.
22. D. S. Hochbaum and P. Baumann, "Sparse computation for large-scale data mining," *IEEE Transactions on Big Data*, vol. 2, no. 2, pp. 151–174, June 2016.
23. V. Bui, L.-C. Chang, D. Li, L.-Y. Hsu, and M. Y. Chen, "Comparison of lossless video and image compression codecs for medical computed tomography datasets," in *2016 IEEE International Conference on Big Data (Big Data)*. IEEE, 2016, pp. 3960–3962.
24. J. Ma, T. Zhang, and M. Dong, "A novel ECG data compression method using adaptive Fourier decomposition with security guarantee in e-health applications," *IEEE Journal of Biomedical and Health Informatics*, vol. 19, no. 3, pp. 986–994, 2015.
25. D. L. Donoho, "Compressed sensing," *IEEE Transactions on Information Theory*, vol. 52, no. 4, pp. 1289–1306, 2006.
26. M. Lustig, D. Donoho, and J. M. Pauly, "Sparse MRI: The application of compressed sensing for rapid MR imaging," *Magnetic Resonance in Medicine*, vol. 58, no. 6, pp. 1182–1195, 2007.
27. H. Mamaghanian, N. Khaled, D. Atienza, and P. Vandergheynst, "Compressed sensing for real-time energy-efficient ECG compression on wireless body sensor nodes," *IEEE Transactions on Biomedical Engineering*, vol. 58, no. 9, pp. 2456–2466, 2011.
28. A. Holzinger, "Interactive machine learning for health informatics: When do we need the human-in-the-loop?" *Brain Informatics*, vol. 3, no. 2, pp. 119–131, 2016.
29. Y. Li, "Deep reinforcement learning: An overview," *arXiv preprint arXiv:1701.07274*, 2017.
30. R. Miotto, F. Wang, S. Wang, X. Jiang, and J. T. Dudley, "Deep learning for healthcare: Review, opportunities and challenges," *Briefings in Bioinformatics*, p. bbx044, 2017.
31. F. Movahedi, J. L. Coyle, and E. Sejdić, "Deep belief networks for electroencephalography: A review of recent contributions and future outlooks," *IEEE Journal of Biomedical and Health Informatics*, 2017.
32. N. Papernot, P. McDaniel, S. Jha, M. Fredrikson, Z. B. Celik, and A. Swami, "The limitations of deep learning in adversarial settings," in *016 IEEE European Symposium on Security and Privacy (EuroS&P)*. IEEE, 2016, pp. 372–387.
33. S. Dodge and L. Karam, "Understanding how image quality affects deep neural networks," in *Eighth International Conference on Quality of Multimedia Experience (QoMEX)*. IEEE, 2016, pp. 1–6.
34. D. I. Shuman, S. K. Narang, P. Frossard, A. Ortega, and P. Vandergheynst, "The emerging field of signal processing on graphs: Extending high-dimensional data analysis to networks and other irregular domains," *IEEE Signal Processing Magazine*, vol. 30, no. 3, pp. 83–98, 2013.
35. L. Stanković, M. Daković, and E. Sejdić, "Vertex-frequency analysis: A way to localize graph spectral components," *IEEE Signal Processing Magazine*, vol. 34, no. 4, pp. 176–182, July 2017.
36. I. Jestrović, J. L. Coyle, and E. Sejdić, "Differences in brain networks during consecutive swallows detected using an optimized vertex-frequency algorithm," *Neuroscience*, vol. 344, pp. 113–123, 2017.
37. I. Jestrović, J. L. Coyle, and E. Sejdić, "A fast algorithm for vertex-frequency representations of signals on graphs," *Signal Processing*, vol. 131, pp. 483–491, 2017.
38. M. Castellaro, G. Rizzo, M. Tonietto, M. Veronese, F. Turkheimer, M. Chappell, and A. Bertoldo, "A variational Bayesian inference method for parametric imaging of PET data," *Neuroimage*, vol. 150, pp. 136–149, 2017.
39. A. Cichocki, D. Mandic, L. De Lathauwer, G. Zhou, Q. Zhao, C. Caiafa, and H. A. Phan, "Tensor decompositions for signal processing applications: From two-way to multiway component analysis," *IEEE Signal Processing Magazine*, vol. 32, no. 2, pp. 145–163, 2015.
40. M. Offroy and L. Duponchel, "Topological data analysis: A promising big data exploration tool in biology, analytical chemistry and physical chemistry," *Analytica Chimica Acta*, vol. 910, pp. 1–11, 2016.
41. O. Sofrygin, Z. Zhu, J. A. Schmittdiel, A. S. Adams, R. W. Grant, M. J. van der Laan, and R. Neugebauer, "Targeted learning with daily EHR data," *arXiv preprint arXiv:1705.09874*, 2017.

2

Collecting and Making Sense of Big Data for Improved Health Care

Thomas R. Clancy

CONTENTS

Introduction 9
I. Sources of Data in Health Care..9
II. Factors Driving the Creation of Big Data in Health Care 12
III. Benefits of Big Data Science in Health Care..14
IV. Big Data and Smart Medical Devices..17
V. Conclusions 18
References 19

Introduction

A number of converging factors are driving the creation of large-scale, complex electronic data repositories in health care today. These repositories may contain clinical data from electronic health records, insurance claims information, streaming data from wearable sensors, patient-reported outcomes from surveys, social media, global positioning systems, genetic sequencing, as well as a host of other sources. The opportunities to mine and discover new knowledge from these emerging electronic data repositories, or *big data*, are immense. From creating improved evidence-based guidelines to treat and diagnose disease conditions to predicting hospital readmissions, big data science has the capacity to transform health care and improve the health and well-being of populations. This chapter will explore the factors driving the creation of large-scale data repositories in health care, the various sources of data upon which they are built, the advanced computational methods used to analyze big data, and the benefits and challenges they create.

I. Sources of Data in Health Care

A. Electronic Health Records

There are multiple sources of data available for research purposes in health care. Electronic health records (EHRs) are one of the primary sources and contain data in either structured or unstructured formats. Structured formats generally include discrete values such as patient vital signs (temperature, pulse, respirations, and blood pressure), medications (drug, dose, and route), diagnosis and procedure codes, lab results, problem lists, demographic data, and so forth. Unstructured data includes clinical documentation (physician, nurse, and other providers), reports (radiology, operative, and consultation), provider orders, discharge summaries, and other types of narrative notes. A general listing of the categories of data in an EHR is listed as follows [1].

- Patient demographics
- Progress notes
- Vital signs
- Medical histories
- Diagnoses
- Medications
- Immunization dates
- Allergies
- Radiology images
- Lab and test results

Electronic health records are utilized in multiple health care settings including acute care hospitals, outpatient clinics, subacute care (long-term care facilities), and home care settings. Each setting may have its own unique documentation requirements. For example, documentation of Medicare patients in home health care must be in OASIS (Outcome and Assessment Information Set), the instrument/data collection tool used to enter and report performance data by home health agencies. Personal health records, often an extension of a hospital's or clinic's EHR, provide home computer portals that allow patients to document their personal outcomes. These data elements may include patients tracking their own fitness, nutrition and diet programs, lab results, medications, and other information.

B. Health Insurance Claims Data

Although originally intended to support reimbursement from insurance providers, claims data is an excellent source for secondary analysis of health care data. Claims data generally contain detailed patient demographics; procedure; diagnosis; facility and provider codes; ancillary service codes for lab, radiology, and pharmacy; reimbursement amounts; insurance coverage; and other types of data required for billing purposes. One advantage of claims data (assuming continuous enrollment in one insurance carrier) is that the data can span the entire continuum of care (acute, outpatient, and home) including ancillary tests, procedures, and pharmacy. This allows investigators to track the total cost of care of individuals and is frequently used in comparative effectiveness and cost/benefit analysis studies used in health policy decisions.

Claims databases can contain a high volume of provider records over many years. For example, the Medicare Claims Public Use File in 2008 contained 38,161,499 beneficiaries enrolled in fee-for-service Medicare programs by gender and age category [2]. The Health Care Cost Institute (HCCI) manages a data set of 40 million patient records of claims data developed by a network of large health care insurance providers [3]. Mini-Sentinel is a data set of 193 million claims data records that is maintained by a network of health insurance providers and supported by the US Food and Drug Administration [4]. Optum Labs™, a unique research collaborative with over 20 partners, supports a data set of 150 million commercial insurance enrollees [5].

C. Publically Supported Databases

There are a number of large databases supported through the US Department of Health and Human Services. Three important databases are the National Institutes of Health's Clinical and Translation Science Award (CTSA) program, the Agency for Healthcare Research and Quality (AHRQ), and the Centers for Disease Control's (CDC's) National Center for Health Statistics (NCHS). The CTSA program is a national network of medical research institutions, known as hubs, that collaborate to improve the translation of research findings into practice. The CTSA is sponsored through the National Institutes of Health, and its network currently includes 62 medical research institutions in 31 states and the District of Columbia [6].

The CTSA Accrual to Clinical Trials (CTSA ACT) program, implemented in 2014, consists of a network of sites that share EHR data to determine the availability of potential research participants for clinical studies [7]. By establishing common standards, data terminology, and shared resources, the CTSA ACT program provides the opportunity to combine patient EHR records from multiple participant sites, which vastly improves the number of patient records available for analysis [7].

The AHRQ [8] hosts a number of databases supported through survey data. These include surveys that meet the AHRQ's goals of

- Supporting researchers and research networks that address concerns of very high public priority, such as health disparities, drugs and other therapeutics, primary care practice, and integrated health care delivery systems
- Supporting projects that test and evaluate successful methods that translate research

into practice to improve patient care in diverse health care settings
- Supporting Evidence-based Practice Centers that review and synthesize scientific evidence for conditions or technologies that are costly, common, or important to the Medicare or Medicaid programs
- Translating the recommendations of the US Preventive Services Task Force into resources for providers, patients, and health care systems

For example, the Medical Expenditure Panel (MEP) Survey is a set of large-scale surveys of families and individuals, their medical providers (doctors, hospitals, pharmacies, etc.), and employers across the United States. The MEP collects data on the specific US health services that include the type, cost, scope, and breadth of health insurance held by and available to US workers [9]. The survey, which began in 1996, has 589,666 sample providers.

Another AHRQ survey, the Healthcare Cost and Utilization Project (HCUP), is a collection of health care databases from the United States. One example, the State Inpatient Database (SID), captures hospital inpatient stays from 48 different states. The SID contains a core set of clinical and nonclinical information on 97% of all community hospital inpatient discharge abstracts in participating states, translated into a uniform format to facilitate multistate comparisons and analyses [10].

Finally, the NCHS is one of 13 principal statistical agencies in the federal government. The center has a number of downloadable public use data files that include national vital statistics (births, deaths, and so forth) and health care surveys on hospitals, clinics, hospices, home care, nutrition, immunizations, family growth, and aging. These surveys have sample sizes as large as 42,000 participants [11].

D. Patient-Reported Outcomes Measures (PROM)
Patient-reported outcomes measures (PROMs) are defined as "any report of the status of a patient's health condition that comes directly from the patient, without interpretation of the patient's response by a clinician or anyone else" [12]. A growing body of research has explored the role of PROMs in assisting patients make better health care decisions by assessing the *patients' personal* impression of the impact of disease or treatment on their physical, psychological, and social functioning and general well-being. Typical PROMs include patient satisfaction with clinical services, clinician decision-making, and clinical outcomes and treatment monitoring [13].

The amount of data collected from PROMs can be extensive. For example, the Patient-Centered Outcomes Research Institute (PCORI), an independent, nonprofit, nongovernmental organization, utilizes patient-powered research networks (PPRNs), which comprise patients and/or caregivers who are providing data for effectiveness research. There are 20 PPRNs listed at the PCORI website, each with participants providing PROM data on specific disease conditions. By combining EHR and PROM data, the PCORI has over 42 million patient-reported outcomes [14].

E. Clinical Registries
A registry is a collection of information about individuals, usually focused around a specific diagnosis or condition. Many registries collect information about people who have a specific disease or condition, while others seek participants of varying health status who may be willing to participate in research about a particular disease. Individuals provide information about themselves to these registries on a voluntary basis. Registries can be sponsored by a government agency, nonprofit organization, health care facility, or private company. The National Institutes of Health has over 50 registries listed on its website and is an excellent source for big data studies of specific disease conditions. For example, the Breast Cancer Family Registry (CFR) includes lifestyle, medical history, and family history data collected from more than 55,000 women and men from 14,000 families with and without breast cancer. The Breast CFR began recruiting families in 1996, and all participants are followed-up 10 years after recruitment to update personal and family histories and expand recruitment if new cases have occurred since baseline [15].

F. Genetic Data
A growing source of health care data today is from sequencing the human genome to discover those at risk for specific disease conditions. Determining the order of DNA building blocks (nucleotides) in an individual's genetic code, called DNA sequencing, has advanced the study of genetics and is one method used to test for

genetic disorders. Finding small variations in the genetic code may help predict a person's risk of particular diseases and response to certain medications. By developing and applying genome-based strategies for early detection and diagnosis, providers can better treat and in some cases prevent the onset of disease. Although the volume of data analyzed for whole-gene sequencing is enormous (6 billion diploid pairs/genome) new methods to generate, store, analyze, and visualize "omics" data have reduced the time and cost of profiling this data by a factor of 1 million [16]. Thus, integrating genomic data will demand high-performance computational environments supported by genome centers that have the capacity to support Exabyte's (10^{18} bytes) of data in the next 5 years.

G. Sensor Data

Improvements in sensor technology have also created new opportunities to generate, store, and analyze health care data. For example, smartphone software applications can now allow individuals to wirelessly monitor their own pulse, respirations, blood pressure, electrocardiogram, ultrasound images, heart sounds, sleep patterns, exercise activity, lab tests, and a host of other physiologic parameters. This data can then be streamed to the *cloud*, where it can be stored for secondary data analysis. Given that there are an estimated 6 billion cellular phones in the world today, the volume, velocity, and variety of data streaming from sensors is enormous. It is estimated that by the year 2019, there will be 2.9 billion sensors integrated with devices such as in-home monitors, smartphones, vehicles, and so forth [17]. As a result of this, scientists utilizing data science approaches are investigating patterns that in combination with data from sensors are providing both clinicians and patients insights that can improve both health and well-being. These patterns may include correlations between chronic disease and the quality and quantity of sleep, activity levels, nutrition, and stress levels. In-home sensors such as motion detectors are now being used to monitor patients with memory loss in their homes. A subtle change in activity, such as frequent trips to the bathroom, can indicate a possible urinary tract infection. Or, a decrease in activity may suggest a progressive problem with ambulation and increased risk of fall.

Traditionally, patients who required physiologic monitoring needed to be close to specialized equipment and clinicians located in hospitals or clinics. However, with improvements in sensor technology, much of this monitoring can now be accomplished outside of the hospital in patient homes. This includes implanted devices such as insulin pumps, pacemakers, and other forms of technology. Thus, the potential for continued growth in streaming health care data from sensors remains very high.

H. Social Media

The use of data generated from social media is an emerging new source of information regarding health care. For example, Facebook, the number one social media site, currently hosts over 1.59 billion active users [18]. Twitter has over 310 million monthly active users [19]. while LinkedIn has 255 million active users [20]. By using natural language processing, text analysis, and computational linguistics, researchers are conducting sentiment analysis studies using data sourced from social media websites. Sentiment analysis focuses on identifying the attitude of an individual or collection of individuals with regard to a specific topic. In health care, this may involve identifying individuals at risk for depression in online support groups or attitudes toward various treatments for a disease condition. Sentiment analysis remains a very active area of research in health care today.

II. Factors Driving the Creation of Big Data in Health Care

A. The HITECH Act, Meaningful Use, and ACOs

In addition to new sources of data as previously described, a number of additional factors have contributed to the increased creation of big data in health care. These factors include implementation of the HITECH Act (Health Information Technology for Economic and Clinical Health), the emergence of accountable care organizations (ACOs), and new database architectures that have the capacity to store and process big data. The accelerated transition of paper to electronic health records as a result of the HITECH Act has had a profound impact on the increased storage of clinical information in a digital format. The HITECH Act, enacted as part of the American Recovery and Reinvestment Act of 2009, was signed into law on February 17, 2009, to promote the adoption and meaningful use of health information technology. Meaningful Use is defined as using certified EHR technology to [21]

- Improve quality, safety, and efficiency, and reduce health disparities
- Engage patients and family
- Improve care coordination, and population and public health
- Maintain privacy and security of patient health information

Meaningful Use's goal is to provide better clinical outcomes through improved population health outcomes, increased transparency and efficiency, empowered individuals, and more robust research data on health systems. The Centers for Medicare and Medicaid Services (CMS) has set specific objectives that eligible professionals and hospitals must achieve to qualify for CMS financial incentive programs. Collectively these incentive programs have accelerated the adoption of EHRs in nonfederal acute care hospitals from 9.4% in 2008 to 96.4% in 2014 [22]. As can be imagined, the conversion of once-paper medical records to a digital format has rapidly expanded the volume and variety of health care data available for knowledge discovery and data mining (KDDM).

At the same time that Meaningful Use has accelerated the adoption of EHRs, the Affordable Care Act has spurred the growth of large integrated health systems across the US. The Affordable Care Act is a federal statute signed into law in March 2010 as a part of the health care reform agenda of the Obama administration. Among its multiple provisions, the Act encouraged the formation of ACOs or networks of providers, hospitals, clinics, and other health care entities to better coordinate care and eliminate unnecessary spending. At least 744 organizations have become ACOs since 2011, serving an estimated 23.5 million Americans [23].

A key requirement of ACOs is information sharing among clinicians, patients, and authorized entities from multiple clinical, financial, operational, and patient-reported sources within and across the system [24]. To do so requires the collection and integration of clinical and administrative data in centralized data repositories. As more and more hospitals and clinics consolidate to grow ever-larger health systems, the size of their data repositories are also expanding. For example, Kaiser Permanente's database of 9 million patients has approximately 30 petabytes of stored data and is adding 2 additional terabytes per day [25].

B. Database Architecture

Traditional health care information system database architecture has relied primarily on structured data that can be stored in connected tables of columns and rows. Because of regulatory policies and billing requirements, the structured data elements most commonly stored in electronic repositories are codes that map to standardized terminologies such as the International Classification of Diseases, Tenth Revision, Clinical Modification (ICD-10-CM), Logical Observation Identifiers Names and Codes (LOINC), Systematized Nomenclature of Medicine—Clinical Terms (SNOMED-CT), Current Procedural Terminology (CPT), and so forth. These standardized terminologies identify diagnostics, procedures, lab tests, radiology tests, medications, clinical documentation of providers, and other data. Standardized terminologies grew out of a need to document medical care by providers in health care facilities for insurance purposes. Much work has gone into the development of standardized terminologies for clinical documentation, and as a result, conducting searches for big data analysis can be accomplished through Structured Query Language (SQL) searches and is, for the most part, straightforward.

However, as the volume, variety, and velocity of data have increased, health care system database architecture has been slow to adapt to new ways of storing and processing data. For example, it is difficult to store streaming physiologic data from sensors in the rows and columns of traditional relational database tables. The data takes up an enormous amount of storage; it is unstandardized and has frequent gaps in it. Thus, preprocessing the data for analysis is very time consuming, and search methods using SQL are problematic. In addition, although the structured data in traditional databases is searchable, it only constitutes about 20% of the actual documentation in EHRs [26]. The remainder is in unstructured data, which is generally unstandardized and difficult to preprocess and search.

With the advent of cloud computing, the increased use of distributed databases, and parallel processing, health care systems are now adapting their database structures to be more in line with social media giants such as Google, Facebook, and LinkedIn. The transition to new data base architectures such as Hadoop, an open-source distributed data storage framework created by Yahoo!, and MapReduce, an algorithm to query the data, developed by Google, is slowly occurring in health systems [27]. These changes are enabling the ongoing growth of big data in health care.

III. Benefits of Big Data Science in Health Care

As described above, it is evident that traditional and new sources of health care data are meeting the four V's of big data: high volume, variety, velocity, and veracity. And although this can provide unique challenges in storing and preprocessing health care data, the opportunities to mine the data for new knowledge are immense. The following are a series of research areas that illustrate some of the potential benefits of big data science in health care today.

A. Precision Medicine

Precision medicine (PM) is an emerging approach for disease treatment and prevention that takes into account individual variability in genes, environment, and lifestyle for each person [28]. Associating an individual's genetic, environmental, and biological factors with specific disease conditions is a key goal of PM. To support research in this area, President Obama in January 2015 announced, during his State of the Union Address, the Precision Medicine Initiative (PMI), which allocates $130 million dollars to the National Institutes of Health and $70 million to the National Cancer Institute to build a national, large-scale research participant cohort group.

The concept of individual variability in the prevention and treatment of disease is not a new concept in medicine. However, the broad application of PM has recently been enabled by the creation of large-scale biologic databases (such as the human genome sequence), powerful methods for characterizing patients, and computational tools for analyzing large sets of data [29]. Add to this EHR, mobile devices, and lifestyle data, and the goals of PM edge closer to reality. A good illustration of the benefits of PM can be seen from potential studies beginning to emerge from the Kavali HUMAN Project (KHP) [30].

The KHP is unique in that researchers plan to measure as many aspects as possible of 10,000 New York City resident volunteers in approximately 2500 households over a 10-year period. The longitudinal study's goal is to provide new insights and understanding into how the interaction of biology, environment, and human behavior determines health. Data sets will be gathered from genome, microbiome, metabolome, epigenetics, medical records, drug and chemical exposure testing, diet assessment, sleep behavior, physical activity levels, and home air and noise quality. Additional data being considered are personality traits, IQ, mental health status, long-term and working memory, social network structure, communication partners and patterns, continuous location capture, educational history, employment data, and detailed financial transaction records.

The KHP offers the opportunity, among many, to better understand the association between genes and addictive behavior such as smoking, alcoholism, and recreational drug use through genome-wide association studies (GWASs). A GWAS is an approach that involves rapidly scanning markers across the genomes of many people to find genetic variations associated with a particular disease [31]. Previous studies on the association between genes and smoking behavior have been challenged by the lack of accurate measurements on various facets of smoking such as initiation, intensity, and cessation. These facets may be affected by different biologic and environmental factors. For instance, there is strong evidence that a set of single-nucleotide polymorphisms (SNPs) located in the chromosome 15 cluster of virtually adjacent nicotinic receptor genes is a risk factor for heaviness of smoking as well as the strongest genetic risk for development of lung cancer [31]. An SNP is a variation in a single nucleotide that occurs at a specific position in the genome and can be used as a biomarker for certain disease conditions. However, current measurements such as the maximum level of smoking at a point in time provide only limited information to researchers.

Through disparate data sources collected through the KHP, researchers have the ability to mesh multiple databases and create behavioral phenotypes that better determine the relationship between biology, environment, and lifestyle on smoking behavior. For example, self-reported smoking quantities can be cross-checked directly through lab tests (saliva cotinine levels) and indirectly through financial records such as credit cards, EHRs, and other data sources. By inclusion of these factors with GWASs, the KHP provides an opportunity to identify the genes that, in combination with environmental and lifestyle factors, account for variation across individuals and can explain how environmental factors can change genetic risk [31].

Although GWASs have vastly improved identification of SNPs, the availability of larger and larger data sets will significantly increase the number of polymorphisms for a specific disease condition. These multiple associations can be regarded as a consequence or as a benefit of big data. Regardless, exciting research investigating the relationship of behavioral, biological, and environmental measures with gene discovery continues to accelerate. The ability to integrate multiple databases regarding an individual's data supports the concept of PM. The KHP project is one emerging example of how big data can make PM possible.

B. Predicting Disease Risk

The ability to study the historic trajectory of a disease can help determine the underlying risk of developing it. The advent of EHRs, with their rich source of clinical data, has allowed investigators to track the clinical signs and symptoms of multiple trajectories for specific disease conditions. Since many diseases follow a similar progression over time, inferences can be made, and predictive models can be developed. By developing predictive models, clinicians can initiate interventions sooner and, in some cases, prevent disease progression. To illustrate this, Wonsuk et al. [32] used a novel method to demonstrate the typical trajectory of type 2 diabetes mellitus (T2DM) and predict disease risk. Using 13 years of EHR follow-up data, the authors were able to study the sequence of comorbidities among multiple patients later diagnosed with T2DM. The three comorbidities tracked were high cholesterol (HDL), hypertension (HTN), and impaired fasting glucose (IFG).

Using a retrospective observational study design, investigators developed phenotyping algorithms to classify whether or not T2DM patients had existing comorbidities at baseline. After inclusion and exclusion criteria were met, investigators tracked the progression of comorbidities of 43,509 patients from baseline till a diagnosis of T2DM was made. For patients who had already developed comorbidities at baseline but fell outside the range of available EHR data, the probability of following a certain sequence of comorbidities was estimated by calculating the probability of a trajectory. To measure the association between different trajectories and the risk of developing T2DM, a multivariate logistic regression model was constructed using demographics, glucose level, staged comorbidities, and three trajectories.

Results of the study provided the likelihood of developing T2DM following a typical versus an atypical trajectory. The most likely sequence of comorbidities, or *typical* trajectory, was HLD, HTN, IFG, and T2DM (27%). The greatest number of patients followed this series of comorbidities prior to a diagnosis of T2DM. However, the study also demonstrated that patients can follow a different sequence, or an *atypical* trajectory, and still be diagnosed with T2DM (15%).

The use of large, complex sets of EHR data is one illustration of how big data science is providing an alternative to traditional research methods such as clinical trials. Small sample sizes, lengthy trial periods, and high costs have limited the number of trials that can be conducted. However, using a data cohort of 43,509 patient EHR records, the authors used a novel method to overcome many of the challenges of retrospective observational studies using EHR data: unreliable diagnostic codes and identification of the onset of a disease condition such as T2DM. Given that T2DM is one of the fastest-growing public health concerns in the nation, identifying phenotypes characteristic of typical and atypical trajectories leading to its diagnosis is a key to early intervention.

C. Coordination of Care

As previously described, health insurance claims repositories are a robust source of data that can be used in retrospective observational research studies. One limitation of claims data is that its primary use is for validating reimbursable services and it does not contain the richness of unstructured clinical progress notes found in EHRs, survey data from public databases, and personal interviews. However, these other sources of data can be used to support inferences from claims data and help explain the outcomes. In their article, "Big data, little data, and care coordination for Medicare beneficiaries with Medigap coverage," Ozminkowski et al. [33] used a combination of claims data, survey data, and personal interviews to identify patients that had a high propensity for successful care coordination. Care coordination involves deliberately organizing patient care activities and sharing information among all of the participants concerned with a patient's care to achieve safer and more effective care [34].

The setting for this study was a high-risk case management (HRCM) care coordination program, named MyCarePath, that included seniors in parts of California, Florida, New York, North Carolina, and Ohio, who participated in the American Association of Retired Persons (AARP) Medigap insurance program [33]. The AARP Medigap insurance program supports about 4 million adults across the US. The MyCarePath program offers individualized care coordination for selected individuals who meet certain criteria such as congestive heart failure, depression, coronary artery disease, diabetes, and other chronic conditions. Using a combination of claims data, health risk appraisal surveys, and telephone interviews, analysts were able to utilize algorithms to classify individuals who meet criteria for the program. Individualized care plans were then developed and informed by indicators of illness and gaps in medical care found in the large data repositories of medical and pharmacy claims.

Although the MyCarePath program provides strategies to obtain higher-quality health care while saving money for both the individual and AARP Medigap, the data indicated that certain indicators predicted a higher success rate or a *propensity to succeed* (PTS). Using three dichotomous variables (patient engagement, financial success, and likelihood of receiving quality care), the authors used logistic regression to estimate participants' PTS scores. Of the approximately 40,000 individuals per month classified as meeting the requirements for the MyCarePath program, 4500 had PTS scores that prompted care coordinators to contact them for outreach programs. Early evaluation of the PTS algorithm has demonstrated an 11% increase in new participants to the program over a 9-month period since it was implemented. This example demonstrates how secondary use of claims big data can be used to create predictive models and aid in clinical decision support.

D. Emergency Department Revisits

Frequent patient revisits to the emergency department (ED) place an excessive burden on hospitals and contribute to the rising cost of health care. Moreover, the Affordable Care Act financially penalizes institutions exceeding set revisit rates [35]. which can have a significant negative impact on hospitals revenue. Therefore it is important to understand the factors that contribute to patient revisits to identify patients at risk and intervene in advance.

The following study investigated patient factors that contributed to frequent 72-hour ED revisits (defined as an ED encounter leading to a subsequent inpatient admission in the past 6 months) and how these factors differ between other ED patients [36]. The data set consisted of 15,000 potential factors such as primary and secondary ICD-9-CM codes, discharge statuses, patient demographics, and treatment dates and times, extracted from 1,149,738 EHR patient files from a single institution, spanning a period of 2 years. Through expert opinion and statistical analysis, the 15,000 factors were mapped to 385 discrete and continuous features. Feature compression methods included lower-dimensional mapping or binning of variables such as primary and secondary ICD-9 diagnosis codes, discharge statuses, and zip codes.

To determine the impact of the 385 factors, log odds ratios were examined on patients with an ED encounter that resulted in a 72-hour revisit versus no resulting 72-hour revisit. The same methods were then applied to those patients that met criteria for infrequent versus frequent 72-hour revisits to compare differences.

Preprocessing and cleaning of the data resulted in 50,127 ED encounters, of which 2735 (5.46%) met the criteria for a frequent 72-hour revisit event (slightly higher than the general revisit rate of 4.51%). To understand the impact of individual features on frequent 72-hour revisit patients, infrequent patients, and patients without a 72-hour revisit, predictive modeling methods (log odds ratios and logistic regression) were performed. The experiments were repeated using a nonlinear classification method, random forests, which resulted in no significant difference in predictive accuracy.

Results of the analysis showed that the highest frequency diagnosis codes for infrequent 72-hour revisits consisted of patients admitted for psychiatric disorders such as alcoholism and drug dependency, as well as skin and subcutaneous tissue infections, complications with pregnancy, leaving against medical advice, and HIV.

Frequent 72-hour revisit patients tended to have three or more diagnoses, with the most frequent related to use of alcohol, substance abuse, mental health, headache, back problems, and HIV.

The study demonstrated significant differences in the patterns of Medicaid patient encounters that result in a 72-hour revisit to the ED versus those that do not. This information is valuable to hospitals as they can identify the factors most likely to result in a readmission to the ED. Such high-risk patients can be scheduled for follow-up postdischarge appointments with medical providers and monitored by care coordinators to ensure they comply with their appointments.

IV. Big Data and Smart Medical Devices

The use of machine learning and other computational methods to classify disease conditions and recommend interventions is a robust area of research today. Patterns or signals discovered in large-scale data sets can be embedded in software as algorithms used for clinical decision support. As the technology for mobile devices advances, the software embedded in them provides the opportunity for patients to receive care once exclusively limited to hospitals or clinics, to be provided in the home. The following section discusses the emerging role of smart medical devices and how big data science is creating artificial intelligence that will transform medical practice in the future.

A. Electronic Stethoscopes

A mainstay tool of clinical providers has been the acoustic stethoscope, which is primarily used to auscultate arterial pressure waves used for measuring blood pressure as well as heart valve sounds and lung sounds. With training and over time, clinicians become experts in recognizing abnormal heart and lung sounds, which can often be the first sign of a progressing disease condition. A major drawback of an acoustic stethoscope is that in certain patients, the sound level is too low to discriminate normal from abnormal sounds. A second challenge is the time it takes to build expertise in accurately assessing abnormal sounds. It can take years of training to accurately assess cardiac and pulmonary problems with an acoustic stethoscope. These problems are being addressed through the emergence of electronic stethoscopes, which use sensors in the diaphragm of the stethoscope to capture heart sounds and convert them into electrical signals, which can then be amplified. Studies have demonstrated that electronic stethoscopes capture heart sounds better than acoustic stethoscopes [37].

A benefit of capturing heart sounds as electronic signals is that the analog signal can be converted to streaming digital data points. Through machine learning approaches such as neural networks and support vector machines, the patterns in the data points can be coded as algorithms and used to classify normal and abnormal heart sounds. The algorithms can then be programmed in microchips located in the stethoscope or an attached smartphone. Although in their early stages, the capacity of electronic stethoscopes to accurately diagnose heart sounds is impressive. Studies have shown that the mean sensitivities and specificities for diagnosing aortic regurgitation, aortic stenosis, mitral regurgitation, and mitral stenosis are 89.8 and 98.0%, 88.4 and 98.3%, 91.0 and 97.52%, and 92.2 and 99.29%, respectively [38]. The use of electronic stethoscopes to diagnose abnormal heart sounds is one example of how big data science will augment clinical providers' cognition and support their clinical decision making.

B. Point-of-Care Lab Testing

Point-of-care (POC) lab testing is the use of mobile devices to analyze blood samples at the bedside, clinic, or home. Advances in microfluidics, a multidisciplinary field that includes engineering, physics, chemistry, biochemistry, nanotechnology, and biotechnology, are increasing both the number and complexity of POC tests available on the market today. Low-volume blood samples in combination with optoelectronic image sensors located in smartphones are dramatically growing the number of POC tests being administered. Cell phones can now support megapixel counts that can act as general-purpose microscopes capable of identifying a single virus on a chip.

POC devices are on the verge of detecting viruses, diagnosing infectious diseases, identifying allergens, detecting protein binding events, analyzing tumors, and providing other tests. The growing global network of connected smartphones will create large-scale data warehouses of lab results that can be shared by researchers. These large data warehouses should allow researchers to use data science methods to quickly classify different pathologies to diagnose and track the evolution of infectious outbreaks and epidemics much faster [39].

C. Activity Monitors

The quantified self-movement, defined as the incorporation of technology for data acquisition on aspects of a person's daily life, has created new sources of real-time data that can be used for analysis of an individual's health. One example is the exponential growth in the use of activity monitors. Improvements in microelectromechanical (MEM) accelerometers and gyro sensors have popularized activity monitors by more accurately measuring the energy expended in such activities as walking and jogging [40]. These sensors can convert the analog signals of motion, velocity, and orientation to streaming digital data points that can be analyzed using data science approaches. The intensity of exercise can then be classified using a variety of methods such as naive Bayes classifiers, Markov chains, and k-means clustering [41]. A number of companies are combining data from multiple monitors that capture diet, medications, sleep, stress, work schedule, and other elements to customize fitness, stress reduction, and diet programs.

D. In-Home Monitors

An emerging trend is the use of in-home monitors to assist caregivers in keeping patients with chronic disease in their homes longer. Sensors placed on chairs, beds, refrigerators, walls, carpets, and other areas allow nurses and family members to monitor the daily habits of seniors with memory loss or at high risk for falls. Streaming data from sensors allow nurses and family caregivers to investigate unusual patterns that might indicate a problem. These problems may include a subtle change in how often a refrigerator door is opening or how long a resident stays in a chair or bed, patterns that may indicate a developing problem with diet or mobility.

The use of big data science methods has played a significant role in aiding nurses to recognize patterns and provide insight into a developing problem. In a recent study [42], investigators were able to extend residents' length of stay (LOS) in an independent living facility, in part, through the use of in-home sensors. The purpose of the study was to compare the LOS of residents living with an environmentally embedded sensor to the LOS of those without the sensors. Through a collaboration with the Missouri University Sinclair School of Nursing and the Center for Eldercare and Rehabilitation Technology (CERT) in the College of Engineering, an embedded sensor system was developed to monitor health status among senior residents in a 54-apartment facility. All sensors were nonwearable and included (1) bed sensors to monitor heart rate, respiratory rate, and nighttime bed restlessness; (2) motion sensors to monitor activity in rooms; and (3) kinect depth images to automatically monitor walking and gait parameters and report falls in real time with alerts e-mailed to direct care staff [42].

Computer algorithms were designed to monitor a rolling 14-day period and compare differences in patterns [43]. These patterns were classified as an early indicator for a new or progressing disease condition. For example, advancing heart failure may show a pattern of an elevated heart rate coupled with increased nighttime bed restlessness; early signs of a urinary tract infection may present as increased nighttime bed restlessness and increased time in the bathroom. A change in activity or sleep patterns may indicate progressing dementia or depression. As unusual patterns were recognized by the computer algorithms, health alerts were automatically sent to nurses to evaluate for early intervention. The computer program was able to successfully classify patterns for multiple illnesses such as pneumonia, upper respiratory infections, congestive heart failure, posthospitalization pain, delirium, hypoglycemia, and urinary tract infections.

Results of the 4.8-year study indicated that there was a significant difference ($p = .0006$) in LOS between the group living with sensors ($n = 52$) and those living without ($n = 81$). The group with sensors had an average LOS of 1557 days (4.3 years), while for the comparison group without sensors, it was 936 days (2.6 years). Groups were comparable based on admission age, gender, number of chronic illnesses, short form 12 (SF12) physical health, SF12 mental health, Geriatric Depression Scale (GDS), activities of daily living, independent activities of daily living, and mini-mental status examination scores. The study highlighted the benefits of advanced sensor technology combined with big data science to allow seniors to safely "age in place."

V. Conclusions

The emergence of large complex data sets, or big data, in health care is being fueled by a number of key factors.

These include the Affordable Care Act, which includes financial incentives for hospitals and clinics to adopt EHRs, cloud computing, improvements in sensor technology and the quantified self-movement, telehealth, advances in gene sequencing, new database architectures and microchip processing speed, and the relentless growth of the Internet. The opportunities to discover patterns or new knowledge hidden in the data are immense. From precision medicine to predictive models to clinical decision support, the success of health care institutions at improving clinical and administrative outcomes will depend, to a large extent, on their capacity to harness big data. Many challenges exist today and include a shortage of data scientists, unstandardized clinical documentation, legacy database architectures, limited interoperability between EHR systems, and privacy and security issues to name a few. However, although these barriers are steep, health care is rapidly transitioning to meet these challenges and discover the immense potential of big data.

References

1. Health IT Government website accessed. https://www.healthit.gov/providers-professionals/meaningful-use-definition-objectives (accessed July 16, 2016).
2. Centers for Medicare and Medicaid website. https://www.cms.gov/Research-Statistics-Data-and-Systems/Statistics-Trends-and-Reports/BSAPUFS/Downloads/2008_Enrollment_and_User_Rates.pdf (accessed June 3, 2016).
3. Health Care Cost Institute website. http://www.healthcostinstitute.org/about (accessed July 27, 2016).
4. Mini-Sentinel website. http://www.mini-sentinel.org/ (accessed July 15, 2016).
5. OptumLabs website. https://www.optumlabs.com/partners.html (accessed June 4, 2016).
6. Clinical Translational Science Award website. https://ctsacentral.org/(accessed July 7, 2016).
7. Clinical Translational Science Award Accrual to Clinical Trials website. http://www.ncats.nih.gov/ctsa/about (accessed July 7, 2016).
8. Agency for Healthcare Research and Quality (AHRQ) website. http://www.ahrq.gov/health-care-information/index.html (accessed July 18, 2016).
9. Medical Expenditure Panel website. https://meps.ahrq.gov/mepsweb/about_meps/survey_back.jsp (accessed July 18, 2016).
10. Healthcare Cost and Utilization Project website. https://www.hcup-us.ahrq.gov/sidoverview.jsp#about (accessed July 25, 2016).
11. National Health Interview Survey website. https://www.cdc.gov/nchs/data_access/ftp_data.htm (accessed June 18, 2016).
12. National Quality Forum website. http://www.qualityforum.org/Projects/n-r/Patient-Reported_Outcomes/Patient-Reported_Outcomes.aspx (accessed July 30, 2016).
13. Locklear, T., Flynn, K., and K. Weinfurt. 2016. Impact of patient-reported outcomes on clinical practice. *NIH Collaboratory*. https://www.nihcollaboratory.org/Products/Impact_of_Patient-Reported_Outcomes_on_Clinical_Practice_Jan_19_2016_Version_1.pdf (accessed July 18, 2016).
14. Patient-Centered Outcomes Research Institute website. http://www.pcori.org/ (accessed July 30, 2016).
15. National Institutes of Health Breast Cancer Family Registry website. https://www.nih.gov/health-information/nih-clinical-research-trials-you/list-registries (accessed July 15, 2016).
16. Costa, F.F. 2016. Big data in genomics: Challenges and solutions. Is life sciences prepared for a big data revolution? *GIT Lab J* 11–12/2012. http://genomicenterprise.com/yahoo_site_admin/assets/docs/Costa_GLJ11-1212_-_3.330103806.pdf (accessed July 18, 2016).
17. Kharif, O. 2013. Trillions of smart sensors will change life. *Bloomberg Technology*. http://www.bloomberg.com/news/articles/2013-08-05/trillions-of-smart-sensors-will-change-life-as-apps-have (accessed July 21, 2016).
18. Statista website. http://www.statista.com/statistics/272014/global-social-networks-ranked-by-number-of-users (accessed July 18, 2016).
19. Statista website/Twitter Users. http://www.statista.com/statistics/282087/number-of-monthly-active-twitter-users/(accessed July 18, 2016).
20. Statista website/LinkedIn Users. http://www.statista.com/statistics/282087/number-of-monthly-active-LinkedIn-users/(accessed July 18, 2016).
21. Health IT Government Providers—FAQs website. https://www.healthit.gov/providers-professionals/faqs/what-information-does-electronic-health-record-ehr-contain (accessed July 16, 2016).
22. Dustin, C., Gabriel, M., and T. Searcy. 2015. Adoption of electronic health record systems among U.S. Non-Federal Acute Care Hospitals: 2008–2014. ONC Data Brief. https://www.healthit.gov/sites/default/files/data-brief/2014HospitalAdoptionDataBrief.pdf (accessed June 4, 2016).
23. Gold, J. 2015. Accountable care organizations explained. *Kaiser Health News*. http://khn.org/news/aco-accountable-care-organization-faq/ (accessed June 8, 2016).
24. Certification Commission for Health Information Technology. 2013. A Health IT Framework for Accountable Care. https://www.healthit.gov/FACAS/sites/faca/files/a_health_it_framework_for_accountable_care_0.pdf (accessed July 1, 2016).
25. Jaret, P. 2013. Mining electronic records for revealing health data. *New York Times Online*. http://www.nytimes.com/2013/01/15/health/mining-electronic-records-for-revealing-health-data.html?_r=2 (accessed July 25, 2016).
26. Wels-Maug, C. 2012. Unlocking the potential of unstructured medical data. Ovum.com. ovum.com/2012/05/11/unlocking-the-potential-ofunstructured-medical-data (accessed June 3, 2016).

27. Crapo, J. 2016. Hadoop in healthcare: A no-nonsense Q and A. Health Catalyst. https://www.healthcatalyst.com/Hadoop-in-healthcare (accessed July 28, 2016).
28. National Institutes of Health Precision Medicine Initiative Cohort Program. https://www.nih.gov/precision-medicine-initiative-cohort-program (accessed July 9, 2016).
29. Collins, H. and H. Varmus. 2015. A new initiative on precision medicine. *N Engl J Med* 372:793–795.
30. Kavli HUMAN Project website. http://kavlihumanproject.org/ (accessed July 30, 2016).
31. Bierut, L. and D. Cesarini. 2015. How genetic and other biological factors impact smoking decisions. *Big Data* 3(3):198–202. doi: 10.1089/big.2015.0013.
32. Oh, K., Kim, E., Castro, R., Caraballo, J., Kumar, V., Steinbach, M.S., and G.J. Simon. 2016. Type 2 diabetes mellitus trajectories and associated risks. *Big Data* 4(1):25–30. doi: 10.1089/big.2015.0029.
33. Ozminkowski, R.J., T.S. Wells, K. Hawkins, and G.R. Bhattarai. 2015. *Big Data* 3(2):114–125 doi: 10.1089/big.2014.0034.
34. Agency for Healthcare Research and Quality (AHRQ) website, Care Coordination. http://www.ahrq.gov/professionals/prevention-chronic-care/improve/coordination/index.html (accessed July 30, 2016).
35. Patient Protection and Affordable Care Act. 2010. 42 USC, 18001 et seq.
36. Ryan, R., Hendler, J. and K.P. Bennett. 2015. Understanding emergency department 72-hour revisits among Medicaid patients using electronic healthcare records. *Big Data* 3(4):238–248. doi: 10.1089/big.2015.0038.
37. Tourtier J.P., N. Libert, P. Clapson, K. Tazarourte, M. Borne, L. Grasser et al. 2011. Auscultation in flight: Comparison of conventional and electronic stethoscopes. *Air Med J* 30(3):158–160. doi: 10.1016/j.amj.2010.11.009.
38. Shuang L., Ru San Tan, K., Tshun C.C., Chao W., Dhanjoo, G., and L. Zhong. 2014. The electronic stethoscope. *Biomed Eng Online* 14(66):1–37. http://gooa.las.ac.cn/external/share/1218610 (accessed July 15, 2016).
39. News Medical website. 2015. New report on global market for point of care diagnostics. http://www.news-medical.net/news/20150626/New-report-on-global-market-for-point-of-care-diagnostics.aspx (accessed June 12, 2016).
40. Ferguson, T., Rowlands, A., Olds, T., and C. Maher. 2015. The validity of consumer-level, activity monitors in healthy adults worn in free-living conditions: A cross-sectional study. *Int J Behav Nutr Phys Act* 12:42. doi: 10.1186/s12966-015-0201-9.
41. Mannini, A. and A. Maria Sabatini. 2010. Machine learning methods for classifying human physical activity from on-body accelerometers. *Sensors* 10:1154–1175. doi: 10.3390/s100201154.
42. Rantz, M., Lane, K., Lorraine, J.P., Despins, L.A., Colleen, G., Alexander, G.L., Koopman, J., Hicks, L., Skubic, M., and S.J. Miller. 2015. Enhanced registered nurse care coordination with sensor technology: Impact on length of stay and cost in aging in place housing. *Nurs Outlook* 6(3):650–655.

3

Big Data Era in Magnetic Resonance Imaging of the Human Brain

Xiaoyu Ding, Elisabeth de Castro Caparelli, and Thomas J. Ross

CONTENTS

3.1 Background on Magnetic Resonance Imaging Techniques 21
    3.1.1 MRI Signal Types 22
        3.1.1.1 Structural MRI 22
        3.1.1.2 Diffusion Tensor Imaging 22
        3.1.1.3 fMRI 22
3.2 Big Data Projects of the Human Brain 23
    3.2.1 Human Connectome Project 23
        3.2.1.1 WU–Minn–Oxford Consortium 23
        3.2.1.2 MGH/Harvard–UCLA Consortium 24
    3.2.2 1000 Functional Connectomes Project (FCP) 24
    3.2.3 Federal Interagency Traumatic Brain Injury Research 25
    3.2.4 National Database for Autism Research 25
    3.2.5 SchizConnect 26
    3.2.6 AD Neuroimaging Initiative 27
    3.2.7 IMAGEN 27
3.3 Machine Learning for Structural and Functional Imaging Data 28
    3.3.1 Statistical Comparison vs. Machine Learning Classification 28
    3.3.2 Feature Extraction and Selection on Brain Imaging Data 28
    3.3.3 Commonly Used Classifiers 30
    3.3.4 Performance Evaluation and Pattern Interpretation 31
    3.3.5 Diagnostic Predictions of Neurological and Psychiatric Diseases 32
3.4 Challenges and Future Directions 36
    3.4.1 Overfitting 36
    3.4.2 Heterogeneity in Patients and Disease Subtype Classification 39
    3.4.3 Multimodal Data Fusion 40
    3.4.4 Translation to Real Clinical Utility 41
Acknowledgements 42
List of Abbreviations in Tables (Sorted by Initial) 42
References 43

3.1 Background on Magnetic Resonance Imaging Techniques

Magnetic resonance imaging (MRI) is a non-invasive imaging technique that is characterized by superior soft-tissue contrast, providing detailed information on the anatomy and physiological processes of the human body in health and disease. Due to its non-ionizing nature, it offers a safe research tool, since experiments can be extensively repeated without side effects. For this reason, this technique has been extensively used in neuroscience research, allowing an indirect marker of neuronal activity to be recorded, non-invasively, while the human brain is involved in either a specific task or in the rest condition. In the next section, we will describe the basic types of MRI signals that have been shown to be useful in machine-learning applications.

3.1.1 MRI Signal Types

One of the main advantages of MRI techniques is the range of signal types that can be obtained even in the absence of exogenous contrast or tracer administration. Functional and anatomical MRI techniques, such as functional MRI (fMRI), structural MRI (sMRI), and diffusion tensor imaging (DTI), have the ability to evaluate different aspects of the living human brain. In particular, fMRI allows one to evaluate brain function, either when the human brain is involved in specific tasks or when it is at rest. Different MRI techniques have been applied in a variety of studies designed to explore cognitive functions and emotional regulation in the brain, to understand brain functional and anatomical connectivity, as well as to evaluate possible functional and anatomical alterations as a consequence of or an antecedent to neurological disorders or pharmacologic manipulations, including drugs of abuse.

3.1.1.1 Structural MRI

sMRI is used to obtain quantitative and qualitative information about the brain structure and anatomy. Contrast among the different brain tissue types can be optimized by modifying imaging parameters; for example, so-called T_1-weighted images will emphasize differences between gray and white matter, whereas T_2-weighted images will highlight differences between brain tissue and cerebrospinal fluid (CSF). Therefore, by using different pulse sequences, different aspects of normal and abnormal brain tissue can be highlighted.

Different characteristics of brain tissue, such as shape, size, and density, are evaluated using structural images. Morphometric techniques, such as volume/density (for example, voxel-based morphometry) [1] and cortical thickness [2], are commonly used to measure the volume or shape of gray matter tissue, while changes in white matter volumes provide some information about possible inflammation, edema, or demyelination. Additional information on the white matter quality is mostly obtained by using diffusion-weighted MRI techniques.

3.1.1.2 Diffusion Tensor Imaging

DTI is an MRI technique used to measure the three-dimensional diffusion of water in the tissue by describing the magnitude, the degree, and the orientation of the diffusion anisotropy, allowing the estimation of the white matter connectivity patterns in the brain [3,4].

In biological organisms, the diffusion of water is limited by the cellular structure of the tissue. It can be anisotropic, as in fibrous tissues, where water diffusion is basically unconstrained in the direction parallel to the fiber orientation, but highly restricted in the directions perpendicular to the fibers, such as in the white matter; or it can be isotropic, where water diffusion is mostly unconstrained in any direction, as in the gray matter. Therefore, DTI techniques can be used to characterize the level of diffusivity in the tissue [5].

In DTI, diffusion is typically characterized through two parameters: the apparent diffusion coefficient (ADC), which is a scalar measure of the total diffusion within a voxel (three-dimensional volume element), and the fractional anisotropy (FA), which measures the degree of anisotropy of a diffusion process in the voxel. FA values range between 0 (diffusion is completely isotropic) and 1 if it is completely constrained to one direction.

DTI is mostly used to evaluate white matter, where the location, orientation, and anisotropy of the tracts can be evaluated. This is because the architecture of the axons, which is in parallel bundles, and their myelin sheaths constrain the diffusion of the water molecules along their main direction. Consequently, the information about the water diffusion in the white matter can be used to perform tractography, allowing an estimation of white-matter connection patterns in three dimensions. Fiber tracking algorithms follow the coherent spatial patterns in the major eigenvectors of the diffusion tensor field to track a fiber along its whole length [4]. The ability of DTI to estimate fiber orientation and strength is increasingly accurate; however, it cannot directly image multiple fiber orientations within a single voxel (usually 3 mm isotropic). To address this limitation, diffusion spectrum imaging (DSI) [6] was developed to reveal the complex distributions of intravoxel fiber orientation, demonstrating the capability to image crossing fibers in neural tissue.

3.1.1.3 fMRI

fMRI, based on blood oxygenation level dependence (BOLD) contrast, uses information generated by the hemodynamic process to study the brain activity. This process is defined as a dynamic regulation of the blood flow and blood oxygenation in the brain, and its connection with neural activity has been known since the end of the nineteenth century [7]. Essentially, when the neuronal cells are active during a particular task, their oxygen consumption increase leads to an increase in the local blood flow, which occurs after a delay of about 1–5 s, rising to a peak after 4–5 s before falling back to baseline [8]. Since this increase in blood supply exceeds that required to offset oxygen consumption, there is a local change in the blood oxygenation level [9], which is detectable by MRI. Essentially, local hemoglobin in blood is acting as an endogenous MRI contrast agent.

Variation in blood flow and oxygenation level leads to change in tissue magnetic permeability, since hemoglobin is diamagnetic when oxygenated (oxyhemoglobin) but paramagnetic when deoxygenated (deoxyhemoglobin) [10], therefore inducing transient changes to the MRI signal. Particularly, the seemingly paradoxical decrease in deoxyhemoglobin concentration (caused by local overperfusion) in activated brain regions leads to a local increase in the MRI signal, which is the BOLD effect. Using fast imaging techniques sensitive to this effect, such as echo planar imaging (EPI) [11], changes in BOLD contrast can be mapped across the entire brain, revealing regional changes in brain activity.

In addition to the use of higher static magnetic fields [12], which enhances BOLD contrast-to-noise ratio [13,14], other recent technological advances have benefitted functional imaging. These include multichannel radio frequency (RF) reception [15] and the implementation of multislices (multiband) techniques [16] that have helped to improve fMRI data by improving spatial and temporal resolution. Currently whole brain functional images can be acquired in as short as a TR (repetition time—the time to acquire an entire volume) of 0.7 s and with a spatial resolution of 2 mm isotropic [17].

BOLD imaging has been extensively used to evaluate brain function either when the human brain is involved in specific tasks or when it is under the rest condition. The latter technique, looking at the spontaneous fluctuations in BOLD signal that reflect the resting brain activity [18], has garnered particular attention, as these scans are easily acquired, tend to be similarly acquired across imaging sites, and have been shown to recapitulate the same networks as task-based imaging [19]. BOLD imaging scans, along with anatomical and DTI scans, are increasingly being contributed to various databases, permitting "big data" analysis.

3.2 Big Data Projects of the Human Brain

The issue of reproducibility, which requires strong statistical power, has become an important issue in neuroimaging research; however, it can be difficult for single studies to achieve the required sample size for this purpose. For this reason, several efforts have been made to create databases where researchers can share their data and have the possibility to perform large-data analysis to address important scientific questions, improving our understanding of the human brain and its relationship to genetics and behavior.

Big data projects, such as the human connectome project (HCP), are building a more comprehensive map of the healthy human brain, while others, such as the 1000 functional connectome project, are addressing issues related to the reproducibility of findings across datasets and individuals. In addition, several data repositories (e.g., the Federal Interagency Traumatic Brain Injury Research [FITBIR] and the National Database for Autism Research [NDAR]) and virtual databases (e.g., the SchizConnect, which was designed to connect different data banks), as well as some longitudinal-multicenter studies (e.g., Alzheimer's Disease Neuroimaging Initiative [ADNI] and IMAGEN), have been initiated to expand collaborative and advanced research in distinct fields in order to improve treatment outcome. In the following, these initiatives are presented in detail.

3.2.1 Human Connectome Project

The HCP is a project sponsored by the National Institutes of Health (NIH) and is divided into two consortia of research institutions: (1) The WU–Minn–Oxford consortium, directed by three centers: Washington University in Saint Louis, the University of Minnesota, and Oxford University; and (2) The MGH/Harvard–UCLA consortium, directed by Harvard University, Massachusetts General Hospital, and University of California Los Angeles. The goal of the HCP is to map the human connectome as accurately as possible in a large number of normal healthy adults to better elucidate the anatomical and functional connectivity within the healthy human brain. These data are freely available to the scientific community in an open-source web-accessible format using a robust as well as user-friendly informatics platform [17,20].

3.2.1.1 WU–Minn–Oxford Consortium

The WU–Minn–Oxford consortium is a multicenter association involving the participation of many institutions in the United States and overseas: Washington University in Saint Louis, Center for Magnetic Resonance Research at the University of Minnesota, Oxford University (United Kingdom), Saint Louis University, Indiana University, University d'Annunzio (Chieti, Italy), Ernst Strungmann Institute (Frankfurt, Germany), Warwick University (United Kingdom), Radboud University Nijmegen (Nijmegen, Netherlands), Duke University, Advanced MRI Technologies, and the University of California at Berkeley [17]. This consortium has collected a large amount of MRI, behavioral, and genetic data on 1200 healthy young adults (twin pairs and their siblings) from 300 families, as well as MRI and behavioral data in healthy volunteers for different ages across the lifespan (HCP Lifespan Project) [21], covering six age groups (4–6, 8–9, 14–15, 25–35, 45–55, 65–75).

The MRI data were collected on three MRI scanners: (1) the "Connectome Skyra," which is a 3-T Skyra (MAGNETOM Skyra Siemens Healthcare) with a customized SC72 gradient insert (maximum gradient strength = 100 mT/m), housed at Washington University in St. Louis [22]; (2) a 3 T Prisma (MAGNETOM Prisma Siemens Healthcare—maximum gradient strength = 80 mT/m); and (3) a 7-T Siemens (maximum gradient strength = 70 mT/m), both housed at the University of Minnesota. These scanners were optimized to provide rapid high-resolution fMRI and DTI, using standardized imaging protocols. For example, by using the multiband EPI sequence [23], fMRI data were generated with a 2-mm isotropic resolution covering the whole brain with a TR of 0.7 s [22]. All 1200 participants from the young adult project were scanned in the Connectome Skyra, where resting-state fMRI (rsfMRI, in which a subject is scanned without performing any explicit task), task-evoked fMRI, and DTI were acquired; a subset of 200 subjects were also scanned in the 7-T scanner. Images for the HCP lifespan pilot project were acquired using 3-T Siemens systems, the Connectome Skyra (83 subjects) and the Prisma (75 subjects), and the 7-T Siemens scanner (51 subjects). Similar scanning protocols to those for the young adult project were used for this project, but with shorter duration [21].

Extensive demographic, behavioral (to assess sensory, motor, and cognitive functions), and genetic (blood samples) data were collected in each of the 1200 subjects scanned. Additionally, magnetoencephalography (MEG) and electroencephalography (EEG) images were acquired on a subset of 100 subjects, including both resting-state MEG/EEG and task-evoked MEG/EEG, using some of the tasks used during the task-evoked fMRI. Thus, data exist yielding information about brain function on a millisecond timescale provided by MEG/EEG as well as at the spatial specificity provided by fMRI.

Data are being analyzed using processing pipelines for maximal comparability and are integrated using a variety of analysis and visualization tools that were refined and optimized to achieve accurate within-subject alignment of datasets of multiple modalities, as well as precise intersubject registration. Results from these data analyses show the anatomical and functional connections between parts of the brain for each individual, and their relation with behavior. By comparing the connectomes and genetic data of identical twins vs. siblings, it will be possible to identify the specific genes that influence an individual pattern of connectivity, therefore revealing the contributions of genes and environment in shaping brain circuitry. The most recent data release for the young adult HCP, in December 2015, included behavioral and 3-T MRI data from over 900 healthy adult participants [24].

3.2.1.2 MGH/Harvard–UCLA Consortium

The MGH/Harvard–UCLA consortium [20] was created with the purpose of optimizing MRI technology to explore finer resolution in neural connectivity using DSI. They modified a 3-T Skyra system (MAGNETOM Skyra Siemens Healthcare) to build the customized MGH Siemens 3-T Connectome scanner, specially developed to improve the spatial resolution and imaging quality in diffusion imaging [25]. The new scanner is housed at the MGH/HST Athinoula A. Martinos Center for Biomedical Imaging, and has a maximum gradient strength of 300 mT/m, with b-values tested up to 20,000 [26]. These values are much higher than those currently commercially available, as a standard gradient strength is about 45 mT/m, with a b-value ranging from 700 to 1000. For example, even the most recent 3-T model from Siemens, the 3-T Prisma, which denotes an important advance in gradient fields for human whole body scans, has a gradient strength of 80 mT/m. Development of new data acquisition protocols, pulse sequences, as well as novel algorithms for detailed analysis of fiber structure and interregional connections, including the development of novel graphical means for interactively navigating brain connectivity, were also carried by this consortium in order to translate the MRI data into connectomic maps detailing the fibrous connections in the brain.

The MGH/Harvard–UCLA Consortium scanned participants at MGH, but data analysis was carried on in both sites using advanced software. Diffusion data with voxel size of 1.5 mm isotropic from 35 healthy adult volunteers, obtained from this consortium, were released in 2014 [25].

3.2.2 1000 Functional Connectomes Project (FCP)

The 1000 FCP is part of the International Neuroimaging Data-sharing Initiative (INDI), which is now sponsored by the Child Mind Institute [27]. This project was launched in December 2009 and was initially designed to gather and share fMRI human data in order to establish the comprehensive mapping of the functional connectome. In a retrospective data sharing, rsfMRI data from over 1400 healthy adult volunteers collected from 35 sites around the world were publicly released [28]. Additionally, the data processing steps used to evaluate the feasibility of this project were also made available on Neuroimaging Informatics Tools and Resources Clearinghouse (NITRC) [29]. As a result, by combining datasets from different centers, the influence of different data acquisition protocols and scanners on resting-state fMRI measures were investigated [30], and the use of the intersubject variance-based method to identify presumed limits between functional networks was explored, finally establishing the presence of a universal functional architecture in the human brain [28].

After achieving the first goal of large-scale sharing of rsfMRI data, an initiative to aggregate and share phenotype datasets to explore brain–behavior relationships in large scale took place. It was achieved by making available comprehensive phenotypic information with imaging datasets to facilitate sophisticated data mining, such as the machine learning approaches described herein. In addition, prospective data sharing was made available by the contributing sites. In this new step, the FCP aimed to establish active data sharing by the neuroimaging community to aid in the concept of modeling large-scale imaging. It was achieved by fulfilling the requirements of prospective, phenotyped rsfMRI data sharing, providing a model for the broader imaging community, while simultaneously creating a public dataset where all datasets are accepted regardless of publication status. The open, prospective data sharing for the neuroimaging community started with eight imaging sites that committed to share data on a pre-established schedule: Baylor College of Medicine, Beijing Normal University, Berlin Mind and Brain Institute, Harvard-MGH, MPI-Leipzig, NKI-Rockland, NYU Institute for Pediatric Neuroscience, and the Valencia node of the Spanish Resting State Network. Additional sites such as Stanford University, Duke University Medical Center, and others have since joined the prospective data-sharing plan [27].

A common protocol for sharing phenotypic/metadata via the 1000 FCP was established to ensure compatibility between the FCP and HCP platforms. For this purpose, the eXtensible Neuroimaging Archiving Toolkit (XNAT—xnat.org) format was selected. Existing datasets on the 1000 FCP website were redistributed as an XNAT virtual machine, including both imaging data and available metadata (including extensive imaging parameters when available).

The FCP is currently sharing data, either retrospective or prospective, from several imaging centers and different neuroimaging modalities, including not only rsfMRI but also task-evoked fMRI, DTI, arterial spin labeling (ASL), and anatomical images (T_1, T_2, MPRAGE). Although most shared imaging data are from various MRI modalities, one site also shared their positron emission tomography (PET) data (retrospective from Dallas Lifespan Brain Study). A variety of data from different populations covering from adult healthy controls (HCs), developing and aging, as well as clinical populations, such as attention deficit hyperactivity disorder (ADHD), epilepsy, autism spectrum disorder (ASD), schizophrenia (SZ), Alzheimer's disease (AD), and cocaine dependents, have been shared and are available on their website [27,29].

3.2.3 Federal Interagency Traumatic Brain Injury Research

The FITBIR is a database created by the NIH, in partnership with the Department of Defense (DoD), sponsored by the US Army Medical Research and Material Command (USAMRMC) through Congressionally Directed Medical Research Programs (CDMRP) and by the National Institute of Neurological Disorders and Stroke (NINDS) and Center for Information Technology (CIT). This initiative seeks to create an infrastructure that integrates heterogeneous datasets to improve collaborative and advanced research in traumatic brain injury (TBI) [31], which is the leading cause of death and disability in young adult patients (under the age of 44) [32]. It aspires to expand the expertise on diagnosing, treating, and improving outcomes in TBI patients by identifying biomarkers to determine either injury progression or recovery, defining a biologically based classification system for the impairment, and establishing comparative research to determine treatment effectiveness for specific cases.

A secure bioinformatics platform was developed by the FITBIR together with Sapient Government Services [32] to share data, methodologies, and associated tools across the TBI research field, facilitating collaboration between laboratories and interconnectivity with other informatics platforms. This informatics system serves as a central data repository and includes advanced tools that facilitate the identification, access, and sharing of the research data. It provides guidelines on data collection for clinical studies, and contains comprehensive and coherent informatics approaches to capture, process, and analyze vast quantities of research data, allowing valid comparison of results of brain injury treatments and diagnoses across studies.

This database contains approximately 200,000 data records from over 4500 patients [31] that were acquired in studies funded by the DoD and NIH. This comprehensive dataset includes demographics, outcome assessments, imaging, and biomarkers. FITBIR is designed to accept human subjects-related imaging, phenotypic, and genomic data from any TBI-related research projects, regardless of funding source and location [31]. Currently, the imaging database is predominantly MRI (87%), but also includes contrast imaging (12%) and PET (1%) [31].

3.2.4 National Database for Autism Research

The formation of the NDAR, in 2007, was an initiative from five institutes and centers at the NIH: The National Institute of Mental Health (NIMH), the National Institute of Child Health and Human Development (NICHD), the NINDS, the National Institute of Environmental Health Sciences (NIEHS), and the CIT. It intends to establish a repository for data acquired in autism research projects, providing a common platform for data collection, retrieval, and archiving. This initiative facilitates data sharing and encourages collaboration among ASD investigators, therefore supporting the aims of the Interagency

Autism Coordinating Committee (IACC) by promoting the improvement on autism research. In order to take advantage of the ASD-related data that were already aggregated in privately funded data repositories, a federation was put in place linking the NDAR with data from the Autism Genetic Resource Exchange (AGRE—genomic and clinical data) [33], the Autism Tissue Program (ATP—a post-mortem brain donation program), and the Interactive Autism Network (IAN—self-report database), all funded by Autism Speaks [34]. In addition to these private databases, NDAR has become the repository for the data acquired in the NIH Study of Normal Brain Development, which enrolled over 500 children to study their development using a variety of clinical and behavioral measurements, as well as different MRI techniques: sMRI, DTI, and MR spectroscopy [35].

The NDAR also developed a set of tools to facilitate the aggregation and organization of vast quantities of heterogeneous data. Specifically, the NDAR Implementation Team (NIT) manages data submission and access, in order to protect the confidentiality of a research subject, and the Autism Informatics Consortium (AIC) is responsible for making informatics tools and resources, establishing common data definitions and specifications, as well as comprehensive and coherent informatics approaches to help autism researchers. The AIC includes Autism Speaks, Kennedy Krieger Institute, Simons Foundation, Prometheus Research, and the NIH. As a result, two main components were developed and form the basis of NDAR: The Global Unique Identifier (GUID) for subjects, and the researcher-defined Data Dictionary to describe the experiments.

The GUID has become the standard as a patient identifier for autism research and is the result of a collaboration between NDAR, the Simons Foundation, and a team of researchers from Columbia University [36]. The Data Dictionary encompasses over 300 definitions for clinical, imaging, and genomic data modalities and was developed with the cooperation of the ASD research community. It includes the genomics definition tool and the imaging tool that makes the process of uploading experimental results straightforward. In addition, a validation tool was developed to ensure that the reported data are consistent with NDAR standards.

This large repository currently comprises data from over 127,000 subjects, including phenotype, genomic (over 25,000), and neuroimaging data (over 9800), with MRI partaking most of the neuroimaging data (over 8000 MRI: DTI, fMRI, sMRI, spectroscopy; 1000 EEG, 600 eye-tracking; 160 MEG; 130 PET) [37].

3.2.5 SchizConnect

SchizConnect is a virtual database that was developed to mediate multiple data repositories in SZ-related neuroimaging studies [38]. The website [39] works as a portal presenting a unified view of the data from the different databases, which remain at the primary sources structured under their original representations. The software that links the databases was developed at the Information Sciences Institute of the University of Southern California (ISI) and was built upon the Biomedical Informatics Research Network (BIRN) [40]. It provides a virtual synchronized outline where, by a given user query, it determines the sources with the appropriate data and translates the user query to the source's format, allowing the access of the data sources in real time [38].

This system provides integrated access to different sources of SZ data that are publicly available, including demographics, cognitive and clinical assessments, and imaging data. These sources have been organized and underwent quality assurance [39]:

1. Function BIRN (FBIRN) Phase II, with data housed at the University of California, Irvine Human Imaging Data (UCI_HID) [41]. This study includes structural and functional MRI from chronic patients with SZ and age-matched controls, which were scanned in two visits each. It is cross-sectional multisite data that were collected on different 1.5- and 3-T scanners; fMRI was performed using different paradigms including a working memory, auditory oddball, breath-holding, and sensorimotor tasks.

2. Northwestern University Schizophrenia Data and Software Tool (NUSDAST) [42] with data held at XNAT_Central and Northwestern University Research Electronic Data Capture (NU_REDCap) [43]. The NUSDAST is a repository of SZ neuroimaging data that contains sMRI collected at the Northwestern 1.5-T Siemens Vision scanner. The data were acquired from individuals with SZ, HCs, and their respective siblings, most with a 2-year longitudinal follow-up. The data are stored in XNAT central, a public repository of neuroimaging and clinical data, hosted at Washington University at Saint Louis, and at NU_REDCap, which is a clinical research database, facilitating the integration of data from clinical and cognitive assessments and facilitating searching across imaging and non-imaging databases for the same subject [44].

3. The Center for Biomedical Research Excellence (COBRE) [45] and Mind Clinical Imaging Consortium (MCICShare) [46], with data stored at the Collaborative Imaging and Neuroinformatics System (COINS) Data Exchange [47]. The COBRE

project contains sMRI and rsfMRI data that were collected on a 3-T scanner, as well as psychiatric, neuropsychological, and genetic testing from people with SZ and controls. The MCICShare project, on the other hand, contains only MRI data (sMRI, rsfMRI, and DTI) from people with SZ and sex and age-matched controls collected on 1.5- and 3-T scanners.

4. Neuromorphometry by Computer Algorithm Chicago (NMorphCH) retained at NU_REDCap and Northwestern University Neuroimaging Data Archive (NUNDA), a data archiving system powered by XNAT [48]. The NMorphCH is a longitudinal study, conducted at Northwestern University, designed to acquire a variety of data from people with SZ and control subjects at three time points: at baseline, and after 2 and 4 years. It includes clinical data, where symptom and mood are assessed, and cognitive data, involving neuropsychological assessments on working memory, episodic memory, and executive function domains, all archived at NU_REDCap. The neuroimaging data, which include sMRI (T_1 and T_2 contrast), DTI, rsfMRI, and task-evoked fMRI [39], are archived at NUNDA.

SchizConnect started in 2014 and already provides access to more than 21,000 images from over 1000 participants, including people with SZ and controls [38]. It is rapidly becoming an important neuroimaging resource for SZ research, in which research outcomes from this mega-dataset is expected to enhance current knowledge about the mechanisms underlying SZ.

3.2.6 AD Neuroimaging Initiative

ADNI is a multicenter program intended to address fundamental questions related to disease initiation and progression of the AD. This longitudinal study aimed to develop clinical, imaging, genetic, and biological markers for the early detection and tracking of AD using a large cohort, in order to establish standardized methods for imaging and biomarker collection and analysis. Therefore, the outcome of this study should help to determine the relationships among the different characteristics of the entire spectrum of AD, providing a broad understanding of the series of pathophysiological events that initiate AD and lead to dementia at the molecular, cellular, brain, and clinical levels [49].

The ADNI project began in 2004 studying changes in cognition, brain structure, and function in elderly controls, subjects with mild cognitive impairment (MCI), and subjects with AD. Its initial phase, known by ADNI1, was a 6-year study (2004–2010) funded by both the public and private sectors, where data from 800 participants were acquired, using sMRI and PET. The second phase of this project was called ADNI GO (2009–2011), which assessed the existing ADNI1 cohort and added 200 participants identified as having early MCI to examine biomarkers in an earlier stage of disease [50]. The third phase, the ADNI2, ran from 2011 to 2015 and was designed to assess the participants from the ADNI1/ADNI GO cohort in addition to a new set of 550 participants including elderly controls and early and late MCI participants [51]. The imaging components used in both latter phases, ADNI GO and ADNI2, were sMRI, DTI, ASL, rsfMRI, and PET [50]. The ADNI study is now in its fourth phase, ADNI3, which proposes to study change in cognition, function, brain structure, and biological markers among the ADNI volunteers [52].

Data from all phases, acquired from 57 sites, are shared through USC's Laboratory of Neuro Imaging's (LONI) Image Data Archive (IDA) [51]. This data repository de-identifies and uploads data from the different ADNI sites, ensures quality control and preprocessing of images, manages their association with the metadata, controls search functions, and supervises user access. The ADNI MRI Core, besides establishing the criterion for imaging preprocessing, has also created standardized analysis sets of data composed of scans that met the quality control criteria. This standardized dataset was created to promote consistency in the data analysis ensuring meaningful direct comparisons of sMRI landmarks [53].

The ADNI program has also extended worldwide (WW-ADNI) forming the umbrella organization for the international neuroimaging initiative, which is carried out by the North American-ADNI, European-ADNI, Japan-ADNI, Australian-ADNI, Taiwan-ADNI, Korea-ADNI, China-ADNI, and Argentina-ADNI [49,51,52]. This worldwide project focuses on evaluating functional and structural changes in the brain of Alzheimer's patients using PET and MRI techniques, as well as on assessing changes in their blood and CSF. This international effort aims to develop effective treatments to slow AD progression, as well as preventive methods to avert the disease.

3.2.7 IMAGEN

IMAGEN is a longitudinal study, initially funded by the European Commission, that involves eight research centers from four European countries: England (King's College London, University of Nottingham), Ireland (Trinity College Dublin), France (The Frederic Joliot Hospital Department of CEA, Paris), and Germany (Charité–University Medicine Berlin, University Medical Center Hamburg-Eppendorf, The Central Institute for

Mental Health in Mannheim, and Dresden University of Technology) [54]. The project aims to understand the consequences of stressors on the brain–behavior system during development by identifying the basis for individual differences in impulsivity, sensitivity to either reward or punishment, and emotional reactivity. The results from this study are expected to aid in understanding the effect of the environment in brain development and its association with mental illness, therefore improving prediction of risk and treatment of psychiatric disorders.

The research teams have been collecting data from 2000 youths from the age of 14, with follow-up assessments at ages 16, 19, and 22. The data collected include sMRI, fMRI, and DTI on 3-T MRI scanners; cognitive and behavioral assessments; self-report questionnaires considering elements such as relationships, feelings, and personality; drug- and alcohol use-related questionnaires; and blood sampling for genetic and biological analyses. Data storage is carried in a central database in France (Neurospin, CEA) using XNAT technology; MRI data are preprocessed centrally using an automated pipeline [55].

Data from this project can be requested by applying a proposal, endorsed by a collaborator in the consortium who will act as a contact person. It must describe the data being requested, details of the planned analyses, and period of the project. Access of the data is granted after the project is approved by the project executive committee (PEC), which is composed of the principle investigators of all associate institutes [54].

The development of these large databases and consortia makes it clear that data aggregation and sharing are paramount in the neuroimaging field. These datasets are sufficiently large to be assessed using big data/machine learning techniques.

3.3 Machine Learning for Structural and Functional Imaging Data

As a method of data analysis that evolved from the study of pattern recognition and computational learning theory in the field of artificial intelligence, machine learning builds analytical models from example inputs in order to make predictions or decisions to unknown data. It not only has given us high-tech industrial products such as self-driving cars, effective Internet search engines, and speech recognition systems, but also has been increasingly applied to healthcare fields such as heartbeat monitoring, medication adherence, and drug development, as well as brain imaging analysis.

3.3.1 Statistical Comparison vs. Machine Learning Classification

Historically, analytical methods applied to structural and functional brain imaging data were dominated by statistical comparisons—for example, between a patient group and a healthy cohort [56–58]. In general, these studies employed mass conventional univariate analysis in which a voxel-by-voxel comparison of brain structural/functional maps is applied to groups of patients and HCs in order to identify regions whose difference reaches statistical significance. While this type of statistical comparison yields valuable characterizations of local abnormalities in brain regions as a function of psychiatric or neurological disease, it is usually insufficient for determining a clinical diagnosis of an individual subject. The main reason is that while the findings derived from a statistical comparison may be significant at the group level, there still may be considerable overlap between the distributions, such that the discriminative power at the individual level is poor. Moreover, for univariate approaches, statistical comparisons treat each voxel independently, which is an overly simplistic assumption of brain organization [59,60]. This inability of mass univariate statistical modeling to make inferences at the individual level has greatly hampered the ability to translate neuroimaging research results into clinical practice.

In contrast to statistical comparison, machine learning-based pattern classification, also known as multivoxel pattern analyses in neuroimaging studies, is a class of multivariate analyses with twofold advantages: first, machine learning methods learn features from the training set and can be applied later at the individual level, therefore generating results with a potentially high level of clinical translation; second, machine learning allows one to detect structural and functional differences between groups with higher accuracy as the multivariate nature of machine learning techniques permits the comparison of spatially distributed patterns of brain structure and activity. In recent years, there has been growing interest within the neuroimaging community in the use of machine learning techniques to analyze structural and functional imaging data to do clinical inferential studies of brain disorders.

3.3.2 Feature Extraction and Selection on Brain Imaging Data

A distinct character of neuroimaging data is that they usually contain a huge number of voxels. Typically, a preprocessed functional brain contains on the order of 100,000 nonzero voxels (with structural images containing more than 10× that number), whereas the sample size in neuroimaging studies is typically fewer than 50.

This common issue is known in the machine learning literature as the small-n-large-p [61] or the curse of dimensionality, where the sample size is much smaller than the number of features (i.e., voxels). In this situation, developing a machine learning model without effectively selecting the most relevant features and discarding other redundant ones will be likely to result in overfitting [62,63], which means that the model fits the training data well but has a poor generalization ability to make reliable predictions on completely novel samples. Therefore, two necessary steps known as "feature extraction" and "feature selection" are frequently performed prior to developing a machine learning model on brain imaging data.

Feature extraction refers to the transformation of the original brain imaging data into a set of representative features that can be used as inputs for the model. For sMRI data, widely used features are gray matter volume/density maps [64–66] and cortical thickness maps [67,68]; whereas for DTI data, features like FA maps [69,70] and mean diffusivity maps [71] are common. A variety of brain activation maps, and thus features used in machine learning models, can be derived from functional and resting state scans, such as contrast images, calculated from task-based functional imaging studies, using the general linear model [69,72,73]; individual component maps derived from independent component analysis [74–76]; and other local features (i.e., amplitude of low-frequency fluctuations, regional homogeneity, functional connectivity strength, and voxel mirrored homotopic connectivity) or global maps (i.e., functional or effective connectivity between anatomical regions) [77–80].

After feature extraction, the features derived from the brain scans will contain redundant information, as there is known to be spatial autocorrelation in neuroimaging data. Feature selection, sometimes known as feature reduction, involves selecting a subset of informative features and removing other less important features to facilitate model training. The rationale for feature selection is threefold. First, it helps to mitigate the small-n-large-p problem mentioned above. Second, it speeds up the learning process by reducing the computational load. Third, it may help elucidate the underlying neurobiological mechanisms of the specific question of interest by localizing, for example, which brain regions contain the most discriminative information.

Feature selection techniques, sometimes in a combination with feature extraction for neuroimaging data, can be categorized into knowledge-driven, model-driven, and data-driven approaches respectively. Knowledge-driven refers to choosing regions of interest (ROIs) based on prior knowledge of a brain atlas [81]. Under an assumption that voxels that are anatomically close respond similarly to a stimuli, brain anatomical maps (e.g., the automated anatomical labeling [AAL] template [82] or templates in their sub-parcellated version [83,84]), treated as *a priori* knowledge of the brain, are applied to feature maps to get regional mean values so as to reduce the feature dimension [65, 77,85,86]. An advanced knowledge-driven approach employed to neuroimaging recently is to apply meta-analysis techniques that have previously been used to report brain activation or group-level differences across neuroimaging studies [87]. Popular meta-analysis techniques include activation likelihood estimation, kernel-density estimation, and multilevel kernel density estimation [88]. These techniques have been used in modeling distributions of reported active regions from previous fMRI studies and the resulting ROIs used as masks to reduce input feature dimension for machine learning analyses [89–91].

Model-driven methods perform statistical tests on the training data and threshold at some significance level, user-defined value, or percentage to obtain a mask that will be applied later to both training and testing samples to select input features for the classifier. In neuroimaging studies, statistical testing techniques such as two-sample t-tests (for two-group classification cases [81,92,93]) and ANOVA (for cases with more than two groups [94–96]) have been extensively used to detect average group-level differences in training samples. Since these methods are simply performing mass-univariate analyses, they suffer from the shortcomings that interactions between multiple ROIs are neglected and that contiguous voxels, which frequently contain redundant information, are selected.

Data-driven approaches are a set of multivariate analysis techniques that aim to optimize a cost function while selecting relevant features. Among these, recursive feature elimination (RFE) [97] is an iterative feature ranking and selection algorithm developed specifically for support vector machines (SVMs). In order to identify the set of features with the greatest discriminative ability, RFE removes uninformative features that would potentially yield the margin between groups increase according to a ranking criterion at each iteration. For a linear kernel SVM, the importance of each feature is directly related to its weight coefficient; features are thus usually sorted according to the absolute or square values of their weight coefficients. A user-defined percentage of the lowest ranking features are removed recursively until the feature set is empty, with the "best performing" (e.g., resulting in the highest accuracy) iteration being selected [65,72,79,98].

RFE is categorized as backward elimination since it starts searching with all features in the training set and excludes uninformative features iteratively. In contrast, forward selection is another category of data-driven methods that begins searching for relevant features with an empty set, and features are iteratively added in until the process meets some stopping criteria [99]. Searchlight is a representative forward selection algorithm that has been developed for neuroimaging data [95,100,101]. First, a 3D volume (e.g., a sphere) with predefined size and shape (called a searchlight) are centered at each voxel in the training data. Next, a machine learning classifier is trained using voxels within the searchlight, and classifier accuracy from the searchlight at every voxel is recorded. Through permutation methods, the searchlight accuracies at every voxel are converted into p-values and thresholded to remove indiscriminative voxels. Since this method assumes that any discriminative information lies within the searchlight volume, it will be deficient at detecting multivariate interactions across distant regions [102].

The aforementioned backward elimination and forward selection approaches both use global search methods starting with either a full set or an empty set of features. A third data-driven category are embedded techniques that add a sparsity penalty directly to a classifier's cost function and combine the feature selection and classifier training steps, which has been successfully applied to neuroimaging data to distinguish patient groups [103–105]. The most popular approaches are the least absolute shrinkage and selection operator (LASSO) [106,107] and Elastic Net [108]. LASSO adds a penalty to the L1-norm of a model's cost function in order to control the sparsity of the model, while Elastic Net adds constraints to both the L2-norm and the L1-norm. Here the L1 penalty promotes sparsity in the solution, whereas the L2 penalty promotes stability of the solution [108].

3.3.3 Commonly Used Classifiers

After feature extraction and selection, prepared features are input to a machine learning classifier. A classifier is essentially a model of the relationship between the input features and the class labels in the training set. Using a mathematical form, given an example x, the classifier is a function f that predicts the label $y = f(x)$. The specific function being learned and the underlying assumption built into it vary between different types of classifiers. Here we review the classifiers that are most commonly used on neuroimaging data.

Machine learning algorithms can be categorized into supervised learning and unsupervised learning depending on whether or not the training data comprise samples along with their corresponding target labels of categories. In the case of two-class classification, supervised learning seeks to develop a function that maps two sets of samples to their given categories; subsequently, the trained mapping function can assign previously unseen new samples to one of the categories with certain accuracy. In contrast, unsupervised learning seeks to learn how the data are organized without *a priori* determined categories. Clustering analysis is a commonly used unsupervised machine learning algorithm [109] that has been applied to fMRI data to divide data into groups [110,111]. It groups a set of samples in such a way that, given some distance measure, samples in the same cluster are closer to each other than to those in other clusters. Two examples are k-means clustering, which aims to partition n samples into k clusters in which each sample belongs to the cluster whose mean value is closest to the sample, and hierarchical clustering where the closest pair of samples is replaced at their mean and the process is repeated until there is a single cluster.

There are a variety of supervised learning classifiers that are applied to brain imaging data; among these, SVM [112] is probably the most popular. It has been used to classify based upon structural and functional MRI data in brain diseases such as AD [113–115], SZ [116–118], and ADHD [119–121]. Given samples that belong to one of two groups, SVM performs pattern classification by finding a separating hyperplane that maximizes the margin between samples in a high dimensional space defined by the features. Those samples closest to the hyperplane are called the support vectors. Frequently the samples are not linearly separable in the original space; thus, there is often an advantage to using a kernel function to map the features into a higher dimensional space for easier separation. The Gaussian radial basis function (RBF) is the most commonly used kernel function on neuroimaging data [74,122,123]. The RBF kernel for two samples x and x' is defined as $K(x,x') = \exp(-\gamma||x-x'||^2)$, where γ is a hyperparameter that is optimized in the model-building stage.

Linear discriminant analysis (LDA) [124], a linear transformation technique, is also a simple and straightforward classifier that has been applied to neuroimaging data [125–127]. In contrast to the kernel SVM that maps data to a higher dimensional space, LDA projects data onto a lower dimensional space with good class separability by computing linear discriminants that will represent the axes that maximize the separation between multiple classes. The general process of performing LDA contains five steps: (1) compute the d-dimensional mean vectors for the different classes of the dataset; (2) compute the between-class and within-class scatter matrices; (3) compute the eigenvectors and corresponding eigenvalues for the scatter matrices; (4) sort the eigenvectors by descending eigenvalues and choose k eigenvectors with the largest eigenvalues to form a $d \times k$ dimensional eigenvector matrix M; (5) use matrix M to transform the samples onto the new subspace.

Deep learning, which is a powerful set of techniques for learning representations in neural networks, has been successfully introduced into the neuroimaging field to distinguish group membership of patients and HCs [85,128–130]. This set of algorithms, also called as deep neural networks (DNNs), attempts to use a cascade of many layers of nonlinear processing units to learn latent feature representations from large-scale unlabeled data. In the nonlinear unsupervised cascade processing, higher-level features are derived from lower-level features to form a hierarchical representation. In contrast to traditional machine learning methods, deep learning is capable of data-driven automatic feature learning, which is an important capability in selecting the relevant features from neuroimaging data as usually many features exist even after feature selection. Another advantage of deep learning is the capability of modeling very complicated data patterns due to its hierarchical structure of the nonlinear layers. To date, stacked auto-encoders (SAE) and restricted Boltzmann machines (RBMs) have been applied to brain imaging data among the variety of DNNs [85,128–130].

An auto-encoder (AE) is composed of input, hidden, and output layers. The output layer is a reconstruction of the input through the activation of the many-fewer hidden layers. It offers an elegant solution to dimension reduction and compression. The output layer can be omitted after training, and the hidden layer can be used as input features for downstream classification. AEs can be stacked one after another in a greedy layer-wise fashion such that the outputs from the hidden units of the lower layers are taken as the input to the upper layers to construct a SAE. After a SAE is trained layer by layer, its output can be used as the input to a supervised classification or regression algorithm. Another DNN that has been applied to neuroimaging data is formed by stacking RBMs. Comparing to the AEs, RBMs also have visible and hidden layers, and are probabilistic models. Instead of deterministic (e.g., logistic) units, RBMs use stochastic units with a particular distribution (e.g., usually Gaussian). The RBM training process consists of several steps of Gibbs sampling that includes repeated propagate (i.e., sampling hidden layers given visible ones) and reconstruct (i.e., sampling visible layers given hidden one) processes, and then adjusting the weights to minimize reconstruction error. This differs from the training of AEs, which simply calculates functions between layers to minimize an error. When trained in an unsupervised way, layers of the RBMs act as feature detectors to find latent representations of the input data. After unsupervised learning, RBMs can be further trained in a supervised way to perform classification, or its output can be used as the input to other supervised algorithms.

Other supervised classifiers have also been utilized in neuroimaging studies. For example, random forest [73,131,132], an ensemble learning method that operates by constructing a multitude of decision trees, is particularly appropriate when the number of predictor variables is much larger than the number of subjects—typical with neuroimaging datasets. Some studies used multikernel learning to do brain disease inference [92,133,134]. Multikernel learning constructs an optimal linear or nonlinear combination of a predefined set of kernel functions. Its main advantage is in combining features from different neuroimaging data phenotypes that have different notions of similarity and may require different kernels for training.

3.3.4 Performance Evaluation and Pattern Interpretation

In many fields using machine learning, such as computer vision, a classifier is trained on a large dataset and tested on a completely independent dataset. Although some neuroimaging studies have validated the trained classifier on a completely independent test set [64,65,135,136], the small sample sizes in most neuroimaging studies do not allow for this. Most neuroimaging studies use n-fold cross-validation (CV) to test the generalizability of the classifier.

In n-fold CV, the whole dataset is randomly partitioned into n equal-sized subsets. One subset is retained as the validation set for testing the classifier, and the remaining $n-1$ subsets are used as training data. Critically, all data transformation (such as scaling/normalizing), feature selection, and model hyperparameter selection are determined exclusively using the training data and applied to the validation data. The CV process is repeated n times until all subsets are used exactly once as the validation set. Ten-fold CV is commonly used, but there is a particular case where n is equal to the number of samples, which is called leave-one-out CV (LOOCV). More complex approaches require nested CV in which the data are partitioned twice: In an outer fold CV, a data partition is left out for testing; the remaining samples are partitioned again in an inner fold CV. The outer loop enables an estimation of the classifier's generalizability, while the inner loop is usually used for optimizing parameters.

The classifier performance can be assessed by several measures such as sensitivity, specificity, and accuracy. In a clinical context, the sensitivity refers to the percentage of people with a disease who are identified as being sick; while the specificity refers to the percentage of HCs who are classified as being healthy. The accuracy represents the total proportion of samples classified correctly; if there is a large imbalance in group size, balanced accuracy (the average of the accuracy calculated separately for each group) is preferred. In addition, the receiver operating characteristic (ROC) is also used to measure the performance of the model. The ROC curve provides information on the balance between the true positive rate (i.e., sensitivity) and the false positive rate (i.e., 1-specificity) across a range of decision thresholds. The area under the ROC curve (AUC) can be interpreted as the probability that the classifier will assign a higher score to a randomly selected positive sample than to a randomly selected negative sample.

The weights determining the classification boundary allow for visualization of the features, which in neuroimaging is in voxel space; this is known as the discriminative map for the classification [137]. However, interpreting the localization of an effect in such multivariate patterns is not straightforward [138]. Some discriminative regions on the map may be linked to clinical or physiological measures via post-hoc regression analyses. However, in general, a multivariate pattern must be considered as a whole. Moreover, for nonlinear classifiers, it is almost impossible to map the weights back to the original feature space. When using a CV procedure, the feature selection step will retain different features from fold to fold, and the overlap between the set of selected features may be very small, which also makes the interpretation difficult. Indeed, this reflects the fact that different multivariate combinations of different features contain sufficient information for classification.

3.3.5 Diagnostic Predictions of Neurological and Psychiatric Diseases

Machine learning-based single subject prediction of neuropsychiatric disorders has gained increased attention in recent years. Compared to a conventional diagnosis for a neuropsychiatric disease that relies on information gathered subjectively during clinical interviews, it objectively seeks out diagnostic differences in patterns across the brain that can be applied to a "new" individual to identify the diagnostic category to which they may belong. This applicability at the level of the single individual makes the machine learning approach a potentially valuable diagnostic and/or prognostic tool. It has been applied to a variety of neuroimaging modalities (e.g., structural, functional, and diffusion MRI) on patients with neurological and psychiatric diseases such as MCI, AD, SZ, bipolar disorder (BPD), major depressive disorder (MDD), ASD, and ADHD. Here we review some of the progress made in these fields.

MCI is a brain disorder that involves cognitive decline in memory, language, thinking, and judgment greater than that expected based on an individual's age and education level. As an intermediate stage between the expected cognitive decline of normal aging and the more-serious decline of dementia, studies suggest that individuals with MCI tend to progress to AD at a rate of approximately 10% to 15% per year [139]. The goal of neuroimaging in AD has thus moved from the diagnosis of advanced AD to the diagnosis of very early AD or MCI at a prodromal stage of MCI [140]. Table 3.1 lists studies that used neuroimaging-based machine learning classification of MCI and AD [141,142]. Since the accuracies in most of the "big data" studies range between 80% and 90%, and are even higher on small datasets, it seems that we are generally able to distinguish both AD and MCI from HC.

SZ, affecting about 24 million people globally, is a chronic and serious mental disorder that interferes with a person's ability to think clearly, manage emotions, and make decisions. Neuroimaging studies in SZ patients have revealed extensive abnormalities, such as altered structural relationships among regional morphology in the thalamus, frontal, temporal, and parietal cortices [143]; altered structural integrity of white matter in frontal and temporal brain regions [144]; and differences in left superior temporal gyrus (STG)–dorsal lateral prefrontal cortex and STG–ventrolateral prefrontal cortex functional connectivity [145]. Compared to other psychiatric disorders, machine learning techniques have been most commonly applied to SZ due to data availability. Studies have shown that this disorder can be accurately predicted using functional and structural MRI data phenotypes (see Table 3.2 [146]).

BPD, formally called manic depression, is a brain disorder that causes extreme mood swings including emotional highs (known as mania) and lows (known as depression). During periods of mania, a patient behaves extremely energetic, happy, or irritable, whereas during periods of depression, the patient shows sadness and a negative outlook on life. In contrast to BPD, MDD is often accompanied by low energy and loss of interest in normally enjoyable activities. Pattern recognition-based BPD and MDD studies have thus far been performed on relatively small- to medium-sized samples (Table 3.3), and thus there is a lack of definitive conclusions.

ASD is a developmental disorder including a wide range of social deficits and communication difficulties. It is characterized by persistent deficits in social communication and interaction across multiple contexts, as well as restricted, repetitive patterns of behavior, interests, and activities. Although the causation of ASD is not clearly known, it is suggested that genetic, developmental, and environmental factors could be combined as possible causal or predisposing effects that develop ASD [147]. Table 3.4 lists papers that applied automatic diagnosis technology to ASD using neuroimaging features and machine learning. These studies were mainly performed on small- or medium-sized data with a wide accuracy range, which indicates that as a disorder, ASD has a huge range of deficits (i.e., it may be a heterogeneous disease).

ADHD is a brain disorder in which an individual has trouble paying attention to tasks and tends to

TABLE 3.1

Summary of Machine Learning-Based MCI/AD Diagnostic Studies (Sorted by Publication Year)

Study	Modality	Sample	Features	Classifiers	Validation	Performance
DeCarli et al. [148]	sMRI	AD=31, HC=29	Brain volume, temporal lobe matter	Discriminant analysis	2-fold validation	Accuracy=100%
Kaufer et al. [149]	sMRI	AD=17, HC=12, FTD=16	Morphometric measures of ROIs	Discriminant analysis	2-fold validation	Accuracy=91%
Frisoni et al. [150]	sMRI	AD=46, HC=31	Linear measurements	LDA	2-fold CV	Sensitivity=81–87%
Freeborough and Fox [151]	sMRI	AD=24, HC=40	Texture features	LDA	2-fold validation	Accuracy=91%
Bottino et al. [152]	sMRI	AD=39, MCI=21, HC=20	Volumetric measures	Discriminant function analysis	2-fold validation	Accuracy=80.5–88.1%
Pennanen et al. [153]	sMRI	AD=48, MCI=65, HC=59	Volumes of hippocampus and ERC	Discriminant function analysis	Step-wise analysis	Accuracy=65.9–90.7%
Li et al. [154]	sMRI	AD=19, HC=20	Surface-based measures	SVM	LOOCV, 3-fold CV	Accuracy=84.6–94.9%
Wang et al. [155]	sMRI	AD=18, HC=26	Hippocampus shape measures	Logistic regression	LOOCV	Accuracy=81.1–84.6%
Arimura et al. [156]	sMRI	AD=29, HC=25	Cortical thickness	SVM	LOOCV	AUC=90.9%
Klöppel et al. [157]	sMRI	AD=85, HC=91, FTD=19	GM maps	SVM	LOOCV	Accuracy=87–96%
Lerch et al. [158]	sMRI	AD=19, HC=17	Cortical thickness	LDA, QDA, logistic regression	LOOCV	Accuracy=90–100%
Vemuri et al. [115]	sMRI	AD=190, HC=190	Tissue densities and genotype	SVM	4-fold CV	Accuracy=85.6–89.3%
Gerardin et al. [159]	sMRI	AD=23, MCI=23, HC=25	Spherical harmonics of hippocampi	SVM	LOOCV	Accuracy=83–94%
Hinrichs et al. [160]	sMRI	Total=183	GM probability maps	Boosting	Leave-many-out CV	AUC=82.0%
Magnin et al. [161]	sMRI	AD=16, HC=22	GM distribution of ROIs	SVM	Bootstrap	Accuracy=94.5%
McEvoy et al. [162]	sMRI	AD=84, MCI=175, HC=139	Volumetric and cortical thickness measures	LDA	LOOCV	Accuracy=89–92%
Young et al. [163]	sMRI	AD=24, HC=23, FTD=19	GM volume	LDA	10-fold CV	Accuracy=81–96%
Zarei et al. [164]	DTI, sMRI	AD=16, VaD=13, HC=22	Transcallosal prefrontal FA and Fazekas score	LDA	LOOCV	Accuracy=87.5%
Haller et al. [165]	DTI	MCI=67, HC=35	FA, longitudinal, radial, and mean diffusivity features	SVM	10-fold CV	Accuracy=91.4–97.5%
Mueller et al. [166]	sMRI	AD=18, MCI=20, HC=53	Hippocampal volumetric measures	LDA	Step-wise analysis	Accuracy=73.7–77.5%
Oliveira et al. [167]	sMRI	AD=14, HC=20	Volumetric measures	SVM	LOOCV	Accuracy=88.2%
Plant et al. [168]	sMRI	AD=32, MCI=24, HC=18	GM and WM maps	SVM, Bayes statistic, and voting	LOOCV	Accuracy=92%
Zhou et al. [126]	rsfMRI	AD=12, HC=12, FTD=12	ROI-based difference	LDA	LOOCV	Accuracy=92%
Abdulkadir et al. [169]	sMRI	AD=91, HC=226	GM probability maps	SVM	LOOCV	Accuracy=87%
Chincarini et al. [170]	sMRI	AD=144, HC=189	Intensity and texture of VOIs	SVMs	20-fold CV	Accuracy=65–94%
Costafreda et al. [171]	sMRI	AD=71, MCI=103, HC=88	3D hippocampal shape morphology	SVM	4-fold CV	Accuracy=80%

(Continued)

TABLE 3.1 (CONTINUED)

Summary of Machine Learning-Based MCI/AD Diagnostic Studies (Sorted by Publication Year)

Study	Modality	Sample	Features	Classifiers	Validation	Performance
Cuingnet et al. [86]	sMRI	AD=137, HC=162	Voxel-based features, cortical thickness features, hippocampus-based features	SVM	LOOCV	Accuracy=81–95%
Graña et al. [172]	DTI	AD=20, HC=25	FA	SVM	10/5/2-fold CV, LOOCV	Accuracy=100%
Hinrichs et al. [173]	sMRI, PET	AD=48, MCI=119, HC=66	GM and WM maps	Multikernel learning	10-fold CV	Accuracy=87.6%
McEvoy et al. [174]	sMRI	AD=164, MCI=317, HC=203	Longitudinal volumetric measures	QDA	LOOCV	Accuracy=85%
Wee et al. [175]	DTI	MCI=10, HC=17	Clustering coefficient of WM connectivity	SVM	LOOCV	Accuracy=88.9%
Yang et al. [176]	sMRI	AD=298, HC=516, MCI=400	Coefficient of ICA	SVM	LOOCV	Accuracy=67.5–99%
Zhang and Davatzikos [177]	sMRI	AD=50, HC=50	ODVBA of RAVENS maps	SVM	5-fold CV	Accuracy=90%
Zhang et al. [92]	sMRI, PET, CSF	AD=51, MCI=99, HC=52	GM, WM and CSF volumes, average intensity	SVM	10-fold CV	Accuracy=76.4–93.2%
Batmanghelich et al. [178]	sMRI	AD=54, HC=63	RAVENS maps	Logistic model trees	10-fold CV	Accuracy=87–89%
Casanova et al. [179]	sMRI	AD=171, MCI=351, HC=205	GM map	SVM, logistic/linear regression	10-fold CV	Accuracy=80–90%
Chu et al. [81]	sMRI	AD=131, MCI=260, HC=188	Hippocampus GM maps	SVM	LOOCV	Accuracy=70–85%
Coupé et al. [180]	sMRI	AD=50, HC=50	Atrophic patterns	QDA	LOOCV	Accuracy=93%
Cui et al. [181]	sMRI, DTI	MCI=79, HC=204	Subcortical volumetric measures, FA	SVM	10-fold CV	Accuracy=71.1%
Dai et al. [77]	sMRI, rsfMRI	AD=16, HC=22	GMV, ALFF, ReHo, FC	LDA	LOOCV	Accuracy=89.5%
Li et al. [182]	sMRI	AD=37, HC=40	Cortical thickness, cortex thinning dynamics, network features	SVM	LOOCV	Accuracy=81.7–96.1%
Mahanand et al. [183]	sMRI	AD=30, HC=30	GM maps	Self-adaptive resource allocation network	Bootstrap	Accuracy=97.1–99.7%
O'Dwyer et al. [184]	DTI	MCI=33, HC=40	FA, DA, DR, and MD	SVM	10-fold CV	Accuracy=93.0%
Polat et al. [185]	sMRI	AD=31, HC=31	VBM measures	SVM	LOOCV	Accuracy=74–79%
Sabuncu et al. [186]	sMRI	AD=150, HC=150	Cortical thickness	RVoxM	5-fold CV	AUC=93.0%
Aguilar et al. [187]	sMRI	AD=116, MCI=119, HC=110	Cortical thickness and volumetric measures	SVM	10-fold CV	Accuracy=88.1%
Cuingnet et al. [188]	sMRI	AD=137, HC=162	Voxel-based and cortical thickness-based schemes	SVM	2-fold validation	Accuracy=83–91%
Dukart et al. [91]	sMRI, PET	AD=49, HC=41	Volumes of interest	SVM	LOOCV	Accuracy=86–100%
Dyrba et al. [71]	DTI, sMRI	AD=137, HC=143	FA, MD, GMD, WMD	SVM	10-fold CV, other CV	Accuracy=63.6–91.1%

(Continued)

TABLE 3.1 (CONTINUED)

Summary of Machine Learning-Based MCI/AD Diagnostic Studies (Sorted by Publication Year)

Study	Modality	Sample	Features	Classifiers	Validation	Performance
Gray et al. [132]	sMRI, PET, CSF, genetics	AD=37, MCI=75, HC=35	ROI-based volumetric measures, voxel-wise intensity	Random forest	Leave-25%-out CV	Accuracy=74.6–89.0%
Lee et al. [189]	DTI	MCI=39, HC=45	FA and the volume of fiber pathways	SVM	10-fold CV	Accuracy=100%
Liu et al. [190]	sMRI	AD=86, HC=137	Volume and cortical thickness	Logistic regression, SVM, LDA	LOOCV	Accuracy=51–89%
Wee et al. [191]	sMRI	AD=198, HC=200	ROI-based correlative features	Multikernel SVM	10-fold CV	Accuracy=79.2–97.4%
Wu et al. [192]	rsfMRI	AD=15, HC=16	Averaged voxel intensities	Multivariate ROC	LOOCV	Accuracy=100%
Yang et al. [193]	sMRI	AD=17, MCI=18, HC=17	Volumetric measures and ventricle shape	SVM	LOOCV	Accuracy=88.9–94.1%
Apostolova et al. [194]	sMRI	AD=95, MCI=182, HC=111	Hippocampus volume and other measures	SVM	LOOCV	Accuracy=64–78%
Chen et al. [195]	sMRI	AD=48, HC=164	SIFT features	Ensemble of SVMs	LOOCV	Accuracy=70–87%
Farhan et al. [196]	sMRI	AD=37, HC=48	GM, WM and CSF volumes	SVM, MLP, decision tree	10-fold CV	Accuracy=93.7%
Hidalgo-Muñoz et al. [197]	sMRI	AD=185, HC=185	WM and GM voxels	SVM-RFE	10-fold CV	Accuracy=94.3–95.1%
Jie et al. [198]	rsfMRI	MCI=12, HC=25	Local connectivity, global topological properties	Multikernel learning	LOOCV	Accuracy=91.9%
Lahmiri and Boukadoum [199]	sMRI	AD=11, MCI=11, HC=11	Hurst's exponents at different scales	SVM	10-fold CV	Accuracy=97.1–97.5%
Lee et al. [200]	sMRI	AD=33, HC=84	Cortical thickness, hippocampus shape	LDA	Online learning	Accuracy=87.5%
Li et al. [201]	sMRI, DTI	AD=21, HC=15	FA and GM volumes	SVM	LOOCV	Accuracy=94.3%
Li et al. [202]	sMRI	AD=80, MCI=141, HC=142	Local binary pattern features	SVM	10-fold CV	Accuracy=61.5–82.8%
Lillemark et al. [203]	sMRI	AD=114, MCI=240, HC=170	Surface connectivity	LDA	LOOCV	AUC=76.6–87.7%
Liu et al. [204]	sMRI, PET	AD=51, MCI=99, HC=52	GM volume from sMRI, average intensity from PET	Multikernel SVM	10-fold CV	Accuracy=78.8–94.8%
Liu et al. [205]	sMRI	AD=198, MCI=225, HC=229	GM maps	SVMs	10-fold CV	Accuracy=85.3–92.0%
Tang et al. [206]	sMRI	AD=175, HC=210	Hippocampus, amygdala, and ventricle shape measures	LDA	LOOCV	Accuracy=86%
Tong et al. [207]	sMRI	AD=198, HC=231	Intensity patches of ROIs	SVM	LOOCV	Accuracy=82.9–89%
Zhang et al. [208]	sMRI, PET, CSF, SNP	AD=49, MCI=93, HC=47	GM volume, average intensity, CSF and SNP features	SVM	10-fold CV	Accuracy=71.0–94.8%
Beheshti et al. [209]	sMRI	AD=130, HC=130	PDF of VOI based on VBM	SVM	10-fold CV	Accuracy=86%
Challis et al. [210]	rsfMRI	AD=27, MCI=50, HC=39	FC among AAL	Bayesian Gaussian process, LR	LOOCV	Accuracy=75–90%
Dyrba et al. [211]	sMRI, DTI, rsfMRI	AD=28, HC=25	GMV, fiber tract integrity, graph-theoretical measure	SVM	LOOCV	AUC=74–85%

(Continued)

TABLE 3.1 (CONTINUED)

Summary of Machine Learning-Based MCI/AD Diagnostic Studies (Sorted by Publication Year)

Study	Modality	Sample	Features	Classifiers	Validation	Performance
Farzan et al. [212]	sMRI	AD=30, HC=30	Percentage of brain volume changes	SVM	LOOCV	Accuracy=91.7%
Goryawala et al. [213]	sMRI	AD=55, MCI=205, HC=125	Volumetric measures	LDA	2-fold CV	Accuracy=90.8–94.5%
Granziera et al. [214]	sMRI	MCI=42, HC=77	Volume, mean T1, MTR and T2*	SVM	LOOCV	Accuracy=75%
Jung et al. [215]	sMRI, DTI	AD=27, MCI=138, HC=27	Cortical thickness, subcortical volume, WM integrity	SVM	10-fold CV	Accuracy=70.5–96.3%
Khazaee et al. [113]	rsfMRI	AD=20, HC=20	Graph measures	SVM	LOOCV	Accuracy=100%
Klöppel et al. [216]	sMRI	AD=483, MCI=418, HC=604	GM and WM maps	SVM	CV	AUC=73–97%
Liu et al. [217]	sMRI	AD=97, HC=128	GM maps	SVMs	10-fold CV	Accuracy=93.8%
Nir et al. [218]	DTI	AD=37, MCI=113, HC=50	FA and MD values	SVM	10-fold CV	Accuracy=68.3–84.9%
Ortiz et al. [219]	sMRI, PET	AD=70, MCI=111, HC=68	Functional and structural connectivity	SVM	10-fold CV	Accuracy=84–92%
Prasad et al. [220]	sMRI, DTI	AD=38, MCI=112, HC=50	Network topology, tractography connectivity	SVM	10-fold CV	Accuracy=59.2–78.2%
Retico et al. [221]	sMRI	AD=144, HC=189	GM maps	SVM	20-fold CV	Accuracy=80%
Salvatore et al. [222]	sMRI	AD=137, HC=162	GM and WM maps	SVM	20-fold CV	Accuracy=72–76%
Xu et al. [223]	sMRI, PET	AD=113, MCI=110, HC=117	Mean volume of GM, SUVr value of PET	Sparse representation-based classifier	10-fold CV	Accuracy=74.5–94.8%
Yun et al. [224]	sMRI, PET	AD=71, MCI=163, HC=85	Cortical and volumetric measures	PLS LDA	LOOCV	Accuracy=76.5–90.1%
Zhang et al. [225]	sMRI	AD=28, HC=98	Eigen brains of key slices	SVM	10-fold CV	Accuracy=92.3%
Zu et al. [226]	sMRI, PET	AD=51, HC=52	ROI-based GM volumes, average intensity	Multikernel SVM	10-fold CV	Accuracy=80.3–95.9%
Yu et al. [227]	sMRI, PET, CSF	AD=50, MCI=97, HC=52	GMV, average intensity, CSF measures	Graph-guided multitask learning	10-fold CV	Accuracy=80.0–92.6%

Source: Modified from M. R. Arbabshirani et al., *Neuroimage*, vol. 145, pp. 137–65, Jan 2017; B. Sundermann et al., *AJNR Am J Neuroradiol*, vol. 35, pp. 848–55, May 2014.

act without thinking. It is one of the most commonly diagnosed functional disorders with approximately 3–10% of school aged children diagnosed with ADHD [228]. Neuroimaging studies have shown substantial evidence of structural and functional alterations in regions related to the frontostriatal circuitry, as well as in the cerebellum and the parietal lobes [229]. In 2011, a global competition called ADHD-200 was held in order to use neuroimaging and phenotypic measures to automatically diagnose ADHD [230]; most of these studies are summarized in Table 3.5.

3.4 Challenges and Future Directions

3.4.1 Overfitting

The goals in machine learning training are twofold: develop a model that describes a set of training data well, yet is able to generalize to novel data. In overfitting, the model describes noise or random error rather than the underlying pattern in the data. Overfitting results in very good performance on the training data but very poor performance on the novel test data. It occurs when

TABLE 3.2

Summary of Machine Learning-Based SZ Diagnostic Studies (Sorted by Publication Year)

Study	Modality	Sample	Features	Classifiers	Validation	Performance
Csernansky et al. [231]	sMRI	SZ=52, HC=65	Hippocampal and thalamic shape eigenvectors	Discriminant function analysis	LOOCV	Accuracy=79%
Davatzikos et al. [232]	sMRI	SZ=69, HC=79	Volumetric measures	Nonlinear classifier	LOOCV	Accuracy=81%
Fan et al. [233]	sMRI	SZ=23, HC=38	Volumetric measures	SVM-RFE	LOOCV	Accuracy=92%
Caan et al. [234]	DTI	SZ=34, HC=24	FA maps	LDA	5-fold CV	Accuracy=75%
Fan et al. [235]	sMRI	SZ=69, HC=79	Volumetric measurements	SVM	LOOCV	Accuracy=91%
Kawasaki et al. [135]	sMRI	SZ=46, HC=46	Mean of eigen image from VBM	Simple thresholding	2-fold validation	Accuracy=80–90%
Caprihan et al. [236]	DTI	SZ=45, HC=45	Discriminant PCA of FA maps	Fisher's linear discriminant	LOOCV	Accuracy=80%
Demirci et al. [237]	fMRI	SZ=57, HC=91	ICA spatial maps	Projection pursuit	LOOCV	Accuracy=80–90%
Yoon et al. [96]	fMRI	SZ=19, HC=15	Active voxels from contrast map	MVPA	CV on different runs	Accuracy=59–72%
Sun et al. [238]	sMRI	SZ=36, HC=36	Cortical GMD	Logistic regression	LOOCV	Accuracy=86%
Shen et al. [110]	rsfMRI	SZ=32, HC=20	Functional connectivity	C-means clustering	LOOCV	Accuracy=86%
Yang et al. [239]	fMRI, SNP	SZ=20, HC=20	Voxels in fMRI activation map, SNPs, ICA maps	Majority voting among 3 SVMs	LOOCV	Accuracy=87%
Ardekani et al. [240]	DTI	SZ=50, HC=50	Voxels of FA and MD maps	LDA	2-fold CV	Accuracy=96%
Castro et al. [241]	fMRI	SZ=52, HC=54	ICA and GLM spatial maps	Recursive composite kernels	Leave-two-out CV	Accuracy=95%
Fan et al. [242]	rsfMRI	SZ=31, HC=31	Functional connectivity	SVMs	LOOCV	Accuracy=85–87%
Karageorgiou et al. [243]	sMRI	SZ=28, HC=47	Volumetric measurements	LDA	LOOCV	Accuracy=72%
Kasparek et al. [244]	sMRI	SZ=39, HC=39	Whole brain voxel intensity value	Maximum-uncertainty LDA	LOOCV	Accuracy=72%
Takayanagi et al. [245]	sMRI	SZ=52, HC=40	Volume and mean cortical thickness of ROIs	Discriminant function analysis	2-fold validation	Accuracy=80%
Bansal et al. [246]	sMRI	SZ=65, HC=40	Surface morphological measures	Hierarchical clustering	LOOCV	Accuracy=94%
Bassett et al. [247]	rsfMRI	SZ=29, HC=29	Size of connected components in graphs	SVM	Bootstrap	Accuracy=75%
Du et al. [248]	fMRI	SZ=28, HC=28	Kernel PCA on ICA maps	Fisher's linear discriminant	LOOCV	Accuracy=93–98%
Greenstein et al. [249]	sMRI	SZ=98, HC=99	Cortical thickness	Random forest	LOOCV	Accuracy=74%
Honorio et al. [250]	fMRI	SZ=13, HC=15	Mean of the largest activation cluster	Majority vote of 3 decision stumps	LOOCV	Accuracy=96%
Nieuwenhuis et al. [64]	sMRI	SZ=283, HC=233	GMD	SVM	LOOCV	Accuracy=71%
Tang et al. [251]	rsfMRI	SZ=22, HC=22	Functional connectivity	SVM	LOOCV	Accuracy=93%
Yoon et al. [127]	fMRI	SZ=51, HC=51	Voxels of left DLPFC in the contrast map	LDA	LOOCV	Accuracy=62%

(Continued)

TABLE 3.2 (CONTINUED)

Summary of Machine Learning-Based SZ Diagnostic Studies (Sorted by Publication Year)

Study	Modality	Sample	Features	Classifiers	Validation	Performance
Anderson and Cohen [252]	rsfMRI	SZ=72, HC=74	Graph metrics based on FNC	SVM	10-fold CV	Accuracy=65%
Arbabshirani et al. [253]	rsfMRI	SZ=29, HC=29	Functional network connectivity	SVM	LOOCV	Accuracy=96%
Cao et al. [254]	fMRI, SNP	SZ=92, HC=116	Sparse representation based variables	Sparse learning	LOOCV	Accuracy=77%
Fekete et al. [255]	rsfMRI	SZ=8, HC=10	Local and global complex network measures	SVM	LOOCV	Accuracy=100%
Iwabuchi et al. [256]	sMRI	SZ=19, HC=20	GM and WM maps	SVM	LOOCV	Accuracy=66.6–77%
Su et al. [118]	rsfMRI	SZ=32, HC=32	Functional connectivity	SVM	LOOCV	Accuracy=83%
Sui et al. [257]	sMRI, rsfMRI, DTI	SZ=35, HC=28	GMD from sMRI, FA from DTI, ALFF from fMRI	SVM	10-fold CV	Accuracy=79%
Yu et al. [258]	rsfMRI	SZ=24, HC=47	Functional connectivity	SVM	LOOCV	Accuracy=62%
Zanetti et al. [259]	sMRI	SZ=62, HC=62	Volumetric measures based on RAVENS	SVM	LOOCV	Accuracy=73%
Zhang and Davatzikos [177]	sMRI	SZ=69, HC=79	Optimally discriminative VBM	SVM	5-fold CV	Accuracy=71%
Anticevic et al. [260]	rsfMRI	SZ=90, HC=90	MVPA based on thalamic connectivity map	SVM	LOOCV	Accuracy=73.9%
Bleich-Cohen et al. [116]	fMRI	SZ=33, HC=20	MVPA on GLM contrast values	SVM	Leave-two-out CV	Accuracy=75–91%
Castro et al. [261]	fMRI	SZ=31, HC=21	ICA spatial maps	Multikernel learning	LOOCV	Accuracy=85%
Guo et al. [117]	rsfMRI	SZ=46, HC=46	fALFF values of the left ITG	SVM	LOOCV	Accuracy=75%
Radulescu et al. [262]	sMRI	SZ=27, HC=24	Texture and volumetric measures	LDA	LOOCV	Accuracy=65.0–72.7%
Watanabe et al. [263]	rsfMRI	SZ=71, HC=74	Functional connectivity	Fused Lasso, GraphNet	5-fold CV	Accuracy=91%
Cetin et al. [264]	fMRI	SZ=27, HC=28	FNC scores	LDA, shapelet-based classifier	CV	Accuracy=72%
Cheng et al. [265]	rsfMRI	SZ=19, HC=29	Graph measure of FC	SVM	LOOCV	Accuracy=80%
Chyzhyk et al. [266]	rsfMRI	SZ=72, HC=74	Correlation from ReHo, (f)ALFF and VMHC	ELMs	10-fold CV	Accuracy=80–91%
Janousova et al. [267]	sMRI	SZ=49, HC=49	MR intensities, GMD, deformation based morphometry	Combination of mMLDA, centroid method, and average linkage	LOOCV	Accuracy=81.6%
Kaufmann et al. [268]	rsfMRI	SZ=71, HC=196	Functional connectivity	Regularized LDA	LOOCV	Accuracy=75–84%
Koch et al. [269]	fMRI	SZ=44, HC=44	MVPA of task activation pattern	Searchlight SVM	LOOCV	Accuracy=93%
Kim et al. [129]	rsfMRI	SZ=50, HC=50	Functional connectivity	Deep neural network	5-fold CV	Accuracy=86%

Sources: Modified from M. R. Arbabshirani et al., *Neuroimage*, vol. 145, pp. 137–65, Jan 2017; B. Sundermann et al., *AJNR Am J Neuroradiol*, vol. 35, pp. 848–55, May 2014; T. Wolfers et al., *Neurosci Biobehav Rev*, vol. 57, pp. 328–49, Oct 2015.

TABLE 3.3

Summary of Machine Learning-Based BPD/MDD Diagnostic Studies (Sorted by Publication Year)

Study	Modality	Sample	Features	Classifiers	Validation	Performance
Gong et al. [270]	sMRI	RDD=23, NDD=23, HC=42	GM and WM densities	SVM	LOOCV	Accuracy=58.7–84.6%
Mourão-Miranda et al. [123]	fMRI	MDD=19, HC=19	Activation maps, ROI-averaged activation features	SVM	LOOCV	Accuracy=63–65.5%
Fang et al. [271]	DTI	MDD=22, HC=26	Anatomical connectivity	SVM	LOOCV	Accuracy=91.7%
Lord et al. [272]	rsfMRI	MDD=21, HC=22	Network-based measures from FC	SVM	Bootstrap	Accuracy=99%
Mwangi et al. [273]	sMRI	MDD=30, HC=32	Feature-based morphometric measures	SVM, RVM	LOOCV	Accuracy=90.3%
Wei et al. [274]	rsfMRI	MDD=20, HC=20	Hurst components of RS networks	SVM	LOOCV	Accuracy=90%
Cao et al. [275]	rsfMRI	MDD=39, HC=37	FC among AAL regions	SVM	LOOCV	Accuracy=76.6%
MacMaster et al. [276]	sMRI	MDD=32, BP=14, HC=22	Volumetric measurements	Discriminant function analysis	2-fold CV	Accuracy=81.0%
Serpa et al. [277]	sMRI	MDD=19, BP=23, HC=71	GM, WM and ventricles volumetric maps	SVM	LOOCV	Accuracy=54.6–66.1%
Zeng et al. [111]	rsfMRI	MDD=24, HC=29	FC maps of sACC	Maximum margin clustering	LOOCV	Accuracy=92.5%
Foland-Ross et al. [278]	sMRI	MDD=18, HC=15	Cortical thickness of ROIs	SVM	10-fold CV	Accuracy=70%
Fung et al. [279]	sMRI	MDD=19, BP=16, HC=29	Cortical thickness and surface area	SVM	10-fold CV	Accuracy=74.3%
Patel et al. [280]	sMRI, rsfMRI, DTI	LLD=33, HC=35	Variety features from each modality	Alternating decision trees	LOOCV	Accuracy=87.3%
Rosa et al. [281]	fMRI	MDD=19, HC=19	Sparse network-based features	SVM	LOSGO-CV	Accuracy=78.9–85.0%
Sacchet et al. [282]	sMRI	MDD=57, RMD=35, BP=40, HC=61	GMV of caudate and ventral diencephalon	SVM	10-fold CV	Accuracy=59.5–62.7%
Sato et al. [283]	fMRI	MDD=25, HC=21	GM maps of PPI analysis	Maximum entropy LDA	LOOCV	AUC=78.1%
Shimizu et al. [284]	fMRI	MDD=25, HC=31	Voxel-wise contrast map	Logistic regression, SVM	10-fold CV	Accuracy=90–95%

Sources: Modified from M. R. Arbabshirani et al., *Neuroimage*, vol. 145, pp. 137–65, Jan 2017; B. Sundermann et al., *AJNR Am J Neuroradiol*, vol. 35, pp. 848–55, May 2014; T. Wolfers et al., *Neurosci Biobehav Rev*, vol. 57, pp. 328–49, Oct 2015.

a model is excessively complex or when the training data have a very small number of samples but a large number of features. Neuroimaging datasets used in pattern recognition studies are based on a small to medium number of samples, which are much less than the number of derived features. Tables 3.1 through 3.5 show that reported accuracies in existing studies tend to decrease with increasing sample size; this trend indicates that many of these studies may suffer from overfitting. Effective feature extraction and selection, as well as proper model choosing and CV, can help control for overfitting. To date, some neuroimaging machine learning studies have validated their model on a completely independent dataset [64,65,135,136]. The decrement in accuracy on the independent test set implies that overfitting still exists in CV, which is likely caused by a limited sample size or allowing the validation samples to influence model hyperparameters or choice (e.g., trying several different machine learning algorithms, using the whole dataset, and choosing the one that worked best). Since the goal of using machine learning techniques on neuroimaging data is often to build computer-assisted systems for clinical diagnoses of neurological and psychiatric diseases, good generalization performance is ultimately required for classification models. Given the reality of overfitting caused by limited sample sizes, the best approach to validate a model's generalizability is testing it on a completely independent dataset that was not used in any step of model construction [285].

3.4.2 Heterogeneity in Patients and Disease Subtype Classification

Neurological and psychiatric disorders may be characterized by different etiologic factors across subjects. Since the same disease may result in different symptoms,

TABLE 3.4

Summary of Machine Learning-Based ASD Diagnostic Studies (Sorted by Publication Year)

Study	Modality	Sample	Features	Classifiers	Validation	Performance
Ecker et al. [66]	sMRI	ASD=22, HC=22	GM and WM maps	SVM	LOOCV	Accuracy=77%
Ecker et al. [286]	sMRI	ASD=20, HC=20	Volumetric and geometric features	SVM	LOOCV	Accuracy=85%
Jiao et al. [67]	sMRI	ASD=22, HC=16	Regional thickness	Logistic model trees	10-fold CV	Accuracy=87%
Anderson et al. [287]	rsfMRI	ASD=40, TDC=40	FC among ROIs	Thresholding	LOOCV	Accuracy=79%
Ingalhalikar et al. [288]	DTI	ASD=45, TDC=30	FA and MD of ROIs	SVM	LOOCV	Accuracy=80%
Uddin et al. [101]	sMRI	ASD=24, TDC=24	Voxel-wise GM and WM maps	SVM	LOOCV	Accuracy=92%
Calderoni et al. [289]	sMRI	ASD=30, TDC=38	GM maps	SVM	Leave-pair-out CV	AUC=80%
Murdaugh et al. [290]	fMRI	ASD=13, TDC=14	AG-, MPFC-, and PCC-based FC maps	Logistic regression	LOOCV	Accuracy=96%
Deshpande et al. [291]	fMRI, DTI	ASD=15, TDC=15	Causal connectivity, FC, FA	SVM	10-fold CV	Accuracy=96%
Uddin et al. [292]	rsfMRI	ASD=20, TDC=20	ICA components	Logistic regression	2-fold validation	Accuracy=78%
Segovia et al. [293]	sMRI	ASD=52, ASD-Sib=40, HC=40	GM volume map	SVM	10-fold CV	Accuracy=80–85%
Wee et al. [294]	sMRI	ASD=58, HC=59	Thickness and volumetric of ROIs	Multikernel SVM	2-fold CV	Accuracy=96.3%
Zhou et al. [295]	sMRI, rsfMRI	ASD=127, TDC=153	Volume of subcortical ROIs, fALFF, number of voxels and z-values of ROIs, VMHC	Random tree classifier	10-fold CV	Accuracy=70%
Chen et al. [296]	rsfMRI	ASD=126, TDC=126	FC among ROIs	Random forest	Bootstrap	Accuracy=91%
Gori et al. [297]	sMRI	ASD=21, HC=20	Morphometric features	SVM	Leave-pair-out CV	AUC=74%
Iidaka [80]	rsfMRI	ASD=312, TDC=328	FC among ROIs	Probabilistic neural network	LOOCV	Accuracy=90%
Libero et al. [298]	sMRI, DTI, MRS	ASD=19, TDC=18	Cortical thickness, FA, neurochemical concentration	Decision tree	LOOCV	Accuracy=91.9%
Plitt et al. [299]	rsfMRI	ASD=148, TDC=148	FC among ROIs	Logistic regression, SVM	LOOCV	Accuracy=76.7%

Sources: Modified from M. R. Arbabshirani et al., *Neuroimage*, vol. 145, pp. 137–65, Jan 2017; T. Wolfers et al., *Neurosci Biobehav Rev*, vol. 57, pp. 328–49, Oct 2015.

different features might be meaningful across patients in the same diagnostic group, making it difficult to generate a common multivariate discriminative pattern to classify a specific diagnostic group. The heterogeneity in patients, as well as other factors such as sample size and scanning parameters, also makes it difficult to compare classifiers' performances across studies.

One approach to solve these problems is to partition patients with the same disease into smaller, less heterogeneous subgroups. Indeed, as it becomes clearer that the clinical diagnoses are quite heterogeneous, there is increased interest in finding subgroups of psychiatric and neurologic disorders [300–302]. This approach aligns with current attempts by the National Institute on Mental Health to characterize disease by Research Domain Criteria—the RDoc initiative [303]. Subdividing data can be done in two ways: by looking at neurophysiologic measures for subgroupings that might identify previously unrecognized clinical subgroups; or by trying to subgroup patients by other measures known to be associated with the disease but not to define it. The biological validity of these subgroups could then be evaluated by applying machine learning algorithms to neuroimaging derived features.

3.4.3 Multimodal Data Fusion

Multimodal neuroimaging data provide new opportunities to systematically characterize human brain structure and function; thus, multimodal data fusion can utilize the strengths of each imaging modality and explore their interrelationships. Multiple multimodal

TABLE 3.5

Summary of Machine Learning-Based ASD Diagnostic Studies (Sorted by Publication Year)

Study	Modality	Sample	Features	Classifiers	Validation	Performance
Zhu et al. [304]	rsfMRI	ADHD=12 HC=12	ReHo maps	PCA-based FDA	LOOCV	Accuracy=85%
Bansal et al. [246]	sMRI	ADHD=41 HC=42	Surface morphometric measures	Hierarchical clustering	LOOCV	Accuracy=91.0%
Bohland et al. [305]	sMRI, rsfMRI	ADHD=285 TDC=491	Anatomical, network, and phenotypic measures	SVM	2-fold CV	AUC=80.0%
Chang et al. [306]	sMRI	ADHD=210 HC=226	Texture features	SVM	10-fold CV	Accuracy=69.9%
Colby et al. [307]	sMRI, rsfMRI	ADHD=285 TDC=491	Morphological measures, FC, power spectra, graph measures	SVM-RFE	10-fold CV	Accuracy=55%
Dai et al. [308]	sMRI, rsfMRI	ADHD=222 TDC=402	Cortical thickness, GM maps, ReHo	SVM, multikernel learning	10-fold CV	Accuracy=61.5%
Fair et al. [309]	rsfMRI	ADHD=192 HC=455	Graph-based features	SVM-based MVPA	LOOCV	Accuracy=63.4–82.7%
Igual et al. [310]	sMRI	ADHD=39 HC=39	Caudate nucleus volumetric measures	Adaboost, SVM	5-fold CV	Accuracy=72.5%
Sato et al. [311]	rsfMRI	ADHD=383 HC=546	ReHo, ALFF, RSN	Logistic regression	Monte Carlo subsampling	Accuracy=54%
Sidhu et al. [312]	rsfMRI	ADHD=239 HC=429	FFT and different variation of PCA on BOLD signals	SVM	10-fold CV	Accuracy=68.86–76%
Peng et al. [313]	sMRI	ADHD=55 HC=55	Cortical thickness measures	ELM	LOOCV	Accuracy=90.2%
Wang et al. [119]	rsfMRI	ADHD=23 HC=23	ReHo maps	SVM	LOOCV	Accuracy=80%
Dey et al. [120]	rsfMRI	ADHD=180 HC=307	Graph-based measures	SVM	2-fold validation	Accuracy=73.5%
Hart et al. [314]	fMRI	ADHD=30 HC=30	GLM coefficient map	Gaussian process	LOOCV	Accuracy=77%
Hart et al. [315]	fMRI	ADHD=20 HC=20	Brain activation map	Gaussian process	LOOCV	Accuracy=75%
Johnston et al. [316]	sMRI	ADHD=34 HC=34	WM maps	SVM	LOOCV	Accuracy=93%
Deshpande et al. [317]	rsfMRI	ADHD=433 TDC=744	Directional connectivity measures	ANN	LOOCV	Accuracy=90%
Iannaccone et al. [121]	sMRI, fMRI	ADHD=18 HC=18	GLM coefficients and GM maps	SVM	LOOCV	Accuracy=61.1–77.8%

Sources: Modified from M. R. Arbabshirani et al., *Neuroimage*, vol. 145, pp. 137–65, Jan 2017; T. Wolfers et al., *Neurosci Biobehav Rev*, vol. 57, pp. 328–49, Oct 2015.

studies indicate that patients with brain disorders exhibit unique functional connectivity patterns and structural characteristics that could not be revealed using separate unimodal analyses as typically performed in traditional neuroimaging experiments [318–320]. Hence, applying machine learning-based pattern classification to these multimodal characteristics could identify specifically useful biomarkers of neurological and psychiatric diseases, so as to facilitate more accurate diagnosis and lead to more appropriate treatments.

3.4.4 Translation to Real Clinical Utility

Valuable findings in machine learning-based neuroimaging studies should ultimately be translated into real clinical applications to diagnose brain disorders. In a typical research study, patients with a single disease diagnosis are included, while subjects with uncertain diagnoses or comorbidities are excluded. Patients and HCs are usually well balanced to train a machine learning model to get an unbiased classifier. However, the diagnostic process in real clinical populations is more complex than in research settings. Patients frequently have multiple comorbid diseases; for example, drug use and abuse is very high in psychiatric populations. Furthermore, populations of psychiatric and neurological patients are much smaller than that of HCs, which means that classifier performance derived from an assumption of equal group sizes—typical in research settings—may have reduced utility for predictions in real clinical cases. Therefore, considerable work remains for translating machine learning-based diagnosis toward clinical practice.

Acknowledgements

This work was supported by the Intramural Research Program of the National Institute on Drug Abuse.

List of Abbreviations in Tables (Sorted by Initial)

AAL	automated anatomical labeling
AD	Alzheimer's disease
ADHD	attention deficit hyperactivity disorder
AG	angular gyrus
ALFF	amplitude of low-frequency fluctuations
ANN	artificial neural network
ASD	autistic spectrum disorder
ASD-Sib	siblings of individuals with ASD
AUC	area under the curve
BOLD	blood-oxygen-level dependent
BP	bipolar depression
CSF	cerebrospinal fluid
CV	cross-validation
DA	axial diffusion
DLPFC	dorsal lateral prefrontal cortex
DR	radial diffusion
DTI	diffusion tensor imaging
ELM	extreme learning machines
ERC	entorhinal cortex
FA	fractional anisotropy
fALFF	fractional amplitude of low-frequency fluctuations
FC	functional connectivity
FDA	Fisher discriminative analysis
FFT	fast Fourier transform
fMRI	functional MRI
FNC	functional network connectivity
FTD	frontotemporal dementia
GLM	general linear model
GM	gray matter
GMD	gray matter density
GMV	gray matter volume
HC	healthy controls
ICA	independent component analysis
ITG	inferior temporal gyrus
Lasso	least absolute shrinkage and selection operator
LDA	linear discriminant analysis
LLD	late-life depression
LOOCV	leave-one-out cross-validation
LOSGO-CV	leave-one-subject-per-group-out cross-validation
LR	logistic regression
MCI	mild cognitive impairment
MD	mean diffusion
MDD	major depressive disorder
MLP	multilayer perceptron
mMLDA	modified maximum uncertainty linear discriminant analysis
MPFC	medial prefrontal cortex
MR	magnetic resonance
MRS	magnetic resonance spectroscopy
MTR	magnetization transfer ratio
MVPA	multivariate pattern analysis
NDD	nonrefractory depressive disorder
ODVBA	optimally discriminative voxel-based analysis
PCA	principle component analysis
PCC	posterior cingulate cortex
PDF	probability distribution function
PET	positron emission tomography
PLS	partial least squares
PPI	psychophysiological interaction
QDA	quadratic discriminant analysis
RAVENS	regional analysis of volumes examined in normalized space
RDD	refractory depressive disorder
ReHo	regional homogeneity
RMD	remitted major depressive disorder
ROC	receiver operational characteristic
ROI	region of interest
RS	resting state
rsfMRI	resting-state fMRI
RSN	resting-state networks
RVM	relevance vector machine
RVoxM	relevance voxel machine
sACC	subgenual anterior cingulate cortex
SIFT	scale invariant feature transform
sMRI	structural MRI
SNP	single-nucleotide polymorphism
SUVr	standard uptake value ratio
SVM	support vector machine
SVM-RFE	support vector machine with recursive feature elimination
SZ	schizophrenia
TDC	typically developing controls
VaD	vascular dementia
VBM	voxel-based morphometry
VMHC	voxel-mirrored homotopic connectivity
VOI	volume of interest
WM	white matter
WMD	white matter density

References

1. J. Ashburner and K. J. Friston, "Voxel-based morphometry—The methods," *Neuroimage*, vol. 11, pp. 805–21, Jun 2000.
2. B. Fischl and A. M. Dale, "Measuring the thickness of the human cerebral cortex from magnetic resonance images," *Proc Natl Acad Sci U S A*, vol. 97, pp. 11050–5, Sep 26, 2000.
3. P. J. Basser, J. Mattiello, and D. Lebihan, "Estimation of the effective self-diffusion tensor from the NMR spin echo," *J Magn Reson B*, vol. 103, pp. 247–54, 1994.
4. P. J. Basser, S. Pajevic, C. Pierpaoli, J. Duda, and A. Aldroubi, "In vivo fiber tractography using DT-MRI data," *Magn Reson Med*, vol. 44, pp. 625–32, Oct 2000.
5. T. L. Chenevert, J. A. Brunberg, and J. G. Pipe, "Anisotropic diffusion in human white matter: Demonstration with MR techniques in vivo," *Radiology*, vol. 177, pp. 401–5, Nov 1990.
6. V. J. Wedeen, P. Hagmann, W.-Y. I. Tseng, T. G. Reese, and R. M. Weisskoff, "Mapping complex tissue architecture with diffusion spectrum magnetic resonance imaging," *Magn Reson Med*, vol. 54, pp. 1377–86, 2005.
7. C. S. Roy and C. S. Sherrington, "On the regulation of the blood-supply of the brain," *J Physiol*, vol. 11, pp. 85–158.17, 1, 1890.
8. M. E. Raichle and M. A. Mintun, "Brain work and brain imaging," *Ann Rev Neurosci*, vol. 29, pp. 449–476, 7, 2006.
9. P. T. Fox and M. E. Raichle, "Stimulus rate determines regional brain blood flow in striate cortex," *Ann Neurol*, vol. 17, pp. 303–5, 3, 1985.
10. L. Pauling and C. D. Coryell, "The magnetic properties and structure of hemoglobin, oxyhemoglobin and carbon monoxyhemoglobin," *Proc Natl Acad Sci U S A*, vol. 22, pp. 210–6, 03, 1936.
11. P. A. Bandettini, E. C. Wong, R. S. Hinks, R. S. Tikofky, and J. S. Hyde, "Time course EPI of human brain function during task activation," *Magn Reson Med*, vol. 25, pp. 390–7, 06, 1992.
12. W. van der Zwaag, S. Francis, K. Head, A. Peters, P. Gowland, P. Morris et al., "fMRI at 1.5, 3 and 7 T: Characterising BOLD signal changes," *Neuroimage*, vol. 47, pp. 1425–34, 10, 2009.
13. J. S. Gati, R. S. Menon, K. Ugurbil, and B. K. Rutt, "Experimental determination of the BOLD field strength dependence in vessels and tissue," *Magn Reson Med*, vol. 38, pp. 296–302, 8, 1997.
14. T. Okada, H. Yamada, H. Ito, Y. Yonekura, and N. Sadato, "Magnetic field strength increase yields significantly greater contrast-to-noise ratio increase: Measured using BOLD contrast in the primary visual area," *Acad Radiol*, vol. 12, pp. 142–7, 2, 2005.
15. K. Pruessmann, M. Weiger, M. Scheidegger, and P. Boesiger, "SENSE: Sensitivity encoding for fast MRI," *Magn Reson Med*, vol. 42, pp. 952–62, 11, 1999.
16. S. Moeller, E. Yacoub, C. A. Olman, E. Auerbach, J. Strupp, N. Harel et al., "Multiband multislice GE-EPI at 7 tesla, with 16-fold acceleration using partial parallel imaging with application to high spatial and temporal whole-brain fMRI," *Magn Reson Med*, vol. 63, pp. 1144–53, May 2010.
17. WU–Minn–Oxford consortium—HCP. Available at https://www.humanconnectome.org/
18. C. Rosazza and L. Minati, "Resting-state brain networks: Literature review and clinical applications," *Neurol Sci*, vol. 32, pp. 773–85, Oct 2011.
19. S. M. Smith, P. T. Fox, K. L. Miller, D. C. Glahn, P. M. Fox, C. E. Mackay et al., "Correspondence of the brain's functional architecture during activation and rest," *Proc Natl Acad Sci U S A*, vol. 106, pp. 13040–5, Aug 2009.
20. The MGH/Harvard–UCLA consortium—HCP. Available at http://www.humanconnectomeproject.org/
21. HCP Lifespan Pilot Project. Available at http://www.lifespan.humanconnectome.org/
22. *WU–Minn HCP 500 Subjects + MEG2 Data Release: Reference Manual*. Available at http://www.humanconnectome.org/documentation/S500/
23. K. Uğurbil, X. Xu, E. J. Auerbach, S. Moeller, A. T. Vu, J. M. Duarte-Carvajalino et al., "Pushing spatial and temporal resolution for functional and diffusion MRI in the Human Connectome Project," *NeuroImage*, vol. 80, pp. 80–104, 2013.
24. 900 Subjects Data Release. Available at https://www.humanconnectome.org/documentation/S900/
25. MGH Adult Diffusion Data. Available at https://www.humanconnectome.org/documentation/MGH-diffusion
26. K. Setsompop, R. Kimmlingen, E. Eberlein, T. Witzel, J. Cohen-Adad, J. A. McNab et al., "Pushing the limits of in vivo diffusion MRI for the Human Connectome Project," *NeuroImage*, vol. 80, pp. 220–33, 2013.
27. 1000 Functional Connectomes Project. Available at http://fcon_1000.projects.nitrc.org/
28. B. B. Biswal, M. Mennes, X. N. Zuo, S. Gohel, C. Kelly, S. M. Smith et al., "Toward discovery science of human brain function," *Proc Natl Acad Sci U S A*, vol. 107, pp. 4734–9, Mar 9, 2010.
29. NITRC. Available at http://www.nitrc.org/
30. D. Tomasi and N. D. Volkow, "Functional connectivity density mapping," *Proc Natl Acad Sci U S A*, vol. 107, pp. 9885–90, May 25, 2010.
31. Federal Interagency Traumatic Brain Injury Research (FITBIR). Available at https://fitbir.nih.gov/
32. Sapient Government Services. Available at http://www.sapientgovernmentservices.com/content/dam/sgs/pdf/OurWork/508/FITBIR.pdf
33. C. M. Lajonchere, "Changing the landscape of autism research: The autism genetic resource exchange," *Neuron*, vol. 68, pp. 187–91, Oct 21, 2010.
34. D. Hall, M. F. Huerta, M. J. McAuliffe, and G. K. Farber, "Sharing heterogeneous data: The national database for autism research," *Neuroinformatics*, vol. 10, pp. 331–9, 2012.
35. A. C. Evans, "The NIH MRI study of normal brain development," *Neuroimage*, vol. 30, pp. 184–202, Mar 2006.

36. S. B. Johnson, G. Whitney, M. McAuliffe, H. Wang, E. McCreedy, L. Rozenblit et al., "Using global unique identifiers to link autism collections," *J Am Med Inform Assoc*, vol. 17, pp. 689–95, Nov–Dec 2010.
37. National Database for Autism Research (NDAR).
38. J. L. Ambite, M. Tallis, K. Alpert, D. B. Keator, M. King, D. Landis et al., "SchizConnect: Virtual data integration in neuroimaging," *Data Integr Life Sci*, vol. 9162, pp. 37–51, Jul 2015.
39. SchizConnect. Available at http://schizconnect.org/
40. K. G. Helmer, J. L. Ambite, J. Ames, R. Ananthakrishnan, G. Burns, A. L. Chervenak et al., "Enabling collaborative research using the Biomedical Informatics Research Network (BIRN)," *J Am Med Inform Assoc*, vol. 18, pp. 416–22, 2011.
41. G. H. Glover, B. A. Mueller, J. A. Turner, T. G. van Erp, T. T. Liu, D. N. Greve et al., "Function biomedical informatics research network recommendations for prospective multicenter functional MRI studies," *J Magn Reson Imaging*, vol. 36, pp. 39–54, Jul 2012.
42. L. Wang, A. Kogan, D. Cobia, K. Alpert, A. Kolasny, M. I. Miller et al., "Northwestern University Schizophrenia Data and Software Tool (NUSDAST)," *Front Neuroinform*, vol. 7, p. 25, 2013.
43. P. A. Harris, R. Taylor, R. Thielke, J. Payne, N. Gonzalez, and J. G. Conde, "Research electronic data capture (REDCap)—A metadata-driven methodology and workflow process for providing translational research informatics support," *J Biomed Inform*, vol. 42, pp. 377–81, 4, 2009.
44. L. Wang, K. I. Alpert, V. D. Calhoun, D. J. Cobia, D. B. Keator, M. D. King et al., "SchizConnect: Mediating neuroimaging databases on schizophrenia and related disorders for large-scale integration," *Neuroimage*, vol. 124, pp. 1155–67, Jan 1, 2016.
45. M. S. Çetin, F. Christensen, C. C. Abbott, J. M. Stephen, A. R. Mayer, J. M. Cañive et al., "Thalamus and posterior temporal lobe show greater inter-network connectivity at rest and across sensory paradigms in schizophrenia," *NeuroImage*, vol. 97, pp. 117–126, 2014.
46. R. L. Gollub, J. M. Shoemaker, M. D. King, T. White, S. Ehrlich, S. R. Sponheim et al., "The MCIC collection: A shared repository of multi-modal, multi-site brain image data from a clinical investigation of schizophrenia," *Neuroinformatics*, vol. 11, pp. 367–88, Jul 2013.
47. A. Scott, W. Courtney, D. Wood, R. de la Garza, S. Lane, M. King et al., "COINS: An innovative informatics and neuroimaging tool suite built for large heterogeneous datasets," *Front Neuroinform*, vol. 5, p. 33, 2011.
48. K. Alpert, A. Kogan, T. Parrish, D. Marcus, and L. Wang, "The Northwestern University Neuroimaging Data Archive (NUNDA)," *NeuroImage*, vol. 124, Part B, pp. 1131–6, 2016.
49. M. W. Weiner, P. S. Aisen, C. R. Jack, Jr., W. J. Jagust, J. Q. Trojanowski, L. Shaw et al., "The Alzheimer's disease neuroimaging initiative: Progress report and future plans," *Alzheimers Dement*, vol. 6, pp. 202–11, e7, May 2010.
50. ADNI. Available at http://adni.loni.usc.edu/
51. M. W. Weiner, D. P. Veitch, P. S. Aisen, L. A. Beckett, N. J. Cairns, J. Cedarbaum et al., "2014 Update of the Alzheimer's Disease Neuroimaging Initiative: A review of papers published since its inception," *Alzheimers Dement*, vol. 11, pp. e1–120, Jun 2015.
52. ADNI-info. Available at http://www.adni-info.org/Home.html
53. B. T. Wyman, D. J. Harvey, K. Crawford, M. A. Bernstein, O. Carmichael, P. E. Cole et al., "Standardization of analysis sets for reporting results from ADNI MRI data," *Alzheimers Dement*, vol. 9, pp. 332–7, May 2013.
54. Imagen-Europe. Available at https://imagen-europe.com/
55. G. Schumann, E. Loth, T. Banaschewski, A. Barbot, G. Barker, C. Buchel et al., "The IMAGEN study: Reinforcement-related behaviour in normal brain function and psychopathology," *Mol Psychiatr*, vol. 15, pp. 1128–39, Dec 2010.
56. M. E. Shenton, C. C. Dickey, M. Frumin, and R. W. McCarley, "A review of MRI findings in schizophrenia," *Schizophr Res*, vol. 49, pp. 1–52, Apr 15, 2001.
57. Y. Liu, K. Wang, C. Yu, Y. He, Y. Zhou, M. Liang et al., Regional homogeneity, functional connectivity and imaging markers of Alzheimer's disease: A review of resting-state fMRI studies, *Neuropsychologia*, vol. 46, pp. 1648–56, 2008.
58. L. Wang, D. F. Hermens, I. B. Hickie, and J. Lagopoulos, "A systematic review of resting-state functional-MRI studies in major depression," *J Affect Disord*, vol. 142, pp. 6–12, Dec 15, 2012.
59. E. Bullmore and O. Sporns, "The economy of brain network organization," *Nat Rev Neurosci*, vol. 13, pp. 336–49, May 2012.
60. Y. He, Z. J. Chen, and A. C. Evans, "Small-world anatomical networks in the human brain revealed by cortical thickness from MRI," *Cereb Cortex*, vol. 17, pp. 2407–19, Oct 2007.
61. G. Fort and S. Lambert-Lacroix, "Classification using partial least squares with penalized logistic regression," *Bioinformatics*, vol. 21, pp. 1104–11, Apr 1, 2005.
62. D. M. Hawkins, "The problem of overfitting," *J Chem Inform Comput Sci*, vol. 44, pp. 1–12, Jan–Feb 2004.
63. E. R. Dougherty, "Feature-selection overfitting with small-sample classifier design," *IEEE Intelligent Syst*, vol. 20, pp. 64–6, Nov–Dec 2005.
64. M. Nieuwenhuis, N. E. van Haren, H. E. Hulshoff Pol, W. Cahn, R. S. Kahn, and H. G. Schnack, "Classification of schizophrenia patients and healthy controls from structural MRI scans in two large independent samples," *Neuroimage*, vol. 61, pp. 606–12, Jul 2012.
65. X. Ding, Y. Yang, E. A. Stein, and T. J. Ross, "Multivariate classification of smokers and nonsmokers using SVM-RFE on structural MRI images," *Hum Brain Mapp*, vol. 36, pp. 4869–79, Dec 2015.
66. C. Ecker, V. Rocha-Rego, P. Johnston, J. Mourao-Miranda, A. Marquand, E. M. Daly et al., "Investigating the predictive value of whole-brain structural MR scans in autism: A pattern classification approach," *Neuroimage*, vol. 49, pp. 44–56, Jan 2010.

67. Y. Jiao, R. Chen, X. Ke, K. Chu, Z. Lu, and E. H. Herskovits, "Predictive models of autism spectrum disorder based on brain regional cortical thickness," *Neuroimage*, vol. 50, pp. 589–99, Apr 2010.
68. J. R. Sato, G. M. de Araujo Filho, T. B. de Araujo, R. A. Bressan, P. P. de Oliveira, and A. P. Jackowski, "Can neuroimaging be used as a support to diagnosis of borderline personality disorder? An approach based on computational neuroanatomy and machine learning," *J Psychiatr Res*, vol. 46, pp. 1126–32, Sep 2012.
69. A. Rizk-Jackson, D. Stoffers, S. Sheldon, J. Kuperman, A. Dale, J. Goldstein et al., "Evaluating imaging biomarkers for neurodegeneration in pre-symptomatic Huntington's disease using machine learning techniques," *Neuroimage*, vol. 56, pp. 788–96, May 2011.
70. E. C. Robinson, A. Hammers, A. Ericsson, A. D. Edwards, and D. Rueckert, "Identifying population differences in whole-brain structural networks: A machine learning approach," *Neuroimage*, vol. 50, pp. 910–9, Apr 15, 2010.
71. M. Dyrba, M. Ewers, M. Wegrzyn, I. Kilimann, C. Plant, A. Oswald et al., "Robust automated detection of microstructural white matter degeneration in Alzheimer's disease using machine learning classification of multicenter DTI data," *PLoS ONE*, vol. 8, p. e64925, 2013.
72. F. De Martino, G. Valente, N. Staeren, J. Ashburner, R. Goebel, and E. Formisano, "Combining multivariate voxel selection and support vector machines for mapping and classification of fMRI spatial patterns," *Neuroimage*, vol. 43, pp. 44–58, Oct 2008.
73. T. M. Ball, M. B. Stein, H. J. Ramsawh, L. Campbell-Sills, and M. P. Paulus, "Single-subject anxiety treatment outcome prediction using functional neuroimaging," *Neuropsychopharmacology*, vol. 39, pp. 1254–61, Apr 2014.
74. F. De Martino, F. Gentile, F. Esposito, M. Balsi, F. Di Salle, R. Goebel et al., "Classification of fMRI independent components using IC-fingerprints and support vector machine classifiers," *Neuroimage*, vol. 34, pp. 177–94, Jan 1, 2007.
75. P. K. Douglas, S. Harris, A. Yuille, and M. S. Cohen, "Performance comparison of machine learning algorithms and number of independent components used in fMRI decoding of belief vs. disbelief," *Neuroimage*, vol. 56, pp. 544–53, May 15, 2011.
76. V. Pariyadath, E. A. Stein, and T. J. Ross, "Machine learning classification of resting state functional connectivity predicts smoking status," *Front Hum Neurosci*, vol. 8, p. 425, 2014.
77. Z. Dai, C. Yan, Z. Wang, J. Wang, M. Xia, K. Li et al., "Discriminative analysis of early Alzheimer's disease using multi-modal imaging and multi-level characterization with multi-classifier (M3)," *Neuroimage*, vol. 59, pp. 2187–95, Feb 2012.
78. E. D. Fagerholm, P. J. Hellyer, G. Scott, R. Leech, and D. J. Sharp, "Disconnection of network hubs and cognitive impairment after traumatic brain injury," *Brain*, vol. 138, pp. 1696–709, Jun 2015.
79. G. Deshpande, Z. Li, P. Santhanam, C. D. Coles, M. E. Lynch, S. Hamann et al., "Recursive cluster elimination based support vector machine for disease state prediction using resting state functional and effective brain connectivity," *PLoS ONE*, vol. 5, p. e14277, 2010.
80. T. Iidaka, "Resting state functional magnetic resonance imaging and neural network classified autism and control," *Cortex*, vol. 63, pp. 55–67, Feb 2015.
81. C. Chu, A. L. Hsu, K. H. Chou, P. Bandettini, C. Lin, and A. s. D. N. Initiative, "Does feature selection improve classification accuracy? Impact of sample size and feature selection on classification using anatomical magnetic resonance images," *Neuroimage*, vol. 60, pp. 59–70, Mar 2012.
82. N. Tzourio-Mazoyer, B. Landeau, D. Papathanassiou, F. Crivello, O. Etard, N. Delcroix et al., "Automated anatomical labeling of activations in SPM using a macroscopic anatomical parcellation of the MNI MRI single-subject brain," *Neuroimage*, vol. 15, pp. 273–89, Jan 2002.
83. Q. Cao, N. Shu, L. An, P. Wang, L. Sun, M. R. Xia et al., "Probabilistic diffusion tractography and graph theory analysis reveal abnormal white matter structural connectivity networks in drug-naive boys with attention deficit/hyperactivity disorder," *J Neurosci*, vol. 33, pp. 10676–87, Jun 2013.
84. A. Zalesky, A. Fornito, I. H. Harding, L. Cocchi, M. Yücel, C. Pantelis et al., "Whole-brain anatomical networks: Does the choice of nodes matter?," *Neuroimage*, vol. 50, pp. 970–83, Apr 2010.
85. H. I. Suk, S. W. Lee, and D. Shen, "Latent feature representation with stacked auto-encoder for AD/MCI diagnosis," *Brain Struct Funct*, vol. 220, pp. 841–59, Mar 2015.
86. R. Cuingnet, E. Gerardin, J. Tessieras, G. Auzias, S. Lehericy, M. O. Habert et al., "Automatic classification of patients with Alzheimer's disease from structural MRI: A comparison of ten methods using the ADNI database," *Neuroimage*, vol. 56, pp. 766–81, May 15, 2011.
87. S. B. Eickhoff, A. R. Laird, C. Grefkes, L. E. Wang, K. Zilles, and P. T. Fox, "Coordinate-based activation likelihood estimation meta-analysis of neuroimaging data: A random-effects approach based on empirical estimates of spatial uncertainty," *Hum Brain Mapp*, vol. 30, pp. 2907–26, Sep 2009.
88. T. D. Wager, M. Lindquist, and L. Kaplan, "Meta-analysis of functional neuroimaging data: Current and future directions," *Soc Cogn Affect Neurosci*, vol. 2, pp. 150–8, Jun 2007.
89. T. Yarkoni, R. A. Poldrack, T. E. Nichols, D. C. Van Essen, and T. D. Wager, "Large-scale automated synthesis of human functional neuroimaging data," *Nat Methods*, vol. 8, pp. 665–70, Aug 2011.
90. T. M. Mitchell, "From journal articles to computational models: A new automated tool," *Nat Methods*, vol. 8, pp. 627–8, Aug 2011.
91. J. Dukart, K. Mueller, H. Barthel, A. Villringer, O. Sabri, M. L. Schroeter et al., "Meta-analysis based SVM classification enables accurate detection of Alzheimer's

disease across different clinical centers using FDG-PET and MRI," *Psychiatry Res*, vol. 212, pp. 230–6, Jun 30, 2013.
92. D. Zhang, Y. Wang, L. Zhou, H. Yuan, and D. Shen, "Multimodal classification of Alzheimer's disease and mild cognitive impairment," *Neuroimage*, vol. 55, pp. 856–67, Apr 2011.
93. C. Y. Wee, P. T. Yap, D. Zhang, K. Denny, J. N. Browndyke, G. G. Potter et al., "Identification of MCI individuals using structural and functional connectivity networks," *Neuroimage*, vol. 59, pp. 2045–56, Feb 1, 2012.
94. S. G. Costafreda, C. H. Fu, M. Picchioni, T. Toulopoulou, C. McDonald, E. Kravariti et al., "Pattern of neural responses to verbal fluency shows diagnostic specificity for schizophrenia and bipolar disorder," *BMC Psychiatr*, vol. 11, p. 18, 2011.
95. M. N. Coutanche, S. L. Thompson-Schill, and R. T. Schultz, "Multi-voxel pattern analysis of fMRI data predicts clinical symptom severity," *Neuroimage*, vol. 57, pp. 113–23, Jul 1, 2011.
96. J. H. Yoon, D. Tamir, M. J. Minzenberg, J. D. Ragland, S. Ursu, and C. S. Carter, "Multivariate pattern analysis of functional magnetic resonance imaging data reveals deficits in distributed representations in schizophrenia," *Biol Psychiatr*, vol. 64, pp. 1035–41, Dec 15, 2008.
97. I. Guyon, J. Weston, S. Barnhill, and V. Vapnik, "Gene selection for cancer classification using support vector machines," *Machine Learning*, vol. 46, pp. 389–422, 2002.
98. R. C. Craddock, P. E. Holtzheimer, 3rd, X. P. Hu, and H. S. Mayberg, "Disease state prediction from resting state functional connectivity," *Magn Reson Med*, vol. 62, pp. 1619–28, Dec 2009.
99. B. Mwangi, T. S. Tian, and J. C. Soares, "A review of feature reduction techniques in neuroimaging," *Neuroinformatics*, vol. 12, pp. 229–44, Apr 2014.
100. N. Kriegeskorte, R. Goebel, and P. Bandettini, "Information-based functional brain mapping," *Proc Natl Acad Sci U S A*, vol. 103, pp. 3863–8, Mar 7, 2006.
101. L. Q. Uddin, V. Menon, C. B. Young, S. Ryali, T. Chen, A. Khouzam et al., "Multivariate searchlight classification of structural magnetic resonance imaging in children and adolescents with autism," *Biol Psychiatr*, vol. 70, pp. 833–41, Nov 2011.
102. J. A. Etzel, J. M. Zacks, and T. S. Braver, "Searchlight analysis: Promise, pitfalls, and potential," *Neuroimage*, vol. 78, pp. 261–9, Sep 2013.
103. R. Casanova, C. T. Whitlow, B. Wagner, J. Williamson, S. A. Shumaker, J. A. Maldjian et al., "High dimensional classification of structural MRI Alzheimer's disease data based on large scale regularization," *Front Neuroinform*, vol. 5, p. 22, 2011.
104. E. Duchesnay, A. Cachia, N. Boddaert, N. Chabane, J. F. Mangin, J. L. Martinot et al., "Feature selection and classification of imbalanced datasets: Application to PET images of children with autistic spectrum disorders," *Neuroimage*, vol. 57, pp. 1003–14, Aug 1, 2011.
105. L. Shen, S. Kim, Y. Qi, M. Inlow, S. Swaminathan, K. Nho et al., "Identifying neuroimaging and proteomic biomarkers for MCI and AD via the elastic net," *Multimodal Brain Image Anal (2011)*, vol. 7012, pp. 27–34, Sep 2011.
106. R. Tibshirani, "Regression shrinkage and selection via the Lasso," *J R Stat Soc B Methodol*, vol. 58, pp. 267–88, 1996.
107. R. Tibshirani, "Regression shrinkage and selection via the lasso: A retrospective," *J R Stat Soc B Stat Methodol*, vol. 73, pp. 273–82, 2011.
108. H. Zou and T. Hastie, "Regularization and variable selection via the elastic net," *J R Stat Soc B Stat Methodol*, vol. 67, pp. 301–20, 2005.
109. A. K. Jain, M. N. Murty, and P. J. Flynn, "Data clustering: A review," *ACM Comput Surveys*, vol. 31, pp. 264–323, Sep 1999.
110. H. Shen, L. Wang, Y. Liu, and D. Hu, "Discriminative analysis of resting-state functional connectivity patterns of schizophrenia using low dimensional embedding of fMRI," *Neuroimage*, vol. 49, pp. 3110–21, Feb 2010.
111. L. L. Zeng, H. Shen, L. Liu, and D. Hu, "Unsupervised classification of major depression using functional connectivity MRI," *Hum Brain Mapp*, vol. 35, pp. 1630–41, Apr 2014.
112. C. J. Burges, "A tutorial on support vector machines for pattern recognition," *Data Mining Knowledge Discov*, vol. 2, pp. 121–67, 1998.
113. A. Khazaee, A. Ebrahimzadeh, and A. Babajani-Feremi, "Identifying patients with Alzheimer's disease using resting-state fMRI and graph theory," *Clin Neurophysiol*, vol. 126, pp. 2132–41, Nov 2015.
114. T. Zhang and C. Davatzikos, "ODVBA: Optimally-discriminative voxel-based analysis," *IEEE Trans Med Imaging*, vol. 30, pp. 1441–54, Aug 2011.
115. P. Vemuri, J. L. Gunter, M. L. Senjem, J. L. Whitwell, K. Kantarci, D. S. Knopman et al., "Alzheimer's disease diagnosis in individual subjects using structural MR images: Validation studies," *Neuroimage*, vol. 39, pp. 1186–97, Feb 1, 2008.
116. M. Bleich-Cohen, S. Jamshy, H. Sharon, R. Weizman, N. Intrator, M. Poyurovsky et al., "Machine learning fMRI classifier delineates subgroups of schizophrenia patients," *Schizophr Res*, vol. 160, pp. 196–200, Dec 2014.
117. W. Guo, Q. Su, D. Yao, J. Jiang, J. Zhang, Z. Zhang et al., "Decreased regional activity of default-mode network in unaffected siblings of schizophrenia patients at rest," *Eur Neuropsychopharmacol*, vol. 24, pp. 545–52, Apr 2014.
118. L. Su, L. Wang, H. Shen, G. Feng, and D. Hu, "Discriminative analysis of non-linear brain connectivity in schizophrenia: An fMRI Study," *Front Hum Neurosci*, vol. 7, p. 702, 2013.
119. X. Wang, Y. Jiao, T. Tang, H. Wang, and Z. Lu, "Altered regional homogeneity patterns in adults with attention-deficit hyperactivity disorder," *Eur J Radiol*, vol. 82, pp. 1552–7, Sep 2013.
120. S. Dey, A. R. Rao, and M. Shah, "Attributed graph distance measure for automatic detection of attention deficit hyperactive disordered subjects," *Front Neural Circuits*, vol. 8, p. 64, 2014.
121. R. Iannaccone, T. U. Hauser, J. Ball, D. Brandeis, S. Walitza, and S. Brem, "Classifying adolescent

121. attention-deficit/hyperactivity disorder (ADHD) based on functional and structural imaging," *Eur Child Adolesc Psychiatr*, vol. 24, pp. 1279–89, Oct 2015.
122. M. Misaki, Y. Kim, P. A. Bandettini, and N. Kriegeskorte, "Comparison of multivariate classifiers and response normalizations for pattern-information fMRI," *Neuroimage*, vol. 53, pp. 103–18, Oct 15, 2010.
123. J. Mourão-Miranda, D. R. Hardoon, T. Hahn, A. F. Marquand, S. C. Williams, J. Shawe-Taylor et al., "Patient classification as an outlier detection problem: An application of the One-Class Support Vector Machine," *Neuroimage*, vol. 58, pp. 793–804, Oct 1, 2011.
124. R. A. Fisher, "The use of multiple measurements in taxonomic problems," *Ann Eugenics*, vol. 7, pp. 179–88, Sep 1936.
125. M. I. Miller, C. E. Priebe, A. Qiu, B. Fischl, A. Kolasny, T. Brown et al., "Collaborative computational anatomy: An MRI morphometry study of the human brain via diffeomorphic metric mapping," *Hum Brain Mapp*, vol. 30, pp. 2132–41, Jul 2009.
126. J. Zhou, M. D. Greicius, E. D. Gennatas, M. E. Growdon, J. Y. Jang, G. D. Rabinovici et al.,"Divergent network connectivity changes in behavioural variant frontotemporal dementia and Alzheimer's disease," *Brain*, vol. 133, pp. 1352–67, May 2010.
127. J. H. Yoon, D. V. Nguyen, L. M. McVay, P. Deramo, M. J. Minzenberg, J. D. Ragland et al., "Automated classification of fMRI during cognitive control identifies more severely disorganized subjects with schizophrenia," *Schizophr Res*, vol. 135, pp. 28–33, Mar 2012.
128. H. I. Suk, S. W. Lee, D. Shen, and A. s. D. N. Initiative, "Hierarchical feature representation and multimodal fusion with deep learning for AD/MCI diagnosis," *Neuroimage*, vol. 101, pp. 569–82, Nov 2014.
129. J. Kim, V. D. Calhoun, E. Shim, and J. H. Lee, "Deep neural network with weight sparsity control and pre-training extracts hierarchical features and enhances classification performance: Evidence from whole-brain resting-state functional connectivity patterns of schizophrenia," *Neuroimage*, vol. 124, pp. 127–46, Jan 1, 2016.
130. H. I. Suk, C. Y. Wee, S. W. Lee, and D. Shen, "State-space model with deep learning for functional dynamics estimation in resting-state fMRI," *Neuroimage*, vol. 129, pp. 292–307, Apr 1, 2016.
131. A. Anderson, J. S. Labus, E. P. Vianna, E. A. Mayer, and M. S. Cohen, "Common component classification: What can we learn from machine learning?," *Neuroimage*, vol. 56, pp. 517–24, May 15, 2011.
132. K. R. Gray, P. Aljabar, R. A. Heckemann, A. Hammers, D. Rueckert, and I. Alzheimer's Disease Neuroimaging, "Random forest-based similarity measures for multi-modal classification of Alzheimer's disease," *Neuroimage*, vol. 65, pp. 167–75, Jan 15, 2013.
133. D. Zhang and D. Shen, "Multi-modal multi-task learning for joint prediction of multiple regression and classification variables in Alzheimer's disease," *Neuroimage*, vol. 59, pp. 895–907, Jan 2012.
134. W. Pettersson-Yeo, S. Benetti, A. F. Marquand, R. Joules, M. Catani, S. C. Williams et al., "An empirical comparison of different approaches for combining multimodal neuroimaging data with support vector machine," *Front Neurosci*, vol. 8, p. 189, 2014.
135. Y. Kawasaki, M. Suzuki, F. Kherif, T. Takahashi, S. Y. Zhou, K. Nakamura et al., "Multivariate voxel-based morphometry successfully differentiates schizophrenia patients from healthy controls," *Neuroimage*, vol. 34, pp. 235–42, Jan 2007.
136. R. Whelan, R. Watts, C. A. Orr, R. R. Althoff, E. Artiges, T. Banaschewski et al., "Neuropsychosocial profiles of current and future adolescent alcohol misusers," *Nature*, vol. 512, pp. 185–9, Aug 2014.
137. J. Mourao-Miranda, A. L. Bokde, C. Born, H. Hampel, and M. Stetter, "Classifying brain states and determining the discriminating activation patterns: Support vector machine on functional MRI data," *Neuroimage*, vol. 28, pp. 980–95, Dec 2005.
138. S. Haufe, F. Meinecke, K. Gorgen, S. Dahne, J. D. Haynes, B. Blankertz et al., "On the interpretation of weight vectors of linear models in multivariate neuroimaging," *Neuroimage*, vol. 87, pp. 96–110, Feb 15, 2014.
139. M. Grundman, R. C. Petersen, S. H. Ferris, R. G. Thomas, P. S. Aisen, D. A. Bennett et al., "Mild cognitive impairment can be distinguished from Alzheimer disease and normal aging for clinical trials," *Arch Neurol*, vol. 61, pp. 59–66, Jan 2004.
140. H. Matsuda, "The role of neuroimaging in mild cognitive impairment," *Neuropathology*, vol. 27, pp. 570–7, Dec 2007.
141. M. R. Arbabshirani, S. Plis, J. Sui, and V. D. Calhoun, "Single subject prediction of brain disorders in neuroimaging: Promises and pitfalls," *Neuroimage*, vol. 145, pp. 137–65, Jan 2017.
142. B. Sundermann, D. Herr, W. Schwindt, and B. Pfleiderer, "Multivariate classification of blood oxygen level-dependent FMRI data with diagnostic intention: A clinical perspective," *AJNR Am J Neuroradiol*, vol. 35, pp. 848–55, May 2014.
143. P. F. Buckley, "Neuroimaging of schizophrenia: Structural abnormalities and pathophysiological implications," *Neuropsychiatr Dis Treat*, vol. 1, pp. 193–204, Sep 2005.
144. H. Cheng, S. D. Newman, J. S. Kent, A. Bolbecker, M. J. Klaunig, B. F. O'Donnell et al., "White matter abnormalities of microstructure and physiological noise in schizophrenia," *Brain Imaging Behav*, vol. 9, pp. 868–77, Dec 2015.
145. R. E. Gur and R. C. Gur, "Functional magnetic resonance imaging in schizophrenia," *Dialogues Clin Neurosci*, vol. 12, pp. 333–43, 2010.
146. T. Wolfers, J. K. Buitelaar, C. F. Beckmann, B. Franke, and A. F. Marquand, "From estimating activation locality to predicting disorder: A review of pattern recognition for neuroimaging-based psychiatric diagnostics," *Neurosci Biobehav Rev*, vol. 57, pp. 328–49, Oct 2015.
147. L. Wing, "The autistic spectrum," *Lancet*, vol. 350, pp. 1761–6, Dec 13, 1997.
148. C. DeCarli, D. G. Murphy, A. R. McIntosh, D. Teichberg, M. B. Schapiro, and B. Horwitz, "Discriminant analysis of MRI measures as a method to determine the presence

of dementia of the Alzheimer type," *Psychiatr Res*, vol. 57, pp. 119–30, Jul 28, 1995.
149. D. I. Kaufer, B. L. Miller, L. Itti, L. A. Fairbanks, J. Li, J. Fishman et al., "Midline cerebral morphometry distinguishes frontotemporal dementia and Alzheimer's disease," *Neurology*, vol. 48, pp. 978–85, Apr 1997.
150. G. B. Frisoni, A. Beltramello, C. Weiss, C. Geroldi, A. Bianchetti, and M. Trabucchi, "Linear measures of atrophy in mild Alzheimer disease," *AJNR Am J Neuroradiol*, vol. 17, pp. 913–23, May 1996.
151. P. A. Freeborough and N. C. Fox, "MR image texture analysis applied to the diagnosis and tracking of Alzheimer's disease," *IEEE Trans Med Imaging*, vol. 17, pp. 475–9, Jun 1998.
152. C. M. Bottino, C. C. Castro, R. L. Gomes, C. A. Buchpiguel, R. L. Marchetti, and M. R. Neto, "Volumetric MRI measurements can differentiate Alzheimer's disease, mild cognitive impairment, and normal aging," *Int Psychogeriatr*, vol. 14, pp. 59–72, Mar 2002.
153. C. Pennanen, M. Kivipelto, S. Tuomainen, P. Hartikainen, T. Hanninen, M. P. Laakso et al., "Hippocampus and entorhinal cortex in mild cognitive impairment and early AD," *Neurobiol Aging*, vol. 25, pp. 303–10, Mar 2004.
154. S. Li, F. Shi, F. Pu, X. Li, T. Jiang, S. Xie et al., "Hippocampal shape analysis of Alzheimer disease based on machine learning methods," *AJNR Am J Neuroradiol*, vol. 28, pp. 1339–45, Aug 2007.
155. L. Wang, F. Beg, T. Ratnanather, C. Ceritoglu, L. Younes, J. C. Morris et al., "Large deformation diffeomorphism and momentum based hippocampal shape discrimination in dementia of the Alzheimer type," *IEEE Trans Med Imaging*, vol. 26, pp. 462–70, Apr 2007.
156. H. Arimura, T. Yoshiura, S. Kumazawa, K. Tanaka, H. Koga, F. Mihara et al., "Automated method for identification of patients with Alzheimer's disease based on three-dimensional MR images," *Acad Radiol*, vol. 15, pp. 274–84, Mar 2008.
157. S. Klöppel, C. M. Stonnington, C. Chu, B. Draganski, R. I. Scahill, J. D. Rohrer et al., "Automatic classification of MR scans in Alzheimer's disease," *Brain*, vol. 131, pp. 681–9, Mar 2008.
158. J. P. Lerch, J. Pruessner, A. P. Zijdenbos, D. L. Collins, S. J. Teipel, H. Hampel et al., "Automated cortical thickness measurements from MRI can accurately separate Alzheimer's patients from normal elderly controls," *Neurobiol Aging*, vol. 29, pp. 23–30, Jan 2008.
159. E. Gerardin, G. Chetelat, M. Chupin, R. Cuingnet, B. Desgranges, H. S. Kim et al., "Multidimensional classification of hippocampal shape features discriminates Alzheimer's disease and mild cognitive impairment from normal aging," *Neuroimage*, vol. 47, pp. 1476–86, Oct 1, 2009.
160. C. Hinrichs, V. Singh, L. Mukherjee, G. Xu, M. K. Chung, S. C. Johnson et al., "Spatially augmented LPboosting for AD classification with evaluations on the ADNI dataset," *Neuroimage*, vol. 48, pp. 138–49, Oct 15, 2009.
161. B. Magnin, L. Mesrob, S. Kinkingnehun, M. Pelegrini-Issac, O. Colliot, M. Sarazin et al., "Support vector machine-based classification of Alzheimer's disease from whole-brain anatomical MRI," *Neuroradiology*, vol. 51, pp. 73–83, Feb 2009.
162. L. K. McEvoy, C. Fennema-Notestine, J. C. Roddey, D. J. Hagler, Jr., D. Holland, D. S. Karow et al., "Alzheimer disease: Quantitative structural neuroimaging for detection and prediction of clinical and structural changes in mild cognitive impairment," *Radiology*, vol. 251, pp. 195–205, Apr 2009.
163. K. Young, A. T. Du, J. Kramer, H. Rosen, B. Miller, M. Weiner et al., "Patterns of structural complexity in Alzheimer's disease and frontotemporal dementia," *Hum Brain Mapp*, vol. 30, pp. 1667–77, May 2009.
164. M. Zarei, J. S. Damoiseaux, C. Morgese, C. F. Beckmann, S. M. Smith, P. M. Matthews et al., "Regional white matter integrity differentiates between vascular dementia and Alzheimer disease," *Stroke*, vol. 40, pp. 773–9, Mar 2009.
165. S. Haller, D. Nguyen, C. Rodriguez, J. Emch, G. Gold, A. Bartsch et al., "Individual prediction of cognitive decline in mild cognitive impairment using support vector machine-based analysis of diffusion tensor imaging data," *J Alzheimers Dis*, vol. 22, pp. 315–27, 2010.
166. S. G. Mueller, N. Schuff, K. Yaffe, C. Madison, B. Miller, and M. W. Weiner, "Hippocampal atrophy patterns in mild cognitive impairment and Alzheimer's disease," *Hum Brain Mapp*, vol. 31, pp. 1339–47, Sep 2010.
167. P. P. Oliveira, R. Nitrini, G. Busatto, C. Buchpiguel, J. R. Sato, and E. Amaro, "Use of SVM methods with surface-based cortical and volumetric subcortical measurements to detect Alzheimer's disease," *J Alzheimers Dis*, vol. 19, pp. 1263–72, 2010.
168. C. Plant, S. J. Teipel, A. Oswald, C. Bohm, T. Meindl, J. Mourao-Miranda et al., "Automated detection of brain atrophy patterns based on MRI for the prediction of Alzheimer's disease," *Neuroimage*, vol. 50, pp. 162–74, Mar 2010.
169. A. Abdulkadir, B. Mortamet, P. Vemuri, C. R. Jack, Jr., G. Krueger, S. Kloppel et al., "Effects of hardware heterogeneity on the performance of SVM Alzheimer's disease classifier," *Neuroimage*, vol. 58, pp. 785–92, Oct 1, 2011.
170. A. Chincarini, P. Bosco, P. Calvini, G. Gemme, M. Esposito, C. Olivieri et al., "Local MRI analysis approach in the diagnosis of early and prodromal Alzheimer's disease," *Neuroimage*, vol. 58, pp. 469–80, Sep 15, 2011.
171. S. G. Costafreda, I. D. Dinov, Z. Tu, Y. Shi, C. Y. Liu, I. Kloszewska et al., "Automated hippocampal shape analysis predicts the onset of dementia in mild cognitive impairment," *Neuroimage*, vol. 56, pp. 212–9, May 1, 2011.
172. M. Graña, M. Termenon, A. Savio, A. Gonzalez-Pinto, J. Echeveste, J. M. Perez et al., "Computer aided diagnosis system for Alzheimer disease using brain diffusion tensor imaging features selected by Pearson's correlation," *Neurosci Lett*, vol. 502, pp. 225–9, Sep 20, 2011.

173. C. Hinrichs, V. Singh, G. Xu, S. C. Johnson, and I. Alzheimers Disease Neuroimaging, "Predictive markers for AD in a multi-modality framework: An analysis of MCI progression in the ADNI population," *Neuroimage*, vol. 55, pp. 574–89, Mar 15, 2011.
174. L. K. McEvoy, D. Holland, D. J. Hagler, Jr., C. Fennema-Notestine, J. B. Brewer, A. M. Dale et al., "Mild cognitive impairment: Baseline and longitudinal structural MR imaging measures improve predictive prognosis," *Radiology*, vol. 259, pp. 834–43, Jun 2011.
175. C. Y. Wee, P. T. Yap, W. Li, K. Denny, J. N. Browndyke, G. G. Potter et al., "Enriched white matter connectivity networks for accurate identification of MCI patients," *Neuroimage*, vol. 54, pp. 1812–22, Feb 1, 2011.
176. W. Yang, R. L. Lui, J. H. Gao, T. F. Chan, S. T. Yau, R. A. Sperling et al., "Independent component analysis-based classification of Alzheimer's disease MRI data," *J Alzheimers Dis*, vol. 24, pp. 775–83, 2011.
177. T. Zhang and C. Davatzikos, "Optimally-discriminative voxel-based morphometry significantly increases the ability to detect group differences in schizophrenia, mild cognitive impairment, and Alzheimer's disease," *Neuroimage*, vol. 79, pp. 94–110, Oct 1, 2013.
178. N. K. Batmanghelich, B. Taskar, and C. Davatzikos, "Generative-discriminative basis learning for medical imaging," *IEEE Trans Med Imaging*, vol. 31, pp. 51–69, Jan 2012.
179. R. Casanova, F. C. Hsu, M. A. Espeland, and Alzheimer's Disease Neuroimaging Initiative, "Classification of structural MRI images in Alzheimer's disease from the perspective of ill-posed problems," *PLoS ONE*, vol. 7, p. e44877, 2012.
180. P. Coupé, S. F. Eskildsen, J. V. Manjon, V. S. Fonov, D. L. Collins, and I. Alzheimer's disease Neuroimaging, "Simultaneous segmentation and grading of anatomical structures for patient's classification: Application to Alzheimer's disease," *Neuroimage*, vol. 59, pp. 3736–47, Feb 15, 2012.
181. Y. Cui, W. Wen, D. M. Lipnicki, M. F. Beg, J. S. Jin, S. Luo et al., "Automated detection of amnestic mild cognitive impairment in community-dwelling elderly adults: A combined spatial atrophy and white matter alteration approach," *Neuroimage*, vol. 59, pp. 1209–17, Jan 16, 2012.
182. Y. Li, Y. Wang, G. Wu, F. Shi, L. Zhou, W. Lin et al., "Discriminant analysis of longitudinal cortical thickness changes in Alzheimer's disease using dynamic and network features," *Neurobiol Aging*, vol. 33, pp. 427, e15–30, Feb 2012.
183. B. S. Mahanand, S. Suresh, N. Sundararajan, and M. Aswatha Kumar, "Identification of brain regions responsible for Alzheimer's disease using a Self-adaptive Resource Allocation Network," *Neural Netw*, vol. 32, pp. 313–22, Aug 2012.
184. L. O'Dwyer, F. Lamberton, A. L. Bokde, M. Ewers, Y. O. Faluyi, C. Tanner et al., "Using support vector machines with multiple indices of diffusion for automated classification of mild cognitive impairment," *PLoS ONE*, vol. 7, p. e32441, 2012.
185. F. Polat, S. O. Demirel, O. Kitis, F. Simsek, D. I. Haznedaroglu, K. Coburn et al., "Computer based classification of MR scans in first time applicant Alzheimer patients," *Curr Alzheimer Res*, vol. 9, pp. 789–94, Sep 2012.
186. M. R. Sabuncu, K. Van Leemput, and I. Alzheimer's Disease Neuroimaging, "The relevance voxel machine (RVoxM): A self-tuning Bayesian model for informative image-based prediction," *IEEE Trans Med Imaging*, vol. 31, pp. 2290–306, Dec 2012.
187. C. Aguilar, E. Westman, J. S. Muehlboeck, P. Mecocci, B. Vellas, M. Tsolaki et al., "Different multivariate techniques for automated classification of MRI data in Alzheimer's disease and mild cognitive impairment," *Psychiatr Res*, vol. 212, pp. 89–98, May 30, 2013.
188. R. Cuingnet, J. A. Glaunes, M. Chupin, H. Benali, O. Colliot, and I. Alzheimer's Disease Neuroimaging, "Spatial and anatomical regularization of SVM: A general framework for neuroimaging data," *IEEE Trans Pattern Anal Mach Intell*, vol. 35, pp. 682–96, Mar 2013.
189. W. Lee, B. Park, and K. Han, "Classification of diffusion tensor images for the early detection of Alzheimer's disease," *Comput Biol Med*, vol. 43, pp. 1313–20, Oct 2013.
190. X. Liu, D. Tosun, M. W. Weiner, N. Schuff, and Alzheimer's Disease Neuroimaging Initiative, "Locally linear embedding (LLE) for MRI based Alzheimer's disease classification," *Neuroimage*, vol. 83, pp. 148–57, Dec 2013.
191. C. Y. Wee, P. T. Yap, D. Shen, and I. Alzheimer's Disease Neuroimaging, "Prediction of Alzheimer's disease and mild cognitive impairment using cortical morphological patterns," *Hum Brain Mapp*, vol. 34, pp. 3411–25, Dec 2013.
192. X. Wu, J. Li, N. Ayutyanont, H. Protas, W. Jagust, A. Fleisher et al., "The Receiver operational characteristic for binary classification with multiple indices and its application to the neuroimaging study of Alzheimer's disease," *IEEE-ACM Trans Comput Biol Bioinform*, vol. 10, pp. 173–180, Jan–Feb 2013.
193. S. T. Yang, J. D. Lee, T. C. Chang, C. H. Huang, J. J. Wang, W. C. Hsu et al., "Discrimination between Alzheimer's disease and mild cognitive impairment using SOM and PSO-SVM," *Comput Math Methods Med*, vol. 2013, p. 253670, 2013.
194. L. G. Apostolova, K. S. Hwang, O. Kohannim, D. Avila, D. Elashoff, C. R. Jack, Jr. et al., "ApoE4 effects on automated diagnostic classifiers for mild cognitive impairment and Alzheimer's disease," *Neuroimage Clin*, vol. 4, pp. 461–72, 2014.
195. Y. Chen, J. Storrs, L. Tan, L. J. Mazlack, J. H. Lee, and L. J. Lu, "Detecting brain structural changes as biomarker from magnetic resonance images using a local feature based SVM approach," *J Neurosci Methods*, vol. 221, pp. 22–31, Jan 15, 2014.
196. S. Farhan, M. A. Fahiem, and H. Tauseef, "An ensemble-of-classifiers based approach for early diagnosis of Alzheimer's disease: Classification using structural features of brain images," *Comput Math Methods Med*, vol. 2014, p. 862307, 2014.

197. A. R. Hidalgo-Muñoz, J. Ramirez, J. M. Gorriz, and P. Padilla, "Regions of interest computed by SVM wrapped method for Alzheimer's disease examination from segmented MRI," *Front Aging Neurosci*, vol. 6, p. 20, 2014.
198. B. Jie, D. Zhang, W. Gao, Q. Wang, C. Y. Wee, and D. Shen, "Integration of network topological and connectivity properties for neuroimaging classification," *IEEE Trans Biomed Eng*, vol. 61, pp. 576–89, Feb 2014.
199. S. Lahmiri and M. Boukadoum, "New approach for automatic classification of Alzheimer's disease, mild cognitive impairment and healthy brain magnetic resonance images," *Healthc Technol Lett*, vol. 1, pp. 32–6, Jan 2014.
200. G. Y. Lee, J. Kim, J. H. Kim, K. Kim, and J. K. Seong, "Online learning for classification of Alzheimer disease based on cortical thickness and hippocampal shape analysis," *Healthc Inform Res*, vol. 20, pp. 61–8, Jan 2014.
201. M. Li, Y. Qin, F. Gao, W. Zhu, and X. He, "Discriminative analysis of multivariate features from structural MRI and diffusion tensor images," *Magn Reson Imaging*, vol. 32, pp. 1043–51, Oct 2014.
202. M. Li, K. Oishi, X. He, Y. Qin, F. Gao, S. Mori et al., "An efficient approach for differentiating Alzheimer's disease from normal elderly based on multicenter MRI using gray-level invariant features," *PLoS ONE*, vol. 9, p. e105563, 2014.
203. L. Lillemark, L. Sorensen, A. Pai, E. B. Dam, M. Nielsen, and I. Alzheimer's Disease Neuroimaging, "Brain region's relative proximity as marker for Alzheimer's disease based on structural MRI," *BMC Med Imaging*, vol. 14, p. 21, Jun 02, 2014.
204. F. Liu, C. Y. Wee, H. Chen, and D. Shen, "Inter-modality relationship constrained multi-modality multi-task feature selection for Alzheimer's disease and mild cognitive impairment identification," *Neuroimage*, vol. 84, pp. 466–75, Jan 1, 2014.
205. M. Liu, D. Zhang, D. Shen, and A. s. D. N. Initiative, "Hierarchical fusion of features and classifier decisions for Alzheimer's disease diagnosis," *Hum Brain Mapp*, vol. 35, pp. 1305–19, Apr 2014.
206. X. Tang, D. Holland, A. M. Dale, L. Younes, M. I. Miller, and I. Alzheimer's Disease Neuroimaging, "Shape abnormalities of subcortical and ventricular structures in mild cognitive impairment and Alzheimer's disease: Detecting, quantifying, and predicting," *Hum Brain Mapp*, vol. 35, pp. 3701–25, Aug 2014.
207. T. Tong, R. Wolz, Q. Gao, R. Guerrero, J. V. Hajnal, D. Rueckert et al., "Multiple instance learning for classification of dementia in brain MRI," *Med Image Anal*, vol. 18, pp. 808–18, Jul 2014.
208. Z. Zhang, H. Huang, D. Shen, and I. Alzheimer's Disease Neuroimaging, "Integrative analysis of multidimensional imaging genomics data for Alzheimer's disease prediction," *Front Aging Neurosci*, vol. 6, p. 260, 2014.
209. I. Beheshti, H. Demirel, and I. Alzheimer's Disease Neuroimaging, "Probability distribution function-based classification of structural MRI for the detection of Alzheimer's disease," *Comput Biol Med*, vol. 64, pp. 208–16, Sep 2015.
210. E. Challis, P. Hurley, L. Serra, M. Bozzali, S. Oliver, and M. Cercignani, "Gaussian process classification of Alzheimer's disease and mild cognitive impairment from resting-state fMRI," *Neuroimage*, vol. 112, pp. 232–43, May 15, 2015.
211. M. Dyrba, M. Grothe, T. Kirste, and S. J. Teipel, "Multimodal analysis of functional and structural disconnection in Alzheimer's disease using multiple kernel SVM," *Hum Brain Mapp*, vol. 36, pp. 2118–31, Jun 2015.
212. A. Farzan, S. Mashohor, A. R. Ramli, and R. Mahmud, "Boosting diagnosis accuracy of Alzheimer's disease using high dimensional recognition of longitudinal brain atrophy patterns," *Behav Brain Res*, vol. 290, pp. 124–30, Sep 1, 2015.
213. M. Goryawala, Q. Zhou, W. Barker, D. A. Loewenstein, R. Duara, and M. Adjouadi, "Inclusion of neuropsychological scores in atrophy models improves diagnostic classification of Alzheimer's disease and mild cognitive impairment," *Comput Intell Neurosci*, vol. 2015, p. 865265, 2015.
214. C. Granziera, A. Daducci, A. Donati, G. Bonnier, D. Romascano, A. Roche et al., "A multi-contrast MRI study of microstructural brain damage in patients with mild cognitive impairment," *Neuroimage Clin*, vol. 8, pp. 631–9, 2015.
215. W. B. Jung, Y. M. Lee, Y. H. Kim, and C. W. Mun, "Automated classification to predict the progression of Alzheimer's disease using whole-brain volumetry and DTI," *Psychiatr Investig*, vol. 12, pp. 92–102, Jan 2015.
216. S. Klöppel, J. Peter, A. Ludl, A. Pilatus, S. Maier, I. Mader et al., "Applying automated MR-based diagnostic methods to the memory clinic: A prospective study," *J Alzheimers Dis*, vol. 47, pp. 939–54, 2015.
217. M. Liu, D. Zhang, E. Adeli, and D. Shen, "Inherent structure-based multiview learning with multitemplate feature representation for Alzheimer's disease diagnosis," *IEEE Trans Biomed Eng*, vol. 63, pp. 1473–82, Jul 2016.
218. T. M. Nir, J. E. Villalon-Reina, G. Prasad, N. Jahanshad, S. H. Joshi, A. W. Toga et al., "Diffusion weighted imaging-based maximum density path analysis and classification of Alzheimer's disease," *Neurobiol Aging*, vol. 36, Suppl 1, pp. S132–40, Jan 2015.
219. A. Ortiz, J. Munilla, I. Alvarez-Illan, J. M. Gorriz, J. Ramirez, and I. Alzheimer's Disease Neuroimaging, "Exploratory graphical models of functional and structural connectivity patterns for Alzheimer's Disease diagnosis," *Front Comput Neurosci*, vol. 9, p. 132, 2015.
220. G. Prasad, S. H. Joshi, T. M. Nir, A. W. Toga, P. M. Thompson, and I. Alzheimer's Disease Neuroimaging, "Brain connectivity and novel network measures for Alzheimer's disease classification," *Neurobiol Aging*, vol. 36 Suppl 1, pp. S121–31, Jan 2015.
221. A. Retico, P. Bosco, P. Cerello, E. Fiorina, A. Chincarini, and M. E. Fantacci, "Predictive models based on support vector machines: Whole-brain versus regional analysis

of structural MRI in the Alzheimer's disease," *J Neuroimaging*, vol. 25, pp. 552–63, Jul–Aug 2015.
222. C. Salvatore, A. Cerasa, P. Battista, M. C. Gilardi, A. Quattrone, I. Castiglioni et al., "Magnetic resonance imaging biomarkers for the early diagnosis of Alzheimer's disease: A machine learning approach," *Front Neurosci*, vol. 9, p. 307, 2015.
223. L. Xu, X. Wu, K. Chen, and L. Yao, "Multi-modality sparse representation-based classification for Alzheimer's disease and mild cognitive impairment," *Comput Methods Programs Biomed*, vol. 122, pp. 182–90, Nov 2015.
224. H. J. Yun, K. Kwak, J. M. Lee, and I. Alzheimer's Disease Neuroimaging, "Multimodal discrimination of Alzheimer's disease based on regional cortical atrophy and hypometabolism," *PLoS ONE*, vol. 10, p. e0129250, 2015.
225. Y. Zhang, Z. Dong, P. Phillips, S. Wang, G. Ji, J. Yang et al., "Detection of subjects and brain regions related to Alzheimer's disease using 3D MRI scans based on eigenbrain and machine learning," *Front Comput Neurosci*, vol. 9, p. 66, 2015.
226. C. Zu, B. Jie, M. Liu, S. Chen, D. Shen, D. Zhang et al., "Label-aligned multi-task feature learning for multimodal classification of Alzheimer's disease and mild cognitive impairment," *Brain Imaging Behav*, vol. 10, pp. 1148–59, Dec 2016.
227. G. Yu, Y. Liu, and D. Shen, "Graph-guided joint prediction of class label and clinical scores for the Alzheimer's disease," *Brain Struct Funct*, vol. 221, pp. 3787–801, Sep 2016.
228. A. De Sousa and G. Kalra, "Drug therapy of attention deficit hyperactivity disorder: Current trends," *Mens Sana Monogr*, vol. 10, pp. 45–69, Jan 2012.
229. M. V. Cherkasova and L. Hechtman, "Neuroimaging in attention-deficit hyperactivity disorder: Beyond the frontostriatal circuitry," *Can J Psychiatr*, vol. 54, pp. 651–64, Oct 2009.
230. H. D. Consortium, "The ADHD-200 Consortium: A model to advance the translational potential of neuroimaging in clinical neuroscience," *Front Syst Neurosci*, vol. 6, p. 62, 2012.
231. J. G. Csernansky, M. K. Schindler, N. R. Splinter, L. Wang, M. Gado, L. D. Selemon et al., "Abnormalities of thalamic volume and shape in schizophrenia," *Am J Psychiatr*, vol. 161, pp. 896–902, May 2004.
232. C. Davatzikos, D. Shen, R. C. Gur, X. Wu, D. Liu, Y. Fan et al., "Whole-brain morphometric study of schizophrenia revealing a spatially complex set of focal abnormalities," *Arch Gen Psychiatr*, vol. 62, pp. 1218–27, Nov 2005.
233. Y. Fan, D. Shen, and C. Davatzikos, "Classification of structural images via high-dimensional image warping, robust feature extraction, and SVM," *Med Image Comput Assist Interv*, vol. 8, pp. 1–8, 2005.
234. M. W. Caan, K. A. Vermeer, L. J. van Vliet, C. B. Majoie, B. D. Peters, G. J. den Heeten et al., "Shaving diffusion tensor images in discriminant analysis: A study into schizophrenia," *Med Image Anal*, vol. 10, pp. 841–9, Dec 2006.
235. Y. Fan, D. Shen, R. C. Gur, R. E. Gur, and C. Davatzikos, "COMPARE: Classification of morphological patterns using adaptive regional elements," *IEEE Trans Med Imaging*, vol. 26, pp. 93–105, Jan 2007.
236. A. Caprihan, G. D. Pearlson, and V. D. Calhoun, "Application of principal component analysis to distinguish patients with schizophrenia from healthy controls based on fractional anisotropy measurements," *Neuroimage*, vol. 42, pp. 675–82, Aug 15, 2008.
237. O. Demirci, V. P. Clark, and V. D. Calhoun, "A projection pursuit algorithm to classify individuals using fMRI data: Application to schizophrenia," *Neuroimage*, vol. 39, pp. 1774–82, Feb 15, 2008.
238. D. Sun, T. G. van Erp, P. M. Thompson, C. E. Bearden, M. Daley, L. Kushan et al., "Elucidating a magnetic resonance imaging-based neuroanatomic biomarker for psychosis: Classification analysis using probabilistic brain atlas and machine learning algorithms," *Biol Psychiatr*, vol. 66, pp. 1055–60, Dec 1, 2009.
239. H. Yang, J. Liu, J. Sui, G. Pearlson, and V. D. Calhoun, "A hybrid machine learning method for fusing fMRI and genetic data: Combining both improves classification of schizophrenia," *Front Hum Neurosci*, vol. 4, p. 192, 2010.
240. B. A. Ardekani, A. Tabesh, S. Sevy, D. G. Robinson, R. M. Bilder, and P. R. Szeszko, "Diffusion tensor imaging reliably differentiates patients with schizophrenia from healthy volunteers," *Hum Brain Mapp*, vol. 32, pp. 1–9, Jan 2011.
241. E. Castro, M. Martínez-Ramón, G. Pearlson, J. Sui, and V. D. Calhoun, "Characterization of groups using composite kernels and multi-source fMRI analysis data: Application to schizophrenia," *Neuroimage*, vol. 58, pp. 526–36, Sep 2011.
242. Y. Fan, Y. Liu, H. Wu, Y. Hao, H. Liu, Z. Liu et al., "Discriminant analysis of functional connectivity patterns on Grassmann manifold," *Neuroimage*, vol. 56, pp. 2058–67, Jun 15, 2011.
243. E. Karageorgiou, S. C. Schulz, R. L. Gollub, N. C. Andreasen, B. C. Ho, J. Lauriello et al., "Neuropsychological testing and structural magnetic resonance imaging as diagnostic biomarkers early in the course of schizophrenia and related psychoses," *Neuroinformatics*, vol. 9, pp. 321–33, Dec 2011.
244. T. Kasparek, C. E. Thomaz, J. R. Sato, D. Schwarz, E. Janousova, R. Marecek et al., "Maximum-uncertainty linear discrimination analysis of first-episode schizophrenia subjects," *Psychiatr Res*, vol. 191, pp. 174–81, Mar 2011.
245. Y. Takayanagi, T. Takahashi, L. Orikabe, Y. Mozue, Y. Kawasaki, K. Nakamura et al., "Classification of first-episode schizophrenia patients and healthy subjects by automated MRI measures of regional brain volume and cortical thickness," *PLoS ONE*, vol. 6, p. e21047, 2011.
246. R. Bansal, L. H. Staib, A. F. Laine, X. Hao, D. Xu, J. Liu et al., "Anatomical brain images alone can accurately diagnose chronic neuropsychiatric illnesses," *PLoS ONE*, vol. 7, p. e50698, 2012.
247. D. S. Bassett, B. G. Nelson, B. A. Mueller, J. Camchong, and K. O. Lim, "Altered resting state complexity in

schizophrenia," *Neuroimage*, vol. 59, pp. 2196–207, Feb 1, 2012.
248. W. Du, V. D. Calhoun, H. Li, S. Ma, T. Eichele, K. A. Kiehl et al., "High classification accuracy for schizophrenia with rest and task FMRI data," *Front Hum Neurosci*, vol. 6, p. 145, 2012.
249. D. Greenstein, J. D. Malley, B. Weisinger, L. Clasen, and N. Gogtay, "Using multivariate machine learning methods and structural MRI to classify childhood onset schizophrenia and healthy controls," *Front Psychiatr*, vol. 3, p. 53, 2012.
250. J. Honorio, D. Tomasi, R. Z. Goldstein, H. C. Leung, and D. Samaras, "Can a single brain region predict a disorder?," *IEEE Trans Med Imaging*, vol. 31, pp. 2062–72, Nov 2012.
251. Y. Tang, L. Wang, F. Cao, and L. Tan, "Identify schizophrenia using resting-state functional connectivity: An exploratory research and analysis," *Biomed Eng Online*, vol. 11, p. 50, Aug 16, 2012.
252. A. Anderson and M. S. Cohen, "Decreased small-world functional network connectivity and clustering across resting state networks in schizophrenia: An fMRI classification tutorial," *Front Hum Neurosci*, vol. 7, p. 520, 2013.
253. M. R. Arbabshirani, K. A. Kiehl, G. D. Pearlson, and V. D. Calhoun, "Classification of schizophrenia patients based on resting-state functional network connectivity," *Front Neurosci*, vol. 7, p. 133, 2013.
254. H. Cao, J. Duan, D. Lin, V. Calhoun, and Y. P. Wang, "Integrating fMRI and SNP data for biomarker identification for schizophrenia with a sparse representation based variable selection method," *BMC Med Genom*, vol. 6, Suppl 3, p. S2, 2013.
255. T. Fekete, M. Wilf, D. Rubin, S. Edelman, R. Malach, and L. R. Mujica-Parodi, "Combining classification with fMRI-derived complex network measures for potential neurodiagnostics," *PLoS One*, vol. 8, p. e62867, 2013.
256. S. J. Iwabuchi, P. F. Liddle, and L. Palaniyappan, "Clinical utility of machine-learning approaches in schizophrenia: Improving diagnostic confidence for translational neuroimaging," *Front Psychiatr*, vol. 4, p. 95, 2013.
257. J. Sui, H. He, Q. Yu, J. Chen, J. Rogers, G. D. Pearlson et al., "Combination of resting state fMRI, DTI, and sMRI data to discriminate schizophrenia by N-way MCCA + jICA," *Front Hum Neurosci*, vol. 7, p. 235, 2013.
258. Y. Yu, H. Shen, H. Zhang, L. L. Zeng, Z. Xue, and D. Hu, "Functional connectivity-based signatures of schizophrenia revealed by multiclass pattern analysis of resting-state fMRI from schizophrenic patients and their healthy siblings," *Biomed Eng Online*, vol. 12, p. 10, Feb 07, 2013.
259. M. V. Zanetti, M. S. Schaufelberger, J. Doshi, Y. Ou, L. K. Ferreira, P. R. Menezes et al., "Neuroanatomical pattern classification in a population-based sample of first-episode schizophrenia," *Prog Neuropsychopharmacol Biol Psychiatr*, vol. 43, pp. 116–25, Jun 3, 2013.
260. A. Anticevic, M. W. Cole, G. Repovs, J. D. Murray, M. S. Brumbaugh, A. M. Winkler et al., "Characterizing thalamo-cortical disturbances in schizophrenia and bipolar illness," *Cereb Cortex*, vol. 24, pp. 3116–30, Dec 2014.
261. E. Castro, V. Gomez-Verdejo, M. Martinez-Ramon, K. A. Kiehl, and V. D. Calhoun, "A multiple kernel learning approach to perform classification of groups from complex-valued fMRI data analysis: Application to schizophrenia," *Neuroimage*, vol. 87, pp. 1–17, Feb 15, 2014.
262. E. Radulescu, B. Ganeshan, S. S. Shergill, N. Medford, C. Chatwin, R. C. Young et al., "Grey-matter texture abnormalities and reduced hippocampal volume are distinguishing features of schizophrenia," *Psychiatr Res*, vol. 223, pp. 179–86, Sep 30, 2014.
263. T. Watanabe, D. Kessler, C. Scott, M. Angstadt, and C. Sripada, "Disease prediction based on functional connectomes using a scalable and spatially-informed support vector machine," *Neuroimage*, vol. 96, pp. 183–202, Aug 1, 2014.
264. M. S. Cetin, S. Khullar, E. Damaraju, A. M. Michael, S. A. Baum, and V. D. Calhoun, "Enhanced disease characterization through multi network functional normalization in fMRI," *Front Neurosci*, vol. 9, p. 95, 2015.
265. H. Cheng, S. Newman, J. Goni, J. S. Kent, J. Howell, A. Bolbecker et al., "Nodal centrality of functional network in the differentiation of schizophrenia," *Schizophr Res*, vol. 168, pp. 345–52, Oct 2015.
266. D. Chyzhyk, A. Savio, and M. Grana, "Computer aided diagnosis of schizophrenia on resting state fMRI data by ensembles of ELM," *Neural Netw*, vol. 68, pp. 23–33, Aug 2015.
267. E. Janousova, D. Schwarz, and T. Kasparek, "Combining various types of classifiers and features extracted from magnetic resonance imaging data in schizophrenia recognition," *Psychiatr Res Neuroimaging*, vol. 232, pp. 237–49, Jun 30, 2015.
268. T. Kaufmann, K. C. Skatun, D. Alnaes, N. T. Doan, E. P. Duff, S. Tonnesen et al., "Disintegration of sensorimotor brain networks in schizophrenia," *Schizophr Bull*, vol. 41, pp. 1326–35, Nov 2015.
269. S. P. Koch, C. Hagele, J. D. Haynes, A. Heinz, F. Schlagenhauf, and P. Sterzer, "Diagnostic classification of schizophrenia patients on the basis of regional reward-related FMRI signal patterns," *PLoS ONE*, vol. 10, p. e0119089, 2015.
270. Q. Gong, Q. Wu, C. Scarpazza, S. Lui, Z. Jia, A. Marquand et al., "Prognostic prediction of therapeutic response in depression using high-field MR imaging," *Neuroimage*, vol. 55, pp. 1497–503, Apr 2011.
271. P. Fang, L. L. Zeng, H. Shen, L. Wang, B. Li, L. Liu et al., "Increased cortical-limbic anatomical network connectivity in major depression revealed by diffusion tensor imaging," *PLoS ONE*, vol. 7, p. e45972, 2012.
272. A. Lord, D. Horn, M. Breakspear, and M. Walter, "Changes in community structure of resting state functional connectivity in unipolar depression," *PLoS ONE*, vol. 7, p. e41282, 2012.
273. B. Mwangi, K. P. Ebmeier, K. Matthews, and J. D. Steele, "Multi-centre diagnostic classification of individual structural neuroimaging scans from patients with major

depressive disorder," *Brain*, vol. 135, pp. 1508–21, May 2012.
274. M. Wei, J. Qin, R. Yan, H. Li, Z. Yao, and Q. Lu, "Identifying major depressive disorder using Hurst exponent of resting-state brain networks," *Psychiatr Res*, vol. 214, pp. 306–12, Dec 30, 2013.
275. L. Cao, S. Guo, Z. Xue, Y. Hu, H. Liu, T. E. Mwansisya et al., "Aberrant functional connectivity for diagnosis of major depressive disorder: A discriminant analysis," *Psychiatr Clin Neurosci*, vol. 68, pp. 110–9, Feb 2014.
276. F. P. MacMaster, N. Carrey, L. M. Langevin, N. Jaworska, and S. Crawford, "Disorder-specific volumetric brain difference in adolescent major depressive disorder and bipolar depression," *Brain Imaging Behav*, vol. 8, pp. 119–27, Mar 2014.
277. M. H. Serpa, Y. Ou, M. S. Schaufelberger, J. Doshi, L. K. Ferreira, R. Machado-Vieira et al., "Neuroanatomical classification in a population-based sample of psychotic major depression and bipolar I disorder with 1 year of diagnostic stability," *Biomed Res Int*, vol. 2014, p. 706157, 2014.
278. L. C. Foland-Ross, M. D. Sacchet, G. Prasad, B. Gilbert, P. M. Thompson, and I. H. Gotlib, "Cortical thickness predicts the first onset of major depression in adolescence," *Int J Dev Neurosci*, vol. 46, pp. 125–31, Nov 2015.
279. G. Fung, Y. Deng, Q. Zhao, Z. Li, M. Qu, K. Li et al., "Distinguishing bipolar and major depressive disorders by brain structural morphometry: a pilot study," *BMC Psychiatr*, vol. 15, p. 298, Nov 21, 2015.
280. M. J. Patel, C. Andreescu, J. C. Price, K. L. Edelman, C. F. Reynolds, 3rd, and H. J. Aizenstein, "Machine learning approaches for integrating clinical and imaging features in late-life depression classification and response prediction," *Int J Geriatr Psychiatr*, vol. 30, pp. 1056–67, Oct 2015.
281. M. J. Rosa, L. Portugal, T. Hahn, A. J. Fallgatter, M. I. Garrido, J. Shawe-Taylor et al., "Sparse network-based models for patient classification using fMRI," *Neuroimage*, vol. 105, pp. 493–506, Jan 15, 2015.
282. M. D. Sacchet, E. E. Livermore, J. E. Iglesias, G. H. Glover, and I. H. Gotlib, "Subcortical volumes differentiate major depressive disorder, bipolar disorder, and remitted major depressive disorder," *J Psychiatr Res*, vol. 68, pp. 91–8, Sep 2015.
283. J. R. Sato, J. Moll, S. Green, J. F. Deakin, C. E. Thomaz, and R. Zahn, "Machine learning algorithm accurately detects fMRI signature of vulnerability to major depression," *Psychiatr Res*, vol. 233, pp. 289–91, Aug 30, 2015.
284. Y. Shimizu, J. Yoshimoto, S. Toki, M. Takamura, S. Yoshimura, Y. Okamoto et al., "Toward probabilistic diagnosis and understanding of depression based on functional MRI data analysis with logistic group LASSO," *PLoS ONE*, vol. 10, p. e0123524, 2015.
285. J. D. Gabrieli, S. S. Ghosh, and S. Whitfield-Gabrieli, "Prediction as a humanitarian and pragmatic contribution from human cognitive neuroscience," *Neuron*, vol. 85, pp. 11–26, Jan 2015.
286. C. Ecker, A. Marquand, J. Mourao-Miranda, P. Johnston, E. M. Daly, M. J. Brammer et al., "Describing the brain in autism in five dimensions—magnetic resonance imaging-assisted diagnosis of autism spectrum disorder using a multiparameter classification approach," *J Neurosci*, vol. 30, pp. 10612–23, Aug 11, 2010.
287. J. S. Anderson, J. A. Nielsen, A. L. Froehlich, M. B. DuBray, T. J. Druzgal, A. N. Cariello et al., "Functional connectivity magnetic resonance imaging classification of autism," *Brain*, vol. 134, pp. 3742–54, Dec 2011.
288. M. Ingalhalikar, D. Parker, L. Bloy, T. P. Roberts, and R. Verma, "Diffusion based abnormality markers of pathology: Toward learned diagnostic prediction of ASD," *Neuroimage*, vol. 57, pp. 918–27, Aug 1, 2011.
289. S. Calderoni, A. Retico, L. Biagi, R. Tancredi, F. Muratori, and M. Tosetti, "Female children with autism spectrum disorder: An insight from mass-univariate and pattern classification analyses," *Neuroimage*, vol. 59, pp. 1013–22, Jan 2012.
290. D. L. Murdaugh, S. V. Shinkareva, H. R. Deshpande, J. Wang, M. R. Pennick, and R. K. Kana, "Differential deactivation during mentalizing and classification of autism based on default mode network connectivity," *PLoS ONE*, vol. 7, p. e50064, 2012.
291. G. Deshpande, L. E. Libero, K. R. Sreenivasan, H. D. Deshpande, and R. K. Kana, "Identification of neural connectivity signatures of autism using machine learning," *Front Hum Neurosci*, vol. 7, p. 670, 2013.
292. L. Q. Uddin, K. Supekar, C. J. Lynch, A. Khouzam, J. Phillips, C. Feinstein et al., "Salience network-based classification and prediction of symptom severity in children with autism," *JAMA Psychiatr*, vol. 70, pp. 869–79, Aug 2013.
293. F. Segovia, R. Holt, M. Spencer, J. M. Gorriz, J. Ramirez, C. G. Puntonet et al., "Identifying endophenotypes of autism: A multivariate approach," *Front Comput Neurosci*, vol. 8, p. 60, 2014.
294. C. Y. Wee, L. Wang, F. Shi, P. T. Yap, and D. Shen, "Diagnosis of autism spectrum disorders using regional and interregional morphological features," *Hum Brain Mapp*, vol. 35, pp. 3414–30, Jul 2014.
295. Y. Zhou, F. Yu, and T. Duong, "Multiparametric MRI characterization and prediction in autism spectrum disorder using graph theory and machine learning," *PLoS ONE*, vol. 9, p. e90405, 2014.
296. C. P. Chen, C. L. Keown, A. Jahedi, A. Nair, M. E. Pflieger, B. A. Bailey et al., "Diagnostic classification of intrinsic functional connectivity highlights somatosensory, default mode, and visual regions in autism," *Neuroimage Clin*, vol. 8, pp. 238–45, 2015.
297. I. Gori, A. Giuliano, F. Muratori, I. Saviozzi, P. Oliva, R. Tancredi et al., "Gray matter alterations in young children with autism spectrum disorders: Comparing morphometry at the voxel and regional level," *J Neuroimaging*, vol. 25, pp. 866–74, Nov–Dec 2015.
298. L. E. Libero, T. P. DeRamus, A. C. Lahti, G. Deshpande, and R. K. Kana, "Multimodal neuroimaging based classification of autism spectrum disorder using anatomical, neurochemical, and white matter correlates," *Cortex*, vol. 66, pp. 46–59, May 2015.
299. M. Plitt, K. A. Barnes, and A. Martin, "Functional connectivity classification of autism identifies highly

predictive brain features but falls short of biomarker standards," *Neuroimage Clin*, vol. 7, pp. 359–66, 2015.
300. K. Iqbal, M. Flory, S. Khatoon, H. Soininen, T. Pirttila, M. Lehtovirta et al., "Subgroups of Alzheimer's disease based on cerebrospinal fluid molecular markers," *Ann Neurol*, vol. 58, pp. 748–57, Nov 2005.
301. K. H. Brodersen, L. Deserno, F. Schlagenhauf, Z. Lin, W. D. Penny, J. M. Buhmann et al., "Dissecting psychiatric spectrum disorders by generative embedding," *Neuroimage Clin*, vol. 4, pp. 98–111, 2014.
302. M. J. Wu, B. Mwangi, I. E. Bauer, I. C. Passos, M. Sanches, G. B. Zunta-Soares et al., "Identification and individualized prediction of clinical phenotypes in bipolar disorders using neurocognitive data, neuroimaging scans and machine learning," *Neuroimage*, vol. 145, pp. 254–64, Jan 2017.
303. B. N. Cuthbert, "Research Domain Criteria: Toward future psychiatric nosologies," *Dialogues Clin Neurosci*, vol. 17, pp. 89–97, Mar 2015.
304. C. Z. Zhu, Y. F. Zang, Q. J. Cao, C. G. Yan, Y. He, T. Z. Jiang et al., "Fisher discriminative analysis of resting-state brain function for attention-deficit/hyperactivity disorder," *Neuroimage*, vol. 40, pp. 110–20, Mar 1, 2008.
305. J. W. Bohland, S. Saperstein, F. Pereira, J. Rapin, and L. Grady, "Network, anatomical, and non-imaging measures for the prediction of ADHD diagnosis in individual subjects," *Front Syst Neurosci*, vol. 6, p. 78, 2012.
306. C. W. Chang, C. C. Ho, and J. H. Chen, "ADHD classification by a texture analysis of anatomical brain MRI data," *Front Syst Neurosci*, vol. 6, p. 66, 2012.
307. J. B. Colby, J. D. Rudie, J. A. Brown, P. K. Douglas, M. S. Cohen, and Z. Shehzad, "Insights into multimodal imaging classification of ADHD," *Front Syst Neurosci*, vol. 6, p. 59, 2012.
308. D. Dai, J. Wang, J. Hua, and H. He, "Classification of ADHD children through multimodal magnetic resonance imaging," *Front Syst Neurosci*, vol. 6, p. 63, 2012.
309. D. A. Fair, J. T. Nigg, S. Iyer, D. Bathula, K. L. Mills, N. U. Dosenbach et al., "Distinct neural signatures detected for ADHD subtypes after controlling for micro-movements in resting state functional connectivity MRI data," *Front Syst Neurosci*, vol. 6, p. 80, 2012.
310. L. Igual, J. C. Soliva, S. Escalera, R. Gimeno, O. Vilarroya, and P. Radeva, "Automatic brain caudate nuclei segmentation and classification in diagnostic of Attention-Deficit/Hyperactivity Disorder," *Comput Med Imaging Graph*, vol. 36, pp. 591–600, Dec 2012.
311. J. R. Sato, M. Q. Hoexter, A. Fujita, and L. A. Rohde, "Evaluation of pattern recognition and feature extraction methods in ADHD prediction," *Front Syst Neurosci*, vol. 6, p. 68, 2012.
312. G. S. Sidhu, N. Asgarian, R. Greiner, and M. R. Brown, "Kernel Principal Component Analysis for dimensionality reduction in fMRI-based diagnosis of ADHD," *Front Syst Neurosci*, vol. 6, p. 74, 2012.
313. X. Peng, P. Lin, T. Zhang, and J. Wang, "Extreme learning machine-based classification of ADHD using brain structural MRI data," *PLoS ONE*, vol. 8, p. e79476, 2013.
314. H. Hart, K. Chantiluke, A. I. Cubillo, A. B. Smith, A. Simmons, M. J. Brammer et al., "Pattern classification of response inhibition in ADHD: Toward the development of neurobiological markers for ADHD," *Hum Brain Mapp*, vol. 35, pp. 3083–94, Jul 2014.
315. H. Hart, A. F. Marquand, A. Smith, A. Cubillo, A. Simmons, M. Brammer et al., "Predictive neurofunctional markers of attention-deficit/hyperactivity disorder based on pattern classification of temporal processing," *J Am Acad Child Adolesc Psychiatr*, vol. 53, pp. 569–78, e1, May 2014.
316. B. A. Johnston, B. Mwangi, K. Matthews, D. Coghill, K. Konrad, and J. D. Steele, "Brainstem abnormalities in attention deficit hyperactivity disorder support high accuracy individual diagnostic classification," *Hum Brain Mapp*, vol. 35, pp. 5179–89, Oct 2014.
317. G. Deshpande, P. Wang, D. Rangaprakash, and B. Wilamowski, "Fully connected cascade artificial neural network architecture for attention deficit hyperactivity disorder classification from functional magnetic resonance imaging data," *IEEE Trans Cybern*, vol. 45, pp. 2668–79, Dec 2015.
318. R. W. McCarley, M. Nakamura, M. E. Shenton, and D. F. Salisbury, "Combining ERP and structural MRI information in first episode schizophrenia and bipolar disorder," *Clin EEG Neurosci*, vol. 39, pp. 57–60, Apr 2008.
319. A. M. Michael, S. A. Baum, J. F. Fries, B. C. Ho, R. K. Pierson, N. C. Andreasen et al., "A method to fuse fMRI tasks through spatial correlations: applied to schizophrenia," *Hum Brain Mapp*, vol. 30, pp. 2512–29, Aug 2009.
320. J. Sui, G. Pearlson, A. Caprihan, T. Adali, K. A. Kiehl, J. Liu et al., "Discriminating schizophrenia and bipolar disorder by fusing fMRI and DTI in a multimodal CCA+ joint ICA model," *Neuroimage*, vol. 57, pp. 839–55, Aug 1, 2011.

Section II

Signal Processing for Big Data

4

Data-Driven Approaches for Detecting and Identifying Anomalous Data Streams

Shaofeng Zou, Yingbin Liang, H. Vincent Poor, and Xinghua Shi

CONTENTS

4.1 Introduction 57
4.2 Two Types of Problems 58
    4.2.1 Detection of Existence 58
    4.2.2 Identification of Anomalous Data Streams 59
4.3 Nonparametric Approaches 59
    4.3.1 Generalized Likelihood-Based Approach 59
    4.3.2 Kernel-Based Approach 60
    4.3.3 KL Divergence-Based Approach 60
4.4 Detection of Existence 61
    4.4.1 Parametric Model 61
    4.4.2 Semiparametric Model 62
    4.4.3 Nonparametric Model 63
4.5 Identification of Anomalous Data Streams 64
    4.5.1 Parametric Model 64
    4.5.2 Semiparametric Model 65
    4.5.3 Nonparametric Model 67
4.6 Applications in Biomedical Signal Processing 70
4.7 Open Problems and Future Directions 71
Acknowledgment 71
References 72

4.1 Introduction

Consider a sensor network, in which all nodes take observations from the environment. If an anomalous event occurs, then a subset of sensors are affected and thus receive data streams with different statistical behavior from that of unaffected sensors. It is typically of interest to detect whether such an anomalous event occurs, and if such an event occurs, to further identify the location of the event. This kind of application gives rise to the problem of detection of anomalous data streams out of a large number of typical data streams based on their statistical properties. Such problems are usually solved in two steps: first detect whether or not there exist anomalous data streams, and then if they exist, identify such anomalous data streams. Statistically, the typical data streams are assumed to be drawn from a certain distribution, and the anomalous data streams are assumed to be drawn from distributions that are different from the typical data streams.

This type of problem models a large number of applications in various domains. For example, in bioinformatics, the data set may contain gene expressions across a group of people with a certain disease. It is anticipated that the genes related to the disease should have a different expression behavior from other irrelevant genes, and hence can be detected and identified. More applications in biomedical signal processing and health care are reviewed in Section 4.1.6. Other applications include detecting virus-infected computers in a network, detecting bank fraud, detecting vacant channels in cognitive radio networks, etc.

The problems that we consider here are fundamentally different from some other outlier/anomaly detection problems from the following perspectives. (1) The aim is to differentiate distributions rather than individual data samples as widely studied in data mining, e.g., Refs. [1,2]. Thus, the problem can generally be viewed as hypothesis testing. (2) The learning is unsupervised, i.e., without the use of labeled training data. Thus, it is different from a common semisupervised problem, e.g., Refs. [3,4,5], in which training samples drawn from a typical distribution are given, and a new sample needs to be tested as to whether it is drawn from the typical distribution.

In order to solve the problem of interest here, most previous studies have focused on parametric scenarios (see, e.g., Refs. [6,7,8,9,10,11]), assuming the distributions are known *a priori*, although practical applications typically provide only raw data. Implicitly, these studies assume that the distributions are learned from the data. However, since the ultimate goal is to detect and identify anomalous data streams, learning the distributions first and then constructing the detection rules may not yield optimal performance. It is thus desirable to design *data-driven nonparametric* tests, which directly accomplish the goal of detection using data without estimating the distributions as an intermediate step. Furthermore, since such tests do not exploit any information about the distributions, they can be designed to provide *universal* performance guarantees for arbitrary distributions.

Several approaches have been developed to solve the nonparametric problem of detecting anomalous data streams, including (1) a generalized likelihood-based approach [12,13], which is applicable to the case with discrete distributions; (2) a kernel-based approach [14], which is applicable to the cases with arbitrary distributions; and (3) a Kullback–Leibler (KL) divergence-based approach [15], which is applicable to the case with continuous distributions. In fact, the KL divergence-based approach can also be applied to discrete distributions if estimators for KL divergence between discrete distributions are applied. The focus of this chapter is to review these detection techniques and their performance characteristics for solving the detection problems. Although our interest is in nonparametric scenarios, we also introduce results for parametric and semiparametric models to develop useful understanding toward nonparametric models.

This chapter is organized as follows. In Section 4.1.2, we introduce two types of problems that have been studied so far: one problem models detection of the existence of anomalous data streams, and the other one models identification of anomalous data streams if they exist. In Section 4.1.3, we review several approaches that have been developed to solve the nonparametric problems of interest here. In Section 4.1.4, we review the results in Ref. [13] on the problem of detecting the existence of anomalous data streams. In Section 4.1.5, we review the results in Refs. [12,14,15] on the problem of identifying anomalous data streams. In Section 4.1.6, we provide several applications in biomedical signal processing. In Section 4.1.7, we conclude this chapter with a few interesting future directions.

4.2 Two Types of Problems

The problem of detecting anomalous data streams naturally comprises two steps: first detecting whether there exists any anomalous data stream, and then further identifying anomalous data streams if existence is confirmed. To define these problems more formally, suppose that there are M data streams denoted by $y^{(i)}$ for $i = 1,2,\ldots,M$. Each data stream consists of n independent and identically distributed (i.i.d.) samples, and we denote the k-th sample in the i-th data stream by $y_k^{(i)}$. It is assumed that different data streams are independent of each other. Among these data streams, a typical data stream contains samples drawn from a distribution π, and an anomalous data stream (if it exists) contains samples drawn from a distribution μ. Both μ and π are supported on the same alphabet $\mathcal{Y}$. We denote the cardinality of $\mathcal{Y}$ as $|\mathcal{Y}|$, if μ and π are discrete. We further assume that $\mu \neq \pi$. We then denote all data streams as $y^{Mn} = (y^{(1)}, y^{(2)}, \ldots, y^{(M)})$.

4.2.1 Detection of Existence

The first problem is to detect the existence of anomalous data streams. This problem is studied in detail in Ref. [13] when both μ and π are discrete distributions with finite alphabet size. This problem is equivalent to distinguishing between the following two hypotheses:

H_0 : All data streams $y^{(i)}$ are typical, for $i = 1, 2, \ldots, M$.

H_1 : There exist s anomalous data streams,

where $s > 0$ is fixed. (4.1)

Under the null hypothesis H_0, all the samples in y^{Mn} are i.i.d. generated by π. However, under the alternative hypothesis H_1, there exist anomalous data streams that contain samples generated i.i.d. by μ, whereas the samples in typical data streams are still generated i.i.d. by π. Hence, y^{Mn} may take multiple distributions depending on which data streams are anomalous. For example, when there is only one anomalous data stream, it can be any of M data streams. For a more general model with $s > 1$ anomalous data streams, the index set S of the anomalous data streams can be any subset of $\mathcal{M} = \{1,\ldots,M\}$ such that the cardinality $|S| = s$. Thus, the Equation 4.1 is a binary composite hypothesis testing problem with multiple subhypotheses under H_1.

A test δ to distinguish between the two hypotheses is defined as a mapping from the set of realizations of data streams y^{Mn} into $\{H_0, H_1\}$, as follows:

$$\delta : \{y^{Mn}\} \to \{H_0, H_1\}. \quad (4.2)$$

The performance of a test δ is characterized by two types of error probabilities. The type I error probability $e_1(\delta)$ is the probability that all data streams are generated by π but δ declares that there are anomalous data streams, and is given by $e_1(\delta) = P_{H_0}(\delta = H_1)$. And the type II error probability $e_2(\delta)$ is the probability that there are anomalous data streams but δ declares that all the data streams are generated by π, and is given by $e_2(\delta) = P_{H_1}(\delta = H_0)$. The following risk function $R(\delta)$ is defined to measure the overall performance of a test for Equation 4.1:

$$R(\delta) = P_{H_0}(\delta = H_1) + \max_{S \subset \mathcal{M}: |S|=s} P_{H_1}(\delta = H_0). \quad (4.3)$$

We are interested in the asymptotic regime, in which $n \to \infty$.

Definition 1 A test δ is said to be consistent if $R(\delta) \to 0$, as $n \to \infty$.

The exponent $E_R(\delta)$ of $R(\delta)$ is further defined as

$$E_R(\delta) = \lim_{n \to \infty} -\frac{1}{n} \log R(\delta). \quad (4.4)$$

Definition 2 A test is said to be exponentially consistent if $E_R(\delta) > 0$, i.e., the risk function $R(\delta)$ converges to zero exponentially fast.

4.2.2 Identification of Anomalous Data Streams

If there exist anomalous data streams among M data streams, it is of practical importance to identify those anomalous data streams. This is a multiple hypothesis testing problem with each hypothesis corresponding to a particular set of data streams being anomalous. This problem is studied in detail for the case when both μ and π are discrete distributions with finite alphabet size in Ref. [12], for the general case that both μ and π are arbitrary and can be either discrete or continuous in Ref. [14], and for the case with both μ and π being continuous in Ref. [15].

In this problem, it is assumed that there are exactly s anomalous data streams among the M data streams. The goal of this problem is to identify which s data streams are anomalous, i.e., to distinguish among multiple hypotheses. The value of s might be known or unknown depending on the applications. Let S denote the set that contains indices of all anomalous data streams with cardinality $|S| = s$. This problem is equivalent to distinguishing among the following hypotheses:

$$\Big\{ H_S : y^{(i)} \text{ is anomalous, for } i \in S, \text{ and } y^{(j)} \text{ is typical,}$$

$$\text{for } i \notin S \Big\} \text{ for } S \subset \mathcal{M}, \text{ and } |S| = s.$$

A test δ to distinguish among the hypotheses is defined as a mapping from the set of realizations of data streams y^{Mn} into $\{H_S : S \subset \mathcal{M}, |S| = s\}$, as follows:

$$\delta : \{y^{Mn}\} \to \{H_S : S \subset \mathcal{M}, |S| = s\}. \quad (4.5)$$

We define the maximum probability of detection error as the performance measure of tests. The maximal error probability for a test δ is defined as follows:

$$R(\delta) = \max_{S: |S|=s} P\{\delta \neq H_S\}. \quad (4.6)$$

The consistency and exponential consistency for a test can be defined in the same way as in Definitions 1 and 2.

4.3 Nonparametric Approaches

In this section, we introduce three nonparametric approaches recently developed to deal with the problem of detecting and identifying anomalous data streams.

4.3.1 Generalized Likelihood-Based Approach

If the distributions μ and π are discrete with finite alphabet size, the generalized likelihood approach can be designed to solve the nonparametric problems [12,13]. Consider the case in which both distribution μ and π are known; then parametric likelihood-based tests can be constructed for both problems. And it can be shown that those parametric likelihood-based tests are optimal. For example, for the problem of identifying anomalous data streams with $s = 1$, the optimal test is given as follows:

$$\begin{aligned}
\delta_{\text{ML}}(y^{Mn}, \pi, \mu) &= \arg\max_{1 \leq i \leq M} P_i(y^{Mn}) \\
&= \arg\max_{1 \leq i \leq M} \prod_{k=1}^{n} \left(\mu(y_k^{(i)}) \prod_{j \neq i} \pi(y_k^{(j)}) \right).
\end{aligned} \quad (4.7)$$

Since the distributions μ and π are discrete, when the number of samples in each data stream is large, the empirical distribution can be used as a good estimate of the underlying true distribution. Hence, if the distributions are unknown, the idea of the generalized likelihood-based approach is to replace the distribution functions μ and π in the parametric likelihood-based tests by their empirical distributions conditioned on each hypothesis.

4.3.2 Kernel-Based Approach

For the general case when both μ and π are arbitrary and can be either discrete or continuous, a kernel-based approach is proposed in Ref. [14], which employs maximum mean discrepancy (MMD) [16] as a distance metric to measure the similarity between the distributions that generate the data streams and further constructs the test.

We next introduce the idea of mean embedding of distributions into a reproducing kernel Hilbert space (RKHS) [17,18] and the metric of MMD [16]. Suppose the set $\mathcal{P}$ includes a class of probability distributions, and suppose $\mathcal{H}$ is the RKHS with an associated kernel $k(\cdot,\cdot)$. A mapping from $\mathcal{P}$ to $\mathcal{H}$ is defined such that each distribution $p \in \mathcal{P}$ is mapped into an element in $\mathcal{H}$ as follows:

$$\eta_p(\cdot) = \mathbb{E}_p[k(\cdot, x)]. \quad (4.8)$$

Here, $\eta_p(\cdot)$ is referred to as the *mean embedding* of the distribution p into the Hilbert space $\mathcal{H}$. Due to the reproducing property of $\mathcal{H}$, it is clear that $\mathbb{E}_p[f] = \langle \eta_p, f \rangle_{\mathcal{H}}$ for all $f \in \mathcal{H}$.

It is desirable that the embedding is *injective* such that each $p \in \mathcal{P}$ is mapped to a unique element $\eta_p \in \mathcal{H}$. It has been shown in Refs. [18,19,20,21] that for many RKHSs such as those associated with Gaussian and Laplacian kernels, the mean embedding is injective. In order to measure the distance between two distributions p and q, Ref. [16] introduced the following quantity based on the mean embeddings η_p and η_q of p and q in the RKHS:

$$\mathrm{MMD}[p,q] := \|\eta_p - \eta_q\|_{\mathcal{H}}. \quad (4.9)$$

Due to the reproducing property of the kernel, it can be shown that

$$\mathrm{MMD}^2[p,q] = \mathbb{E}_{x,x'}[k(x,x')] - 2\mathbb{E}_{x,y}[k(x,y)] + \mathbb{E}_{y,y'}[k(y,y')],$$

where x and x' are independent but have the same distribution p, and y and y' are independent but have the same distribution q. An unbiased estimate of $\mathrm{MMD}^2[p,q]$ based on n samples of x and m samples of y is given by

$$\mathrm{MMD}_u^2[X,Y] = \frac{1}{n(n-1)} \sum_{i=1}^n \sum_{j \neq i}^n k(x_i, x_j)$$
$$+ \frac{1}{m(m-1)} \sum_{i=1}^m \sum_{j \neq i}^m k(y_i, y_j) \quad (4.10)$$
$$- \frac{2}{nm} \sum_{i=1}^n \sum_{j=1}^m k(x_i, y_j).$$

And it can be show that $\mathbb{E}[\mathrm{MMD}_u^2[X,Y]] = \mathrm{MMD}^2[p,q]$.

From the above, it can be seen that MMD measures the distance between two distributions, and it can be easily estimated from samples using the unbiased estimator (Equation 4.10). Thus, MMD can be applied as a distance metric to construct tests for the nonparametric detection problem. The MMD distance metric has several advantages. For example, (1) MMD can be easily estimated based on samples with exponentially fast convergence rate, which leads to low-complexity tests, and (2) MMD-based approaches do not require estimation of probability density functions as an intermediate step, but directly estimate the distance between distributions to build tests, thus avoiding error propagation.

4.3.3 KL Divergence-Based Approach

Similarly to MMD, KL divergence can also be used as a distance metric to solve the detection problems here. For nonparametric scenarios, estimators of KL divergence can be used to approximate the test statistic of the parametric maximum likelihood test. Such a KL divergence-based approach has been proposed in Ref. [15] for continuous μ and π, which utilizes the KL divergence estimator based on data-dependent partitions developed in Ref. [22] and introduced below.

Consider a simple case, in which the distribution p is *unknown* and the distribution q is known, and both p and q are continuous. A sequence of i.i.d. samples $Y \in \mathbb{R}^n$ is generated from p. The goal is to estimate the KL divergence between p and q, which is defined as follows:

$$D(p\|q) \triangleq \int dp \log \frac{dp}{dq}.$$

Denote the order statistics of Y by $\{Y_{(1)}, Y_{(2)}, \ldots, Y_{(n)}\}$ where $Y_{(1)} \leq Y_{(2)} \leq \ldots \leq Y_{(n)}$. The real line is partitioned into empirically equiprobable segments as follows:

$$\{I_t^n\}_{t=1,\ldots,T_n}$$
$$= \{(-\infty, Y_{(\ell_n)}], (Y_{(\ell_n)}, Y_{(2\ell_n)}], \ldots, (Y_{(\ell_n(T_n-1))}, \infty)\}, \quad (4.11)$$

where $\ell_n \in \mathbb{N} \leq n$ is the number of points in each interval except possibly the last one, and $T_n = \lfloor n/\ell_n \rfloor$ is the number of intervals. A divergence estimator between the sequence $Y \in \mathbb{R}^n$ and the distribution q was proposed in Ref. [22], which is given by

$$\hat{D}_n(Y\|q) = \sum_{t=1}^{T_n-1} \frac{\ell_n}{n} \log \frac{\ell_n/n}{q(I_t^n)} + \frac{\epsilon_n}{n} \log \frac{\epsilon_n/n}{q(I_{T_n}^n)}, \quad (4.12)$$

where $\epsilon_n = (n - \ell_n(T_n - 1))$ is the number of points in the last segment.

The consistency of such an estimator is shown in Ref. [22]. The convergence rate is further characterized in Ref. [15] by introducing the following boundedness condition on the density ratio between p and q:

$$0 < K_1 \leq \frac{dp}{dq} \leq K_2, \quad (4.13)$$

where K_1 and K_2 are positive constants. In practice, such a boundedness condition is often satisfied, for example, for truncated Gaussian distributions. The above consistency properties are very useful for analyzing the performance of KL divergence-based approaches in the detection problems described in Section 4.1.2. We further note that various other estimators [23,24,25,26,27] can be similarly applied to solve nonparametric detection problems.

4.4 Detection of Existence

In this section, we consider the problem of detecting the existence of anomalous data streams. This problem has been studied in Ref. [13] for the case when both μ and π are discrete with finite alphabet size. We first introduce the results in Ref. [13] on the parametric model with both μ and π known, which serves as a benchmark to gauge the performance of the proposed tests for the semiparametric case with μ unknown and π known and the nonparametric case with neither μ nor π known.

4.4.1 Parametric Model

Suppose both π and μ are known, and assume that the two distributions are discrete. First consider the simple scenario with at most one anomalous data stream, i.e., there exists only one anomalous data stream under H_1. Under H_0, the probability distribution of y^{Mn} is given by

$$P_0(y^{Mn}) = L_0(y^{Mn}, \pi) = \prod_{k=1}^{n}\prod_{j=1}^{M} \pi(y_k^{(j)})$$

$$= \exp\left\{-n\sum_{j=1}^{M} H(\gamma_j) - n\sum_{j=1}^{M} D(\gamma_j \| \pi)\right\}, \quad (4.14)$$

where γ_j denotes the empirical distribution of $y^{(j)}$, and $D(p\|q)$ denotes the KL divergence between two discrete distributions p and q defined as follows:

$$D(p \| q) = \sum_{y \in \mathcal{Y}} p(y) \log \frac{p(y)}{q(y)}. \quad (4.15)$$

And, under H_1 with the i-th data stream being anomalous, the probability of y^{Mn} is given by

$$P_i(y^{Mn}) = L_i(y^{Mn}, \pi, \mu) = \prod_{k=1}^{n}\left[\mu(y_k^{(i)})\prod_{j\neq i}\pi(y_k^{(j)})\right]$$

$$= \exp\left\{-n\sum_{j=1}^{M} H(\gamma_j) - nD(\gamma_i \| \mu) - n\sum_{j\neq i} D(\gamma_j \| \pi)\right\}. \quad (4.16)$$

For this setting, Ref. [13] constructed the following generalized likelihood ratio test (GLRT):

$$\delta(y^{Mn}) : \frac{1}{n}\log\frac{\max_i P_i(y^{Mn})}{P_0(y^{Mn})} \overset{H_1}{\underset{H_0}{\gtrless}} \tau,$$

where τ is used to denote a generic decision threshold here and in subsequent tests. Further simplification based on Equations 4.14 and 4.16 yields the following equivalent test:

$$\max_i [D(\gamma_i \| \pi) - D(\gamma_i \| \mu)] \overset{H_1}{\underset{H_0}{\gtrless}} \tau. \quad (4.17)$$

Under H_0, all the samples are generated from π. Hence, γ_i converges to π almost surely (a.s.) as $n\to\infty$, for $i = 1,\ldots,M$. Thus the test value in Equation 4.17 converges to $-D(\pi\|\mu)$ as $n\to\infty$. Under H_1, if $y^{(j)}$ is generated from μ, then γ_j converges to μ a.s., and hence $D(\gamma_j\|\pi)-D(\gamma_j\|\mu)$ converges to $D(\mu\|\pi)$. For $i\neq j$, $y^{(i)}$ is generated from π, and γ_i converges to π. Therefore, $D(\gamma_i\|\pi)-D(\gamma_i\|\mu)$ converges to $-D(\pi\|\mu)$. Thus, the test value converges to $D(\mu\|\pi)$ under H_1. In order to distinguish between H_0 and H_1, τ should be chosen between $D(\mu\|\pi)$ and $-D(\pi\|\mu)$. It can be shown that by setting $\tau=0$, the GLRT (Equation 4.17) is exponentially consistent and achieves the optimal exponent of the risk function $R(\delta)$ given by

$$E_R(\delta) = C(\mu, \pi), \quad (4.18)$$

where $C(p,q)$ denotes the Chernoff distance between distributions p and q given by

$$C(p,q) = \max_{0\leq\lambda\leq 1} -\log\left(\sum_y p(y)^\lambda q(y)^{1-\lambda}\right). \quad (4.19)$$

Such an idea can be easily extended to the more general scenario with multiple anomalous data streams. Assume that each data stream $y^{(i)}$ takes a different anomalous distribution μ_i if it is anomalous, where μ_i for $i=1,\ldots,M$ are not necessarily the same. All the typical data streams are generated by the same distribution π. Assume that there are $s \geq 1$ anomalous data streams under H_1, and s is fixed and known.

The GLRT can be constructed similarly as follows:

$$\delta(y^{Mn}) : \frac{1}{n}\log\frac{\max_{S\subseteq\mathcal{M},|S|=s} P_S(y^{Mn})}{P_0(y^{Mn})} \overset{H_1}{\underset{H_0}{\gtrless}} \tau,$$

which is equivalent to

$$\max_{S\subset\mathcal{M},|S|=s}\sum_{j\in S}\left[D(\gamma_j\|\pi)-D(\gamma_j\|\mu_j)\right] \overset{H_1}{\underset{H_0}{\gtrless}} \tau. \quad (4.20)$$

Similarly to the case with $s=1$, τ is set to 0 in order to distinguish between H_0 and H_1. The following theorem characterizes the optimality of the GLRT.

Theorem 1

Consider the problem (Equation 4.1) with s anomalous data streams under H_1, where s is fixed and known. Suppose both π and μ_j for $j = 1, 2, \ldots, M$ are known. The GLRT (Equation 4.20) with the threshold $\tau = 0$ is exponentially consistent and achieves the optimal exponent of the risk function $R(\delta)$ given by

$$E_R(\delta) = \min_{S \subset \mathcal{M}, |S|=s} C(\mu_S, \pi_s), \quad (4.21)$$

where $\mu_S = \prod_{i \in S} \mu_i$, and $\pi_s = \pi^s$.

4.4.2 Semiparametric Model

Consider the semiparametric model, in which π is known but μ is unknown. We first introduce the simple model with at most one anomalous data stream and then present the general results with multiple anomalous data streams.

First consider the case with $s = 1$. Without the knowledge of μ, the GLRT (Equation 4.17) is not applicable. For this setting, Ref. [13] constructed the following GLRT by replacing μ in Equation 4.17 with its maximum likelihood estimate $\hat{\mu}_i = \gamma_i$:

$$\max_i D(\gamma_i \| \pi) \underset{H_0}{\overset{H_1}{\gtrless}} \tau. \quad (4.22)$$

In order to distinguish between the two hypotheses, τ should be chosen between 0 and $D(\mu \| \pi)$, which requires the knowledge of the value of $D(\mu \| \pi)$. If $D(\mu \| \pi)$ is known, then the GLRT (Equation 4.17) with the threshold $\tau \in (0, D(\mu \| \pi))$ is exponentially consistent and achieves the exponent of the risk function $R(\delta)$ given by

$$E_R(\delta) = \min \left\{ \tau, \min_{q : D(q\|\pi) \leq \tau} D(q\|\mu) \right\}. \quad (4.23)$$

It is clear that the exponent $E_R(\delta)$ varies for different threshold values $\tau \in (0, D(\mu \| \pi))$. It can be shown that $E_R(\delta)$ achieves the optimal error exponent $C(\pi, \mu)$ if $\tau = C(\mu, \pi)$. This is the optimal error exponent for the parametric model and is hence optimal here.

Without the knowledge of $D(\mu \| \pi)$ and $C(\pi, \mu)$, it is impossible to construct any universally exponentially consistent test over all π and μ. A natural question is whether it is still possible to distinguish H_0 and H_1 even without knowledge of $D(\mu \| \pi)$ and $C(\pi, \mu)$. In order to distinguish between H_0 and H_1, τ should be set between 0 and $D(\mu \| \pi)$. One possible way is to set τ to satisfy $\tau_n \to 0$, such that for large enough n, τ falls between 0 and $D(\mu \| \pi)$. In particular, Ref. [13] sets τ as follows:

$$\tau_n > \frac{|\mathcal{Y}| \log(n+1)}{n}. \quad (4.24)$$

With this choice, the GLRT is universally consistent but not exponentially consistent. Furthermore, the type II error probability converges to zero exponentially fast for any μ and π.

It is clear that choosing τ within $(0, D(\mu \| \pi))$ is necessary for the GLRT to be consistent. Moreover, a large distance between τ and 0 guarantees a small type I error probability, and a large distance between τ and $D(\mu \| \pi)$ guarantees a small type II error probability. Since $D(\mu \| \pi)$ is unknown, a diminishing τ_n eventually falls in the range $(0, D(\mu \| \pi))$ for large enough n by sacrificing exponential consistency of type I error while still keeping exponential consistency of type II error. Furthermore, τ_n cannot converge to zero faster than the test value (i.e., the left-hand side of Equation 4.22) under H_0, which is guaranteed by the condition in Equation 4.24.

We now consider the more general case with s anomalous data streams and assume that $y^{(i)}$ takes the distribution μ_i if it is anomalous for $i = 1, \ldots, M$. Assume that s is fixed and known. Following a similar idea of using the maximum likelihood estimate γ_i of μ_i, for $i \in S$, to replace μ_i in Equation 4.20, the GLRT is given as follows:

$$\max_{S \subset \mathcal{M}, |S|=s} D(\gamma_S \| \pi_s) \underset{H_0}{\overset{H_1}{\gtrless}} \tau. \quad (4.25)$$

In order to distinguish between H_0 and any subhypothesis associated with S under H_1, τ should be chosen between 0 and $\min_{S \subset \mathcal{M}, |S|=s} D(\mu_S \| \pi_s)$.

Theorem 2

Consider the problem (Equation 4.1) with s anomalous data streams, where s is fixed and known. Suppose π is known but μ_j for $j = 1, 2, \ldots, M$ are unknown. Further assume that $\min_{S \subset \mathcal{M}, |S|=s} D(\mu_S \| \pi_s)$ is known. Then the GLRT in Equation 4.25 with the threshold $0 < \tau < \min_{S \subset \mathcal{M}, |S|=s} D(\mu_S \| \pi_s)$ is exponentially consistent and achieves the risk decay exponent given by

$$\min\{\alpha(\delta), \beta(\delta)\}, \quad (4.26)$$

where $\alpha(\delta) = \tau$ and $\beta(\delta)$ is given by

$$\min_{\substack{S \subset \mathcal{M} \\ s.t. \, |S|=s}} \min_{q_{\mathcal{M}} = \{q_1, \ldots, q_M\}} D(q_S \| \mu_S) + D(q_{S^c} \| \pi_{M-s})$$

$$s.t. \, D(q_{S'} \| \pi_s) \leq \tau \text{ for all } S', |S'| = s. \quad (4.27)$$

In the above theorem, $\alpha(\delta)$ and $\beta(\delta)$ respectively correspond to the exponents of the type I error and type II error probabilities, and are characterized based on Sanov's theorem (see Ref. [13] for further details). It can be seen that the GLRT with the specified τ is exponentially consistent without knowing the exact anomalous distributions $\mu_1,\ldots,\mu_M$ but only the distance $\min_{S \subset \mathcal{M}, |S|=s} D(\mu_S \| \pi_s)$. Although the exponent given as the solution to the convex optimization problem (Equation 4.27) does not have an explicit form, it can be solved using numerical methods efficiently. It can also be verified that the optimization problem (Equation 4.27) reduces to that for the model with one anomalous data stream by setting $s = 1$ and $\mu_j = \mu$ for $j = 1,2,\ldots,M$.

By setting $\tau = \min_{S \subset \mathcal{M}, |S|=s} C(\mu_S, \pi_s)$, the risk exponent equals $\min_{S \subset \mathcal{M}, |S|=s} C(\mu_S, \pi_s)$, which is the same as the optimal error exponent of the parametric model, and thus is optimal for the semiparametric model.

For the case without any knowledge about the anomalous distributions $\mu_1,\ldots,\mu_M$ (or even any information about distance between π and $\mu_1,\ldots,\mu_M$), then there does not exist any universally exponentially consistent test. In such a case, it is still possible to construct a consistent test. Set the threshold τ in the GLRT to satisfy $\tau_n \to 0$ and $\tau_n > \frac{s|\mathcal{Y}| \log(n+1)}{n}$. Then the GLRT is universally consistent. Furthermore, the type II error probability converges to zero exponentially fast.

4.4.3 Nonparametric Model

In this section, we consider the nonparametric model when neither π nor μ is known. Again, we start with the simpler model with $s = 1$.

Conditioned on H_0, the maximum likelihood estimate of π is $\hat{\pi}_0 = \frac{\sum_{i=1}^M \gamma_i}{M}$. Conditioned on the i-th data stream being anomalous, the maximum likelihood estimate of μ is γ_i, and the maximum likelihood estimate of π is $\hat{\pi}_i = \frac{\sum_{j \neq i} \gamma_j}{M-1}$. By replacing the distributions with their maximum likelihood estimates, the GLRT can be constructed as follows:

$$\max_i \left[D(\gamma_i \| \hat{\pi}_0) + \sum_{j \neq i} (D(\gamma_j \| \hat{\pi}_0) - D(\gamma_j \| \hat{\pi}_i)) \right] \overset{H_1}{\underset{H_0}{\gtrless}} \tau. \quad (4.28)$$

In order to distinguish between H_0 and H_1, the threshold τ should be set between 0 and $D(\mu \| \frac{1}{M}\mu + \frac{M-1}{M}\pi)$. Assume that $D(\mu \| \frac{1}{M}\mu + \frac{M-1}{M}\pi) := d$ is known. Then the GLRT (Equation 4.28) with the threshold $\tau \in (0,d)$ is shown in Ref. [13] to be exponentially consistent and achieves the exponent

$$E_R(\delta) = \min\{\alpha(\delta), \beta(\delta)\} \quad (4.29)$$

where

$$\alpha(\delta) = \min_{q_j, j=1,\ldots,M} D(q_1 \| \pi) + D(q_2 \| \pi) + \ldots + D(q_M \| \pi)$$

$$\text{s.t. } D(q_1 \| \bar{q}) + \sum_{j \neq 1} \left[D(q_j \| \bar{q}) - D(q_j \| \bar{q}_{-1}) \right] \geq \tau \quad (4.30)$$

and

$$\beta(\delta) = \min_{q_j, j=1,\ldots,M} D(q_1 \| \mu) + \sum_{j \neq 1} D(q_j \| \pi)$$

$$\text{s.t. } D(q_k \| \bar{q}) + \sum_{j \neq k} \left[D(q_j \| \bar{q}) - D(q_j \| \bar{q}_{-k}) \right] \leq \tau \quad (4.31)$$

for $k = 1, 2, \ldots, M$.

In the above definitions for $\alpha(\delta)$ and $\beta(\delta)$, $\bar{q} = \sum_{j=1}^M q_j / M$ and $\bar{q}_{-i} = \sum_{j \neq i} \frac{q_j}{M-1}$ for $i = 1,\ldots,M$.

It can be shown that $\alpha(\delta) > 0$ and $\beta(\delta) > 0$. Hence, the GLRT is exponentially consistent. Such exponential consistency does not exploit the full knowledge of the distributions but only the KL divergence $D(\mu \| \frac{1}{M}\mu + \frac{M-1}{M}\pi)$ to set the threshold. For large M, $D(\mu \| \frac{1}{M}\mu + \frac{M-1}{M}\pi)$ is close to $D(\mu \| \pi)$. And as $M \to \infty$, the exponent of $R(\delta)$ converges to $C(\pi, \mu)$ if τ is set to be $C(\mu, \pi)$, which is the optimal risk exponent for the parametric model, and thus is optimal.

If no information about the distance between the distributions is known, then there does not exist any universally exponentially consistent test. And if the threshold τ is set to satisfy $\tau_n \to 0$ and $\tau_n > \frac{M|\mathcal{Y}| \log(n+1)}{n}$, then the GLRT is universally consistent. Furthermore, the type II error is universally exponentially consistent. For large enough n, a diminishing τ_n eventually falls into the desirable range of τ that can distinguish between H_0 and H_1.

We further consider the case with s anomalous data streams. Suppose that $y^{(i)}$ is generated by μ_i, if it is anomalous, for $i = 1,\ldots,M$. Assume that s is fixed and known, and both π and $\{\mu_i\}_{i=1}^M$ are unknown.

Under H_0, we replace π with its maximum likelihood estimate $\hat{\pi}_0 = \frac{\sum_{i=1}^M \gamma_i}{M}$. Under H_1, and conditioned on the case with data streams supported on S being anomalous, we use the maximum likelihood estimate $\hat{\pi}_{-S} = \frac{\sum_{j \notin S} \gamma_j}{M-s}$ to replace π, and $\hat{\mu}_i = \gamma_i$ to replace μ_i, for $i \in S$. Thus, the GLRT is given by

$$\max_{S, |S|=s} \left[\sum_{j \in S} D(\gamma_j \| \hat{\pi}_0) + \sum_{j \notin S} (D(\gamma_j \| \hat{\pi}_0) - D(\gamma_j \| \hat{\pi}_{-S})) \right] \overset{H_1}{\underset{H_0}{\gtrless}} \tau. \quad (4.32)$$

In order to distinguish between H_0 and H_1, the threshold τ should be chosen between 0 and $\min_{S \subset \mathcal{M}, |S|=s}$

$$D\left(\mu_S \left\| \prod_{i=1}^s \left(\frac{1}{M} \sum_{j \in S} \mu_j + \frac{M-s}{M} \pi \right) \right. \right).$$

Theorem 3

Consider the problem (Equation 4.1) with s anomalous data streams. Assume that s is fixed and known, and π is known but μ_j for $j = 1,2,\ldots,M$ are unknown. Further assume that

$$\min_{S \subset \mathcal{M}, |S|=s} D\left(\mu_S \parallel \prod_{i=1}^{s}\left(\frac{1}{M}\sum_{j \in S}\mu_j + \frac{M-s}{M}\pi\right)\right) := d \text{ is}$$

known. Then the GLRT (Equation 4.32) with the threshold $\tau \in (0,d)$ is exponentially consistent and achieves the risk decay exponent $R(\delta)$ given by

$$E_R(\delta) = \min\{\alpha(\delta), \beta(\delta)\} \tag{4.33}$$

where

$$\alpha(\delta) = \min_{\{q_j\}_{j=1}^{M}} \sum_{j=1}^{M} D(q_j \parallel \pi)$$

$$\text{s.t.} \sum_{j=1}^{M} D\left(q_j \parallel \frac{\sum_{i=1}^{M}q_i}{M}\right) - \sum_{j \notin S} D\left(q_j \parallel \frac{\sum_{i \notin S}q_i}{M-s}\right) \geq \tau$$

for any $S \subset \mathcal{M}$ s.t. $|S| = s$ (4.34)

$$\beta(\delta) = \min_{S:|S|=s} \min_{\{q_j\}_{j=1}^{M}} \sum_{j \in S} D(q_j \parallel \mu_j) + \sum_{j \notin S} D(q_j \parallel \pi)$$

$$\text{s.t.} \sum_{j=1}^{M} D\left(q_j \parallel \frac{\sum_{i=1}^{M}q_i}{M}\right) - \sum_{j \notin S'} D\left(q_j \parallel \frac{\sum_{i \notin S'}q_i}{M-s}\right) \leq \tau$$

for all $S' \subset \mathcal{M}$ s.t. $|S'| = s$ (4.35)

It can be shown that $\alpha(\delta) > 0$ and $\beta(\delta) > 0$, such that the exponential consistency is guaranteed. Furthermore, if $M \to \infty$, $s/M \to 0$, and τ is set to be $\min_{S,|S|=s} C(\mu_S, \pi_s)$, then the exponent of $R(\delta)$ converges to $\min_{S,|S|=s} C(\mu_S, \pi_s)$ which is the optimal value. The condition $s/M \to 0$ guarantees that there are not too many anomalous data streams so that the estimate of the typical distribution π can be accurate enough.

For the case without any knowledge of the distance between the distributions (including no information about the distance between π and μ_j for $j = 1,2,\ldots,M$ is known), then there does not exist any universally exponentially consistent test. In such a case, it is still possible to construct a consistent test. Set the threshold τ in the GLRT to satisfy $\tau_n \to 0$ and $\tau_n > \frac{M|\mathcal{Y}|\log(n+1)}{n}$, where $|Y|$ is the cardinality of the support set of π and μ. Then the GLRT is universally consistent. Furthermore, the type II error probability converges to zero exponentially fast. This implies that without knowing the distance between π and μ_j for $j = 1,2,\ldots,M$, a diminishing τ_n helps to keep the type II error probability exponentially decaying to zero while keeping the type I error probability decaying to zero although not exponentially.

4.5 Identification of Anomalous Data Streams

In this section, we consider the problem of identifying anomalous data streams if they exist. This problem has been studied in Ref. [12] for discrete distributions, in Ref. [15] for continuous distributions, and in Ref. [14] for arbitrary distributions. We start with the parametric case and then the semiparametric and nonparametric cases.

4.5.1 Parametric Model

The parametric model with both μ and π known has been studied in Ref. [12]. First consider the simple case with one anomalous data stream. If the i-th data stream is anomalous, the joint distribution of all samples is given by

$$P_i(y^{Mn}) = \prod_{k=1}^{n}(\mu(y_k^{(i)})\prod_{j \neq i}\pi(y_k^{(j)})) \tag{4.36}$$

$$\triangleq L_i(y^{Mn}, \mu, \pi).$$

The test that achieves the optimal error exponent for the problem is the maximum likelihood test [12]:

$$\delta_{ML}(y^{Mn}, \pi, \mu) = \arg\max_{1 \leq i \leq M} P_i(y^{Mn}). \tag{4.37}$$

And the optimal error exponent is equal to $2B(\mu, \pi)$.

These results are further generalized to the case with multiple anomalous data streams in Ref. [12]. In this model, most of the data streams are generated by the typical distribution π, whereas only a small subset $S \subset \mathcal{M}$ of data streams are anomalous. Again, we denote the number of anomalous data streams by $s = |S|$. The data stream i is generated by π if it is typical and is generated by μ_i if it is anomalous. For the hypothesis with the data streams in the set S being anomalous, the joint distribution of y^{Mn} is as follows:

$$P_S(y^{Mn}) = \prod_{k=1}^{n}\left\{\prod_{i \in S}\mu_i(y_k^{(i)})\prod_{j \notin S}\pi(y_k^{(j)})\right\} \tag{4.38}$$

$$\triangleq L_S(y^{Mn}, \{\mu_i\}_{i \in S}, \pi).$$

It is shown in Ref. [12] that the test that achieves the optimal error exponent for this problem is the maximum likelihood test:

$$\delta_{ML}(y^{Mn}, \pi, \mu) = \arg\max_{S:|S|=s} P_S(y^{Mn}). \tag{4.39}$$

Then the error exponent of the test (Equation 4.39) was characterized in Ref. [12] in the following theorem.

Theorem 4

Consider the problem of identifying anomalous data streams with s anomalous data streams. Assume that π, μ_i for $i = 1,\ldots, M$, and s are known. The optimal error exponent is equal to

$$\min_{1 \leq i < j \leq M} C(\mu_i(y)\pi(y'), \pi(y)\mu_j(y'))). \quad (4.40)$$

Furthermore, when $\mu_i = \mu$, for $i = 1,\ldots,M$, the optimal error exponent is equal to $2B(\mu,\pi)$.

4.5.2 Semiparametric Model

In this section, we consider the semiparametric model, in which π is known but μ is unknown. This model has been studied via various approaches, which are applicable to different cases. We next describe these studies.

For the case when both π and μ are discrete, the GLRT has been developed in Ref. [12]. First consider the case with $s = 1$. Conditioned on the i-th data stream being anomalous, the maximum likelihood estimate of μ is γ_i. Then the GLRT is constructed in Ref. [12] by replacing the unknown μ with its maximum likelihood estimate γ_i in L_i as follows:

$$\delta(y^{Mn}) = \arg\max_{i=1,\ldots,M} L_i(y^{Mn}, \gamma_i, \pi). \quad (4.41)$$

It is shown in Ref. [12] that the GLRT (Equation 4.41) achieves the optimal error exponent $2B(\mu,\pi)$. Thus, even without knowing μ, the optimal error exponent for the parametric case with μ and π known can still be achieved.

The authors in Ref. [12] further generalized these results to the case with multiple anomalous data streams. Suppose each data stream i takes distribution μ_i if it is anomalous. Consider the case with s being known. When π is known and μ_i for $i = 1,\ldots,M$ are unknown, the GLRT is constructed in Ref. [12] by replacing μ_i with its maximum likelihood estimate as follows:

$$\delta(y^{Mn}) = \arg\max_{S \subset \{1,\ldots,M\}, |S|=s} L_S(y^{Mn}, \{\gamma_i\}_{i \in S}, \pi). \quad (4.42)$$

The performance of the GLRT (Equation 4.42) is characterized by Ref. [12] in the following theorem.

Theorem 5

Consider the problem of identifying anomalous data streams with s anomalous data streams. Assume that π and s are known, and μ_i for $i = 1,\ldots,M$ are unknown. The error exponent achieved by the GLRT (Equation 4.42) is equal to

$$\min_{1 \leq i \leq M} 2B(\mu_i,\pi). \quad (4.43)$$

Furthermore, when $\mu_i = \mu$, for $i = 1,\ldots,M$, the error exponent is equal to $2B(\mu,\pi)$, which is optimal.

For the case when μ and π are continuous, the KL divergence-based test has been developed in Ref. [15]. First consider the case with $s = 1$. The test is motivated as follows. If both μ and π are known, the optimal test for the problem is the maximum likelihood test in Equation 4.37. By normalizing $P_i(y^{Mn})$ with $\pi(y^{Mn})$, Equation 4.37 is equivalent to

$$\delta_{ML}(y^{Mn},\pi,\mu) = \arg\max_{1 \leq i \leq M} \frac{P_i(y^{Mn})}{\pi(y^{Mn})}$$

$$= \arg\max_{1 \leq i \leq M} \left\{ \frac{1}{n} \sum_{k=1}^{n} \log \frac{\mu(y_k^{(i)})}{\pi(y_k^{(i)})} \right\}$$

$$= \arg\max_{1 \leq i \leq M} \hat{L}_i.$$

where

$$\hat{L}_i \triangleq \frac{1}{n} \sum_{k=1}^{n} \log \frac{\mu(y_k^{(i)})}{\pi(y_k^{(i)})}. \quad (4.44)$$

It can be shown that the test (Equation 4.37) achieves the optimal error exponent $2B(\pi,\mu)$.

If $y^{(i)}$ is anomalous, $\hat{L}_i$ is an empirical estimate of the KL divergence between μ and π, and then $\hat{L}_i \to D(\mu\|\pi)$ as $n \to \infty$, by the law of large numbers. If $y^{(i)}$ is typical, $\hat{L}_i \to -D(\pi\|\mu)$ as $n \to \infty$. These observations motivate the design of the KL divergence-based test.

Now consider the KL divergence estimator $\hat{D}_n(y^{(i)}\|\pi)$. It is clear that if $y^{(i)}$ is anomalous, then $\hat{D}_n(y^{(i)}\|\pi)$ is a good estimator of $D(\mu\|\pi)$, which is a positive constant. On the other hand, if $y^{(i)}$ is typical, $\hat{D}_n(y^{(i)}\|\pi)$ should be close to $D(\pi\|\pi) = 0$. Based on such understanding, the following test is constructed in Ref. [15] for the problem of identifying anomalous data streams:

$$\delta_{KL}(y^{Mn}) = \arg\max_{1 \leq i \leq M} \hat{D}_n(y^{(i)}\|\pi). \quad (4.45)$$

The test is designed using the KL divergence estimator based on data-dependent partitions. The following theorem provides a lower bound on the error exponent of δ_{KL}, which further implies that δ_{KL} is universally exponentially consistent. Such performance of the test is mainly determined by the properties of the KL divergence estimator.

Theorem 6

If the density ratio between μ and π satisfies Equation 4.13, then δ_{KL} defined in Equation 4.45 is exponentially consistent, and the error exponent is lower-bounded by $\frac{1}{32}(\frac{K_1}{K_1+K_2})^2 D^2(\mu\|\pi)$.

We note that such a KL divergence-based approach can also be generalized to the case with s anomalous data streams. Furthermore, by using appropriate KL divergence estimators, such an approach can be generalized to the case with discrete μ and π, and for the nonparametric case with μ and π unknown.

For the cases when μ and π are arbitrary (either continuous or discrete), the kernel-based approach has been studied in Ref. [15]. First consider the simple case $s = 1$.

The motivation of the kernel-based approach is as follows. It is clear that if $y^{(i)}$ is anomalous, $\text{MMD}_u^2[y^{(i)},\pi]$ is a good estimator of $\text{MMD}^2[\mu,\pi]$, which is a positive constant. On the other hand, if $y^{(i)}$ is typical, $\text{MMD}_u^2[y^{(i)},\pi]$ should be a good estimator of $\text{MMD}^2[\pi,\pi]$, which is zero. Based on the above understanding, the following test is constructed in Ref. [15]:

$$\delta_{\text{MMD}} = \arg\max_{1 \leq i \leq M} \text{MMD}_u^2[y^{(i)},\pi]. \quad (4.46)$$

In Ref. [15], a lower bound on the error exponent of δ_{MMD} is provided, and it is further demonstrated that the test δ_{MMD} is universally exponentially consistent. This result is given in the following theorem.

Theorem 7

Consider the problem of identifying an anomalous data stream with $s = 1$. Suppose δ_{MMD} defined in Equation 4.46 applies a bounded kernel with $0 \leq k(x,y) \leq K$ for any (x,y). Then, the error exponent is lower-bounded by $\dfrac{\text{MMD}^4[\mu,\pi]}{9K^2}$.

Next, we show the comparison of the performance of the KL divergence-based approach and kernel-based approach as in Ref. [15]. Set the number of sequences $M = 5$, choose the typical distribution $\pi = \mathcal{N}(0,1)$, and choose the outlier distribution $\mu = \mathcal{N}(0,0.2)$, $\mathcal{N}(0,1.2)$, and $\mathcal{N}(0,1.8)$, respectively. In Figures 4.1–4.3 the logarithm $logR$ of the probability of error is plotted as a function of the sample size n.

It can be seen that for both tests, the probability of error converges to zero as the sample size increases. Furthermore, $logR$ decreases with n linearly, which demonstrates the exponential consistency of both the KL divergence-based test and the kernel-based test.

By comparing the four figures, it can be seen that as the variance of μ deviates from the variance of π, δ_{KL} has a better performance than δ_{MMD}. The numerical results

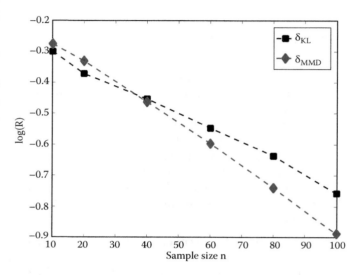

FIGURE 4.2
Comparison of the performance between KL divergence- and kernel-based tests with $\pi = \mathcal{N}(0,1)$ and $\mu = \mathcal{N}(0,1.2)$.

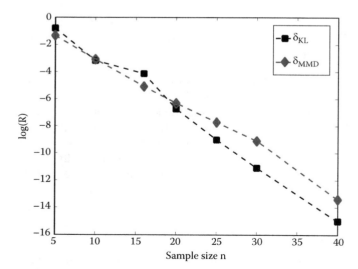

FIGURE 4.1
Comparison of the performance between KL divergence- and kernel-based tests with $\pi = \mathcal{N}(0,1)$ and $\mu = \mathcal{N}(0,0.2)$.

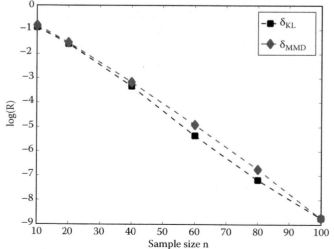

FIGURE 4.3
Comparison of the performance between KL divergence- and kernel-based tests with $\pi = \mathcal{N}(0,1)$ and $\mu = \mathcal{N}(0,1.8)$.

and theoretical lower bounds on error exponents give us some intuitions to identify regimes in which one test outperforms the other. It can be seen from the figures that when the distributions μ and π become more different from each other, δ_{KL} outperforms δ_{MMD}. The reason is that for any pair of distributions, MMD is bounded between $[0,2K]$, while the KL divergence is not bounded. As the distributions become more different from each other, the KL divergence increases, and the KL divergence-based test thus has a larger error exponent than the kernel-based test.

4.5.3 Nonparametric Model

In this section, we consider the nonparametric model, in which both μ and π are unknown.

For the case when both μ and π are discrete, the generalized likelihood-based approach is studied in Ref. [12]. First consider the case $s = 1$. It is interesting that for the semiparametric model, even without knowing μ, the optimal error exponent for the parametric case with μ and π known can still be achieved when $s = 1$. Given such an assertion, it is of interest to determine whether or not it is possible to design a test that achieves the optimal error exponent $2B(\mu,\pi)$ when neither μ nor π is known. In fact, an example given in Ref. [12, Example 1] shows that when neither μ nor π is known, it is impossible to design a test that achieves the optimal error exponent $2B(\mu,\pi)$ universally (i.e., for all μ and π).

Although the optimal error exponent for the parametric case is not achievable universally, a universally exponentially consistent test can still be constructed. Conditioned on the i-th data stream being anomalous, the maximum likelihood estimate of π is $\frac{\sum_{j\neq i}\gamma_j}{M-1}$. The GLRT is constructed by further replacing π with its maximum likelihood estimate in L_i as follows:

$$\delta(y^{(Mn)}) = \arg\max_{i=1,\ldots,M} L_i\left(y^{(Mn)}, \gamma_i, \frac{\sum_{j\neq i}\gamma_j}{M-1}\right). \quad (4.47)$$

The following error exponent for the GLRT (Equation 4.47) is characterized in Ref. [12] via Sanov's theorem:

$$\min_{q_1,\ldots,q_M} D(q_1\|\mu) + \sum_{i=2}^{M} D(q_i\|\pi)$$

$$\text{s.t.} \sum_{j\neq 1} D\left(q_j \;\|\; \frac{\sum_{k\neq 1}q_k}{M-1}\right) \geq \sum_{j\neq 2} D\left(q_j \;\|\; \frac{\sum_{k\neq 2}q_k}{M-1}\right). \quad (4.48)$$

The authors in Ref. [12] further show that the GLRT (Equation 4.47) achieves the optimal error exponent $2B(\mu,\pi)$, as $M \to \infty$. Note that for any $M \geq 3$ and $\epsilon > 0$, it holds that

$$\lim_{n\to\infty} P_i\left\{ \left\|\frac{\sum_{j=1}^{M}\gamma_j}{M} - \left(\frac{\mu}{M} + \frac{(M-1)\pi}{M}\right)\right\|_1 > \epsilon \right\} = 0, \quad (4.49)$$

for any anomalous data stream, where $\|\cdot\|_1$ denotes the l_1 norm. Furthermore, $\frac{\mu}{M} + \frac{(M-1)\pi}{M} \to \pi$ as $M \to \infty$. Based on these two facts, a consistent estimate of π can be obtained by using the empirical distribution of y^{Mn} for large enough M. Such an observation gives an intuitive explanation of the result that the GLRT (Equation 4.47) achieves the optimal error exponent $2B(\mu,\pi)$, as $M \to \infty$.

These results are further generalized to the case with multiple anomalous data streams in Ref. [12]. Suppose that the i-th data stream takes distribution μ_i if it is anomalous, for $1 \leq i \leq M$. When none of π or μ_i for $i = 1,\ldots,M$ are known, the GLRT is constructed by further replacing π with its maximum likelihood estimate as follows:

$$\delta(y^{Mn}) = \arg\max_{S \subset \{1,\ldots,M\}, |S|=s} L_S\left(y^{Mn}, \{\gamma_i\}_{i\in S}, \frac{\sum_{k\notin S}\gamma_k}{M-s}\right). \quad (4.50)$$

The performance of the GLRT (Equation 4.50) is characterized in Ref. [12] in the following theorem.

Theorem 8

Consider the problem of identifying anomalous data streams with s anomalous data streams. Assume that s is known, and π and μ_i for $i = 1,\ldots,M$ are unknown. The GLRT (Equation 4.50) is exponentially consistent, and the error exponent achieved is characterized as follows:

$$\alpha = (\delta) = \min_{\substack{S,S'\subset\{1,\ldots,M\} \\ |S|=|S'|=s}} \min_{q_1,\ldots,q_M} \left(\sum_{i\in S} D(q_i\|\mu_i) + \sum_{j\notin S} D(q_j\|\pi)\right) \quad (4.51)$$

$$\text{s.t.} \sum_{i\notin S} D\left(q_i \;\|\; \frac{\sum_{k\notin S}q_k}{M-s}\right) \geq \sum_{i\notin S'} D\left(q_i \;\|\; \frac{\sum_{k\notin S'}q_k}{M-s}\right). \quad (4.52)$$

Furthermore, when $\mu_i = \mu$, for $i = 1,\ldots,M$, as $M \to \infty$, the error exponent is equal to $2B(\mu,\pi)$, which is optimal.

The authors in Ref. [12] further studied the case when s is unknown. Assume that for a fixed number of anomalous data streams $k = 0,1,\ldots,M/2$, $\mathcal{S}$ either contains all hypotheses with k anomalous data streams or none of them. First consider the case when all anomalous data streams are identically distributed, i.e., $\mu_i = \mu$ for $i = 1,\ldots,M$, and neither π nor μ is known. The GLRT is constructed as follows:

$$\delta(y^{Mn}) = \arg\max_{S \in \mathcal{S}} L_S\left(y^{Mn}, \frac{\sum_{i \in S} y_i}{|S|}, \frac{\sum_{j \notin S} y_j}{M - |S|}\right). \quad (4.53)$$

It was shown in Ref. [12] that the GLRT (Equation 4.53) is universally consistent for every hypothesis set. If the anomalous data streams are not necessarily identically distributed, Ref. [12] showed that there does not exist any universally exponentially consistent test.

We next consider the case when μ and π can be either discrete or continuous, for which the kernel-based approach is developed in Ref. [14]. Furthermore, the authors in Ref. [14] focused on the regime of massive data sets, in which the number M of data streams and the number s of anomalous data streams possibly scale to infinity, which is different from the preceding scenarios with fixed M and s. This is motivated by applications in which the problem dimensions (i.e., M and s) are large, which is very common in biomedical signal processing. It is clear that as M (and possibly s) becomes large, it is increasingly challenging to consistently identify anomalous data streams. It then requires the number of samples n in each data stream to increase with M and s in order to guarantee consistent identification. Hence, the focus here is on how the sample size n should scale with M and s to guarantee the consistency of tests.

First consider the case with s known. We start with the simple case with $s = 1$, and then consider a more general model with $s \geq 1$ anomalous data streams, in which $\frac{s}{M} \to \lambda$ as $M \to \infty$, where $0 \leq \lambda \leq 1$.

For each data stream $y^{(i)}$, denote the $(M-1)n$-dimensional sequence that stacks all other data streams together as $\overline{y^{(i)}}$, which is given by

$$\overline{y^{(i)}} = \left\{y^{(1)}, \ldots, y^{(i-1)}, y^{(i+1)}, \ldots, y^{(M)}\right\}.$$

The authors in Ref. [27] developed the following test based on the MMD:

$$\delta(y^{Mn}) = \arg\max_{1 \leq i \leq M} \mathrm{MMD}_u^2[y^{(i)}, \overline{y^{(i)}}]. \quad (4.54)$$

The intuition behind the test (Equation 4.54) is as follows. If $y^{(i)}$ is anomalous, then $\overline{y^{(i)}}$ is fully composed of typical data streams. Hence, $\mathrm{MMD}_u^2[y^{(i)}, \overline{y^{(i)}}]$ is a good estimate of $\mathrm{MMD}^2[\mu, \pi]$, which is a positive constant. On the other hand, if $y^{(i)}$ is typical, $\overline{y^{(i)}}$ is composed of $M-2$ data streams generated by μ and only one data stream generated by π. As M increases, the impact of the anomalous data stream in $\overline{y^{(i)}}$ is negligible, and $\mathrm{MMD}_u^2[y^{(i)}, \overline{y^{(i)}}]$ should be asymptotically close to zero. Based on the above understanding, the test (Equation 4.54) is able to identify the anomalous data stream.

The authors in Ref. [14] showed that if the test (Equation 4.54) applies a bounded kernel with $0 \leq k(x,y) \leq K$ for any (x,y), then the maximum error probability is upper-bounded as follows:

$$e(\delta) \leq \exp\left(\log M - \frac{n(\mathrm{MMD}^2[\mu,\pi] - \xi)^2}{16K^2(1 + \Theta(\frac{1}{M}))}\right), \quad (4.55)$$

where ξ is a constant arbitrarily close to zero. And the error exponent is lower-bounded by

$$\frac{(\mathrm{MMD}^2[\mu,\pi] - \xi)^2}{16K^2(1 + \Theta(\frac{1}{M}))} - \frac{\log M}{n}. \quad (4.56)$$

Thus, the test (Equation 4.54) is exponentially consistent if

$$n \geq \frac{16K^2(1 + \eta)}{\mathrm{MMD}^4[\mu,\pi]} \log M, \quad (4.57)$$

where η is any positive constant. For the scenario with one anomalous data stream, the order of $\log M$ samples is sufficient to guarantee exponentially consistent detection.

The authors in Ref. [14] further generalized the test (Equation 4.54) to the case with arbitrary $s \geq 1$. Assume that $\frac{s}{M} \to \lambda$ as $M \to \infty$, where $0 \leq \lambda < \frac{1}{2}$. Although it is assumed that $\lambda < \frac{1}{2}$, the case with $\lambda > \frac{1}{2}$ can also be solved, with the roles of π and μ being exchanged. The test is constructed by choosing the data streams with the largest s values of $\mathrm{MMD}_u^2[y^{(i)}, \overline{y^{(i)}}]$ as follows:

$$\delta(y^{Mn}) = \{i : \mathrm{MMD}_u^2[y^{(i)}, \overline{y^{(i)}}]$$

is among the s largest values of $\mathrm{MMD}_u^2[y^{(k)}, \overline{y^{(y)}}]$ $\quad (4.58)$

for $k = 1, \ldots, n\}$

The authors in Ref. [14] characterize the following condition under which the above test is consistent.

Theorem 9

Consider the problem of identifying anomalous data streams with s anomalous data streams, where $\frac{s}{M} \to \lambda$ as $M \to \infty$ and $0 \leq \lambda < \frac{1}{2}$. Assume that the value of s is known. Further assume that the test (Equation 4.58) applies a bounded kernel with $0 \leq k(x,y) \leq K$ for any (x,y). Then the maximum error probability is upper-bounded as follows:

$$e(\delta) \leq \exp\left(\log((M-s)s) - \frac{n((1-2\lambda)\mathrm{MMD}^2[\mu,\pi] - \xi)^2}{16K^2(1 + \Theta(\frac{1}{M}))}\right), \quad (4.59)$$

where ξ is a constant that can be picked arbitrarily close to zero, and $f(n) = \Theta(g(n))$ implies that there exist $k_1, k_2, n_0 > 0$ s.t. for all $n > n_0$, $k_1 g(n) \leq f(n) \leq k_2 g(n)$. The error exponent is lower-bounded as follows:

$$\alpha(\delta) \geq \frac{((1-2\lambda)\text{MMD}^2[\mu,\pi] - \xi)^2}{16K^2(1+\Theta(\frac{1}{M}))} - \frac{\log((M-s)s)}{n}. \quad (4.60)$$

Furthermore, the test (Equation 4.58) is exponentially consistent for any μ and π if

$$n \geq \frac{16K^2(1+\eta)}{(1-2\lambda)^2 \text{MMD}^4[\mu,\pi]} \log(s(M-s)), \quad (4.61)$$

where η is any positive constant.

It can be seen that $\log((M-s)s)$ is of the order of $\log M$, for $1 \leq s < M$. Hence, Theorem 9 implies that even with s anomalous data streams, the test (Equation 4.58) requires only the order of $\log M$ samples in each data stream in order to guarantee consistency of the test. Hence, the increase of s does not affect the order-level requirement on the sample size n. Theorem 9 is also applicable to the case in which $\lambda > \frac{1}{2}$ simply with the roles of μ and π exchanged.

Note that Theorem 9 characterizes the conditions to guarantee consistency for a pair of fixed but unknown distributions μ and π. Hence, the condition (Equation 4.61) depends on the underlying distributions μ and π. In fact, such a condition implies that the test (Equation 4.58) is universally consistent for any arbitrary pair of μ and π, if $n = \omega(\log M)$, where $f(n) = \omega(g(n))$ means that for all $k > 0$, there exists $n_0 > 0$ s.t. for all $n > n_0$, $|f(n)| \geq k|g(n)|$.

Further consider the case that the value of s is unknown. Assume that $\frac{s}{n} \to 0$, as $n \to \infty$, which means that the majority of the data streams are typical, and only a diminishing portion of them are anomalous. This includes two cases: (1) s is fixed, and (2) $s \to \infty$ and $\frac{s}{n} \to 0$ as $n \to \infty$. Without knowledge of s, the test in Equation 4.58 is not applicable anymore, because it depends on the value of s.

For $i = 1,...,M$, although $y^{(i)}$ contains mixed samples from π and μ, it is dominated by samples from π due to the assumption that $s/M \to 0$. Thus, for large enough M and n, $\text{MMD}_u^2[y^{(i)}, \overline{y^{(i)}}]$ should be close to zero if $y^{(i)}$ is drawn from π, and should be far away enough from zero (in fact, close to $\text{MMD}^2[\pi,\mu]$) if $y^{(i)}$ is drawn from μ. Based on this understanding, the following test is constructed:

$$\delta(y^{Mn}) = \{i : \text{MMD}_u^2[y^{(i)}, \overline{y^{(i)}}] > \delta_M\}, \quad (4.62)$$

where $\delta_M \to 0$ and $\frac{s^2}{M^2 \delta_M} \to 0$ as $M \to \infty$. The reason for the condition $\frac{s^2}{M^2 \delta_M} \to 0$ is to guarantee that δ_M converges to zero more slowly than $\text{MMD}_u^2[y^{(i)}, \overline{y^{(i)}}]$ with $y^{(i)}$ drawn from π so that as M goes to infinity, δ_M asymptotically falls between $\text{MMD}_u^2[y^{(i)}, \overline{y^{(i)}}]$ with $y^{(i)}$ drawn from π and $\text{MMD}_u^2[y^{(i)}, \overline{y^{(i)}}]$ with $y^{(i)}$ drawn from μ. We note that the scaling behavior of s as M increases needs to be known in order to choose δ_M for the test, which can be obtained based on prior knowledge of the problem.

The authors in Ref. [14] characterize the following condition under which the test (Equation 4.62) is consistent.

Theorem 10

Consider the problem of identifying anomalous data streams, where $\frac{s}{M} \to 0$, as $M \to \infty$. Assume that s is unknown in advance. Further assume that the test (Equation 4.62) adopts a threshold δ_M such that $\delta_M \to 0$ and $\frac{s^2}{M^2 \delta_M} \to 0$, as $M \to \infty$, and the test applies a bounded kernel with $0 \leq k(x,y) \leq K$ for any (x,y). Then the maximum error probability is upper-bounded as follows:

$$e(\delta) \leq \exp\left(\log s - \frac{n(\text{MMD}^2[\mu,\pi] - \delta_M)^2}{16K^2(1+\Theta(\frac{1}{M}))}\right) \\ + \exp\left(\log(M-s) - \frac{n(\delta_M - \mathbb{E}[\text{MMD}_u^2[Y_k, \overline{Y_k}]])^2}{16K^2(1+\Theta(\frac{1}{M}))}\right). \quad (4.63)$$

Furthermore, the test (Equation 4.62) is consistent if

$$n \geq 16(1+\eta)K^2 \\ \max\left\{\frac{\log(\max\{1,s\})}{(\text{MMD}^2[\mu,\pi] - \delta_M)^2}, \frac{\log(M-s)}{(\delta_M - \mathbb{E}[\text{MMD}_u^2[Y,\overline{Y}]])^2}\right\} \quad (4.64)$$

where η is any positive constant. In the above equation, $\mathbb{E}[\text{MMD}_u^2[Y,\overline{Y}]]$ is a constant, where Y is a data stream generated by π and $\overline{Y}$ is a stack of $(n-1)$ data streams with s data streams generated by μ and the remaining data streams generated by π.

Note that Theorem 10 is also applicable to the case with $s = 0$, i.e., the null hypothesis when there is no anomalous data stream. This solves the problems of detecting the existence of anomalous data streams and identifying them simultaneously. But the test (Equation 4.62) is not exponentially consistent. In fact, when there is no null hypothesis (i.e., $s \geq 1$), an exponentially consistent test can be built as follows. For each subset S of $\{1,...,M\}$, we compute $\text{MMD}_u^2[y^S, \overline{y^S}]$, and the test finds the set of indices corresponding to the largest average value. However, for such a test to be consistent, n needs to scale linearly with M, which is not desirable.

Theorem 10 implies that n should be in the order $\omega(\log M)$ to guarantee consistency, because $\frac{s}{M} \to 0$ and $\delta_M \to 0$ as $M \to \infty$. Compared to the case with s known (for which it is sufficient for n to scale on the order $\log M$), the threshold on n has an order-level increase due to lack of the knowledge of s. Furthermore, the above understanding on the order-level condition on n also implies that the test (Equation 4.62) is universally consistent for any arbitrary pair of π and μ, if $n = \omega (\log M)$.

Comparison between the two universal consistency results implies that the knowledge of s does not affect the order-level sample complexity to guarantee a test to be universally consistent. Furthermore, the authors in Ref. [14] also characterized the conditions under which no test is universally consistent for arbitrary μ and π, given in the following proposition.

Proposition 1

Consider the problem of identifying anomalous data streams with s anomalous data streams. Suppose that the sample size n satisfies

$$n = O\left(\frac{\log \frac{M}{s}}{s}\right), \quad (4.65)$$

where $f(n) = O(g(n))$ means that there exist $k, n_0 > 0$ s.t. for all $n > n_0$, $|f(n)| \leq k|g(n)|$. Then there exists no test that is universally consistent for arbitrary distribution pairs π and μ.

The sufficient and necessary conditions on sample complexity establish the following performance optimality for the kernel-based test.

Theorem 11

Consider the problem of identifying anomalous data streams with $s \geq 1$. For s being known and unknown, the kernel-based test (Equation 4.58, under the conditions in Theorem 9) and the test (Equation 4.62, under the conditions in Theorem 10) are respectively order-level optimal in sample complexity required to guarantee universal consistency for arbitrary π and μ.

4.6 Applications in Biomedical Signal Processing

The problems of detecting and identifying anomalous data streams have wide applications in biomedical signal processing and health care. These anomaly detection and identification methods can be applied to large biomedical data sets at various molecular and phenotypic layers including DNA, RNA, protein, metabolite, and epigenetic sequence data for biomedical research and clinical diagnosis. These methods can also be applied to analyze medical records, treatment and supervision logs, and traces from health trackers and mobile services. Since these methods do not rely on prior knowledge of anomaly, they are suitable for solving numerous problems in biomedical research and health care.

It is well known that biological and medical data are corrupted by noises. This situation brings about a challenge of efficiently detecting meaningful signals from background noise when analyzing biomedical data. One problem that these anomaly detection approaches can solve is to detect sample biases, and reduce sampling biases by removing sequences with significantly differently distributed sequences [28]. Another example of such applications is similar to the work in Ref. [28], where anomaly detection and identification methods can be applied to improving labeled disordered protein sequence data sets by discovering protein sequences that are substantially different from any known disordered proteins. Since it is usually costly and time consuming to sample and label biological data sets, using anomaly detection and identification methods can greatly improve data quality, increase data set size, and help identify novel examples of new patterns (e.g., novel examples of disordered proteins).

More generally, these methods can be used to detect and alleviate batch effects by removing batch biases in many experimental data sets on various platforms. Applied to data generated from sequencing and microarray technologies, these anomaly detection and identification methods can be used to identify and remove sequencing or assay biases from different batches, lanes, and platforms. Such analysis can be imposed at many different layers including DNA, RNA, metabolite, protein, and epignetic factors.

Particularly, these methods can also be applied to identify anomalous sequences that help in distinguishing disease populations from normal controls, dissecting subtypes of diseases, and detecting various pathological stages or treatment response groups by mining the underlying genetic, epigenetic, and clinical data. For example, these methods can be applied to a genetic data set composed of the DNA sequences or genotype profiles of a set of individuals, with a goal of detecting and identifying individuals belonging to different phenotypic groups or subtypes. Such knowledge will be of great help for precision medicine where personalized prognosis and treatment plans can be developed.

Another straightforward application of these anomaly detection methods is to detect whether a particular patient care flow trace is anomalous comparing with normal traces in clinical processes and practices [29]. In typical clinical settings, clinical practices and processes are generally dynamic, complex, and potentially ad hoc [29]. Hence, traces in clinical processes and practices are usually flexible, and could be unknown a priori, unexpected, or unpredictable [29]. Therefore, the proposed nonparametric anomaly detection and identification methods are well posed to analyze health logs and care flow traces in clinical practices and processes. Anomalies identified in care flow traces may have insightful clinical significance, since these anomalies could represent unexpected treatment activities, personalized treatments or responses to particular drugs, and/or treatment/response patterns occurring at unexpected or different time points. With the ubiquitous availability of electronic medical records (EMRs), these approaches can be applied to a large set of EMR, with a goal of mining anomalous patterns to improve clinical practices and processes, enhance the quality of and access to health care, and reduce health care costs and patient burden.

In addition to mining health records, these anomaly detection and identification methods can be widely applied to emerging data from mobile health trackers (e.g., Fitbit). The traces and logs of these mobile health trackers can track a set of health-related information about basic human behavioral factors and surveillance of human health (e.g., diet, weight, heart rate, blood pressure, sleep pattern, pain level, etc.). For example, these methods can detect and report anomalous events in blood pressure and spikes of pain levels. These reports can be used to provide better health surveillance and timely communications with health care providers for general populations, especially for elders and individuals with chronic diseases.

With the big data deluge in biomedicine and health, we witness a rapid accumulation of biomedical and health data at many different layers and from various perspectives on human individuals and populations. New analytical methods are thus called upon to analyze such data, including the methods we described above that aim at detecting and identifying anomalous data streams. We expect to see that these anomaly detection methods have wide applications in a variety of scenarios and have significant potentials to advance biomedical research and health care.

4.7 Open Problems and Future Directions

In this chapter, we have reviewed recent advances in detection and identification of anomalous data streams in nonparametric scenarios. Although there has been substantial progress in this area, there are still important scenarios that have not been addressed yet. For example, for the problem of detecting the existence of anomalous data streams, only the case when the distributions are discrete has been studied in Ref. [13]. It is of great practical interest to further investigate this problem under continuous distributions.

Existing studies have established three approaches: (1) the generalized likelihood-based approach [12,13]; (2) the kernel-based approach [14]; and (3) the KL divergence-based approach [15]. It will be of great interest to comprehensively compare the performances of these approaches as well as advantages/disadvantages of competitive approaches. Furthermore, for each type of approach, there can be various implementations. For example, for the kernel-based approach, estimators of MMD and kernel functions can be chosen differently, which can greatly affect the performance. For the KL divergence-based approach, estimators of KL divergence can also be chosen differently. It is important to address these issues in order to provide useful guidelines for applying these approaches in practice.

It occurs very often in practice that the problem dimensions (the numbers of data streams and anomalous data streams) are very large. In such regimes, the computational complexity of the three nonparametric approaches can be significant, although these approaches achieve good error performance. However, reducing complexity of detection tests can sacrifice performance. It is important to design nonparametric tests that have low computational costs but still achieve acceptable performance.

Furthermore, this chapter has focused on an offline setting, in which decisions are made after receiving all the samples. In practice, samples are often received in a sequential manner. For example, in the problem here, samples of each data stream may arrive sequentially over time. It is possible that with sufficient samples in each data stream, an accurate decision can already be made, and there is no need to wait until all samples arrive. Such problems have been explored in Ref. [30] for the problem of identifying anomalous data streams under discrete distributions; it is interesting to study other online scenarios as well. The major issue here is to design stopping rules to balance the trade-off between the sample size and the detection error.

Another related topic is to explore the change-point problem as in Ref. [31]. For example, samples of all data streams can initially be generated from the same typical distribution, and samples of some data streams may be generated from a different distribution from a certain time onward. Such a change can happen simultaneously for these data streams or can occur at different time instances. It is interesting to detect the changes as soon as possible, and furthermore, to identify which data streams experience changes, subject to a tolerable false alarm rate. Such a problem has wide applications in control and monitoring systems.

Acknowledgment

The work of S. Zou was supported in part by the U.S. National Science Foundation under Grant CCF 16-18658. The work of Y. Liang was supported in part by the U.S. National Science Foundation under Grants CCF-1801855 and ECCS-1818904. The work of H. V. Poor was supported in part by the U.S. National Science Foundation under Grant CNS-1456793. The work of X. Shi was supported in part by the U.S. National Science Foundation under Grant IIS-1502172.

References

1. V. Chandola, A. Banerjee, and V. Kumar. Anomaly detection: A survey. *ACM Comput. Surv. (CSUR)*, 41(3):15, 2009.
2. A. Patcha and J.-M. Park. An overview of anomaly detection techniques: Existing solutions and latest technological trends. *Comput. Netw.*, 51(12):3448–3470, 2007.
3. A. O. Hero. Geometric entropy minimization (GEM) for anomaly detection and localization. In *Proc. Advances in Neural Information Processing Systems (NIPS)*, pages 585–592, 2006.
4. A. O. Hero and O. J. J. Michel. Asymptotic theory of greedy approximations to minimal k-point random graphs. *IEEE Trans. Inform. Theory*, 45(6):1921–1938, 1999.
5. M. Zhao and V. Saligrama. Anomaly detection with score functions based on nearest neighbor graphs. In *Proc. Advances in Neural Information Processing Systems (NIPS)*, pages 2250–2258, 2009.
6. G. Fellouris and A. Tartakovsky. Unstructured sequential testing in sensor networks. In *52nd IEEE Conference on Decision and Control*, pages 4784–4789, 2013.
7. F. Grubbs. Procedures for detecting outlying observations in samples. *Technometrics*, 11(1):1–21, 1969.
8. L. Lai, H. V. Poor, Y. Xin, and G. Georgiadis. Quickest search over multiple sequences. *IEEE Trans. Inform. Theory*, 57(8):5375–5386, 2011.
9. J. Laurikkala, M. Juhola, and E. Kentala. Informal identification of outliers in medical data. In *International Workshop on Intelligent Data Analysis in Medicine and Pharmacology*, pages 20–24, 2000.
10. A. Tajer and H. V. Poor. Quick search for rare events. *IEEE Trans. Inform. Theory*, 59(7):4462–4481, 2013.
11. A. G. Tartakovsky, X. R. Li, and G. Yaralov. Sequential detection of targets in multichannel systems. *IEEE Trans. Inform. Theory*, 49(2):425–445, 2003.
12. Y. Li, S. Nitinawarat, and V. V. Veeravalli. Universal outlier hypothesis testing. *IEEE Trans. Inform. Theory*, 60(7):4066–4082, July 2014.
13. W. Wang, Y. Liang, and H. V. Poor. Nonparametric composite outlier detection. In *Proc. Asilomar Conference on Signals, Systems, and Computers*, 2016.
14. S. Zou, Y. Liang, H. V. Poor, and X. Shi. Nonparametric detection of anomalous data streams. *IEEE Trans. Signal Process.*, 2017, to be published.
15. Y. Bu, S. Zou, Y. Liang, and V. V. Veeravalli. Universal outlying sequence detection for continuous observations. In *Proc. IEEE International Conference on Acoustics, Speech and Signal Processing (ICASSP)*, pages 4254–4258. IEEE, 2016.
16. A. Gretton, K. Borgwardt, M. Rasch, B. Schölkopf, and A. Smola. A kernel two-sample test. *J. Mach. Learn. Res.*, 13:723–773, 2012.
17. A. Berlinet and C. Thomas-Agnan. *Reproducing Kernel Hilbert Spaces in Probability and Statistics*. Kluwer, 2004.
18. B. Sriperumbudur, A. Gretton, K. Fukumizu, G. Lanckriet, and B. Schölkopf. Hilbert space embeddings and metrics on probability measures. *J. Mach. Learn. Res.*, 11:1517–1561, 2010.
19. K. Fukumizu, A. Gretton, X. Sun, and B. Schölkopf. Kernel measures of conditional dependence. In *Proc. Advances in Neural Information Processing Systems (NIPS)*, 2008.
20. K. Fukumizu, B. Sriperumbudur, A. Gretton, and B. Schölkopf. Characteristic kernels on groups and semigroups. In *Proc. Advances in Neural Information Processing Systems (NIPS)*, 2009.
21. B. Sriperumbudur, A. Gretton, K. Fukumizu, G. Lanckriet, and B. Schölkopf. Injective Hilbert space embeddings of probability measures. In *Proc. Annual Conference on Learning Theory (COLT)*, 2008.
22. Q. Wang, S. R. Kulkarni, and S. Verdú. Divergence estimation of continuous distributions based on data-dependent partitions. *IEEE Trans. Inform. Theory*, 51(9):3064–3074, 2005.
23. Y. Bu, S. Zou, Y. Liang, and V. V. Veeravalli. *Estimation of KL divergence: Optimal minimax rate*. ArXiv e-prints 1607.02653, July 2016.
24. H. Cai, S. R. Kulkarni, and S. Verdú. Universal estimation of entropy and divergence via block sorting. In *Proc. IEEE Int. Symp. Information Theory (ISIT)*, page 433, 2002.
25. H. Cai, S. R. Kulkarni, and S. Verdú. Universal divergence estimation for finite-alphabet sources. *IEEE Trans. Inform. Theory*, 52(8):3456–3475, 2006.
26. Q. Wang, S. R. Kulkarni, and S. Verdú. Divergence estimation for multidimensional densities via nearest-neighbor distances. *IEEE Trans. Inform. Theory*, 55(5):2392–2405, 2009.
27. Z. Zhang and M. Grabchak. Nonparametric estimation of Küllback–Leibler divergence. *Neural Comput.*, 26(11):2570–2593, 2014.
28. S. Vucetic, D. Pokrajac, H. Xie, and Z. Obradovic. Detection of underrepresented biological sequences using class-conditional distribution models. In *Proceedings of the SIAM International Conference on Data Mining*, page 5, 2003.
29. Z. Huang, X. Lu, and H. Duan. Anomaly detection in clinical processes. *AMIA Annu Symp Proc.*, pages 370–379, November 2012.
30. Y. Li, S. Nitinawarat, and V. V. Veeravalli. Universal sequential outlier hypothesis testing. In *48th Asilomar Conference on Signals, Systems and Computers*, pages 281–285. IEEE, 2014.
31. S. Nitinawarat and V. V. Veeravalli. Universal quickest outlier detection and isolation. In *Proc. IEEE Int. Symp. Information Theory (ISIT)*, pages 770–774, June 2015.

5

Time–Frequency Analysis for EEG Quality Measurement and Enhancement with Applications in Newborn EEG Abnormality Detection: Multichannel EEG Enhancement and Classification for Newborn Health Outcome Prediction

Boualem Boashash, Samir Ouelha, Mohammad Al-Sa'd, Ayat Salim, and Paul Colditz

CONTENTS

5.1 Introduction: Big Data and Information Retrieval	74
5.1.1 DSP Approach and Engineering Rationale	74
5.1.2 EEG and Its Clinical Prognostic Value	74
5.1.3 EEG Clinical Rationale and the Problem of Artefacts	75
5.1.4 Study Tasks and Organization of the Chapter	76
5.2 Review of State-of-the-Art of EEG Signal	76
5.2.1 EEG and Neurodevelopmental Outcomes	76
5.2.2 Newborn EEG Abnormalities	77
5.2.3 Artefacts in the EEG	77
5.2.3.1 Main Types of EEG Artefacts	78
5.2.3.2 Standard Methods for Handling Artefacts	78
5.2.4 Features for Detecting EEG Abnormalities and Artefacts	79
5.3 Methodology for EEG Enhancement and Classification	79
5.3.1 Data Acquisition and Labeling	79
5.3.2 Pre-Processing: Artefact Detection and Removal	80
5.3.3 Signal Representation Using High-Resolution TF Techniques	80
5.3.3.1 Formulation of Quadratic TF Distributions	80
5.3.3.2 Design of High-Resolution TFDs for EEG Data Processing	81
5.3.3.3 Selection of High-Resolution TFDs	81
5.3.3.4 Multidirectional Distribution (MDD)	82
5.3.4 TF Features for the Detection of EEG Abnormalities and Artefacts	83
5.3.4.1 Inherent (t, f) Features	83
5.3.4.2 Translation Approach-Based TF Features	85
5.3.4.3 Feature Selection	85
5.3.4.4 Feature Fusion	87
5.3.5 TF Matched Filter Approach for Newborn Abnormality Detection	88
5.3.6 Classification	89
5.3.7 BSS Algorithms	90
5.4 TF Modelling of Neonatal Multichannel Non-Stationary EEG and Artefacts	90
5.4.1 Modelling Newborn Mono-Channel Non-Stationary EEG	90
5.4.1.1 The *IF-Based* Approach	91
5.4.2 Newborn Head Model	92
5.4.2.1 Four-Sphere Head Model	93
5.4.3 EEG Multichannel Propagation	93
5.4.3.1 Modelling Multichannel Attenuation Matrix	95

    5.4.3.2 Modelling Multichannel Translation Matrix 95
    5.4.3.3 Solving for Multichannel Path Lengths..96
    5.4.3.4 Generating Multichannel Background and Seizure Patterns...96
    5.4.3.5 Assumptions and Functionality of the Model 97
    5.4.3.6 Validating the Multichannel EEG Model..98
  5.4.4 EEG Quality Measure..98
    5.4.4.1 Validation of Real EEG with Simulated Artefacts..99
    5.4.4.2 Model of Artefact..100
    5.4.4.3 Validation for Real EEG with Real Artefact 100
5.5 Discussion of Results for Stage 1 Experiments...101
  5.5.1 Newborn EEG Abnormality Detection Using TF-Based Features...101
    5.5.1.1 Selected Features...101
    5.5.1.2 Seizure Detection Results 101
  5.5.2 Newborn EEG Abnormality Detection Using (t, f) Matched Filters 102
5.6 Results for Stage 2 Experiments..104
  5.6.1 High-Resolution TFDs: Multidirectional Distributions 104
  5.6.2 Machine Learning 104
  5.6.3 BSS-Based Artefact Removal 108
  5.6.4 Overall System 108
5.7 Conclusions and Perspectives 108
Acknowledgments..110
References 110

5.1 Introduction: Big Data and Information Retrieval

5.1.1 DSP Approach and Engineering Rationale

Recent advances in automated analyses of the electroencephalogram (EEG) indicate that their clinical implementation is feasible in principle and getting closer in practice [1]. Given the amount of EEG data to be analyzed, it is important to develop an automatic EEG analysis system in order to deal with the large amounts of EEG data requiring analysis. An important aspect of such automated EEG analysis is the ability to identify the EEG that is of poor quality and improve its quality by first removing artefacts if required. This ensures that automated newborn EEG methods are robust to the deficiencies of the recording environment. The combination of newborn EEG artefact detection, quality assessment, and improvement can increase the confidence of any decision-making process (such as classification and diagnosis) based on manual or automated analysis of the EEG. Contamination and distortion of EEG recordings, by various biological and extracerebral sources (artefacts), remains a leading obstacle in the practical deployment of any automated EEG classification algorithms. EEG artefact research to date has centered primarily on, prior to this study, removing ocular artefacts. These artefacts prove most troubling in clinical settings where the more intrusive head movements and facial gestures cannot be minimized. In real-world settings, movement and other artefacts prove severely problematic in terms of newborn EEG contamination, yet they have received very little attention in the literature, which is so far dominated by the study of adult EEG. This is a problem as the amount and type of artefact seen on newborn EEG are very different from those observed in adult EEG, primarily due to the intensive care setting of the newborn EEG and the desire for long duration recording. The lower frequency content of newborn EEG, in addition to significant differences in its patterns, compared to adult EEG also indicates that methods designed for the removal of adult EEG artefacts must be modified when applied to newborn EEG. Given the above setting, the modern engineering approach to deal effectively with the artefact problem is to design special signal processing and machine learning algorithms that, combined, allow the detection and removal of artefacts, hence improving newborn EEG quality [2, Chapter 16].

5.1.2 EEG and Its Clinical Prognostic Value

The qualitative and quantitative improvements of neonatal and obstetric care have resulted in a significant decrease in the perinatal mortality rate. The improvement in newborn survival rate experienced in the last few decades has not, in many conditions, resulted in a reduction in the incidence of adverse neurological sequelae in survivors. Recent studies have shown that

25–35% of newborns who experience hypoxia–ischaemia during their delivery and who have an encephalopathy have long-term neurodevelopmental disabilities that include cerebral palsy, mental retardation, and seizures [3]. Similar studies have also shown that even children who did not exhibit clearly identifiable perinatal difficulties are at risk of developing neurodevelopment problems later in life due to injuries contracted before or during birth [4]. Follow-up studies indicate that up to 50% show variable degrees of long-term sequelae ranging from hearing or visual impairment, a delay in psychomotor development, or neurological damage (see Ref. [5] and references therein). These observations identify a clear need for more sensitive diagnostic and prognostic tools for the evaluation of the central nervous system (CNS) to complement clinical observations and permit the development of improved early intervention strategies focused on prevention of disability or death.

In recent years, brain imaging techniques have radically changed the way major structural disorders of the newborn brain are diagnosed. Techniques such as ultrasonography and magnetic resonance imaging (MRI) allow early detection of brain lesions. These imaging modalities are, however, not suitable when the brain abnormalities are expressed in functional terms with or without demonstrable structural correlates. Although functional MRI (fMRI) is capable of registering functional and physiological changes in the brain, its widespread use is hampered by a number of factors such as [6, p. 14] (a) the very low time resolution (approximately 2 frames/s), (b) the high cost, and (c) the fact that many types of brain disorders have low effect on the level of oxygenated blood. For such cases and when a continuous long-term recording is required, the EEG offers a valuable non-invasive screening and an ideal tool for (1) monitoring the function of the CNS and (2) improving diagnosis and prognosis of at-risk infants (see Figure 5.1).

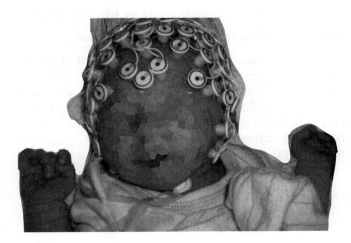

FIGURE 5.1
Procedure for recording newborn EEG signals using a cap.

In practice, as reported in Ref. [7], one can say that "the EEG remains the only bedside neurodiagnostic procedure that provides a continuous record of cerebral function over time."

Although most newborn EEG patterns are non-specific and, therefore, cannot help provide the diagnosis when used alone, they can point the care providers to specific neuroimaging tests. For example, interhemispheric or regional asymmetric EEG can possibly suggest issues such as perinatal stroke, brain genetic malformations, or an inborn error of metabolism [7]. The EEG is also ideal for identifying subclinical seizures and avoiding misdiagnosis of seizures, a particular problem in the newborn brain where only about 50% of seizures are clinically manifest and not all events that clinically appear to be a seizure have an EEG seizure correlation. However, the newborn EEG's greatest value is its potential to predict short- and long-term outcomes.

5.1.3 EEG Clinical Rationale and the Problem of Artefacts

The desired outcome of neonatal intensive care is to achieve the survival and development of healthy children who can thrive, learn, and develop normally, so that they can then fulfill their maximum potential as members of their families and society. Unfortunately, neurologically intact survival is still currently an elusive goal for both neonatologists and parents. Lifelong disability due to brain injury continues to afflict many who are born too soon, are too small, or are subject to a disruption in the blood or oxygen supply (hypoxia ischaemia [HI]) during late pregnancy, labor, or delivery). Newborns who become ill for other reasons, such as heart or lung disease, are also at risk of HI injury. Newborns can develop seizures (also called fits) because of HI injury and for many other reasons such as stroke, meningitis, or inborn errors of metabolism. Accurate, rapidly available and reliable information regarding brain function is essential if intensive care therapies are to be adapted to achieve the best possible outcomes.

Continuous cot-side EEG brain monitoring currently offers the best window into brain function; EEG can correctly identify newborns who have suffered a significant HI injury and are likely to benefit from neuroprotective therapies (e.g., hypothermia*). The EEG can also be used to diagnose seizures, monitor the effect of treatment, and uncover the effect of episodes of hypoxia, hypotension, intracranial bleeding, and

---
* Hypothermia is the condition that happens when a person's body loses heat faster than it can produce it. Therapeutic hypothermia can be induced by cooling the baby to a core temperature of about 33°C and is an effective therapy to reduce the degree of brain injury after moderate or severe hypoxia ischaemia.

hypoglycaemia on brain function. Continuous EEG monitoring provides information about evolving neurological conditions in newborns in the neonatal intensive care unit (NICU), which can indicate when immediate adjustment of medical interventions is required. At the time of publication, neonatal EEG is still, however, difficult to interpret except by neurophysiologists or paediatric neurologists with extensive newborn experience (and these are scarce). In many NICUs, it may be possible to realize a short EEG recording, but it could take hours or even days to get a useful report. Given the urgent need to identify newborns who need treatment for seizures or are suitable for therapeutic hypothermia in the immediate postnatal period, interpretation of the EEG may be needed rapidly and often late at night or in the early morning. Neurophysiologists worldwide struggle to provide a 24-hour emergency EEG service for NICUs. Therefore, the most efficient way of dealing with the large datasets generated by multichannel EEG is an automated real-time system. From a conceptual point of view, it is not difficult to design an expert-based digital system that would provide the required analysis. However, a significant obstacle in the development of such analyses is the presence of artefacts within the newborn EEG. Such artefacts can be present in a significant part of newborn EEG recordings [8]. The detection of artefacts in clinical practice currently requires close and time-consuming analysis of multiple channels of EEG and related signals by an experienced paediatric neurologist. By obscuring or distorting the newborn EEG, artefacts can affect the interpretation of the signals, often making it difficult to detect, interpret, and localize events in time. This study results in the design of a robust and reliable system for monitoring brain function in newborns as described in the rest of this chapter.

5.1.4 Study Tasks and Organization of the Chapter

The neurological impairment in patients resulting from congenital or acquired injuries contracted before or during birth incurs a large social and financial cost to the community throughout the patient's life. A more widespread use of EEG in the NICU is limited by several factors. In particular, the visual interpretation of long-term EEG recordings is an expensive, time-consuming, and subjective process that is restricted to a small number of highly trained medical staff.

This study addresses the problem of automated classification of newborn EEG abnormalities for neurodevelopmental outcome prediction. Its main objectives are to

- Design accurate automated systems for detecting newborn EEG abnormalities using advanced signal processing methods adapted to match the characteristics of the signals
- Design and implement various signal processing algorithms that would allow the realization of the above according to specifications
- Utilize a labeled newborn EEG database for testing, validating, and fine-tuning the above-mentioned algorithms.

The objectives of this study were met by executing the following tasks:

1. Multichannel newborn EEG acquisition, labeling, and pre-processing
2. Creation and analysis of a database of neonatal EEG with annotated artefacts
3. Analysis and design of algorithms for time–frequency (TF) distributions
4. Artefact detection and removal
5. Feature extraction, fusion, and selection for classification
6. System integration and implementation
7. EEG quality metric (develop a statistic that incorporates the results of the artefact detection and rejection process to determine the overall quality of the EEG recording)

5.2 Review of State-of-the-Art of EEG Signal

5.2.1 EEG and Neurodevelopmental Outcomes

Previous studies have concluded that the EEG is an excellent prognostic tool for both preterm and term infants [3]. Neurodevelopmental outcomes are measured using different assessment methods and criteria, and at different ages from 12 months onwards. Studies have used structured examinations [9] and/or general neurological examinations to define *normal* or *abnormal* neurodevelopmental outcomes. A few studies used their own follow-up index to define normal or abnormal outcomes [10,11]. After the review of previous studies, we chose to consider normal, minor, or mildly abnormal outcome as a normal outcome with moderate or severely abnormal or death as an abnormal outcome. In short, a normal EEG is thus highly predictive of a normal outcome, whereas the various abnormal EEG features have been consistently associated with unfavorable neurodevelopmental outcomes or death [5,11].

5.2.2 Newborn EEG Abnormalities

Newborn EEG abnormalities have been classified in previous studies as follows:

A1: *Amplitude-based abnormalities* are characterized by a significant decrease in amplitude. There are two prominent members of this category. (1) Electrocerebral inactivity (ECI) consists of EEG activity below 10 μV that is continuously present throughout the record [7]; and (2) low amplitude activity is characterized by a voltage of 5–15 μV during wakefulness and active sleep and a voltage of 10–25 μV in quiet sleep [7].

A2: *Discontinuity-based abnormalities* are characterized by repeated bursts of high-amplitude activity followed by periods of low-amplitude activity. There are two important patterns in this class. (1) Burst-suppression (B-S) is characterized by a burst of high-voltage activity lasting 1–10 seconds and composed of various waveforms followed by periods of marked attenuation (<5 μV) lasting more than 20 seconds [7,12]. This pattern is different from both the normal trace alternant and trace discontinue patterns [7]. (2) Hypsarrhythmia is defined as asynchronous bursts of high-amplitude slow activity embedded with multifocal spikes and sharp waves. It differs from B-S by its high-amplitude activity (typically > 1 mV) [7].

A3: *Asymmetric/asynchronous-based abnormalities* are based on relative measures between *homologous* regions of each hemisphere of the brain and are of two types: (1) interhemispheric asymmetry, which is defined as persistent voltage asymmetries of 50% or more in all EEG states [7]; and (2) interhemispheric asynchrony, which is defined as the temporal separation of morphologically similar waveforms of more than 1.5 seconds with less than 25% synchrony.

A4: *Frequency-based abnormalities* are characterized by excessively slow or fast patterns in EEG signals. One of such abnormalities is the slow wave pattern characterized by diffuse delta activity mainly in the band 0.5–1 Hz with an amplitude range of 20–100 μV and little activity in the theta band (frequency range from 4 to 7 Hz) [13,14].

A5: *Transient-based abnormalities*: The random occurrence of sharp transients, such as spikes and sharp waves, in newborn EEG is widely accepted as normal [7]. Sharp transients occurring over frontal, rolandic, and temporal areas are, however, considered abnormal if they are excessively frequent, appear in short runs, are consistently unilateral, and are polyphasic [7].

A6: *EEG seizures* exhibit variations in voltage, duration, frequency content, and waveform shape. There are two common patterns [15]. The first one consists of trains of rhythmic spikes or sharp waves with a frequency of 4–10 Hz. The second pattern is composed of stereotyped, repetitive, rhythmic paroxysmal discharges consisting of sharp broad-based waves that typically occur with a frequency of 0.5–1 Hz.

A7: *Sleep state-based abnormalities*: Sleep states are determined using EEG patterns and other physiological information. EEG recordings, in newborns older than 30 weeks postconceptional age, in which no recognizable sleep states can be found, are considered abnormal [7,16]. Typical values of sleep-state cycling for a term newborn are 50% active sleep and 30–35% quiet sleep, and large alterations of this pattern indicate an abnormality of prognostic significance [7].

A8: *Maturation-based abnormalities*: Newborn EEG is considered dysmature if it appears more than 2 weeks younger than the actual postconceptional age [7,17].

5.2.3 Artefacts in the EEG

As explained in Section 5.1, EEG recordings can often be contaminated by different types of artefacts, such as those due to electrode displacement, motion, ocular, and electromyogram (EMG) artefacts from muscle activity. There are two types of artefacts: (1) internal and (2) external. Internal artefacts are due to physiological activities of the subject and its movement. Some artefacts may be present in several neighboring channels (global), while some of them can be found in only a single channel (local). Therefore, artefact detection and removal is required for neural information processing applications especially when dealing with newborns. The variety of artefacts and their overlap with signals of interest in both spectral and temporal domains make it difficult for simple signal pre-processing techniques to identify them within the EEG. In addition, there is currently no universal standard quantitative metric available for performance evaluation of existing artefact removal methods.

5.2.3.1 Main Types of EEG Artefacts

Artefacts are undesired contaminating signals that may introduce changes in the measurements and affect the signal of interest. They can be coarsely separated into physiological and nonphysiological origin. The latter can be reduced, e.g., (1) by proper electrodes placement, (2) by recording in a controlled environment, etc. [18]. However, the former cannot be avoided and most algorithms developed for EEG artefact processing are intended to deal with physiological artefacts. The most common artefacts are described below [19–21]:

- *Muscle artefacts*: This artefact is typical of awake patient and occurs when the baby is, e.g., swallowing. The shapes and amplitudes of the artefacts depend on the degree of muscle contraction and on the type of muscle. In addition, muscle artefacts have a wide spectral range and therefore contaminate and distort all standard EEG bands.
- *Cardiac activity*: The amplitude of the cardiac activity component on the scalp is usually low; however, it depends on the electrode positions, electrode impedance, and instrumentation characteristics. The electrocardiogram (ECG) has a very characteristic repetitive and regular pattern, which may be mistaken for seizure [22]. Pulse artefacts are also related to cardiac activity, and they occur when an electrode is placed near a pulsating vessel; this generates slow periodic waves that may be similar to EEG activity [23].
- *Eye movements and blinks*: the strength of these artefacts depends on the closeness of the electrodes to the eyes and the direction of eye movement. Blinking causes generally more abrupt change than eye movement. In addition, the amplitude of the blinking artefact is generally much larger than the EEG background activity [20].

5.2.3.2 Standard Methods for Handling Artefacts

5.2.3.2.1 Artefact Detection

Identifying artefacts is the first and most important step for handling them. Artefacts, often, overlap with EEG signals in both spectral and temporal domains such that it becomes difficult to use simple filters or standard signal processing techniques. Some methods adopted the idea of machine learning for artefact separation from useful EEG signals by training a classifier like artificial neural network (ANN) [24,25] or support vector machines (SVMs) [26]. Once artefactual epochs are identified by applying a machine learning algorithm, such epochs are either highlighted as artefacts or can be rejected before being examined by a clinician. The drawback of rejecting contaminated artefact epochs is to also remove important EEG information, resulting in a possible loss of information [27]. Therefore, there is a need to develop more advanced processing techniques for artefact removal.

5.2.3.2.2 Artefact Removal Using Automated Component Separation

This approach aims to correct the signal contaminated by artefacts without distorting the signal of interest. It is primarily done in two ways: either by filtering and regression or by decomposing the EEG data into other transform domains. Different types of techniques used in previous studies are described below:

1. *Blind source separation (BSS)*: This is one of the most common artefact removal methods. It extracts individual sources from their mixtures. It can also estimate the unknown mixing medium; this is done by using only the information within the observed mixtures from the output of each channel without any knowledge about the source signals and mixing matrix [28,29]. Specific techniques include (1) independent component analysis (ICA), (2) canonical correlation analysis (CCA), and morphological component analysis (MCA). They are described below.
   a. ICA requires manual intervention to reject independent components (ICs) with visually detected artefacts after decomposition.
   b. CCA uses a weaker condition than the statistical independence sought by the ICA

algorithm because it is looking for uncorrelated components [30].

  c. MCA: This method's efficiency depends on the available artefact-template database [31].

2. *Principal components analysis (PCA)*: This technique was used for artefact removal in Refs. [32,33]. One important limitation of PCA is that it fails to separate ocular or similar artefacts from EEG when amplitudes are comparable.

3. *Empirical mode decomposition (EMD)*: this is a data-driven method with non-stationary signals. The theory behind EMD is not complete; therefore, it is difficult to predict its robustness for all EEG signals [34].

4. *Hybrid methods*: such combinations include

  a. *Wavelet-BSS*: Multichannel datasets are transformed into ICs or CCs (components for CCA algorithm), and then the artefactual component is identified and decomposed by wavelet transform. After that, the artefactual coefficients are denoised by thresholding; this is intended to eventually preserve the residual neural signals of low amplitude after thresholding the higher artefactual segment. Wavelet-ICA was used in Refs. [27,35] and Wavelet-CCA in Ref. [36].

  b. *EMD-BSS*: The first stage is to decompose the signal into intrinsic mode functions (IMFs) by EMD, or an enhanced version (EEMD), and then apply BSS on the IMFs to identify the artefactual component and reject it before reconstruction [37–39].

  c. *BSS-SVM*: The idea is to apply ICA or CCA and then extract features from each component and feed them into a SVM classifier. The SVM then identifies artefact components and rejects them [40].

  d. Other hybrid methods have been proposed in previous studies using different methods like the discrete wavelet transform (DWT) with ANN [41] or adaptive noise cancellation [42].

5.2.3.2.3 EEG Signal Enhancement
TF matching pursuit can be used to remove spike-like short duration artefacts from EEG signals. This approach uses a dictionary of redundant functions to iteratively decompose a signal [43]. It has been shown that the use of artefact removal algorithms as a pre-processing stage can improve the total accuracy of a TF matched filtering-based signal classification algorithm by 5% [44].

5.2.4 Features for Detecting EEG Abnormalities and Artefacts

Most features used for the detection and classification of EEG abnormalities can be categorized into four main groups, namely temporal features, spectral features, TF features, and time-scale (TS) features.

- Temporal features are extracted from the signal representation in the time domain (t-domain) and include features based on the amplitude, non-linear and chaos analyses, and autoregressive modeling of EEG signals [45,46].

- Spectral features are extracted from the signal representation in the frequency domain (f-domain) and are usually derived from the fast Fourier transform (FFT) coefficients [45,46].

- TF features are extracted from the representation of EEG signals in the (t, f) domain and are based on the energy distribution of the signals in the (t, f) plane, singular value decomposition (SVD) of TFDs, TF divergence measures, and matched filtering methods and image features [45,46].

- TS features are extracted from the coefficients of the DWT and have been widely used for detecting seizures in EEG signals [47].

As EEG signals are non-stationary, TF and TS methods seem naturally more suitable for the detection and classification of newborn EEG abnormalities. This study focuses on TF features because they exploit the additional information provided by the signal variations in terms of non-stationarities. Novel features are developed and extracted from the TF representation of EEG signals for the purpose of detecting EEG abnormalities and artefacts.

5.3 Methodology for EEG Enhancement and Classification

5.3.1 Data Acquisition and Labeling

This study used a large newborn EEG database that includes 20-channel continuous EEG (cEEG) signals. When one uses 20-channel signals, it results in manageable dataset sizes, unless recording periods are very long. However, if more EEG channels are collected, (as required if spatial localization of the origin of the EEG signals is to be determined), then the amounts of data collected rapidly increases, e.g., with 64 channels and the sampling frequency equal to fs = 256 Hz. As the recordings are continuous, each day, it will produce ≈500 million samples. The signals were recorded according to the 10–20 international electrode placement system using bipolar montage (see Figure 5.11 in Section 5.4), using a Medelec Profile system (Medelec, Oxford Instruments, Old Woking, UK) at f_s = 256 Hz sampling rate. EEG recordings are from 36 sick term newborns admitted to the NICU at the Royal Brisbane and Women's Hospital, Brisbane, Australia and labeled for seizure, artefacts, and normal background patterns by an expert pediatric neurologist.

5.3.2 Pre-Processing: Artefact Detection and Removal

For the detection and removal of artefacts from newborn EEG signals used in this study, two different techniques based on matched filtering and TF feature extraction were developed [44,45]. The matched filtering technique presented in Ref. [44] uses a dictionary populated with atoms defined as scaled, translated, and modulated replicas of an elementary Gabor function and targets spike-like and short-duration artefacts caused by eye blinks, electrode pop-ups, sleep waves, and body movements. On the other hand, the method presented in Ref. [45] uses TF features to discriminate between artefact and non-artefact segments. These two techniques are described in the next sections, and details of those techniques can be found in Refs. [44,45].

As mentioned earlier, artefacts cause a major problem in the implementation of fully automated EEG signal classification systems, e.g., respiratory artefact may look like a seizure signal and can be misinterpreted by the automatic abnormality detection system, therefore resulting in false alarms. Hence, successful removal of the artefacts is important, and there are two basic approaches for this:

1. Use machine learning technique to detect and reject EEG segments corrupted by artefact [46].
2. Correct EEG segments corrupted by artefacts; some artefacts can be corrected by simple filtering in a frequency domain, e.g., notch filter can be used to remove 50-Hz noise, but most artefacts need to be corrected using more advanced methods than simple filtering.

Once data are cleaned and enhanced, one needs to use high precision and resolution techniques to extract relevant information needed for decision making. Given the findings in previous studies that EEG data are non-stationary, a suitable approach is to use TF methods that have the property of high resolution; this is presented in the next section.

5.3.3 Signal Representation Using High-Resolution TF Techniques

5.3.3.1 Formulation of Quadratic TF Distributions

Let us consider a real signal $x(t)$, with analytic associate $z(t) = x(t) + j\mathcal{H}\{x(t)\}$ (with $\mathcal{H}\{.\}$ being the Hilbert transform); it can be transformed to the TF domain using a quadratic TF distribution (QTFD) $\rho_z(t,f)$, which can be expressed as [2, Chapters 2 and 3]:

$$\rho_z(t,f) = \gamma(t,f) \underset{tf}{**} W_z(t,f), \quad (5.1)$$

where $\gamma(t,f)$ is the (t,f) kernel associated with the desired QTFD, $\underset{tf}{**}$ is a *double convolution*, and $W_z(t,f)$ is the Wigner–Ville distribution (WVD) defined as

$$W_z(t,f) = \int_{-\infty}^{\infty} z\left(t+\frac{\tau}{2}\right) z^*\left(t-\frac{\tau}{2}\right) e^{-j2\pi f\tau} d\tau, \quad (5.2)$$

where * stands for complex conjugation.

QTFDs represent a class of TF methods widely used in a wide range of practical applications for representing, analyzing, and processing non-stationary signals [1,46]. The discrete form of a QTFD can be expressed as [2, Chapter 6]:

$$\rho_z[n,k] = \underset{m \to k}{\text{DFT}} \left\{ G[n,m] \underset{n}{*} (z[n+m]z^*[n-m]) \right\}, \quad (5.3)$$

where $G[n,m]$ is defined as the discrete time-lag kernel of the TFD, and the symbol $\underset{n}{*}$ denotes discrete convolution in time. For an N-point signal $x[n]$, $\rho_z[n,k]$ becomes an $N \times M$ matrix ρ_z where M is the number of FFT points used in calculating the TFD.

Different kernels, $G[n,m]$ in Equation 5.3, define different TFDs, which can be specifically adapted to selected classes of signals [46]. For the analysis of multicomponent signals, such as some types of newborn EEGs, it is generally expected that one needs to utilize TFDs that reduce cross-terms while retaining a good auto-term resolution. These TFDs are often labeled as reduced interference TFDs (RI-TFDs). In this study, five RI-TFDs are mainly considered in the analysis of EEG signals, namely modified B-distribution (MBD), smoothed Wigner–Ville distribution (SWVD), Choi–Williams distribution (CWD), spectrogram (SPEC), and extended MBD (EMBD) [1,46]. In addition, the standard WVD is also included in the comparison as previous findings indicated that under some circumstances, the cross-terms of the WVD can be useful for classification [48]. The formulas for the time-lag kernels $G[n,m]$ of those selected TFDs are shown in Table 5.1. Other QTFDs with high resolution are introduced in more detail in the next section.

TABLE 5.1

The Time-Lag Kernels of a Few QTFDs Used in this Chapter

Distribution	$G[n,m]$
WVD	$\delta[n]$
SWVD	$\delta[n]w[m]$
CWD	$\dfrac{\sqrt{\pi\sigma}}{2|m|}\exp\left(\dfrac{-\pi^2\sigma n^2}{4m^2}\right)$
SPEC	$w[n+m]w[n-m]$
MBD	$\dfrac{\cosh^{-2\beta}n}{\sum_n \cosh^{-2\beta}n}$
EMBD	$\dfrac{\cosh^{-2\beta}n}{\sum_n \cosh^{-2\beta}n}\dfrac{\cosh^{-2\alpha}m}{\sum_n \cosh^{-2\alpha}m}$

Note: The parameters α, β, and σ are real and positive; N is the length of the signal under analysis; and $w[n]$ represents the window function used in both SWVD and SPEC kernels. For a complete list, see Ref. [2, Chapters 2 and 3].

5.3.3.2 Design of High-Resolution TFDs for EEG Data Processing

The WVD is an efficient tool for mono-component signals, because of its high (t,f) "resolution". The WVD's weakness is introducing cross-terms between multiple signal components. The occurrence of such cross-terms in QTFDs is the fundamental reason that prevents more widespread use of WVD-based TF signal analysis in most real-life applications dealing with multicomponent non-stationary signals.

Let us consider designing separable kernel RI-TFDs, i.e., the Doppler-lag kernel of the TFD can be written as $G[n,m] = g_1[n]g_2[m]$ [46]. The requirement is to find suitable shapes and widths for the window functions $g_1[n]$ and $g_2[m]$ such that the TFD with a time-lag kernel $G[n,m] = g_1[n]g_2[m]$ provides a good energy concentration for each component and a good suppression of cross-terms. Results presented in Ref. [46] show that different window functions can be used for $g_1[n]$ and $g_2[m]$; some are listed in Table 5.2. An extended methodology for designing high-resolution TFDs with compact-support kernels can be found in Ref. [46], and it is summarized in the next section.

5.3.3.3 Selection of High-Resolution TFDs

Separable kernel TFDs are conceived to mitigate the resolution limitation of the spectrogram. This is done by adding an additional degree of freedom to separately control the smoothing requirements along both f and t axes. Let us now consider two cases of high-resolution

TABLE 5.2

Different Window Functions for $g_1[n]$ and $g_2[m]$

Function Type	$g_1[n]$	$g_2[m]$
Hyperbolic	$\dfrac{\cosh^{-2\beta}[n]}{\sum \cosh^{-2\beta}[n]}$	$\cosh^{-2\alpha}[m]$
Gaussian	$\dfrac{e^{-(\pi n\beta)^2}}{\sum e^{-(\pi n\beta)^2}}$	$e^{-(\pi m\alpha)^2}$
sinc	$\dfrac{\mathrm{sinc}^2(\beta n)}{\sum \mathrm{sinc}^2(\beta n)}$	$\mathrm{sinc}^2(\alpha m)$
Cauchy	$\dfrac{(\beta^2+n^2)^{-1}}{\sum(\beta^2+n^2)^{-1}}$	$\alpha^2(\alpha^2+m^2)^{-1}$

separable kernel TFDs; these are the S-method [49] and the compact-support kernel distribution (CKD) [45]. These TFDs are selected for further description as they are recognized as top performing methods [2].

5.3.3.3.1 S-Method (SM): An Enhanced Spectrogram

The S-Method [49] is defined as

$$SM_z(t,f) = 2\int_{-\infty}^{\infty} P(v)F_z^w(t,f+v)F_z^{w*}(t,f-v)dv, \quad (5.4)$$

where $F_z^w(t,f)$ is the short-time Fourier transform (STFT) of $z(t)$, obtained with a window w, and $P(v)$ is a concentrated narrow-band spectral window (e.g., Hamming or Hanning window). The length of $P(v)$ is chosen to control the cross-term suppression as well as the auto-term resolution characteristics of the enhanced spectrogram. The ambiguity domain (v,τ) kernel of the SM is given by [45]

$$g(v,\tau) = P\left(\dfrac{v}{2}\right) *_v \int_{-\infty}^{\infty} w\left(u+\dfrac{\tau}{2}\right)w^*\left(u-\dfrac{\tau}{2}\right)e^{-j2\pi vu}du. \quad (5.5)$$

The shape of this ambiguity domain SM kernel depends on the shapes and sizes of $w(t)$ and $p(t)$ (which is the inverse FT (IFT) of $P(v)$). Previous studies indicated that with a correct selection of the length of $P(v)$, the SM can combine key advantages of both spectrogram and WVD [49].

5.3.3.3.2 Compact Kernel Distribution (CKD)

Compact-support kernels are designed with the constraint of finite support in a region of the ambiguity domain, i.e., they are required to vanish outside a selected range in the (v,τ) domain. These TFDs are designed with an extra degree of freedom to control both the length and the shape of the ambiguity domain filter. They can then obtain high-energy concentration for auto-terms with the required reduction of cross-terms. The CKD kernel can be expressed as [45]

$$g(v, \tau) = \begin{cases} e^{2c} e^{\frac{cD^2}{v^2-D^2}} e^{\frac{cE^2}{\tau^2-E^2}} & \text{if } |v| < D, |\tau| < E \\ 0 & \text{otherwise} \end{cases}. \quad (5.6)$$

The parameters D and E specify the cut-off of the ambiguity domain filters along the v and τ axes, while c controls the shape of the smoothing kernel.

5.3.3.4 Multidirectional Distribution (MDD)

A new kernel TFD has been defined based on directional information, which can represent newborn EEG signals more accurately and effectively than standard methods. Previous kernels did not take into account the directional information and made some rough approximations about the location of auto-terms and cross-terms in the ambiguity domain.

5.3.3.4.1 Analysis for a Piecewise Linear Frequency Modulated (LFM) Model

Let us consider the exact location of auto-terms and cross-terms for a basic piecewise LFM (PW-LFM) signal defined by

$$x(t) = \Pi_T\left(t - \frac{T}{2}\right) e^{\left(2j\pi\left(\frac{\alpha}{2}t^2 + f_0 t\right) + j\phi\right)}, \quad (5.7)$$

where $\Pi_T(t)$ is defined by

$$\Pi_T(t) = \begin{cases} 1 & \text{if } |t| \leq T/2 \\ 0 & \text{elsewhere} \end{cases}$$

Let us further consider a noise-free multicomponent PW-LFM $z(t)$ expressed as

$$z(t) = \sum_{i=1}^{M} z_i(t) = \sum_{i=1}^{M} \sum_{k=1}^{N_i} z_{ki}(t), \quad (5.8)$$

where z_{ki} is the kth branch of the ith signal component z_i, M is the number of components, and N_i is the number of segments for the ith component; z_{ki} is expressed by

$$z_{ki}(t) = \begin{cases} e^{\left(2j\pi\left(\frac{\alpha_{ki}}{2}t^2 + f_{0ki}t\right) + j\phi_{0ki}\right)} & t_{0ki} < t < t_{0ki} + T_{ki} \\ 0 & \text{elsewhere} \end{cases}, \quad (5.9)$$

where $\alpha_{ki}, f_{0ki}, \phi_{0ki}, t_{0ki},$ and T_{ki} are the frequency rate, the initial frequency, the initial phase, the birth (or start) time, and the duration of $z_{ki}(t)$, respectively. The derivation of the ambiguity domain of $z(t)$ highlighted the different properties of the auto- and cross-terms location and behavior. One of the main properties is that the auto-terms fall around straight lines passing through the origin in the (t, f) plane. But they are not located only in the vicinity of the origin. Other properties and more details can be found in Ref. [1].

5.3.3.4.2 Design of a New Kernel

The above properties yield a new kernel design that should have the following characteristics: (1) It should be a multidirectional kernel, i.e., it should have parameters that allow (and control) filtering along multiple directions θ_i in the ambiguity domain. (2) It should have parameters to adjust the support along both the major and minor axes of auto-terms. To account for the fact that real signals are multicomponent and can have several directions of energy concentration in the (t, f) domain, an MDD has been defined [1], which is formulated as

$$\rho_z(t,f) = W_z(t,f) \underset{t,f}{**} \gamma_\theta(t,f), \quad (5.10)$$

where γ_θ is a directional non-separable kernel in the (t, f) domain, dependent on the set $\theta = (\theta_1, \theta_2, \ldots, \theta_P)$ of direction angles of the signal*; these angles are defined as the angles between the lag axis and the auto-term directions in the ambiguity domain. In the ambiguity domain, this expression is given by

$$A_z(v, \tau) = A_z(v, \tau) g_\theta(v, \tau), \quad (5.11)$$

where $g_\theta(v, \tau)$ is similarly a non-separable kernel in the Doppler-lag domain, also dependent on the set of signal direction angles θ. This kernel is defined as

$$g_\theta(v, \tau) = \frac{e^{c_0+c}}{P} \sum_{i=1}^{P} F_{\theta_i}(v, \tau) \; h_{\theta_i}(v, \tau), \quad (5.12)$$

where the factor e^{c_0+c}/P in front of the summation is a normalization coefficient, c_0 and c are slope-adjustment parameters, and θ_i is related to the frequency rate α_i by $\alpha_i = tan(\theta_i)$. The factor $F_{\theta_i}(v, \tau)$ is defined as

$$F_{\theta_i}(v, \tau) = \begin{cases} e^{\left(\frac{x_{\theta_i}(v,\tau)}{D_i}\right)^2 - 1} & , |x_{\theta_i}(v, \tau)| < D_i \\ 0, & \text{otherwise} \end{cases} \quad (5.13)$$

where $x_{\theta_i}(v, \tau) = cos(\theta_i)v - sin(\theta_i)\tau$, D_i is the half-support of $g_\theta(v, \tau)$ along the direction perpendicular to the ith branch of the multidirectional kernel (MDK), and $h_{\theta_i}(v, \tau)$ is defined as

$$h_{\theta_i}(v, \tau) = \begin{cases} e^{\left(\frac{y_{\theta_i}(v,\tau)}{E_i}\right)^2 - 1} & , |y_{\theta_i}(v, \tau)| < E_i \\ 0, & \text{otherwise} \end{cases}$$

---
* The set of angles θ is not known *a priori*.

where $y_{\theta_i}(v,\tau) = sin(\theta_i)v + cos(\theta_i)\tau$ and E_i is related to either the time duration of LFM components or bandwidth of spike components. Then, in order to automatically set the parameters of the MDD, one can apply a radon transform (RT) to the squared modulus of the ambiguity function (AF), and then a detrending method. After detecting the peaks, one can find the angles and the associated parameters of each branch of the kernel [1].

In an automatic classification system, a key step is to extract discriminating features; the next section presents TF features that can be extracted from the above presented TFDs.

5.3.4 TF Features for the Detection of EEG Abnormalities and Artefacts

The design of an automated abnormality pattern recognition system requires defining representations that can show these abnormal patterns, as well as feature extraction and selection. For newborn EEG abnormality detection, features extracted from the signal TF representation can result in reliable and accurate classification systems as they can account for signals with non-stationary characteristics (using the extra information available in the (t,f) domain). TF-based features suitable for the classification of non-stationary signals were developed and used to detect newborn EEG abnormalities [2,45,46,50]. Such features can be classified in two groups as described next.

5.3.4.1 Inherent (t, f) Features

Inherent (t,f) features are exclusive to the (t,f) domain and do not necessarily have an obvious direct counterpart in either the t-domain or the f-domain. A few such (t,f) features are listed below, and more details can be found in Refs. [45,46,50]. Note that in the following equations, $\rho_{z_x}[n,k]$ is the TFD of the analytic associate of a given real N-point discrete–time signal $x[n]$ given in Equation 5.3. The resulted TFD can be represented as an $N \times M$ matrix ρ_{z_x}.

1. *Instantaneous frequency (IF)–based features* are derived from statistics of the signal IF such as its mean, deviation, skewness, and kurtosis.
2. *Instantaneous amplitude (IA)–based features* are derived from the signal IA statistics.
3. *Matrix decomposition-based features* are derived from a decomposition of the TFD matrix ρ_{z_x}. Two matrix decomposition methods are commonly used, namely SVD and non-negative matrix factorization (NMF) as described below.

a. The *SVD* divides the TFD matrix ρ_{z_x} into two subspaces of the form [45]

$$\rho_{z_x} = USV^H, \quad (5.14)$$

where U is a real matrix, S is an $N \times M$ diagonal matrix with non-negative real numbers ($S_i, i = 1,2,\cdots,N$), and V^H (the conjugate transpose of V) is a real unitary matrix. The diagonal entries of S represent the singular values of ρ_{z_x}. Features are then extracted from the singular values of the matrix ρ_{z_x}. Such SVD-based features include the maximum and the variance of the singular values. Another possible feature that is based on the SVD is the (t,f) complexity measure, denoted by $CM(x)$, which is derived from the Shannon entropy of the singular values of ρ_{z_x} and describes the magnitude and the number of estimated non-zero singular values for a TFD [46]. It can be expressed as

$$CM(x) = -\sum_{i=1}^{N} \bar{S}_i \log_2 \bar{S}_i, \quad (5.15)$$

where $\bar{S}_i$ is the ith normalized singular value, i.e., $\bar{S}_i = \dfrac{S_i}{\sum_i S_i}$.

b. The *NMF* is another matrix decomposition method that has the advantage of preserving the non-negativity of the entries, which is important to obtain meaningful physical interpretation under some circumstances. The NMF factorizes the TFD matrix ρ_{z_x} into two non-negative matrices as [51]

$$\rho_{z_x} \approx W_{N \times R} H_{R \times M} = \sum_{r=1}^{R} w_r h_r, \quad (5.16)$$

where the columns of the matrices W and H^T are referred to as base and coefficient vectors, respectively, and R ($R \ll \min(M, N)$) is the decomposition parameter that is usually application-dependent. Note that the decomposition of ρ_{z_x} using the NMF requires the matrix to be non-negative [51].

The base vectors $w_r, r = 1,\ldots,R$, can be interpreted as characteristic frequency structures, whereas the coefficient vectors $h_r, r = 1,...,R$, can be seen as the temporal location of these structures [52]. Based on the decomposition in Equation 5.16, the sparsity of the base and coefficient vectors are defined respectively as

$$S_{w_r} = \frac{1}{\sqrt{N}}\left(\sqrt{N} - \frac{\sum_{l=1}^{N} w_r(l)}{\sqrt{\sum_{l=1}^{N} w_r^2(l)}}\right), \quad (5.17)$$

$r = 1,...,R,$

$$S_{h_r} = \frac{1}{\sqrt{M}}\left(\sqrt{M} - \frac{\sum_{l=1}^{M} h_r(l)}{\sqrt{\sum_{l=1}^{M} h_r^2(l)}}\right), \quad (5.18)$$

$r = 1,...,R,$

and their statistics can be considered as NMF-based features.

4. *TFD concentration measures* can also be used for the classification of non-stationary signals. Different measures were presented in Ref. [2, Section 7.3] and their properties were discussed. One such measure is

$$M_p(x) = \left(\sum_{n=1}^{N}\sum_{k=1}^{M}|\rho_{z_x}[n,k]|^{\frac{1}{p}}\right)^p, \quad (5.19)$$

with $p > 1$. The advantage of this measure is that it is not sensitive to small values in $\rho_{z_x}[n,k]$. Note that signals with power distributed all over the (t,f) plane yield a larger $M_p(x)$, while those with power concentrated in certain areas have a smaller $M_p(x)$. In this study, the definition given in Ref. [53] was used with $p = 2$ yielding the following expression:

$$\mathcal{M} = \left(\sum_{n=1}^{N}\sum_{k=1}^{M}|\rho_{z_x}[n,k]|^{\frac{1}{2}}\right)^2. \quad (5.20)$$

5. The (t,f) *Renyi entropy* provides a measure of complexity of a non-stationary signal in the (t,f) plane and is given by

$$R_\alpha(x) = \frac{1}{1-\alpha}\log\left(\sum_{n=1}^{N}\sum_{k=1}^{M}\frac{\rho_{z_x}^\alpha[n,k]}{\sum_n\sum_k\rho_{z_x}[n,k]}\right), \quad (5.21)$$

with $\alpha > 2$. The advantage of (t,f) Renyi entropy over other expressions of entropies is that it can also be applied to all TFDs, which may assume negative values (see Ref. [2, Section 7.3]).

6. (t,f) *image-based features* are image descriptors extracted from the TFD matrix ρ_{z_x} considered as an image. First a segmentation technique, e.g., water-shed, is applied on ρ_{z_x} to detect regions where most signal information appears and a binary segmented image, denoted by $\rho_{z_x}^{seg}[n,k]$, is generated [45]. Figure 5.2 shows an example of (t,f) image of newborn EEG signal obtained using the EMBD and its binary-segmented image obtained using the methodology proposed in Ref. [45]. The segmented image allows the selection of regions where the normal and abnormal patterns appear.

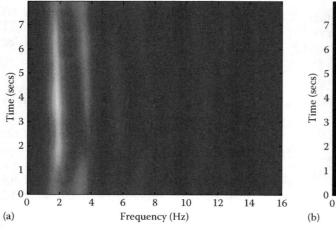
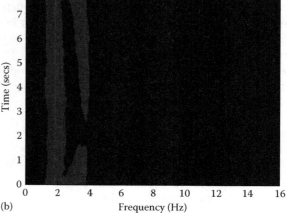

FIGURE 5.2
(a) Example of the TFD of an EEG signal using the EMBD, and (b) binary-segmented image. The lighter gray part of the binary-segmented image is the selected region after thresholding step.

Morphometric features are then extracted from the moments of $\rho_{z_x}^{seg}[n,k]$ defined as

$$m_{pq} = \sum_{n=1}^{N}\sum_{k=1}^{M} n^p k^q \rho_{z_x}^{seg}[n,k], \quad p,q = 0,1,2,\ldots$$

Such features include [45, 46]

a. Area: $A(x) = m_{00}$,
b. Compactness: $C(x) = \frac{P(x)^2}{A(x)}$ where $P(x) = (m_{30} + m_{12})^2 + (m_{03} + m_{21})^2$,
c. Coordinates of the centroid for the segmented region: $CC_X(x) = \frac{m_{10}}{m_{00}}$ and $CC_Y(x) = \frac{m_{01}}{m_{00}}$,
d. Rectangularity: $R(x) = (m_{20} - m_{02})^2 + 4m_{11}^2$,
e. Aspect ratio: $AR(x) = m_{20} - m_{02}$.

5.3.4.2 Translation Approach-Based TF Features

A process for defining new TF features by extending t-domain and f-domain features to the joint (t,f) domain is described in Refs. [2,45,46].

Some statistical features that can be extracted from the t-domain representation are listed in Table 5.3. In order to exploit the additional information provided by TFDs, the t-domain features can be extended to the (t,f) features by simply substituting the 1D t-domain moments with the corresponding 2D (t,f) domain moments as shown in Table 5.3.

New (t,f) features can also be derived by extending f-domain features such as the spectral flux, spectral entropy, and spectral flatness to the (t,f) domain. Table 5.4 lists the (t,f) representations of the extended features.

More details about (t,f)-based features and their application to newborn EEG abnormality detection are provided in Refs. [2,45,46,50]. In Section 5.5.1, the results of using such features for the purpose of newborn EEG abnormality detection are presented.

5.3.4.3 Feature Selection

Feature selection/reduction is an important step in classification as a large number of features lead to overtraining of classifier and significantly higher computational cost [2, Section 15.6]. Two families of feature selection methods are presented in the next subsections.

TABLE 5.3

TF Extension of t-Domain Features

Feature	t-Domain Representation	(t,f) Extension
Mean	$m_{(t)} = \frac{1}{N}\sum_n x[n]$	$m_{(t,f)} = \frac{1}{NM}\sum_n\sum_k \rho[n,k]$
Variance	$\sigma_{(t)}^2 = \frac{1}{N}\sum_n (x[n] - m_{(t)})^2$	$\sigma_{(t,f)}^2 = \frac{1}{NM}\sum_n\sum_k (\rho_{z_x}[n,k] - m_{(t,f)})^2$
Skewness	$\gamma_{(t)} = \frac{1}{N\sigma_{(t)}^3}\sum_n (x[n] - m_{(t)})^3$	$\gamma_{(t,f)} = \frac{1}{(NM-1)\sigma_{(t,f)}^3}\sum_n\sum_k (\rho_{z_x}[n,k] - m_{(t,f)})^3$
Kurtosis	$k_{(t)} = \frac{1}{N\sigma_{(t)}^4}\sum_n (x[n] - m_{(t)})^4$	$k_{(t,f)} = \frac{1}{(NM-1)\sigma_{(t,f)}^4}\sum_n\sum_k (\rho_{z_x}[n,k] - m_{(t,f)})^4$
Coefficient of variation	$c_{(t)} = \frac{\sigma_{(t)}}{m_{(t)}}$	$c_{(t,f)} = \frac{\sigma_{(t,f)}}{m_{(t,f)}}$

TABLE 5.4

TF Extension of f-Domain Features

Feature	f-Domain Notation	(t,f) Extension
Spectral flux	$\mathcal{FL}_{(f)}$	$\mathcal{FL}_{(t,f)} = \sum_{n=1}^{N-l}\sum_{k=1}^{M-m} \left\| \rho_{z_x}[n+l,k+m] - \rho_{z_x}[n,k] \right\|$
Spectral entropy	$\mathcal{SE}_{(f)}$	$RE_{(t,f)} = \frac{1}{1-\alpha}\log_2 \sum_{n=1}^{N}\sum_{k=1}^{M} \left(\frac{\rho_{z_x}[n,k]}{\sum_n\sum_k \rho_{z_x}[n,k]} \right)^\alpha$
Spectral flatness	$\mathcal{SF}_{(f)}$	$\mathcal{SF}_{(t,f)} = MN \frac{\prod_n\prod_k \rho_{z_x}[n,k]}{\sum_n\sum_k \rho_{z_x}[n,k]}$

5.3.4.3.1 Filter Methods

For this type of methods, the criterion to rank each feature is based on the property of the dataset. Four popular approaches are presented below:

- *Fisher criterion* [54]: it measures the separability between data clusters as

$$F(d) = \frac{\sum_{c=1}^{C} \left(n_c (\mu_c^d - \mu^d)^2 \right)}{\sum_{c=1}^{C} \left(n_c (\sigma_c^d)^2 \right)}, \quad (5.22)$$

where n_c is the number of elements in class c, μ_c^d and σ_c^d represent the mean and the standard deviation, respectively, of the dth feature in the class c, and μ^d is the overall average of the dth feature.

- *Maximal marginal diversity* (MMD): this criterion for feature selection is based on the Kullback–Liebler distance (dKL) [2]. The selected best solution for the problem (of selecting the most discriminating features) is to choose the axes that maximize the MMD. More details can be found in Ref. [55].
- *Minimum redundancy and maximum relevance*: this measure combined two criteria based on (1) maximum relevance and (2) minimum redundancy; more details can be found in Ref. [55].
- *Receiver operating characteristics (ROC) analysis*: it is chosen in this study to evaluate the performance of features and select good performing (t, f) features when used to detect an abnormality or artefacts in newborn EEG signals. In this approach, the performance of each feature (in detecting an abnormality or artefacts) was evaluated by doing an ROC analysis of the feature values extracted for EEG segments corresponding to different states (e.g., seizure and non-seizure). For a particular feature, the ROC curve of a binary classifier based on that feature was assessed as its discrimination threshold changed. The area under the resulting ROC, i.e., area under the ROC curve (AUC), which has a range of 0 to 1, was computed and utilized as a summary of the ROC curve and a measure of how well a feature can effectively discriminate between two groups. In other words, a classifier with an AUC value of 1 is considered perfect, and an AUC value of 0.5 is in effect a random-guessing classifier [56].

5.3.4.3.2 Wrapper Methods

Wrapper methods use classifiers as a black box to score subsets of features that maximize the classifier performance [57]. The overall block diagram for the wrapper algorithm is illustrated in Figure 5.3. Wrapper methods are designed to automatically define the number of features needed to maximize the accuracy [57], which is not the case for filter-based methods.

5.3.4.3.3 Feature Search Techniques

To select a subset for evaluation at each step, in the case of wrapper methods, the approaches listed below are relevant:

- *Exhaustive search*: this intuitive method returns the best feature subset for a given criterion. A weakness of the method is that its computational complexity is too large. For example, an exhaustive search with a feature set of D features requires $(2^D - 1)$ evaluations of the criterion. When D becomes very large, the exhaustive search becomes too costly.
- *Smart search algorithms*: they iteratively add or remove features from the set depending on the particular search algorithm. Three considered main types of search procedures are listed below [58]:
  1. *Sequential forward selection* (SFS): starting from an empty set, this algorithm adds features iteratively until a selected stopping condition is reached.

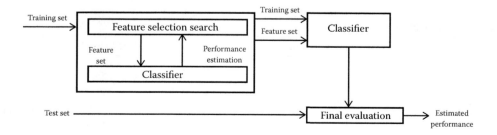

FIGURE 5.3
Wrapper-based method. The best subset is selected w.r.t. the classification accuracy. With this kind of method, the number of features needed is automatically set [45], which is not the case for filter-based methods.

2. *Sequential backward selection* (SBS): This approach starts with the set of all features; then selected features are iteratively removed from the set until convergence of the algorithm is obtained.
3. *Sequential forward floating selection* (SFFS): The SFS and SBS have a tendency to become stuck in local minima and have a risk of overfitting. The SFFS algorithm avoids this issue by combining both SFS and SBS. The resulting combined SFFS includes and removes features based on the prevailing search direction.

Other feature search algorithms can be found in Ref. [59]. In this study, the SFFS method is selected with wrapper as it is the best suboptimal search method [59].

5.3.4.4 Feature Fusion

When dealing with the classification of multichannel data, one needs to merge the information extracted from all the channels to obtain a single decision. This information fusion can be performed at different levels, i.e., at the channel level, feature level, or decision level, as discussed below.

5.3.4.4.1 Fusion at the Channel Level

In this method, the acquired multichannel signals are converted into a single channel obtained by taking a spatial average of the multichannel data [45]. It is expressed as

$$x[n] = \frac{1}{M}\sum_{p=1}^{M} x_p[n], \qquad (5.23)$$

where M represents the number of channels, x_p denotes the time-domain signal recorded by the pth channel, and n represents the sample (or time) index of the signal x_p.

The weakness of this approach is that some useful signal components can get eliminated, reduced, or distorted by the averaging, but it is computationally efficient (Figure 5.4). Another related technique is the feature concatenation, which uses all the extracted features from each channel; this is not used in this study as the dimensionality problem becomes too large when using this method.

5.3.4.4.2 Fusion at the Features Level

This approach first extracts features from each independent channel separately and then merges these extracted features by forming a new single feature set that is then used at the classification stage as shown in Figure 5.5. It is defined as

$$F_j[n_{seg}] = \frac{1}{M}\sum_{p=1}^{M} F_{jp}[n_{seg}], \qquad (5.24)$$

where F_{jp} is the jth feature for the pth channel, and the index n_{seg} represents the segment index.

5.3.4.4.3 Decision Level Fusion

This approach detects abnormality in each channel separately by independently extracting features from each channel followed by training of a classifier for each channel independently as illustrated in Figure 5.6. The independent decisions obtained from each channel can then be combined using, e.g., regional correlation [44]. This approach is most effective as it allows both detection and localization of abnormality, but it requires the channel-by-channel annotation of the acquired database.

The EEG application described in this study uses only the first two feature fusion methods because independent decisions could not be obtained from each channel (the reason is that the EEG database available for this study does not have channel-by-channel annotations).

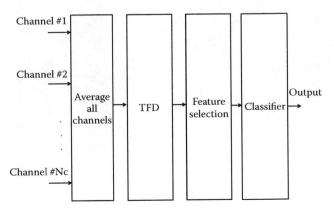

FIGURE 5.4
Multichannel feature extraction: channel fusion approach.

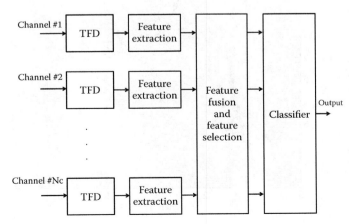

FIGURE 5.5
Multichannel feature extraction: feature fusion approach.

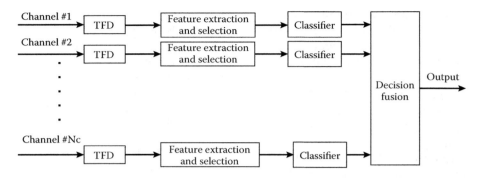

FIGURE 5.6
Multichannel feature extraction: decision fusion approach.

5.3.5 TF Matched Filter Approach for Newborn Abnormality Detection

The problem of newborn EEG abnormality detection can be considered as a special case of the classical detection problem in which there is a need to detect the presence of a non-stationary signal $s(t)$ that may exhibit certain unknown parameters such as unknown time and/or frequency shifts from a noisy measured signal $x(t)$ of duration T, representing an epoch of the EEG signal under analysis. Following the classical detection problem, the two hypothesis on $x(t)$ are

$$\begin{aligned}\mathcal{H}_0 &: x(t) = n(t), \quad &\text{signal absent,}\\ \mathcal{H}_1 &: x(t) = s(t;\Theta) + n(t), \quad &\text{signal present,}\end{aligned} \quad (5.25)$$

where $n(t)$ is the additive noise and Θ represents the unknown parameters of the known deterministic signal $s(t)$. In one such scenario, the signal-to-detect is a (t, f) shifted version of $s(t)$, i.e., $s(t;\Theta) = s(t-t')e^{j2\pi ft}$.

For example, the problem of newborn seizure detection can be formulated as Equation 5.25 in which $s(t)$ is a piecewise LFM signal with unknown time delay and/or frequency shift. This is based on the finding that newborn EEG seizures can be described as piecewise LFM signals with harmonics (the number of LFM pieces depending on the duration of the EEG seizure epoch) [44,46]. The background patterns, on the other hand, usually exhibit irregular activities with no clear consistent behavior [46]. However, as the parameters of the LFM signals (i.e. time duration and slope of each piece of $s(t)$) are not known and may be different for each EEG epoch, a template set $r(t)$ composed of M piecewise LFM signals can be used, i.e., $r(t) = \{r_j(t)\}_{j=1}^{M}$. Also, since the joint probability density function (pdf) of time delay and frequency shift of $s(t)$ is not known either, the kernel of the TFD needs to be found adaptively. Figure 5.7 illustrates the different (t, f) characteristics observed in newborn EEG seizure and background.

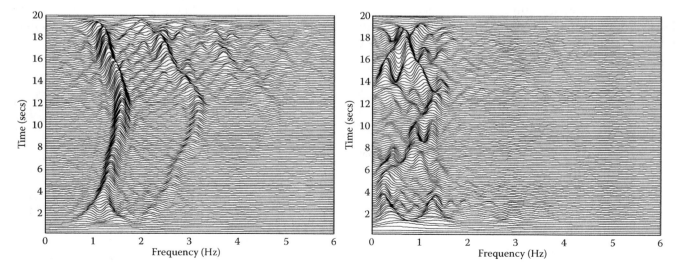

FIGURE 5.7
Example of (t, f) representations of EEG signals acquired from a newborn during seizure (left) and non-seizure activities (right) [2, Section 12.5].

The optimum decision strategy for determining the correct hypothesis in the detection problem formulated in Equation 5.25 involves finding the test statistic

$$\eta = \max_{\Theta}\left\{\int_{(T)} x(t)s^*(t;\Theta)(t)\,dt\right\}, \quad (5.26)$$

where * stands for complex conjugation, and comparing it with a predefined threshold value. For cases where there is no unknown parameter, the signal $s(t)$ is real, and the noise $n(t)$ is a zero-mean white Gaussian process, the conventional time-domain matched filter with the test statistic

$$\eta = \int_{(T)} x(t)s^*(t)\,dt \quad (5.27)$$

is the optimum detector. Based on the inner-product invariance property of the WVD [2, Chapter 2], the test statistic of the optimum detector in the (t,f) domain can be expressed as

$$\eta_{QMF} = |\eta|^2 = \iint_{(T)} W_x(t,f)W_s(t,f)\,dt\,df. \quad (5.28)$$

The (t,f) matched filter (TFMF) with the test statistic given in Equation 5.28 is known as the quadrature matched filter (QMF) [60]. Alternatively, based on the formulation of TFDs in the ambiguity domain (presented in Section 3.2.2), η_{QMF} can be written as

$$\eta_{QMF} = \iint A_x(\nu,\tau)A_s^*(\nu,\tau)\,d\nu\,d\tau, \quad (5.29)$$

where $A_s(\nu,\tau)$ is the symmetrical AF of $s(t)$ [2].

In most real-life applications, the QMF is not optimal as the signal-to-detect is neither deterministic nor known completely. One such case is when $s(t;\Theta) = s(t-t')e^{j2\pi f't}$, where $s(t)$ is assumed known and deterministic while t' and f' are random variables with joint pdf $\rho_{t'f'(t',f')}$. For this scenario, the optimum solution is provided by a TFMF with the test statistic given in Equation 5.30 with $\gamma(t,f) = \rho_{t'f'}(t'=t, f'=f)$ as the (t,f) kernel of the TFD [60]. When the time delay t' and the frequency shift f' are statistically independent, the optimal kernel will be a separable one, i.e., $g(\nu,\tau) = G_1(\nu)g_2(\tau)$, that is, a TFD with separable kernel. For such cases, detectors based on the general formulation of TFMFs developed in Refs. [44,60] should be used.

Based on Equations 5.28 and 5.29, the test statistic of the general formulation of the WVD-based TFMFs in the (t,f) and ambiguity domains can be written as

$$\eta_{TF}^{(WV)} = \iint_{(T)} W_x(t,f)\rho_s(t,f)\,dt\,df \quad (5.30a)$$

$$= \iint A_x(\nu,\tau)A_s^*(\nu,\tau)g^*(\nu,\tau)\,d\nu\,d\tau, \quad (5.30b)$$

where $g(\nu,\tau)$ is the Doppler-lag kernel of the TFD. The choice of different kernels $g(\nu,\tau)$ in Equations 5.30a and

TABLE 5.5

Test Statistics Resulting from the Choice of Doppler-Lag Kernel of the TFD in Equations 5.30a nd 5.30b

Test Statistic	Doppler-Lag Kernel in Equation 5.30b
$\iint_{(T)} W_x(t,f)W_s(t,f)\,dt\,df$	1
$\iint_{(T)} W_x(t,f)\rho_s(t-t_0,f-f_0)\,dt\,df$	$g(\nu,\tau)e^{-j2\pi t_0 \nu}e^{j2\pi f_0 \tau}$
$\iint_{(T)} \rho_x(t,f)\rho_s(t-t_0,f-f_0)\,dt\,df$	$|g(\nu,\tau)|^2 e^{-j2\pi t_0 \nu}e^{j2\pi f_0 \tau}$

5.30b results in different test statistics. Table 5.5 lists a few known test statistics as special cases of Equations 5.30a and 5.30b with different Doppler-lag kernels for the TFD used in the equations.

By replacing $W_x(t,f)$ in Equation 5.30a with the XWVD of the signals $x(t)$ and $s(t)$, the test statistic of the general XWVD-based formulation of TFMFs can be defined as

$$\eta_{TF}^{(XWV)} = \iint_{(T)} W_{xs}(t,f)\rho_s(t,f)\,dt\,df. \quad (5.31)$$

Details about TFMFs and their application to newborn EEG abnormality detection appear in Refs. [44,60,61]. Some results appear in Section 5.5.2.

5.3.6 Classification

SVMs are used to evaluate the overall detection performance of the (t,f) features developed in this project for detecting newborn EEG artefacts as well as abnormalities. The choice of SVMs is justified by previous findings reported in Refs. [45,62]. The SVM decision is, e.g., whether the EEG signal $x[n]$ is seizure or non-seizure. The performance of the proposed methodology for the detection of artefacts and abnormalities in newborn EEG signals is estimated using standard statistical parameters, i.e., sensitivity (SEN), specificity (SPE), positive predictive value (PPV), negative predictive value (NPV), and total accuracy (ACC), as expressed below.

$$SEN = \frac{number\ of\ true\ positives}{number\ of\ true\ positives\ +\ number\ of\ false\ negatives}, \quad (5.32a)$$

$$SPE = \frac{number\ of\ true\ negatives}{number\ of\ true\ negatives\ +\ number\ of\ false\ positives}, \quad (5.32b)$$

$$PPV = \frac{number\ of\ true\ positives}{number\ of\ true\ negatives\ +\ number\ of\ false\ positives}, \quad (5.32c)$$

$$NPV = \frac{number\ of\ true\ negatives}{number\ of\ true\ negatives\ +\ number\ of\ false\ negatives},\quad(5.32\text{d})$$

$$ACC = \frac{number\ of\ true\ positives\ +\ number\ of\ true\ negatives}{number\ of\ positives\ +\ number\ of\ negatives}.\quad(5.32\text{e})$$

5.3.7 BSS Algorithms

A BSS algorithm has been developed for multichannel signals. This algorithm is based on three key points:

1. Use of high-resolution TFD: the MDD; this QTFD is defined in Section 5.3.3.4.
2. A robust noise thresholding step based on statistical test given by

$$\begin{aligned}\mathcal{H}_0 &: \mu M - \varrho_0 \le Y[n,k] \le \mu M + \varrho_0 \\ \mathcal{H}_1 &: otherwise,\end{aligned}\quad(5.33)$$

where $Y[n,k] = \sum_{i=1}^{M} \rho_{z_i,z_i}[n,k]$, M is the number of sensors, $\mu = \sigma_0^2 g[0,0]$ (g is the Doppler-lag kernel used for the TFD and σ_0 is the standard deviation of the noise), and ϱ_0 is the solution of the following equation [63]:

$$\frac{(-1)^M\sqrt{\pi}\ {}_1\frac{\varrho^2}{4b^2}M_1F_2\left(M,[M+\frac{1}{2},M+1],\frac{\varrho^2}{4b^2}\right)}{\Gamma\left(M+\frac{1}{2}\right)\Gamma(M+1)}$$

$$-\frac{(-1)^M\sqrt{\pi}\ {}_1F_2\left(\frac{1}{2},[\frac{3}{2},\frac{3}{2}-M],\frac{\varrho^2}{4b^2}\right)\varrho}{b\ \Gamma(M)\Gamma\left(\frac{3}{2}-M\right)} = 1-\xi,$$

(5.34)

where ξ is the probability of false alarm, ${}_pF_q$ is the generalized hypergeometric function, and $b = \frac{\sigma_0^2}{\sqrt{2}}\sqrt{\sum_n\sum_m|G[n,m]|^2}$.

3. IF estimation algorithm based on image processing techniques. More precisely, the following steps are undertaken:

   a. Creation of a binary mask
   b. Creation of the mask skeleton
   c. Detection of the cross (t,f) points using an 8-neighboring set
   d. Thresholding w.r.t. the number of pixels of the selected segments
   e. Dilatation of the skeleton by convolving with a 3×3 matrix containing only ones

These steps enable us to improve the mixing matrix estimation. The separation source problem is rewritten to use the least absolute shrinkage and selection operator (LASSO) algorithm to improve the BSS. More details can be found in Ref. [63].

5.4 TF Modelling of Neonatal Multichannel Non-Stationary EEG and Artefacts

Current techniques for newborn seizure detection lack the required robustness that is expected for clinical implementation, due to variability of EEG characteristics in time and among patients [64], and scarcity of databases for validation and verification [65]. EEG variability is unavoidable, as background and seizure patterns change significantly among different patients, and is often significantly affected by the conceptional age* (CA) of the patient [66]. In addition, obtaining large annotated databases requires sustained effort, and data protection policies may restrain independent scientists from sharing and comparing their results on a common ground basis.

In this section, a modelling approach for multichannel newborn EEG is introduced to provide a common ground validation source. This model can be adopted for different applications, such as using the model parameters as new features for classification algorithms, testing artefact detection and removal, quantifying EEG quality, and calibrating new EEG machines, and it can be used as an alternative when patient confidentiality and ethical privacy laws forbid the use of real EEG [67].

5.4.1 Modelling Newborn Mono-Channel Non-Stationary EEG

An approach, namely the *IF-Based*, is presented in Ref. [68]. Background epochs are modelled using random processes with a time varying spectrum, while a multi-component signal with a piecewise linear IF law is used to model seizure waveforms. In this model, the statistical

---

* Age of infants can be defined in terms of gestational age or conceptional age. Gestational age (GA) is the time in weeks from conception, or the time after the last menstrual cycle to the date of birth. Conceptional age (CA) is defined as gestational age plus chronological age (weeks of life after birth).

properties of all parameters are known; consequently, it is capable of simulating the newborn background and seizure EEG patterns. Background epochs, generated by the *IF-based* approach, are further improved by using a band-limited fractional Brownian process with time-varying Hurst exponent (the *FBM* approach). Another technique in newborn EEG modelling appeared in Ref. [67]. This model is based on a Duffing oscillator, hence its name—the *Duffing* approach. The model is driven by a non-stationary impulse train to simulate seizure epochs and by white Gaussian noise that simulates background EEG. The advantages of this model are as follows: (1) reduction in the required number of parameters and (2) more realistic EEG when compared with previously proposed models.

5.4.1.1 The IF-Based Approach

The newborn mono-channel non-stationary EEG model is constructed using the *IF-based* approach described in Refs. [2, Chapter 16.2; 68,69]. It consists of two simulators: background and seizure EEG. Both simulators produce good approximations as they are correlated with real newborn EEG epochs (0.817 for background and 0.901 for seizure in the (t,f) domain [68]). Figures 5.8 and 5.9 illustrate generated newborn epochs composed of background and seizure patterns, respectively, where N, F_s, N_{FFT}, l, and t_r denote the signal length, sampling frequency, FFT length, lag window length, and time resolution, respectively. Figure 5.8 depicts the random nature of normal newborn EEG, while Figure 5.9 shows

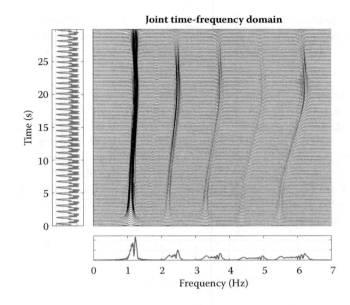

FIGURE 5.9
Simulated mono-channel newborn seizure EEG (EMBD, $N = 720$, $F_s = 24$ Hz, $N_{FFT} = 1024$, lag $l = N-1$, $\alpha = 0.025$, $\beta = 0.5$, $t_r = 1$).

the poly-harmonic nature of seizure EEG. Seizure harmonics illustrate the multipath reception of LFMs propagating through a non-linear frequency shifting material. Seizure power decays as one moves across the harmonic number; however, frequency deviation and distortion is increasing. Such observations construct the following statement: the fundamental harmonic represents a seizure event that propagates the least distance to reach an EEG electrode, while other harmonics represent scattered versions of the fundamental harmonic. Consequently, the *IF-based* mono-channel EEG model considers the possible multipath reception of newborn EEG.

5.4.1.1.1 Formulation of Background EEG Simulator

Let us consider the fractal dimension (*FD*) of real EEG signals. Synthetic background EEG generation can be obtained from the relationship between *FD* and the spectrum power law index [68]. The power spectrum of background EEG approximately follows a power law such that

$$S(f) \approx \frac{c}{|f|^{\gamma}}, \tag{5.35}$$

where c is a constant, f is the frequency, and γ is the power law exponent* [68]. The non-stationary behavior is

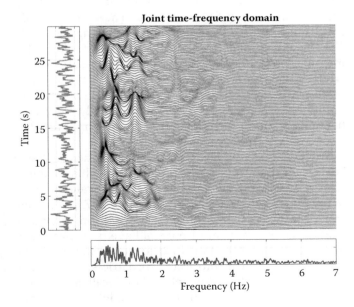

FIGURE 5.8
Simulated mono-channel newborn background EEG (EMBD, $N = 720$, $F_s = 24$ Hz, $N_{FFT} = 1024$, lag $l = N/4-1$, $\alpha = 0.025$, $\beta = 0.5$, $t_r = 1$).

---

* Note that $\gamma > 0$ and $f \in \mathbb{R} - \{0\}$.

modelled using a time-varying power law exponent γ_i; therefore, the aforementioned equation is approximated as

$$S_i(f) = \frac{c}{|f|^{\gamma_i}}, \quad (5.36)$$

where $S_i(f)$ is the power spectrum of the ith epoch with duration T. Let us assume that the EEG is quasi-stationary in every duration T; therefore, γ_i is constant in an epoch, but varies from epoch to epoch. The parameter γ_i is estimated using its linear relationship with the FD, $\gamma_i = 5 - 2FD$, where it follows a Beta distribution with parameters α and β valued at 7.82 and 7.44, respectively [68]. Furthermore, the power spectrum $S_i(f)$ can be described as

$$S_i(f) = X_i(f) X_i^*(f), \quad (5.37)$$

$$X_i(f) = \frac{\sqrt{c}}{|f|^{\gamma_i/2}} \exp[j\theta_i(f)], \quad (5.38)$$

where $X_i(f)$ is the FT of the ith epoch, and $\theta_i(f)$ is the phase spectrum, which is assumed to be a random process with uniform distribution having a range between 0 and 2π [70]. Synthesis of $x_i(t)$ is done by taking the IFT of $X_i(f)$; however, $X_i(f)$ has a smooth power law, contradicting real background EEG, which shows random fluctuations around the power law. Thus, 15 sub-epochs with the same power law but with different random phase spectra can be added together to mimic real random fluctuations [68].

5.4.1.1.2 Formulation of Seizure EEG Simulator

It is based on the (t, f) non-stationary signal model expressed as

$$s(t) = \sum_{q=1}^{Q} a_q(t) \cos\left[2\pi \int_0^t f_q(\tau)\,d\tau + \theta_q\right], \quad (5.39)$$

where $a_q(t)$, $f_q(t)$, and θ_q represent the amplitude modulation, time-varying IF function, and initial phase for the qth signal component, respectively. Equation 5.39 accounts for all important (t, f) characteristics of newborn EEG seizures such as multiple components, piecewise IF laws, and harmonic amplitude modulation [2, Chapter 16.2;69].

The IF function $f_q(t)$ is modelled as a piecewise linear function. The general form of a piecewise LFM function $f(t)$, with N_{LFM} pieces, is

$$f(t) = \sum_{n_{lfm}=1}^{N_{LFM}} \left[(\xi_{n_{lfm}} t + C_{n_{lfm}}) \operatorname{rec}\left(\frac{t - 0.5(B_{n_{lfm}+1} - B_{n_{lfm}})}{B_{n_{lfm}+1} - B_{n_{lfm}}}\right)\right],$$

$$C_{n_{lfm}} = \begin{cases} f_{st} & n_{lfm} = 1 \\ B_{n_{lfm}}(\xi_{n_{lfm}-1} - \xi_{n_{lfm}}) + C_{n_{lfm}-1} & n_{lfm} \geq 2, \end{cases}$$

$$(5.40)$$

where $\xi = [\xi_1, \xi_2, \cdots, \xi_{N_{LFM}}]$ are the gradients in Hz/s, $C_{n_{lfm}}$ is the alignment intercept that ensures continuity, $B = [B_1, B_2, \cdots, B_{N_{LFM}}, B_{N_{LFM}}+1]$ are the turning points in seconds*, and rec is the rectangular function, e.g., $\operatorname{rec}(\frac{t-\alpha}{\tau})$ is centered at α with a duration τ [68]. The multiple harmonics of the newborn EEG seizure are related to the fundamental frequency by $f_q(t) = q f_1(t)$.

The amplitude modulation function $a_q(t)$ of each harmonic q is parametrized by

$$a_q(t) = \varphi(R_q, V_n, P; t), \quad (5.41)$$

where R_q is the harmonic ratio, V_n is the normalized variation with a mean of 0.33, and P is the number of turning points. The component amplitude modulation function is computed by a cubic spline interpolation of P random turning points with amplitudes described as $a_q = R_q(0.67 + V_n)$. Locations of the turning points can be computed using $w = \frac{N(p + X)}{P}$, where $p = [0, 1, \cdots, P-1]$, N is the number of samples, and X is a uniform random process ranging between 0 and 1. Finally, the initial phase θ_q of the qth harmonic is assumed to be random with a uniform pdf on $[-\pi, \pi)$.

The complexity of the aforementioned model can be reduced by setting Q and N_{LFM} to 5 and 3, respectively, assuming B as a uniformly distributed random process, and limiting P to a maximum value of 8. Estimates for all newborn EEG seizure model parameters are listed in Refs. [2, Chapter 16.2;68].

5.4.2 Newborn Head Model

The newborn head consists basically of a scalp, skull, cerebrospinal fluid (CSF), and a brain. The scalp is the soft tissue enveloping the cranial vault containing the skull, CSF, and the brain. The skull is the bony structure the mainly forming the head and protects inner soft tissues, while the CSF is the clear, colorless body fluid seen in the brain and spine; it provides the basic mechanical and immunological protection to the brain. Finally, the brain, the main organ of the human nervous system, provides a coherent control over the body actions.

Newborn head models, e.g., realistic or spherical, have been often used in various applications such as solving EEG forward and inverse problems, and validation of brain optical tomographic images [66,71–75]. Various spherical models have been proposed in previous studies. Four sphere head models have been proposed and adopted in Refs. [66,71,72] to mimic the newborn head by dividing it into four concentric spheres. This methodology is utilized in Ref. [66] for localizing seizure dipole sources using a realistic head model to investigate the utility of EEG source imaging in newborns with post-

---

* Note that, $(\forall n_{LFM})\{ n_{lfm} \in N \mid n_{lfm} \leq N_{LFM} \} \exists B_{n_{lfm}} + 1 : B_{n_{lfm}} + 1 \geq B_{n_{lfm}}$.

asphyxia seizures. In addition, it is used in Ref. [71] to compare several spherical and realistic head modelling techniques to produce an EEG forward solution from current dipole sources. Finally, it appeared in Ref. [72] to estimate the sensitivity of impedance measurement configurations to bleeding in the brains of premature newborns. In addition, other approaches, e.g., five-sphere and isotropic/anisotropic multisphere models, have been proposed in Refs. [73–75].

5.4.2.1 Four-Sphere Head Model

The "four-sphere" approach, described in Ref. [72], is utilized to build the newborn head model. It is adopted due to its reliability in mimicking the true newborn head. All dimensions were selected from the inspection of archived newborn MRI models. The four-sphere head model divides the newborn head into four concentric spheres, namely, scalp, skull, CSF, and brain, such that $S^2_{brain} \subseteq S^2_{csf} \subseteq S^2_{skull} \subseteq S^2_{scalp}$, where S^2 is a 2D manifold*, centered at zero, with a radius of R.

Figure 5.10 depicts a basic head model to illustrate the viewing perspective of front, back, and side views. Furthermore, Figure 5.11a illustrates the head model in 3D and in 2D cross sections, showing the different head regions with their dimensions shown in different shades of grey scalp, skull, CSF, and finally the brain. Moreover, Figure 5.11b shows 21 EEG electrodes placed on the scalp surface of the four-sphere head model in accordance with the international 10–20 system. It also depicts the right and left hemisphere electrodes, the mid-line electrodes, and the front and back electrodes.

5.4.3 EEG Multichannel Propagation

EEG propagation models describe how EEG signals disperse and evolve through the different brain regions. They can attenuate, scatter, and/or modulate EEG source signals. Constructing such models requires calculating the electrode potentials for a given source (forward problem), or finding dipole parameters that best describe measured potentials at the scalp electrodes (inverse problem). The forward problem has been investigated using various methods such as combining both EEG and MRI to localize seizures in 3D space [66]; using particle numerical models [74]; using electromagnetic source localization [75]; and using boundary element method, finite-element method, and finite difference method [71,76–78]. All the aforementioned methodologies have common pitfalls such as

---

* A 2D manifold S^2 is a 3D sphere, such that $S^2 = \{(x,y,z):x^2 + y^2 + z^2 = R^2\}$, where $(x,y,z) \in \mathbb{R}^3$.

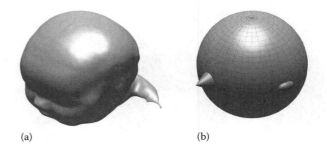

FIGURE 5.10
(a) Basic newborn head model (meshing resolution for each manifold utilized in the (b) simulated newborn head is 32 × 32).

- Need for high computational power [71].
- The need for dedicated software, as standard commercial programs, such as MATLAB, may not solve 3D numerical problems, as required.
- The need for accurate conductivity measurement of all head regions and precise placement of EEG electrodes [66].
- The need for detailed head modelling, e.g., in including or excluding the fontanel [77,78].

Utilizing newborn mono-channel simulators simplifies the forward problem to become a matter of assigning relative amplitudes and delays to synthetic waveforms appearing on the four-sphere head model. Nevertheless, this has to be done by considering the location of an EEG source signal in a way that relative amplitudes and delays can be affected by the depth and properties of the brain structures through which the signal propagates. Constructing the EEG multichannel propagation model assumes that

1. Locations of electrodes and EEG source events are precisely known.
2. Head regions in the four-sphere head model, S^2_{brain}, S^2_{csf}, S^2_{skull}, and S^2_{scalp}, have homogeneous stationary propagation properties.
3. EEG events are represented by point sources inside the brain region S^2_{brain}.
4. Background EEG patterns are stochastic; thus, they appear as random independent and identically distributed (i.i.d.) processes on all channels.
5. Seizure EEG patterns are not stochastic; consequently, the same pattern can appear on all channels, but with different amplitudes and delays correlated with the EEG event location in time and space.

Let us consider a single source signal $s(t)$ having a stationary spatial location in 3D. The source signal energy

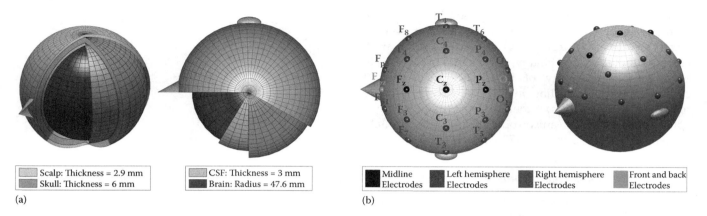

FIGURE 5.11
Newborn four-sphere head model (meshing resolution for each manifold utilized in the four-sphere head is 32 × 32) (a) Cross sections in 3D and 2D top view (b) Placement of 21 EEG electrodes.

propagates through all brain regions to be picked up by an electrode $x(t, r)$ arbitrarily located at r, where r is the relative position of the electrode with respect to the source signal position. $r = (x, y, z)$, where $(x, y, z) \in S^2_{scalp}$. $x(t,r)$ and $s(t)$ are related by the convolution operation, such that $x(t, r) = h(t, r) * s(t)$, as expressed by the following equation:

$$x(t, r) = \int_{-\infty}^{\infty} h(t - \tau, r) s(\tau) \, d\tau, \quad (5.42)$$

where $h(t,r)$ is the mono-channel impulse response that is responsible for two operations: attenuation using $\lambda(r)$ and translation using $\delta(t-\phi(r))$ as described by the following equation:

$$h(t, r) = \sum_{i=1}^{E} \lambda_i(r) \, \delta(t - \phi_i(r)), \quad (5.43)$$

where $\delta(t)$ is the Dirac function that is defined as $\delta(t) = 0$ for $t \ne 0$ and $\int_{-\infty}^{\infty} \delta(t) \, dt = 1$, while $\phi(r)$ is the time delay exerted on $s(t)$ when received by $x(t, r)$, and E is the number of received reflections. Assuming only line-of-Sight propagation, $E = 1$, Equation 5.43 is simplified into

$$h(t, r) = \lambda(r) \, \delta(t - \phi(r)). \quad (5.44)$$

This is a valid assumption as propagation in biological tissues damps most reflections, e.g., multiple reflections in ultrasound. Finally, note that $h(t, r)$ is a linear time-invariant space-variant system; thus, conventional signal processing techniques are adequate to be used for such propagation model.

Equations 5.42 and 5.44 only describe the reception on one electrode, and thus an extension to the multichannel case needs to be formulated. Let us consider a multi-channel EEG acquisition system that has M electrodes, attached on S^2_{scalp} according to the 10–20 international standard, describing the propagation of n source signals inside S^2_{brain}. $x(t) = [x_1(t), x_2(t), \cdots, x_M(t)]T$ represents the set of signals recorded by the M electrodes, and $s(t) = [s_1(t), s_2(t), \cdots, s_n(t)]T$ represents the set of n EEG source signals. Note that r is omitted from the vector $x(t)$ as all M electrodes have stationary spatial locations described by the subscript number. $x(t)$ and $s(t)$ are related by the convolution operation, such that $x(t) = H(t, R) * s(t)$, as shown below:

$$x(t) = \int_{-\infty}^{\infty} H(t - \tau, R) \times \mathbf{s}(\tau) \, d\tau, \quad (5.45)$$

$H(t,R)$ is the multichannel impulse response matrix, mixing matrix, described as

$$H(t, R) = \Lambda(R) \cdot \Phi(t, R), \quad (5.46)$$

where R is a matrix containing the relative position vectors of all electrodes with respect to all source signals, i.e., the coefficient r_{ij} is the relative position of the ith electrode with respect to the jth EEG source signal position. $\Lambda(R)$ is the multichannel attenuation matrix, while $\Phi(t, R)$ is the multichannel translation matrix; these two matrices can be expressed by

$$\Lambda(R) = \begin{bmatrix} \lambda(r_{11}) & \lambda(r_{12}) & \cdots & \lambda(r_{1n}) \\ \lambda(r_{21}) & \lambda(r_{22}) & \cdots & \lambda(r_{2n}) \\ \vdots & \vdots & \ddots & \vdots \\ \lambda(r_{M1}) & \lambda(r_{M2}) & \cdots & \lambda(r_{Mn}) \end{bmatrix}, \quad (5.47)$$

$$\Phi(t, R) = \begin{bmatrix} \delta(t - \phi(r_{11})) & \delta(t - \phi(r_{12})) & \cdots & \delta(t - \phi(r_{1n})) \\ \delta(t - \phi(r_{21})) & \delta(t - \phi(r_{22})) & \cdots & \delta(t - \phi(r_{2n})) \\ \vdots & \vdots & \ddots & \vdots \\ \delta(t - \phi(r_{M1})) & \delta(t - \phi(r_{M2})) & \cdots & \delta(t - \phi(r_{Mn})) \end{bmatrix}. \quad (5.48)$$

$H(t,R)$ is related to the lead-field matrix presented in multiple previous studies such as in Ref [2, Chapter 8.1] and Refs. [79,80]. The lead-field mixing matrix is composed of two matrices that describe the source dipoles location and orientation. Thus, both matrices carry mixed information for attenuation and delay. However, the mixing matrix $H(t,R)$ separates such information into two matrices. Thus, with respect to information, $H(t,R)$ can be considered as a separable version of the lead-field matrix.

5.4.3.1 Modelling Multichannel Attenuation Matrix

The multichannel attenuation matrix $\Lambda(R)$ can be modelled using the radiation transport equation (RTE), which represents the dispersion and decay in light intensity caused by absorption, scattering, and reflection during propagation through biological tissues [73,81–84]. Although RTE is conventionally utilized for light and EEG is an electrical measurement, both signals obey the same fundamental laws of physics, e.g., absorption, scattering, reflection, and the inverse square law [85]. Consequently, EEG power can be assumed to decay and disperse according to the head tissues optical properties.

In Ref. [81], RTE is used to verify reported tissue optical properties, while in Ref. [82], it is used to discuss both strongly scattering tissues and weakly scattering high-transparent tissues. Moreover, in Refs. [83,84], RTE is adopted to investigate the effects of the newborn fontanel by predicting the photon propagation using a Monte Carlo approach. Another utilization of RTE is in Ref. [73], where authors have used absorption coefficients, reduced scattering coefficients, and refractive indices from Ref. [83] to model the photon transport.

Elements of the multichannel attenuation matrix $\Lambda(R)$ are given in Equation 5.49 by adopting RTE for K head regions.

$$\lambda(r_{ij}) = \sqrt{I_o \prod_{k=1}^{K-1}\left(1 - R_k^{(k+1)}\right) \exp\left(-\sum_{k=1}^{K} \mu^{(k)} \| r_{ij}^{(k)} \|\right)}, \quad (5.49)$$

where

$$R_k^{(k+1)} = \left(\frac{n_k - n_{k+1}}{n_k + n_{k+1}}\right)^2, \quad (5.50)$$

I_o is the initial signal intensity; $R_k^{(k+1)}$ is the Fresnel reflection coefficient at the normal beam incidence between regions k and $k+1$; $\mu^{(k)}$ is the absorption coefficient of region k; $\| r_{ij}^{(k)} \|$ is the signal path length when propagating from the jth source signal to the ith electrode through the kth head region; and finally n_k and n_{k+1} are the refractive indices of regions k and $k+1$, respectively. RTE generally describes the drop in intensity

TABLE 5.6
Optical Properties of the Head Model Regions

k	Tissue Type	$\mu^{(k)}(cm^{-1})$	n_k
1	Brain	0.425	1.3
2	CSF	0.041	1.3
3	Skull	0.16	1.3
4	Scalp	0.18	1.3

μ = absorption coefficient, n = refractive index, k = region number. These values are verified and validated in Ref. [83].

(power transferred per unit area); therefore, a square root is added to reflect the drop in amplitude.

Table 5.6 summarizes the optical properties of the newborn four-sphere head model. These values are used in Ref. [73], while some of them are also used in previous studies [86–88]. Table 5.6 values simplify Equation 5.49 by omitting the reflection terms (Equation 5.50), and by assuming a unit initial intensity, the segmented propagation path lengths $\| r_{ij}^{(k)} \|$ become the only unknowns.

5.4.3.2 Modelling Multichannel Translation Matrix

The multichannel translation matrix $\Phi(t,R)$ can be calculated by knowing the signal travelled distance through each head region, and its propagation speed. As newborn EEG propagation speed is not precisely known (to the authors' best knowledge at the time of the publication), adult EEG propagation speed is used: one of its characteristics is that it varies over several orders of magnitude [89–91]. A previous study proposed that EEG propagation speed varies uniformly from 0.1 to 100 mm/s [89]. Another study suggested that afterdischarges can propagate with relative uniform speed and are independent of the speed of the EEG waveform (20–100 mm/s) [90]. Another experiment showed how anxiety level can change the EEG propagation speed in Ref. [91].

Elements of the multichannel translation matrix $\Phi(t,R)$ are given for all K head regions as

$$\delta(t - \phi(r_{ij})) = \delta\left(t - \sum_{k=1}^{K}\left[\frac{\| r_{ij}^{(k)} \|}{v^{(k)}(r)}\right]\right), \quad (5.51)$$

where $v^{(k)}(r)$ is the EEG propagation speed at an arbitrarily r in the kth head region. Calculations are carried out using the mean propagation speed* of EEG in Ref. [90], which is chosen as 0.6 mm/s. Hence, $\| r_{ij}^{(k)} \|$ are now the only unknowns in Equation 5.51, matching the unknowns of the multichannel attenuation matrix $\Lambda(R)$ in Equation 5.49.

---

* Since all the head model regions have the same refraction index n (Table 5.6), EEG propagation speed $v^{(k)}(r)$ is independent of head regions and space, such that $v^{(k)}(r) = v$.

5.4.3.3 Solving for Multichannel Path Lengths

Computing the multichannel attenuation and translation matrices, $\Lambda(R)$ and $\Phi(t,R)$, requires calculating the segmented propagation path lengths $\|r_{ij}^{(k)}\|$ in 3D. The problem needs first to be simplified by reducing its column space to 2D and the number of regions to two ($K = 2$). After solving this simplified case, the solution can then be extended to the general 3D case with K head regions.

Figure 5.12 depicts a methodology for solving the segmented propagation path lengths problem in 2D. It illustrates a seizure event occurring inside two concentric disks, composed of different materials. The seizure event energy spreads out radially and is picked up by a single electrode attached at the surface of region 2. The following steps summarize the solution approach:

1. Create an arbitrary line that passes through the EEG event location A. This will be the reference for all angle calculations.
2. Create a vector AC originating from A and ending up at the receiving electrode location C.
3. Create an infinite line L, where $AC \subseteq L$.
4. Find all intersection points V between L and the two regions circular surfaces $V = (B, C, \widetilde{B}, \widetilde{C})$.
5. Create "ghost" vectors originating from A and ending up at V (AB, AC, $A\widetilde{B}$, $A\widetilde{C}$).
6. Keep all intersection points in V that creates a ghost vector in the direction of AC (B, C). These points are ghost electrodes that represent projections of the receiving electrode on every inner region.
7. Solve for $\|r_{ij}^{(1)}\|$ by calculating the magnitude of region 1 ghost vector $\|AB\|$, and for $\|r_{ij}^{(2)}\|$ by calculating the magnitude difference between regions 2 and 1 ghost vectors $\|AC\|-\|AB\|$.

The aforementioned procedure can be used to simultaneously calculate all $\|r_{ij}^{(k)}\|$ for any number of concentric circular regions K. Furthermore, the solution space can be extended to 3D by replacing the line-and-circle intersection method with line and sphere, as the problem would consist of concentric spheres rather than circles. In addition, it can be used to solve $\|r_{ij}^{(k)}\|$ in the multichannel case by solving each relative distance independently, as if it is a mono-sensor case.

5.4.3.4 Generating Multichannel Background and Seizure Patterns

Background EEG waveforms are independent multichannel stochastic processes, i.e., $s(t) = [S_1, S_2, \cdots, S_n]^T$, where S_i is a random process appearing on the ith channel. Consequently, in the case of background EEG simulation, Equation 5.47 is adjusted, to respect the i.i.d. property, by equating the number of source signals n to the number of sensors M, and by omitting all interdependencies in the multichannel attenuation matrix $\Lambda(R)$ such that

$$\Lambda_B(R) = \begin{bmatrix} \lambda(r_{11}) & 0 & \cdots & 0 \\ 0 & \lambda(r_{22}) & \cdots & 0 \\ \vdots & \vdots & \ddots & \vdots \\ 0 & 0 & \cdots & \lambda(r_{Mn}) \end{bmatrix}. \quad (5.52)$$

Moreover, relative delays are not needed when simulating background source signals, as they are i.i.d. random processes. Thus, the multichannel translation matrix $\Phi(t,R)$ (Equation 5.48) is modified to account for this variation as shown in the following:

$$\Phi_B(t, R) = \begin{bmatrix} \delta(t) & 0 & \cdots & 0 \\ 0 & \delta(t) & \cdots & 0 \\ \vdots & \vdots & \ddots & \vdots \\ 0 & 0 & \cdots & \delta(t) \end{bmatrix}. \quad (5.53)$$

On the other hand, seizure EEG signals are not stochastic [66]; consequently, the same pattern is imposed on most channels, but with different amplitudes and delays ($n = 1$), i.e., $s(t) = [s_1(t)]$. Thus, in the case of seizure simulation, the multichannel attenuation $\Lambda(R)$ and translation $\Phi(t,R)$ matrices are modified to contain one

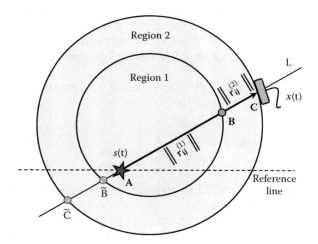

FIGURE 5.12
Solving for $\|r_{ij}^{(k)}\|$ in two concentric disks. A, B, and C are respectively the seizure event, ghost electrode, and receiving electrode locations; $\widetilde{B}$ and $\widetilde{C}$ are the locations of irrelevant intersection points; $\|r_{ij}^{(1)}\|$ and $\|r_{ij}^{(2)}\|$ are the signal propagation path lengths in regions 1 and 2; $s(t)$ is the EEG source signal; and $x(t)$ is the received signal.

steering vector as expressed by

$$\Lambda_S(R) = \begin{bmatrix} \lambda(\mathbf{r}_{11}) \\ \lambda(\mathbf{r}_{21}) \\ \vdots \\ \lambda(\mathbf{r}_{M1}) \end{bmatrix}, \quad (5.54)$$

$$\Phi_S(t,R) = \begin{bmatrix} \delta(t - \phi(r_{11})) \\ \delta(t - \phi(r_{21})) \\ \vdots \\ \delta(t - \phi(r_{M1})) \end{bmatrix}. \quad (5.55)$$

The aforementioned characteristics imply that the *IF-based* background simulator has to be excited M times to generate M stochastic EEG source signals, while the seizure simulator has to be excited only once. The

originating location of background source signals, r_b, has to be random, but confined within S^2_{brain}. This is important to ensure its spatial stochasticity as well as its broad dispersion of energy through all channels [66]. On the other hand, the originating location of the seizure source signals r_s has to be within the cortex, away from the center of the brain and its edges. This is crucial to ensure its directional dispersion of energy and localized oscillatory behavior [66].

Figure 5.13 depicts location permutations for both background and seizure source signals. They are characterized by a northern hemispheric 3D uniform distribution centered at zero with a radius of 3.76 cm (1 cm less than the brain outer shell radius). Background events locations are confined to an inner hemispheric region with a 1-cm radius, while seizure event locations are located in the complementary region of the total hemisphere.

Amplitude spatial distribution of nine randomly located background and seizure epochs are illustrated in Figure 5.14. These distributions are calculated by computing the attenuation matrix factor $\Lambda(R)$ on every point located on S^2_{scalp}. Note that the depicted distributions are not visually distorted by projection. They illustrate the top view of a 4D solution, where the fourth dimension (amplitude) is mapped by color. Results confirm the direction of amplitude dispersion for both background and seizure epochs, where background amplitudes are stochastic, while seizure amplitudes are directional.

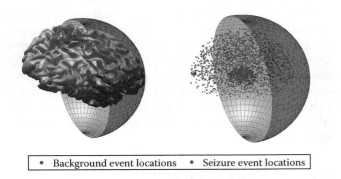

FIGURE 5.13
Normal and abnormal source signals locations. Real brain atlas is shown on the left to facilitate visualizing the location of background and seizure source signals.

5.4.3.5 Assumptions and Functionality of the Model

Constructing a complete multichannel EEG waveform composed of background and seizure epochs requires a

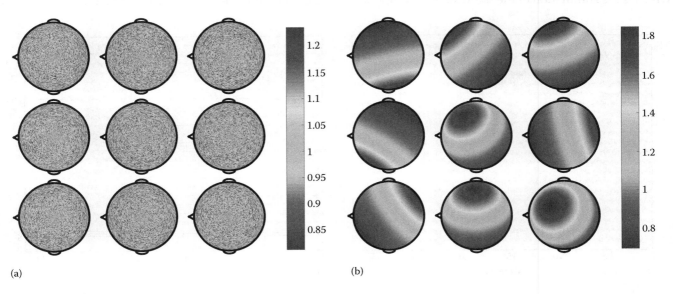

FIGURE 5.14
Amplitude spatial distribution of different EEG epochs (a) Background EEG (b) Seizure EEG.

time-domain binary mask, containing ones when seizure is present, and zeros otherwise. A user-defined mask is utilized to produce EEG patterns matching their preference (w.r.t. seizure or background patterns). Construction of the multichannel EEG model assumes the following about the supplied mask:

1. Seizure "birth" or start is determined by the start of the first seizure pattern that shows on any of the channels.
2. Seizure "death" or end is determined by the end of the last seizure pattern occurring on any of the channels.
3. The birth and death of a background epoch is determined by the observed absence of a seizure pattern on all the EEG channels simultaneously.
4. Seizure source signal location changes at the beginning of every indicated seizure.
5. The locations of the background sources are random and change every 1 second, ensuring its stochasticity.

Example: Figure 5.15 depicts 32 seconds of simulated multichannel background and seizure epochs, received on 21 electrodes (corresponding to 20 channels), on the four-sphere head model, respectively. Results in Figure 5.15a validate the stochastic behavior of background EEG, as the resultant waveform varies across electrodes, through time, and through space. On the other hand, Figure 5.15b depicts the attenuation and delay across every channel, and the different seizure epochs generated according to the supplied EEG mask. Furthermore, Figure 5.16a shows a complete multichannel EEG signal, which is the summation of the multichannel components presented in Figure 5.15. The duration, start, and end of all seizure epochs are illustrated by highlighting the multichannel waveform, while their location in space is depicted by the two amplitude distributions attached on Figure 5.16b.

5.4.3.6 Validating the Multichannel EEG Model

The generated multichannel EEG signals can be validated by exploiting EEG cross-channel properties. Previous studies have shown that EEG cross-channel correlation and cross-channel causality measurements can be utilized as classifying features [2, Chapter 16.4]. Figure 5.17 demonstrates (t, f) correlation matrices of four different EEG epochs. The first column (plots a and c) depicts the correlation matrices of simulated and real background epochs. They illustrate a random behavior, which corresponds to the i.i.d. property of background EEG. Moreover, the second column (plots b and d) illustrates the correlation matrices of real and simulated seizure epochs. They clearly show a distinct pattern, when compared to background EEG, that implies cross-channel dependencies. Note that, in the proposed model, simulated seizures are random; thus, these results do not imply an exact correspondence with real EEG, but rather a clear connection between the two cross-channel properties, and emphasizing their validity for different applications.

5.4.4 EEG Quality Measure

The development of an EEG quality measure is important because it can provide objective quantitative information regarding the quality of the newborn EEG. This measure should be able to be interpreted by

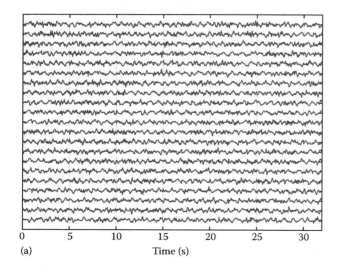
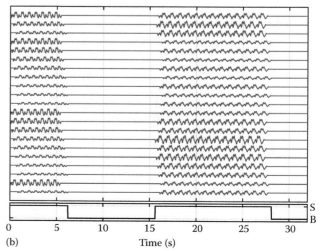

FIGURE 5.15
Simulated multichannel newborn EEG components. EEG mask is shown at the bottom of (b) showing where seizures start and finish.

Time–Frequency Analysis for EEG Quality Measurement and Enhancement

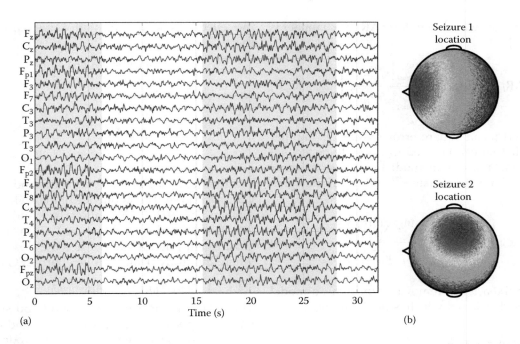

FIGURE 5.16
(a) Simulated multichannel newborn EEG. Seizure epochs are highlighted. The first highlighted part corresponds to Seizure 1 and the second highlighted part to Seizure 2. The two plots in (b) show the location of each seizure, respectively (standard minimum duration of a seizure: 10 s).

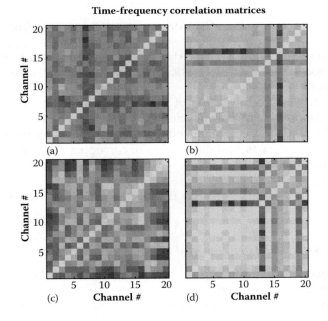

FIGURE 5.17
TF correlation matrices for simulated and real multichannel newborn EEG. Note that electrode F_{pz} is omitted from the simulation to match the real data. (a) Simulated Background, (b) Simulated Seizure, (c) Real Background, (d) Real Seizure.

neurophysiologists and automated newborn EEG analysis, and should be able to differentiate between clean EEG and contaminated EEG. Let us present in this section a model of artefacts to validate two different quality measures.

5.4.4.1 Validation of Real EEG with Simulated Artefacts

It is quite difficult to compare different artefact removal methods based on their efficiency in removing artefacts since there is no current standard and very few quantitative evaluations have been reported in previous studies. Most of the published articles evaluate them in terms of qualitative plots. In addition, very few of them quantify the distortion to desired EEG signals due to removal effect. The most difficult point for evaluating the performance of algorithms for artefact removal purpose is that the clean EEG signal is not known *a priori*. One solution to overcome this problem is to create simulated artefacts to be added to clean real or simulated EEG data. In Ref. [92], the authors proposed a model of EEG signal. From another point of view, it is better to use real EEG signal with simulated artefacts because it is harder to mimic all the behaviors of EEG signal. The most used metric to assess artefact removal algorithms is the signal-to-artefact ratio (SAR). Let x_l be the signal recorded from the lth sensor such that $x_l = [x_l(1),\ldots,x_l(N)]$; then the EEG data matrix X is given by $X = [x_1^T x_2^T \ldots x_M^T]^T$, where N is the number of samples and M is the number of channels. This matrix is considered as a superposition of the signal of interest $X^{(s)}$ and the artefact components $X^{(a)}$ such as

$$X = X^{(s)} + X^{(a)}, \qquad (5.56)$$

where X is defined such that each row corresponds to one measured EEG channel. $X^{(s)}$ and $X^{(a)}$ are then defined in a similar way. Then, the SAR for channel l is defined by [93]

$$\text{SAR}_l = 10 \log_{10} \left(\frac{\sum_{n=1}^{N} \left| \mathbf{x}_l^{(s)}[n] \right|^2}{\sum_{n=1}^{N} \left| \mathbf{x}_l^{(a)}[n] \right|^2} \right). \quad (5.57)$$

The normalized mean square error (NMSE) is another popular metric that can be used to assess the effectiveness of an artefact removal method; it measures the error η between the channel estimate $\hat{\mathbf{x}}_l^{(s)}$ and the clean signal $\mathbf{x}_l^{(s)}$ [94,95]:

$$\eta_i = 10 \log_{10} \left(\mathbb{E} \left[\frac{\sum_{n=1}^{N} \left| \hat{\mathbf{x}}_l^{(s)} - \mathbf{x}_l^{(s)} \right|^2}{\sum_{n=1}^{N} \left| \mathbf{x}_l^{(s)}[n] \right|^2} \right] \right). \quad (5.58)$$

To evaluate the artefact removal algorithms, a model of artefact was developed which is described next.

5.4.4.2 Model of Artefact

Newborn EEG artefacts can be modelled by combining the various different contaminating signals that corrupt multichannel EEG waveforms. Blood vessel pulsation (BVP), ECG spikes (ECGS), and short-time high-amplitude (STHA) artefacts are chosen as the contamination signals. These physiological artefacts are chosen because of their unique behavior in mimicking EEG seizure patterns [30,96].

5.4.4.2.1 Blood Vessel Pulsation

This corresponds to a pulsation effect induced on electrodes that are close to a blood vessel. It can be represented by a continuous oscillatory Gaussian signal, highly correlated in time, but uncorrelated to other components in the EEG signal [96,97]. The BVP can be modelled as a sine wave with a Gaussian additive noise, and a frequency close to the heart rate of newborns [98], such that

$$\text{BVP}(t) = \sin(2\pi f_{\text{BVP}} t) + n(t), \quad (5.59)$$

where f_{BVP} is the pulsation frequency, and $n(t)$ is a white Gaussian noise. The additive noise $n(t)$ forces the probability distribution of BVP(t) to take a bell shape and to simulate the noisy nature of the BVP artefactual signal. The pulsation frequency f_{BVP} of the sine wave is set to 2 Hz, as recommended in Refs. [97,98].

5.4.4.2.2 ECG Spikes

This represents the corrupted QRS complexes of the ECG as picked up by an electrode when placed in the vicinity of a blood vessel. It appears as a train of sharp pulses, which is not strictly periodic but correlated with the newborn heart rate [97,98]. The ECGS artefact can be modelled as a spike train with Gaussian pdf [99], such that

$$\text{ECGS}(t) = g(t) + n(t), \quad (5.60)$$

where $g(t)$ is defined by

$$g(t) = \begin{cases} 1 ; & t = \dfrac{n}{f_e} \\ 0 ; & \text{otherwise,} \end{cases} \quad (5.61)$$

f_e is the frequency of the spike train, and $n(t)$ is a white Gaussian noise. The frequency of the spike train is set to 1 Hz in Ref. [97], 2.5 Hz in Ref. [98], and 1.2 Hz in Ref. [99]. In this work, f_e is chosen as 1 Hz, as Ref. [97] is the most recent publication and it is the closest to Ref. [99], where the study dealt with newborn EEG artefacts.

5.4.4.2.3 Short-Time High-Amplitude

The STHA artefact can correspond to burst suppression patterns appearing in recorded EEG, and can also correspond to the patient and/or electrode movements [99,100]. STHA is characterized by periods of high electrical activities alternating with periods of no activities; thus, it is modelled using a heavy-tailed noise [101], such that

$$\text{STHA}(t) = n_h(t), \quad (5.62)$$

where $n_h(t)$ is a Levy stable symmetrical stochastic process with characteristic function parameters taken from Ref. [102].

Automatic generation of real EEG corrupted by simulated artefacts was developed, where one can control the number of contaminated segments, the time duration of the artefacts, and the number of corrupted electrodes, such that

$$x_i^{(a)} = \text{BVP}_i + \text{ECGS}_i + \text{STHA}_i. \quad (5.63)$$

5.4.4.3 Validation for Real EEG with Real Artefact

From a classification point of view, one can use a metric based on the machine learning outputs. After classifying the segment into one of the following categories: (1) background, (2) seizure, or (3) artefact, one can assess the quality of EEG recordings in a database using the following ratio:

$$Q = \frac{N_b + N_s}{N_a}, \quad (5.64)$$

where N_b is the number of segments classified as background, N_s is the number of segments classified as seizure, and finally N_a is the number of segments classified as artefacts.

5.5 Discussion of Results for Stage 1 Experiments

5.5.1 Newborn EEG Abnormality Detection Using TF-Based Features

This section presents and discusses the results of utilizing selected TF features for the detection of artefacts and abnormalities in newborn EEG signals. In order to minimize the length of the chapter, only the results for EEG seizure detection are presented. More details can be found in Ref. [45].

5.5.1.1 Selected Features

Among the (t, f) features presented in Section 5.3.4, the selected subset of features is listed in Table 5.7. The (t, f) features were extracted from the images of EEG signals formed by QTFDs. The list of the TFDs used in these experiments and the values of their parameters are given in Table 5.8. Note that the selected values for the

TABLE 5.7

TF Feature Set Used for Automatic Detection of Artefacts and Seizures in Newborn EEG Signals

Class	Feature Name	Formula
Extended f-domain features	(t, f) flux	$TF_1 = \mathcal{FL}_{(t,f)}$
	(t, f) flatness	$TF_2 = S\mathcal{F}_{(t,f)}$
	Renyi entropy	$TF_3 = RE_{(t,f)}$
Extended t-domain features	Mean	$TF_4 = m_{(t,f)}$
	Variance	$TF_5 = \sigma^2_{(t,f)}$
	Skewness	$TF_6 = \gamma_{(t,f)}$
	Kurtosis	$TF_7 = k_{(t,f)}$
	Coefficient of variation	$TF_8 = c_{(t,f)}$
Inherent (t, f) features	Mean of the IF	$TF_9 = \frac{1}{N}\sum_{n=1}^{N} f_{z_x}[n]$
	Deviation of the IF	$TF_{10} = max(f_{z_x}[n]) - min(f_{z_x}[n])$
	Maximum of singular values	$TF_{11} = max(\bar{s}_1,...,\bar{s}_N)$
	Complexity measure	$TF_{12} = CM$
	Mean of S_{w_r}	$TF_{13} = \frac{1}{R}\sum_{r=1}^{R} S_{w_r}$
	Standard deviation of S_{w_r}	$TF_{14} = std(S_{w_1},...,S_{w_r})$
	Mean of S_{h_r}	$TF_{15} = \frac{1}{R}\sum_{r=1}^{R} S_{h_r}$
	Standard deviation of S_{h_r}	$TF_{16} = std(S_{h_1},...,S_{h_r})$
	TFD concentration measure	$TF_{17} = \mathcal{M}$

Note: TF_i is the ith (t, f) feature; $f_{z_x}[n]$ is the IF of $x[n]$; $\bar{s}_1,...,\bar{s}_N$ are the normalized singular values of the matrix ρ_{z_x}; and S_{w_r} and S_{h_r} are the sparsity of the base and coefficient vectors of ρ_{z_x}, respectively.

TABLE 5.8

TFDs Used in the Experiments and Their Parameters (See Also Table 5.1)

Distribution	Parameters
WVD	N/A
SWVD	$w[n]$:Hamming, $\frac{N}{4}$ samples long
CWD	$\sigma = 5$
SPEC	$w[n]$:Hamming, $\frac{N}{4}$ samples long
MBD	$\beta = 0.01$
EMBD	$\alpha = 0.9, \beta = 0.01$

TABLE 5.9

t-Domain and f-Domain Features Used for Automatic Detection of Artefacts and Seizures in Newborn EEG Signals

Class	Feature Name	Formula
f-Domain	Spectral flux	$F_1 = \mathcal{FL}_{(f)}$
	Spectral flatness	$F_2 = S\mathcal{F}_{(f)}$
	Spectral entropy	$F_3 = SE_{(f)}$
t-Domain features	Mean	$T_1 = m_{(t)}$
	Variance	$T_2 = \sigma^2_{(t)}$
	Skewness	$T_3 = \gamma_{(t)}$
	Kurtosis	$T_4 = k_{(t)}$
	Coefficient of variation	$T_5 = c_{(t)}$

Note: T_i and F_i mean are the ith t-domain and f-domain features, respectively.

parameters of TFDs are the standard ones for which these TFDs demonstrated good performances in EEG signals analysis [1,2,46].

In order to compare their performance with their t-domain and f-domain feature counterparts, the features listed in Table 5.9 were also extracted from the EEG signals.

5.5.1.2 Seizure Detection Results

5.5.1.2.1 Data Acquisition and Pre-Processing

This experiment used the database described in Section 5.3.1 using a subset of five newborns. The EEG signals were first inspected visually by an EEG expert to remove highly artefactual segments. The channel fusion methodology is used. A set of 80 non-overlapping seizure segments and 80 non-overlapping non-seizure segments of length 8 seconds were selected and extracted randomly.

5.5.1.2.2 ROC Analysis

All the features listed in Tables 5.7 and 5.9 were extracted from the TFD of $x[n]$ obtained using channel fusion, and for each feature, ROC analysis was performed and AUC values were calculated as summarized in Table 5.10.

TABLE 5.10

ROC Analysis Results of the (t, f), t-Domain, and f-Domain Features for Seizures Detection in Newborn EEG Signals

(t, f) Feature (See Table 5.7)	WVD	SWVD	CWD	SPEC	MBD	EMBD	Original t- or f-Domain Feature (See Table 5.9)
TF_1	0.67	0.64	0.70	0.67	0.65	0.73	0.54 (F_1)
TF_2	0.67	0.67	0.71	0.74	0.62	0.60	0.54 (F_2)
TF_3	0.79	0.85	0.80	0.85	0.88	0.52	0.90 (F_3)
TF_4	0.66	0.66	0.66	0.66	0.66	0.66	0.53 (T_1)
TF_5	0.64	0.62	0.61	0.61	0.62	0.60	0.66 (T_2)
TF_6	0.79	0.75	0.75	0.80	0.80	0.52	0.53 (T_3)
TF_7	0.79	0.78	0.77	0.79	0.79	0.51	0.65 (T_4)
TF_8	0.73	0.92	0.83	0.86	0.92	0.52	0.51 (T_5)
TF_9	0.52	0.92	0.62	0.70	0.83	0.59	N/A
TF_{10}	0.51	0.53	0.51	0.57	0.52	0.52	
TF_{11}	0.57	0.55	0.57	0.59	0.55	0.60	
TF_{12}	0.64	0.78	0.83	0.83	0.79	0.83	
TF_{13}	0.89	0.93	0.91	0.90	0.93	0.89	
TF_{14}	0.51	0.53	0.68	0.53	0.52	0.58	
TF_{15}	0.60	0.79	0.83	0.85	0.83	0.83	
TF_{16}	0.54	0.79	0.55	0.69	0.62	0.62	
TF_{17}	0.64	0.54	0.59	0.59	0.57	0.61	

It can be observed that the feature TF_{13}, i.e., the average of the base vectors of ρ_{z_x}, extracted from the SWVD outperforms other features. The results show that most extended (t, f) features give better performance than their t-domain or f-domain counterparts. For illustration, box plots of the (t, f) features TF_1 and TF_6 and their t-domain or f-domain counterparts (i.e., F_1 and T_3, respectively) for seizure and non-seizure segments are shown in Figure 5.18. The box plots show that the features TF_1 and TF_6 allow for better discrimination between the two classes. The AUC scores in Table 5.10 also show that some features, e.g., TF_{10} (deviation of the IF), fail to discriminate between seizure and non-seizure segments. The results indicate that the selection of the best performing (t, f) features depends on the choice of TFD used to represent the signal $x[n]$ in the (t, f) domain. For example, if one chooses the SWVD for transforming the signal to the (t, f) domain, then the features TF_8, TF_9, and TF_{13} are the best performing ones with AUCs ≥ 0.92.

5.5.1.2.3 Classification

Three feature sets, $FV_1 = \{F_i\}_{i=1}^{3} \cup \{T_i\}_{i=1}^{5}$, $FV_2 = \{TF_i\}_{i=1}^{8}$, and $FV_3 = \{TF_i\}_{i=9}^{17}$, were used to train three SVMs. A leave-one-out cross-validation was used to evaluate the performance of the classifiers. Table 5.11 shows the values of the statistical parameters of the SVM-based classifier used for detecting newborn EEG seizures for the different feature sets and for different TFDs.

The results show that SVM with the combined feature set FV_1 gives a total accuracy of 86.88%. This is mainly explained by the presence of the signal power and its spectral entropy (i.e., T_2 and F_3, respectively) in the feature set, i.e., the features that are very discriminative between seizure and non-seizure classes [103]. However, the use of the combined (t, f) feature set FV_2 yields a better performance (up to 7% higher total accuracy) compared to FV_1. The results also show that the best performing classifier yields a total accuracy of 93.75% and high SEN and SPE, and it uses the (t, f) feature vector FV_2 extracted from the SWVD of EEG signals.

5.5.2 Newborn EEG Abnormality Detection Using (t, f) Matched Filters

The methodology described in Section 5.3.5 can be used for the detection and classification of different newborn EEG abnormalities such as seizures, burst and suppression patterns, as well as artefacts. However, this needs a thorough data analysis in order to find the templates that best represent those patterns. In this section, only the results of a TFMF-based approach for newborn EEG seizure detection are presented. More details can be found in Ref. [60].

Based on the findings of previous studies, the problem of newborn seizure detection can be formulated as Equation 5.25 in which $x(t)$ is the EEG signal and $s(t)$ is a

Time–Frequency Analysis for EEG Quality Measurement and Enhancement

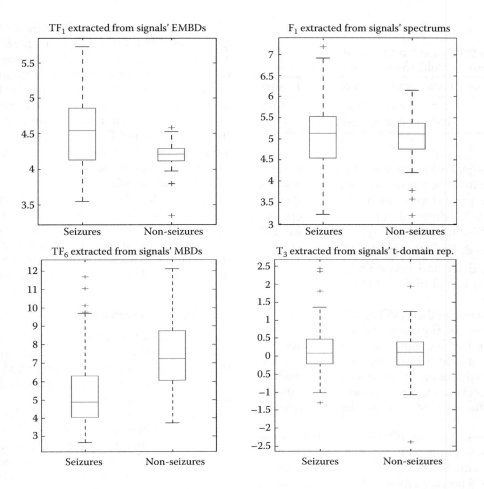

FIGURE 5.18
Box plots of four chosen features showing improved discrimination between seizure and non-seizure segments using the extended t-domain and f-domain features. Top: (left) TF_1 extracted from the EMBD and (right) F_1 extracted from the signal spectrum for seizure and non-seizure segments. Bottom: (left) TF_6 extracted from the MBD and (right) T_3 extracted from the signal t-domain representation for seizure and non-seizure segments. TF_1 is the extended version of F_1, and TF_6 is the extended version of T_3.

TABLE 5.11

Newborn EEG Seizure Detection Results in the Case of SVMs Trained with the (t, f), t-Domain, and f-Domain Features (See Tables 5.7 and 5.9 for the List of Features)

Feature Vector	TFD	SEN	SPE	PPV	NPV	ACC
t- and f-domains features, $FV_1 = \{F_i\}_{i=1}^{3} \cup \{T_i\}_{i=1}^{5}$	N/A	92.50	81.25	85.07	92.58	86.88
(t,f) extended t- and f-domains features, $FV_2 = \{TF_i\}_{i=1}^{8}$	WVD	83.75	92.50	92.50	86.35	88.13
	SWVD	95.00	92.50	93.24	95.19	93.75
	CWD	86.25	97.50	97.57	89.30	91.88
	SPEC	86.25	95.00	95.42	89.42	90.63
	MBD	91.25	91.25	92.09	92.06	91.25
	EMBD	57.50	76.25	66.77	67.62	66.88
Complementary (t, f) features, $FV_3 = \{TF_i\}_{i=9}^{17}$	WVD	86.25	83.75	84.23	87.93	85.00
	SWVD	93.75	90.00	91.67	94.62	91.88
	CWD	95.00	90.00	91.46	95.67	92.50
	SPEC	93.75	92.50	92.92	94.09	93.13
	MBD	93.75	87.50	89.68	94.62	90.63
	EMBD	98.75	76.25	82.02	98.33	87.50

TABLE 5.12

AUC Scores of the WVD- and XWVD-Based TFMFs in Detecting Seizures in Multichannel Newborn EEG

Method	TFD Kernel	WVD	SWVD	SPEC	MBD	EMBD
WVD-based		0.87	0.89	0.89	0.89	0.67
XWVD-based		0.88	0.94	0.94	0.95	0.88

piecewise LFM signal with unknown time delay and/or frequency shift. Since the parameters of the LFM signals (i.e., time duration and slope of each piece of $s(t)$) are not known and may be different for each EEG epoch, a template set $r(t)$ composed of N_{LFM} piecewise LFM signals can be used, i.e., $r(t) = \{r_j(t)\}_{j=1}^{N_{LFM}}$. Also, since the joint pdf of time delay and frequency shift of $s(t)$ is not known either, the kernel of the TFD needs to be found adaptively.

Two TFMFs based on the WVD and XWVD (with the test statistics given in Equations 5.30a and 5.31) were used with a template set composed of only three LFM signals ($N_{LFM} = 3$), and a multichannel decision fusion was achieved by combining the independent decisions of the different EEG channels. The performance of the two TFMFs was evaluated using the database described in Section 5.5.1.2.1.

The AUC scores for the WVD- and XWVD-based detectors with different (t, f) kernels are given in Table 5.12. Note that with the WVD as the (t, f) kernel, the WVD-based TFMF becomes the QMF, which has similar performance as the time-domain matched filter.

The results imply that the XWVD-based TFMF (with highest AUC score of 0.95) outperforms the one based on the WVD (with highest AUC score of 0.89) and the QMF (with the AUC score of 0.87). The MBD kernel shows higher detection accuracy than other kernels.

5.6 Results for Stage 2 Experiments

5.6.1 High-Resolution TFDs: Multidirectional Distributions

Let us consider a four-component signal composed of two tones and two parallel LFM signals with different durations, births and deaths. Figure 5.19 shows that for such signal, the MDD (Figure 5.19f) outperforms all the other considered TFDs. The MDD kernel performs better in this case because it can follow the directions of auto-terms (as indicated by Equation 5.12). The CKD does not give a high concentration of auto-terms energy without being affected by cross-terms, because the auto-term components have different directions.

Let us now consider an EEG seizure signal with both spike and LFM (almost tone) characteristics that is sampled at $f_s = 32$ Hz. The MDD here again outperforms in terms of auto-term energy resolution and cross-term suppression. Figure 5.20 shows that the EMBD and CKD cannot concentrate the energy of the auto-terms for both tones and spikes after suppression of the cross-terms. The radially kernel Gaussian [104] method gives high-energy concentration for tones but fails to represent the auto-term energy for spikes.

Results on simulated signals show that the MDD outperforms the state of the art w.r.t. subjective visualization and Boashash–Susic criterion (for more details, see Ref. [2, Section 7.4]) and also in terms of visual perception. Results on real EEG signals also validate the improved performance as the artefactual cross-terms are attenuated while maintaining a good resolution.

5.6.2 Machine Learning

This study assessed three separate classifiers using all features*: (1) SVM, (2) ANN, and (3) random forest (RF). The outcome resulted in selecting SVM because it gives the best performance with respect to accuracy and its generalization ability [105]. Then, two different multiclass strategies are compared, the one vs. one and the one vs. all methodologies [106]. The two parameters for an RBF kernel, C and σ, are selected using a grid-search strategy [105,106]. The parameter values that gave the highest accuracy are chosen. So, for each configuration, the set parameters C and σ are different.

This experiment uses an SVM classifier with RBF kernel. The parameters C and σ of this kernel have been optimized using a grid-search methodology [105,106]. Finally in the case of multiclass problem, one vs. one and one vs. all methodologies are compared [106]. The classifier performance is assessed using SEN, SPE, and ACC. These quality measures are computed using the leave-one-patient-out cross-validation approach by using segments from one patient for testing, and training is carried out using the segment from other patients. This process is repeated for each patient in the database so that the segments corresponding to each patient are used once for validation. Feature fusion and/or channel fusion approaches are compared. For both feature extraction methods, TFDs used for signal representation are the EMBD with parameters $\alpha = 0.05$ and $\beta = 0.05$; CKD with parameters $c = 1$, $D = 0.04$, and $E = 0.04$; the spectrogram with Hamming window length 65 samples (2 seconds); the SM with hamming window length 127 samples (3.94 seconds) and a rectangular window P of length 3; and the WVD. The selected TFD parameters are

---

* This is because the feature selection step is a very time-consuming task, and it is not needed to apply this method to all the classifiers.

Time–Frequency Analysis for EEG Quality Measurement and Enhancement

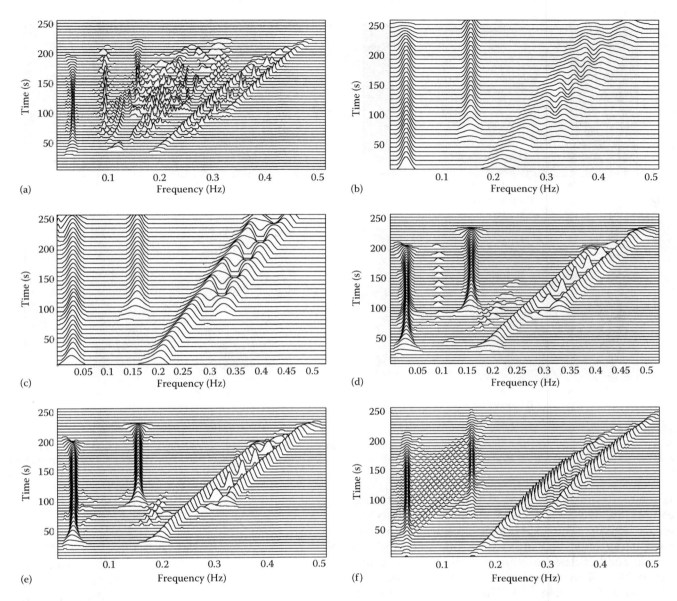

FIGURE 5.19
TFDs of a simulated FM multicomponent signal: (a) WVD, (b) spectrogram (hamming window of length 85), (c) S-method (Hanning window, $L = 6$, $N = 85$, overlap $N = 84$, $FFT_N = 512$), (d) EMBD ($\alpha = 0.18$ and $\beta = 0.24$), (e) CKD ($c = 1$, $D = 0.19$, $E = 0.12$), and (f) MDD. A threshold of 5% of the maximum of each TFD is applied for a better visualization.

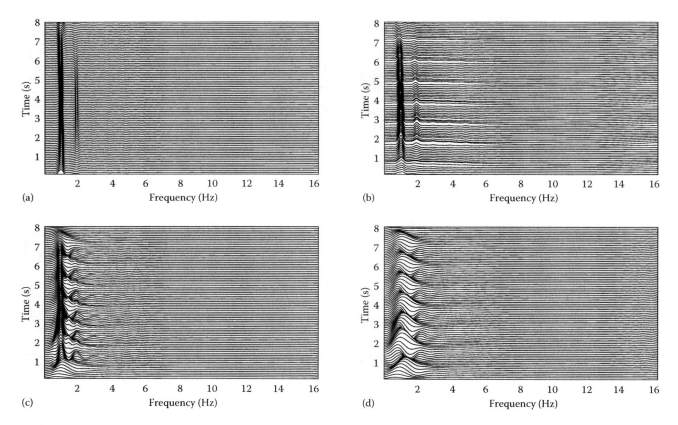

FIGURE 5.20
TF plot of a real newborn EEG seizure signal composed of a tone (pseudo-sinusoid) and spikes. (a) Radially kernel Gaussian method ($\alpha = 2$), (b) MDD, (c) EMBD ($\alpha = 0.1$, $\beta = 0.2$), and (d) CKD ($c = 1$, $D = 0.075$, $E = 0.075$). A threshold of 5% of the maximum of each TFD is applied for a better visualization.

chosen to maximize the mean AUC [2,45]. The extracted features are fed to the SVM classifier for detecting seizure activity. This experiment considers both combined background and artefacts as one class and seizures as another class (resulting in a binary classification approach). Different set of features are used and are described in Table 5.13. The classification results, reported in Table 5.14, show that

1. For all TFDs considered, signal-related features result in a better performance than image or statistical features.
2. Among all TFDs considered, considering all features, the CKD yields the best classification performance with total accuracy 82.02%.
3. The feature fusion approach systematically results in better classification results for all considered TFDs and all selected sets of TF features.

A second multiclass approach is assessed to detect seizures in the presence of both background and artefacts; it labels seizures, artefacts, and background as different classes. Multiclass signal classification

TABLE 5.13

Definition of the Abbreviation Use in This Section

Class	Feature Abbreviation	Feature Name
Signal-related features	S_1	TF flux ($l = 0$, $m = 1$)
	S_2	TF flux ($l = 1$, $m = 0$)
	S_3	TF flux ($l = 1$, $m = 1$)
	S_4	TF flatness
	S_5	Energy concentration
	S_6	Normalized Renyi entropy
	IF_1 (IA_1)	Mean of the IF (IA)
	IF_2 (IA_2)	Variance of the IF (IA)
	IF_3 (IA_3)	Skewness of the IF (IA)
	IF_4 (IA_4)	Kurtosis of the IF (IA)
Statistical features	T_1	Mean of the TF plane
	T_2	Variance of the TF plane
	T_3	Skewness of the TF plane
	T_4	Kurtosis of the TF plane
	T_5	Coefficient of variation
Image features	I_1	Area
	I_2	Centroid along t-axis
	I_3	Centroid along f-axis
	I_4	Perimeter
	I_5	Compactness

TABLE 5.14

Seizure Detection Experimental Results in the Case of Binary Classification

TFD	Set	SEN	SPE	ACC
EMBD	1	85.12% (71.42%)	73.89% (73.02%)	**80.93%** (72.60%)
	2	84.45% (75.80%)	61.26% (69.52%)	67.45% (71.20%)
	3	83.52% (84.14%)	69.26% (58.92%)	73.06% (65.65%)
	4	79.20% (72.30%)	80.67% (78.89%)	80.27% (77.13%)
WVD	1	85.12% (74.67%)	71.69% (75.35%)	75.27% (75.16%)
	2	87.02% (87.13%)	44.79%(37.50%)	56.06% (50.74%)
	3	74.82% (62.77%)	35.07% (63.06%)	45.67% (62.98%)
	4	80.69% (75.13%)	73.81% (76.23%)	**75.65%** (75.93%)
Spec	1	83.83% (76.57%)	76.19% (73.92%)	78.23% (74.63%)
	2	85.89% (87.90%)	64.07% (36.12%)	69.89% (49.93%)
	3	83.06% (72.55%)	55.94% (56.03%)	63.17% (60.44%)
	4	79.51% (65.50%)	80.12% (79.36%)	**79.96%** (75.66%)
CKD	1	85.48% (79.09%)	76.23% (70.40%)	78.70% (72.72%)
	2	85.02% (85.12%)	64.61% (54.35%)	70.05% (62.55%)
	3	81.87% (81.46%)	65.68% (55.25%)	70.00% (62.24%)
	4	80.79% (67.46%)	82.47% (81.92%)	**82.02%** (78.06%)
SM	1	81.82% (75.13%)	77.07% (72.56%)	78.34% (73.24%)
	2	84.86% (87.44%)	39.30% (37.52%)	51.46% (50.84%)
	3	78.99% (77.86%)	62.74% (52.81%)	67.07% (59.49%)
	4	79.40% (74.87%)	80.67% (73.47%)	**80.33%** (73.85%)

Note: The numbers without brackets are for feature fusion and the results between brackets are for channel fusion. 1 corresponds to signal-related features; 2 corresponds to image features; 3 corresponds to statistical features; and 4 corresponds to all features.

TABLE 5.15

Seizure Detection Experimental Results Obtained Using the CKD by Taking a Spatial Averaging of Features with Multiclass Classification and Sequential Forward Feature Selection

Rule	Method	SEN	SPE	ACC
ACC	Binary	76.31%	89.12%	85.70%
	1 vs 1	76.93%	90.13%	86.61%
	1 vs all	74.43%	90.48%	85.93%
Eq.(1.65)	Binary	84.6%	84.39%	84.45%
	1 vs 1	84.09%	84.25%	84.20%
	1 vs all	85.12%	84.36%	84.56%

Note: Two separate criteria are considered for feature selection. The first one maximizes the total accuracy while the second is given in Equation 5.69.

TABLE 5.16

Selected Features for the Six Cases Defined in Table 5.15

Rule	Method	Selected Features
ACC	Binary	$S_1,S_3,S_5,S_6,T_1,T_2,T_4,T_5,I_1,I_2,IF_1,IA_1,IF_2,IA_2$
	1 vs. 1	$S_1,S_2,S_3,S_6,S_7,T_1,T_3,T_4,I_1,I_2,IF_1,IA_1,IF_2,IA_2,IA_4$
	1 vs. all	$S_1,S_2,S_3,S_5,S_6,S_7,T_3,T_5,I_1,I_2,IF_1,IA_1,IF_2,IA_2$
Eq.(1.65)	Binary	$S_1,S_2,S_3,S_5,S_6,S_7,T_1,T_2,T_3,T_5,I_1,I_4,I_5,IF_1,IA_1,IA_4$
	1 vs. 1	$S_1,S_3,S_6,S_7,T_1,T_2,T_3,T_4,T_5,I_1,I_2,I_3,I_4,I_5,IF_1,IA_1,IA_2,IA_4$
	1 vs. all	$S_1,S_3,S_6,T_2,I_1,IF_1,IA_1,IA_4$

strategies (including one vs. one and one vs. all) are utilized to detect both artefacts and seizures in the EEG data to reduce the chances of misinterpreting artefacts as seizure.

SEN, SPE, and ACC are utilized, however, by treating seizures as one class, and both artefacts and background as another class. The wrapper-based SFFS is chosen to select the best performing features for both approaches, i.e., binary class and multiclass. Two different optimization criteria are used: (1) the accuracy and (2) the criterion defined by

$$S_{opt} = \arg\max_{S} (ACC - |SPE - SEN|) \qquad (5.65)$$

where S_{opt} represents the selected features subset and S is the set of all possible subsets.

The criterion in Equation 5.65 and the total accuracy are combined to select optimal sets of TF features for each multiclass classification strategy (see Tables 5.14 and 5.15). Table 5.16 presents the *selected* features for each combination of criteria and classification strategy. The method to extract features uses the feature fusion approach because its performance is better compared to the channel fusion approach as shown earlier in Table 5.14. Similarly, the CKD is used for feature extraction because its classification performance is better considering all features, compared to other TFDs (see Table 5.14). The results presented indicate that

1. Feature selection using the total accuracy as performance measure suggests that one can improve the classification performance with the CKD. The total classification accuracy obtained for the binary classification problem is 85.70% (see Table 5.15), which is 3.68% more than the classification accuracy achieved by using all TF features (see Table 5.14).

2. The features selected using the criterion of total accuracy have also high SPE but poor SEN. On the other hand, the features obtained using the new criterion, given by Equation 5.65, have higher SEN with a slight reduction in SPE and ACC. For example, for the CKD, referring to Table 5.15 (binary rows 1 and 4), SEN, SPE, and ACC assessments using the new criterion yield 84.60%, 84.39%, and 84.45%, respectively, whereas SEN, SPE, and ACC obtained with the total accuracy as a performance measure are 76.31%, 89.12%, and 85.70%, respectively.

3. The multiclass strategy results in a better classification outcome than the binary classification approach. For example, for the CKD, the one vs. one multiclass strategy with the criterion of maximum accuracy for feature selection gives a total accuracy of 86.61%, (this is 0.91% over the binary approach). In this experiment, in cases where there is indetermination for the one vs. one strategy, the segment is classified as "non-seizure" because it is the most likely case in real life.

4. The wrapper feature selection with SFFS selects different TF features for each case because this approach conducts a search for the optimal subset using the classifier algorithm as a black box; so that in such a case, the selected features depend on the classifier.

5.6.3 BSS-Based Artefact Removal

Let us consider in this experiment three sensors and five sources. The sources are two LFMs and three constant FMs. The LFMs are defined by $[f_{0_1}, f_{0_4}] = [0.05, 0.4]$ and $[\alpha_1, \alpha_4] = [14 \times 10^{-4}, 14 \times 10^{-4}]$, while the constant FMs are defined by $[f_{0_2}, f_{0_3}, f_{0_5}] = [0.225, 0.45, 0.03]$. For all the sources, the sampling frequency is equal to 1 Hz. This signal has a (t, f) point with three intersecting sources (i.e., M sources), and other (t, f) points contain $M-1$ intersecting sources.

This section compares the subspace projection-based method presented in Ref. [2, Chapter 8] with the LASSO method* [63]. Both methods are applied to the same set of selected (t, f) points. Figure 5.21 shows a comparison of the separation performance for the two methods. It is observed that the LASSO method provides much better separation results than those obtained by the subspace-based method; in terms of NMSE, the improvement is about 1.5 dB for different SNRs. Figure 5.22 illustrates the efficiency of the LASSO by showing the original and reconstructed sources in the time and (t, f) domains.

This illustrative example shows that this new algorithm can be applied for artefact removal to discard all sources identified as artefacts. This allows us then to reconstruct a clean EEG neonatal signal.

5.6.4 Overall System

Previous sections described algorithms and methods developed for automated artefact removal in newborn EEG recordings. These consist of (1) high-resolution TFDs that are adapted to various EEG signals, (2) a variety of features that are distinctive among artefacts,

---
* This is a method of linear model approximation.

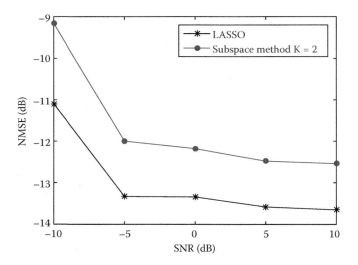

FIGURE 5.21
Comparison between the LASSO method and the subspace-based algorithm: normalized MSE (NMSE) vs. SNR for five FM sources and three sensors. (From S. Ouelha, A. Aissa-El-Bey, and B. Boashash, "Improving DOA estimation algorithms using high-resolution quadratic time-frequency distributions." *IEEE Trans*. SP; vol. 65, no. 19, pp. 5179–5190, 2017.)

background, and abnormal EEG patterns, and (3) an artefact removal strategy. Integration of these algorithms and methods yields a system that can process EEG data from raw signals to signals that are artefact-free or maximally attenuated. Relevant software/codes have been developed with effectiveness and efficiency being taken into consideration and tested using an adequate database of clinical newborn EEG data.

5.7 Conclusions and Perspectives

The contents of this chapter suggest a number of concluding remarks and reflections, as listed below:

1. A TF-based approach allows for accurate detection and classification of different newborn EEG abnormalities, in the presence of artefact and noise, and results in a system that is potentially useful in clinical practice [45].

2. In many applications, TF features give better performance as compared to time-only and frequency-only features [45].

3. The actual performance of the developed system can be further improved by using data-dependent TF distributions and optimizing the parameters of their kernels, as well as by using more computationally efficient algorithms for implementation [1].

Time–Frequency Analysis for EEG Quality Measurement and Enhancement

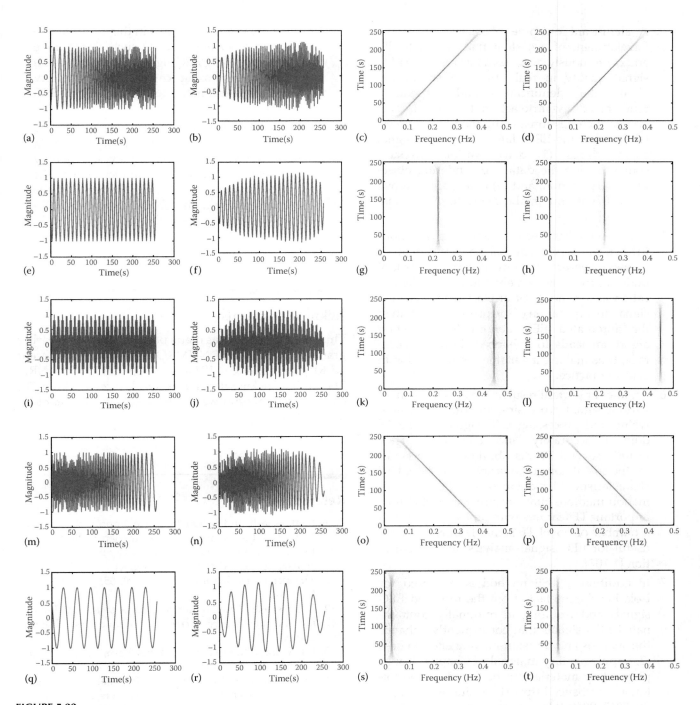

FIGURE 5.22
Source recovery illustration on a simulated example in the case of two LFMs and three constant FMs; three sensors are used. Each row represents one source and in the first and third columns, the t-domain and (t,f) domain of the original source are represented, respectively, while in the second and fourth columns, the t-domain and (t,f) domain of the reconstructed signal are shown. (From S. Ouelha, A. Aissa-El-Bey, and B. Boashash, "Improving DOA estimation algorithms using high-resolution quadratic time-frequency distributions." *IEEE Transactions on Signal Processing*, vol. 65, no. 19, pp. 5179–5190, 2017.)

4. A significant outcome of this study is the development of a system that automatically and continuously provides a report on the EEG signal quality in real time and specifically allows the identification and subsequent removal of any artefact in the signal. This system includes three components as follows: (1) segment the EEG data with similar signal characteristics, (2) detect various artefacts contaminating the data and hindering objective interpretation, and (3) detect and remove artefacts with minimal changes to the original EEG signals.

5. The system has been tested using newborn EEG from sick term babies with hypoxic ischaemic encephalopathy at very high risk of brain injury. It accurately allows the identification and removal of artefact from the EEG signal. Incorporation of the proposed system in the automated EEG abnormality detection algorithm leads to enhanced detection accuracy, thus making it suitable for application in clinical practice.

6. Another important result is the development of various signal processing algorithms for analyzing and processing EEG signals in the TF domain accurately and effectively. They include several TF distributions that account for the signal non-stationarity and different TF characteristics, instantaneous frequency estimation methods that capture one of the most important TF features with high accuracy, and a number of novel TF signal features that are useful for EEG signal analysis and classification [1,107].

7. In addition, a BSS method is proposed for isolating the artefacts from the recorded EEG signals and resulting in minimally contaminated EEG signals that consequently enhance the analysis and decision-making effectiveness [2, Supplementary material of Chapter 8; supplementary material can be found on the following website: http://booksite.elsevier.com/9780123984999/supplementary.php].

8. Finally, the overall system is mainly composed of (1) an artefact detection system based on advanced machine learning techniques; (2) an artefact removal system based on innovative BSS methods, and (3) an EEG quality measure stage; all these components improve the overall performance of an automated EEG analysis system.

9. The TF features developed in this study can be used for other applications such as cognitive monitoring using EEG signals in which a user's cognitive processing capacity and memory workload is estimated using changes in EEG signals.

10. The developed software-based system described above is ready to be tested on a large cohort in order to examine its practicability and identify the potential barriers in implementing the system for daily monitoring. Elements of the software can be found on the companion website of Ref. [2].

Acknowledgments

This research study was made possible by several earlier grants from ARC and NHMRC as well as more recent NPRP grants 09-465-2-174 and 4-1303-2-517 from QNRF. The authors acknowledged feedback received from Dr. Ghasem Azemi who was employed as a consultant on this project.

References

1. B. Boashash and S. Ouelha, "An improved design of high-resolution quadratic time-frequency distributions for the analysis of nonstationary multicomponent signals using directional compact kernels," *IEEE Transactions on Signal Processing*, vol. 65, pp. 2701–2713, May 2017.
2. B. Boashash, *Time Frequency Signal Analysis and Processing: A Comprehensive Reference*. Academic Press: Elsevier, 2nd ed., 2015.
3. J. J. Volpe, "Neurology of the newborn," *Elsevier Health Sciences*, vol. 899, 2008.
4. M. G. Rosen and C. J. Hobel, "Prenatal and perinatal factors associated with brain disorders," *Obstetrics & Gynecology*, vol. 68, no. 3, pp. 416–421, 1986.
5. M. A. Awal, M. M. Lai, G. Azemi, B. Boashash, and P. B. Colditz, "EEG background features that predict outcome in term neonates with hypoxic ischaemic encephalopathy: A structured review," *Clinical Neurophysiology*, vol. 127, no. 1, pp. 285–296, Jan. 2016.
6. S. Sanei and J. A. Chambers, *EEG Signal Processing*. Wiley, 2008.
7. M. S. Scher, Electroencephalography of the newborn: normal and abnormal features, in

Electroencephalography; Basic Principles, pp. 896–946. Lippincott Williams & Wilkins, 1999.
8. C. F. Hagmann, N. J. Robertson, and D. Azzopardi, "Artifacts on electroencephalograms may influence the amplitude-integrated EEG classification: a qualitative analysis in neonatal encephalopathy," *Pediatrics*, vol. 118, no. 6, pp. 2552–2554, 2006.
9. P. Snyder, J. M. Eason, D. Philibert, A. Ridgwa, and T. McCaughey, "Concurrent validity and reliability of the Alberta infant motor scale in infants at dual risk for motor delays," *Physical & Occupational Therapy in Pediatrics*, vol. 28, no. 3, pp. 267–282, 2008.
10. N. N. Finer, C. M. Robertson, K. L. Peters, and J. H. Coward, "Factors affecting outcome in hypoxic-ischemic encephalopathy in term infants," *American Journal of Diseases of Children*, vol. 137, no. 1, pp. 21–25, 1983.
11. S. Hamelin, N. Delnard, F. Cneude, T. Debillon, and L. Vercueil, "Influence of hypothermia on the prognostic value of early EEG in full-term neonates with hypoxic ischemic encephalopathy," *Clinical Neurophysiology*, vol. 41, no. 1, pp. 19–27, 2011.
12. A. L. Rose and C. T. Lombroso, "Neonatal seizure states: A study of clinical, pathological, and electroencephalographic features in 137 full-term babies with long term follow-up," *Pediatrics*, vol. 45, no. 3, pp. 404–425, 1970.
13. N. Monod, N. Pajot, and S. Guidasci, "The neonatal EEG: Statistical studies and prognostic value in full-term and pre-term babies," *Electroencephalography and Clinical Neurophysiology*, vol. 32, pp. 529–544, May 1972.
14. B. Tharp, F. Cukier, and N. Monod, "The prognostic value of the electroencephalogram in premature infants," *Electroencephalography and Clinical Neurophysiology*, vol. 51, pp. 219–236, Mar. 1981.
15. H. J. Niemarkt, P. Andriessen, J. Pasman, J. S. Vles, L. J. Zimmermann, and S. B. Oetomo, "Analyzing EEG maturation in preterm infants: The value of a quantitative approach," *Medicine*, vol. 1, pp. 131–144, 2008.
16. B. R. Tharp, "Electrophysiological brain maturation in premature infants: an historical perspective," *Journal of Clinical Neurophysiology*, vol. 7, no. 3, pp. 302–314, 1990.
17. C. T. Lombroso, "Neonatal polygraphy in full-term and premature infants: a review of normal and abnormal findings," *Journal of Clinical Neurophysiology*, vol. 2, pp. 105–55, Apr. 1985.
18. P. Anderer, S. Roberts, A. Schlögl, G. Gruber, G. Klösch, W. Herrmann, P. Rappelsberger, O. Filz, M. J. Barbanoj, G. Dorffner et al., "Artifact processing in computerized analysis of sleep EEG—a review," *Neuropsychobiology*, vol. 40, no. 3, pp. 150–157, 1999.
19. G. Gratton, "Dealing with artifacts: The EOG contamination of the event-related brain potential," *Behavior Research Methods, Instruments, & Computers*, vol. 30, no. 1, pp. 44–53, 1998.
20. R. J. Croft and R. J. Barry, "Removal of ocular artifact from the EEG: a review," *Neurophysiologie Clinique/Clinical Neurophysiology*, vol. 30, no. 1, pp. 5–19, 2000.
21. I. Daly, N. Nicolaou, S. J. Nasuto, and K. Warwick, "Automated artifact removal from the electroencephalogram: a comparative study," *Clinical EEG and Neuroscience*, vol. 44, no. 4, pp. 291–306, 2013.
22. S. R. Benbadis and D. Rielo, "EEG artifacts," *eMedicine Neurology*, Mar. 2010.
23. B. J. Fisch and R. Spehlmann, "Fisch and Spehlmann's EEG primer: basic principles of digital and analog EEG," *Elsevier Health Sciences*, 1999.
24. C. Burger and D. J. van den Heever, "Removal of eog artefacts by combining wavelet neural network and independent component analysis," *Biomedical Signal Processing and Control*, vol. 15, pp. 67–79, 2015.
25. A. Jafarifarmand and M. A. Badamchizadeh, "Artifacts removal in EEG signal using a new neural network enhanced adaptive filter," *Neurocomputing*, vol. 103, pp. 222–231, 2013.
26. S.-Y. Shao, K.-Q. Shen, C. J. Ong, E. P. Wilder-Smith, and X.-P. Li, "Automatic EEG artifact removal: a weighted support vector machine approach with error correction," *IEEE Transactions on Biomedical Engineering*, vol. 56, no. 2, pp. 336–344, 2009.
27. N. Mammone and F. C. Morabito, "Enhanced automatic wavelet independent component analysis for electro-encephalographic artifact removal," *Entropy*, vol. 16, no. 12, pp. 6553–6572, 2014.
28. M. A. Klados, C. Papadelis, C. Braun, and P. D. Bamidis, "REG-ICA: a hybrid methodology combining blind source separation and regression techniques for the rejection of ocular artifacts," *Biomedical Signal Processing and Control*, vol. 6, no. 3, pp. 291–300, 2011.
29. S.-C. Ng and P. Raveendran, "Enhanced rhythm extraction using blind source separation and wavelet transform," *IEEE Transactions on Biomedical Engineering*, vol. 56, no. 8, pp. 2024–2034, 2009.
30. K. T. Sweeney, T. E. Ward, and S. F. McLoone, "Artifact removal in physiological signals—practices and possibilities," *IEEE transactions on Information Technology in Biomedicine*, vol. 16, no. 3, pp. 488–500, 2012.
31. X. Yong, R. K. Ward, and G. E. Birch, "Artifact removal in EEG using morphological component analysis," in *2009 IEEE International Conference on Acoustics, Speech and Signal Processing*, pp. 345–348, IEEE, 2009.
32. A. Turnip, "Automatic artifacts removal of EEG signals using robust principal component analysis," in *Technology, Informatics, Management, Engineering, and Environment (TIME-E), 2014 2nd International Conference on*, pp. 331–334, IEEE, 2014.
33. A. Turnip and E. Junaidi, "Removal artifacts from EEG signal using independent component analysis and principal component analysis," in *Technology, Informatics, Management, Engineering, and Environment (TIME-E), 2014 2nd International Conference on*, pp. 296–302, IEEE, 2014.
34. M. K. I. Molla, M. R. Islam, T. Tanaka, and T. M. Rutkowski, "Artifact suppression from EEG signals using data adaptive time domain filtering," *Neurocomputing*, vol. 97, pp. 297–308, 2012.
35. R. Mahajan and B. I. Morshed, "Unsupervised eye blink artifact denoising of EEG data with modified multiscale

sample entropy, kurtosis, and Wavelet-ICA," *IEEE Journal of Biomedical and Health Informatics*, vol. 19, no. 1, pp. 158–165, 2015.
36. C. Zhao and T. Qiu, "An automatic ocular artifacts removal method based on wavelet-enhanced canonical correlation analysis," in *2011 Annual International Conference of the IEEE Engineering in Medicine and Biology Society*, pp. 4191–4194, IEEE, 2011.
37. X. Chen, C. He, and H. Peng, "Removal of muscle artifacts from single-channel EEG based on ensemble empirical mode decomposition and multiset canonical correlation analysis," *Journal of Applied Mathematics*, vol. 2014, 2014.
38. K. T. Sweeney, S. F. McLoone, and T. E. Ward, "The use of ensemble empirical mode decomposition with canonical correlation analysis as a novel artifact removal technique," *IEEE Transactions on Biomedical Engineering*, vol. 60, no. 1, pp. 97–105, 2013.
39. H. Zeng, A. Song, R. Yan, and H. Qin, "EOG artifact correction from EEG recording using stationary subspace analysis and empirical mode decomposition," *Sensors*, vol. 13, no. 11, pp. 14839–14859, 2013.
40. L. Shoker, S. Sanei, and J. Chambers, "Artifact removal from electroencephalograms using a hybrid BSS-SVM algorithm," *IEEE Signal Processing Letters*, vol. 12, no. 10, pp. 721–724, 2005.
41. H.-A. T. Nguyen, J. Musson, F. Li, W. Wang, G. Zhang, R. Xu, C. Richey, T. Schnell, F. D. McKenzie, and J. Li, "EOG artifact removal using a wavelet neural network," *Neurocomputing*, vol. 97, pp. 374–389, 2012.
42. H. Peng, B. Hu, Q. Shi, M. Ratcliffe, Q. Zhao, Y. Qi, and G. Gao, "Removal of ocular artifacts in EEG: An improved approach combining DWT and ANC for portable applications," *IEEE Journal of Biomedical and Health Informatics*, vol. 17, no. 3, pp. 600–607, 2013.
43. S. G. Mallat and Z. Zhang, "Matching pursuits with time-frequency dictionaries," *IEEE Transactions on Signal Processing*, vol. 41, pp. 3397–3415, Dec. 1993.
44. M. S. Khlif, P. B. Colditz, and B. Boashash, "Effective implementation of time–frequency matched filter with adapted pre and postprocessing for data-dependent detection of newborn seizures," *Medical Engineering & Physics*, vol. 35, no. 12, pp.1762–1769, 2013.
45. B. Boashash and S. Ouelha, "Automatic signal abnormality detection using time-frequency features and machine learning: a newborn EEG seizure case study," *Knowledge-Based Systems*, vol. 106, pp. 38–50, 2016.
46. B. Boashash, G. Azemi, and J. O' Toole, "Time-frequency processing of nonstationary signals: advanced TFD design to aid diagnosis with highlights from medical applications," *Signal Processing Magazine, IEEE*, vol. 30, no. 6, pp. 108–119, 2013.
47. A. Subasi, "EEG signal classification using wavelet feature extraction and a mixture of expert model," *Expert Systems with Applications*, vol. 32, no. 4, pp. 1084–1093, 2007.
48. B. D. Forrester, *Time–Frequency Signal Analysis: Methods and Applications*, chapter 18, pp. 406–423. Longman-Cheshire/Wiley, 1992.
49. L. Stanković, "A method for time-frequency analysis," *IEEE Transactions on Signal Processing*, vol. 42, no. 1, pp. 225–229, 1994.
50. B. Boashash, H. Barki, and S. Ouelha, "Performance evaluation of time–frequency image feature sets for improved classification and analysis of non-stationary signals: application to newborn EEG seizure detection," *Knowledge-Based Systems*, 2017.
51. M. W. Berry, M. Browne, A. N. Langville, V. P. Pauca, and R. J. Plemmons, "Algorithms and applications for approximate nonnegative matrix factorization," *Computational Statistics & Data Analysis*, vol. 52, no. 1, pp. 155–173, 2007.
52. B. Ghoraani and S. Krishnan, "Time–frequency matrix feature extraction and classification of environmental audio signals," *IEEE Transactions on Audio, Speech, and Language Processing*, vol. 19, no. 7, pp. 2197–2209, 2011.
53. L. Stanković, "A measure of some time–frequency distributions concentration," *Signal Processing*, vol. 81, no. 3, pp. 621–631, 2001.
54. T. S. Furey, N. Cristianini, N. Duffy, D. W. Bednarski, M. Schummer, and D. Haussler, "Support vector machine classification and validation of cancer tissue samples using microarray expression data," *Bioinformatics*, vol. 16, no. 10, pp. 906–914, 2000.
55. B. Boashash and S. Ouelha, "Designing high-resolution time-frequency and time-scale distributions for the analysis and classification of non stationary signals: a tutorial review with features performance comparison," *Digital Signal Processing*, 2017.
56. T. Fawcett, "An introduction to roc analysis," *Pattern Recognition Letters*, vol. 27, pp. 861–874, Jun. 2006.
57. I. Guyon and A. Elisseeff, "An introduction to variable and feature selection," *The Journal of Machine Learning Research*, vol. 3, pp. 1157–1182, 2003.
58. Z. M. Hira and D. F. Gillies, "A review of feature selection and feature extraction methods applied on microarray data," *Advances in Bioinformatics*, vol. 2015, 2015.
59. A. Jain and D. Zongker, "Feature selection: evaluation, application, and small sample performance," *IEEE Transactions on Pattern Analysis and Machine Intelligence*, vol. 19, no. 2, pp. 153–158, 1997.
60. B. Boashash and G. Azemi, "A review of time–frequency matched filter design with application to seizure detection in multichannel newborn EEG," *Digital Signal Processing*, vol. 28, pp. 28–38, 2014.
61. J. M. O. Toole and B. Boashash, "time–frequency detection of slowly varying periodic signals with harmonics: methods and performance evaluation," *EURASIP Journal on Advances in Signal Processing*, vol. 2011, pp. 5:1–5:16, Jan. 2011.
62. F. Lotte, M. Congedo, A. Lécuyer, F. Lamarche, and B. Arnaldi, "A review of classification algorithms for EEG-based brain–computer interfaces," *Journal of Neural Engineering*, vol. 4, Jun. 2007.
63. S. Ouelha, A. Aissa-El-Bey, and B. Boashash, "Improving DOA estimation algorithms using high-resolution quadratic time-frequency distributions." *IEEE Transactions on Signal Processing*, vol. 65, no. 19, pp. 5179–5190, 2017.

64. L. Orosco, A. G. Correa, and E. Laciar, "Review: A survey of performance and techniques for automatic epilepsy detection," *Journal of Medical and Biological Engineering*, vol. 33, no. 6, pp. 526–537, 2013.
65. A. Schulze-Bonhage, H. Feldwisch-Drentrup, and M. Ihle, "The role of high-quality EEG databases in the improvement and assessment of seizure prediction methods," *Epilepsy & Behavior*, vol. 22, Supplement 1, pp. S88–S93, 2011.
66. I. Despotovic, P. J. Cherian, M. De Vos, H. Hallez, W. Deburchgraeve, P. Govaert, M. Lequin, G. H. Visser, R. M. Swarte, E. Vansteenkiste, S. Van Huffel, and W. Philips, "Relationship of EEG sources of neonatal seizures to acute perinatal brain lesions seen on mri: A pilot study," *Human Brain Mapping*, vol. 34, no. 10, pp. 2402–2417, 2013.
67. N. J. Stevenson, M. Mesbah, G. B. Boylan, P. B. Colditz, and B. Boashash, "A nonlinear model of newborn EEG with nonstationary inputs," *Annals of Biomedical Engineering*, vol. 38, no. 9, pp. 3010–3021, 2010.
68. L. Rankine, N. Stevenson, M. Mesbah, and B. Boashash, "A nonstationary model of newborn EEG," *IEEE Transactions on Biomedical Engineering*, vol. 54, pp. 19–28, Jan 2007.
69. B. Boashash, "Time-frequency signal analysis," in *Advances in Spectrum Analysis and Array Processing* (S. Haykin, ed.), vol. 1, chapter 9, pp. 418–517, Prentice-Hall, Englewood Cliffs, NJ, 1991.
70. L. Rankine, H. Hassanpour, M. Mesbah, and B. Boashash, "Newborn EEG simulation from nonlinear analysis," in *ISSPA 2005: The 8th International Symposium on Signal Processing and its Applications, Vols 1 and 2, Proceedings*, vol. 1, pp. 191–194, IEEE, 2005.
71. F. Vatta, F. Meneghini, F. Esposito, S. Mininel, and F. D. Salle, "Realistic and spherical head modeling for EEG forward problem solution: a comparative cortex-based analysis," *Computational Intelligence and Neuroscience*, vol. 2010, pp. 13:3–13:3, Jan. 2010.
72. R. J. Sadleir and T. Tang, "Electrode configurations for detection of intraventricular haemorrhage in the premature neonate," *Physiological Measurement*, vol. 30, no. 1, p. 63, 2009.
73. S. Brigadoi, P. Aljabar, M. Kuklisova-Murgasova, S. R. Arridge, and R. J. Cooper, "A 4d neonatal head model for diffuse optical imaging of pre-term to term infants," *NeuroImage*, vol. 100, pp. 385–394, 2014.
74. G. Ala and E. Francomano, "A multi-sphere particle numerical model for non-invasive investigations of neuronal human brain activity," *Progress in Electromagnetics Research Letters*, vol. 36, pp. 143–153, 2013.
75. D. Hyde, F. Duffy, and S. Warfield, "Anisotropic partial volume CSF modeling for EEG source localization," *NeuroImage*, vol. 62, no. 3, pp. 2161–2170, 2012.
76. P. Gargiulo, P. Belfiore, E. Frigeirsson, S. Vanhatalo, and C. Ramon, "The effect of fontanel on scalp EEG potentials in the neonate," *Clinical Neurophysiology*, vol. 126, no. 9, pp. 1703–1710, 2015.
77. B. Lanfer, M. Scherg, M. Dannhauer, T. Knsche, M. Burger, and C. Wolters, "Influences of skull segmentation inaccuracies on EEG source analysis," *NeuroImage*, vol. 62, no. 1, pp. 418–431, 2012.
78. S. Lew, D. D. Sliva, M. sun Choe, P. E. Grant, Y. Okada, C. H. Wolters, and M. S. Hmlinen, "Effects of sutures and fontanels on MEG and EEG source analysis in a realistic infant head model," *NeuroImage*, vol. 76, pp. 282–293, 2013.
79. R. Grech, T. Cassar, J. Muscat, K. P. Camilleri, S. G. Fabri, M. Zervakis, P. Xanthopoulos, V. Sakkalis, and B. Vanrumste, "Review on solving the inverse problem in EEG source analysis," *Journal of NeuroEngineering and Rehabilitation*, vol. 5, no. 1, p. 25, 2008.
80. K. Sekihara, S. S. Nagarajan, D. Poeppel, S. Miyauchi, N. Fujimaki, H. Koizumi, and Y. Miyashita, "Estimating neural sources from each time-frequency component of magnetoencephalographic data," *IEEE Transactions on Biomedical Engineering*, vol. 47, pp. 642–653, May 2000.
81. S. L. Jacques, "Optical properties of biological tissues: a review," *Physics in Medicine and Biology*, vol. 58, no. 11, p. R37, 2013.
82. V. V. Tuchin, "Light scattering study of tissues," *Physics-Uspekhi*, vol. 40, no. 5, pp. 495–515, 1997.
83. M. Dehaes, K. Kazemi, M. Pélégrini-Issac, R. Grebe, H. Benali, and F. Wallois, "Quantitative effect of the neonatal fontanel on synthetic near infrared spectroscopy measurements," *Human Brain Mapping*, vol. 34, no. 4, pp. 878–889, 2013.
84. M. Jäger and A. Kienle, "Non-invasive determination of the absorption coefficient of the brain from time-resolved reflectance using a neural network," *Physics in Medicine and Biology*, vol. 56, no. 11, p. N139, 2011.
85. F. T. Ulaby and U. Ravaioli, eds., *Fundamentals of Applied Electromagnetics*, 7th ed., Pearson, 2015.
86. A. Custo and D. A. Boas, "Comparison of diffusion and transport in human head," in *Biomedical Topical Meeting*, p. WF31, Optical Society of America, 2004.
87. J. Heiskala, T. Neuvonen, P. E. Grant, and I. Nissilä, "Significance of tissue anisotropy in optical tomography of the infant brain," *Applied Optics*, vol. 46, pp. 1633–1640, Apr 2007.
88. J. Heiskala, M. Pollari, M. Metsäranta, P. E. Grant, and I. Nissilä, "Probabilistic atlas can improve reconstruction from optical imaging of the neonatal brain," *Optics Express*, vol. 17, pp. 14977–14992, Aug. 2009.
89. A. J. Trevelyan, D. Sussillo, and R. Yuste, "Feedforward inhibition contributes to the control of epileptiform propagation speed," *Journal of Neuroscience*, vol. 27, no. 13, pp. 3383–3387, 2007.
90. A. J. Trevelyan, T. Baldeweg, W. van Drongelen, R. Yuste, and M. Whittington, "The source of after discharge activity in neocortical tonic–clonic epilepsy," *Journal of Neuroscience*, vol. 27, no. 49, pp. 13513–13519, 2007.
91. T. Asakawa, A. Muramatsu, T. Hayashi, T. Urata, M. Taya, and Y. Mizuno-Matsumoto, "Comparison of EEG propagation speeds under emotional stimuli on smartphone between the different anxiety states," *Frontiers in Human Neuroscience*, vol. 8, p. 1006, 2014.

92. L. Rankine, N. Stevenson, M. Mesbah, and B. Boashash, "A nonstationary model of newborn EEG," *IEEE Transactions on Biomedical Engineering*, vol. 54, no. 1, pp. 19–28, 2007.
93. S. Romero, M. A. Mañanas, and M. J. Barbanoj, "A comparative study of automatic techniques for ocular artifact reduction in spontaneous EEG signals based on clinical target variables: a simulation case," *Computers in Biology and Medicine*, vol. 38, no. 3, pp. 348–360, 2008.
94. D. Safieddine, A. Kachenoura, L. Albera, G. Birot, A. Karfoul, A. Pasnicu, A. Biraben, F. Wendling, L. Senhadji, and I. Merlet, "Removal of muscle artifact from EEG data: comparison between stochastic (ICA and CCA) and deterministic (EMD and wavelet-based) approaches," *EURASIP Journal on Advances in Signal Processing*, vol. 2012, no. 1, pp. 1–15, 2012.
95. M. T. Akhtar, W. Mitsuhashi, and C. J. James, "Employing spatially constrained ICA and wavelet denoising, for automatic removal of artifacts from multichannel EEG data," *Signal Processing*, vol. 92, no. 2, pp. 401–416, 2012.
96. M. De Vos, W. Deburchgraeve, P. Cherian, V. Matic, R. Swarte, P. Govaert, G. H. Visser, and S. Van Huffel, "Automated artifact removal as preprocessing refines neonatal seizure detection," *Clinical Neurophysiology*, vol. 122, no. 12, pp. 2345–2354, 2011.
97. A. Janardhan and K. K. Rao, "Application of signal separation algorithms for artifact removal from EEG signals," *International Journal of Modern Communication Technologies & Research*, vol. 3, Jan. 2015.
98. V. Matic, W. Deburchgraeve, and S. Van Huffel, "Comparison of ICA algorithms for ECG artifact removal from EEG signals," in *Proc. of the 4th Annual symposium of the IEEE-EMBS Benelux Chapter. (IEEE-EMBS)*, pp. 137–140, 2009.
99. M. S. Khlif, M. Mesbah, B. Boashash, and P. Colditz, "Influence of EEG artifacts on detecting neonatal seizure," in *Information Sciences Signal Processing and their Applications (ISSPA), 2010 10th International Conference on*, pp. 500–503, IEEE, 2010.
100. M. Zima, P. Tichavsky, K. Paul, and V. Krajča, "Robust removal of short-duration artifacts in long neonatal EEG recordings using wavelet-enhanced ICA and adaptive combining of tentative reconstructions," *Physiological Measurement*, vol. 33, no. 8, p. N39, 2012.
101. J. Brotchie, L. Rankine, M. Mesbah, P. Colditz, and B. Boashash, "Robust time-frequency analysis of newborn EEG seizure corrupted by impulsive artefacts," in *2007 29th Annual International Conference of the IEEE Engineering in Medicine and Biology Society*, pp. 11–14, IEEE, 2007.
102. Y. Liang and W. Chen, "A survey on computing Lévy stable distributions and a new MATLAB toolbox," *Signal Processing*, vol. 93, no. 1, pp. 242–251, 2013.
103. B. R. Greene, S. Faul, W. P. Marnane, G. Lightbody, I. Korotchikova, and G. B. Boylan, "A comparison of quantitative EEG features for neonatal seizure detection," *Clinical Neurophysiology*, vol. 119, no. 6, pp. 1248–1261, 2008.
104. R. G. Baraniuk and D. L. Jones, "Signal-dependent time-frequency analysis using a radially Gaussian kernel," *Signal Processing*, vol. 32, no. 3, pp. 263–284, 1993.
105. C. J. Burges, "A tutorial on support vector machines for pattern recognition," *Data Mining and Knowledge Discovery*, vol. 2, no. 2, pp. 121–167, 1998.
106. C.-W. Hsu and C.-J. Lin, "A comparison of methods for multiclass support vector machines," *IEEE Transactions on Neural Networks*, vol. 13, no. 2, pp. 415–425, 2002.
107. M. A. Awal, S. Ouelha, S. Dong, and B. Boashash, "Robust time–frequency representation based on the local optimization of the short-time fractional Fourier transform," *Digital Signal Processing*, vol. 70, pp. 125–144, Nov. 2017.

6

Active Recursive Bayesian State Estimation for Big Biological Data*

Mohammad Moghadamfalahi, Murat Akcakaya, and Deniz Erdogmus

CONTENTS

6.1 Introduction 115
6.2 Active Recursive Bayesian State Estimation 116
    6.2.1 Active Learning for RBSE 118
    6.2.2 Submodular-Monotone Set Functions for Set Optimization Problems 119
    6.2.3 On the Objective Functions for Query Optimization 120
        6.2.3.1 A Solution for the Proposed Objective Function 121
    6.2.4 Combinatorial Optimization 122
6.3 Illustrative BCI Design Example 123
    6.3.1 ERP-Based BCI for Letter-by-Letter Typing 124
        6.3.1.1 Presentation Component 124
    6.3.2 Decision-Making Component 125
        6.3.2.1 EEG Feature Extraction and Classification 125
        6.3.2.2 Language Model 126
    6.3.3 Experimental Results and Discussions 126
        6.3.3.1 RSVP Paradigm 127
        6.3.3.2 Matrix-Based Presentation Paradigm with Overlapping Trials 127
        6.3.3.3 Matrix-Based Presentation Paradigm with Single-Character Trials ...129
6.4 Open Problems and Future Directions 129
References 130

6.1 Introduction

The spring of technology and availability of large electronic storage have presented scientists with the benefits and challenges of large volumes of data. In the field of life sciences, data about genome, transcriptome, epigenome, proteome, metabolome, molecular imaging, molecular pathways, different populations of people, and clinical/medical records has already been stored in volumes of petabytes and exabytes. This data is not only large but complex and dynamic. These characteristics well fit the definitions of *big data* by TechAmerica Foundation:

"Big data is a term that describes large volumes of high velocity, complex and variable data that require advanced techniques and technologies to enable the capture, storage, distribution, management, and analysis of the information" [1].

Most traditional machine learning techniques that are well grounded on clear mathematical definition and assumptions are cost inefficient for analyzing big data. Hence, more than ever, we need some tools that can handle all aspects of big data, namely, large volume, high velocity, and variability. These tools when applied on biological data would offer rich information and good understanding of biological dynamics and biomedical basics, extracted from massive amounts of raw data that can be transferred and used in developing personalized medicine and inference from limited or noisy measurements. Moreover, this information can be used to administer data collection when data velocity is high and analysis time and computational capacity are limited.

Interactive machine learning applications with high velocity of data flow are rapidly becoming ubiquitous with increased demand for personalized computational solutions in both social and professional settings.

---

* This work was supported by NSF CNS-1136027, IIS1149570; NIH 2R01DC009834-06A1; and NIDRR H133E140026.

For instance, personalized music or movie recommendations by services such as Pandora, Spotify, and Netflix; personalized advertising by social and professional networking services such as Facebook and LinkedIn; and personalized health care at multiple spatial scales (e.g., monitoring daily activity and health records at a large scale or personalized gene therapies at a small scale) are all high-impact domains where interactive machine learning and traditional active learning (AL) promise to deliver solid theoretical frameworks for highly desirable solutions.

AL is a subfield of machine learning that attempts to determine a series of queries to be answered by an oracle (labeler) in order to learn a classification or regression model. Significant advances have been made in the AL area with the primary focus having been on efficiently learning stationary model parameters with one or more ideal or noisy labelers [2]. Existing work has largely been applied assuming the availability of the entire data set of interest, using a stationary data model for efficient model learning [3–12]. In contrast, most human–computer interaction settings are dynamic and online, where latent states and observations are dependent on the users' intent and history, implying the need for a time-recursive approach. Note that AL can help us to significantly reduce the volume of noninformative and redundant data recording.

In this chapter, we present a state estimation framework that can use previously obtained information from a large volume of data to administer high-velocity data collection and inference in real time, and is able to adapt to variability presented in time by the source of data. In Section 6.2, we introduce the theoretical basis for our framework, *active recursive Bayesian state estimation (Active-RBSE)*. Next we explain the application of Active-RBSE in the brain/body computer interface (BBCI) as a toy example in Section 6.3. And we conclude this chapter in Section 6.4.

6.2 Active Recursive Bayesian State Estimation

We represent our fusion and joint inference architecture as a hidden Markov model of order n (HMM-n). In this dynamic system setup, recursive Bayesian state estimation (RBSE) is used to extract information about parameters, or states, of the system in real time given the noisy measurements of the system output.

Through this section, we build a probabilistic graphical model (PGM) of our system in four steps. At each step, based on the structure of the problem, we impose a set of assumptions on HMM-n that make the model more restricted and specific to Active-RBSE. First we

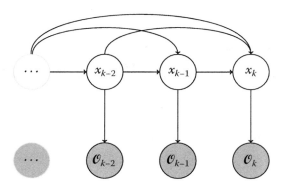

FIGURE 6.1
Hidden Markov model of order n (HMM-n).

start with the abstract PGM of the proposed HMM-n as shown in Figure 6.1.

In this figure, x_k represents the system state at time k, and $\mathcal{O}_k$ is the system output measurement. Here, we assume that system state belongs to a finite discrete space while we let the measurement space be continuous. $\mathcal{O}_k$ is multimodal evidences, such that $\mathcal{O}_k = \{O_k^1, O_k^2, \cdots, O_k^m\}$ where m represents the number of measurement modalities. The evidence that we consider here can be divided into two types:

1. **Type I, abbreviated as T-I throughout this chapter,** is the set of measurements that are generated and received by the system, and we have no control to guide the data flow. Two examples of this type of evidence are

   - Heart rate of a patient in an intensive care unit
   - Volitional cortical potentials (VCPs) in electroencephalography (EEG) signal

2. **Type II, abbreviated as T-II throughout this chapter,** is the set of measurements that are generated in response to a set of stimuli (or questions) presented to the user. Two examples for this type of evidence are

   - Mouse click in an interactive image segmentation scenario
   - Event-related potentials (ERPs) in EEG

Consequently, system output measurements can be partitioned into two sets of $\mathcal{O}_{1,k} = \{O_{1,k}^1, O_{1,k}^2, \cdots, O_{1,k}^{m_1}\}$ for T-I and $\mathcal{O}_{2,k} = \{O_{2,k}^1, O_{2,k}^2, \cdots, O_{2,k}^{m_2}\}$ for T-II evidences, such that $\mathcal{O}_k = \mathcal{O}_{1,k} \cup \mathcal{O}_{2,k}$, thus $m = m_1 + m_2$. Next we describe our assumptions on the probabilistic relationships among different measurements originating from various modalities. Accordingly, the abstract graphical model presented in Figure 6.1 is detailed in the PGM as illustrated in Figure 6.2 to represent our assumptions on the interdependency among observations.

Active Recursive Bayesian State Estimation for Big Biological Data

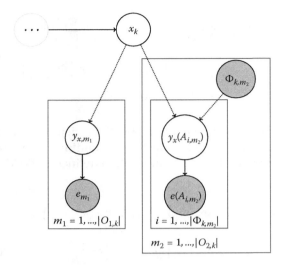

FIGURE 6.2
Probabilistic graphical model of the system during inference cycle k.

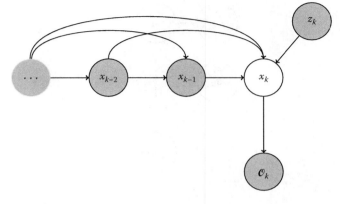

FIGURE 6.3
HMM-n while prior states are observed.

In this chapter, using the assumptions imposed by the graphical model in Figure 6.2, we employ a maximum a posteriori (MAP) inference method to estimate the system state. To compute the posterior probability mass function (PMF) over the state space, we use the Bayes rule of *posterior* $\propto$ *prior* $\times$ *likelihood*. But according to the PGM shown in Figure 6.2, for a given state value, likelihoods corresponding to different modalities can be calculated, up to a normalization factor, independently from each other. Hence for the rest of the chapter, we will focus on estimating the posterior from one type of evidence as other likelihoods can be calculated likewise and be fused with each other easily.

The PGMs illustrated in Figures 6.1 and 6.2 correspond to real-time casual systems, and they are designed to infer system state and execute certain tasks when a confidence threshold is attained. In this chapter, we refer to the time window in which the system reaches a confident decision as an *epoch*. In this setup, when the output measurements up to epoch k are observed, the goal is to estimate the current system state. Note that this setup represents a dynamic system in which the state dynamics might change during the operation of the system. For instance, in a letter-by-letter typing brain–computer interface, the state at epoch k represents only a character, while the user needs to type a sequence of characters to form words and sentences and eventually communicate the desired massage. As a result, it seems inevitable to adaptively update the (probabilistic or deterministic) system belief about the state space throughout time. Upon state estimation, the system will take an action, and hence, the system state at all past epochs can be assumed as observed. In most applications, state value at epoch k is a function of previously estimated system states, and the oracle* in the loop acts as a controller of this closed-loop system to perform a task. Moreover, the system state can be affected based on some external factors z_k. Then the high-level graphical model in Figure 6.1 can be updated as in Figure 6.3.

In the remainder of this chapter, we will focus on the state inference mechanism from T-II-based measurement outputs. In the presence of noisy measurements, the system queries the user with a set of questions to achieve a more confident estimation. These queries may contain sets of state values presented to the oracle, and it responds to these questions with a yes/no answer through some voluntarily or involuntarily generated physiological evidence. Let us define $C_k = \{x_{k-n}, \ldots, x_{k-1}, z_k\}$ and the query set as Φ_k; then we update the PGM as in Figure 6.4. Throughout this chapter, we will refer to C_k as the context information.

In Figure 6.4, A_i is a subset of state space $\mathcal{V}$, which we call a *trial*, the $y_x(A_i) \in \{0,1 \mid 1 = \text{yes}, 0 = \text{no}\}$, and $e(A_i)$ represents the physiological measurement in response to $A_{i,j}$, the jth element of the subset A_i. But to keep the framework general, we assume measurements are noisy, and the system might need to query the agent with multiple *sequences* of trials to obtain a confident estimation.

In this setup, the system queries the oracle iteratively and updates the state space posterior PMF until the probability of the most likely state value reaches a predefined confidence threshold. The updated PGM that allows for multiple sequences is shown in Figure 6.5.

In this model (see Figure 6.5), we have set an upper bound on the number of sequences m_s within each epoch to discard the possibility of extremely long decision cycles. This setup allows us to update the posterior PMF

---

* The oracle can be a human user or an automatic controller that is aimed at reducing estimation error.

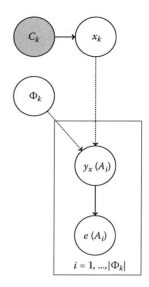

FIGURE 6.4
PGM of the system at epoch k.

recursively, after every sequence. Assume $1 \leq s \leq m_s$ sequences have been shown to the agent and define $\mathcal{E}^s = \{E^i\}_{i=1}^s$, where $E^i = \{e(A^i_j)|j = 1, \cdots, |\Phi^i_k|\}$. Similarly, take $Y^i_{x_k} = \{y_{x_k}(A^i_j)|j = 1, \cdots, |\Phi^i_k|\}$, then define $\mathcal{Y}^s_{x_k} = \{Y^i_{x_k}\}_{i=1}^s$. The MAP framework estimates the system state by solving the following optimization problem:

$$\hat{x}_k = \arg\max_x P\left(x_k = x | \mathcal{E}^s, C, \{\Phi^i_k\}_{i=1}^s\right). \quad (6.1)$$

The posterior probability defined in Equation 6.1 can be factorized in terms of likelihood and context prior using the assumptions imposed in Figure 6.5. We have

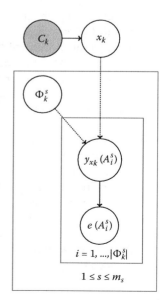

FIGURE 6.5
HMM-n while prior states are observed.

$$P(x_k = x | \mathcal{E}^s, C, \{\Phi^i_k\}_{i=1}^s) = \frac{p(x_k = x, \mathcal{E}^s | C, \{\Phi^i_k\}_{i=1}^s)}{p(\mathcal{E}^s | C, \{\Phi^i_k\}_{i=1}^s)}$$
$$\propto p\left(\mathcal{E}^s | x_k = x, \{\Phi^i_k\}_{i=1}^s\right). \quad (6.2)$$
$$P(x_k = x | C),$$

but for a given x_k, the $\mathcal{Y}^s_{x_k}$ for $\{\Phi^i_k\}_{i=1}^s$ is deterministically defined. Hence, according to the conditional independence of $\mathcal{E}^s$ and context information defined in PGM, we obtain

$$p\left(\mathcal{E}^s | x_k = x; \{\Phi^i_k\}_{i=1}^s\right) = p\left(\mathcal{E}^s | \mathcal{Y}^s_{x_k}; \{\Phi^i_k\}_{i=1}^s\right) =$$
$$\prod_{\substack{i=1,\ldots,s \\ j=1,\ldots,|\Phi^i_k|}} p\left(e(A^i_j) | y_{x_k}(A^i_j); \{\Phi^i_k\}_{i=1}^s\right). \quad (6.3)$$

Then we can rewrite Equation 6.1 as

$$P\left(x_k = x | \mathcal{E}^s, C, \{\Phi^i_k\}_{i=1}^s\right) \propto$$
$$\prod_{\substack{i=1,\ldots,s \\ j=1,\ldots,|\Phi^i_k|}} p\left(e(A^i_j) | y_{x_k}(A^i_j); \{\Phi^i_k\}_{i=1}^s\right) \cdot P(x_k = x | C) \propto$$
$$\prod_{\substack{i=1,\ldots,s \\ \{j | y_{x_k}(A^i_j) = 1\}}} \frac{p\left(e(A^i_j) | y_{x_k}(A^i_j) = 1; \{\Phi^i_k\}_{i=1}^s\right)}{p\left(e(A^i_j) | y_{x_k}(A^i_j) = 0; \{\Phi^i_k\}_{i=1}^s\right)} \cdot P(x_k = x | C).$$
$$(6.4)$$

6.2.1 Active Learning for RBSE

T-II-based measurements for the proposed system are obtained in response to a set of labeling questions queried from the oracle. For a small set of actions, it is possible to query the agent for the label of all possible state values at every sequence. However, if the state space is large, depending on the level of abstraction offered by questions, this can potentially lead to long sequences with a large amount of noninformative multivariate measurements to be processed. In addition, obtaining expert (i.e., oracle) answers to these questions can be very expensive in terms of time and cognitive frustration especially when the oracle is a human. To mitigate these problems, one can propose to select a random subset of state space to be presented to the agent. But instead, we propose to use AL, to intelligently select samples for annotation that enables efficiently learning an accurate posterior with as few questions as possible. Here, the implicit assumptions are that the labeling costs (in terms of time or cognitive load) are the same for all of the queries and also significantly larger than the computational cost of the querying algorithms. The latter assumption leads us toward AL algorithms that generate a suboptimal batch of queries, even though they might not be the optimal solutions of defined objective functions. Later in this chapter, through experimental results of a toy example, we show that this mechanism outperforms the cheap passive learning with random queries.

Accordingly, we define the Active-RBSE for inference and query optimization. Within this framework, a generic AL and MAP inference loop will iterate by alternating between the following two steps:

Query: $\widehat{\Phi}_k^{s+1} = \arg\max_{\Phi_k^{s+1}} g(\Phi_k^{s+1})$ s.t $\Phi_k^{s+1} \in \mathcal{F}_k \subseteq 2^\mathcal{V}$. (6.5)

Inference: $\hat{x}_k = \arg\max_x P\left(x_k = x | \mathcal{E}^s, C, \left\{\Phi_k^i\right\}_{i=1}^s\right)$. (6.6)

Here, Φ_k^{s+1} is a potential query set restricted to the set of feasible queries at time k, $\mathcal{F}_k$, which is a subset of all possible queries, $2^\mathcal{V}$, the power set of $\mathcal{V}$. The quality of a query from the perspective of AL is measured by the set function g.

An AL setting typically starts with an initial model (which we obtain from the context information); then samples are selected for label querying. Performing AL in batch mode (sequence by sequence) introduces new challenges. Since we need to select a *set* of queries, one should also make sure that the samples are nonredundant to maximize the amount of information that they provide. Another related challenge is that optimally selecting a subset of samples based on a given objective function defined over sets is, in general, an NP-hard combinatorial optimization problem, and it can easily lead to intractable solutions.

6.2.2 Submodular-Monotone Set Functions for Set Optimization Problems

Submodular set functions offer various mathematical properties that can be exploited to define tractable solutions in combinatorial optimization problems. Submodular set functions are discrete analogs of concave or convex real-valued functions [13]. Next we introduce certain definitions and theorems about the submodular functions.

Definition 1 (discrete derivative)

Assume a set function $f: 2^\mathcal{W} \to \mathbb{R}$, $B \subseteq \mathcal{W}$, and $w \in \mathcal{W}$, then $\Delta_f(w|B) := f(B \cup \{w\}) - f(B)$ is discrete derivative of f at B with respect to w.

Now we can define a *submodular* function as follows.

Definition 2 (submodular set function)

A function $f: 2^\mathcal{W} \to \mathbb{R}$ is submodular if for every $B_1 \subseteq B_2 \subseteq \mathcal{W}$ and $w \in \mathcal{W} \setminus B_2$,

$$\Delta(w|B_1) \geq \Delta(w|B_2)$$

or equivalently, the function: $2^\mathcal{W} \to \mathbb{R}$ is submodular if for every $B_1, B_2 \subseteq \mathcal{W}$

$$f(B_1 \cap B_2) + f(B_1 \cup B_2) \leq f(B_1) + f(B_2).$$

In particular, one can use a greedy forward algorithm to find a solution within a guaranteed bound around the global optimum when the objective is a monotone submodular set function [14]. To provide the proof, first we need to define *monotone* set functions.

Definition 3 (monotone set function)

A set function $f: 2^\mathcal{W} \to \mathbb{R}$ is monotone if for every $B_1 \subseteq B_2 \subseteq \mathcal{W}$, we get $f(B_1) \leq f(B_2)$.

In a maximization problem, the greedy forward algorithm starts with $B_0 = \emptyset$, and iteratively adds the elements that maximize the discrete derivative of the function at the set from prior iteration, with respect to that element. Accordingly, the subproblem for iteration i is

$$B_i = B_{i-1} \cup \left\{\arg\max_w \Delta(w|B_{i-1})\right\}. \quad (6.7)$$

Theorem 1 [13]

Assume a nonnegative monotone submodular set function $f: 2^\mathcal{W} \to \mathbb{R}_+$. Also define $\{B_i\}_{i \geq 0}$ to be the greedily selected sets according to Equation 6.7. Then for all positive integers k and l, we have

$$f(B_l) \geq \left(1 - e^{\frac{-l}{k}}\right) \max_{B: |B| \leq k} f(B).$$

Proof. Fix k and l and get $B^* \in \arg\max\{f(B) : |B| \leq k\}$ as an optimal set with $|B^*| \leq k$. Since the function f is a monotone set function, we can assume $|B^*| = k$ without loss of generality and define $B^* = \{w_1^*, w_2^*, \cdots, w_k^*\}$. Then for all $i \leq l$

$$f(B^*) \leq f(B^* \cup B_i)$$

$$= f(B_i) + \sum_{j=1}^k \Delta\left(w_j^* | B_i \cup \{w_1^*, \cdots, w_{j-1}^*\}\right)$$

$$\leq f(B_i) + \sum_{w^* \in B^*} \Delta(w^* | B_i)$$

$$\leq f(B_i) + \sum_{w^* \in B^*} (f(B_{i+1}) - f(B_i))$$

$$\leq f(B_i) + k(f(B_{i+1}) - f(B_i)).$$

Hence we have
$$f(B^*) - f(B_i) \le k(f(B_{i+1} - B_i)).$$

Now define $s_i = f(B^*) - f(B_i)$; then we get
$$s_i \le (s_i - s_{i+1}) \Rightarrow s_{i+1} \le \left(1 - \frac{1}{k}\right)s_i \Rightarrow s_l \le \left(1 - \frac{1}{k}\right)^l s_0.$$

We know that $s_0 = f(B^*) - f(\emptyset)$ since f is nonnegative by assumption. Consequently, by use of the well-known inequality $1 - x \le e^{-x}$, $\forall x \in \mathbb{R}$, we get
$$s_l \le \left(1 - \frac{1}{k}\right)^l s_0 \le e^{\frac{-l}{k}} f(B^*) \Rightarrow f(B^*) - f(B_l) \le e^{\frac{-l}{k}} f(B^*)$$
$$\Rightarrow f(B_l) \ge \left(1 - e^{\frac{-l}{k}}\right) f(B^*).$$
□

More interestingly, for a modular-monotone objective function, the greedy forward algorithm leads to the global optimum solution. The proof of this proposition follows easily from the following definition as the contribution of each element does not depend on the set size.

Definition 4 (modular set function)

A function $f: 2^{\mathcal{W}} \to \mathbb{R}$ is modular if for every $B_1 \subseteq B_2 \subseteq \mathcal{W}$ and $w \in \mathcal{W} \setminus B_2$,
$$\Delta(w|B_1) = \Delta(w|B_2),$$
or equivalently, the function $f: 2^{\mathcal{W}} \to \mathbb{R}$ is modular if for every $B_1, B_2 \subseteq \mathcal{W}$,
$$f(B_1 \cap B_2) + f(B_1 \cup B_2) = f(B_1) + f(B_2).$$

6.2.3 On the Objective Functions for Query Optimization

System parameter/state learning can be more efficient if we can query the oracle, to obtain the labels of state values that convey the most salient information. Such querying can be achieved through careful selection of objective functions in the Active-RBSE inference framework. It is important to note here that for efficient solution of the subset selection through a greedy forward optimization in an online setting, in addition to being informative about the state estimation, either the objective functions need to be monotone-modular set functions or they should be upper- or lower-bounded by such set functions.

Here, we consider $g(.)$ to be used in the Active-RBSE framework, as specified in Equation 6.5. Let us assume that the actual state value for the current epoch (i.e., epoch k) is given as x_k^*, and s sequences of questions have already been presented to the oracle. The goal is to optimize the query set for sequence $s + 1$ with the assumption that s prior sequences have not led to a confident decision. Then although the measurements for that sequence are not observed yet, one can estimate a prediction of x_k^* posterior probability by introducing and marginalizing the random variable for measurements, when Φ_k^{s+1} is given. We define a function $g: \mathcal{V}, 2^{2^{\mathcal{V}}} \to \mathbb{R}$ as

$$\begin{aligned}
g(\mathbf{x},\Phi_k^{s+1}) &= P(x_k = \mathbf{x}|\mathcal{E}^s, C, \{\Phi_k^i\}_{i=1}^{s+1}, x_k^* = \mathbf{x}) \\
&= \int_{\tilde{E}^{s+1}} P(x_k = \mathbf{x}, \tilde{E}^{s+1}|\mathcal{E}^s, C, \{\Phi_k^i\}_{i=1}^{s+1}, x_k^* = \mathbf{x}) d(\tilde{E}^{s+1}) \\
&= \mathbb{E}_{\tilde{E}^{s+1}|\Phi_k^{s+1}, x_k^*}\left[P\left(x_k = \mathbf{x}|\tilde{E}^{s+1}, \mathcal{E}^s, C, \{\Phi_k^i\}_{i=1}^{s}, x_k^*\right)\right] \\
&= \mathbb{E}_{\tilde{E}^{s+1}|\Phi_k^{s+1}, x_k^*}\left[\frac{\Pi^{s+1}(\mathbf{x}) \cdot p(\tilde{E}^{s+1}|x_k = \mathbf{x}, \Phi_k^{s+1})}{\sum_{\mathbf{v}\in\mathcal{V}} \Pi^{s+1}(\mathbf{v}) \cdot p(\tilde{E}^{s+1}|x_k = \mathbf{v}, \Phi_k^{s+1})}\right],
\end{aligned}$$
(6.8)

where $\Pi^{s+1}(\mathbf{x}) = P(x_k = \mathbf{x}|\mathcal{E}^s, C, \{\Phi_k^i\}_{i=1}^{s})$ represents the prior probability of $\mathbf{x} \in \mathcal{V}$ before observing sequence $s + 1$. Moving from the third line to the fourth of Equation 6.8, we use

$$P\left(x_k = \mathbf{x}|\tilde{E}^{s+1}, \mathcal{E}^s, C, \{\Phi_k^i\}_{i=1}^{s+1}\right) = \frac{\Pi^{s+1}(\mathbf{x}) \cdot p(\tilde{E}^{s+1}|x_k = \mathbf{x}, \Phi_k^{s+1})}{\sum_{\mathbf{v}\in\mathcal{V}} \Pi^{s+1}(\mathbf{v}) \cdot p(\tilde{E}^{s+1}|x_k = \mathbf{v}, \Phi_k^{s+1})}, \quad (6.9)$$

for which the denominator is the normalization constant.

Note that $g(\mathbf{x},\Phi_k^{s+1})$ computes the posterior probability of a hypothesized target for a particular Φ_k^{s+1} given previously observed measurements and context information. But note that during the current epoch, x is yet to be estimated; hence, it is not known. Consequently, we can marginalize out the dependency on this unobserved random variable by computing the expected value of $g(\mathbf{x},\Phi_k^{s+1})$ with respect to the most recent estimate of state space posterior PMF, $\Pi^{s+1}(\mathbf{x})$.

Accordingly, the objective function for query set selection is then defined as

$$\widehat{\Phi}_k^{s+1} = \arg\max_{\Phi_k^{s+1}} \mathbb{E}_{\Pi^{s+1}(\mathbf{x})}\left[g(\mathbf{x},\Phi_k^{s+1})\right]. \quad (6.10)$$

The function defined in Equation 6.10 is the expected value of the predicted target posterior probability with respect to current probability distribution over the state space that was obtained through the evidence gained until the sequence $s + 1$. Optimizing this function in terms of Φ_k^{s+1} exploits our current knowledge to minimize the uncenrtanity about the targeted but unknown state x_k.

6.2.3.1 A Solution for the Proposed Objective Function

Considering the proposed graphical model and as we computed the equations for inference (see Equation 6.4), in Equation 6.9, we can define

$$p\left(\tilde{E}^{s+1}|x_k = x, \Phi_k^{s+1}\right)$$
$$\propto \prod_{\{j|y_{x_k}(A_j^i)=1\}} \frac{p\left(e\left(A_j^{s+1}\right)\Big|y_{x_k}\left(A_j^{s+1}\right) = 1; \Phi_k^{s+1}\right)}{p\left(e\left(A_j^{s+1}\right)\Big|y_{x_k}\left(A_j^{s+1}\right) = 0; \Phi_k^{s+1}\right)}. \quad (6.11)$$

In the function presented in Equation 6.8, the argument inside the expectation is only a function of $\tilde{E}^{s+1}$ when x and Φ_k^{s+1} are fixed. Hence, we define $\tilde{\sigma} = [\tilde{\sigma}(A_1^{s+1}), \cdots, \tilde{\sigma}(A_{|\Phi_k^{s+1}|}^{s+1})]$, where

$$\tilde{\sigma}\left(A_j^i\right) = \frac{p\left(\tilde{e}\left(A_j^i\right)\Big|1\right)}{p\left(\tilde{e}\left(A_j^i\right)\Big|0\right)}.$$

Then we define a new function $\mathcal{F}: \mathbb{R}^{|\Phi_k^{s+1}|} \to \mathbb{R}$ using Equations 6.4 and 6.11 as

$$\mathcal{F}(\tilde{\sigma}) = \frac{\Pi^{s+1}(x) \cdot \prod_{\{j|y_x(A_j^{s+1})=1\}} \tilde{\sigma}\left(A_j^{s+1}\right)}{\sum_{v \in \mathcal{V}} \Pi^{s+1}(v) \cdot \prod_{\{j|y_v(A_j^{s+1})=1\}} \tilde{\sigma}\left(A_j^{s+1}\right)}. \quad (6.12)$$

To obtain a practical algorithm for processing the big data, we need a time-efficient optimization mechanism. For that, we simplify the problem by approximating the $g(x, \Phi_k^{s+1})$ using the Taylor series expansion of the function defined in Equation 6.12. Namely,

$$g\left(x, \Phi_k^{s+1}\right) = \mathbb{E}_{\tilde{E}^{s+1}|\Phi_k^{s+1}, x_k^*}[\mathcal{F}(\tilde{\sigma})] =$$
$$\mathbb{E}_{\tilde{E}^{s+1}|\Phi_k^{s+1}, x_k^*}[\mathcal{F}(\mu_\sigma) + (\tilde{\sigma} - \mu_\sigma) \cdot \nabla \mathcal{F}(\mu_\sigma) + \cdots]. \quad (6.13)$$

In Equation 6.13, $\mu_\sigma = \mathbb{E}_{\tilde{E}^{s+1}|\Phi_k^{s+1}, x_k^*}[\tilde{\sigma}]$. Now we use Equation 6.13 to define a substitute objective function as in Equation 6.14, which is the locally suboptimal linear approximation around the μ_σ of the original objective function. This type of approximation is commonly used in the field of signal processing [15], especially for distributions with negligible higher-order central moments. This is an important assumption that needs to be considered when this particular solution is used.

$$g\left(x, \Phi_k^{s+1}\right) \approx \hat{g}\left(x, \Phi_k^{s+1}\right) =$$
$$\mathcal{F}(\mu_\sigma) + \mathbb{E}_{\tilde{E}^{s+1}|\Phi_k^{s+1}, x_k^*}[(\tilde{\sigma} - \mu_\sigma)] \cdot \nabla \mathcal{F}(\mu_\sigma) = \mathcal{F}(\mu_\sigma) \quad (6.14)$$

The Taylor expansion in Equation 6.13 is done around $\mu_\sigma = \mathbb{E}_{\tilde{E}^{s+1}|\Phi_k^{s+1}, x_k^*}[\tilde{\sigma}]$; hence the second term in Equation 6.14 is zero. Now we need to compute $\mu_\sigma = \mathbb{E}_{\tilde{E}^{s+1}|\Phi_k^{s+1}, x_k^*}[\tilde{\sigma}]$. Here we use the conditional independence of trials as presented in the proposed graphical model in Figure 6.5, for a given $y_{x_k^*}^{s+1}$. This means that $\tilde{\sigma}(A_i^{s+1})$ is independent from $\tilde{\sigma}(A_j^{s+1})$ $\forall i,j = 1, \ldots, |\Phi_k^{s+1}|$ for all $i \neq j$. Note that, $\tilde{\sigma}(A_j^{s+1})$ is calculated at samples drawn from

$$\begin{cases} \tilde{e}\left(A_j^{s+1}\right) \sim p(e(.)|1), & \text{if } x_k^* \in A_j^{s+1} \\ \tilde{e}\left(A_j^{s+1}\right) \sim p(e(.)|0), & \text{if } x_k^* \notin A_j^{s+1} \end{cases}$$

distributions.

Accordingly, we define $\mu_\sigma = [\hat{\sigma}(A_1^{s+1}), \cdots, \hat{\sigma}(A_{|\Phi_k^{s+1}|}^{s+1})]$, such that

$$\hat{\sigma}\left(A_j^{s+1}\right) = \begin{cases} \widehat{\sigma^+} = \mathbb{E}_{e(.)|1}\left[\frac{p\left(e\left(A_j^{s+1}\right)\big|1\right)}{p\left(e\left(A_j^{s+1}\right)\big|0\right)}\right], & \text{if } x_k^* \in A_j^{s+1} \\[2em] \widehat{\sigma^-} = \mathbb{E}_{e(.)|0}\left[\frac{p\left(e\left(A_j^{s+1}\right)\big|1\right)}{p\left(e\left(A_j^{s+1}\right)\big|0\right)}\right], & \text{if } x_k^* \notin A_j^{s+1} \end{cases}$$

$$(6.15)$$

Then $\hat{g}(x, \Phi_k^{s+1})^*$ can be defined as

$$\hat{g}\left(x, \Phi_k^{s+1}\right) = \frac{\Pi^{s+1}(x) \cdot \widehat{\sigma^+}^{c_{x,x}^+(\Phi_k^{s+1})}}{\sum_{v \in \mathcal{V}} \Pi^{s+1}(v) \cdot \widehat{\sigma^+}^{c_{x,v}^+(\Phi_k^{s+1})} \cdot \widehat{\sigma^-}^{c_{x,v}^-(\Phi_k^{s+1})}}, \quad (6.16)$$

where

$$c_{x,v}^+\left(\Phi_k^{s+1}\right) = \sum_{j=1}^{|\Phi_k^{s+1}|} y_x\left(A_j^{s+1}\right) \cdot y_v\left(A_j^{s+1}\right) \quad \text{and}$$

$$c_{x,v}^-\left(\Phi_k^{s+1}\right) = \sum_{j=1}^{|\Phi_k^{s+1}|} \left(1 - y_x\left(A_j^{s+1}\right)\right) \cdot y_v\left(A_j^{s+1}\right)$$

for $y_x(A_j^{s+1}) \in \{0,1\}$ (i.e., target and nontarget classes). Then we can use the approximate objective function $\hat{g}(x, \Phi_k^{s+1})$ and redefine the optimization problem as

$$\widehat{\Phi}_k^{s+1} = \arg\max_{\Phi_k^{s+1}} \mathbb{E}_{\Pi^{s+1}(x)}\left[\hat{g}\left(x, \Phi_k^{s+1}\right)\right]$$
$$= \arg\max_{\Phi_k^{s+1}} \log\left(\mathbb{E}_{\Pi^{s+1}(x)}\left[\hat{g}\left(x, \Phi_k^{s+1}\right)\right]\right). \quad (6.17)$$

Here we propose to optimize the logarithm of the objective function, as the solution does not change due to this monotonically increasing transformation. To solve the problem defined in Equation 6.17, we use Jensen's inequality to define a lower bound of the objective function, namely,

---

* Note that the approximation in Equation 6.14 also corresponds to defining a point estimate of the evidence scores by calculating their mean value as computed in Equation 6.15.

$$\log\left(\mathbb{E}_{\Pi^{s+1}(x)}\left[\hat{g}(\mathbf{x},\Phi_k^{s+1})\right]\right) \geq \mathbb{E}_{\Pi^{s+1}(x)}\left[\log(\hat{g}(\mathbf{x},\Phi_k^{s+1}))\right] =$$

$$\mathbb{E}_{\Pi^{s+1}(x)}\left[\log\left(\frac{\Pi^{s+1}(\mathbf{x}) \cdot \widehat{\sigma^+}^{c_{\mathbf{x},\mathbf{x}}^+(\Phi_k^{s+1})}}{\sum_{v \in \mathcal{V}} \Pi^{s+1}(v) \cdot \widehat{\sigma^+}^{c_{\mathbf{x},v}^+(\Phi_k^{s+1})} \cdot \widehat{\sigma^-}^{c_{\mathbf{x},v}^-(\Phi_k^{s+1})}}\right)\right]$$

$$= \mathbb{E}_{\Pi^{s+1}(x)}\left[\log(\Pi^{s+1}(\mathbf{x})) + c_{\mathbf{x},\mathbf{x}}^+(\Phi_k^{s+1})\log(\widehat{\sigma^+})\right] -$$

$$\mathbb{E}_{\Pi^{s+1}(x)}\left[\log\left(\sum_{v \in \mathcal{V}} \Pi^{s+1}(v) \cdot \widehat{\sigma^+}^{c_{\mathbf{x},v}^+(\Phi_k^{s+1})} \cdot \widehat{\sigma^-}^{c_{\mathbf{x},v}^-(\Phi_k^{s+1})}\right)\right].$$

(6.18)

Here, we propose that the class-conditional probability density functions (PDFs) are well separated to assume, $\widehat{\sigma^+} > 1$ and $\widehat{\sigma^-} < 1$. Note that if the optimal threshold for discrimination between two classes is not centered around zero, then one can always use kernel density estimation (KDE) or a shift of variable to make this assumption reasonable (for more on this, please see Section 6.3). In this example we proposed an upper bound $|\Phi_k^{s+1}| \leq m_t$ as the limit on the number of sequences in each epoch to prevent extremely long state estimation cycles. We have

$$\widehat{\sigma^+}^{c_{\mathbf{x},v}^+(\Phi_k^{s+1})} \leq \left(\widehat{\sigma^+}\right)^{m_t} \text{ and } \widehat{\sigma^-}^{c_{\mathbf{x},v}^-(\Phi_k^{s+1})} \leq \widehat{\sigma^-}^0 = 1$$

This leads to $\widehat{\sigma^+}^{c_{\mathbf{x},v}^+(\Phi_k^{s+1})} \cdot \widehat{\sigma^-}^{c_{\mathbf{x},v}^-(\Phi_k^{s+1})} \leq (\widehat{\sigma^+})^{m_t}$. Finally,

$$\mathbb{E}_{\Pi^{s+1}(x)}\left[\log(\Pi^{s+1}(\mathbf{x})) + c_{\mathbf{x},\mathbf{x}}^+(\Phi_k^{s+1})\log(\widehat{\sigma^+})\right] -$$

$$\mathbb{E}_{\Pi^{s+1}(x)}\left[\log\left(\sum_{v \in \mathcal{V}} \Pi^{s+1}(v) \cdot \widehat{\sigma^+}^{c_{\mathbf{x},v}^+(\Phi_k^{s+1})} \cdot \widehat{\sigma^-}^{c_{\mathbf{x},v}^-(\Phi_k^{s+1})}\right)\right] \geq$$

$$\mathbb{E}_{\Pi^{s+1}(x)}\left[\log(\Pi^{s+1}(\mathbf{x})) + c_{\mathbf{x},\mathbf{x}}^+(\Phi_k^{s+1})\log(\widehat{\sigma^+})\right] -$$

$$\mathbb{E}_{\Pi^{s+1}(x)}\left[\log\left(\left(\widehat{\sigma^+}\right)^{m_t}\right)\right]. \quad (6.19)$$

Now we can exclude the terms that are independent of Φ_k^{s+1} in Equation 6.19, and use Equation 6.17 to define the optimization problem as

$$\widehat{\Phi}_k^{s+1} \approx \arg\max_{\Phi_k^{s+1}} Q = \mathbb{E}_{\Pi^{s+1}(x)}\left[c_{\mathbf{x},\mathbf{x}}^+(\Phi_k^{s+1})\log(\widehat{\sigma^+})\right].$$

(6.20)

Through this simplification, we maximize the lower bound on the original cost function, and we show next that such a simplification leads to a time-efficient solution to the optimization problem.

6.2.4 Combinatorial Optimization

The approximated objective function defined in Equation 6.20 is a modular and monotonic set function $Q: 2^{2^{\mathcal{V}}} \to \mathbb{R}$; therefore, the optimization defined in Equation 6.20 has guaranteed convergence properties [14]. Here, we prove that Q is a monotone modular set function.

Lemma 1

Take $\mathcal{W} = 2^{\mathcal{V}}$; then the function $Q: 2^{\mathcal{W}} \to \mathbb{R}$ as defined in Equation 6.20 is a modular set function.

Proof. Assume $\Phi_1 \subseteq \Phi_2 \subseteq 2^{\mathcal{V}}$ and $A \in 2^{\mathcal{V}} \setminus \Phi_2$; then

$$\Delta_Q(A|\Phi_1) =$$

$$\mathbb{E}_{\Pi^{s+1}(x)}\left[c_{\mathbf{x},\mathbf{x}}^+(\Phi_1 \cup \{A\})\log(\widehat{\sigma^+})\right] - \mathbb{E}_{\Pi^{s+1}(x)}\left[c_{\mathbf{x},\mathbf{x}}^+(\Phi_1)\log(\widehat{\sigma^+})\right] =$$

$$\mathbb{E}_{\Pi^{s+1}(x)}\left[c_{\mathbf{x},\mathbf{x}}^+(\Phi_1 \cup \{A\})\log(\widehat{\sigma^+}) - c_{\mathbf{x},\mathbf{x}}^+(\Phi_1)\log(\widehat{\sigma^+})\right].$$

Since $A \notin \Phi_1$, we use the definition of $c_{\mathbf{x},\mathbf{x}}^+(.)$ to write

$$c_{\mathbf{x},\mathbf{x}}^+(\Phi_1 \cup \{A\}) = c_{\mathbf{x},\mathbf{x}}^+(\Phi_1) + c_{\mathbf{x},\mathbf{x}}^+(\{A\}) \Rightarrow \Delta_Q(A|\Phi_1) =$$

$$\mathbb{E}_{\Pi^{s+1}(x)}\left[(c_{\mathbf{x},\mathbf{x}}^+(\Phi_1) + c_{\mathbf{x},\mathbf{x}}^+(\{A\}))\log(\widehat{\sigma^+}) - c_{\mathbf{x},\mathbf{x}}^+(\Phi_1)\log(\widehat{\sigma^+})\right] =$$

$$\mathbb{E}_{\Pi^{s+1}(x)}\left[c_{\mathbf{x},\mathbf{x}}^+(\{A\})\log(\widehat{\sigma^+})\right].$$

Similarly as $A \notin \Phi_2$, we have

$$\Delta_Q(A|\Phi_2) = \mathbb{E}_{\Pi^{s+1}(x)}\left[c_{\mathbf{x},\mathbf{x}}^+(\{A\})\log(\widehat{\sigma^+})\right] \Rightarrow$$

$$\Delta_Q(A|\Phi_1) = \Delta_Q(A|\Phi_2).$$

□

Lemma 2

Take $\mathcal{W} = 2^{\mathcal{V}}$; then the function $Q: 2^{\mathcal{W}} \to \mathbb{R}$ as defined in Equation 6.20 is a monotone set function.

Proof. Assume $\Phi_1 \subseteq \Phi_2 \subseteq 2^{\mathcal{V}}$, and define $\Phi_3 = \Phi_2 \setminus \Phi_1$; then $\Phi_3 \cup \Phi_1 = \Phi_2$, and we can write

$$Q(\Phi_2) = \mathbb{E}_{\Pi^{s+1}(x)}\left[c_{\mathbf{x},\mathbf{x}}^+(\Phi_1 \cup \Phi_3)\log(\widehat{\sigma^+})\right].$$

Moreover, $\Phi_3 \cap \Phi_1 = \emptyset$; then according to the definition of $c_{\mathbf{x},\mathbf{x}}^+(.)$, we have

$$Q(\Phi_2) = \mathbb{E}_{\Pi^{s+1}(x)}\left[(c_{\mathbf{x},\mathbf{x}}^+(\Phi_1) + c_{\mathbf{x},\mathbf{x}}^+(\Phi_3))\log(\widehat{\sigma^+})\right] =$$

$$\mathbb{E}_{\Pi^{s+1}(x)}\left[c_{\mathbf{x},\mathbf{x}}^+(\Phi_1)\log(\widehat{\sigma^+})\right] + \mathbb{E}_{\Pi^{s+1}(x)}\left[c_{\mathbf{x},\mathbf{x}}^+(\Phi_3)\log(\widehat{\sigma^+})\right] =$$

$$Q(\Phi_1) + \mathbb{E}_{\Pi^{s+1}(x)}\left[c_{\mathbf{x},\mathbf{x}}^+(\Phi_3)\log(\widehat{\sigma^+})\right].$$

Based on our assumption, $\widehat{\sigma^+} \geq 1 \Rightarrow \log(\widehat{\sigma^+}) \geq 0$. Also due to definition, $c_{\mathbf{x},\mathbf{x}}^+(.) \geq 0$. Hence

$$\mathbb{E}_{\Pi^{s+1}(x)}\left[c_{\mathbf{x},\mathbf{x}}^+(\Phi_3)\log(\widehat{\sigma^+})\right] \geq 0 \Rightarrow$$

$$\mathbb{E}_{\Pi^{s+1}(x)}\left[c_{\mathbf{x},\mathbf{x}}^+(\Phi_3)\log(\widehat{\sigma^+})\right] + Q(\Phi_1) \geq 0 + Q(\Phi_1) \Rightarrow$$

$$Q(\Phi_2) \geq Q(\Phi_1).$$

□

As shown in Section 6.2.2, a greedy forward algorithm can provide a good approximation of the solution for an NP-hard set optimization problem when the objective function is a submodular-monotone set function. It is easy to see that this algorithm provides the global optimum if the objective function is a modular-monotone set function. Here we assume that the number of trials within each sequence is fixed and equal to N_t. Accordingly, the *deterministic greedy algorithm* is described in Algorithm 1. The selected subset according to this algorithm is the global optimum of the optimization problem defined in Equation 6.20. Next, we illustrate the usage of this objective function in stimuli subset selection for a language-model-assisted BCI for letter-by-letter typing. In the following subsections, after a short introduction of the typing Brain Computer Interface (BCI), we approximate the proposed objective function with a modular-monotone set function and provide the algorithm for stimuli subset selection. Then through an experimental study, we demonstrate the benefit of the AL component of Active-RBSE in terms of typing speed and accuracy.

Algorithm 1 Greedy algorithm for maximization of Q

Input: The size of sequence set N_t
Output: Estimated sequence set $\widehat{\Phi}_k^{s+1}$

/* Initializations */

1 $\widehat{\Phi}_k^{s+1} \leftarrow \emptyset$

/* Starting the Iterations */

2 **for** $i = 1 \to N_t$ **do**

  /* Adding a the next optimal $A \in 2^{\mathcal{V}} | \hat{\Phi}_k^{s-1}$ */

3 $\widehat{\Phi}_k^{s+1} \leftarrow \widehat{\Phi}_k^{s+1} \cup \{\arg\max_A \Delta_Q(A|\widehat{\Phi}_k^{s+1})\}$ where $A \in 2^{\mathcal{V}} \setminus \widehat{\Phi}_k^{s+1}$.
  end
4 **return** $\widehat{\Phi}_k^{s+1}$

6.3 Illustrative BCI Design Example

A safe and portable set of BCIs utilize noninvasively recorded EEG for inference. Among many, a class of these BCIs employ external cues to induce an ERP in response to user intent. For example, most commonly, ERP-based BCI systems that rely on visual stimulation can utilize various presentation paradigms. The pioneering example of these systems is the matrix speller from Farwell and Donchin, which demonstrates how to design a presentation paradigm for inducing P300* in response to user intent as a control signal for BCI-based communication [16]. In this study, the subjects observe a 6 × 6 matrix of letters in the English alphabet, numbers from 1 to 9, and a space symbol distributed on the screen. While the user is focused on the intent character, rows and columns of the matrix are flashed randomly. This work led to extensive efforts for designing different configurations or algorithms to improve the communication speed and accuracy with the matrix speller, as well as other audio, visual, and tactile stimulus presentation techniques for eliciting P300 responses. In the following, we will review some of these stimulus presentation techniques.

Visuospatial presentation: We categorize different visuospatial presentation techniques into the following groups:

Matrix presentation: Generally the matrix spellers use an $R \times C$ matrix of symbols with R rows and C columns. Traditionally, in these systems, each row and column of the matrix is intensified in a pseudo-random fashion, while the participants count the number of highlighted rows or columns (or, in general, subsets) that include the desired symbol. Among all rows and columns in the matrix, only two contain the target symbol; hence it is proposed that they will induce a P300 response. By detecting this signature in the EEG, the BCI system can identify the target letter to enable typing.

The accuracy of BCIs highly depends on signal-to-noise ratio (SNR). Consequently, due to low SNR of EEG, matrix speller systems require sacrificing the speed by cuing the user with multiple sequences of flashes to achieve an acceptable accuracy. It was demonstrated that the matrix speller can achieve 7.8 characters/minute with 80% accuracy, using bootstrapping and averaging the trials in different sequences [17]. Many signal processing and machine learning techniques have been proposed by researchers in the field, to improve the matrix speller performance in terms of speed and accuracy [18–40].

Considering the target population, matrix-based presentation paradigms might not offer a suitable solution for BCIs since they performs well in overt attention mode; however, in covert attention mode, its performance degrades significantly [41]. BCI researchers have proposed minimally gaze-dependent stimulus presentation techniques such as rapid serial visual presentation and balanced-tree visual presentation to overcome such performance drops.

Rapid serial visual presentation (RSVP): RSVP is a technique in which stimuli are presented one at a time at a fixed location on the screen, in pseudo-random order, with a short time gap in between. Within a sequence of RSVP stimuli, each symbol is shown only once; hence, the user intent is considered as rare event that can induce an ERP containing the P300 wave as consequence of the

---

* P300 is an ERP that is elicited in response to a desired unpredictable and rare event. This evidence is characterized as a positive peak around 300 ms after the onset of desired stimuli.

target matching process that takes place in the brain. RSVP aims to be less dependent on gaze control, by utilizing temporal separation of symbols instead of spatial separation as in the matrix speller [42–46].

Usually, in RSVP-based BCIs, the inference speed is lower than matrix spellers, as the binary tree that leads to symbol selections in a matrix speller could reduce expected bits to select a symbol (determined by entropy), by exploiting the opportunity of highlighting a subset of symbols, while RSVP is constrained to a highly structured right-sided binary tree that can only offer a larger expected bits per symbol. Letter-by-letter typing RSVP-BCIs designed by Berlin BCI and RSVP Keyboard™ groups have achieved up to 5 characters/minute [42–44,46]. Utilization of color cues and language models has offered some enhancements in typing speeds with RSVP [43,46].

Balanced-tree visual presentation paradigms: In the balanced-tree visual presentation technique, visual stimuli are distributed spatially into multiple presentation groups with balanced numbers of elements. For example, in a system from Berlin BCI known as Hex-o-Spell, a set of 30 symbols is distributed among six presentation groups each containing five symbols. Presentation groups are flashed in a random fashion to induce an ERP in response to the group that contains the intended symbol. Upon selection of a group, the symbols of that set are distributed individually to different presentation groups, typically with one group containing a command symbol for moving back to the first presentation stage. Then the system utilizes the same flash paradigm to decide on the user desired symbol [41,47]. In a similar system known as Geospell, 12 groups of 6 symbols, corresponding to rows and columns of a 6 × 6 matrix speller, are arranged in a circular fashion [48,49]. In another study, this subset of 12 overlapping symbols are presented in RSVP manner [50]. In these systems, the intersection of the selected groups gives the desired symbol.

In this section, we introduce a language-model-assisted EEG-based BCI that can utilize any of two well-known matrix presentation paradigms (matrix row and column [RCP] and matrix single-character presentation [SCP]) and RSVP paradigm to cue the user for the inference of the desired symbol.

6.3.1 ERP-Based BCI for Letter-by-Letter Typing

Figure 6.6 represents the complete flow chart of the BCI in this example. The system can be segmented in three main components: (I) *a presentation component* that controls the presentation scheme, (II) *a feature extraction component* that extracts the likelihoods from raw EEG evidence for Bayesian fusion, and (III) *a decision-making component* that combines the EEG (physiology) and context information to estimate the user intent. In the next subsection, these components are described in more detail.

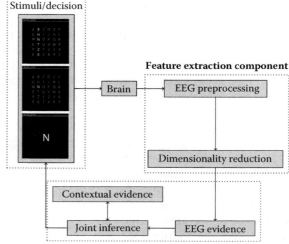

FIGURE 6.6
Typical BCI block diagram. (From M. Moghadamfalahi et al., "Language-model assisted brain computer interface for typing: A comparison of matrix and rapid serial visual presentation," IEEE Transactions on Neural Systems and Rehabilitation Engineering, vol. 23, no.5, pp. 910–920, Sept 2015.)

6.3.1.1 Presentation Component

Definitions: Let $V = \{x_1, x_2, x_3, \ldots, x_{|V|}\}$ be the vocabulary set, i.e., the state space. In this example, for a letter-by-letter typing application, V consists of letters in the (English) alphabet, numerical symbols, and space and backspace symbols (represented here by _ and < respectively). Define $2^V = \{A_1, A_2, \ldots, A_{2^{|V|}}\}$ as the power set of V; $A_i \subset V$.

As a reminder, we define a *trial* as a subset A_i, which is highlighted during the presentation. In RCP paradigm, each trial consists of multiple characters, i.e., $|A_i| \geq 1$, but in RSVP and SCP paradigms each trial is a singleton; i.e., $|A_i| = 1$. A *sequence* is a series of consecutive flashes of trials with a predefined short intertrial interval (ITI) in between. Among many definitions, here we use the ITI as the time gap between the onset of two consecutive trials in a sequence. After every sequence, the system fuses the likelihoods obtained from recorded EEG, in response to that sequence, to compute the posterior PMF over the vocabulary set V and tries to estimate the user intent through MAP inference. However, a final decision is not made until a predefined confidence level is reached.* Therefore, the system may need to query the user with multiple sequences before commiting to a

---
* In the current implementation, confidence is measured by the maximum posterior probability over V; this corresponds to using Rényi entropy of order ∞ as the measure of uncertainty. Other entropy definitions such as Shannon's could also be used.

decision. In this chapter a set, of sequences that leads to a decision is referred to as an *epoch*.

In matrix-based presentation paradigms, symbols are spatially distributed on an $R \times C$ grid with R number of rows and C number of columns [52]. To cue the user for the inference of the desired character, subsets of these symbols are intensified typically in pseudo-random order.

In every sequence, usually trials $A_1, A_2, \ldots, A_n$ are selected such that $\cup_{i=1}^{n} A_i = \mathcal{V}$. RCP is a paradigm in which a trial A_i is constrained to contain exactly all the symbols in a row or a column of the matrix of symbols with $n = (R + C)$ [16]. Accordingly, in RCP, every symbol in $\mathcal{V}$ is flashed twice in every sequence since, $|A_i \cap A_j| \leq 1, i \neq j$. For this example, to obtain the best coverage of the wide-screen monitors used in our experiments, we utilize a matrix of size 4×7. Researchers in the field have suggested that a probability of less than 25% can lead to a detectable P300 wave in response to a desired symbol [16]. In our setup for RCP, each sequence contains 11 flashes, among which only 2 contain the target symbol, and hence, the probability of each symbol trial in a sequence is $\frac{2}{11} \approx 0.18 \leq 0.25$.

In contrast to RCP, SCP paradigm was shown to increase the P300 signal quality, which led to more accurate target detection [53]. In SCP, each trial is a singleton, i.e., $|A_i| = 1$. Moreover, for the trials in a sequence, it is required that, $A_i \cap A_j = \emptyset; i \neq j$. We choose a large enough number of flashes ($n \geq 5$) in a sequence to reach a target probability less than 25%.

Similar to SCP, each trial in RSVP includes only a single symbol. It has been shown that RSVP-BCI systems that present all 28 symbols in every sequence, can only achieve a speed of 5 symbols/minute [42–44,46]. Instead, one may choose to present a subset of vocabulary in each sequence to improve typing speed/accuracy. Active-RBSE offers a principled mechanism for selecting this subset not only for RSVP but also for RCP and SCP paradigms to achieve an optimal solution based on an objective function that is associated with gaining information about the user intent inference, more specifically for state estimation as described in Section 6.2.1. The results from an experimental study as described in Section 6.3.3 show that this framework can improve both the typing speed and accuracy significantly.

6.3.2 Decision-Making Component

The inference engine of the BCI in this example uses language structure information and provides context prior to be fused with EEG likelihoods. The joint decision from context information and EEG evidence is estimated through MAP inference. System parameters are optimized for each user individually, with the data obtained from a calibration session, collected in a supervised fashion.

6.3.2.1 EEG Feature Extraction and Classification

More reliable inference from EEG can be achieved by preprocessing the data and enhancing the signal quality. The processed signal is used in this system to form EEG feature vectors, and the details of this procedure are provided later in Section 6.3.3. These feature vectors are computed by applying certain linear operations on EEG time series. EEG is widely considered as a Gaussian process in the field [51,54,55]. Hence, it seems natural to use quadratic discriminant analysis (QDA)* to project these vectors onto a one-dimensional space with minimum expected classification risk. But QDA requires invertible class-conditional covariance matrix estimates, which are not feasible in the practical usage of the BCI due to the high dimensionality of the EEG feature vectors and low number of calibration samples in each class.[†] This problem can be mitigated by employing regularized discriminant analysis (RDA), which provides full-rank class-conditional covariance estimates [56].

Rank deficient covariance matrices are obtained using maximum likelihood estimation from calibration data. Then RDA converts these estimates to full-ranked covariance matrices using shrinkage and regularization. Shrinkage is defined as a convex combination of each class-covariance matrix and the overall class-mean-subtracted covariance. Define $\mathbf{f}_i \in \mathbb{R}^p$ as a p-dimensional feature vector, and $l \in \{0, 1\}$ as class label where 0 and 1 represent nontarget and target classes respectively; then the maximum likelihood estimator for mean and covariance matrices are

$$\boldsymbol{\mu}_l = \frac{1}{N_l} \sum_{i=1}^{N} \mathbf{f}_i \delta(y_i, l)$$

$$\Sigma_l = \frac{1}{N_l} \sum_{i=1}^{N} (\mathbf{f}_i - \boldsymbol{\mu}_l)(\mathbf{f}_i - \boldsymbol{\mu}_l)^T \delta(y_i, l). \quad (6.21)$$

Here N_l is the number of trials in class l; thus $N = N_0 + N_1$ is the total number of feature vectors, and $\delta(.,.)$ is the Kronecker-δ. The shrinkage step of RDA is defined as

$$\hat{\Sigma}_l(\lambda) = \frac{(1-\lambda)N_l \Sigma_l + (\lambda)\sum_{j=0}^{1} N_j \Sigma_j}{(1-\lambda)N_l + (\lambda)\sum_{j=0}^{1} N_j}, \quad (6.22)$$

where $\lambda \in [0, 1]$ is known as the shrinkage parameter. Note that, $\lambda = 1$ provides equal class-conditional covariance matrices and converts the problem to linear discriminant analysis (LDA).

---

* The Gaussian distribution assumption here is a direct consequence of the assumption that filtered EEG is a Gaussian random process.
† In BCI design, a limited number of calibration samples are obtained in order to keep the calibration sessions short.

The regularization step of RDA is defined as

$$\hat{\Sigma}_l(\lambda,\gamma) = (1 - \gamma)\hat{\Sigma}_l(\lambda) + (\gamma)\frac{1}{p}tr[\hat{\Sigma}_l(\lambda)]\mathbf{I}_p. \quad (6.23)$$

In this equation, we use $tr[.]$ as the trace operator, a $p \times p$ identity matrix is shown by $\mathbf{I}_p$, and $\gamma \in [0, 1]$ is the regularization parameter. The regularization step corresponds to diagonal loading on these matrices to make them invertible. The discriminant scores of RDA are then estimated similar to QDA but with regularized and shrinked class-conditional covariance matrices.

$$e = \log\frac{f_N(\mathbf{f};\boldsymbol{\mu},\hat{\Sigma}_1(\lambda,\gamma))\hat{\pi}_1}{f_N(\mathbf{f};\boldsymbol{\mu}_0,\hat{\Sigma}_0(\lambda,\gamma))\hat{\pi}_0} \quad (6.24)$$

In Equation 6.24, f_N (f; μ, Σ) serves as a Gaussian PDF when $f \sim \mathcal{N}(\mu, \Sigma)$, and the the prior probability of class l is shown as $\hat{\pi}_l$. Here, with no knowledge about a trial, we use $\hat{\pi}_1 = \hat{\pi}_0$. The EEG scores e obtained from RDA, corresponding to the evidences defined in Figure 6.5, are then treated as random variables for which the class-conditional PDF are computed using KDE as in Equation 6.25 [46].

$$P\left(\mathbf{f} = \hat{\mathbf{f}}\middle| l = \hat{l}\right) = f_{KDE}(e = \hat{e}|l = \hat{l}) = \frac{1}{N_{\hat{l}}}\sum_{i=1}^{N}\mathcal{K}_{h_l}(\hat{e},e_i)\delta_{l_i,\hat{l}}$$
(6.25)

In this equation, $\mathcal{K}_{h_k}(.,.)$ is a suitable kernel function with bandwidth h_l. In this example, we use a Gaussian kernel and we estimate the kernel bandwidth h_l for each class using the Silverman rule of thumb [57], applied over the RDA scores for the corresponding class.

6.3.2.2 Language Model

In a letter-by-letter typing scenario, a letter n-gram language model can be used to obtain a prior PMF over the state space of user intent. This information can be used both for sequence optimization and for inference. A letter n-gram model corresponds to a Markov mode of order $n - 1$, which estimates the conditional probability of every letter in the alphabet based on $n - 1$ previously typed letters [58].

According to the Bayes rule and conditional independence provided by an $n - 1$ Markov model, the conditional probability of each character is computed as

$$p(x_k = x|\mathbf{x}_{k-1} = \mathbf{x}) = \frac{p(x_k = x, \mathbf{x}_{k-1} = \mathbf{x})}{p(\mathbf{x}_{k-1} = \mathbf{x})}. \quad (6.26)$$

As defined in Section 6.2, x_k is the system state (yet) to be estimated during epoch k, and the string of previously written $n - 1$ symbols are denoted by $\mathbf{x}_{k-1}$. In this particular example, we use a 6-gram letter model. This model is trained on the *New York Times* portion of the English Gigaword corpus [58].

6.3.3 Experimental Results and Discussions

In this example, we employed a set of supervised data, collected from 12 healthy participants following an Institutional Review Board (IRB) approved protocol [51]. In this experiment, we used 16 EEG electrode locations of Fp1, Fp2, F3, F4, Fz, Fc1, Fc2, Cz, P1, P2, C1, C2, Cp3, Cp4, P5, and P6 according to the International 10/20 configuration. The EEG was recorded at a sampling rate of 256 Hz utilizing g.USBamp biosignal amplifier with active g.Butterfly electrodes. These data were collected in three separate days each for one presentation paradigm of RCP, SCP, and RSVP at an ITI of 150 ms. The order of sessions was distributed uniformly among participants to exclude the effect of learning or frustration from statistical analysis on the results. Each calibration session contained 100 sequences with 10 trials per sequence among which one is the target. Prior to each sequence, we showed the target to the user to make the calibration task a supervised data collection session. These data sets were used to obtain class-conditional PDFs and other system parameters for each user and presentation paradigm combinations. These PDFs were then used in Monte Carlo simulations of the system in an online typing scenario.

Twenty Monte Carlo simulations of the system were executed using the samples drawn from estimated class-conditional PDFs. During each simulation, the system types missing words in 10 different phrases. Note that the prior suggested by the context information is not always helping the user. For instance, the user might need to type a that which is not common in the English language. To consider these cases in our analysis, we define 5 different difficulty levels,* with 1 as the easiest and 5 as the most difficult words to type. These phrases in these simulations are selected uniformly across these five difficulty levels. Here we compare the results for simulated online performance of our system under the Active-RSBE framework and baseline methods that either use the entire vocabulary at every sequence or perform random stimuli subset selections.

The performance of the system is measured in terms of (I) total typing duration (TTD) for typing 10 phrases, which is inversely proportional to typing speed, and (II) probability of phrase completion (PPC), which is a measure of typing accuracy. Next, we present the simulation results for different presentation paradigms and for different users with various BCI usage accuracy levels. These accuracy levels are represented by the area under the receiver operating curve (AUC) values for

---
* Lower levels consist of copying phrases that have letters that are assigned high probabilities by the language model. As the level increases, the language model probabilities become increasingly adversarial. Level 3 is neutral on average.

each user. These AUC values are obtained through performing cross-validation over the supervised data collected at calibration sessions.

6.3.3.1 RSVP Paradigm

To assess the effect of the proposed query set optimization on online system performance, we conducted two sets of Monte Carlo simulations (1) with random trial selection and (2) with optimal query selection. Based on an earlier experimental study, the upper bound on the number of sequences in each epoch was set to $m_t = 8$, and number of trials per sequence was selected as $k = 14$ [51,59]. The scatterplot in Figure 6.7 shows the TTD for active RSVP (ARSVP) vs. the random RSVP paradigm. In this figure, the horizontal and vertical axes represent TTD for random RSVP and ARSVP respectively. In this plot, the width and the height of the box around each data point represent the standard deviation of TTD from 20 Monte Carlo simulations for each case in the corresponding dimensions. This figure shows that 9 out of 12 users could benefit from optimal sequence selection and achieve higher typing speed. To quantify these results, we used a Wilcoxon signed-rank sum test, which showed a statistically significant improvement with $P < 0.03$ under the assumption of a significance level of $\alpha = 0.05$.

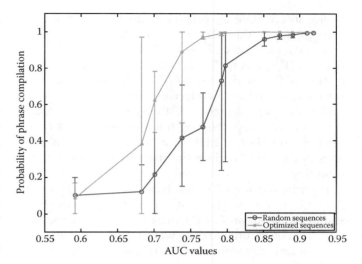

FIGURE 6.8
Average probability of phrase completion with 90% confidence intervals for RSVP and ARSVP paradigm. (From M. Moghadamfalahi et al., "Active learning for efficient querying from a human oracle with noisy response in a language-model assisted brain computer interface," in 2015 IEEE 25th International Workshop on Machine Learning for Signal Processing (MLSP). IEEE, 2015, pp.1–6.)

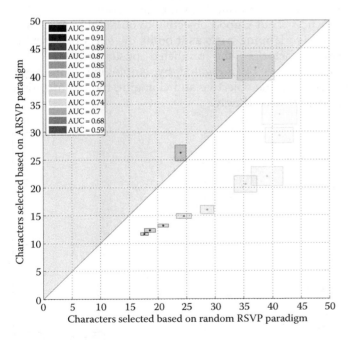

FIGURE 6.7
Scatterplot of average TTD in minutes from 20 Monte Carlo simulations for RSVP and ARSVP. (From M. Moghadamfalahi et al., "Active learning for efficient querying from a human oracle with noisy response in a language-model assisted brain computer interface," in 2015 IEEE 25th International Workshop on Machine Learning for Signal Processing (MLSP). IEEE, 2015, pp.1–6.)

In these typing scenarios, we mark a phrase as incomplete if the system cannot type the correct phrase in a predefined duration or if more than five consecutive mistakes occur. Consequently, we define the PPC as the ratio of number of completed phrases to the total number of phrases. The estimated PPCs from the simulation sets are presented in Figure 6.8. In this figure, the AUC values (which are a measure of EEG classification performance for each user) are mapped on the x-axis, and the PPCs are presented on the y-axis. The averaged PPCs of 20 Monte Carlo simulations are shown as "*" points in green for ARSVP and as "o" in red for random RSVP. The 90% area under a beta distribution fitted to PPCs for each parameter set is obtained for both cases, which are shown as error bars in corresponding colors around the averaged PPCs.

These results suggest that the optimal query strategy improves the typing accuracy especially for the AUCs $\in$ [0.7, 0.9], which is typically the range that includes most of the users in the healthy population. Wincoxon signed-rank sum hypothesis testing results ($P < 0.003$) applied on averaged PPCs from both conditions show a significant improvement in typing accuracy due to optimizing query sets for each sequence according to proposed objective function.

6.3.3.2 Matrix-Based Presentation Paradigm with Overlapping Trials

In contrast to RSVP and SCP, trials in RCP are not constrained to be singletons. This can potentially lead to higher typing speed. Here we propose to define a more relaxed search space for the optimization problem at hand.

Define a function $\mathbf{c} : 2^{\mathcal{V}} \to \{1, 0\}^{\mathcal{V}}$ where $\mathbf{c}(A_i) = [1\{\mathcal{V}_1 \in A_i\}, \cdots, 1\{\mathcal{V}_{|\mathcal{V}|} \in A_i\}]^T$. Then, define a $|\mathcal{V}| \times k$ binary code matrix as $\mathbf{C} = [\mathbf{c}(A_1), \cdots, \mathbf{c}(A_k)]$. Each row of this code matrix $\mathbf{C}$ associates a code word of zeros and ones to each state space element, i.e., the symbols in the vocabulary. In this setup, the length of the code words represents the number of trials in a sequence. The value of ones in each row defines the trials in which the corresponding character is flashed.

In particular for RCP, the trials have overlaps with $A_i \cap A_j \leq 1$, $\forall\, i,j \in \{1, \cdots, |\Phi_k^{s+1}|\}$, $i \neq j$. Then under these assumptions, for an $\mathcal{R} \times \mathcal{C}$ matrix of characters, the RCP paradigm offers unique code words of length $\mathcal{R} + \mathcal{C}$ with two nonzero elements. The visual presentation component of the system used in this example utilizes a 4×7 background matrix. Hence the length of code words in RCP is $4 + 7 = 11$.

In such an RCP setup, we propose to define the $\mathbf{C}$ such that each letter is assigned with a unique code word. For application in ERP-BCIs, we need to impose certain constraints on the search space to control for the frequency of letter presentation in each sequence to induce ERP in response to intent. But it is important to note that, unlike RSVP-based paradigms, matrix-based presentation paradigms can benefit from visual evoked potentials (VEPs). Hence, we can relax the constraints and allow for more frequent flashes of the same character in each sequence [61]. This can lead to improved typing speed by reducing the sequences' length. More specifically, we define the feasible set such that there exists a unique code word for each symbol while each symbol is presented with a frequency less than 0.5 in each sequence.

Based on these propositions, we define the length of the code words to be equal to 6, to get at least as many code words as the size of our vocabulary set with three or less nonzero elements. With this number of trials, we can produce $\binom{6}{3} + \binom{6}{2} + \binom{6}{1} = 41$ unique code words to be assigned to each character. The scatterplot of TTD for standard RCP (using 11 flashes corresponding to all the rows and columns in each sequence) and AL-based presentation (ALP) paradigms are presented in Figure 6.9. This plot suggests that every user can achieve shorter TTD with the ALP paradigm. This effect is more clear for the lower AUCs that are represented by the points toward the center of the figure. Also, the statistical hypothesis testing result confirms a significant improvement ($P < 0.0005$). But on the other hand, as shown in Figure 6.10, the improvement on the PPCs is not as significant. Here note that in the RCP paradigm, the EEG classification AUCs are generally high, and this leads to high PPCs (above 95%) with no sequence optimization. Thus, as expected, we don't have a statistically significant improvement in PPCs ($P > 0.75$). But in sum, from the results presented in Figures 6.9 and

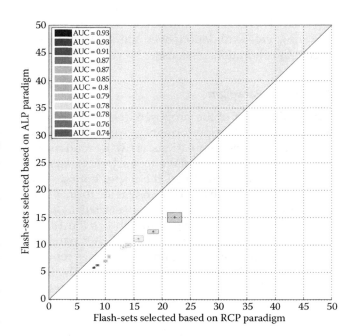

FIGURE 6.9
Scatterplot of total typing duration of 10 phrases in terms of minutes for RCP and ALP.

6.10, we infer that ALP can significantly reduce the TTD while preserving the PPC.

Here we should mention that the proposed monotone-modular lower bound on the approximated objective function is not a good objective to obtain the query sets with overlapping trials as it omits the effect of normalization factor in the predicted posterior PMF. This problem can be mitigated by employing more tight bounds on the objective and changing the optimization approach accordingly.

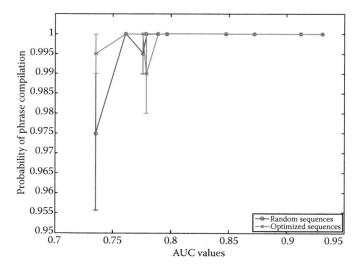

FIGURE 6.10
Average probability of phrase completion with 90% confidence intervals for RCP and ALP.

6.3.3.3 Matrix-Based Presentation Paradigm with Single-Character Trials

For the RSVP paradigm, it has been shown that the best typing performance can be achieved when not all letters but a subset of vocabulary is presented in each sequence [59]. Similar to RSVP, in SCP, each trial consists of a single letter, and each letter is presented at most once in each sequence. Accordingly, we use the results of an RSVP study to propose that the best typing performance for SCP can be achieved with sequences of length 14 [59]. Hence, we constrain the search space of our objective function to sequence sets of length $k = 14$ while each trial is a singleton, i.e., $|A_i| = 1$. Here we compare our methods to a typical SCP paradigm in which all the vocabulary elements are presented in random order at every sequence.

The results are summarized in Figures 6.11 and 6.12. The scatterplot of TTD is shown in Figure 6.11. In this figure, the horizontal axis shows the TTD value for the standard SCP paradigm and the vertical axis shows the TTD when the Active-SCP (ASCP) paradigm is used. The results here suggest that the typing speed of an SCP paradigm can be significantly improved when optimized sequence sets of length 14 are used ($P < 0.01$). Moreover, as the results show in Figure 6.12, users can achieve a significantly improved ($P < 0.008$) typing accuracy when the ASCP paradigm is used rather than the standard SCP paradigm that uses all the symbols in the vocabulary at every sequence.

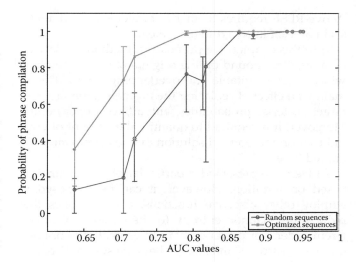

FIGURE 6.12
Average probability of phrase completion with 90% confidence intervals for SCP and ASCP.

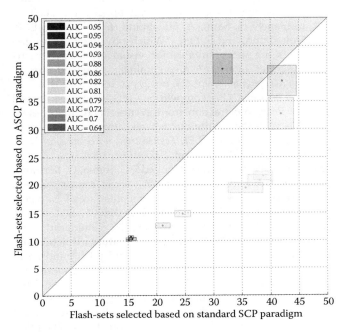

FIGURE 6.11
Scatterplot of TTD of 10 phrases in terms of minutes for SCP and ASCP.

6.4 Open Problems and Future Directions

In this chapter, we presented the Active-RBSE framework for system state inference in the presence of high-velocity multimodal data. Active-RBSE imposes conditional independence among different physiological measurements; hence, different modalities can be processed in parallel, and extracted evidence then will be fused in a probabilistic fashion for joint inference. Also, inspired by cognitive systems, in this framework, the system state is defined as a random variable tailored by context information, which defines a prior probability distribution over the state space. The RBSE mechanism employed in this system offers recursive update rules for estimated posterior distribution after observing a set of measurements that reduces both memory and processing demand during online operation.

Moreover, Active-RBSE employs the AL concept to exploit the information extracted from the context and previously observed evidence to optimize query sets offered to the oracle. This can potentially reduce the processing time. To illustrate the importance of information exploitation in query set design recursively, we used 36 supervised data sets collected from 12 healthy participants who utilized an EEG-based language-model-assisted letter-by-letter typing interface with three different presentation paradigms of RSVP, SCP, and RCP with optimized and random query sets. A thorough assessment of the Active-RBSE framework through Monte Carlo simulations demonstrated that this framework offers a significant improvement in terms of typing speed and accuracy over well known existing presentation paradigms. The presented version of

Active-RBSE requires further improvements and analysis in at least in two different categories.

(I) In this chapter, we proposed a method for solving an alternative bound on the original objective function, which is not suitable for overlapping trials. This is mainly an effect of replacing the normalization factor of target posterior probabilities with a fixed upper bound. Moreover, it is required to quantify the amount of error induced on the optimal solution due to approximations offered here.

(II) Here we presented a particular objective function based on intuition. However, it can be proposed to employ other objective functions, e.g., the ones that reduce the Shannon entropy (or Rényi entropy of different orders) of posterior PMF to estimate the optimal query set. Also it of interest to allow the system to define the length of each sequence for achieving the optimal solution.

References

1. Federal Big Data Commission. "Demystifying big data: A practical guide to transforming the business of Government." Retrieved from https://bigdatawg.nist.gov/_uploadfiles/M0068_v1_3903747095.pdf, TechAmerica Foundation's, 2012.
2. B. Settles, "Active learning," *Synthesis Lectures on Artificial Intelligence and Machine Learning*, vol. 6, no. 1, pp. 1–114, 2012.
3. B. Anderson and A. Moore, "Active learning for hidden Markov models: Objective functions and algorithms," in *Proceedings of the 22nd International Conference on Machine Learning*. ACM, 2005, pp. 9–16.
4. T. Scheffer, C. Decomain, and S. Wrobel, "Active hidden Markov models for information extraction," in *Advances in Intelligent Data Analysis*. Springer, 2001, pp. 309–318.
5. T. Scheffer and S. Wrobel, "Active learning of partially hidden Markov models," in *Proceedings of the ECML/PKDD Workshop on Instance Selection*. Citeseer, 2001.
6. E. Ringger, P. McClanahan, R. Haertel, G. Busby, M. Carmen, J. Carroll, K. Seppi, and D. Lonsdale, "Active learning for part-of-speech tagging: Accelerating corpus annotation," in *Proceedings of the Linguistic Annotation Workshop*. Association for Computational Linguistics, 2007, pp. 101–108.
7. J. F. Bowring, J. M. Rehg, and M. J. Harrold, "Active learning for automatic classification of software behavior," in *ACM SIGSOFT Software Engineering Notes*, vol. 29, no. 4. ACM, 2004, pp. 195–205.
8. B. Settles and M. Craven, "An analysis of active learning strategies for sequence labeling tasks," in *Proceedings of the Conference on Empirical Methods in Natural Language Processing*. Association for Computational Linguistics, 2008, pp. 1070–1079.
9. A. Culotta, T. Kristjansson, A. McCallum, and P. Viola, "Corrective feedback and persistent learning for information extraction," *Artificial Intelligence*, vol. 170, no. 14, pp. 1101–1122, 2006.
10. M.-Y. Chen, M. Christel, A. Hauptmann, and H. Wactlar, "Putting active learning into multimedia applications: Dynamic definition and refinement of concept classifiers," in *Proceedings of the 13th Annual ACM International Conference on Multimedia*. ACM, 2005, pp. 902–911.
11. H. Cao, D. Jiang, J. Pei, E. Chen, and H. Li, "Towards context-aware search by learning a very large variable length hidden Markov model from search logs," in *Proceedings of the 18th International Conference on World Wide Web*. ACM, 2009, pp. 191–200.
12. S. Ji, B. Krishnapuram, and L. Carin, "Variational Bayes for continuous hidden Markov models and its application to active learning," *Pattern Analysis and Machine Intelligence, IEEE Transactions on*, vol. 28, no. 4, pp. 522–532, 2006.
13. A. Krause and D. Golovin, "Submodular function maximization," *Tractability: Practical Approaches to Hard Problems*, vol. 3, p. 19, 2012.
14. G. L. Nemhauser, L. A. Wolsey, and M. L. Fisher, "An analysis of approximations for maximizing submodular set functions I," *Mathematical Programming*, vol. 14, no. 1, pp. 265–294, 1978.
15. S. Kay, Fundamentals of Statistical Signal Processing, Volume II: Detection Theory. 2008.
16. L. Farwell and E. Donchin, "Talking off the top of your head: Toward a mental prosthesis utilizing event-related brain potentials," *Electroencephalography and Clinical Neurophysiology*, vol. 70, pp. 510–523, 1988.
17. E. Donchin, K. M. Spencer, and R. Wijesinghe, "The mental prosthesis: Assessing the speed of a p300-based brain–computer interface," *IEEE Transactions on Rehabilitation Engineering*, vol. 8, no. 2, pp. 174–179, 2000.
18. L. Bianchi, S. Sami, A. Hillebrand, I. P. Fawcett, L. R. Quitadamo, and S. Seri, "Which physiological components are more suitable for visual erp based brain–computer interface? A preliminary meg/eeg study," *Brain Topography*, vol. 23, no. 2, pp. 180–185, 2010.
19. V. Bostanov, "Bci competition 2003—Data sets ib and iib: Feature extraction from event-related brain potentials with the continuous wavelet transform and the t-value scalogram," *IEEE Transactions on Biomedical Engineering*, vol. 51, no. 6, pp. 1057–1061, 2004.
20. H. Cecotti, B. Rivet, M. Congedo, C. Jutten, O. Bertrand, E. Maby, and J. Mattout, "A robust sensor-selection method for p300 brain–computer interfaces," *Journal of Neural Engineering*, vol. 8, no. 1, p. 016001, 2011.
21. A. Combaz, N. V. Manyakov, N. Chumerin, J. A. Suykens, and M. M. Van Hulle, "Feature extraction and classification of eeg signals for rapid p300 mind spelling," in *International Conference on Machine Learning and Applications, 2009. ICMLA'09*. IEEE, 2009, pp. 386–391.
22. A. Combaz, N. Chumerin, N. V. Manyakov, A. Robben, J. A. Suykens, and M. M. Van Hulle, "Error-related potential recorded by eeg in the context of a p300 mind speller brain–computer interface," in *2010 IEEE*

International Workshop on Machine Learning for Signal Processing. IEEE, 2010, pp. 65–70.
23. A. Combaz, N. Chumerin, N. V. Manyakov, A. Robben, J. A. Suykens, and M. M. Van Hulle, "Towards the detection of error-related potentials and its integration in the context of a p300 speller brain–computer interface," *Neurocomputing*, vol. 80, pp. 73–82, 2012.
24. M. Kaper, P. Meinicke, U. Grossekathoefer, T. Lingner, and H. Ritter, "Bci competition 2003—Data set iib: Support vector machines for the p300 speller paradigm," *IEEE Transactions on Biomedical Engineering*, vol. 51, no. 6, pp. 1073–1076, 2004.
25. P.-J. Kindermans, D. Verstraeten, and B. Schrauwen, "A Bayesian model for exploiting application constraints to enable unsupervised training of a p300-based bci," *PloS One*, vol. 7, no. 4, p. e33758, 2012.
26. P.-J. Kindermans, H. Verschore, D. Verstraeten, and B. Schrauwen, "A p300 bci for the masses: Prior information enables instant unsupervised spelling," in *Advances in Neural Information Processing Systems*, 2012, pp. 710–718.
27. D. J. Krusienski, E. W. Sellers, F. Cabestaing, S. Bayoudh, D. J. McFarland, T. M. Vaughan, and J. R. Wolpaw, "A comparison of classification techniques for the p300 speller," *Journal of Neural Engineering*, vol. 3, no. 4, p. 299, 2006.
28. D. J. Krusienski, E. W. Sellers, D. J. McFarland, T. M. Vaughan, and J. R. Wolpaw, "Toward enhanced p300 speller performance," *Journal of Neuroscience Methods*, vol. 167, no. 1, pp. 15–21, 2008.
29. Y. Li, C. Guan, H. Li, and Z. Chin, "A self-training semi-supervised SVM algorithm and its application in an EEG-based brain computer interface speller system," *Pattern Recognition Letters*, vol. 29, no. 9, pp. 1285–1294, 2008.
30. D. J. McFarland, W. A. Sarnacki, and J. R. Wolpaw, "Should the parameters of a BCI translation algorithm be continually adapted?" *Journal of Neuroscience Methods*, vol. 199, no. 1, pp. 103–107, 2011.
31. R. C. Panicker, S. Puthusserypady, and Y. Sun, "Adaptation in p300 brain–computer interfaces: A two-classifier cotraining approach," *IEEE Transactions on Biomedical Engineering*, vol. 57, no. 12, pp. 2927–2935, 2010.
32. A. Rakotomamonjy, V. Guigue, G. Mallet, and V. Alvarado, "Ensemble of SVMs for improving brain computer interface p300 speller performances," in *International Conference on Artificial Neural Networks*. Springer, 2005, pp. 45–50.
33. A. Rakotomamonjy and V. Guigue, "Bci competition III: Dataset II—Ensemble of SVMs for BCI p300 speller," *IEEE Transactions on Biomedical Engineering*, vol. 55, no. 3, pp. 1147–1154, 2008.
34. B. Rivet, A. Souloumiac, V. Attina, and G. Gibert, "xdawn algorithm to enhance evoked potentials: Application to brain–computer interface," *IEEE Transactions on Biomedical Engineering*, vol. 56, no. 8, pp. 2035–2043, 2009.
35. D. B. Ryan, G. Frye, G. Townsend, D. Berry, S. Mesa-G, N. A. Gates, and E. W. Sellers, "Predictive spelling with a p300-based brain–computer interface: Increasing the rate of communication," *International Journal of Human–Computer Interaction*, vol. 27, no. 1, pp. 69–84, 2010.
36. H. Serby, E. Yom-Tov, and G. F. Inbar, "An improved p300-based brain–computer interface," *IEEE Transactions on Neural Systems and Rehabilitation Engineering*, vol. 13, no. 1, pp. 89–98, 2005.
37. Y. Shahriari and A. Erfanian, "Improving the performance of p300-based brain–computer interface through subspace-based filtering," *Neurocomputing*, vol. 121, pp. 434–441, 2013.
38. W. Speier, C. Arnold, J. Lu, R. K. Taira, and N. Pouratian, "Natural language processing with dynamic classification improves p300 speller accuracy and bit rate," *Journal of Neural Engineering*, vol. 9, no. 1, p. 016004, 2012.
39. M. Spüler, W. Rosenstiel, and M. Bogdan, "Online adaptation of a C-VEP brain–computer interface (BCI) based on error-related potentials and unsupervised learning," *PloS One*, vol. 7, no. 12, p. e51077, 2012.
40. M. Thulasidas, C. Guan, and J. Wu, "Robust classification of EEG signal for brain–computer interface," *IEEE Transactions on Neural Systems and Rehabilitation Engineering*, vol. 14, no. 1, p. 24, 2006.
41. M. S. Treder and B. Blankertz, "Research (C) overt attention and visual speller design in an ERP-based brain–computer interface," *Behavioral & Brain Functions*, vol. 6, 2010.
42. L. Acqualagna, M. S. Treder, M. Schreuder, and B. Blankertz, "A novel brain–computer interface based on the rapid serial visual presentation paradigm," in *Engineering in Medicine and Biology Society (EMBC), 2010 Annual International Conference of the IEEE*. IEEE, 2010, pp. 2686–2689.
43. L. Acqualagna and B. Blankertz, "Gaze-independent BCI-spelling using rapid serial visual presentation (RSVP)," *Clinical Neurophysiology*, vol. 124, no. 5, pp. 901–908, 2013.
44. U. Orhan, K. E. Hild, D. Erdogmus, B. Roark, B. Oken, and M. Fried-Oken, "RSVP keyboard: An EEG based typing interface," in *IEEE International Conference on Acoustics, Speech and Signal Processing (ICASSP)*, 2012, pp. 645–648, 2012.
45. U. Orhan, D. Erdogmus, B. Roark, B. Oken, S. Purwar, K. E. Hild, A. Fowler, and M. Fried-Oken, "Improved accuracy using recursive Bayesian estimation based language model fusion in ERP-based BCI typing systems," in *Engineering in Medicine and Biology Society (EMBC), 2012 Annual International Conference of the IEEE*. IEEE, 2012, pp. 2497–2500.
46. U. Orhan, D. Erdogmus, B. Roark, B. Oken, and M. Fried-Oken, "Offline analysis of context contribution to ERP-based typing BCI performance," *Journal of Neural Engineering*, vol. 10, no. 6, p. 066003, 2013.
47. M. S. Treder, N. M. Schmidt, and B. Blankertz, "Gaze-independent brain–computer interfaces based on covert attention and feature attention," *Journal of Neural Engineering*, vol. 8, no. 6, p. 066003, 2011.
48. P. Aricò, F. Aloise, F. Schettini, A. Riccio, S. Salinari, F. Babiloni, D. Mattia, and F. Cincotti, "Geospell: An alternative p300-based speller interface towards no

eye gaze required," *International Journal of Bioelectromagnetism*, vol. 13, no. 3, pp. 152–153, 2011.
49. F. Schettini, F. Aloise, P. Arico, S. Salinari, D. Mattia, and F. Cincotti, "Control or no-control? reducing the gap between brain–computer interface and classical input devices," in *2012 Annual International Conference of the IEEE Engineering in Medicine and Biology Society*. IEEE, 2012, pp. 1815–1818.
50. Y. Liu, Z. Zhou, and D. Hu, "Gaze independent brain–computer speller with covert visual search tasks," *Clinical Neurophysiology*, vol. 122, no. 6, pp. 1127–1136, 2011.
51. M. Moghadamfalahi, U. Orhan, M. Akcakaya, H. Nezamfar, M. Fried-Oken, and D. Erdogmus, "Language-model assisted brain computer interface for typing: A comparison of matrix and rapid serial visual presentation," *IEEE Transactions on Neural Systems and Rehabilitation Engineering*, vol. 23, no. 5, pp. 910–920, Sept 2015.
52. M. Akcakaya, B. Peters, M. Moghadamfalahi, A. Mooney, U. Orhan, B. Oken, D. Erdogmus, and M. Fried-Oken, "Noninvasive brain computer interfaces for augmentative and alternative communication," *IEEE Reviews in Biomedical Engineering*, vol. 7, no. 1, pp. 31–49, 2014.
53. C. Guan, M. Thulasidas, and J. Wu, "High performance p300 speller for brain–computer interface," in *IEEE International Workshop on Biomedical Circuits and Systems, 2004*. IEEE, 2004, pp. S3–S5.
54. S. Faul, G. Gregorcic, G. Boylan, W. Marnane, G. Lightbody, and S. Connolly, "Gaussian process modeling of EEG for the detection of neonatal seizures," *IEEE Transactions on Biomedical Engineering*, vol. 54, no. 12, pp. 2151–2162, 2007.
55. M. Zhong, F. Lotte, M. Girolami, and A. Lécuyer, "Classifying EEG for brain computer interfaces using gaussian processes," *Pattern Recognition Letters*, vol. 29, no. 3, pp. 354–359, 2008.
56. J. H. Friedman, "Regularized discriminant analysis," *Journal of the American Statistical Association*, vol. 84, no. 405, pp. 165–175, 1989.
57. B. W. Silverman, Density Estimation for Statistics and Data Analysis, vol. 26. CRC Press, Boca Raton, FL, 1986.
58. B. Roark, J. D. Villiers, C. Gibbons, and M. Fried-Oken, "Scanning methods and language modeling for binary switch typing," in *Proceedings of the NAACL HLT 2010 Workshop on Speech and Language Processing for Assistive Technologies*. Association for Computational Linguistics, 2010, pp. 28–36.
59. M. Moghadamfalahi, P. Gonzalez-Navarro, M. Akcakaya, U. Orhan, and D. Erdogmus, "The effect of limiting trial count in context aware BCIs: A case study with language model assisted spelling," in *Foundations of Augmented Cognition*. Springer, 2015, pp. 281–292.
60. M. Moghadamfalahi, J. Sourati, M. Akcakaya, H. Nezamfar, M. Haghighi, and D. Erdogmus, "Active learning for efficient querying from a human oracle with noisy response in a language-model assisted brain computer interface," in *2015 IEEE 25th International Workshop on Machine Learning for Signal Processing (MLSP)*. IEEE, 2015, pp. 1–6.
61. S. Chennu, A. Alsufyani, M. Filetti, A. M. Owen, and H. Bowman, "The cost of space independence in P300-BCI spellers," *Journal of Neuroengineering and Rehabilitation*, vol. 10, no. 82, pp. 1–13, 2013.

7

Compressive Sensing Methods for Reconstruction of Big Sparse Signals

Ljubiša Stanković, Miloš Daković, and Isidora Stanković

CONTENTS

7.1 Introduction 133
7.2 Basic Definitions 134
    7.2.1   Conditions for Reconstruction 135
    7.2.2   Common Measurement Matrices 137
7.3 Norm-Zero-Based Reconstruction 137
    7.3.1   Reconstruction with Known/Estimated Positions 138
    7.3.2   Position Estimation and Reconstruction 138
    7.3.3   Coherence 139
    7.3.4   Block and One-Step Reconstruction 139
7.4 Norm-One-Based Reconstruction Algorithms 141
    7.4.1   LASSO Minimization 142
    7.4.2   Signal Reconstruction with a Gradient Algorithm 143
    7.4.3   Total Variations 144
7.5 Bayesian-Based Reconstruction 145
7.6 Applications in Biomedical Signal Processing 148
7.7 Open Problems and Future Directions 148
7.8 Conclusions 149
References 149

7.1 Introduction

A signal is sparse if the number of its nonzero coefficients is much smaller than the total number of signal samples. Discrete-time signals can be transformed and represented in various transformation domains. Some signals that assume nonzero values in most of the discrete-time domain could be sparse in a transformation domain. They may have just a few nonzero transform coefficients. These signals are then sparse in the respective transformation domain. An observation or measurement is a linear combination of signal coefficients in their sparsity domain. The weights of this linear combination can be random numbers. If a signal is sparse in a transformation domain, then its samples, being linear combinations of transform coefficients, can be considered as measurements as well. Knowing that a signal is sparse, one can represent and reconstruct it using a reduced set of measurements. The number of these measurements/observations is much smaller than the total number of samples [1–10].

A reduced set of measurements can be a result of an aspiration to sense and represent a sparse signal with the lowest possible number of measurements (compressive sensing). This is of great importance in the case of big data setups when a large number of signal elements is considered. A reduced set of measurements can also be a result of a physical unavailability to take a complete set of measurements/samples. It can also happen that some arbitrarily positioned samples of a signal are heavily corrupted by disturbances that should rather be omitted and considered as unavailable. Although the reduced set of measurements/samples in the first case appears as a result of user strategy to compress the information, while in the later two cases, the reduced set of samples is not a result of the user intention, all of them can be considered within the unified compressive sensing theoretical framework.

The reconstruction of big sparse signals from a reduced set of measurements is the topic of this chapter. The chapter is an adaption for big data setups of the chapter presented in Ref. [11]. The conditions and few algorithms for the signal/data reconstruction are presented. Under some conditions, a full (exact) reconstruction of a big sparse signal can be achieved from a reduced set of measurements.

7.2 Basic Definitions

A big set of discrete-time data $x(n), n = 0,1,\ldots,N-1$, with a large number of samples N, is considered. Its coefficients in a representation domain are denoted as

$$\mathbf{X} = [X(0), X(1), \ldots, X(N-1)]^T,$$

where T represents the transpose operation. We consider a signal to be sparse in this representation domain if the number of nonzero coefficients K is much smaller than the number of the original signal samples N, i.e., if $\mathbf{X}(k) = 0$ for $k \notin \mathbb{K} = \{k_1, k_2,\ldots,k_K\}$ and $K \ll N$. The number of nonzero coefficients is commonly denoted by $\|\mathbf{X}\|_0$

$$\|\mathbf{X}\|_0 = \text{card}\{\mathbb{K}\} = K,$$

where $\text{card}\{\mathbb{K}\}$ is the cardinality of set $\mathbb{K}$. It is equal to the number of elements in $\mathbb{K}$. It is called the ℓ_0-norm (norm-zero) or the ℓ_0-pseudo-norm of vector $\mathbf{X}$ although it does not satisfy the norm properties.

The observations/measurements are defined as linear combinations of signal coefficients in the sparsity domain

$$y(m) = \sum_{k=0}^{N-1} X(k)\psi_k(m), \quad (7.1)$$

where $m = 0,1,\ldots,M-1$ is the measurement index and $\psi_k(m)$ are the weighting coefficients. The vector form of the measurement signal is denoted by $\mathbf{y}$

$$\mathbf{y} = [y(0), y(1), \ldots, y(M-1)]^T.$$

The measurements defined by Equation 7.1 can be written as a system of M equations as

$$\begin{bmatrix} y(0) \\ y(1) \\ \vdots \\ y(M-1) \end{bmatrix} = \begin{bmatrix} \psi_0(0) & \psi_1(0) & \cdots & \psi_{N-1}(0) \\ \psi_0(1) & \psi_1(1) & \cdots & \psi_{N-1}(1) \\ \vdots & \vdots & \ddots & \vdots \\ \psi_0(M-1) & \psi_1(M-1) & \cdots & \psi_{N-1}(M-1) \end{bmatrix}$$

$$\times \begin{bmatrix} X(0) \\ X(1) \\ \vdots \\ X(N-1) \end{bmatrix}$$

or using vector notation

$$\mathbf{y} = \mathbf{AX}$$

where $\mathbf{A}$ is the measurement matrix of size $M \times N$.

The fact that the signal is sparse with $X(k) = 0$ for $k \notin \mathbb{K} = \{k_1, k_2,\ldots,k_K\}$ is not included in the measurement matrix $\mathbf{A}$ since the positions of the nonzero values are unknown. If the knowledge that $X(k) = 0$ for $k \notin \mathbb{K}$ were included, then a reduced system would be obtained as

$$\begin{bmatrix} y(0) \\ y(1) \\ \vdots \\ y(M-1) \end{bmatrix} = \begin{bmatrix} \psi_{k_1}(0) & \psi_{k_2}(0) & \cdots & \psi_{k_K}(0) \\ \psi_{k_1}(1) & \psi_{k_2}(1) & \cdots & \psi_{k_K}(1) \\ \vdots & \vdots & \ddots & \vdots \\ \psi_{k_1}(M-1) & \psi_{k_2}(M-1) & \cdots & \psi_{k_K}(M-1) \end{bmatrix}$$

$$\times \begin{bmatrix} X(k_1) \\ X(k_2) \\ \vdots \\ X(k_K) \end{bmatrix}$$

with a reduced $M \times K$ measurement matrix $\mathbf{A}_K$ defined as

$$\mathbf{y} = \mathbf{A}_K \mathbf{X}_K. \quad (7.2)$$

Matrix $\mathbf{A}_K$ would be formed if we assumed/knew the positions of nonzero samples $k \in \mathbb{K}$. It would follow from the measurement matrix $\mathbf{A}$ by omitting the columns corresponding to the zero-valued coefficients in $\mathbf{X}$.

In compressive sensing, the weighting coefficients $\psi_k(m)$ are commonly random numbers, such as Gaussian distributed values. In the case of signal processing, these coefficients usually correspond to the signal transform coefficients. Common measurement matrices will be discussed later in this section.

We will illustrate the basic definitions in a simple example of when only two coefficients in vector $\mathbf{X}$ are nonzero (i.e., the sparsity is $K = 2$). We do not know either their positions or their values. The aim is to find the positions and the values of these coefficients.

The nonzero values will be denoted by $X(i)$ and $X(l)$. A direct way to find the positions i and l would be to analyze the complete set of N independent measurements and to get all values of $X(k)$, $k = 0,1,...,N-1$. However, for big data signals, N is very large, and there are only $K = 2 \ll N$ nonzero coefficients; we can get the result with a very reduced set of observations/measurements.

Consider one measurement

$$y(0) = \sum_{k=0}^{N-1} X(k)\psi_k(0). \tag{7.3}$$

This equation represents an N-dimensional hyperplane in the space of variables $X(0)$, $X(1)$, ... , $X(N-1)$. In order to be able to calculate two coefficients, we should have at least two equations, with one more measurement

$$y(1) = \sum_{k=0}^{N-1} X(k)\psi_k(1). \tag{7.4}$$

Assuming that any two coefficients are nonzero, we will get a possible solution from the system

$$\begin{aligned} y(0) &= X(i)\psi_i(0) + X(l)\psi_l(0) \\ y(1) &= X(i)\psi_i(1) + X(l)\psi_l(1) \end{aligned}. \tag{7.5}$$

Assuming that all weighting coefficients are such that

$$\psi_i(0)\psi_l(1) - \psi_i(1)\psi_l(0) = \det\begin{bmatrix} \psi_i(0) & \psi_l(0) \\ \psi_i(1) & \psi_l(1) \end{bmatrix} \neq 0 \tag{7.6}$$

for any i and l, then the solution will exist for any pair of i and l. Thus, with two measurements, we will get $\binom{N}{2}$ possible solutions with nonzero value in $\mathbf{X}$ at two positions. As expected, two measurements are not enough to be able to solve the problem and to find the positions and the values of two nonzero coefficients.

If we perform two more measurements $y(2)$ and $y(3)$ with other sets of weighting coefficients $\psi_k(2)$ and $\psi_k(3)$, for $k = 0,1,...,N-1$, the result will be an additional two hyperplanes

$$\begin{aligned} y(2) &= \sum_{k=0}^{N-1} X(k)\psi_k(2) = X(i)\psi_i(2) + X(l)\psi_l(2) \\ y(3) &= \sum_{k=0}^{N-1} X(k)\psi_k(3) = X(i)\psi_i(3) + X(l)\psi_l(3). \end{aligned} \tag{7.7}$$

Assuming again that any two coefficients are nonzero, we will get $\binom{N}{2}$ possible solutions as in the case of the first two measurements. If these two sets of solutions produce only one common value for only one pair of positions (i,l), then it is the solution of our problem.

In a matrix form, these four measurements can be written as

$$\mathbf{y} = \mathbf{AX}$$

where $\mathbf{A}$ is the matrix of transform coefficients called the measurement matrix,

$$\mathbf{A} = \begin{bmatrix} \psi_0(0) & \psi_1(0) & \cdots & \psi_{N-1}(0) \\ \psi_0(1) & \psi_1(1) & \cdots & \psi_{N-1}(1) \\ \psi_0(2) & \psi_1(2) & \cdots & \psi_{N-1}(2) \\ \psi_0(3) & \psi_1(3) & \cdots & \psi_{N-1}(3) \end{bmatrix},$$

and $\mathbf{y} = [y(0)\ y(1)\ y(2)\ y(3)]^T$ are the measurements/observations of sparse variable $\mathbf{X} = [X(0)X(1)...X(N-1)]^T$.

A common solution for two pairs of measurements is unique if

$$\det\begin{bmatrix} \psi_i(0) & \psi_l(0) & \psi_p(0) & \psi_q(0) \\ \psi_i(1) & \psi_l(1) & \psi_p(1) & \psi_q(1) \\ \psi_i(2) & \psi_l(2) & \psi_p(2) & \psi_q(2) \\ \psi_i(3) & \psi_l(3) & \psi_p(3) & \psi_q(3) \end{bmatrix} \neq 0 \tag{7.8}$$

for any i,l,p,q. This will be proven in the next section.

7.2.1 Conditions for Reconstruction

The analysis of a signal with an arbitrary sparsity K is similar to the analysis as in the example in the previous section. To get the first set of possible solutions for K nonzero coefficients (i.e., sparsity), we need K measurements. For any combination of K (out of N) nonzero coefficients $X(k)$, $k \in \{k_1, k_2,...,k_K\}$, we will get a possible solution. There exist $\binom{N}{K}$ such possible combinations/solutions (binomial coefficient). Additional K measurements will be used to produce another set of $\binom{N}{K}$ possible solutions. The intersection of these two sets is then the solution of our problem.

The solution is unique if the determinant of all $\mathbf{A}_{2K}$ submatrices of matrix $\mathbf{A}$ are different from zero. This statement will be proven by contradiction. Assume that $M = 2K$ measurements are available within the vector $\mathbf{y}$. Assume that two different solutions for $\mathbf{X}$ of sparsity K exist. Denote the nonzero parts of the solutions by $\mathbf{X}_K^{(1)}$ and $\mathbf{X}_K^{(2)}$. Both of them satisfy the measurements equation,

$$\mathbf{A}_K^{(1)} \mathbf{X}_K^{(1)} = \mathbf{y}$$

and

$$\mathbf{A}_K^{(2)} \mathbf{X}_K^{(2)} = \mathbf{y},$$

where $\mathbf{A}_K^{(1)}$ and $\mathbf{A}_K^{(2)}$ are two different submatrices of matrix $\mathbf{A}$ of size $M \times K$ corresponding to the elements in

$\mathbf{X}_K^{(1)}$ and $\mathbf{X}_K^{(2)}$. If we rewrite these equations by adding zeros

$$\left[\mathbf{A}_K^{(1)} \mathbf{A}_K^{(2)}\right] \begin{bmatrix} \mathbf{X}_K^{(1)} \\ \mathbf{0}_K \end{bmatrix} = \mathbf{y} \text{ and } \left[\mathbf{A}_K^{(1)} \mathbf{A}_K^{(2)}\right] \begin{bmatrix} \mathbf{0}_K \\ \mathbf{X}_K^{(2)} \end{bmatrix} = \mathbf{y} \quad (7.9)$$

and subtract them, we get

$$\left[\mathbf{A}_K^{(1)} \mathbf{A}_K^{(2)}\right] \begin{bmatrix} \mathbf{X}_K^{(1)} \\ -\mathbf{X}_K^{(2)} \end{bmatrix} = \mathbf{0}. \quad (7.10)$$

There are no nonzero solutions for $\mathbf{X}_K^{(1)}$ and $\mathbf{X}_K^{(2)}$ if the determinant of matrix $\mathbf{A}_{2K} = [\mathbf{A}_K^{(1)} \mathbf{A}_K^{(2)}]$ is nonzero. If all possible submatrices $\mathbf{A}_{2K}$ (including all lower-order submatrices) of measurement matrix $\mathbf{A}$ are nonsingular, then two solutions of sparsity K cannot exist, and the solution is unique. Note that there are $\binom{N}{2K}$ submatrices $\mathbf{A}_{2K}$.

Based on the previous analysis, the solution for a K sparse problem is unique if

$$\text{spark}\{\mathbf{A}\} > 2K,$$

where spark$\{\mathbf{A}\}$ is defined as the smallest number of linearly dependent columns, i.e., the smallest submatrix $\mathbf{A}_K$ with $\det(\mathbf{A}_K^H \mathbf{A}_K) = 0$ (singular submatrix). For example, if any column of matrix $\mathbf{A}$ is all zero-valued, then spark$\{\mathbf{A}\} = 1$; if there is no zero column in $\mathbf{A}$ and there exists a singular $\mathbf{A}_2^H \mathbf{A}_2$, then spark$\{\mathbf{A}\} = 2$; and so on. It means that if spark $\{\mathbf{A}\} > 2K$, then all $2K \times 2K$ (and lower-order) submatrices $\mathbf{A}_{2K}^H \mathbf{A}_{2K}$ are nonsingular. This has already been established as the uniqueness condition.

Note that for any square matrix, its determinant is equal to the product of its eigenvalues

$$\det\{\mathbf{A}_{2K}\} = d_1 d_2 \cdot \ldots \cdot d_{2K}.$$

The uniqueness condition can be rewritten as

$$\min_i |d_i| > 0$$

for all submatrices $\mathbf{A}_{2K}$ of the measurement matrix $\mathbf{A}$ with eigenvalues d_i.

In numerical and practical applications, we would not be satisfied if any of the determinants is very close to zero. In this case, the theoretical condition for a unique solution would be satisfied; however, the analysis and possible inversion would be highly sensitive to any kind of noise in measurements [12,13]. Thus, a practical requirement is that the determinant is not just different from zero, but that it sufficiently differs from zero so that an inversion stability and noise robustness are achieved.

From the matrix theory, it is known that the norm of a matrix $\mathbf{A}_{2K}$ satisfies

$$\lambda_{\min} \leq \frac{\|\mathbf{A}_{2K} \mathbf{X}_{2K}\|_2^2}{\|\mathbf{X}_{2K}\|_2^2} = \frac{\mathbf{X}_{2K}^T \mathbf{A}_{2K}^T \mathbf{A}_{2K} \mathbf{X}_{2K}}{\mathbf{X}_{2K}^T \mathbf{X}_{2K}} \leq \lambda_{\max}, \quad (7.11)$$

where $\lambda_{\min}$ and $\lambda_{\max}$ are the minimal and the maximal eigenvalues of the (Gram) matrix $\mathbf{A}_{2K}^T \mathbf{A}_{2K}$ (squared absolute singular values of $\mathbf{A}_{2K}$) and $\|\mathbf{X}\|_2^2 = |X(0)|^2 + \ldots |X(N-1)|^2$ is the squared ℓ_2-norm (norm-two).

The isometry property for a linear transformation matrix $\mathbf{A}$ holds if

$$\|\mathbf{AX}\|_2^2 = \|\mathbf{X}\|_2^2 \quad \text{or} \quad \frac{\|\mathbf{AX}\|_2^2}{\|\mathbf{X}\|_2^2} = 1.$$

The restricted isometry property (RIP) for a matrix $\mathbf{A}_{2K}$ and a $2K$-sparse vector $\mathbf{X}_{2K}$ holds if

$$1 - \delta_{2K} \leq \frac{\|\mathbf{A}_{2K} \mathbf{X}_{2K}\|_2^2}{\|\mathbf{X}_{2K}\|_2^2} \leq 1 + \delta_{2K}, \quad (7.12)$$

where $0 \leq \delta_{2K} < 1$ is the isometric constant [14,15]. From Equations 7.11 and 7.12, we can write

$$\delta_{2K} = \max\{1 - \lambda_{\min}, \lambda_{\max} - 1\}.$$

Commonly, isometric constant is defined by $\lambda_{\max} - 1$, and it is calculated as maximal eigenvalue of matrix $\mathbf{A}_{2K}^T \mathbf{A}_{2K} - \mathbf{I}$. Normalized energies of the columns of matrix $\mathbf{A}$ (diagonal elements of $\mathbf{A}_{2K}^T \mathbf{A}_{2K}$) are assumed. Otherwise, the normalization factors should be added. For complex-valued matrices, Hermitian transpose should be used in $\mathbf{A}_{2K}^H \mathbf{A}_{2K}$.

For a K-sparse vector $\mathbf{X}$ and a measurement matrix $\mathbf{A}$, the RIP is satisfied if Equation 7.12 holds for all submatrices $\mathbf{A}_K$ with $0 \leq \delta_K < 1$. The solution for K-sparse vector is unique if the measurement matrix satisfies the RIP for $2K$-sparse vector $\mathbf{X}$ with $0 \leq \delta_{2K} < 1$.

Note that if the RIP is satisfied, then $\lambda_{\min} > 0$.

Restricted isometry property for small δ_{2K} is closer to the isometry property and improves the solution stability. It can be related to the matrix conditional number. The conditional number of a matrix $\mathbf{A}_{2K}^T \mathbf{A}_{2K}$ is defined as the ratio of its maximal and minimal eigenvalues

$$\text{cond}\{\mathbf{A}_{2K}^T \mathbf{A}_{2K}\} = \frac{\lambda_{\max}}{\lambda_{\min}}.$$

If a matrix A_{2K} satisfies the RIP with δ_{2K}, then

$$\text{cond}\{\mathbf{A}_{2K}^T \mathbf{A}_{2K}\} \leq \frac{1 + \delta_{2K}}{1 - \delta_{2K}}.$$

With small values of δ_{2K}, the conditional number is close to 1, meaning stable invertibility and low sensitivity to the input noise (small variations of the input signal [measurements] do not cause large variations of the result).

7.2.2 Common Measurement Matrices

Some common measurement matrices used in practical applications and theoretical considerations will be presented here.

Randomness of measurement matrices is a favorable property in compressive sensing, and matrices with random elements are often used. The most common is the measurement matrix with zero-mean unity variance Gaussian distributed numbers as elements

$$\phi_k(n) = \frac{1}{\sqrt{M}} \mathcal{N}(0,1)$$

normalized with $1/\sqrt{M}$ so that the energy of each column is 1.

In signal processing, the most common transform is the Discrete Fourier transform (DFT). The coefficients of its direct transform matrix Φ are defined as

$$\phi_k(n) = \exp(-j2\pi nk/N).$$

The inverse DFT matrix coefficients are $\psi_k(n) = \frac{1}{N}\exp(j2\pi nk/N)$. Commonly, the measurements are the signal samples $y(m-1) = x(n_m)$ for $m = 1,\ldots,M$ where

$$n_m \in \mathbb{M} = \{n_1, n_2, \ldots, n_M\} \subset \{0, 1, \ldots, N-1\},$$

and

$$y(m-1) = x(n_m) = \frac{1}{N} \sum_{k=0}^{N-1} X(k) e^{j2\pi n_m k/N}.$$

Therefore, the measurement matrix is obtained by keeping the rows of the inverse DFT matrix corresponding to the samples at $n_m \in \{0,1,\ldots,N-1\}$, for the measurements $m = 1,2,\ldots,M$,

$$\mathbf{A} = \frac{1}{N} \begin{bmatrix} 1 & e^{j2\pi n_1/N} & \cdots & e^{j2\pi n_1(N-1)/N} \\ 1 & e^{j2\pi n_2/N} & \cdots & e^{j2\pi n_2(N-1)/N} \\ \vdots & \vdots & \ddots & \vdots \\ 1 & e^{j2\pi n_M/N} & \cdots & e^{j2\pi n_M(N-1)/N} \end{bmatrix}. \quad (7.13)$$

This is a partial inverse DFT matrix. In compressive sensing theory, it is common to normalize the measurement matrix so that the energy of its columns (diagonal elements of $\mathbf{A}^H\mathbf{A}$ matrix) is equal to 1. Then the factor $1/N$ in $\mathbf{A}$ should be replaced by $1/\sqrt{M}$.

In order to increase randomness in the Fourier transform matrix, the measurements may be taken at any random instant. Then the measurement vector elements are $y(m-1) = x(t_m)$ where t_m, $m = 1,2,\ldots,M$ are random instants within the considered time interval T. The measurement matrix follows then from the Fourier series definition $x(t) = \sum_{k=0}^{N-1} X(k) \exp(j2\pi kt/T)$. It has been assumed that the Fourier series coefficients are within $0 \le k \le N-1$. The measurement matrix is

$$\mathbf{A} = \begin{bmatrix} 1 & e^{j2\pi t_1/T} & \cdots & e^{j2\pi t_1(N-1)/T} \\ 1 & e^{j2\pi t_2/T} & \cdots & e^{j2\pi t_2(N-1)/T} \\ \vdots & \vdots & \ddots & \vdots \\ 1 & e^{j2\pi t_M/T} & \cdots & e^{j2\pi t_M(N-1)/T} \end{bmatrix} \quad (7.14)$$

with a possible normalization factor $1/\sqrt{M}$. This measurement matrix is a partial random inverse Fourier transform matrix.

Other signal transforms and other random distributions can be used to form an appropriate measurement matrix. For example, if a signal is sparse in the discrete cosine transform (DCT) domain (or in the discrete sine transform [DST] domain), then this transform will be used to perform measurements and form the corresponding measurement matrix.

Next we will present reconstruction algorithms with three different approaches. The first group of algorithms is based on the direct problem solution with a minimal number of nonzero coefficients, avoiding the direct search combinatorial approach. In the second group of algorithms, the convex relaxation of the ℓ_0-norm to the ℓ_1-norm is used with appropriate iterative solvers. The third group of algorithms is based on the Bayesian approach to the sparse signals' reconstruction.

7.3 Norm-Zero-Based Reconstruction

For a given set of measurements defined by $\mathbf{y}$, the reconstruction problem can be observed as finding the vector $\mathbf{X}$ with the smallest number $K = \|\mathbf{X}\|_0$ of nonzero coefficients satisfying $\mathbf{y} = \mathbf{A}\mathbf{X}$,

$$\min \|\mathbf{X}\|_0 \text{ subject to } \mathbf{y} = \mathbf{A}\mathbf{X}. \quad (7.15)$$

The ℓ_0-norm cannot be used in common minimization methods. However, a class of algorithms is based on the solution of the measurement equation $\mathbf{y} = \mathbf{A}\mathbf{X}$ with a minimal possible number of the nonzero coefficients. Therefore, these algorithms are based on the minimization of the number of coefficients $K = \|\mathbf{X}\|_0$ in an implicit way.

Direct search reconstruction is an example of such algorithms. Commonly $M > 2K$ is used in order to get a solution with sufficient stability. Then the measurement matrix $\mathbf{A}$ is of an $M \times N$ order, while the matrix corresponding to a K sparse vector $\mathbf{X}_K$ is $\mathbf{A}_K$. It is of an $M \times K$ order. Note that the product $\mathbf{A}_K^T \mathbf{A}_K$ is of $K \times K$ order. Thus, all previous conclusions are valid for this kind of matrix as well.

Consider a measurement matrix $\mathbf{A}$ that we want to use for M measurements of K-sparse vector $\mathbf{X}$. This measurement matrix $\mathbf{A}$ will produce a unique solution for any K-sparse vector if all its submatrices $\mathbf{A}_{2K}$ satisfy spark or RIP. To check any of these properties, we have to make $\binom{N}{2K}$ calculations for all possible $\mathbf{A}_{2K}$ submatrices of matrix $\mathbf{A}$.

For each possible arrangement of K nonzero coefficients in $\mathbf{X}$ denoted by $\mathbf{X}_K$, the system of M equations with $K < M$ unknowns $\mathbf{y} = \mathbf{A}_K \mathbf{X}_K$ is solved in the least-mean-square sense (details will be presented later). There are $\binom{N}{K}$ such systems. The arrangement whose solution produces zero-valued mean square error is the solution of our problem.

For any reasonable N and K, the number of combinations $\binom{N}{K}$ and $\binom{N}{2K}$ is extremely large and computationally not feasible. This combinatorial problem is an non-deterministic polynomial-time (NP) hard problem (with nonpolynomial number of combinations).

Computational complexity will be significantly reduced if we are able to estimate positions of the nonzero coefficients in the solution $\mathbf{X}$ and to solve the problem with the minimal possible number of nonzero coefficients [16,17].

7.3.1 Reconstruction with Known/Estimated Positions

Let us firstly assume that the positions of nonzero coefficients are already estimated and known. We will denote the set of the known positions as

$$\mathbb{K} = \{k_1, k_2, \ldots, k_K\}.$$

The values of the coefficients at these positions are unknown. The goal is to reconstruct them. We will denote them as

$$\mathbf{X}_K = [X(k_1), X(k_2), \ldots, X(k_K)]^T.$$

In a matrix form, the system $\mathbf{y} = \mathbf{A}\mathbf{X}$ reduces to

$$\mathbf{y} = \mathbf{A}_K \mathbf{X}_K, \quad (7.16)$$

where $\mathbf{A}_K$ is an $M \times K$ matrix obtained from the measurement matrix $\mathbf{A}$ with columns corresponding to the nonzero transform coefficients. The matrix can be defined as

$$\mathbf{A}_K = \begin{bmatrix} \psi_{k_1}(0) & \psi_{k_2}(0) & \cdots & \psi_{k_K}(0) \\ \psi_{k_1}(1) & \psi_{k_2}(1) & \cdots & \psi_{k_K}(1) \\ \vdots & \vdots & \ddots & \vdots \\ \psi_{k_1}(M-1) & \psi_{k_2}(M-1) & \cdots & \psi_{k_K}(M-1) \end{bmatrix}. \quad (7.17)$$

Equation 7.16 is used with the relation $K < M \leq N$. Its solution, in the least-mean-square sense, follows from the minimization of difference of the available signal values and the values produced by the reconstructed coefficients $X(k)$, defined by

$$e^2 = (\mathbf{y} - \mathbf{A}_K \mathbf{X}_K)^H (\mathbf{y} - \mathbf{A}_K \mathbf{X}_K)$$
$$= \|\mathbf{y}\|_2^2 - 2\mathbf{X}_K^H \mathbf{A}_K^H \mathbf{y} + \mathbf{X}_K^H \mathbf{A}_K^H \mathbf{A}_K \mathbf{X}_K, \quad (7.18)$$

where H denotes the Hermitian transpose conjugate. Using symbolic derivation over the vector of unknowns, we get the minimum of error e^2 from

$$\frac{\partial e^2}{\partial \mathbf{X}_K^H} = -2\mathbf{A}_K^H \mathbf{y} + 2\mathbf{A}_K^H \mathbf{A}_K \mathbf{X}_K = 0.$$

The solution of the reconstruction is then

$$\mathbf{X}_K = \left(\mathbf{A}_K^H \mathbf{A}_K\right)^{-1} \mathbf{A}_K^H \mathbf{y}. \quad (7.19)$$

7.3.2 Position Estimation and Reconstruction

In this section, we will present a procedure to estimate the set of positions $\mathbb{K}$ of the nonzero coefficients in $\mathbf{X}$. This is the *matching pursuit* (MP) approach to the sparse signal reconstruction.

We have a set of equations represented in the measurement matrix form as $\mathbf{y} = \mathbf{A}\mathbf{X}$. For the initial estimation, we will use

$$\mathbf{X}_0 = \mathbf{A}^H \mathbf{y} = \mathbf{A}^H \mathbf{A} \mathbf{X}. \quad (7.20)$$

In the ideal case, matrix $\mathbf{A}^H \mathbf{A}$ would be the identity matrix, and the initial estimate would fully correspond to the solution. However, with a reduced number of measurements, this cannot be achieved. Still, the requirement is that diagonal elements of $\mathbf{A}^H \mathbf{A}$ are larger than the other elements. The properties related to the elements of $\mathbf{A}^H \mathbf{A}$ will be discussed later.

The first element of K is found as the position of the maximum of $\mathbf{X}_0$,

$$k_1 = \arg\max |\mathbf{A}^H \mathbf{y}|. \quad (7.21)$$

The reduced system (Equation 7.16) is solved using the solution in Equation 7.19 with the position $\mathbb{K} = \{k_1\}$. In this way, we obtain $\mathbf{X}_1$. Then $\mathbf{y}_1 = \mathbf{A}_1 \mathbf{X}_1$ is calculated. If $\mathbf{y}_1 = \mathbf{y}$, the signal is of sparsity $K = 1$, and $\mathbf{X}_1$ is the problem solution. If this is not the case, the estimated component is removed from $\mathbf{y}$, and the signal $\mathbf{e}_1 = \mathbf{y} - \mathbf{y}_1$ is formed. Then the second nonzero position is estimated as

$$k_2 = \arg\max |\mathbf{A}^H \mathbf{e}_1|,$$

and the set $\mathbb{K} = \{k_1, k_2\}$ is formed. Equation 7.16 is solved again with the new set, and $\mathbf{X}_2$ and $\mathbf{y}_2$ are calculated. Next we form the vector $\mathbf{e}_2 = \mathbf{y} - \mathbf{y}_2$. If it is a zero vector, the solution is found as $\mathbf{y} = \mathbf{y}_2$. If this is not the case, we will continue to estimate k_3 using $\mathbf{e}_2$ and to find $\mathbb{K} = \{k_1, k_2, k_3\}$, followed by $\mathbf{X}_3$ and $\mathbf{y}_3$ calculation. This procedure is continued until zero (or acceptable) error is achieved. The algorithm for MP reconstruction is presented in Algorithm 1.

Algorithm 1 Norm-zero-based MP reconstruction

Input:

- Measurements vector **y**
- Measurement matrix **A**
- Number of selected coefficients in each iteration r
- Required precision ε

1: $\mathbb{K} \leftarrow \varnothing$
2: $\mathbf{e} \leftarrow \mathbf{y}$
3: **while** $\|\mathbf{e}\|_2 > \varepsilon$ **do**
4: $(k_1, k_2, \ldots, k_r) \leftarrow$ positions of r highest values in $\mathbf{A}^H \mathbf{e}$
5: $\mathbb{K} \leftarrow \mathbb{K} \cup \{k_1, k_2, \ldots, k_r\}$
6: $\mathbf{A}_K \leftarrow$ columns of matrix **A** selected by set $\mathbb{K}$
7: $\mathbf{X}_K \leftarrow \text{pinv}(\mathbf{A}_K)\mathbf{y}$
8: $\mathbf{y}_K \leftarrow \mathbf{A}_K \mathbf{X}_K$
9: $\mathbf{e} \leftarrow \mathbf{y} - \mathbf{y}_K$
10: **end while**
11: $\mathbf{X} \leftarrow \begin{cases} 0 & \text{for positions not in } \mathbb{K} \\ \mathbf{X}_K & \text{for positions in } \mathbb{K} \end{cases}$

Output:

- Reconstructed signal coefficients **X**

7.3.3 Coherence

Now we will analyze the elements of $\mathbf{A}^H \mathbf{A}$. The diagonal elements are

$$a_{kk} = \sum_{i=0}^{M-1} \psi_k(i) \psi_k^*(i) = \langle \psi_k, \psi_k^* \rangle,$$

while the elements outside the diagonal are

$$a_{mk} = \sum_{i=0}^{M-1} \psi_m(i) \psi_k^*(i) = \langle \psi_m, \psi_k^* \rangle, \quad m \neq k,$$

where $\langle \psi_m, \psi_k^* \rangle$ denotes a scalar product of $\psi_m(i)$ and $\psi_k^*(i)$.

The coherence index of a matrix **A** is defined as a maximal absolute value of the normalized scalar product of its two columns

$$\mu = \max|\mu(m, k)|, \text{ for } m \neq k$$

where

$$\mu(m, k) = \frac{a_{mk}}{a_{kk}} = \frac{\sum_{i=0}^{M-1} \psi_m(i) \psi_k^*(i)}{\sum_{i=0}^{M-1} |\psi_k(i)|^2} = \frac{\langle \psi_m, \psi_k^* \rangle}{\langle \psi_k, \psi_k^* \rangle} \quad (7.22)$$

and ψ_k is the kth column of matrix **A**. This index plays an important role in the analysis of measurement matrices. Smaller values of coherence index are preferred, since the matrix $\mathbf{A}^H \mathbf{A}$ is closer to the identity matrix. However, the coherence index cannot be arbitrarily small for an $M \times N$ matrix **A** ($M < N$). The Welch upper bound relation holds,

$$\mu \geq \sqrt{\frac{N-M}{M(N-1)}}. \quad (7.23)$$

Consider a K-sparse signal. The worst-case influence to the strongest component occurs when the remaining $K-1$ components are equally strong (with equal amplitudes). Then the amplitude of the strongest component is 1 (assuming normalized measurement matrix with $a_{kk} = 1$ and the normalized signal component amplitude). The worst case for the detection of this component is when all other components maximally reduce the value of the considered component in the initial estimate. Influence of the kth component to the mth position is $\sum_{i=0}^{M-1} \psi_m(i) \psi_k^*(i)$. Its maximal possible value is μ. The worst-case amplitude of the considered component is then $1 - (K-1)\mu$. At the position where there is no component, in the worst case, the maximal possible contributions of all K components (being equal to μ) sum up in phase to produce maximal possible disturbance $K\mu$. Detection of the component is always successful if the worst possible amplitude, at the considered component position, is greater than the largest possible disturbance [18],

$$1 - (K-1)\mu > K\mu,$$

producing

$$K < \frac{1}{2}\left(1 + \frac{1}{\mu}\right). \quad (7.24)$$

After the first component is successfully detected, then the previous relation will hold for $K - 1$ sparsity, guaranteeing the exact (unique) solution. Relation between K, M and N can be derived from Equations 7.22 and 7.24.

The RIP constant can be related to the coherence index as

$$\delta_K \leq (1 - K)\mu.$$

7.3.4 Block and One-Step Reconstruction

For the iterative procedure calculation, we need matrix inversion in Equation 7.19 for each iteration (in addition to the other, less complex operations). For sparsity K, the order of calculation complexity is $1^3 + 2^3 + \ldots + K^3 \sim K^4$. The calculation complexity will be reduced if we are able to estimate nonzero positions in blocks in each iteration.

The most efficient way is to estimate all positions in a single iteration, with calculation complexity of K^3 order.

It is interesting to note that in the reconstructions based on the position estimation, if a few false positions are detected, in general, the algorithm will produce correct solutions of all nonzero coefficients $X(k)$, including these few zero values of the coefficients [19]. The maximal number of equations that we can use in Equation 7.19 is equal to the number of measurements M. Sparsity of the result has to meet the reconstruction condition.

Example 1 Signal

$$x(t) = \sin\left(12\pi t + \frac{\pi}{3}\right) + 0.6\cos\left(32\pi t + \frac{\pi}{4}\right)$$

is considered within $0 \le t < 1$. This signal is uniformly sampled at $t_n = n\Delta t = n/N$, with $N = 64$. A random subset $\mathbf{y}$ of $M = 32$ samples at $n_i, i = 1, 2, \ldots, M$ is available, Fig. 7.1 (first row, left). The measurement matrix $\mathbf{A}$ is a partial DFT matrix corresponding to the samples defined by time indices $n_i, i = 1, 2, \ldots, M$.

The one-step procedure for nonzero coefficient indices' determination is used. The recovered signal is calculated for detected DFT positions. The recovered DFT values are denoted as $X_r(k)$ and presented in Fig. 7.1 (second row, right). Its values are the same as in the DFT of the original signal, Fig. 7.1 (first row, right). A simple and efficient uniqueness check in this case is proposed [20].

The reconstruction of a sparse signal with the desired sparsity K can be modified and obtained in an iterative way. The algorithm is called compressive sampling matching pursuit (CoSaMP), and it is introduced in Ref. [21]. The measurement vector $\mathbf{y}$ is projected onto columns of the measurement matrix $\mathbf{A}$, and $2K$ positions with the highest projection magnitudes are selected. The set of the selected positions is expanded with nonzero positions in the current estimate of sparse signal $\mathbf{X}$. A solution in the least-square sense is found, and K coefficients with the highest magnitudes are selected as the reconstructed signal $\mathbf{X}$. The measurements vector is adjusted by subtracting the current solution, and the iterative procedure is repeated. The CoSaMP reconstruction procedure is presented in Algorithm 2.

The stopping criterion of the algorithm can be a predefined number of iterations or the norm of the adjusted measurement vector $\mathbf{e}$.

Algorithm 2 CoSaMP Reconstruction Algorithm

Input:

- Measurements vector $\mathbf{y}$
- Measurement matrix $\mathbf{A}$
- Desired sparsity K

1: $\mathbf{X} \leftarrow \mathbf{0}_{N\times 1}$
2: $\mathbf{e} \leftarrow \mathbf{y}$
3: **repeat**
4: $\mathbb{T}_1 \leftarrow$ positions of $2K$ highest values in $\mathbf{A}^H \mathbf{e}$
5: $\mathbb{T}_2 \leftarrow$ positions of nonzero coefficients in $\mathbf{X}$
6: $\mathbb{T} \leftarrow \mathbb{T}_1 \cup \mathbb{T}_2$
7: $\mathbf{A}_T \leftarrow$ columns from matrix $\mathbf{A}$ selected by set $\mathbb{T}$
8: $\mathbf{B} \leftarrow \text{pinv}(\mathbf{A}_T)\mathbf{y}$

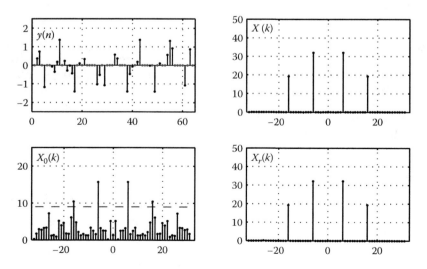

FIGURE 7.1
One-step signal recovery: available samples $y(n)$; exact DFT coefficients $X(k)$; initial estimation of the DFT coefficients $X_0(k)$; the estimated DFT coefficients $X_r(k)$.

9: Put K coefficients with highest magnitude from **B** to the corresponding positions in **X** and set remaining coefficients to zero.
10: $\mathbf{e} \leftarrow \mathbf{y} - \mathbf{AX}$ stopping criterion is satisfied.

Output:

- Reconstructed K-sparse signal vector **X**

7.4 Norm-One-Based Reconstruction Algorithms

The ℓ_0-norm-based minimization is not a convex problem and cannot be solved using well-developed iterative and linear programming methods. Significant efforts have been made to reformulate the ℓ_0-norm-based minimization problem into a convex form, and the closest convex form is the ℓ_1-norm (norm-one). It has been shown that, under certain conditions, minimization of the ℓ_1-norm produces the same solution as the minimization of the ℓ_0-norm [1–3,22].

For the ℓ_1-norm-based reconstructions, the problem is reformulated as

$$\min \| \mathbf{X} \|_1 \text{ subject to } \mathbf{y} = \mathbf{AX}$$

where

$$\| \mathbf{X} \|_1 = \sum_{k=0}^{N-1} |X(k)|.$$

This is the so-called *basis pursuit* (BP) approach to the sparse signal reconstruction.

Let us consider a signal with sparsity $K = 1$, dimension $N = 3$, and two measurements $M = 2$ described by

$$y(0) = X(0)\psi_0(0) + X(1)\psi_1(0) + X(2)\psi_2(0)$$
$$y(1) = X(0)\psi_0(1) + X(1)\psi_1(1) + X(2)\psi_2(1). \quad (7.25)$$

Each of the equations is a plane in space of variables $X(0)$, $X(1)$, and $X(2)$, while the two equations represent a line in this space, as illustrated in Fig. 7.2. The "ball" with a constant norm is presented for the ℓ_1-norm and the $\ell_{1/4}$-norm that is close to the ℓ_0-norm. The solution of the minimization problem is the intersection of the "measurement line" with a minimal possible ball. We can conclude from both both cases that $X(0)$ is indicated as a nonzero coefficient.

The solution is the same as long as the measurement line does not penetrate through the minimal ball. For the ℓ_0-norm case, the measurement line should not have zeros in the direction vector. For this illustrative example, this condition is equivalent to the condition that all submatrices of the measurement matrix of order 2×2 have a nonzero determinant (since these determinants are equal to the direction coefficients of the measurement line). For the ℓ_1-norm case, the condition for measurement line direction is obviously more restrictive since it must not penetrate through the ℓ_1-ball.

Generally, the RIP is the condition that should be satisfied to give equivalent solutions between ℓ_0-norm and the ℓ_1-norm. For a K-sparse vector and a measurement matrix **A**, the solution of the ℓ_0-norm minimization problem is the same as the solution of the corresponding ℓ_1-norm-based minimization process if the measurement matrix satisfies

$$1 - \delta_{2K} \leq \frac{\| \mathbf{A}_{2K} \mathbf{X}_{2K} \|_2^2}{\| \mathbf{X}_{2K} \|_2^2} \leq 1 + \delta_{2K}, \quad (7.26)$$

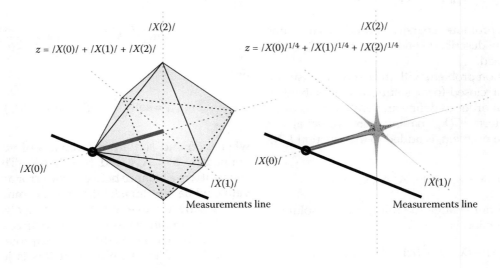

FIGURE 7.2
Illustration of the solution with the ℓ_1-norm (left) and the $\ell_{1/4}$-norm (right; close to the ℓ_0-norm) for a three-dimensional case.

where $0 \leq \delta_{2K} < \sqrt{2} - 1$ for all submatrices of order $2K$ of the measurement matrix $\mathbf{A}$ [14]. Note that for the ℓ_0-norm, the restricted isometry constant range was $0 \leq \delta_{2K} < 1$.

The minimization problem based on the ℓ_1-norm can be reformulated in various ways. One example is its Lagrangian formulation

$$F(\mathbf{X}) = \| \mathbf{y} - \mathbf{AX} \|_2^2 + \lambda \| \mathbf{X} \|_1,$$

where $F(\mathbf{X})$ is the function to be minimized.

The problem reformulated in a constrained form is

$$\min \| \mathbf{X} \|_1 \text{ subject to } \| \mathbf{y} - \mathbf{AX} \|_2^2 < \varepsilon,$$

where ε is a sufficiently small parameter. This problem is the minimization of the ℓ_1-norm with quadratic constraint.

There are many ways to solve the stated problem, based on the constrained or Lagrangian form [22]. Many of them are developed within the regression theory. We will present in detail just one of them, based on the least absolute selection and shrinkage operator (LASSO) formulation and Lagrangian minimization form [23].

7.4.1 LASSO Minimization

The ℓ_1-norm-based minimization can be formulated as the minimization of the error $\mathbf{y} - \mathbf{AX}$ with a condition imposed on $\mathbf{X}$. The cost function

$$F(\mathbf{X}) = \| \mathbf{y} - \mathbf{AX} \|_2^2 + \lambda \| \mathbf{X} \|_1$$
$$= \| \mathbf{y} \|_2^2 - \mathbf{X}^T \mathbf{A}^T \mathbf{y} - \mathbf{y}^T \mathbf{AX} + \mathbf{X}^T \mathbf{A}^T \mathbf{AX} + \lambda \mathbf{X}^T \text{sign}\{\mathbf{X}\}$$

is used here. The LASSO minimization problem is formulated as

$$\mathbf{X} = \arg \min_{\mathbf{X}} \{ \| \mathbf{y} - \mathbf{AX} \|_2^2 + \lambda \| \mathbf{X} \|_1 \}.$$

Function $\|X\|_1$ promotes sparsity. It produces the same results (under the described conditions) as if $\|X\|_p$, with p close to 0, is used.

The minimization problem with the ℓ_1-norm constraint does not have a closed-form solution. It is solved in iterative ways. In order to define an iterative procedure, a nonnegative term $G(\mathbf{X})$, having zero value at the solution $\mathbf{X}_s$ of the problem, is added to the function $F(\mathbf{X})$. It is defined as

$$G(\mathbf{X}) = (\mathbf{X} - \mathbf{X}_s)^T (\alpha \mathbf{I} - \mathbf{A}^T \mathbf{A})(\mathbf{X} - \mathbf{X}_s).$$

This term will not change the minimization solution. The new cost function is

$$H(\mathbf{X}) = F(\mathbf{X}) + (\mathbf{X} - \mathbf{X}_s)^T (\alpha \mathbf{I} - \mathbf{A}^T \mathbf{A})(\mathbf{X} - \mathbf{X}_s),$$

where α is a constant so that the added term is always nonnegative. It means that $\alpha > \lambda_{\max}$, where $\lambda_{\max}$ is the largest eigenvalue of $\mathbf{A}^T \mathbf{A}$. The gradient of $H(\mathbf{X})$ is

$$\nabla H(\mathbf{X}) = \frac{\partial H(\mathbf{X})}{\partial \mathbf{X}^T}$$
$$= -2\mathbf{A}^T \mathbf{y} + 2\mathbf{A}^T \mathbf{AX} + \lambda \text{sign}\{\mathbf{X}\} + 2(\alpha \mathbf{I} - \mathbf{A}^T \mathbf{A})(\mathbf{X} - \mathbf{X}_s).$$

The solution of $\nabla H(\mathbf{X}) = 0$ is

$$-\mathbf{A}^T \mathbf{y} + \frac{\lambda}{2} \text{sign}\{\mathbf{X}\} - (\alpha \mathbf{I} - \mathbf{A}^T \mathbf{A})\mathbf{X}_s + \alpha \mathbf{X} = 0$$

$$\mathbf{X} + \frac{\lambda}{2\alpha} \text{sign}\{\mathbf{X}\} = \frac{1}{\alpha} \mathbf{A}^T (\mathbf{y} - \mathbf{AX}_s) + \mathbf{X}_s.$$

The corresponding iterative relation is of the form

$$\mathbf{X}_{s+1} + \frac{\lambda}{2\alpha} \text{sign}\{\mathbf{X}_{s+1}\} = \frac{1}{\alpha} \mathbf{A}^T (\mathbf{y} - \mathbf{AX}_s) + \mathbf{X}_s.$$

The soft-thresholding rule is used as a solution of the scalar equation

$$x + \lambda \text{sign}(x) = y.$$

It is defined by the function $\text{soft}(y, \lambda)$ as

$$x = \text{soft}(y, \lambda) = \begin{cases} y + \lambda & \text{for} \quad y < -\lambda \\ 0 & \text{for} \quad |y| \leq \lambda \\ y - \lambda & \text{for} \quad y > \lambda \end{cases},$$

or

$$\text{soft}(y, \lambda) = \text{sign}(y) \max\{0, |y| - \lambda\}.$$

The same rule can be applied to each coordinate of vector $\mathbf{X}_{s+1}$,

$$\mathbf{X}_{s+1} = \text{soft}\left(\frac{1}{\alpha} \mathbf{A}^T (\mathbf{y} - \mathbf{AX}_s) + \mathbf{X}_s, \frac{\lambda}{2\alpha} \right) \quad (7.27)$$

or

$$X(k)_{s+1} = \text{soft}\left(\frac{1}{\alpha}(a(k) - b(k)) + X(k)_s, \frac{\lambda}{2\alpha} \right)$$

where $a(k)$ and $b(k)$ are coordinates of vectors $\mathbf{a}$ and $\mathbf{b}$ defined by $\mathbf{a} = \mathbf{A}^T \mathbf{y}$ and $\mathbf{b} = \mathbf{A}^T \mathbf{AX}_s$. The Lagrangian constant λ is a balance between the error and the ℓ_1-norm value, while $\alpha = 2\max\{\text{eig}\{\mathbf{A}^T \mathbf{A}\}\}$ is commonly used.

This is the iterative soft-thresholding algorithm (ISTA) for LASSO minimization. It is summarized in Algorithm 3. It can be easily modified to improve convergence to fast ISTA (FISTA). Note that this is just one of the possible solutions of the minimization problem with the ℓ_1-norm.

One of the most popular software tools for the ℓ_1-norm-based signal reconstruction is the ℓ_1-magic [24]. It is based on the primal-dual algorithm for linear programming [25].

Algorithm 3 LASSO–ISTA reconstruction

Input:

- Measurements vector **y**
- Measurement matrix **A**
- Regularization parameter α
- Sparsity promotion parameter λ

1: $\mathbf{X} \leftarrow \mathbf{0}_{N \times 1}$
2: **repeat**
3: $\mathbf{s} \leftarrow \frac{1}{\alpha} \mathbf{A}^T (\mathbf{y} - \mathbf{A}\mathbf{X}) + \mathbf{X}$
4: **for** $k \leftarrow 1$ to N **do**
5: $X(k) \leftarrow \begin{cases} s(k) + \lambda & \text{for } s(k) < -\lambda \\ 0 & \text{for } |s(k)| \leq \lambda \\ s(k) - \lambda & \text{for } s(k) > \lambda \end{cases}$
6: **end for**
7: **until** stopping criterion is satisfied

Output:

- Reconstructed signal coefficients **X**

Example 2

Measurement matrix **A** is formed as a Gaussian random matrix of the size 40 × 100. Since there are 40 measurements, the random variable $N(0, \sigma^2)$ with $\sigma^2 = 1/40$ is used. The original sparse signal of the total length $N = 100$ is $X(k) = \delta(k - 3) + 0.7\delta(k - 17) - 0.5\delta(k - 35) + 0.7\delta(k - 41)$ in the transformation domain. It is measured with a matrix **A** with 40 measurements stored in vector **y**. All 60 signal values are reconstructed using these 40 measurements **y** and the matrix **A**, in 1000 iterations. In the initial iteration, $X_0 = 0$ is used. Then, for each next s, the new values of **X** are calculated using Equation 7.27, given data **y** and matrix **A**. The value of $\alpha = 2\max\{\text{eig}\{\mathbf{A}^T\mathbf{A}\}\}$ is used. The results for $\lambda = 0.1$ and $\lambda = 0.001$ are presented in Fig. 7.3. For very small $\lambda = 0.001$, the result is not sparse, since the constraint is too weak.

7.4.2 Signal Reconstruction with a Gradient Algorithm

We already mentioned that a sparse signal **X** can be reconstructed from a reduced set of $M < N$ measurements. We will introduce here a different approach of using the ℓ_1-minimization for reconstruction of the signal. Here, we assume that we have some missing measurements or some of the measurements in the signal are highly corrupted and need to be reconstructed. This algorithm is a kind of modified direct search method. The advantage of the gradient-based algorithm over the direct search is that when the number of missing samples is large, then direct search cannot be used due to its high calculation complexity.

We know that the vector of available measurements **y** is obtained by using/selecting the available M measurements from the full set **x**, i.e.,

$$\mathbf{y} = [x(n_1), x(n_2), \ldots, x(n_M)]^T.$$

The positions of the selected (available) measurements will be denoted by $\mathbb{M}$ and the set of the positions of the

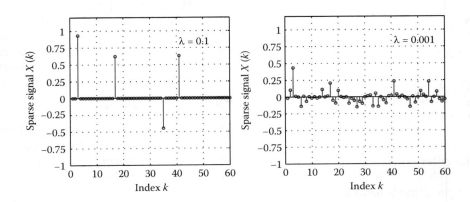

FIGURE 7.3
A sparse signal with $N = 100$ and $K = 4$ reconstructed using a reduced set of $M = 40$ measurements and LASSO iterative algorithm. The results for $\lambda = 0.1$ and $\lambda = 0.001$ are presented.

remaining (unavailable) ones by $\mathbb{Q}$ The vector of the corrupted/unavailable measurements $\mathbf{y}_c$ is declared

$$\mathbf{y}_c = [y_c(0), y_c(1), \ldots, y_c(N - M - 1)]^T$$

$$= [x(n_{M+1}), x(n_{M+1}), \ldots, x(n_{N-1})]^T.$$

Now the standard formulation of the sparse signal reconstruction

$$\min \|\mathbf{X}\|_1 \quad \text{subject to } \mathbf{y} = \mathbf{AX}$$

can be reformulated for the gradient as follows. The unavailable/missing measurements are found by minimization of $\|\mathbf{X}\|_1 = \|\Phi\mathbf{x}\|_1$ where $\mathbf{x}$ is a complete vector of measurements with elements $x(n_i) = y(i)$ for $n_i \in \mathbb{M}$ and $x(n_i) = y_c(i)$ for $n_i \in \mathbb{Q}$ The simplest way to solve this problem is by changing all the missing samples $x(n_i)$, $n_i \in \mathbb{M}$ within the range of their possible values and then to select the combination of their values that produced the minimal sparsity measure. The reconstruction procedure is presented in Algorithm 4.

Algorithm 4 Gradient-based reconstruction procedure

Input:

- Set of missing/omitted sample positions $\mathbb{Q}$
- Set of available sample positions $\mathbb{M}$
- Available samples (measurements) $\mathbf{y}$
- Transformation matrix Φ
- Step α

1: $m \leftarrow 0$
2: Set initial estimate signal vector $x^{(0)}$ as $x^{(0)}(n_i) = y(i)$ for $n_i \in \mathbb{M}$ and $x^{(0)}(n_i) = 0$ for $n_i \in \mathbb{Q}$
3: $\Delta \leftarrow \max_n |x^{(0)}(n)|$
4: **repeat**
5: **repeat**
6: $\mathbf{x}^{(m+1)} \leftarrow \mathbf{x}^{(m)}$
7: **for** $n_i \in \mathbb{Q}$ **do**
8: $\mathbf{z}_1 \leftarrow \mathbf{x}^{(m)}$
9: $z_1(n_i) \leftarrow z_1(n_i) + \Delta$
10: $\mathbf{z}_2 \leftarrow \mathbf{x}^{(m)}$
11: $z_2(n_i) \leftarrow z_2(n_i) - \Delta$
12: $g(n_i) \leftarrow \|\Phi \mathbf{z}_1\|_1 - \|\Phi \mathbf{z}_2\|_1$
13: $x^{(m+1)}(n_i) \leftarrow x^{(m)}(n_i) - \alpha g(n_i)$
14: **end for**
15: $m \leftarrow m+1$
16: **until** stopping criterion is satisfied
17: $\Delta \leftarrow \Delta/3$
18: **until** required precision is achieved
19: $\mathbf{x} \leftarrow \mathbf{x}^{(m)}$
20: $\mathbf{X} \leftarrow \Phi \mathbf{x}$

Output:

- Reconstructed signal vector $\mathbf{X}$
- Full set of measurements $\mathbf{x}$

As the initial iteration, we can use zero values for missing measurements, i.e., $\mathbf{y}_c^{(0)} = \mathbf{0}$. For each missing sample $n_i, i = 1, 2, \ldots, M$, two full sets of measurements are considered, where the value of the i-th missing sample is increased and decreased by some constant Δ

$$z_1(n) = \begin{pmatrix} x^{(m)}(n) & \text{for } n \neq n_i \\ x^{(m)}(n) + \Delta & \text{for } n = n_i \in \mathbb{Q} \end{pmatrix}$$

$$z_2(n) = \begin{pmatrix} x^{(m)}(n) & \text{for } n \neq n_i \\ x^{(m)}(n) - \Delta & \text{for } n = n_i \in \mathbb{Q} \end{pmatrix}$$

in the m-th iteration. The n_i-th coefficient of gradient vector $\mathbf{g}^{(m)}$ is estimated as

$$g^{(m)}(n_i) = \frac{\|\Phi \mathbf{z}_1\|_1 - \|\Phi \mathbf{z}_2\|_1}{2\Delta}, n_i \in \mathbb{Q}.$$

The gradient for available samples remains unchanged, $g^{(m)}(n_i) = 0$, $n_i \in \mathbb{M}$.

Usually the value Δ used is of the order of the maximal amplitude of available samples in the initial iteration, i.e., $\Delta = \max\{\mathbf{y}\}$. When the algorithm reaches a stationary point, with a given Δ, the mean squared error will assume almost constant value, and values of missing samples will oscillate. The fact that the reconstructed values oscillate when the stationary point is reached may be used as an indicator to reduce the step Δ, in order to approach the signal true value with a given precision. The oscillations of the solution are detected by measuring the angle between two successive gradient vectors. If the angle is greater than, for example 170°, the calculation is continued with a reduced Δ. Otherwise the Δ stays the same. This is also a possible stopping criterion in Algorithm 4. More details on the value Δ can be found in Ref. [26]. The iterative algorithm should be stopped when the change of reconstructed missing samples between two consecutive Δ reductions is bellow the desired precision.

7.4.3 Total Variations

The presented ℓ_1-norm-based algorithms can be used to solve other convex problems. One of the examples is when the vector $\mathbf{X}$ is not sparse, but instead its first-order finite difference $Z(k)$ is sparse,

$$Z(k) = X(k) - X(k - 1)$$

$$\mathbf{Z} = \text{diff}\{\mathbf{X}\}.$$

A simple example is $X(k) = 1$ for $0 \le k < N/2$ and $X(k) = -1$ for $N/2 \le k < N$. In this case, the minimization problem is formulated as

$$\min \|Z\|_1 \quad \text{subject to} \quad y = AX,$$

since the first difference vector Z is sparse. This is the minimization of total variations in the space of X. The result is the value of Z with smallest possible number of nonzero elements. The value of X is then with minimal variations (maximally flat). Its values follow from $X(0) = Z(0), X(1) = Z(1) + X(0)\ldots, X(N-1) = Z(N-1) + X(N-2)$. Total variations can be extended to two-dimensional signals in order to get maximally flat images.

7.5 Bayesian-Based Reconstruction

In supervised learning, a set of the output values and a set of the input values are given. The idea is to learn the model describing the relation between the input and the output data and to use it for the prediction of next input values. In this sense, the measurement relation $y = AX$ can be understood as a linear combination of basis funtions in A with adjustable parameters X as inputs to the model. The output of this linear combination is given by y, whose elements are

$$y(m) = \sum_{k=0}^{N-1} X(k)\psi_k(m). \quad (7.28)$$

The task here is to describe a Bayesian model for the estimation of linear combination parameters X, with the a priori knowledge that they are sparse (only a few of them will assume nonzero values) [27,28].

For a noisy measurement,

$$y = AX + \varepsilon,$$

where ε is a vector of Gaussian zero-mean noise with variance σ^2. The probability density function of the measurement error $\varepsilon = y - AX$ is Gaussian. The Gaussian likelihood model is

$$p(y|X,\sigma) = \frac{1}{(\sigma\sqrt{2\pi})^M} e^{-\frac{1}{2\sigma^2}\|y-AX\|_2^2}. \quad (7.29)$$

The sparsity promotion prior condition on X should be added. A common sparsity promotion on X in Bayesian formulation is the Laplacian density function for these coefficients, corresponding to the ℓ_1-norm,

$$p(X|\lambda) = (\frac{\lambda}{2})^N e^{-\lambda \|X\|_1}. \quad (7.30)$$

However, rather than imposing a Laplace prior on the values of X, in the relevance vector machine (RVM) approach, a hierarchical prior has been used. It exhibits similar properties. The zero-mean Gaussian prior distribution for X is assumed with hyperparemeters d_i

$$p(X|D) = \prod_{i=0}^{N-1} \frac{1}{\sqrt{2\pi d_i^{-1}}} e^{-X(i)d_i X^*(i)}$$

$$= e^{-X^T D X} \prod_{i=0}^{N-1} \frac{1}{\sqrt{2\pi d_i^{-1}}}, \quad (7.31)$$

where $D = diag(d_0, d_1, \ldots, d_{N-1})$ is a diagonal matrix of hyperparameters. Their role is to promote sparsity.

Posterior distribution of coefficients X, denoted by $p(X, D, \sigma^2 | y)$, could be calculated from

$$p(X,D,\sigma^2|y)p(y) = p(y|X,D,\sigma^2)p(X,D,\sigma^2). \quad (7.32)$$

However, the calculation of probabilities in Equation 7.32 is not possible in an analytic way. Some approximations are done. The posterior for X is decomposed as

$$p(X,D,\sigma^2|y) = p(X|y,D,\sigma^2)p(D,\sigma^2|y), \quad (7.33)$$

and the corresponding probabilities are calculated. First

$$p(X|y,D,\sigma^2) = \frac{p(y|X,\sigma^2)p(X|D)}{p(y|D,\sigma^2)} \quad (7.34)$$

is calculated using the probabilities defined by Equations 7.29 and 7.31. This is done from

$$p(X|y,D,\sigma^2)p(y|D,\sigma^2) = p(y|X,\sigma^2)p(X|D) \quad (7.35)$$

by grouping the exponential terms on the right side of Equation 7.35 with Equation 7.29 and Equation 7.31. After some matrix transformations, we can conclude that $p(X|y,D,\sigma^2)$ is Gaussian with covariance and mean-value matrices defined by [27,28]

$$\Sigma = (A^T A/\sigma^2 + D)^{-1}, \quad (7.36)$$

$$V = \Sigma A^T y/\sigma^2. \quad (7.37)$$

It is interesting to note that for small noise variance σ^2, the solution for the mean value has the form of Equation 7.19 since

$$\mathbf{V} = \Sigma \mathbf{A}^T \mathbf{y}/\sigma^2 = (\mathbf{A}^T\mathbf{A} + \sigma^2 \mathbf{D})^{-1} \mathbf{A}^T \mathbf{y}$$

$$\approx (\mathbf{A}^T\mathbf{A})^{-1} \mathbf{A}^T \mathbf{y}. \quad (7.38)$$

Obviously the next step is to promote sparsity by detecting candidates for zero-valued coefficients and omitting them from $\mathbf{X}$ and $\mathbf{A}$, until the K-sparse vector $\mathbf{X}_K$ is achieved. Corresponding matrix $\mathbf{A}_K$, resulting in $(\mathbf{A}_K^T \mathbf{A}_K)^{-1} \mathbf{A}_K^T \mathbf{y}$ as in Equation 7.19, is used. In the earlier presented algorithm, the sparsity is promoted starting from its lowest order, by taking the strongest component, and then the sparsity is increased until the model and the true sparsity are achieved. Here the process runs in the opposite direction. The sparsity will be reduced from its maximal value by using hyperparameters.

In order to calculate the function $p(\mathbf{D},\sigma^2|y)$ and promote sparsity, we have to make approximations. One is to replace this function with its most probable value. With this approximation, the search for the hyperparameters $\mathbf{D}$ reduces to the maximization of $p(\mathbf{D},\sigma^2|y) = p(\mathbf{y}|\mathbf{D},\sigma^2)p(\mathbf{D})p(\sigma^2)$. For uniform hyperparameters only, $p(\mathbf{y}|\mathbf{D},\sigma^2)$ should be minimized. Its value follows in analytic form, as the marginal value

$$p(\mathbf{y}|\mathbf{D},\sigma^2) = \int p(\mathbf{y}|\mathbf{X},\sigma^2) p(\mathbf{X}|\mathbf{D}) d\mathbf{X} \quad (7.39)$$

$$= \frac{1}{(2\pi)^{M/2}} \frac{1}{\sqrt{|\sigma^2 \mathbf{I} + \mathbf{A}\mathbf{D}^{-1}\mathbf{A}^T|}}$$

$$\times \exp\left(-\frac{1}{2} \mathbf{y}^H (\sigma^2 \mathbf{I} + \mathbf{A}\mathbf{D}^{-1}\mathbf{A}^T)^{-1} \mathbf{y}\right). \quad (7.40)$$

This is known as the marginal likelihood, and its maximization is type-II maximum likelihood method. Values of parameters $\mathbf{D}$ and σ^2 that maximize this probability cannot be obtained in a closed form. An iterative procedure is defined by [27,28]

$$d_i^{new} = \frac{\gamma_i}{V_i^2},$$

where V_i are the elements of posterior mean vector $\mathbf{V}$ defined by Equation 7.37 and γ_i is related to the diagonal elements Σ_{ii} of covariance matrix Σ, defined by Equation 7.36, and the sparsity hyperparameters d_i as

$$\gamma_i = 1 - d_i \Sigma_{ii}.$$

The noise variance is reestimated as

$$(\sigma^2)^{new} = \frac{\|\mathbf{y} - \mathbf{A}\mathbf{V}\|^2}{M - \sum_i \gamma_i}.$$

In this way, we get the coefficients as the resulting mean values of the coefficients, and also we get the variances of the estimated values from the covariance matrix. The previous relation defines iterative procedure, which is done until a required precision is achieved.

During the iterations, many parameters d_i will tend to infinity, indicating that the corresponding coefficients $X(i)$ tend to zero. These coefficients d_i and corresponding $X(i)$ are omitted from their matrices in next iterations. Only a small set of finite values d_i will remain at the end of iterations, indicating the nonzero values of $X(i)$ and producing vector $\mathbf{X}_K$ with corresponding $\mathbf{A}_K$. Note that as a result of this procedure, we will get mean values of nonzero coefficients in vector $\mathbf{V}$,

$$\mathbf{X}_K = \mathbf{V} = \Sigma_K \mathbf{A}_K^T \mathbf{y}/\sigma^2,$$

given by Equations 7.37 and 7.38. In addition, we will obtain variances of coefficients $\mathbf{X}_K$. Note that if a component position is wrongly detected and omitted in one iteration, the algorithm cannot recover it later (unrecoverable error).

The Bayesian-based algorithm is summarized in Algorithm 5.

Algorithm 5 Bayesian-based reconstruction

Input:

- Measurements vector $\mathbf{y}_{M \times 1}$
- Measurement matrix $\mathbf{A}_{M \times N}$

1: $\alpha_i \leftarrow 1$ ▷ For $i=1,2,\ldots,N$
2: $\sigma^2 \leftarrow 1$ ▷ Initial estimate
3: $T_h = 10^2$ ▷ Threshold
4: $\mathbf{p} = [1,2,\ldots,N]^T$
5: **repeat**
6: $\mathbf{D} \leftarrow$ diagonal matrix with d_i values
7: $\Sigma \leftarrow (\mathbf{A}^T\mathbf{A}/\sigma^2 + \mathbf{D})^{-1}$
8: $\mathbf{V} \leftarrow \Sigma \mathbf{A}^T \mathbf{y}/\sigma^2$
9: $\gamma_i \leftarrow 1 - d_i \Sigma_{ii}$ ▷ For each i
10: $d_i \leftarrow \gamma_i/V_i$ ▷ For each i
11: $\sigma^2 \leftarrow \dfrac{\|\mathbf{y}-\mathbf{A}\mathbf{V}\|^2}{M - \sum_i \gamma_i}$
12: $\mathbb{R} \leftarrow \{i : |d_i| > T_h\}$
13: Remove columns from matrix $\mathbf{A}$ selected by $\mathbb{R}$
14: Remove elements from array d_i selected by $\mathbb{R}$
15: Remove elements from vector $\mathbf{p}$ selected by $\mathbb{R}$
16: **until** stopping criterion is satisfied
17: Reconstructed vector $\mathbf{X}$ nonzero coefficients are in vector $\mathbf{V}$ with corresponding positions in vector $\mathbf{p}$, $X_{p_i} = V_i$

Output:

- Reconstructed signal vector $\mathbf{X}$

Example 3. A Signal Sparse in the Discrete Sine Transform (DST) Domain

$$x(n) = \sqrt{\frac{2}{N}} \sum_{i=1}^{K} X(k_i) \sin\left(\frac{2\pi(2n+1)k_i}{4N}\right) + \varepsilon(n), k_i \neq 0$$

with $N = 128$, $K = 10$, and noise variance $\sigma_\varepsilon^2 = 0.01$ is considered. The coefficients $X(k_i)$ values and positions are presented in Fig. 7.4a. They are considered as unknown, and they should be found from the available measurements. Assume that only $M = N/2 = 64$ randomly positioned measurements $\mathbf{y} = [x(n_1), x(n_1), \ldots, x(n_M)]^T$ are available (Fig. 7.4b). Observation matrix $\mathbf{A}$ is obtained from the full DST matrix keeping the rows corresponding to the measurement instants $n_1, n_2, \ldots, n_M$ only. In order to start the iterative algorithm, the initial values $\mathbf{D} = \mathbf{I}$ and $\sigma = 0.1$ are assumed. The assumed threshold for considering

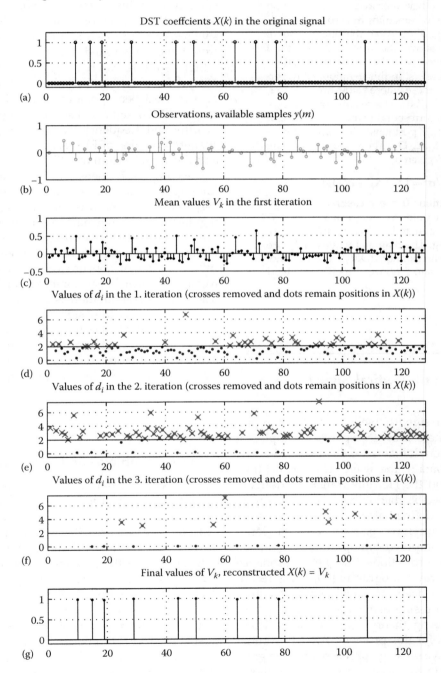

FIGURE 7.4
Bayesian reconstruction of a signal sparse in the DST domain. (a) Signal coefficients in the DST domain. (b) Available samples/measurements. (c) Distribution mean values (estimation of coefficients) in the initial iteration. (d–f) Hyperparameters in the first, second, and third iteration with a threshold. (g) The final mean value (the final estimated DST coefficients) at the positions of remaining hyperparameters from (f).

hyperparameters extremely large is $T_h = 100$. Hyperparameters above this threshold are omitted from calculation (along with the corresponding values in X, A, D, and V). The results for estimated mean value V in the first iteration are shown in Figure 7.4c, along with the values of hyperparameters V in Fig. 7.4d. The hyperparameters whose value is above T_h are omitted (pruned) along with the corresponding values at the same positions in all other matrices. In the second iteration, the values of remaining hyperparameters V are shown in Fig. 7.4e. After the elimination of hyperparameters above the threshold, the third iteration is calculated with the remaining positions of the hyperparameters. In this iteration, all hyperparameters, except those whose values are close to 1, are eliminated (Fig. 7.4f). The remaining positions, after this iteration, correspond to the nonzero coefficients $X(k_i), i = 1,2,\ldots,K$ positions, with corresponding pruned matrices Σ_K, A_K, D_K. The values of $X(k_i)$ are estimated using V_i given by

$$V_K = \Sigma_K A_K^T y / \sigma^2 = (A_K^T A_K + \sigma^2 D_K)^{-1} A_K^T y$$

in the final iteration. If the measurements were noise-free, this would be the exact recovery. The values of estimated $X(k_i), i = 1,2,\ldots,K$ are shown in Fig. 7.4g. The diagonal values of Σ_K are the variances of the estimated $X(k_i)$. The parameters d_i for the estimated nonzero coefficients tend to zero.

7.6 Applications in Biomedical Signal Processing

For most biomedical signals, we can define a representation domain where they are sparse. Examples of common representation domains are the discrete Fourier domain, the DCT domain, the wavelet, the short-time Fourier transform, and the Hermite transform domain. For some signals, specific dictionaries for their sparse representation can be defined as well. One of the initial successes in the compressive sensing was the long-studied problem of computed tomography (CT) image reconstructions from projections (magnetic resonance imaging [MRI] and x-ray tomography [30]). This reconstruction problem had been intensively studied within biomedical signal processing using various approaches. The compressive sensing theory provided an efficient and superior solution for this problem. After this initial success, the compressive sensing approach attracted great interest of researchers in various engineering applications, including many other biomedical applications.

The compressive sensing approach has been adapted to various tomography area problems. In addition to CT and MRI, it has been used in diffuse optical tomography (DOT) [31] and positron emission tomography (PET) [32].

Reconstruction of electrocardiography (ECG) signals is another challenging area in the compressive sensing applications. Various sparsity domains of the ECG signals are used, starting from the time domain (since the ECG signal has wide low-amplitude intervals). Other sparsity domains of the ECG are the Hermite and the wavelet transformation domains [33–35]. The compressive sensing approach is used to reduce the number of stored and transmitted data in this case.

Significant research efforts have been done in the compressive sensing formulation and application in the area of electroencephalogram (EEG) and electromyography (EMG) signals. The EMG signals can be sparse in time and frequency domains [35,36]. The common sparsity domains of EEG signals are the time-frequency and the wavelet domain [37].

The compressive sensing (CS) techniques are successfully applied to wireless body area networks (WBANs) [34]. In this case, many biomedical signals are monitored regularly, so the big data problem arises. CS methods can lower energy consumption and communication volume. In general, whenever an observed signal has a small number of nonzero values in some domain, it is worthy to form linear combinations of signal samples (each linear combination is a single measurement). In this manner, very long signals can be represented with a small number of measurements. Expected benefits are power efficiency, storage, and transmission efficiency. Additionally, more efficient signal processing methods could be derived under the signal sparsity assumption.

7.7 Open Problems and Future Directions

Signal reconstruction uniqueness is important and a theoretically well-studied topic in CS. However, for practical applications, these results are often very pessimistic, and these results could be improved based on the used signal sparsity domain. The main question here is to determine the smallest possible number of measurements that will guarantee unique signal reconstruction for the whole class of considered sparse signals in a specific application and its sparsity domain.

Since the introduction of compressive sensing, there exist many reconstruction algorithms. Some of them are highly accurate and efficient but require an increased number of measurements. When the number of measurements is close to the lower limit (determined by uniqueness conditions), efficiency of the reconstruction algorithms decreases. Efficient algorithms that can produce accurate results with the theoretical lower limit of measurements are an open research topic. Faster and more efficient results could be obtained by the parallelization of existing algorithms.

Biomedical signals are often measured with several sensors, resulting in a complex, multichannel data. In this case, more sophisticated methods than individual channel processing could be developed. For example, multichannel data could share the same or similar support in the sparsity domain. Then we could find the support for one channel and efficiently process other channels with assumed support using a lower number of measurements.

7.8 Conclusions

Big data, which are sparse in one of the representation domains, can be reconstructed from a reduced set of measurements/observations. A reduced set of measurements is interesting in practice from several points of view. For big data, it can be a desire to sense sparse signals with the lowest possible number of measurements (compressive sensing). It can also be a result of a physical unavailability to take a complete set of measurements/samples or of some signal samples being heavily corrupted by disturbances that it is better to omit them and consider them as unavailable in the analysis. Then the unavailable samples are reconstructed to the full signal using the sparsity property. In this way, an efficient filtering and disturbance removal can be achieved. The practical part of sparse signal processing is not limited to big data or any particular application and can be found in many recent papers.

In this chapter, we describe the conditions as well as five reconstruction algorithms. The algorithms belong in three groups of methods: norm-zero- and norm-one-based algorithms, and the Bayesian approach.

References

1. D. L. Donoho, "Compressed sensing," *IEEE Transactions on Information Theory*, vol. 52, no. 4, 2006, pp. 1289–1306.
2. E. J. Candès, J. Romberg, T. Tao, "Robust uncertainty principles: Exact signal reconstruction from highly incomplete frequency information," *IEEE Transactions on Information Theory*, vol. 52, no. 2, 2006, pp. 489–509.
3. E.J. Candès, M. Wakin, "An Introduction to Compressive Sampling," *IEEE Signal Processing Magazine*, vol. 25, no. 2, March 2008, pp. 21–30.
4. M. Fornsaier, H. Rauhut, "Iterative thresholding algorithms," *Applied and Computational Harmonic Analysis*, vol. 25, no. 2, Sept. 2008, pp. 187–208.
5. I. Daubechies, M. Defrise, C. De Mol, "An iterative thresholding algorithm for linear inverse problems with a sparsity constraint," *Communications on Pure and Applied Mathematics*, vol. 57, no. 11, Nov. 2004, pp. 1413–1457.
6. M. G. Amin, *Compressive Sensing for Urban Radar*, CRC Press, Boca Raton, FL, 2014.
7. E. Sejdic, "Time-frequency compressive sensing," in *Time-Frequency Signal Analysis and Processing*, ed. B. Boashash, Academic Press, Nov. 2015. pp. 424–429.
8. M. Elad, *Sparse and Redundant Representations: From Theory to Applications in Signal and Image Processing*, Springer, 2010.
9. Y. C. Eldar, G. Kutyniok, *Compressed Sensing: Theory and Applications*, Cambridge University Press, 2012.
10. R. Baraniuk, "Compressive sensing," *IEEE Signal Processing Magazine*, vol. 24, no. 4, 2007, pp. 118–121.
11. L. Stanković, M. Daković, S. Stanković, I. Orović, "Reconstruction of sparse signals—Introduction," *Wiley Encyclopedia of Electrical and Electronic Engineering*, Wiley, 2017.
12. L. Stanković, S. Stanković, M. G. Amin, "Missing samples analysis in signals for applications to L-estimation and compressive sensing," *Signal Processing*, Elsevier, vol. 94, Jan. 2014, pp. 401–408.
13. L. Stanković, "A measure of some time-frequency distributions concentration," *Signal Processing*, vol. 81, 2001, pp. 621–631.
14. E. J. Candès, "The restricted isometry property and its implications for compressed sensing," *Comptes Rendus Mathematique*, vol. 346, nos. 9–10, May 2008, pp. 589–592.
15. D. L. Donoho, M. Elad, V. Temlyakov, "Stable recovery of sparse overcomplete representations in the presence of noise," *IEEE Transactions on Information Theory*, vol. 52, 2006, pp. 6–18.
16. S. G. Mallat, Z. Zhang, "Matching pursuits with time-frequency dictionaries," *IEEE Transactions on Signal Processing*, vol. 41, no. 12, 1993, pp. 3397–3415.
17. J.A. Tropp, A.C. Gilbert, "Signal recovery from random measurements via orthogonal matching pursuit," *IEEE Transactions on Information Theory*, vol. 53, no. 12, Dec. 2007, pp. 4655–4666.
18. L. Stanković, *Digital Signal Processing with Applications: Adaptive Systems, Time-Frequency Analaysis, Sparse Signal Processing*, CreateSpace Independent Publishing Platform, 2015.
19. S. Stanković, I. Orović, E. Sejdić, *Multimedia Signals and Systems: Basic and Advanced Algorithms for Signal Processing*, 2nd edition. Springer International Publishing, 2015.
20. L. Stanković, M. Daković, "On the uniqueness of the sparse signals reconstruction based on the missing samples variation analysis," *Mathematical Problems in Engineering*, vol. 2015, Article ID 629759, 14 pages, doi:10.1155/2015/629759
21. D. Needell, J. A. Tropp, "CoSaMP: Iterative signal recovery from incomplete and inaccurate samples," *Communication of the ACM*, vol. 53, no. 12, Dec. 2010, pp. 93–100.
22. M. A. Figueiredo, R. D. Nowak, S. J. Wright, "Gradient projection for sparse reconstruction: Application to

compressed sensing and other inverse problems," *IEEE Journal of Selected Topics in Signal Processing*, vol. 1, no. 4, 2007, pp. 586–597.
23. R. Tibshirani, M. Saunders, S. Rosset, J. Zhu, K. Knight, "Sparsity and Smoothness via the Fused lasso," *Journal of the Royal Statistical Society. Series B (Statistical Methodology)*, vol. 67, no. 1, 2005, pp. 91–108.
24. E. Candès, J. Romberg, "ℓ_1-magic: Recovery of sparse signals via convex programming," Caltech. Available at http://users.ece.gatech.edu/justin/l1magic/downloads/l1magic.pdf, Oct. 2005.
25. S. Boyd, L. Vandenberghe, *Convex Optimization*, Cambridge University Press, 2004.
26. L. Stanković, M. Daković, S. Vujović, "Adaptive variable step algorithm for missing samples recovery in sparse signals," *IET Signal Processing*, vol. 8, no. 3, 2014, pp. 246–256.
27. M. Tipping, "Sparse Bayesian learning and the relevance vector machine," *Journal of Machine Learning Research*, vol. 1, 2001, pp. 211–244.
28. S. Ji, Y. Xue, L. Carin, "Bayesian compressive sensing," *IEEE Transactions on Signal Processing*, vol. 56, no. 6, June 2008, pp. 2346–2356.
29. G. Wang, Y. Bresler, V. Ntziachristos, "Compressive sensing for biomedical imaging," Guest Editorial, *IEEE Transactions on Medical Imaging*, vol. 30, no. 5, May 2011.
30. M. Lustig, D. Donoho, J. Santos, J. Pauly, "Compressed sensing MRI," *IEEE Signal Processing Magazine*, March 2008.
31. M. Süzen, A. Giannoula, T. Durduran, "Compressed sensing in diffuse optical tomography," *Optics Express*, vol. 18, no. 23, pp. 23676–23690, 2010.
32. M. Valiollahzadeh, "Compressive sensing in positron emission tomography (PET) imaging," PhD dissertation, Rice University, 2015.
33. M. Brajović, I. Orović, M. Daković, and S. Stanković, "Gradient-based signal reconstruction algorithm in the Hermite transform domain," *Electronics Letters*, vol. 52, no. 1, pp. 41–43, 2016.
34. R. Carrillo, A. Ramirez, G. Arce, K. Barner and B. Sadler, "Robust compressive sensing of sparse signals: A review," *EURASIP Journal on Advances in Signal Processing*, doi: 10.1186/s13634-016-0404-5, 2016.
35. A. Dixon, E. Allstot, D. Gangopadhyay, D. Allstot, "Compressed sensing system considerations for ECG and EMG wireless biosensors," *IEEE Transactions on Biomedical Circuits and Systems*, vol. 6, no. 2, April 2012.
36. Y. Liu, I. Gligorijevic, V. Matic, M. De Vos, S. Van Huffel, "Multi-sparse signal recovery for compressive sensing," *34th Annual International Conference of the IEEE EMBS*, San Diego, California, 2012.
37. Y. Liu, M. De Vos, S. Van Huffel, "Compressed sensing of multichannel EEG signals: The simultaneous cosparsity and low-rank optimization," *IEEE Transactions on Biomedical Engineering*, vol. 62, no. 8, August 2015.

8

Low-Complexity DCT Approximations for Biomedical Signal Processing in Big Data

Renato J. Cintra, Fábio M. Bayer, Yves Pauchard, and Arjuna Madanayake

CONTENTS

8.1 Introduction 152
8.2 Mathematical Preliminaries and Definitions 153
    8.2.1 KLT and DCT 153
    8.2.2 Fast Algorithms 154
    8.2.3 Approximations 154
8.3 DCT Approximations 155
    8.3.1 Classical Multiplierless Transforms 156
        8.3.1.1 Hadamard Transform 156
        8.3.1.2 Walsh–Hadamard Transform 156
    8.3.2 DCT Approximation Based on Integer Functions 156
        8.3.2.1 Signed DCT 156
        8.3.2.2 Rounded DCT 157
        8.3.2.3 Collection of Integer DCT Approximations 157
    8.3.3 DCT Approximations by Inspection 158
        8.3.3.1 Modified RDCT 158
        8.3.3.2 BAS Series of Approximations 158
        8.3.3.3 SPM Transform 158
        8.3.3.4 Signed SPM Transform 159
    8.3.4 DCT Approximations Based on Computational Search 159
        8.3.4.1 DCT Approximations for RF Imaging 160
        8.3.4.2 Improved 8-Point DCT Approximation 160
        8.3.4.3 DCT Approximation for IR 161
    8.3.5 Parametric DCT Approximations 161
        8.3.5.1 Level 1 Approximation 161
        8.3.5.2 Parametric BAS 162
        8.3.5.3 FW Class of DCT Approximations 162
8.4 Performance Comparison 163
    8.4.1 Performance Measures 163
        8.4.1.1 Transform Distortion 163
        8.4.1.2 Total Error Energy 165
        8.4.1.3 Mean Square Error 166
        8.4.1.4 Unified Transform Coding Gain 166
        8.4.1.5 Transform Efficiency 166
    8.4.2 Comparison and Discussion 166

8.5　Applications in Biomedical Signal Processing 167
　　8.5.1　Image Compression 167
　　8.5.2　Image Registration 169
8.6　Open Problems and Future Directions 172
Acknowledgments 172
References 172

8.1 Introduction

Contemporary health technology generates very large volumes of medical data [1,2]. Such health-care data should be locally stored, processed, analyzed, transmitted, and shared. In order to properly manage this huge amount of medical information, the investigation of efficient signal processing tools plays an essential role [2]. Indeed, the medical big data is a good example of how *the 3 Vs* of big data—volume, variety, and velocity [3]—are an intrinsic problem in this field [4].

Efficient data compression is a pivotal step in several biomedical data processing [1,2]. In particular, medical imagery shares the common big data characteristics [5]. However, the proposition of good and efficient compression methods is not a trivial task [1]. Popular image compression algorithms operate over the spectral domain according to tailored transform operators [1,2,6]. Orthogonal discrete transforms are widely used in digital signal processing field [7,8]. In particular, the Karhunen–Loève transform (KLT) is the optimal transform for data compression [8–13] and presents a large number of possible applications, such as in image compression [14–16], image filtering [17,18], speech processing [19,20], pattern recognition [21,22], communications [23,24], and several other biomedical applications [25–27].

In this sense, the discrete cosine transform (DCT) is widely regarded as a key operation in image compression [9,10,12]. In fact, the DCT is the asymptotic configuration of the well-known KLT [9–12,28]. When high correlated first-order Markov signals are considered [10,12]—such as natural images [29]—the DCT can closely emulate the KLT [9].

A number of signal processing methods [30] for biomedical data considers and effectively adopts the KLT and DCT [1,2,26,27]. In fact, the DCT is the central mathematical operation for the following standards of image and video coding: JPEG [31], MPEG-1 [32], MPEG-2 [33], H.261 [34], H.263 [35], H.264 [36], and the recent HEVC [37,38]. These coding standards are commonly found in a multitude of biomedical applications related to image and video compression [2,39–41]. All above standards employ the DCT or related DCT-based integer transforms with blocklength set to 8 and, possibly, 16.

Implantable and miniaturized biomedical devices possess severe restrictions on available resources, especially in terms of computational and power capabilities. Such constraints impose significant challenges to the design of energy efficient methods for signal processing [42]. In particular, image coding procedures must be tailored in order to have their energy consumption significantly reduced [43].

Thus, developing fast algorithms for the efficient evaluation of the 8- and 16-point DCT is a main task in the circuits, systems, and signal processing communities, especially focusing in biomedical applications [43]. Archived literature contains a multitude of fast algorithms for these particular blocklengths [44,45]. Remarkably extensive reports have been generated amalgamating scattered results for the standard DCT [10,12]. Among the most popular techniques, we mention the following algorithms: Chen algorithm [46], Wang factorization [47], Lee DCT for power-of-two blocklengths [48], Arai DCT scheme [49], Loeffler algorithm [50], Vetterli–Nussbaumer algorithm [45], Hou algorithm [44], and Feig–Winograd (FW) factorization [51]. These are traditional methods in the field and have been considered for practical applications [32,52]. For instance, the Arai DCT scheme was employed in various recent hardware implementations of the DCT [53–55].

Naturally, DCT fast algorithms that result in major computational savings compared to direct computation of the DCT were already developed decades ago. In fact, the intense research in the field has led to methods that are very close to the minimum theoretical complexity of DCT [29,49,50,56,57]. Thus, the computation of the exact DCT is a task with very little room for major improvements in terms of minimization of computational complexity by means of standard methods.

Consequently, DCT approximations—operations that closely emulate the DCT—are mathematical tools that can furnish an alternative venue for DCT evaluation. Effectively, DCT approximations have already been considered in a number of works [58–74]. Moreover, although usual fast algorithms can reduce the computational complexity significantly, they still require floating-point operations [10]. In contrast, approximate DCT methods can be tailored to require very low arithmetic complexity. This way, these methods can significantly increase the velocity of biomedical signal processing.

A comprehensive list of approximate methods for the DCT is found in Ref. [10]. Prominent techniques include the signed DCT (SDCT) [69], the binDCT [29], the level 1 approximation by Lengwehasatit–Ortega [71], the Bouguezel–Ahmad–Swamy (BAS) series of algorithms [59–64], the DCT round-off approximation [65], the modified DCT round-off approximation [58], the DCT approximation for image registration (IR) [72], the collection of integer approximations [66], and the FW class of DCT approximations [74]. Some works focusing on DCT approximations with other lengths are Refs. [67,70,75–77].

The goal of this chapter is to review low-complexity transforms useful for biomedical applications. Several classical and state-of-the-art DCT approximations are cataloged and compared. Their utilities in biomedical image compression and IR are then demonstrated.

8.2 Mathematical Preliminaries and Definitions

8.2.1 KLT and DCT

Let $\mathbf{x} = [x_0\ x_2\ \ldots\ x_{N-1}]^T$ be an N-point random vector with associate covariance matrix given by

$$\mathbf{R}_x = E\{(\mathbf{x} - E\{\mathbf{x}\})(\mathbf{x} - E\{\mathbf{x}\})^T\}.$$

The rows of the $N \times N$ KLT matrix $\mathbf{K}_N$ are the eigenvectors of the symmetric positive definite covariance matrix $\mathbf{R}_x$. The KLT transform maps $\mathbf{x}$ into the N-point vector $\mathbf{X} = [X_0\ X_1\ \cdots\ X_{N-1}]^T$ given by

$$\mathbf{X} = \mathbf{K}_N \cdot \mathbf{x}.$$

Vector $\mathbf{X}$ contains the uncorrelated transform coefficients. In statistical terminology, the elements of $\mathbf{X}$ are the principal components [78].

Because the KLT matrix depends on the input signal $\mathbf{x}$, it is not trivial to derive fast algorithms to compute the associate transform coefficients [7,13,79]. There is no simple way to compute the eigenvectors, rendering the application range of the KLT severely limited due to the high computational requirements [10]. However, it is possible to derive simple expressions for the KLT, when the input signal is constrained to known classes of stationary processes. In fact, if the signal $\mathbf{x}$ is modeled after a first-order Markov process, then its covariance matrix entries are given by [10]

$$r^{(x)}_{i,j} = \rho^{|i-j|},$$

where $0 < \rho < 1$ is the correlation coefficient and $i, j = 1, 2, \ldots, N$. For this particular class of stationary process, the (i, j)th elements of the matrix $\mathbf{K}_N$, for a fixed value of ρ, are given by

$$u_{ij} = \sqrt{\frac{2}{N + \lambda_j}} \sin\left[\omega_j\left(i - \frac{N-1}{2}\right) + \frac{(j+1)\pi}{2}\right],$$

where the eigenvalues of the covariance matrix $\mathbf{R}_x$ are

$$\lambda_j = \frac{1 - \rho^2}{1 + \rho^2 - 2\rho\cos\omega_j}.$$

The quantities $\omega_1, \omega_2, \ldots, \omega_N$ are the N solutions of the following equation:

$$\tan N\omega = \frac{-(1-\rho^2)\sin\omega}{(1+\rho^2)\cos\omega - 2\rho}.$$

In image and video compression, adjacent pixels are highly correlated [12]. The supposition that natural images are well described by first-order Markov process highly correlated is widely accepted [10]. In particular, a common choice of correlation coefficient is $\rho = 0.95$.

For example, the 8-point KLT matrix for first-order Markovian data with $\rho = 0.95$ is given by

$$\mathbf{K}_8^{(0.95)} = \begin{bmatrix} 0.338 & 0.351 & 0.360 & 0.364 & 0.364 & 0.360 & 0.351 & 0.338 \\ 0.481 & 0.420 & 0.286 & 0.101 & -0.101 & -0.286 & -0.420 & -0.481 \\ 0.467 & 0.207 & -0.179 & -0.456 & -0.456 & -0.179 & 0.207 & 0.467 \\ 0.423 & -0.085 & -0.487 & -0.278 & 0.278 & 0.487 & 0.085 & -0.423 \\ 0.360 & -0.347 & -0.356 & 0.351 & 0.351 & -0.356 & -0.347 & 0.360 \\ 0.283 & -0.488 & 0.094 & 0.415 & -0.415 & -0.094 & 0.488 & -0.283 \\ 0.195 & -0.462 & 0.460 & -0.190 & -0.190 & 0.460 & -0.462 & 0.195 \\ 0.100 & -0.279 & 0.416 & -0.490 & 0.490 & -0.416 & 0.279 & -0.100 \end{bmatrix}.$$

Asymptotically when $\rho \to 1$, the KLT converges to the DCT. The (i, j)th elements of the N-point DCT matrix is given by [10]

$$c_{i,j} = \alpha_{i-1}\cos\left(\frac{\pi(i-1)(2(j-1)+1)}{2N}\right),$$

where $\alpha_0 = \frac{1}{\sqrt{N}}$, $\alpha_i = \sqrt{\frac{2}{N}}$ for $i > 0$, and $i, j = 1, 2, \ldots, N$.

In particular, for $N = 8$, the DCT matrix is

$$\mathbf{C}_8 = \begin{bmatrix} 0.354 & 0.354 & 0.354 & 0.354 & 0.354 & 0.354 & 0.354 & 0.354 \\ 0.490 & 0.416 & 0.278 & 0.098 & -0.098 & -0.278 & -0.416 & -0.490 \\ 0.462 & 0.191 & -0.191 & -0.462 & -0.462 & -0.191 & 0.191 & 0.462 \\ 0.416 & -0.098 & -0.490 & -0.278 & 0.278 & 0.490 & 0.098 & -0.416 \\ 0.354 & -0.354 & -0.354 & 0.354 & 0.354 & -0.354 & -0.354 & 0.354 \\ 0.278 & -0.490 & 0.098 & 0.416 & -0.416 & -0.098 & 0.490 & -0.278 \\ 0.191 & -0.462 & 0.462 & -0.191 & -0.191 & 0.462 & -0.462 & 0.191 \\ 0.098 & -0.278 & 0.416 & -0.490 & 0.490 & -0.416 & 0.278 & -0.098 \end{bmatrix}.$$

The one-dimensional (1-D) DCT transform of $\mathbf{x}$ is the N-point vector $\mathbf{X} = [X_0 \; X_1 \; \cdots \; X_{N-1}]^T$ given by

$$\mathbf{X} = \mathbf{C}_N \cdot \mathbf{x}.$$

Because $\mathbf{C}_N \cdot \mathbf{C}_N^T$ is the identity matrix, the inverse transformation can be written according to

$$\mathbf{x} = \mathbf{C}_N^T \cdot \mathbf{X}.$$

Let $\mathbf{A}$ and $\mathbf{B}$ be square matrices of size N. For two-dimensional (2-D) signals, such as medical images, we have the following expressions that relate the forward and inverse 2-D DCT operations, respectively:

$$\mathbf{B} = \mathbf{C}_N \cdot \mathbf{A} \cdot \mathbf{C}_N^T \quad \text{and} \quad \mathbf{A} = \mathbf{C}_N^T \cdot \mathbf{B} \cdot \mathbf{C}_N. \quad (8.1)$$

8.2.2 Fast Algorithms

Because the DCT matrix does not depend on the input signal, substituting the DCT for the KLT effects a significant reduction in computational cost. Nevertheless, the DCT itself still poses a noticeable computational complexity, which requires dedicated fast algorithms for its efficient calculation. This issue becomes particularly pressing when huge amounts of data are expected to be processed, such as it is typical in big data and biomedical contexts.

Thus, a multitude of efficient implementations of DCT is archived in the literature. Some of the most popular algorithms for the 8-point DCT computation are presented in Table 8.1 along with their arithmetic complexity. These algorithms employ real-valued arithmetic and are based on matrix factorization methods, which can reduce the number of multiplications for the DCT computation. Indeed, DCT fast algorithms result in important computational savings compared to direct computation of the DCT.

TABLE 8.1

Computational Complexity of Some 8-Point DCT Fast Algorithms

Algorithm	Multiplications	Additions
DCT by definition	64	56
Chen et al. [46]	16	26
Lee [48]	12	29
Vetterli–Nussbaumer [45]	12	29
Wang [47]	13	29
Suehiro–Hatori [80]	13	29
Hou [44]	12	29
Arai et al. [49]	13	29
Loeffler et al. [50]	11	29
Feig–Winograd [51]	22	28
Yuan et al. [81]	12	29

8.2.3 Approximations

Although usual fast algorithms can reduce the computational complexity significantly, they still need floating-point operations [10]. Due to the intense research in the field, algorithms with very close computational cost to the minimum theoretical complexity were already developed.

Currently, the computation of the exact DCT is a task with very little room for major improvements in terms of minimization of computational complexity. In this scenario, the interest in DCT approximations has increased. In fact, approximate DCT methods can be tailored to require very low arithmetic complexity and deliver good performance. Several works present fast algorithms for DCT approximations where multiplications or bit-shifting operations are absent [10].

In general terms, a DCT approximation is a transformation $\hat{\mathbf{C}}$ that—according to some specified metric—behaves similarly to the *exact* DCT matrix $\mathbf{C}$. An approximation matrix $\hat{\mathbf{C}}$ is usually based on a transformation matrix $\mathbf{T}$ of low computational complexity. Indeed, matrix $\mathbf{T}$ is the key component of a given DCT approximation.

Often the elements of the transformation matrix $\mathbf{T}$ possess null multiplicative complexity. For instance, this property can be satisfied by restricting the entries of $\mathbf{T}$ to the set of powers of two $\{0, \pm 1, \pm 2, \pm 4, \pm 8 \ldots\}$. In fact, multiplications by such elements are trivial and require only bit-shifting operations.

In this work, we adopt the following terminology. A matrix $\mathbf{A}$ is orthogonal if $\mathbf{A} \cdot \mathbf{A}^T$ is a diagonal matrix. In particular, if $\mathbf{A} \cdot \mathbf{A}^T$ is the identity matrix, then $\mathbf{A}$ is said to be orthonormal.

Approximations for the DCT can be classified into two categories depending on whether $\hat{\mathbf{C}}$ is orthonormal or not. In principle, given a low-complexity matrix $\mathbf{T}$, it is possible to derive an orthonormal matrix $\hat{\mathbf{C}}$ based on $\mathbf{T}$ by means of the polar decomposition [82,83]. Indeed, if $\mathbf{T}$ is a full rank real matrix, then the following factorization is uniquely determined:

$$\hat{\mathbf{C}} = \mathbf{S} \cdot \mathbf{T},$$

where $\mathbf{S}$ is a symmetric positive definite matrix [83, p. 348]. Matrix $\mathbf{S}$ is explicitly related to $\mathbf{T}$ according to the following relation:

$$\mathbf{S} = \sqrt{\left(\mathbf{T} \cdot \mathbf{T}^T\right)^{-1}},$$

where $\sqrt{\cdot}$ denotes the matrix square root operation [84,85]. Being orthonormal, such type of approximation satisfies $\hat{\mathbf{C}}^{-1} = \hat{\mathbf{C}}^T$. Therefore, we have that

$$\hat{\mathbf{C}}^{-1} = \mathbf{T}^T \cdot \mathbf{S}.$$

As a consequence, the inverse transformation $\hat{\mathbf{C}}^{-1}$ inherits the same computational complexity of the forward transformation.

From the computational point of view, it is desirable that $\mathbf{S}$ be a diagonal matrix. In this case, the computational complexity of $\hat{\mathbf{C}}$ is the same as that of $\mathbf{T}$, except for the scale factors in the diagonal matrix $\mathbf{S}$. Moreover, depending on the considered application, even the constants in $\mathbf{S}$ can be disregarded in terms of computational complexity assessment. This occurs when the involved constants are trivial multiplicands, such as the powers of two. Another more practical possibility for neglecting the complexity of $\mathbf{S}$ arises when it can be absorbed into other sections of a larger procedure. This is the case in JPEG-like compression, where the quantization step is present [31]. Thus, matrix $\mathbf{S}$ can be incorporated into the quantization matrix [58,59,61,63,65,71,75]. In terms of the inverse transformation, it is also beneficial that $\mathbf{S}$ is diagonal, because the complexity of $\hat{\mathbf{C}}^{-1}$ becomes essentially that of $\mathbf{T}^T$.

In order that $\mathbf{S}$ be a diagonal matrix, it is sufficient that $\mathbf{T}$ satisfies the orthogonality condition

$$\mathbf{T} \cdot \mathbf{T}^T = \mathbf{D}, \qquad (8.2)$$

where $\mathbf{D}$ is a diagonal matrix [82].

If Equation 8.2 is not satisfied, then $\mathbf{S}$ is not a diagonal and the advantageous properties of the resulting DCT approximation are, in principle, lost. In this case, the off-diagonal elements contribute to a computational complexity increase and the absorption of matrix $\mathbf{S}$ cannot be easily done. However, at the expense of not providing an orthogonal approximation, one may consider approximating $\mathbf{S}$ itself by replacing the off-diagonal elements of $\mathbf{D}$ by zeros. Thus, the resulting matrix $\hat{\mathbf{S}}$ is given by

$$\hat{\mathbf{S}} = \sqrt{\left[\mathrm{diag}(\mathbf{T} \cdot \mathbf{T}^T)\right]^{-1}},$$

where $\mathrm{diag}(\cdot)$ returns a diagonal matrix with the diagonal elements of its matrix argument. Thus, the non-orthogonal approximation is furnished by

$$\tilde{\mathbf{C}} = \hat{\mathbf{S}} \cdot \mathbf{T}.$$

Matrix $\tilde{\mathbf{C}}$ can be a meaningful approximation if $\hat{\mathbf{S}}$ is, in some sense, close to $\mathbf{S}$, or, alternatively, if $\mathbf{T}$ is almost orthogonal.

From the algorithm designing perspective, proposing non-orthogonal approximations may be a less demanding task, since Equation 8.2 is not required to be satisfied. However, since $\tilde{\mathbf{C}}$ is not orthogonal, the inverse transformation must be cautiously examined. Indeed, the inverse transformation does not employ directly the low-complexity matrix $\mathbf{T}$ and is given by

$$\tilde{\mathbf{C}}^{-1} = \mathbf{T}^{-1} \cdot \hat{\mathbf{S}}^{-1}.$$

Even if $\mathbf{T}$ is a low-complexity matrix, it is not guaranteed that $\mathbf{T}^{-1}$ also possesses low computational complexity figures. Nevertheless, it is possible to obtain non-orthogonal approximations whose both direct and inverse transformation matrices have low computational complexity. Two prominent examples are the SDCT [69] and the BAS approximation described in Ref. [59].

8.3 DCT Approximations

In this section, we present a review of the literature. We present a multitude of 8-point approximate DCT categorizing by obtaining method. For the sake of notation, hereafter the matrices $\mathbf{C}_8$ and $\hat{\mathbf{C}}_8$ are referred to as $\mathbf{C}$ and $\hat{\mathbf{C}}$. In the following, in Section 8.4, we evaluate these discussed transforms in terms of computational complexity and performance measures.

8.3.1 Classical Multiplierless Transforms

8.3.1.1 Hadamard Transform

Finding several applications in coding theory [86, p. 117], the Hadamard transform employs the Hadamard matrix, which is composed of elements −1 and +1 [87]. For power-of-two blocklengths, the Hadamard matrix can be recursively derived according to the Sylvester construction [87]:

$$\mathbf{H}_{2^n} = \begin{bmatrix} \mathbf{H}_{2^{n-1}} & \mathbf{H}_{2^{n-1}} \\ \mathbf{H}_{2^{n-1}} & -\mathbf{H}_{2^{n-1}} \end{bmatrix}, \quad n = 2, 3, \ldots$$

where $\mathbf{H}_2 = \frac{1}{2}\begin{bmatrix} 1 & 1 \\ 1 & -1 \end{bmatrix}$. In order to ensure that the transform is orthogonal, a scaling factor $1/\sqrt{N}$ is included. Thus, for $N = 8$, we have the following transformation matrix:

$$\mathbf{H}_8 = \frac{1}{\sqrt{8}} \cdot \begin{bmatrix} 1 & 1 & 1 & 1 & 1 & 1 & 1 & 1 \\ 1 & -1 & 1 & -1 & 1 & -1 & 1 & -1 \\ 1 & 1 & -1 & -1 & 1 & 1 & -1 & -1 \\ 1 & -1 & -1 & 1 & 1 & -1 & -1 & 1 \\ 1 & 1 & 1 & 1 & -1 & -1 & -1 & -1 \\ 1 & -1 & 1 & -1 & -1 & 1 & -1 & 1 \\ 1 & 1 & -1 & -1 & -1 & -1 & 1 & 1 \\ 1 & -1 & -1 & 1 & -1 & 1 & 1 & -1 \end{bmatrix}.$$

8.3.1.2 Walsh–Hadamard Transform

The Walsh–Hadamard transform (WHT) is related to the HT according to a particular column permutation based on the binary expansion of the column index [88]. The WHT is usually considered as a low-complexity transformation method for image processing [6]. For $N = 8$, it is given by [6]

$$\mathbf{W}_8 = \frac{1}{\sqrt{8}} \cdot \begin{bmatrix} 1 & 1 & 1 & 1 & 1 & 1 & 1 & 1 \\ 1 & 1 & 1 & 1 & -1 & -1 & -1 & -1 \\ 1 & 1 & -1 & -1 & -1 & -1 & 1 & 1 \\ 1 & 1 & -1 & -1 & 1 & 1 & -1 & -1 \\ 1 & -1 & -1 & 1 & 1 & -1 & -1 & 1 \\ 1 & -1 & -1 & 1 & -1 & 1 & 1 & -1 \\ 1 & -1 & 1 & -1 & -1 & 1 & -1 & 1 \\ 1 & -1 & 1 & -1 & 1 & -1 & 1 & -1 \end{bmatrix}.$$

The complexity of the WHT direct implementation is in $O(N^2)$. Fast algorithms can reduce the complexity to only $N \log_2(N)$ additions.

8.3.2 DCT Approximation Based on Integer Functions

Approximations archived in the literature often possess transformation matrices with entries defined on the set $\{0, \pm1, \pm2\}$ [58,63–65,69]. Thus such transformations exhibit null multiplicative complexity, because the required arithmetic operations can be implemented exclusively by means of additions and bit-shifting operations. Some of these DCT approximations have been proposed considering integer functions [89, Cap. 3]. Here an integer function is understood as a function whose values are integers. These functions can be employed to map the entries of the DCT matrix into integer quantities. The resulting matrices are approximates of the DCT.

8.3.2.1 Signed DCT

In Ref. [59], Haweel introduced a simple approach for designing a DCT approximation. This DCT approximation was termed the signed DCT (SDCT) and was defined as follows [69]:

$$\text{sign}(\mathbf{C}_N),$$

where $\text{sign}(\cdot)$ is the signum function, which is applied to each entry of $\mathbf{C}_N$ and is given by

$$\text{sign}(x) = \begin{cases} +1, & \text{if } x > 0, \\ 0, & \text{if } x = 0, \\ -1, & \text{if } x < 0. \end{cases}$$

The SDCT can be regarded as a seminal work in the field of DCT approximations.

Especially for $N = 8$, the SDCT matrix is given by

$$\hat{\mathbf{C}}_{\text{SDCT}} = \frac{1}{\sqrt{8}} \cdot \begin{bmatrix} 1 & -1 & -1 & -1 & -1 & -1 & -1 & -1 \\ 1 & -1 & -1 & -1 & -1 & -1 & -1 & -1 \\ 1 & -1 & -1 & -1 & -1 & -1 & -1 & -1 \\ 1 & -1 & -1 & -1 & -1 & -1 & -1 & -1 \\ 1 & -1 & -1 & -1 & -1 & -1 & -1 & -1 \\ 1 & -1 & -1 & -1 & -1 & -1 & -1 & -1 \\ 1 & -1 & -1 & -1 & -1 & -1 & -1 & -1 \\ 1 & -1 & -1 & -1 & -1 & -1 & -1 & -1 \end{bmatrix}.$$

8.3.2.2 Rounded DCT

Additionally, in Ref. [65], an 8-point low-complexity DCT approximation was proposed based on the following matrix:

$$\text{round}(2 \cdot \mathbf{C}),$$

where round($\cdot$) is the entrywise rounding-off function as implemented in MATLAB programming environment [90]. Thus, the resulting matrix of the rounded DCT (RDCT) is shaped as follows:

$$\hat{\mathbf{C}}_{\text{RDCT}} = \mathbf{S}_{\text{RDCT}} \cdot \begin{bmatrix} 1 & 1 & 1 & 1 & 1 & 1 & 1 & 1 \\ 1 & 1 & 1 & 0 & 0 & -1 & -1 & -1 \\ 1 & 0 & 0 & -1 & -1 & 0 & 0 & 1 \\ 1 & 0 & -1 & -1 & 1 & 1 & 0 & -1 \\ 1 & -1 & -1 & 1 & 1 & -1 & -1 & 1 \\ 1 & -1 & 0 & 1 & -1 & 0 & 1 & -1 \\ 0 & -1 & 1 & 0 & 0 & 1 & -1 & 0 \\ 0 & -1 & 1 & -1 & 1 & -1 & 1 & 0 \end{bmatrix},$$

where $\mathbf{S}_{\text{RDCT}} = \text{diag}\left(\frac{1}{2\sqrt{2}}, \frac{1}{\sqrt{6}}, \frac{1}{2}, \frac{1}{\sqrt{6}}, \frac{1}{2\sqrt{2}}, \frac{1}{\sqrt{6}}, \frac{1}{2}, \frac{1}{\sqrt{6}}\right)$.

The good features of RDCT could enable to outperform the SDCT in any compression ratio [65].

In terms of complexity assessment, matrix $\mathbf{S}_{\text{RDCT}}$ may not introduce an additional computational overhead. For image compression, the DCT operation is a preprocessing step for a subsequent coefficient quantization procedure. Therefore, the scaling factors in the diagonal matrix $\mathbf{S}_{\text{RDCT}}$ can be merged into the quantization step. The RDCT was chosen as the basis transform for the scalable orthogonal approximation proposed by Jridi et al. [70].

8.3.2.3 Collection of Integer DCT Approximations

In Ref. [66], a method for deriving low-complexity DCT approximations based on integer functions is proposed. As a result, a collection of 8-point DCT approximations is derived, capable of encompassing several approximation methods, including the SDCT and RDCT.

As a venue to design DCT approximations, the following general mapping was considered:

$$\mathbb{R} \to \mathcal{M}_8(\mathbb{Z})$$
$$\alpha \mapsto \text{int}(\alpha \cdot \mathbf{C}),$$

where $\mathcal{M}_8(\mathbb{Z})$ is the space of 8 × 8 matrices over the set of integers $\mathbb{Z}$, and int($\cdot$) is a prototype integer function [89, p. 67]. Function int($\cdot$) operates entrywise over its matrix argument. Parameter α is termed the expansion factor and scales the exact DCT matrix allowing a wide range of possible integer mappings [91].

Particular examples of integer functions are the floor, the ceiling, the truncation (round towards zero), and the round-away-from-zero function. These functions are defined, respectively, as follows:

$$\text{floor}(x) = \lfloor x \rfloor = \max\{m \in \mathbb{Z} | m \leq x\},$$
$$\text{ceil}(x) = \lceil x \rceil = \min\{n \in \mathbb{Z} | n \geq x\},$$
$$\text{trunc}(x) = \text{sign}(x) \cdot \lfloor |x| \rfloor,$$
$$\text{round}_{\text{AFZ}}(x) = \text{sign}(x) \cdot \lceil |x| \rceil,$$

where $|\cdot|$ returns the absolute value of its argument.

Another particularly useful integer function is the round to nearest integer function [92, p. 73]. This function possesses various definitions depending on its behavior for input arguments whose fractional part is exactly 1/2. Thus, we have the following rounding-off functions: round-half-up, round-half-down, round-half-away-from-zero, round-half-towards-zero, round-half-to-even, and round-half-to-odd function. These different nearest integer functions are, respectively, given by

$$\text{round}_{\text{HU}}(x) = \left\lfloor x + \frac{1}{2} \right\rfloor,$$
$$\text{round}_{\text{HD}}(x) = \left\lceil x - \frac{1}{2} \right\rceil,$$
$$\text{round}_{\text{HAFZ}}(x) = \text{sign}(x) \cdot \left\lfloor |x| + \frac{1}{2} \right\rfloor,$$
$$\text{round}_{\text{HTZ}}(x) = \text{sign}(x) \cdot \left\lceil |x| - \frac{1}{2} \right\rceil,$$
$$\text{round}_{\text{EVEN}}(x) = \begin{cases} \left\lfloor x - \frac{1}{2} \right\rfloor, & \text{if } \frac{2x-1}{4} \in \mathbb{Z}, \\ \left\lfloor x + \frac{1}{2} \right\rfloor, & \text{otherwise}, \end{cases}$$
$$\text{round}_{\text{ODD}}(x) = \begin{cases} \left\lfloor x + \frac{1}{2} \right\rfloor, & \text{if } \frac{2x-1}{4} \in \mathbb{Z}, \\ \left\lfloor x - \frac{1}{2} \right\rfloor, & \text{otherwise}. \end{cases}$$

The round-half-away-from-zero function is the implementation employed in the round function in MATLAB/Octave. The international technical standard ISO/IEC/IEEE 60559:2011 recommends $\text{round}_{\text{EVEN}}(\cdot)$ as the nearest integer function of choice [93]. This latter implementation is adopted in the scientific computation software Mathematica [94].

The above approach returns a large number of candidate approximations. After screening for the approximations with best performance and complexity trade-off, the following matrices were found:

$$\mathbf{T}_4 = \mathbf{S}_{\mathbf{T}_4} \cdot \begin{bmatrix} 1 & 1 & 1 & 1 & 1 & 1 & 1 & 1 \\ 1 & 1 & 1 & 0 & 0 & -1 & -1 & -1 \\ 1 & 1 & -1 & -1 & -1 & -1 & 1 & 1 \\ 1 & 0 & -1 & -1 & 1 & 1 & 0 & -1 \\ 1 & -1 & -1 & 1 & 1 & -1 & -1 & 1 \\ 1 & -1 & 0 & 1 & -1 & 0 & 1 & -1 \\ 1 & -1 & 1 & -1 & -1 & 1 & -1 & 1 \\ 0 & -1 & 1 & -1 & 1 & -1 & 1 & 0 \end{bmatrix},$$

$$\tilde{\mathbf{T}}_1 = \mathbf{S}_{\mathbf{T}_1} \cdot \begin{bmatrix} 1 & 1 & 1 & 1 & 1 & 1 & 1 & 1 \\ 1 & 1 & 0 & 0 & 0 & 0 & -1 & -1 \\ 1 & 0 & 0 & -1 & -1 & 0 & 0 & 1 \\ 1 & 0 & -1 & 0 & 0 & 1 & 0 & -1 \\ 1 & -1 & -1 & 1 & 1 & -1 & -1 & 1 \\ 0 & -1 & 0 & 1 & -1 & 0 & 1 & 0 \\ 0 & -1 & 1 & 0 & 0 & 1 & -1 & 0 \\ 0 & 0 & 1 & -1 & 1 & -1 & 0 & 0 \end{bmatrix},$$

$$\tilde{\mathbf{T}}_3 = \mathbf{S}_{\mathbf{T}_3} \cdot \begin{bmatrix} 1 & 1 & 1 & 1 & 1 & 1 & 1 & 1 \\ 2 & 2 & 1 & 1 & -1 & -1 & -2 & -2 \\ 2 & 1 & -1 & -2 & -2 & -1 & 1 & 2 \\ 2 & -1 & -2 & -1 & 1 & 2 & 1 & -2 \\ 1 & -1 & -1 & 1 & 1 & -1 & -1 & 1 \\ 1 & -2 & 1 & 2 & -2 & -1 & 2 & -1 \\ 1 & -2 & 2 & -1 & -1 & 2 & -2 & 1 \\ 1 & -1 & 2 & -2 & 2 & -2 & 1 & -1 \end{bmatrix},$$

where $\mathbf{S}_{\mathbf{T}_4} = \mathrm{diag}\left(\frac{1}{2\sqrt{2}}, \frac{1}{\sqrt{6}}, \frac{1}{2\sqrt{2}}, \frac{1}{\sqrt{6}}, \frac{1}{2\sqrt{2}}, \frac{1}{\sqrt{6}}, \frac{1}{2\sqrt{2}}, \frac{1}{\sqrt{6}}\right)$, $\mathbf{S}_{\mathbf{T}_1} = \mathrm{diag}\left(\frac{1}{2\sqrt{2}}, \frac{1}{2}, \frac{1}{2}, \frac{1}{2}, \frac{1}{2\sqrt{2}}, \frac{1}{2}, \frac{1}{2}, \frac{1}{2}\right)$, and $\mathbf{S}_{\mathbf{T}_3} = \mathrm{diag}\left(\frac{1}{2\sqrt{2}}, \frac{1}{2\sqrt{5}}, \frac{1}{2\sqrt{5}}, \frac{1}{2\sqrt{5}}, \frac{1}{2\sqrt{2}}, \frac{1}{2\sqrt{5}}, \frac{1}{2\sqrt{5}}, \frac{1}{2\sqrt{5}}\right)$.

Matrix $\mathbf{T}_4$ is orthogonal, whereas $\tilde{\mathbf{T}}_1$ and $\tilde{\mathbf{T}}_3$ are non-orthogonal.

8.3.3 DCT Approximations by Inspection

Several approximations in the literature are proposed without a clearly identifiable methodology. Some of these matrices are simple solutions by inspection based on the exact DCT matrix or some other DCT approximation.

8.3.3.1 Modified RDCT

After judiciously replacing elements of the RDCT matrix with zeros, the modified RDCT (MRDCT) is given by the following orthogonal matrix [58]:

$$\hat{\mathbf{C}}_{\mathrm{MRDCT}} = \mathbf{S}_{\mathrm{MRDCT}} \cdot \begin{bmatrix} 1 & 1 & 1 & 1 & 1 & 1 & 1 & 1 \\ 1 & 0 & 0 & 0 & 0 & 0 & 0 & -1 \\ 1 & 0 & 0 & -1 & -1 & 0 & 0 & 1 \\ 0 & 0 & -1 & 0 & 0 & 1 & 0 & 0 \\ 1 & -1 & -1 & 1 & 1 & -1 & -1 & 1 \\ 0 & -1 & 0 & 0 & 0 & 0 & 1 & 0 \\ 0 & -1 & 1 & 0 & 0 & 1 & -1 & 0 \\ 0 & 0 & 0 & -1 & 1 & 0 & 0 & 0 \end{bmatrix},$$

where $\mathbf{S}_{\mathrm{MRDCT}} = \mathrm{diag}\left(\frac{1}{\sqrt{8}}, \frac{1}{\sqrt{2}}, \frac{1}{2}, \frac{1}{\sqrt{2}}, \frac{1}{\sqrt{8}}, \frac{1}{\sqrt{2}}, \frac{1}{2}, \frac{1}{\sqrt{2}}\right)$.

8.3.3.2 BAS Series of Approximations

In Refs. [59–62], Bouguezel, Ahmad, and Swamy (BAS) introduced a series of low-complexity transforms to replace the DCT in image compression application. Table 8.2 shows some of these transforms.

In Ref. [63], another BAS transform is introduced based on a given parameterization of the transform matrix. This approximation is discussed in Section 8.3.5.

8.3.3.3 SPM Transform

In Ref. [95], Senapati, Pati, and Mahapatra (SPM) proposed a transform matrix by appropriately replacing particular elements of the SDCT matrix for zeros or $\pm\frac{1}{2}$. The SPM matrix is given by

$$\hat{\mathbf{C}}_{\mathrm{SPM}} = \frac{\mathbf{S}_{\mathrm{SPM}}}{2\sqrt{2}} \cdot \begin{bmatrix} 1 & 1 & 1 & 1 & 1 & 1 & 1 & 1 \\ 1 & 1 & 0 & 0 & 0 & 0 & -1 & -1 \\ 1 & 0.5 & -0.5 & -1 & -1 & -0.5 & 0.5 & 1 \\ 0 & 0 & -1 & 0 & 0 & 1 & 0 & 0 \\ 1 & -1 & -1 & 1 & 1 & -1 & -1 & 1 \\ 1 & -1 & 0 & 0 & 0 & 0 & 1 & -1 \\ 0.5 & 0 & 0 & -0.5 & -0.5 & 0 & 0 & 0.5 \\ 0 & 0 & 0 & -1 & 1 & 0 & 0 & 0 \end{bmatrix},$$

TABLE 8.2
BAS Series of Transforms

Transform Matrix	Orthogonal?	Description
$\begin{bmatrix} 1 & 1 & 1 & 1 & 1 & 1 & 1 & 1 \\ 1 & 1 & 0 & 0 & 0 & 0 & -1 & -1 \\ 1 & \frac{1}{2} & -\frac{1}{2} & -1 & -1 & -\frac{1}{2} & \frac{1}{2} & 1 \\ 0 & 0 & -1 & 0 & 0 & 1 & 0 & 0 \\ 1 & -1 & -1 & 1 & 1 & -1 & -1 & 1 \\ 1 & -1 & 0 & 0 & 0 & 0 & 1 & -1 \\ \frac{1}{2} & -1 & 1 & -\frac{1}{2} & -\frac{1}{2} & 1 & -1 & \frac{1}{2} \\ 0 & 0 & 0 & -1 & 1 & 0 & 0 & 0 \end{bmatrix}$	Yes	Introduced in Ref. [59]
$\begin{bmatrix} 1 & 1 & 1 & 1 & 1 & 1 & 1 & 1 \\ 1 & 1 & 1 & 0 & 0 & -1 & -1 & -1 \\ 1 & 1 & -1 & -1 & -1 & -1 & 1 & 1 \\ 1 & 0 & -1 & 0 & 0 & 1 & 0 & -1 \\ 1 & -1 & -1 & 1 & 1 & -1 & -1 & 1 \\ 1 & -1 & 1 & 0 & 0 & -1 & 1 & -1 \\ 1 & -1 & 1 & -1 & -1 & 1 & -1 & 1 \\ 1 & -1 & -1 & 1 & -1 & 1 & 1 & 1 \end{bmatrix}$	No	Introduced in Ref. [60]
$\begin{bmatrix} 1 & 1 & 1 & 1 & 1 & 1 & 1 & 1 \\ 1 & 1 & 0 & 0 & 0 & 0 & -1 & -1 \\ 1 & 1 & -1 & -1 & -1 & -1 & 1 & 1 \\ 0 & 0 & -1 & 0 & 0 & 1 & 0 & 0 \\ 1 & -1 & -1 & 1 & 1 & -1 & -1 & 1 \\ 1 & -1 & 0 & 0 & 0 & 0 & 1 & -1 \\ 1 & -1 & 1 & -1 & -1 & 1 & -1 & 1 \\ 0 & 0 & 0 & -1 & 1 & 0 & 0 & 0 \end{bmatrix}$	Yes	Introduced in Ref. [61]
$\begin{bmatrix} 1 & 1 & 1 & 1 & 1 & 1 & 1 & 1 \\ 1 & 1 & 1 & 1 & -1 & -1 & -1 & -1 \\ 2 & 1 & -1 & -2 & -2 & -1 & 1 & 2 \\ 2 & 1 & -1 & -2 & 2 & 1 & -1 & -2 \\ 1 & -1 & -1 & 1 & 1 & -1 & -1 & 1 \\ 1 & -1 & -1 & 1 & 1 & 1 & -1 \\ 1 & -2 & 2 & -1 & -1 & 2 & -2 & 1 \\ 1 & -2 & 2 & -1 & 1 & -2 & 2 & -1 \end{bmatrix}$	Yes	Introduced in Ref. [62]

(Continued)

TABLE 8.2 (CONTINUED)
BAS Series of Transforms

Transform Matrix	Orthogonal?	Description
$\begin{bmatrix} 1 & 1 & 1 & 1 & 1 & 1 & 1 & 1 \\ 1 & 1 & 1 & 1 & -1 & -1 & -1 & -1 \\ 1 & 1 & -1 & -1 & 1 & 1 & -1 & -1 \\ 1 & -1 & -1 & 1 & -1 & 1 & -1 & -1 \\ 1 & -1 & 1 & -1 & 1 & -1 & 1 & -1 \\ 1 & -1 & 1 & -1 & -1 & 1 & -1 & 1 \\ 1 & -1 & -1 & 1 & 1 & -1 & -1 & 1 \\ 1 & 1 & -1 & -1 & -1 & -1 & 1 & 1 \end{bmatrix}$	Yes	Introduced in Ref. [64]

where $\mathbf{S}_{\text{SPM}} = \text{diag}(1, \sqrt{2}, 2\sqrt{2/5}, 2, 1, \sqrt{2}, 2\sqrt{2/5}, 2)$. This is a non-orthogonal approximation.

8.3.3.4 Signed SPM Transform

By applying the signum function operator to the SPM transform matrix [95], Haweel et al. [68] proposed a new approximation matrix. It consists of a quasi-orthogonal transform given by

$$\hat{\mathbf{C}}_{\text{SSPM}} = \mathbf{S}_{\text{SSPM}} \cdot \begin{bmatrix} 1 & 1 & 1 & 1 & 1 & 1 & 1 & 1 \\ 1 & 1 & 0 & 0 & 0 & 0 & -1 & -1 \\ 1 & 1 & -1 & -1 & -1 & -1 & 1 & 1 \\ 0 & 0 & -1 & 0 & 0 & 1 & 0 & 0 \\ 1 & -1 & -1 & 1 & 1 & -1 & -1 & 1 \\ 1 & -1 & 0 & 0 & 0 & 0 & 1 & -1 \\ 1 & 0 & 0 & -1 & -1 & 0 & 0 & 1 \\ 0 & 0 & 0 & -1 & 1 & 0 & 0 & 0 \end{bmatrix},$$

where $\mathbf{S}_{\text{SSPM}} = \text{diag}\left(\frac{1}{\sqrt{8}}, \frac{1}{2}, \frac{1}{\sqrt{8}}, \frac{1}{\sqrt{2}}, \frac{1}{\sqrt{8}}, \frac{1}{2}, \frac{1}{2}, \frac{1}{\sqrt{2}}\right)$.

8.3.4 DCT Approximations Based on Computational Search

Some DCT approximations are proposed by solving an optimization problem in the following format:

$$\hat{\mathbf{C}}^* = \arg \min_{\hat{\mathbf{C}}} \text{approx}(\hat{\mathbf{C}}, \mathbf{C}),$$

where $\text{approx}(\hat{\mathbf{C}}, \mathbf{C})$ is a relevant objective function—such as proximity measures presented in Section 8.4. Usually the solution is subject to a variety of constraints such as low-arithmetic cost or orthogonality.

8.3.4.1 DCT Approximations for RF Imaging

In Ref. [96], a DCT approximation tailored to radio-frequency (RF) multi-beamforming applications was proposed. This particular low-complexity transform was derived by means of a multivariate non-linear optimization approach with the following objective function:

$$\text{approx}(\hat{\mathbf{C}}, \mathbf{C}) = \epsilon(\hat{\mathbf{C}}),$$

where $\epsilon(\hat{\mathbf{C}})$ is the total error energy discussed in the next section. Additionally, the following constraints were adopted:

i. Elements of matrix must be in {0, ±1, ±2} to ensure that resulting multiplicative complexity is null.
ii. In order to maintain the symmetry of the DCT matrix, the structure of the candidate matrices was restricted to the following:

$$\begin{bmatrix} a_0 & a_0 & a_0 & a_0 & a_0 & a_0 & a_0 & a_0 \\ a_1 & a_2 & a_3 & a_4 & -a_4 & -a_3 & -a_2 & -a_1 \\ a_5 & a_6 & -a_6 & -a_5 & -a_5 & -a_6 & a_6 & a_5 \\ a_7 & -a_8 & -a_9 & -a_{10} & a_{10} & a_9 & a_8 & -a_7 \\ a_0 & -a_0 & -a_0 & a_0 & a_0 & -a_0 & -a_0 & a_0 \\ a_{11} & -a_{12} & a_{13} & a_{14} & -a_{14} & -a_{13} & a_{12} & -a_{11} \\ a_{15} & -a_{16} & a_{16} & -a_{15} & -a_{15} & a_{16} & -a_{16} & a_{15} \\ a_{17} & -a_{18} & a_{19} & -a_{20} & a_{20} & -a_{19} & a_{18} & -a_{17} \end{bmatrix}$$

where $a_i \in \{0, 1, 2\}$, for $i = 1, \ldots, 20$.
iii. The candidate matrices must be orthogonal.

The above optimization problem is algebraically intractable. However, exhaustive computational search can lead to solutions. As a result, the following DCT approximation was found:

$$\hat{\mathbf{C}}_{\text{RF}} = \mathbf{S}_{\text{RF}} \cdot \begin{bmatrix} 1 & 1 & 1 & 1 & 1 & 1 & 1 & 1 \\ 2 & 1 & 1 & 0 & 0 & -1 & -1 & -2 \\ 2 & 1 & -1 & -2 & -2 & -1 & 1 & 2 \\ 1 & 0 & -2 & -1 & 1 & 2 & 0 & -1 \\ 1 & -1 & -1 & 1 & 1 & -1 & -1 & 1 \\ 1 & -2 & 0 & 1 & -1 & 0 & 2 & -1 \\ 1 & -2 & 2 & -1 & -1 & 2 & -2 & 1 \\ 0 & -1 & 1 & -2 & 2 & -1 & 1 & 0 \end{bmatrix}$$

where $\mathbf{S}_{\text{RF}} = \frac{1}{2} \cdot \text{diag}\left(\frac{1}{\sqrt{2}}, \frac{1}{\sqrt{3}}, \frac{1}{\sqrt{5}}, \frac{1}{\sqrt{3}}, \frac{1}{\sqrt{2}}, \frac{1}{\sqrt{3}}, \frac{1}{\sqrt{5}}, \frac{1}{\sqrt{3}}\right)$.

8.3.4.2 Improved 8-Point DCT Approximation

An improved approximate DCT for image and video compression was proposed in Ref. [73], which proved to be a good candidate for reconfigurable video standards such as HEVC [38]. Such transformation was obtained by solving an optimization problem related to the following objective function:

$$\text{approx}(\hat{\mathbf{C}}, \mathbf{C}) = \text{cost}(\hat{\mathbf{C}}),$$

where $\text{cost}(\hat{\mathbf{C}})$ returns the arithmetic complexity of $\hat{\mathbf{C}}$. Additionally, the following constraints were adopted:

i. Elements of matrix must be in {0, ±1, ±2} to ensure that resulting multiplicative complexity is null.
ii. The structure of the candidate matrices was restricted to the following form:

$$\begin{bmatrix} a_3 & a_3 & a_3 & a_3 & a_3 & a_3 & a_3 & a_3 \\ a_0 & a_2 & a_4 & a_6 & -a_6 & -a_4 & -a_2 & -a_0 \\ a_1 & a_5 & -a_5 & -a_1 & -a_1 & -a_5 & a_5 & a_1 \\ a_2 & -a_6 & -a_0 & -a_4 & a_4 & a_0 & a_6 & -a_2 \\ a_3 & -a_3 & -a_3 & a_3 & a_3 & -a_3 & -a_3 & a_3 \\ a_4 & -a_0 & a_6 & a_2 & -a_2 & -a_6 & a_0 & -a_4 \\ a_5 & -a_1 & a_1 & -a_5 & -a_5 & a_1 & -a_1 & a_5 \\ a_6 & -a_4 & a_2 & -a_0 & a_0 & -a_2 & a_4 & -a_6 \end{bmatrix},$$

where $a_i \in \{0, 1, 2\}$, for $i = 0, 1, \ldots, 6$.
iii. All rows of the matrix are non-null.
iv. The candidate matrix must be orthogonal.

Constraint (ii) was required to preserve the DCT-like matrix structure.

As a result of the above optimization problem, eight candidate matrices were found, including the transform matrix proposed in Ref. [58]. Among these minimal cost matrices, the authors separated the matrix that presents the best performance in terms of image quality of compressed images according the JPEG-like technique. The resulting matrix is given by

$$\hat{\mathbf{C}}_{\text{Improve}} = \mathbf{S}_{\text{Improve}} \cdot \begin{bmatrix} 1 & 1 & 1 & 1 & 1 & 1 & 1 & 1 \\ 0 & 1 & 0 & 0 & 0 & 0 & -1 & 0 \\ 1 & 0 & 0 & -1 & -1 & 0 & 0 & 1 \\ 1 & 0 & 0 & 0 & 0 & 0 & 0 & -1 \\ 1 & -1 & -1 & 1 & 1 & -1 & -1 & 1 \\ 0 & 0 & 0 & 1 & -1 & 0 & 0 & 0 \\ 0 & -1 & 1 & 0 & 0 & 1 & -1 & 0 \\ 0 & 0 & 1 & 0 & 0 & -1 & 0 & 0 \end{bmatrix},$$

where $\mathbf{S}_{\text{Improve}} = \text{diag}\left(\frac{1}{\sqrt{8}}, \frac{1}{\sqrt{2}}, \frac{1}{2}, \frac{1}{\sqrt{2}}, \frac{1}{\sqrt{8}}, \frac{1}{\sqrt{2}}, \frac{1}{2}, \frac{1}{\sqrt{2}}\right)$.

8.3.4.3 DCT Approximation for IR

With the objective to reduce the computation time of the residual complexity similarity metric employed in biomedical IR [97], in Ref. [72], a multiplier-free DCT approximation was proposed. Its derivation stems from a parametric-based optimization approach similar to the method described in Refs. [73,96]. In fact, the objective function, based on the product of two transformation performance measures, was given by

$$\text{approx}(\hat{\mathbf{C}}, \mathbf{C}) = \left[\epsilon(\hat{\mathbf{C}}) \times d_2(\hat{\mathbf{C}})\right],$$

where $\epsilon(\hat{\mathbf{C}})$ and $d_2(\hat{\mathbf{C}})$ are, respectively, the total error energy and the transform distortion, measures to be detailed in Section 8.4.1. Additionally, the following constraints were adopted:

i. Elements of matrix must be in $\{0, \pm 1\}$ to ensure that resulting null multiplicative complexity.
ii. The additive complexity of the candidate matrices must be smaller than the additive complexity of the SDCT, RDCT, HT, and WHT.
iii. In order to maintain the symmetry of the DCT matrix, the structure of the candidate matrices was restricted to the following:

$$\gamma \cdot \begin{bmatrix} a_3 & a_3 & a_3 & a_3 & a_3 & a_3 & a_3 & a_3 \\ a_0 & a_2 & a_4 & a_6 & -a_6 & -a_4 & -a_2 & -a_0 \\ a_1 & a_5 & -a_5 & -a_1 & -a_1 & -a_5 & a_5 & a_1 \\ a_2 & -a_6 & -a_0 & -a_4 & a_4 & a_0 & a_6 & -a_2 \\ a_3 & -a_3 & -a_3 & a_3 & a_3 & -a_3 & -a_3 & a_3 \\ a_4 & -a_0 & a_6 & a_2 & -a_2 & -a_6 & a_0 & -a_4 \\ a_5 & -a_1 & a_1 & -a_5 & -a_5 & a_1 & -a_1 & a_5 \\ a_6 & -a_4 & a_2 & -a_0 & a_0 & -a_2 & a_4 & -a_6 \end{bmatrix},$$

where $a_i \in \{0, 1\}$, for $i = 0, \ldots, 6$ and the scaling factor γ ranging from 0.1 to 0.9 by 0.01 step.

An exhaustive computational search returned the following transform:

$$\hat{\mathbf{C}}_{\text{IR}} = 0.47 \cdot \begin{bmatrix} 1 & 1 & 1 & 1 & 1 & 1 & 1 & 1 \\ 1 & 1 & 0 & 0 & 0 & 0 & -1 & -1 \\ 1 & 0 & 0 & -1 & -1 & 0 & 0 & 1 \\ 1 & 0 & -1 & 0 & 0 & 1 & 0 & -1 \\ 1 & -1 & -1 & 1 & 1 & -1 & -1 & 1 \\ 0 & -1 & 0 & 1 & -1 & 0 & 1 & 0 \\ 0 & -1 & 1 & 0 & 0 & 1 & -1 & 0 \\ 0 & 0 & 1 & -1 & 1 & -1 & 0 & 0 \end{bmatrix} \cdot$$

We note that the low-complexity matrix associated to this transform is the same in $\tilde{\mathbf{T}}_1$ [66]. However, in Ref. [66], it was considered a diagonal matrix for quasi-orthogonalization and in Ref. [72] was considered just a scalar. In the biomedical application in Ref. [72], it was important to consider transform matrices without diagonal adjustment matrices.

8.3.5 Parametric DCT Approximations

Some DCT approximations can be embedded into the same framework, inducing a class of transformations. In these cases, depending on the value of a free parameter, or a vector of parameters, the parametric transform can lead to some special—known or unknown—matrices.

8.3.5.1 Level 1 Approximation

DCT approximations were proposed in Ref. [71] where rational multipliers were approximated by parameters that require additions and bit-shifting operations only. Such parameters were proposed in five levels of precision. Considering the coarsest approximation level, the resulting transform matrix possesses entry values in $\{0,1/2,1\}$. This multiplierless "Level 1" approximation is given by [71]

$$\hat{\mathbf{C}}_{\text{Level 1}} = \mathbf{S}_{\text{Level 1}}$$

$$\cdot \begin{bmatrix} 1 & 1 & 1 & 1 & 1 & 1 & 1 & 1 \\ 1 & 1 & 1 & 0 & 0 & -1 & -1 & -1 \\ 1 & 1/2 & -1/2 & -1 & -1 & -1/2 & 1/2 & 1 \\ 1 & 0 & -1 & -1 & 1 & 1 & 0 & -1 \\ 1 & -1 & -1 & 1 & 1 & -1 & -1 & 1 \\ 1 & -1 & 0 & 1 & -1 & 0 & 1 & -1 \\ 1/2 & -1 & 1 & -1/2 & -1/2 & 1 & -1 & 1/2 \\ 0 & -1 & 1 & -1 & 1 & -1 & 1 & 0 \end{bmatrix},$$

where $\mathbf{S}_{\text{Level 1}} = \dfrac{1}{2\sqrt{2}} \text{diag}(1.0, 1.1162, 1.2617, 1.1162, 1.0, 1.1162, 1.2617, 1.1162)$.

8.3.5.2 Parametric BAS

Proposed in 2011 by Bouguezel et al. [63], the parametric BAS transform is an 8-point orthogonal transform containing a single parameter a in its transformation matrix $\mathbf{C}_{(a)}$. It is given as follows:

$$\hat{\mathbf{C}}_{(a)} = \mathbf{S}_{(a)} \cdot \begin{bmatrix} 1 & 1 & 1 & 1 & 1 & 1 & 1 & 1 \\ 1 & 1 & 0 & 0 & 0 & 0 & -1 & -1 \\ 1 & a & -a & -1 & -1 & -a & a & 1 \\ 0 & 0 & 1 & 0 & 0 & -1 & 0 & 0 \\ 1 & -1 & -1 & 1 & 1 & -1 & -1 & 1 \\ 0 & 0 & 0 & 1 & -1 & 0 & 0 & 0 \\ 1 & -1 & 0 & 0 & 0 & 0 & 1 & -1 \\ a & -1 & 1 & -a & -a & 1 & -1 & a \end{bmatrix},$$

where $\mathbf{S}_{(a)} = \text{diag}\left(\dfrac{1}{\sqrt{8}}, \dfrac{1}{2}, \dfrac{1}{\sqrt{4+4a^2}}, \dfrac{1}{\sqrt{2}}, \dfrac{1}{\sqrt{8}}, \dfrac{1}{\sqrt{2}}, \dfrac{1}{2}, \dfrac{1}{\sqrt{4+4a^2}}\right)$. Usually the parameter a is selected as a small integer in order to minimize the complexity of $\hat{\mathbf{C}}_{(a)}$. In Ref. [63], suggested values are $a \in \{0, 1/2, 1\}$.

8.3.5.3 FW Class of DCT Approximations

In Ref. [74], a new class of matrices based on a parameterization of the FW [51] DCT factorization is proposed. Such parameterization induces a matrix subspace, which unifies a number of existing methods for DCT approximation. By solving a multicriteria optimization problem, several new DCT approximations were proposed.

In Ref. [51], Feig and Winograd introduced a fast algorithm whose factorization can be given by

$$\mathbf{C}_8 = \dfrac{1}{2} \cdot \mathbf{P} \cdot \mathbf{K} \cdot \mathbf{B}_1 \cdot \mathbf{B}_2 \cdot \mathbf{B}_3,$$

where $\mathbf{P}$ is a signed permutation matrix, $\mathbf{K}$ is a multiplicative matrix, and $\mathbf{B}_1$, $\mathbf{B}_2$, and $\mathbf{B}_3$ are symmetric additive matrices. These matrices are given by

$$\mathbf{B}_1 = \text{bdiag}\left(\begin{bmatrix} 1 & 1 \\ 1 & -1 \end{bmatrix}, \mathbf{I}_6\right),$$

$$\mathbf{B}_2 = \text{bdiag}\left(\begin{bmatrix} 1 & 0 & 0 & 1 \\ 0 & 1 & 1 & 0 \\ 0 & 1 & -1 & 0 \\ 1 & 0 & 0 & -1 \end{bmatrix}, \mathbf{I}_4\right),$$

$$\mathbf{B}_3 = \begin{bmatrix} \mathbf{I}_4 & \bar{\mathbf{I}}_4 \\ \bar{\mathbf{I}}_4 & -\mathbf{I}_4 \end{bmatrix}, \quad \mathbf{P} = \begin{bmatrix} 1 & 0 & 0 & 0 & 0 & 0 & 0 & 0 \\ 0 & 0 & 0 & 0 & -1 & 0 & 0 & 0 \\ 0 & 0 & 1 & 0 & 0 & 0 & 0 & 0 \\ 0 & 0 & 0 & 0 & 0 & -1 & 0 & 0 \\ 0 & 1 & 0 & 0 & 0 & 0 & 0 & 0 \\ 0 & 0 & 0 & 0 & 0 & 0 & 0 & -1 \\ 0 & 0 & 0 & 1 & 0 & 0 & 0 & 0 \\ 0 & 0 & 0 & 0 & 0 & 0 & 1 & 0 \end{bmatrix},$$

and

$$\mathbf{K} = \begin{bmatrix} \gamma_3 & 0 & 0 & 0 & 0 & 0 & 0 & 0 \\ 0 & \gamma_3 & 0 & 0 & 0 & 0 & 0 & 0 \\ 0 & 0 & \gamma_5 & \gamma_1 & 0 & 0 & 0 & 0 \\ 0 & 0 & -\gamma_1 & \gamma_5 & 0 & 0 & 0 & 0 \\ 0 & 0 & 0 & 0 & -\gamma_6 & -\gamma_4 & -\gamma_2 & -\gamma_0 \\ 0 & 0 & 0 & 0 & \gamma_4 & \gamma_0 & \gamma_6 & -\gamma_2 \\ 0 & 0 & 0 & 0 & -\gamma_0 & \gamma_2 & -\gamma_4 & \gamma_6 \\ 0 & 0 & 0 & 0 & -\gamma_2 & -\gamma_6 & \gamma_0 & -\gamma_4 \end{bmatrix},$$

where $\gamma_k = \cos(2\pi(k+1)/32)$, $\mathbf{I}_l$ and $\bar{\mathbf{I}}_l$ denote the identity and counter-identity matrices of size l, respectively, and bdiag($\cdot$) is the block diagonal operator.

Above factorization circumscribes the entire multiplicative complexity of the DCT into the block diagonal matrix $\mathbf{K}$. Indeed, the seven distinct non-null elements of $\mathbf{K}$, namely γ_i, $i = 0, 1, \ldots, 6$, are the only non-trivial quantities in FW algorithm. This factorization paves the way for defining a class of 8×8 matrices generated according to the following mapping [74]:

$$\text{FW}: \mathbb{R}^7 \to \mathcal{M}(8) \qquad (8.3)$$
$$\boldsymbol{\alpha} \mapsto \mathbf{P} \cdot \mathbf{K}_\alpha \cdot \mathbf{B}_1 \cdot \mathbf{B}_2 \cdot \mathbf{B}_3,$$

where $\boldsymbol{\alpha} = [\alpha_0 \ \alpha_1 \ \cdots \ \alpha_6]^T$ is a 7-point parameter vector, $\mathcal{M}(8)$ is the 8×8 matrix space over the real numbers, and

$$\mathbf{K}_\alpha = \begin{bmatrix} \alpha_3 & 0 & 0 & 0 & 0 & 0 & 0 & 0 \\ 0 & \alpha_3 & 0 & 0 & 0 & 0 & 0 & 0 \\ 0 & 0 & \alpha_5 & \alpha_1 & 0 & 0 & 0 & 0 \\ 0 & 0 & -\alpha_1 & \alpha_5 & 0 & 0 & 0 & 0 \\ 0 & 0 & 0 & 0 & -\alpha_6 & -\alpha_4 & -\alpha_2 & -\alpha_0 \\ 0 & 0 & 0 & 0 & \alpha_4 & \alpha_0 & \alpha_6 & -\alpha_2 \\ 0 & 0 & 0 & 0 & -\alpha_0 & \alpha_2 & -\alpha_4 & \alpha_6 \\ 0 & 0 & 0 & 0 & -\alpha_2 & -\alpha_6 & \alpha_0 & -\alpha_4 \end{bmatrix}.$$

The image of the multivariate function FW(·) is a subset of $\mathcal{M}(8)$. It is straightforward to verify that such subset is also closed under the operations of addition and scalar multiplication. Therefore, this subset is a matrix subspace, which is referenced as the FW matrix subspace [74].

Mathematically, the FW factorization induces a matrix subspace by allowing its multiplicative constants to be treated as parameters. Thus, an appropriate parameter selection may result in suitable approximations, which are denoted by $\mathbf{T}_\alpha = \text{FW}(\alpha)$.

Considering the mapping described in Equation 8.3, for any choice of α, $\mathbf{T}_\alpha$ satisfies the FW factorization. Thus, all matrices in this particular subspace possess the same general fast algorithm structure.

To ensure orthogonality or near orthogonality, FW transforms are given by

$$\hat{\mathbf{C}}_\alpha = \mathbf{S}_\alpha \cdot \mathbf{T}_\alpha$$

where $\mathbf{S}_\alpha = \text{diag}(s_0, s_1, s_2, s_1, s_0, s_1, s_2, s_1)$ if matrix $\mathbf{T}_\alpha$ is orthogonal, for $s_0 = 1/(2^{3/2}\alpha_3)$, $s_1 = 1/\sqrt{2(\alpha_6^2 + \alpha_4^2 + \alpha_2^2 + \alpha_0^2)}$, $s_2 = 1/(2\sqrt{\alpha_5^2 + \alpha_1^2})$; and $\hat{\mathbf{S}}_\alpha = \sqrt{[\text{diag}(\mathbf{T}_\alpha \cdot \mathbf{T}_\alpha^T)]^{-1}}$ if it is not orthogonal.

The FW formalism generalizes dozens of transforms under the same framework. Based on different values of the parameter vector α, possibly distinct FW approximation matrices can be obtained. However, with the goal of finding good DCT approximations, the authors in Ref. [74] solved the following multicriteria optimization problem [98,99]:

$$\arg \min_\alpha \left(\epsilon(\hat{\mathbf{C}}_\alpha), \text{MSE}(\hat{\mathbf{C}}_\alpha), -C_g(\hat{\mathbf{C}}_\alpha), -\eta(\hat{\mathbf{C}}_\alpha), \mathcal{A}(\hat{\mathbf{C}}_\alpha), \mathcal{S}(\hat{\mathbf{C}}_\alpha) \right), \quad (8.4)$$

where $\mathcal{A}(\hat{\mathbf{C}}_\alpha)$ and $\mathcal{S}(\hat{\mathbf{C}}_\alpha)$ are the addition and bit-shifting counts, respectively, and $\epsilon(\hat{\mathbf{C}}_\alpha)$, $\text{MSE}(\hat{\mathbf{C}}_\alpha)$, $C_g(\hat{\mathbf{C}}_\alpha)$, and $\eta(\hat{\mathbf{C}}_\alpha)$ are performance measures described in Section 8.4.1. To address the above optimization problem, the authors identified a search space. With the goal of finding low-complexity approximations, the candidate solutions α were restricted by $\alpha_k \in \mathcal{P}$, $k = 0, 1, \ldots, 6$, where $\mathcal{P} = \{0, \pm 1/2, \pm 1, \pm 2\}$. Thus, the search space for Equation 8.4 is the set $\mathcal{P}^7$. As a result of the computational search, nine efficient solutions were found, namely, $\mathbf{T}_8^{(1)}, \mathbf{T}_8^{(2)}, \mathbf{T}_8^{(3)}, \mathbf{T}_8^{(4)}, \mathbf{T}_8^{(5)}, \mathbf{T}_8^{(6)}, \mathbf{T}_8^{(7)}, \mathbf{T}_8^{(8)}$, and $\mathbf{T}_8^{(16)}$. Among these efficient transforms, the matrices $\mathbf{T}_8^{(1)}, \mathbf{T}_8^{(2)}, \mathbf{T}_8^{(3)}$, and $\mathbf{T}_8^{(16)}$ are already known in the literature, being the Level 1 approximation [71], RDCT [65], MRDCT [58], and DCT approximation for IR [72], respectively. The other five transforms were new. These efficient transform matrices are presented in Table 8.3.

8.4 Performance Comparison

In practice, the characteristics of decorrelation and compaction of DCT approximations are very similar to the characteristics of DCT. Figure 8.1 illustrates these properties from a biomedical image considering the DCT, SDCT, and RDCT. However, it is important to consider some quantitative measures to assess (i) the proximity of each DCT approximation to the exact DCT and (ii) their coding capability. Section 8.4.1 presents the main performance measures considered in the literature. In Section 8.4.2, all above discussed transforms are assessed by means of arithmetic complexity evaluation and performance measures.

8.4.1 Performance Measures

In order to evaluate the DCT approximations, some figures of merit can be considered. The following measures are usually assessed: (i) the transform distortion [100]; (ii) the total error energy [65]; (iii) the mean square error (MSE) [101]; (iv) the unified coding gain [10,102,103]; and (v) the transform efficiency [10]. The transform distortion, total error energy, and MSE are employed to quantify how close a given DCT approximation $\hat{\mathbf{C}}_N$ is to the exact DCT matrix $\mathbf{C}_N$. To maintain the compatibility between the approximation and the *exact* DCT outputs, these similarity measures should be minimized [10,29]. The coding gain and transform efficiency capture the coding performance of a given transformation [10]. Approximations exhibiting high transform coding gains and transform efficiency can compact more energy into few coefficients [29].

For coding performance evaluation, it is assumed that the signals are occurrences of a first-order Markov process with zero-mean, unit variance, and a correlation coefficient equal to $\rho = 0.95$ [10,29,103], as described in Section 8.2.1. Below we briefly describe each of these performance measures.

8.4.1.1 Transform Distortion

The transform distortion [100] is a figure of merit that measures the difference percentage between the exact matrix and its approximation. Being a similarity measure, we adopted the exact DCT $\mathbf{C}_N$ as reference framework. The transform distortion measure is given by

$$d_2(\hat{\mathbf{C}}_N) = 1 - \frac{1}{N} \cdot \left\| \text{diag}(\mathbf{C}_N \cdot \hat{\mathbf{C}}_N^T) \right\|_2^2,$$

where $\|\cdot\|_2$ is the Euclidean norm [104].

TABLE 8.3
FW DCT Approximations

Transform	Parametric Vector α	Matrix	Orthogonal?
$\mathbf{T}_8^{(4)}$	$[1\ 2\ 0\ 1\ 0\ 1\ 0]^T$	$\begin{bmatrix} 1 & 1 & 1 & 1 & 1 & 1 & 1 & 1 \\ 1 & 0 & 0 & 0 & 0 & 0 & 0 & -1 \\ 2 & 1 & -1 & -2 & -2 & -1 & 1 & 2 \\ 0 & 0 & -1 & 0 & 0 & 1 & 0 & 0 \\ 1 & -1 & -1 & 1 & 1 & -1 & -1 & 1 \\ 0 & -1 & 0 & 0 & 0 & 0 & 1 & 0 \\ 1 & -2 & 2 & -1 & -1 & 2 & -2 & 1 \\ 0 & 0 & 0 & -1 & 1 & 0 & 0 & 0 \end{bmatrix}$	Yes
$\mathbf{T}_8^{(5)}$	$[0\ 1\ 1\ 1\ 1\ 0\ 0]^T$	$\begin{bmatrix} 1 & 1 & 1 & 1 & 1 & 1 & 1 & 1 \\ 0 & 1 & 1 & 0 & 0 & -1 & -1 & 0 \\ 1 & 0 & 0 & -1 & -1 & 0 & 0 & 1 \\ 1 & 0 & 0 & -1 & 1 & 0 & 0 & -1 \\ 1 & -1 & -1 & 1 & 1 & -1 & -1 & 1 \\ 1 & 0 & 0 & 1 & -1 & 0 & 0 & -1 \\ 0 & -1 & 1 & 0 & 0 & 1 & -1 & 0 \\ 0 & -1 & 1 & 0 & 0 & -1 & 1 & 0 \end{bmatrix}$	Yes
$\mathbf{T}_8^{(6)}$	$[0\ 2\ 1\ 1\ 1\ 1\ 0]^T$	$\begin{bmatrix} 1 & 1 & 1 & 1 & 1 & 1 & 1 & 1 \\ 0 & 1 & 1 & 0 & 0 & -1 & -1 & 0 \\ 2 & 1 & -1 & -2 & -2 & -1 & 1 & 2 \\ 1 & 0 & 0 & -1 & 1 & 0 & 0 & -1 \\ 1 & -1 & -1 & 1 & 1 & -1 & -1 & 1 \\ 1 & 0 & 0 & 1 & -1 & 0 & 0 & -1 \\ 1 & -2 & 2 & -1 & -1 & 2 & -2 & 1 \\ 0 & -1 & 1 & 0 & 0 & -1 & 1 & 0 \end{bmatrix}$	Yes
$\mathbf{T}_8^{(7)}$	$[0\ 2\ 2\ 1\ 1\ 1\ 0]^T$	$\begin{bmatrix} 1 & 1 & 1 & 1 & 1 & 1 & 1 & 1 \\ 0 & 2 & 1 & 0 & 0 & -1 & -2 & 0 \\ 2 & 1 & -1 & -2 & -2 & -1 & 1 & 2 \\ 2 & 0 & 0 & -1 & 1 & 0 & 0 & -2 \\ 1 & -1 & -1 & 1 & 1 & -1 & -1 & 1 \\ 1 & 0 & 0 & 2 & -2 & 0 & 0 & -1 \\ 1 & -2 & 2 & -1 & -1 & 2 & -2 & 1 \\ 0 & -1 & 2 & 0 & 0 & -2 & 1 & 0 \end{bmatrix}$	Yes

(Continued)

TABLE 8.3 (CONTINUED)
FW DCT Approximations

Transform	Parametric Vector α	Matrix	Orthogonal?
$\mathbf{T}_8^{(8)}$	$[2\ 2\ 0\ 1\ 0\ 1\ 1\ 1/2]^T$	$\begin{bmatrix} 1 & 1 & 1 & 1 & 1 & 1 & 1 & 1 \\ 2 & 0 & 0 & 1/2 & -1/2 & 0 & 0 & -2 \\ 2 & 1 & -1 & -2 & -2 & -1 & 1 & 2 \\ 0 & -1/2 & -2 & 0 & 0 & 2 & 1/2 & 0 \\ 1 & -1 & -1 & 1 & 1 & -1 & -1 & 0 \\ 0 & -2 & 1/2 & 0 & 0 & -1/2 & 2 & 0 \\ 1 & -2 & 2 & -1 & -1 & 2 & -2 & 1 \\ 1/2 & 0 & 0 & -2 & 2 & 0 & 0 & -1/2 \end{bmatrix}$	Yes

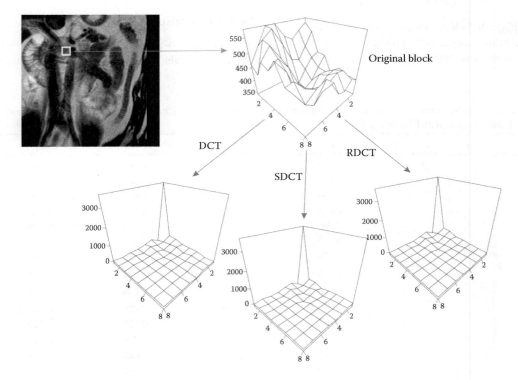

FIGURE 8.1
Energy concentration.

8.4.1.2 Total Error Energy

Each row of a given transform matrix can be understood as finite impulse response filter with associate transfer functions [38,105]. Based on Ref. [69], the magnitude of the difference between the transfer functions of the DCT and the SDCT was advanced in Ref. [65] as a similarity measure. Such measure, termed total error energy, was further employed as a proximity measure between several other approximations [65,75]. Although originally defined on the spectral domain [75], the total error energy can be given a simple matrix form by means of the Parseval theorem [106, p. 18]. The total error energy ϵ for a given DCT approximation matrix $\hat{\mathbf{C}}_N$ is furnished by

$$\epsilon(\hat{\mathbf{C}}_N) = \pi \cdot \|\mathbf{C}_N - \hat{\mathbf{C}}_N\|_F^2,$$

where $\|\cdot\|_F$ denotes the Frobenius norm [104, p. 115].

8.4.1.3 Mean Square Error

The MSE for an approximation matrix $\hat{\mathbf{C}}_N$ is defined as [10,29]

$$\mathrm{MSE}(\hat{\mathbf{C}}_N) = \frac{1}{N} \cdot \mathrm{tr}\left((\mathbf{C}_N - \hat{\mathbf{C}}_N) \cdot \mathbf{R}_x \cdot (\mathbf{C}_N - \hat{\mathbf{C}}_N)^T\right),$$

where $tr(\cdot)$ is the trace function [107].

8.4.1.4 Unified Transform Coding Gain

For non-orthogonal transformations, the unified coding gain generalizes the usual coding gain [103]. Let $\mathbf{h}_k$ and $\mathbf{g}_k$ be the kth row of $\hat{\mathbf{C}}_N$ and $\hat{\mathbf{C}}_N^{-1}$, respectively. Thus, the coding gain of $\hat{\mathbf{C}}_N$ is given by

$$C_g(\hat{\mathbf{C}}_N) = 10 \cdot \log_{10}\left[\prod_{k=1}^N \frac{1}{(A_k \cdot B_k)^{1/N}}\right] \quad \text{(in dB)},$$

where $A_k = \mathrm{su}[(\mathbf{h}_k^T \cdot \mathbf{h}_k) \circ \mathbf{R}_x]$, $\mathrm{su}(\cdot)$ returns the sum of the elements of its matrix argument [107], operator $\circ$ is the element-wise matrix product [83], and $B_k = \|\mathbf{g}_k\|_2^2$.

The transform coding gain for the KLT and 8-point DCT are 8.8462 and 8.8259, respectively [10].

8.4.1.5 Transform Efficiency

Let $r_{m,n}^{(X)}$ be the (m, n)-th entry of the covariance matrix of the transformed signal $\mathbf{X}$, which is given by $\mathbf{R}_X = \hat{\mathbf{C}}_N \cdot \mathbf{R}_x \cdot \hat{\mathbf{C}}_N^T$. The transform efficiency is defined as [10,108]

$$\eta(\hat{\mathbf{C}}_N) = \frac{\sum_{m=1}^N |r_{m,m}^{(X)}|}{\sum_{m=1}^N \sum_{n=1}^N |r_{m,n}^{(X)}|} \cdot 100.$$

The transform efficiency η measures the decorrelation ability of the transform [10]. The optimal KLT converts signals into completely uncorrelated coefficients and has a transform efficiency equal to 100.

8.4.2 Comparison and Discussion

Among all 8-point approximations, measurements shown in Table 8.4 revealed that the transform $\tilde{\mathbf{T}}_3$ [66]

TABLE 8.4

Coding and Similarity Performance Assessment: Transform Distortion (d_2), Total Error Energy (ϵ), MSE, Unified Transform Coding Gain (C_g), and Transform Efficiency (η)

Transform	d_2	ϵ	MSE	C_g	η
8-point DCT	0.000	0.000	0.000	8.826	93.991
HT	0.835	47.613	0.224	7.946	85.314
WHT	0.184	5.049	0.025	7.946	85.314
SDCT [69]	0.126	3.316	0.021	6.026	82.619
RDCT [65]	0.069	1.794	0.010	8.183	87.432
T_4 [66]	0.069	1.794	0.010	8.183	87.157
$\tilde{T}_1$ [66]	0.126	3.316	0.021	6.046	83.081
$\tilde{T}_3$ [66]	0.023	0.577	0.004	8.438	89.706
MRDCT [58]	0.296	8.659	0.059	7.333	80.897
BAS-2008 [60]	0.152	4.188	0.019	6.268	83.173
BAS-2008 [59]	0.205	5.929	0.024	8.119	86.863
BAS-2009 [61]	0.240	6.854	0.028	7.913	85.380
BAS-2010 [62]	0.148	4.093	0.021	8.325	88.218
BAS-2013 [64]	0.739	35.064	0.102	7.946	85.314
SPM [95]	0.311	9.792	0.044	4.865	81.772
Signed SPM [68]	0.328	10.255	0.044	6.438	82.288
DCT approximation for RF imaging [96]	0.034	0.870	0.006	8.344	88.059
Improved DCT approximation [73]	0.364	11.313	0.079	7.333	80.897
DCT approximation for IR [72]	0.007	3.866	0.116	5.672	88.760
Level 1 approximation [71]	0.066	0.856	0.006	8.388	88.972
BAS-2011 with $a = 1$ [63]	0.437	26.864	0.071	7.913	85.380
FW $\mathbf{T}_8^{(4)}$ [74]	0.261	7.734	0.056	7.540	81.985
FW $\mathbf{T}_8^{(5)}$ [74]	0.296	8.659	0.059	7.369	81.179
FW $\mathbf{T}_8^{(6)}$ [74]	0.261	7.734	0.055	7.576	82.275
FW $\mathbf{T}_8^{(7)}$ [74]	0.255	7.532	0.054	7.555	82.701
FW $\mathbf{T}_8^{(8)}$ [74]	0.252	7.414	0.053	7.575	83.085

presents superior performance according to all performance measures. Nevertheless, its high computational complexity shown in Table 8.5 makes it less attractive in extremely low-power applications. On the other hand, transforms MRDCT [58] and improved DCT approximation [73] require just 14 additions for their computation. These transforms are the fastest DCT approximations. In general, BAS matrices possess good coding measures (C_g and η) but poor similarity measures (d_2, ϵ, and MSE).

We note that there is a trade-off between computational cost and performance measures. To analyze this balance, it is necessary to compare the results in Tables 8.4 and 8.5. Among all 8-point DCT approximations, the efficient solutions of the FW class of transforms in Ref. [74] are highlighted, namely, Level 1 approximation [71], RDCT [65], MRDCT [58], $\mathbf{T}_8^{(4)}$, $\mathbf{T}_8^{(5)}$, $\mathbf{T}_8^{(6)}$, $\mathbf{T}_8^{(7)}$, $\mathbf{T}_8^{(8)}$, and DCT approximation for IR [72]. These transforms are a solution of a multicriteria optimization problem, presenting a good balance between computational cost and performance measures.

8.5 Applications in Biomedical Signal Processing

In this section, the above DCT approximation schemes are applied to biomedical scenarios. There are numerous big data biomedical applications. Two representative applications were selected to demonstrate the potential of DCT approximations in this field. The first application targets image compression in medical imaging, where DCT approximations are employed to reduce data complexity in a computationally efficient manner. In the second application, medical images are aligned into a common frame of reference utilizing IR algorithms. Approximate transforms are employed to reduce computational complexity.

8.5.1 Image Compression

In the context of biomedical application, low-complexity image compression techniques are important tasks [1,2,6].

TABLE 8.5

Comparison of Computational Complexities

Transform	Mult	Add	Shifts	Total
Exact 8-point DCT	64	56	0	120
Exact 8-point DCT with fast algorithm [50]	11	29	0	40
HT	0	24	0	24
WHT	0	24	0	24
SDCT [69]	0	24	0	24
RDCT [65]	0	22	0	22
T_4 [66]	0	24	0	24
$\tilde{T}_1$ [66]	0	18	0	18
$\tilde{T}_3$ [66]	0	28	10	38
MRDCT [58]	0	14	0	14
BAS-2008 [60]	0	21	0	21
BAS-2008 [59]	0	18	2	20
BAS-2009 [61]	0	18	0	18
BAS-2010 [62]	0	24	4	28
BAS-2011 with $a = 1$ [63]	0	18	0	18
BAS-2013 [64]	0	24	0	24
SPM [95]	0	17	2	19
Signed SPM [68]	0	17	0	17
DCT approximation for RF imaging [96]	0	24	6	30
Improved DCT approximation [73]	0	14	0	14
DCT approximation for IR [72]	0	18	0	18
Level 1 approximation [71]	0	24	2	26
FW $\mathbf{T}_8^{(4)}$ [74]	0	16	2	18
FW $\mathbf{T}_8^{(5)}$ [74]	0	18	0	18
FW $\mathbf{T}_8^{(6)}$ [74]	0	20	2	22
FW $\mathbf{T}_8^{(7)}$ [74]	0	20	6	26
FW $\mathbf{T}_8^{(8)}$ [74]	0	20	10	30

In medical imaging, data volume, storage, and transfer increase in fast pace requiring optimized methods. For instance, medical facilities that used to produce 5.5 TB/day of medical images in 2005 produce 35.5 TB/day in 2010 [109], which is consistent with Moore's law. Such increase in data volume can be attributed to improvements in scanning technology with higher image resolution, the availability of new three-dimensional and four-dimensional image acquisition devices, and increased diagnostic value, resulting in a higher number of image acquisitions.

The industry standard medical image storage format Digital Imaging and Communications in Medicine (DICOM) has long supported both lossless and lossy compression including JPEG and JPEG-2000. In applications such as teleradiology, the potential of lossy image compression was assessed to reduce the amount of data to transfer [110]. However, the fear of "throwing away" diagnostically relevant image information has slowed adoption of lossy image compression for data storage in routine medical practice. Only in the past years radiology societies are starting to release guidelines defining specific compression ratios to be used with lossy image compression for various imaging modalities and applications (see, e.g., Europe [109] and North America [111]). Despite the slow adoption, lossy image compression is routinely used when generating diagnostic reports to produce supporting screen-shots of image slices, three-dimensional computer renderings, and other animations. Therefore, reducing complexity of compression algorithms is an important task.

In fact, all previously discussed transformations can be considered as tools for JPEG-like image compression. Given a digital image, this compression method can be summarized by the following algorithm:

1. The image is subdivided in $N \times N$ blocks.
2. Each block is submitted to a 2-D transformation similar to Equation 8.1, where the exact DCT matrix is replaced by a selected DCT approximation.
3. The resulting N^2 coefficients in transform domain are ordered in the standard zigzag sequence [31], and only the r initial coefficients in each block are retained.
4. The inverse 2-D transform was applied to the truncated transform domain coefficients and the compressed block is obtained.
5. Each compressed block is reallocated into the original place.

Figure 8.2 depicts the above-described JPEG-like procedure. This type of computational experiment is commonly employed to assess the compression capabilities of discrete transforms [58,59,65,66,69].

For a comparative analysis, we selected DCT approximations with favorable characteristics. To this end, all

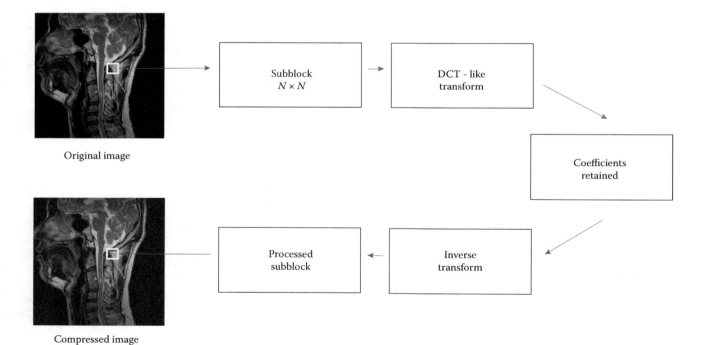

FIGURE 8.2
JPEG-like compression scheme.

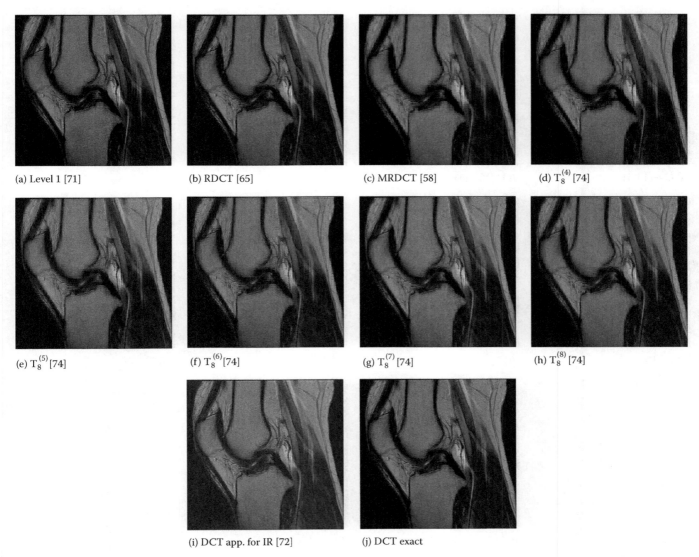

FIGURE 8.3
JPEG-like compressed "knix" image using several transforms, for $r = 15$.

the efficient solutions of the FW class of transforms in Ref. [74] were considered. Figures 8.3 and 8.4 show "Knix" and "Spine" images after being submitted to the JPEG-like compression experiment for $r = 15$ and $r = 25$, respectively. The "Knix" image was accessed from a web archive in DICOM format [112]; the "Spine" image is part of the R language oro.dicom package [113]. While differences such as small block effects are visible for some approximate transforms, e.g., Figure 8.3c, image quality in the context of medical image compression would need to be assessed carefully for each application with input from end-users, in particular radiologists.

8.5.2 Image Registration

In medical image analysis, besides standard image processing techniques, such as noise reduction, filtering, etc., there are two large application areas: image segmentation—utilized to extract anatomical objects from images; and IR. IR solves the task of finding a geometric transformation to align a fixed and a moving image [114]. These two images might represent the same patient at a different time point or different patients. Images might come from the same imaging modality, e.g., X-ray to X-ray registration, or might come from different modalities, e.g., computed tomography to magnetic resonance imaging registration. There are multiple ways to find the geometric transformation: aligning fiducial points, aligning anatomical surfaces, and matching pixel/voxel intensity information. In Ref. [115], a comprehensive overview is available. Intensity-based IR is a popular choice, as minimal pre-processing is needed with high potential for automating.

In this intensity-based IR approach, parameters of a geometric transformation are determined using an

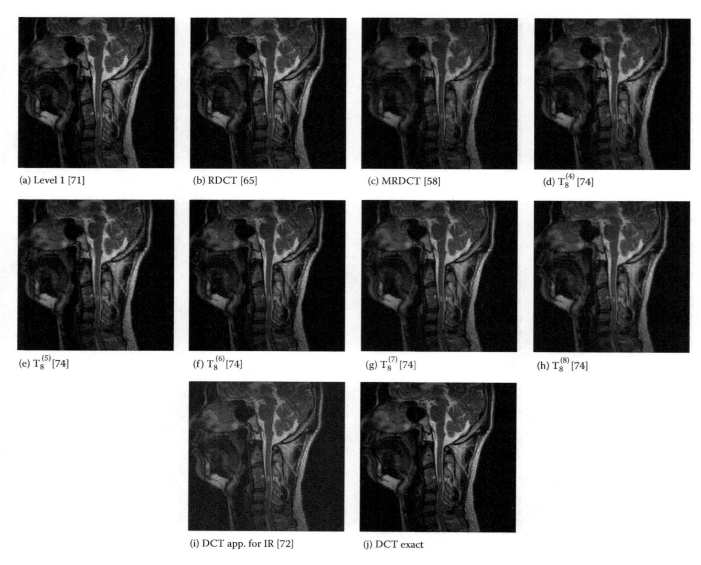

FIGURE 8.4
JPEG-like compressed "Spine1" image using several transforms, for $r = 25$.

iterative optimization algorithm to minimize (or maximize) a cost function, the so-called image similarity metric. The choice of the image similarity metric depends on the kind of images to be registered, noise expected, and computation time available. For example, when registering images from different imaging modalities, voxel intensities of similar structures might possess complex relationships in the two images. Mutual information is a similarity metric able to handle such cases and is often used in other situations, due to the possibility of random sub-sampling resulting in a reduction in computation time. Because IR is an iterative procedure, the similarity metric must be repeatedly evaluated for each iteration, rendering the operation computationally expensive.

Recently, a novel similarity metric termed residual complexity has been proposed [97]. In this metric, the difference between the fixed and transformed moving image is decomposed with the DCT, and the DCT coefficients are combined into the metric to penalize presence of high-frequency components arising from misalignment of object edges. This metric is especially suitable for registering images that differ by an additive low-frequency intensity distortion in addition to a geometric transform and Gaussian noise. These types of images are found in magnetic resonance imaging [116], X-ray images [117], confocal microscopy [118], retina images, and 3-D echocardiography [97].

In Ref. [72], a residual complexity similarity metric based on low-complexity transforms was introduced. The DCT needed in the computation of the similarity measure is replaced with multiplier-free low-complexity approximate transforms. The computational complexity of the original method was reduced by a factor of 8–9 being the similarity computation performed with additions only. In this sense, low-power hardware-based IR of medical images [119–121] and real-time registration of confocal microscopy images [122] can be efficiently produced.

Let $\mathbf{I}$ and $\mathbf{J}$ be two $N_T \times N_T$ images to be aligned. A popular similarity metric is the sum of squared intensity differences [123,124] and intensity-based IR would tune a geometric transformation $\mathcal{T}$ to operate on $\mathbf{J}$ in such a way that the following difference

$$\mathbf{r} = \mathbf{I} - \mathcal{T}\{\mathbf{J}\}$$

is minimized in least-squared sense. Instead of analyzing the difference image as part of the similarity metric, residual complexity aims at analyzing the image difference in the 2-D DCT domain. The 2-D DCT of $\mathbf{r}$ is given by

$$\mathbf{R} = \mathbf{C}_{N_T} \cdot \mathbf{r} \cdot \mathbf{C}_{N_T}^T.$$

In order to apply sub-block processing, Ref. [72] introduces a new approach for residual complexity computation. Instead of subjecting the $N_T \times N_T$ residual image $\mathbf{r}$ to the N-point 2-D transform, they proposed to split such image into square sub-blocks of size $N \times N$. Following Ref. [72], the block-wise residual complexity is given by

$$E_T = \sum_{k=1}^{(N_T/N)^2} E_k,$$

where E_k is the residual complexity of the kth block, given by [97]

$$E_k = \sum_{m=1}^{N} \sum_{n=1}^{N} \log \left(\frac{R_{m,n}^2}{\alpha} + 1 \right),$$

where $R_{m,n}$ is the (m, n) th entry of each $\mathbf{R}_k = \mathbf{C}_N \cdot \mathbf{r}_k \cdot \mathbf{C}_N^T$ and α is a trade-off parameter. In Ref. [97], it was established that useful values of α are in the range of $[0.01, 3]$.

This clock-wise procedure has the advantage of possessing a significantly lower computational complexity. In addition, aiming at a further minimization of the computational cost of the residual complexity calculation, low-complexity transforms can be employed instead of the exact DCT. A suitable low-complexity transform for IR was introduced in Ref. [72].

In full body examinations, multiple X-ray images have to be combined due to finite detector size. This so-called stitching of two (or more) X-ray images was selected as a practical application in Ref. [72]. In this problem, two 2-D X-ray images depicting sections of a larger object with an overlapping region are to be aligned [117]. The two images potentially possess different illumination in addition to noise and would, therefore, benefit from the residual complexity similarity metric. While intensity-based IR has been applied to this problem [117], utilization of residual complexity similarity metric has not been reported. X-ray images of the head and torso were re-sampled to the same pixel resolution. Subsequently, images were automatically aligned using the method summarized above to form a combined (stitched) image as shown in Figure 8.5. The original images are available online [112].

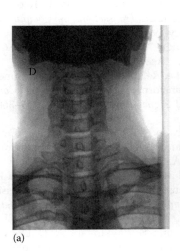
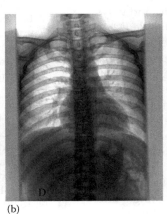
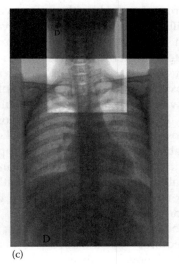

(a)   (b)   (c)

FIGURE 8.5
X-ray stitching example: (a) original image of the head, (b) original image of the torso, and (c) stitched images using residual complexity metric with DCT approximation for IR [72] and $\alpha = 0.01$.

8.6 Open Problems and Future Directions

The intense generation of biomedical imagery and medical data in general calls for correspondingly efficient computational methods to facilitate the storage, as well as processing and analysis of such large volumes of data. A possible approach is to revisit the fundamental signal processing building blocks aiming at alternative computation methods. We suggest approximate methods as a viable tool for significantly reducing the complexity of transform and spectral estimation evaluation. By employing low-complexity approximations, existing algorithms can be immediately re-used after substituting exact computing blocks with their approximate counterparts.

In this chapter, we advanced the use of approximate discrete transforms as a venue for addressing the huge computational demands related to big data processing. In particular, we focused on the DCT because of its very well-documented energy compaction properties and ubiquitous applications in image processing, including biomedical data. Several approximations were discussed and compared. We hope that this review provides a useful resource for selecting approximate transforms for specific applications with specific demands. Moreover, we demonstrated in actual biomedical image processing contexts the effectiveness of approximate computation. In particular, compression of medical data to reduce storage requirements and allow for efficient processing will remain an important topic. While development of algorithms, such as the methods presented, is important, tailoring methods to different applications is pivotal. For example, compression might be tailored to storage and re-use by an automated processing system or alternatively to medical professionals. In either case, working across disciplines will be necessary to develop relevant solutions.

To unleash the full potential of approximate transforms presented, further research is needed to find ways to implement and make these methods available. Most often, approximate transforms are implemented in custom hardware using field programmable gate arrays (FPGAs) [53]. Depending on the application, this might not be the most practical approach to make these capabilities available. In medical image compression and registration, it might be more useful to implement on readily available consumer hardware, either personal computers or mobile devices. Progress in this direction has been made recently [68].

Applying biomedical signal processing to big data problems requires optimization of existing and development of novel analysis methods. The use of approximate transforms offers an avenue for optimization and novel development in tasks aimed at reducing data complexity in a computationally efficient manner.

Acknowledgments

This work was partially supported by CNPq, CAPES, FACEPE, and FAPERGS.

References

1. N. Karimi, S. Samavi, S. M. Reza Soroushmehr, S. Shirani, and K. Najarian. Toward practical guideline for design of image compression algorithms for biomedical applications. *Expert Systems with Applications*, 56:360–367, 2016.
2. A. Nait-Ali and C. Cavaro-Ménard. *Compression of Biomedical Images and Signals*. John Wiley & Sons, 2008.
3. P. C. Zikopoulos, C. Eaton, D. deRoos, T. Deutsch, and G. Lapis. *Understanding Big Data: Analytics for Enterprise Class Hadoop and Streaming Data*. McGraw-Hill, 2012.
4. A. Belle, R. Thiagarajan, S. M. R. Soroushmehr, F. Navidi, D. A. Beard, and K. Najarian. Review article big data analytics in healthcare. *BioMed Research International*, 2015, 2015.
5. W. Raghupathi and V. Raghupathi. Big data analytics in healthcare: Promise and potential. *Health Information Science and Systems*, 2(3), 2014.
6. R. C. Gonzalez and R. E. Woods. *Digital Image Processing*. Prentice Hall, 3rd edition, 2008. p. 854.
7. M. Unser. On the approximation of the discrete Karhunen–Loève transform for stationary processes. *Signal Processing*, 7(3):231–249, 1984.
8. R. Wang. *Introduction to Orthogonal Transforms: With Applications in Data Processing and Analysis*. Cambridge University Press, ISBN 9780521516884, ICCN 2012405448, 2012. Cambridge Books Q7 Online. URL https://books.google.ca/books?id=4KEKGjaiJn0C.
9. N. Ahmed, T. Natarajan, and K. R. Rao. Discrete cosine transform. *IEEE Transactions on Computers*, C-23(1):90–93, January 1974.
10. V. Britanak, P. Yip, and K. R. Rao. *Discrete Cosine and Sine Transforms*. Academic Press, 2007.
11. R. J. Clarke. Relation between the Karhunen-Loève and cosine transforms. *IEEE Proceedings F Communications, Radar and Signal Processing*, 128(6):359–360, November 1981.
12. K. R. Rao and P. Yip. *Discrete Cosine Transform: Algorithms, Advantages, Applications*. Academic Press, San Diego, CA, 1990.
13. I. S. Reed and L.-S. Lan. A fast approximate Karhunen–Loève transform (AKLT) for data compression. *Journal of Visual Communication and Image Representation*, 5(4):304–316, 1994.
14. N. Kambhatla, S. Haykin, and R. D. Dony. Image compression using KLT, wavelets and an adaptive mixture of principal components model. *Journal of VLSI Signal Processing Systems for Signal, Image and Video Technology*, 18(3):287–296, 1998.

15. Y. Shi, Z. Tao, and P. Gong. An improved generalized K-L transform and applications in image compression. *Journal of Information and Computational Science*, 10 (1):167–173, 2013.
16. D. Zhang and S. Chen. Fast image compression using matrix KL transform. *Neurocomputing*, 68:258–266, 2005.
17. Y. Ding, Y.-C. Chung, S. V. Raman, and O. P. Simonetti. Application of the Karhunen–Loève transform temporal image filter to reduce noise in real-time cardiac cine MRI. *Physics in Medicine and Biology*, 54(12):3909, 2009.
18. G. Mihai, Y. Ding, H. Xue, Y.-C. Chung, S. Rajagopalan, J. Guehring, and O. P. Simonetti. Non-rigid registration and KLT filter to improve SNR and CNR in GRE-EPI myocardial perfusion imaging. *Journal of Biomedical Science and Engineering*, 5(12A):871–877, 2012.
19. Y. Nagata, K. Mitsubori, T. Kagi, T. Fujioka, and M. Abe. Fast implementation of KLT-based speech enhancement using vector quantization. *IEEE Transactions on Audio, Speech, and Language Processing*, 14(6):2086–2097, 2006.
20. A. Rezayee and S. Gazor. An adaptive KLT approach for speech enhancement. *IEEE Transactions on Speech and Audio Processing*, 9(2):87–95, 2001.
21. P. Barat and A. Roy. Modification of Karhunen–Loève transform for pattern recognition. *Sadhana*, 23(4):341–350, 1998.
22. Y. Yamashita, Y. Ikeno, and H. Ogawa. Relative Karhunen–Loève transform method for pattern recognition. In *Proceedings. Fourteenth International Conference on Pattern Recognition*, volume 2, pages 1031–1033, Aug 1998.
23. M. Gastpar, P. L. Dragotti, and M. Vetterli. The distributed Karhunen–Loève transform. *IEEE Transactions on Information Theory*, 52(12):5177–5196, 2006.
24. G. Z. Karabulut, D. Panario, and A. Yongacoglu. Integer to integer Karhunen–Loéve transform over finite fields. In *Acoustics, Speech, and Signal Processing, 2004. Proceedings. (ICASSP '04). IEEE International Conference on*, volume 5, pages V–213–16, vol. 5, May 2004.
25. S. Olmos, M. Millan, J. Garcia, and P. Laguna. ECG data compression with the Karhunen–Loève transform. In *Computers in Cardiology*, pages 253–256, Sept. 1996.
26. D. C. Reddy. *Biomedical Signal Processing: Principles and Techniques*. Tata McGraw-Hill Education, 2005. p. 411.
27. F. J. Theis and A. Meyer-Base. *Biomedical Signal Analysis: Contemporary Methods and Applications*. MIT Press, 2010. p. 415.
28. J. Huang and Y. Zhao. A DCT-based fast signal subspace technique for robust speech recognition. *IEEE Transactions on Speech and Audio Processing*, 8(6):747–751, 2000.
29. J. Liang and T. D. Tran. Fast multiplierless approximation of the DCT with the lifting scheme. *IEEE Transactions on Signal Processing*, 49:3032–3044, 2001.
30. V. Bhaskaran and K. Konstantinides. *Image and Video Compression Standards*. Kluwer Academic Publishers, Boston, 1997.
31. G. K. Wallace. The JPEG still picture compression standard. *IEEE Transactions on Consumer Electronics*, 38(1):xviii–xxxiv, 1992.
32. N. Roma and L. Sousa. Efficient hybrid DCT-domain algorithm for video spatial downscaling. *EURASIP Journal on Advances in Signal Processing*, 2007(2):30–30, 2007.
33. International Organisation for Standardisation. Generic coding of moving pictures and associated audio information—Part 2: Video. ISO/IEC JTC1/SC29/WG11—coding of moving pictures and audio, ISO, 1994.
34. International Telecommunication Union. ITU-T recommendation H.261 version 1: Video codec for audiovisual services at $p \times 64$ kbits. Technical report, ITU-T, 1990.
35. International Telecommunication Union. ITU-T recommendation H.263 version 1: Video coding for low bit rate communication. Technical report, ITU-T, 1995.
36. Joint Video Team. Recommendation H.264 and ISO/IEC 14 496-10 AVC: Draft ITU-T recommendation and final draft international standard of joint video specification. Technical report, ITU-T, 2003.
37. M. T. Pourazad, C. Doutre, M. Azimi, and P. Nasiopoulos. HEVC: The new gold standard for video compression: How does HEVC compare with H.264/AVC? *IEEE Consumer Electronics Magazine*, 1(3):36–46, July 2012.
38. G. J. Sullivan, J. Ohm, Woo-Jin Han, and T. Wiegand. Overview of the high efficiency video coding (HEVC) standard. *IEEE Transactions on Circuits and Systems for Video Technology*, 22(12):1649–1668, 2012.
39. A. Chaabouni, Y. Gaudeau, J. Lambert, J.-M. Moureaux, and P. Gallet. H.264 medical video compression for telemedicine: A performance analysis. *IRBM*, 37(1):40–48, 2016. Medical Image Analysis for Computer Aided Diagnosis.
40. S. Gujjunoori and B. B. Amberker. DCT based reversible data embedding for MPEG-4 video using HVS characteristics. *Journal of Information Security and Applications*, 18 (4):157–166, 2013.
41. K. Najarian and R. Splinter. *Biomedical Signal and Image Processing*. CRC Press, Boca Raton, FL, 2005. p. 448.
42. T. Sheltami, M. Musaddiq, and E. Shakshuki. Data compression techniques in wireless sensor networks. *Future Generation Computer Systems*, in press, 64:151–162, 2016. ISSN 0167-739X. doi: https://doi.org/10.1016/j.future.2016.01.015, URL http://www.sciencedirect.com/science/article/pii/S0167739X16000285.
43. A. Madanayake, R. J. Cintra, V. Dimitrov, F. Bayer, K. A. Wahid, S. Kulasekera, A. Edirisuriya, U. Potluri, S. Madishetty, and N. Rajapaksha. Low-power VLSI architectures for DCT/DWT: Precision vs approximation for HD video, biomedical, and smart antenna applications. *IEEE Circuits and Systems Magazine*, 15(1):25–47, First quarter 2015.
44. H. S. Hou. A fast recursive algorithm for computing the discrete cosine transform. *IEEE Transactions on Acoustic, Signal, and Speech Processing*, 6(10):1455–1461, 1987.
45. M. Vetterli and H. Nussbaumer. Simple FFT and DCT algorithms with reduced number of operations. *Signal Processing*, 6:267–278, August 1984.
46. W. H. Chen, C. Smith, and S. Fralick. A fast computational algorithm for the discrete cosine transform. *IEEE Transactions on Communications*, 25(9):1004–1009, 1977.
47. Z. Wang. Fast algorithms for the discrete W transform and for the discrete Fourier transform. *IEEE Transactions*

on *Acoustics, Speech and Signal Processing*, ASSP-32:803–816, August 1984.
48. B. G. Lee. A new algorithm for computing the discrete cosine transform. *IEEE Transactions on Acoustics, Speech and Signal Processing*, ASSP-32:1243–1245, December 1984.
49. Y. Arai, T. Agui, and M. Nakajima. A fast DCT-SQ scheme for images. *Transactions of the IEICE*, E-71(11):1095–1097, November 1988.
50. C. Loeffler, A. Ligtenberg, and G. Moschytz. Practical fast 1D DCT algorithms with 11 multiplications. In *Proceedings of the International Conference on Acoustics, Speech, and Signal Processing*, pages 988–991, 1989.
51. E. Feig and S. Winograd. Fast algorithms for the discrete cosine transform. *IEEE Transactions on Signal Processing*, 40(9):2174–2193, 1992.
52. B. Vasudev and N. Merhav. DCT mode conversions for field/frame coded MPEG video. In *IEEE Second Workshop on Multimedia Signal Processing*, pages 605–610, December 1998.
53. A. Edirisuriya, A. Madanayake, V. Dimitrov, R. J. Cintra, and J. Adikari. VLSI architecture for 8-point AI-based Arai DCT having low area-time complexity and power at improved accuracy. *Journal of Low Power Electronics and Applications*, 2(2):127–142, 2012.
54. H. L. P. A. Madanayake, R. J. Cintra, D. Onen, V. S. Dimitrov, and L. T. Bruton. Algebraic integer based 8 × 8 2-D DCT architecture for digital video processing. In *IEEE International Symposium on Circuits and Systems (ISCAS)*, pages 1247–1250, 2011.
55. N. Rajapaksha, A. Edirisuriya, A. Madanayake, R. J. Cintra, D. Onen, I. Amer, and V. S. Dimitrov. Asynchronous realization of algebraic integer-based 2D DCT using Achronix Speedster SPD60 FPGA. *Journal of Electrical and Computer Engineering*, 2013:1–9, 2013.
56. M. T. Heideman and C. S. Burrus. *Multiplicative Complexity, Convolution, and the DFT*. Signal Processing and Digital Filtering. Springer-Verlag, 1988.
57. A. Madanayake, A. Edirisuriya, R. J. Cintra, V. S. Dimitrov, and N. T. Rajapaksha. A single-channel architecture for algebraic integer based 8 × 8 2-D DCT computation. *IEEE Transactions on Circuits and Systems for Video Technology*, PP(99):1–1, 2013.
58. F. M. Bayer and R. J. Cintra. DCT-like transform for image compression requires 14 additions only. *Electronics Letters*, 48(15):919–921, 19 2012.
59. S. Bouguezel, M. O. Ahmad, and M. N. S. Swamy. Low-complexity 8 × 8 transform for image compression. *Electronics Letters*, 44(21):1249–1250, September 2008.
60. S. Bouguezel, M. O. Ahmad, and M. N. S. Swamy. A multiplication-free transform for image compression. In *2nd International Conference on Signals, Circuits and Systems*, pages 1–4, November 2008.
61. S. Bouguezel, M. O. Ahmad, and M. N. S. Swamy. A fast 8 × 8 transform for image compression. In *International Conference on Microelectronics (ICM)*, pages 74–77, December 2009.
62. S. Bouguezel, M. O. Ahmad, and M. N. S. Swamy. A novel transform for image compression. In *53rd IEEE International Midwest Symposium on Circuits and Systems (MWSCAS)*, pages 509–512, August 2010.
63. S. Bouguezel, M. O. Ahmad, and M. N. S. Swamy. A low-complexity parametric transform for image compression. In *IEEE International Symposium on Circuits and Systems (ISCAS)*, 2011.
64. S. Bouguezel, M. O. Ahmad, and M. N. S. Swamy. Binary discrete cosine and Hartley transforms. *IEEE Transactions on Circuits and Systems I: Regular Papers*, 60(4):989–1002, 2013.
65. R. J. Cintra and F. M. Bayer. A DCT approximation for image compression. *IEEE Signal Processing Letters*, 18(10):579–582, October 2011.
66. R. J. Cintra, F. M. Bayer, and C. J. Tablada. Low-complexity 8-point DCT approximations based on integer functions. *Signal Processing*, 99:201–214, 2014.
67. T. L. T. da Silveira, R. S. Oliveira, F. M. Bayer, R. J. Cintra, and A. Madanayake. Multiplierless 16-point DCT approximation for low-complexity image and video coding. *Signal, Image and Video Processing*, 11(2):227–233, 2017.
68. R. T. Haweel, W. S. El-Kilani, and H. H. Ramadan. Fast approximate DCT with GPU implementation for image compression. *Journal of Visual Communication and Image Representation*, in press, 40:357–365, 2016. ISSN 1047-3203. doi: https://doi.org/10.1016/j.jvcir.2016.07.003, URL http://www.sciencedirect.com/science/article/pii/S1047320316301298.
69. T. I. Haweel. A new square wave transform based on the DCT. *Signal Processing*, 82:2309–2319, 2001.
70. M. Jridi, A. Alfalou, and P. K. Meher. A generalized algorithm and reconfigurable architecture for efficient and scalable orthogonal approximation of DCT. *IEEE Transactions on Circuits and Systems—Regular paper*, 62(2):449–457, 2015.
71. K. Lengwehasatit and A. Ortega. Scalable variable complexity approximate forward DCT. *IEEE Transactions on Circuits and Systems for Video Technology*, 14(11):1236–1248, November 2004.
72. Y. Pauchard, R. J. Cintra, A. Madanayake, and F. M. Bayer. Fast computation of residual complexity image similarity metric using low-complexity transforms. *IET Image Processing*, 9(8):699–708, 2015.
73. U. S. Potluri, A. Madanayake, R. J. Cintra, F. M. Bayer, S. Kulasekera, and A. Edirisuriya. Improved 8-point approximate DCT for image and video compression requiring only 14 additions. *IEEE Transactions on Circuits and Systems I: Regular Papers*, 61(6):1727–1740, June 2014.
74. C. J. Tablada, F. M. Bayer, and R. J. Cintra. A class of DCT approximations based on the Feig Winograd algorithm. *Signal Processing*, 113:38–51, 2015.
75. F. M. Bayer, R. J. Cintra, A. Edirisuriya, and A. Madanayake. A digital hardware fast algorithm and FPGA-based prototype for a novel 16-point approximate DCT for image compression applications. *Measurement Science and Technology*, 23(8):114010, 2012.
76. F. M. Bayer, R. J. Cintra, A. Madanayake, and U. S. Potluri. Multiplierless approximate 4-point DCT VLSI architectures for transform block coding. *Electronics Letters*, 49(24):1532–1534, 21 2013.

77. T. L. T. da Silveira, F. M. Bayer, R. J. Cintra, S. Kulasekera, A. Madanayake, and A. J. Kozakevicius. An orthogonal 16-point approximate DCT for image and video compression. *Multidimensional System and Signal Processing*, 27(1):87–104, 2016.
78. R. A. Johson and D. W. Wichern. *Applied Multivariate Statistical Analysis*. Pearson, 6th edition, 2007.
79. R. Kouassi, P. Gouton, and M. Paindavoine. Approximation of the Karhunen Loève transformation and its application to colour images. *Signal Processing: Image Communication*, 16(6):541–551, 2001.
80. N. Suehiro and M. Hateri. Fast algorithms for the DFT and other sinusoidal transforms. *IEEE Transactions on Acoustic, Signal, and Speech Processing*, 34(6):642–644, 1986.
81. W. Yuan, P. Hao, and C. Xu. Matrix factorization for fast DCT algorithms. In *IEEE International Conference on Acoustic, Speech, Signal Processing (ICASSP)*, volume 3, pages 948–951, 2006.
82. R. J. Cintra. An integer approximation method for discrete sinusoidal transforms. *Journal of Circuits, Systems, and Signal Processing*, 30(6):1481–1501, 2011.
83. G. A. F. Seber. *A Matrix Handbook for Statisticians*. John Wiley & Sons, Inc, 2008.
84. N. J. Higham. Computing real square roots of a real matrix. *Linear Algebra and its Applications*, 88/89:405–430, 1987.
85. MATLAB. *Version 8.1 (R2013a) Documentation*. The MathWorks Inc., Natick, Massachusetts, 2013.
86. I. S. Reed and X. Chen. *Error-Control Coding for Data Networks*. Kluwer Academic Publishers, Boston, MA, 1999.
87. M. R. Schroeder. *Number Theory in Science and Communication*. Springer, Berlin, 2009.
88. K. J. Horadam. *Hadamard Matrices and Their Applications*. Princeton University Press, 2007.
89. R. L. Graham, D. E. Knuth, and O. Patashnik. *Concrete Mathematics*. Addison-Wesley, Upper Saddle River, NJ, 2nd edition, 2008.
90. MathWorks. *MATLAB*. The MathWorks, Inc., Natick, MA, 2011.
91. G. Plonka. A global method for invertible integer DCT and integer wavelet algorithms. *Applied and Computational Harmonic Analysis*, 16:90–110, 2004.
92. K. Oldham, J. Myland, and J. Spanier. *An Atlas of Functions*. Springer, 2nd edition, 2008.
93. International Organization for Standardization. ISO/IEC/IEEE 60559:2011, 2011.
94. Wolfram Research. Round—nearest integer function. http://functions.wolfram.com/IntegerFunctions/Round/27/01/01/01/, September 2013.
95. R. K. Senapati, U. C. Pati, and K. K. Mahapatra. A low complexity orthogonal 8 × 8 transform matrix for fast image compression. In *Proceeding of the Annual IEEE India Conference (INDICON)*, pages 1–4, Kolkata, India, 2010.
96. U. S. Potluri, A. Madanayake, R. J. Cintra, F. M. Bayer, and N. Rajapaksha. Multiplier-free DCT approximations for RF multi-beam digital aperture-array space imaging and directional sensing. *Measurement Science and Technology*, 23(11):114003, 2012.
97. A. Myronenko and X. Song. Intensity-based image registration by minimizing residual complexity. *IEEE Transactions on Medical Imaging*, 29(11):1882–1891, November 2010.
98. M. Ehrgott. *Multicriteria Optimization*. Lecture Notes in Economics and Mathematical Systems. Springer-Verlag GmbH, 2000.
99. K. Miettinen. *Nonlinear Multiobjective Optimization*. International series in operations research and management science. Kluwer Academic Publishers, 1999.
100. C.-K. Fong and W.-K. Cham. LLM integer cosine transform and its fast algorithm. *IEEE Transactions on Circuits and Systems for Video Technology*, 22(6):844–854, 2012.
101. Z. Wang and A. C. Bovik. Mean squared error: Love it or leave it? A new look at signal fidelity measures. *IEEE Signal Processing Magazine*, 26(1):98–117, January 2009.
102. V. K. Goyal. Theoretical foundations of transform coding. *IEEE Signal Processing Magazine*, 18(5):9–21, 2001.
103. J. Katto and Y. Yasuda. Performance evaluation of subband coding and optimization of its filter coefficients. *Journal of Visual Communication and Image Representation*, 2(4):303–313, 1991.
104. D. S. Watkins. *Fundamentals of Matrix Computations*. Pure and Applied Mathematics: A Wiley Series of Texts, Monographs and Tracts. Wiley, 2004.
105. G. Strang. The discrete cosine transform. *SIAM Review*, 41(1):135–147, March 1999.
106. A. K. Jain. *Fundamentals of Digital Image Processing*. Prentice-Hall, Inc., Upper Saddle River, NJ, USA, 1989.
107. J. K. Merikoski. On the trace and the sum of elements of a matrix. *Linear Algebra and its Applications*, 60:177–185, 1984.
108. W. K Cham. Development of integer cosine transforms by the principle of dyadic symmetry. In *IEE Proceedings I Communications, Speech and Vision*, volume 136, pages 276–282, 1989.
109. D. Pinto dos Santos, F. Jungmann, C. Friese, C. Dber, and P. Mildenberger. Irreversible Bilddatenkompression in der Radiologie. *Der Radiologe*, 53(3):257–260, March 2013.
110. J. R. Varela, P. G. Tahoces, M. Souto, and J. J. Vidal. A compression and transmission scheme of computer tomography images for telemedicine based on JPEG2000. *Telemedicine Journal and E-Health: The Official Journal of the American Telemedicine Association*, 10 Suppl 2:S–40–44, 2004.
111. D. A. Koff and H. Shulman. An overview of digital compression of medical images: Can we use lossy image compression in radiology? *Canadian Association of Radiologists Journal = Journal l'Association Canadienne Des Radiologistes*, 57(4):211–217, October 2006.
112. Osirix-Viewer: DICOM sample image sets. http://www.osirix-viewer.com/datasets/. Accessed 2016-07-15.
113. Brandon Whitcher, Volker J. Schmid, and Andrew Thornton. Working with the DICOM and NIfTI data standards in R. *Journal of Statistical Software*, 44(6):1–28, 2011.
114. S. Klein, M. Staring, K. Murphy, M. A. Viergever, and J. P. W. Pluim. Elastix: A toolbox for intensity-based medical image registration. *IEEE Transactions on Medical Imaging*, 29(1):196–205, January 2010.

115. J. B. Maintz and M. A. Viergever. A survey of medical image registration. *Medical Image Analysis*, 2(1):1–36, March 1998.
116. M. Styner, C. Brechbuhler, G. Szckely, and G. Gerig. Parametric estimate of intensity inhomogeneities applied to MRI. *IEEE Transactions on Medical Imaging*, 19(3):153–165, 2000.
117. A. Gooßen, M. Schlüter, T. Pralow, and R.-R. Grigat. A stitching algorithm for automatic registration of digital radiographs. In A. Campilho and M. Kamel, editors, *Image Analysis and Recognition*, number 5112 in Lecture Notes in Computer Science, pages 854–862. Springer, Berlin, January 2008.
118. S. Gopinath, Q. Wen, N. Thakoor, K. Luby-Phelps, and J. X. Gao. A statistical approach for intensity loss compensation of confocal microscopy images. *Journal of Microscopy*, 230:143–159, April 2008.
119. C. R. Castro-Pareja, J. M. Jagadeesh, and R. Shekhar. FAIR: A hardware architecture for real-time 3-D image registration. *IEEE Transactions on Information Technology in Biomedicine*, 7(4):426–434, December 2003.
120. O. Dandekar and R. Shekhar. FPGA-accelerated deformable image registration for improved target-delineation during CT-guided interventions. *IEEE Transactions on Biomedical Circuits and Systems*, 1(2):116–127, June 2007.
121. R. Shekhar, V. Zagrodsky, C. R. Castro-Pareja, V. Walimbe, and J. M. Jagadeesh. High-speed registration of three- and four-dimensional medical images by using voxel similarity. *Radiographics*, 23(6):1673–1681, November 2003.
122. S. E. Budge, A. M. Mayampurath, and J. C. Solinsky. Real-time registration and display of confocal microscope imagery for multiple-band analysis. In *Conference Record of the Thirty-Eighth Asilomar Conference on Signals, Systems and Computers, 2004*, volume 2, pages 1535–1539, 2004.
123. D. L. G. Hill, P. G. Batchelor, M. Holden, and D. J. Hawkes. Medical image registration. *Physics in Medicine and Biology*, 46:R1–R45, 2001.
124. P. Thevenaz, Urs E. Ruttimann, and M. Unser. A pyramid approach to subpixel registration based on intensity. *IEEE Transactions on Image Processing*, 7(1):27–41, 1998.

9
Dynamic Processes on Complex Networks

June Zhang and José M.F. Moura

CONTENTS

9.1 Introduction 177
9.2 Network Process Dynamics: The Scaled SIS Process 178
9.3 Continuous-Time Markov Process 179
9.4 Scaled SIS Process: Equilibrium Distribution 180
9.5 Scaled SIS Process: Parameter Regimes 181
9.6 Application: Vulnerable Communities 182
9.7 Scaled SIS Process: MPCP 183
9.8 Regime: (I) Healing Dominant and (IV) Infection Dominant 183
9.9 Regime II: Endogenous Infection Dominant 183
    9.9.1 Equilibrium Distribution and Induced Subgraphs 184
    9.9.2 Most-Probable Configuration and Induced Subgraphs 185
    9.9.3 Most-Probable Configuration and the Densest Subgraph 186
9.10 Examples 186
9.11 Regime III: Exogenous Infection Dominant 187
9.12 Application in Biomedical Signal Processing 189
9.13 Open Problems and Future Direction 189
9.14 Conclusions 189
References 190

9.1 Introduction

The Internet has led to an explosion of user-generated content especially from social networks. Next-generation sequencing has provided much data in biology at the cellular scale. Advances in brain sensing technologies have led to increasing interest in studying the brain as a connected system. In these complex systems, the interactions between components—individual users, genetic sequences, neurons—give rise to surprising macroscopic behavior.

Networks are used as abstract representations of pairwise interactions between components in complex systems. Complex systems can be studied from two perspectives. One is *synthesis*, in which the underlying network structure is inferred from multiple measurements: for example, inferring the functional connectivity of the brain (called the Connectome [1]) from various measurements (i.e., EEG, MRI) of different spatial regions in the brain; and inferring the transmission network of infections from genetic sequences in epidemiology [2,3]. In these types of applications, advances in sensing and measurements have given rise to large amounts of data, which requires novel analytical methods to derive the underlying interactions [4,5].

In the other direction, *analysis*, the network is given, and questions are asked about the effect of the network structure on the behavior of the system. For example, network structure affects how cascading failures spread in the power grid [6,7] and how networked systems are vulnerable to attacks [8]. Network-based epidemics models consider the impact of the topology of the contact network on the dynamics and extent of infection spreading [9–11]: a densely connected contact network facilitates the spread of infection, while a sparsely connected network may impede the spread of infection. Analysis provides insights on the interdependence between the underlying network structure and the observable network processes.

In this chapter, we present and study the *scaled SIS process*, a stochastic network process that models how infections or failures spread through a population due to neighbor-to-neighbor contagion via a contact network. Unlike classical epidemics models, which assumes homogeneous mixing (i.e., individuals are all in contact with one another), we consider that the contact network is explicitly represented by an undirected graph, which can be highly asymmetric in that some individuals have much larger number of contacts than others (i.e., potential super spreaders). An edge in the contact network denotes substantial pairwise contact through which infection *may* be transmitted. Under the assumption that the dynamics of the spreading process are much faster than the evolution of the network topology, we can assume that the network structure remains static while the state of each node changes according to the dynamics described by the scaled susceptible-infected-susceptible (SIS) process. For example, while contacts between individuals in a population can be highly fluid, contacts between *individual populations*, such as through major flight or shipping routes, are fairly static. The scaled SIS process assumes that each node can be in one of two states (e.g., infected vs. healthy) and that the structure of the network is static. It is similar to the class of network process models in Refs. [12–19], but differs from these in three important aspects:

First, for a given range of model parameters, the scaled SIS process mimics the behavior that individuals prefer the healthy state but interactions spread the infection. But for another choice of parameters, we can also model the less intuitive but also equally important phenomenon where interactions can also impede the spread of infection (i.e., agents seek measures to protect their health when they observe that many of their neighbors become infected). Second, the scaled SIS process leads to a reversible Markov process, which makes it more amenable to analysis. Third, in contrast with other network processes, we are able to derive the precise detailed equilibrium distribution of the scaled SIS process.

While there is an intuition that the densely connected populations are more susceptible to contagion, the scaled SIS process formalizes the relationship between dynamics and connectivity. The chapter shows how we use the scaled SIS process to identify individuals in the population that are vulnerable to infection by finding the configuration with the maximum equilibrium probability; we call this the most-probable configuration. The solution to the most-probable configuration problem depends not only on the relative strengths of infection rate to healing rate but also on the topological connectivity in the underlying contact network. We show that there is a strong connection between subgraphs and subgraph densities (i.e., average degree) to identifying vulnerable communities.

9.2 Network Process Dynamics: The Scaled SIS Process

In this section, we introduce the dynamics of the scaled SIS process; see Refs. [20,21]. In the scaled SIS process, contacts between N agents (i.e., individuals, components, or populations depending on applications) are described by an unweighted, undirected, simple graph $G(V,E)$, where V is the set of vertices representing individuals, and E is the set of edges representing contacts between individuals. The structure of the network is characterized by its adjacency matrix, $\mathbf{A}$, where $A_{ij} = A_{ji} = 1$ if there is substantial contact between i and j that may lead to contagion if one individual is infected while the other is susceptible, and $A_{ij} = A_{ji} = 0$ otherwise.

The state of the ith vertex in the network is denoted by x_i. When $x_i = 0$, the individual is healthy but susceptible to infection. When $x_i = 1$, the individual is infected and capable of spreading the infection to others. The state of the network or of the population is the N-tuple collecting the states of all the vertices,

$$\mathbf{x} = [x_1, x_2, \ldots, x_N]^T$$

We refer to $\mathbf{x}$ the *configuration*. The set of all possible configurations is $\mathcal{X} = \{\mathbf{x}\} = \{0,1\}^N$. Since each node can be in one of two states, the cardinality of $\mathcal{X}$ is

$$|\mathcal{X}| = 2^N$$

The scaled SIS process, $\{X(t), t \geq 0\}$, models the evolution of the network state by a stochastic process on the network $G(V,E)$. The dynamics of the scaled SIS process is based on the susceptible-infected-susceptible (SIS) epidemics framework, which assumes that infected individuals can heal and healthy individuals will become infected in a random amount of time. As a result, the process will transition from one configuration to another configuration as individuals heal or become infected. Since infected individuals can pass the virus to their healthy neighbors, how long an individual remains healthy depends on the state of their neighbors; for example, it is more likely that a healthy individual with many infected neighbors will be infected by proxy than a healthy individual with few infected neighbors. The scaled SIS process assumes the following dynamics rules:

Dynamic Processes on Complex Networks

- Only one individual can change its state at any time.
- Assuming that it was infected, the time it takes for the ith individual to heal is exponentially distributed with rate

$$\mu > 0.$$

The parameter μ is the *healing rate*. From the properties of the exponential distribution, the expected time it takes for an infected individual to heal is

$$\overline{T}_h = \frac{1}{\mu}.$$

The scaled SIS process assumes that all the infected individuals in the network have the same expected healing time $\overline{T}_h$. For example, the expected healing time for V_2 in Figure 9.1a and for V_3 in Figure 9.1b are the same.

- Assuming that it was healthy, the time it takes for the ith individual to become infected is exponentially distributed with rate

$$\lambda \gamma^{m_i} > 0,$$

where m_i is the total number of infected neighbors of the ith individual.

  - The rate $\gamma > 0$ is the *endogenous infection rate* or the contagion rate (technically, γ is unitless so it is a factor rather than a rate). It characterizes contagion effect. When $\gamma = 1$, the infection rate does not depend on the number of infected neighbors; the structure does not affect the dynamics. When $\gamma > 1$, the infection rate increases as the number of infected neighbors increases, which models a typical epidemics scenario.

  - The rate $\lambda > 0$ is the *exogenous infection rate* or the spontaneous infection rate. It is the infection rate at which a healthy individual becomes infected from sources outside the network (i.e., *not* by its infected neighbors).

The expected time it takes for the ith individual to become infected is

$$\overline{T}_i = \frac{1}{\lambda \gamma^{m_i}}.$$

Healthy individuals with the same number of infected neighbors, m_i, have the same expected infection time $\overline{T}_i$. The expected infection time for individual V_3 in Figure 9.2a is not the same as the expected infection time for individual V_5 in Figure 9.2b since V_3 has three infected neighbors, while V_5 has no infected neighbors. Since the total number of infected neighbors, m_i, of a node is upper-bounded by its total number of neighbors, d_i, the infection rate depends on the underlying network topology.

9.3 Continuous-Time Markov Process

The scaled SIS process, $\{X(t), t \geq 0\}$, is a stationary, homogenous, irreducible, finite-size continuous-time Markov process [22]. Each state of the Markov process corresponds to a configuration, $\mathbf{x}$. The state space of the scaled SIS process, X, is the space of all possible configurations. The possible transitions of the scaled SIS process are as follows.

1. Consider the configuration

$$\mathbf{x} = [x_1, x_2, \ldots, x_j = 1, x_k, \ldots x_N]^T.$$

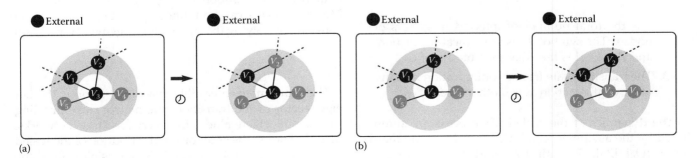

FIGURE 9.1
Examples of healing transitions (dark gray = infected, medium gray = healthy). (a) individual V_2 Heals (b) individual V_3 Heals.

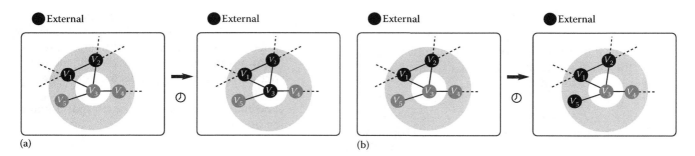

FIGURE 9.2
Examples of infection transitions (dark gray = infected, medium gray = healthy). (a) individual V_3 becomes infected (b) indvidual V_5 becomes infected.

Let x' denote the configuration identical to x, except that the jth individual heals:

$$x' = [x_1, x_2, \ldots, x_j = 0, \ldots, x_N]^T$$

The transition rate $q(x, x')$ of the scaled SIS process from x to x' is

$$q(\mathbf{x}, \mathbf{x}') = \mu \quad (9.1)$$

2. Consider the configuration

$$\mathbf{x} = [x_1, x_2, \ldots, x_j, x_k = 0, \ldots x_N]^T$$

Let x' denote the configuration identical to x, except that the kth individual becomes infected:

$$x' = [x_1, x_2, \ldots, x_k = 1, \ldots, x_N]^T$$

The transition rate $q(x, x')$ of the scaled SIS process from x to x' is

$$q(\mathbf{x}, \mathbf{x}') = \lambda \gamma^{m_k}, \quad (9.2)$$

where

$$m_k = \sum_{j=1}^{N} \mathbb{1}(x_j = 1) A_{jk}$$

is the total number of infected neighbors of node k. The symbol $\mathbb{1}(\cdot)$ is the indicator function, and $A = [A_{jk}]$ is the adjacency matrix of G.

3. The transition rate from a configuration x to any other configuration x' is 0 otherwise.

The dynamics of the scaled SIS process are summarized by the asymetric, $2^N \times 2^N$ transition rate matrix, $Q = [Q_{ij}]$ [23]. The ith row and jth column of $\mathbf{Q}$ correspond to the decimal representations of different configurations, $x, x' \in X$, respectively. Depending on the type of transition, the value of Q_{ij}, $i \neq j$ is determined by Equation 9.1, 9.2, or is 0. The diagonal elements $Q_{ii} = -\sum_{j=1}^{N} Q_{ij}$.

The transition rate matrix specifies the master equation [12]

$$\frac{d}{dt} \mathbf{P}(t) = \mathbf{P}(t) \mathbf{Q}, \quad (9.3)$$

where the i,jth entry of $\mathbf{P}(t)$ denotes the probability that the scaled SIS process is in some configuration x' (represented by j) at time $t > 0$ given that it started in some configuration $\mathbf{x}$ (represented by i) at time $t = 0$.

9.4 Scaled SIS Process: Equilibrium Distribution

The probability distribution for which

$$\frac{d}{dt} \mathbf{P}(t) = 0$$

is the equilibrium distribution. Adopting a convention from Markov process literature [22], we denote the equilibrium distribution as $\pi(x)$, which is a probability mass function (PMF) over the configuration space X. The equilibrium distribution for a finite-size, continuous-time Markov process can be found, up to the normalization constant, by solving the eigenvalue–eigenvector problem

$$\pi \mathbf{Q} = 0$$

The unnormalized equilibrium distribution is the left eigenvector of the transition rate matrix corresponding to the null eigenvalue. However, for large networks, directly finding the equilibrium distribution of the scaled SIS process by this method is infeasible since the dimension of the $\mathbf{Q}$ matrix is exponential in the size of the network. For $N = 100$, the transition rate matrix is

$2^{100} \times 2^{100}$. Solving this eigenvalue–eigenvector problem is computationally intractable for all but small size networks. We proved that

Theorem 9.4.1

The scaled SIS process, $\{X(t), t \geq 0\}$, is a reversible Markov process and the equilibrium distribution is

$$\pi(\mathbf{x}) = \frac{1}{Z}\left(\frac{\lambda}{\mu}\right)^{\mathbf{1}^T\mathbf{x}} \gamma^{\frac{\mathbf{x}^T A \mathbf{x}}{2}}, \quad \mathbf{x} \in \mathcal{X} \qquad (9.4)$$

where Z is the partition function defined as

$$Z = \sum_{\mathbf{x}\in\mathcal{X}}\left(\frac{\lambda}{\mu}\right)^{\mathbf{1}^T\mathbf{x}} \gamma^{\frac{\mathbf{x}^T A \mathbf{x}}{2}}. \qquad (9.5)$$

The proof of Theorem 9.4.1 is in Ref. [20]. Briefly, the proof is based on the scaled SIS process being a reversible Markov process [22], for which the equilibrium distribution also satisfies the detailed balance condition in addition to the full balance condition. The equilibrium distribution characterizes the behavior of the process at steady state. It shows that at steady state, the probability of the system being in a configuration x is determined by only two statistics of the configuration:

$$\mathbf{1}^T\mathbf{x} = \sum_{i=1}^N x_i,$$

the total number of infected nodes in the network configuration, and

$$\frac{\mathbf{x}^T A \mathbf{x}}{2} = \frac{1}{2}\sum_{i=1}^N \sum_{j=1}^N A_{ij} x_i x_j,$$

the total number of edges where both end nodes are infected. Note that $0 \leq \mathbf{1}^T\mathbf{x} \leq N$ and $0 \leq \frac{\mathbf{x}^T A \mathbf{x}}{2} \leq |E|$, where N is the total number of individuals in the network and $|E|$ is the total number of edges.

The equilibrium distribution of the scaled SIS process is a Gibbs distribution [24]. An alternate formulation of Equation 9.4 is

$$\pi(\mathbf{x}) = \frac{1}{Z} e^{H(\mathbf{x})} \qquad (9.6)$$

with

$$H(\mathbf{x}) = \mathbf{1}^T\mathbf{x} \log\left(\frac{\lambda}{\mu}\right) + \frac{\mathbf{x}^T A \mathbf{x}}{2} \log(\gamma) \qquad (9.7)$$

$$= \sum_{i=1}^N x_i \log\left(\frac{\lambda}{\mu}\right) + \sum_{i=1}^N \sum_{j=1}^N A_{ij} x_i x_j \log(\gamma). \qquad (9.8)$$

In statistical mechanics, $H(\mathbf{x})$ is called the *Hamiltonian* and is considered the energy, or potential, of the network configuration $\mathbf{x}$. By the Hammersley–Clifford theorem, the distribution $\pi(x)$, being a Gibbs distribution, is also a Markov random field (MRF, also known as undirected Markov network in the probabilistic graphical model literature [25]). Using MRF terminology, we will refer to $x_i \log\left(\frac{\lambda}{\mu}\right)$ as the *unary potential* and $A_{ij} x_i x_j \log(\gamma)$ as the **pairwise** potential.

9.5 Scaled SIS Process: Parameter Regimes

According to Equation 9.4, the equilibrium behavior of the scaled SIS process, $\{X(t), t \geq 0\}$, depends on the underlying network, $\mathbf{A}$, and the dynamics parameters: $\frac{\lambda}{\mu}$ and γ. The parameter $\frac{\lambda}{\mu}$ is the ratio of the exogenous infection rate and the healing rate; it is intuitive that in an epidemic process, there is competition between infection and healing. The parameter γ is the endogenous infection rate. The evolution of the scaled SIS process is driven by two processes: one that is controlled by γ, that we refer to as the *topology-dependent process*, since it depends on the network structure through the number of infected neighbors, and one that is controlled by $\frac{\lambda}{\mu}$, which we refer to the *topology-independent process*.

The parameter space of the scaled SIS process can be divided into four regimes, as shown in Table 9.1.

The parameter $\frac{\lambda}{\mu}$ models the preferences of an individual or node in the network.

TABLE 9.1

Parameter Regimes

I) Healing Dominant	II) Endogenous Infection Dominant
$0 < \frac{\lambda}{\mu} \leq 1$	$0 < \frac{\lambda}{\mu} \leq 1$
$0 < \gamma \leq 1$	$\gamma > 1$
III) Exogenous Infection Dominant	IV) Infection Dominant
$\frac{\lambda}{\mu} > 1$	$\frac{\lambda}{\mu} > 1$
$0 < \gamma \leq 1$	$\gamma > 1$

- When $\frac{\lambda}{\mu} > 1$, on average, an individual is in the infected state longer than it is in the healthy state. When $\frac{\lambda}{\mu}$ falls in this parameter range, individuals prefer the infected state ($x_i = 1$) to the healthy state ($x_i = 0$).
- When $0 < \frac{\lambda}{\mu} < 1$, on average, an individual is in the healthy state longer than it is in the infected state. When $\frac{\lambda}{\mu}$ falls in this parameter range, individuals *prefer* the healthy state ($x_i = 0$) to the infected state ($x_i = 1$).
- When $\frac{\lambda}{\mu} = 1$, individuals have equal preference for the infected and healthy state.

The parameter γ models how an individual is affected by its neighborhood. This parameter couples the underlying network topology to the dynamics of the process. It determines if the network structure will facilitate or impede the spread of infection.

- When $\gamma > 1$, additional infected neighbors of a healthy individual will increase the infection rate, thereby making the individual more vulnerable to infection. As the number of infections increases, the population will be more vulnerable to the epidemics as a single infection may quickly lead to additional infections.
- When $0 < \gamma < 1$, additional infected neighbors of a healthy individual will decrease the infection rate, thereby making the individual *less* vulnerable to infection. This means that additional infections will actually strengthen the population. For example, in a system with active countermeasures, increasing the number of infected neighbors is a signal to boost the susceptible individual's defenses.
- When $\gamma = 1$, the state of an individual is unaffected by the states of its neighbors. In this case, the underlying network topology does not affect the dynamics of the scaled SIS process.

The behavior modeled by the four regimes are as follows:

(I) *Healing dominant:* $0 < \frac{\lambda}{\mu} \leq 1$ and $0 < \gamma \leq 1$. Individuals prefer the healthy state. In addition, individuals adopt defensive mechanisms so that the infection rate decreases with increasing number of infected neighbors; the network also helps to impede the infection. In this regime, the topology-dependent process supports the topology-independent process.

(II) *Endogenous infection dominant:* $0 < \frac{\lambda}{\mu} \leq 1$ and $\gamma > 1$. Individuals prefer the healthy state. However, the infection rate increases with the increasing number of infected neighbors; the network helps to spread the infection. This regime models the behavior of traditional epidemics. In this regime, the topology-independent process opposes the topology-dependent process.

(III) *Exogenous infection dominant:* $\frac{\lambda}{\mu} > 1$ and $0 < \gamma \leq 1$. Individuals prefer the infected state. However, the infection rate decreases with the increasing number of infected neighbors; the network helps to impede the infection. In this regime, the topology-independent process opposes the topology-dependent process.

(IV) *Infection dominant:* $\frac{\lambda}{\mu} > 1$ and $\gamma > 1$. Individuals prefer the infected state. The infection rate also increases with the increasing number of infected neighbors; the network helps to spread the infection. In this regime, the topology-dependent process supports the topology-independent process.

9.6 Application: Vulnerable Communities

Epidemics models characterize the spread of infection in a population. Network-based epidemics models, as they account for contact network structures, can be used to determine which individuals in the population are more vulnerable to infection. Generally, infections spread more easily in well-connected communities than in sparsely connected communities. As a result, many approaches in finding vulnerable communities are related to the graph theoretic problem of clustering—finding subsets of nodes that are more connected than others. These problems include (1) densest subgraph problem, (2) K-densest subgraph problem, and (3) maximum clique problem; some of these problems have been solved while others are provably NP-hard and remain open research questions [26,27].

However, graph-based techniques only consider the network structure and not the dynamical process of infection spreading. The scaled SIS process accounts for both the network structure *and* infection and healing dynamics. We use the equilibrium distribution of the scaled SIS process and the parameter regimes to identify vulnerable communities in the network using the most-probable configuration, x^*, the network configuration with maximum equilibrium probability. An individual is vulnerable to infection if it is infected in the most-probable configuration.

Finding the most-probable configuration means solving the most-probable configuration problem (MPCP), a combinatorial optimization problem. This may be infeasible to find for large networks since the state space of configurations grows exponentially with the size of the network. We showed in Ref. [21] that for certain regime of parameter values, MPCP can be solved in polynomial time. We can then use MPCP to identify vulnerable communities in networks of thousands of individuals. Intuitively, these vulnerable individuals do belong to well-connected communities; however, the size and composition of vulnerable communities (i.e., the solution of MPCP) change depending on the infection and healing rates.

9.7 Scaled SIS Process: MPCP

The MPCP solves for the configuration in X with the maximum probability

$$x^* = \arg \max_{x \in \mathcal{X}} \pi(x) = \arg \max_{x \in \mathcal{X}} \left\{ \left(\frac{\lambda}{\mu}\right)^{1^T x} \gamma^{\frac{x^T A x}{2}} \right\} \quad (9.9)$$

The MPCP 1.9 is a combinatorial optimization problem as agents can only be in one of two states. Its solution space depends on both the network topology, A, and the infection and healing rates, λ, γ, μ. For large networks, it is infeasible to iterate through all 2^N configurations of the scaled SIS process to find the most-probable configuration. The solutions of MPCP depend on the four parameter regimes identified in Section 9.5 as well as the underlying contact network structure. We consider the ith individual to be vulnerable if it is infected in the most-probable configuration (i.e., $x_i^* = 1$).

9.8 Regime: (I) Healing Dominant and (IV) Infection Dominant

The MPCP is trivial for regimes I and IV as the individual preferences of the agents (i.e., topology-independent process), controlled by $\frac{\lambda}{\mu}$, are also supported by the network effect (i.e., topology-dependent process), controlled by γ.

In regime I, *healing dominant*, $0 < \frac{\lambda}{\mu} \leq 1, 0 < \gamma \leq 1$, both the topology-independent process and the topology-dependent process favor the healthy state. The most-probable configuration in equilibrium is $x^* = x^0 = [0,0,\ldots,0]^T$, the configuration where all the agents are healthy. In regime I, the healing rate (μ) dominates both the endogenous and exogenous infection rates (γ and λ) so no individual in the population is considered vulnerable to infection. This holds for any network topology, $G(V,E)$.

In regime IV, *infection dominant*, $\frac{\lambda}{\mu} > 1, \gamma > 1$, both the topology-independent process and the topology-dependent process favor infection. The most-probable configuration is $x^* = x^N = [1,1,\ldots,1]^T$, the configuration where all the agents are infected. In regime IV, the infection rates (γ and λ) dominate over the healing rate (μ) so all individuals are vulnerable to infection. This holds for any network topology, $G(V,E)$.

9.9 Regime II: Endogenous Infection Dominant

Unlike regimes I and IV, in regime II, the topology-independent process, controlled by the rate $\frac{\lambda}{\mu}$, opposes the topology-dependent process, controlled by the contagion rate γ. This introduces an additional complexity in solving MPCP. The solution space of MPCP in regime II exhibits *phase transition*; the solution of Equation 9.9 changes depending on the parameter values. It is the case that in this parameter regime and depending on the network topology, some individuals may be vulnerable while others are not.

With $0 < \frac{\lambda}{\mu} \leq 1$, the exogenous infection rate, λ, is smaller than or equal to the healing rate, μ. If the exogenous infection rate is ignored (i.e., $\gamma = 1$), then the most-probable network configuration is $x^0 = [0,0,\ldots 0]^T$. However, since $\gamma > 1$, additional infected agents *increase* the infection rate. The process utilizes the network structure to spread the infection to healthy agents. Regime II models the behavior of standard epidemics.

In regime II, the topology-dependent process favors maximizing the number of infected edges (i.e., edges where both end nodes are infected), whereas the topology-independent process wants to minimize the number of infected agents. Intuitively, if $\frac{\lambda}{\mu}$ is very small and γ is close to 1, then the behavior of the process will be dictated by the topology-independent process; a good guess for x^* is $x^0 = [0,0,\ldots 0]^T$. When $\gamma \gg 1$ and $\frac{\lambda}{\mu}$ is close to 1, the process behavior will be dictated by the topology-dependent process; a good guess for x^* is $x^N = [1,1,\ldots 1]^T$.

When neither the topology-independent process nor the topology-dependent dominates the dynamic, the most-probable configuration may be a configuration neither x^0 nor x^N; we call such solutions *non-degenerate* configurations.

We showed in Ref. [21] that in regime II, solving the MPCP is equivalent to minimizing a submodular function [28]. Ref. [29] proved that the minimization of a pseudo-Boolean function that is submodular can be solved in polynomial time. Using the Max-Flow/Min-Cut algorithm proposed in Ref. [30], we can find the most-probable configuration for a network consisting of 4941 nodes, meaning that the number of configurations is 2^{4941}, in less than 0.1 sec on a desktop with 3.7 GHz Quad Core Xeon processor and 16 GB of RAM.

In addition to being readily obtainable by numerical methods, non-degenerate most-probable configurations also give insights on how the underlying network structure affects the epidemics process. In the next section, we discuss how non-degenerate most-probable configurations relate to dense subgraphs in the underlying network structure.

9.9.1 Equilibrium Distribution and Induced Subgraphs

First, we will define the graph theoretic terms used in this section.

Definition 9.9.1

(From Ref. [31]) *The graph F is an induced subgraph of G if two vertices in F are connected if and only if they are connected in G and the vertex set and edge set of F are subsets of the vertex set and edge set of G.*

$$V(F) \subseteq V(G), E(F) \subseteq E(G)$$

Definition 9.9.2

The graph F(x) is an induced subgraph of configuration $x = [x_1, x_2, \ldots, x_N]^T$ if the nodes/edges in the subgraph are the infected nodes/edges in x.

$$V(F(\mathbf{x})) = \{v_i \in V(G) | x_i = 1\} \quad (9.10)$$

$$E(F(\mathbf{x})) = \{(i,j) \in E(G) | x_i = 1, x_j = 1\} \quad (9.11)$$

By definition, $|V(F(x))| = \mathbf{1}^T\mathbf{x}$ and $|E(F(\mathbf{x}))| = \dfrac{\mathbf{x}^T A \mathbf{x}}{2}$. Figures 9.3 and 9.4 show two network configurations and their corresponding induced subgraphs. We proved

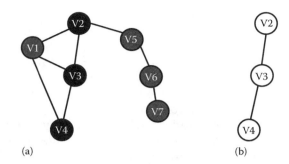

FIGURE 9.3
(a) Configuration $x_1 = [0,1,1,1,0,0,0]^T$. (b) Induced subgraph $F(x_1) = F_1$.

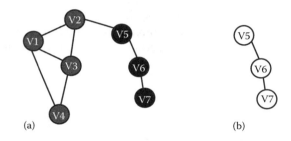

FIGURE 9.4
(a) Configuration $x_2 = [0,0,0,0,1,1,1]^T$. (b) Induce subgraph $F(x_2) = F_2$.

in Ref. [32] that configurations whose induced subgraphs are isomorphic are equally probable. Unless we need to refer explicitly to the underlying network configuration x, for notational simplicity, we will write F to denote an induced subgraph instead of writing F(x).

Definition 9.9.3

The set of all possible induced subgraphs of G is $\mathcal{F} = \{F(x)\}, \forall x \in X$.

The set $\mathcal{F}$ includes the empty graph that is induced by the configuration $x^0 = [0,0,\ldots,0]^T$, and G that is the subgraph induced by the configuration $x^N = [1,1,\ldots,1]^T$.

Definition 9.9.4

(From Ref. [26]) *The density of a graph G is*

$$d(G) = \frac{|E(G)|}{|V(G)|}$$

There is an alternative definition for graph density that is the number of edges divided by the total number of possible edges [33]. These two definitions of density are not equivalent.

Dynamic Processes on Complex Networks

We will refer to the density of the entire network, $d(G) = d(F(x^N))$, as the *network density* and the density of an induced subgraph of G as the *subgraph density*. The density of the empty graph, $d(F(x^0))$, is 0 by definition. The subgraphs in $\mathcal{F}$ can be partially ordered by their density. There may be many subgraphs with the same density. A special induced subgraph in $\mathcal{F}$ is the densest subgraph.

Definition 9.9.5

Let $\overline{F}$ be the densest subgraph in G. Then

$$d(\overline{F}) \geq d(F), \quad \forall F \in \mathcal{F}.$$

Finding $\overline{F}$ is known as the *densest subgraph problem*. It is known that, for undirected, unweighted graphs, $\overline{F}$ can be found in polynomial time [34].

Since there is a one-to-one relationship between the network configuration x and its induced subgraph $F(x)$, we can rewrite the equilibrium distribution (Equation 9.4) of the scaled SIS process in terms of the induced subgraph density and the size of the induced subgraph:

$$\pi(F) = \frac{1}{Z}\left(\left(\frac{\lambda}{\mu}\right)\gamma^{d(F)}\right)^{|V(F)|}, \quad F \in \mathcal{F}, \quad (9.12)$$

where $d(F)$ is the density of the induced subgraph F and Z is the partition function.

Rewriting the equilibrium distribution (Equation 9.4) as a function of induced subgraphs (Equation 9.12) allows us to see that, when the induced subgraphs of two configurations are isomorphic (e.g., configurations x_1 in Figure 9.3 and x_2 in Figure 9.4), then the configurations are equally probable at equilibrium for the scaled SIS process.

9.9.2 Most-Probable Configuration and Induced Subgraphs

Using (Equation 9.12), the MPCP (Equation 9.9) is then also an optimization problem over all the possible induced subgraphs in G:

$$F(x^*) = \arg\max_{F \in \mathcal{F}}\left(\left(\frac{\lambda}{\mu}\right)\gamma^{d(F)}\right)^{|V(F)|}. \quad (9.13)$$

The subgraph induced by the most-probable configuration, $F(x^*)$, is the *most-probable subgraph*, which is *not* necessarily the same subgraph as the densest subgraph, $\overline{F}$.

Assuming the following conditions hold, in Ref. [21], we proved the following theorems.

Assumption 9.1

The scaled SIS process operates in regime II, *endogenous infection dominant*. This limits the dynamics parameters to the range $0 < \frac{\lambda}{\mu} \leq 1$ and $\gamma > 1$.

Assumption 9.2

The underlying network G is a simple, undirected, unweighted, and connected graph.

Theorem 9.9.1

The most-probable configuration $x^* \neq x^0$ if and only if there exists at least one induced subgraph $F \in \mathcal{F}$ with density $d(F)$ for which $\lambda\gamma^{d(F)} > \mu$.

Theorem 9.9.2

The most-probable configuration $x^* \neq x^N$ if and only if there exists at least one induced subgraph $F \in \mathcal{F}\backslash G$ with density $d(F) = E'\frac{}{N'}$ for which

$$\frac{\log\left(\frac{\lambda}{\mu}\gamma^{d(G)}\right)}{\log\left(\frac{\lambda}{\mu}\gamma^{d(F)}\right)} < \frac{N'}{N}. \quad (9.14)$$

Combining Theorems 9.9.1 and 9.9.2, we can obtain the following corollary regarding the non-degenerate most-probable configurations.

Corollary 9.9.1

Let the density of the network be $d(G) = \frac{E}{N}$. Then, the most-probable configuration is a non-degenerate configuration, $x^* \in X\backslash\{x^0, x^N\}$, if and only if there exists at least one induced subgraph $F \in \mathcal{F}$ with density $d(F) = E'\frac{}{N'}$ for which $\lambda\gamma^{d(F)} > \mu$, and

$$\frac{\log\left(\frac{\lambda}{\mu}\gamma^{d(G)}\right)}{\log\left(\frac{\lambda}{\mu}\gamma^{d(F)}\right)} < \frac{N'}{N}.$$

In regime II, individual agents have a preference for being healthy, but the epidemics might spread to other agents through neighbor-to-neighbor contagion. Under the scaled SIS process, the subgraph density $d(F)$ scales the exogenous infection rate γ, thereby affecting the overall infection rate. Theorem 9.9.1 states that, if the network contains *dense-enough* subgraphs, then even when the effective exogenous infection rate, $\frac{\lambda}{\mu}$, is small (i.e., $0 < \frac{\lambda}{\mu} \ll 1$), the exogenous infection rate, γ, can leverage dense subgraphs to spread the infection in the network. The vulnerable individuals are those that belong in these dense subgraphs.

On the other hand, if the endogenous infection rate, γ, is large (i.e., $\gamma \gg 1$), then most certainly the epidemics will spread throughout the entire network and all individuals are vulnerable to infection. Theorem 9.9.2 states when this does not happen. Furthermore, Theorem 9.9.2 shows that it is important to consider if the densest subgraph in the network is the entire network or a smaller subgraph. Corollary 9.9.1 proves that the existence of the non-degenerate configurations is related to the existence of subgraphs with density larger than the network density. The existence of these subgraphs that are *denser-than G* is crucial to the existence of non-degenerate configurations (i.e., different from x^0 and x^N) as solutions to the MPCP; when the most-probable configuration is a non-degenerate configuration, agents belonging to denser subgraphs are more vulnerable to the epidemics.

In network science, dense clusters of agents have often been identified as either the network *core* or *community* [35–37]. Solving for the non-degenerate configuration is an alternative method for determining these network structures. Previous works in core/community detection are algorithmic and do not consider the dynamical process on the network. The scaled SIS process show that, what is considered a *community* changes depending on the parameters of the dynamical process: the most-probable configuration changes depending on the dynamics parameters.

9.9.3 Most-Probable Configuration and the Densest Subgraph

We showed that the most-probable configuration is related to the density of induced subgraphs in the network. The densest subgraph, $\overline{F}$, is a special induced subgraph. In this section, we focus specifically on the relationship between the most-probable configuration and the densest subgraph.

Corollary 9.9.2.

The most-probable configuration $x^* = x^0$ if and only if $\lambda \gamma^{d(\overline{F})} \leq \mu$.

Corollary 9.9.2 follows the result of Theorem 9.9.1. If the densest subgraph in the network is not *dense enough* to overcome individual preferences for being healthy, then the endogenous infection rate γ will not be able to drive the most-probable configuration away from x^0.

9.10 Examples

The most-probable configuration changes depending on the dynamics parameters λ, μ, γ. When the healing rate μ dominates the infection rates λ and γ, $x^* = x^0$; this means that the epidemics is not severe. When the infection rates dominate over the healing rate, $x^* = x^N$; this means that the epidemics is severe. When x^* is a non-degenerate configuration (i.e., $x^* \neq x^0, x^N$), this indicates that sets of agents in the network are more vulnerable than others to the epidemics. We illustrate this by solving for the most-probable configuration under different $\left(\frac{\lambda}{\mu}, \gamma\right)$ parameters for two networks: a social network generated from interview data [38] and a large artificially generated R-MAT network [39].

The network shown in Figure 9.5 is a 193 node, 273 edge social network of drug users in Hartford, CT. The network was determined through interviews. Ref. [40] looked for influential agents in the network by considering it as a graph connectivity problem. However, they did not consider a dynamical model of influence. Assuming that we can model drug habits as an epidemics (i.e., there is a social contagion aspect to the behavior), we applied the scaled SIS process to this network and solved for the most-probable configuration under different parameters to find influential network structures.

We show the resultant most-probable configurations in Figure 9.5a–d as we change $\left(\frac{\lambda}{\mu}, \gamma\right)$. We can see from these results that there is a small community of users who are infected when others are healthy. The size of this community increases or decreases depending on the parameters. If there is a social contagion component to drug usage, then these agents may be more vulnerable to the social contagion component of drug usage and therefore more likely to persist in their habit.

The network shown in Figure 9.6a and b is an artificially generated R-MAT network using Ref. [41]. It contains 3215 nodes and 6605 edges. R-MAT graphs have been shown to demonstrate real-world characteristics such as power law degree distribution and small diameter. We use R-MAT to model a contact network where nodes represent individuals and edges represent significant contacts. Figure 9.6a and b show the most-probable configuration when the scaled SIS process parameters are $\left(\frac{\lambda}{\mu} = 0.1, \gamma = 2.5\right)$ and $\left(\frac{\lambda}{\mu} = 0.1, \gamma = 5\right)$.

We can see that, for the same $\frac{\lambda}{\mu}$, as γ increases, thereby increasing the infectiousness of contagion, the size of the vulnerable communities increases. This is intuitive since, for large γ, the epidemics is severe, and the most-probable configuration is driven toward x^N, the configuration where all the components are infected. Moreover, these

Dynamic Processes on Complex Networks

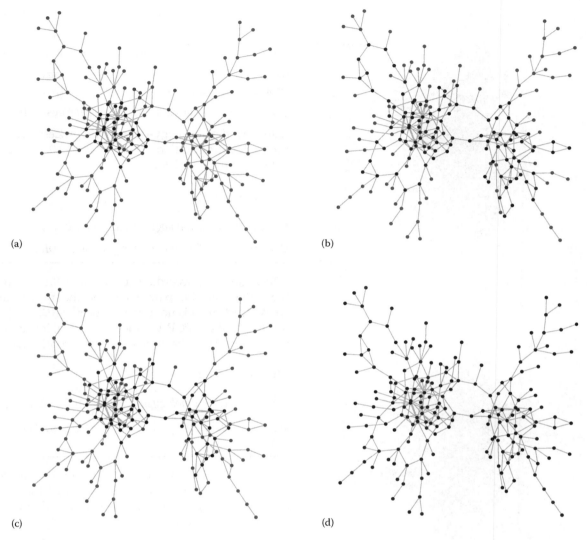

FIGURE 9.5

Susceptible individuals: most-probable configuration x^* under different $\left(\frac{\lambda}{\mu},\gamma\right)$ parameters (dark gray = infected, medium gray = healthy). (a) $\frac{\lambda}{\mu} = 0.208, \gamma = 2.421$; (b) $\frac{\lambda}{\mu} = 0.269, \gamma = 2.94$; (c) $\frac{\lambda}{\mu} = 0.3, \gamma = 2.053$; (d) $\frac{\lambda}{\mu} = 0.3, \gamma = 3.71$.

most-probable configurations are both non-degenerate configurations. From Theorem 9.9.2, we have the additional insight that these components are more vulnerable to failures because they belong to a dense subgraph that satisfies a relationship between subgraph densities and infection and healing rates.

9.11 Regime III: Exogenous Infection Dominant

Unlike regimes I and IV, in regime III, the topology-independent process, controlled by $\frac{\lambda}{\mu}$, opposes the topology-dependent process, controlled by γ. It will also be the case that in this parameter regime and depending on the network topology, some individuals will be vulnerable while others are not.

With $\frac{\lambda}{\mu} > 1$, the exogenous infection rate, λ, is larger than the healing rate, μ. If the exogenous infection rate is ignored (i.e., we take $\gamma = 1$), then the most-probable network configuration is $x^* = x^N = [1,1,\ldots 1]^T$. However, since $0 < \gamma \leq 1$, additional infected nodes *decrease* the infection rate. This implies that healthy agents adopt a defense mechanism in response to having infected neighbors.

In regime III, the topology-dependent process favors minimizing the number of infected edges (i.e., edges where both end nodes are infected), whereas the topology-independent process favors maximizing the number of

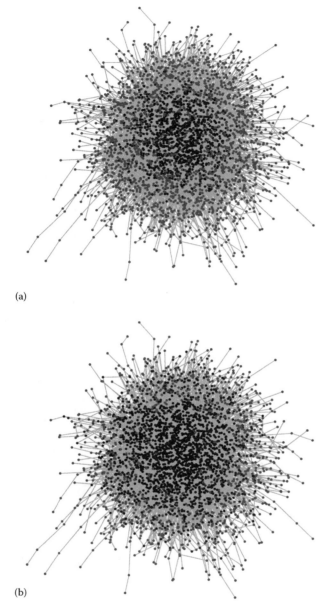

FIGURE 9.6
Susceptible individuals: most-probable configuration x^* under different $\left(\frac{\lambda}{\mu}, \gamma\right)$ parameters (dark gray = infected, medium gray = healthy).
(a) $\frac{\lambda}{\mu} = 0.1, \gamma = 2.5$; (b) $\frac{\lambda}{\mu} = 0.1, \gamma = 5$.

infected agents. In other words, these processes favor isolated infected nodes. The solution space of the MPCP is related to the graph theoretic concept of *independent sets*. An independent set is a subset of nodes such that the induced subgraph is composed entirely of isolated nodes. The *maximum independent set* is the largest possible independent set for a given graph [42]. The maximum independent set is also the largest maximal independent set (i.e., an independent set that is not a subset of any other independent set).

Intuitively, if $\frac{\lambda}{\mu} \gg 1$ and γ is close to 1, then the epidemics will be dominated by the topology-independent process; a good guess for x^* is $x^N = [1,1,\ldots1]^T$. When $0 < \gamma \ll 1$ and $\frac{\lambda}{\mu}$ is close to 1, the epidemics will be dominated by the topology-dependent process. The most-probable configuration cannot be $x^0 = [0,0,\ldots0]^T$ as shown below (proved in [20]).

Theorem 9.11.1

For any network topology, when $\frac{\lambda}{\mu} > 1, 0 < \gamma \le 1$, then $x^* \ne x^0 = [0,0,\ldots0]^T$ for any feasible parameter values.

In regime III, according to Theorem 9.11.1, no matter the values of the parameters or the underlying network structure, it is not possible for $x^0 = [0,0,\ldots,0]^T$ to be a solution of the MPCP. On the other hand, it is possible for $x^N = [1,1,\ldots,1]^T$ to be a most-probable configuration.

Theorem 9.11.2

For any network topology and when $\frac{\lambda}{\mu} > 1, 0 < \gamma \le 1$, sort the degrees so that $k_1 \ge k_2 \ge \ldots \ge k_N$. If $\lambda \gamma^{k_1} > \mu$, then $x^* = x^N = [1,1,\ldots 1]^T$.

Theorem 9.11.2 shows that the only consideration necessary for $x^* = x^N$ are the values of the dynamics parameters λ, γ, μ, and the maximum degree of the underlying network.

Theorem 9.11.3

For any network topology and when $\frac{\lambda}{\mu} > 1, 0 < \gamma \le 1$, if in addition $\lambda \gamma < \mu$, then the most-probable configuration, x^*, is the configuration with the maximum number of infected nodes and 0 infected edges; this is also the maximum independent set [43].

Theorem 9.11.3 shows that, in the parameter regime III, $\frac{\lambda}{\mu} > 1, 0 < \gamma \le 1$, and $\lambda \gamma < \mu$, the most-probable configuration x^* of the scaled SIS process at equilibrium is the maximum independent set. There exist polynomial-time algorithms to find this set for a special class of graphs called *perfect graphs* [44]. Well-known examples of perfect graphs are complete graphs, bipartite graphs, and chordal graphs. Unfortunately, finding the maximum independent set is NP-hard for general graph topologies [42]. As a result, it is in general infeasible to analyze the equilibrium behavior for other large, real-world networks when the dynamics parameters are in regime III.

9.12 Application in Biomedical Signal Processing

Classical compartmental epidemics models (SIS, SIR, etc.) are based on the thermodynamic principles of *full mixing* and *mean field*, which assumed an infinitely large population where each individual can potentially come in contact with every other individual [45]. In these models, the susceptibility of the population to infection depends *only* on the dynamics parameters of infection and healing. In contrast, network processes explicitly model the dynamics of microscopic interactions, accounting, in particular, for asymmetry in interactions as represented by a network structure.

The scaled SIS process is a network process model that focuses on population-scale analysis. It is also possible to consider network process models at the molecular scale. For example, the standard model of HIV dynamics describes the evolution of three variables: T, the total number of uninfected target cells; V, the total number of free virions (i.e., infective form of virus outside a host cell); and T^*, the total number of infected target cells [46] by

$$\frac{dT}{dt} = \lambda - dT - kTV, \text{ and } \frac{dT^*}{dt}$$

$$= kTV - \mu T^*, \text{ and } \frac{dV}{dt} = pT^* - cV, \infty \qquad (9.15)$$

where λ and d are the birth and death rate of uninfected target cells, respectively, and k is the infection rate of the uninfected cell by the virions. Infected cells produce new virions at rate p and are killed with death rate μ. The dynamics of this model are similar to the classical epidemics models, which were also based on thermodynamic interactions and model mean-field dynamics by assuming homogeneous mixing. In reality, there are many sources of heterogeneity that affect viral evolution, so network process models will give different insights than Equation 9.15. Ref. [47] has used network processes, called Boolean networks, to study cell regulation. Network analysis, especially network process models, offers new insights to problems involving complex interactions in system biology.

Many system biology applications can be studied using network abstractions. From an application perspective, there are many open questions regarding how to construct these network representations from observations such as what scale of measurements to consider, what assumptions to retain, etc. A major open question remains in how to integrate all of these scales together. At what scale of analysis are network perspectives necessary? Much like the separation of time scale, for example, at the scale of analyzing human-to-human interaction, the dynamics of gene-to-gene interaction will be too fast to be considered, whereas if the time scale is at the gene-to-gene interaction, then human-to-human interaction can be neglected since the time scale of human interaction will be much too slow.

From an analysis perspective, much of the work has focused on analyzing the topology structure of *static, unweighted* graph. Future direction of network-based analysis is not only to develop analytical tools for larger classes of graphs but also to explicitly model the network process: dynamic process associated with the time-varying network structure. In this chapter, we considered a model for modeling epidemics in which the network structure (i.e., interactions) remains static but the values associated with each node vary according to a random process. Alternatively, a network process may consider a model in which the edges of the network vary according to some dynamical process. The major challenge of analyzing network processes is due to the combinatorial complexity introduced by the network structure. As we showed with the scaled SIS process, combinatorial complexity may be avoidable using certain assumptions and mathematical properties (e.g., reversibility). Additionally, instead of dealing with any heterogeneous network structure, analysis showed that network density is the statistic of interest. Furthermore, under certain conditions, the network structure may not impart any insight to the behavior of the process (e.g., regimes I and IV of the most-probable configuration).

Network-based analysis offers new insight to classical approaches to system analysis. Its value to modeling complex interactive systems such as biological systems is clear. However, there remain many challenges and open questions to both analysis and applications of such an approach, making this an interesting direction for interdisciplinary future research.

9.13 Open Problems and Future Direction

Network-based analyses have been increasingly adapted to analyze complex biological systems such as the connectome (i.e., brain network) [1], protein–protein interaction networks [48], and viral quasi-species networks [49].

9.14 Conclusions

In this chapter, we discussed a simple network process, the scaled SIS process, for modeling the spread of infection in a population where the interactions between individuals are represented as a contact network; we showed that the structure of the contact network, in addition to the infection and healing rates, has an impact on the susceptibility of the population, specifically that, for certain dynamics parameters and contact network structure, a subset of nodes (i.e., subgraphs with high density) are more vulnerable to infection. We used the most-probable configuration to show that subgraph density is the topological property related to infection vulnerability.

References

1. Olaf Sporns, Giulio Tononi, and Rolf Ktter. The human connectome: A structural description of the human brain. *PLOS Computational Biology*, 1(4), 09 2005.
2. Thibaut Jombart, Anne Cori, Xavier Didelot, Simon Cauchemez, Christophe Fraser, and Neil Ferguson. Bayesian reconstruction of disease outbreaks by combining epidemiologic and genomic data. *PLOS Computational Biology*, 10(1):e1003457, 2014.
3. Pavel Skums, Olga Glebova, David S. Campo, Nana Li, Zoya Dimitrova, Seth Sims, Leonid Bunimovich, Alex Zelikovsky, and Yury Khudyakov. Algorithms for prediction of viral transmission using analysis of intra-host viral populations. In Computational Advances in Bio and Medical Sciences (ICCABS), 2015 IEEE 5th International Conference on, pages 1–1. IEEE, 2015.
4. Jonathan Mei and José M.F. Moura. Signal processing on graphs: Causal modeling of unstructured data. *IEEE Transactions on Signal Processing*, 65(8):2077–2092, 2017.
5. Christopher J. Quinn, Ali Pinar, and Negar Kiyavash. Bounded-degree connected approximations of stochastic networks. *IEEE Transactions on Molecular, Biological and Multi-Scale Communications*, 2017.
6. Ian Dobson, Benjamin A. Carreras, Vickie E. Lynch, and David E. Newman. Complex systems analysis of series of blackouts: Cascading failure, critical points, and self-organization. *Chaos: An Interdisciplinary Journal of Nonlinear Science*, 17(2):026103, 2007.
7. David L. Pepyne. Topology and cascading line outages in power grids. *Journal of Systems Science and Systems Engineering*, 16(2):202–221, 2007.
8. L. Blume, D. Easley, J. Kleinberg, R. Kleinberg, and É. Tardos. Which networks are least susceptible to cascading failures? In Foundations of Computer Science (FOCS), 2011 IEEE 52nd Annual Symposium on, pages 393–402. IEEE, Palm Springs, 2011.
9. Leon Danon, Ashley P. Ford, Thomas House, Chris P. Jewell, Matt J. Keeling, Gareth O. Roberts, Joshua V. Ross, and Matthew C. Vernon. Networks and the epidemiology of infectious disease. *Interdisciplinary Perspectives on Infectious Diseases*, 2011.
10. Matt J. Keeling and Ken T.D. Eames. Networks and epidemic models. *Journal of the Royal Society Interface*, 2(4):295–307, 2005.
11. Romualdo Pastor-Satorras and Alessandro Vespignani. Epidemic spreading in scale-free networks. *Physical Review Letters*, 86(14):3200, 2001.
12. Alain Barrat, Marc Barthelemy, and Alessandro Vespignani. *Dynamical Processes on Complex Networks*, volume 1. Cambridge University Press, 2008.
13. Deepayan Chakrabarti, Yang Wang, Chenxi Wang, Jurij Leskovec, and Christos Faloutsos. Epidemic thresholds in real networks. *ACM Transactions on Information and System Security*, 10(4):1:1–1:26, 2008.
14. Chi-Kin Chau, Vasileios Pappas, Kang-Won Lee, and Asser Tantawi. Self-organizing processes of collective behavior in computer networks. 2009. https://pdfs.semanticscholar.org/5cd7/54007fcd05b0431dd92e87d6224738ea8c9b.pdf.
15. Moez Draief, Ayalvadi Ganesh, and Laurent Massoulié. Thresholds for virus spread on networks. In Proceedings of the 1st International Conference on Performance Evaluation Methodologies and Tools, valuetools '06, New York, NY, USA, 2006. ACM.
16. Arvind Ganesh, Laurent Massoulié, and Don Towsley. The effect of network topology on the spread of epidemics. In Proceedings of the Annual Joint Conference of the IEEE Computer and Communications Societies, pages 1455–1466, vol. 2, Miami, USA, March 2005.
17. James P. Gleeson. Binary-state dynamics on complex networks: Pair approximation and beyond. *Physical Review X*, 3(2):021004, 2013.
18. Yang Wang, Deepayan Chakrabarti, Chenxi Wang, and Christos Faloutsos. Epidemic spreading in real networks: An eigenvalue viewpoint. In Proceedings of 22nd International Symposium on Reliable Distributed Systems, pages 25–34, Oct 2003.
19. June Zhang and José M.F. Moura. Accounting for topology in spreading contagion in non-complete networks. In 2012 IEEE International Conference on Acoustics, Speech and Signal Processing (ICASSP), pages 2681–2684, March 2012.
20. June Zhang and José M.F. Moura. Diffusion in social networks as SIS epidemics: Beyond full mixing and complete graphs. *IEEE Journal of Selected Topics in Signal Processing*, 8(4):537–551, Aug 2014.
21. June Zhang and José M.F. Moura. Role of subgraphs in epidemics over finite-size networks under the scaled SIS process. *Journal of Complex Networks*, 3(4):584–605, 2015.
22. Frank P. Kelly. *Reversibility and Stochastic Networks*. Cambridge University Press, 2011.
23. James R. Norris. *Markov Chains*. Cambridge University Press, 1998.
24. Alexander K. Hartmann and Martin Weigt. *Phase Transitions in Combinatorial Optimization Problems: Basics, Algorithms and Statistical Mechanics*. John Wiley & Sons, 2006.
25. Daphne Koller and Nir Friedman. *Probabilistic Graphical Models: Principles and Techniques*. MIT Press, 2009.
26. Samir Khuller and Barna Saha. On finding dense subgraphs. In *Automata, Languages and Programming*, pages 597–608. Springer, 2009.

27. Victor E. Lee, Ning Ruan, Ruoming Jin, and Charu Aggarwal. A survey of algorithms for dense subgraph discovery. In *Managing and Mining Graph Data*, pages 303–336. Springer, 2010.
28. László Lovász. Submodular functions and convexity. In *Mathematical Programming the State of the Art*, pages 235–257. Springer, 1983.
29. Martin Grötschel, László Lovász, and Alexander Schrijver. The ellipsoid method and its consequences in combinatorial optimization. *Combinatorica*, 1(2):169–197, 1981.
30. Vladimir Kolmogorov and Ramin Zabin. What energy functions can be minimized via graph cuts? *IEEE Transactions on Pattern Analysis and Machine Intelligence*, 26(2):147–159, 2004.
31. Gordon Royle and Chris Godsil. *Algebraic Graph Theory*. Springer-Verlag, 2001.
32. June Zhang and José M.F. Moura. Subgraph density and epidemics over networks. In 2014 IEEE International Conference on Acoustics, Speech and Signal Processing (ICASSP), pages 1125–1129, May 2014.
33. Stanley Wasserman. *Social Network Analysis: Methods and Applications*, volume 8. Cambridge University Press, 1994.
34. Andrew V. Goldberg. *Finding a Maximum Density Subgraph*. University of California Berkeley, CA, 1984.
35. Stephen P. Borgatti and Martin G. Everett. Models of core/periphery structures. *Social Networks*, 21(4):375–395, 2000.
36. Ulrik Brandes, Jürgen Pfeffer, and Ines Mergel. *Studying Social Networks: A Guide to Empirical Research*. Campus Verlag, 2012.
37. Peter Csermely, András London, Ling-Yun Wu, and Brian Uzzi. Structure and dynamics of core/periphery networks. *Journal of Complex Networks*, 1(2):93–123, 2013.
38. Margaret R. Weeks, Scott Clair, Stephen P. Borgatti, Kim Radda, and Jean J. Schensul. Social networks of drug users in high-risk sites: Finding the connections. *AIDS and Behavior*, 6(2):193–206, 2002.
39. Deepayan Chakrabarti, Yiping Zhan, and Christos Faloutsos. R-MAT: A recursive model for graph mining. In Proceedings of the 2004 SIAM International Conference on Data Mining, pages 442–446. SIAM, 2004.
40. Stephen P. Borgatti. The key player problem. In *Dynamic Social Network Modeling and Analysis: Workshop Summary and Papers*, page 241. National Academies Press, 2003.
41. Farzad Khorasani, Rajiv Gupta, and Laxmi N. Bhuyan. Scalable simd-efficient graph processing on gpus. In Proceedings of the 24th International Conference on Parallel Architectures and Compilation Techniques, PACT '15, pages 39–50, 2015.
42. James Abello, Sergiy Butenko, Panos M. Pardalos, and Mauricio G.C. Resende. Finding independent sets in a graph using continuous multivariable polynomial formulations. *Journal of Global Optimization*, 21(2):111–137, 2001.
43. West, Douglas Brent *Introduction to Graph Theory*, volume 2. Prentice Hall, Upper Saddle River, 2001.
44. Yoshio Okamoto, Takeaki Uno, and Ryuhei Uehara. Linear-time counting algorithms for independent sets in chordal graphs. In *Graph-Theoretic Concepts in Computer Science*, pages 433–444. Springer, 2005.
45. Matthew O. Jackson. *Social and Economic Networks*. Princeton University Press, 2008.
46. Colleen L. Ball, Michael A. Gilchrist, and Daniel Coombs. Modeling within-host evolution of HIV: Mutation, competition and strain replacement. *Bulletin of Mathematical Biology*, 69(7):2361–2385, 2007.
47. Sui Huang. Genomics, complexity and drug discovery: Insights from Boolean network models of cellular regulation. *Pharmacogenomics*, 2(3):203–222, 2001.
48. Jean-François Rual, Kavitha Venkatesan, Hao Tong, Tomoko Hirozane-Kishikawa, et al. Towards a proteome-scale map of the human protein-protein interaction network. *Nature*, 437(7062):1173, 2005.
49. David S. Campo, Guo-Liang Xia, Zoya Dimitrova, Yulin Lin, Joseph C. Forbi, Lilia Ganova-Raeva, Lili Punkova, Sumathi Ramachandran, Hong Thai, Pavel Skums, Inna Rytsareva, Gilberto Vaughan, Ha-Jung Roh, Michael A. Purdy, Amanda Sue, and Yury Khudyakov. Accurate genetic detection of hepatitis c virus transmissions in outbreak settings. *Journal of Infectious Diseases*, 213(6):957–965, 2016.

10
Modeling Functional Networks via Piecewise-Stationary Graphical Models*

Hang Yu and Justin Dauwels

CONTENTS

10.1 Introduction 193
10.2 Preliminaries 194
    10.2.1 Undirected Graphical Models and GGMs 194
    10.2.2 Copulas and Gaussian Copulas 195
10.3 Copula GGMs 196
    10.3.1 Standard Copula GGMs 196
    10.3.2 Hidden Variable Copula GGMs 197
        10.3.2.1 Hidden Variable GGM 197
        10.3.2.2 Copula GGMs with Hidden Variables 198
10.4 Change Point Detection 198
10.5 Regularization Selection 200
    10.5.1 Regularization Selection for Change Point Detection 200
    10.5.2 Regularization Selection for Graphical Model Inference 200
        10.5.2.1 Standard Copula GGMs 201
        10.5.2.2 Hidden Variable Copula GGMs 202
10.6 Numerical Results 203
    10.6.1 Synthetic Data 203
    10.6.2 Real Data 204
10.7 Possible Applications in Biomedical Signal Processing 207
10.8 Conclusion, Open Problems, and Future Directions 207
Acknowledgment 207
References 207

10.1 Introduction

Approximately 50 million people worldwide exhibits symptoms of epilepsy [2], a neurological disorder of the brain characterized by sudden and unpredictable seizures. Seizures typically lead to a lapse of attention or a convulsion of the entire body, thus posing a serious risk of physical injuries or even death to the patient. In order to treat epileptic seizures, it is essential to understand how the cerebral disorder propagates.

In this chapter, we analyze multichannel scalp electroencephalogram (EEG) signals and characterize the evolution of brain states through piecewise-stationary functional brain networks. Note that scalp EEG recording systems are inexpensive and (potentially) mobile in comparison with other measures such as functional magnetic resonance imaging (fMRI) [3]. Furthermore, they are not invasive. Due to these advantages, EEG has the potential of being a commonly used tool for seizure detection.

Here, we learn functional brain networks using the framework of graphical models. Graphical models provide sparse networks to represent the statistical relations between numerous variables, resulting in

---

* A preliminary version of this work was presented at the International Conference on Acoustics, Speech and Signal Processing (ICASSP), Florence, Italy, May 2014 [1].

compact representations and efficient inference algorithms [4,5]. Specifically, when tackling EEG signals, we resort to copula Gaussian graphical models (GGMs). Such models allow us to tie different kinds of marginal distributions (Gaussian, non-Gaussian, or even nonparametric) together to form a graphical model [6–11]. As a result, copula GGMs have been applied in such diverse areas as computational biology [6,7] and neuroscience [8], geophysics [9], extreme events analysis [10], and sociology [11].

So far, most literature on network inference via graphical models focuses on stationary data. However, during epileptic seizures, functional brain networks are shown to evolve through a sequence of distinct topologies [2]. Therefore, statistical models designed for stationary data may not yield accurate results for these problems. Inferring such evolving networks in the framework of graphical models has received little attention until now. A reasonable approach is to detect change points, and then infer graphical models in the stationary segments between the change points. There is a handful of works concerning identifying change points in multivariate time series, and we review them below. Xuan and Murphy [12] employed the Bayesian change point detection approaches: they adopt a geometric prior on the time segment lengths, and then iterate between MAP segmentation and graphical model inference. The main restriction, however, is that the graph for all segments must be decomposable. On the other hand, a greedy binary segmentation scheme is proposed in Ref. [13]. A change point is inserted such that the Bayesian information score (BIC) of the two graphical models of the data before and after the change point is minimized; this procedure is repeated until no further splits reduce the BIC. Unfortunately, besides the high computational complexity, the binary segmentation can be misleading and overestimate the number of change points, as pointed out in Ref. [14]. To address this concern, dynamic programming is applied in Ref. [15], resulting in joint estimation of all the change points. Unfortunately, the method has computational complexity of order $\mathcal{O}(n^3)$ in the number of fixed points n, which is impractical for most real-life time series with tens or hundreds of change points. Another limitation of the aforementioned methods is the assumption of Gaussian distributed data, which is not always fulfilled in practice.

In this work, our objective is to establish piecewise-stationary copula GGMs for EEG recordings during seizure. In order to reduce the computational complexity, we disentangle the process of change point detection and graphical model inference. Specifically, we first detect the change points by minimizing a cost function defined on covariance matrix using low-complexity pruned exact linear time (PELT) method [16], and next learn the graphical model based on the covariance of the data in each stationary time segment. The procedure also infers the number of change points as well as the graphical model structure in an automated fashion without tuning any parameters. Numerical results for both synthetic and real data show that the proposed method provides an effective and efficient tool to identify change points and infer networks.

This chapter is structured as follows. In this section, we briefly introduce Gaussian copula graphical models (with hidden variables) for stationary data. In Section 10.2, we extend those models to piecewise stationary data. In Section 10.3, we present results for synthetic and real data, validating the proposed model. We offer concluding remarks in Section 10.4.

10.2 Preliminaries

10.2.1 Undirected Graphical Models and GGMs

In this section, we first give a brief introduction to undirected graphical models. We then focus on GGMs in which all variables follow Gaussian distributions.

A undirected graphical model (i.e., a Markov random field [MRF]) can be defined as a multivariate probability distribution $p(x)$ that factorizes according to a graph $\mathcal{G}$, which consists of nodes $\mathcal{V}$ and edges $\mathcal{E}$. Concretely, each node $i \in \mathcal{V}$ is associated with a random variable x_i. An edge $(i, j) \in \mathcal{E}$ is absent if and only if the corresponding two variables x_i and x_j are conditionally independent: $p(x_i, x_j \mid x_{\mathcal{V} \mid i,j}) = p(x_i \mid x_{\mathcal{V} \mid i,j}) p(x_j \mid x_{\mathcal{V} \mid i,j})$, where $\mathcal{V} \mid i,j$ denotes all the nodes except i and j. Hammersley and Clifford [17] further summarize the relation between the Markov properties implied by $\mathcal{G}$ and the factorization of the probability distribution $p(x)$ in the following theorem:

Theorem 10.1 Hammersley–Clifford Theorem

Define a clique C as a fully connected subgraph, that is, a subgraph in which each node is a neighbor of every other node. If $p(x) > 0$ for all x, and $p(x)$ is Markov with respect to the graph $\mathcal{G}$, then $p(x)$ can be expressed as a product of factors corresponding to cliques of $\mathcal{G}$:

$$p(\mathbf{x}) = \frac{1}{Z} \prod_{C \in \mathcal{C}} \psi_C(\mathbf{x}_C), \tag{10.1}$$

where $\psi_C(x_C)$ is a compatibility function defined on a clique C, $\mathcal{C}$ is the set of all cliques in $\mathcal{G}$, and Z is a normalization term called the partition function. Conversely, if $p(x)$ can be factorized as in Equation 10.1 with $\psi_C(x_C) \geq 0$, then $p(x)$ is Markov with respect to the corresponding graph $\mathcal{G}$.

In this chapter, we restrict our attention to pairwise MRFs in which the cliques are chosen as the nodes and edges in the graph. The resulting probability density function (PDF) can be written as

$$p(\mathbf{x}) = \frac{1}{Z} \prod_{i \in \mathcal{V}} \psi_i(x_i) \prod_{(i,j) \in \mathcal{E}} \psi_{ij}(x_i, x_j), \quad (10.2)$$

where the node potential is $\psi_i(x_i)$ and the edge potential is $\psi_{ij}(x_i, x_j)$.

If the random variables x corresponding to the nodes in G are jointly Gaussian, then the undirected graphical model is called a GGM (or Gauss–Markov random field [GMRF]). Let $x \sim \mathcal{N}(\mu, \Sigma)$ with mean vector μ and positive-definite covariance matrix Σ. The GGM can be written equivalently as $\mathcal{N}(K^{-1}h, K^{-1})$ with a precision matrix $K = \Sigma^{-1}$ and a potential vector $h = K\mu$. The resulting PDF can be expressed as

$$p(x) \propto \exp\left\{-\frac{1}{2}x^T K x + h^T x\right\}. \quad (10.3)$$

The corresponding node and edge potentials are

$$\psi_i(x_i) = \exp\left\{-\frac{1}{2}K_{ii}x_i^2 + h_i x_i\right\}, \quad (10.4)$$

$$\psi_{ij}(x_i, x_j) = \exp\left\{-x_i K_{ij} x_j\right\}. \quad (10.5)$$

Obviously, $K_{ij} = 0$ implies that x_i and x_j are conditionally independent in a GGM. We emphasize that $K_{ij} = 0$ does not mean x_i and x_j are uncorrelated. Instead, Σ_{ij} usually is nonzero, since the inverse of a sparse precision matrix is often a full covariance matrix.

Since the structure of a GGM is well characterized by the precision matrix, we only need to infer the precision matrix when learning the GGM. Moreover, a sparse graphical model is usually preferable in practice. We therefore aim to learn a sparse precision matrix given the N multivariate Gaussian distributed observations $x^{(1:N)}$. The resulting optimization problem can be formulated as [18]

$$\hat{K} := \underset{K \succ 0}{\operatorname{argmin}} \; \operatorname{tr}(SK) - \log \det K + \lambda \| K \|_1, \quad (10.6)$$

where tr represents the trace of a matrix, and S is the empirical covariance matrix defined as

$$S = \frac{1}{N} \sum_{i=1}^{N} x^{(i)} \left(x^{(i)}\right)^T, \quad (10.7)$$

$\|\cdot\|_1$ is the ℓ_1 norm (i.e., the sum of the absolute value of all the elements in the matrix), and λ is the regularization or penalty parameter. Note that the term $\operatorname{tr}(SK) - \log \det K$ can be interpreted as the divergence between the estimated distribution and the empirical distribution or the negative log-likelihood of the observed data $x^{(1:N)}$. As such, the problem minimizes the divergence between the estimated and original distribution with an ℓ_1-norm penalty on the elements of the precision matrix K, resulting in a sparse GGM approximation. It is a trade-off between the data fidelity and the model sparsity, which is determined by the regularization parameter λ. As shown in Ref. [19], if λ is small, then $\hat{K}$ tends to be dense; otherwise, $\hat{K}$ tends to be sparse. To recover the correct matrices K, the parameter λ needs to be chosen appropriately, which is a critical issue that will be addressed in the following sections. After selecting a proper λ, Equation 10.6 can be solved efficiently using interior-point methods, block coordinate descent methods, or the graphical lasso (glasso) algorithm [19]. It has been proven in Ref. [20] that solving Equation 10.6 can consistently yield the true graphical model structure (i.e., the sparsity pattern of the precision matrix) under certain conditions.

10.2.2 Copulas and Gaussian Copulas

A copula $C(u_1,\ldots,u_p)$ can be defined as a distribution function mapping from the unit P-cube $[0,1]^P$ to the unit interval $[0,1]$ (i.e., a distribution with uniform marginals), satisfying the following conditions [21]:

1. $C(1,\ldots,1,a_i,1,\ldots,1) = a_i$ for every $i \leq P$ and all a_i in $[0,1]$.
2. $C(a_1,\ldots,a_P) = 0$ if $a_i = 0$ for any $i \leq P$.
3. C is P-increasing.

The term copula was first introduced by Sklar [22]. Thus, we next present the Sklar's theorem:

Theorem 10.2 Sklar's Theorem

Suppose that $x_1,\ldots,x_P$ are random variables with marginal cumulative distribution functions (CDFs) $F_1,\ldots,F_P$ and joint CDF F. Then there exists a copula C such that for all $x = [x_1,\ldots,x_P]^T$

$$F(x_1,\ldots,x_P) = C(F_1(x_1),\ldots,F_P(x_P)). \quad (10.8)$$

Furthermore, if the marginals $F_1,\ldots,F_P$ are continuous, then the copula C can be uniquely determined. Conversely, given any marginals $F_1,\ldots,F_P$ and an arbitrary copula C, F defined through Equation 10.8 is a P-dimensional distribution function with marginals $F_1,\ldots,F_P$.

Note that $u_i = F_i(x_i)$ follows a uniform distribution in $[0,1]$. According to Sklar's theorem, the joint distribution can be specified via its marginals and a copula that "glues" the marginals together. It is worthy noticing that copulas separate the marginal specification from dependence modeling. On the other hand, the intricate dependencies between numerous variables can be captured effectively and efficiently by graphical models (cf. Section 10.2.1). Therefore, as demonstrated in this chapter, copula-based graphical models can tackle data with various families of marginal distributions in a compact and flexible fashion.

Assuming that the partial derivatives of C exist, the PDF of the joint distribution (Equation 10.8) can be written as [21]

$$f(x_1,\ldots,x_p) = \frac{\partial^p F(x_1,\ldots,x_p)}{\partial x_1 \cdots \partial x_p} \qquad (10.9)$$

$$= c\big(F_1(x_1),\ldots,F_p(x_p)\big)\prod_{i=1}^{p} f_i(x_i), \qquad (10.10)$$

where c is the copula density function.

It follows from Equation 10.8 that a copula can be derived from a closed-form multivariate distribution as

$$C(u_1,\ldots,u_P) = F\big(F_1^{-1}(u_1),\ldots,F_P^{-1}(u_P)\big). \qquad (10.11)$$

As such, Gaussian copulas have the form

$$C(u_1,\ldots,u_P) = \Phi\big(\Phi^{-1}(u_1),\ldots,\Phi^{-1}(u_P);\Sigma\big), \qquad (10.12)$$

where $\Phi(\cdot,\Sigma)$ represents a zero-mean multivariate Gaussian distribution with covariance matrix Σ (with normalized diagonal), while Φ is the standard normal distribution. The Gaussian copula can also be constructed by introducing a vector of latent Gaussian variables $z = [z_1,\ldots,z_P]^T$ that are related to the observed variables $x = [x_1,\ldots,x_P]^T$. The two-layer model can be formulated as

$$z \sim N(0,\Sigma) \qquad (10.13)$$

$$x_i = F_i^{-1}(\Phi(z_i)). \qquad (10.14)$$

The correlation matrix Σ is the dependence parameter for Gaussian copulas; it can easily be determined once the latent variables z are inferred. The flexibility and analytical tractability of Gaussian copulas make them a handy tool in many applications. Their popularity is due to the fact that they describe dependence between variables in much the same way that Gaussian distributions do.

10.3 Copula GGMs

10.3.1 Standard Copula GGMs

We denote the observed non-Gaussian variables and latent Gaussian variables as $x = [x_1,\ldots,x_P]^T$ and $z = [z_1,\ldots,z_P]^T$, respectively. Following from the two-layer construction of Gaussian copulas (cf. Equations 10.13 and 10.14), a standard Copula GGM can be defined as [23]

$$z \sim \mathcal{N}(0,K^{-1}) \qquad (10.15)$$

$$x_i = F_i^{-1}(\Phi(z_i)), \qquad (10.16)$$

where K is the precision matrix whose inverse (the covariance matrix) has normalized diagonal, Φ is the CDF of the standard Gaussian distribution, and F_i is the CDF of x_i. The latter is often approximated by the empirical distributions $\hat{F}_i$. The pseudo-inverse F_i^{-1} of F_i is defined as

$$F_i^{-1}(y) = \inf_{x_i \in \mathcal{X}}\{F_i(x_i) \geq y\}, \qquad (10.17)$$

where inf denotes infimum, and $\mathcal{X}$ is the domain of x_i such that $F_i(x_i) \leq 1$ for all $x_i \in \mathcal{X}$. Note that we use a precision matrix as the dependence parameter of the Gaussian copula instead of the correlation matrix in Equation 10.13. Next, we prove that for a CGGM, the conditional independence between variables is specified by the zero elements in the precision matrix K [6].

Proposition 10.1

Define $z_i = h_i(x_i) = \Phi^{-1}(\hat{F}_i(x_i))$ and let K be the precision matrix of the latent Gaussian variables z. If the functions $h_i(x_i)$ are differentiable with respect to x_i, then $p(x_i,x_j | x_{\mathcal{V}\setminus i,j}) = p(x_i | x_{\mathcal{V}\setminus i,j})p(x_j | x_{\mathcal{V}\setminus i,j})$ if and only if $K_{ij} = 0$.

Proof. If the functions $h_i(x_i)$ are differentiable, it follows from Equations 10.9 and 10.12 that the PDF of the CGGM is given by

$$p(x) \propto \exp\left\{-\frac{1}{2}h(x)^T K h(x)\right\}\prod_{i=1}^{P} h_i'(x_i). \qquad (10.18)$$

The first factor corresponds to a GGM, and thus can be further decomposed as the product of node and edge potentials as defined in Equations 10.4 and 10.5. Thus, the CGGM only modifies the node potentials of the corresponding GGM. By applying the Hammersley–Clifford theorem (i.e., Theorem 10.1), we can find that the conditional dependence between variables is characterized by the precision matrix.

As a result, we aim to estimate a sparse precision matrix (so as to learn a sparse graphical model) from the observed non-Gaussian data x. To move forward this objective, we first map the non-Gaussian variables x to the latent Gaussian layer:

$$z_i = \Phi^{-1}\big(\hat{F}_i(x_i)\big). \qquad (10.19)$$

We then learn the sparse precision matrix based on the latent Gaussian variables z by solving the ℓ_1-norm penalized optimization problem introduced in Section 10.2.1 [6]:

$$\hat{K} := \underset{K \succ 0}{\operatorname{argmin}}\ \operatorname{tr}(SK) - \log \det K + \lambda \|K\|_1, \quad (10.20)$$

where S is the empirical covariance of z.

10.3.2 Hidden Variable Copula GGMs

One typically constructs a sparse graphical model by discovering the most important interactions between observed variables [19], as shown in Figure 10.1a. Such graphical models, however, fail to deal with the case where there exist hidden variables. Now suppose that the light gray nodes in Figure 10.1a represent hidden variables; we only observe samples of the observed variables (medium gray nodes in Figure 10.1a), and no information is provided about the hidden variables. Under such a scenario, approaches to inferring standard graphical models would yield a dense graph (see Figure 10.1b), including both interactions originally between observed variables as well as those coming from hidden variables. Instead, we may consider the effect of hidden variables and obtain a sparse graph that only characterizes the direct interdependencies between observed variables. In practice, it is quite common that data are unavailable for some relevant variables. For instance, when inferring functional brain networks given brain signals, the signals may only be measured from some specific brain areas (e.g., cortex in the case of scalp EEGs). However, those signals may be affected by brain areas from which no measurements are available (e.g., deeper areas such as hippocampus). The latter may then be treated as hidden variables in a statistical model.

To address the above problem, we propose a novel hidden variable copula GGM (HVCGGM). Before presenting the proposed model, we first introduce hidden variable GGMs.

10.3.2.1 Hidden Variable GGM

Here, we consider the simplest case of hidden variable graphical models, where both the observed variables and hidden variables are Gaussian. Suppose we have observable variables z_O (medium gray nodes in Figure 10.1a) and hidden variables z_H (light gray nodes in Figure 10.1a), which are jointly Gaussian distributed. The joint precision matrix of $z_O \cup z_H$, $K_{O \cup H}$, which characterizes the graphical model in Figure 10.1a, can be expressed as

$$K_{O \cup H} = \begin{bmatrix} K_{OO} & K_{OH} \\ K_{HO} & K_{HH} \end{bmatrix}. \quad (10.21)$$

Then according to the Schur complement [24], the marginalized precision matrix $\tilde{K}_{OO}$ of z_O can be written as

$$\tilde{K}_{OO} = K_{OO} - K_{OH} K_{HH}^{-1} K_{HO} = K_{OO} - L, \quad (10.22)$$

where $L = K_{OH} K_{HH}^{-1} K_{HO}$. Note that given the joint covariance matrix $\Sigma_{O \cup H}$, the marginal precision matrix of observed variables $\tilde{K}_{OO} = ([\Sigma_{O \cup H}]_{OO})^{-1}$.

The two components of $\tilde{K}_{OO}$ have their own properties [25]. K_{OO} is the conditional precision matrix of z_O, conditioned on z_H (see Figure 10.1c). It is supposed to be sparse as it describes the interactions only between observed variables. L summarizes the effect of marginalization over the hidden variables. The rank of L is equal to the number of hidden variables, and it is assumed to be low since the number of hidden variables is supposed to be small. Since z_H are connected to many of z_O, K_{OH} and K_{HO} are not sparse, thus making the product matrix dense. Resulting from the subtraction, $\tilde{K}_o$ is also dense, as shown in Figure 10.1b. CGGMs cannot yield a sparse graph in this case because they can only estimate $\tilde{K}_o$.

Given i.i.d. samples of z_O, our objective is to estimate K_{OO} and L; we are especially interested in the rank of L since it equals the number of hidden variables. Those matrices may be recovered by solving the convex relaxation [25]

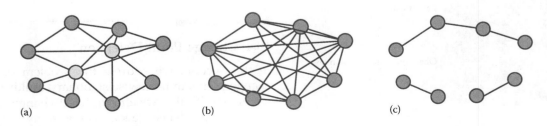

FIGURE 10.1
Graphical model with hidden variables (light gray nodes): (a) the joint graphical model; (b) the marginal graphical model of observed variables (medium gray nodes); and (c) the conditional graphical model of observed variables.

$$\left(\hat{K}_{OO}, \hat{L}\right) = \underset{K_{OO} \geq 0, L \pm 0}{\operatorname{argmin}} \operatorname{tr}((K_{OO} - L)S_{OO})$$

$$- \log \det(K_{OO} - L)$$

$$+ \lambda(\gamma \| K_{OO} \|_1 + \operatorname{tr}(L)), \quad (10.23)$$

where $\hat{K}_{OO}$ and $\hat{L}$ are estimates of K_{OO} and L, respectively, and S_{OO} is the empirical marginal covariance of z_O.

Note that similar to Equation 10.20, $\operatorname{tr}((K_{OO} - L)S_{OO}) - \log \det(K_{OO} - L)$ is the divergence between the observed data and the estimated model. The two regularization parameters λ and γ can be interpreted as follows. The product of λ and γ is the regularization parameter of the ℓ_1 norm; it controls the trade-off between the sparsity of K_{OO} and the fidelity of the data. On the other hand, λ alone is the regularization parameter of the nuclear norm (which reduces to the trace norm for symmetric, positive-semidefinite matrices), and therefore it penalizes the rank of L. In addition, γ alone balances the trade-off between the rank of L and the sparsity.

The convex problem 10.23 can be solved efficiently by the Newton-CG primal proximal point algorithm [26]. To recover the correct K_{OO} and L, the parameters λ and γ need to be chosen properly, and this will be discussed in Section 10.5.2.

10.3.2.2 Copula GGMs with Hidden Variables

We now present the proposed hidden variable copula GGM. The observed (continuous) variables x (dark gray nodes in Figure 10.2) are non-Gaussian, and each of them is associated with a Gaussian distributed latent variable z_{Oi} (medium gray nodes in Figure 10.2), as in the standard copula GGM. However, besides the latent variables z_O, there exist several hidden variables z_H (light gray nodes in Figure 10.2) that are not associated with observed variables. In the corresponding graphical model, the nodes z_H are only connected to latent variables z_O; they are not connected to observed variables x.

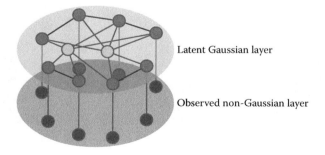

FIGURE 10.2
Copula GGM with hidden variables: dark gray nodes denote observed non-Gaussian variables, medium gray nodes denote latent Gaussian variables, and light gray nodes denote hidden variables in the latent layer.

In other words, the variables x and z_O constitute a Gaussian copula, while the variables z_O and z_H form a hidden variable GGM. Together, the variables (x, z_O, z_H) form a hidden variable copula GGM with associated conditional precision matrix K_{OO} and product matrix L (cf. (10.21)).

Given i.i.d. samples of the non-Gaussian variables x, we wish to infer the conditional precision matrix K_{OO} of z_O conditioned on z_H (corresponding to the dark gray edges in Figure 10.2), and the product matrix L, associated with the HVCGGM.

As a first step, we transform the non-Gaussian observed variables x into Gaussian distributed hidden variables z_O associated with the observed variables x:

$$z_{Oi} = \Phi^{-1}\left(\hat{F}_i(x_i)\right), \quad (10.24)$$

where Φ is the CDF of the standard Gaussian distribution and $\hat{F}_i$ is the empirical CDF of x_i. As a result, we are dealing with Gaussian variables z_O which together with z_H constitute a HVGGM.

In the second step, we follow the procedure of Equation 10.23 to infer the sparse conditional precision matrix K_{OO} of z_O and the low-rank product matrix L:

$$\left(\hat{K}_{OO}, \hat{L}\right) = \underset{K_{OO} \succ 0, L \geq 0}{\operatorname{argmin}} \operatorname{tr}((K_{OO} - L)S_{OO})$$

$$- \log \det(K_{OO} - L)$$

$$+ \lambda(\gamma \| K_{OO} \|_1 + \operatorname{tr}(L)). \quad (10.25)$$

The theoretical computational complexity of the proposed method is $\mathcal{O}(P^3)$, where P is the number of variables. The computational bottleneck stems from solving the above convex problem. In practice, the algorithm proposed in Ref. [26] can efficiently deal with problems of dimension up to several thousands. Note that the selection of regularization parameters λ and γ in Equation 10.25 needs special attention, as will be discussed in Section 10.5.2.

10.4 Change Point Detection

In this section, we address the problem of detecting change points in piecewise stationary multivariate time series. Specifically, we aim to detect changes in the statistical dependence (a.k.a. connectivity or network structure) among the time series. The data between the change points are assumed to be stationary and will be modeled by copula GGMs.

Let us assume that we have an ordered sequence of N samples for each of the P variables, $x_i^{(n)}$, where $n = 1, \cdots, N$ and $i = 1, \cdots, P$. We wish to infer an unknown number M of change points $\tau_{1:M} = (\tau_1, \cdots, \tau_M)$. Each change point is an integer between ℓ and $N - \ell$, where ℓ is the minimum length of one segment. We further define $\tau_0 = 0$ and $\tau_{M+1} = n$, and thus the M change points will split the data into $M + 1$ segments, where the kth segment is given by $\mathbf{x}_{1:P}^{(\tau_{k-1}+1:\tau_k)}$.

As a first step, we transform the non-Gaussian observed variables x_i into Gaussian latent variables z_i (associated with the observed variables x_i), i.e., $z_i = \Phi^{-1}(\hat{F}_i(x_i))$.

We next solve the problem of identifying all the change points together in the Gaussian latent layer. Concretely, we minimize a cost function with a penalty on the number of change points, as suggested in the literature [16,27]:

$$\sum_{k=1}^{M+1} L_k\left(\mathbf{z}_{1:P}^{(\tau_{k-1}+1:\tau_k)}\right) + \beta M, \quad (10.26)$$

where β is the cost associated with each change point, in order to limit overfitting. The negative log-likelihood L_k is defined as

$$L_k\left(\mathbf{z}_{1:P}^{(\tau_{k-1}+1:\tau_k)}\right) = \frac{\tau_k - \tau_{k-1}}{n} \log \det S_k, \quad (10.27)$$

where S_k is the empirical covariance of segment k. We apply the PELT method [16] to efficiently find the global minimum of Equation 10.26. Specifically, let $F_M(t)$ denote the global minimum of Equation 10.26 for data $\mathbf{z}_{1:P}^{(1:t)}$ and let $T_t = \{\boldsymbol{\tau}: 0 = \tau_0 < \tau_1 < \cdots < \tau_M < \tau_{M+1} = t\}$ be the set of candidate change points at time t. It therefore follows that [28]

$$F_M(t) = \min_{\boldsymbol{\tau} \in T_t} \sum_{k=1}^{M+1}\left[L\left(\mathbf{z}_{1:P}^{(\tau_{k-1}+1:\tau_k)}\right) + \beta\right] \quad (10.28)$$

$$= \min_s \left\{ \min_{\boldsymbol{\tau} \in T_s} \sum_{k=1}^{M}\left[L\left(\mathbf{z}_{1:P}^{(\tau_{k-1}+1:\tau_k)}\right) + \beta\right] + L\left(\mathbf{z}_{1:P}^{(s+1:t)}\right) + \beta \right\}$$

$$(10.29)$$

$$= \min_s \left[F_{M-1}(s) + L\left(\mathbf{z}_{1:P}^{(s+1:t)}\right) + \beta\right]. \quad (10.30)$$

The interpretation is that the optimal value $F_M(t)$ of the cost function at time t can be derived using the optimal partition of the data prior to the last change point plus the cost for the segment from the last change point to t. In other words, the expression 10.30 offers a recursion expressing the minimum for data $\mathbf{z}_{1:P}^{(1:t)}$ in terms of the minimum for $\mathbf{z}_{1:P}^{(1:s)}$ for $s < t$. We then set $F(0) = -\beta$ so that the penalty equals βM in Equation 10.28, and Equation 10.30 can be solved in turn for $t = \ell + 1, \cdots, N$ by finding the most recent change point s before t. The computational cost of solving (Equation 10.30) is linear in t, since we need to consider all candidate positions of the last change point before time t, namely, $s \in \{\ell + 1, \ell + 2, \cdots, t - \ell\}$. Therefore, the overall computational complexity of finding the optimal partition is $\mathcal{O}(N^2)$.

The computational complexity can be further reduced by pruning the candidate set of last change points at time t as t increases. Such pruning can be done by removing those candidates that can never be change points after t. The following theorem asserts a condition to determine which time points to be pruned:

Theorem 10.3

[16] *We assume that when introducing a change point into a sequence of observations, the cost L of the sequence reduces. More formally, for all $r < s < t$,*

$$L\left(\mathbf{z}^{(r+1:s)}\right) + L\left(\mathbf{z}^{(s+1:t)}\right) \leq L\left(\mathbf{z}^{(r+1:t)}\right). \quad (10.31)$$

Therefore, if it holds that

$$F(r) + L\left(\mathbf{z}^{(r+1:s)}\right) \geq F(s), \quad (10.32)$$

then at a future time $t > s$, r can never be the optimal last change point prior to t.

By discarding time points satisfying the above condition in each iteration, such procedure successfully removes computations that are irrelevant to obtaining the final set of change points and accelerates the algorithm. The resulting computational cost of the PELT method can be reduced to be linear under certain conditions as proven in Ref. [16]. In summary, the PELT method is proceeded as follows:

1. Initialize $F(0) = -\beta$. The set $cp(t)$ of previous change points at $t \leq \ell$ is initialized as $cp(t) = \emptyset$, and the set $R_{\ell+1}$ of candidate change points at $t = \ell + 1$ is initialized as $R_{\ell+1} = \{0\}$.

2. Compute $F(t) = \min_{\tau \in R_t}[F(\tau) + L(\mathbf{z}_{1:P}^{(\tau+1:t)}) + \beta]$ and let $\tau^* = \operatorname{argmin}_{\tau \in R_t} F(t)$.

3. Update the set $cp(t)$ of previous change points at time t: $cp(t) = cp(\tau^*) \cup \tau^*$.

4. Prune the set R_t by removing $\{\tau \in R_t : F(\tau) + L(\mathbf{z}_{1:P}^{(\tau+1:t)}) > F(t)\}$. If $t \geq 2\ell - 1$, then update R_{t+1} of candidate change points at the next time position $t + 1$: $R_{t+1} = R_t \cup \{t + 1 - \ell\}$.

5. Return to Step 2 if $t < N$ and increase t by 1.

The resulting $cp(N)$ is the optimal set of change points.

10.5 Regularization Selection

10.5.1 Regularization Selection for Change Point Detection

Another issue with solving Equation 10.26 is the selection of regularization parameter β, which determines the final number of change points. There is no uniform rule to compute β for all kinds of data. An adaptive method to determine the number of change points is introduced in Ref. [27], which keeps increasing the number of change points until the negative log-likelihood of the entire time series ceases to decrease significantly. In the example of Figure 10.3a, the optimal number of change points is three, as the negative log-likelihood starts decreasing more slowly for larger number of change points. Since we implicitly infer the number of change points by tuning the value of β, we modify the abovementioned approach to choose β in the same spirit.

Let $L = \sum_{k=1}^{M+1} L_k$ and define a general penalty $f(M)$, which equals M in our case. It is apparent that the estimated number of change points $\hat{M}(\beta)$ is a piecewise constant function of β. As such, if $\hat{M}(\beta) = a$, $L(a) + \beta f(a) < \min_{b \neq a}(L(b) + \beta f(b))$. Therefore, β satisfies the following condition [27]:

$$\max_{b>a} \frac{L(a) - L(b)}{f(b) - f(a)} < \beta < \min_{b<a} \frac{L(b) - L(a)}{f(a) - f(b)}. \quad (10.33)$$

In the example of Figure 10.3, for β such that $\hat{m}(\beta) = 3$, $L(3) - L(4) < \beta < L(2) - L(3)$. There hence are ordered sequences $1 = m_1 < m_2 < \ldots$ and $\infty = \beta_0 > \beta_1 \ldots$ defined as [27]

$$\beta_i = \frac{L(m_i) - L(m_{i+1})}{f(m_{i+1}) - f(m_i)}, \quad i \geq 1, \quad (10.34)$$

such that $\hat{m}(\beta) = m_i, \forall \beta \in [\beta_i, \beta_{i-1})$ (as marked by the medium gray circles in Figure 10.3b). To find the m for which L ceases to decrease significantly, we need to look for a *break* in the slope of the function $L(f)$, which is defined by β according to Equation 10.34. Therefore, the change of the slope is determined by the length h_{m_i} of the interval $[\beta_i, \beta_{i-1})$. As a result, the regularization selection procedure can be executed as follows:

1. Select an arithmetic sequence of $\beta = (\beta_1, \beta_2, \cdots)$ (denoted by black plus signs in Figure 10.3b). In practice, we set the interval between two β values to be

$$c \log(N) P(P+1)/4N, \quad (10.35)$$

   where $c = 0.02$ is a user-defined constant and $\log(N)P(P+1)/4N$ is the regularization parameter when the Bayesian information criterion (BIC) is employed for regularization selection.

2. For each β_i, estimate the number of change points $\hat{M}(\beta_i)$ by solving Equation 10.26 using the PELT method.

3. Compute the length h_{M_j} of the interval by counting the number of β_i that generates M_j change points, which is represented by the length between two circles with the same number of change points in Figure 10.3b.

4. Choose the smallest value of β such that $h_{M_j} \gg h_{M_k}$, for $k > j$. In Figure 10.3b, as an example, β corresponding to $h_{\hat{M}(\beta)=3}$ is optimal since $h_{\hat{M}(\beta)=3} \gg h_{\hat{M}(\beta)=4}$.

10.5.2 Regularization Selection for Graphical Model Inference

Once we obtain the optimal segmentation of the time series, we can learn graphical models based on the empirical covariance S_k for each time segment. Here we consider copula GGMs without and with

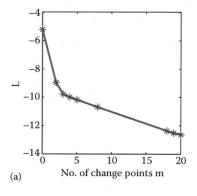
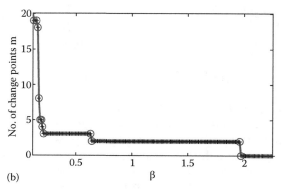

FIGURE 10.3
Adaptive regularization selection. (a) L vs. m. (b) m vs. β.

hidden variables (i.e., CGGMs and HVCGGMs), which are inferred by solving Equations 10.20 and 10.25, respectively.

A suitable choice of regularization parameters λ and γ in Equations 10.20 and 10.25 can recover the underlying true graphical model. Standard approaches for regularization selection, including cross validation (cv), the Akaike information criterion (AIC), and the BIC, are known to overfit the data, and they typically result in graphs that are too dense [29]. As an alternative, we circumvent the delicate issue of regularization selection by learning the graph structure via stability selection [30,31]. The objective of stability selection is to estimate a stable graph from the data. It generates multiple sample sets by bootstrapping the original data, and then determines the graphical model that appears consistently over those sample sets. Next, we present the stability selection process for copula GGMs without and with hidden variables separately at length.

10.5.2.1 Standard Copula GGMs

Suppose we have a P-dimensional dataset S with sample size N. The method presented below is referred to as the bootstrap inference for network construction (BINCO) method [31].

1. We randomly draw M sample sets $S_1, S_2, \ldots, S_M$ without replacement from the dataset S, each of size $N/2$. In our case, we choose $M = 100$.
2. We select a candidate set of λ. Now let us concentrate on one candidate λ in that set. For each sample set S_m (for $m = 1, \ldots, M$), we estimate one precision matrix K_m for each sample set S_m by solving Equation 10.20, resulting in M precision matrices.
3. For each element (i, j) in the matrix K_m, the number of times it is nonzero for each sample set S_m is counted and divided by M. As a result, we obtain the selection frequency (or stability) $\hat{p}_{ij}$ of the corresponding edge associated with λ. Note that $\hat{p}_{ij}$ can only take values in $\{0,1/M,2/M, \ldots,1\}$. Typically, a proper selection of λ will result in a U-shaped empirical density function $f^\lambda(x)$ of selection frequencies of all the candidate edges, as illustrated in Figure 10.4a. The selection frequencies fall into two categories, i.e., "true" or "null," depending on whether the edge exists in the true graphical model. Consequently, the density function $f^\lambda(x)$ can be decomposed as

$$f^\lambda(x) = (1-\pi)f_0^\lambda(x) + \pi f_1^\lambda(x),$$
$$x \in \left\{0, \frac{1}{M}, \frac{2}{M}, \ldots, 1\right\}, \quad (10.36)$$

where π is the proportion of true edges, and f_0^λ and f_1^λ are the density functions of selection frequencies when the edge belongs to the "null" and "true" category, respectively. As illustrated in Figure 10.4b, the distribution of f_0^λ is monotonically decreasing and it almost equals f^λ when the selection frequencies are small. In other words, we can obtain the following proper condition:

Proper Condition 10.1

[31] *There exists V_1 and V_2, $0 < V_1 < V_2 < 1$, such that*

a. $f_1^\lambda \to 0$ on $(V_1, V_2]$.
b. f_0^λ is monotonically decreasing on $(V_1, 1]$.

This proper condition is satisfied by a class of procedures as described in the following theorem:

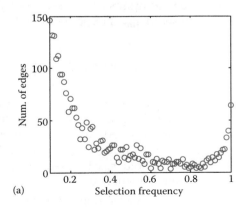
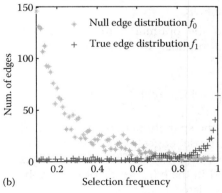

FIGURE 10.4
Distribution of selection frequencies based on a simulated dataset. (a) All the edges; (b) null edges and true edges separately.

Theorem 10.4

[31] *A selection procedure satisfies the proper condition if, as the sample size increases, $\hat{p}_{ij}$ tends to one uniformly for all true edges and has a limit superior strictly less than one for all null edges.*

It has been verified in Ref. [31] that all consistent procedures for graphical model selection (e.g., the glasso algorithm) satisfy the condition in Theorem 10.4.

4. The proper condition motivates us to estimate f_0^λ and π using the empirical distribution of selection frequencies f^λ in the region $(V_1, V_2]$. Specifically, we posit a parametric model over f_0^λ [31]:

$$f_0^\lambda = \int_0^1 b_1(x|\tau) b_2(\tau|a,b) d\tau, \quad (10.37)$$

where $b_1(x \mid \tau)$ is a binomial distribution, and $b_2(\tau \mid a, b)$ is a beta distribution that is conjugate to the binomial distribution. It is intuitive to model the selection frequency by a rescaled binomial distribution, since whether an edge is selected can be described by a binomial distribution, and a beta distribution can have the shape with f_0^λ in the region $(V_1, 1]$. We then estimate the parameters a, b, and π by minimizing the Kullback–Leibler (KL) divergence between $(1 - \pi) f_0^\lambda$ and f^λ in the region $(V_1, V_2]$.

5. We extrapolate the parametric model 10.37 of f_0^λ to the region $(V_2, 1]$, and compute the false detection error (FDR) for possible threshold $c \in (V_2, 1]$ as

$$\text{FDR}^\lambda(c) = \frac{\sum_{x \geq c} f_0^\lambda(x)}{\sum_{x \geq c} f^\lambda(x)}. \quad (10.38)$$

For each λ, we choose the threshold c that minimizes the FDR.

6. We then find the optimal λ that maximizes the estimated number of true edges whose definition is

$$N_e(\lambda) = \left(1 - \text{FDR}^\lambda(c)\right) \sum_{x \geq c} f^\lambda(x). \quad (10.39)$$

Eventually, the sparsity pattern K is obtained by retaining those edges with selection frequencies above threshold c. The selected edges are then stored in the edge set $\mathcal{E}_K$ of the graphical model associated with K.

7. After learning the structure of the graphical model, the problem of learning the nonzero elements in the precision matrix $K^{(\kappa+1)}$ can be formulated as

$$K^{(\kappa+1)} = \underset{K \succ 0}{\arg\max} \log \det K - \text{tr}\left(S^{(\kappa)} K\right), \quad (10.40)$$
$$\text{s.t} \quad K_{ij} = 0 \quad \forall\, (i,j) \notin \mathcal{E}_K.$$

This parameter learning problem can be solved efficiently by iterative proportional fitting [32].

10.5.2.2 Hidden Variable Copula GGMs

Here, we denote the undirected graphical model associated with K_{OO} as $\mathcal{G}_{OO} = (\mathcal{V}_{OO}, \mathcal{E}_{OO})$. We refer to the stability selection method below as the stability surface (SS) method.

1. The first set is the same with that for standard copula GGMs. We randomly bootstrap M sample sets from the original data.

2. We select a range of λ and γ (cf. Equation 10.25). Now let us concentrate on one pair of parameters (λ, γ) in that range. For each sample set S_m (for $m = 1, \ldots, M$), we estimate one conditional precision matrix K_{OO} using Equation 10.25, resulting in M precision matrices $K_1, \ldots, K_M$. For each element (i, j) in the matrix K_m, the number of times it is nonzero ($[K_m]_{ij} \neq 0$) among the M matrices is counted and divided by M; as a result, we obtain the probability (referred to as "stability") that this edge exists in the graphical model associated with (λ, γ). By varying λ and γ through the chosen range, we can draw a surface of the stability for each edge.

3. We include edge (i, j) in the edge set $\mathcal{E}_{OO}$, if the probability of that edge, for at least one pair (λ, γ) in the selected range, is larger than a threshold π_{thr}. It has been proven in Ref. [30] that given the expected number of falsely selected edges E, the threshold π_{thr} is upper-bounded by

$$\pi_{\text{thr}} \leq \frac{\bar{p}^2}{P(P-1)E} + 0.5 \quad (10.41)$$

The parameter $\bar{p}$ is the *average* number of edges in the graphs associated with each pair (λ, γ) in the selected range, inferred from the entire data set S through Equation 10.25. In practice, the threshold π_{thr} is often set to be the upper bound (Equation 10.41).

4. With the learned sparsity pattern of K_{OO}, we proceed to estimate K_{OO} and L. Specifically, we solve a problem similar to Equation 10.25,

subject to the learned graphical model structure $\mathcal{E}_{OO}$:

$$\left(\hat{K}_{OO}, \hat{L}\right) = \underset{K_{OO} \succ 0, L \succ 0}{\mathrm{argmin}}\ \log \det(K_{OO} - L)$$
$$-\mathrm{tr}((K_{OO} - L)S_{OO}) + \lambda\, \mathrm{tr}(L)$$
$$\text{s.t.}\quad [K_{OO}]_{ij} = 0 \quad \forall\ (i,j) \notin \mathcal{E}_{OO}. \tag{10.42}$$

The parameter λ is selected as the mean of the values of λ in the chosen range in the second step of stability selection. The number of hidden variables can be estimated easily as the rank of L. The convex problem 10.42 can be solved using the solver LogdetPPA [26].

Note that we employ the SS method [30] to learn the structure of HVCGGMs, since it can deal with multiple regularization parameters. However, the SS method often leads to a graphical model that is too sparse for problems with one penalty parameter (Equation 1.20), as demonstrated in Refs. [29,30]. As a consequence, we exploit the BINCO method [31] for CGGMs with one single regularization parameter, since it determines the edge selection threshold more reasonably by minimizing the false detection rate.

10.6 Numerical Results

In this section, we apply the proposed model to both synthetic data and real scalp EEG data during seizures.

10.6.1 Synthetic Data

The piecewise stationary non-Gaussian data are generated as follows:

1. We generate $M + 1$ random precision matrices K_k ($k = 1, \cdots, m + 1$) with $P = 25$ as in previous chapters, where $M = 3$ is the predefined number of change points. Specifically, first we uniformly sample $x_1, \ldots, x_n$ from a unit square. The precision matrix is initialized as a unit matrix. Next, we set the element $[K_k]_{ij} = [K_k]_{ji}$ of precision matrix equal to $\rho = 0.245$ with probability $(\sqrt{2\pi})^{-1} \exp(-4\|x_i - x_j\|^2)$, and equal to zero otherwise. For assessing the hidden variable graphical models (cf. Section 10.3.2), we add a few variables and connect each of them to at least 80% of other variables; the corresponding elements in precision matrix are nonzero.

2. From each precision matrix K_k, we generate N_k Gaussian samples. Consequently, N_k is the length of the kth segment. We then concatenate the Gaussian data corresponding to all the precision matrices. For assessing the hidden variable graphical models, we discard all the samples of variables added in Step 1 (hidden variables).

3. Transform the P Gaussian variables to continuous non-Gaussian variables following Equation 10.16, where F_i can be any non-Gaussian continuous marginal. Normalize the non-Gaussian data set to have unit variance, so that its empirical covariance becomes more tractable.

We generate 100 100-dimensional data sets with 9000 non-Gaussian samples for each variable. The common true values of change points are (2221, 4301, 6751). In order to assess both the copula GGMs with and without hidden variables, the first half of 100 data sets does not contain hidden variables while the other half does.

We first test the accuracy of change point detection. Specifically, we benchmark the proposed copula PELT method with adaptive regularization selection (denoted as "CPELT-A") against three other approaches: the copula PELT method with the regularization parameter selected via the BIC (denoted as "CPELT-B"), the original PELT method with adaptive regularization selection ("PELT-A"), and the original PELT method with the BIC ("PELT-B"). The results are listed in Table 10.1, where we show the distribution of the detected number of change points. We also report the mean absolute error (MAE) between the estimated position of change points and the ground truth, for the cases where the correct number of change points is inferred. Clearly, the proposed CPELT-A method greatly outperforms other methods, reliably identifying the number and the position of change points. On the other hand, CPELT-B and PELT-B seriously underestimate the number of change points; the BIC does not seem to be suitable here for regularization selection. The PELT-A method can often detect the correct number of change points (in 68% of the cases),

TABLE 10.1

Comparison of Different Methods for Accuracy of Change Point (cp) Detection

Methods	\multicolumn{4}{c}{Distribution of Detected cp No.}	MAE (cp No. = 3)
	0	1	2	3	
CPELT-A	0%	0%	0%	100%	1.7
CPELT-B	100%	0%	0%	0%	
PELT-A	0%	33%	2%	65%	3.2
PELT-B	92%	8%	0%	0%	

TABLE 10.2

Comparison of Different Methods for Accuracy of Graphical Model Inference

Models	CGGMs		HVCGGMs	
Methods	BIC	BINCO	BIC	SS
Precision	0.22	0.74	0.18	0.79
Recall	0.99	0.76	0.97	0.73
F_1-score	0.36	0.75	0.30	0.76

but the inferred position of change points is not as accurate compared to CPELT-A.

We also validated the graphical models inferred from the stationary segments between change points obtained from CPELT-A. We compare the stability selection-based methods introduced in Section 10.5.2 with BIC for regularization selection. To evaluate the performance of these two methods, we consider three criteria: precision, recall, and F_1-score. Precision is defined as the proportion of correctly estimated edges to all the edges in the estimated graph; recall is defined as the proportion of successfully estimated edges to all the edges in the true graph. Moreover, F_1-score = 2·precision·recall/(precision+recall) is a weighted average of the precision and recall. The results are listed in Table 10.2. We can tell from the table that BIC favors graphs that are much denser than the ground truth, as pointed out in Ref. [29]. However, it is apparent that the stability selection-based method (i.e., BINCO and SS) outperforms BIC and can reliably infer the structure of graphical models.

10.6.2 Real Data

We now apply the proposed model to scalp EEG recorded during epilepsy seizures [33]. The EEG time series were collected from 23 pediatric patients (5 males, ages 3–22; and 17 females, ages 1.5–19) with intractable seizures, using the international 10–20 system at a sampling frequency of 256 Hz. During the recordings, the patients experienced 198 events that were judged to be clinical seizures by experts. The pediatric EEG data used in this chapter are contained within the CHB-MIT database, which can be downloaded from the PhysioNet website: http://www.physionet.org/pn6/chbmit/. Here, we select 140 events that are longer than 20 seconds and extract EEG data starting 30 seconds before seizure onset and ending 30 seconds after seizure termination. We then downsample the signal to retain one fourth of all the samples, so as to further limit the computational complexity. The EEGs are further band-pass-filtered with a digital third-order Butterworth filter between 0.5 and 25 Hz, since most seizure activities fall within this range [34].

First, we evaluate the performance of the change-point detection algorithm: we check whether we can identify a change point when the seizure starts or ends. Specifically, we calculate the relative error, which is the ratio between the distance from the onset or termination of the seizure to a closest change point and the time length of the seizure, and then depict the histogram of the relative error in Figure 10.5. We can find that most of the errors are very small, indicating that the proposed algorithm can reliably detect the beginning and the end of seizures. More precisely, the relative error is smaller than 0.1 for above 76% of all selected seizure events when identifying the onset and for 90% when identifying the termination.

Next, we infer copula GGMs without and with hidden variables for all stationary segments. We randomly choose 16 seizure events and plot the number of edges in the graphical models as a function of time in Figure 10.6. The results for other seizure events have similar patterns. We first focus on CGGMs without hidden variables. It can be observed from Figure 10.6 that the networks resulting from this approach typically become sparser at the onset

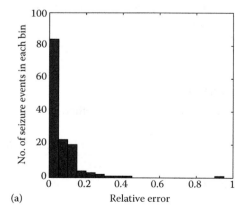
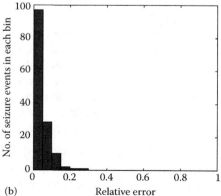

FIGURE 10.5
Histogram of relative error for change point detection w.r.t. to the onset and termination of seizures

Modeling Functional Networks via Graphical Models

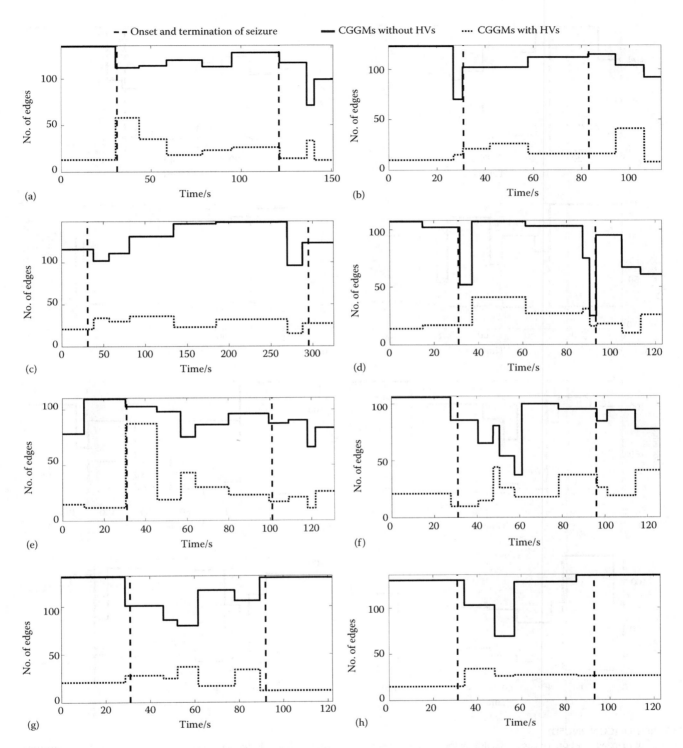

FIGURE 10.6
Results of piecewise-stationary copula GGMs without (solid line) and with (dotted line) hidden variables for scalp EEG data during seizures.
(Continued)

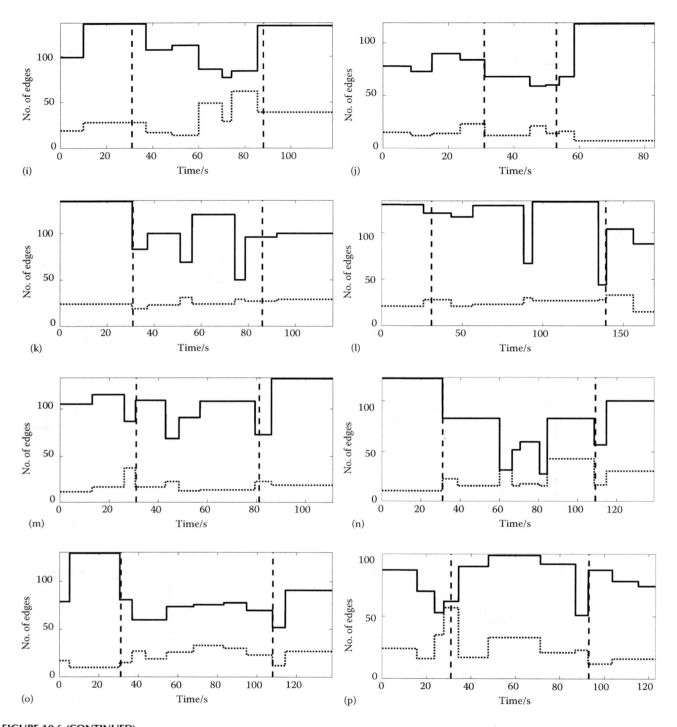

FIGURE 10.6 (CONTINUED)
Results of piecewise-tationary copula GGMs without (solid line) and with (dotted line) hidden variables for scalp EEG data during seizures.

and termination of seizures than those during inter-ictal periods (see black solid lines in Figure 10.6b–p). However, in the middle portion of seizures, the networks may possess many more edges than those at the beginning and end of seizures (see Figure 10.6c, d, f, g, i, k, l, n, and o), and can even be denser than networks during inter-ictal periods.

On the other hand, CGGMs with hidden variables yield networks that are dense at the onset and termination of seizures but sparse in the middle portion of seizures. Networks during inter-ictal periods are also sparse. Interestingly, Kramer et al. [2] analyzed the dynamics of functional networks based on cross-correlation through the entire seizure in intracranial electrocorticogram (ECoG) recordings. They found that the networks are dense at seizure onset and termination, but sparse during the middle portion of the seizure. Here, although scalp EEG is noisier than intracranial ECoG recordings, the proposed method can find the same pattern as observed in Ref. [2]. It is noteworthy that, in contrast to Ref. [2], the proposed method allows us to infer the number and location of change points as well as the structure of graphical models in an automated fashion.

10.7 Possible Applications in Biomedical Signal Processing

The proposed method for estimating abruptly changing networks can find broad applications in the field of biomedical signal processing. For example, over the course of a cellular process (e.g., a cell cycle or an immune response), the functionalities of molecules and the relation of the molecules to each other may vary across different stages. Detecting the change points and inferring the functional networks between the molecules can offer us a deeper insight in the underlying biological mechanisms. In a similar fashion, modeling the evolution of protein networks during the progression of a disease may help us to better understand the mechanisms underlying that disease and may shed light upon the treatment of the disease. More generally, the proposed statistical method may serve as a powerful tool for modeling biological processes that undergo distinct changes in behavior, describing them in terms of networks that change at specific points in time.

10.8 Conclusion, Open Problems, and Future Directions

An effective method is proposed to infer abruptly changing functional networks from piecewise-stationary brain recordings. A low-complexity PELT-based algorithm is introduced to detect change points. Next graphical models (with and without hidden variables) are inferred for each stationary time segment. Numerical results for synthetic and real data demonstrate the utility of the proposed model. As an illustration, we have shown that the proposed model can be an effective tool to determine "when" and "how" the functional brain networks change during epileptic seizures, which may lead to novel insights in the phenomenology of seizures.

Although the proposed method is reliable and tuning-free, the techniques for regularization selection typically require running the algorithm multiple times, which is obviously computationally demanding. One appealing approach for solving this problem is to exploit the Bayesian framework in which the regularization parameters are random variables and their posterior distributions can be inferred from the observed data along with other parameters [35,36]. It is of great interest to further speed up the algorithm by following such Bayesian approach.

Another problem with the present algorithm is that the computational complexity of inferring graphical models is $O(P^3)$, making it prohibitive to high-dimensional problems such as modeling fMRI data. Considerable efforts have been undertaken to scale the algorithm up to 1 million variables by means of quadratic approximation and parallelization [37]. As an alternative, we plan to reduce the computational complexity of the algorithm directly, probably with the help of stochastic gradient methods [38].

Acknowledgment

This project is supported by the Singapore Ministry of Education (Tier 2 project ARC5/14).

References

1. H. Yu, C. Li, and J. Dauwels, "Network inference and change point detection for piecewise-stationary time series," *Proceedings of ICASSP 2014*, pp. 4498–4502, 2014.
2. M. A. Kramer, U. T. Eden, E. D. Kolaczyk, R. Zepeda, E. N. Eskandar, and S. S. Cash, "Coalescence and fragmentation of cortical networks during focal seizures," *Journal of Neuroscience*, vol. 30, no. 30, pp. 10076–10085, 2010.
3. J. Dauwels, F. Vialatte, A. Cichocki, "Diagnosis of Alzheimer's disease from EEG signals: Where are we standing?" *Current Alzheimer Research*, vol. 7, no. 6, pp. 487–505, 2010.
4. A. T. Ihler, S. Krishner, M. Ghil, A. W. Robertson, and P. Smyth, "Graphical models for statistical inference and data assimilation," *Physica D*, vol. 230, pp. 72–87, 2007.
5. H.-A. Loeliger, J. Dauwels, J. Hu, S. Korl, P. Li, and F. Kschischang, "The factor graph approach to model-based signal processing," *Proceedings of the IEEE*, vol. 95, no. 6, pp. 1295–1322, 2007.
6. H. Liu, J. Lafferty, and L. Wasserman, "The non-paranormal: Semiparametric estimation of high dimensional undirected graphs," *Journal of Machine Learning Research*, vol. 10, pp. 2295–2328, 2010.
7. H. Yu, J. Dauwels, and X. Wang, "Copula Gaussian graphical models with hidden variables," *Proceedings of ICASSP 2012*, pp. 2177–2180, 2012.

8. J. Dauwels, H. Yu, X. Wang, F. Vialatte, C. Latchoumane, J. Jeong, and A. Cichocki, "Inferring brain networks through graphical models with hidden variables," *Machine Learning and Interpretation in Neuroimaging, Lecture Notes in Computer Science*, Berlin Heidelberg: Springer, pp. 194–201, 2012.
9. H. Yu, J. Dauwels, X. Zhang, S. Y. Xu, and W. I. T. Uy, "Copula Gaussian multiscale graphical models with application to geophysical modeling," *Proceedings of 15th International Conference on Information Fusion*, pp. 1741–1748, 2012.
10. H. Yu, Z. Choo, W. I. T. Uy, J. Dauwels, and P. Jonathan, "Modeling extreme events in spatial domain by copula graphical models," *Proceedings of 15th International Conference on Information Fusion*, pp. 1761–1768, 2012.
11. J. Dauwels, H. Yu, S. Y. Xu, and X. Wang, "Copula Gaussian graphical model for discrete data," *Proceedings of ICASSP 2013*, pp. 2177–2180, 2013.
12. X. Xuan, and K. Murphy, "Modeling changing dependency structure in multivariate time series," *Proceedings of the 24th ICML*, 2007.
13. I. Cribben, R. Haraldsdottir, L. Y. Atlas, T. D. Wager, and M. A. Lindquist, "Dynamic connectivity regression: Determining state-related changes in brain connectivity," *NeuroImage*, vol. 61, pp. 907–920, 2012.
14. M. Lavielle, and G. Teyssière, "Adaptive detection of multiple change-points in asset price volatility," in: G. Teyssière and A. Kirman (Eds), *Long-Memory in Economics*, Berlin Heidelberg: Springer, pp. 129–156, 2005.
15. D. Angelosante, and G. B. Giannakis, "Sparse graphical modeling of piecewise-stationary time series," *Proceedings of ICASSP 2011*, pp. 1960–1963, 2011.
16. R. Killick, P. Fearnhead, I. A. Eckley, "Optimal detection of changepoints with a linear computational cost," *Journal of the American Statistical Association*, vol. 107, no. 500, pp. 1590–1598, 2012.
17. J. M. Hammersley and P. E. Clifford, "Markov fields on finite graphs and lattices," unpublished manuscript, 1971.
18. O. Banerjee, L. E. Ghaoui, and A. d'Aspremont, "Model selection through sparse maximum likelihood estimation," *Journal of Machine Learning Research*, vol. 9, pp. 485–516, 2008.
19. J. Friedman, T. Hastie, and R. Tibshirani, "Sparse inverse covariance estimation with the graphical lasso," *Biostatistics*, vol. 9, pp. 432–441, 2008.
20. P. Ravikumar, M. J. Wainwright, G. Raskutti, and B. Yu, "Model selection in Gaussian graphical models: High-dimensional consistency of ℓ_1-regularized MLE," In *Advances in Neural Information Processing Systems 22*, pp. 1329–1336, 2009.
21. P. K. Trivedi, and D. M. Zimmer, *Copula Modeling: An Introduction for Practitioners*. Hanover, MA, USA: Now Publishers Inc, 2007.
22. A. Sklar, "Fonctions de répartition à n dimensions et leurs marges," *Publications de l'Institut de Statistique de L'Université de Paris 8*, pp. 229–231, 1959.
23. A. Dobra and A. Lenkoski, "Copula Gaussian graphical models and their application to modeling functional disability data," *Annals of Applied Statistics*, vol. 5, No. 2A, pp. 969–993, 2011.
24. R. A. Horn and C. R. Johnson, *Matrix Analysis*, Cambrige, UK: Cambridge University Press, 1990.
25. V. Chandrasekaran, P. A. Parrilo, and A. S. Willsky, "Latent variable graphical model selection via convex optimization," *The Annals of Statistics*, vol. 40, no. 4, pp. 1935–1967, 2012.
26. C. Wang, D. Sun, and K. C. Toh, "Solving log-determinant optimization problems by a Newton-CG primal point algorithm," *Society for Industrial and Applied Mathematics*, vol. 20, pp. 2994–3013, 2009.
27. M. Lavielle and G. Teyssière, "Detection of multiple change-points in multivariate time series," *Lithuanian Mathematical Journal*, vol. 46, pp. 287–306, 2006.
28. B. Jackson, J. D. Scargle, D. Barnes, S. Arabhi, A. Alt, P. Gioumousis, E. Gwin, P. Sangtrakulcharoen, L. Tan, and T. T. Tsai, "An algorithm for optimal partitioning of data on an interval," *IEEE Signal Processing Letters*, vol. 12, no. 2, pp. 105–108, 2005.
29. H. Liu, K. Roeder, and L. Wasserman, "Stability approach to regularization selection (StARS) for high dimensional graphical models," *Advances in Neural Information Processing Systems*, pp. 1432–1440, 2010.
30. N. Meinshausen and P. Bühlmann, "Stability selection," *Journal of the Royal Statistical Society*, vol. 72, Series B, pp. 417–473, 2010.
31. S. Li, L. Hsu, J. Peng, and P. Wang, "Bootstrap inference for network construction with an application to a breast cancer microarray study," *Annals of Applied Statistics*, vol. 7, no. 1, pp. 391–417, 2013.
32. P.-F. Xu, J. Guo, and M.-L. Tang, "An improved HaraTakamura procedure by sharing computations on junction tree in Gaussian graphical models," *Statistics and Computing*, vol. 22, no. 5, pp. 1125–1133, 2012.
33. A. L. Goldberger, L. A. N. Amaral, L. Glass, J. M. Hausdorff, P. C. Ivanov, R. G. Mark, J. E. Mietus, G. B. Moody, C. K. Peng, H. E. Stanley, "PhysioBank, PhysioToolkit, and PhysioNet: Components of a new research resource for complex physiologic signals," *Circulation*, vol. 101, pp. 215–220, 2000.
34. A. Shoeb and J. Guttag, "Application of machine learning to epileptic seizure onset detection," in *Proceedings of 27th International Conference on Machine Learning (ICML)*, 2010.
35. H. Yu and J. Dauwels, "Variational inference for graphical models of multivariate piecewise-stationary time series," in *Proceedings of 18th International Conference on Information Fusion*, pp. 808–813, 2015.
36. H. Yu and J. Dauwels, "Variational Bayes learning of graphical models with hidden variables," in *Proceedings of MLSP*, 2015.
37. C.-J. Hsieh, M. A. Sustik, I. S. Dhillon, P. Ravikumar, and R. A. Poldrack, "BIG & QUIC: Sparse inverse covariance estimation for a million variables," *Advances in Neural Information Processing Systems*, 2013.
38. H. Robbins and S. Monro, "A stochastic approximation method," *The Annals of Mathematical Statistics*, vol. 22, no. 3, pp. 400–407, 1951.

11
Topological Data Analysis of Biomedical Big Data

Angkoon Phinyomark, Esther Ibáñez-Marcelo, and Giovanni Petri

CONTENTS

11.1 Introduction 209
11.2 Time-Series TDA Processing Pipeline 210
    11.2.1 Biomedical Time-Series Data 211
    11.2.2 Data Segmentation 212
    11.2.3 Input Data for TDA Techniques 213
11.3 Persistent Homology 214
    11.3.1 Simplicial Complexes 214
        11.3.1.1 Čech and Rips–Vietoris Complexes 215
        11.3.1.2 Clique Complex 216
    11.3.2 Topological Invariants 216
        11.3.2.1 An Example with Interpretation of Topological Invariants 216
        11.3.2.2 Mathematical Formalism 218
    11.3.3 Feature Extraction–Based Topological Summaries 219
    11.3.4 Properties of Persistent Homology for Big Data Analysis 219
        11.3.4.1 Stability of Persistent Homology 219
        11.3.4.2 Bottleneck Matching 221
        11.3.4.3 Auction Algorithm for Wasserstein Distance 221
        11.3.4.4 Complexity of Simplicial Complex Constructions and Persistent Homology 222
11.4 Topological Machine Learning 223
    11.4.1 Traditional Machine Learning Algorithms 223
    11.4.2 Topological Kernel and Kernel-Based Learning Methods 224
11.5 Topological Simplification 225
    11.5.1 Cluster Analysis–Based Topological Network 225
    11.5.2 Feature Selection-Based Topological Network 227
11.6 Applications in Biomedical Signal Processing 227
    11.6.1 EEG Signal 227
    11.6.2 EMG Signal 228
    11.6.3 Human Motion Capture Data 229
11.7 Open Problems and Future Directions 229
References 230

11.1 Introduction

To capture and describe the variability and complexity of biosignals and images acquired from human systems for biomedical applications, a massive amount of information is necessary. The collection of big volumes of biosignal and image data is therefore the first, crucial step in modern science. Thanks to recent developments in low-cost commercial products of wireless and wearable biosignal devices (e.g., EMOTIV* for recording brain activity signal and Myo[†] for recording muscle activity signal) as well as public big biosignal and image

---
* https://www.emotiv.com
[†] https://www.myo.com

resources (e.g., TUH-EEG database [1], which comprises approximately 22,000 electroencephalography (EEG) records from 15,000 patients, or the HCP database [2] which consists of 76 terabytes of behavioral and magnetic resonance imaging data from 1200 healthy subjects), we are being ushered into the era of Big Data. To translate this huge amount of information into a better understanding of the basic biomedical mechanisms and to further biomedical applications, analytic tools, and techniques to analyze Big Data are needed. The name "Big Data" itself only contains a term related to the volume of data while there are other important features of Big Data such as variability of data sources, veracity of the data quality, and velocity of processing the data [3]. These hallmarks of Big Data need to be characterized by special analytic tools and techniques as well. This additional signal processing and classification stage is very important to turn any collected large data set into meaningful biomedical applications.

Recently a number of methods rooted in algebraic topology have been successfully adopted as novel tools for data analysis in biological and neurological contexts (e.g., [4–6]). The fundamentally new character of these tools, collectively referred to as *Topological Data Analysis (TDA)*, stems from abandoning the standard measures between data points as the fundamental building blocks, and focusing on extracting and understanding the "shape" of data at the mesoscopic scale. This set of tools allows for the extraction of relevant insights from complex data with high dimensionality, high variability, time dependence, low signal-to-noise ratio, as well as missing values, which are prominent challenges in the analysis of biomedical Big Data. Moreover, these approaches come with guarantees on their robustness to perturbations in data and have been recently shown to scale well with data set size. For example, one TDA technique, Mapper [7], provides a separation of the global clustering problem into a set of many smaller problems using local clusterings, reducing the necessity of relying on poor estimates of whole metric spaces and making the algorithm immediately amenable to parallelization. Another TDA technique, persistent homology, is able to provide a multiscale summary of a whole data set, but it is computationally cumbersome if computed naively. However, recent algorithmic advances have significantly reduced its complexity, and parallel algorithms have become available (such as a spectral sequence algorithm [8] and a chunk algorithm [9]). As a result, TDA can be now used to approach very large, high-dimensional data sets. In addition, the robustness of TDA outputs originates from its capacity to capture the global topological structure of point cloud data rather than their local geometric behavior, thereby making the summaries robust to missing data and small errors. For example, the persistent homology of a point cloud is strongly robust to subsampling and coarse-graining of the points, a fact that is used in practice to both reduce the complexity of the computational problem and infer with confidence a system's underlying topology also in cases where there is no natural common metric associated with the data set's features.

These few examples of TDA properties clearly show the potential of different TDA techniques to be used as Big Data analytic tools. In the rest of the chapter, we delve more into the properties of TDA that make it a promising paradigm to handle Big Data.

This chapter begins with a comprehensive introduction to a time-series processing pipeline involving TDA techniques, followed by a discussion of how these techniques are suitable for biomedical signal processing in Big Data. Three major TDA techniques are presented in this chapter, which can be considered as potential Big Data analytics: (1) persistent homology, (2) topological machine learning, and (3) topological simplification, as exemplified by the Mapper algorithm. Next, an overview of existing methods in the application of biomedical signal processing is presented. Finally, concluding remarks and future research directions are given.

11.2 Time-Series TDA Processing Pipeline

A pipeline (or workflow) is a set of interconnected data processing steps, where the output of one step is the input of the following step. There are a number of general TDA pipelines that have been applied to different types of data across many fields. However, biomedical time-series data differ in many ways from input data for general TDA and thus require a newly developed time-series TDA processing pipeline [10]. The overall workflow of TDA for the analysis of time-series data involving biomedical data can be divided into several steps, as shown in Fig. 11.1. The first step is to acquire an initial raw time-series data and then partition the time series into a sequence of discrete segments. The next step is to create the input data for TDA, for example, by constructing point clouds via embedded signals using a time-delay embedding technique or by calculating a similarity matrix via correlation or distance measures from each of the given input temporal segments. The final step is to analyze the transformed data using techniques from topology (i.e., persistent homology, topological machine learning, and topological simplification). Each TDA technique also includes several pipeline steps such as constructing a simplicial complex and computing its topological invariants. A detailed description of each is provided in the following sections.

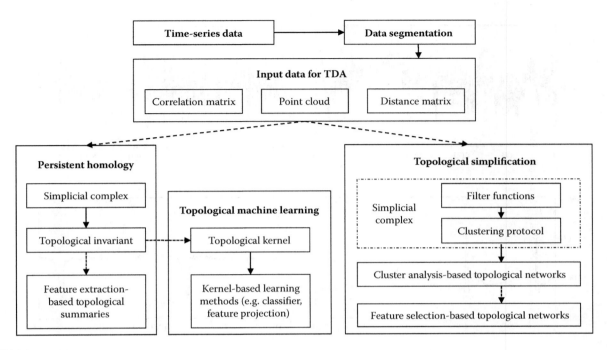

FIGURE 11.1
An overview of the time-series TDA processing pipeline.

11.2.1 Biomedical Time-Series Data

The human body consists of many systems, such as the nervous system, the cardiovascular system, and the musculoskeletal system. Each system also consists of several subsystems that carry on many physiological processes, which are accompanied by or manifest themselves as *signals*. Such signals can be electrical, physical, or biochemical, and can be measured using several types of sensors over time. For example, electrophysiological techniques are quantitative measurements of the electrical activity of biological systems. Recording of spontaneous electrical activity of the brain is called electroencephalography (EEG) (Fig. 11.2a) [1], while electromyography (EMG) [11] is the recording of electrical activity produced by skeletal muscles (Fig. 11.2b). Other common electrophysiological recordings include electrocardiography (ECG) (Fig. 11.2c) [12] and electro-oculography (EOG) (Fig. 11.2d) [13]. Medical imaging time series from a medical scanner is another important source of biomedical data; for instance, it is possible to explore functional connectivity in the human brain by measuring the level of coactivation of brain regions via resting-state functional magnetic resonance imaging (fMRI) time series [14]. We can also analyze microscopic images used in medical diagnostics in the form of one-dimensional (1-D) sequences [15]. These signals are useful in the diagnosis and treatment of patients due to the fact that physiological processes associated with diseases (or pathological processes) typically generate signals that are different from the corresponding signals originating from normal physiological processes. Typically, these signals are recorded continuously as a function of time and refered to as *a time series*, i.e., a sequence of data points consisting of successive measurements made over a time interval. In general, the data points or samples refer to the rows in the matrix, while dimensions or variables indicate the columns. After a matrix of raw time-series data is prepared, a number of data preprocessing steps (or *a preprocessing pipeline*) needs to be applied to reduce the influence of artifacts and noise, which could compromise the interpretation. This processing step is used to prepare initial input data for further processing such as to study both healthy biological functions and mechanisms of disease [16] or to classify signal patterns for controlling assistive and rehabilitation devices [17]. It is important to note that each type of biomedical signal requires a specific preprocessing pipeline, e.g., EEG [18] and fMRI [19], and it has been widely shown that different data preprocessing steps can change the classification performance and patient-control group differences for traditional statistics and machine learning techniques [20]. Therefore, a better understanding of the effects of different preprocessing choices on the classification and detection performance of TDA is warranted. Potential applications of TDA to a number of biomedical signals (i.e., EEG signals, EMG signals, and human motion capture data) and a discussion of their properties for Big Data analysis are presented in Section 11.6.

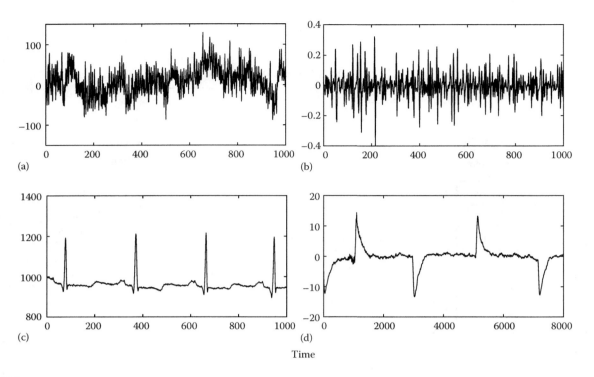

FIGURE 11.2
Examples of biomedical signals: (a) EEG signal, (b) EMG signal, (c) ECG signal, (d) EOG signal.

11.2.2 Data Segmentation

This processing step comprises various techniques to handle preprocessed time-series data before applying TDA and is referred to as data segmentation, i.e., a time-series analysis method used to divide an initial input time series into a series of temporal segments of finite length. This step is necessary due to the fact that biomedical signals obtained as time series in a time–amplitude domain are nonstationary, or exhibit "nonstationarity." The term *stationarity* means that statistical properties of the signal do not vary in time. If the properties of the signal source change over time, then the process is nonstationary. Since electrophysiological signals (like EEG or EMG signals) are recorded from a human body part and a human being performs different activities, the brain or muscles cannot be considered stationary in time. Conversely, many data analysis methods are not designed to reliably quantify a nonstationary signal. Thus, partitioning a longer time signal into sufficiently shorter time segments is necessary to determine the properties of a signal as it changes over time, and those signals could be considered as a weaker form of stationarity [21]. As a result, we can apply general TDA to each segment and can improve accuracy and response time for specific biomedical applications of signal and image processing.

However, biomedical signals could claim some form of stationarity under some specific conditions. For instance, when EEG recordings are obtained for a short period of time while subjects are instructed to stay at rest with their eyes closed, the brain as a process could be assumed to be stationary, and then those signals could be considered weak- or wide-sense stationary. EMG recordings could be considered stationary as well during a short-term isometric muscle contraction [21]. Unfortunately, limiting any application to utilize signals acquired under a specific condition would limit the benefits of biomedical applications. Therefore, dividing a longer time signal into shorter segments of appropriate length is still necessary to involve different states of the signal in the analysis and increase the utility of the system.

Moreover, many biomedical applications require real-time computing. For myoelectric control systems, as an example, a segment length plus the processing time should be equal to or less than 300 ms (0.3 s) due to real-time constraints [22]. Although a small segment length is required for the stationarity of time-series data and real-time computing of the system, a trade-off in response time and accuracy exists, and thus, the segment length should be carefully selected. In general, the bias and variance of features increase as the segment length decreases and consequently degrade the classification performance [23]. Also, a sufficient number of data points is required to obtain topological features (e.g., persistent diagrams) with reasonable accuracy. Since the results of TDA could change due to the segment length,

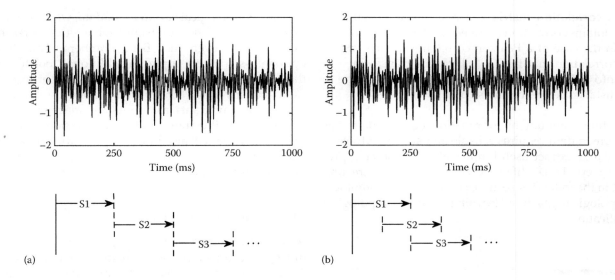

FIGURE 11.3
Data windowing techniques: (a) adjacent windowing and (b) overlapping windowing.

the optimal segment length needs to be investigated for each specific data type [10]. For instance, the optimal size of segmentation is 4 s for detecting epileptic EEG patterns using persistent homology [24], while the time segment of 0.3 s was used to develop brain–computer interfaces (BCIs) using topological simplification [25].

Another important point in data segmentation is *data windowing*. There are two major techniques: adjacent windowing and overlapping windowing. In adjacent windowing, as shown in Fig. 11.3a, adjacent disjoint segments with a predefined length are used. On the other hand, in overlapping windowing the new segment slides over the current segment with an increment time less than the segment length (Fig. 11.3b). In addition, several biomedical signals could be partitioned into segments that lie in between important events such as ECG signals using QRS peaks [12] and joint kinematic data during walking and running using gait events [26,27]. For long-term monitoring, we can apply onset detection techniques to identify resting state and activity state of several biomedical signals involving EEG and EMG signals [28].

11.2.3 Input Data for TDA Techniques

To extract the global shape of high-dimensional preprocessed data obtained from the previous step, suitable low-dimensional data that can represent the global structure is needed. There are many types of data where global features are present. The most usual types of input data for TDA techniques include the following:

- A **point cloud** is a set of data points $\{x^s\}_{s \in S}$ in a given coordinate system, usually $\mathbb{R}^n$. Although this notion originates from a set of points in the usual three-dimensional (3-D) system, a point cloud is a more general notion. Each data point can be thought of as a measurement of several variables where each variable would correspond to a different coordinate. There are several ways to compute point cloud data when we cannot obtain this type of data directly from the high-dimensional preprocessed data. The most frequently used technique for biomedical signals (e.g., [10,29–31]) is the so-called *time-delay embedding technique* [32,33]. This technique transforms a given time-series data into a point cloud in a (typically) lower-dimensional space in such a way that the periodicity of the signal corresponds to the appearance of a cycle in the point cloud. Inspired by this technique, for example, Perea et al. [34,35] proposed a method called SW1PerS (Sliding Windows and 1-Dimensional Persistence Scoring) to discover periodicity in gene expression time-series data.

- A **distance matrix** is constructed from a given distance function by specifying all pairwise distances between points. *Euclidean distance* and its variance normalized version are the most common use of distance (e.g., [5,36,37,38]). The original version is suitable for data that is not directly comparable, while the standardized version is able to give better performance when data contains heterogeneous scale variables. Other distance measures, which can be applied to biomedical time-series data, include Manhattan distance, Minkowski distance, and cosine similarity.

- A **correlation matrix** is a pairwise matrix that contains correlations between each pair of points. In the case of a Pearson correlation matrix, the correlation matrix $C = \{c_{ij}\}$, can be transformed into a distance matrix $D = \{d_{ij}\}$ by defining the distance between points i and j as $d_{ij} = 1 - c_{ij}$.

This huge amount of information contained in this transformed data requires reliable and sufficient tools to analyze and extract useful information in the next processing step. Three different TDA techniques are presented in the following sections: (1) persistent homology, (2) topological machine learning, and (3) topological simplification.

11.3 Persistent Homology

Homology is a topological invariant that classifies a topological space depending on some characteristics of its shape. It is called an invariant because when two topological spaces can be transformed into each other via homeomorphisms (continuous maps with continuous inverse), the homology of the two spaces is the same. Roughly speaking, homology counts the number of holes in different dimensions, that is, how many connected components, cycles, voids, and so on there are in a given topological space. On the other hand, *persistent homology* studies the multiscale shape of data by computing a graded variant of homology across a range of scales.

The pipeline to compute persistent homology is composed of three steps. The first step is transforming a point cloud to a sequence of simplicial complexes using a family of Čech or Vietoris–Rips complexes. This skeletal structure is dependent on some parameter, i.e., the radius r of the neighborhoods of points. By varying r, we obtain a filtration built from the point cloud, i.e., a collection of nested simplicial complexes. The second step is computing the persistent homology of the filtration, yielding a topological summary of data, which can come in a number of ways, for example, barcodes, persistence diagrams, or even just Betti numbers. The third—optional—step is to apply an additional tool to study these topological invariants for a specific purpose such as to classify data. This can be done by leveraging topological invariants as inputs for machine learning techniques, as illustrated in the next section. Finally, in this section, we will also briefly discuss the properties of TDA tools and indicators, e.g., stability and computational complexity, that are relevant for Big Data analysis.

11.3.1 Simplicial Complexes

In order to compute homology from a point cloud, one must produce a simplicial approximation to the data set. That is, one needs to convert the original data set into a specific type of topological space, a *simplicial complex*, which allows homology to be readily computed. A simplicial complex, X, is a topological space constructed from the union of simple elements, the n-simplices. A simplex $\sigma = \{p_0, \ldots, p_{k-1}\}$ of dimension k is defined by its vertices, and geometrically, it can be thought of as the convex hull of the vertices defining it. If we imagine them embedded in space, these simplices can be visualized as points (we call this set X_0), lines (we call this set X_1), triangles (set X_2), and their n-dimensional counterparts (set X_n) (see example in Fig. 11.4). A simplicial complex K is then a set of simplices respecting a few additional conditions: each face δ' of a simplex $\sigma \in K$, i.e., each subset of the vertices forming σ ($\delta \subset \sigma$) belongs to K; and the intersection of any two simplices $\sigma, \sigma' \in K$ is another simplex $\delta = \sigma \cap \sigma'$ and belongs to K. In this chapter, we focus on two main types of input from which we are able to construct simplicial complexes: (1) data points in a metrical space, which is the most common situation when dealing with biomedical data, and (2) networks, which have been found to very useful in cases where the data does not come with a well-defined (or reliable) metrical structure.

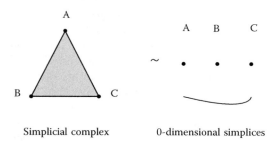
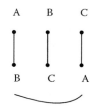
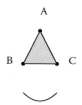

FIGURE 11.4
An example of simplicial complex and its pieces. A full triangle is composed of three 0-simplices (points), three 1-simplices (segments) and a 2-simplex (a full triangle).

11.3.1.1 Čech and Rips–Vietoris Complexes

We focus first on how to construct a simplicial complex, or a sequence of them, starting from a set of data points equipped with metrical information. The most common constructions are the *Čech* and the *Rips–Vietoris simplicial complexes* [39,40]. Both leverage metrical information to construct a simplicial complex. In both cases, simplices are defined in terms of overlapping neighborhoods of the data points. However, for the Čech complex, given a radius $r \in \mathbb{R}^+$, one considers the neighborhood of radius r of each point in the data set; then, whenever two neighborhoods overlap, one adds the 1-simplex (a line) formed by the corresponding two points; when three neighborhoods overlap, one adds the corresponding 2-simplex (a full triangle); and so on for higher-dimension simplices. In short, in principle, one needs to check all the possible combinations of k points for every k up to the number of points in the data set. The problem naturally is that for large point clouds, the Čech complex becomes computationally intractable. The Rips–Vietoris simplicial complex addresses this problem by reducing the computational cost and is the most used in practice. In fact, for Rips–Vietoris complexes, simplices are constructed in the same way as for the Čech complex, but instead of adding a $(k-1)$-simplex when the intersection of the k neighborhoods is nonempty, we add it to all the neighborhoods that have nonempty pairwise intersections. This procedure is still computationally cumbersome, but it already reduces the complexity mightily with respect to the Čech complex.

We encounter now a new problem: if we believe that the topology of the point data set can give us information about the systems the data comes from, we need this topological information to be well defined; however, it is immediately evident that different values of the radius r of the neighborhoods of the data points can produce simplicial complexes with radically different topologies (e.g., consider the Rips–Vietoris complexes built from $r = 0$ and from $r = d$, where d is the diameter of the point cloud). Indeed, in most applications, there is no *a priori* way to pick the value of r. One way to solve this apparent conundrum is to consider properties that persist across changes in the choice of r. Persistent homology takes its name from exactly this idea. Indeed, usually one does not consider a single (Čech) Rips–Vietoris complex K, but rather, a family of (Čech) Rips–Vietoris complexes $\{K_r\}$ parametrized by the radius r. These families have the property that for two radii r,r', such that $r' < r$, the corresponding complexes are such that $K_{r'} \cap K_r$, and the whole family constitutes a *filtration*, the key element in computing persistent homology. In other words, persistent homology is obtained by computing homology along each r in the filtration.

Figs. 11.5 and 11.6 illustrate the construction of a Čech simplicial complex from a set of data points and a Čech filtration, respectively. In Fig. 11.5, from a given data point and a radius r, we create as many disks as there are points, of radius r, each disk centered in the point. When two disks overlap, we add a line, 1-simplex (dark gray line), but when three disks overlap each with both neighbors, then a 2-simplex is created (full triangle). In this case, two 2-simplices are generated (medium gray lines), besides six 1-simplices (dark gray lines) plus each segment that forms triangles, and finally, six 0-simplices represented by the points. In Fig. 11.6, we show an example of filtration via Čech complexes. From step 1 to

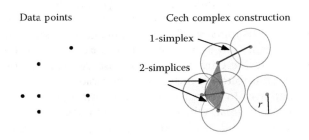

FIGURE 11.5
From data points to a Čech simplicial complex.

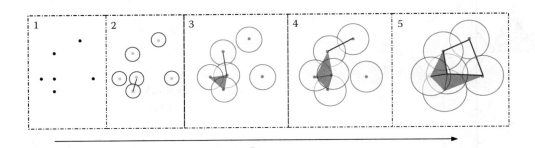

FIGURE 11.6
Filtration via Čech complex.

step 5, the disk radius r grows, and it creates more overlaps among neighbors. At step 1, we only have six 0-simplices because the neighborhood radius is very small, but, as r increases, higher-dimensional simplices begin to appear.

11.3.1.2 Clique Complex

A second way to get a simplicial complex is from networks. In this case, networks are transformed to complexes via clique complexes, e.g., real-world complex networks, weighted networks obtained from repeated spreading or cascading processes, or structural and functional brain networks. A *clique complex* is fully defined by its underlying 1-skeleton, i.e., the set of its 1-simplices. The 1-simplices can be thought of as edges of a network. The higher-dimension simplices are defined by the cliques of the 1-skeleton. A clique is a subset of vertices such that they induce a complete subgraph. That is, every two distinct vertices in the clique are adjacent. So to obtain a clique complex from a network, we map k-cliques to $(k-1)$-simplices. Interestingly, although all the information is encoded in the pairwise interactions, converting a graph to a simplicial complex reveals a mesoscopic organizational structure that is not appreciable at the network level, thanks to the nonlocality of the topological invariants of the simplicial complex, similar to what happens with Rips–Vietoris complexes in metrical spaces. In this case, the filtration used for the persistent homology often consists in the family of clique complexes obtained from the progressive thresholding of a weighted network. We illustrate this for a simple network in Fig. 11.7: 1-cliques correspond to nodes, 2-cliques correspond to edges, and three-cliques to triangles (2-simplices); in this way, it is easy to see how we obtain a clique complex by promoting the network's clique to its corresponding simplex.

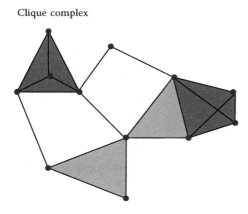

FIGURE 11.7
Clique complex from a network.

11.3.2 Topological Invariants

Coming back to the selection of radius r in the construction of a simplicial complex, *persistent homology* [41,42] solves that problem using each r to obtain different simplicial complexes, and studying homology and its related features (intervals) throughout them. That is, persistent homology, by design, is focused on the features that live across intervals of values and assigns an importance to them proportional to the length of such intervals, with the obvious implication that features surviving across many scales are more meaningful than those that live only for short intervals, which are usually considered to represent topological noise. Via persistent homology, we are able to compute the homology of a filtration. This gives us some topological invariants that later will be the features used to classify data sets through machine learning techniques.

Betti numbers are the corresponding dimensions of the homology. In other words, the kth Betti number, b_k, refers to the number of k-dimensional holes on a topological surface. The first few Betti numbers have the following definitions: b_0 is the number of 0-dimensional holes, that is, the number of connected components; b_1 is the number of one-dimensional or circular holes; and, b_2 is the number of two-dimensional voids or cavities.

The form of a parameterized version of a Betti number is its graphical representation called a *barcode* [43]. A *barcode* represents each persistent generator with a horizontal line, as an interval, beginning at the first filtration level i where it appears and ending at the filtration level j where it disappears. That is a collection of horizontal line segments (i,j) in a plane whose horizontal axis corresponds to the parameter and whose vertical axis represents an (arbitrary) ordering of homology generators. Another way to represent the same information is the *persistence diagram*, which is a multiset consisting of the pairs of birth and death indices for all generators. It can also be readily visualized in the plane by adding a point for each generator with the birth time i as its x-coordinate and the death time j as its y-coordinate. Both barcodes and persistence diagrams are well defined for all dimensions. Note that barcodes and persistence diagrams are equivalent and represent the same information, converting interval (i,j) in the barcode to the 2-D point with coordinates (i,j).

11.3.2.1 An Example with Interpretation of Topological Invariants

A complete example of topological invariants is shown in Fig. 11.8. These are the results of computing the persistent homology of the filtration in Fig. 11.6. Starting from the top, it represents the barcode, for 0-homology H_0 (top, black lines), which counts connected

Topological Data Analysis of Biomedical Big Data

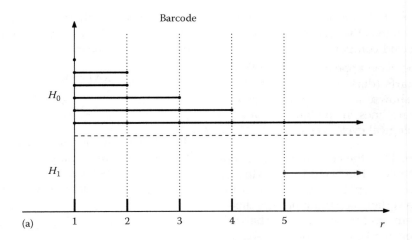

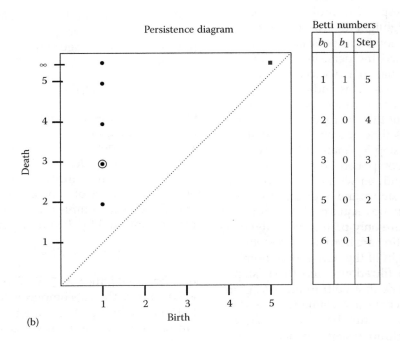

FIGURE 11.8
Topological invariants: Barcode, persistence diagram, and Betti numbers from Čech filtration in Fig. 11.6. Round points: 0-dimensional homology generators (connected components); square point: one-dimensional homology generator (1-D cycle).

components, and 1-homology H_1 (bottom, dark gray line), which counts cycles. In the barcode plot, the x-axis represents each filtration's step given by the growing of the radius disk r:

- In step 1, we have six different connected components, the points: for this reason, we have six different bars that appear at 1.
- From step 1 to step 2, two connected components merge into one. The number of connected components (intervals in H_0) reduces from six to five different connected components. The five separated components are represented in the barcode by the five surviving lines, while the component that merged is represented by the line disappearing at step 2.
- At step 3, two different connected components merge again, reducing the number of surviving lines for H_0. This makes for finishing two intervals, which have started from the beginning and finished at step 3.
- In order to finish the process for H_0, at step 4, another connected component disappears, and at step 5, all the initial points become connected

in only one connected component. The only surviving interval line is the one corresponding to the final connected component.

- Finally, in H_1, one cycle appears at step 5; then a new interval starts (dark gray line) until the cycle dies (not shown in the figure). Before step 5, there are no lines in the barcode of H_1 because no homological cycle was present.

At the bottom of Fig. 11.8, the corresponding persistence diagram is represented. The x-axis is the birth step, and the y-axis is the death step. Squares represent that the object was born at a certain step but it has not died at the last step and so it remains forever. As in the barcode, black shade is for H_0 elements, and dark gray shade is for H_1 elements. Points in the persistence diagram surrounded by a circle correspond to multiple generators. We use this notation to highlight that two generators can have the same representation in the persistence diagram. Let us explain how to read the persistence diagram in more detail. For birth equal to 1, we can see different points along the death axis. This indicates that such points, representing different connected components, were all born at step 1, but they merged into other components at different steps (one at step 2, two at step 3, one at step 4, another at step 5), finally all merging into the complete connected component, represented by the square. On the other hand, as we have shown in the barcode, a cycle is born at step 5, and it remains forever. It is represented by the dark gray square. Note that in persistence diagrams, there are only points above the diagonal, because generators first are born and then they die. Finally, on the bottom right of the figure, Betti numbers are represented for each filtration's step (from step 0, bottom, to step 5, top). The first Betti number b_0 corresponds to the number of connected components and the second Betti number b_1 to the number of cycles. As you can see, b_0 goes down as filtration step goes ahead, and b_1 leaves to be zero at the last step when a cycle is born.

11.3.2.2 Mathematical Formalism

Some technical notions are necessary in the context of persistent homology: chain complex, boundary map, and the formal definition of homology [39,40].

The set of n-dimensional chains $C_n(X)$ of a simplicial complex X is the formal sums of n-simplices, formally written as

$$C_n(X) = \{r_1\sigma_1 + r_2\sigma_2 + \ldots | r_i \in \mathbb{Z}, \sigma_i \in X_n\}.$$

Then we can define a map between n-dimensional chains $C_n(X)$ and n−1-dimensional chains $C_{n-1}(X)$. This map is going to correspond to our intuitive notion of boundary. For example, if we have a simplex of dimension 2 (a full triangle), it will be converted to its boundary, that is, three concatenated edges (simplices of dimension 1). This is called *boundary map* ∂_n and is defined as

$$\partial_n : C_n(X) \to C_{n-1}(X)$$

$$[v_1, \ldots, v_n] \mapsto \sum_{i=0}^{n}(-1)^i[v_0, \ldots, \hat{v}_i, \ldots, v_n]$$

where the hat denotes the omission of the vertex. That is, the boundary of a simplex is the alternating sum of restrictions to its faces. This map satisfies $\partial_n \partial_{n+1} = 0 \; \forall n$.

A simplicial complex X induces the *chain complex*,

$$\cdots \xrightarrow{\partial_{n+2}} C_{n+1} \xrightarrow{\partial_{n+1}} C_n \xrightarrow{\partial_n} C_{n-1} \xrightarrow{\partial_{n-1}} \cdots,$$

often written as $(C_\bullet, \partial_\bullet)$.

The *n-homology* of this complex is defined by the quotient of two vector spaces, the kernel of the map ∂_n quotiented by the image of the boundary map one upper dimension, ∂_{n+1},

$$H_n(X) = \ker \partial_n / \operatorname{im} \partial_{n+1},$$

where n indicates the dimension of the generators in the homology group. We call the kernel ker ∂n the nth cycle module, denoted Z_n, and the image im ∂_n the nth boundary module, denoted B_n.

Formally, given a simplicial complex X, a filtration is a totally ordered set of subcomplexes $X_i \subset X$ that starts with the empty complex and ends with the complete complex, indexed by the nonnegative integers,

$$\emptyset = X_0 \subseteq X_1 \subseteq \ldots \subseteq X_m = X,$$

such that if $i \leq j$, then $X_i \subseteq X_j$. The total ordering itself is called a filter. The subcomplexes X_i are the analog of the sublevel sets in the Morse function setting [44].

In order to define *persistent homology*, we use superscripts to denote the index in a filtration. The ith simplicial complex X_i in a filtration gives rise to its own chain complex $(C_\bullet^i, \partial_\bullet^i)$, and the kth chain, cycle, boundary, and homology modules are denoted by C_k^i, Z_k^i, B_k^i, and H_k^i, respectively.

For a positive integer p, *the p-persistent kth homology module of X_i* is

$$H_k^{i,p} = Z_k^i / \left(B_k^{i+p} \cap Z_k^i\right).$$

The form of $H_k^{i,p}$ should seem similar to the formula for H_k^i, except that instead of characterizing the k-cycles in X_i that do not come from a (k + 1)-chain in X_i, it characterizes the k-cycles in the X_i subcomplex that are not the boundary of any (k + 1)-chain from the larger complex X_{i+p}. Put another way, $H_k^{i,p}$ characterizes the k-dimensional holes in X_{i+p} created by the subcomplex X_i. These holes exist for all complexes X_j in the filtration with index

$i \leq j \leq i + p$. Equivalently, it is possible to define persistent homology using the fact that the inclusion $X_i \hookrightarrow X_j$ induces a homomorphism $f_p^{i,j}: H_p(X_i) \to H_p(X_j)$ on the simplicial homology groups for each dimension p. The pth persistent homology groups are the images of these homomorphisms, and the pth persistent Betti numbers $\beta_p^{i,j}$ are the ranks of those groups [8,42].

11.3.3 Feature Extraction–Based Topological Summaries

It is possible to derive other topological features (or summaries) from Betti numbers (or barcodes or persistence diagrams) that can be easier to combine with tools from statistics and machine learning. Three topological summaries are briefly presented in this chapter involving persistence landscapes, vineyards, and persistent entropy.

- **Persistence landscapes** have been introduced by Bubenik [45] as a new summary of a persistence diagram that is transformed through certain functions $\lambda: \mathbb{N} \times \mathbb{R} \to \bar{\mathbb{R}}$, where $\bar{\mathbb{R}}$ is the extended real numbers, $[-\infty, \infty]$. As a statistical descriptor, a persistence landscape has the advantage that it is a function, so we can use the vector space structure of its underlying function space as a domain for statistical approaches. Given a barcode $B = \{[b_i, d_i]: 1 \leq i \leq n\}$ and performing the following transformation, we obtain a persistence landscape:

  $\lambda_k(t) = k$th largest value of min $(t - b_i, d_i - t)_+$,

  where + denotes max(•,0). The properties of persistence landscapes are discussed in detail in Section 11.4.2, while an implementation of this feature is discussed in detail in Ref. [46].

- **Vineyards** are time-varying persistence diagrams [47,48]. They consist in following point trajectories along a time-varying family of persistence diagrams. Take a dynamic point cloud $X(t) = x_1(t), \ldots, x_n(t)$, that is, a point cloud moving continuously for a finite amount of time. Then, there is the corresponding set of persistence diagrams $D(X(t))$ for each time t. This 1-parameter family of persistence diagrams is called the *vineyard*. Each off-diagonal point p in $D(X(t))$ moves in time, tracing out a curve, is a *vine*. An application of this feature is shown in Ref. [49]. Analyzing dynamic brain connectivity of resting and gaming states, it is possible to determine temporally dynamic properties of the brain in a threshold-free and robust manner.

- **Persistent entropy**: Given a persistence barcode $B = \{[b_i, d_i]: 1 \leq i \leq n\}$, let $L = \{l_i = d_i - b_i : 1 \leq i \leq n\}$. The persistent entropy H_L [50,51] is

$$H_L = -\sum_{i=1}^{n} \frac{l_i}{S_L} \log \frac{l_i}{S_L},$$

where $S_L = \sum_{i \in I} l_i$. Note that the maximum persistent entropy would correspond to the situation in which all the intervals in the barcode are of equal length. More concretely, if B has n intervals, the possible values of the persistent entropy H_L associated with the barcode B lie in the interval $[0, \log(n)]$. The idea is that given a filtration of a simplicial complex, persistence entropy helps to find a *proper* filter preserving the partial ordering imposed by the filtration trying to minimize the number of long-life homological classes that are associated with significant intervals in the persistence barcodes. This topological feature has been successful, for example, in discovering hidden patterns among antibodies in the immune system [51,52] as well as in discriminating real long-length noisy signals of DC electrical motors [53]. This feature can also be used as a measure for separating topological noise (k-dimensional holes with short lifetime) from topological features (those with a long lifetime) [54].

In future studies, a comprehensive comparison between the performances of these topological features should be performed using both simulation data and real-world data, while new topological summaries should also be proposed.

11.3.4 Properties of Persistent Homology for Big Data Analysis

11.3.4.1 Stability of Persistent Homology

In previous sections, we saw that persistent homology describes the multiscale properties of a data set by capturing the birth and death times of topological features. These properties are summarized by the persistence diagrams, multisets of points in the $\mathbb{R}^2$. For such diagrams to be relevant and interpretable for data analysis and applications, they must be *robust*: small perturbations of the data should correspond to small alterations in the corresponding persistence diagrams. Fortunately, the stability of persistence diagrams is one of the key features of the study of persistent homology and has been shown in a number of different works.

Fig. 11.9 shows the properties of topological information: first, significant structures, in this case the large circle, are well reconstructed and persist for long intervals along the filtration; secondly, small perturbations, for example, due to noise or measurement errors, do not affect significantly the topological information (i.e., the

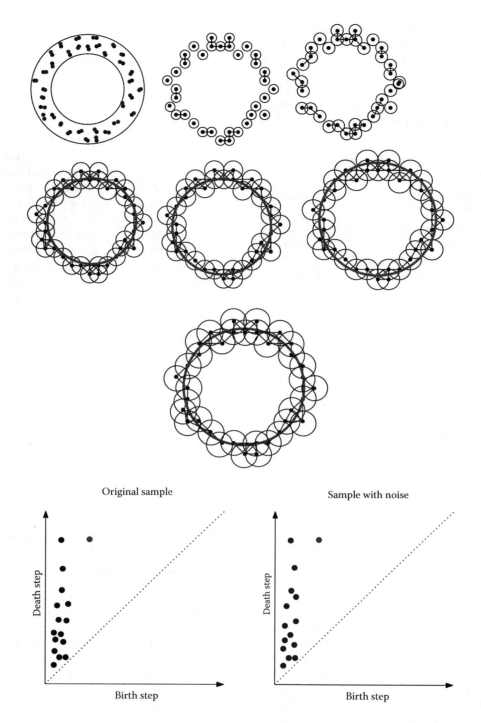

FIGURE 11.9
Stability of persistence diagrams and how a main shape lives across many scales of the filtration.

persistence diagram only shows minor changes). On the top left of the figure, black dots represent the original sample, while black dots show the perturbed sample. Regarding persistence diagrams, the persistence diagram on the left is given by the original sample (black points, 0-homology; dark gray point, 1-homology), while the persistence diagram on the right corresponds to the perturbed sample (black points, 0-homology; dark gray point, 1-homology).

The first proof of the stability of persistence diagrams was given by Cohen-Steiner et al. [55]. It relies on showing that the distance between two topological spaces constrains that between the corresponding persistence diagrams in all dimensions. In particular, they focused on the case in which the diagram encodes the sublevel set homology of a function defined on a topological space. They choose the distance between two functions, f and g, to be the L_∞-norm, and that between the corresponding persistence diagrams, $D(f)$ and $D(g)$, to be the bottleneck distance. Using these distances, the stability theorem bounds the diagrams' distance by the functions as follows:

Theorem 11.3.1

Let X be a topological space with tame functions $f,g:X\to\mathbb{R}$. Then for each dimension p, the bottleneck distance between the persistence diagrams is bounded from above by the difference between the functions,

$$d_B(D(f), D(g)) \leq \|f - g\|_\infty$$

The assumptions of Cohen-Steiner et al. [55] are satisfied, among others, by Morse functions on compact manifolds and piecewise linear functions on simplicial complexes (which include the general weighting schemes adopted in Refs. [4,56]). The bottleneck distance is based on the set of all bijections γ between points in the two diagrams,

$$d_B(D(f), D(g)) = \inf_\gamma \sup_x \|x - \gamma(x)\|_\infty,$$

and is therefore always at least the Hausdorff distance between the two diagrams. Note that the bottleneck distance is a special case of the qth Wasserstein distance

$$W_q(D(f), D(g)) = \left[\inf_{\gamma:D(f)\to D(g)} \sum_{x\in D(f)} \|x - \gamma(x)\|_\infty^q\right]^{1/q}$$

for $q\to\infty$.

The theoretical guarantees on the results of persistent homology do not require the actual calculation of the bottleneck distance value between diagrams. However, for practical applications (e.g., to produce a clustering of a set of diagrams), its calculation becomes necessary, and it is easily shown to be very cumbersome: one must consider all possible mappings between diagram points. Formally, this yields an assignment problem over the two sets of points in $\mathbb{R}^2$ [41]. To address this issue, a number of algorithms have been devised, the two main ones being the *Hopcroft–Karp algorithm for bottleneck matchings* and *Bertsekas's auction algorithm for Wasserstein distances*, which we briefly illustrate in the following section.

11.3.4.2 Bottleneck Matching

Hopcroft and Karp's bottleneck algorithm is a variant of the maximum matching algorithm. Following Ref. [57], consider a weighted graph $G = (E,V)$ (E set of edges, V set of vertices) and the subgraph $G[r]$ containing only the edges with weight $\omega < = r$. In this setting, one can write that the bottleneck distance of G has the minimal r such that $G[r]$ contains a perfect matching for G, which for a general graph has a complexity $O(V^2E)$. Since the bottleneck cost for G must be equal to the weight of one of the edges, we can find it exactly by combining a test for a perfect matching with a binary search on the edge weights. For a given graph $G[r]$, the algorithm computes a maximum matching, i.e., a matching of maximal cardinality. $G[r]$, with $2n$ vertices, has a perfect matching if and only if its maximum matching has n edges. The algorithm starts from an empty matching M. Edges are then added to M via the augmenting paths, i.e., a path composed by alternative edges inside and outside M. The edges in the path are then state-switched, those in M are removed from M, and those outside M are added to M, resulting in an increase of the size of M by 1. This procedure is repeated until there are no more augmenting paths or all the nodes are matched. The algorithm in this form has an overall running time of $O(m\sqrt{n}) = O(n^{2.5})$.

The contribution of Efrat et al. [58] is the realization that it is possible to avoid the explicit construction of $G[r]$ by using a near-neighbor search data structure $D_r(S)$, which stores a point set S and a radius r, and can be constructed in the case of $\mathbb{R}^2$ in $O(n \log n)$ preprocessing time, finally making the Hopcroft–Karp algorithm cost only $O(n^{1.5} \log n)$. Taking advantage of the mapping between the persistence diagram distance and the matching, it is immediate to see that the bottleneck distance between persistence diagrams can be computed in $O(n^{1.5} \log n)$. A similar implementation proposed by Kerber et al. [57] substitutes the D_r structure with a k-d tree, or k-dimensional tree, to explore the local point structure, showing a reduction of the complexity to $O(\log n)$.

11.3.4.3 Auction Algorithm for Wasserstein Distance

The Wasserstein matching has complexity problems similar to those of the bottleneck distance. To improve the performance, an algorithm was proposed in Ref. [57] based on Bertsekas's auction algorithm [59], an asymmetric approach to identifying the maximum weight matching. The algorithm works by considering half of the nodes as bidders and half as objects, and allowing for iterative multiple bidding on the objects, in such a way as to find the maximum auction value across the possible bidder–object matchings. The most cumbersome part of the algorithm is the phase in which bidders select the object of maximum value. The brute-force approach requires a search over all objects by each bidder, resulting in a quadratic running time per iteration of the auction algorithm. However, the same result can be obtained also via local searches on a k-d tree, which is based on the geometry of the actual data set. In this way, significant speedups were observed, with speedup factors between 50 and 400 depending on the size of the problem instance.

11.3.4.4 Complexity of Simplicial Complex Constructions and Persistent Homology

In the previous section, we discussed how expensive it is to compute the distance between the topological summaries of two spaces. However, one of the long-standing problems of simplicial methods is that realistic simplicial complexes easily become very large, making it difficult to even calculate the persistence diagram. It is important to understand how such complexes can be approximated or shrunk to manageable sizes. The most famous one, the Čech complex, is built by checking the intersection of every collection of simplices, making it extremely cumbersome computationally. Indeed, building a Čech complex from N points up to it highest dimension n can cost $O(N^2 + N^{2n})$, making the direct calculation absolutely unfeasible in most cases [60]. The standard alternative is the Rips–Vietoris complex, which can be defined as an abstract simplicial complex defined on the basis of a metric space M with distance d by constructing a simplex for each set of points with diameter at most δ, or equivalently, as the clique complex associated to a metrical graph. The construction is rather straightforward, but the Rips–Vietoris complex has a major drawback: it is very large; for example, the k-skeleton alone (the simplices up to dimension k) grows as $O(n^{k+1})$ for a simplicial complex with n vertices. It it therefore important to be able to produce smaller-sized simplicial complexes that approximate the properties of the original complex. Luckily, a number of controlled approaches have been proposed to address this.

Zomorodian [61] introduced the tidy set: a minimal simplicial set that captures the topology of a simplicial complex. The advantage of this approach is particularly evident in the calculation of the homology of clique complexes and weak witness complexes [62], because it relies on avoiding the explicit construction of the full clique complex, focusing in particular on high-dimensional simplices (since the number of faces grows exponentially with the simplex dimension). The construction produces sets that are significantly smaller than the original clique complex, providing a performance increase not only to the simplex construction but also in the homology computation.

A second approach to the sparsification or the Rips–Vietoris complex and filtration was proposed in Ref. [63] that is able to reduce the complexity of a filtered simplicial complex on an n-point metric space to $O(n)$ by using weighted metric perturbations of the original metric space and adopting a hierarchical representation of the original point cloud via net-trees [63,64]. Moreover, the filtration has a time complexity scaling as $O(n \log n)$ and has tight bounds on the resulting persistent homology, allowing one to significantly scale up the possible data set dimension.

Another classic reduction technique for simplicial complexes based on point sets is the *witness complex* [62]. Its inspiration comes from the observation that standard constructions (Rips, Čech, ...) are very redundant since the same homology (and homotopy) is often realizable on smaller node sets. This can be done by selecting a subset of vertices, called *landmark points*, and then constructing the witness complex on such set in a similar way to the Delaunay complex in Euclidean space. Effectively, the nonlandmark data points become witnesses to the existence of edges or simplices described by combinations of landmark points. This approach is able to reduce strongly the size of the final simplicial complex and comes with strong bounds on the changes in topology between the original complex and the reduced witness complex. The reduced number of simplices often also makes it easier to identify the homology classes because it both reduces the topological noise that can arise in very large complexes and allows one to identify with more accuracy the localization of the homology class.

In addition to the difficulties of obtaining a simplicial complex, or a whole filtration, starting from a point cloud, it is important to turn our attention to what calculating its (persistent) homology entails. The classical persistent homology algorithm relies on direct reduction of the boundary matrix; hence it has the complexity of Gaussian elimination, $O(n)$, where n is the number of simplices. Milosavljevic et al. [65] proposed a persistent homology algorithm that takes matrix multiplication time $O(n\omega)$ with a current best estimation of $\omega = 2.376$, which depends on the algorithm for matrix multiplication. Chen and Kerber [66] contributed a randomized algorithm with a complexity that depends on the number of persistence pairs living longer than a given threshold. They also enhanced the matrix reduction of the standard persistent homology algorithm, obtaining a complexity of $O(n \log n)$ for cubical complexes. More recently, Bauer and collaborators [9,67] proposed and implemented a parallelizable computation of persistent homology that works by dividing the simplicial complex to analyze in chunks and applying two optimization steps before the actual persistent homology calculation, referred to as *clean* and *compress*. It is able to achieve a complexity of $O(ml^3 + gln + g^3)$, where n is the number of simplices, m the number of chunks, l the maximum chunk-size, and g is the number of columns not paired during the compression phase. The resulting complexity bound is still cubic, because g is $O(n)$ in the worst-case scenario, but in practice, the calculations become significantly easier to perform thanks to the partition of the original complex. Despite these worst-case complexity results, recently a number of algorithmic advances have reduced the number of redundant steps in Rips–Vietoris persistent homology [68], improving memory and time efficiency; compressed the data structures used for storing simplices

[69–71]; and provided new matroid-based formulations [72] that make the calculation of persistent homology of large data sets feasible in practice.

11.4 Topological Machine Learning

Machine learning (ML) is the term that describes a set of algorithmic techniques that try to understand data and create models from them, in order to predict the correct classification when presented with new data. Nowadays, standard ML algorithms include an improvement, *kernel trick*, which provides embeddings of data in higher-dimensional spaces in such a way as to be able to split and classify data clouds that could otherwise not be linearly split. However, for kernel tricks to work, the input data space needs to satisfy some requirements, which will be discussed later in this section. On the other hand, persistent homology creates outputs that can be used to differentiate groups and classify data. From this perspective, topological ML can be seen as a bridge between persistent homology outputs and ML inputs. In this section, we briefly introduce different ways to use topological outputs to classify data, and we describe an appropriate kernel to apply ML algorithms using persistence diagrams as inputs.

11.4.1 Traditional Machine Learning Algorithms

Machine learning is a field related to computer science whose goal is the creation of algorithms that are able to learn from data in order to be later used to make predictions on new unknown data, or to discover hidden underlying structures [73]. ML techniques have found applications in a very diverse set of fields, ranging from web page classification, email classification (e.g., spam or not spam), and speller correction, to the screening of large molecule and drug databases and the selection of targets in recommendation systems (e.g., movies in Netflix, books and items in Amazon, etc.). The main subcategories of ML algorithms are as follows:

- *Supervised learning*: Given a set of inputs with the corresponding outputs (labels), the goal is to learn a general rule that maps inputs to outputs. Supervised learning approaches and algorithms include artificial neural network, linear discriminant analysis, decision tree learning and random forests, and so on.
- *Unsupervised learning*: Without giving corresponding outputs to the inputs, but given a set of constraints on the data structure (e.g., number of clusters), the goal is to find the best instance of such structure in the input set. Unsupervised learning approaches and algorithms include hierarchical clustering, *k*-means clustering, independent component analysis, and so on.

Two very well-known algorithms are *Support Vector Machine (SVM)* in the group of supervised learning, and *Principal Component Analysis (PCA)* in the group of unsupervised learning. Both start as linear methods, but they can act as nonlinear methods using a *kernel trick*, that is, by moving to higher-dimensional spaces, where it becomes easier to split data.

SVMs are models associated with supervised learning that analyze data used for classification and regression analysis. Like any supervised learning method, they use a set of training examples to create a model to classify new data. The learned models are usually tested on another part of the data set to measure their performances. The simplest and usual classification is to split data in two categories (classes or labels). SVM performs classification by finding the hyperplane that maximizes the margin between the two classes. The vectors (instances) that define the hyperplane are the support vectors. Then, new samples are mapped onto that same space and predicted to belong to a category based on which side of the gap they fall on. SVM is a nonprobabilistic linear classifier.

PCA, on the other hand, is a statistical procedure that identifies a reduced number of linearly uncorrelated variables, called principal components, from a large set of data in order to reduce the dimensionality, and hence the complexity, of the input data set. The goal of PCA is to explain the maximum amount of variance with the fewest number of principal components. The resulting vectors of PCA transformation are an uncorrelated orthogonal basis set.

A positive kernel function, usually simply called *kernel function K*, is a function of similarity given two points, x and x', in an input space $\mathcal{X}$.

$$K : \mathcal{X} \times \mathcal{X} \to \mathbb{R}$$

Typically, problems in ML need a kernel written in the form of feature map $\phi : \mathcal{X} \to \mathcal{V}$ that satisfies $K(x,x') = \langle \phi(x), \phi(x') \rangle_\mathcal{V}$. However, it is not needed to know explicitly the ϕ map, and this is the *trick*. The key restriction is that the inner product $\langle \cdot, \cdot \rangle_\mathcal{V}$ must be a proper inner product. That is, $\mathcal{V}$ must be a *Hilbert space*. Remember that a Hilbert space is a real or complex inner product space that is also a complete metric space with respect to the distance function induced by the inner product [74]. Thus, a kernel K effectively computes dot products in a higher-dimensional space $\mathcal{V}$ than the original space $\mathcal{X}$.

Therefore, the *kernel trick* approach consists in using kernel functions to embed points in higher spaces and classify them without ever computing the coordinates of the data in $\mathcal{V}$ space, but rather by simply computing the inner products between the images of all pairs of data in the feature space. This approach gives the opportunity to transform linear methods like SVM and PCA to nonlinear ones via a kernel trick. In fact, given N points, they usually cannot be linearly separated in $d < N$ dimensions, but they become almost always linearly separated in $d \geq N$ when using a kernel trick.

Here are some usual used kernels [75]:

$$K(\vec{x}_i, \vec{x}_j) = \begin{cases} \vec{x}_i \cdot \vec{x}_j & \text{Linear} \\ (\gamma \vec{x}_i \cdot \vec{x}_j + C)^d & \text{Polynomial} \\ exp(-\gamma|\vec{x}_i - \vec{x}_j|^2) & \text{Radial Basis Function} \\ & \text{(RBF)} \\ tanh(-\gamma \vec{x}_i \cdot \vec{x}_j + C) & \text{Sigmoid} \end{cases}$$

where $K(\vec{x}_i, \vec{x}_j) = \phi(\vec{x}_i) \cdot \phi(\vec{x}_j)$ without knowing ϕ. That is, the kernel function represents a dot product of input data points mapped into the higher-dimensional feature space by transformation ϕ.

11.4.2 Topological Kernel and Kernel-Based Learning Methods

TDA can capture characteristics of the data that other methods often fail to provide. The study of persistent homology is a popular method for TDA. As illustrated in Section 11.3, it catches information about birth and death times of topological features for different dimensions. Moreover, it is done over multiple scales, and the information is captured by the persistence diagram (see Fig. 11.8). Topological ML tries to create a bridge between TDA, which will provide data inputs, and ML, which is able to classify in groups the input features of TDA. The main problem lies in defining a suitable kernel for persistence diagrams. To apply SVM or PCA, we require a Hilbert space, and the persistence diagrams equipped with the Wasserstein distance only form a metric space, i.e., only a distance is defined but not an inner product.

A potential solution to this problem is given by Reininghaus et al. [76]. They proposed and described a persistence scale-space kernel stable for the 1-Wasserstein distance, with a scale parameter $\sigma > 0$. Note that the stability of a kernel is very important for classification problems [55]. Essentially, if there exists a hyperplane H that divides a set of points into two classes with margin m, and if the data points are perturbed by some $\epsilon < m/2$, then a stable kernel H still correctly splits the data into two classes with margin $m - 2\epsilon$.

The multiscale kernel k_σ proposed in Ref. [76,77] can be evaluated directly given two persistence diagrams, F and G, using the following expression:

$$k_\sigma(F,G) = \frac{1}{8\pi\sigma} \sum_{p \in F, q \in G} e^{-\frac{\|p-q\|}{8\sigma}} - e^{-\frac{\|p-\bar{q}\|}{8\sigma}},$$

where $\bar{q} = (q_2, q_1)$ if $q = (q_1, q_2)$ is the mirrored point at the diagonal. Computationally speaking, the proposed kernel can be computed in $\mathcal{O}(|F| \cdot |G|)$, where $|F|$ and $|G|$ denote the cardinality of the multisets F and G, respectively.

Other proposals that try to sidestep this problem exist, the most relevant for our discussion being the work by Bubenik et al. [45]. They adopt a statistical approach to the feature map for persistent diagrams, converting them through certain functions $\lambda : \mathbb{N} \times \mathbb{R} \to \mathbb{R}$ to a new object, the persistence landscape.

As a descriptor, the persistence landscape has the advantage that it is a function. We can then use the vector space structure of its underlying function space as opposed to the geometry of the space of persistence diagrams, which instead presents a number of statistical and interpretational problems. As an example, sets of persistence diagrams need not have a unique mean. However, we can obtain unique means and generally do statistics in the space of persistence landscapes. This is possible thanks to the fact that this function space is "well-behaved": it is a separable Banach space, and consequently, we can apply statistics there, i.e., we can compute means, create confidence intervals, and make hypotheses tests. For example, we can check whether two distributions are equal, $\mu_X = \mu_Y$, where X and Y are distributions.

Note that an individual persistence landscape has a corresponding barcode and persistence diagram, but the mean persistence landscape does not. It was proved that persistence landscapes are stable with respect the supremum norm and the p-landscape distance defined in Ref. [45]. The advantages of using the multiscale kernel defined in Ref. [76] rather than others derived from the persistence landscapes are shown in Ref. [45] for shape classification data and texture image classification, using persistence diagrams for dimensions 0 and 1.

A different approach to the use of persistence diagrams for classification was proposed by Pachauri et al. [78]. There, persistence diagrams are first rasterized on a regular grid, then a probability density function (PDF) is computed choosing a suitable kernel-density estimate (KDE), and eventually the vectorized discrete PDF is used as a feature vector to train SVM using standard kernels for $\mathbb{R}^n$. However, this approach does not make clear whether stability is affected and how it behaves with respect to the existing metrics (bottleneck or Wasserstein distance) in the resulting kernel-induced distance. This method was used to separate subjects with and without Alzheimer's disease, via an SVM trained with the estimated PDFs of 0-dimensional persistence diagrams obtained from cortical thickness data.

Regarding the use of topological invariants as features for classification, in Li et al. [79], a higher recognition accuracy was obtained when both bag-of-features and persistence diagrams were taken into account together. In short, using structural data properties provided by 0-dimensional homology and its corresponding persistence landscapes as complementary features, which captures the distribution values of a given function, yields significantly better results in object recognition than applying both methods individually.

11.5 Topological Simplification

Topological simplification refers to a set of techniques that are able to extract a topological backbone from an unstructured data cloud. If a metrical or a similarity structure is available, it is possible to produce controlled simplications of the data set by means of a series of local clusterings in overlapping regions of the data space and by successively linking together clusters that share common data points. This is the basis of the Mapper algorithm, the most well-known technique originally introduced by Singh et al. [7].

Although topological simplification keeps the essential features of topology, the pipeline to compute the Mapper algorithm is different from conventional TDA, which is composed of four steps (see Fig. 11.10 for an illustration). The first step is transforming raw data into a point cloud either using *filter functions* as a standalone, or using intrinsic information to the *metric* as the input of filter functions. The second step is defining *a resolution* by segmenting point cloud data in the low-dimensional space obtained from the previous step into intervals of length L overlapped with percentage O. The third step is applying any standard clustering algorithm to create clusters from each sub-data set corresponding to the defined intervals, i.e., *a clustering protocol*. The clusters in the intervals will become the *nodes* of a simplicial complex approximating the data. The fourth step is constructing the Mapper graph by connecting a pair of nodes that share data points, i.e., creating *edges* or *connections* of the simplicial complex. Note that by considering sets of clusters/nodes larger than two, one can define higher-dimension simplices in a natural way, similarly to the Čech complex construction. Any node that contains no shared points remains a singleton. The constructed simplicial complex, often called *topological network*, is then ready for further analysis. It is important to note that the Mapper algorithm is based on topological ideas in the sense that the described construction tends to preserve the notion of nearness in data cloud data but discard the effects of large distances, which often carry little meaning or are scarcely reliable in applications.

The selection of filter functions is the most important step in the Mapper algorithm since the filter functions are used to define a set of overlapping regions that slice the point cloud data and guide the clustering method. There are several types of filter: summary statistics (e.g., mean, max, min, variance, kurtosis, n-moment), geometric filter (e.g., Gaussian kernel density, L1 or L-infinity centrality, curvature, harmonic cycles), and feature projection filter (e.g., the first two components of singular value decomposition or PCA, autoencoders, Isomap, multidimensional scaling, t-Distributed Stochastic Neighbor Embedding, two-dimensional embedding of the k-nearest neighbor graph, SVM distance from hyperplane) [7]. Since each filter function extracts relevant information from the noisy input data in a different manner, we can use more than one filter (or multiple filters) at the same time. It should be noted that filters do not need to be continuous and can map to space other than $\mathbb{R}$. Since the Mapper algorithm is driven by the filter selected, it is less sensitive to the selection of the metric than persistent homology.

In the case of further visualization of topological networks, additional visual properties such as *color* or *size* of the nodes can be added. Commonly, one can define the color to the nodes to represent the average value (for the continuous values) of either any meaningful filter function used for the low-dimensional mapping or any target output. Fig. 11.10 shows the variation of color spectrum ranging from red to blue, where a red node represents higher average values, while a blue node represents lower average values. For the categorical values, color directly represents a value concentration. Similar color patterns (a set of nodes) over the topological network then represent highly correlated variables and can help us to identify which shape contains more information than other shapes. Another approach is to define the size of the nodes to represent the proportion of points belonging to each cluster with respect to the whole data set. Consequently, we can use this information for further exploration. In practice, there are two main purposes for using topological networks: (1) cluster analysis and (2) feature selection (more details in the subsections to follow).

11.5.1 Cluster Analysis–Based Topological Network

The output of this topological simplification approach has been used in a number of studies as cluster analysis to extract nontrivial qualitative information from

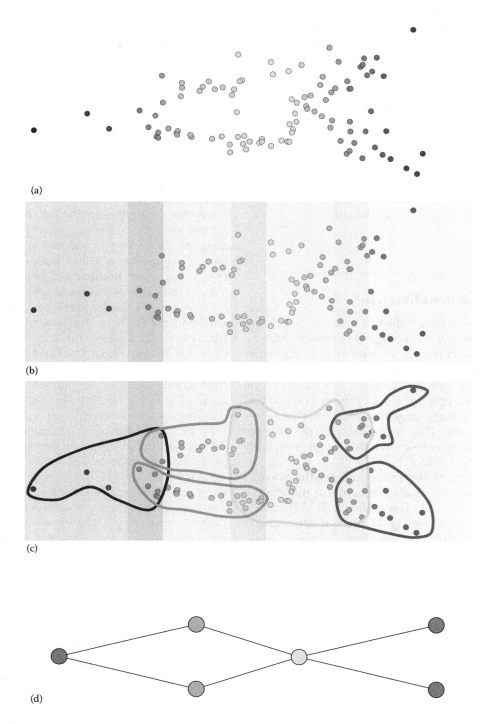

FIGURE 11.10
An overview of the Mapper algorithm pipeline. (a) Point cloud data, (b) Overlapping regions of the data space, (c) Local clustering of points within regions, (d) Mapper graph.

large data sets that was hard to discern when studying the data set globally. An interesting shape of a graph consists of subgroups that display distinct patterns of the data such as "loops" (continuous circular segments) and "flares" (long linear segments), as opposed to a straight line. To avoid a biased comparison, an interesting subgroup should contain a sufficient number of samples (e.g., >30 samples). The fundamental subgroups can also discriminate from artifacts if the shapes of the subgroup persist over a large-scale change by tuning the resolution parameters across a range of scales. For instance, Nicolau et al. [80] detected a previously unknown subtype of breast cancer characterized by a greatly reduced mortality, while Lum et al. [5] and Nielson et al. [38] identified a number of new subgroups in data sets as diverse as genomic data, NBA players, and spinal cord injuries.

11.5.2 Feature Selection-Based Topological Network

We can further use topological networks for the selection of relevant features. The principle of applying the Mapper algorithm to feature selection is to recognize that the shape of the graph encodes essential structural information of the data. Specifically, after the fundamental and interesting topological subgroups were defined in the previous step, we are able to select features that best discriminate them (either between the interesting subgroups [36] or between any interesting subgroup and the rest of the structure [5]) based on statistical tests (typically a Kolmogorov–Smirnov [K-S] test). A set of features with either high K-S scores or low p-values (i.e., the top-ranked features) can then be used as input for any classification or predictive model. For example, Guo and Banerjee [36] selected, topologically, a set of key process variables in a manufacturing pipeline that affected the final product yield. Using this selected feature set as an input to three different predictive models, they obtain the same prediction accuracy as by using all the process variables, hence providing a way to reduce costs in monitoring and control of the manufacturing process. Lum et al. [5] also performed the K-S test to identify a list of genes that best differentiated the interesting subgroup from the rest of the structure.

The aforementioned approaches have shown potential and are therefore good candidates as a way to build simplicial complexes used to compute persistent homology that convey local summaries of the data set's features. However, one of the main limitations of the Mapper algorithm is that it requires a specific choice of scales, in the definitions of the bins, of their overlap, and also in the choice of the underlying filter function. This limitation is typically dealt with by swiping across a range of parameters and ensuring that the result is stable, but there is no formal way to choose the optimal parameter set. This problem is a common one for tools that are based on set theoretic concepts because in most applications, the sets need to be defined, and that entails a choice of parameters. Persistent homology, however, turns this limitation upside down by embedding this scaling problem in its definition. On the other hand, topological simplification is more intuitive than persistent homology and does not require a deeper understanding of mathematics.

11.6 Applications in Biomedical Signal Processing

Machine learning approaches involving feature extraction, dimensionality reduction, and learning algorithms still remain largely unknown for Big Data application in biomedical research. Regardless, TDA would largely benefit big biomedical data analysis and should be investigated in future studies. We can use TDA either instead of or together with existing ML methods. Recently, there is much interest in finding new applications that build upon the strong theoretical foundations of TDA. In this chapter, biomedical applications of three different proposed TDA techniques to three different data types are presented: EEG data, EMG data, and human motion capture data.

11.6.1 EEG Signal

EEG signals are valuable indicators of neural activity widely used in medical settings (e.g., to diagnose epilepsy) and BCIs [81]. EEG signals are most often used to diagnose epilepsy. Researchers in the topology community then have selected EEG signals collected from epileptic brains as a first case study [24,37,82,83]. The purpose of this study case is to determine whether topological features of EEG can identify the preictal state, i.e., the state before the onset of an epileptic seizure, or the ictal state, as well as detect the phase transition between the two states. Merelli et al. [24] analyzed the EEG signals of the PhysioNet database* by representing brain complex systems with multivariate time series and transforming signals into complex networks that are analyzed by TDA and information theory. Specifically, the initial input time-series data were divided into a series of temporal segments, and the Pearson (partial) correlation coefficient matrices were computed together with applying a threshold. The threshold matrices were used as weighted edgelists, and then simplicial complexes were characterized by *persistent homology*, i.e., their clique weight rank persistent homology [56], providing a new entropy measure called the weighted persistent entropy [51]. The transient preictal to ictal can be detected by observing the number of connected components, i.e., the number of connected ones tends to be 1 (all the features are persistent) during a phase transition, while this number is higher before and after the period. In support of the finding of Merelli et al. [24], Piangerelli et al. [37] show that the persistent entropy can be used to discriminate the epileptic state from nonepileptic states (p-value = 1.8346e−36; area under the curve = 0.972).

---

* http://www.physionet.org/physiobank/database/

The topological entropy features also achieved higher accuracy than the sample entropy features. Instead of computing persistent homology from complex networks, a piecewise linear function combined with a lower-star filtration was used. They also found that Vietoris–Rips filtration can help to improve the understanding of which region plays the role of trigger for an epileptic seizure (as an EEG channel selection method). Furthermore, Wang et al. [82] studied the persistent homology of EEG signals, smoothed by a weighted Fourier series estimator, using the persistence landscape. The 2-persistence landscape distance, which is the measure of difference between two persistence landscapes, was used as a feature to identify differences between preseizure and seizure patterns. The results show that the proposed topological features were able to identify the significant site T3 (p-value = 0.0005), where the patient's epileptic seizure originates, among eight channels without any prior information (in an unsupervised learning way). We can use these topological features to understand the underlying transition mechanisms and the ictogenesis, which are necessary in developing an automatic system to predict the onset of epileptic seizures. It is of great practical interest to further investigate the potential of topological features for other medical and research uses of EEG such as to monitor the depth of anesthesia or Alzheimer's disease developments.

For BCI research, Perez-Guevara [25] proposed a framework to explore the features of EEG signals based on personalized unsupervised learning using *topological simplification*. Data were taken from a bigger data set of Blankertz et al. [84] in which subjects were asked to move their hand or foot as well as to be at rest. Data segmentation was applied, and then a 36-dimensional vector consisting of the average power spectrum features for three different frequency bands and 12 different electrode locations in the scalp was created for each segment. Two different dimensionality reduction and clustering methods were employed for the Mapper algorithm: the SIGFRIED methodology with a Gaussian mixture model (GMM) clustering [85] and the Fisher's information measure with a single-linkage clustering [86]. Topological networks obtained from the first method show patterns and structures that can discriminate the rest from nonrest tasks as well as meaningful differences between the nonrest tasks, i.e., moving hand or foot, in an unsupervised way. This indicates the potential of the Mapper algorithm to characterize EEG features to build personalized BCIs. Several necessary points have also been suggested for further development such as to employ a two-dimensional meaningful mapping on the Mapper algorithm instead of the current one-dimensional mapping [25]. Other EEG signals that have been studied are such as EEG data capturing eye state (open or closed) [10].

11.6.2 EMG Signal

Similar to EEG signal, EMG can be used as a diagnostics tool to identify neuromuscular diseases [16] or a control signal for prosthetic and assistive devices, known as a myoelectric control system or muscle–computer interface [22]. Either to interpret EMG signals for translational or personalized medicine or to translate EMG signals into control signals, pattern recognition approaches need to be applied, and the success of these approaches depends almost entirely on the selection and extraction of relevant features [87,88]. Despite the growing number of studies related to EMG feature extraction, and the good performance of EMG features reported in the literature, there still exist considerable challenges before these developments can be translated from the laboratory to clinical practice. One of such challenges is to develop EMG features that can achieve a high degree of reliability and robustness. Although to our knowledge, no prior studies have applied TDA to EMG research, some of the previous limitations related to reliability and robustness could be overcome using TDA techniques. For example, several robust EMG features (such as Willison amplitude, zero crossing, and slope sign change features) use a threshold to reduce the effect of background noise [89,90]. While the threshold value has a strong impact on the feature space [89,90], the optimum threshold value of these features is highly dependent on data and subject (i.e., it is not consistent between two different data sets) and thus does not generalize well [91]. Conversely, since TDA is a coordinate-free approach, the TDA features may provide consistent results between different EMG data sets (i.e., different platforms or different coordinate systems) [5]. Another important issue is the development of EMG features able to analyze the nonlinear and nonstationary properties of EMG [92,93]. Similar to other manifold learning approaches [88,94], TDA can unfold and capture nonlinear structures that are not well described by linear extraction methods [95]. Topological features extracted from the TDA via a technique of persistent homology (e.g., persistence barcode lengths) with an unsupervised ML approach have shown very good performance and results in recognizing different data classes for high-dimensional time series from nonlinear dynamic systems (see, for example, the study of Berward et al. [29]).

Further, pattern recognition approaches for myoelectric control systems have been developed based on the assumption that there exist distinguishable and repeatable EMG patterns originated by muscle activities. It means that data classes (i.e., specific motions) are defined *a priori* to consequently select the features. Due to the fact that signal characteristics of surface EMG are subject-specific and change over time [96], the choice of feature and muscle location that has been optimized for specific motions is likely to be suboptimal [21,91,97]. Since there is no strong *a priori* basis for selecting muscle activities and EMG features, one of the possible approaches is to self-label a muscle activity (i.e, a combined set of motions and muscle locations that share similar feature spaces [98] and/or statistical properties [99]) using an unsupervised ML method (such as cluster analysis). We can also apply TDA as an approach to group EMG features into several types since most of the studied EMG features often contain information that is redundant [98].

11.6.3 Human Motion Capture Data

Seversky et al. [10] evaluated their proposed time-series TDA processing pipeline on several motion-based time-series data sets. For each data set, the initial input time-series data were divided into a series of temporal segments, and then the authors computed delay coordinate embeddings and generated persistence diagrams. Wasserstein and bottleneck distances between all segment pairs and configurations as well as the scale-space persistence kernel were computed to explore the discriminating and learning potential of the topological information. For example, 3-D accelerometer data [100] capturing general daily life activities (such as walking, standing, climbing stairs, and working seated) were recorded and used to develop a user verification system based on an automatic daily life physical activities recognition subsystem and a walking-based authentication subsystem. The results [10] show that topological features of activities like standing and working at computer tend to be uniform (i.e., the distances are close to zero), while the distances for activities like walking are more varied. Therefore, larger variability in the topology of segments within the walking class could be used as a biometric unobtrusive pattern for a user's authentication and verification. This indicates the potential to use topological information obtained from *persistent homology* as biometric measures. It is also possible to couple topological features obtained from persistent homology with other traditional ML approaches to gain deeper insights into big gait data for clinical application. For example, Venkataraman et al. [31] proposed the use of persistence diagrams as feature extraction and a nearest neighbor algorithm with the 1-Wasserstein distance measure as classifier to discriminate five human actions (i.e., dance, jump, run, sit, and walk) using 3-D motion capture sequences of body joints [101]. The proposed topological method achieved 96.48% accuracy, while traditional chaotic invariants only achieved 52.44% accuracy.

We suggest applying *topological simplification* as a cluster analysis method to overcome the limitations of traditional clustering techniques (such as hierarchical clustering and *k*-means clustering) in order to identify the distinct subgroups of subjects with and without running-related injuries using 3-D motion capture data, and then clinicians can use this information to subtype patients and provide more appropriate and targeted interventions [102]. Further, an open-access data set containing multimodal measures of human tasks involved in cooking and food preparation (e.g., 3-D motion capture and accelerometer data) [103] was used to compare the classification performance of the chaotic invariant feature and the scale-space topological kernel with kernel SVM classifier, i.e., *topological ML*. On average, the topological kernel method achieved higher classification accuracy than the chaotic invariant method [10]. Using the same topological kernel-based method on another benchmark multimodal data set [104], topological information extracted from motion-based time-series data (which are physically and trajectory based) also achieved relatively good classification accuracy as compared to other data types (such as nonphysical based).

In summary, these aforementioned initial findings clearly show the potential of different TDA techniques in biomedical applications of various types of multivariate time-series data. The TDA techniques have also been applied to other types of biomedical signals and images as well different biomedical applications. For example, Kyeong et al. [105] applied the Mapper algorithm to analyze resting-state functional network data in assessing attention-deficit/hyperactivity disorder symptom severity, while Emrani et al. [30] applied persistent homology to analyze breathing sound signals for wheeze detection.

11.7 Open Problems and Future Directions

In this chapter, we have reviewed and discussed a recent advanced analytics tool for *Big Data* called *Topological Data Analysis*. Although this set of tools allows for the extraction of relevant insights from complex data with high dimensionality, high variability, time dependence, low signal-to-noise ratio, as well as missing values, which are prominent challenges in the analysis of biomedical Big Data, there are still open problems that have not been addressed yet, particularly the analysis of Big Data in practice. The future challenges for this set of tools will mainly come from the area of personalized data streams requiring localized topological information. As biosensors become progressively cheaper and more widespread, real-time streaming data of biomechanical and biological signals will present a great opportunity for personalized diagnosis and monitoring. Streaming data, however, present unique challenges for topological tools, as they effectively require us to be able to track the evolution of complex shapes in high-dimensional spaces and time. The first steps in this direction—at both the algorithmic and conceptual levels—have been already taken (see the work of Ulrich Bauer on streaming graphs, for example [68], and that of Cohen on temporal homology [47]), and a principled approach to these data types appears to be possible. A second research direction that will present interesting problems is the localization of homology. Indeed, using topological information to identify specific features within the data requires the possibility to accurately localize such information within the simplicial representations. However, this problem is known to be NP-hard in general, and the approximation schemes introduced so far will likely require significant computational advancements before becoming directly applicable to large data sets. Other future challenges include improving the scalability of the persistent homology algorithms in such a way as be able to adapt to larger and larger data sets; building efficient algorithms together with user-friendly algorithms for multipersistence [106,107], i.e., persistent homology with multiple filtration parameters, which, for example, would allow looking at temporal data sets in a natural way; and clarifying the role of topological structures in the evolution of dynamical systems taking place on them, for example, via the study of combinatorial Laplacians [108].

In conclusion, there is much interest in finding new applications that build upon the strong theoretical foundations of TDA involving applications in *biomedical signal processing*. Three different TDA techniques, persistent homology, topological ML, and topological simplification, proposed in this chapter, can be applied as potential tools not only in biomedical signal processing but also in ML and data mining (e.g., feature extraction, feature and channel selection, clustering, and so on). As examples, this chapter has presented the potential of different TDA techniques in the analysis of only three different types of biomedical signals. However, it is important to emphasize that this set of techniques can also be applied to various types of biomedical signals and images. Since each type of biomedical signal and image has different characteristics and properties, it will be of great interest to comprehensively investigate and compare the performances of these TDA techniques for each data type and application.

References

1. I. Obeid and J. Picone. The Temple University Hospital EEG Data Corpus. *Frontiers in Neuroscience*, 10:196, 2016.
2. D. C. Van Essen, S. M. Smith, D. M. Barch, T. E. J. Behrens, E. Yacoub, and K. Ugurbil. The WU-Minn Human Connectome Project: An overview. *NeuroImage*, 80:62–79, 2013.
3. Y. Demchenko, Z. Zhao, P. Grosso, A. Wibisono, and C. de Laat. Addressing big data challenges for scientific data infrastructure. In *4th IEEE International Conference on Cloud Computing Technology and Science*, pages 614–617, Dec. 2012.
4. G. Petri, P. Expert, F. Turkheimer, R. Carhart-Harris, D. Nutt, P. J. Hellyer, and F. Vaccarino. Homological scaffolds of brain functional networks. *Journal of the Royal Society Interface*, 11(101):20140873, 2014.
5. P. Y. Lum, G. Singh, A. Lehman, T. Ishkanov, M. Vejdemo-Johansson, M. Alagappan, J. Carlsson, and G. Carlsson. Extracting insights from the shape of complex data using topology. *Scientific Reports*, 3, 2013.
6. J. M. Chan, G. Carlsson, and R. Rabadan. Topology of viral evolution. *Proceedings of the National Academy of Sciences*, 110(46):18566–18571, 2013.
7. G. Singh, F. Mémoli, and G. E. Carlsson. Topological methods for the analysis of high dimensional data sets and 3D object recognition. In *SPBG*, pages 91–100, 2007.
8. H. Edelsbrunner and J. Harer. *Computational Topology: An Introduction*. American Mathematical Soc., Providence, Rhode Island, USA, 2010.
9. U. Bauer, M. Kerber, and J. Reininghaus. Clear and compress: Computing persistent homology in chunks. In *Topological Methods in Data Analysis and Visualization III*, pages 103–117. Springer, Cham, Switzerland, 2014.
10. L. M. Seversky, S. Davis, and M. Berger. On time-series topological data analysis: New data and opportunities. In *29th IEEE Conference on Computer Vision and Pattern Recognition*, page 5967, June 2016.
11. M. B. I. Reaz, M. S. Hussain, and F. Mohd-Yasin. Techniques of EMG signal analysis: Detection, processing, classification and applications. *Biological Procedures Online*, 8(1):11–35, 2006.
12. M. Elgendi, B. Eskofier, S. Dokos, and D. Abbott. Revisiting QRS detection methodologies for portable, wearable, battery-operated, and wireless ECG systems. *PLoS ONE*, 9(1):1–18, 01 2014.
13. P. Phukpattaranont, S. Aungsakul, A. Phinyomark, and C. Limsakul. Efficient feature for classification of eye movements using electrooculography signals. *Thermal Science*, 20(suppl. 2):563–572, 2016.
14. M. P. van den Heuvel and H. E. Hulshoff Pol. Exploring the brain network: A review on resting-state fMRI functional connectivity. *European Neuropsychopharmacology*, 20(8):519–534, 2010.
15. A. Phinyomark, S. Jitaree, P. Phukpattaranont, and P. Boonyapiphat. Texture analysis of breast cancer cells in microscopic images using critical exponent analysis method. *Procedia Engineering*, 32:232–238, 2012.
16. A. Subasi. Classification of EMG signals using PSO optimized SVM for diagnosis of neuromuscular disorders. *Computers in Biology and Medicine*, 43(5):576–586, 2013.
17. A. Phinyomark, P. Phukpattaranont, and C. Limsakul. A review of control methods for electric power wheelchairs based on electromyography signals with special emphasis on pattern recognition. *IETE Technical Review*, 28(4):316–326, 2011.
18. N. Bigdely-Shamlo, T. Mullen, C. Kothe, K. M. Su, and K. A. Robbins. The PREP pipeline: Standardized preprocessing for large-scale EEG analysis. *Frontiers in Neuroinformatics*, 9:16, 2015.

19. M. F. Glasser, S. N. Sotiropoulos, J. A. Wilson, T. S. Coalson, B. Fischl, J. L. Andersson, J. Xu, S. Jbabdi, M. Webster, J. R. Polimeni, D. C. Van Essen, and M. Jenkinson. The minimal preprocessing pipelines for the Human Connectome Project. *NeuroImage*, 80:105–124, 2013.
20. Victor M. Vergara, Andrew R. Mayer, Eswar Damaraju, Kent Hutchison, and Vince D. Calhoun. The effect of preprocessing pipelines in subject classification and detection of abnormal resting state functional network connectivity using group ICA. *NeuroImage*, 145:365–376, 2017.
21. S. Thongpanja, A. Phinyomark, F. Quaine, Y. Laurillau, B. Wongkittisuksa, C. Limsakul, and P. Phukpattaranont. Effects of window size and contraction types on the stationarity of biceps brachii muscle EMG signals. In *Proceedings of the 7th International Convention on Rehabilitation Engineering and Assistive Technology*, i-CREATe '13, pages 44:1–44:4, Kaki Bukit TechPark II, Singapore, 2013. Singapore Therapeutic, Assistive & Rehabilitative Technologies (START) Centre.
22. M. A. Oskoei and H. Hu. Myoelectric control systems—A survey. *Biomedical Signal Processing and Control*, 2(4):275–294, 2007.
23. M. A. Oskoei and H. Hu. Support vector machine–based classification scheme for myoelectric control applied to upper limb. *IEEE Transactions on Biomedical Engineering*, 55(8):1956–1965, Aug. 2008.
24. E. Merelli, M. Piangerelli, M. Rucco, and D. Toller. A topological approach for multivariate time series characterization: The epileptic brain. In *Proceedings of the 9th EAI International Conference on Bio-inspired Information and Communications Technologies*, BICT'15, pages 201–204, ICST, Brussels, Belgium, Belgium, 2016. ICST (Institute for Computer Sciences, Social-Informatics and Telecommunications Engineering).
25. M. Perez-Guevara. Exploratory framework on EEG signals for the development of BCIs, 2010.
26. A. Phinyomark, S. T. Osis, B. A. Hettinga, D. Kobsar, and R. Ferber. Gender differences in gait kinematics for patients with knee osteoarthritis. *BMC Musculoskeletal Disorders*, 17(1):157, https://www.ncbi.nlm.nih.gov/pmc/articles/PMC4830067/. 2016.
27. A. Phinyomark, S. Osis, B. A. Hettinga, R. Leigh, and R. Ferber. Gender differences in gait kinematics in runners with iliotibial band syndrome. *Scandinavian Journal of Medicine & Science in Sports*, 25(6):744–753, 2015.
28. G. Staude, C. Flachenecker, M. Daumer, and W. Wolf. Onset detection in surface electromyographic signals: A systematic comparison of methods. *EURASIP Journal on Advances in Signal Processing*, 2001(2):867853, 2001.
29. J. Berwald, M. Gidea, and M. Vejdemo-Johansson. Automatic recognition and tagging of topologically different regimes in dynamical systems. *ArXiv e-prints*, December 2013.
30. S. Emrani, T. Gentimis, and H. Krim. Persistent homology of delay embeddings and its application to wheeze detection. *IEEE Signal Processing Letters*, 21(4):459–463, Apr. 2014.
31. V. Venkataraman, K. N. Ramamurthy, and P. Turaga. Persistent homology of attractors for action recognition. In *IEEE International Conference on Image Processing*, pages 4150–4154, Sept. 2016.
32. F. Takens. *Detecting Strange Attractors in Turbulence*, pages 366–381. Springer, Berlin, 1981.
33. N. H. Packard, J. P. Crutchfield, J. D. Farmer, and R. S. Shaw. Geometry from a time series. *Physical Review Letters*, 45:712–716, Sept. 1980.
34. J. A. Perea and J. Harer. Sliding windows and persistence: An application of topological methods to signal analysis. *Foundations of Computational Mathematics*, 15(3):799–838, 2015.
35. J. A. Perea, A. Deckard, S. B. Haase, and J. Harer. SW1PerS: Sliding windows and 1-persistence scoring; discovering periodicity in gene expression time series data. *BMC Bioinformatics*, 16(1):257, 2015.
36. W. Guo and A. G. Banerjee. Toward automated prediction of manufacturing productivity based on feature selection using topological data analysis. In *Proceedings of IEEE International Symposium on Assembly and Manufacturing*, 2016.
37. M. Piangerelli, M. Rucco, and E. Merelli. Topological classifier for detecting the emergence of epileptic seizures. *ArXiv e-prints*, November 2016.
38. J. L. Nielson, J. Paquette, A. W. Liu, C. F. Guandique, C. A. Tovar, T. Inoue, K. A. Irvine, J. C. Gensel, J. Kloke, T. C. Petrossian, P. Y. Lum, G. E. Carlsson, G. T. Manley, W. Young, M. S. Beattie, J. C. Bresnahan, and A. R. Ferguson. Topological data analysis for discovery in preclinical spinal cord injury and traumatic brain injury. *Nature Communications*, 6:8581, 2015.
39. J. R. Munkres. *Elements of Algebraic Topology*, volume 2. Addison-Wesley, Menlo Park, 1984.
40. A. Hatcher. *Algebraic Topology*. Cambridge University Press, Cambridge, UK, 2002.
41. H. Edelsbrunner and J. Harer. Persistent homology—A survey. *Contemporary Mathematics*, 453:257–282, 2008.
42. A. Zomorodian and G. Carlsson. Computing persistent homology. *Discrete & Computational Geometry*, 33(2):249–274, 2005.
43. R. Ghrist. Barcodes: The persistent topology of data. *Bulletin of the American Mathematical Society*, 45(1):61–75, 2008.
44. J. Milnor. *Morse Theory. (AM-51)*, volume 51. Princeton University Press, Princeton, New Jersey, USA, 2016.
45. P. Bubenik. Statistical topological data analysis using persistence landscapes. *Journal of Machine Learning Research*, 16(1):77–102, 2015.
46. P. Bubenik and P. Dotko. A persistence landscapes toolbox for topological statistics. *Journal of Symbolic Computation*, 78:91–114, 2017.
47. D. Cohen-Steiner, H. Edelsbrunner, and D. Morozov. Vines and vineyards by updating persistence in linear time. In *Proceedings of the Twenty-second Annual Symposium on Computational Geometry*, pages 119–126. ACM, 2006.
48. D. Morozov. *Homological Illusions of Persistence and Stability*. ProQuest, Duke University, Durham, North Carolina, USA, 2008.

49. Y.M. Ahn J. Yoo, E.Y. Kim and Y. Jong Chul. Topological persistence vineyard for dynamic functional brain connectivity during resting and gaming stages. *Journal of Neuroscience Methods*, 267:1–13, 2016.
50. H. Chintakunta, T. Gentimis, R. Gonzalez-Diaz, M. J. Jimenez, and H. Krim. An entropy-based persistence barcode. *Pattern Recognition*, 48(2):391–401, 2015.
51. E. Merelli, M. Rucco, P. Sloot, and L. Tesei. Topological characterization of complex systems: Using persistent entropy. *Entropy*, 17(10):6872–6892, 2015.
52. M. Rucco, F. Castiglione, E. Merelli, and M. Pettini. *Characterisation of the Idiotypic Immune Network Through Persistent Entropy*, pages 117–128. Springer International Publishing, Cham, 2016.
53. M. Rucco, R. Gonzalez-Diaz, M. J. Jimenez, N. Atienza, C. Cristalli, E. Concettoni, A. Ferrante, and E. Merelli. A new topological entropy-based approach for measuring similarities among piecewise linear functions. *Signal Processing*, 134:130–138, 2017.
54. N. Atienza, R. Gonzalez-Diaz, and M. Rucco. *Separating Topological Noise from Features Using Persistent Entropy*, pages 3–12. Springer International Publishing, Cham, 2016.
55. D. Cohen-Steiner, H. Edelsbrunner, and J. Harer. Stability of persistence diagrams. *Discrete & Computational Geometry*, 37(1):103–120, 2007.
56. G. Petri, M. Scolamiero, I. Donato, and F. Vaccarino. Topological strata of weighted complex networks. *PloS ONE*, 8(6):e66506, 2013.
57. M. Kerber, D. Morozov, and A. Nigmetov. Geometry helps to compare persistence diagrams. *Journal of Experimental Algorithmics*, 22(1.4):1.4:1–1.4:20, 2017. ACM, New York, NY, USA, Sep 2017. ISSN 1084–6654, URL http://doi.acm.org/10.1145/3064175, doi: 10.1145 /3064175. Assignment problems, bipartite matching, k-d tree, persistent homology.
58. A. Efrat, A. Itai, and M. J. Katz. Geometry helps in bottleneck matching and related problems. *Algorithmica*, 31(1):1–28, 2001.
59. D. P. Bertsekas. The auction algorithm: A distributed relaxation method for the assignment problem. *Annals of Operations Research*, 14(1):105–123, 1988.
60. N.K. Le, P. Martins, Decreusefond L., and A. Vergne. Construction of the generalized czech complex. In *Vehicular Technology Conference (VTC Spring), 2015 IEEE 81st*, pages 1–5. IEEE, 2015.
61. A. Zomorodian. The tidy set: A minimal simplicial set for computing homology of clique complexes. In *Proceedings of the twenty-sixth annual symposium on Computational geometry*, pages 257–266. ACM, 2010.
62. V. de Silva and G. Carlsson. Topological estimation using witness complexes. In *Proceedings of the First Eurographics Conference on Point-Based Graphics*, SPBG'04, pages 157–166, Aire-la-Ville, Switzerland, Switzerland, 2004. Eurographics Association.
63. D. R. Sheehy. Linear-Size Approximations to the Vietoris–Rips Filtration. *ArXiv e-prints*, March 2012.
64. T. K. Dey, F. Fan, and Y. Wang. Computing topological persistence for simplicial maps. In *Proceedings of the Thirtieth Annual Symposium on Computational Geometry*, SOCG'14, pages 345:345–345:354. ACM, New York, NY, USA, 2014.
65. N. Milosavljević, D. Morozov, and P. Skraba. *Zigzag Persistent Homology in Matrix Multiplication Time*. ACM, New York, NY, USA, Jun. 2011.
66. C. Chen and M. Kerber. An output-sensitive algorithm for persistent homology. *Computational Geometry*, 46 (4):435–447, 2013.
67. U. Bauer, M. Kerber, and J. Reininghaus. *Distributed Computation of Persistent Homology*, Society for USA, Proceedings of the Meeting on Algorithm Engineering & Experiments, Society for Industrial and Applied Mathematics, pp. 8, 31–38, acmid 2790178, URL http:// dl.acm.org/citation.cfm?id=2790174.2790178, Portland, OR, Philadelphia, PA, USA, 2014.
68. U. Bauer. Ripser, 2016.
69. J. D. Boissonnat, K. C. S., and S. Tavenas. Building efficient and compact data structures for simplicial complexes. *Algorithmica*, Sept. 2016.
70. J. D. Boissonnat and D. Mazauric. On the complexity of the representation of simplicial complexes by trees. *Theoretical Computer Science*, 617:17, Feb. 2016.
71. J. D. Boissonnat, T. K. Dey, and C. Maria. The compressed annotation matrix: An efficient data structure for computing persistent cohomology. *Algorithmica*, 1–13, Apr. 2015.
72. G. Hensenlman. Eirene, 2016.
73. T. M. Mitchell. *Machine Learning*, vol. 45, page 37. McGraw Hill, Burr Ridge, IL, 1997.
74. W. Rudin. *Real and Complex Analysis*. Tata McGraw-Hill Education, New York City, New York, USA, 1987.
75. B. Scholkopf and A. J. Smola. *Learning with Kernels: Support Vector Machines, Regularization, Optimization, and Beyond*. MIT Press, Cambridge, Massachusetts, USA, 2001.
76. J. Reininghaus, S. Huber, U. Bauer, and R. Kwitt. A stable multi-scale kernel for topological machine learning. In *2015 IEEE Conference on Computer Vision and Pattern Recognition (CVPR)*, pages 4741–4748, Jun. 2015.
77. R. Kwitt, S. Huber, M. Niethammer, W. Lin, and U. Bauer. Statistical topological data analysis—A kernel perspective. In *Advances in Neural Information Processing Systems*, pages 3070–3078, 2015.
78. D. Pachauri, C. Hinrichs, M. K. Chung, S. C. Johnson, and V. Singh. Topology-based kernels with application to inference problems in Alzheimer's disease. *IEEE Transactions on Medical Imaging*, 30(10):1760–1770, Oct. 2011.
79. C. Li, M. Ovsjanikov, and F. Chazal. Persistence-based structural recognition. In *2014 IEEE Conference on Computer Vision and Pattern Recognition*, pages 2003–2010, Jun. 2014.
80. M. Nicolau, A. J. Levine, and G. Carlsson. Topology based data analysis identifies a subgroup of breast cancers with a unique mutational profile and excellent survival. *Proceedings of the National Academy of Sciences*, 108(17):7265–7270, 2011.
81. F. Lotte, M. Congedo, A. Lcuyer, F. Lamarche, and B. Arnaldi. A review of classification algorithms for EEG-based brain–computer interfaces. *Journal of Neural Engineering*, 4(2):R1–R13, 2007.

82. Y. Wang, H. Ombao, and M. K. Chung. Persistence landscape of functional signal and its application to epileptic electroencaphalogram data. Paper submitted for ENAR Distinguished Student Paper Awards. Available online at http://pages.stat.wisc.edu/.~mchung/papers/wang.2014.ENAR.pdf, 2014.

83. Y. Wang, H. Ombao, and M. K. Chung. Topological data analysis of single-trial electroencephalographic signals, *The Annals of Applied Statistics*, 2016.

84. B. Blankertz, G. Dornhege, M. Krauledat, K. R. Mller, and G. Curio. The non-invasive berlin brain–computer interface: Fast acquisition of effective performance in untrained subjects. *NeuroImage*, 37(2):539–550, 2007.

85. G. Schalk, P. Brunner, L.A. Gerhardt, H. Bischof, and J.R. Wolpaw. Brain–computer interfaces (BCIs): Detection instead of classification. *Journal of Neuroscience Methods*, 167(1):51–62, 2008.

86. D. Lowe, C. J. James, and R. Germuska. *Tracking complexity characteristics of the wake brain state*, Papadourakis, G.M. (ed.) In *Proceedings of the Fourth International Conference on Neural Networks and Expert Systems in Medicine and Healthcare*. Technological Educational Institute of Crete, pages 318–323, 2001.

87. A. Phinyomark, F. Quaine, S. Charbonnier, C. Serviere, F. Tarpin-Bernard, and Y. Laurillau. EMG feature evaluation for improving myoelectric pattern recognition robustness. *Expert Systems with Applications*, 40(12):4832–4840, 2013.

88. A. Phinyomark, P. Phukpattaranont, and C. Limsakul. Investigation long-term effects of feature extraction methods for continuous EMG pattern classification. *Fluctuation and Noise Letters*, 11(4):1250028, 2012.

89. A. Phinyomark, C. Limsakul, and P. Phukpattaranont. EMG feature extraction for tolerance of white Gaussian noise. In *International Workshop and Symposium Science Technology*, pages 178–183, 2008.

90. A. Phinyomark, C. Limsakul, and P. Phukpattaranont. EMG feature extraction for tolerance of 50 Hz interference. In *4th PSU-UNS International Conference on Engineering Technologies*, pages 289–293, 2009.

91. E. N. Kamavuako, E. J. Scheme, and K. B. Englehart. Determination of optimum threshold values for EMG time domain features; a multi-dataset investigation. *Journal of Neural Engineering*, 13(4):046011, 2016.

92. A. Phinyomark, P. Phukpattaranont, and C. Limsakul. Fractal analysis features for weak and single-channel upper-limb EMG signals. *Expert Systems with Applications*, 39(12):11156–11163, 2012.

93. A. Phinyomark, F. Quaine, S. Charbonnier, C. Serviere, F. Tarpin-Bernard, and Y. Laurillau. Feature extraction of the first difference of EMG time series for EMG pattern recognition. *Computer Methods and Programs in Biomedicine*, 117(2):247–256, 2014.

94. A. Phinyomark, H. Hu, P. Phukpattaranont, and C. Limsakul. Application of linear discriminant analysis in dimensionality reduction for hand motion classification. *Measurement Science Review*, 12(3):8289, 2012.

95. V. de Silva, D. Morozov, and M. Vejdemo-Johansson. Persistent cohomology and circular coordinates. *Discrete & Computational Geometry*, 45(4):737–759, 2011.

96. A. Phinyomark, F. Quaine, S. Charbonnier, C. Serviere, F. Tarpin-Bernard, and Y. Laurillau. A feasibility study on the use of anthropometric variables to make muscle–computer interface more practical. *Engineering Applications of Artificial Intelligence*, 26(7):1681–1688, 2013.

97. S. Thongpanja, A. Phinyomark, F. Quaine, C. Limsakul, and P. Phukpattaranont. The choice of muscles impacts the linear relationship between the probability density function of surface EMG signal and the level of contraction force. In *Proceedings of the 7th PSU-UNS International Conference on Engineering and Technology*, pages 92–95, 2015.

98. A. Phinyomark, P. Phukpattaranont, and C. Limsakul. Feature reduction and selection for EMG signal classification. *Expert Systems with Applications*, 39(8):7420–7431, 2012.

99. A. Phinyomark, F. Quaine, Y. Laurillau, S. Thongpanja, C. Limsakul, and P. Phukpattaranont. EMG amplitude estimators based on probability distribution for muscle–computer interface. *Fluctuation and Noise Letters*, 12(03):1350016, 2013.

100. P. Casale, O. Pujol, and Petia P. *Radeva*. Personalization and user verification in wearable systems using biometric walking patterns. *Personal and Ubiquitous Computing*, 16(5):563–580, 2012.

101. S. Ali, A. Basharat, and M. Shah. Chaotic invariants for human action recognition. In *2007 IEEE 11th International Conference on Computer Vision*, pages 1–8, Oct. 2007.

102. A. Phinyomark, E. Ibá nez Marcelo, and G. Petri. Analysis of big data in gait biomechanics: Current trends and future directions. *Journal of Medical and Biological Engineering*, 38(2):244–260, 2018.

103. A. Bargteil X. Martin J. Macey A. Collado F. de la Torre, J. Hodgins and P. Beltran. Guide to the Carnegie Mellon University multimodal activity (CMU-MMAC) database. In *Tech. report CMU-RI-TR-08-22, Robotics Institute, Carnegie Mellon University*, Apr. 2008.

104. A. Reiss and D. Stricker. Introducing a new benchmarked dataset for activity monitoring. In *2012 16th International Symposium on Wearable Computers*, pages 108–109, Jun. 2012.

105. S. Kyeong, S. Park, K. A. Cheon, J. J. Kim, D. H. Song, and E. Kim. A new approach to investigate the association between brain functional connectivity and disease characteristics of attention-deficit/hyperactivity disorder: Topological neuroimaging data analysis. *PLoS ONE*, 10(9):1–15, Sept. 2015.

106. W. Chacholski, M. Scolamiero, and F. Vaccarino. Combinatorial resolutions of multigraded modules and multipersistent homology. *ArXiv e-prints*, June 2012.

107. M. Lesnick and M. Wright. Interactive visualization of 2-D persistence modules. *ArXiv e-prints*, Dec. 2015.

108. O. Parzanchevski and R. Rosenthal. Simplicial complexes: Spectrum, homology and random walks. *Random Structures & Algorithms*, 2016.

12

Targeted Learning with Application to Health Care Research

Susan Gruber

CONTENTS

12.1 Introduction 235
    12.1.1 TL Estimation Roadmap 236
12.2 Super Learning 237
    12.2.1 SL Algorithm 238
    12.2.2 SL in Practice 238
    12.2.3 A Note on Reproducible Research 239
12.3 Targeted Minimum Loss-Based Estimation 240
    12.3.1 TMLE for Estimating a Marginal Mean Outcome Under Missingness 241
    12.3.2 TMLE for Point Treatment Effect Estimation 243
        12.3.2.1 Point Treatment Estimation When Some Outcomes Are Not Observed 244
        12.3.2.2 Additional Remarks 244
    12.3.3 TMLE for Longitudinal Treatment Regimes 245
        12.3.3.1 Longitudinal TMLE to Estimate the Conditional Mean Outcome of a Multiple Time Point Intervention 246
        12.3.3.2 Two Time-Point Example 247
        12.3.3.3 Extensions: Dynamic Regimes and Marginal Structural Models 248
12.4 Applications in Health Care Research 250
12.5 Summary 251
12.6 Open Problems and Future Directions 251
References 252

12.1 Introduction

The increasing availability of electronic data provides an unprecedented opportunity to gain insight into the world around us. Learning from data requires care and expertise, yet careful analysis can meet the information need, whether the underlying goal is prediction, classification, estimation, hypothesis testing, or hypothesis generation (data mining). Statistical analyses provide evidence of real-world associations in the form of estimates and measures of uncertainty. Obtaining an unbiased and efficient estimate of a statistical parameter of interest necessitates accounting for potential bias introduced through sources such as model misspecification, informative treatment assignment, and missingness in the outcome data.

Targeted learning (TL) is a paradigm for transforming data into reliable, actionable knowledge [96]. TL relies on two core analytic methodologies, *super learning* (SL), a data-adaptive approach to predictive modeling, and *targeted minimum loss-based estimation* (TMLE), an efficient double-robust semi-parametric substitution estimator. The combination of SL and TMLE produces parameter estimates that have a clear interpretation and relevance for addressing the motivating question.

TL was developed by Mark van der Laan and colleagues at the University of California at Berkeley, beginning with a seminal 2006 paper on targeted maximum likelihood estimation [97]. Multi-disciplinary research provided the early foundation for TL, including groundbreaking work in statistical causal inference by Robins and colleagues [6,31,57,62,64,76,95] and Pearl

[46], in semi-parametric efficiency theory [7,19,30,33], and in machine learning [2,8,86,98,100].

The goal of this chapter is to introduce key concepts in TL. Although the mathematical foundations of TL apply universally, the majority of published applications of TL address issues in health-related research. Terminology and motivating examples in this chapter are drawn from the biostatistics literature. TL has also been applied to problems in economics, web-based advertising, and online music sales [35,38,74,83,96].

12.1.1 TL Estimation Roadmap

The TL estimation roadmap provides a systematic guide to extracting knowledge from data (Figure 12.1) [96]. We view observed data as a random draw from some underlying distribution, P_0. The knowledge we seek corresponds to parameters of P_0. A crucial first step in learning from data is understanding the goal(s) of the research. After articulating a precise substantive question, the next task is translating it into a statistical question that can be informed by data. This process ensures that evaluating a well-defined parameter of the data distribution provides information relevant to the substantive goal. Posing and refining these questions may be an iterative activity that depends on the suitability of the available data. Once these elements are in place, estimating the target parameter becomes a matter of applying the appropriate tools and methodology.

Step 1 of the roadmap is defining the collection of possible probability distributions of the data that delineate the solution space. This is known as the statistical model $\mathcal{M}$. It is logical to define $\mathcal{M}$ in a way that ensures P_0 is a member; however, this is not standard statistical practice. $\mathcal{M}$ is typically chosen by default based on the type of outcome. Conventions differ across disciplines. Econometricians favor probit regression when modeling binary outcomes, while epidemiologists use logistic regression. There is no real reason to prefer one of these models over the other, and of course neither might be correct. In general, an estimator that relies on a misspecified model produces asymptotically biased results. This means that as more and more data are available to fit the model, the estimate converges to the wrong number. A dramatic consequence is that as more data become available, a test of a true null hypothesis will be rejected with probability tending to 1. Despite this, in many fields, an applied statistical analysis consists of fitting parameters of a parametric model from a well-studied family.

The effects of model misspecification are attenuated when estimator variance overshadows bias. However, in this era of increasingly large-scale data, bias can more often swamp variance. Naive analyses will produce a narrow confidence interval around a biased point estimate that is erroneously interpreted as the truth. This is strong motivation to pursue better alternatives. The TL framework respects statistical and domain knowledge by defining $\mathcal{M}$ such that it contains the true P_0. In the absence of true knowledge, $\mathcal{M}$ can be defined nonparametrically, assuming independence among units of observation. When true domain knowledge exists, restricting $\mathcal{M}$ based on this knowledge makes sense. For example, when the outcome is known to respect global bounds or subject-level conditional bounds, constraining the set of possible probability distributions under consideration improves finite sample bias and variance [4,22,23].

Step 2 of the roadmap conceptually defines a target parameter, ψ_0^{full}, in terms of the distribution of full data, $X^{full} \sim P_0^{full}$, which is only partially observable. This obliges us to precisely define ψ_0^{full} non-parametrically, independent of the choice of model or estimator. To make this concrete, consider evaluating the impact of a

---

Estimation roadmap

Step 1. Define a statistical model, $\mathcal{M}$, that contains the true distribution of the data, P_0.

Step 2. Define the target parameter of interest, Ψ_0^{full}, as a feature of a full data distribution P_0^{full}.

Step 3. Specify a mapping from the observed data to the full data, $\Psi: \mathcal{M} \to \mathbf{R}^d$ such that under explicitly stated identifying assumptions $\Psi_0^{full} = \Psi(P_0)$.

Step 4. Estimation and inference of statistical parameter $\Psi_0 = \Psi(P_0)$ using super learning and targeted minimum loss based estimation.

Step 5. Provide a considered interpretation of the result.

FIGURE 12.1
TL estimation roadmap.

treatment, A, on an outcome Y. Under the Neyman–Rubin counterfactual framework, the potential outcomes under each possible level of treatment are presumed to exist [73]. These potential (or counterfactual) outcomes Y_0 and Y_1 correspond to the values that would have been observed had we intervened to withhold treatment from the entire population, or alternatively exposing everyone to treatment, respectively. The full data contain information on both counterfactual outcomes, $X^{full} = (W, Y_0, Y_1)$. The marginal additive treatment effect (ATE) is easily defined as the expectation of the difference in counterfactual outcomes, $E(Y_1 - Y_0)$.

Another way to conceptualize the full data is in terms of a non-parametric structural equation model (NPSEM) [44]. A NPSEM is a system of equations describing the true data-generating mechanism. The NPSEM describes the structure of relationships in the data without imposing *a priori* parametric assumptions. For example, consider the following NPSEM: $W = f_W(U_W)$, $A = f_A(W, U_A)$, $Y = f_Y(W, A, U_Y)$, where U_W, U_A, and U_Y are exogenous error terms. This specification indicates that while W is a possible cause of A, Y is certainly not. It also indicates that both A and W are possible causes of Y. Counterfactual outcomes Y_0 and Y_1 correspond to values obtained by evaluating the function $f_Y(W, A, U_Y)$ with A set to the appropriate value. This paradigm also allows us to express our statistical question directly in terms of a parameter of P_0^{full}.

The data to which we have access is a coarsened version of the full data that contains less information. In real life, each of our n subjects is either treated or not at a particular point in time. The single outcome recorded in the data for each subject arises subsequent to the observed treatment. Observed data thus consist of n independent and identically distributed observations, $O_1, \ldots, O_n$ of $O = (W, A, Y) \sim P_0$, where W is a vector of baseline covariates, A is a binary indicator of exposure to a treatment or intervention of interest, and Y is the outcome observed under the recorded exposure. P_0 is the underlying joint distribution of the observed data.

Step 3 of the road map requires the analyst to explicitly establish links between the full and observed data. These *identifying assumptions* are conditions under which the target parameter can be estimated from data under the assumed statistical model. Identifying assumptions establish the correspondence between a parameter of the distribution of the observed data, $P_0(O)$, and the target parameter defined in terms of the full data. For example, under causal assumptions described in Section 12.3, the ATE parameter is equivalent to the statistical parameter $\psi_0 = \Psi_0(O) = E_W[E(Y|A=1,W) - E(Y|A=0,W)]$.

Step 4 of the road map is concerned with efficient unbiased estimation of the statistical parameter. SL is used for data-driven predictive modeling. SL can be used to model the regression of Y on A and W, and any of the functions comprising the NPSEM. When the goal is estimating a causal contrast or variable importance measure, SL and TMLE are combined for robust, efficient estimation of the target parameter. SL is used on its own when the end goal is purely individual-level prediction or classification. The remainder of this chapter describes the fundamentals of SL and TMLE in detail, and illustrates their use in practice.

Step 5 is the culmination of the process. Analytical results obtained after following Steps 1–4 have a clear statistical interpretation and relevance to the substantive question. The validity of a causal interpretation rests on the extent to which the identifying assumptions hold.

In summary, the roadmap teaches us to formulate a statistical question, define a parameter of the full data distribution that would successfully address this question, and state the assumptions under which this parameter can be identified from data. Together with the statistical model, this defines our statistical estimand. SL and TMLE provide efficient unbiased estimation of these statistical parameters. Following the roadmap facilitates communication among multi-disciplinary collaborators, who can weigh in at each step in the process. In this way the data analysis is directly tied to the substantive goals of the study.

12.2 Super Learning

SL provides a flexible approach to predictive modeling and classification that places fewer restrictions on $\mathcal{M}$ than traditional parametric modeling [94]. If we knew the true functional form of the the relationship between the outcome and the predictors, this is the model we would fit. However, because the true P_0 is always unknown, the optimal modeling procedure cannot be specified in advance. SL is a data-adaptive stacked ensemble machine learning algorithm. In lieu of imposing a single parametric regression model, or even a single data-adaptive procedure, SL provides a way to explore a larger portion of the space of possible probability distributions of the data.

The machine learning literature teaches that an alternative to relying on a single parametric model is to combine predictions from multiple models [8,86,100], or more generally from multiple predictive algorithms [2,51]. SL is an example of the latter. The analyst assembles a collection of prediction algorithms known as a *library* and uses SL to estimate the conditional mean outcome (or classification). SL predictions are a weighted combination of the predictions from each algorithm in the library. Alternatively, predicted SL values can be set equal to those produced by the single best performing algorithm in the library. This is known as *discrete SL*.

SL relies on proven properties of V-fold cross-validation to produce asymptotically optimal predictions, with respect to the collection available in a specified prediction algorithm library. SL converges to the true data model when the library algorithms search over the correct portion of the solution space. If no combination of candidates is consistent, SL will converge to the minimizer of the cross-validated risk, defined as the expectation of a loss function, $\mathcal{L}(O)$, such as the negative log likelihood [52,94,98]. If one of the prediction algorithms converges at a parametric rate to the truth, SL will converge at almost the parametric rate of $\log(n)/n$. If none of the prediction algorithms converge at a parametric rate to the true model, SL will be asymptotically equivalent to an oracle algorithm that can accurately evaluate the loss function with respect to the truth [3,90,94].

12.2.1 SL Algorithm

Given data consisting of n observations $O = (Y, X)$, SL provides an estimate of the conditional expectation, $E[Y \mid X]$. Continuing the example from above, Y might be a health outcome of interest and the set of predictors X include treatment indicator A and covariates W. The ensemble SL uses the following algorithm [52].

Initializations: Define the ensemble library $\{Alg_1, \ldots, Alg_K\}$, where K is the number of algorithms in the library.

Step 1: Train each algorithm $(Alg_1, \ldots, Alg_K)$ using all n observations, and obtain predicted values,

$$Z' = \begin{bmatrix} \hat{Y}_{11} & \cdots & \hat{Y}_{1K} \\ \vdots & \ddots & \vdots \\ \hat{Y}_{n1} & \cdots & \hat{Y}_{nK} \end{bmatrix}_{n \times K}.$$

Step 2: Create V equal-sized partitions of the data indexed by $v \in 1, \ldots, V$. For each v, this defines two disjoint sets: training set $Tr(v)$ contains all observations *except* those in partition v, and validation set $Val(v)$, containing observations in partition v.

Step 3: For each v, train each algorithm on $Tr(v)$ and obtain predicted values for observations in $Val(v)$, yielding a matrix of predictions containing one predicted value per observation for each algorithm,

$$Z = \begin{bmatrix} \hat{Y}_{11}^{Val} & \cdots & \hat{Y}_{1L}^{Val} \\ \vdots & \ddots & \vdots \\ \hat{Y}_{n1}^{Val} & \cdots & \hat{Y}_{nK}^{Val} \end{bmatrix}_{n \times K}.$$

Step 4: Propose a family of weighted combinations of candidate estimators indexed by α, $m(z|\alpha) = \sum_{k=1}^{K} \alpha_k \hat{Y}_k$, subject to the constraints $\alpha_k \geq 0 \ \forall k$ and $\sum_{k=1}^{K} \alpha_k = 1$. Calculate the contribution of each algorithm to the final prediction $(\hat{\alpha}_1, \ldots, \hat{\alpha}_K)$ by determining $\hat{\alpha}$ that minimizes the cross-validated risk, such as

$$\hat{\alpha} = \arg\min_{\alpha} \sum_{i=1}^{n} (Y_i - m(z_i|\alpha))^2,$$

subject to the constraints, $\alpha_k \geq 0 \ \forall k$, and $\sum_{k=1}^{K} = 1$, where z_i is the vector of predicted values $(\hat{Y}_{i1}^{Val}, \ldots, \hat{Y}_{iK}^{Val})$ for subject i from the ith row of prediction matrix Z.

Step 5: Evaluate the SL predictions

$$\hat{Y}_{SL} = \sum_{k=1}^{K} \hat{\alpha}_k z'_k,$$

where z'_k is column k of the Z' prediction matrix obtained in Step 1.

Discrete SL involves only a slight modification to Step 4. The discrete SL is evaluated by setting $\hat{\alpha}_k = 0$ for all but a single algorithm, Alg_{k^*}, that minimizes the cross-validated risk, e.g., $k^* = \arg\min_k \sum_{i=1}^{n}(Y_i - Z[k,i])^2$. Ties can be decided arbitrarily.

Cross-validation is the key to avoiding the general tendency of data-adaptive procedures to overfit the data. Overfitting occurs when the model fit responds to noise in the data beyond the true signal. Overfitting worsens predictive accuracy for novel inputs. Cross-validated risk is based on the loss for observations that were excluded from data used to fit the model. An algorithm that overfits to the data may minimize empirical risk, but will have poor cross-validated risk. This performance degradation will be reflected in the weights calculated in Step 4. An algorithm that overfits might contribute little or nothing to SL predictions.

12.2.2 SL in Practice

A data analyst can harness the power of SL by using a flexible, extensible R package, *SuperLearner* [53,54]. Perhaps the biggest decision facing a data analyst is deciding what algorithms to include in the SL library. Theory encourages a rich library containing many algorithms (polynomial in sample size). Ideally, these algorithms would be capable of modeling distributions P spanning all of $\mathcal{M}$. In practice, theoretical and practical considerations will inform the decision, and perhaps restrict the search to a portion of the solution space.

Algorithms that run quickly and are easy to interpret are attractive candidates, but a more diverse library will typically offer a better trade-off between search quality and computation time. A rich library of prediction algorithms weakens modeling assumptions. This should improve the fit of the model as long as measured covariates in the data include important predictors of the outcome. In published SL applications, the library typically contains a mixture of non-parametric, parametric, and semi-parametric procedures, some of which are themselves data-adaptive [50,70]. Algorithms applied in the literature include logistic regression, penalized regression, neural nets, k-nearest neighbor algorithms, and boosted or Bayesian classification and regression trees and generalized additive models [10,13,17,29,39,56,99]. When the response variable is a highly non-linear function of the covariates, procedures that avoid modeling the functional form directly might achieve more success than those that do not [28,36].

In large datasets there is tension between carrying out a computationally intensive search over the solution space and the feasibility of such a search in practice. Computation time and space increase with the dimension of the data, the size of the dataset, and the number of algorithms in the library. Several alternatives to pruning the library exist when the empirical distribution of a subset of the data closely resembles that of all the data, as in many of today's big data problems. One obvious approach is to obtain fitted values for all observations from a super learner trained on only a subset of observations [14,75]. Another is to train the models on one subset of the data and evaluate fit on the remaining hold-out test set [20]. The *SuperLearner* package also offers a built-in parallel processing option that speeds execution time when multiple computing cores are available.

A second choice facing the analyst is how to choose the number of cross-validation folds, V. Heuristics suggest V should be set somewhere between 3 and 20, with larger values being more appropriate when the size of the dataset is small [1,91]. Increasing V increases the number of observations in the training sample, which in turn stabilizes the model fits across folds, leading to improved performance. The commensurate decrease in the size of the validation set is less problematic in practice. Of course, when the number of observations in the dataset is quite large, this is a moot point, since the true distribution of the data is well captured by the empirical distribution of the data in both the training and validation sets. When the dataset is small relative to its inherent variability, increasing V will noticeably improve performance.

A valid loss function is one that is minimized at the truth. The scientific goal can be taken into account. For example, cross-validated squared error loss rewards predictive accuracy. Cross-validated area under the receiver operating curve rewards discrimination, so is appropriate for a classification task.

Cross-validated risk provides a sound basis for comparing the performance of individual algorithms in the library, with larger values indicating poorer fit. The weights calculated in Step 4 cannot be interpreted as measures of relative performance. To see this, consider an extreme example where the library consisted of exactly two algorithms, Alg_1 and Alg_2, that produce identical predictions. In Step 4, Alg_1 would receive a weight of 1 while Alg_2 would receive a weight of 0. Clearly, Alg_2 is in no way inferior to Alg_1. Rather, the weight of 0 reflects the fact that no combination of Alg_1 and Alg_2 predictions can improve the fit relative to predictions from Alg_1 alone. The set of weights is a function of the data and all the algorithms in the SL library collectively. The weights do not reflect the goodness of fit of each individual algorithm.

12.2.3 A Note on Reproducible Research

Randomness in SL stems from the inclusion of data-adaptive algorithms in the SL library and from the construction of cross-validation splits. For this reason, SL results typically vary from one run to the next. For analysts accustomed to running parametric regressions that always produce the same answer when presented with the same data, this behavior can be unnerving. If repeated SL runs produce different results, they wonder how to decide which result to report. Instead of viewing this as a drawback of machine learning, we can view it as a wake-up call. Variability is inherent in statistical analyses of data sampled from a population. We expect to get slightly different parameter estimates if we used two separate independent samples drawn from the same population to fit the model. Neither fit is more or less correct than the other. They provide complementary information that can be considered jointly, or summarized through meta-analysis or pooling the data. In practice, predictions can be averaged over multiple SL runs until stability is achieved.

Analytical results will not change substantially from one SL run to the next when there is sufficient information in the data. If there are important inconsistencies in performance, then there is a lack of information in the data for predicting the outcome. In a parametric setting, the sparsity of information may be less apparent. Nevertheless, in these situations, modeling assumptions are driving the analysis, not data. Stability is a function of the underlying distribution of the data, sample size, and characteristics of the algorithm such as its own level of data adaptivity [2,9]. SL's degree of data adaptivity is largely a function of the composition of the library. Data-adaptive methodologies by their very nature respond to information in the data. Their use encourages us to explicitly acknowledge the variability inherent in reasoning from incomplete information.

A core tenet of reproducible research is that study results should be verifiable and replicable [18,47]. Reproducibility and verifiability of SL results are accomplished by setting a random seed prior to invoking the function, reporting this value, and citing all software environments and packages used in the analysis, including version numbers. These measures promote transparency and scientific integrity.

12.3 Targeted Minimum Loss-Based Estimation

TMLE is a procedure for estimating any pathwise differentiable parameter of a distribution of the data, such as casual effect or statistical association [97]. Recall from the roadmap that the parameter is defined as a mapping from data to a scalar, or to a higher dimensional real number. Although SL predictions could be directly plugged into this parameter mapping, applying TMLE will often reduce bias and variance in the target parameter estimate. This is particularly true in high dimensional settings when our target parameter is a low-dimensional feature of the data. While SL and other maximum likelihood estimation–based procedures optimize a global bias variance trade-off, TMLE aims to make a more favorable trade-off with respect to the target statistical parameter, ψ_0. TMLE also provides insurance against bias due to model misspecification if despite our best efforts SL is somewhat misspecified.

The distribution of the data, P_0, can be expressed as $P_0 = (Q_0, g_0)$, where Q_0 denotes the components required to evaluate the defined mapping and g_0 denotes the remaining components of the likelihood. Restated, we define Q so that $\Psi(P_0) = \Psi_1(Q_0(P))$ for some Ψ_1, i.e., knowledge about Q_0 is sufficient for evaluating ψ_0. For example, the ATE parameter is given by $\psi_0 = E_W[E(Y|A=1,W) - E(Y|A=0,W)]$. For this parameter, Q_0 corresponds to the distribution of the outcome conditional on A and W, and the distribution of W, while g refers to the distribution of A given W. If we knew Q_0, we would know the mean outcome for each subject under both levels of treatment and could evaluate the parameter directly. Since we do not know Q_0, we must estimate it from data.

We use SL to model the conditional mean outcome $E(Y|A,W)$. This would give us a model of the conditional mean outcome, $\bar{Q}_n(A,W)$ from which we could evaluate predicted values under both levels of exposure, $\bar{Q}_n(0,W)$ and $\bar{Q}_n(1,W)$. One option is to plug these values directly into the mapping. The parameter estimate, ψ_n, is evaluated by taking the mean difference over all observations in the data (n signifies a quantity estimated from data). However, if SL is misspecified, ψ_n will be a biased estimate of ψ_0. TMLE can make use of information in g to reduce this bias. In fact, TMLE will be consistent if either Q or g is correct. This property is known as *double robustness (DR)*.

The key to DR lies in solving the efficient influence curve estimating equation. Theory tells us that an efficient regular asymptotically linear RAL estimator solves the efficient influence curve equation for the target parameter up to a second-order term [7]. An influence curve, or influence function, describes the behavior of an estimator under slight perturbations in the distribution of the data. Hampel [27] described the influence curve as the first derivative of an estimator, viewed as a functional, at some distribution in an infinite-dimensional space. Operationally, the influence function can be used as an estimating equation to evaluate a statistical parameter of interest [95]. Different estimators of the same quantity may be characterized by different influence functions. Unique among these influence functions is the one with the smallest variance. This is known as the efficient influence function, $D^*(P)$.

Robins and Rotnitzky developed the first DR estimator, an augmented inverse probability weighted (A-IPW) estimator that uses $D^*(P)$ as an estimating equation by setting its empirical mean to 0 and solving for ψ_n. We write this as $P_n(D^*(P)) = 0$, where P_n denotes $1/n \sum_{i=1}^{n}$. This approach requires that such an estimating function representation exists, while TMLE does not [64,67,68]. Like TMLE, A-IPW is consistent if either Q or g is correctly specified, and locally efficient when both models are correct. Unlike TMLE, it is not a substitution estimator, and in some circumstances will produce estimates outside the bounds of the statistical model.

TMLE views the efficient influence curve as a path instead of an estimating equation [97]. This allows TMLE to be applied for estimation of parameters where the estimating equation is not solvable in ψ. Formal proofs, derivations, and instructions on how to construct a TMLE for a novel parameter are presented elsewhere [72,96]. The focus here is on how to use TMLE to estimate parameters that are often of interest to health care researchers.

A series of examples is provided that illustrates the application of TMLE in increasingly complex settings. The core principles of TMLE remain the same in each setting. TMLE consists of two stages. Stage 1 is concerned with obtaining the initial estimate of Q, denoted as Q_n^0. Stage 2 produces a targeted estimate, Q_n^*, that uses information in g to update the initial Q_n^0. The parameter is evaluated by plugging Q_n^* into the mapping. More formally, the general requirements for TMLE are (1) an initial estimate of the relevant Q portion of P_0 needed to evaluate the target parameter mapping; (2) a loss function $\mathcal{L}(Q)$ that is minimized at the truth, $Q_0 = \underset{Q \in \mathcal{Q}}{\operatorname{argmin}}\ E\mathcal{L}(Q)(O)$; and (3) a parametric submodel, $Q_{ng}^0(\epsilon)$, with parameter ϵ. ϵ satisfies $\dfrac{d}{d\epsilon}\mathcal{L}(Q_{ng}^0(\epsilon))(O)\Big|_{\epsilon=0} = D^*(Q_n^0, g)(O)$, where $D^*(Q_0, g_0)$ (equivalently, $D^*(P_0)$) is the efficient influence curve of Ψ: $\mathcal{M} \to I\mathbb{R}$ at P_0.

12.3.1 TMLE for Estimating a Marginal Mean Outcome Under Missingness

A problem that frequently arises when analyzing study data is that the outcome may not have been ascertained for some study members. A naive estimator that considers only complete cases is inefficient and will be biased when missingness is informative. When information is available in the measured covariates, W, TMLE can reduce bias and variance. The causal inference literature distinguishes among three forms of missingness. Missingness that is independent of the outcome is known as missing completely at random (MCAR). Missingness that is a function of only measured covariates is known as missing at random (MAR). Missingness that is a function of some unmeasured covariates is known as missing not at random (MNAR) [37]. A TMLE for estimating parameters when data are MNAR uses an instrumental variables approach [87]. This chapter focuses on applications of TMLE to data that are MCAR and MAR. This is also known as data that are coarsened at random (CAR) [19,33].

The full data are given by $X^{full} = (Y_{\Delta=1}, W)$, where $Y_{\Delta=1}$ is the value of the outcome when, perhaps contrary to fact, the outcome is observed. Our parameter of interest is the marginal mean outcome defined in terms of full data by the mapping $\Psi(P_0^{full}) = E(Y_{\Delta=1})$. Access to X^{full} would allow us to evaluate the marginal mean outcome by simply taking the mean of Y over all observations. Because we only have access to the observed data, this estimator is not available in practice, since in the observed data, some outcomes are missing. The observed data consist of n independent and identically distributed (i.i.d.) observations of $O = (\Delta Y, \Delta, W) \sim P_0$, where Δ is a binary missingness indicator that takes on the value 1 when the outcome is observed. The full data parameter is identifiable when we assume the data are CAR, such that $\Delta \perp X^{full} | W$. This *conditional exchangeability* assumption states that there are no unmeasured causes of both missingness and the outcome. Identifiability also requires a positivity assumption stating that there is a positive probability of observing the outcome within every strata defined by W, $P(\Delta = 1 | W) > 0, \forall W$. A final consistency assumption states that the observed Y equals the counterfactual value $Y_{\Delta=1}$ that would have been observed had there been no missingness in the data. Under these assumptions, the statistical parameter corresponding to our causal parameter of interest is defined as $\psi = E_W [E (\Delta Y | \Delta = 1, W)]$ [45].

The observed data likelihood factorizes as $P_0(O) = P_0 (\Delta Y | \Delta = 1, W) P_0 (\Delta | W) P_0 (W)$. The first and third factors of the likelihood are required to evaluate the parameter, and thus $Q = (Q_Y, Q_W)$. Nuisance parameter g is the second factor of the likelihood.

The first ingredient for a TMLE is an initial estimate of the components of Q. Although Q_Y refers to the entire density of Y, we see that only a consistent estimate of the conditional mean outcome under no missingness is required to evaluate this parameter. This conditional mean is denoted by $\bar{Q}(W) = E(Y | \Delta = 1, W)$. Although many analysts would use linear or logistic regression to model $\bar{Q}(W)$, the road map promotes the use of SL to reduce the opportunity for model misspecification bias. Q_W is consistently estimated non-parametrically by the empirical distribution of covariates W. Stage 1 of TMLE is complete upon obtaining these initial estimates, $(\bar{Q}_n(W), Q_{W_n})$.

Stage 2 requires us to specify a parametric submodel and loss function that is minimized at the truth. The goal of this second stage is to fluctuate the initial estimate in a manner that reduces bias in the parameter estimate. This fluctuation must be carefully designed to ensure that the empirical mean of the efficient influence curve for the target parameter is equal to 0. The efficient influence function for this parameter is given by

$$D^*(P) = H(\Delta, W)[\Delta Y - \bar{Q}(W)] + \bar{Q}(W) - \psi, \quad (12.1)$$

where $H(\Delta, W) = (\Delta) / g(\Delta, W)$, missingness mechanism $g(\Delta, W) = P(\Delta = 1 | W)$ is the conditional probability of observing the outcome, $\bar{Q}(W)$ is the mean outcome conditional on W, $\bar{Q}(W) = E(Y | \Delta = 1, W)$, and ψ is the parameter of interest [64]. We define a submodel logit $\bar{Q}_n^*(W) = \operatorname{logit} \bar{Q}_n^0(W) + \epsilon H(\Delta, W)$. This submodel specifies that we will fluctuate the initial estimate of the conditional mean outcome, $\bar{Q}_n^0(W)$, on the logit scale by adding a term that is a function of the nuisance parameter, g. Fluctuation parameter ϵ will be fit by maximum likelihood, i.e., by minimizing the negative log likelihood loss function.

Sample R code shows how these ingredients can be used to estimate the marginal mean outcome when some outcomes are missing (Figure 12.2). This code is deliberately kept simple. A more sophisticated implementation that incorporates SL is available in an open source R package, *tmle*. The package also provides

```
1.  n <- 1000
2.  W1 <- rnorm(n)
3.  W2 <- rnorm(n)
4.  W3 <- rnorm(n)
5.  Delta <- rbinom(n, 1, plogis(-0.5 + 0.2 * W1 + 0.5 * W2))
6.  Y <- rbinom(n, 1, plogis(-1 + 0.1 * W1 + 0.5 * W2 - 0.3 * W3))
7.  Y[Delta == 0] <- NA
8.  m.mis <- glm(Y ~ W1 + W3, family = binomial, subset = Delta == 1)
9.  Q.mis <- predict(m.mis, newdata = data.frame(W1, W3))
10. g <- glm(Delta ~ W1 + W2, family = binomial)
11. g1W <- predict(g, type = "response")
12. h <- 1/g1W
13. epsilon <- coef(glm(Y ~ -1 + h + offset(Q.mis), family = binomial,
        subset = Delta == 1))
14. Qstar <- plogis(Q.mis + epsilon * h)
15. psi <- mean(Qstar)
```

FIGURE 12.2
R code for using TMLE to estimate the marginal mean outcome when some outcomes are missing.

TMLEs for estimating a variety of point treatment parameters [24,25].

The data for this example are generated such that the outcome is a function of (W_1, W_2, W_3), and missingness is a function of W_1 and W_2 (lines 1–7). Estimation of $\bar{Q}_n^0$ (line 8) is deliberately based on a misspecified outcome regression model (line 8). This was done in order to demonstrate the ability of TMLE's Stage 2 to reduce bias. A preliminary evaluation of the plug-in estimator of ψ_0 based on these initial predicted values would be biased, since the model omits confounder W_2. The covariate $H(\Delta, W) = 1 / g(1, W)$ is evaluated based on the estimated missingness probabilities $g(1, W) = P(\Delta = 1 | W)$ (lines 10–12). SL could be used in place of logistic regression on line 10. ε is fit using a univariate logistic regression of ΔY on $H(\Delta, W)$ with offset logit $\bar{Q}_n^0(W)$ and no intercept. The model is fit on observations where the outcome is observed, $\Delta = 1$ (line 13). The magnitude of $\hat{\varepsilon}$ reflects the degree of residual bias in the target parameter estimate. For example, when $\bar{Q}_n^0$ is correct, $\hat{\varepsilon}$ is essentially 0; however, even this small fluctuation can reduce variance if the initial estimator of $\bar{Q}_0$ was not efficient. It is important to avoid overfitting $\bar{Q}_n^0$, as this attenuates the signal in the residuals that is needed for bias reduction. Targeted estimates of the conditional mean outcome, $\bar{Q}_n^*(W)$, are obtained on line 14. The TMLE is evaluated by plugging $\bar{Q}_n^*$ into the mapping for ψ (line 15). The expectation in the mapping is approximated by the empirical mean over all observations in the data.

The covariate H is parameter-specific. Recall that the efficient influence function for this target parameter is given by $D^*(P) = H(\Delta, W)[\Delta Y - \bar{Q}(W)] + \bar{Q}(W) - \psi$ (Equation 12.1). The maximum likelihood procedure to fit ε solves score equation $\sum H(\Delta, W)[\Delta Y - \bar{Q}(W)] = 0$, ensuring the first term in Equation 12.1 has mean 0. By construction, $\bar{Q}_n^*(W) - \psi_n$ has mean 0. Thus, convergence when fitting ε implicitly guarantees the TMLE solves the empirical efficient influence curve estimating equation $P_n(D^*(P)) = 0$, which confers DR.

When both Q and g are correctly specified, TMLE is locally efficient. The variance of the parameter estimate can be estimated analytically as $var(D^*(P))/n$, and used to construct a confidence interval or Wald-type test statistic for hypothesis testing. In a repeated measures dataset where data are not i.i.d., the variance is calculated as the mean with respect to the number of independent units, where each unit's contribution is its mean contribution to the IC among observations sharing a common group identifier. Although SL was not used in this example, note that even when SL is used to estimate the outcome regression, the analytic variance is valid because there is no contribution to the influence function from the estimation of Q.

The true marginal mean outcome under the data generating mechanism shown Figure 12.2 (lines 1–7) is $\psi_0 = 0.2832$. In a Monte Carlo simulation study with the random seed set to 10, the average TMLE over 1000 replicates was $\psi_n = 0.2833$. In contrast, the average estimate based on the misspecified outcome regression model, ψ_{mis} = expit (Q.mis), was 0.3108. Mean squared error (MSE) for the biased estimator was more than double that of the TMLE (0.0013 versus 0.0005). These simulation study results demonstrate the advantage of using a DR estimator that can remain unbiased when g is correctly specified, despite utilizing a misspecified initial $\bar{Q}_n^0$.

12.3.2 TMLE for Point Treatment Effect Estimation

A slightly more complex example involves comparing the impact of one treatment versus another (or no treatment) on an outcome. In this section, we consider TMLEs to estimate these point treatment effects. For simplicity, the discussion focuses on binary treatments, but TMLEs for categorical and continuous treatments have also been developed [12,24].

Marginal treatment effects can be expressed as causal contrasts of marginal mean outcomes under different levels of exposure. Let $\mu_1 = E_0(Y_1)$ and $\mu_0 = E_0(Y_0)$ be the true marginal mean outcomes under exposure to the treatment of interest, $A = 1$, and no exposure to the treatment of interest, $A = 0$. The ATE parameter previously discussed can be expressed as $\mu_1 - \mu_0$. The relative risk (RR = μ_1/μ_0,) and odds ratio (OR = $\mu_1 (1 - \mu_0)/ [\mu_0 (1 - \mu_1)]$) are defined for binary outcomes. These parameters are easily evaluated when μ_1 and μ_0 are known. Identifying assumptions will allow us to equate these parameters with statistical parameters that can be estimated from observed data.

The observed data for this type of problem consist of n i.i.d. copies of $O = (Y, A, W)$. Outcome Y that arises under exposure A is observed when $\Delta = 1$. A factorization of the likelihood is given by $P(O) = P(Y | A, W) P(A | W) P(W)$. As in the previous example, $Q = (Q_Y, Q_W)$. In this context, $g = P(A | W)$, the conditional probability of receiving treatment, also known as the propensity score [71].

Assumptions of consistency, conditional exchangeability, and positivity allow us to define a two-dimensional statistical estimand, $\psi = (\psi_{Y_1}, \psi_{Y_0})$, where $\psi_{Y_1} = E(Y | A = 1, W)$ and $\psi_{Y_0} = E(Y | A = 0, W)$. The consistency assumption states that the outcome in the data for a subject with exposure $A = a$ corresponds to the counterfactual outcome Y_a. Conditional exchangeability is an assumption of no unmeasured confounding that is satisfied when data are CAR, $A, \Delta \perp X^{full} | W$. These two assumptions are untestable from data. The positivity assumption requires that within strata defined by W, there is a non-zero probability of receiving treatment at all levels, $a \in 0, 1$. Conceptually, this requirement guards against a complete lack of support in the data for estimating the desired causal effect.

A TMLE to estimate the effect of a binary point treatment begins by obtaining Stage 1 estimates of counterfactual outcomes under $A = 0$ and $A = 1$ for all observations, $\bar{Q}_n^0(0, W)$ and $\bar{Q}_n^0(1, W)$, respectively. SL is ideally suited for this task. In Stage 2, we will target μ_0 and μ_1 simultaneously. This is accomplished by defining a submodel with a two-dimensional parameter, $\epsilon = (\epsilon_0, \epsilon_1)$. The TMLE targeting step for updating $\bar{Q}_n^0$ with respect to $(E(Y_1), E(Y_0))$ is as follows:

$$\text{logit}\,\bar{Q}_n^*(A, W) = \text{logit}\,\bar{Q}_n^0(A, W) + \hat{\epsilon}_0 H_0(A, W)$$
$$+ \hat{\epsilon}_1 H_1(A, W),$$

where $H_0(A, W) = \dfrac{I(A = 0)}{g(0|W)}$, and $H_1(A, W) = \dfrac{I(A = 1)}{g(1|W)}$. A side effect of fitting ϵ by maximum likelihood is solving the efficient influence curve estimating equation for each component of our two-dimensional ψ. Thus, the TMLE will be consistent if either Q or g is correctly specified. Targeted estimates of the counterfactual outcomes for each subject are obtained by evaluating

$$\text{logit}\,\bar{Q}_n^*(0, W) = \text{logit}\,\bar{Q}_n^0(1, W) + \hat{\epsilon}_0 H_0(0, W),$$

$$\text{logit}\,\bar{Q}_n^*(1, W) = \text{logit}\,\bar{Q}_n^0(0, W) + \hat{\epsilon}_1 H_1(1, W).$$

Finally, we set $\psi_{nY_0} = P_n(\bar{Q}_n^*(0, W))$, and $\psi_{nY_1} = P_n(\bar{Q}_n^*(1, W))$. These estimates of μ_0 and μ_1 are then plugged into the appropriate formula for the parameter of interest, ATE, RR, or OR.

It is also possible to directly target a specific causal contrast by specifying an appropriate parameter-specific submodel and loss function. Consider the ATE parameter whose efficient influence function is given by

$$D^*(P) = H(A, W)[Y - \bar{Q}(A, W)] + \bar{Q}(1, W)$$
$$- \bar{Q}(0, W) - \psi, \qquad (12.2)$$

with $H(A, W) = \frac{A}{g(1, W)} - \frac{1 - A}{1 - g(1, W)}$ [97]. Our goal is to define a submodel with a single fluctuation parameter, ε, such that the $\hat{\varepsilon}$ minimizing the empirical loss function also solves the empirical efficient influence function equation, $P_n(D^*(P)) = 0$. We will do this by ensuring that two separate components of Equation 12.2 that are summed together each has mean 0. Because TMLE is a substitution estimator, it is the case that $\psi_n = P_n(\bar{Q}_n^*(1, W) - \bar{Q}_n^*(0, W)$. This indicates that the last three terms on the right-hand side of Equation 12.2 have mean 0. By arranging the first term in Equation 12.2 to also have mean 0, we guarantee the entire equation has mean 0. We arrange this by defining the submodel logit $\bar{Q}_n^*(A, W) =$ logit $\bar{Q}_n^0(A, W) + \varepsilon H(A, W)$, and fitting ε by maximum likelihood. The score equation solved by fitting a logistic regression model to the data corresponds to the first term in Equation 1.2. Convergence ensures this term has mean zero at the updated $\bar{Q}_n^*$. The update is given by $\bar{Q}_n^*(0, W) = \text{expit} \{\text{logit } \bar{Q}_n^0(0, W) - \hat{\varepsilon}/[1 - g(1, W)]\}$, and $\bar{Q}_n^*(1, W) = \text{expit}\{\text{logit } \bar{Q}_n^0(1, W) + \hat{\varepsilon}/g(1, W)\}$. This example demonstrates that the TMLE for the ATE parameter is double robust. One could instead target the RR or OR parameter directly. The influence functions for these parameters are slightly more complicated. Mathematical convergence requires multiple iterations of the targeting step, but is certainly feasible [24].

12.3.2.1 Point Treatment Estimation When Some Outcomes Are Not Observed

The next example combines the previous two scenarios by showing how TMLE can be used to estimate a point treatment effect when some outcomes are missing. The observed data for this type of problem consist of n i.i.d. copies of $O = (\Delta Y, \Delta, A, W)$. Outcome Y that arises under exposure A is observed only when $\Delta = 1$. The distribution of the data factorizes as $P(O) = P(Y | \Delta = 1, A, W) P(\Delta | A, W) P(A | W) P(W)$. As in the earlier examples, $Q = (Q_Y, Q_W)$, but g now has two components, $g = (g_\Delta, g_A)$, corresponding to the conditional distribution of censoring and the conditional distribution of treatment, respectively.

Identifiability rests on an extended conditional exchangeability assumption that includes no unmeasured confounders of the association between missingness and the outcome, $A, \Delta \perp X^{full} | W$. The positivity assumption is also extended to require a non-zero probability of the outcome being observed within all levels of (A, W). With these assumptions in place, our focus becomes estimating the statistical parameter, $\psi = (\psi_{Y_1}, \psi_{Y_0})$, where $\psi_{Y_1} = E(Y | \Delta = 1, A = 1, W)$ and $\psi_{Y_0} = E(Y | \Delta = 1, A = 0, W)$.

For Stage 1, we can use SL to obtain initial estimates of the conditional mean outcomes under treatment and no treatment for each subject, $\bar{Q}_n^0(1, W)$ and $\bar{Q}_n^0(0, W)$. The models are fit using observations where $\Delta = 1$. We obtain predicted values $\bar{Q}_n^0(1, W)$ from this model for all subjects when exposure is set to $A = 1$. Next, we obtain predicted values $\bar{Q}_n^0(0, W)$ by setting exposure $A = 0$ for all subjects.

In Stage 2, ψ_{Y_1} and ψ_{Y_0} can be targeted simultaneously.

$$\text{logit } \bar{Q}_n^*(A, W) = \text{logit } \bar{Q}_n^0(A, W) + \epsilon_0 H_0(\Delta, A, W)$$
$$+ \epsilon_1 H_1(\Delta, A, W),$$

where $H_0(\Delta, A, W) = \frac{\Delta}{P(\Delta = 1 | A, W)} \times \frac{I(A = 0)}{g(0 | W)}$, and $H_1(\Delta, A, W) = \frac{\Delta}{P(\Delta = 1 | A, W)} \times \frac{I(A = 1)}{g(1 | W)}$. The regression is fit using observations where $\Delta = 1$. Targeted estimates are obtained by evaluating

$$\text{logit } \bar{Q}_n^*(0, W) = \text{logit } \bar{Q}_n^0(1, W) + \hat{\epsilon}_0 H_0(1, 0, W),$$

$$\text{logit } \bar{Q}_n^*(1, W) = \text{logit } \bar{Q}_n^0(0, W) + \hat{\epsilon}_1 H_1(1, 1, W).$$

Consistency of the TMLE is predicated on consistent estimation of either both Q factors or both g factors.

12.3.2.2 Additional Remarks

Near violations of the positivity assumption. If there is some combination of characteristics in W for which treatment is very likely (or very unlikely), then there will be some strata of W for which the propensity score $g(1, W)$ is close to 0 or 1. Because both $g(1, W)$ and $(1 - g(1, W))$ are in the denominator of fluctuation covariate H, this de-stabilizes the estimator. This near violation of the positivity assumption signals a sparsity of information for evaluating the desired causal contrast. Practitioners sometimes choose to deal with this issue by trimming the population to exclude observations in these problematic strata. This re-frames the question in terms of a different target population having a different distribution of (A, W). Another approach is to artificially truncate the estimated probabilities to ensure being bounded away from $(0,1)$. This typically improves finite sample bias and variance, but introduces asymptotic bias. The literature provides guidance on diagnosing and responding to violations in the positivity assumption [48].

A collaborative TMLE (C-TMLE) was developed in response to the problem of how to best estimate nuisance parameter g in the context of double-robust estimation. C-TMLE data-adaptively selects a subset of covariates $\tilde{W} \subseteq W$ sufficient to address bias in ψ_n that remains after the completion of Stage 1. It does this by exploiting a collaborative DR property of DR estimators that guarantees consistency under certain complementary forms of dual misspecification [21,26,92]. The Stage 1 outcome regression smooths over areas of little support in the data. If this extrapolation is unsuccessful, C-TMLE's Stage 2 provides an opportunity to selectively model g in order to reduce the MSE of the estimate. As originally proposed, the C-TMLE uses a greedy stepwise forward selection search strategy to incorporate covariates into the model for g whose impact on bias reduction outweighs its contribution to the variance of the overall estimator. This can be a computationally intensive procedure. A recent scalable C-TMLE designed specifically for large-scale datasets reduces this computational burden. It uses a slightly less data-adaptive algorithm than the original C-TMLE that is still directed towards targeting residual bias [34]. These C-TMLEs address violations of the positivity assumption by making a favorable finite-sample bias/variance trade-off in response to signal in the data.

Continuous outcomes. Although it seems natural to carry out the targeting step on the linear scale instead of the logit scale when the outcome is continuous, this approach is not always successful. While targeting on the linear scale will work for many data distributions, fluctuations on the linear scale may violate known bounds on the outcome, particularly when there are near violations of the positivity assumption. Recall that the fluctuation of the initial density estimate is required to be a *submodel* of the observed data model, $\mathcal{M}$. Unlike a logistic fluctuation, a linear fluctuation provides no assurance that the targeted estimate of the conditional mean remains within the defined model space. The TL solution is to define a simple transformation of the data that allows us to carry out the fluctuation on the logit scale even when Y is continuous. The only requirement is the existence of known finite bounds on Y [22]. The negative log likelihood for binary outcomes is a valid loss function for continuous outcomes bounded between 0 and 1, and provides a procedure for mapping outcome Y, bounded by (a, b), into Y^*, a continuous outcome bounded by $(0,1)$: $Y^* = (Y - a)/(b - a)$. Values on the Y^* scale are easily mapped to their counterparts on the original scale:

$$E_W(Y_0) = E_W(Y_0^*(b-a) + a),$$

$$E_W(Y_1) = E_W(Y_1^*(b-a) + a).$$

Initial estimates $\bar{Q}_n^0(A, W)$ are transformed in this way, fluctuated, and back-transformed to the original scale. This procedure ensures that the Stage 2 parametric submodel remains within $\mathcal{M}$. When bounds are known to vary within strata of W imposing conditional bounds $(a(W), b(W))$ can improve MSE [23].

Variable importance. The validity of a causal interpretation of the statistical parameter rests on assumptions that are not testable from data. For some classes of problems, these statistical parameter estimates have utility, even if causal assumptions are not met. Variable importance measures are non-causal statistical association parameters that can be used to identify strong predictors of risk or to rank variables according to the strength of their impact upon the outcome. SL and TMLE have been used to calculate variable importance measures to rank predictors of mortality in hospital intensive care units (ICUs) [32,50]. Utilizing TMLE to target each variable importance prediction was shown to have an impact on bias and reduce variance.

12.3.3 TMLE for Longitudinal Treatment Regimes

The parameters considered up until now are of interest when there is only a single opportunity for right censoring or treatment. TMLE can also be used to estimate the impact of a series of treatment decisions made over time. Consider the task of estimating a marginal mean outcome under a specific treatment regime administered over K time points. In an ideal randomized control trial (RCT), one would randomly assign subjects to this regime, ensure complete adherence, measure the outcome with no loss to follow-up, and calculate the mean. In a realistic RCT, lack of adherence or informative loss to follow-up may bias this naive estimator. Observational studies are affected by these same sources of bias, as well as by selection bias due to non-random treatment assignment. A longitudinal TMLE (L-TMLE) can be used to estimate the marginal mean outcome of a multiple time-point intervention, and to evaluate causal contrasts that are functions of these intervention-specific means. As a TMLE, L-TMLE is also an efficient double robust semi-parametric substitution estimator, and is RAL.

Time-dependent confounding is a problem that is unique to longitudinal data analyses. In the longitudinal setting, a covariate that is affected by prior treatment may also confound the association between future treatment and the outcome. Because we are interested in estimating the impact of the entire series of treatments, this poses a dilemma. In order to control for confounding, one would like to include this time-varying covariate in the outcome regression model. However, because it is a downstream effect of prior treatment, including it in the model would bias the contribution to the estimated effect stemming from prior treatment [31]. This problem was identified by Robins three decades ago. Robins also developed the necessary mathematical foundation for unbiased estimation of causal effects, the G-computation formula [57–59,61]. Nevertheless, including time-varying covariates in a Cox proportional hazards model remains standard practice. This is despite the fact that the Cox model assumes uninformative right censoring and inadequately handles time-dependent confounding [11].

Robins and colleagues have also proposed a variety of estimators derived from the G-computation formula that appropriately address time-dependent confounding, including a double robust estimator proposed in Bang and Robins [6] (see also Refs. [31,41,60,62,63,65,66,69]). L-TMLE confers an advantage in practice over this earlier DR approach by being a substitution estimator that respects bounds on $\mathcal{M}$ and allows for flexible modeling of the Q portion of the likelihood.

12.3.3.1 Longitudinal TMLE to Estimate the Conditional Mean Outcome of a Multiple Time Point Intervention

Suppose we are interested in understanding the impact of a daily dose of an anti-retroviral drug on nausea over the course of one week. The full data parameter corresponding to this conditional mean outcome of a multiple time point intervention is expressed as $E(Y_{\bar{a}})$, where $\bar{a}$ corresponds to a specified sequence of treatments. In this example, the series of seven treatments is written as $\bar{a} = \{1,1,1,1,1,1,1\}$, and $E(Y_{\bar{a}})$ represents the proportion of patients who experience the outcome. The parameter is defined by the mapping $\Psi: \mathcal{M} \to \mathbb{R}$, such that $\Psi(P_{\bar{a}}) = E(Y_{\bar{a}})$.

The observed data consist of n i.i.d. observations $O = (L_0, A_0, L_1, A_1, \ldots, L_{K+1} = Y)$. This structure describes baseline covariate vector L_0, exposures A_k at time points 0 through K, and time-varying covariates, $L_1, \ldots L_K$. L_k may also contain an outcome or event indicator at time k. L_{K+1} contains the outcome of interest. In our notation, $\bar{L}_k$ refers to the entire covariate history through time point k, and we will use Y and L_{K+1} interchangeably. $\bar{A}_k$ refers to the entire treatment history through time k. $\bar{a}_K$ refers to a specific sequence of treatment settings corresponding to a longitudinal treatment regime of interest, $\bar{a}_K = (a_0, \ldots a_K)$.

Analogous to the point treatment setting, we can map questions about the effect of a longitudinal treatment regime, $\bar{a}$, to parameters of the distribution of the data under the regime of interest by estimating the relevant components of $\bar{L}_{K+1}$ under a hypothetical intervention that sets $\bar{A}_K = \bar{a}_k$. We can also encode loss to follow-up by more broadly defining A_k as a vector of exposure and censoring indicators, (A_{k_T}, A_{k_C}), at each time point. These are collectively known as "intervention nodes." The counterfactual outcome of interest is uncensored, so our hypothetical intervention also sets $\bar{A}_{K_C} = 0$. Following Robins [57], the distribution of the data under an intervention that sets $\bar{A}_K = \bar{a}_K$ is given by $P_{\bar{a}} = \prod_{k=0}^{K+1} Q_{L_k}^a(\bar{l}_k)$. Under appropriate causal identifying assumptions, the statistical parameter is given by the mapping $\Psi: \mathcal{M} \to \mathbb{R}$, such that $\Psi(P_{\bar{a}}) = E(Y_{\bar{a}})$.

Bang and Robins (BR) [6] showed that under a set of identifying assumptions, the G-computation formula can be re-written as a sequence of nested conditional means. The law of iterated expectations tells us that $E(Y) = E_x(E[Y|X])$. The double robust estimator presented by BR evaluates $\psi_{nY_{\bar{a}}}$ by applying this law repeatedly.

From the TL perspective, the Q portion of the likelihood corresponds to the conditional distribution of $\bar{L}_{K+1}$ given the prior covariate history at each time point. The g portion of the likelihood corresponds to the conditional distribution of treatment and right censoring at each time point. The calculation starts by estimating the expected value of the outcome at time $K+1$ conditional on the entire past, $\bar{Q}_{L_{K+1}}^a = E\{Y_{\bar{A}}|\bar{L}_K, \bar{A}_K = \bar{a}_K\}$ (innermost term; Figure 12.3). Taking the expectation of $\bar{Q}_{L_{K+1}}^a$ with respect to the distribution of covariates measured at time point K under treatment $\bar{A}_K = \bar{a}_K$ yields the next term in the sequence, $\bar{Q}_{L_K}^a = E[E\{Y_{\bar{A}}|\bar{L}_K, \bar{A}_K = \bar{a}_K\}|\bar{L}_{K-1}, \bar{A}_{K-1} = \bar{a}_{K-1}]$. This process iterates over all remaining time points. Ultimately, the expectation is taken with respect to the empirical distribution of baseline covariates L_0. The L-TMLE mapping for the target parameter is a function of an iteratively defined sequence of *targeted* conditional means, $\Psi(\bar{Q}^{a*})$, where $\bar{Q}^{a*} = (\bar{Q}_Y^{a*}, \bar{Q}_{L_K}^{a*}, \ldots, \bar{Q}_{L_0}^{a*})$ are estimated flexibly, then targeted to reduce bias in the parameter of interest [93]. Other longitudinal TMLEs (L-TMLEs) utilizing a different choice of submodel and loss function have been developed, but will not be discussed further [84,88].

The first required ingredient for the L-TMLE is the initial estimate, $\bar{Q}_{L_k}^a$. As in the point treatment setting, this can be modeled using parametric regression, or more flexibly using SL. The influence function for $\Psi(P_{\bar{a}}) = E(Y_{\bar{a}})$ is given by $D^* = \sum_{k=0}^{K+1} D_k^*$, the sum of the time point-specific influence functions given in Ref. [93].

$$E(Y_{\bar{a}}) = E(E\{\ldots E[\underbrace{E\{Y_{\bar{a}}|\bar{L}_K, \bar{A}_K = \bar{a}_K\}}_{\bar{Q}_{L_{K+1}}^a}|\bar{L}_{K-1}, \bar{A}_{K-1} = \bar{a}_{K-1}]\ldots|L_0\})$$

FIGURE 12.3
Iterated conditional means approach to estimating the marginal mean outcome under longitudinal treatment strategy $\bar{a}$.

The submodel and loss function are given by

$$\text{logit}\bar{Q}_{L_k}^{a,*}(\epsilon_k) = \text{logit}\bar{Q}_{L_k}^a + \epsilon_k \frac{1}{g_{0:k-1}},$$

$$\mathcal{L}_{k,\bar{Q}_{L_{k+1}}^a,g}(\bar{Q}_{L_k}^a) = -\frac{I(\bar{A}_{k-1} = \bar{a}_{k-1})}{g_{0:k-1}}$$
$$\times \{\bar{Q}_{L_{k+1}}^a \log \bar{Q}_{L_k}^a + (1 - \bar{Q}_{L_{k+1}}^a) \log(1 - \bar{Q}_{L_k}^a)\},$$

where $g_k = P(A_k = 1|\bar{L}_k, \bar{A}_{k-1})$, and $g_{0:k-1}$ is the cumulative product of g at time points $0, \ldots, k-1$. The parameter is defined by the mapping $\Psi(\bar{Q}_n^{a,*}) = \bar{Q}_{L_0,n}^{a,*} = \frac{1}{n}\sum_{i=1}^n \bar{Q}_{1,n}^{a,*}(L_{0i})$. Estimation proceeds backwards in time from $k = K+1$ to 0. At each step, the initial estimate is targeted by setting $\bar{Q}_{L(k)}^{a,*}$

$= \bar{Q}_{L(k)}^a + \epsilon_k \dfrac{1}{g_{0:k}}$. Each ϵ_k is fit by minimizing the loss function, using observations having $\bar{A}_k = \bar{a}_k$. As in the point treatment setting, fitting ϵ_k at each step ensures the contribution to the efficient influence curve estimating equation at time point k has mean 0, $P_n(D_k^*) = 0$. As a consequence, since $D^*(P)$ is the sum of these contributions over all time points, $P_n(D^*) = 0$. Thus, L-TMLE is double robust. If all components of Q or all components of g are modeled correctly, then the estimator is unbiased.

12.3.3.2 Two Time-Point Example

An example helps to illustrate how this L-TMLE algorithm works. Suppose there is a new vaccine on the market that is supposed to be administered in two doses six months apart. We are interested in understanding the effect of the vaccine on muscle weakness. If we were to design a hypothetical RCT to compare the impact of two doses of the vaccine versus two doses of an inactive comparator, we would randomize patients to one of these two treatment arms, and at the end of follow-up one year later compare the proportion of vaccinated versus unvaccinated subjects who experienced muscle weakness. In a real-world RCT, this evaluation plan is complicated by the fact that some patients do not return for the second dose, and others will be lost to follow-up before the outcome is measured. An analysis of observational data is further complicated by non-random treatment assignment. L-TMLE can be used to obtain an unbiased estimate of the causal effect when causal assumptions hold.

The time ordering of the covariates in this scenario is given in Figure 12.4. L_0 includes baseline measures of static and time-varying covariates. A_{0_T} is baseline treatment. Censoring indicator A_{0_C} captures loss to follow-up prior to measuring time-varying covariates L_1. A_{1_T} indicates treatment assignment at time point 1. Further loss to follow-up is captured by A_{1_C}. Outcome Y is

FIGURE 12.4
Time ordering of intervention and non-intervention nodes, baseline covariates L_0, treatment nodes (A_{0_T}, A_{1_T}), censoring nodes (A_{0_C}, A_{1_C}), time-dependent covariate L_1, and outcome $Y = L_2$.

measured at time point 2. Intervention nodes are $\{A_{0_T}, A_{0_C}, A_{1_T}, A_{1_C}\}$. The statistical question concerns the additive effect of this two time-point intervention on treatment in a study subject to right censoring.

We are interested in comparing the effect of exposure at both time points versus no exposure. Counterfactual outcomes are defined in terms of the full data where there is no censoring. The target ATE parameter is given by $\psi = E\{Y_{(1,0,1,0)} - Y_{(0,0,0,0)}\}$. The strategy for estimating ψ will be to estimate the mean outcome under each intervention separately, then evaluate the difference. The algorithm is as follows.

1. Set all intervention nodes to the desired values for regime 1, $\bar{a} = \{1,0,1,0\}$.

2. Obtain an initial estimate of $\bar{Q}_{L_2}^{\bar{a}} = E(Y|\bar{L}_1, \bar{A}_1 = \bar{a}_1)$.
3. Target this estimate to reduce bias.

$$\bar{Q}_{L_2}^{*\bar{a}} = \text{expit}\left(\text{logit}\bar{Q}_{L_2}^{0\bar{a}} + \hat{\epsilon} H_{L_2}\right),$$

where $H_{L_2} = 1/(g_{A_{0:1}}g_{C_{0:1}})$.

4. Obtain an initial estimate of $\bar{Q}_{L_1}^{\bar{a}}$ that integrates over covariates measured at L_1.

$$\bar{Q}_{L_1}^{0\bar{a}} = E\left(\bar{Q}_{L_2}^{*\bar{a}}|L_0, A_0 = a_0\right)$$

5. Target this estimate to reduce bias.

$$\bar{Q}_{L_1}^{*\bar{a}} = \text{expit}\left(\text{logit}\bar{Q}_{L_1}^{0\bar{a}} + \hat{\epsilon} H_{L_1}\right),$$

where $H_{L_1} = 1/(g_{A_0}g_{C_0})$.

6. Evaluate the mapping for the intervention-specific mean, $\psi_{\bar{a}}$, by taking the average over all observations.

$$\psi_{n\bar{a}} = \bar{Q}_{L_0}^{*\bar{a}} = P_n(\bar{Q}_{L_1}^{*\bar{a}}).$$

7. Set $\bar{a} = (0,0,0,0)$, then repeat steps 1–6.
8. Evaluate $\psi_n = \psi_{n\bar{a}_{1,0,1,0}} - \psi_{n\bar{a}_{0,0,0,0}}$.

A previously published Monte Carlo simulation study compares the performance of the TMLE and the untargeted G-computation estimator [93]. Data for the study were generated as follows.

$$W_1, W_2 \sim \text{Bernoulli}(0.5)$$
$$W_3 \sim N(4,1)$$
$$g_{0_A}(1, L_0) = P_0(A_{0_T} | L_0) = 0.5$$
$$g_{0_C}(0, L_0, A_{0_T}) = P_0(A_{0_C} = 0 | L_0, A_{0_T})$$
$$= \text{expit}(0.1 + 0.5 W_1 + W_2 - 0.1 W_3)$$
$$L_1 = 3 + A_{0_T} - 0.5 W_1 W_3 - 0.5 W_3 + \epsilon_1$$
$$g_{1_A}(1, \bar{L}_1, A_{0_T}) = P_0(A_{1_T} = 1 | A_{0_T}, \bar{L}_1)$$
$$= \text{expit}(-1.2 - 0.2 W_2 + 0.1 W_3 + 0.4 L_1)$$
$$g_{1_C}(0, A_{1_T}, A_{0_T}) = P_0(A_{1_C} = 0 | A_{0_T}, A_{1_T}, \bar{L}_1)$$
$$= \text{expit}(2 - 0.05 W_3 - 0.4 L_1)$$
$$Y = \text{expit}(3 - 0.3 A_{0_T} + 0.1 W_2$$
$$- 0.5 L_1 - 0.5 A_{1_T} + \epsilon_2)$$

with $\epsilon_1, \epsilon_2 \sim_{i.i.d.} N(0,1)$. Under this data generating distribution, $\psi_0 = -0.160$ is the true value of the ATE comparing regimes $\bar{a}_{K_1} = \{1, 0, 1, 0\}$ versus $\bar{a}_{K_2} = \{0, 0, 0, 0\}$. The semi-parametric efficiency bound on the variance of the estimator is $\sigma^2 = 0.39/n$. Datasets analyzed for the simulation study were right-censored at time point k when loss of follow-up indicator $A_{k_C} = 1$. Observations where the observed treatment at k was not consistent with the target regime, $A_{k_T} \ne a_{k_T}$, do not contribute to the model fits at time points $k \le t \le K$.

L-TMLE performance is compared with the untargeted G-computation estimator under correct and misspecified models for the Q and g factors of the likelihood (Table 12.1). Results labeled Q_c were obtained using a set of logistic regression models that includes all terms used to generate the data at each covariate and event node. Using these models to estimate conditional means $\bar{Q}^a_{L(k)}$ gives practically unbiased estimation of ψ_0. Q_{m_1} is a set of mildly misspecified logistic regression models that include main term baseline covariates only. Q_{m_2} is a set of more severely misspecified intercept-only logistic regression models for the outcome. The use of misspecified models for $\bar{Q}^a_k$ illustrates the value of using a DR estimator when the true distribution is unknown.

In order to illustrate L-TMLE's DR, two approaches were used to estimate $g_{0_A}, g_{0_C}, g_{1_A}, g_{1_C}$: the initial treatment assignment probabilities, censoring (loss to follow-up) after the first treatment was received, intermediate switching from treatment to control, and loss to follow-up after the second treatment was received. Targeting was carried out once using correctly specified models for g, and a second time using misspecified main terms logistic regression models that included only baseline covariates, $L_0 = (W_1, W_2, W_3)$. These are labeled as correct and misspecified models for g in Table 12.1.

Empirical bias, percent bias relative to the true parameter value, variance, and MSE of the Monte Carlo estimates are reported in Table 12.1. Results confirm that all estimators are unbiased under correct parametric model specification. As expected, untargeted G-computation estimates are biased under misspecified models Q_{m_1} and Q_{m_2}. Stage 2 targeting using correctly specified models for g greatly reduced bias and had little impact on variance. Under severe misspecification (Q_{m_2}), utilizing a misspecified g provided limited bias reduction.

12.3.3.3 Extensions: Dynamic Regimes and Marginal Structural Models

Up until now, we have considered only static longitudinal regimes in which a pre-defined sequence of treatments is fixed at baseline. However, medical decisions often unfold over time in response to changes in a subject's health status. A dynamic longitudinal treatment regime is one that defines a strategy or decision rule $d(t)$ at baseline. Although the rule is fixed, the actual treatment prescribed at time t depends on values of covariates that arise in the course of treatment, and in response to prior treatment choices. For example, the Centers for Disease Control recommends that adults who are

TABLE 12.1

Empirical Bias, Percent Bias Relative to the True Parameter Value, Variance, and MSE of L-TMLE and an Untargeted G-Computation Estimator of the ATE of a Two Time-Point Intervention ($n = 1000, 500$ Monte Carlo Replicates, $\psi_0 = -0.160$)

		g Correctly Specified				g Misspecified			
		%Rel Bias	Bias	Var	MSE	%Rel Bias	Bias	Var	MSE
Q_c	Untargeted	−0.82	0.001	0.0003	0.0004	−0.82	0.001	0.0003	0.0004
	L-TMLE	−0.60	0.001	0.0003	0.0003	−0.58	0.001	0.0004	0.0004
Q_{m_1}	Untargeted	6.21	−0.010	0.0004	0.0005	6.21	−0.010	0.0004	0.0005
	L-TMLE	1.81	−0.003	0.0004	0.0004	6.15	−0.010	0.0004	0.0005
Q_{m_2}	Untargeted	12.61	−0.020	0.0005	0.0009	12.61	−0.020	0.0005	0.0009
	L-TMLE	3.13	−0.005	0.0004	0.0005	7.26	−0.012	0.0005	0.0006

infected with the human immunodeficiency virus (HIV) be treated with antiretroviral therapy consisting of a combination of drugs [42]. Different drug cocktails are recommended for patients based on medical characteristics, affordability, ease of adherence, and efficacy. Over time, treatment may be altered in response to monitored changes in any of these factors. For example, a lab test might show that the virus has developed resistance to one of the current drugs, with consequences for patient health and mortality. In this case, a physician might follow a dynamic treatment rule such as *begin treatment at baseline and monthly monitoring. Switch to an alternate drug if monitoring shows evidence of drug resistance.* While the physician will not know in advance what drug will be prescribed at each time point, the strategy itself is well defined at baseline.

The mean outcome had, contrary to fact, everyone followed the same dynamic treatment rule can be evaluated using the same L-TMLE algorithm described above for static regimes. However, whether the question is about the effects of static or dynamic regimes, in observational studies subjects will be observed to follow many treatment strategies over time and few subjects may be observed to be following the specific regimes of interest. The data may therefore contain little information for estimating the effects of the entire longitudinal treatment. Marginal structural models (MSMs) were proposed to smooth over areas of little support in the data [31]. An MSM imposes parametric or semiparametric constraints on the counterfactual conditional mean outcome expressed as a function of the treatment strategy over time. An L-TMLE for estimating the parameters of an MSM was presented in Ref. [49] and is described next in the context of dynamic treatment regimes. MSMs are also used to estimate the effects of static regimes.

Consider a logistic MSM for estimating the counterfactual mean outcome under and a set of treatment strategies of interest, $d \in D$. An MSM can be used to model the marginal mean outcome under rule d adjusting for time and any arbitrary function of treatment over time, $f(d, t)$. Baseline covariates can be included in the MSM, but for simplicity are omitted here. The MSM is given by logit $M_\beta(d, t) = \beta_1 + \beta_2 t + \beta_3 f(d, t)$. We are interested in estimating $\beta = (\beta_1, \beta_2, \beta_3)$. Recall that the TL estimation road map dictates defining the target parameter as a feature of the underlying data distribution, P_0. If the MSM is not correctly specified, then there is no direct correspondence between model parameters and features of P_0. We can, however, define ψ in terms of the projection of P_0 onto the MSM, and view the MSM itself as a working model. The target parameter is defined as the minimizer of the loss function rather than as a mapping $\Psi(P_0)$. The statistical target parameter is defined as

$$\psi = \operatorname*{argmin}_{\beta} -E_0 \sum_{t \in \tau} \sum_{d \in D} \{Y_d(t) \log m_\beta(d, t) + (1 - Y_d(t)) \log(1 - m_\beta(d, t))\}.$$

The estimand $\psi_0 = \beta$ solves the equation

$$0 = E_0 \sum_{t \in \tau} \sum_{d \in D} \frac{\frac{d}{d\beta} m_\beta(d, t)}{m_\beta(1 - m_\beta)} \left(E_0\left(Y^d(t) | L(0)\right) - m_\beta(d, t) \right).$$

A stratified longitudinal TMLE algorithm using the BR iterated conditional means approach proceeds by updating an initial estimate of the Q portion of the likelihood to produce a targeted estimate, $\bar{Q}_{L_0}^{d,t*}$, for each time point, t, and strategy $d \in D$. Targeting a multidimensional parameter requires a multi-dimensional fluctuation at each step, $\epsilon_k = (\epsilon_{1_k}, \epsilon_{2_k}, \epsilon_{3_k})$.

$$\bar{Q}_{L_k}^{d*} = \bar{Q}_{L_k}^{d} + \epsilon_k \frac{h_1(d, t)}{g_{0:k-1}},$$

with $h_1(d, t) = \dfrac{\frac{d}{d\beta} m_\beta(d, t)}{m_\beta(1 - m_\beta)}$. ϵ is fit by minimizing the loss function (e.g., the negative log likelihood), using observations where $\bar{A}_{k-1} = \bar{a}_{k-1}$. Estimates of the MSM parameters are obtained by regressing $\bar{Q}_{L_0}^{d,t*}$ onto appropriate covariates in the model $(1, t, f(d, t))$, in a dataset where observations at all time points are stacked [78].

A general L-TMLE for estimating parameters of an arbitrary MSM was developed by Petersen et al. [49]. This L-TMLE modifies the procedure just described by allowing data to be pooled over treatment regimes when estimating fluctuation parameter ε. The advantage of this is that pooling stabilizes the estimates at time points where there may be sufficient support in the data for some regimes but not others. The pooled L-TMLE stacks the datasets for all strategies to estimate a single (multidimensional) ε_k at each time point. The dataset has $n \times |D|$ observations. An alternative pooled TMLE pools over time points as well as strategies. Initial estimates of $\bar{Q}_{L_k}^d$ are obtained for all k from $K + 1$ to 0, then targeted simultaneously. The targeting step is iterated until convergence. This dataset has $n \times |D| \times K + 1$ observations. Targeting using data pooled over time ensures $P_n(D^*) = 0$ without requiring $P_n(D_k^*) = 0$ at all time points. An open source R package, *ltmle*, implements all L-TMLEs discussed in this chapter [80].

12.4 Applications in Health Care Research

TL has widespread applicability in health care research to improve patient and population morbidity and mortality. Utilizing SL for risk score prediction can help identify and tailor care for high-risk patients. Risk scores in common use may be simple summary measures, such as the Apgar score applied to newborn infants [16], or predictions from logistic regression models. SL may be able to provide more accurate predictions by better capturing the functional form of covariate–outcome relationships.

Consider hospital ICU mortality, where risk scores typically used to identify patients at high risk of mortality in hospital ICUs include the Simplified Acute Physiology Score (SAP, SAP-II) and the Acute Physiology and Chronic Health Evaluation (APACHE, APACHE-II) scores [50]. A study of 24,508 patients admitted to a Boston, Massachusetts hospital ICU compared SL with these standard approaches. The SL library included neural networks, random forest, Bayesian additive regression trees, penalized regression, logistic regression, and bagging and boosting machine learning algorithms. Performance was measured as the cross-validated area under the receiver operating curve (cv-AUC), which evaluates a predictor's ability to discriminate between higher and lower risk patients. SL slightly increased cv-AUC and more accurately predicted subsequent mortality than the widely used scoring algorithms (SL cv-AUC = 0.85, new SAPS cv-AUC = 0.83). SL's performance improved even more when the individual components of composite covariates used by the SAPS algorithm were provided (cv-AUC = 0.88).

TMLE can be combined with SL to identify and rank covariates according to their predictive strength. Hubbard et al. [32] employed TL to obtain variable importance rankings for predicting time-dependent mortality after major trauma. Following the roadmap, they defined the parameter of interest as the impact of a clinically meaningful change in the level of each covariate on the risk difference (ATE) parameter. To assess how the level of risk for each factor changed over time, they divided the outcome into three non-overlapping risk windows. SL was used to fit the initial regression of the outcome on covariates measured on 980 patients in the Prospective Observational Multicenter Major Trauma Transfusion study [55]. TMLE was used to target the effect estimate for each of 23 pre-selected risk factors in turn during each risk window. This was done by viewing each risk factor as the treatment of interest in turn, and obtaining a targeted window-specific ATE estimate that adjusted for measured confounders. This provides nuanced, interpretable insight into how the importance of different risk factors can rise and fall in the hours following a major trauma.

When causal assumptions are met, TL's parameter estimates have a valid causal interpretation. For example, in a randomized controlled trial where patients are closely monitored, investigators may feel confident that informative loss to follow-up is well explained by measured covariates. TL can also produce valid casual effect estimates from observational data when sufficient information on confounders exists.

TMLE was applied in an observational study to assess the impact of statin use on one-year risk of mortality following an acute myocardial infarction (MI) [43]. Electronic health records on adults with no prior MI and no prior use of statins were extracted from the Clinical Practice Research Datalink and linked with hospitalization records (n = 32,792). Because treatment was not randomized, it was necessary to adjust for selection bias. A high-dimensional propensity score (hdPS) procedure offers a pragmatic approach to identifying likely confounders among a set of diagnosis codes, procedure codes, and constructed correlated proxies of unmeasured confounders from these high-dimensional data [77].

The study investigated the performance of TMLE under different specifications of outcome regression models and propensity score models in comparison with IPTW. As a substitution estimator that respects bounds on $\mathcal{M}$, TMLE was able to provide more stable estimates despite near-positivity violations in the data. TMLE's marginal odds ratio estimate of approximately 0.7 was stable across variations in the specification of the propensity score model when the outcome regression was adjusted for all confounders. This suggests that bias reduction was largely taking place in Stage 1 of the TMLE procedure. RCTs and meta-analyses of the effect of statins on mortality have found a protective effect in a range of patient sub-populations, with similar estimated odds ratios [15,82].

TL is also useful for evaluating the impact of a series of treatment decisions made over time in the presence of time-dependent confounding, and provides valid inference even when data are clustered. Schnitzer et al. [78] used re-analyzed data collected in the PROmotion of Breastfeeding Intervention Trial (PROBIT) in order to evaluate the effect of breastfeeding on gastrointestinal infection in infants. Because the duration of breastfeeding varies across mother–infant pairs, the researchers defined a set of questions and target parameters corresponding to different lengths of time of breastfeeding. Data on 17,036 mother–infant pairs clustered within 31 hospitals included demographic information, mother characteristics (age, smoking status, number of previous children, etc.), infant characteristics (gestational age, height, weight, etc.), and geographic region. Additional data were collected longitudinally at six scheduled follow-up visits during the infant's first year. Intermediate infections were a suspected source of time-dependent confounding. L-TMLE with and without SL was compared with IPTW and G-computation. All estimators found a protective effect of breastfeeding for longer versus shorter durations. For two out of three target parameters, TMLE + SL estimated effect sizes were slightly larger than the other estimators, with narrower width confidence intervals. TL's smaller standard errors were presumably due to a combination of SL's ability to better explain variation in the outcome and further reducing variance in the parameter estimate itself through TMLE's targeting step.

12.5 Summary

TL provides a sound foundation for learning from data. It is built upon a theoretical foundation that incorporates machine learning into statistical analysis while preserving the ability to quantify uncertainty in the parameter estimate. The estimation roadmap stresses the importance of clearly stating the scientific question and translating that into a statistical question that can be answered from data. TL practice is to start by defining a statistical model $\mathcal{M}$ that reflects what is truly known about the possible distribution of the data. Formulating the question is a separate activity from choosing an estimation approach. With few exceptions, the parameter of interest is defined as a feature of the unknown underlying probability distribution, P_0, rather than as a coefficient in a parametric model. Identifying assumptions link this parameter with a feature of observed data. At this point, many estimation procedures are available. TL encourages flexible modeling through the use of SL, and double robust targeting of the parameter of interest using TMLE. As a substitution estimator, TMLE is designed to respect known bounds on $\mathcal{M}$. Practical benefits include limiting bias amplification due to model misspecification, and improved finite sample performance.

Large observational datasets are increasingly being used to address research, policy, and business concerns. When data are high-dimensional, the number of potential main terms and higher order interactions describing the joint distribution will be explosive. Although n is large in these settings, there will still be sparsity in the data. While a parametric model will smooth over areas of little support in the data, parametric modeling simply does not scale to high-dimensional settings. It is not possible to *a priori* specify a model that accurately captures the complexity of the relationships among high-dimensional covariates. In practice, such an estimator will inevitably be biased, producing results driven by assumptions of convenience rather than by data. Because variance shrinks as the number of observations grows, the biased estimate will be bracketed by a narrow confidence interval. This analysis will provide an inaccurate result, and misleading assurance regarding its reliability. Contrast this with TL's data adaptive semi-parametric approach. While TL cannot overcome a complete lack of information in the data, the combination of SL and TMLE is designed to optimally exploit the information relevant for estimating the low-dimensional parameter of interest.

Appropriately adjusting for sources of bias requires insight into mechanisms underlying potential baseline and time-dependent confounding. While using sophisticated methodologies does not guarantee success, it does avoid falling prey to common data analysis pitfalls. Publicly available software enables analysts to effectively use SL and TMLE in practice. Sound statistical principles, explicit statement of assumptions, and transparent rationale for modeling choices remain the guiding principles for extracting knowledge from data.

12.6 Open Problems and Future Directions

TL will evolve as applied researchers tackle ever more complex estimation problems. In areas such as infectious disease research, obesity research, and social epidemiology, some subjects' treatments and outcomes will affect the outcomes of other subjects. Methodology for estimating causal effects when these relationships exists is on the cutting edge of causal inference research. The dependence structure is captured by a network that may contain clusters of nodes that are closely related, and more loosely connected to neighboring clusters. TMLE for network analysis has been described and published as an open source package [79,81,89] with limited application to date [5].

A second compelling open problem involves causal effect estimation in the presence of unmeasured confounding. There has been an increasing interest in estimating causal effects when some confounders are not measured due to the increasing availability of large administrative health care datasets re-purposed for secondary analysis. In the absence of a strong instrumental variable, causal effect estimation is feasible only by making additional assumptions. The hdPS approach described earlier assumes that composite covariates can be constructed to serve as correlated proxies of unmeasured confounders. We have demonstrated that combining hdPS with a collaborative TMLE that data-adaptively models the propensity score can improve bias and variance of causal effect estimates obtained using electronic health records data, but needs to further develop our understanding of the trade-offs [34]. A second innovative approach to causal effect estimation when there is unmeasured confounding requires the availability of data on known negative controls [40,85]. These are drugs known not to be causally associated with the outcome under study, and outcomes known not to be causally associated with the treatment of interest. Successful application of TL in this context is an important area of future study.

References

1. S. Arlot. V-fold cross-validation improved: V-fold penalization. *arXiv preprint arXiv:0802.0566*, 2008.
2. S. Arlot and A. Celisse. A survey of cross-validation procedures for model selection. *Statistic Surveys*, 4:40–79, 2010.
3. S. Arlot and M. Lerasle. V-fold cross-validation and V-fold penalization in least-squares density estimation. *eprint arXiv:1210.5830*, 2012.
4. L. Balzer, J. Ahern, S. Galea, and M.J. van der Laan. Estimating effects with rare outcomes and high dimensional covariates: Knowledge is power. *Epidemiologic Methods*, 2015.
5. L. Balzer, P. Staples, J.P. Onnela, and V. DeGruttola. Using a network-based approach and targeted maximum likelihood estimation to evaluate the effect of adding pre-exposure prophylaxis to an ongoing test-and-treat trial. *Clinical Trials*, 2017.
6. H. Bang and J.M. Robins. Doubly robust estimation in missing data and causal inference models. *Biometrics*, 61:962–72, 2005.
7. P.J. Bickel, C.A.J. Klaassen, Y. Ritov, and J. Wellner. *Efficient and Adaptive Estimation for Semiparametric Models*. Springer-Verlag, 1997.
8. L. Breiman. Stacked regression. *Machine Learning*, 24:49–64, 1996.
9. L. Breiman. Statistical modeling: The two cultures. *Statistical Science*, 16(3):199–215, 2001.
10. L. Breiman, J.H. Friedman, R. Olshen, and C.J. Stone. *Classification and Regression Trees*. The Wadsworth Statistics/Probability Series. Chapman and Hall, New York, 1984.
11. N.E. Breslow. Analysis of survival data under the proportional hazards model. *International Statistical Review/Revue Internationale de Statistique*, 43(1):45–57, 1975.
12. A. Chambaz, P. Neuvial, and M.J. van der Laan. Estimation of a non-parametric variable importance measure of a continuous exposure. *Electronic Journal of Statistics*, 6:1059, 2012.
13. H.A. Chipman, E.I. George, and R.E. McCulloch. Bart: Bayesian additive regression trees. *The Annals of Applied Statistics*, 4(1):266–298, 2010.
14. M.L. Petersen E. Ledell, M.J. van der Laan. *Case-Control Subsampling for Super-Learner*, 2014. R package.
15. S. Ebrahim, F. Taylor, K. Ward, A. Beswick, M. Burke, and G. Davey Smith. Multiple risk factor interventions for primary prevention of coronary heart disease. *Cochrane Database of Systematic Reviews*, 2011.
16. M. Finster and M. Wood. The Apgar score has survived the test of time. *Anesthesiology*, 102:855–857, 2005.
17. J. Friedman, T. Hastie, and R. Tibshirani. Regularization paths for generalized linear models via coordinate descent. *Journal of Statistical Software*, 33(1):1–22, 2010.
18. M. Fuentes. Reproducible research in JASA. *AMSTAT News*, July 2016.
19. R.D. Gill, M.J. van der Laan, and J.M. Robins. Coarsening at random: Characterizations, conjectures and counter-examples. In D.Y. Lin and T.R. Fleming, editors, *Proceedings of the First Seattle Symposium in Biostatistics*, pages 255–294. Springer-Verlag, New York, 1997.
20. S. Gruber, R.W. Logan, I. Jarrín, S. Monge, and M.A. Hernán. Ensemble learning of inverse probability weights for marginal structural modeling in large observational datasets. *Statistics in Medicine*, 34(1):106–117, 2015.
21. S. Gruber and M.J. van der Laan. An application of collaborative targeted maximum likelihood estimation in causal inference and genomics. *The International Journal of Biostatistics*, 6(1), 2010.
22. S. Gruber and M.J. van der Laan. A targeted maximum likelihood estimator of a causal effect on a bounded continuous outcome. *The International Journal of Biostatistics*, 6(1), 2010.
23. S. Gruber and M.J. van der Laan. Targeted minimum loss based estimation of a causal effect on an outcome with known conditional bounds. *The International Journal of Biostatistics*, 8, 2012.
24. S. Gruber and M.J. van der Laan. tmle: An R package for targeted maximum likelihood estimation. *Journal of Statistical Software*, 51, November 2012.
25. S. Gruber and M.J. van der Laan. *tmle: Targeted Maximum Likelihood Estimation*, 2014. R package version 1.2.0-4, http://CRAN.R-project.org/package=tmle.
26. S. Gruber and M.J. van der Laan. Consistent causal effect estimation under dual misspecification and implications for confounder selection procedures. *Statistical Methods in Medical Research*, 24(6):1003–1008, 2015.
27. F.R. Hampel. The influence curve and its role in robust estimation. *Journal of the American Statistical Association*, 69(346):383–393, 1974.
28. T. Hastie, R. Tibshirani, and J. Friedman. *The Elements of Statistical Learning*. Springer Series in Statistics. Springer, 2002.
29. T.J. Hastie. *gam: Generalized Additive Models*, 2015. R package version 1.12.
30. D.F. Heitjan and D.B. Rubin. Ignorability and coarse data. *The Annals of Statistics*, 19(4):2244–2253, December 1991.
31. M.A. Hernan, B. Brumback, and J.M. Robins. Marginal structural models to estimate the causal effect of zidovudine on the survival of HIV-positive men. *Epidemiology*, 11(5):561–570, 2000.

32. A. Hubbard, I. Diaz Munoz, A. Decker, J.B. Holcomb, M.A. Schreiber, E.M. Bulger, K.J. Brasel, E.E. Fox, D.J. Del Junco, C.E. Wade et al. Time-dependent prediction and evaluation of variable importance using super learning in high dimensional clinical data. *The Journal of Trauma and Acute Care Surgery*, 75(1 Suppl 1):S53–S60, 2013.
33. M. Jacobsen and N. Keiding. Coarsening at random in general sample spaces and random censoring in continuous time. *The Annals of Statistics*, 23:774–786, 1995.
34. C. Ju, S. Gruber, S.D. Lendle, J.M. Franklin, R. Wyss, S. Schneeweiss, and M.J. van der Laan. *Scalable collaborative targeted learning for large scale and high-dimensional data*. Technical report, Division of Biostatistics, University of California, Berkeley, 2016.
35. N. Kreif, R. Grieve, R. Radice, S. Gruber, and J.S. Sekhon. Comparing targeted maximum likelihood estimation and bias-corrected matching for average treatment effects when using data-adaptive estimation methods. *Statistical Methods in Medical Research*, 2014.
36. B.K. Lee, J. Lessler, and E.A. Stuart. Improved propensity score weighting using machine learning. *Statistics in Medicine*, 29:337–346, 2009.
37. R.J.A. Little and D. B. Rubin. *Statistical Analysis with Missing Data*. Wiley Series in Probability and Mathematical Statistics. Wiley, 1987.
38. S. McBride. Determination of rates and terms for digital performance in sound recordings and ephemeral recordings (web IV), testimony of Stephan McBride, 2014.
39. D. McCaffrey, L. Burgette, B.A. Griffin, and C. Martin. *Toolkit for Weighting and Analysis of Nonequivalent Groups: A Tutorial for the Twang SAS Macro*, 2014.
40. W. Miao and E.J. Tchetgen Tchetgen. Bias attenuation and identification of causal effects with multiple negative controls. *American Journal of Epidemiology*, 185(10):950–953, 2017.
41. S.A. Murphy, M.J. van der Laan, and J.M. Robins. Marginal mean models for dynamic treatment regimens. *Journal of the American Statistical Association*, 96:1410–1424, 2001.
42. Panel on Antiretroviral Guidelines for Adults, Adolescents. Department of Health, and Human Services. Guidelines for the use of antiretroviral agents in HIV-1-infected adults and adolescents, 2016.
43. M. Pang, T. Schuster, K.B. Filion, M. Eberg, and R.W. Platt. Targeted maximum likelihood estimation for pharmacoepidemiologic research. *Epidemiology*, 27:570–577, 2016.
44. J. Pearl. *Causality: Models, Reasoning, and Inference*. Cambridge University Press, Cambridge, 2nd edition, 2000.
45. J. Pearl. Causal inference in statistics: An overview. *Statistics Surveys*, 3:96–146, 2009.
46. J. Pearl. An introduction to causal inference. *The International Journal of Biostatistics*, 6(2), 2010.
47. R.D. Peng. Reproducible research in computational science. *Science*, 334(6060):1226–1227, 2011.
48. M.L. Petersen, K.E. Porter, S. Gruber, Y. Wang, and M.J. van der Laan. Diagnosing and responding to violations in the positivity assumption. *Statistical Methods in Medical Research*, published online October 28 (doi: 10.1177/0962280210386207), 2010.
49. M.L. Petersen, J. Schwab, S. Gruber, N. Blaser, M. Schomaker, and M.J. van der Laan. Targeted maximum likelihood estimation for dynamic and static longitudinal marginal structural working models. *Journal of Causal Inference*, 2(2):147–185, 2014.
50. R. Pirracchio, M.L. Petersen, M. Carone, M.R. Rigon, S. Chevret, and M.J. van der Laan. Mortality prediction in intensive care units with the Super ICU learner algorithm (SICULA): A population-based study. *The Lancet Respiratory Medicine*, 3(1):42–52, 2015.
51. R. Polikar. Ensemble learning. *Scholarpedia*, 4(1):2776, 2009.
52. E.C. Polley and M.J. van der Laan. *Super learner in prediction*. Technical Report 200, U.C. Berkeley Division of Biostatistics Working Paper Series, 2010.
53. E.C. Polley and M.J. van der Laan. *SuperLearner: Super Learner Prediction*, 2016. R package version 2.0-19, http://CRAN.R-project.org/package=SuperLearner.
54. R Core Team. *R: A Language and Environment for Statistical Computing*. R Foundation for Statistical Computing, Vienna, Austria, 2016.
55. M.H. Rahbar, E.E. Fox, D.J. del Junco, B.A. Cotton, J.M. Podbielski, N. Matijevic, M.J. Cohen, M.A. Schreiber, J. Zhang, P. Mirhaji et al., and PROMMTT Investigators. Coordination and management of multicenter clinical studies in trauma: experience from the PRospective Observational Multicenter Major Trauma Transfusion (PROMMTT) study. *Resuscitation*, 83:459–464, 2012.
56. G. Ridgeway, D. McCaffrey, A. Morral, L. Burgette, and B.A. Griffin. *Toolkit for Weighting and Analysis of Nonequivalent Groups: A Tutorial for the Twang Package*, 2013. R package.
57. J.M. Robins. A new approach to causal inference in mortality studies with sustained exposure periods—Application to control of the healthy worker survivor effect. *Mathematical Modelling*, 7:1393–1512, 1986.
58. J.M. Robins. The analysis of randomized and non-randomized aids treatment trials using a new approach in causal inference in longitudinal studies. In L. Sechrest, H. Freeman, and A. Mulley, editors, *Health Service Methodology: A Focus on AIDS*, pages 113–159. U.S. Public Health Service, National Center for Health Services Research, Washington DC, 1989.
59. J.M. Robins. The control of confounding by intermediate variables. *Statistics in Medicine*, 8:679–701, 1989.
60. J.M. Robins. Correcting for non-compliance in randomized trials using structural nested mean models. *Communications in Statistics*, 23:2379–2412, 1994.
61. J.M. Robins. Causal inference from complex longitudinal data. In M. Berkane, editor, *Latent Variable Modeling and Applications to Causality*, pages 69–117. Springer-Verlag, New York, 1997.
62. J.M. Robins. Marginal structural models. 1997 Proceedings of the American Statistical Association Section on Bayesian Statistical Science, pages 1–10, 1997.
63. J.M. Robins. Structural nested failure time models. In P. Armitage, T. Colton, P.K. Andersen, and N. Keiding, editors, *The Encyclopedia of Biostatistics*. John Wiley and Sons, Chichester, UK, 1997.

64. J.M. Robins. Robust estimation in sequentially ignorable missing data and causal inference models. In Proceedings of the American Statistical Association: Section on Bayesian Statistical Science, pages 6–10, 2000.
65. J.M. Robins and M.A. Hernan. Estimation of the causal effects of time-varying exposures. In Fitzmaurice G., Davidian M., Verbeke G., and Molenberghs G., editors, *Advances in Longitudinal Data Analysis*, pages 553–599. Chapman and Hall/CRC Press, Boca Raton, FL, 2009.
66. J.M. Robins, M.A. Hernan, and B. Brumback. Marginal structural models and causal inference in epidemiology. *Epidemiology*, 11(5):550–560, 2000.
67. J.M. Robins and A. Rotnitzky. Semiparametric efficiency in multivariate regression models with missing data. *Journal of the American Statistical Association*, 90(429):122–9, March 1995.
68. J.M. Robins and A. Rotnitzky. Comment on the Bickel and Kwon article, 'Inference for semiparametric models: Some questions and an answer.' *Statistica Sinica*, 11(4):920–936, 2001.
69. J.M. Robins and A. Rotnitzky. Estimation of treatment effects in randomized trials with noncompliance and a dichotomous outcome using structural nested mean models. *Biometrika*, 91(4):763–783, 2004.
70. S Rose. Mortality risk score prediction in an elderly population using machine learning. *American Journal of Epidemiology*, 177(5):443–452, 2013.
71. P.R. Rosenbaum and D.B. Rubin. The central role of the propensity score in observational studies for causal effects. *Biometrika*, 70:41–55, 1983.
72. M. Rosenblum and M.J. van der Laan. *Simple examples of estimating causal effects using targeted maximum likelihood estimation*. Technical report 262, Division of Biostatistics, University of California, Berkeley, 2010.
73. D.B. Rubin. Estimating causal effects of treatments in randomized and nonrandomized studies. *Journal of Educational Psychology*, 64:688–701, 1974.
74. K.E. Rudolph and M.J. van der Laan. *Double robust estimation of encouragement-design intervention effects transported across sites*. Technical report 335, U.C. Berkeley Division of Biostatistics Working Paper Series, 2015.
75. S. Sapp, M.J. van der Laan, and J. Canny. Subsemble: An ensemble method for combining subset-specific algorithm fits. *Journal of Applied Statistics*, 41:1247–1259, 2014.
76. D.O. Scharfstein, A. Rotnitzky, and J.M. Robins. Adjusting for non-ignorable drop-out using semiparametric nonresponse models, (with discussion and rejoinder). *Journal of the American Statistical Association*, 94(448):1096–1120 (1121–1146), 1999.
77. S. Schneeweiss, J.A. Rassen, R.J. Glynn, J. Avorn, H. Mogun, and M.A. Brookhart. High-dimensional propensity score adjustment in studies of treatment effects using health care claims data. *Epidemiology*, 20, 2009.
78. M.E. Schnitzer, M.J. van der Laan, E.E.M. Moodie, and R.W. Platt. Effect of breastfeeding on gastrointestinal infection in infants: A targeted maximum likelihood approach for clustered longitudinal data. *The Annals of Applied Statistics*, 8(2):703–725, 2014.
79. J. Schwab, S. Lendle, M. Petersen, and M.J. van der Laan. tmlenet: *Targeted Maximum Likelihood Estimation for Network Data*, 2015. R package version 0.1.0, https://CRAN.R-project.org/package=tmlenet.
80. J. Schwab, S. Lendle, M. Petersen, and M.J. van der Laan. ltmle: *Longitudinal Targeted Maximum Likelihood Estimation*, 2016. R package version 0.9-8, https://cran.r-project.org/web/package=ltmle.
81. O. Sofrygin and M.J. van der Laan. *Semi-parametric estimation and inference for the mean outcome of the single time-point intervention in a causally connected population*. Technical Report 344, UC Berkeley Division of Biostatistics Working Paper Series, December 2015.
82. U. Stenestrand, L. Wallentin, and Swedish Register of Cardiac Intensive Care (RIKS-HIA). Early statin treatment following acute myocardial infarction and 1-year survival. *Journal of the American Medical Association*, 285:430–436, 2001.
83. O. Stitelman, B. Dalessandro, C. Perlich, and F. Provost. Estimating the effect of online display advertising on browser conversion. *Data Mining and Audience Intelligence for Advertising*, 8, 2011.
84. O.M. Stitelman, V. De Gruttola, and M.J. van der Laan. *A general implementation of TMLE for longitudinal data applied to causal inference in survival analysis*. Technical report 281, Division of Biostatistics, University of California, Berkeley, April 2011.
85. E.J. Tchetgen Tchetgen. The control outcome calibration approach (COCA) for causal inference with unobserved confounding. *American Journal of Epidemiology*, 179:633–640, 2014.
86. K.M. Ting and I.H. Witten. Issues in stacked generalization. *Journal of Artificial Intelligence Research*, 10:271–289, 1999.
87. B. Toth and M.J. van der Laan. *TMLE for marginal structural models based on an instrument*. Technical Report 350, U.C. Berkeley Division of Biostatistics Working Paper Series, June 2016.
88. M.J. van der Laan. Targeted maximum likelihood based causal inference: Part I. *The International Journal of Biostatistics*, 6(2):Article 2, 2010.
89. M.J. van der Laan. Causal inference for a population of causally connected units. *Journal of Causal Inference*, 2:1374, 2014.
90. M.J. van der Laan and S. Dudoit. *Unified cross-validation methodology for selection among estimators and a general cross-validated adaptive epsilon-net estimator: Finite sample oracle inequalities and examples*. Technical report 130, Division of Biostatistics, University of California, Berkeley, November 2003.

91. M.J. van der Laan, S. Dudoit, and A.W. van der Vaart. *The cross-validated adaptive epsilon-net estimator.* Technical report 142, Division of Biostatistics, University of California, Berkeley, February 2004.
92. M.J. van der Laan and S. Gruber. Collaborative double robust penalized targeted maximum likelihood estimation. *The International Journal of Biostatistics*, 6(1), January 2010.
93. M.J. van der Laan and S. Gruber. Targeted minimum loss based estimation of causal effects of multiple time point interventions. *The International Journal of Biostatistics*, 8, 2012.
94. M.J. van der Laan, E. Polley, and A. Hubbard. Super Learner. *Statistical Applications in Genetics and Molecular Biology*, 6(25), 2007.
95. M.J. van der Laan and J.M. Robins. *Unified Methods for Censored Longitudinal Data and Causality.* Springer-Verlag, New York, 2003.
96. M.J. van der Laan and S. Rose. *Targeted Learning: Prediction and Causal Inference for Observational and Experimental Data.* Springer, New York, 2011.
97. M.J. van der Laan and D. Rubin. Targeted maximum likelihood learning. *The International Journal of Biostatistics*, 2(1), 2006.
98. A.W. van der Vaart, S. Dudoit, and M.J. van der Laan. Oracle inequalities for multi-fold cross-validation. *Statistics and Decisions*, 24(3):351–371, 2006.
99. W.N. Venables and B.D. Ripley. *Modern Applied Statistics with S.* Springer, New York, 4th edition, 2002.
100. D.H. Wolpert. Stacked generalization. *Neural Networks*, 5:241–259, 1992.

Section III

Applications of Signal Processing and Machine Learning for Big Biomedical Data

13

Scalable Signal Data Processing for Measuring Functional Connectivity in Epilepsy Neurological Disorder

Arthur Gershon, Samden D. Lhatoo, Curtis Tatsuoka, Kaushik Ghosh, Kenneth Loparo, and Satya S. Sahoo

CONTENTS

13.1 Introduction 259
    13.1.1 Functional Networks in the Brain 259
    13.1.2 Related Work 260
    13.1.3 Outline 261
13.2 Methods 261
    13.2.1 Recording of Brain Electrical Activity 261
    13.2.2 Managing Signal Data: The CSF 261
    13.2.3 Processing Signal Data to Compute Functional Connectivity Measures 263
    13.2.4 Computational Pipeline for Analyzing Signal Data 263
13.3 Results 265
13.4 Discussion 266
    13.4.1 Developing an Accurate Measure of Signal Correlation in Neurological Disorders 266
    13.4.2 Improving the Performance of the Computational Pipeline 266
    13.4.3 Use of Ontologies for Standardizing Terminology in Signal Data Analysis 267
13.5 Conclusion 267
Acknowledgement 267
References 268

13.1 Introduction

13.1.1 Functional Networks in the Brain

The human brain is one of the most important organs in the body; it is responsible for many critical functions, including cognition, memory, language, and execution [1]. The brain has been the subject of a large body of research; nevertheless, we still have a rather limited understanding of various aspects of the brain and its constituents, as well as the interactions between brain structures, especially in the context of neurological disorders. Previous research efforts, which were focused solely on understanding the complexities of the brain's structure and functions, have identified numerous substructures in various regions of the brain that are responsible for a wide variety of different cognitive and physiological functions [2,3].

In addition to understanding the functions of each substructure as an individual unit, it is important to determine how distinct regions in the brain interact and work together. Physical connections between distinct brain regions form *structural networks*, whereas associations that are formed over time and developed during processes such as speech, memorization, or other physiological events are referred to as *functional networks* [4]. *Functional connectivity measures* are computed by evaluating statistical correlations of physiological signals recorded from different brain regions [2]. Analyses of these latter connections have proven to be quite approachable using signal processing methods.

There are several important applications of functional connectivity analysis; for example, it can be used toward the goal of better understanding the mechanisms that underpin various capabilities of the human brain. The measurement of functional connectivity in the brain can also be used to study neurological disorders, such as epilepsy. Epilepsy is one of the most common serious neurological disorders, affecting more than 65 million individuals worldwide in various forms [5].

Epileptic disorders are characterized by seizures resulting from the generation and propagation of abnormal electrical activity in the brain [6,7]. These characteristic phenomena of epilepsy are commonly recorded and observed through the use of electroencephalograms (EEGs), and functional connectivity analysis can be applied to EEG data to aid in the understanding of the development and progression of epilepsy. There has been extensive research focused on the use of signal processing and data mining methods on intracranial recordings to determine case-specific collections of intracranial sites involved in seizure onset and that constitute the *epileptogenic zone* [2,8]. Similarly, other research has focused on computing functional connectivity measures between different brain regions that constitute an epileptogenic network [8].

Several different categories of correlation measures have been used to evaluate functional connectivity [9,10]. One example is the *linear* correlation measures that include cross-correlation, Pearson's correlation coefficient, and coherence. These measures quantify the *linear dependence* of the time series, assuming that information propagates directly from one site to another without interference from ambient noise. By contrast, functional connectivity measures such as the average amount of mutual information (AAMI) index and other dynamical system analysis approaches do take into account interference from other signals, and thus do not make any direct assumption on how signal information propagates as a function of time. Measures in this category are often referred to as *non-linear* measures of functional connectivity [2,10]. Non-linear functional connectivity measures have been shown to provide more accurate results vis-à-vis linear approaches [11]. It is important to note, however, that existing non-linear correlation techniques are based on the temporal characteristics of the signal (e.g., the signal amplitude) instead of frequency domain features, such as those derived from the Fourier transform [8], whereas coherence, a linear correlation measure, uses signal frequency values for analysis [8,10].

In addition to the development of new theoretical models to correlate signal data, the massive volume, wide variety, and rapid rate of signal data generation require the development of highly scalable computational tools and platforms [1,12,13]. The computational challenges associated with this "Big Data" in neuroscience require the development of algorithms and data structures that can leverage high-performance distributed computing resources (e.g., cloud platforms) to store, analyze, and visualize large-scale datasets [1]. In particular, neuroscience Big Data requires the development of new multi-modal data representation formats that can effectively address the limitations of existing file formats and leverage distributed computing infrastructure for scalability and efficient analysis.

It is important to address the limitations of present data formats, such as the European Data Format (EDF) that is widely used for storing physiological signal data [14,15]. Files using EDF consist of two major components: (1) a header containing patient data and recording metadata, including the names of each channel, recording times, and units of measurement, represented as ASCII strings; and (2) the data record with a list of all signal values, stored in a binary format, corresponding to each time sample of the data. EDF files store signal data as a collection of data recordings organized by the time of recording, and although this significantly reduces the overall size of the file it makes, it can be difficult to extract and analyze recordings from individual channels, e.g., in the computation of functional connectivity measures.

There has been a significant amount of work in the development of neuroscience data storage formats to address the limitations of EDF and other existing file formats. In the next section, we describe various approaches used to manage neuroscience data and the computation of functional connectivity measures derived from the signal data.

13.1.2 Related Work

Although we focus on computing functional connectivity measures derived from signal data in this chapter, there are a variety of other approaches used to determine functional networks in the brain. For example, instead of using EEG electrical signal data, a substantial amount of neurological research uses blood oxygen level dependency signals, measured using functional magnetic resonance imaging (fMRI), to derive functional connectivity measures both in general neuroscience [2] and specifically in epilepsy research [16,17].

There has also been extensive work in the development of file formats to address the challenges in storing and analyzing neurological data [18]. For example, the Neuroscience Electrophysiology Object (NEO) format is an object-oriented file format based on the Python programming language. The NEO format is proposed as a natural method for storing neurological data due to its object-oriented nature, which makes it suitable for use across computing platforms [19]. Similarly, the Hierarchical Data Format (HDF5) has been developed as a general scientific data storage format with implementations in a variety of programming environments. The HDF5 project has developed various optimization techniques for data storage and access [20], which has made it popular as a file format for storing neurological data.

13.1.3 Outline

The remainder of this chapter is structured as follows. In Section 13.2, we describe the computation of functional connectivity measures in epilepsy using techniques of signal analysis, with a particular focus on various statistical models used to derive correlations among signal data recorded from different channels. In addition, we introduce a novel computational pipeline that uses a new data format, called the *Cloudwave Signal Format (CSF)*, to process and analyze signal data using a nonlinear correlation technique. In Section 13.3, we give a broad overview of the results we have obtained from the use of this pipeline. Then, in Section 13.4, we describe the broader application of our techniques and tools used to compute functional connectivity in epilepsy patients, and discuss proposed enhancements. We conclude in Section 13.5 with a summary of our work.

13.2 Methods

In this section, we give an overview of the CSF and its role in enabling the distributed storage of signal data, we describe techniques of signal data analysis used to derive functional connectivity measures, and finally we introduce the multi-step computational pipeline we have developed that implements the CSF and the above data analytic techniques for measuring functional connectivity.

13.2.1 Recording of Brain Electrical Activity

The study of electrical activity in the brain has been of fundamental importance in neurology since Galvani's experiments on electrical activity in frogs and the subsequent development of an electronic theory of the nervous system [21]. We recall that the brain is composed of special cells known as *neurons* that regulate various processes according to location, and a cognitive function is understood to result from the transmission of information between two neurons [22]. There are a number of ways in which signals can be relayed among neurons, including the use of chemicals such as neurotransmitters for processes that require a lossless transmission of information [23].

Electrical signals comprise another category of methods for neuronal communication. They are used to rapidly convey information among different regions of the brain in the execution of reactive and motor skills, and in the synchronization of cognitive functions that form the basis of processes such as learning and perception [23]. The science of *electroencephalography* (EEG) encompasses a variety of methods of gathering data on intracranial electrical activity, including the use of scalp electrodes and magnets. *Niedermeyer's Electroencephalography* [6] provides a comprehensive overview of the subject.

EEG signal data are recorded using electrodes placed on the scalp according to the 10–10 system of placing electrodes at 10% intervals. The 10–10 placement scheme is a standard developed by the American Electroencephalographic Society and can be viewed as an amelioration of the International Standard 10–20 system that instead uses 20% spacing [24]. We note that there is ongoing research related to the optimal placement of electrodes to record brain electrical activity [25]. In contrast to scalp electrodes, depth electrodes are implanted in the brain (penetrating gray matter), and signals are recorded by one or more electrical contacts on each electrode. The specific number of contacts on the electrode depends on the position of the electrode and the depth of its implant [26,27].

Depth electrodes are often implanted using a stereotactic approach, and the corresponding method of recording signal data is called *stereotactic electroencephalography* or *SEEG* [28]. Although SEEG is an invasive recording technique, the quality of data is robust with brain electrical activity recorded at a high resolution. The analysis of SEEG data is therefore used as a gold standard in the diagnosis and treatment of epilepsy [29].

13.2.2 Managing Signal Data: The CSF

The storage and the management of signal data are significant challenges in brain connectivity research since effective analysis of data requires the use of both data and essential contextual metadata, for example instrument parameters, sampling rate, and study protocol. EDF is one of the most widely used signal storage formats in neuroscience applications [15]; however, it is not well suited for developing efficient data integration and analysis techniques. In addition, the EDF does not support the FAIR principles that facilitate data sharing and reusability [30]. The FAIR principles allow efficient discovery of and access to scientific data using the following properties associated with datasets:

- *Findable*: Data should be easy to locate and easy to identify through the use of persistent identifiers and appropriate contextual metadata.
- *Accessible*: Data should be easy to access using existing network protocols and associated metadata information.
- *Interoperable*: Data should be annotated using a standard ontology term that allows easy sharing and analysis of data aggregated from different sources.

- *Reusable*: Data and any associated metadata should be made available with a clearly defined license to allow secondary use of datasets.

EDF files have limited to no support for these FAIR principles; for example, it is difficult to locate specific segments of signal data in an EDF file due to the format's minimal use of metadata information and lack of semantic annotations using standardized terminology. Similarly, the storage of signal data in EDF files as collection of temporally ordered recordings makes it difficult to analyze channel-specific signals over a period of time. In particular, the retrieval of channel-specific signal data for a specific time period in an EDF file requires multiple "look-ups" for each timestamp, significantly increasing the number of computations required for time-series analysis.

To address these limitations of the EDF, we have developed the CSF that allows for the efficient storage, retrieval, and processing of signal data [31,32]; Figure 13.1 illustrates the overall structure of a CSF file. The CSF has been developed using the JavaScript Object Notation (JSON) framework that associates "values," such as text data, numerical data, or other JSON objects, with textual strings known as "keys" [33]. EDF files can be transformed into CSF files without any loss of information or any other difficulty in the reusability of the signal data. On the contrary, the CSF enables significant improvements over the EDF in terms of signal data accessibility and interoperability. For example, we recall how the signal processing of data in EDF files requires several steps, each involving some computation of byte offset values in order to access the data. By contrast, CSF files can be easily processed with a single invocation of an appropriate value retrieval function in a programming language (e.g., a "getter" function in Java) with the associated key string as the function's input.

```
{
    "Header": {
        "firstFragment": A,
        "lastFragment": B,
        "epochDuration": 30.0,
        "fragmentNoA": {
            "fragmentNo": A,
            "startDate": "MM.DD.YY",
            "startTime": "HH.MM.SS"
        },
        "fragmentNo(A+1)": {
            "fragmentNo": A+1,
            "startDate": "MM.DD.YY",
            "startTime": "HH.MM.SS" // 30 seconds later
        },
        ...,
        "fragmentNoB": {
            "fragmentNo": B,
            "startDate": "MM.DD.YY",
            "startTime": "HH.MM.SS" // 30*(B-A) seconds later
        }
    },
    "studyMetadata": {
        "edfFileName": "eegRecord.edf",
        "dataFormatVersion": 0,
        "localPatientID": "patientIDString",
        "recordingStartDate": "11.22.33",
        "dateFormat": "MM.DD.YY",
        "recordingStartTime": "12.34.56",
        "recordingStartTimeFormat": "HH.MM.SS",
        "numberHeaderBytes": "56064",
        "numberDataRecords": "14400",
        "dataRecordDuration": "0.1",
        "dataRecordDurationUnit": "seconds",
        "numberSignals": "N+1"
    },
    "clinicalAnnotationList": {
        "timestamp_1": "Annotation_1",
        "timestamp_2": "Annotation_2",
        ...
    },
    ...
    "channelMetadata": {
        "channelName_0": {
            "channelNumber": "0",
            ... // other signal metadata listed in the EDF header
        },
        "channelName_1": {
            "channelNumber": "1",
            ... // other signal metadata listed in the EDF header
        },
        ...
        "channelName_N": {
            "channelNumber": "N",
            ... // other signal metadata listed in the EDF header
        },
        "channelList": "[channelName_0, channelName_1,
                        ...,channelName_N]"
    },
    "dataRecords": {
        "fragmentNumberA": {
            "channelName_0": "[ ... ]", // values are decimal arrays
            "channelName_1": "[ ... ]",
            ... ,
            "channelName_N": "[ ... ]"
        },
        "fragmentNumber(A+1)": {
            "channelName_0": "[ ... ]",
            "channelName_1": "[ ... ]",
            ... ,
            "channelName_N": "[ ... ]"
        },
        ... ,
        "fragmentNumberB": {
            "channelName_0": "[ ... ]",
            "channelName_1": "[ ... ]",
            ... ,
            "channelName_N": "[ ... ]"
        }
    }
}
```

FIGURE 13.1
Example of the structure of the CSF. Each value is associated to a plaintext key that can be used to easily retrieve data.

In addition to greater accessibility, CSF files also support the interoperability of signal data generated from different sources through the use of ontology terms for data annotation. This feature allows for the reconciliation of data heterogeneity and improves the integration of data to allow researchers to query and analyze large repositories of signal data. In addition, the use of ontology terms for the annotation of signal data in CSF also enables the greater reusability of data and supports the creation of efficient indices for data segments. Although the use of descriptive "keys" and "values" in the CSF leads to an increase in the storage size of the resulting files, we believe that the increasing availability of cheap data storage infrastructures will address this challenge and allow CSF files to be used in practical data management systems.

13.2.3 Processing Signal Data to Compute Functional Connectivity Measures

The CSF supports the efficient computation of functional connectivity measures between different channels using various statistical techniques. The Pearson linear regression coefficient is a common statistical technique used to measure correlation between two datasets [34]. However, measures of *linear* correlation assume that electrical signals in the brain propagate as a *linear* function of time, which is not corroborated by clinical data [8]. To address this limitation, *non-linear* regression techniques have been explored in the brain connectivity research community, such as the measurement of the AAMI shared by the signals. These regression techniques have led to the creation of non-linear functional connectivity measures that address certain limitations present in linear regression techniques [11].

A correlation metric developed by Pijn and Lopes da Silva [11] called the *non-linear correlation coefficient* has been found to be useful for computing functional connectivity in epilepsy patients [8]. This non-linear correlation measure views discretely recorded signals as continuous functions of time and uses the well-known mathematical fact that any continuous function can be approximated by a piece-wise linear function, where the error in the approximation is controlled by both the number and the locations of the endpoints of each linear piece [35]. The correlation coefficient of Pijn and Lopes da Silva uses linear regression on each (linear) piece of the approximation, and the average of the corresponding linear correlation coefficients is computed as an approximation to a "true" correlation.

The non-linear correlation coefficient generates accurate results with respect to correlation measures for signal data as demonstrated by Pijn and Lopes da Silva [11] and Wendling et al. [8]. In addition, the non-linear correlation coefficient is applicable in scenarios where the signals are linearly correlated as a function of time; in such cases, the value of the measure identically matches the value of Pearson's linear correlation coefficient. Moreover, the proposed non-linear correlation coefficient is *asymmetric*; that is, the value $h^2(X, Y)$ of the non-linear correlation coefficient comparing signal X to signal Y may differ from the value $h^2(Y, X)$ used to correlate signal Y to signal X. (The notation h^2 for, and indeed the name of, the non-linear correlation coefficient is meant to be analogous to the notation r^2 for, and the name of, Pearson's linear correlation coefficient [8].) This asymmetric property of the correlation coefficient measure introduces a notion of *directionality* and allows us to evaluate if the activity at location X influences the activity at location Y, if

$$h^2(X, Y) \geq h^2(Y, X) \tag{13.1}$$

if conversely activity at Y influences activity at X, if

$$h^2(X, Y) \leq h^2(Y, X) \tag{13.2}$$

or if the activities at each site have some mutual influence on one another [11], if

$$h^2(X, Y) \approx h^2(Y, X). \tag{13.3}$$

We note, however, that the statistical measures described above address the issue of correlating signal amplitude values only, whereas clinicians often use signal frequency analysis to identify epilepsy-related events [36]. Although linear functional connectivity measures of signal frequencies such as coherence are used in the research community, there are no non-linear functional connectivity measures available to correlate frequencies of signal data [8].

13.2.4 Computational Pipeline for Analyzing Signal Data

In the previous section, we described various components of the signal analysis pipeline we have developed in order to analyze neurological signal data. We now describe our implementation and use of this signal processing pipeline to evaluate functional connectivity.

In the first phase, de-identified signal data recorded from an epilepsy patient and stored as a collection of EDF files are processed and transformed into CSF files. This process involves subdividing the entire duration of the signal data recording into smaller segments (typically 30 seconds in duration). For each segment, data are extracted from the EDF files by parsing the file and computing the byte location of each data element as described by the EDF specifications, and the extracted data are stored in an intermediate data structure.

A predetermined number of segments (provided as user input) are aggregated in a CSF file, which is generated using the Java JSON Application Programming Interface (API) [37].

The process of rewriting EDF files as CSF files involves a transformation of the layout of signal data from a collection of signals recorded during a given time period into a collection of time series data corresponding to each recording channel. In effect, this is a "transposition" of the time value-recording site matrix to a channel-oriented layout by measuring the byte offset to locate and extract the information. We present a schematic diagram of the transformation process in Figure 13.2.

The computational pipeline takes as input a list of user-defined parameters including start and end time stamps for a seizure event (or ictal period) under investigation, and a list of recording channels. In the next step, the tool iterates over pairs of signal channels listed in the input parameters, and extracts the data for each pair of signals over the given ictal period. The channel-oriented

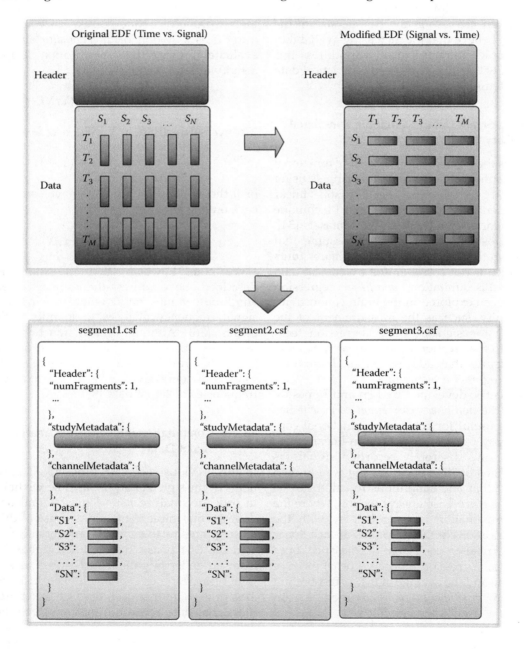

FIGURE 13.2
Schematic diagram of the signal processing pipeline. The first arrow represents the transposition of time/site data as written in EDF to site/time data stored in an intermediate object. The second arrow denotes the fragmentation of the data into multiple CSF files for use in distributed computing environment.

layout of the signal data in the CSF files facilitates the retrieval of relevant signal data during this step. In the next phase, the pipeline computes Pijn's non-linear correlation coefficient for each pair of signal channel using the extracted data. We note that this step is performed for all pairs of signal recording channels, which ensures that we perform all relevant computations in both directions of signal propagation (as discussed earlier in Section 13.2.3).

The output of the computational pipeline is a two-dimensional matrix $\{h^2(X, Y)\}$ of non-linear correlation coefficient values for a given ictal period. In the next step, the matrix values are analyzed to qualify the correlation between different channels of signal data. We evaluate the relative strengths of each correlation during the ictal period by computing the average value, denoted by μ, and the standard deviation, denoted by σ, of all of the values in the matrix. These values are used to compute $N_\sigma(X, Y) = (h^2(X, Y) - \mu)/\sigma$ for each pair of signals, a value that measures the number of standard deviations between a specific (i.e., local) correlation coefficient $h^2(X, Y)$ and the global average μ (a similar method is used by Wendling et al. [8]).

The final output of the computational pipeline is a visualization of the data as a network graph with a set of vertices corresponding to the set of signal recording sites in the brain and edges corresponding to the matrix of N_σ values. It is common practice in statistics to characterize those values in a given set that differ from the average of all the values in that set by more than two standard deviations as *statistically significant*, as this behavior is observed for at most 5% of the values in a set with a Gaussian distribution [38]. While EEG signal data do not typically satisfy a Gaussian distribution, a similar proportion of signal values is observed to lie more than two standard deviations from the mean [8]. In accordance with these observations, we have therefore added directed edges in our output graph from vertex X to vertex Y for each pair of signals (X, Y) with $N_\sigma(X, Y) \geq 2$.

13.3 Results

In this section, we describe the preliminary results from analyzing de-identified signal data of an epilepsy patient using the computational pipeline described in Section 13.2. Figure 13.3 provides a schematic representation of the computational pipeline used to generate the matrix of h^2 values, the matrix of N_σ values, and the corresponding directed network graph. A preliminary review of the results shows that correlation of signals is not transitive. For example, given three electrode contacts A, B, and C, if activity at A is correlated with activity at

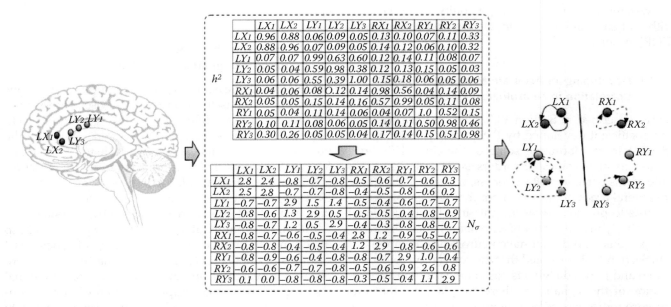

FIGURE 13.3
Conceptual overview of the data acquisition, processing, and analysis phases of the pipeline are illustrated. The diagram on the left displays an example placement of depth electrodes within the brain (Brain image created by N. Byrd [39], "A 2D vector drawing of the brain sliced down the center viewed from the side. Created using Adobe Illustrator CS3," Wikimedia Commons, 2014.). The middle shows the resulting matrix of h^2 correlation values and its conversion to a matrix of N_σ values. Finally, the right-most image gives an example of the network graph output created from the N_σ matrix. The notation X_i and Y_i ($i = 1, 2, 3$) represent electrode contacts with LX_i and RX_i corresponding to placement of the electrode in the left and right hemisphere, respectively. The directed edges connecting the electrode contact nodes represent correlation measures computed by the computational pipeline with solid lines and dashed lines used to differentiate between different correlation measure values.

B and activity at B is correlated with activity at C, then it is not necessarily true that A is correlated with activity at C (in either direction). We propose to investigate the underlying cause for this result in collaboration with clinical researchers as part of our ongoing research.

Furthermore, we noted in our analysis that correlated sites of activity are more likely to be located on the same electrode, whereas correlated activity rarely occurs between contacts on different electrodes, especially across the two brain hemispheres. We found that our conclusions concur with those obtained in a clinical setting using evoked potentials [27]. We emphasize, however, that our results are computed using data from a single patient, and additional analysis is required to understand the underlying causes for these characteristics in the signal data.

13.4 Discussion

In this section, we discuss some of the limitations and proposed improvements to the signal data correlation coefficient measure proposed by Pijn and Lopes da Silva. In addition, we discuss the use of parallel and distributed computing techniques for the goal of improving the performance of our computational pipeline. Finally, we describe the importance of using common terminological systems to facilitate interoperability of signal data with patient data stored in electronic health record (EHR) systems.

13.4.1 Developing an Accurate Measure of Signal Correlation in Neurological Disorders

As we described in Section 13.3, the non-linear correlation coefficient developed by Pijn and Lopes da Silva is effective in corroborating some of the clinical findings related to neurological disorders such as epilepsy. In spite of this perceived effectiveness, we believe that there are several areas of improvement that will enable signal analysis to provide better insights into brain functional connectivity in both patients with neurological disorders and persons who do not have neurological disorders.

In Section 13.2, we noted that the correlation coefficient of Pijn and Lopes da Silva is based on a discrete approximation of the signal; it is therefore plausible that this measure represents a discrete approximation of a more accurate connectivity measure. Thus, an area of potential improvement lies in determining how to compute this "true" correlation measure using different techniques, such as through the use of some type of limiting process. Such a development could resolve some of the issues described in Section 13.2, such as an accurate determination of those pairs of signals for which correlation is statistically significant.

The results of our evaluation agree with previous findings by Wendling et al. [8] that the non-linear measure developed by Pijn and Lopes da Silva [11] effectively measures functional connectivity. This suggests that intracranial signals propagate in a fashion that is *non-linear* with regard to time, which in turn implies the existence of some kind of signal interference that influences the transmission of electrical signals during epileptic events. However, the non-linear correlation coefficient cannot accurately determine the nature of this interference. Further investigation into this matter may require the incorporation of other techniques, such as dynamical system analysis.

Finally, we note that our current use of the non-linear correlation coefficient does allow us to determine the direction of influence among pairs of signals due to the inherent asymmetric properties of the measure. However, this correlation measure only computes *instantaneous* correlation; that is, we do not know how long it takes for a signal to reach some other site in the brain. The incorporation of additional features in the correlation coefficient measure for signal data, such as a method to compute any *time lag*, will significantly help advance our understanding of functional connectivity in epilepsy.

13.4.2 Improving the Performance of the Computational Pipeline

The statistical measures used to compute correlations within signal datasets require the pairwise processing of data recorded from different locations. With rapid technological advances in recording brain activities, the number of data points that are available to be processed for the computation of functional connectivity measures has increased dramatically in the past few years. For example, current SEEG recording techniques can record data at a rate of 10 kHz from 256 electrode contacts. In addition, the volume of signal data is expected to keep increasing. Although the processing of such vast amounts of data is useful for advancing functional connectivity research, it presents significant computational challenges. The current implementation of our computational pipeline, for example, requires several hours to process data for an ictal event lasting only 30 seconds. As signal recording technology continues to improve and the volume of data correspondingly increases, there is a clear need to develop efficient computational approaches to analyze signal data on a large scale.

The use of high-performance parallel and distributed computing approaches, including the use of a cloud computing infrastructure, will allow us to improve the performance of the computational pipeline used to derive functional connectivity measures. In particular, the use of Apache Hadoop [40] or Apache Spark [41] will allow multiple ictal periods to be analyzed simultaneously. Apropos, the inherent ability of CSF to fragment and store signal data across multiple sites is ideally suited for use with a cloud computing infrastructure. We successfully developed a proof-of-concept implementation of our computational pipeline using Apache Pig that processed 750 GB of EDF file data into CSF files using a 31-node Hadoop cluster [42]. Following this pilot implementation, we plan to develop an Apache Spark-based implementation of our computational pipeline to significantly improve the performance time for large-scale signal data processing.

13.4.3 Use of Ontologies for Standardizing Terminology in Signal Data Analysis

Terminological heterogeneity in data generated from multiple sources arises due to the use of disparate terms to describe similar physiological events (e.g., signal complexes in EEG recordings), and it represents a key challenge in integrating large-scale neuroscience data [1]. To address this critical challenge, we use terms modeled in existing biomedical ontologies to annotate signal data in CSF files as part of the computational pipeline. Ontologies are knowledge models that represent terms in a domain of discourse using formal knowledge representation languages, such as the description logic-based Web Ontology Language (OWL2) [43]. Biomedical ontologies have been widely adopted and used to reconcile data heterogeneity and support data integration and querying. For example, Gene Ontology (GO) is widely used to annotate genomic data to facilitate the use of common terminology across different data sources and also enable users to easily query the integrated data [44].

The National Center for Biomedical Ontologies lists more than 500 open source biomedical ontologies that can be used for a semantic annotation of biomedical data and an automated reconciliation of heterogeneous terms used to describe similar data values [45]. At present, we are building on our experience in the development and application of a domain ontology for epilepsy called the Epilepsy and Seizure Ontology (EpSO) [46] to integrate additional neuroscience-specific ontologies in the computational pipeline for a semantic annotation of signal data. These semantic annotations are expected to significantly improve the integration and retrieval of signal data aggregated from multiple sources, including data generated in multi-center research studies. In addition, the use of terminological systems such as the Systematized Nomenclature of Medicine Clinical Terms (SNOMED CT) for semantic annotation is also expected to facilitate the interoperability of signal data with related clinical data stored in EHR systems. This will enable neuroscience researchers to perform clinical research studies.

13.5 Conclusion

The determination of dynamic properties of functional networks in the brain, in both healthy individuals and persons suffering from neurological disorders, is an important and challenging research problem. Visualizing the brain as an interactive and interconnected network of structures, we can create maps of functionally connected brain regions by observing the generation and propagation of electrical activity. In this chapter, we have outlined the use of statistical correlation techniques to compute functional connectivity measures from SEEG signal data. We have also described a computational pipeline that incorporates the new CSF signal data representation format, along with other data processing and signal analysis functionalities. We expect that our pipeline will help to analyze signal data on a large scale, and thereby potentially advance our understanding of complex neurological disorders such as epilepsy. Our computational pipeline makes effective use of the novel CSF file format for signal data representation and storage. The CSF has been designed to effectively support time-series signal analysis and parallel processing techniques. We believe that integrating new functionalities and improving correlation measures for signal data will allow us to effectively leverage the growing volume of signal data for further research in neurological disorders.

Acknowledgement

This work is supported in part by the NIH-NIBIB Big Data to Knowledge (BD2K) 1U01EB020955 grant and NSF grant# 1636850.

References

1. C. Bargmann, Newsome, W., Anderson, D. et al., "BRAIN 2025: A scientific vision.," in *Brain Research through Advancing Innovative Neurotechnologies (BRAIN) Working Group Report to the Advisory Committee to the Director NIH*," US National Institutes of Health. 2014.
2. K. J. Friston, "Functional and effective connectivity: A review," *Brain Connect*, vol. 1, no. 1, pp. 13–36, 2011.
3. S. Seung, *Connectome: How the Brain's Wiring Makes Us Who We Are*. Boston: Houghton Mifflin Harcourt, 2012.
4. O. Sporns, "Structure and function of complex brain networks," *Dialogues in Clinical Neuroscience*, vol. 15, no. 3, pp. 247–262, 2013.
5. P. O. Shafer, *About Epilepsy: The Basics*. Available at http://www.epilepsy.com/start-here/about-epilepsy-basics (accessed on January 15, 2015).
6. D. L. Schomer and F. H. Lopes da Silva, *Niedermeyer's Electroencephalography: Basic Principles, Clinical Applications, and Related Fields*. Philadelphia: Lippincott Williams & Wilkins, 2011.
7. P. O. Shafer, What happens during a seizure? Available at http://www.epilepsy.com/start-here/about-epilepsy-basics/what-happens-during-seizure (accessed on January 15, 2015).
8. F. Wendling, F. Bartolomei, and L. Senhadji, "Spatial analysis of intracerebral electroencephalographic signals in the time and frequency domain: identification of epileptogenic networks in partial epilepsy," *Philosophical Transactions of the Royal Society A: Mathematical, Physical and Engineering Sciences*, vol. 367, no. 1887, pp. 297–316, Jan 28, 2009.
9. O. David, D. Cosmelli, and K. J. Friston, "Evaluation of different measures of functional connectivity using a neural mass model," *Neuroimage*, vol. 21, no. 2, pp. 659–673, Feb 2004.
10. K. Ansari-Asl, L. Senhadji, J.-J. Bellanger, and F. Wendling, "Quantitative evaluation of linear and nonlinear methods characterizing interdependencies between brain signals," *Physical Review. E, Statistical, Nonlinear, and Soft Matter Physics*, vol. 74, no. 3 Pt 1, pp. 31916, Sept 26, 2006.
11. J. P. Pijn and F. Lopes da Silva, "Propagation of electrical activity: Nonlinear associations and time delays between EEG signals," in *Basic Mechanisms of the EEG*, S. Zschocke and E.-J. Speckmann, Eds. Boston: Birkhauser, 1993.
12. S. S. Sahoo, C. Jayapandian, G. Garg, F. Kaffashi, S. Chung, A. Bozorgi, C. Chen, K. Loparo, S. D. Lhatoo, G. Q. Zhang, "Heartbeats in the cloud: Distributed analysis of electrophysiological "big data" using cloud computing for epilepsy clinical research," *Journal of American Medical Informatics Association (Special Issue on Big Data)*, vol. 21, no. 2, pp. 263–271, 2014.
13. Editorial-Introduction, "Challenges and opportunities," *Science*, vol. 331, no. 6018, p. 692, 2011.
14. B. Kemp, A. Värri, A. C. Rosa, K. D. Nielsen, and J. Gade, "A simple format for exchange of digitized polygraphic recordings," *Electroencephalography and Clinical Neurophysiology*, vol. 82, no. 5, pp. 391–393, 5, 1992.
15. B. Kemp and J. Olivan, "European data format 'plus' (EDF+), an EDF alike standard format for the exchange of physiological data," *Clinical Neurophysiology*, vol. 114, no. 9, pp. 1755–1761, Sept 2003.
16. C. Kesavadas and B. Thomas, "Clinical applications of functional MRI in epilepsy," *Indian Journal of Radiology and Imaging*, vol. 18, no. 3, pp. 210–217, Aug 2008.
17. L. Maccotta et al., "Impaired and facilitated functional networks in temporal lobe epilepsy," *NeuroImage: Clinical*, vol. 2, pp. 862–872, 2013.
18. A. Schlögl, "An overview on data formats for biomedical signals," in *World Congress on Medical Physics and Biomedical Engineering, September 7–12, 2009, Munich, Germany: Vol. 25/4 Image Processing, Biosignal Processing, Modelling and Simulation, Biomechanics*, O. Dössel and W. C. Schlegel, Eds. Berlin: Springer, 2010, pp. 1557–1560.
19. S. Garcia et al., "Neo: An object model for handling electrophysiology data in multiple formats," *Frontiers in Neuroinformatics*, vol. 8, p. 10, 2014.
20. M. T. Dougherty et al., "Unifying biological image formats with HDF5," *Communications of the ACM*, vol. 52, no. 10, pp. 42–47, 2009.
21. E. Niedermeyer and D. L. Schomer, "Historical aspects of EEG," in *Niedermeyer's Electroencephalography: Basic Principles, Clinical Applications, and Related Fields*, D. L. Schomer and F. H. Lopes da Silva, Eds. Philadelphia: Lippincott Williams & Wilkins, 2011.
22. I. Singh, *Textbook of Human Neuroanatomy*, 7th ed. New Delhi: Jaypee Brothers Medical Publishers, 2006.
23. S. G. Hormuzdi, M. A. Filippov, G. Mitropoulou, H. Monyer, and R. Bruzzone, "Electrical synapses: A dynamic signaling system that shapes the activity of neuronal networks," *Biochimica et Biophysica Acta (BBA)—Biomembranes*, vol. 1662, nos. 1–2, pp. 113–137, Mar 23, 2004.
24. J. Malmivuo and R. Plonsey, "Electroencephalography," in *Bioelectricomagnetism: Principles and Applications of Bioelectric and Biomagnetic Fields*, J. Malmivuo and R. Plonsey, Eds. New York: Oxford University Press, 1995.
25. V. Jurcak, D. Tsuzuki, and I. Dan, "10/20, 10/10, and 10/5 systems revisited: Their validity as relative head-surface-based positioning systems," *Neuroimage*, vol. 34, no. 4, pp. 1600–1611, Feb 15, 2007.
26. J. Gonzalez-Martinez et al., "Stereotactic placement of depth electrodes in medically intractable epilepsy," *Journal of Neurosurgery*, vol. 120, no. 3, pp. 639–644, Mar 2014.
27. N. Lacuey et al., "Homotopic reciprocal functional connectivity between anterior human insulae," *Brain Structure and Function*, vol. 221, no. 5, pp. 2695–2701, Jun 2016.
28. H. Lüders, J. Engel Jr., and C. Munari, "General principles," in *Surgical Treatment of the Epilepsies*, J.J. Engel, Ed. New York: Raven Press, 1993, pp. 137–153.
29. D. Cosandier-Rimélé, J.-M. Badier, P. Chauvel, and F. Wendling, "Modeling and interpretation of scalp-EEG and depth-EEG signals during interictal activity," *Conference Proceedings*, vol. 1, pp. 4277–4280, 2007.
30. M. D. Wilkinson et al., "The FAIR Guiding Principles for scientific data management and stewardship," *Scientific Data*, Comment vol. 3, p. 160018, online, Mar 15, 2016.

31. C. P. Jayapandian, Chen, C. H., Bozorgi, A., Lhatoo, S. D., Zhang, G. Q., Sahoo, S. S., "Cloudwave: Distributed Processing of "Big Data" from Electrophysiological Recordings for Epilepsy Clinical Research Using Hadoop," in *American Medical Informatics Association (AMIA) Annual Symposium*, Washington, DC, 2013, pp. 691–700: AMIA.
32. C. Jayapandian, Wei, A., Ramesh, P., Zonjy, B., Lhatoo, S. D., Loparo, K., Zhang, G. Q, Sahoo, S. S., "A scalable neuroinformatics data flow for electrophysiological signals using MapReduce," *Frontiers in Neuroinformatics*, vol. 9, no. 4, 2015.
33. ECMA International, "The JSON Interchange Format," no. ECMA-404, Available at http://www.ecma-international.org/publications/files/ECMA-ST/ECMA-404.pdf (accessed on January 15, 2015).
34. S. W. Scheff, *Fundamental Statistical Principles for the Neurobiologist: A Survival Guide*. London: Academic Press, 2016.
35. E. Süli and D. F. Mayers, *An Introduction to Numerical Analysis*. Cambridge, UK: Cambridge University Press, 2003.
36. M. Zijlmans, P. Jiruska, R. Zelmann, F. S. Leijten, J. G. Jefferys, and J. Gotman, "High-frequency oscillations as a new biomarker in epilepsy," *Annals of Neurology*, vol. 71, no. 2, pp. 169–178, Feb 2012.
37. *Java API for JSON Processing (JSON-P)*. Available at http://json-processing-spec.java.net/ (accessed on January 15, 2015).
38. J. Rosenblatt, *Basic Statistical Methods and Models for the Sciences*. Boca Raton, FL: Chapman & Hall, 2002.
39. N. Byrd, "A 2D vector drawing of the brain sliced down the center viewed from the side. Created using Adobe Illustrator CS3," ed: Wikimedia Commons, 2014.
40. The Apache Software Foundation. *Welcome to Apache Hadoop!* Available at http://hadoop.apache.org/ (accessed on January 15, 2015).
41. The Apache Software Foundation. *Apache Spark™—Lightning-Fast Cluster Computing*. Available at http://spark.apache.org/ (accessed on January 15, 2015).
42. S. S. Sahoo et al., "NeuroPigPen: A scalable toolkit for processing electrophysiological signal data in neuroscience applications using Apache Pig," *Frontiers in Neuroinformatics*, vol. 10, p. 18, 2016.
43. P. Hitzler, M. Krötzsch, B. Parsia, P. F. Patel-Schneider, and S. Rudolph, "OWL 2 web ontology language primer," *W3C Recommendation*, vol. 27, no. 1, p. 123, 2009.
44. M. Ashburner et al., "Gene Ontology: Tool for the unification of biology," *Nature Genetics*, vol. 25, no. 1, pp. 25–29, 2000.
45. M. A. Musen et al., "The national center for biomedical ontology," *Journal of the American Medical Informatics Association*, vol. 19, no. 2, pp. 190–195, 2012.
46. S. S. Sahoo, Lhatoo, S. D., Gupta, D. K., Cui, L., Zhao, M., Jayapandian, C., Bozorgi, A., Zhang, G. Q., "Epilepsy and seizure ontology: Towards an epilepsy informatics infrastructure for clinical research and patient care," *Journal of American Medical Informatics Association*, vol. 21, no. 1, pp. 82–89, 2014.

14
Machine Learning Approaches to Automatic Interpretation of EEGs

Iyad Obeid and Joseph Picone

CONTENTS

14.1 Introduction 271
14.2 Big Data Issues in Manual Interpretation of EEGs 273
    14.2.1 Waveform Displays Are Still a Primary Visualization Tool 273
    14.2.2 Signal Conditioning Enhances Interpretation 275
    14.2.3 Locality Is an Extremely Important Feature 276
    14.2.4 Annotations Play a Critical Role in Machine Learning Systems 276
    14.2.5 Inter-Rater Agreement Is Low 277
    14.2.6 Evaluation Metrics Are Important 279
    14.2.7 Decision Support Systems Can Enhance Interpretation 280
    14.2.8 Unbalanced Data Problem Makes Machine Learning Challenging 280
14.3 TUH EEG Corpus 282
    14.3.1 Digitization and Signal Processing 282
    14.3.2 EEG Report Pairing 283
    14.3.3 De-Identification of the Data 284
    14.3.4 Basic Descriptive Statistics 284
    14.3.5 Structure of the Released Data 286
14.4 Automatic Interpretation of EEGs 287
    14.4.1 Machine Learning Approaches 287
    14.4.2 Typical Static Classification Systems 288
    14.4.3 Feature Extraction 288
    14.4.4 Hidden Markov Models 291
    14.4.5 Normal/Abnormal Detection 293
    14.4.6 Seizure Detection 295
14.5 Summary and Future Directions 295
Acknowledgments 296
References 297

14.1 Introduction

An EEG measures the spontaneous electrical activity of the brain, as shown in Figure 14.1 [1]. A routine EEG typically lasts 20 to 30 minutes and is used in clinical circumstances where a short measurement is sufficient (e.g., distinguishing epileptic seizures from other types of seizures). Routine EEGs often include standard activation procedures that increase the chance of capturing seizure-like discharges or even seizures (e.g., hyperventilation and photic stimulation). The entire session for a routine EEG, including the time required to affix sensors to a patient's scalp, requires one to two hours. Patients are asked to lie still in a prone position and are periodically requested to perform limited movements (e.g., breath, blink).

Long-term monitoring (LTM) is useful in a variety of situations in which patients have intermittent disturbances that are difficult to capture during routine EEG

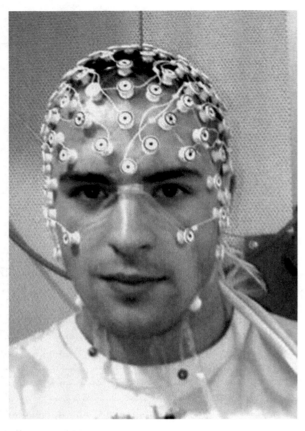

FIGURE 14.1
EEG measures electrical activity (ionic currents) along the scalp using gold-plated or silver/silver chloride electrodes. The signals, typically in the microvolt range, are very noisy and must be viewed by examining differential voltages between and electrode and a ground point such as an electrode connected to the left ear. (Photo adapted from Electroencephalography, April 19, 2017.)

sessions. Patients who experience medical conditions such as epilepsy or stroke are often subjected to LTM in a critical care setting such as an epilepsy monitoring unit (EMU). In such cases, recordings can last several hours or several days, generating a large amount of data that need to be reviewed by a clinician. Advances in digital technology have greatly enhanced the ability to acquire, store, and review large amounts of clinical data, which typically now include electrical signals, video, and other vital signs. Clinical practice is struggling to keep pace with the vast amount of patient data being collected. This has created an opportunity for automated computer-based processing and interpretation of EEGs.

The signals measured along the scalp can be correlated with brain activity, which makes it a primary tool for diagnosis of brain-related illnesses [2,3]. The electrical signals are digitized and presented in a waveform display as shown in Figure 14.2. EEG specialists review these waveforms and develop a diagnosis. The output of the process is a physician's EEG report, as shown in Figure 14.3.

It is important to note that a vast majority of all routine EEGs conducted are inconclusive [4]: "In healthy adults with no declared history of seizures, the incidence of epileptiform discharge in routine EEG was 0.5%. A slightly higher incidence of 2–4% is found in healthy children and in nonepileptic patients referred to hospital EEG clinics. The incidence increases substantially to 10–30% in cerebral pathologies such as tumor, prior head injury, cranial surgery, or congenital brain injury." Patients experiencing epilepsy rarely seize during a routine EEG session, even though audio visual stimulation such as rhythmic stimulation using light and sound is often applied to induce seizures. In recent years, with the advent of wireless technology, LTM occurring over a period of days to weeks has become possible. Ambulatory data collections, in which untethered patients are continuously monitored, are becoming increasingly popular due to their ability to capture seizures and other critical infrequently occurring events [5]. Unfortunately, the data collected under such conditions are often sufficiently noisy and poses serious challenges for automated analysis systems.

EEGs traditionally have been used to diagnose epilepsy and strokes [2] although other common clinical diagnoses include coma [6], encephalopathies [7], brain death [8], and sleep disorders [9]. EEGs and other forms of brain imaging such as fMRI are increasingly being used to diagnose conditions such as head-related trauma injuries, Alzheimer's disease, posterior reversible encephalopathy syndrome, and middle cerebral artery infarction. Computerized EEG signal processing applications have included predicting dementia in Parkinson's disease patients [10,11], stroke volume measurement and outcome prediction [12,13], psychosis evaluation in high-risk patients [14], and assessment of traumatic brain injury severity.

The increasing scope of conditions addressable by an EEG suggest that there is a growing need for expertise to interpret EEGs and, equally importantly, research to understand how various conditions manifest themselves in an EEG signal. Computer-generated EEG interpretation and identification of critical events can therefore be expected to significantly increase the quality and efficiency of a neurologist's diagnostic work. Clinical consequences include regularization of reports, real-time feedback, and decision-making support to physicians. Computerized EEG assessment can therefore potentially alleviate the bottleneck of inadequate resources to monitor and interpret these tests.

The primary focus of this chapter is to introduce readers to the process of developing machine learning technology for automatic interpretation of EEGs. Readers interested in the fundamentals of the electrophysiology or neuroscience on which an EEG is based are encouraged to read one of many excellent discourses [2,15] on the topic. Here we

Machine Learning Approaches to Automatic Interpretation of EEGs 273

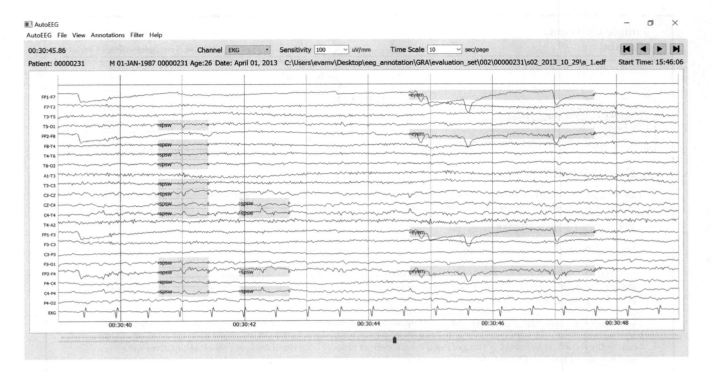

FIGURE 14.2
EEG signal is a multi-channel signal typically consisting of 22 channels. Neurologists often visualize the data using a montage—a set of differential voltages designed to accentuate anomalous behavior like spikes and sharp waves. Event-based annotations, which are created on a per-channel basis, are shown.

focus less on basic EEG science and more on the issues that impact machine learning systems. We also describe the challenges in providing reliable real-time alerts, which is an important capability for LTM, and a critical gap preventing EEGs from becoming more useful and pervasive. This chapter begins with an overview of the challenges in manual interpretation of an EEG in Section 14.2. We then introduce the TUH EEG Corpus (TUH-EEG) and discuss the important role big data is playing in the development of automatic interpretation technology in Section 14.3. Next, in Section 14.4, we introduce machine learning systems that can detect critical EEG events such as spikes with performance close to human experts. We then conclude with a discussion of emerging directions in high-performance classification using deep learning in Section 14.5. The latter is enabled by the existence of large corpora of labeled data such as that described in Section 14.3.

14.2 Big Data Issues in Manual Interpretation of EEGs

The information yielded by an EEG channel is essentially the difference of electrical activity between two electrodes. Figure 14.4 shows a standard electrode mapping for a 10/20 configuration as recommended by the American Clinical Neurophysiology Society (ACNS) [16]. Because conduction of electricity along the scalp is a nonlinear process, changes in the electrode locations on the scalp often cause significant variations in the signals observed, as does the reference point used to measure scalp voltages. Grounding also plays a critically important role in the quality of the observed signals.

14.2.1 Waveform Displays Are Still a Primary Visualization Tool

Manual interpretation of an EEG is a subtle process that can require extensive knowledge of the patient's medical history, medication history, and alertness, as well as the duration and morphology of EEG signal events. EEG signals are often analyzed based on their temporal properties, e.g., amplitude, shape, and frequency. The latter is typically measured by counting peaks in the time domain. No single feature or collection of features identify an EEG as normal. It is the overall orderly progression of signal over time that best represents a normal pattern [15]. Essential features of a normal EEG include frequency content, polarity of spikes in the signal, symmetry of transient behavior, and perhaps most importantly, the locality of an event (e.g., whether an event is observed across all channels or only a few channels that correspond to a particular brain region). Frequency

Temple University Hospital

Clinical Neurophysiology Center

Temple University Health System	3509 Broad Street 5th Floor, Boyer Pavilion Philadelphia, PA 19140	Tel (215) 707-4523

EEG REPORT

PATIENT NAME: Smith, John	DOB: 10/09/1979
DATE: 04/01/2013	MR: 12345678
ACCT: 123456789012	OP/RM#
EEG: 13-528	REFERRING PHYSICIAN: Daniel Jones/Rodriguez

REASON FOR STUDY: Migrains.

CLINICAL HISTORY: This is a 33-year-old female with a history of migraines using Fioricet. She has a past medical history of hypertension, gastric bypass, obesity, and migraines.

TECHNICAL DIFFICULTIES: None.

MEDICATIONS: Fioricet, guaifenesin, Paxil, amlodipine, Reglan, Carafate, Flonase, omeprazole, Topamax, and vitamins.

INTRODUCTION: A routine EEG was performed using standard 10–20 electrode placement system with the addition of anterior temporal and EKG electrode. The patient was recorded in wakefulness and stage I and stage II sleep. Activating procedures included hyperventilation and photic stimulation.

DESCRIPTION OF THE RECORD: The record opens to a well-defined posterior dominant rhythm that reaches 9–10 Hz which is reactive to eye opening. There is normal frontocentral beta. The patient reached stage I and stage II sleep. She also during the recording had short periods of rapid eye movement noted. Hyperventilation and photic stimulation were performed and produced no abnormal discharges.

ABNORMAL DISCHARGES: None.

HEART RATE: 60.

SEIZURES: None.

IMPRESSION: Normal awake and sleep EEG.

CLINICAL CORRELATION: This is a normal awake and sleep EEG. No seizures or epileptiform discharges were seen. Please note that the findings of REM during a routine EEG could be suggestive or indicative of sleep disorder and sleep consultation could be helpful.

Camilo Gutierrez, MD

MedQ 557391452/559219
DD:04/01/2013 13:56:56
DT:04/01/2013 15:10:37

FIGURE 14.3
Example of a typical EEG Report. The format of the report is fairly standard across institutions and contains a brief medical history, medication history, a description of the neurologist's interpretation of the EEG signal, and the implications of these findings (clinical correlation).

content in the alpha, beta, delta, and theta bands, which are not measured directly but can be derived from the channels available in a standard EEG recording, can often be used to detect anomalous behavior. Time-based waveform displays are still the most popular means by which a neurologist interprets an EEG, although frequency domain visualizations have become more popular for rapidly scanning large amounts of data to locate regions of interest [17,18].

Neurologists can often determine whether an EEG is normal or abnormal with high reliability by examining the first few minutes of an EEG session. Not surprisingly, machine learning systems can approach human performance on this task operating on similar amounts of data [19,20]. Neurologists can also identify what are often referred to as benign variants with high reliability—behaviors that on the surface might be considered anomalous, but that are not indicative of a medical condition or are simply inconclusive. For example, patient movement or eye blinks can often cause spikes in the signal that can be easily misclassified by machine learning systems. These benign variants contribute to the high false alarm rate from which most commercial systems suffer [21]. Therefore, interactive tools used by neurologists typically include many digital signal processing options that accentuate certain behaviors in the signal [18,22], including low pass, high pass, bandpass, and notch filters.

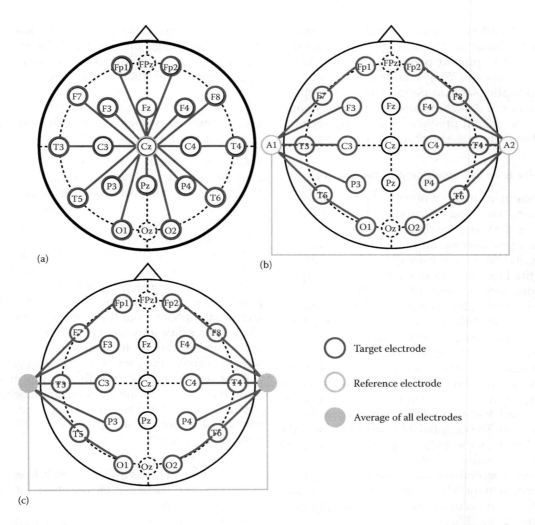

FIGURE 14.4
Electrode locations are shown for three common referential montages for a standard 10/20 configuration: (a) the common vertex reference (C_z), (b) the linked ears reference (LE), and (c) the average reference (AR).

14.2.2 Signal Conditioning Enhances Interpretation

Because typical electrical voltage ranges for EEG signals are in the tens of microvolts and extremely noisy, EEG signals are typically visualized using a differential view, known as a montage, that consists of signal differences from various pairs of electrodes (e.g., Fp1-F7). ACNS recognizes that there are a great variety of montage styles in use among EEG practitioners, and has proposed guidelines for a minimum set of montages [16]. Neurologists are often particular about the specific montage used when interpreting an EEG. At Temple University Hospital (TUH), for example, the Temporal Central Parasagittal (TCP) montage [23] is often used, as it accentuates spike behavior. Despite ACNS guidelines, several voltage reference sites are still used during EEG recordings depending on the purpose of the EEG recording [24].

Some commonly used reference schemes, which are depicted in Figure 14.4, include the following:

- Common vertex reference (Cz): uses an electrode in the middle of the head
- Linked ears reference (A1+A2, LE, RE): based on the assumption that sites like the ears and mastoid bone lack electrical activity, often implemented using only one ear
- Average reference (AR): uses the average of a finite number of electrodes as a reference

The robustness of a state-of-the-art machine learning system that decodes EEG signals depends highly on the ability of the system to maintain its performance when there are variations in the recording conditions. The specific montage of a recording could potentially affect the operation of such systems in a negative way, which constitutes a fundamental problem, given the fact that EEG signals tend to present high variability in clinical settings. In the work of [25], it was observed that there are some systematic biases in performance that depend on the source of the data, but these are relatively small compared to the overall problem of detecting seizures, spikes, or other such transient phenomena.

14.2.3 Locality Is an Extremely Important Feature

The spatial locality of an event often plays a major role in its classification. Since each electrode is tied to a particular location on the scalp, the channels in which an event occurs prominently become an important key for classification. Frontal lobes, which are defined as the area at the front of the brain behind the forehead, are responsible for voluntary movement, conscious thought, learning, and speech [26]. Temporal lobes, which are defined as the areas of the brain at the side of the head above the ears, are responsible for memory and emotions. Parietal lobes are the area of the brain at the top of your head behind your frontal lobes and control cognitive functions such as how we process sensory input, how we judge spatial relationships, and coordination. Our ability to read, write, and do math is also tied to this region of the brain. The occipital lobes are the area at the back of the brain at the back of your head and are responsible for our sense of sight.

Conditions such as epilepsy are caused by disruptions in normal brain activity. There are several key types of disruptions that can occur. These can be broadly clustered into two classes [27]:

- *Partial (focal) seizures*: seizures that happen in, and affect, only part of the brain. The signatures of these types of seizures depend on which part of the brain is affected.
- *Generalized seizures*: seizures that happen in, and affect, both sides of the brain. The patient is often unconscious during this type of seizure and will not remember the seizure itself. The most well-known category for this type of seizure is a tonic clonic (convulsive) seizure.

A seizure causes a change in the EEG, so detecting changes from normal patterns becomes an important first step in the process. Observation of an abnormal EEG does not prove that the patient has epilepsy. EEGs must be used alongside other tests to conclusively diagnose a condition. For example, video of the patient is often examined along with an EEG, and MRIs are increasingly being used to confirm diagnoses. The combination of the montage, used to accentuate spike or transient behavior, and the nature of the locality are important cues for manual interpretation of an EEG. Clinicians also visually adapt to the background channel behavior of a patient's data before they can identify cues such as spikes that lead to the identification of a seizure.

14.2.4 Annotations Play a Critical Role in Machine Learning Systems

There are generally two approaches to developing machine learning technology to automatically interpret EEGs. The first approach, which relies on expert knowledge, requires a deeper understanding of how EEGs are manually interpreted and the translation of this process into an algorithm description. Low-level events, such as spikes, are detected, and then a higher level of logic is applied to map sequences of these events to diseases or outcomes. This is analogous to the process of recognizing phonemes in speech recognition and then building word-level transcriptions from these phoneme hypotheses [28]. This requires data annotated in such a way that low-level event models can be trained, which, in turn, requires some agreement on a set of low-level labels.

After several iterations with a group of expert neurologists, and following popular approaches found in the literature [29], we have developed the following six-way classification for a segment of an EEG, which we refer to as an epoch:

1. *Spike and/or sharp wave (SPSW)*: epileptiform transients that are typically observed in patients with epilepsy.
2. *Periodic lateralized epileptiform discharges (PLED)*: EEG abnormalities consisting of repetitive spike or sharp wave discharges, which are focal or lateralized over one hemisphere and that recur at almost fixed time intervals.
3. *Generalized periodic epileptiform discharges (GPED)*: periodic short-interval diffuse discharges, periodic long-interval diffuse discharges, and suppression-burst patterns according to the interval between the discharges. Triphasic waves (diffuse and bilaterally synchronous spikes with bifrontal predominance, typically periodic at a rate of 1–2 Hz) are included in this class.
4. *Artifacts (ARTF)*: recorded electrical activity that is not of cerebral origin, such as those due to equipment or environment.
5. *Eye movement (EYEM)*: common events that can often be confused for a spike.
6. *Background (BCKG)*: all other signals.

These classes are very similar to what others have used [30,31] to perform stroke and epilepsy detection.

Epochs are usually 1 second in duration and are further subdivided in time for signal conditioning and analysis.

Examples of the SPSW and EYEM classes are shown in Figure 14.2. We typically annotate data in a channel-dependent manner since we need to establish the locality of an event. We refer to such annotations as event-based annotations since they identify the start and stop times of events on specific channels. A summary judgment for each epoch is then made based on the channel-dependent annotations. We refer to these annotations as term-based, following a convention used in other research communities [32]. We generate these automatically using a majority voting scheme that looks across all channels. In cases where the outcome of a majority vote is not clear, we resolve ambiguity manually.

Identification of these six events is important towards making a final classification of a section of data as constituting a seizure event. The first three classes are information bearing in that they describe events that are critical in manual interpretation of an EEG. What primarily distinguishes these three classes is the degree of periodicity and the extent to which these events occur across channels. Neurologists can identify PLEDs and GPEDs with a high degree of accuracy since these events are distinctive because of their long-term repetitive behavior. Accuracy for manually detecting spikes, however, is more problematic [21].

The last three classes are used to improve our ability to model the background channel. Background modeling is an important part of any machine learning system that attempts to model the temporal evolution of a signal (e.g., hidden Markov models [HMMs], deep learning). We let the system automatically perform background/non-background classification as part of the modeling process rather than use a heuristic preprocessing algorithm to detect signals of interest. This is described in more detail in Section 14.4. This follows a very successful approach that we have used in speech recognition [33].

Artifacts, eye blinks, and eye-related muscle movements occur frequently enough that they merit separate classes. These events appear as transient events in the signal and to an untrained eye can be misinterpreted as spike behavior. The rest of the events that do not match the first five classes are lumped into the background class. Hence, it is important that the background class model be robust and powerful, because this is the primary way the false alarm rate is minimized. Further, the critical aspects of performance are related to the sensitivity and specificity of the first three classes since these are the events neurologists will key on to interpret a session. Hence, as discussed in Section 14.2.6, we often adjust our scoring criteria to give proper weight to these classes.

The second machine learning approach, which is embodied in many deep learning-based systems that are so popular today, is to let the system learn the underlying structure of the data. In this approach, we provide manual annotations of seizure events that simply indicate the start and stop time of a seizure, and optionally the type of seizure. For this approach to work, a large archive of data is needed, and hence the focus on big data resources described in Section 14.3. A waveform display for a typical seizure event, along with the corresponding spectrogram for a selected number of channels, is shown in Figure 14.5. The TUH EEG Seizure Corpus [34], which is a subset of TUH-EEG, provides a large annotated corpus of seizure events that can be used to develop such technology.

Our seizure event annotations include start and stop times; localization of a seizure (e.g., focal, generalized) with the appropriate channels marked; type of seizure (e.g., simple partial, complex partial, tonic–clonic, gelastic, absence, atonic); and the nature of the seizure (e.g., convulsive). The non-seizure event annotations include artifacts, which could be confused with seizure-like events such as ventilatory artifacts and lead artifacts; non-epileptiform activity that may resemble epileptiform discharges, such as psychomotor variant, mu, breach rhythms, and positive occipital sharp transients of sleep (POSTS); abnormal background that could be confused with seizure-like events (e.g., triphasics); and interictal and postictal states. The types of features are important when manually interpreting an EEG and determining how a seizure manifests itself. We use these finer categorizations of seizures to build models specific to these events, which also helps reduce the false alarm rate.

14.2.5 Inter-Rater Agreement Is Low

A board-certified EEG specialist currently is required by law to interpret an EEG and produce a diagnosis. It takes several years of additional training post-medical school for a physician to qualify as a clinical specialist. However, despite completing a rigorous training process, there is only moderate inter-rate agreement (IRA) in EEG interpretation [35,36] for low-level events such as spikes. In the work of Halford et al. [37], it was noted that IRA among experts was significantly higher for identification of electrographic seizures compared to periodic discharges. In the work of Swisher et al. (17), it was noted that even augmenting traditional waveform displays with a display based on advanced "brain mapping" analytics, known as a quantitative EEG (qEEG), was not adequate as the sole method for reviewing continuous EEG data. Kappa statistics [38] in the range of 0.6 are common for manual interpretations and drop considerably as the data become more challenging or are collected under typical clinical conditions.

What makes this problem so challenging is that an EEG signal is a very low voltage signal (e.g., microvolts).

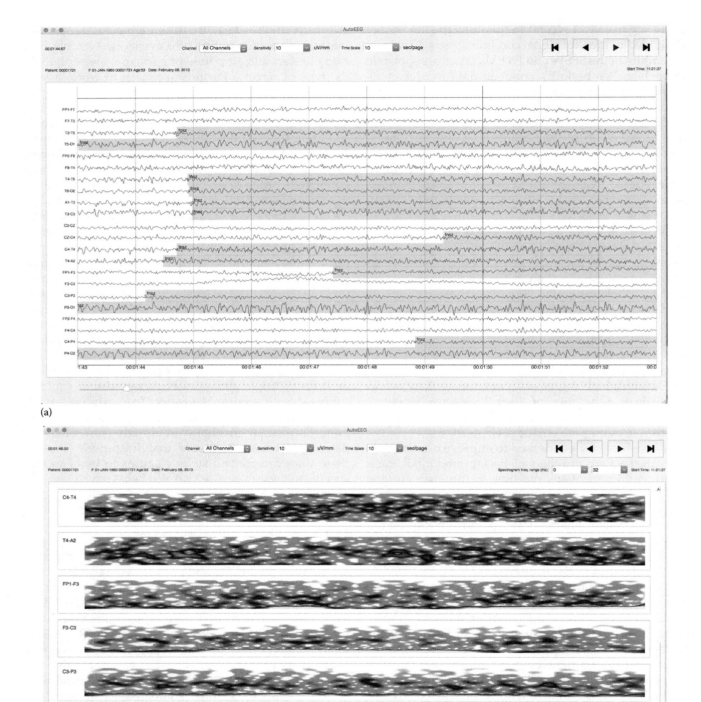

FIGURE 14.5
Typical seizure event is shown in (a) a waveform plot for a subset of the channels; (b) a spectrogram plot for a subset of the channels. Note that the seizure begins on a few channels (e.g., T5-O1) and then spreads to adjacent channels.

The slightest disturbances, such as simply pressing on the electrical connections, cause large deflections in the waveforms. There are many anomalies that produce spike-like behavior in the signal. An example is shown in Figure 14.2, where we see an SPSW event on the left side of the image and an EYEM event on the right side of the image. Video is often used concurrently with an EEG to characterize paroxysmal clinical events that might be seizures, including grimacing, chewing, or nystagmoid eye movements; abrupt and otherwise unexplained changes in pulse, blood pressure, or respiratory pattern; or abrupt deterioration in conscious level. Ideally, video and the EEG signals should be recorded concurrently. Accurate recognition and distinction of benign variants in an EEG are essential to avoid overinterpretation. The range of benign variants include highly confusable events such as small sharp spikes (often referred to as BSSS or BSST for benign small sleep transients), 14 and 6 positive spikes, 6-Hz "phantom" spike and wave, and subclinical rhythmic EEG discharge [39]. These often require additional input beyond EEG waveforms (e.g., video). Therefore, it is not surprising that IRA is fairly low even among experts, particularly on clinical data in which patient behavior is not well controlled.

14.2.6 Evaluation Metrics Are Important

Annotations play a key role in most machine learning applications where supervised training [33] is used. Accurate system evaluation, however, also represents a challenge in itself. Researchers typically report performance in terms of sensitivity and specificity [40] of epochs in biomedical research applications [41]. Each epoch is considered as a separate testing example, even though EEG events can span multiple epochs. The results of the classifier are presented in a confusion matrix, which gives a very useful overview of performance. For example, for a two-class problem such as seizure detection, a confusion matrix has the following categories:

- *True positives (TP)*: the number of epochs identified as a seizure in the reference annotations, and were correctly labeled as a seizure
- *True negatives (TN)*: the number of epochs correctly identified as non-seizures
- *False positives (FP)*: the number of epochs incorrectly labeled as seizure
- *False negatives (FN)*: the number of epochs incorrectly labeled as non-seizure

Sensitivity (TP/TP+FN) and specificity (TN/TN+FP) are derived from these quantities. A precision–recall (PR) curve is an alternative method of scoring [42] in which precision is the percentage of correctly detected seizure divided by predicted seizure epochs (TP/TP+FP), while recall is called sensitivity.

However, sensitivity can often be increased arbitrarily if one is willing to tolerate a poor specificity or a high false alarm rate. Interviews conducted with many clinicians have indicated that the primary reason commercially available technology is not used in clinical settings is due to the high false alarm rate [43,44]. This is perhaps the single most important metric today in guiding machine learning research applications in critical care. Critical care units are overwhelmed with the number of false positives that automated event detection equipment generates. To put this in perspective, one false alarm per bed per hour in a 12-bed ICU generates 12 interrupts per hour that must be serviced. This can easily overwhelm healthcare providers. Since many types of automated monitoring equipment are used in an ICU setting, each with significant false alarm issues, the number of false alarms that must be serviced by healthcare providers is overwhelming [45]. As a result, clinicians report that in practice they simply ignore these systems.

Of course, one must balance sensitivity, specificity, and false alarms. This has been studied extensively in other communities focused on event-spotting technology such as spoken term detection in voice signals [46]. A measure that we have borrowed from this research community is the term-weighted value (TWV) [32], which is based on the notion of a detection error tradeoff (DET) curve [47]. A DET curve is very similar to a receiver operating characteristic originally developed to assess the performance of a communications system [48]. TWV essentially assigns an application-dependent reward to each correct detection and a penalty to each incorrect detection. TWV is one minus the weighted sum of the term-weighted probability of missed detection and the term-weighted probability of false alarms. The actual TWV (ATWV) is performance measured for a specific decision threshold—essentially establishing a specific operating point on the DET curve. This measure is useful when it is preferred to compare two systems based on a single number, though it is always better to compare DET curves over a range of operating characteristics. ATWV and DET curves are our recommended way to evaluate EEG interpretation systems.

To use ATWV, however, you need what are referred to as term-based annotations of the data, and you need to tune the weights assigned to various error modalities [32,49]. Epoch-based annotations are defined as those in which each frame is labeled. Researchers often choose to evaluate their systems on a subset of the epochs available in a database because the overwhelming majority of epochs are assigned to a background, non-seizure, or the equivalent. Since the data are dominated by the presence of events assigned to the background class, performance and evaluation will be biased towards background events if proper normalizations are not considered. In a typical clinical corpus, seizure events account for less than 0.01% of the data (as measured by the cumulative number of seconds seizure events exist). Hence, optimization of a system based on such a metric will focus on non-seizure events, which is not desirable in practice.

Therefore, we annotate the data using term-based annotations, which simply denote the start and stop time of specific events such as a seizure. ATWV allows one to tune the tradeoffs between various types of error classes. It is ideal for these types of applications because it adequately weights false alarms, which is crucial to this application. ATWV ranges from $[-\infty, 1]$, with a score greater than 0.5 being indicative of a system that is performing well. Negative ATWV scores are typically indicative of systems with high false alarm rates. ATWV software is available from the National Institute of Standards and Technology [50] as part of the Open Keyword Search Evaluation package.

14.2.7 Decision Support Systems Can Enhance Interpretation

Decision support systems in healthcare, which can greatly improve manual interpretation, can leverage vast archives of electronic medical records (EMRs) if high-performance automated data wrangling can be achieved [51–53]. EMRs can include unstructured text, temporally constrained measurements (e.g., vital signs), multichannel signal data (e.g., EEGs), and image data (e.g., MRIs). Clinicians who specialize in visual interpretation of data often require second opinions, consult data banks of reference samples, or even reference textbooks for difficult cases that are outside of their normal daily experiences. Medical students specializing in neurology spend years shadowing clinicians while they learn how to read EEGs through experiential training.

One application of automatic interpretation technology that integrates high-performance classification with big data is cohort retrieval. This application can positively impact clinical work and medical student training. When observing an event of interest, such as a seizure, it is desirable to locate other similar examples of such signals, either from previous sessions from the same or similar patients. Information from EEG reports and EEG signals can be mined in such a way that database queries to locate such events can be performed on the aggregated data. Clinical consequences include regularization of EEG reports, real-time feedback and decision-making support, and enhanced training for young neurophysiologists.

One of the challenges in this task is that the EEG reports, such as the one shown in Figure 14.3, are captured as unstructured text in most clinical environments. Therefore, natural language processing is required to identify key medical concepts in the reports. Identification of the type and temporal location of EEG signal events such as spikes or generalized periodic epileptiform discharges in the EEG signal is critical to the interpretation of an EEG. Cohort retrieval systems allow users to query such information using natural language (e.g., "Show me all similar young patients with focal cerebral dysfunction who were treated with Topamax"). In Figure 14.6, we show an example of one such system [51] based on TUH-EEG (described in Section 14.3).

14.2.8 Unbalanced Data Problem Makes Machine Learning Challenging

Finally, as mentioned previously in Section 14.2.6, something that makes this problem additionally challenging is the large imbalance between events of interest (e.g., seizures, spikes) and non-events (e.g., background). This is often referred to as the imbalanced data problem [54]. The amount of time that a patient experiences a seizure is typically less than 0.01% of the overall data in a clinical setting. Clinicians must sift through vast amounts of data in real time to diagnose a disorder. For example, a patient is often admitted to an EMU or ICU for continuous monitoring. Seizure events will occur optimistically only a few times per day. Nevertheless, all data must be manually reviewed. Further, patients now can also use ambulatory EEG systems that allow continuous data collection from their normal living environments, further amplifying the amount of data that must be reviewed.

Direct training on this type of data poses several challenges for traditional machine learning approaches. Even the best deep learning systems will ignore such infrequently occurring events in their efforts to optimize a performance metric. In such situations where one class is significantly more probable than another, the obvious solution is to always guess the most probable class. Though overall performance will appear to be good, performance on events of interest is very poor. This is a common problem in this application space, and another reason a weighted metric like ATWV becomes important. As we will see in the next section, big data plays a critical role in this field, because we need large amounts of these infrequently occurring events to train high-performance statistical models. Techniques such as

Machine Learning Approaches to Automatic Interpretation of EEGs 281

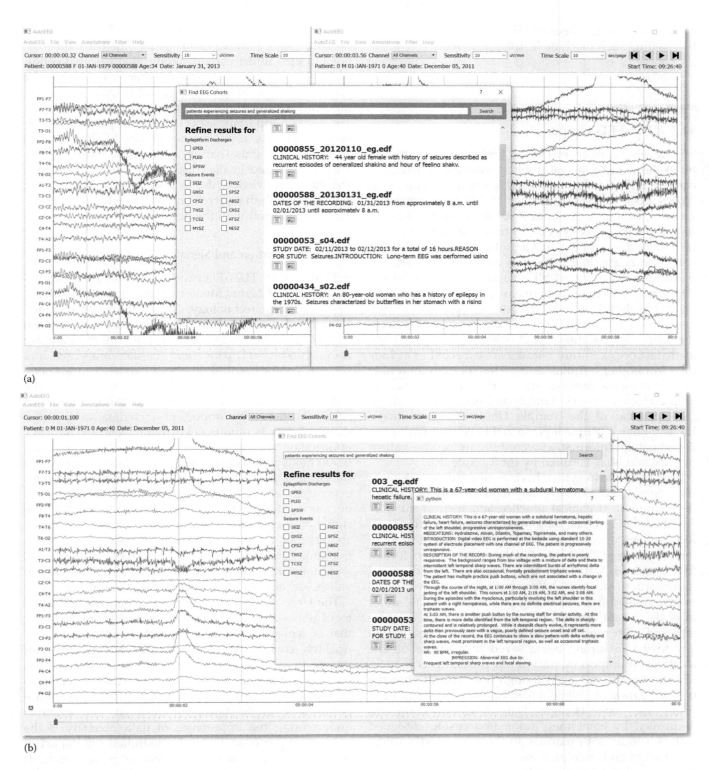

FIGURE 14.6
Cohort retrieval system can provide relevant decision support that improves a neurologist's ability to manually interpret an EEG. (a) The user interface allows signal and text data to be simultaneously searched using unconstrained text queries. (b) The system returns links to signal data (EDF files) and EEG reports (plain text files) for those patient records that match the query.

cross-validation and boosting [55] play an important role in avoiding such machine learning problems.

14.3 TUH EEG Corpus

The development of TUH-EEG [56] was an attempt to build the world's largest publicly available database of clinical EEG data. It currently comprises more than 30,000 EEG records from over 16,000 patients and is growing at a rate of about 3,000 sessions per year. It represents the collective output from Temple University Hospital's Department of Neurology since 2002. Data collection began in 2013 and is an ongoing effort. We have currently released all data from 2002 to 2013 in v0.0.6 and will continue to release additional data on an annual basis (v1.0.0 was released in Spring 2017 and contains data through 2015). The data are available from the Neural Engineering Data Consortium (http://www.nedcdata.org/). Future releases are expected to include data from other hospitals and metadata from a number of collaborative projects. All work was performed in accordance with the Declaration of Helsinki and with the full approval of the Temple University Institutional Review Board (IRB). All personnel in contact with privileged patient information were fully trained on patient privacy and were certified by the Temple IRB.

Because of the long time horizon of the data collection, the original data exist in many data formats that reflect the evolution of clinical practice and instrumentation. Archival EEG signal data were recovered from CD-ROMs. Files were converted from their native proprietary file format (Nicolet's NicVue) to an open format EDF standard. Data were then rigorously de-identified to conform to the HIPAA Privacy Rule by eliminating 18 potential identifiers including patient names and dates of birth. Patient medical record numbers (MRNs) were replaced with randomized database identifiers, with a key to that mapping being saved to a secure off-line location. An important part of our process was to identify similar patients even though they appeared in the original data with different MRNs, name spellings, or in some cases name changes. Data de-identification was performed by combining automated custom-designed software tools with manual editing and proofreading. All storage and manipulation of source files was conducted on dedicated non-network connected computers that were physically located within the TUH Department of Neurology.

There are two distinguishing aspects of these data. First and foremost, it cannot be overemphasized that the data were collected in a live clinical setting. TUH is the public hospital for Philadelphia and serves a diverse population. The EEGs were collected from adults ranging in age from 16 to 90+ years old. The data were not collected under carefully controlled research conditions. This becomes apparent when we discuss the challenges that EEG signal events such as patient movements pose in terms of robust pattern recognition. The second important aspect of this corpus is that it is openly available. Users are not required to be added to an IRB or sign data-sharing agreements. Further, users are not restricted in their use of the data, though they should acknowledge the source of the data in all publications. Both research and commercialization work can be conducted with the data.

14.3.1 Digitization and Signal Processing

EEG signals in TUH-EEG were recorded using several generations of Natus Medical Incorporated's Nicolet™ EEG recording technology. The raw signals obtained from the studies consist of recordings that vary between 20 and 128 channels sampled at a minimum of 250 Hz using a 16-bit A/D converter. The data are stored in a proprietary format that has been exported to an open standard, the European Data Format (EDF) [57], with the use of NicVue v5.71.4.2530. These EDF files contain an ASCII header with important metadata information distributed in 24 unique fields that contain the patient's information and the signal's condition. There are additional fields that describe signal conditions, such as the maximum amplitude of the signals, which are stored for every channel. A complete description of this header can be found at the TUH-EEG project website [58]. The signal data are stored in an uncompressed format using 16 bits per sample. The data are normalized by a minimum and maximum value to maximize precision and minimize quantization effects. These normalization values are stored in the EDF header for each file.

The large variability among EEG channels and montages utilized in clinical EEGs is not usually something that is represented in data collected under controlled research conditions. However, this is an important practical issue present in clinical data. For example, in TUH-EEG, there are over 40 different channel configurations and at least four different types of reference points used in the EEGs administered. One example underscoring the importance of data diversity is the work of Lopez et al. [25], which studied the impact of sensor configuration on classification performance. It was determined that there was a statistically significant degradation in performance when reference channel conditions were mismatched (e.g., training on Average Reference data and evaluating on Linked Ears data). Attempts to mitigate this using standard approaches such as Cepstral Mean Subtraction (CMS) [59] were not successful, indicating that further study of this problem is necessary. It is unclear whether these data can be modeled by a single statistical model, or whether special measures must be taken to account for this variability. Research fields such as speech recognition have dealt with this problem for many years using technologies such as speaker and channel adaptation [59], but these technologies have yet to be explored in EEG research.

Although there are many unique sensor configurations, about 50% of the data follow the 10/20 convention shown in Figure 14.4 closely. The remaining 50% of the data can be mapped onto this configuration using a combination of channel selection and spatial interpolation. We have developed signal processing software to abstract these details from the user so that the data can be easily processed using typical machine learning paradigms. In our preliminary experiments described in Section 14.4, we will focus on data adequately modeled using a 10/20 configuration since this is the most prevalent configuration for an EEG. Neurologists can manually interpret EEGs collected using this configuration with relatively high accuracy, and there is no compelling evidence that higher resolution EEGs improve human performance. However, higher resolution EEGs are still useful for techniques such as localization of seizures and brain mapping [60].

The EEG data archived by a hospital are, unfortunately, not the entire signal of record. Because these are multichannel signals, the amount of data storage required for a single EEG would exceed the capacity of a DVD, which are still the primary way these data are archived. For example, a 22-channel signal digitized at 16 bits/sample for 20 minutes totals about 132 MB. An LTM EEG, which can last for 72 hours or more, can grow in size to several gigabytes of data. At the time these digital systems were designed, this was deemed excessive. Therefore, during a session, technicians mark sections of interest in the signal. When the data are archived to disk, only these sections are retained. This is a process referred to as EEG pruning [61]. The EEGs in TUH-EEG are all pruned. When these EEGs are exported to an EDF file, they are split into multiple files corresponding to each segment of interest. Start and stop times of these segments relative to the original signal are retained in the EDF file so that some parts of the original timeline can be reconstructed.

14.3.2 EEG Report Pairing

For every EEG, there is also a report, such as the one shown in Figure 14.3, that was generated by a board-certified neurologist. This report contains a summary of the physician's findings (e.g., clinical correlation sections) as well as information such as the patient's history and medications. These reports are generated by the neurologist after analyzing the EEG scan and are the official hospital summary of the clinical impression. They are generated anywhere from a few hours to a few days after the patient has been treated depending on the particular workflow for the neurologist and the hospital. These reports are composed of unstructured text that describes the patient, relevant history, medications, and clinical impression. The report also includes information about the location of the session (e.g., inpatient or outpatient), the type of EEG test (e.g., LTM or standard), and the protocol invoked for the test (e.g., the type of stimulation used).

These reports typically have four sections relevant to our research: clinical history, medications, introduction, description of the record, and clinical correlations. The last section is perhaps the most important for machine learning research since it contains a summary of the findings. Not surprisingly, the language used in these reports is intentionally measured, which often makes it difficult to automatically interpret using natural language processing techniques. Due to the limitations of information technology (IT) systems at many hospitals, reports are not often easily matched with the EEG signals. Reports are often stored in a separate IT system, and their connection to the actual EEG data is lost over time. Therefore, we manually paired each retrieved EEG with its corresponding clinician report. Reports were mined from the hospital's central EMR archives and typically consisted of image scans of printed reports. Various levels of image processing were employed to improve the image quality before applying optical character recognition (OCR) to convert the images into text. A combination of software and manual editing was used to scrub protected health information (PHI) from the reports and to correct errors in OCR transcription. Only sessions with both an EEG and a corresponding clinician report were included in the final corpus. The unpaired data are still available and are useful for studies where the lack of an EEG report or a summary of the findings is not an issue (e.g., clustering, unsupervised training, or self-organizing data analysis).

The pairing process can often be challenging due to a number of complicating factors. A patient can often have multiple MRNs either due to practical issues such as clerical errors (e.g., misspelled names or a typo in an MRN), a change in medical insurance, the use of another patient's medical insurance (not uncommon in public hospitals), and changes in marital status, which trigger a name change. The timestamp is also a valuable key in pairing reports, but even that can be misleading. A patient can receive multiple EEGs in the same day in some cases, or often receives a standard 20-minute baseline EEG before beginning an LTM session. We were able to resolve these issues using a combination of automated software that detects potential conflicts and manual review. In extreme cases, the EEG report must be consulted to make sure the patient's medical history matches demographic data collected by the technician at the time the EEG is administered. The latter is logged as part of the data collection system and is available in our private, unreleased version of the database that we refer to when we need to disambiguate data. We have been able to manually pair over 95% of the data with reports using a combination of MRNs, timestamps, and patient demographic information.

14.3.3 De-identification of the Data

The EEG reports in TUH-EEG have been manually de-identified so that a patient's identity remains anonymous. This is extremely important because HIPAA compliance [62] requires a patient's identity be kept anonymous. Patient information appears in several places in the original data: the EDF header, annotation channels where technicians make comments during the session, EEG reports, and some auxiliary files generated by the NicVue software to document the recording session (e.g., the Impedance Report). De-identification, also referred to as anonymization, requires removal of this information. We decided to release only two types of data—an EEG report as a plain text file and the EEG signals as a collection of pruned EDF files.

The EDF files were redacted in order to ensure the patients' anonymity. This process included modifying the MRNs, names, exact dates of birth, and study numbers in the ASCII EDF header. Patients were assigned a randomized ID, which can be mapped back to the original MRN through a mapping file stored in a secure, off-line location. Technician comments, which are time-aligned with the data and show up in a binary format as additional channels of the signal, were stripped since these tended to be unproductive for our research needs. These can at times contain patient names or other sensitive information. The manual effort required to redact these was not considered cost-effective relative to our machine learning research goals.

Redaction of personal information in the EEG reports was a much more sensitive issue and required intensive amounts of manual review. Information relevant to the outcome and interpretation of the EEGs, such as gender, age, medical history, and medications, was retained. Selected fields from this header that contain important metadata are shown in Table 14.1.

Though the EEG reports appear to have a semantic structure, the reports are created typically in Microsoft Word as flat unstructured files. Extracting useful information from these reports can only be done using natural language processing techniques. More research is needed on the organization and representation of these reports, but we are making progress in parsing and representing these data in a related research project [63,64] and plan to release these metadata soon (see http://www.nedcdata.org for further details on its release). The Word versions of the EEG reports have been manually redacted, and then automatically converted to flat text files. These data were reviewed multiple times by several data entry specialists and run through a series of text filters designed to spot thousands of special cases indicating incorrect redaction.

14.3.4 Basic Descriptive Statistics

The first release of TUH-EEG, referred to a v0.0.6, was made in 2015. Approximately 75% of the sessions are standard EEGs less than one hour in duration, while the remaining 25% are from LTM sessions. A distribution of the number of records per year is presented in the upper right of Figure 14.7. To put the size of this corpus in perspective, the total EEG signal data collected thus far, including all unreleased data, require over 2 TB of storage with a median file size of 20 MB. Though the EEG signal data are pruned, the amount of data is staggering. For example, if we treat each channel of data as an independent signal, there is over 1 billion seconds of data. Though this might seem huge at first, the events we are interested in are relatively rare, often occupying less than 0.01% of the recording duration. The number of patients experiencing seizures during a session is on the order of several hundred. When these sessions are cross-referenced by patient medical histories, even this huge amount of data appears small.

TABLE 14.1

Selected Fields from an EDF Header That Contain De-Identified Patient Information and Key Signal Processing Parameters Are Shown

Field	Description	Example
1	Version Number	0
2	Patient ID	TUH123456789
4	Gender	M
6	Date of Birth	57
8	Firstname_Lastname	TUH123456789
11	Startdate	01-MAY-2010
13	Study Number/Tech. ID	TUH123456789/TAS X
14	Start Date	01.05.10
15	Start time	11.39.35
16	Number of Bytes in Header	6400
17	Type of Signal	EDF+C
19	Number of Data Records	207
20	Dur. of a Data Record (Secs)	1
21	No. of Signals in a Record	24
27	Signal Prefiltering	HP:1.000 Hz LP:70.0 Hz N:60.0
28	No. Signal Samples/Rec.	250

The completed v0.0.6 corpus comprises 16,986 sessions from 10,874 unique subjects. Each of these sessions contains at least one EDF file (more in the case of LTM sessions that were broken into multiple files) and one physician report. Corpus metrics are summarized in Figure 14.7. Subjects were 51% female and ranged in age from less than 1 year to over 90 years (average 51.6, stdev 55.9; see Figure 14.7 bottom left). The average number of sessions per patient was 1.56, although as many as 37 EEGs were recorded for a single patient over an eight-month period (Figure 14.7, top left). The number of sessions per year varies from approximately 1,000 to 2,500 (except for years 2000–2002 and 2005, in which limited numbers of complete reports were found in the various EMR archives; see Figure 14.7, top right).

There was a substantial degree of variability with respect to the number of channels included in the corpus (see Figure 14.7, bottom right). The corpus is relatively evenly split between AR reference data (51.5%) and LE referenced data (48.5%). (There are a very small number of reference schemes that do not conform to these two standards). EDF files typically contained both EEG-specific channels as well as supplementary channels such as detected bursts, EKG, EMG, and photic stimuli. The most common number of EEG-only channels per EDF file was 31, although there were cases with as few as 20. A majority of the EEG data were sampled at 250 Hz (87%) with the remaining data being sampled at 256 Hz (8.3%), 400 Hz (3.8%), and 512 Hz (1%).

An overview of the distribution of signal events described in Section 14.2.4 is shown in Table 14.2 for a subset of the data we have used to develop baseline technology. We see that all classes except SPSW occur frequently enough that robust statistical models can be built. A subset of the SPSW events will ultimately correspond to seizures, so we see how infrequently seizure events actually occur. We estimate that about 5% of the sessions contain actual seizure events, and less than 0.01% of the recorded data contain an actual seizure event. PLEDs and GPEDs can be located with relative ease. ARTF and EYEM events also can be relatively easily identified using a variety of heuristic or statistical methods.

An initial analysis of the physician reports reveals a wide range of medications and medical conditions. Unsurprisingly, the most commonly listed medications were anti-convulsants such as Keppra and Dilantin, as well as blood thinners such as Lovenox and heparin. Approximately 87% of the reports included the text string "epilep," and about 12% included "stroke." Only 48 total reports included the string "concus." The EEG

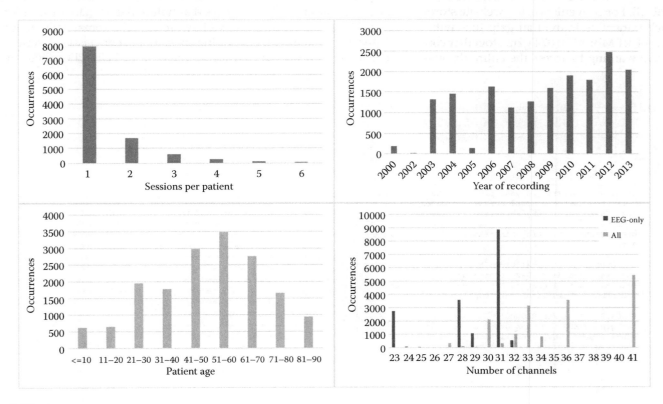

FIGURE 14.7
Metrics describing the TUH EEG Corpus: (top left) histogram showing number of sessions per patient; (top right) histogram showing number of sessions recorded per calendar year; (bottom left) histogram of patient ages; (bottom right) histogram showing number of EEG-only channels (black) and total channels (gray).

TABLE 14.2

Overview of the Distribution of Events in the Subset of the TUH EEG Corpus Used to Develop Our Baseline Technology for the Six-Event Classification Task

Event	Train No.	% (CDF)	Eval No.	% (CDF)
SPSW	645	0.8% (1%)	567	1.9% (2%)
GPED	6184	7.4% (8%)	1,998	6.8% (9%)
PLED	11,254	13.4% (22%)	4,677	15.9% (25%)
EYEM	1,170	1.4% (23%)	329	1.1% (26%)
ARTF	11,053	13.2% (36%)	2,204	7.5% (33%)
BCKG	53,726	63.9% (100%)	19,646	66.8% (100%)
Total:	84,032	100.0% (100%)	29,421	100.0% (100%)

reports contain a total of 3.5M words, which makes it an interesting corpus for natural language processing research.

14.3.5 Structure of the Released Data

The TUH EEG Corpus v0.0.6 has been released and is freely available online at http://www.nedcdata.org. Users must register with a valid email address. The uncompressed EDF files and reports together comprise 572 GB. For convenience, the website stores all data from each patient as individual gzip files with a median file size of 4.1 MB; all 10,874 gzips together comprise 330 GB. Users wanting to access the entire database are encouraged to physically mail a USB hard drive to the authors in order to avoid the downloading process.

The corpus was defined with a hierarchical Unix-style file tree structure. The top folder, edf, contains 109 numbered folders, each of which contains numbered folders for up to 100 patients. Each of these patient folders contains sub-folders that correspond to individual recording sessions. Those folder names reflect the session number and date of recording. Finally, each session folder includes one or more EEG (.edf) data files as well as the clinician report in .txt format. Figure 14.8 summarizes the corpus file structure and gives examples of text and signal data.

We also have released subsets of the data of interest to the community. TUH EEG Epilepsy Corpus (v0.0.1) is a subset of the TUH EEG Corpus that contains 100 subjects with and without epilepsy. The TUH EEG Seizure Corpus [34] is a subset designed to support seizure detection experiments. This corpus contains a 50-patient evaluation set and a 250-patient training set. The evaluation set has been manually annotated for seizure events by a panel of expert neurologists. The training set has been annotated by a team of experienced annotators and validated using a variety of commercial systems. The TUH EEG Abnormal EEG Corpus is a subset in which each EEG has been manually classified as normal or abnormal by our expert annotation team [20]. In addition to these subsets, we have released automatically generated alignments of the six signal event classes described in Section 14.2.4 for all the data in TUH-EEG. Over the

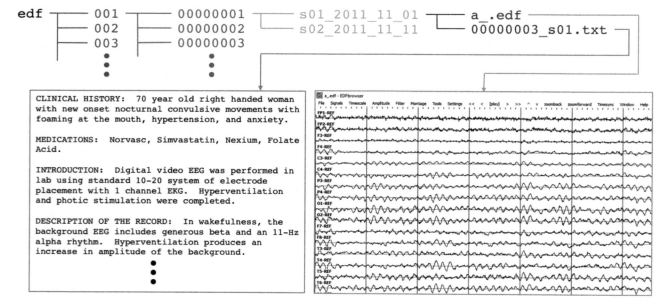

FIGURE 14.8
Directory and file structure of TUH-EEG is shown. Data are organized by patient and then by session. Each session contains one or more signal files (edf) and a physician's report (txt). To accommodate filesystem management issues, patients are grouped into sets of about 100.

next few years, we also expect to expand the corpus to include data from other hospitals.

14.4 Automatic Interpretation of EEGs

Automatic interpretation of EEGs essentially involves detecting epileptiforms and then classifying their nature. For example, if they are persistent, an event will be classified as a GPED. If it is an isolated even, it is classified as SPSW. Disambiguating spikes from background noise is a significant problem, especially for signals such as EEGs where many types of artifacts produce spike-like behavior. Most spike detection systems use one of two approaches: (1) heuristic waveform detection techniques or (2) static frame classification techniques. The performance of commercial tools, which are often based on heuristics, has proven to be unacceptable for clinical use [17,21].

14.4.1 Machine Learning Approaches

There have been historically three general machine-learning approaches to the problem of EEG classification: (1) a single classification of an entire EEG, (2) segment-level classification, and (3) sequential decoding. Single classification of an entire signal represents the simplest approach in which an entire EEG segment is classified by learning a mapping of a single aggregate feature vector [65,66]. Lopez et al. [20] demonstrated the classification of abnormal EEGs using this approach. Nonlinear statistical models such as neural networks (NNs), support vector machines (SVMs), and even visual pattern recognition techniques are typically used. Such systems often make a binary decision (e.g., normal/abnormal), but do not identify critical events within the signal.

A second variation on this approach is to segment each channel of the multi-channel signal into frames, analyze and classify each frame, and then perform some overall classification of a segment or an entire file based on these event probabilities [67,68]. These two-level hierarchical systems can be viewed as static classifiers since each frame of data is judged independently of the others. These approaches often use feature extraction techniques based on wavelets [69,70] or other similar time/frequency representations of the signal, and often concatenate feature vectors in time to increase temporal context. They do not make use of the temporal sequence of EEG events.

Both these approaches can be viewed as a bottom-up approach because the statistical models that transform segments of the signal to probabilities are not generally informed by preceding or following context. When the low-level signal is highly ambiguous, as is the case in speech recognition or EEG event detection, these techniques fail because the event probabilities generated are not adequately distinct for high-performance inference. This was a lesson learned many years ago in speech recognition when expert systems [71] and event spotting [72] were abandoned in favor of statistical models based on Bayes' rule and exhaustive search [73].

The third approach, segment decoding, has been popular in applications such as speech recognition, where language provides an overall structure to the signal. We can exploit sequential relationships between words and phonemes to improve the accuracy of this low-level event detection (known as acoustic decoding in speech recognition) [33,74]. Feature extraction and the transformations used to postprocess features, which we refer to as signal models, learn context-dependent statistics [75] and use these to better calibrate the frequency domain and time domain predictions of the signal. This type of top-down processing is critical to achieving high performance in speech recognition, but has yet to be applied to EEG analysis, because the decision tree (DT) or logic behind classifying an EEG based on expert knowledge has not been adequately codified.

Most previous studies have focused on small numbers of patients (typically less than 20) and have not demonstrated robust performance. For example, the Seizure Detection Challenge (https://www.kaggle.com/c/seizure-detection) [76] was based on "prolonged intracranial EEG recorded from four dogs with naturally occurring epilepsy and from 8 patients with medication resistant seizures during evaluation for epilepsy surgery." Most research focused on EEG classification has concentrated on static classification of the data. An example of this type of approach was used by Bao et al. [68], in which 94% classification accuracy was achieved for epilepsy detection on six normal and six epileptic patients. The signal was segmented into 20-second non-overlapping intervals (4096 samples at 200 Hz), and converted to a feature vector using a collection of heterogeneous features that included power spectral intensity, fractal dimension, and other measures of nonlinearity. A probabilistic neural network (PNN) was applied to each channel, and the individual channel outputs were combined using a voting system.

Although the feature extraction process was manually optimized by using bandpass filters and increasing the segment length to 40 seconds, automated optimization could have been exploited if more data were available and discriminative training of features could be employed. Our experience in other classification problems, such as speech recognition [77], is that feature extraction plays a relatively insignificant role in overall system performance if the higher-level classification system is suitably powerful. "recent years, deep learning systems are circumventing a model-based feature extraction process (e.g., cepstral features) and operating directly on samples of the signal [78] using an approach called representation learning.

The drawbacks of many of the static classification approaches being used are that the spectral behavior of the signal is coarsely quantized, and hence the channel-specific classifiers do not account for the temporal behavior of the signal. The analysis windows typically used are on the order of 1 to 2 seconds, which give poor temporal resolution. In addition, the sample size is typically too small to achieve true statistical significance. In such limited studies, there is always the potential for the classifier keying in on artifacts of the data, such as the background noise or the quality of the transducer conduction.

14.4.2 Typical Static Classification Systems

As machine learning research has evolved, interest in detecting low-level events has been growing. For example, Wulsin et al. [67] defined a detection problem based on the six signal events described in Section 14.2.4. Their recall (0.22) and precision (0.19) rates were respectable, but their study was conducted on only 11 patients. A small portion of the data were selected that contained significant EEG events, and these data were hand-labeled by a clinical epileptologist. For feature extraction, temporal features, such as mean energy, zero crossings, and average peak/valley amplitude, and spectral features, such as frequency band power and wavelet energy, were combined. A variety of classifiers were evaluated including SVMs and deep belief networks (DBNs; a particular form of a deep learning–based system).

False positive rates were in the range of 4 to 25 false alarms per hour [67]. A mean F-score was used as an overall performance metric. Best performance was obtained using a principal components analysis (PCA) of the features followed by a DT classifier. More interestingly, the performance of DTs, SVMs, and DBNs was comparable and slightly better than the k-nearest neighbor (kNN). We conjecture that this is a byproduct of insufficient training data and less mature deep learning technology. Head-to-head comparisons of deep learning systems on speech recognition tasks with large amounts of training data [79, 80] have shown two things: (1) for the same amount of training data, deep learning systems can exceed the performance of more conventional HMMs if properly configured, and (2) the performance of HMMs can often approach the performance of deep learning systems if additional training data are provided. Bengio et al. (2013) also demonstrated that the feature extraction process could be automatically learned using a deep learning–based system.

We have replicated the Wulsin results on a publicly available seizure detection task [81] using a relatively simple standard HMM-based [82,83] approach as a proof of concept. We achieved a sensitivity of 96.5% and a false alarm rate of 3.8/hour. It was possible to build this system quickly and efficiently because time-aligned markers for the seizure events were provided with the data. It has been more challenging to reach our goal of 95% sensitivity on TUH-EEG due to the lack of time-aligned markers for events. This has necessitated investigation into active learning approaches that can be used to bootstrap systems from small amounts of data [84].

14.4.3 Feature Extraction

There are two general approaches to contemporary signal processing—functional or model-based features (e.g., cepstral coefficients) or statistically based features that are embedded in some larger machine learning system. The latter appear to work well when there are ample amounts of data and offer the benefit of integrating key algorithmic enhancements such as discriminative training [85]; the former is more traditional and has worked well in a variety of recognition applications (e.g., speech recognition) where there is significant subject matter expertise. Model-based features tend to work well for EEG analysis, although statistically based features are slowly emerging as more data become available. In this section, we describe a standard model-based approach.

Neurologists most often review EEGs in 10-second windows when doing fine-grained analysis. Larger time intervals are used to triage data and locate sections of interest, but detailed interpretations are most often performed on 10-second windows. The time resolution used in these assessments is often on the order of 1 second. Therefore, we decompose the signal into fixed intervals typically of 1 to 2 seconds, which we refer to as an epoch. We have experimentally adjusted these parameters and found that a 1-second epoch is an appropriate tradeoff between time resolution, computational complexity, and performance [86]. We further subdivide this interval into 0.1-second frames and use an overlapping window approach to compute features, as shown in Figure 14.9.

We use a standard cepstral coefficient-based feature extraction approach similar to the linear frequency cepstral coefficients (LFCCs) used in speech recognition [75,87]. Though popular alternatives to LFCCs in EEG processing include wavelets, which are used by many commercial systems, our experiments with such features have shown very little advantage over LFCCs on TUH-EEG. Unlike speech recognition, which uses a mel scale for reasons related to speech perception, we use a linear frequency scale for EEGs, since there is no physiological evidence that a log scale is meaningful [15].

It is common in the LFCC approach to compute cepstral coefficients by computing a high-resolution fast Fourier transform, downsampling this representation using an oversampling approach based on a set of overlapping bandpass filters, and transforming the output into the cepstral domain using a discrete cosine transform [59,75]. The zeroth-order cepstral term is typically discarded and replaced with an energy term as described below.

There are two types of energy terms that are often used: time domain and frequency domain. Time domain energy is a straightforward computation using the log of the sum of the squares of the windowed signal:

$$E_t = \log\left(\frac{1}{N}\sum_{n=0}^{N-1}|x(n)|^2\right). \quad (14.1)$$

We use an overlapping analysis window (a 50% overlap was used here) to ensure a smooth trajectory of this feature. The energy of the signal can also be computed in the frequency domain by computing the sum of squares of the oversampled filter bank outputs after they are downsampled:

$$E_f = \log\left(\sum_{k=0}^{N-1}|X(k)|^2\right). \quad (14.2)$$

This form of energy is commonly used in speech recognition systems because it provides a smoother, more stable estimate of the energy that leverages the cepstral representation of the signal. However, the virtue of this approach has not been extensively studied for EEG processing.

To improve differentiation between transient pulse-like events (e.g., SPSW events) and stationary background noise, we have introduced a differential energy term that attempts to model the long-term change in energy. This term examines energy over a range of M frames centered about the current frame, and computes the difference between the maximum and minimum over this interval:

$$E_d = \max_m\left(E_f(m)\right) - \min_m\left(E_f(m)\right). \quad (14.3)$$

We typically use a 0.9-second window for this calculation. This simple feature has proven to be surprisingly effective.

The final step to note in our feature extraction process is the familiar method for computing derivatives of features using a regression approach [59,79,88]:

$$d_t = \frac{\sum_{n=1}^{N}n(c_{t+n}-c_{t-n})}{2\sum_{n=1}^{N}n^2}, \quad (14.4)$$

where d_t is a delta coefficient, from frame t computed in terms of the static coefficients c_{t+n} to c_{t-n}. A typical value for N is 9 (corresponding to 0.9 second) for the first derivative in EEG processing, and 3 for the second derivative. These features, which are often called deltas because they measure the change in the features over times, are one of the most well-known features in speech recognition [59]. We typically use this approach to compute the derivatives of the features and then apply this approach again to those derivatives to obtain an estimate of the second derivatives of the features, generating what are often called delta-deltas. This triples the size of the feature vector (adding deltas and delta-deltas), but is well-known to deliver improved performance. This approach has not been extensively evaluated in EEG processing.

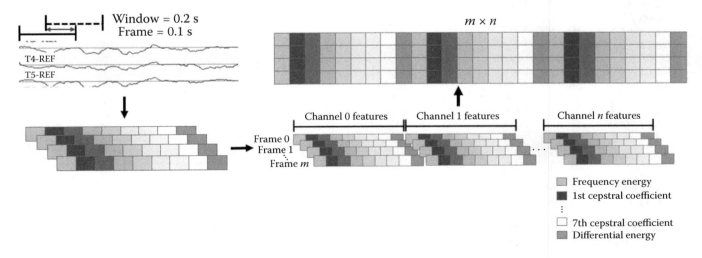

FIGURE 14.9
Illustration of the frame-based analysis that is used to extract features, and how features are stacked so that both temporal and spatial context can be exploited.

Dimensionality is something we must always pay attention to in classification systems since our ability to model features is directly related to the amount of training data available. The use of differential features raises the dimension of a typical feature vector from 9 (e.g., 7 cepstral coefficients, frequency domain energy, and differential energy) to 27. There must be sufficient training data to support this increase in dimensionality or any improvements in the feature extraction process will be masked by poor estimates of the model parameters (e.g., Gaussian means and covariances). The impact of differential energy is shown in Figure 14.10.

We have used a subset of TUH-EEG that has been manually labeled for the six types of events described in Section 14.2.4 to tune feature extraction [86]. We use the HMM system described in Section 14.4.4 as the classification system. The training set contains segments from 359 sessions, while the evaluation set was drawn from 159 sessions. No patient appears more than once in the entire subset, which we refer to as the TUH EEG Short Set. The training set was designed to provide a sufficient number of examples to train statistical models. We refer to the 6sixclasses shown in Section 14.2.4 as the six-way classification problem. This is not necessarily the most informative performance metric. It makes more sense to collapse the three background classes into one category. We refer to this second evaluation paradigm as a four-way classification task: SPSW, GPED, PLED, and BACKG. The latter class contains an enumeration of the three background classes. Finally, in order that we can produce a DET curve [47], we also report a two-way classification task in which we collapse the data into a target class (TARG) and a background class (BCKG).

DET curves are generated by varying a threshold typically applied to likelihoods to evaluate the tradeoff between detection rates and false alarms. However, it is also instructive to look at specific numbers in table form. Therefore, all experiments reported in the tables use a scoring penalty of 0, which essentially means we are evaluating the raw likelihoods returned from the classification system. In virtually all cases, the trends shown in these tables hold up for the full range of the DET curve.

The first series of experiments was run on a simple combination of features. A summary of these experiments is shown in Table 14.3, where error rates are given on the six-way, four-way, and two-way classification tasks described above. Cepstral-only features were compared with several energy estimation algorithms. It is clear that the combination of frequency domain energy and differential energy, system no. 5 in Table 14.3, provides a substantial reduction in performance. However, note that differential energy by itself (system no. 4)

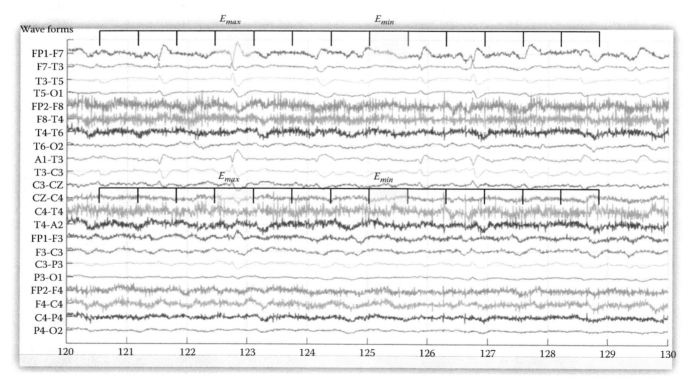

FIGURE 14.10
Illustration of how the differential energy term accentuates the differences between spike-like behavior and noise-like behavior. Detection of SPSW events is critical to the success of the overall system.

TABLE 14.3

Performance on the TUH EEG Short Set of the Base Cepstral Features Augmented with an Energy Feature. System No. 5 Uses Both Frequency Domain and Differential Energy Features. Note That the Results Are Consistent Across All Classification Schemes

No.	System Description	Dims.	6-Way	4-Way	2-Way
1	Cepstral	7	59.3%	33.6%	24.6%
2	Cepstral + E_f	8	45.9%	33.0%	24.0%
3	Cepstral + E_t	8	44.9%	33.7%	24.8%
4	Cepstral + E_d	8	55.2%	32.8%	24.3%
5	Cepstral + E_f + E_d	9	39.2%	30.0%	20.4%

produces a noticeable degradation in performance. Frequency domain energy clearly provides information that complements differential energy. The improvements produced by system no. 5 hold for all three classification tasks. Though this approach increases the dimensionality of the feature vector by one element, the value of that additional element is significant and not replicated by simply adding other types of signal features [86].

A second set of experiments was run to evaluate the benefit of using differential features. These experiments are summarized in Table 14.4, where we again show six-way, four-way, and two-way classification error rates as described above. The addition of the first derivative adds about 7% absolute in performance (e.g., system no. 6 vs. system no. 1). However, when differential energy is introduced, the improvement in performance drops to only 4% absolute.

The story is somewhat mixed for the use of second derivatives. On the base cepstral feature vector, second derivatives reduce the error rate on the six-way task by 4% absolute (systems nos. 6 and 11). However, the improvement for a system using differential energy is much less pronounced (system no. 5 in Table 14.3, systems nos. 10 and 15 in Table 14.4). In fact, it appears that differential energy and derivatives do something very similar. Therefore, we evaluated a system that eliminates the second derivative for differential energy. This system is labeled no. 16 in Table 14.4. We obtained a small but significant improvement in performance over system no. 10. The improvement on four-way classification was larger, which indicates more of an impact on differentiating between PLEDs, GPEDs, and SPSW vs. background. This is satisfying since this feature was designed to address this problem.

The results shown in Tables 14.3 and 14.4 hold up under DET curve analysis as well. DET curves for system nos. 1, 5, 10, and 15 are shown in Figure 14.11. We can see that the relative ranking of the systems is comparable over the range of the DET curves. First derivatives deliver a measurable improvement over absolute features (system no. 10 vs. no. 5). Second derivatives do not provide as significant an improvement (system no. 15 vs. no. 10). Differential energy provides a substantial improvement over the base cepstral features.

14.4.4 Hidden Markov Models

An overview of a generic system to automatically interpret EEGs is shown in Figure 14.12. This system incorporates a signal event detector that operates on

TABLE 14.4

Impact of Differential Features on Performance Is Shown. For the Overall Best Systems (nos. 10 and 15), Second Derivatives Do Not Help Significantly. Differential Energy and Derivatives Appear to Capture Similar Information

No.	System Description	Dims.	6-Way	4-Way	2-Way
6	Cepstral + Δ	14	56.6%	32.6%	23.8%
7	Cepstral + E_f + Δ	16	43.7%	30.1%	21.2%
8	Cepstral + E_t + Δ	16	42.8%	31.6%	22.4%
9	Cepstral + E_d + Δ	16	51.6%	30.4%	22.0%
10	Cepstral + E_f + E_d + Δ	18	35.4%	25.8%	16.8%
11	Cepstral + Δ + $\Delta\Delta$	21	53.1%	30.4%	21.8%
12	Cepstral + E_f + Δ + $\Delta\Delta$	24	39.6%	27.4%	19.2%
13	Cepstral + E_t + Δ + $\Delta\Delta$	24	39.8%	29.6%	21.1%
14	Cepstral + E_d + Δ + $\Delta\Delta$	24	52.5%	30.1%	22.6%
15	Cepstral + E_f + E_d + Δ + $\Delta\Delta$	27	35.5%	25.9%	17.2%
16	(15) but no $\Delta\Delta$ for E_d	26	35.0%	25.0%	16.6%

each channel using channel-independent models, and two stages of postprocessing to produce epoch labels. An N-channel EEG is transformed into N independent feature streams using a standard sliding window–based approach. These features are then transformed into EEG signal event hypotheses using a standard HMM recognition system [83]. These hypotheses are postprocessed by examining temporal and spatial context to produce epoch labels. The system detects three events of clinical interest (SPSW, PLED, and GPED) and three events that map to background (ARTF, EYEM, and BCKG) as described in Section 14.2.4.

A multichannel EEG signal is input to the system, typically as an EDF file. A subset of the channels corresponding to a standard 10/20 EEG are selected. The signal is converted to a sequence of feature vectors as previously described. A group of frames are classified into an event on a per-channel basis using an HMM-based [33,89] classifier. This approach, which we borrow heavily from speech recognition [28,83], uses a left-to-right HMM topology [33] to encode the temporal evolution of the signal. Though there is no direct physiological or neurological motivation for this topology, experiments on alternate topologies have not proven to result in a significant gain on EEG or speech recognition experiments. The standard three-state model works surprisingly well across a wide range of applications.

HMMs are trained for each of the six classes using data for all channels (channel-specific HMMs have not proven to provide a significant gain in performance).

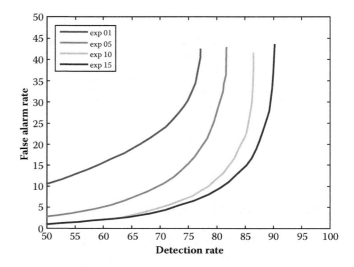

FIGURE 14.11
DET curve analysis of feature extraction systems that compares absolute and differential features. The addition of first derivatives provides a measurable improvement in performance while second derivatives are less beneficial.

Each incoming epoch for each channel is processed through the system, resulting in a likelihood vector that models the probability that the epoch could have been generated from the corresponding model. The event label for a channel is selected based on the most probable class—a forced-choice hypothesis test.

The second level of the system essentially examines multiple adjacent epochs in time, which we refer to as

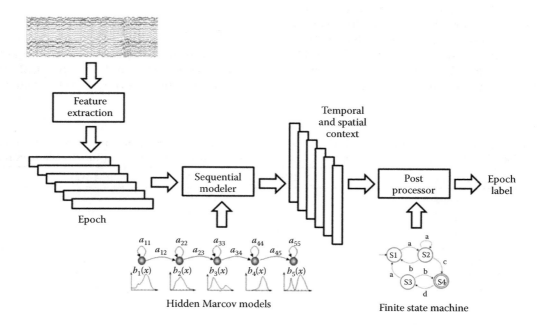

FIGURE 14.12
Two-level architecture for automatic interpretation of EEGs that integrates hidden Markov models for sequential decoding of EEG events with deep learning for decision-making based on temporal and spatial context.

temporal context, and multiple channels, which we refer to as spatial context since each channel is associated with a location of an electrode on a patient's head. There are a wide variety of algorithms that can be used to produce a decision from these inputs. For example, early work in EEG interpretation used a majority vote [90]. We have explored three approaches: (1) a simple heuristic mapping that makes decisions based on a predefined order of preference (e.g., SPWS > PLED > GPED > ARTF > EYEM > BCKG); (2) application of a random forest classification tree approach [91] that we have used successfully for a number of other applications; and (3) a stacked denoising autoencoder [92,93] that has been successfully used in many deep learning systems. Performance of these three approaches is summarized in Table 14.5.

The detection rate (DET) is defined as the percentage of correct recognitions for the classes (SPSW, GPED, PLED). The false alarm rate (FA) is defined as the percentage of incorrect recognitions for (BCKG, ARTF, EYEM)—the number of times these classes are detected as one of the three non-background classes (SSW, GPED, PLED) divided by the number of times they occur. The error rate (ERR) is defined as the number of times any class is incorrectly detected divided by the total number of epochs. Applying a machine learning component as a postprocessor to the event detection level achieves our goal of maximizing the DET rate and minimizing the FA rate.

Further, the errors that the current system makes are not "mission critical." For example, it is not critical that the system detect every spike accurately. Distinguishing spikes from background signal is very hard even for a trained neurologist. However, alerting a neurologist that an EEG has spikes in it, and showing the approximate locations of these spikes, is of great value. The neurologist can then manually review only the areas of the signal that have events of interest such as spikes.

One can often argue that it is relatively straightforward to build an automatic labeling system if fully annotated data are available. This is a great oversimplification since most realistic applications require a significant amount of engineering. For the EEG problem, the situation is more complex as the agreement between transcribers is relatively low. We have developed a highly effective active learning approach to training models that requires a very small amount of transcribed data [84]. An overview of the process is shown in Figure 14.13. We use a small amount of manually transcribed data to seed the models. We then use these models to automatically label the data and produce likelihoods that each epoch could have been one of the six classes. We continue by using these new labels to sort the data and retrain the models. We only consider labels for data for which we are confident that the labels are correct, and

TABLE 14.5

Comparison of Performance for Three Postprocessing Algorithms for the Detection Rate (DET), False Alarm Rate (FA), and Classification Error Rate (Error). The FA Rate Is the Most Critical to This Application

System	DET	FA	Error
1: Simple Heuristics	99%	64%	74%
2: Random Forests	85%	6%	37%
3: Autoencoder	84%	4%	37%

for data of interest such as SPSW events. We typically set a confidence threshold of about 80%. We continue iterating on this process until classification performance on the training data and/or development test data is adequate. In practice, convergence occurs quickly and even after only three iterations, models have improved significantly.

14.4.5 Normal/Abnormal Detection

A similar version of this system can be used to do normal/abnormal classification [19]. The automated classification of an EEG record as normal or abnormal represents a significant step for the reduction of the visual bias intrinsic to the subjectivity of the record's interpretation. The main characteristics of an adult normal EEG are as follows [15]:

- Reactivity: Response to certain physiological changes or provocations
- Alpha rhythm: Waves originated in the occipital lobe (predominantly), between 8 and 13 Hz and 15 to 45 μV
- Mu rhythm: Central rhythm of alpha activity commonly between 8 and 10 Hz visible in 17% to 19% of adults
- Beta activity: Activities in the frequency bands of 18–25, 14–16, and 35–40 Hz
- Theta activity: Traces of 5–7 Hz activity present in the frontal or frontocentral regions of the brain

Neurologists follow procedures similar to the one summarized in Figure 14.14 and can usually make this determination by examining the first few minutes of a recording. Hence, we focused on examining the first 60 seconds of an EEG to calibrate the difficulty of the task.

We selected a demographically balanced subset of TUH-EEG through manual review that consisted of 202 normal EEGs and 200 abnormal EEGs. These sets were further partitioned into a training set (102 normal/100 abnormal), development test set (50 normal/50 abnormal), and an evaluation set (50 normal/50 abnormal).

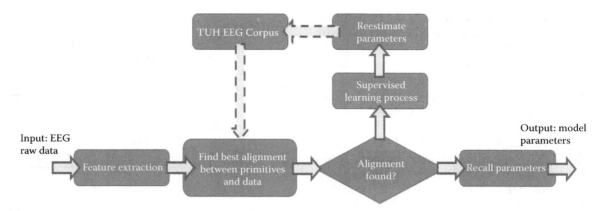

FIGURE 14.13
Overview of our iterative HMM training procedure is shown. An active learning approach is used to bootstrap the system to handle large amounts of data.

To create an appropriate experimental paradigm, only one EEG channel was selected for consideration. Examination of manual interpretation techniques practiced by experts revealed that the most promising channel to explore was the differential measurement T5-O1, which is part of the popular TCP montage. This channel represents the difference between two electrodes located in the left temporal and occipital lobes. The first 60 seconds of each recording was used to extract signal features using the process described in Section 14.4.3. The feature vectors (60 seconds × 10 frames/second = 600 vectors) were concatenated into a supervector of dimension 600 × 27 = 16,200. The dimensionality of the supervector was reduced using class-dependent PCA in which we retained the N most significant eigenvectors of the covariance matrix.

Two standard algorithms were explored: kNN (Duda et al., 2001) and random forest ensemble learning (RF) [91]. We conducted additional searches for an optimal set of parameters for each system. In Table 14.6, we compare performance of these two systems to a baseline based on random guessing. The first system is random guessing based on priors. The second system is kNN with $k = 20$ and a PCA dimension of 86. The third system is RF with $N_t = 50$ and a PCA dimension of 86. The tuned kNN and RF systems outperform random guessing based on priors, which is a promising outcome for these experiments. However, there is a high confusion rate for

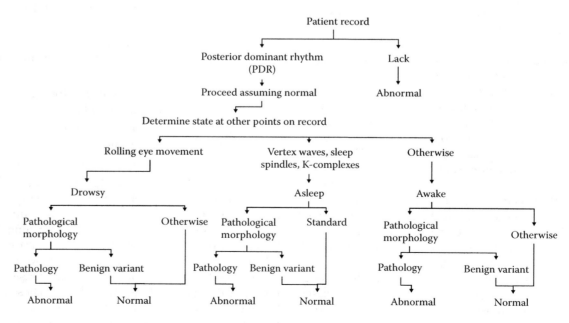

FIGURE 14.14
General process for identifying an abnormal EEG depends heavily on the observation of the posterior dominant rhythm (PDR).

normal EEGs. The dominant error is a normal EEG classified as abnormal. This could be explained by the presence of benign variants, or electroencephalographic patterns that resemble abnormalities, but do not qualify as events that would be of significance for the abnormal classification of a record.

Next, we developed a variant of the HMM system depicted in Figure 14.12. We used standard three-state HMMs with three Gaussian mixture components per state as before. Feature vectors were computed every 0.1 second, generating a total of 600 feature vectors over the first 60 seconds of the EEG signal. Two approaches were followed. First, as discussed before, we built a supervector for each frame consisting of a concatenation of feature vectors for each channel. This supervector was reduced to a dimension of 20 using PCA. This system, shown in Table 14.6 as PCA-HMM, reduced the error rate from 32% to 26%. The goal in this experiment was to evaluate whether PCA could adequately model localization of the event since it has access to data from all channels.

The second approach selected a single channel and applied the raw feature vector to the same three-state HMM. This system, referred to as GMM-HMM, further reduced the error rate to 17%, which is approaching human performance on this task (when neurologists are constrained to look at the first 60 seconds of data). In Table 14.7, we explore performance as a function of the channel used. We verified that performance for T5-O1 was, in fact, optimal, as predicted from neuroscience considerations.

14.4.6 Seizure Detection

We have also trained the system depicted in Figure 14.12, using a basic three-state HMM topology and SdA postprocessing, to perform a seizure/no-seizure binary decision. Here, we use a subset of TUH-EEG that was specifically labeled using term-based labels for seizures by a series of experts [34]. The data were carefully annotated by a team of students who have been trained

TABLE 14.6

Comparison of Performance for Abnormal Classification of an EEG for Several Standard Classification Approaches Including kNN, RF, and Two Variants of a System

No.	System Description	Error
1	Random Guessing	49.8%
2	kNN (k = 20)	41.8%
3	RF (N_t = 50)	31.7%
4	PCA-HMM (#GM = 3 #HMM States = 3)	25.6%
5	GMM-HMM (#GM = 3 #HMM States = 3)	17.0%

TABLE 14.7

Comparison of Performance on an Abnormal EEG Classification Task as a Function of the Channel Selected for Analysis. Neuroscience Considerations Support the Use of T5-O1, Which Delivers the Lowest Error Rate

Channel	Error (%)
FP1-F7	19.8%
T5-O1	17.0%
F7-T3	19.8%
C3-Cz	20.7%
P3-O1	23.6%

to annotate seizures. Their work has been evaluated against expert neurologists who marked a portion of the same data and shown to have an IRA that exceeds that for the expert neurologists. Each event in the evaluation data has been reviewed by at least five different annotators.

Performance was as follows: sensitivity: 29.2%; specificity: 66.7%; false alarms/24 hours: 78; ATWV: −0.42. This is the first significant benchmark on seizure detection for TUH-EEG, and the first benchmark in this application space that uses ATWV. Commercially available tools tend to perform at a sensitivity of around 30% [21] with a high false alarm rate on this task based on other clinical evaluations. Our internal evaluations of these commercial systems indicate that our HMM baseline system delivered performance competitive to these systems. However, the ATWV score is extremely poor for these systems largely due to the large emphasis this metric places on false alarms. Hence, it is understandable that these systems fail to perform well in live clinical settings. The main challenge on this task is to accurately identify short seizures and the start time of slowly evolving seizures.

14.5 Summary and Future Directions

Biomedicine is entering a new age of data-driven discovery driven by ubiquitous computing power, a machine learning revolution, and high-speed Internet connections. Access to massive quantities of properly curated data is now the critical bottleneck to advancement in many areas of biomedical research. Ironically, clinicians generate enormous quantities of data daily, but that information is almost exclusively sequestered in secure archives where it cannot be used for research by the biomedical research community. The quantity, quality, and variability of such data represent a significant unrealized potential, which is doubly unfortunate considering that the cost of generating those data has already been borne.

In this chapter, we have introduced the problem of automatic interpretation of EEGs and described a paradigm that can be used to develop high-performance technology. We began by articulating the problem in terms that machine learning research can understand. We described the development of a big data corpus, TUH-EEG, that is enabling the application of state-of-the-art statistical models. We introduced a baseline system that integrates data-driven model-based parameterizations and subject matter expertise to achieve high performance in EEG signal event detection. This system was based on HMMs because of their proven ability to model sequential data.

Future research is now focused on applying deep learning methodologies to these problems. There are a wide variety of deep learning technologies available that support the integration of temporal and spatial constraints and can learn autonomously from the data. Initial experiments using networks based on long short-term memory (LSTM) [94,95] and convolutional neural networks (CNN) [96] have shown modest improvements in performance in terms of sensitivity and specificity, but comparable false alarm rates. Experiments in which cepstral-based features are replaced by several additional layers of a deep learning system are also showing comparable performance but no significant gains yet. This is more than likely due to a lack of a large amount of annotated data, as well as the lack of maturity of our pre-training and unsupervised learning approaches in deep learning. Such systems more than likely will require one or two orders of magnitude more data than existing HMM-based systems, and such data resources are under development at the Neural Engineering Data Consortium (http://www.nedcdata.org).

However, if we have significantly more data, there are some fundamental challenges that need to be addressed. Incorporating better features into the system will be critical. These features need to expose the deep learning systems to similar information that neurologists use to interpret EEGs, so both spatial and temporal contexts are required. This greatly increases the complexity of the networks and raises some computational issues. It also further underscores the importance of more data, since the dimensionality of these systems gets quite large. Similarly, the imbalanced nature of the data must be dealt with in a fundamental manner. The high false alarm rate is also a fundamental barrier to the acceptance of the technology. Future research will address these problems using a range of approaches that include transfer learning [97], confidence measures [98], and discriminative training [99]. Our ultimate research goal is the prediction of seizures 30 to 60 minutes before they occur, and research is underway to address this challenge as well.

Acknowledgments

Research reported in this publication was most recently supported by the National Human Genome Research Institute of the National Institutes of Health under award number U01HG008468. The content is solely the responsibility of the authors and does not necessarily represent the official views of the National Institutes of Health. This material is also based in part upon work supported by the National Science Foundation under Grant No. IIP-1622765. Any opinions, findings, and conclusions or recommendations expressed in this material are those of the author(s) and do not necessarily reflect the views of the National Science Foundation. The TUH EEG Corpus work was funded by (1) the Defense Advanced Research Projects Agency (DARPA) MTO under the auspices of Dr. Doug Weber through the Contract No. D13AP00065; (2) Temple University's College of Engineering; and (3) Temple University's Office of the Senior Vice-Provost for Research.

In any project of this nature, there are many people who have contributed knowledge, data, and technology. We owe an enormous debt to Dr. Mercedes Jacobson, MD, professor of neurology, Lewis Katz School of Medicine, who has been our colleague and mentor, and was responsible for our access to Temple Hospital's data. We are particularly grateful to Dr. Steven Tobochnik, MD, currently with New York–Presbyterian Hospital–Columbia University for his willingness to review data and answer many questions about the science. We have been fortunate to benefit from a group of over 135 neurologists that have advised us on some portion of this project.

Without the dedication of many graduate and undergraduate students at the Neural Engineering Data Consortium, none of these resources would exist. Amir Harati, Meysam Golmohammadi, and Silvia Lopez were the lead developers of the technology. Vinit Shah, Eva von Weltin, and James Riley McHugh were primarily responsible for the data development. Saeedeh Ziyabari was responsible for the development of the evaluation software. Scott Yang developed the active learning technology. Elliott Krome was responsible for the development of interactive demonstration tools. Over 20 undergraduates have participated in data entry work for the TUH EEG Corpus.

References

1. Electroencephalography. (2017, April 19). In *Wikipedia, The Free Encyclopedia*. (Retrieved 16:57, April 16, 2017, from https://en.wikipedia.org/w/index.php?title=Electroencephalography&oldid=774613669.)
2. Tatum, W., Husain, A., Benbadis, S., & Kaplan P. (2007). *Handbook of EEG Interpretation*. (Kirsch, Ed.). New York City, New York, USA: Demos Medical Publishing.
3. Yamada, T., & Meng E. (2009). *Practical Guide for Clinical Neurophysiologic Testing: EEG*. Philadelphia, Pennsylvania, USA: Lippincott Williams & Wilkins.
4. Smith, S. (2005). EEG in the diagnosis, classification, and management of patients with epilepsy. *Journal of Neurology, Neurosurgery, and Psychiatry*, 76(Suppl 2), ii2–ii7.
5. Dash, D., Hernandez-Ronquillo, L., Moien-Afshari, F., & Tellez-Zenteno J. (2012). Ambulatory EEG: A cost-effective alternative to inpatient video-EEG in adult patients. *Epileptic Disorders*, 14(3), 290–297.
6. Ardeshna, N. I. (2016). EEG and coma. *The Neurodiagnostic Journal*, 56(1), 1–16. http://doi.org/10.1080/21646821.2015.1114879
7. Sutter, R., Kaplan, P. W., Valença, M., & De Marchis G. M. (2015). EEG for diagnosis and prognosis of acute nonhypoxic encephalopathy: History and current evidence. *Journal of Clinical Neurophysiology*. United States: By the American Clinical Neurophysiology Society.
8. Ercegovac, M. (2010). Importance of EEG in brain death diagnosis. *Clinical Neurophysiology*. Elsevier Ireland Ltd.
9. Rudrashetty, S. M., Pyakurel, A., Karumuri, B., Liu, R., Vlachos, I., & Iasemidis L. (2015). Differential diagnosis of sleep disorders based on EEG analysis. *Journal of the Mississippi Academy of Sciences*. Mississippi Academy of Sciences.
10. Klassen, B. T., Hentz, J. G., Shill, H. A., Driver-Dunckley, E., Evidente, V. G. H., Sabbagh, M. N., ... Caviness, J. N. (2011). Quantitative EEG as a predictive biomarker for Parkinson disease dementia. *Neurology*, 77(2), 118–124.
11. Al-Qazzaz, N. K., Ali, S. H. B. M., Ahmad, S. A., Chellappan, K., Islam, M. S., & Escudero J. (2014). Role of EEG as biomarker in the early detection and classification of dementia. *The Scientific World Journal*.
12. Sheorajpanday, R. V. A., Nagels, G., Weeren, A. J. T. M., & De Deyn P. P. (2011). Quantitative EEG in ischemic stroke: Correlation with infarct volume and functional status in posterior circulation and lacunar syndromes. *Clinical Neurophysiology*, 122(5), 884–890.
13. Finnigan, S., & van Putten M. J. a M. (2013). EEG in ischaemic stroke: quantitative EEG can uniquely inform (sub-)acute prognoses and clinical management. *Clinical Neurophysiology*, 124(1), 10–19.
14. van Tricht, M. J., Ruhrmann, S., Arns, M., Müller, R., Bodatsch, M., Velthorst, E., ... Nieman, D. H. (2014). Can quantitative EEG measures predict clinical outcome in subjects at Clinical High Risk for psychosis? A prospective multicenter study. *Schizophrenia Research*, 153(1–3), 42–47.
15. Ebersole, J. S., & Pedley T. A. (2014). *Current Practice of Clinical Electroencephalography* (4th ed.). Philadelphia, Pennsylvania, USA: Wolters Kluwer.
16. Jurcak, V., Tsuzuki, D., & Dan I. (2007). 10/20, 10/10, and 10/5 systems revisited: Their validity as relative head-surface-based positioning systems. *NeuroImage*, 34(4), 1600–1611.
17. Swisher, C. B., White, C. R., Mace, B. E., & Dombrowski K. E. (2015). Diagnostic accuracy of electrographic seizure detection by neurophysiologists and non-neurophysiologists in the adult ICU using a panel of quantitative EEG trends. *Journal of Clinical Neurophysiology*, 32(4), 324–330.
18. Thiess, M., Krome, E., Golmohammadi, M., Obeid, I., & Picone J. (2016). Enhanced visualizations for improved real-time EEG monitoring. In *2016 IEEE Signal Processing in Medicine and Biology Symposium* (SPMB) (p. 1).
19. Lopez, S. (2017). *Automated Identification of Abnormal EEGs*. Temple University. (Available at http://www.isip.piconepress.com/publications/ms_theses/2017/abnormal.)
20. Lopez, S., Suarez, G., Jungries, D., Obeid, I., & Picone J. (2015). Automated identification of abnormal EEGs. In *IEEE Signal Processing in Medicine and Biology Symposium* (pp. 1–4). Philadelphia, Pennsylvania, USA.
21. Scheuer, M. L., Bagic, A., & Wilson S. B. (2017). Spike detection: Inter-reader agreement and a statistical Turing test on a large data set. *Clinical Neurophysiology*, 128(1), 243–250.
22. van Beelen, T. (2013). *EDFbrowser*. (Retrieved from http://www.teuniz.net/edfbrowser/.)
23. Acharya, J., Hani, A., Thirumala, P., & Tsuchida T. (2016). American Clinical Neurophysiology Society Guideline 3: A proposal for standard montages to be used in clinical EEG. *Journal of Clinical Neurophysiology: Official Publication of the American Electroencephalographic Society*, 33 (4).
24. Harati, A., Lopez, S., Obeid, I., Jacobson, M., Tobochnik, S., & Picone J. (2014). The TUH EEG corpus: A big data resource for automated eeg interpretation. In *Proceedings of the IEEE Signal Processing in Medicine and Biology Symposium* (pp. 1–5). Philadelphia, Pennsylvania, USA.
25. Lopez, S., Gross, A., Yang, S., Golmohammadi, M., Obeid, I., & Picone J. (2016). An analysis of two common reference points for EEGs. In *IEEE Signal Processing in Medicine and Biology Symposium* (pp. 1–4). Philadelphia, Pennsylvania, USA.
26. Nolte, J., & Sundsten J. W. (2015). *The Human Brain: An Introduction to Its Functional Anatomy* (7th ed.). St. Louis, Missouri: Mosby.
27. Misulis, K. E., & Abou-Khalil B. (2014). *Atlas of EEG and Seizure Semiology and Management* (2nd ed., p. 384). Oxford: Oxford University Press.
28. Deshmukh, N., Ganapathiraju, A., & Picone J. (1999). Hierarchical search for large-vocabulary conversational speech recognition: Working toward a solution to the decoding problem. *IEEE Signal Processing Magazine*, 16(5), 84–107.

29. Baldassano, S., Wulsin, D., Ung, H., Blevins, T., Brown, M.-G., Fox, E., & Litt B. (2016). A novel seizure detection algorithm informed by hidden Markov model event states. *Journal of Neural Engineering*, 13(3), 36011.

30. Waterstraat, G., Burghoff, M., Fedele, T., Nikulin, V., Scheer, H. J., & Curio G. (2015). Non-invasive single-trial EEG detection of evoked human neocortical population spikes. *NeuroImage*, 105, 13–20.

31. Wulsin, D., Blanco, J., Mani, R., & Litt B. (2010). Semi-supervised anomaly detection for EEG waveforms using deep belief nets. In *International Conference on Machine Learning and Applications (ICMLA)* (pp. 436–441). Washington, D.C., USA.

32. Doddington, G. R., Przybocki, M. A., Martin, A. F., & Reynolds D. A. (2000). The NIST speaker recognition evaluation—Overview, methodology, systems, results, perspective. *Speech Communication*, 31(2), 225–254. http://doi.org/10.1016/S0167-6393(99)00080-1.

33. Picone, J. (1990). Continuous speech recognition using hidden Markov models. *IEEE ASSP Magazine*, 7(3), 26–41.

34. Golmohammadi, M., Shah, V., Lopez, S., Ziyabari, S., Yang, S., Camaratta, J., ... Picone, J. (2017). The TUH EEG Seizure Corpus. In *American Clinical Neurophysiology Society* (p. 1). Phoenix, Arizona, USA.

35. Stroink, H., Schimsheimer, R.-J., de Weerd, A. W., Geerts, A. T., Arts, W. F., MC, E., ... van Donselaar, C. A. (2006). Interobserver reliability of visual interpretation of electroencephalograms in children with newly diagnosed seizures. *Developmental Medicine & Child Neurology*, 48(5), 374–377.

36. van Donselaar, C., Schimsheimer, R., AT, G., & Declerck A. (1992). Value of the electroencephalogram in adult patients with untreated idiopathic first seizures. *Archives of Neurology*, 49(3), 231–237.

37. Halford, J. J., Shiau, D., Desrochers, J. A., Kolls, B. J., Dean, B. C., Waters, C. G., ... LaRoche, S. M. (2015). Inter-rater agreement on identification of electrographic seizures and periodic discharges in ICU EEG recordings. *Clinical Neurophysiology*, 126(9), 1661–1669. http://doi.org/10.1016/j.clinph.2014.11.008.

38. Nizam, A., Chen, S., & Wong S. (2013). Best-case Kappa scores calculated retrospectively from EEG report databases. *Journal of Clinical Neurophysiology*. United States: Copyright American Clinical Neurophysiology Society.

39. Britton, J. W., Frey, L. C., Hopp, J. L., Korb, P., Koubeissi, M., Lievens, W., ... St. Louis, E. K. (2016). *Electroencephalography (EEG): An Introductory Text and Atlas of Normal and Abnormal Findings in Adults, Children, and Infants [Internet]*. (E. K. St. Louis & L. C. Frey, Eds.) American Epilepsy Society (1st ed.). Chicago, Illinois, USA: National Institutes of Health. (Retrieved from https://www.ncbi.nlm.nih.gov/books/NBK390352/.)

40. Japkowicz, N., & Shah M. (2011). *Evaluating Learning Algorithms: A Classification Perspective*. Cambridge; New York; Cambridge University Press.

41. Altman, D. G., & Bland J. M. (1994). Diagnostic Tests 1: Sensitivity and specificity. *BMJ: British Medical Journal*. England: British Medical Association.

42. Manning, C. D., Raghavan, P., & Schütze H. (2008). *Introduction to Information Retrieval*. Cambridge, UK: Cambridge University Press.

43. Hu, P. (2015). Reducing False Alarms in Critical Care. In *Working Group on Neurocritical Care Informatics, Neurocritical Care Society Annual Meeting*. Scottsdale, Arizona, USA.

44. Obeid, I., & Picone J. (2015). *NSF ICORPS Team: AutoEEG*. Philadelphia, Pennsylvania, USA. (Retrieved from https://www.isip.piconepress.com/publications/reports/2016/nsf/icorps/report_v01.pdf.)

45. Christensen, M., Dodds, A., Sauer, J., & Watts N. (2014). Alarm setting for the critically ill patient: A descriptive pilot survey of nurses' perceptions of current practice in an Australian Regional Critical Care Unit. *Intensive & Critical Care Nursing*. Netherlands: Elsevier B.V.

46. Mandal, A., Prasanna Kumar, K. R., & Mitra P. (2014). Recent developments in spoken term detection: A survey. *International Journal of Speech Technology*, 17(2), 183–198.

47. Martin, A., Doddington, G., Kamm, T., Ordowski, M., & Przybocki M. (1997). The DET curve in assessment of detection task performance. In *Proceedings of Eurospeech* (pp. 1895–1898). Rhodes, Greece.

48. Jacobs, I. M., & Wozencraft J. M. (1965). *Principles of Communication Engineering* (1st ed.). Long Grove, Illinois, USA: Waveland Pr Inc.

49. Ziyabari, S., Golmohammadi, M., Obeid, I., & Picone J. (2017). An analysis of objective performance metrics for automatic interpretation of EEG signal events. Under Review. (Available at http://www.isip.piconepress.com/publications/unpublished/journals/2017/neural_engineering/metrics.)

50. NIST. (2010). Retrieved from http://www.itl.nist.gov/iad/mig//tools/.

51. Picone, J., & Obeid I. (2016). Fundamentals in data science: Data wrangling, normalization, preprocessing of physiological signals. In M. Dunn (Ed.), *The BD2K Guide to the Fundamentals of Data Science* (p. 1). Bethesda, Maryland, USA: National Institutes of Health. (Retrieved from http://www.bigdatau.org/data-science-seminars.)

52. Harabagiu, S. (2016). Active deep learning-based annotation of electroencephalography reports for patient cohort identification. In M. Dunn (Ed.), *The BD2K Guide to the Fundamentals of Data Science* (p. 1). Bethesda, Maryland, USA: National Institutes of Health. (Retrieved from http://www.bigdatau.org/data-science-seminars.)

53. Picone, J., Obeid, I., & Harabagiu S. (2015). Automatic discovery and processing of EEG cohorts from clinical records. In Big Data to Knowledge All Hands Grantee Meeting (p. 1). Bethesda, Maryland, USA. (Retrieved from http://www.isip.piconepress.com/publications/conference_presentations/2015/nih_bd2k/cohort/.)

54. He, H., Ma, Y., & Obooks W. O. L. U. A. (2013). *Imbalanced Learning: Foundations, Algorithms, and Applications* (Vol. 1). Hoboken, New Jersey: John Wiley & Sons, Inc.

55. Duda, R. O., Hart, P. E., & Stork D. G. (2001). *Pattern Classification*. (2nd, Ed.). New York City, New York, USA: John Wiley & Sons, Inc.

56. Obeid, I., & Picone J. (2016). The Temple University Hospital EEG Data Corpus. *Frontiers in Neuroscience, Section Neural Technology*, 10, 196.

57. Kemp, R. (2013). European Data Format. (Retrieved March 7, 2017, from http://doi.org/http://www.edfplus.info.)
58. Bergey, S., & Picone J. (2017). A description of the EDF header. (Retrieved March 7, 2017, from https://www.isip.piconepress.com/projects/tuh_eeg/doc/edf/.)
59. Huang, X., Acero, A., & Hon H.-W. (2001). *Spoken Language Processing: A Guide to Theory, Algorithm and System Development*. Upper Saddle River, New Jersey, USA: Prentice Hall.
60. Michel, C. M., & Murray M. M. (2012). Towards the utilization of EEG as a brain imaging tool. *NeuroImage*, 61(2), 371.
61. LaRoche, S., & Collection E. A. C. S. (2013). *Handbook of ICU EEG Monitoring*. New York, New York, USA: Demos Medical Pub.
62. Brzezinski, R. (2016). *HIPAA Privacy and Security Compliance—Simplified: Practical Guide for Healthcare Providers and Managers 2016 Edition* (3rd ed.). Seattle, Washington, USA: CreateSpace Independent Publishing Platform.
63. Harabagiu, S., Goodwin, T., Maldonado, R., & Taylor S. (2016). Active deep learning-based annotation of electroencephalography reports for patient cohort retrieval. In *Big Data to Knowledge All Hands Grantee Meeting* (p. 1). Bethesda, Maryland, USA: National Institutes of Health.
64. Obeid, I., Picone, J., & Harabagiu S. (2016). Automatic discovery and processing of EEG cohorts from clinical records. In *Big Data to Knowledge All Hands Grantee Meeting* (p. 1). Bethesda, Maryland, USA: National Institutes of Health.
65. Chandran, V. (2012). Time-varying bispectral analysis of visually evoked multi-channel EEG. *EURASIP Journal on Advances in Signal Processing*, 2012(1), 1–22.
66. Wang, C., Hu, X., Yao, L., Xiong, S., & Zhang J. (2011). Spatio-temporal pattern analysis of single-trial EEG signals recorded during visual object recognition. *Science China Information Sciences*, 54(12), 2499–2507.
67. Wulsin, D. F., Gupta, J. R., Mani, R., Blanco, J. A., & Litt B. (2011). Modeling electroencephalography waveforms with semi-supervised deep belief nets: Fast classification and anomaly measurement. *Journal of Neural Engineering*, 8(3), 36015.
68. Bao, F. S., Gao, J.-M., Hu, J., Lie, D. Y.-C., Zhang, Y., & Oommen K. J. (2009). Automated epilepsy diagnosis using interictal scalp EEG. In *International Conference of the IEEE Engineering in Medicine and Biology Society* (EMBC) (pp. 6603–6607). Minneapolis, Minnesota, USA.
69. Lin, E.-B., & Shen X. (2011). Wavelet analysis of EEG signals. In Proceedings of the IEEE National Aerospace and Electronics Conference (NAECON) (pp. 105–110). Dayton, Ohio, USA.
70. Rosso, O. A., Martin, M. T., Figliola, A., Keller, K., & Plastino A. (2006). EEG analysis using wavelet-based information tools. *Journal of Neuroscience Methods*, 153(2), 163–182.
71. Lowerre, B. (1980). The HARPY speech understanding system. In W. Lea (Ed.), *Trends in Speech Recognition* (pp. 576–586). Englewood Cliffs, New Jersey, USA: Prentice Hall.
72. Erman, L. D., Hayes-Roth, F., Lesser, V. R., & Reddy D. R. (1980). The Hearsay-II Speech-Understanding System: Integrating knowledge to resolve uncertainty. *ACM Computing Surveys*, 12(2), 213–253.
73. Lee, K.-F., & Hon H.-W. (1989). Speaker-independent phone recognition using hidden Markov models. *IEEE Transactions on Acoustics, Speech, and Signal Processing*, 37(11), 1641–1648.
74. Rabiner, L. (1989). A tutorial on hidden Markov models and selected applications in speech recognition. *Proceedings of the IEEE*, 77(2), 257–286.
75. Picone, J. (1993). Signal modeling techniques in speech recognition. *Proceedings of the IEEE*, 81(9), 1215–1247.
76. The America Epilepsy Society, (NINDS), N. I. of H., Pennsylvania, U. of, & Mayo Clinic. (2014). UPenn and Mayo Clinic's Seizure Detection Challenge. (Retrieved from https://www.kaggle.com/c/seizure-detection.)
77. May, D., Srinivasan, S., Ma, T., Lazarou, G., & Picone J. (2008). Continuous speech recognition using nonlinear dynamical invariants. In *Proceedings of the IEEE International Conference on Acoustics, Speech and Signal Processing*. Las Vegas, Nevada, USA.
78. Bengio, Y., Courville, A., & Vincent P. (2013). Representation learning: A review and new perspectives. *IEEE Transactions on Pattern Analysis and Machine Intelligence*.
79. Xiong, W., Droppo, J., Huang, X., Seide, F., Seltzer, M., Stolcke, A., ... Zweig, G. (2016). *Achieving human parity in conversational speech recognition*. Ithaca, New York, USA: Cornell University Library (arvix.org). (Retrieved from https://arxiv.org/abs/1610.05256.)
80. Hinton, G., Deng, L., Yu, D., Dahl, G., Mohammed, A., Jaitly, N., ... Kingsbury, B. (2012). Deep Neural Networks for Acoustic Modeling in Speech Recognition. *IEEE Signal Processing Magazine*, 29(6), 83–97. (Retrieved from http://ieeexplore.ieee.org/xpl/RecentIssue.jsp?punumber=79.)
81. Goldberger, A. L., Amaral, L. A. N., Glass, L., Hausdorff, J. M., Ivanov, P. C., Mark, R. G., & Stanley H. E. (2000). Physiobank, physiotoolkit, and physionet components of a new research resource for complex physiologic signals. *Circulation*, 101(23), e215–e220. (Retrieved from http://circ.ahajournals.org/content/101/23/e215.short.)
82. Lu, S., & Picone J. (2013). Fingerspelling Gesture Recognition Using A Two-Level Hidden Markov Model. In Proceedings of the International Conference on Image Processing, Computer Vision, and Pattern Recognition (ICPV) (pp. 538–543). Las Vegas, Nevada, USA. http://www.isip.piconepress.com/publications/conference_proceedings/2013/ipcv/asl_hmm/.
83. Huang, K., & Picone J. (2002). Internet-Accessible Speech Recognition Technology. In Proceedings of the IEEE Midwest Symposium on Circuits and Systems (p. III-73-III-76). Tulsa, Oklahoma, USA. https://doi.org/10.1109/MWSCAS.2002.1186973.
84. Yang, S., López, S., Golmohammadi, M., Obeid, I., & Picone J. (2016). Semi-automated annotation of signal events in clinical EEG data. In *Proceedings of the IEEE Signal Processing in Medicine and Biology Symposium* (pp. 1–5). Philadelphia, Pennsylvania, USA.

85. Povey, D., Kanevsky, D., Kingsbury, B., Ramabhadran, B., Saon, G., & Visweswariah K. (2008). Boosted MMI for model and feature-space discriminative training. *Proceedings of the IEEE International Conference on Acoustics, Speech and Signal Processing*. Las Vegas, Nevada, USA.
86. Harati, A., Golmohammadi, M., Lopez, S., Obeid, I., & Picone J. (2015). Improved EEG event classification using differential energy. In 2015 *IEEE Signal Processing in Medicine and Biology Symposium (SPMB)* (pp. 1–4). Philadelphia: IEEE.
87. Davis, S., & Mermelstein P. (1980). Comparison of parametric representations for monosyllabic word recognition in continuously spoken sentences. *IEEE Transactions on Acoustics, Speech, and Signal Processing*, 28(4), 357–366.
88. Furui, S. (1986). Speaker-independent isolated word recognition using dynamic features of speech spectrum. *IEEE Transactions on Acoustics, Speech and Signal Processing*, 34(1), 52–59.
89. Juang, B.-H., & Rabiner L. (1991). Hidden Markov Models for Speech Recognition. *Technometrics*, 33(3), 251–272. (Retrieved from http://www.leonidzhukov.net/hse/2012/stochmod/papers/HMMs_speech_recognition.pdf.)
90. Ahangi, A., Karamnejad, M., Mohammadi, N., Ebrahimpour, R., & Bagheri N. (2013). Multiple classifier system for EEG signal classification with application to braincomputer interfaces. *Neural Computing and Applications*, 23(5), 1319–1327.
91. Breiman, L. (2001). Random forests. *Machine Learning*, 45(1), 5–32.
92. Bengio, Y., Lamblin, P., Popovici, D., & Larochelle H. (2007). Greedy layer-wise training of deep networks. In Advances in Neural Information Processing System (pp. 153–160). Vancouver, B.C., Canada.
93. Vincent, P., Larochelle, H., Bengio, Y., & Manzagol P.-A. (2008). Extracting and composing robust features with denoising autoencoders. In *Proceedings of the 25th international conference on Machine learning* (pp. 1096–1103). New York, NY, USA.
94. Graves, A., Mohamed, A., & Hinton G. (2013). Speech recognition with deep recurrent neural networks. *International Conference on Acoustics, Speech, and Signal Processing (ICASSP)*, (3), 6645–6649.
95. Hochreiter, S., & Schmidhuber J. (1997). Long short-term memory. *Neural Computation*, 9(8), 1735–80.
96. Abdel-Hamid, O., Mohamed, A. R., Jiang, H., & Penn G. (2012). Applying convolutional neural networks concepts to hybrid NN-HMM model for speech recognition. In *ICASSP, IEEE International Conference on Acoustics, Speech and Signal Processing - Proceedings* (pp. 4277–4280).
97. Taylor, M. E. (2009). Transfer in reinforcement learning domains. In *Studies in Computational Intelligence* (vol. 216, p. 218). Berlin, Germany: Springer.
98. Yu, D., Li, J., & Deng L. (2011). Calibration of confidence measures in speech recognition. *Audio, Speech, and Language Processing, IEEE Transactions on*, 19(8), 2461–2473.
99. Manohar, V., Povey, D., & Khudanpur S. (2015). Semi-supervised maximum mutual information training of deep neural network acoustic models. In *Proceedings of INTERSPEECH* (pp. 2630–2634). Dresden, Germany: ISCA.

15

Information Fusion in Deep Convolutional Neural Networks for Biomedical Image Segmentation*

Mohammad Havaei, Nicolas Guizard, Nicolas Chapados, and Yoshua Bengio

CONTENTS

15.1 Introduction 301
15.2 Background: Deep CNNs 303
15.3 Method 305
    15.3.1 Hetero-Modal Image Segmentation 305
    15.3.2 Pseudo-Curriculum Training Procedure 306
    15.3.3 Interpretation as an Embedding 307
15.4 Data and Implementation Details 307
    15.4.1 MS Datasets 307
    15.4.2 Brain Tumors (BRATS) 307
    15.4.3 Pre-Processing and Implementation Details 307
15.5 Experiments and Results 308
15.6 Conclusion 310
References 311

15.1 Introduction

Among clinical datasets, medical images are perhaps the largest and most complex to analyze. Typical individual computed tomography (CT) scans and magnetic resonance images (MRIs) range from a few tens of megabytes to multi-gigabytes per study. Moreover, with the greater availability of equipment, the number of scans performed per year has been steadily increasing, both in the United States and worldwide [1]. Figure 15.1 illustrates the increase in per-capita CT and MRI scans in the United States over the last 20 years. In addition, new screening guidelines coming into force for cancers such as lung, in countries such as the United States [2] and Canada [3], imply that a great increase in scans is to be further expected, leading to a bottleneck in the availability of radiologists to analyze all these data. Hence, tools to automate some of this analysis are beneficial. This chapter proposes a technique based on deep convolutional neural networks (CNNs) to carry out important analytical tasks—such as image segmentation—given a varying set of available imaging modalities.

In medical image analysis, image segmentation is a common subtask to achieve a number of clinical goals, such as lesion detection and characterization, and is primordial to visualize and quantify the severity of the pathology in clinical practice. Multi-modality imaging provides complementary information to discriminate specific tissues, anatomies, and pathologies. However, manual segmentation is long, painstaking, and subject to human variability. In the last decades, numerous automatic approaches have been developed to speed up medical image segmentation. These methods can be grouped into two categories: The first, *multi-atlas approaches* estimate online intensity similarities between the subject being segmented and multi-atlases (images with expert labels). These techniques have shown excellent results in structural segmentation when using non-linear registration [4] when combined with non-local approaches, they have proven effective in segmenting diffuse and sparse pathologies (i.e., multiple sclerosis [MS] lesions [6] as well as more complex

---
* Portions of this work were previously published in Havaei et al. [5].

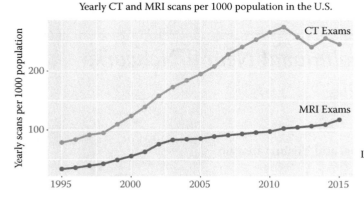

FIGURE 15.1
Yearly CT and MRI scans per 1000 population in the United States. These numbers have been steadily increasing for the past 20 years with the greater availability of equipment, making automated analysis tools ever more desirable.

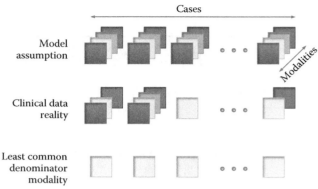

FIGURE 15.2
Most multi-modal computer vision and machine learning models assume that all modalities are available for all patients (cases), illustrated on top; however, clinical datasets available in practice will generally omit one or more modalities from most cases (middle), often reducing model scope to the least common denominator modality (bottom).

multi-label pathology, e.g., in glioblastoma [7]. Multi-atlas methods rely on image intensity and spatial similarity, which can be difficult to be fully described by the atlases and heavily dependent on the image preprocessing. In contrast, *model-based approaches* are typically trained offline to identify a discriminative model of image intensity features. These features can be predefined by the user (e.g., with random forests; [8] or extracted and learned hierarchically directly from the images [9].

Both strategies are typically optimized for a specific set of multi-modal images and usually require these modalities to be available. In clinical settings, image acquisition and patient artifacts, among other hurdles, make it difficult to fully exploit all the modalities; as such, it is common to have one or more modalities to be missing for a given instance, a problem illustrated in Figure 15.2. This problem is not new, and the subject of missing data analysis has spawned an immense literature in statistics (see, e.g., Van Buuren, [10]). In medical imaging, a number of approaches have been proposed, some of which require to re-train a specific model with the missing modalities or to synthesize them [11]. Synthesis can improve multi-modal classification by adding information of the missing modalities in the context of a simple classifier such as random forests [12]. Approaches to imitate with fewer features a classifier trained with a complete set of features have also been proposed [13]. Nevertheless, it should stand to reason that a more complex model should be capable of extracting relevant features from just the available modalities without relying on artificial intermediate steps such as imputation or synthesis.

This chapter introduces a deep learning framework (called *Heteromodal Image Segmentation*, or HeMIS) that can segment medical images from incomplete multi-modal datasets. Deep learning [14] has shown an increasing popularity in medical image processing for segmenting but also to synthesize missing modalities [15]. Here, the proposed approach learns, separately for each modality, an embedding of the input image into a latent space. In this space, arithmetic operations (such as computing first and second moments of a collection of vectors) are well defined and can be taken over the different modalities available at inference time. These computed moments can then be further processed to estimate the final segmentation; an overview of the approach is shown in Figure 15.3. This approach presents the advantage of being robust to any combinatorial subset of available modalities provided as input, without the need to learn a combinatorial number of imputation models.

After presenting a background refresher on deep CNNs (Section 15.2), we describe the method (Section 15.3), follow with a description of the datasets (Section 15.4)

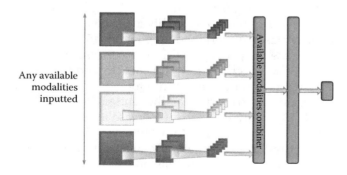

FIGURE 15.3
High-level view of the HeMIS deep CNN structure, wherein a separate projection to a common embedding is learned for each possible modality, which are then combined and further processed before a final output is produced. The following sections detail this model.

and experiments (Section 15.5), and finally offer concluding remarks (Section 15.6).

15.2 Background: Deep CNNs

CNNs are a type of neural network adopted for spatially ordered input. The main building block used to construct a CNN architecture is the *convolutional layer*. As in a regular neural network, several convolutional layers can be stacked on top of each other forming a hierarchy of features. Each layer can be understood as extracting features from its preceding layer in the hierarchy. A single convolutional layer takes as input a stack of input planes and produces as output some number of output planes or *feature maps*. Each feature map can be thought of as a topologically arranged map of responses of a particular spatially local non-linear feature extractor (the parameters of which are learned), applied identically to each spatial neighborhood of the input planes in a sliding window fashion. In the case of a first convolutional layer, the individual input planes correspond to different input channels. In the case of MRI, it can be single slices of different MR imaging modalities,* and in the case of color images, it can be different color channels. In subsequent layers, the input planes typically consist of the feature maps of the previous layer. Computing a feature map in a convolutional layer (see Figure 15.4) consists of the following three steps:

1. *Convolution of kernels (filters):* Each feature map O_s of the layer is associated with one kernel (or several, in the case of Maxout[†]). The feature map O_s is computed as follows:

$$O_s = b_s + \sum_r W_{sr} * X_r \quad (15.1)$$

where X_r is the rth input channel, W_{sr} is the sub-kernel for that channel, * is the convolution operation, and b_s is a bias term.[‡] In other words, the affine operation being performed for each feature map is the *sum* of the application of R

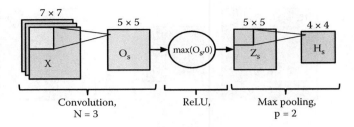

FIGURE 15.4
Single convolution layer block showing computations for a single feature map. The input patch (here 7 × 7) is convolved with a series of kernels (here 3 × 3) followed by ReLU and max-pooling.

different two-dimensional $N \times N$ convolution filters (one per input channel/modality), plus a bias term that is added pixel-wise to each resulting spatial position. The convolutional operation of image X and kernel W is computed as

$$C_{ij} = (W * X)_{ij} = \sum_m \sum_n X_{i+m, j+n} W_{-m, -n}. \quad (15.2)$$

In the above equation, the region in matrix X that is used in computation of C_{ij} is referred to as the *local receptive field* for C_{ij}, and so C_{ij} is only connected to its receptive field, rather than the whole image as it is the case with standard multi-layer perceptrons (MLPs). This greatly reduces the number of parameters of the model. This receptive field is slid across the entire image, a process illustrated in Figure 15.5 [15]. For each receptive field, there is a different hidden neuron (i.e., $C_{s,ij}$). However, the weights to compute every hidden neuron are shared. This further reduces the parameters of the model by a factor of the number of neurons in that feature map. Intuitively, the reason for sharing parameters is that each kernel can be thought of as a feature detector that tries to identify that particular feature at different spatial positions in the image. Also, by sharing parameters, we can greatly reduce the parameters of the model and reduce risk of overfitting. The combination of several of those is illustrated in Figure 15.6 [15].

Whereas traditional image feature extraction methods rely on a fixed recipe (sometimes taking the form of convolutions with a linear filter bank), the key to the success of CNNs is their ability to learn the weights and biases of individual feature maps, giving rise to data-driven, customized, task-specific dense feature extractors. These parameters are learned via stochastic gradient descent on a surrogate loss function, with gradients computed efficiently via the backpropagation algorithm.

---

* In this work, we assume 2D convolutional networks, although the formulation generalize straightforwardly to 3D and higher dimensionalities.
† Maxout will be discussed later in this chapter.
‡ Since the convolutional layer is associated to R input channels, X contains $M \times M \times R$ gray-scale values and thus each kernel W_s contains $N \times N \times R$ weights. Accordingly, the number of parameters in a convolutional block, consisting of S feature maps, is equal to $R \times M \times M \times S$.

FIGURE 15.5
Computing the output values of a discrete convolution. (From Dumoulin, V. and F. Visin (2016, March). A guide to convolution arithmetic for deep learning. *ArXiv e-prints*. With permission.)

Special attention must be paid to the treatment of border pixels by the convolution operation. One option is to employ the so-called *valid* mode convolution, meaning that the filter response is not computed for pixel positions that are less than $\lfloor N/2 \rfloor$ pixels away from the image border. An $M \times M$ input convolved with an $N \times N$ filter patch will result in a $Q \times Q$ output, where $Q = M - N + 1$. In Figure 15.4, $M = 7$, $N = 3$, and thus $Q = 5$. Note that the size (spatial width and height) of the kernels are hyper-parameters that must be specified by the user. One can apply the convolutions in the *same* mode to preserve the input size. In this mode, zero padding is applied around the input prior to the convolution operation.

2. *Non-linear activation function:* To obtain features that are non-linear transformations of the input, an elementwise non-linearity is applied to the result of the kernel convolution. There are multiple choices for this non-linearity, such as the sigmoid, hyperbolic tangent, and rectified linear functions [16,17] or maxout [18]. The rectified linear function, or ReLU, is defined elementwise simply as

$$\mathrm{ReLU}(\cdot) = \max(0, \cdot),$$

whereas maxout features are associated with multiple kernels $\mathbf{W}_s$, wherein maxout map $\mathbf{Z}_s$ is associated with K feature maps: $\{\mathbf{O}_{Ks}, \mathbf{O}_{Ks+1}, ..., \mathbf{O}_{Ks+K-1}\}$. Maxout features correspond to taking the max over the feature maps $\mathbf{O}$, individually for each spatial position,

$$Z_{s,i,j} = \max\{O_{Ks,i,j}, O_{Ks+1,i,j}, ..., O_{Ks+K-1,i,j}\}, \quad (15.3)$$

where i, j are spatial positions. Maxout features are thus equivalent to using a convex activation function, but whose shape is adaptive and depends on the values taken by the kernels. The ReLU function can be considered a special form of Maxout where the max operation is taken over every feature map and a zero matrix of the same size for each spatial position (i.e. $\max(\mathbf{O}_s, \mathbf{0})$),

$$Z_{s,i,j} = \max\{O_{s,i,j}, 0_{i,j}\}. \quad (15.4)$$

Note that in Figure 15.4, the ReLU activation function is used.

Information Fusion in Deep CNNs for Biomedical Image Segmentation   305

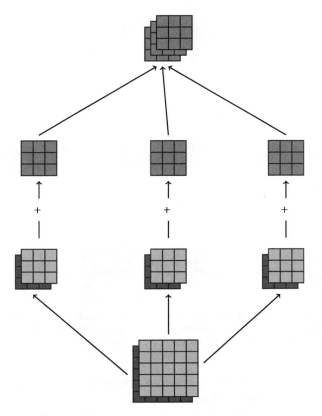

FIGURE 15.6
Convolution mapping from two input feature maps to three output feature maps using a 3 × 2 × 3 × 3 collection of kernels **w**. In the left pathway, input feature map 1 is convolved with kernel $\mathbf{w}_{1,1}$ and input feature map 2 is convolved with kernel $\mathbf{w}_{1,2}$, and the results are summed together elementwise to form the first output feature map. The same is repeated for the middle and right pathways to form the second and third feature maps, and all three output feature maps are grouped together to form the output. (From Dumoulin, V. and F. Visin (2016, March). A guide to convolution arithmetic for deep learning. ArXiv e-prints. With permission.)

3. *Max-pooling:* This operation consists of taking the maximum feature (neuron) value over subwindows within each feature map. This can be formalized as follows:

$$H_{s,i,j} = \max_p Z_{s,Si+p,Sj+p}, \qquad (15.5)$$

where p determines the max-pooling window size, and S is the stride value that corresponds to the horizontal and vertical increments at which pooling sub-windows are positioned. Depending on the stride value, the sub-windows can be overlapping or not (Figure 15.4 shows an overlapping configuration). The max-pooling operation shrinks the size of the feature map. This is controlled by the pooling size p and the stride hyper-parameter. Let $Q \times Q$ be the shape of the feature map before max-pooling. The output of the max-pooling operation would be of size $D \times D$, where $D = (Q - p) / S + 1$.[*] In Figure 15.4, since $Q = 5$, $p = 2$, $S = 1$, the max-pooling operation results into a $D = 4$ output feature map. The motivation for this operation is to introduce invariance to local translations. This subsampling procedure has been found beneficial in other applications [19].

15.3 Method

15.3.1 Hetero-Modal Image Segmentation

Typical CNN architectures take a multiplane image as input and process it through a sequence of convolutional layers (followed by nonlinearities such as ReLU), alternating with optional pooling layers, to yield a per-pixel or per-image output [14]. In such networks, every input plane is assumed to be present within a given instance: since the very first convolutional layer mixes input values coming from all planes, any missing plane introduces a bias in the computation that the network is not equipped to deal with.

We propose an approach wherein each modality is initially processed by its own convolutional pipeline, independently of all others. After a few independent stages, feature maps from all available modalities are *merged* by computing mapwise statistics such as the mean and the variance, quantities whose expectation does not depend on the number of terms (i.e., modalities) that are provided. After merging, the mean and variance feature maps are concatenated and fed into a final set of convolutional stages to obtain network output. This is illustrated in Figure 15.7.

In this procedure, each modality contributes a separate term to the mean and variance; in contrast to a vanilla CNN architecture, a missing modality does not throw this computation off: the mean and variance terms simply become estimated with larger uncertainty. In seeking to be robust to any subset of missing modalities, we call this approach *hetero-modal* rather than multimodal, recognizing that in addition to taking advantage of several modalities, it can take advantage of a diverse, instance-varying, set of modalities. In particular, it does not require that a "least common denominator" modality be present for every instance, as sometimes needed by common imputation methods.

Let $k \in K \subseteq \{1, \ldots, N\}$ denote a modality within the set of available modalities for a given instance, and M_k

---
[*] Note that values p and S should be chosen in a way that the pooling window fits the feature map (i.e. D should be an integer). Alternatively, we can zero pad the feature map Q accordingly.

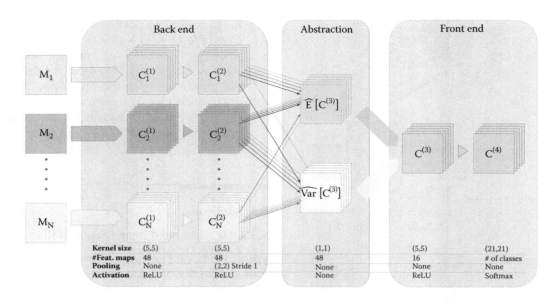

FIGURE 15.7
Illustration of the Hetero-Modal Image Segmentation architecture. Modalities available at inference time, M_k, are provided to independent modality-specific convolutional layers in the *back-end*. Feature maps statistics (first and second moments) are computed in the *abstraction layer*, which after concatenation are processed by further convolutional layers in the *front-end*, yielding pixelwise classification outputs.

represent the image of the kth modality. For simplicity, in this work we assume 2D data (e.g., a single slice of a tomographic image), but it can be extended in an obvious way to full 3D sections. As shown in Figure 15.7, HeMIS proceeds in three stages:

1. *Back-end*: In our implementation, this consists of two convolutional layers with ReLU, the second followed with a (2, 2) max-pooling layer, denoted $C_k^{(1)}$ and $C_k^{(2)}$, respectively. To ensure that the output layer consists of the same number of pixels as the input image, the convolutions are zero-padded and the stride for all operations (including max-pooling) is 1. In particular, pooling with a stride of 1 *does not downsample*, but simply "thickens" the feature maps; this is found to add some robustness to the results. The number of feature maps in each layer is given in Figure 15.7. Let $C_{k,\ell}^{(j)}$ be the ℓth feature map of $C_k^{(j)}$.

2. *Abstraction layer*: Modality fusion is computed here, as first and second moments across available modalities in $C^{(2)}$, separately for each feature map ℓ,

$$\widehat{E}_\ell\left[C^{(2)}\right] = \frac{1}{|\mathcal{K}|}\sum_{k\in\mathcal{K}}C_{k,\ell}^{(2)} \quad \text{and} \quad \widehat{Var}_\ell\left[C^{(2)}\right]$$
$$= \frac{1}{|\mathcal{K}|-1}\sum_{k\in\mathcal{K}}\left(C_{k,\ell}^{(2)} - \widehat{E}_\ell\left[C^{(2)}\right]\right)^2,$$

with $\widehat{Var}_\ell[C^{(2)}]$ defined to be zero if $|\mathcal{K}| = 1$ (a single available modality).

3. *Front-end*: Finally the front-end combines the merged modalities to produce the final model output. In our implementation, we concatenate all $\widehat{E}[C^{(2)}]$ and $\widehat{Var}[C^{(2)}]$ feature maps, pass them through a convolutional layer $C^{(3)}$ with ReLU activation, to finish with a final layer $C^{(4)}$ that has as many feature maps as there are target segmentation classes. The pixelwise posterior class probabilities are given by applying a softmax function across the $C^{(4)}$ feature maps, and a full image segmentation is obtained by taking the pixelwise most likely posterior class. No further postprocessing on the resulting segment classes (such as smoothing) is done.

15.3.2 Pseudo-Curriculum Training Procedure

To carry out segmentation efficiently, the model is trained fully convolutionnally to minimize a pixelwise class cross-entropy loss, in the spirit of Long et al. [20]. It has long been known that noise injection during training is a powerful technique to make neural networks more robust, as shown among others with denoising autoencoders [21], and dropout and related procedures [22]. Here, we make the HeMIS architecture robust to missing modalities by randomly dropping any number for a given training example. Inspired by previous works on curriculum learning [23]—where the model starts learning from easy scenarios before turning to more difficult ones—we used a pseudo-curriculum learning scheme where after a few warm-up epochs where all modalities are shown to the model, we start randomly dropping modalities, ensuring a higher probability of dropping zero or one modality only.

15.3.3 Interpretation as an Embedding

An embedding is a mapping from an arbitrary source space to a target real-valued vector space of fixed dimensionality. In recent years, embeddings have been shown to yield unexpectedly powerful representations for a wide array of data types, including single words [24], variable-length word sequences [25] and images [26], and more.

In the context of HeMIS, the back-end can be interpreted as learning to separately map each modality into an *embedding common to all modalities*, within which vector algebra operations carry well-defined semantics. As such, computing empirical moments to carry out modality fusion is sensible. Since the model is trained entirely end-to-end with backpropagation, the key aspect of the architecture is that this embedding only needs to be defined implicitly as that which minimizes the overall training loss. Cross-modality interactions can be captured within specific embedding dimensions, as long as there are a sufficient number of them (i.e., enough feature maps within $C^{(2)}$), as they can be combined by $C^{(3)}$.

With this interpretation, the back-end consists of a modular assembly of operators, viewed as reusable building blocks that may or may not be needed for a given instance, each computing the embedding from its own input modality. These projections are summarized in the abstraction layer (with a mean and variance, although additional summary statistics are simple to entertain), and this summary further processed in the front-end to yield the final model output.

15.4 Data and Implementation Details

We studied the HeMIS framework on two neurological pathologies: MS with the MS Grand Challenge (MSGC) and a large relapsing–remitting MS (RRMS) cohort, as well as glioma with the brain tumor segmentation (BRATS) dataset [27].

15.4.1 MS Datasets

MSGC: The MSGC dataset [28] provides 20 training MR cases with manual ground truth lesion segmentation and 23 testing cases from the Boston Children's Hospital (CHB) and the University of North Carolina (UNC). We downloaded* the co-registered T1W, T2W, FLAIR images for all 43 cases as well as the ground truth lesion mask images for the 20 training cases. While lesion masks for the 23 testing cases are not available for download, an automated system is available to evaluate the output of a given segmentation algorithm.

RRMS: This dataset is obtained from a multi-site clinical study with 300 RRMS patients (mean age 37.5 years, SD 10.0 years). Each patient underwent an MRI that included sagittal T1W, T2W, and T1 post-contrast (T1C) images. The MRI data were acquired on 1.5 T scanners from different manufacturers (GE, Philips, and Siemens).

15.4.2 Brain Tumors (BRATS)

The *BRATS-2015* dataset contains 220 subjects with high-grade and 54 subjects with low-grade tumors. Each subject contains four MR modalities (FLAIR, T1W, T1C, and T2) and comes with a voxel level segmentation ground truth of five labels: *healthy*, *necrosis*, *edema*, *non-enhancing tumor*, and *enhancing tumor*. As done by Menze et al. [27], we transform each segmentation map into three binary maps, which correspond to three tumor categories, namely, *Complete* (which contains all tumor classes), *Core* (which contains all tumor subclasses except "edema"), and *Enhancing* (which includes the "enhanced tumor" subclass). For each binary map, the Dice similarity coefficient (DSC) is calculated [27].

BRATS-2013 contains two test datasets; Challenge and Leaderboard. The Challenge dataset contains 10 subjects with high-grade tumors while the Leaderboard dataset contains 15 subjects with high-grade tumors and 10 subjects with low-grade tumors. There is no ground truth provided for these datasets and thus quantitative evaluation can be achieved via an online evaluation system [27]. In our experiments, we used Challenge and Leaderboard datasets to compare the HeMIS segmentation performance to the state-of-the-art, when trained on all modalities.

15.4.3 Pre-Processing and Implementation Details

Before being provided to the network, bias field correction [29], and intensity normalization with a zero mean, truncation of 0.001 quantile and unit variance is applied to the image intensity. The multi-modal images are co-registered to the T1W and interpolated to 1 mm isotropic resolution.

We used Keras library [30] for our implementation. To deal with class imbalance, we adopt the patch-wise training procedure described by Havaei et al. [31].

---

* http://www.nitrc.org/projects/msseg/

We first train the model with a balanced dataset that allows learning features that are agnostic to the class distribution. In a second phase, we train only the final classification layer with a distribution close to the ground truth. This ensures that we learn good features yet keep the correct class priors. The method was trained using an Nvidia TitanX GPU, with a stochastic gradient learning rate of 0.001, decay constant of 0.0001, and Nesterov momentum coefficient of 0.9 [32]. For both BRATS-2015 and MS, we split the dataset into three separate subsets—train, valid, and test—with ratios of 70%, 10%, and 20% respectively. To avoid over-fitting, we used early stopping on the validation set.

15.5 Experiments and Results

The main segmentation evaluation metric that we use is the DSC [33], which represents a normalized measure of overlap between the proposed segmentation and the ground truth, defined as

$$DSC(P, T) = \frac{|P_1 \wedge T_1|}{\frac{1}{2}(|P_1| + |T_1|)},$$

where $P \in \{0, 1\}$ is the mask volume of the predictions made by the model, $T \in \{0, 1\}$ is the mask volume of the ground truth segmentation, P_1 and T_1 respectively represent the set of voxels with $P = 1$ and $T = 1$, $|\cdot|$ is the set cardinality operator, and $\wedge$ is the logical "and" operator. These regions are illustrated in Figure 15.8.

We first validate HeMIS performance against state-of-the-art segmentation methods on the two challenge datasets: MSGC and BRATS. Since the test data and the ranking table for BRATS 2015 are not available, we submitted results to BRATS 2013 challenge and leaderboard. These results are presented in Table 15.1.* As we observe, HeMIS outperforms Tustison et al. [34], the winner of the BRATS 2013 challenge, on most tumor region categories.

The MSGC dataset illustrates a direct application of HeMIS flexibility as only three modalities (T1W, T2W, and FLAIR) are provided for a small training set. Therefore, given the small number of subjects, we first trained HeMIS on RRMS dataset with four modalities and fine-tuned on MSGC. Our results were submitted to the MSGC website,[†] with a results summary appearing in Table 15.2. The MSGC segmentation results include three other supervised approaches: a Gaussian mixture model winner of the MSGC 2008 [35], a random decision forest classifier [8], and a more similar approach to ours using a deep convolutional encoder [9]. When compared to them, HeMIS obtains highly competitive results with a combined score of 83.2%, where 90.0% would represent human performance given inter-rater variability. This score is higher than the Gaussian mixture (80.0%) and the random forest (82.1%) a somewhat lower than the auto-encoder approach (84.0%).

For each rater (CHB and UNC), we provide the volume difference (VD), surface distance (SD), true positive rate (TPR), false positive rate (FPR), and the method's score as in Styner et al. [28].

The main advantage of HeMIS lies in its ability to deal with missing modalities, specifically when different subjects are missing different modalities. To illustrate the model's flexibility in such circumstances, we compare HeMIS performance to two common approaches to deal with random missing modalities. The first, *mean filling*, is to replace a missing modality by the modality's mean value. In our case, since all means are zero by construction, replacing a missing modality by zeros can be viewed as imputing with the mean. The second approach is to train an MLP to predict the expected value of specific missing modality given the available ones. Since neural networks are generally trained for a unique task, we need to train 28 different MLPs (one for each ∘ in Table 15.3 for a given dataset) to account for different possibilities of missing modalities. We used the same MLP architecture for all these models, which consists of 2 hidden layers with 100 hidden units each, trained to minimize the mean squared error. Figure 15.9 shows an example of predicted modalities for an MS patient.

The table shows the DSC for all possible configurations of MRI modalities being either absent (∘) or present (•), in order of FLAIR (F), T1W (T_1), T1C (T_1c), and

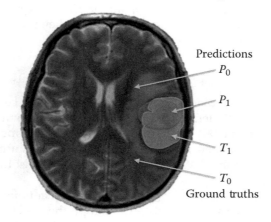

FIGURE 15.8
Illustration of the "Ground truths" and "Predictions" regions that are used in the computation of the DSC.

---
* Note that the results mentioned in Table 15.1 are from methods competing in the BRATS 2013 challenge for which a static table is provided at https://www.virtualskeleton.ch/BraTS/StaticResults 2013. Since then, other methods have been added to the scoreboard but for which no reference is available.
† http://www.ia.unc.edu/MSseg

TABLE 15.1
Comparison of HeMIS When Trained on All Modalities against BRATS-2013 Leaderboard and Challenge Winners, in Terms of Dice Similarity

	Leaderboard					Challenge
Method	Complete	Core	Enhancing	Complete	Core	Enhancing
Tustison et al. [34]	79	65	53	87	**78**	**74**
Zhao et al. [36]	79	59	47	84	70	65
Menze et al. [27]	72	60	53	82	73	69
HeMIS	**83**	**67**	**57**	**88**	75	**74**

Source: Scores from Menze, B. et al., (2015, Oct), *IEEE TMI* 34(10), 1993–2024.

TABLE 15.2
Results of the Full Dataset Training on the MSGC

Method	Rater	VD (%)	SD (mm)	TPR (%)	FPR (%)	Score
Souplet et al. [35]	CHB	86.4	8.4	58.2	70.6	80.0
	UNC	57.9	7.5	49.1	76.3	
Geremia et al. [8]	CHB	**52.4**	**5.4**	59.0	71.5	82.1
	UNC	**45.0**	5.7	51.2	76.7	
Brosch et al. [9]	CHB	63.5	7.4	47.1	**52.7**	**84.0**
	UNC	52.0	6.4	**56.0**	49.8	
HeMIS	CHB	127.4	7.5	**66.1**	55.3	83.2
	UNC	68.2	6.6	52.3	61.3	

TABLE 15.3
DSC Results on the RRMS and BRATS Test Sets (%) When Modalities Are Dropped

				RRMS			BRATS							
Modalities				Lesion			Complete			Core			Enhancing		
F	T_1	T_1c	T_2	HeMIS	Mean	MLP	HeMIS	Mean	MLP	HeMIS	Mean	MLP	HeMIS	Mean	MLP
○	○	○	●	1.74	2.66	**12.77**	58.48	2.70	**61.50**	40.18	4.00	37.32	**20.31**	6.25	18.62
○	○	●	○	2.67	0.00	**3.51**	33.46	23.11	2.04	44.55	23.90	17.70	**49.93**	30.02	32.92
○	●	○	○	3.89	0.00	**6.64**	33.22	0.00	2.07	17.42	0.00	10.52	4.67	6.25	**10.78**
●	○	○	○	34.48	9.77	**38.46**	71.26	**72.30**	63.81	37.45	0.00	34.26	5.57	6.25	**15.90**
○	○	●	●	27.52	4.31	25.83	67.59	35.01	64.97	**63.39**	30.92	49.38	**65.38**	39.00	60.30
○	●	●	○	8.21	0.00	**8.26**	45.93	23.63	1.99	**55.06**	41.89	26.55	**62.40**	43.80	40.93
●	○	●	○	38.81	11.62	**39.15**	80.28	75.58	78.13	49.52	0.00	48.97	22.26	6.25	**25.18**
●	○	○	●	**31.25**	8.31	29.39	69.56	1.77	66.88	47.26	2.63	43.66	23.56	6.25	**26.37**
○	●	○	●	**39.64**	33.31	38.55	**82.1**	81.01	81.35	53.42	25.94	52.41	23.19	6.25	**25.01**
●	●	○	○	**41.38**	6.42	39.33	79.8	45.97	**81.13**	66.12	29.85	65.51	**67.12**	35.14	66.19
●	●	●	○	**41.97**	9.00	40.63	80.88	81.57	**82.19**	69.26	62.13	**69.34**	**71.30**	67.13	70.93
●	●	○	●	**46.6**	41.12	41.83	**83.87**	77.84	80.40	57.76	20.66	53.46	28.46	6.25	28.34
○	●	●	●	**41.90**	38.95	41.47	82.78	64.19	**83.37**	70.62	42.36	70.45	70.52	49.62	**70.56**
●	○	●	●	**34.98**	5.78	29.46	**70.98**	30.86	67.85	66.60	45.79	55.40	**67.84**	50.21	64.81
●	●	●	●	**48.66**	43.48	43.48	**83.15**	82.43	82.43	**72.5**	71.46	71.46	**75.37**	72.08	72.08
# Wins / 15				9	0	6	10	1	4	14	0	1	9	0	6

T2W (T_2). Results are reported for HeMIS, mean (mean filling), and the imputation MLP (MLP).

Table 15.3 shows the DSC for this experiment on the test set.

On the BRATS dataset, for the Core category, HeMIS achieves the best segmentation in almost all cases (14 out of 15), and for the Complete and Enhancing categories, it leads in most cases (10 and 9 cases out of 15, respectively). Also, the mean-filling approach hardly outperforms HeMIS or MLP imputation. These results are consistent with the MS lesion segmentation dataset, where HeMIS outperforms other imputation approaches in 9 out of 15 cases. In scenarios where only one or two modalities are missing, while both HeMIS and MLP imputation obtain good results, HeMIS outperforms the latter in most cases on both datasets.

In cases where two modalities are missing, HeMIS performance is either close to or much higher than that of MLP across all categories.

On BRATS, when missing three out of four modalities, HeMIS outperforms the MLP in a majority of cases. Moreover, whereas the HeMIS performance only gradually drops as additional modalities become missing, the performance drop for MLP imputation and mean filling is much more severe. On the RRMS cohort, the MLP imputation appears to obtain slightly better segmentations when only one modality is available.

Although it is expected that tumor sub-label segmentations should be less accurate with fewer modalities, we should still hope for the model to report a sensible characterization of the tumor "footprint." While MLP and mean filling fail in this respect, HeMIS quite well achieves this goal by outperforming in almost all cases of the Complete and Core tumor categories. This can also be seen in Figure 15.10 where we show how adding modalities to HeMIS improves its ability to achieve a more accurate segmentation. From Table 15.3, we can also infer that the FLAIR modality is the most relevant for identifying the Complete tumor, while T1C is the most relevant for identifying Core and Enhancing tumor categories. On the RRMS dataset, HeMIS results are also seen to degrade slower than the other imputation approaches, preserving good segmentation when modalities go missing. Indeed, as seen in Figure 15.10, even though with FLAIR alone HeMIS already produces good segmentations, it is capable of further refining its results when adding modalities by removing false positives and improving outlines of the correctly identified lesions or tumor.

15.6 Conclusion

We have proposed a new fully automatic segmentation framework for heterogenous multi-modal MRI using a specialized convolutional deep neural network. The embedding of the multi-modal CNN back-end allows to train and segment datasets with missing modalities. We carried out an extensive validation on MS and glioma and achieved state-of-the art segmentation results on two challenging neurological pathology image processing tasks. Importantly, we contrasted the graceful performance degradation of the proposed approach as modalities go missing, compared with other popular imputation approaches, which it achieves without requiring training specific models for every potential missing modality combination.

This type of model offers a direction to fuse data from multiple imaging modalities, making it a step to deal with the information glut arising from the rapid increase in scans performed worldwide.

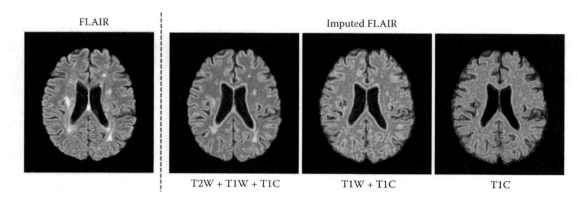

FIGURE 15.9
MLP-imputed FLAIR for an MS patient. The figure shows from left to right the original modality and the predicted FLAIR given other modalities.

Information Fusion in Deep CNNs for Biomedical Image Segmentation 311

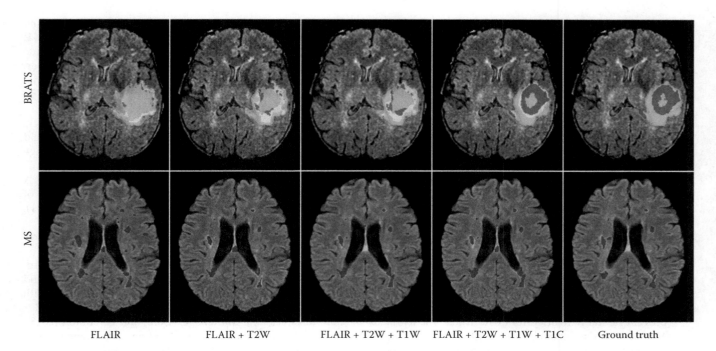

FIGURE 15.10
Example of HeMIS segmentation results on BRATS and MS subjects for different combinations of input modalities. For both cohorts, an axial FLAIR slice of a subject is overlaid with the results where for BRATS (first row) the segmentation colors describe necrosis (gray), non-enhancing (white), active core (light gray), and edema (medium gray). For the MS case, the lesions are highlighted in dark gray. The columns present the results for different combinations of input modalities, with ground truth in the last column.

References

1. OECD (2016). Health at a glance, 2015.
2. U.S. Preventive Services Task Force (2014). Screening for lung cancer: U.S. preventive services task force recommendation statement. *Annals of Internal Medicine* 160(5), 330–338.
3. Canadian Task Force on Preventive Health Care (2016). Recommendations on screening for lung cancer. *Canadian Medical Association Journal*.
4. Iglesias, J. E. and M. R. Sabuncu (2015). Multi-atlas segmentation of biomedical images: A survey. *Medical Image Analysis* 24(1), 205–219.
5. Havaei, M., N. Guizard, N. Chapados, and Y. Bengio (2016). *HeMIS: Hetero-Modal Image Segmentation*, pp. 469–477. Springer International Publishing.
6. Guizard, N., P. Coupé, V. S. Fonov, J. V. Manjón, D. L. Arnold, and D. L. Collins (2015). Rotation-invariant multi-contrast non-local means for ms lesion segmentation. *NeuroImage: Clinical* 8, 376–389.
7. Cordier, N., H. Delingette, and N. Ayache (2016). A patch-based approach for the segmentation of pathologies: Application to glioma labelling. *IEEE TMI* PP(99), 1.
8. Geremia, E., B. H. Menze, and N. Ayache (2013). Spatially adaptive random forests. Biomedical Imaging (ISBI), *2013 IEEE 10th International Symposium on*, pp. 1344–1347.
9. Brosch, T., Y. Yoo, L. Y. W. Tang, D. K. B. Li, A. Traboulsee, and R. Tam (2015). *MICCAI Proceedings*, Chapter "Deep convolutional encoder networks for multiple sclerosis lesion segmentation," pp. 3–11. Springer.
10. Van Buuren, S. (2012). *Flexible Imputation of Missing Data*. CRC Press, Boca Raton, FL.
11. Hofmann, M., F. Steinke, V. Scheel, G. Charpiat, J. Farquhar, P. Aschoff, M. Brady, B. Schölkopf, and B. J. Pichler (2008). MRI-based attenuation correction for PET/MRI: A novel approach combining pattern recognition and atlas registration. *Journal of Nuclear Medicine* 49 (11), 1875–1883.
12. Tulder, G. and M. Bruijne (2015). *MICCAI Proceedings*, Chapter, "Why does synthesized data improve multi-sequence classification?" pp. 531–538. Springer.
13. Hor, S. and M. Moradi (2015). Scandent tree: A random forest learning method for incomplete multimodal datasets. In *MICCAI 2015*, pp. 694–701. Springer.
14. Goodfellow, I., Y. Bengio, and A. Courville (2016). *Deep Learning*. Book in preparation for MIT Press. URL http://www.deeplearningbook.org.
15. Dumoulin, V. and F. Visin (2016, March). A guide to convolution arithmetic for deep learning. *ArXiv e-prints*.
16. Jarrett, K., K. Kavukcuoglu, M. Ranzato, and Y. LeCun (2009, Sept). What is the best multi-stage architecture for object recognition? In *Computer Vision, 2009 IEEE 12th International Conference on*, pp. 2146–2153.

17. Glorot, X., A. Bordes, and Y. Bengio (2011). Domain adaptation for large-scale sentiment classification: A deep learning approach. In *Proceedings of the 28th International Conference on Machine Learning (ICML-11)*, pp. 513–520.
18. Goodfellow, I. J., D. Warde-Farley, M. Mirza, A. Courville, and Y. Bengio (2013). Maxout networks. In *ICML*.
19. Krizhevsky, A., I. Sutskever, and G. E. Hinton (2012). Imagenet classification with deep convolutional neural networks. In *NIPS*, Volume 1, pp. 1097–1105. NIPS.
20. Long, J., E. Shelhamer, and T. Darrell (2015). Fully convolutional networks for semantic segmentation. In *Proceedings of the IEEE CVPR Conference on Computer Vision and Pattern Recognition*, pp. 3431–3440.
21. Vincent, P., H. Larochelle, Y. Bengio, and P.-A. Manzagol (2008). Extracting and composing robust features with denoising autoencoders. In *Proceedings of the 25th International Conference on Machine Learning*, pp. 1096–1103. ACM.
22. Srivastava, N., G. Hinton, A. Krizhevsky, I. Sutskever, and R. Salakhutdinov (2014). Dropout: A simple way to prevent neural networks from overfitting. *The Journal of Machine Learning Research* 15(1), 1929–1958.
23. Bengio, Y., J. Louradour, R. Collobert, and J. Weston (2009). Curriculum learning. In *Proceedings of the 26th Annual International Conference on Machine Learning*, pp. 41–48. ACM.
24. Mikolov, T., I. Sutskever, K. Chen, G. S. Corrado, and J. Dean (2013). Distributed representations of words and phrases and their compositionality. In *NIPS*, pp. 3111–3119.
25. Sutskever, I., O. Vinyals, and Q. V. Le (2014). Sequence to sequence learning with neural networks. In *Advances in Neural Information Processing Systems*, pp. 3104–3112.
26. Xu, K., J. Ba, R. Kiros, K. Cho, A. Courville, R. Salakhudinov, R. Zemel, A. Bengio, Yoshuaand Courville, R. Salakhudinov, R. Zemel, and Y. Bengio (2015). Show, attend and tell: Neural image caption generation with visual attention. In D. Blei and F. Bach (Eds.), *Proceedings of the 32nd International Conference on Machine Learning (ICML-15)*, pp. 2048–2057. JMLR Workshop and Conference Proceedings.
27. Menze, B., A. Jakab, S. Bauer, J. Kalpathy-Cramer, K. Farahani, and J. e. a. Kirby (2015, Oct). The multimodal brain tumor image segmentation benchmark (BRATS). *IEEE TMI* 34(10), 1993–2024.
28. Styner, M., J. Lee, B. Chin, M. Chin, O. Commowick, H. Tran, S. Markovic-Plese, V. Jewells, and S. Warfield (2008). 3D segmentation in the clinic: A grand challenge ii: Ms lesion segmentation. *MIDAS* 2008, 1–6.
29. Sled, J. G., A. P. Zijdenbos, and A. C. Evans (1998). A nonparametric method for automatic correction of intensity nonuniformity in MRI data. *IEEE TMI* 17(1), 87–97.
30. Chollet, F. (2015). Keras. URL https://github.com/keras-team/keras.
31. Havaei, M., A. Davy, D. Warde-Farley, A. Biard, A. Courville, Y. Bengio, C. Pal, P.-M. Jodoin, and H. Larochelle (2015). Brain tumor segmentation with deep neural networks. *arXiv preprint arXiv:1505.03540*.
32. Sutskever, I., J. Martens, G. Dahl, and G. Hinton (2013). On the importance of initialization and momentum in deep learning. In *ICML-13*, pp. 1139–1147.
33. Dice, L. R. (1945). Measures of the amount of ecologic association between species. *Ecology* 26(3), 297–302.
34. Tustison, N. J., K. Shrinidhi, M. Wintermark, C. R. Durst, B. M. Kandel, J. C. Gee, M. C. Grossman, and B. B. Avants (2015). Optimal symmetric multimodal templates and concatenated random forests for supervised brain tumor segmentation (simplified) with ANTsR. *Neuroinformatics* 13(2), 209–225.
35. Souplet, J., C. Lebrun, N. Ayache, and G. Malandain (2008, 07). An automatic segmentation of T2-FLAIR multiple sclerosis lesions.
36. Zhao, L., W. Wu, and J. J. Corso (2013). *MICCAI Proceedings*, Chapter, "Semi-automatic brain tumor segmentation by constrained MRFs using structural trajectories," pp. 567–575. Springer.

16
Automated Biventricular Cardiovascular Modelling from MRI for Big Heart Data Analysis

Kathleen Gilbert, Xingyu Zhang, Beau Pontré, Avan Suinesiaputra, Pau Medrano-Gracia, and Alistair Young

CONTENTS

16.1 Background 313
16.2 Left-Ventricular Analysis 315
    16.2.1 Signal Representation of Shapes 315
    16.2.2 A Finite-Element Model of the Left Ventricle 315
    16.2.3 Building an Atlas of the Heart 315
    16.2.4 Statistical Analysis of Shape 317
    16.2.5 Atlas-Based Quantification of Diseased Shapes 317
    16.2.6 Classification 318
    16.2.7 Regional Analysis 320
16.3 Congenital Heart Disease 322
    16.3.1 Biventricular Finite-Element Model 322
    16.3.2 Guide-Point Modelling 323
    16.3.3 Guide-Point Modelling vs. Manual Contouring 324
16.4 Summary 325
References 325

16.1 Background

In the last few years, there have been rapid advances in the capture and analysis of big data in the healthcare sector, as a result of technology improvement in hospital information technology systems. Patient health records are now stored electronically with clinical diagnosis and monitoring tools interconnected via computer networks. This allows data exchange between hospitals and healthcare institutions. Analysis of healthcare data is expected to have a large impact on the understanding and treatment of several domains (e.g. brain or heart), and this is reflected by the integration of data across cohorts or population studies [1]. There is a growing number of population studies that employ imaging acquisition methods, which have increased the amount of data by several orders of magnitude. This has revolutionized medical image analysis with big data analytics, such as in digital pathology [2], neuroimaging [3] and heart disease [4–6].

This chapter focuses on the application of big data analysis to cardiovascular disease (CVD), which is the world's largest cause of morbidity and mortality. This data analytics allows for a significant and controlled dimension reduction from millions of pixels down to a few hundred cardiac shape parameters for each patient. These parameters are exploited in the application of atlas-based cardiac imaging, modelling and analytics from population studies. Table 16.1 shows major cardiovascular epidemiological studies around the world that employ image acquisitions in their protocols. Cardiac magnetic resonance imaging (MRI) is the most commonly used imaging modality and is considered the gold standard to assess cardiac volume, mass and functions [14,15]. Unlike its counterparts, computed tomography and echocardiography, cardiac MRI has the ability to image the whole heart with a range of contrast mechanisms without the use of ionising radiation [16].

The first major cardiac population study that employed MRI was the Multi-Ethnic Study of Atherosclerosis

TABLE 16.1

Major Cardiovascular Epidemiological Studies with the Inclusion of Imaging Data and More Than 1000 Cohort Size

Study	Country	Modality	Cohort Size	Population Groups	Age Range
Framingham Offspring [7]	USA	MRI	1707	Multi	NA
Jackson Heart Study (JHS) [8]	USA	ECG, CT, MRI	5302	African-American	21–84
Multi-Ethnic Study of Atherosclerosis (MESA) [9]	USA	ECG, CT, MRI	6814	Hispanic, Chinese, White, African-American	45–84
EuroCMR [10]	Europe	MRI	27,781	Multi	47–70
Swedish Cardiopulmonary BioImage Study (SCAPIS) [11]	Sweden	MRI, CT	30,000	Multi	50–64
German National Cohort (GNC) [12]	Germany	MRI	30,000	Multi	20–69
UK Biobank [13]	UK	MRI, ECG, DEXA	100,000	Multi	40–69

Abbreviations: CT = computed tomography, DEXA = dual-energy x-ray absorptiometry, ECG = echocardiography, MRI = magnetic resonance imaging.

(MESA) [9]. MESA has collected and studied 6814 participants from four centers across the United States. MESA participants were initially free from CVD symptoms at the time of recruitment, because the main objective of MESA was to discover new image-derived risk factors that identify the manifestation of subclinical to clinical CVD before symptoms occur. The follow-up MESA study after 10 years of around 3000 participants has just recently been completed [17], which has demonstrated new knowledge about longitudinal changes in heart function in a large cohort [18,19].

The Jackson Heart Study (JHS) [8] is a US-based population study that has specifically targeted people at high-risk CVD. This study was designed to determine the root cause of CVD in 5302 African-American individuals who live in an area known to have economic, sociocultural, behavioural, dietary and physical activity disadvantages. JHS provides a rare opportunity to uncover hidden interactions between cardiac function derived from images with non-medical records.

The UK Biobank is a prospective cohort study of 500,000 aged 40–69 years, designed to identify the determinants of disease over time [13]. Over 10,000 participants have now completed comprehensive imaging examinations, including cardiac MRI for ventricular function, aortic compliance and flow. UK Biobank will complete imaging examinations in 100,000 participants by 2021, including cardiac MRI, abdominal MRI, brain MRI, 3D carotid ultrasound, dual-energy x-ray absorptiometry (DEXA) and 12 lead electrocardiogram (ECG).

Other cardiovascular epidemiological studies such as Canadian Partnership for Tomorrow Project (CPTP) [20], Iceland Myocardial Infarction study [21], Framingham Offspring Study [7] and The Swedish CArdioPulmonary BioImage Study (SCAPIS) [11] are also contributing to big data research in heart disease. EuroCMR [10] and Global Cardiac Magnetic Resonance (CMR) registries have not been specifically designed as clinical studies, but they aim to evaluate the prognostic potential of cardiac MRI as well as its cost-effectiveness. Images and associated clinical information are available for cardiac research. In general, big data analytics is only possible if data are available for research. In the field of cardiovascular research, data sharing has been pioneered by the Cardiac Atlas Project (CAP) [22]. The CAP was established to facilitate data sharing and atlas-based analysis of cardiovascular imaging examinations. One major contributing study is MESA, which is described above. Another major contributing study was the Defibrillators to Reduce Risk by Magnetic Resonance Imaging Evaluation (DETERMINE) study [23], which comprised patients with coronary artery disease and myocardial infarction (MI). Software was developed to facilitate labeling, ontological classification and databasing. All software developed in CAP is open source and available to the research community. Over 3000 de-identified cardiac imaging examinations are currently available in the CAP database. Apart from establishing data sharing infrastructure, the CAP also contributes to research in heart disease. Atlas-based methods applied to large datasets collated from existing population studies have been used to understand cardiac remodelling processes, as described below.

Cardiac remodelling is the term given to the processes by which the heart continuously adapts and remodels its three-dimensional shape and function in response to genetic, environmental and disease processes [24]. These changes in heart shape and function provide important information about the status and progression of cardiac disease, particularly in the transition from pre-clinical (asymptomatic) to clinical (symptomatic) disease [25–27]. Clinically studied changes in morphology have typically been characterized due to vascular events [28], hypertensive [29] and idiopathic cardiomyopathies [30], and these have been shown to be important markers for disease progression.

Pre-clinical remodelling also occurs in asymptomatic individuals prior to the establishment of clinical symptoms, in response to exposure to risk factors and genetic interactions. These have also been associated with long-term adverse outcomes. For example, larger left ventricular (LV) chamber dimension was associated with adverse events in the Framingham Heart Study [24]. Lower systolic dimension change [31] and hypertrophy [32] have also been associated with increased adverse cardiac events in Framingham.

Studies from MESA have also reported pre-clinical remodelling patterns. End-diastolic (ED) volume and mass were found to be predictive of future heart failure [33], and both LV mass and concentric hypertrophy were predictive of stroke [34]. Volume and mass were also shown to be altered with standard risk factors of smoking, hypertension and diabetes in the MESA cohort [35,36]. LV mass and mass-to-volume ratio (a measure of concentric remodelling) were positively associated with measures of obesity in MESA [37]. Over a 10-year follow-up period, mass-to-volume ratio increased in both men and women, which was associated with increases in systolic blood pressure and body mass index [19]. However, such simple shape indices do not capture the full range of shape adaptations in clinical or pre-clinical populations. Current clinical assessment of remodelling is limited to simple measures of size and length ratios, which ignore a large amount of information available in modern medical imaging examinations.

In the following sections, we explore the analysis of cardiac shape and function from MRI using LV shape models (Section 16.2) and statistical applications thereof (including variance, classification and regional analysis), before showing how these methods can be extended to biventricular (left and right ventricular) analyses (Section 16.3).

16.2 Left-Ventricular Analysis

16.2.1 Signal Representation of Shapes

Shapes have traditionally been computationally encoded as a vector or a matrix containing (1) the coordinates of vertices and (2) the topology or relationships of these vertices into faces or patches (mathematically, these are often differentiable two-dimensional manifolds). Such representation of shape as a signal requires a coordinate system and a mathematical basis describing these relationships. In our group, we prefer finite-element modelling (FEM), a mathematical framework that usually maps unity squares in two dimensions to surface patches in three.

The foundations of FEM were established in the 1940s initially for structural analysis [38], but this approach has been extended more generally to numerical modelling [39] with applications in bioengineering found as early as [40]. FEM is now a well-established method in cardiac mechanics.

FEM provides a direct quantitative approach to encoding shape and shape change over time in a continuous fashion (the elements are usually C^k-continuous parametric continuity across the element boundaries in the kth derivative) in a discrete framework (the number of elements is finite) and is typically used for solving systems of differential equations. For example, a FEM has been used to describe the shape and function of the LV [41].

16.2.2 A Finite-Element Model of the Left Ventricle

The LV model typically used in our group comprises two surface manifolds that represent the endocardium and epicardium enclosing the myocardium tissue volume. Each surface was modelled with 16 bicubic elements with C^1 continuity defined in a prolate spheroidal coordinate system. A schematic representation of these manifolds can be seen in Figure 16.1. Each element on the model has its own local domain, represented by ξ. Note that the nodes are numbered in the circumferential (ξ_1) direction, then longitudinal (ξ_2) and lastly transmural (ξ_3).

Fiducial landmarks were manually defined at the hinge points of the mitral valve on the long axis images, and at the endocardial insertions of the right ventricular (RV) free wall into the inter-ventricular septum. These were used to define the coordinate system and position of the model control points in consistent positions registered to the anatomy of the heart. A fully detailed explanation can be found in Ref. [42].

16.2.3 Building an Atlas of the Heart

This consistent representation of shape enables comparison across different subjects, generating statistical population models. However, it is often required to remove differences in orientation and scaling between shapes, using the Procrustes alignment [43]. If there are no corresponding landmarks between two or more objects, this is known as the generalized Procrustes alignment problem, and other algorithms such as the coherent point drift [44] can be applied. When there is point correspondence between two shapes, and they are expressed in the same coordinate system, this is a much simpler problem known as Procrustes superimposition, which can be readily solved by the algorithm developed by Kabsch [45]. This algorithm finds the optimal rotation matrix and translation vector, which minimise the overall distance between two sets of points with respect to the Euclidean norm.

In our case, the LV models have had a cardiac coordinate system defined individually, which guarantees a certain level of anatomical correspondence across subjects. In order to eliminate any residual bias (jitter) due to this procedure, LV models can be sampled and

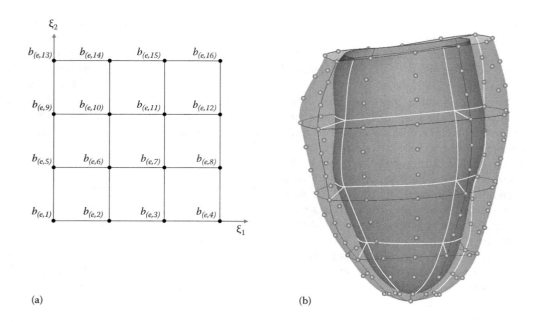

FIGURE 16.1
(a) Diagram of one element with the 16 Bézier control points in element space (ξ_1,ξ_2). (b) 3D rendering of the LV model with control points (medium gray) for the epicardial surface (light gray). The endocardium is shown in dark gray. Elements are delineated in black for the epicardial surface and white for the endocardial. (Reproduced from Pau Medrano-Gracia. *Shape and Function in Large Cardiac MRI Datasets*. PhD thesis, Auckland Bioengineering Institute, Auckland, 11 2014. https://researchspace.auckland.ac.nz/handle/2292/20497. With permission.)

aligned to their mean shape. To investigate the normal distributions of shape of the LV, we generated an atlas from nearly 2000 cases using FEM analysis of the CMR cine images; 1991 de-identified CMR cases were retrieved from the CAP database. The cases were from the MESA study, which had informed participant consent compatible with sharing of de-identified images and Local Review Board approval. The images were de-identified in a HIPAA compliant manner previous to upload, annotated using standard ontological schema, stored in a web-accessible picture archiving and communication system (PACS) database and analyzed using atlas-based techniques [22].

The overall construction process for the atlas is shown in Figure 16.2. Fiducial landmarks were manually placed at the centroid of the LV cavity on the apical and basal

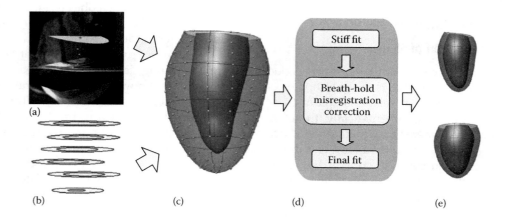

FIGURE 16.2
Flow chart of the atlas construction: (a) Fiducial landmarks defined at ED on short and long axis images (3D view from anterior). Dark gray markers denote the mitral valve and medium gray markers denote the intersections of the RV free wall and the septum. The base plane is drawn as a white disc. (b) Contours drawn on short-axis slices by the core lab. Individual breath-holds for each 2D slice result in mis-alignment between slices. (c) 3D finite-element model showing epicardial control points (model shape parameters) and element boundaries after an initial fit. (d) The process for correcting for mis-aligned slices caused by breath-hold misregistration. It results in a corrected 3D finite-element model. (e) Principal component analysis of atlas shape variation. Upper and lower panels show 2 standard deviations in the principal component shape. (Reproduced from Pau Medrano-Gracia et al., *Journal of Cardiovascular Magnetic Resonance*, 16(1):1, 2014. With permission.)

ED short-axis images, the hinge points of the mitral valve in the ED long axis image 1.2a and at the insertions of the RV free wall into the inter-ventricular septum in the ED short-axis images. These were used to define a patient-specific coordinate system that although individualized, was generally aligned in the same way for all patients, and initialize the position of the model, as described in detail in Ref. [46].

The model coordinates were used to provide the atlas coordinates of the LV: each point was assumed to be in the same anatomical location, and this allowed alignment of the hearts of all patients [47].

16.2.4 Statistical Analysis of Shape

Principal component analysis (PCA) was used to find the most important global shape variations in the atlas [48]. PCA finds the smallest number of mathematically independent shapes, (or components), which explain as much of the global shape variation as possible [49]. The shapes at ED and ES were analyzed separately, creating two atlases. An advantage of PCA is that traditional measures of heart function, such as volume, thickness and dimensional shape change, are inherently included in the PCA modes. The shape components are ordered by the amount of variance they explain, with the first component explaining the largest amount of variance. Figure 16.3 shows the first and second components of both the ED and ES atlas. The first component (Figure 16.3 top left) of the ED atlas can be seen as variation in heart size and accounts for 32% of the total variation in the population. We therefore call this component *size* for convenience; however, it is not pure scaling. The second component (Figure 16.3 top right) at ED is called *sphericity* as it is associated with the height-to-width ratio of the ventricle. This mode accounts for 13% of the total variation in the atlas.

The first mode of the ES atlas (Figure 16.3 bottom left) is similar to the *size* mode seen in the ED atlas. This mode explains 30% of the total variation in the atlas. The second component of the ES atlas (Figure 16.3 bottom right) is called *concentricity* as it describes the ratio of cavity volume to wall volume. Higher amounts of this shape at ES lead to a more concentric shape of the ventricle. The second component accounts for 10% of the variation in the ES atlas. The third component (9%, not shown) was associated with ventricular sphericity at ES.

16.2.5 Atlas-Based Quantification of Diseased Shapes

In Ref. [50], we also included a sample of shapes from patients with MI provided by the DETERMINE study [23]. The asymptomatic cases did not have any clinical symptoms of CVD at the time of recruitment and were regarded as the control group. Patients in the DETERMINE study were taller and heavier than the control group and more likely to be male. The patients also had larger LV mass, end-diastolic volume (EDV), end-systolic volume (ESV) and blood pressures. The DETERMINE study used steady-state free precession (SSFP) imaging for the CMR protocol, whereas MESA used GRE. For this study, all LV models from the patient and control

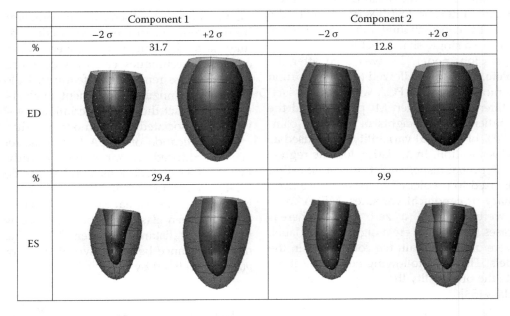

FIGURE 16.3
First and second principal components of variation in the atlas ($N = 1991$) for ED and ES. For each component, the left and right shapes represent the mean 2 std. dev. in the component distribution. Viewpoint is from the septum, posterior wall on the right. (Reproduced from Pau Medrano-Gracia et al., *Journal of Cardiovascular Magnetic Resonance*, 16(1):1, 2014. With permission.)

groups were aligned to their mean shape using translation and rotation. Scale was removed from the model, but height and weight were included in regression models.

The MESA cohort was acquired using a different imaging protocol (GRE) to the DETERMINE cohort (SSFP). It is known that these two protocols result in small differences in the placement of inner and outer surfaces of the heart. SSFP typically gives rise to larger estimates of LV cavity volume and smaller estimates of LV mass than GRE. The shape bias has been shown to be regionally variable and can be effectively removed using a maximum likelihood correction algorithm [51]. Briefly, a transformation between GRE models and SSFP models was learned using data from 40 asymptomatic individuals who were scanned using both protocols. The optimal transformation was found using maximum likelihood methods and was validated previously [51]. All MESA shape models were then transformed using this method, with the transformed shapes then being directly comparable to SSFP models. The coordinates (x,y,z) of the surface sampling points were concatenated into a shape vector. Shape vectors from all cases were formed into a matrix. The eigenvectors of the covariance matrix formed the principal component modes, and their corresponding eigenvalues indicate the proportion of the total variation explained by each mode. Selecting the number of PCA modes to retain in subsequent analysis is contingent on the application. In this paper, enough modes were retained to explain 90% of the total variance. Three PCA cases were considered: the first using only shape vectors at ED, the second using shape vectors at ES and the third using a combination of ED and ES (ED&ES). The ED&ES PCA was formed by concatenating the shape vectors from ED and ES into a single shape vector.

A logistic regression model [52] was used after the PCA was complete. The model allowed for identification of which shape modes from the PCA were most associated with the differences between MI patients and the asymptomatic patients. The weights of the PCA components (up to 90% of the total variability) were used as predictors for classification. In statistics, logistic regression is a type of probabilistic, statistical classification model, which is used to predict a binary response from continuous, binary or canonical variables. MESA cases (non-patients) were assigned a zero label, whereas DETERMINE cases (patients) were assigned a one label. These values were used to obtain the coefficients in the regression models. Thus, the following equation can be used to calculate the probability that a new case belongs to the patient class [53]:

$$P = \frac{1}{1 + \exp(-\beta_0 + \sum \beta_i X_i)}, \quad (16.1)$$

where P is the probability of a certain case belonging to the MI set, X_i are the values of the predictors, which in our case represent the PCA modes, β_i are the coefficient terms of X_i and β_0 is the intercept. The β terms were found by maximum likelihood estimation. The goodness of fit of the resulting model can be examined to determine how well the regression model distinguishes between non-patients and patients. Three common statistics used to quantify the goodness of fit of the model are deviance, Akaike information criterion (AIC) and Bayesian information criterion (BIC) [54]:

$$\text{Deviance} = -2\log(L), \quad (16.2)$$

$$\text{AIC} = -2\log(L) + 2k, \quad (16.3)$$

$$\text{BIC} = -2\log(L) + 2k\log(n), \quad (16.4)$$

where L represents the log-likelihood of the model (i.e., the value that is maximized by computing the maximum likelihood value for the parameters), k is the number of estimated parameters and n is the sample size. In all three measures, a lower number is indicative of a better model. In addition to these three measures, we also evaluated the area under the curve (AUC) of the receiver operating characteristic (ROC) curves (Figure 16.4), since this is also an overall measure of goodness of fit (better models having values closer to 1.0).

16.2.6 Classification

PCA as an unsupervised feature extraction technique has been introduced in the above section, but the method may lack discriminatory power. Supervised feature extraction techniques can provide more efficient and representative remodelling features. In Ref. [55], information maximizing component analysis (IMCA) was used to extract the most discriminatory shape features, which is associated with remodelling due to MI.

The application of IMCA to cardiac remodelling has been introduced in our previous study [55]. Briefly, IMCA views each class as a probability density function (PDF) on a statistical manifold, which can be projected into a low-dimensional Euclidean space [56]. The similarity between classes can be described with the Fisher information distance between PDFs. The Fisher information distance between two distributions $p(x;\sigma_1)$ and $p(x;\sigma_2)$ is defined by

$$DF(\theta_1,\theta_2) = \min_{\theta:\theta(0)=\theta_1,\theta(1)=\theta_2} \int \sqrt{\left(\frac{d\theta}{dt}\right)^T [I(\theta)] \left(\frac{d\theta}{dt}\right)} dt, \quad (16.5)$$

Automated Biventricular Cardiovascular Modelling from MRI for Big Heart Data Analysis

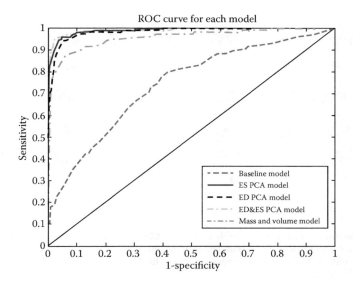

FIGURE 16.4
ROC curve for the logistic regression classification for each model. (Reproduced from Xingyu Zhang et al., *PLoS One*, 9(10):e110243, 2014. With permission.)

where σ_1 and σ_2 are the parameters corresponding to the two PDFs, $\sigma(t)$ is the parameter path along the manifold and is the Fisher information matrix whose elements are defined as

$$[I(\theta)] = \int f(X;q) \frac{\partial \log f(X;\theta)}{\partial \theta^i} \frac{\partial \log f(X;\theta)}{\partial \theta^j} dX \quad (16.6)$$

While the Fisher information distance cannot be exactly computed without knowing the parameterization of the manifold, it can be approximated by the Kullback–Leibler divergence [56], denoted $D_{KL}(p_i, p_j)$. The IMCA projection is defined as one that maximizes the Fisher information distance between classes. Specifically, let $\chi = \{X_1, X_2\}$ be a family of datasets where X_1 corresponds to samples from MESA and X_2 corresponds to samples from DETERMINE, estimating the PDF of X_i as p_i. Following Ref. [55], we refer to $D_{KL}(p_i, p_j)$ as $D_{KL}(X_i, X_j)$ with the knowledge that the divergence is calculated with respect to PDFs, not realizations. We wish to find a single orthonormal projection matrix A such that

$$A = \arg \max_{A:A^TA=I} \| D_{KL}(AX_i, AX_j), \|_F, \quad (16.7)$$

where I is the identity matrix and D_{KL} is the 2 × 2 matrix of Kullback–Leibler divergences. We used the gradient descent algorithm to find the optimal solution. IMCA can be viewed as a generalized and orthogonal version of LDA, which does not make assumptions on the class distributions.

CMR images from 300 cases with MI and 1991 asymptomatic volunteers (AVs) were obtained from the CAP. Finite-element models were customized to model the shape and function of each case using a standardized procedure. IMCA was used to identify global modes of shape variation across all cases, which best discriminate the two groups.

Figure 16.5 shows how these new indices of global remodelling create a continuum where cases can be scored according to their degree of severity; in particular, it shows that the IMCA ED&ES mode captures the larger size and more spherical shape, bulging of the apex and thinner wall thickness, which are known clinically to be associated with remodelling after MI. This new remodelling index enables precise quantification of the amount of remodelling and the effect of treatments designed to reverse remodelling effects. The distribution of IMCA scores at ED&ES between MESA and DETERMINE is shown in Figure 16.6. The asymptomatic group and the MI group were well classified with IMCA scores.

In 2015, CAP ran a challenge as part of the Statistical Atlases and Computational Models of the Heart workshop, in conjunction with the MICCAI conference. The aim of this challenge was to capture the state of the art of automatic MI classification methods based purely on LV anatomical shape.

We therefore designed a challenge to test classification performance in MI given a set of three-dimensional LV surface points. The training set comprised 100 MI patients and 100 AVs.

We provided the training set with labels to participants, and they were asked to submit the likelihood that an LV shape is MI from a different (validation) set of 200 cases (100 AV and 100 MI). Sensitivity, specificity,

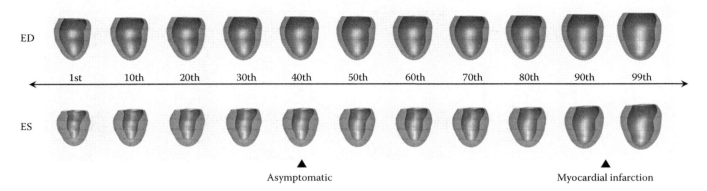

FIGURE 16.5
Derived shape indices allow for a continuous representation of disease remodelling. In the figure, the corresponding shapes from the percentiles of the IMCA ED&ES index are shown. Mean values (black triangles) for the asymptomatic (MESA) and myocardial infarct group (DETERMINE) show over 50 percentiles of separation for this index (in correspondence with Figure 16.6). (Reproduced from Xingyu Zhang et al., *Journal of Translational Medicine*, 13(1):1–9, 2015. With permission.)

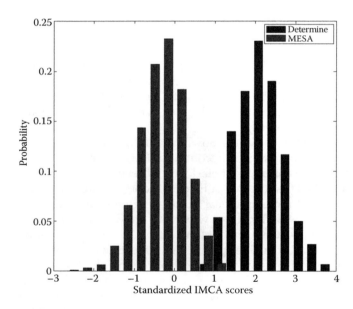

FIGURE 16.6
Distribution of IMCA Scores of MESA and DETERMINE for the best case (ED&ES). ED and ES figures do not show perceivable differences in their equivalent plots and are therefore omitted. (Reproduced from Xingyu Zhang et al., *Journal of Translational Medicine*, 13(1):1–9, 2015. With permission.)

accuracy and area under the ROC curve (AUC) were used as the classification outcome measures. The goals of this challenge were to (1) establish a common dataset for benchmarking classification algorithms and (2) test whether statistical shape modelling provides additional benefits to increase the accuracy of MI detection compared to standard clinical measurements. Eleven groups with a wide variety of classification and feature extraction approaches participated in this challenge.

All methods achieved excellent classification results with accuracy ranges from 0.83 to 0.98. The AUC values were above 0.90. However, only two methods consistently performed better than traditional indices on all scores.

16.2.7 Regional Analysis

It is routine clinical practice to use visual wall motion and infarct scoring from MRI when assessing ventricular function and regions of scar in patients with ischemic heart disease [57]. The guidelines set by the American Heart Association (AHA) state that the myocardium should be partitioned into 17 segments for regional analysis [58]. In Ref. [59], we provided a framework for automatic scoring to alert the diagnostician to potential regions of abnormality (Figure 16.7).

We investigated different shape and motion configurations of a finite-element cardiac atlas of the LV. Again, two patient populations were used: 300 AVs and 105 patients with MI, both randomly selected from the CAP database. Al-Jarrah et al. [60] comment that support vector machines are a common tool in machine learning, allowing the extraction of useful predictors from complex large systems, thus cementing their importance in big data analysis. Support vector machines [61] were employed to estimate the boundaries between the asymptomatic control and patient groups for each of the 16 standard anatomical regions in the heart.

Ground truth visual wall motion scores from standard cines and infarct scoring from late enhancement were provided by experienced observers. From all configurations, ES shape best predicted wall motion abnormalities (global accuracy 78%, positive predictive value 85%, specificity 91%, sensitivity 60%) and infarct scoring (74%, 72%, 91%, 44%), as shown in Table 16.2.

Automated Biventricular Cardiovascular Modelling from MRI for Big Heart Data Analysis 321

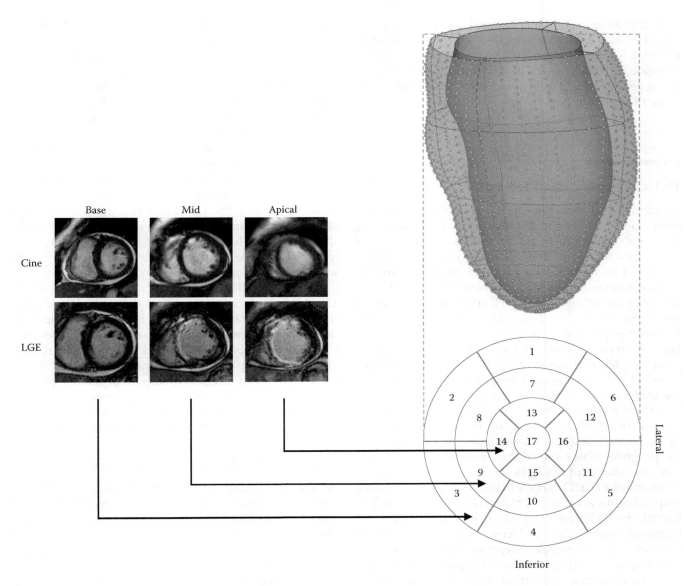

FIGURE 16.7
Clinical workflow for regional wall motion abnormalities (RWMA) and late gadolinium enhancement (LGE) scoring. MR images are acquired at different levels of the heart with and without contrast. Cine images are used to create a 3D LV finite-element model. Cine images and 3D models are mapped onto AHA regions (*bull's eye* plot) and scored visually by a specialist.

TABLE 16.2

Global Summary Statistics for RWMA Prediction and LGE Inference (across All Regions and Cross-Validation Experiments) for the Five Atlas Configurations

Atlas	Accuracy		PPV		Specificity		Sensitivity	
	RWMA	LGE	RWMA	LGE	RWMA	LGE	RWMA	LGE
ED	61.1	65.1	70.0	55.3	92.4	94.5	22.2	12.1
ES	77.5	73.9	84.9	72.0	91.4	90.5	60.2	44.1
EDES	76.6	73.6	84.6	72.1	91.5	90.8	58.1	42.6
ESW	71.0	71.5	74.2	63.6	85.1	84.8	53.4	47.6
EDESW	71.3	70.8	79.0	63.7	89.7	86.4	48.3	42.9

Abbreviation: PPV = Positive predictive value.

In conclusion, computer-assisted wall motion and infarct scoring has the potential to provide robust identification of those segments requiring further clinical attention; in particular, the high specificity and relatively low sensitivity could help avoid unnecessary late gadolinium rescanning of patients.

16.3 Congenital Heart Disease

Congenital heart disease (CHD) is the most common birth defect occurring in 75 out of every 1000 births [62], with moderate to severe malformations affecting 6 out of every 1000 infants [63,64]. Some defects are life-threatening and need immediate surgical treatment following birth. Improved diagnosis and treatment of CHD mean that the population of adult CHD patients is growing at approximately 5% per year and is now larger than the pediatric population [65]. Many of these patients, particularly those with tetralogy of Fallot, functional single right ventricle and transposition of the great vessels, are at risk of RV dilatation and dysfunction with associated morbidity and mortality. RV function measurement is an important prognostic marker, and RV size and function indices are used as indications for intervention [66]. These patients must be imaged repeatedly to determine the progression of remodelling. However, the quantitative assessment of changes in shape and function is problematic in CHD, largely because there is no detailed map of normal and abnormal hearts for comparison [67]. Recent model-based analysis of RV shape identified increased eccentricity and decreased systolic function in patients with pulmonary hypertension [68]. Another recent statistical shape analysis in repaired tetralogy of Fallot patients finds correlation of RV dilatation, outflow tract bulging and apical dilatation with the presence of pulmonary regurgitation [69].

MRI is the gold standard for assessing cardiac structure and function in CHD patients [16]. The use of a nonionising modality is ideal, as patients must be imaged repeatedly over their life in order to determine the progression of remodelling. Current methods for the analysis of biventricular remodelling and function in CHD patients are time-consuming and prone to error. Manual analysis of CMR images is considered the gold standard at present. This involves an expert manually contouring the endocardium and epicardium on all short-axis slices at both ED and ES. However, this method has been shown to have high inter- and intra-observer error for the right ventricle [70]. Figure 16.8 shows four chamber slices from three different CHD pathologies with the location of the short-axis slices for each case.

While most MR-based cardiac analysis focuses on the LV, the importance of the RV in determining the prognosis of CHD patients requires that both ventricles be assessed. For this reason, an analysis tool based on a biventricular model is required. Further, for routine clinical use, the software will need to rapidly fit the model to a set of cine MR images to generate a patient-specific model with minimal user interaction. To meet this purpose, the software needs to allow for the efficient manipulation of large amounts of imaging data.

16.3.1 Biventricular Finite-Element Model

As the RV is also of clinical importance in CHD, we developed a biventricular finite-element model. The initial model was an 88 element finite-element model that was created from an ex vivo porcine heart for biomechanical applications. The model was simplified by reducing collapsing nodes on the valves and removing several elements and converting the model parameterization

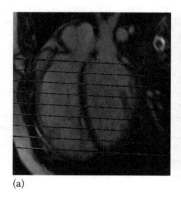
(a)

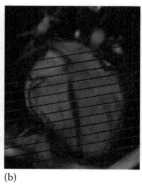
(b)

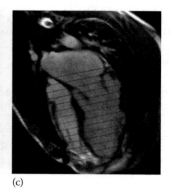
(c)

FIGURE 16.8
Four-chamber ED cine frame with the location of the short-axis slices in three patients with CHD. Patient (a) has tetralogy of Fallot, (b) has transposition of the great arteries where a mustard procedure has been performed and (c) has Marfan's syndrome where a Bentalls procedure and a mitral valve repair have been performed.

Automated Biventricular Cardiovascular Modelling from MRI for Big Heart Data Analysis

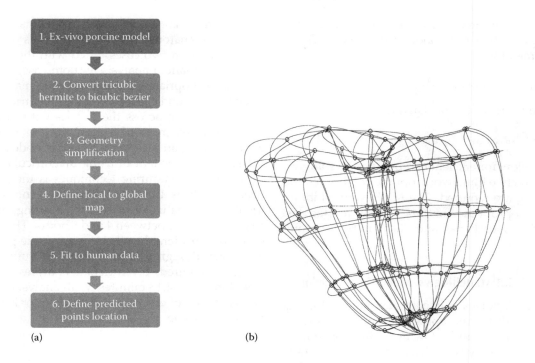

FIGURE 16.9
Workflow to develop a biventricular finite-element model. The grey boxes show model simplification steps and the black boxes show the steps required for it to be a good prior for guide-point modelling. (a) Workflow and (b) initial predicted points, shown on the 82-element model.

from Hermite to Beziér. The result was an 82-element finite-element model with 597 parameters in each of the x, y and z directions. Figure 16.9 details the process undertaken and shows the final 82-element model.

16.3.2 Guide-Point Modelling

Guide-point modelling is a technique that allows iterative updating of a finite-element model thus creating a patient-specific model [71]. The model parameters are optimised by least squares minimisation of an objective function quantifying the difference in displacements between the model and the data in terms of model coordinates (ξ):

$$\mathbf{E}(\mathbf{u}) = \sum_{d=1}^{D} \| \mathbf{w}_d((x^{\text{final}}(\xi_d) - x^{\text{initial}}(\xi_d)) - (\mathbf{h}_d - x^{\text{initial}}(\xi_d))) \|^2 + E_s(\mathbf{u}) \quad (16.8)$$

where D is the total number of data points, $\mathbf{w}_d$ is the weight of the data point (d) and $x^{\text{final}}(\xi_d)$ and $x^{\text{initial}}(\xi_d)$ are the initial and final positions of the model points corresponding to the data point $\mathbf{h}_d$, respectively. The data points included sparse guide points placed by the user, and densely sampled contour points generated by the non-rigid registration image tracking process [71].

$E_S(u)$ is chosen to impose a standard Sobolev regularisation:

$$\mathbf{E}_s(\mathbf{u}) = \int_\Omega \zeta_1 \left\| \frac{\partial \mathbf{u}}{\partial \xi_1} \right\|^2 + \zeta_2 \left\| \frac{\partial \mathbf{u}}{\partial \xi_2} \right\|^2 + \zeta_3 \left\| \frac{\partial \mathbf{u}}{\partial \xi_3} \right\|^2$$
$$+ \zeta_4 \left\| \frac{\partial^2 \mathbf{u}}{\partial \xi_1^2} \right\|^2 + \zeta_5 \left\| \frac{\partial^2 \mathbf{u}}{\partial \xi_2^2} \right\|^2 + \zeta_6 \left\| \frac{\partial^2 \mathbf{u}}{\partial \xi_1 \xi_2} \right\|^2 \quad (16.9)$$
$$+ \zeta_7 \left\| \frac{\partial^2 \mathbf{u}}{\partial \xi_1 \xi_3} \right\|^2 + \zeta_8 \left\| \frac{\partial^2 \mathbf{u}}{\partial \xi_2 \xi_3} \right\|^2 d\Omega$$

The first three terms penalize deviation from the shape in the circumferential, longitudinal and transmural directions. Terms 4 and 5 penalize changes in curvature and terms 6–9 penalize changes in the surface area of the element. The smoothing weights, ζ_i, need to be set so that the optimisation is sufficiently regularised but the smoothing does not dominate the model fit.

The optimisation problem can be represented in the form of $\mathbf{A}x = \mathbf{b}$, where A is a matrix formed by

$$\mathbf{A} = \mathbf{A}_{\text{data}} + \mathbf{A}_{\text{smoothing}} \quad (16.10)$$

Equation 16.10 is solved using a preconditioned conjugate gradient where the preconditioner is the inverse of $A_{\text{smoothing}}$. However, since the smoothing matrix is singular, it does not provide an adequate preconditioner. In order to better constrain the model and improve conditioning, a set of predictor points are required, with one point representing each node in the model. Therefore, **A** is formed by

$$\mathbf{A} = \mathbf{A}_{data} + \mathbf{A}_{smoothing} + \mathbf{A}_{predicted\ points} \quad (16.11)$$

and the preconditioner is calculated as

$$\Xi^{-1} = (A_{smoothing} + A_{predicted\ points})^{-1} \quad (16.12)$$

In order to allow for efficient computation of solutions inside the interactive framework, optimised libraries and multi-threading are required. The Math Kernel Library from Intel is appropriate for this task as it used fast vector matrix operations and has in-built multi-threading.

16.3.3 Guide-Point Modelling vs. Manual Contouring

Manual analysis involves an expert manually contouring both the endocardium and epicardium on a number of short-axis slices at ED and ES. Owing to time constraints, manual contouring may typically be performed on only a small subset of slices from the cardiac MRI dataset. The sparse spatial sampling of the myocardium along the long axis of the heart can introduce inaccuracies in cardiac function measures [72]. Figure 16.8 reinforces the importance of long axis slices in evaluation of cardiac function. In Figure 16.8a, there is a large volume in the RV that would be excluded if the long axis slices were not analyzed. Further, the use of only two time points during the cardiac cycle can affect accuracy of cardiac function. In comparison, guide-point modelling considers anatomical features across the entire cardiac MRI dataset, up to 15 slices. Also, since the whole cardiac cycle can be considered, more detailed assessment of cardiac function rather than measures based on changes between ES and ED is possible.

We have performed a comparison of guide-point modelling against manual analysis in CHD patients with a wide variety of congenital defects [73]. The number of guide points required to fit a model is dependent on how different the anatomy is from the initial model. For CHD cases, there will be significant differences in anatomy, requiring the models to be appropriately customized for each individual patient. On average, 318.1 ± 45.7 guide points were required to fit a case. The biventricular modelling method shows good agreement with the manual gold standard for clinically relevant cardiac function measures in both ventricles including ES and ED volumes, ejection fraction and mass.

Although there is agreement between the two methods, manual contouring can be more reliable and robust in CHD cases than the biventricular model. This difference is especially seen in patients with complex and unique anatomy. Figure 16.10 shows one short-axis cine frame from two cases, fitted with the biventricular model and manual analysis. In both cases, the myocardium is appropriately contoured using manual analysis (Figure 16.10a and c). However, the biventricular analysis is less consistent across these cases with Figure 16.10b showing an example of a poor fit, and Figure 16.10d demonstrating an acceptable fit to the model.

The primary benefit of using biventricular modelling over manual contouring techniques is the reduction in overall analysis time. To perform a complete biventricular analysis using manual contouring, experienced analysts can take between 4 and 5 hours. The time can be reduced considerably when using guide-point modelling. Even though a large number of guide points are typically required, biventricular analyses can be completed in 19.4 ± 4.18 minutes. With analysis times in this range, the use of biventricular analysis for large datasets becomes feasible.

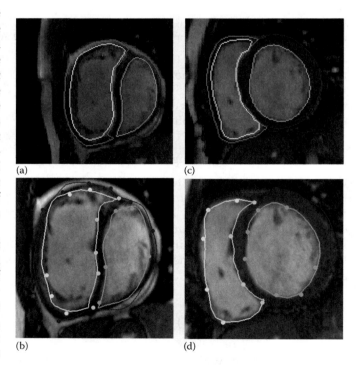

FIGURE 16.10

Short-axis ED frames with contours drawn by manual analysis (left) and contours, which have been interactively fitted by the biventricular modelling tool (right). (a and b) From a participant with transposition of the great arteries who had a mustard procedure. (c and d) From a participant with coarctation of the aorta and a repaired ventricular septal defect. The yellow contour notates the RV endocardium, the green is the LV endocardium and the blue is the biventricular epicardium.

16.4 Summary

As cardiac imaging reaches more patients and large databases are established, there arises the challenge of exploiting this wealth of information into meaningful outputs.

In Section 16.1, we have shown several examples of large-scale studies including cardiac imaging. Analyses of these data are shedding light onto important research questions, such as the risk factors of adverse events. However, most studies tend to discard information embedded in medical images because of the complexity in deriving useful measurements.

We have demonstrated in Section 16.2 that mathematical modelling of the cardiac shape over the cardiac cycle can generate useful shape indices, and how these compare to traditional clinical indices, such as ejection fraction. Shape and function indices can accurately identify disease in patients with MI both at the patient and regional level, thus quantifying image information in a meaningful fashion.

The extension of these tools to both ventricles and beyond (Section 16.3) will be useful for analysis of large cardiac datasets and will enable the creation of biventricular atlases.

We expect that these novel shape indices will contribute to future association studies, advancing and augmenting current diagnostic practices.

References

1. Ivo D Dinov. Volume and value of big healthcare data. *Journal of Medical Statistics and Informatics*, 4, 2016.
2. Anant Madabhushi and George Lee. Image analysis and machine learning in digital pathology: Challenges and opportunities. *Medical Image Analysis*, Jul 2016.
3. Sandeep R Panta, Runtang Wang, Jill Fries, Ravi Kalyanam, Nicole Speer, Marie Banich, Kent Kiehl, Margaret King, Michael Milham, Tor D Wager, Jessica A Turner, Sergey M Plis, and Vince D Calhoun. A tool for interactive data visualization: Application to over 10,000 brain imaging and phantom MRI data sets. *Frontiers in Neuroinformatics*, 10:9, 2016.
4. John S Rumsfeld, Karen E Joynt, and Thomas M Maddox. Big data analytics to improve cardiovascular care: Promise and challenges. *Nature Reviews Cardiology*, 13(6):350–9, Jun 2016.
5. Avan Suinesiaputra, Andrew D McCulloch, Martyn P Nash, Beau Pontre, and Alistair A Young. Cardiac image modelling: Breadth and depth in heart disease. *Medical Image Analysis*, Jun 2016.
6. Avan Suinesiaputra, Pau Medrano-Gracia, Brett R Cowan, and Alistair A Young. Big heart data: Advancing health informatics through data sharing in cardiovascular imaging. *IEEE Journal of Biomedical and Health Informatics*, 19(4):1283–90, Jul 2015.
7. Michael L Chuang, Philimon Gona, Gilion L T F Hautvast, Carol J Salton, Marcel Breeuwer, Christopher J O'Donnell, and Warren J Manning. CMR reference values for left ventricular volumes, mass, and ejection fraction using computer-aided analysis: the Framingham Heart Study. *Journal of Magnetic Resonance Imaging*, 39(4):895–900, Apr 2014.
8. Herman A Taylor, J.R., James G Wilson, Daniel W Jones, Daniel F Sarpong, Asoka Srinivasan, Robert J Garrison, Cheryl Nelson, and Sharon B Wyatt. Toward resolution of cardiovascular health disparities in African Americans: Design and methods of the Jackson Heart Study. *Ethnicity & Disease*, 15(4 Suppl 6):S6-4-17, 2005.
9. Diane E Bild, David A Bluemke, Gregory L Burke, Robert Detrano, Ana V Diez Roux, Aaron R Folsom, Philip Greenland, David R Jacob, Jr., Richard Kronmal, Kiang Liu, Jennifer Clark Nelson, Daniel O'Leary, Mohammed F Saad, Steven Shea, Moyses Szklo, and Russell P Tracy. Multi-ethnic study of atherosclerosis: objectives and design. *American Journal of Epidemiology*, 156(9):871–81, Nov 2002.
10. Oliver Bruder, Anja Wagner, Massimo Lombardi, Jürg Schwitter, Albert van Rossum, Günter Pilz, Detlev Nothnagel, Henning Steen, Steffen Petersen, Eike Nagel, Sanjay Prasad, Julia Schumm, Simon Greulich, Alessandro Cagnolo, Pierre Monney, Christina C Deluigi, Thorsten Dill, Herbert Frank, Georg Sabin, Steffen Schneider, and Heiko Mahrholdt. European Cardiovascular Magnetic Resonance (EuroCMR) registry—Multi national results from 57 centers in 15 countries. *Journal of Cardiovascular Magnetic Resonance*, 15:9, 2013.
11. Göran Bergström, Göran Berglund, Anders Blomberg, J Brandberg, Gunnar Engström, Jan Engvall, M Eriksson, U Faire, A Flinck, Mats G Hansson et al. The Swedish cardiopulmonary bioimage study: Objectives and design. *Journal of Internal Medicine*, 278(6):645–59, 2015.
12. German National Cohort (GNC) Consortium. The German National Cohort: Aims, study design and organization. *European Journal of Epidemiology*, 29(5):371–82, May 2014.
13. Steffen E Petersen, Paul M Matthews, Fabian Bamberg, David A Bluemke, Jane M Francis, Matthias G Friedrich, Paul Leeson, Eike Nagel, Sven Plein, Frank E Rademakers, Alistair A Young, Steve Garratt, Tim Peakman, Jonathan Sellors, Rory Collins, and Stefan Neubauer. Imaging in population science: cardiovascular magnetic resonance in 100,000 participants of UK Biobank—Rationale, challenges and approaches. *Journal of Cardiovascular Magnetic Resonance*, 15:46, 2013.
14. Helene Childs, Lucia Ma, Michael Ma, James Clarke, Myra Cocker, Jordin Green, Oliver Strohm, and Matthias G Friedrich. Comparison of long and short axis quantification of left ventricular volume parameters by cardiovascular magnetic resonance, with ex-vivo

validation. *Journal of Cardiovascular Magnetic Resonance*, 13:40, 2011.
15. Andre La Gerche, Guido Claessen, Alexander Van de Bruaene, Nele Pattyn, Johan Van Cleemput, Marc Gewillig, Jan Bogaert, Steven Dymarkowski, Piet Claus, and Hein Heidbuchel. Cardiac MRI: A new gold standard for ventricular volume quantification during high-intensity exercise. *Circulation: Cardiovascular Imaging*, 6(2):329–38, Mar 2013.
16. Arno AW Roest and Albert De Roos. Imaging of patients with congenital heart disease. *Nature Reviews Cardiology*, 9(2):101–15, 2012.
17. Diane E Bild, Robyn McClelland, Joel D Kaufman, Roger Blumenthal, Gregory L Burke, J Jeffrey Carr, Wendy S Post, Thomas C Register, Steven Shea, and Moyses Szklo. Ten-year trends in coronary calcification in individuals without clinical cardiovascular disease in the multi-ethnic study of atherosclerosis. *PLoS One*, 9(4):e94916, 2014.
18. Bharath Ambale Venkatesh, Gustavo J Volpe, Sirisha Donekal, Nathan Mewton, Chia-Ying Liu, Steven Shea, Kiang Liu, Gregory Burke, Colin Wu, David A Bluemke, and João A C Lima. Association of longitudinal changes in left ventricular structure and function with myocardial fibrosis: The Multi-Ethnic Study of Atherosclerosis study. *Hypertension*, 64(3):508–15, Sep 2014.
19. John Eng, Robyn L McClelland, Antoinette S Gomes, W Gregory Hundley, Susan Cheng, Colin O Wu, J Jeffrey Carr, Steven Shea, David A Bluemke, and Joao A C Lima. Adverse left ventricular remodeling and age assessed with cardiac MR imaging: The multi-ethnic study of atherosclerosis. *Radiology*, 278(3):714–22, Mar 2016.
20. Marilyn J Borugian, Paula Robson, Isabel Fortier, Louise Parker, John McLaughlin, Bartha Maria Knoppers, Karine Bédard, Richard P Gallagher, Sandra Sinclair, Vincent Ferretti, Heather Whelan, David Hoskin, and John D Potter. The Canadian Partnership for Tomorrow Project: Building a pan-Canadian research platform for disease prevention. *CMAJ*, 182(11):1197–201, Aug 2010.
21. Erik B Schelbert, Jie J Cao, Sigurdur Sigurdsson, Thor Aspelund, Peter Kellman, Anthony H Aletras, Christopher K Dyke, Gudmundur Thorgeirsson, Gudny Eiriksdottir, Lenore J Launer, Vilmundur Gudnason, Tamara B Harris, and Andrew E Arai. Prevalence and prognosis of unrecognized myocardial infarction determined by cardiac magnetic resonance in older adults. *JAMA*, 308(9):890–6, Sep 2012.
22. Carissa G Fonseca, Michael Backhaus, David A Bluemke, Randall D Britten, Jae Do Chung, Brett R Cowan, Ivo D Dinov, J Paul Finn, Peter J Hunter, Alan H Kadish, Daniel C Lee, Joao A C Lima, Pau Medrano-Gracia, Kalyanam Shivkumar, Avan Suinesiaputra, Wenchao Tao, and Alistair A Young. The Cardiac Atlas Project—An imaging database for computational modeling and statistical atlases of the heart. *Bioinformatics*, 27(16):2288–95, Aug 2011.
23. Alan H Kadish, David Bello, J Paul Finn, Robert O Bonow, Andi Schaechter, Haris Subacius, Christine Albert, James P Daubert, Carissa G Fonseca, and Jeffrey J Goldberger. Rationale and design for the Defibrillators to Reduce Risk by Magnetic Resonance Imaging Evaluation (DETERMINE) trial. *Journal of Cardiovascular Electrophysiology*, 20(9):982–7, Sep 2009.
24. Ramachandran S Vasan, Martin G Larson, Emelia J Benjamin, Jane C Evans, and Daniel Levy. Left ventricular dilatation and the risk of congestive heart failure in people without myocardial infarction. *New England Journal of Medicine*, 336(19):1350–5, May 1997.
25. Babak A Vakili, Peter M Okin, and Richard B Devereux. Prognostic implications of left ventricular hypertrophy. *American Heart Journal*, 141(3):334–41, Mar 2001.
26. Harvey D White, Robin M Norris, Michael A Brown, PW Brandt, RM Whitlock, and Christopher J Wild. Left ventricular end-systolic volume as the major determinant of survival after recovery from myocardial infarction. *Circulation*, 76(1):44–51, Jul 1987.
27. Selwyn P Wong, John K French, Anna-Maria Lydon, Samuel O M Manda, Wanzhen Gao, Noel G Ashton, and Harvey D White. Relation of left ventricular sphericity to 10-year survival after acute myocardial infarction. *American Journal of Cardiology*, 94(10):1270–5, Nov 2004.
28. GF Mitchell, GA Lamas, DE Vaughan, and MA Pfeffer. Left ventricular remodeling in the year after first anterior myocardial infarction: A quantitative analysis of contractile segment lengths and ventricular shape. *Journal of the American College of Cardiology*, 19(6):1136–44, May 1992.
29. Liqi Li, Yuji Shigematsu, Mareomi Hamada, and Kunio Hiwada. Relative wall thickness is an independent predictor of left ventricular systolic and diastolic dysfunctions in essential hypertension. *Hypertension Research*, 24(5):493–9, Sep 2001.
30. G. William Dec and V Fuster. Idiopathic dilated cardiomyopathy. *New England Journal of Medicine*, 331(23):1564–75, Dec 1994.
31. MS Lauer, JC Evans, and D Levy. Prognostic implications of subclinical left ventricular dilatation and systolic dysfunction in men free of overt cardiovascular disease (the Framingham heart study). *American Journal of Cardiology*, 70(13):1180–4, Nov 1992.
32. William B Kannel, Daniel Levy, and L Adrienne Cupples. Left ventricular hypertrophy and risk of cardiac failure: insights from the Framingham study. *Journal of Cardiovascular Pharmacology*, 10(Suppl 6):S135–40, 1987.
33. David A Bluemke, Richard A Kronmal, João A C Lima, Kiang Liu, Jean Olson, Gregory L Burke, and Aaron R Folsom. The relationship of left ventricular mass and geometry to incident cardiovascular events: The mesa (multi-ethnic study of atherosclerosis) study. *Journal of the American College of Cardiology*, 52(25):2148–55, Dec 2008.
34. Aditya Jain, Robyn L McClelland, Joseph F Polak, Steven Shea, Gregory L Burke, Diane E Bild, Karol E Watson, Matthew J Budoff, Kiang Liu, Wendy S Post, Aaron R Folsom, João A C Lima, and David A Bluemke. Cardiovascular imaging for assessing cardiovascular risk in asymptomatic men versus women: The multi-ethnic study of atherosclerosis (mesa). *Circulation: Cardiovascular Imaging*, 4(1):8–15, Jan 2011.
35. Ola Gjesdal, David A Bluemke, and Joao A Lima. Cardiac remodeling at the population level—risk factors,

screening, and outcomes. *Nature Reviews Cardiology*, 8(12):673–85, Dec 2011.
36. Susan R Heckbert, Wendy Post, Gregory D N Pearson, Donna K Arnett, Antoinette S Gomes, Michael Jerosch-Herold, W Gregory Hundley, Joao A Lima, and David A Bluemke. Traditional cardiovascular risk factors in relation to left ventricular mass, volume, and systolic function by cardiac magnetic resonance imaging: The multiethnic study of atherosclerosis. *Journal of the American College of Cardiology*, 48(11):2285–92, Dec 2006.
37. Evrim B Turkbey, Robyn L McClelland, Richard A Kronmal, Gregory L Burke, Diane E Bild, Russell P Tracy, Andrew E Arai, João A C Lima, and David A Bluemke. The impact of obesity on the left ventricle: The multi-ethnic study of atherosclerosis (mesa). *JACC Cardiovascular Imaging*, 3(3):266–74, Mar 2010.
38. Ray W Clough. Original formulation of the finite element method. *Finite Elements in Analysis and Design*, 7(2):89–101, 1990.
39. Gilbert Strang and George J Fix. *An Analysis of the Finite Element Method*, volume 212. Prentice-Hall, Englewood Cliffs, NJ, 1973.
40. Robert Friedenberg. Direct analysis or finite element analysis in biology: A new computer approach. *Biosystems*, 3(2):89–94, 1969.
41. Alistair A Young, Brett R Cowan, Steven F Thrupp, Warren J Hedley, and Louis J DellItalia. Left ventricular mass and volume: fast calculation with guide-point modeling on MR images 1. *Radiology*, 216(2):597–602, 2000.
42. Pau Medrano-Gracia. *Shape and Function in Large Cardiac MRI Datasets*. PhD thesis, Auckland Bioengineering Institute, Auckland, 11 2014. https://researchspace.auckland.ac.nz/handle/2292/20497.
43. David G Kendall. A survey of the statistical theory of shape. *Statistical Science*, 87–99, 1989.
44. Pau Medrano-Gracia, John Ormiston, Mark Webster, Susann Beier, Chris Ellis, Chunliang Wang, Alistair Young, and Brett Cowan. A statistical model of the main bifurcation of the left coronary artery using coherent point drift. In *Joint MICCAI workshops on Computing and Visualisation for Intravascular Imaging and Computer-Assisted Stenting*, 2015.
45. Wolfgang Kabsch. A solution for the best rotation to relate two sets of vectors. *Acta Crystallographica Section A: Crystal Physics, Diffraction, Theoretical and General Crystallography*, 32(5):922–3, 1976.
46. Pau Medrano-Gracia, Brett R Cowan, Bharath Ambale-Venkatesh, David A Bluemke, John Eng, John Paul Finn, Carissa G Fonseca, Joao AC Lima, Avan Suinesiaputra, and Alistair A Young. Left ventricular shape variation in asymptomatic populations: the multi-ethnic study of atherosclerosis. *Journal of Cardiovascular Magnetic Resonance*, 16(1):1, 2014.
47. Alistair A Young, David J Crossman, Peter N Ruygrok, and Mark B Cannell. Mapping system for coregistration of cardiac MRI and ex vivo tissue sampling. *Journal of Magnetic Resonance Imaging*, 34(5):1065–71, 2011.
48. Alistair A Young and Alejandro F Frangi. Computational cardiac atlases: From patient to population and back. *Experimental Physiology*, 94(5):578–96, 2009.
49. Espen W Remme, Kevin F Augenstein, Alistair A Young, and Peter J Hunter. Parameter distribution models for estimation of population based left ventricular deformation using sparse fiducial markers. *IEEE Transactions on Medical Imaging*, 24(3):381–8, 2005.
50. Xingyu Zhang, Brett R Cowan, David A Bluemke, J Paul Finn, Carissa G Fonseca, Alan H Kadish, Daniel C Lee, Joao AC Lima, Avan Suinesiaputra, Alistair A Young, et al. Atlas-based quantification of cardiac remodeling due to myocardial infarction. *PLoS One*, 9(10):e110243, 2014.
51. Pau Medrano-Gracia, Brett R Cowan, David A Bluemke, J Paul Finn, Alan H Kadish, Daniel C Lee, Joao AC Lima, Avan Suinesiaputra, and Alistair A Young. Atlas-based analysis of cardiac shape and function: correction of regional shape bias due to imaging protocol for population studies. *Journal of Cardiovascular Magnetic Resonance*, 15(1):1, 2013.
52. David W Hosmer and Stanley Lemeshow. Introduction to the logistic regression model. *Applied Logistic Regression, Second Edition*, pages 1–30, 2000.
53. S James Press and Sandra Wilson. Choosing between logistic regression and discriminant analysis. *Journal of the American Statistical Association*, 73(364):699–705, 1978.
54. Jerald B Johnson and Kristian S Omland. Model selection in ecology and evolution. *Trends in Ecology & Evolution*, 19(2):101–8, 2004.
55. Xingyu Zhang, Bharath Ambale-Venkatesh, David A. Bluemke, Brett R. Cowan, J. Paul Finn, Alan H. Kadish, Daniel C. Lee, Joao A. C. Lima, William G. Hundley, Avan Suinesiaputra, Alistair A. Young, and Pau Medrano-Gracia. Information maximizing component analysis of left ventricular remodeling due to myocardial infarction. *Journal of Translational Medicine*, 13(1):1–9, 2015.
56. Kevin M Carter, Raviv Raich, William G Finn, and Alfred O Hero. Information preserving component analysis: Data projections for flow cytometry analysis. *IEEE Journal of Selected Topics in Signal Processing*, 3(1):148–58, 2009.
57. Gautham P Reddy, Sandra Pujadas, Karen G Ordovas, and Charles B Higgins. MR imaging of ischemic heart disease. *Magnetic Resonance Imaging Clinics of North America*, 16(2):201–12, May 2008.
58. Manuel D Cerqueira, Neil J Weissman, Vasken Dilsizian et al. Standardized myocardial segmentation and nomenclature for tomographic imaging of the heart. *Circulation*, 105(4):539–42, Jan 2002.
59. Pau Medrano-Gracia, Avan Suinesiaputra, Brett Cowan, David Bluemke, Alejandro Frangi, Daniel Lee, João Lima, and Alistair Young. An atlas for cardiac MRI regional wall motion and infarct scoring. In *International Workshop on Statistical Atlases and Computational Models of the Heart*, pages 188–197. Springer, 2012.
60. Omar Y Al-Jarrah, Paul D Yoo, Sami Muhaidat, George K Karagiannidis, and Kamal Taha. Efficient machine learning for big data: A review. *Big Data Research*, 2(3):87–93, 2015.

61. Rong-En Fan, Kai-Wei Chang, Cho-Jui Hsieh, Xiang-Rui Wang, and Chih-Jen Lin. Liblinear: A library for large linear classification. *The Journal of Machine Learning Research*, 9:1871–4, 2008.
62. Mark A Fogel. *Ventricular Function and Blood Flow in Congenital Heart Disease*. John Wiley & Sons, 2005.
63. Julien IE Hoffman and Samuel Kaplan. The incidence of congenital heart disease. *Journal of the American College of Cardiology*, 39(12):1890–900, 2002.
64. Ariane J Marelli, Andrew S Mackie, Raluca Ionescu-Ittu, Elham Rahme, and Louise Pilote. Congenital heart disease in the general population changing prevalence and age distribution. *Circulation*, 115(2):163–172, 2007.
65. Julien Guihaire, Francois Haddad, Olaf Mercier, Daniel J Murphy, Joseph C Wu, and Elie Fadel. The right heart in congenital heart disease, mechanisms and recent advances. *Journal of Clinical & Experimental Cardiology*, 8(10):1, 2012.
66. David M Harrild, Charles I Berul, Frank Cecchin, Tal Geva, Kimberlee Gauvreau, Frank Pigula, and Edward P Walsh. Pulmonary valve replacement in tetralogy of Fallot impact on survival and ventricular tachycardia. *Circulation*, 119(3):445–51, 2009.
67. Georgios Giannakoulas, Konstantinos Dimopoulos, and X Yun Xu. Modelling in congenital heart disease. Art or science? *International Journal of Cardiology*, 133(2):141–4, Apr 2009.
68. Peter J Leary, Christopher E Kurtz, Catherine L Hough, Mary-Pierre Waiss, David D Ralph, and Florence H Sheehan. Three-dimensional analysis of right ventricular shape and function in pulmonary hypertension. *Pulmonary Circulation*, 2(1):34–40, 2012.
69. Benedetta Leonardi, Andrew M Taylor, Tommaso Mansi, Ingmar Voigt, Maxime Sermesant, Xavier Pennec, Nicholas Ayache, Younes Boudjemline, and Giacomo Pongiglione. Computational modelling of the right ventricle in repaired tetralogy of Fallot: Can it provide insight into patient treatment? *European Heart Journal—Cardiovascular Imaging*, 14(4):381–6, Apr 2013.
70. Lucy E Hudsmith, Steffen E Petersen, Jane M Francis, Matthew D Robson, and Stefan Neubauer. Normal human left and right ventricular and left atrial dimensions using steady state free precession magnetic resonance imaging. *Journal of Cardiovascular Magnetic Resonance*, 7(5):775–782, 2005.
71. Bo Li, Yingmin Liu, Christopher J Occleshaw, Brett R Cowan, and Alistair A Young. In-line automated tracking for ventricular function with magnetic resonance imaging. *JACC: Cardiovascular Imaging*, 3(8):860–6, 2010.
72. Thomas L Gentles, Brett R Cowan, Christopher J Occleshaw, Steven D Colan, and Alistair A Young. Midwall shortening after coarctation repair: The effect of through-plane motion on single-plane indices of left ventricular function. *Journal of the American Society of Echocardiography*, 18(11):1131–6, 2005.
73. Kathleen Gilbert, H-I Lam, Beau Pontré, Brett R Cowan, Christopher J Occleshaw, JY Liu, and Alistair A Young. An interactive tool for rapid biventricular analysis of congenital heart disease. *Clinical Physiology and Functional Imaging*, 2015.

17
Deep Learning for Retinal Analysis

Henry A. Leopold, John S. Zelek, and Vasudevan Lakshminarayanan

CONTENTS

17.1 Introduction: Background and Context 330
    17.1.1 Chapter Overview..331
17.2 What Is a Neuron, Really? 331
    17.2.1 Visual Inputs...332
    17.2.2 Deep Feedforward Networks 333
        17.2.2.1 Applying What We Know About the V1...333
        17.2.2.2 The Gabor Function 337
    17.2.3 Recurrent Neural Networks 337
17.3 Major Diseases of the Retina...337
    17.3.1 Diabetic Retinopathy 338
        17.3.1.1 Proliferative Diabetic Retinopathy 338
        17.3.1.2 Microaneurysms 338
        17.3.1.3 Diabetic Macular Edema...338
    17.3.2 Glaucoma 339
        17.3.2.1 Optic Nerve Head 339
        17.3.2.2 Peripapillary Atrophy 339
        17.3.2.3 Retinal Nerve Fibre Layer 340
17.4 Computer-Aided Diagnostics 340
    17.4.1 Mimicking Experts..341
    17.4.2 Feature Discovery 342
    17.4.3 Traditional CAD of Retinal Images..343
    17.4.4 Image Processing Fundamentals 343
        17.4.4.1 Global Histogram Enhancement...343
        17.4.4.2 Local Histogram Enhancement 344
        17.4.4.3 Super-Resolution Techniques 344
        17.4.4.4 Canny Edge Detectors 344
        17.4.4.5 Watershed Methods...344
    17.4.5 Common Data Sets 345
        17.4.5.1 Drive 345
        17.4.5.2 Stare...345
        17.4.5.3 Diaret 345
        17.4.5.4 Messidor 345
        17.4.5.5 Kaggle/EyePACS..345
    17.4.6 Why Performance Is Better Than Accuracy 345
    17.4.7 Evaluating Performance..347
17.5 Pixel Sampling for Segmenting Retinal Morphology...348
    17.5.1 Traditional Retinal Segmentation 348
    17.5.2 Pixel Sampling..348
    17.5.3 The Deep Learning Approach to Retinal Morphology Segmentation 350
        17.5.3.1 Adam Optimization...350

17.5.4 Performance Comparison: Vessel Segmentation and Pixel Classification 350
  17.5.4.1 Traditional Algorithms 350
  17.5.4.2 Deep Algorithms 352
  17.5.4.3 Comparing Performances 353
17.6 Superpixel Sampling for Microaneurysm Detection in the Classification of Diabetic Retinopathy 354
  17.6.1 Traditional Microaneurysm Detection 354
  17.6.2 Superpixel Sampling 356
    17.6.2.1 Selective Sampling 356
    17.6.2.2 Sampling with Foveation 356
  17.6.3 The Deep Learning Approach to Microaneurysm Detection 356
  17.6.4 Performance Comparison: MA Detection and Region Classification 357
    17.6.4.1 Traditional Algorithms 357
    17.6.4.2 Deep Expert Mimicry 359
    17.6.4.3 Deep Feature Discovery 360
    17.6.4.4 Comparing Performances 360
  17.6.5 Performance Comparison: Diabetic Retinopathy and Image Classification 361
    17.6.5.1 Traditional Algorithms 361
    17.6.5.2 Deep Algorithms 362
    17.6.5.3 Comparing Performances 363
17.7 Outlook 363
References 363

17.1 Introduction: Background and Context

Ophthalmology can be defined as basic and clinical studies of the eye and vision. It is widely accepted the origins of this field of medicine date back at least 2500 years to Alcmaeon of Cotona, according to the Greek anatomist and physician Galen of Pergamon, who defined the field [1]. Ophthalmology built upon Galen's writings and was incorporated in studies of optics, most notably by Ibn al Haytham. His magnum opus *Kitab al Manazir* (*The Book of Optics*), was published in 1015 AD and had a great influence on scientists in opthalmology [2]. A pivotal breakthrough in studying the eye was the ophthalmoscope—also referred to as the funduscope—developed by the physicist and ophthalmologist Hermann Von Helmholtz in 1851, which made study of the retina possible. Today, ophthalmoscopes are an essential part of every doctors' arsenal in ocular health diagnostics. Modern technological innovations within this space have been in the apparatus, with the most common being the fundus camera [3]. Optical coherence tomography (OCT) is a relatively new technology, first proposed in 1991 [4]. Compared to fundus images, the advantage for OCT imaging is its ability to extract 3-D data on the topology of the eye, enabling a more detailed view of retinal morphology; the disadvantage of OCT is its cost and time to capture images [5]. The methods described herein focus on the use of images collected from such devices.

Health information and data standards are far from uniform, resulting in widespread variation in usability and quality, thus impeding the development of meaningful health information systems. Patient care settings are typically overburdened by the demands of the health system, resulting in a higher risk for diagnostic mistakes that will further encumber practise and degrade the quality of care. These variations in information standards make it incredibly difficult to deliver effective care, especially in remote settings and public health surveillance [6]. This lack of health information standardization is most commonly understood to be within medical records; however, it is much worse with regard to retinal fundus images. Image quality varies due to camera devices, lighting conditions, clinician experience, and patient ethnicity, all in addition to photoparameters such as aperture, shutter speed, etc. All the inconsistency and variability in data creates an enormous feature space that must be generalized by algorithms driving diagnostic support systems, if one wants to improve the quality of care. The miniaturization of fundus cameras has lead to smartphone-based ophthalmology [7], enabling the adoption of computer-aided diagnostic (CAD) systems in remote communities and developing world settings; miniaturization and smartphone use only exaggerate the problems in data collection and consistency experiences in clinic.

Machine learning and pattern recognition techniques drive CAD systems and research-based informatics; however, traditional methods are unable to generalize information disparities in text-based data within medicine, nor are they able to efficiently process medical image data. These traditional methods must be carefully designed for specific problems and data sets, prohibiting reuse and commercialization. Computational methods for image analysis are divided into supervised and unsupervised techniques. Prior to deep learning, supervised methods encompass pattern recognition algorithms, such as k-nearest neighbours, decision trees, and support vector machines (SVMs). Unsupervised methods utilize a series of rule-based processes, such as filters, gradients, and thresholds, whereas supervised methods use pixel-based feature maps as training materials for an algorithmic classifier.

Deep learning methods are powerful supervised and unsupervised techniques that generate abstractions from the combination of multiple nonlinear transformations on a data set [8]. They are a type of representation learning method that is able to automatically derive the representations necessary for processing and relating disparate raw data [9]. Prior to the advent of heterogeneous, graphic processing unit (GPU) based-processing in 2011, the implementation of deep neural networks (DNNs) on large data sets required months to years to train [10]. Currently, GPU implementation is 50 times faster than its CPU equivalent, enabling faster training on larger data sets, resulting in more accurate classifiers [11].

Applications of these techniques for object recognition in images first appeared in 2006 during the Mixed National Institute of Standards and Technology database (MNIST) digit image classification problem, of which convolutional neural networks (CNNs) currently hold the highest accuracy [12]. In the case of retinal image analysis, the algorithms utilize a binary system, learning to differentiate morphologies based on performance masks manually delineated from the images. There are a number of different unsupervised methods that detect disease pathologies and image quality through pixel-based point sampling; mathematical extrapolation based on *a priori* morphological structures; and other techniques combining linear approximations, snakes, and region growing. The current limitation with most unsupervised methods is that they utilize a set of predefined linear kernels to convolve the images or templates that are sensitive to variations in image quality and fundus morphologies [13]. Deep learning approaches overcome these limitations and have been shown to outperform shallow methods for screening and other tasks in diagnostic retinopathy [9,14].

17.1.1 Chapter Overview

The content throughout this chapter draws parallels between the biological and computational elements being discussed, weaving together applicable concepts in a meaningful way. The reader should be able to come away from this chapter understanding the deep learning methodology and their applications with an appreciation of the biological counterparts from which the computational systems have been derived. While there have been many innovations in retinal imaging in the last 20 years, this chapter focuses on *fundus images*, which are retinal images captured utilizing a fundus camera—a widely used method in clinical optometry and ophthalmology. As such, *retinal images* will always refer to *fundus images* herein.

This chapter can be largely divided into four parts: A primer on the human visual system and deep algorithms; major diseases of the eye; CAD; and deep learning applications in retinal image analysis. The chapter begins by first providing context for deep learning in retinal analysis by describing the human visual system in parallel with DNNs, before providing a fundamental overview of convolutional and recurrent neural networks (RNNs). The next section provides a brief overview of diabetic retinopathy and glaucoma as examples of diseases of the eye. The third section on CAD methods sets the stage for deep methods, describing how systems are designed and how they are evaluated through the use of several performance measures.

Since there are many volumes on the CAD technology as well as the anatomy and diseases of the eye, these sections are meant only as a brief introduction in order that a new entrant may better appreciate the applications of deep networks in retinal image analysis that are presented in the final part of this chapter. There are three accompanying performance sections, which delve into the efficacy of traditional and deep methods. The first performance section investigates methods for vessel segmentation as a real-world example for pixel-based classification. The second performance section investigates methods for MA detection as a real-world example of region classification and pathology detection. The final performance section investigates methods for diabetic retinopathy detection as a real-world example of image classification and CAD systems.

17.2 What Is a Neuron, Really?

Throughout this chapter, two varieties of *neurons* are discussed: the biological and those that exist within the principle algorithms involved with deep learning—neural networks. While our focus will be on the latter, biological neurons have played a critical role in the development of artificial neural networks. The study of the mammalian visual system conducted in Stephen Kuffler's lab at Johns Hopkins and later at Harvard

University by David Hubel and Torsten Wiesel and colleagues revealed how neurons in the visual cortex process images. The study found that neurons interpret patterns as they relate to areas of the receptive fields within the retina, characterizing how the brain handles such information [15,16,17]. For brevity, only a simplified version of a visual system that relates to neural networks and image processing will be considered. The remainder of this section is divided into three sections discussing visual inputs, feedforward networks, and feedback networks. For more detailed information on visual systems and neuroscience, refer to Ref. [18].

17.2.1 Visual Inputs

One way to think about the retina is as an automated light sensor, responsible for detecting, selecting, and communicating this information to the brain for inference and internal representation [19]. The primary role of the retina is to efficiently transmit data to the brain; but what *is* efficient?

A typical adult human retina is made up of more than 120 million photoreceptors, which consist of two types—cones (20 million) and rods (100 million)—which are responsible for converting light into biological (electrochemical) signals through a process called photo-transduction [19]. Rather than transmitting all of this information to the brain, the retina preprocesses the information, selecting only salient features to be passed through to the primary visual cortex (V1). Saliency refers to variations in light intensity within a given scene, such as edges or borders, changes in colour, brightness or contrast, and spatiotemporal changes. From a computational perceptive, salient features can be extracted from images through the use of various edge detectors, segmentation methods, and convolution operations. Essentially, the retina only transmits changes in light level, position, etc. to the rest of the visual system; if there is no change, the system adapts prior information. This filtering mechanism is hard-wired into the retina and ensures static information is not redundantly processed while changes in the environment are rapidly assessed.

The macula is responsible for fine central vision and refers to the central region of the retina. The area outside of this region is referred to as the peripheral retina. Within the centre of the macula resides the fovea, a small depressed segment of the retina with the greatest visual acuity—a measure of spatial resolution. The distribution of photoreceptors and ganglion cells within the retina is uneven, with the highest density within the fovea, tapering off roughly in proportion to distance. There is essentially a one-to-one synaptic connection between photoreceptors in the fovea and the ganglion cells, whereas in the periphery, many photoreceptors multiplex into a single ganglion cell. The fovea limits primate eyes' area of focus, requiring the eye be in constant motion to acquire data so that the brain can reconstruct the environment. This process is driven by subconscious micromovements called saccades, tremors, and drifts, which enable continuous generation of object saliency within a scene and are required to prevent neural adaptation [20]. Interestingly, primates are the only mammal with a fovea, though other species possess different variations of the fovea, such as birds and reptiles [19].

While its significance can be argued, the fovea serves to increase saliency, thereby optimizing the quality of information delivered to the V1. This process is known as *foveation* and has been incorporated in image processing techniques to increase computational efficiency, as shown in Section 17.6.2. The same principles have been applied to deep learning in the development of "attention mechanisms," which are actively being researched [21]. The retina must be supplied with an ample blood supply, resulting in the vasculature growing throughout the space around the fovea. Vessels are only able to enter the eye and access the retina via the optic nerve head (ONH), with vessel diameter and concentration decreasing further from the ONH. Figure 17.1 illustrates the structure and basic function of the eye as well as the fundus, including examples of two prominent disease indicators: microaneurysms and exudates. Figure 17.2 provides a complementary OCT cross-section of the retina.

The visual cortex (V1) is the main area of the brain responsible for processing visual inputs received by the retina. Specifically, light stimulates the retina, which transmits the information via electrical impulse through the optic nerve to the lateral geniculate nucleus (LGN), and then to the V1. Here, we consider the retina and V1 to be the main areas of interest, with everything in between simplified to a single data pipeline. The V1 is integrated with other areas of the brain including higher brain areas such as V2, V3, middle temporal Gyrus, etc., which are responsible for memory, association with other senses, and additional transformations/processes via a complex network of layered neurons. For our purposes, we consider the inferotemporal cortex (IT) to be the brain region responsible for memory as it bares the closest resemblance to the decision layers within neural networks, as shown in Figure 17.3. Studies into the function of simple cortical cells in the V1 uncovered a curious duality in their operation, which could be modelled mathematically. The cells' response to visual stimuli is able to be described as a simultaneous combination of spatial domain and spatial-frequency domain representations; this is a fundamental property for many variations of Gaussian kernel functions and most notably Gabor functions, due proximity of function with established V1 models [22].

Deep Learning for Retinal Analysis

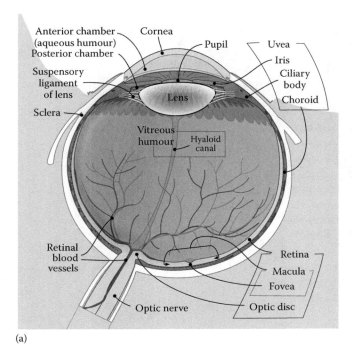

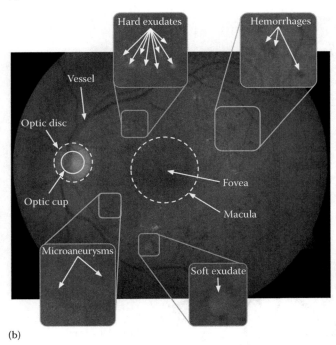

FIGURE 17.1
Key concepts of human vision and retinal image analysis. (a) The basic structure of the human eye. (b) Sample fundus image with the main morphology and common pathologies labelled.

17.2.2 Deep Feedforward Networks

Feedforward networks, also called multilayer perceptrons (MLPs), are the fundamental deep learning model, utilizing a series of neural network layers to transform input data before passing the information through to the final output layer—commonly referred to as the decision or classification layer. MLPs are architected modularly with a series of layers selected to address different classification problems. A layer comprises an input, output, size (number of neurons), and varying set of operating parameters/hyperparameters.

Historically, neural networks are fully connected, meaning all neurons in one layer pass data to all neurons of the next. Unsurprisingly, this has high computational cost that is exacerbated when dealing with high-dimensional data, such as images. These feedforward architectures are the underpinnings for CNNs, an example of which is shown in Figure 17.4. Keep in mind, this is a simplified representation that does not include feedback mechanisms present in biological systems. The inclusion of feedback mechanisms and memory further increases the complexity of neural networks and has resulted in various feedback architectures, typically categorized as RNNs [9].

17.2.2.1 Applying What We Know About the V1

Here is where we begin to draw on the beauty of primate visual systems! The visual system consists of multiple interconnected layers of neurons that sequentially process vision inputs from the retina. Within the V1, neuronal activations resemble the spatial orientation of activated photoreceptors within the retina. This behaviour of sharing spatial parameters between visual system structures dates back to early connectionist models of vision by Marr and has been crucial for optimizing deep networks [23,24]. Similarly, the features of convolutional networks are represented by multi-dimensional arrays resembling the spatial properties of the input or output of the prior layer.

The retina utilizes multiple filters to simultaneously extract numerous features from a scene before streaming the information in parallel to the brain, as illustrated in Figure 17.5 [25,26]. These behaviours implicitly describe how information is preprocessed and passed between layers within a CNN. Given that CNNs are loosely based on the primate visual system, they are unsurprisingly well suited for image analysis and other grid-based information structures. Typically, CNNs comprise 5 to 25 distinct layers where at least 1 layer is a convolutional layer, as seen in Figure 17.4. Following all convolutional operations, the network converges into a decision algorithm, most commonly a fully connected artificial neural network with one or more layers [27]. Alternatively, most classification or regression algorithms may be used at the decision stage, such as SVMs, decision trees, or even other DNNs! The most common layers of a CNN can be described as follows:

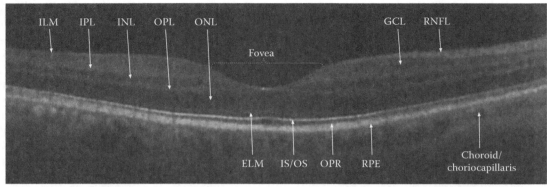

ILM: Inner limiting membrane
IPL: Inner plexiform layer
INL: Inner nuclear layer
OPL: Outer plexiform layer
ONL: Outer nuclear layer
ELM: External limiting membrane
IS/OS: Junction of inner and outer photoreceptor segments
OPR: Outer segment PR/RPE complex
RNFL: Retinal nerve fibre layer
GCL: Ganglion cell layer
RPE: Retinal pigment epithelium

FIGURE 17.2
Cross-sectional OCT image of the human retina.

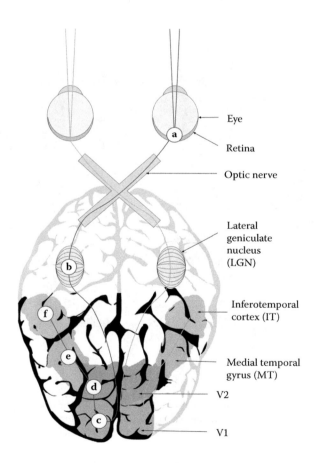

(a) The retina converts an input image into electrochemical signals and sends them to the brain.

(b) The LGN processes the impulses from the eye on the opposite side via the optic nerve and is made up of six neuronal layers.

(c) The visual cortex receives the signals from the LGN. The V1 extracts features from the data and sends them to the V2.

(d) The V2 and other higher layers of the visual cortex further abstract and interpret the image data.

(e) The MT is an area thought to interpret the highly abstracted visual data from the visual cortex.

(f) The IT is believed to be responsible for memory and final categorization of visual stimuli.

FIGURE 17.3
Simplified model of the human visual pathway. The retina (a) assimilates visual data and sends it over the optic nerve in multiple parallel streams to the LGN (b) on the opposite side of the brain from the eye. (c–f) There, streams are processed and forwarded to a series of visual and temporal cortices for interpretation.

Deep Learning for Retinal Analysis

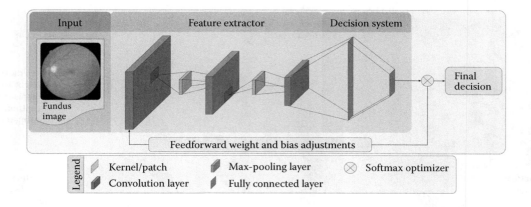

FIGURE 17.4
A simple convolutional neural network. A fundus image is passed through three convolutions, two max pooling operations, two fully connected layers, and a softmax optimizer. The optimizer is only used during training, resulting in fixed weights and bias at test time. The final decision is case specific and kept general in this illustration.

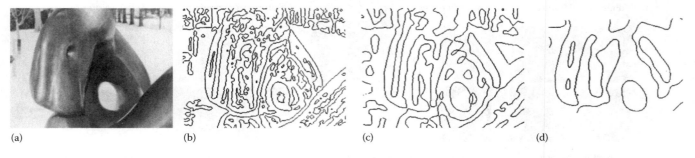

FIGURE 17.5
The neuroscience analog of the Gabor filter is the receptive field description of the centre-surround organization of retinal and cortical cells (see Ref. [25] for example). The original image (a) is filtered at three scales that approximately mimic the human fovea (b–d). Adapted from D. Marr and E. Hildreth, Theory of Edge Detection, *Proceedings of the Royal Society of London B: Biological Sciences*, 207(1167):187–217, 1980.

i. *Convolutional layer (CL)*

Convolution layers extract features from a sliding window on the previous layer. The first convolution of the input image extracts low-level features, such as edges, while subsequent convolutions extract higher level features, as illustrated in Figure 17.6. Each input window is convolved using multiple kernels; the results from each are passed through a nonlinear operation before reaching to the next layer. Nonlinear operations are essential for CNN operation as they are the main method for processing the features. Rectified linear units are the most common choice due to their ease of use and efficiency, best described by the simple maximization function they implement: $y = max(x, 0)$.

ii. *Pooling layer (PL)*

Pooling layers are used to reduce noise and distortion by reducing the resolution of the input. The most common pooling methods are average pooling and max pooling, the latter being more effective when processing fundus images [11].

iii. *Fully connected layer (FC)*

Fully connected layers are classical neural networks and are more computationally taxing than their sparse convolutional counterparts. They are typically used within the final layers of a CNN and take all the features from the prior layer as an input. These layers can be used to merge the outputs of separate CNNs.

A key concept gleaned from the neuronal structure of the primary cortex is *sparsity*—the frequency by which neurons physically connect with (and share information with) other neurons. Sparsely connected neurons are a defining property of the convolutional layers within a CNN, whereby neurons are not connected to every neuron in subsequent layers (as is the case for MLPs and traditional artificial neural networks utilizing only fully

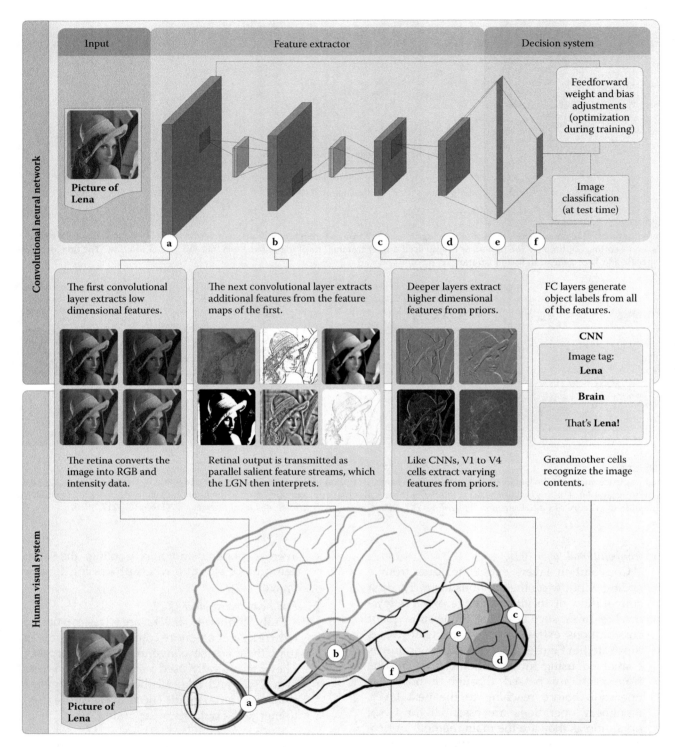

FIGURE 17.6
Comparing the human visual system with a simple CNN. (a): The first convolutional layer extracts low-dimensional features, similar to how the retina converts the image into red green blue (RGB), intensity, and saliency data. (b–d): The LGN and visual cortices extract features from streams in a similar manner to deeper convolutional layers in the DNN. (e–f): The FC layers compress features into tag probabilities, returning a vector and selecting the highest as the likely object, similar to how memory cells activate in the brain.

connected layers). Further, the cells within the V1 have varying roles with regard to information processing, ranging in operational complexity.

Simple processes can be represented mathematically as linear operations on spatial features. These operations have been applied as activation functions within convolutional and fully connected layers. The information output from the cells in the V1 have been well approximated by Gabor filters, which are further described in Section 17.2.2 [28]. More complex processes tend to be invariant to changes in image features, such as orientation and position. These aspects have been modelled as a wide variety of pooling and dropout layers, further reducing the information burden for increasingly deep networks.

Compared to our simplified model of the primate visual system, fully connected layers mimic the so-called grandmother cells, which are believed to activate only when exposed to a specific object, person, or stimulus, regardless to how it is transformed, cropped, or illuminated [17]. This same behaviour is represented within deep layers of a network and relates to performance measures when comparing networks of varying depth and complexity [21]. Figure 17.6 relates the concepts discussed here, using the features and foundations for CNNs and the human visual system illustrated in Figures 17.4 and 17.3.

17.2.2.2 The Gabor Function

Gabor functions are linear filter functions most commonly used for edge detection. They are named after Dennis Gabor, who received a Nobel Prize in Physics for inventing holography in 1971. When utilized in 2-D, Gabor functions are referred to as *kernels* and operate by modulating frequency and spatial orientation parameters within a set of functions. In the case of an image $I(x, y)$, the kernel is centred on pixel x, y within a cell defined by a set of coordinates $(\mathbb{X}, \mathbb{Y})$ and location-dependent weights defining the properties of the kernel. From this, the impulse response of the kernel on the image can be represented by the convolution of the Gabor kernel and the image

$$s(I) = \sum_{x \in \mathbb{X}} \sum_{y \in \mathbb{Y}} g(x, y) \otimes I(x, y)$$

The Gabor function can be further expressed as

$$g(x, y; \alpha, \beta_x, \beta_y, f, \phi, x_0, y_0, \tau)$$
$$= \alpha \exp\left(-\beta_x x'^2 - \beta_y y'^2\right) \cos(fx' + \phi)$$

where

$$x' = (x - x_0)\cos(\tau) + (y - y_0)\sin(\tau)$$
$$y' = -(x - x_0)\sin(\tau) + (y - y_0)\cos(\tau)$$

The Gaussian element $\alpha \exp(-\beta_x x'^2 - \beta_y y'^2)$ restricts the range of the filter to only values close to the kernel's receptive field, denoted by x' and y'. The scaling factor α controls the effect of the kernel, while β_x and β_y control rate of decay for the influence of the kernel. The cosine component $\cos(fx' + \phi)$ defines the kernel's impulse response to brightness along the x' axis of an image—f controls the consine frequency and ϕ its phase offset. In essence, this function enhances edges by manipulating the brightness on small scales, amplifying bright regions to be brighter and dark regions darker. The design of the kernel ensures only lines, and the immediate neighbouring pixels are enhanced [29–31].

17.2.3 Recurrent Neural Networks

One limitation of CNNs is the inability to efficiently process large sets of sequential data. While we have drawn parallels to neurons and some concepts of recognition that relate to memory, we have yet to broach that specific property of biological neurons, such as those in the temporal lobes and cortex. *Recurrent neural networks* (RNNs) are neural networks that specialize in processing sequential data [32]. The main difference between CNNs and RNNs resides in how they implement and share parameters to handle differences in a sequence such as time; a sentence is another example of a sequence. To exemplify the difference between the two, if we use a CNN to classify a key term in a sentence, it would have to be trained against all word orders—it attempts to classify the sentence as a single string, even though the sentence is a sequence of words. An RNN is able to look at each word, even if it has been trained on sentences—this is the property CNNs do not possess. RNNs have recurrent connections between hidden units and are able to read an entire sequence before producing a single output.

17.3 Major Diseases of the Retina

Modern medicine has provided patients with diabetes a greatly increased life span. Like with any chronic condition, the longer one is afflicted, the higher the risk of complications implicit to that condition or comorbidities. For most retinal conditions, early diagnosis and prevention are often the most effective method for reducing patient risk through modifications to lifestyle, medication, and acute monitoring [3]. For example, abnormalities of retinal vessels have been correlated with coronary heart disease, stroke, and cardiovascular mortality [33]. Similarly, the same information, this time gleaned from youth, can be used as indicators in the prediction of those individuals' health later in life [34]. A good example is diabetic retinopathy (DR), which is one of the leading causes for blindness, occuring as a complication of diabetes. The symptoms for DR are changes in retinal vessel morphology such as vessel thickness, leaks, and rapid proliferation. Another leading cause for blindness is that of glaucoma [35]. During the early stages of glaucoma, patients will hardly notice any changes to their vision. Not until the conditions becomes severe and they begin to go irreparably blind are patients able to recognize the symptoms from their vision.

For the purpose of this chapter, the diseases that we will focus on are DR and glaucoma. This choice was made due to the similarity between techniques for identifying glaucoma and DR and those for age-related macular degeneration. Refer to Ref. [5] for more details into the various diseases of the eye and respective imaging modalities.

17.3.1 Diabetic Retinopathy

Diabetic retinopathy (DR) is the most common complication for diabetic patients [36] and a leading cause of blindness globally. DR is characterized by abnormal fractal growth within the vascularization of the retina known as proliferative DR (PDR), accompanied by retinal lesions that may be associated with PDR as well as nonproliferative DR (NPDR) [37]. The disorder causes angiogenesis of aberrant venous tissues and is assessed by the appearance of lesions and neovascularization within the fundus. DR lesions are summarized in Table 17.1 [38]. More than half of those suffering from long-term diabetes mellitus develop such vascular traits [39]. Treatments for DR are highly effective; however, accessibility is the limiting factor with 60–90% coverage in developed countries and significantly lower rates in developing countries [35]. The most common method for assessing the progression of DR is by comparing the morphology of vessel branches, such as the difference in size before and after a vessel branches from within a fundus image; this is shown alongside other DR pathologies in Figure 17.7 [40]. More details on DR pathologies are discussed herein.

17.3.1.1 Proliferative Diabetic Retinopathy

Proliferative diabetic retinopathy (PDR) specifically refers to DR cases with a noticeable increase in venous tissue—known as neovascularization—which is often accompanied by hemorrhages and lesions. Lesions are due to changes in retinal pathology causing an increase in capillary permeability and capillary closures, resulting in fluid accumulation within the retina and cell dealth due to the cessation of axoplasmic flow in the nerve fibre layer. Venous tissue proliferation is accompanied by fibrous

TABLE 17.1

Types of Diabetic Retinopathic Lesions

Microaneurysms
Dot-hemorrhages
Hard exudates
Blotchy hemorrhages
Intraretinal microvascular abnormalities
Venous bleeding
Cotton wool spots
New vessels on or within one disc diameter of the disc
New vessels elsewhere
Fibrous proliferation
Preretinal hemorrhage
Vitreous hemorrhage
Traction retinal detachment
Rhegmatogenous retinal detachment
Macular retinal thickening (edema)
Clinically significant macular edema

Source: Modified from H. Jelinek and M. J. Cree, *Automated Image Detection of Retinal Pathology*, CRC Press, 2009, ISBN 978-0-8493-7556-9, doi: 10.1201/9781420037005, URL http://dx.doi.org/10.1201/9781420037005.

tissue growth with an increasing proportion of fibrous to venous tissue over time. In many cases, the vascular tissue regresses, leaving a buildup of fibrous tissue that exerts mechanical stresses on the retina. These stresses may cause a variety of pathological changes in the retina, most notably edema, striations, and retinal detachment.

17.3.1.2 Microaneurysms

Microaneurysms (MAs) (saccular outpouchings of the capillary wall) are often the earliest indicator of DR [41], enabling detection of DR before morphological dimensions of the branches become noticeable. MAs cause the capillaries in the retina to swell, potentially bursting (at which point they are classified as hemorrhages), resulting in the appearance of red dots in fundus images. MAs are a major source of retinal bleeding and capillary permeability and, for the purposes of clinical diagnosis, often also refer to resulting hemorrhages of equivalent size. Any bleeding in the retina increases risk for complications, such as the development of exudates or severe hemorrhages. MAs develop continuously, making them good indicators for specific retinal regions at risk of complication. The number of MAs is a good indicator for the severity of DR during the early stages of the disease [42].

17.3.1.3 Diabetic Macular Edema

When retinopathic edema develop over or close to the macula, it is referred to as diabetic macular edema (DME). However, DMEs are not the only macular abnormalities

Deep Learning for Retinal Analysis

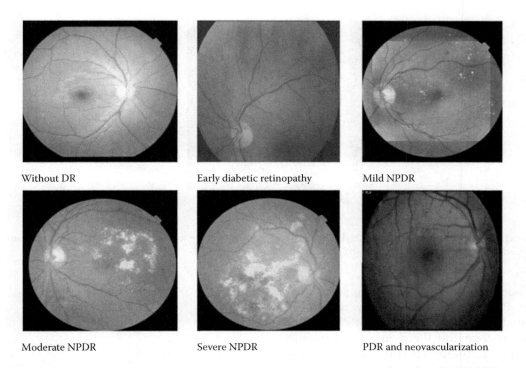

FIGURE 17.7
Fundus images of normal, mild NPDR, moderate NPDR, severe NPDR, PDR, and PDR with neovascularization. Modified from M. F. El-Bab, N. Shawky, A. Al-Sisi, and M. Akhtar, Retinopathy and Risk Factors in Diabetic Patients from al-Madinah al-Munawarah in the Kingdom of Saudi Arabia, *Clinical Ophthalmology (Auckland, N.Z.)*, 6:269–276, 2012, doi: 10.2147/OPTH.S27363, URL http://www.ncbi.nlm.nih.gov/pmc/articles/PMC3284208/.

associated with DR but often are used as an umbrella term for all such pathologies. These macular pathologies include capillary nonperfusion, intraretinal hemorrhage, preretinal hemorrhage, macular surface traction, and macular holes [38]. The techniques used to segment or classify the morphological markers present in the macula with DR could also be applied to similar conditions, such as age-related macular degeneration.

17.3.2 Glaucoma

Glaucoma is the second leading cause of blindness, characterized by long-term degeneration of the optic nerve. Primary open-angle glaucoma (POAG) and angle-closure glaucoma (ACG) are the two most prominent types of glaucoma. POAG is the most common form of glaucoma, accounting for at least 90% of all glaucoma cases [43]. The progression of glaucoma is measured by alterations in the ONH, neuroretinal rim, and vasculature structure. From these structures, four characteristic measures are used by ophthalmologists and medical professionals, and have been incorporated into retinal image analytics: ONH variance, neuroretinal rim loss, defects in the retinal nerve fibre layer, and peripapillary atrophy.

17.3.2.1 Optic Nerve Head

The ONH is the location the optic nerve that enters the back of the eye, the bright circle/ellipse in Figure 17.8a. Since the ONH is essentially nerve fibres and vasculature, it lacks photoreceptors and is referred to as the blind spot. The cup-to-disc ratio (CDR) is a major and common technique for measuring the progression of glaucoma [44]. As a patient's condition worsens, the optic nerve fibres disappear, causing the optic cup to stretch while the optic disc deforms only in the slightest, resulting in an increased CDR. Ophthalmologists and optometrists regularly measure the CDR subjectively, though such measurements are very variable, even between experts of equal experience [45]. In healthy eyes, the neuroretinal rim obeys the "INST rule." The INST rule refers to size of the rim, in particular the relative sizes of the inferior, nasal, superior, and temporal quadrants, as shown in Figure 17.8b. INST can be used as an indicator of glaucoma but is not a consistent measure between patients [46].

17.3.2.2 Peripapillary Atrophy

Peripapillary atrophy (PPA) is the degeneration of the retinal pigment epithelial layer, photoreceptors, and

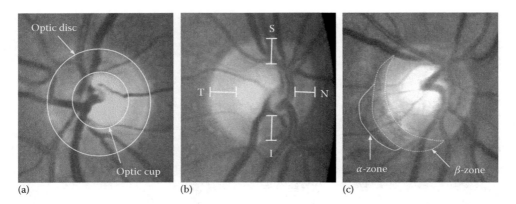

FIGURE 17.8
Elements of the ONH and retinal nerve fibre layer (RNFL). (a) Fundus image illustrating the optic disc and optic cup. (b) Illustration of the INST rule. (c) An ONH with peripapillary atrophy, showing the α-zone and β-zone.

underlying choriocapillaris in the region surrounding the ONH [46]. PPA can occur for various reasons and, if left untreated, results in glaucoma. PPA is subdivided into the α-zone and the β-zone. The α-zone is indicative of the chorioretinal tissue layer thinning and characterized by hyperpigmentation, while the β-zone is characterized by hypopigmentation; the various retinal layers can be seen in Figure 17.2. The β-zone is often larger and its appearance more frequent/obvious than the α-zone, as shown in Figure 17.8c [47].

17.3.2.3 Retinal Nerve Fibre Layer

Defects found in the retinal nerve fibre layer (RNFL) are the earliest sign of glaucoma [48]. Defects cause the disappearance of RNFL striations from the temporal inferior region of the fundus, which can be detected in the green channel of fundus images. When considering defects caused by glaucoma, there are two types:

i. *Localized RNFL defects* extending from the optic disc, which result in structural deformations that appear as dark regions within fundus images, cross-sectional illustrations of which are shown in Figure 17.9 [49] and fundus images in Figure 17.10c

ii. *Diffuse RNFL defects* caused by ganglion cell death, as shown in Figure 17.10 [50].

17.4 Computer-Aided Diagnostics

Computer vision/analysis provides a robust alternative to direct ophthalmology by a medical specialist, providing opportunities for more comprehensive analysis through techniques such as batch image analysis [13].

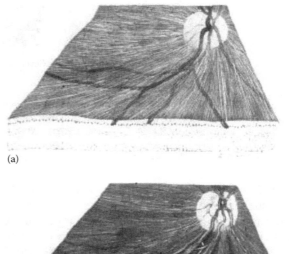

FIGURE 17.9
Healthy and defective retinal nerve fibre layers. (a) Cross-sectional view of normal RNFL. (b) Cross-sectional view of RNFL defect. Adapted from Y. Fengshou, *Extraction of Features from Fundus Images for Glaucoma Assessment*, PhD thesis, National University of Singapore, 2011.

As such, much research has gone into automatically measuring retinal morphology, traditionally utilizing images captured via fundus cameras. Accurate fundus and/or retinal vessel maps give rise to longitudinal studies able to utilize multimodal image registration and disease/condition status measurements, as well as applications in surgery preparation and biometrics [51]. The impacts on clinical efficiency can be substantial but will not be covered in this chapter [52].

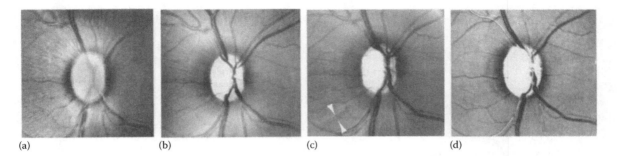

(a) (b) (c) (d)

FIGURE 17.10
Progressive loss of the RNFL. (a) Healthy fundus. (b) Early-stage RNFL deformation and atrophy. (c) Severe RNFL atrophy, indicated by loss of all striations. Arrows mark a dark band indicative of localized RNFL defects. (d) Complete atrophy and loss of nerve fibres. Adapted from H. A. Quigley, R. M. Hohman, E. M. Addicks, and W. R. Green, Blood Vessels of the Glaucomatous Optic Disc in Experimental Primate and Human Eyes, *Investigative Ophthalmology & Visual Science*, 25(8):918–931, 1984.

Computer-aided diagnostics (CAD) can be broken down into two main groups, *computer-aided detection* (CADe) and *computer-aided diagnosis* (CADx), based upon what information the system provides. CADe refers to a software tool that assists an end user in identifying regions of interest (RoIs) within an image. In case of retinal images, these would be in identifying exudates or MAs, or in assessing the size of vessels or the ONH. For CADe, this information is relayed to the user in a manner that facilitates an expert's diagnosis, whereas CADx systems further analyze this information before relaying a diagnostic prediction to the end user. This distinction may seem subtle but is a critical consideration when planning to commercialize research or develop a software product. Simply, regulatory bodies such as the FDA have clear medical gradings for different types of hardware, wet-ware, and software [53]. At the time of writing, CADe systems are at least one "medical device grade" lower than CADx in all first-world settings. This means that the barriers to commercialization, time to market, and penalty for error are far stricter for CADx than CADe, requiring substantially greater fidelity, accuracy, and usability.

The main challenge when processing retinal images is the presence of distortions, such as retinal capillaries, underlying choroidal vessels, and reflection artifacts that may be confused with disease pathologies, such as DR lesions. Distortions are often the result of human elements such as movement causing blur, as well ocular factors like peripheral optical aberrations, cataracts, intraocular scattering, and retinal scarring [54]. Broadly speaking, distortions can be compensated for algorithmically. Only in exceptional cases of poor image quality are images rendered unusable for further analysis. Ideally, CAD systems are able to distinguish healthy from disease states on a morphological level (CADe) or patient level (CADx), measure the progression of a disease between clinical visits, and even predict the progression. Each of these requires a different set of algorithms and interface design; however, there are two approaches for developing such systems, referred to as *mimicry* and *feature discovery*.

17.4.1 Mimicking Experts

Mimicry refers to using clinical standards as the basis for algorithm design, such that the system replicates the approach utilized by clinicians in assessing the retinal image. In line with the applications discussed at the end of this chapter, such a system may evaluate an image for the presence and quantity of MAs or hemorrhages in assessing the presence/absence or severity of DR. Mimicry uses the same methodology as a clinician would, making the results much more digestable for end uses, since they will easily understand how the system arrived at its conclusion. The process is illustrated in Figure 17.11, where a series of fundus images are preprocessed and then analyzed by several feature and disease detection algorithms.

Though it is not explicitly covered within this chapter, many of the same methods used to assess image quality are able to be used in cataract detection and analysis when image quality is known [55].

Success of these techniques depends heavily on optimization carried out against manually delineated images where at least one clinician must go through a training set of images and colour the various regions/morphologies by hand. This process is incredibly time consuming as each type of morphology requires its own delineation and prohibits reuse of the system for analyzing different types of data. Without the use of deep learning, traditional algorithms do not reliably generalize across retinal image data sets, meaning that the algorithms would require vast amounts of training data to compensate for real-world variations in image quality, luminance, and contrast; this is an unrealistic problem to address!

In many cases—even in preliminary evaluations of CDR—experts seldom agree with one another, leading

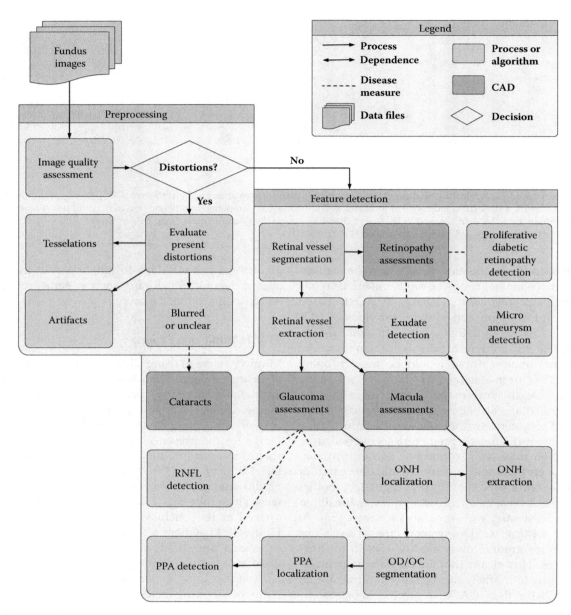

FIGURE 17.11
Summary flow diagram of traditional retinal health assessment algorithm structures. Fundus images are passed into a series of preprocessing steps and then feature detection processes. First, the image quality is evaluated, and the images are routed accordingly. If the distortions can be removed and the quality corrected, the images are passed to the suite of *feature detection* algorithms. A unique case is cataracts, where the quality assessment can be indicative of disease severity. Otherwise, a series of algorithms are often necessary in the evaluation of various retinal diseases.

to a bias in the training data. In one example, six experts were each tasked with creating a set of manually delineated retinal images, which, when compared, found the experts only in agreement on 60–75% of cases [45]. When training an algorithm against all of this data, even if it is using unsupervised learning methods, accuracy becomes less about the actual image in question, but how much the algorithm agrees with the clinicians. A simple way of understanding this phenomenon is that the algorithms will be understandable by clinicians but also imprinted with human bias, resulting in a system that repeats the clinical experts' mistakes.

17.4.2 Feature Discovery

The feature discovery approach eliminates human bias from the system, whereby diagnostic methods used in clinical practise are ignored, forcing the system to learn the features that correlate the condition labels (i.e., healthy vs. disease present) with the retinal images/medical data.

This approach is traditionally presented as an intractable problem, requiring far too much time and computational power to reasonably train a system. CNNs not only solve this problem, but outperform traditional methods at test time. Since the system learns features autonomously, these techniques are able to discover new features, even to the end user. The discovery approach is still limited to the data set; however, it is able to be reapplied across modalities and against different problems, such as measuring the severity of a disease versus detecting multiple diseases. Further, such systems can be used in situations or for modalities outside preexisting clinical expertise. If the final interface is able to share these discoveries with clinicians in a meaningful way, clinicians will learn to recognize these features and plan accordingly, improving their practise and leading to a higher quality of care.

As with all deep learning techniques, data variation becomes the limiting factor when building a system able to generalize across differing data sets. As such, a drawback for using feature discovery is the lack of transparency in how it selects key decision criteria during the training process. These features can be explored following the training process; however, there is always a need to validate the system against variations in data sets to assure confidence in the results. Another obscurity-related limitation is how the results of such a system can be used to measure disease progress (or forecast it) in a meaningful way. When using mimicry, it is much more straightforward to measure the dimensions and morphological changes in fundus images. This is not to say it is not possible, only that it is less obvious. It is worth noting that these approaches are very complementary, able to appropriately address each others' strengths and weaknesses.

17.4.3 Traditional CAD of Retinal Images

The analysis of retinal images has traditionally followed a mimicry approach, beginning with a process to normalize the properties of training images before selectively extracting and analyzing retinal features independently, as illustrated in Figure 17.14 and Figure 17.16. Many of the traditional methods first discard image channels that are not relevant to the objective of a particular task, utilizing a monochromatic image for feature extraction; there have been a number of investigations into the value of each channel. Most healthy retinal tissues do not readily absorb red light, with the exception of the ONH, making it a useful channel for exudate and ONH detection, whereas it is relatively unhelpful for edema, MA, and vessel-related methods, which benefit most from the green channel. The green channel provides the clearest features for detecting vessels and MA, due to hemoglobin's absorption pattern, which peaks for green light. Violet and blue light is highly scattered and partially absorbed before the retina, causing the blue channels to blur [56].

Variations in melanin concentration due to patient ethnicity have been shown to alter light absorption in retinal tissue. Traditional methods cannot readily generalize for these variances, making monochromatic methods less effective for commercial CAD systems without significant image preprocessing; patient subgroups would be subject to variational inaccuracies requiring the clinician to conduct a manual investigation regardless [57]. For example, it has been shown that the blue and red colour channels have beneficial features for MA detection; however, when the method was applied against a testing data set, it falsely detected MA due to features within the test set it was not trained on; this inability to generalize is the greatest limitation for traditional methods [58]. Different morphologies, such as the ONH, macula, and exudates require careful segmentation prior to independent analysis for disease classification. Ethnic differences in tissue colouration result in variances of the colour exudates, lesions, and other disease indicators, making such tasks very challenging on raw images; such pathologies may appear to be dimmer than fundus in colour, resulting in misclassification of both morphologies. For traditional methods, colour normalization is a necessary preprocessing step to match the test image colour distribution with that of the training set.

17.4.4 Image Processing Fundamentals

This section highlights key image processing techniques often used during preprocessing and image analysis, some examples of which may be found in Figure 17.14. Gaussian filter functions, like those discussed in Section 17.2.2, are a prominent underpinning of many image normalization and edge detection techniques, such as image blurring or Canny edge detectors. There are a myriad of preprocessing techniques that can be used in many different sequences, including colour normalization, deblurring/contrast enhancement, edge detection, morphological reconstruction, and phase transformations. The most common methodology begins with single channel normalization through the application of histogram and/or median filters; which channel depends on the morphology of interest. The following steps usually focus on enhancing the image contrast with a combination of deblurring and adaptive edge enhancement techniques. For more information on retinal image normalization, please refer to Ref. [59].

17.4.4.1 Global Histogram Enhancement

Global histogram enhancement is a method for adjusting the pixel values within an image based on all the pixel intensities within an image. Colour histograms are used throughout image processing as they play a large role in image enhancement and object recognition. Global histogram methods for contrast enhancement consider the value of all pixels within an image and then redistribute the overall intensity throughout the entire image to normalize the pixel values. Such methods have limited uses as they can worsen image quality when used in misaligned cases, such as when images have poor colour depth or capture artifacts, such as glare. Fundus images have poor colour depth and often contain capture artifacts. Global methods are useful for thresholding images whereby pixels within or outside specific pixel intensity values are set to a fixed value corresponding to an object. Such methods are used for basic morphological segmentation, such as segmenting the ONH from a fundus image based on its brightness. Unfortunately, while these thresholding methods are computationally cheap, they are subject to the same limitations as the global contrast enhancement techniques, for the same reasons.

17.4.4.2 Local Histogram Enhancement

While more computationally intensive, local histogram enhancement methods are able to improve the enhanced image quality and contrast where global methods fail. Rather than sampling all pixels within an image once, histograms are generated for subsections of the image, each of which is normalized. Windows need to overlap so subsection boundaries are normalized accordingly, resulting in a the spike in computational power. One limitation for local methods is the risk of enhancing noise within the image. Contrast-limited adaptive histogram equalization (CLAHE) is one method that overcomes this limitation. CLAHE limits the maximum pixel intensity peaks within a histogram, redistributing the values across all intensities prior to histogram equalization [60]. This is the contrast enhancement method used during preprocessing in Figure 17.14. More advanced local enhancement techniques exist and often require some calibration for specific problems, such as rank enhancement where pixel values are adjusted based on their relative distance from histogram minima/maxima.

17.4.4.3 Super-Resolution Techniques

Super-resolution techniques are methods for taking low-resolution or low-quality images and enhancing them to high quality through a number of methods including multiresolution techniques and pixel-energy-based quality normalization functions [61]. Applying these methods to retinal images during initial preprocessing or downstream feature extraction steps can greatly improve the the system's ability to detect and properly classify morphology [62]. A review of these techniques in retinal fundus applications can be found in Ref. [63] and nonretinal applications in Ref. [64]. Specific traditional techniques for analyzing different retinal morphologies are covered in subsequent sections alongside their deep learning counterparts.

17.4.4.4 Canny Edge Detectors

Canny edge detectors are a multistep edge detection algorithm, named after John Canny, who developed them in 1986 [65]. The steps are as follows:

1. *Noise reduction*: Noise can have a large impact on edge detector efficiency; Gaussian smoothing is often used to reduce noise within the image.
2. *Ensemble intensity gradient detection*: Canny edge detection methods use four filters with different sigma values within an edge detection operator, combining the results to discern the edge gradients within the image.
3. *Edge thinning*: The results of step 2 are refined with a nonmaximum suppression technique, which erodes the edges, making the stronger results more prominent, while reducing noise.
4. *Thresholding*: A double threshold is applied to select high-intensity pixels and low intensity pixels, with the highest-intensity pixels expected to represent edges. If a pixel's intensity falls between the thresholds, it is classified as a *weak edge*, and if it is less than the low threshold, it is considered to be a nonedge.
5. *Edge refinement*: The final step of the Canny edge detector further evaluates the *weak edges* with a region growing algorithm or similar method. If weak edges are contiguous with a strong edge pixel, the *weak edge* is retained in the final result.

17.4.4.5 Watershed Methods

Another prominent edge detector is the watershed transformation. To Canny edge detectors, the first step typically involves Gaussian smoothing to reduce global noise. Following, these methods utilize a variant form of region growing, most resembling level set methods, which find the gradients of all regions based on intensity maxima or minima, by incrementally lowering or raising (respectively) the intensity threshold. Watershed methods keep track of each region, and rather than combining them when they overlap, as with level set methods, designate the overlapping boundary as an edge. At their simplest, watershed techniques are susceptible to oversampling, leading to the development of semisupervised and hierarchal techniques. The former require the human to label RoIs within an image, and the former keep track of the thresholds from every increment and use subsequent algorithms to decide on the final edges. Region growing, either simple level set or watershed, is used in pathology detection, such as MA, which is described in Section 17.6.1.

17.4.5 Common Data Sets

The most well known methods and data sets for deep learning image analysis are from the MNIST, CIFAR10, and ImageNet competitions. Many of the algorithms used in these competitions are applicable to any other kind of image, including retinal fundus images, albeit with relevant data sets and some architectural adjustments. The most commonly used benchmarking data sets for researching algorithms that analyze the retinal fundus images are: the Digital Retinal Images for Vessel Extraction (DRIVE) database; the Structured Analysis of the Retina (STARE) database; and the Standard Diabetic Retinopathy Database (Diaret). A large clinical data set curated by the UK Biobank has recently become available, offering complete records for more than 500,000 participants. It is not publicly available; however, it offers an unprecedented opportunity in the development of cutting-edge algorithms, CAD systems, and patient programs, and is worth mentioning.*

17.4.5.1 DRIVE

The Digital Retinal Images for Vessel Extraction database is a standardized set of fundus images commonly used to gauge the effectiveness of classification algorithms. The images are 8 bits per red green blue alpha (RGBA) channel with a 565 × 584 pixel resolution. The data set comprises 20 training images with manually delineated performance masks and 20 test images with two sets of manually delineated performance masks by the first and second human observers. The images were collected for a DR screening program in the Netherlands using a Canon CR5 nonmydriatic three charge coupled device (3CCD) camera with a 45-degree field of view [66].[†]

17.4.5.2 STARE

The Structured Analysis of the Retina database has 400 retinal images, which were acquired using the TopCon TRV-50 retinal camera with a 35-degree field of view and pixel resolution of 700 × 605. The database was populated and funded through the US National Institutes of Health [67].*

17.4.5.3 Diaret

The Standard Diabetic Retinopathy Database (Diaret) has two parts, calibration (referred to as DIARETDB0) consisting of 130 colour fundus images, 20 normal and 110 with signs of DR, and the other database is associated with an evaluation protocol (referred to as DIARETDB1) that consists of 84 images with DR and 4 normal images. Images were captured with the ZEISS FF450plus digital camera, possessing a 50-degree field of view. Importantly, the evaluation database includes manual delineations from four experts annotating MAs, hemorrhages, and hard and soft exudates [68].[†]

17.4.5.4 Messidor

The Messidor database contains 1200 losslessly compressed images from three different sites and three different resolutions: 1440 × 960, 2240 × 1488, or 2304 × 1536 pixels. Images were acquired using a colour video 3CCD camera on a Topcon TRC NW6 nonmydriatic retinograph with a 45-degree field of view [69].[‡]

17.4.5.5 Kaggle/EyePACS

The Kaggle dataset is a subset of the EyePACS database, comprising of 35,126 training images graded into five DR severity levels and 53,576 unlabelled test images. Images were acquired using multiple fundus cameras and different field of view, with varying lighting conditions and quality. The images were provided by EyePACS through the California Healthcare Foundation; the specific details about the devices used and locations of acquisition are unknown [70].[§]

17.4.6 Why Performance Is Better Than Accuracy

In machine learning and data science, *accuracy* is a very ambiguous term. In this chapter, *accuracy* is sometimes used synonymously with *classifier performance*, which is calculated utilizing a number of different metrics for evaluating CAD systems and algorithms, referred to collectively as *key performance indicators* (KPIs). Within this section, *accuracy* (Acc) is a straightforward KPI referring to the "rate of truth" for the method in question, in other words, the frequency with which the

---

* http://www.ukbiobank.ac.uk/
[†] http://www.isi.uu.nl/Research/Databases/DRIVE/

* http://cecas.clemson.edu/.11ex~ahoover/stare/
[†] http://www2.it.lut.fi/project/imageret/#DOWNLOAD
[‡] http://www.adcis.net/en/Download-Third-Party/Messidor.html
[§] https://www.kaggle.com/c/diabetic-retinopathy-detection

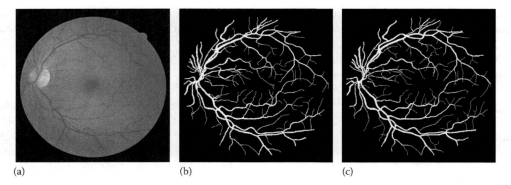

FIGURE 17.12
Sample set of the DRIVE data set. (a) Fundus image. (b) First manual delineation, used as the ground truth. (c) Second manual delineation, referred to as the second human observer and used as the human performance benchmark. Modified from J. Staal, M. D. Abràmoff, M. Niemeijer, M. A. Viergever, and B. van Ginneken, Ridge-Based Vessel Segmentation in Color Images of the Retina, *IEEE Transactions on Medical Imaging*, 23(4):501–509, April 2004, ISSN 0278-0062, doi: 10.1109/TMI.2004.825627.

system classifies a datum as positive within the data set. KPIs for CAD techniques may be scored in a number of ways, most typically by comparing the classification error of pixels or images with ground truth data. The ground truth data for pixel classification techniques tend to be manual delineations created by one or more experts as in the DRIVE data set (see Section 17.4.5), where two ophthalmologists each created a manual delineation of every fundus image, as shown in Figure 17.12 [66].

In cases where there are multiple morphologies present—such as a combination of MAs, exudates, and vessel labels for DR—a separate set of manual delineations must be made for each. One such data set is Diaret (see Section 17.4.5), where the four types of lesions each have their own mask for every image. Another technique is to transform the labels into a database or series of text files that explicitly state the position and label for every positive pixel or pixel region, where the centre pixel and radius of every pathology is defined. Proper conversion of the data set from images and label files rarely results in loss of data quality due to a particular storage format.

There are four potential classification outcomes that are used to generate KPIs:

i. *True positive* (TP): when a positive sample is classified correctly
ii. *False positive* (FP): when a positive sample is classified incorrectly
iii. *True negative* (TN): when a negative sample is classified correctly
iv. *False negative* (FN): when a negative sample is classified incorrectly

The classification outcomes are used to derive KPIs that gauge system performance, the most common of which are mathematically defined in Table 17.2.

TABLE 17.2

Key Performance Indicators

KPI	Value
True positive rate (TPR)	$\dfrac{TP}{\text{Vessel pixel count}}$
False positive rate (FPR)	$\dfrac{FP}{\text{Nonvessel pixel count}}$
Accuracy (Acc)	$\dfrac{TP + TN}{\text{Total pixel count}}$
Sensitivity aka recall (SN)	TPR or $\dfrac{TP}{TP + FN}$
Specificity (SP)	$1 - $ FPR or $\dfrac{TN}{TN + FP}$

Sensitivity (SN) is the proportion of true positive results detected by the classifier; sometimes SN is referred to as the *true positive fraction* or a classifier's *recall*. Specificity (SP) is the proportion of negative samples properly classified by the system. SN and SP are two of the most important KPIs to consider when developing a classification system as they are both representations of the "truth condition" and are thereby a far better performance measure than Acc. In an ideal system, both SN and SP will be 100%; however, this is rarely the case. Systems must make a trade-off between false positives and false negatives; in health care, this is a very tricky question: *which type of false alarm is better?* False positives ensure fewer cases are missed but incur additional resource allocations and associated expenses, whereas false negatives place less burden on the system but result in more missed cases or misdiagnoses.

Assessing the performance of CAD systems by use of a *receiver operating characteristic* (ROC) is one way to develop an optimal design that balances SN and SP. ROC is often visualized as a curve or presented as a metric derived from the *area under the curve* (AUC),

Deep Learning for Retinal Analysis

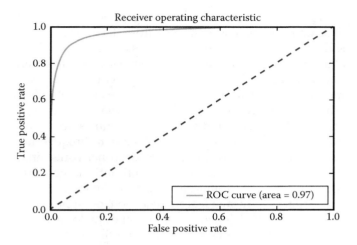

FIGURE 17.13
Receiver operating characteristic (ROC) curve. The dashed line marks the 0.5 value set, representing the minimum value for ROC curves, whereby AUC and ROC become uninformative KPIs. Modified from H. A. Leopold, J. Orchard, J. Zelek, and V. Lakshminarayanan, Segmentation and Feature Extraction of Retinal Vascular Morphology. *Proc. SPIE Medical Imaging*, 10133, 2017b, doi: 10.1117/12.2253744, URL http://dx.doi.org/10.1117/12.2253744.

whereby a perfect AUC score is 1 within the range from 0.5 to 1. The most common ROC curve utilizes SN in the y-axis and the *false positive fraction* in the x-axis such that the graph grows toward the top left corner of the figure with increasing performance, as shown in Figure 17.13 [71]. AUC scores are frequently used when reporting the performance of deep networks as well as CAD systems. AUC has been used herein as the primary evaluation metric for algorithms in the sections on performance (Sections 17.5.4, 17.6.4, and 17.6.5). These methods fall into the area of *signal detection theory*, which is widely used in clinical medicine [72].

17.4.7 Evaluating Performance

Comparison with a second observer's data set is the standard baseline utilized when assessing retinal fundus images classifiers. This method entails that a minimum of two experts have evaluated the same data set, manually delineating the images and providing applicable labels regarding the contents—in this case, health and conditions present within the retina. Experts must complete the entire data set based on the same criteria, such as painting the retinal vessels, circling MA, or labelling disease severity. When collected, each expert's label-set must retain a confidence level of at least 70%, or else be rejected [73]. For supervised algorithms, the first complete label-set is incorporated into the training data set and used to train a classifier. The second expert's labels are used as the human performance benchmark when calculating a classifier's KPIs. Understandably, they are referred to as the *second observer*; see Figure 17.12.

There are two archetypal methods for assessing the performance of a pixel-level classification system, and a third method for image-level systems:

i. *Pixel based (pixel level)*

Pixel-based methods derive system measures from the *truth* of each pixel classification against the ground truth. DRIVE (see Section 17.4.5) is specialized for these types of assessments, as the data set includes retinal images, masks, and manually delineated vessel maps. These maps are mostly a continuous single region without other markers, making pixel-based classification far more useful than region-based classification.

ii. *Region based (pixel level)*

Rather than only measuring the pixel truth, region-based measures evaluate the system's ability to classify each morphological feature in an image. This too is accomplished by training the algorithm against manually delineated ground truths. Many algorithms are able to classify pixels; however, even the best region assessment techniques can struggle when trying to differentiate small, closely packed regions, such as exudates and MA. Classification errors are therefore measured against each RoI instead of against each pixel. As discussed later in the chapter, traditional methods require many different region proposals for preprocessing retinal images for automated assessments. Often, these measures are paired with image quality and overall image classification measures to improve the representation of a system. For instance, a classifier that has an SP and SN of 60% for a low-quality, low-resolution image may in fact be far superior to another classifier with 90% SP and SN on a high-quality/resolution data set.

Image-level classification methods are typically the final stage of the CAD system, whereby the system utilizes the information gleaned from prior segmentation, classification, and KPIs to determine the condition of the retina within the image: if it is healthy or diseased; if so, which diseases; and in well-trained settings, what the severity of the disease(s) may be. The ground truth for image-based classification is often a list of indicators in a data table, which specify the health/disease(s)/severity or morphological features. The outcome measures are determined on the image as a whole, which will limit the size of the training set in many cases. STARE (see Section 17.4.5) is a useful data set for learning image classification as it has 400 labelled images of diverse pathologies and severities. While 400 images is a large size for retinal image data sets, it is far from sufficient when training a deep network, requiring the utilization of optimization methods to artificially expand the data set.

17.5 Pixel Sampling for Segmenting Retinal Morphology

Retinal vessel segmentation from fundus images plays a key role in practically all retinal image analysis, either in the assessment of vasculature or in the removal of vessels prior to the analysis of other morphologies, such as the ONH and macula. For this reason, it is the most crucial step of practically all traditional computer-based analyses of the fundus [73]. Varying quality of fundus images makes automatic segmentation of the vasculature and other retinal morphologies quite challenging, mainly due to inconsistencies in noise, hue, and brightness [13].

17.5.1 Traditional Retinal Segmentation

This method begins with image preprocessing and data set normalization. Figure 17.14 shows the process for pixel-based morphology recognition utilizing manual delineations to segment retinal vessels. While the focus of the original work was on vessel segmentation, such techniques may be applied to any morphology so long as there is a pixel-based ground truth available for training and comparison [74]. Morphological classification is conducted against pixel colouration for each type of tissue, commonly using a k-nearest neighbour method or Gaussian mixture model. System performance is assessed against the second observer, and baselines are most commonly drawn from the algorithm's application on the STARE and DRIVE data sets (see Section 17.4.5). Once the pixels have been classified, image processing techniques may be applied to enhance the results, further analyze either set of pixels, and/or segment the original image as a prerequisite for another algorithm. The positive set of pixels—those representing the vessels, may be assessed for growth patterns indicative of DR, whereas the negative set of pixels representing other morphologies may be passed to another classifier able to extract a different tissue structure or localize on a specific retinal region, such as in Section 17.6.1. The latter case is quite common in traditional CAD systems (see Figure 17.11), where retinal vessel extraction is necessary in ONH localization for glaucoma assessment and similarly, the detection of MA.

17.5.2 Pixel Sampling

Fundus images typically contain 500 × 500 to 2000 × 2000 pixels, making the training of a classifier a time consuming and computationally expensive ordeal. Rather than processing the entire image, *megapixels* (also referred to as image *patches*) are sampled from the image using a sliding window. The simplest pixel sampling algorithms algorithmically select and extract megapixels from an image, using the label corresponding to the centre pixel; megapixel refers to the patch's centre pixel and its label, while considering other pixels within the patch as class features. Patch size and mini-batch volumes are determined on a case-by-case basis. These techniques are particularly useful for optimizing deep networks since they allow for the creation of significantly larger data sets of megapixels from a small number images.

Sampling is a form a feature detection, often preceded by preprocessing, image masking, and/or localization (as described in Section 17.4.3). For this reason, pixel sampling can be considered part of batch preparation and is sometimes referred to as *mini-batch preparation*, since a given batch of images is further subsampled into mini-batches of megapixels. It is worth noting that this optimization technique for artificially increasing data set size is limited to pixel classification; however, algorithms for disease detection are able to be built on top of algorithms trained on megapixels. There are three main types of sampling methods that are used to create a batch of data:

i. *Random pixel sampling*
   A mini-batch of data with n samples is populated with megapixels randomly extracted from an image. Depending on the application and user preference, the algorithm may sample from the retina as well as the surrounding black border.

ii. *Stratified pixel sampling*
   Similar to random sampling, this method extracts a mini-batch of samples; however, the application first assesses every pixel within a batch before selecting an equal distribution of all classes to train the system. Since the algorithm indexes the entire sample, it can become very memory intensive. Depending on the task, number of mini-batch samples, data storage methods, and overall software architecture, it may be more computationally effective to use random sampling.

iii. *Regional pixel sampling*
   This technique obeys the same tenets as stratified sampling. The difference between them is that regional sampling creates a distribution of samples for each segment of an image, such as exudates, ONH, vessels, and fundus, rather than a strata of all pixel values within each label type. Regional sampling typically uses random sampling within each

Deep Learning for Retinal Analysis

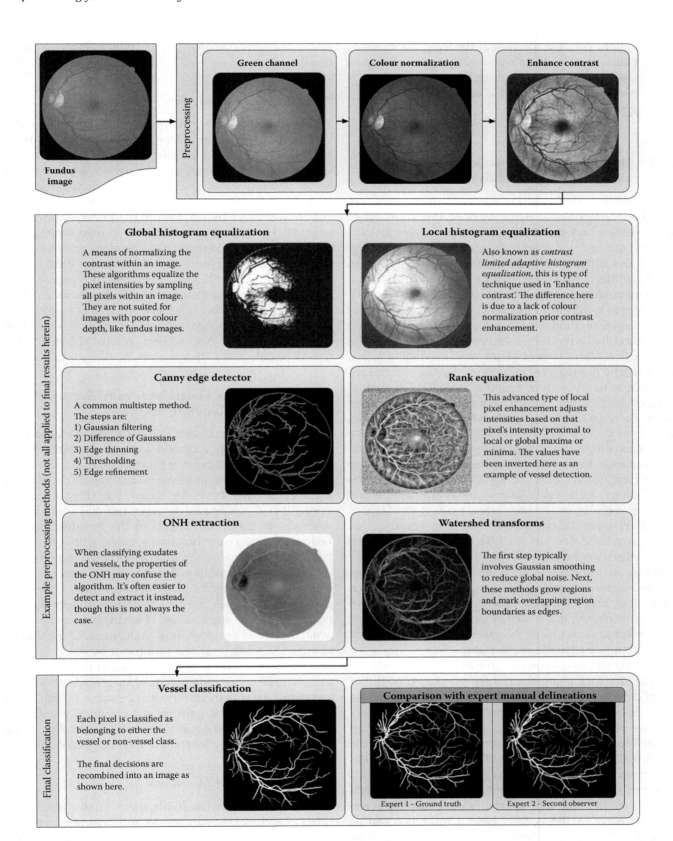

FIGURE 17.14
Traditional pixel-based retinal vessel segmentation.

region, running the risk of drawing duplicate values in the case of small sample sizes. It can also use stratified sampling; however, this may compound computational requirements on the system.

17.5.3 The Deep Learning Approach to Retinal Morphology Segmentation

Deep learning methods for retinal segmentation are based on the techniques that have been applied to image segmentation in other fields, particularly vehicle data sets and microscopy. Most of these methods are based on LeNet and utilize stochastic gradient descent (SGD) to optimize the network [9]. Recent work into stochastic gradient-based optimization has incorporated adaptive estimates of lower-order moments, resulting in the *Adam* optimization method, which is further described below [75]. Recently, Adam was successfully applied to the problem of retinal vessel segmentation—that work is what will be discussed herein; the network architectures for these networks can be found in Refs. [71,74]. A flow diagram of this method is shown in Figure 17.15. Deep networks often utilize at least two fully connected layers to make a final pixel classification, as the feature dimensionality must be greatly reduced from hundreds of thousands to a single class vector of size 2 for any single morphology, plus one variable for each additional morphology. In this case, the vector is size 2 with one class for each of vessel or nonvessel pixel. If exudates or the ONH were included in the training set, the final vector would be size 3 or 4, with a new variable for each morphology respectively.

While these methods can be executed without image enhancement during preprocessing, proper use of such techniques may be beneficial for accuracy and computational efficiency. In the case of retinal vessels and MAs, deep methods can utilize the green channel exclusively to make highly accurate predictions, while reducing the computational requirements due to a greatly reduced sample size (one channel vs. three or four) [71,74]. For ONH detection and extraction, the red channel must also be included. Contrast enhancement techniques may also be utilized [76]. Stratified pixel sampling is then utilized to create a multibatch set of megapixels from the training data set, transforming 20 retinal fundus images (i.e., from DRIVE) into tens to hundreds of thousands of samples. The samples are then used to train a CNN over many (at least 100) epochs, unless training is terminated early by the system. Early termination can be built into the system as an optimization method whereby the network recognizes when the change in accuracy between epochs is significantly low for a statistically significant amount of time, and ends the process assuming future training steps will follow the same trend.

17.5.3.1 Adam Optimization

Adam (from "adaptive moment estimation") is a stochastic gradient optimization algorithm, differentiated from others by its use of only low-order partial derivatives. Typically, only first-order derivatives are required to outperform other stochastic optimization methods in terms of accuracy as well as computational cost. Adam is driven by an adaptive algorithm primarily controlled by three parameters: α representing the learning rate (also referred to as step size), and B_1 and B_s, which represent exponential decay rates for the moment estimates. ε represents a fuzz factor and is typically set to 10^{-8}. In the literature, it has been shown that Adam outperforms all other methods (AdaGrad, RMSProp, SGD, AdaDelta amongst others) for image classification tasks utilizing CNNs, testing against MNIST, and CIFAR10 data sets [75].

17.5.4 Performance Comparison: Vessel Segmentation and Pixel Classification

The first performance section investigates nondeep (traditional) and deep methods for vessel segmentation as a real-world example for pixel-based classification. Table 17.3 takes an unconventional approach wherein algorithms have been organized into the aforementioned types, as opposed to *supervised* and *unsupervised* methods, as commonly found in the literature; sibling tables in the other performance sections follow the same pattern. The secondary purpose for this section is to describe the computational costs at test time for these algorithms. Later performance sections do not reinvestigate execution times as they remain relatively proportional between applications. Herein, performance is evaluated using the algorithms' AUC score for the DRIVE and STARE data sets. The traditional methods selected present a wide variety of approaches to the problem of vessel segmentation, whereas there are considerably fewer archetypal deep methods for doing the same. In both cases, the techniques are organized chronologically with the oldest methods at the top and were chosen due to the unique methodology inherent to their implementation.

17.5.4.1 Traditional Algorithms

Within the traditional methods, there are two prominent types of techniques: supervised and unsupervised, differentiable by the use of a classifier to select pixels. Much of the early work in nondeep methods used supervised machine learning approaches, such as SVMs and decision trees (*2-D Gabor wavelet and Bayesian*

Deep Learning for Retinal Analysis

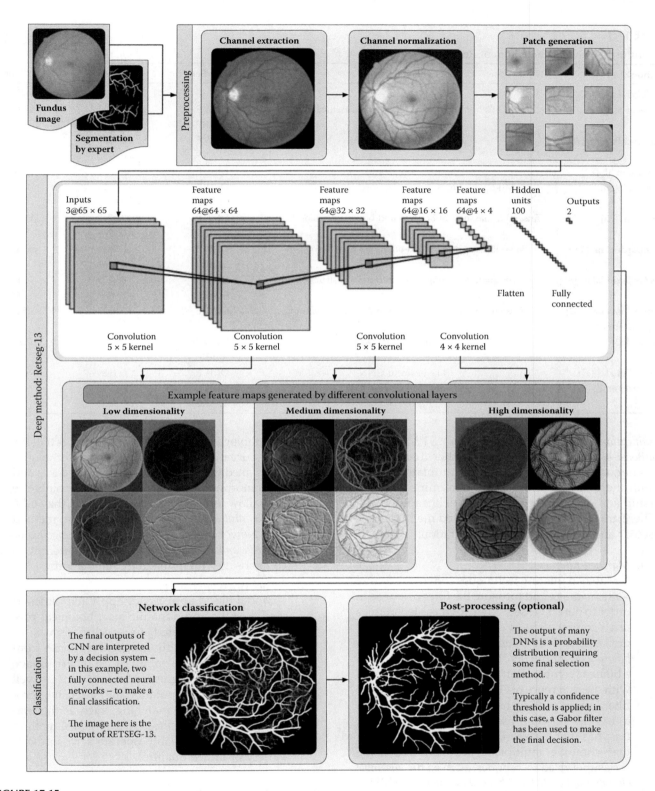

FIGURE 17.15
Flow diagram of the Adam optimized deep learning method for retinal segmentation. Images adapted from H. A. Leopold, J. Orchard, J. Zelek, and V. Lakshminarayanan, Segmentation and Feature Extraction of Retinal Vascular Morphology. *Proc. SPIE Medical Imaging*, 10133, 2017b, doi: 10.1117/12.2253744, URL http://dx.doi.org/10.1117/12.2253744.

TABLE 17.3

Algorithms for the Segmentation of Retinal Vessels

Authors	Method	Data Set	AUC	Test Time
Traditional Methods				
Soares et al. [77]	2-D Gabor wavelet and Bayesian classification	DRIVE	0.9614	180s
		STARE	0.9671	
Ricci and Perfetti [78]	Line operators and SVM	DRIVE	0.9633	N/A
		STARE	0.9680	
Lupascu et al. [79]	AdaBoost-based classification	DRIVE	0.9561	125s
Lam et al. [80]	Multiplanar multiscale concavity combination	DRIVE	0.9614	780s
		STARE	0.9738	
Fraz et al. [73]	Ensemble classifier of boosted and bagged decision trees	DRIVE	0.9747	120s
		STARE	0.9768	
Azzopardi et al. [13]	Bar-selective combination of shifted filter responses	DRIVE	0.9614	10s
		STARE	0.9563	
Kovács and Hajdu [81]	Template matching and contour reconstruction	DRIVE	0.9722	780s
		STARE	0.9893	
Annunziata et al. [82]	Hessian eigenvalue analysis and exudate inpainting	STARE	0.9655	25s
Deep Methods				
Li et al. [83]	Denoising autoencoder	DRIVE	0.9738	72s
		STARE	0.9879	
Leopold et al. [84]	13-layer linear CNN	DRIVE	0.9689	87s
Liskowski et al. [77]	Channel separated CNN	DRIVE	0.9710	92s
		STARE	0.9880	

classification [77], *line operators and SVM* [78], and *AdaBoost-based classification* [79]). Rather than training a classifier, unsupervised techniques stimulate a response within the pixels of an image to determine class membership and do not require manual delineations.

There are three types of unsupervised methods used in vascular and morphological segmentation:

i. *Matched filters*

Matched filtering techniques convolve a kernel with an image and measure the filter response in order to detect vessels. Many of these techniques use a combination or derivative of either Gabor or Gaussian filters, such as the *bar-selective combination of shifted filter responses* method [13] with a bilinear combination of a filter bank populated with the difference of Gaussian kernels. Matched filters work well in healthy images, but suffer a drop in SP when used on pathological cases; this has led to work on ensemble methods that combine multiple types of filters using semisupervised techniques, such as the *ensemble classifier of boosted and bagged decision trees* [73].

ii. *Morphology tracking*

Morphology tracking methods implement edge or watershed techniques to segment the image, which may use greyscale intensity or known mathematical properties of different morphologies, such as vessel shape, tortuosity, and curvature. Historically, these techniques have failed to accurately capture thin vessels and other morphologies that may lack seed points due to shallow intensity variations, as with the *multiplanar multiscale concavity combination* method [80]. Newer methods, such as *Hessian eigenvalue analysis and exudate inpainting* [82], attempt to compensate with multiscale combinations.

iii. *Model-based approaches*

Model-based approaches apply templates as well as static and dynamic morphological profiles that are catered to specific tissues or pathologies to reliably induce a response from medical images, even in cases where image quality is poor. Higher-order functions as well as multistep, computationally intensive processes for template matching, such as the *template matching and contour reconstruction* method [81], are able to reliably operate across a wide variety of image qualities.

17.5.4.2 Deep Algorithms

While there are many emerging methods for deep learning in medicine, three predominant types of DNNs have been used for morphological segmentation, all of which are considered supervised learning methods:

i. *Linear CNNs*

Linear CNNs process an image either extracting a single channel from the image or treating the channels as an additional feature space for the final pixel classification. Figure 17.4 is an example of a linear CNN. These are the most fundamental examples of CNNs, which are utilized in parallel or in sequence in the DNNs discussed throughout this chapter. The *13-layer linear CNN* method [74] is shown in Figure 17.15 [71], which illustrates the following:

1. *Image enhancement*: Fundus image channels are separated and undergo colour normalization and contrast enhancement.
2. *Mini-batch creation*: Next, stratified pixel sampling is used to generate an even number of vessel and nonvessel patches as the training data set.
3. *Training*: The patches are fed through the CNN and regenerated at every epoch.
4. *Testing*: At test time, the resulting pixel classifications are used to generate probability heat maps of the entire fundus.
5. *Final classification*: The heat maps are then postprocessed by thresholding to reach a final binary classification for vessels and nonvessels.

ii. *Multichannel CNNs*

As the name implies, these CNNs split off into multiple parallel and identical architectures for each channel within an image. To reduce computational complexity, equivalent layers share parameters, leading to joint training across all image channels while each stream learns its own set of ideal features for classifying pixels within its channel. The final layer of each channel stream is joined by flattening and fully connected layers in the same way as linear CNNs, producing the same type of pixel probability vector as the output, as in the *channel separated CNN* (CSCNN) method [84].

The benefit for analyzing multiple channels within a fundus image is described in Section 17.4.3; the gains in system performance should come as no surprise when comparing multichannel CNNs with linear CNNs that only use a grey or green channel image for training and classification.

iii. *Autoencoders*

An autoencoder is a neural network trained to copy its input to its output [83]. In the case of vessel segmentation, it is trained to convert a fundus image into a manual delineated image or *vice versa*. The simplest autoencoders are made up of an encoder and a decoder. The encoder is the feature extractor component of a linear CNN, and the decoder is an identical but transposed CNN implemented as the decision system, rather than fully connected layers. The decoder comprises deconvolutional and upsampling layers to match the convolutional and pooling layers in the network's encoder. It is worth noting that the feature extractor and decision system need not be perfect reflections of one another. As an example, a network could combine a multichannel CNN as the feature extractor, but a single channel deconvolutional network as the decision system, so long as they have the same depth and size of layers. In the *denoising autoencoder* method [76], fully connected layers are inserted in between the feature extractor and decision system to denoise the data, leading to an improvement in results. Compared to the prior two types of CNNs, autoencoders are able to output entirely segmented patches at test time, reducing the computational requirements of the system.

17.5.4.3 Comparing Performances*

Amongst the deep methods, the denoising autoencoder (DAE) is the best performer for DRIVE, while the CSCNN is barely the best for segmenting vessels in STARE. Upon more careful study of the KPIs, it seems network depth improves SN, as the linear CNN outperforms the others; it has eight layers for feature extraction deep, while the others have only five, leading to a raw SN of 0.9568 [71], whereas the DAE has 0.7569 [83] and the CSCNN 0.7520 [84]. The reason the final AUC score of the linear CNN is lower than the others is due to SP, which is roughly 0.03 lower, meaning that the final classification of the linear CNN does not agree with the ground truth, not that the network is necessarily performing poorly. Reviewing test time run speed is nonintuitive. At first glance, it appears the DAE runs the most efficiently, requiring only 72s to process a single test image. In actuality, the linear CNN is the most efficient method at 87s, as it is processing larger image patches, which inherently consumes more computation power. If they were both processing the same-sized kernels, the DAE would be much

---

* Computational times discussed herein have been extracted from literary sources. Due to variations in computer architecture and processing power, the reader should take the metrics as only a rough approximation.

slower(16 × 16 pixels per image patch would increase to 65 × 65 pixels).

When looking at the traditional (nondeep) methods described in Table 17.3, there appear to be two different technique philosophies emerging: minimizing run times, as with the *bar-selective combination of shifted filter responses* method [13], which only requires 10s to compute; and maximizing performance, as in the *ensemble classifier of boosted and bagged decision trees* [73] as well as the *template matching and contour reconstruction* methods [81]. The latter respectively boast the best AUC score for DRIVE and STARE amongst all methods presented in this section at the expense computational requirements at test time—2 min and 13 min compared to the 72s of the DAE, which has comparable AUC for both DRIVE and STARE as the superior method of the two. From this, a conclusion can be drawn: In the case of traditional and deep methods for vessel and morphological segmentation, deep learning approaches outperform in terms of accuracy as well as computational efficiency at test time.

17.6 Superpixel Sampling for Microaneurysm Detection in the Classification of Diabetic Retinopathy

Microaneurysm detection algorithms are perhaps one of the most useful CAD systems for augmenting standard practise. Detection of MAs is like any other form of regional morphology detection, such as ONH, exudate, and lesion extraction; however, it can be significantly more challenging due to their size. MAs are quite small at 10–100 μm in diameter, equivalent to a small red dot in a retinal fundus image. Due to the continuous proliferation of MA, scarring, hemorrhages, lesions, and other exudates can obfuscate the real value, even from clinical experts. Prior to recent developments in computational power and deep learning, the only realistic solution for conducting improved early screening was with novel imaging technologies, such as autofluorescence fundus imaging techniques and OCT. Image enhancement techniques, such as multiresolution techniques, are able to enhance normal fundus images to a quality equivalent to autofluorescence imaging [62].

To use pathological morphologies in patient diagnoses or measure disease progression over time requires painstaking effort to find, mark, and count all morphologies in both eyes at every encounter. A system for automating detection and counting is necessary to complete patient ocular health assessments the same day as a patient's appointment, if not in real time. The tools and principles described in this section focus on MA detection; however, they can be applied to other pathologies. This problem is similar to *vessel segmentation*; however, methods are discussed that improve computational efficiency and enhance the system's ability to differente MA from other vessel tissues. These problems differ in that MA detection is a form of *region-based classification*, whereas vessel segmentation is one of *pixel-based classification* (see Section 17.4.7), as evidenced by the image labels utilized within these methods. Diaret (see Section 17.4.5) is commonly used to benchmark these tasks. This is done as there are considerably fewer positives than with vessels or larger morphologies; it is much more efficient and effective to store these labels as text files than as whole images.

17.6.1 Traditional Microaneurysm Detection

Recent work evaluated the efficacy of various traditional MA detection methods utilizing a novel performance algorithm, scoring the *Waikato Microaneurysm Detector* best overall [85]. Many algorithms have adapted the Waikato architecture, making modifications to individual stages since it was first published [86]. More information regarding the Waikato detector can be found in Ref. [87]. Generally, MA detectors build on the preprocessing steps for colour and shade normalization, while utilizing retinal vessel segmentation techniques as described in Section 17.5.1. Figure 17.16 [86] shows a generalized traditional method for MA detection. The process is as follows:

i. *Preprocessing*: The green channel is extracted from the fundus image and preprocessed with colour normalization and contrast enhancement techniques.

ii. *Morphological mask generation*: A region growing algorithm (such as watershed methods) is used to find large morphological components and generate an image mask from the grey image. The algorithm would evaluate only regions larger than MA. At the same time, a vessel segmentation algorithm is used to generate a separate pixel mask from the grey image.

iii. *Mask subtraction*: After a fundus image has been preprocessed, a linear filter is applied to the vessel segments and large morphologies such that only elements larger than MA are subtracted from the fundus image; MA and other small nonvessel structures are left over following this subtraction, while larger structures known to cause false positives have been removed.

iv. *MA candidate detection*: At this stage, pixel intensities are used to seed potential RoIs,

Deep Learning for Retinal Analysis

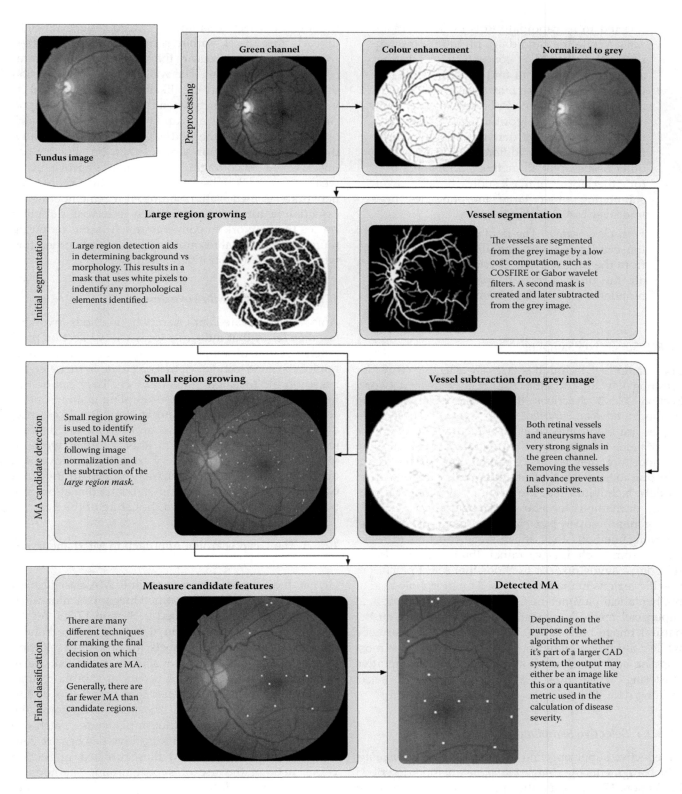

FIGURE 17.16
Flow diagram of a traditional automated microaneurysm detection method. Images adapted from S. A. Ali Shah, A. Laude, I. Faye, and T. B. Tang, Automated Microaneurysm Detection in Diabetic Retinopathy Using Curvelet Transform, *Journal of Biomedical Optics*, 21(10):101404, 2016, doi: 10.1117/1.JBO.21.10.101404, URL http://dx.doi.org/10.1117/1.JBO.21.10.101404.

referred to as candidate RoIs or as the *candidates*. While the candidates provide some representation of MA concentrations, further analysis is required to find and count the true RoIs—failing to do so may greatly decrease SN and SP. The candidate RoIs are expanded with a region growing algorithm and passed to a feature extractor that derives various intensity features that aid in differentiating MA from other tissues. Not all methods implement both (or either) forms of mask subtraction; however, subtracting other large morphologies does improver MA detector accuracy.

v. *Final classification*: The final classifier evaluates the candidates, utilizing the peak responses from the *MA candidate detection* stage to generate the final measures representing the total number of MAs within a fundus image.

17.6.2 Superpixel Sampling

Supersampling techniques are an effective way for selecting key RoIs to efficiently train a deep network. Section 17.5.2 describes three basic pixel sampling methods: *random, stratified*, and *regional*. These techniques are made into superpixel methods through the use of batch sample extraction. *Advanced* supersampling methods are defined as computational strategies for selecting megapixels that utilize complex mathematical models, techniques that are inspired by biological processes, or methods for assessing image saliency.

In general, superpixel sampling methods are pixel sampling methods that utilize some function to select pixels from each image, rather than only randomly selecting a single pixel or methods that select patches in specific regions in the image, based on morphological or mathematical parameters. Compared to pixel sampling, superpixel methods generally subsample based on the natural image boundaries, bolstering local consistency while collecting a higher-fidelity data set [88]. The most popular superpixel method is simple linear iterative clustering, which uses a k-means algorithm to select training data [89].

17.6.2.1 Selective Sampling

An effective supersampling method is *selective sampling*, which augments the *stratified sampling* method described in Section 17.5.2. After the image patches have been extracted in bulk from a data set, positive and negative samples (MA and non-MA in this case) are separated into a representative data set at each epoch. Stratified sampling uses this same methodology, randomly sampling the corpus of megapixels at each epoch to create an training set. Selective sampling instead assigns a weight to each megapixel within some if not all classes—computational resources are the deciding factor for number of classes to weigh. The weights influence selection probability, dictating the likelihood a megapixel is included in a mini-batch when they are generated upon each epoch. Weights are updated following network optimization, before the next batch is created. Typically, these algorithms favour samples that the network has misclassified, encouraging the system improve in its weakest areas. A subset of samples will be those the network classified well, resulting in it remembering all essential features. To summarize, selective sampling causes the network to converge much faster, requiring more computational resources during sample generation when compared to stratified sampling.

17.6.2.2 Sampling with Foveation

Anisotropic foveated supersampling methods involve a *foveation kernel* that induces a local acuity effect on pixel patches at time of sampling [90]. These are nonadaptive, nonlocal means techniques that calculate kernel weights based on the foveated patch distance. They can be utilized in conjunction with other sampling methods, often as an alternative to purely random sampling. In essence, a foveation bias is induced across the image, limiting sampling further from known RoIs. The bias within a given kernel may be varied by adjusting the foveation operators; however, they generally utilize an invariant Gaussian blur with increasing standard deviation proportional to the distance from the centre of the RoI.

17.6.3 The Deep Learning Approach to Microaneurysm Detection

Herein, the deep learning approach to MA classification is based on the CNN and stratified megapixel mini-batch generation processes described in Section 17.5.3. As the pixel and MA segmentation data sets utilize fundamentally different types of labels, the process of selecting megapixel regions must be slightly modified. Early optimization of the network occurs at sampling time, requiring the mini-batch generation process to be tailored for the detection of MA. The method recognizes the size distribution of MA and thus implements a patch size large enough to capture them. Positive patches are considerably more sparse than those utilized in Section 17.5.3, making the use of supersampling methods in optimizing network training comparatively more beneficial given the incurred computational burden.

Selective sampling improves the rate of network convergence, as described in Section 17.6.2. If the base sampling methods described in Section 17.5.2 were utilized instead, it would be much more difficult for the

CNN to differentiate MA from retinal vasculature. In a practical sense, the deep network would still converge to a similar performance with either type of sampling method; the difference is the number of epochs required to do so. The CNN would eventually generate a similar set of features, weights, and biases; however, the rate at which such parameters are selected may be more than 300% longer [91].

Figure 17.17 [76,91] shows an example architecture for deep MA detection, which analyzes supersampled megapixels, classifying each pixel as either MA or non-MA before grouping regions of positive pixels into individual lesions. Similar to the deep method for vessel segmentation discussed in Section 17.5.3, MA classification begins by passing fundus images along with expert-generated labels through preprocessing algorithms. As mentioned in the introduction to this section, the labels are usually text based. As such, "Segmentation by expert" within Figure 17.17 has been created solely for the sake of example. Preprocessing for this method is simply contrast enhancement by convolution with a Gaussian filter, which affects each channel independently. Such techniques consistently improve image quality, contrast, and sharpness in fundus images. Supersampled megapixels are populated at each epoch.

The network's training progress at 5, 15, 25, and 60 epochs is shown, where the coloured mask illustrates the probability for pixels being MA using a heat map spanning from blue (low probability) to red regions (with the highest probability). Upon reaching the classification stage, the system is able to confidently extract MA candidate pixels and present them as a probability heat map. However, the actual shape and size of each individual lesion remain to be determined. At test time, the network must convert the heat maps into MA candidates, similar to the traditional method. A form of region growing is implemented to extract the final candidates, while ensuring regions are not tagged more than once by combining instances into single pathological patches. Determining potential region seed points can be challenging. One method for selecting such sites first blurs the image with a Gaussian kernel prior to selection of seed points based on local probability maximums from the then-smoothed heat maps [92]. Following, images can be scored based on the presence and quantity of lesions and MA as an indicator of DR. Performance for MA detection is compared in Table 17.4, whereas performance for DR detection (image score) is shown in Table 17.5.

17.6.4 Performance Comparison: MA Detection and Region Classification

The second performance section investigates traditional and deep methods for MA detection as a real-world example of region classification and pathology detection. Table 17.4 describes a few of these methods with the best performance as reported in the literature. Larger lesions and exudates are able to be found and analyzed with techniques described in Section 17.5. Detection of MA is a very niche method group, which may experience a decline in explicit usage to focus on assessing disease progression. In general, there are fewer studies into MA detection compared with vessel segmentation or DR analysis. In the latter case, both traditional and deep methods are gravitating toward disease classification through feature discovery rather than expert mimicry, as further discussed in Section 17.6.5. For this reason, the best MA detectors have been developed as part of DR detection systems, resulting in the majority of techniques being discussed in this section seeing application classifying DR in the subsequent section on image classification. Similar to Section 17.5.4, performance is evaluated using the algorithms' AUC score and organized chronologically with the oldest methods at the top. The databases most commonly used in these studies are Diaret and Messidor.

17.6.4.1 Traditional Algorithms

i. *Visual word dictionaries*

Visual word dictionary methods are semi-supervised multistep classification techniques that learn to detect points of interest (PoIs) within manually denoted RoIs of fundus images. Pixels are not explicitly painted or labelled. RoIs are manually delineated by experts who also label the region by which type of morphology or pathology is present within a region, but not count or exact location. Expert labelling becomes much easier as the experts are only outlining key areas of the fundus as opposed to painting every pathological pixel. The RoIs and associated labels are used to train feature detectors for each type of morphology independent of the rest, thereby building a set of *visual word dictionaries*, each containing a set of features specialized for a specific pathology, such as building a separate dictionary for each type of lesion in Table 17.1.

ii. It is possible to implement a multidimensional *Speeded Up Robust Features* (SURF) algorithm to extract and characterize PoIs, feeding the results into a k-means clustering SVM to develop a set of dictionaries for DR classification; MAs are one such dictionary. Using the dictionaries, each image is converted into a projected feature mapping representing the image as a frequency distribution of pathological points/regions of interest [93]. The distributions are used to train a k-means clustering

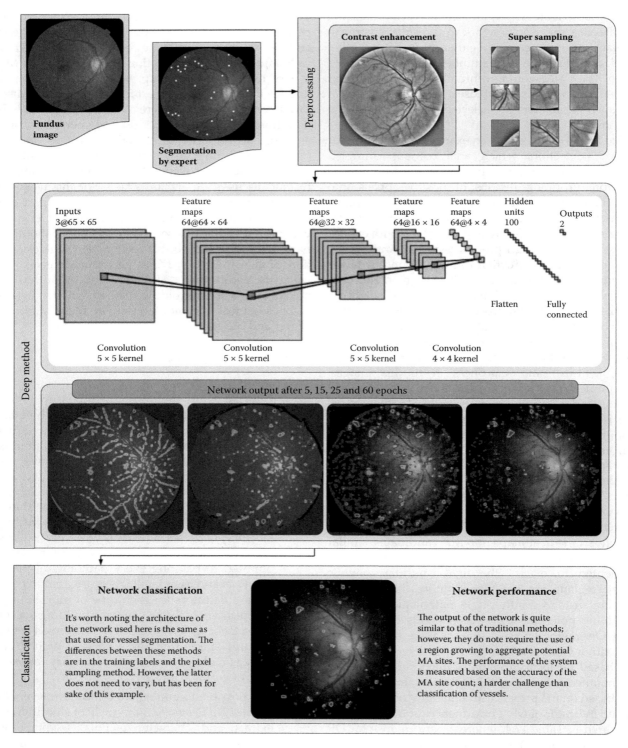

FIGURE 17.17
Flow diagram of the deep learning method for microaneurysm detection. The images have their contrast enhanced by convolution with a Gaussian filter prior to superpixel sampling. Training progress is shown, as are the final probability heat maps for pixels containing lesions. Images modified from M. J. van Grinsven, B. van Ginneken, C. B. Hoyng, T. Theelen, and C. I. Sánchez, Fast Convolutional Neural Network Training Using Selective Data Sampling: Application to Hemorrhage Detection in Color Fundus Images, *IEEE Transactions on Medical Imaging*, 35 (5):1273–1284, May 2016, ISSN 0278-0062, doi: 10.1109/TMI.2016.2526689, and H. A. Leopold, J. Orchard, J. Zelek, and V. Lakshminarayanan, Use of Gabor Filters and Deep Networks in the Segmentation of Retinal Vessel Morphology, *Proceedings of SPIE Biomedical Optics*, 10068, 2017a, doi: 10.1117/12.2252988, URL http://dx.doi.org/10.1117/12.2252988.

TABLE 17.4

Algorithms for Microaneurysm Detection

Authors	Method	Data Set	AUC
Traditional Methods			
Rocha et al. [93]	Visual word dictionary used to train an SVM	Diaret Messidor	0.933
Srivastava et al. [94]	Multiple-kernel learning using radial basis functions	Diaret Messidor	0.973
Deep Methods			
Grinsven et al. [91]	Selective sampling 9-layer linear CNN	Kaggle Messidor	0.894 0.972
Quellec et al. [95]	Sensitivity criterion derivation from a 25-layer linear CNN	Diaret	0.978

TABLE 17.5

Algorithms for DR Detection

Authors	Method	Data Set	AUC
Human [96]	Two experts evaluated the data set manually	Messidor	0.865–0.922
Traditional Methods			
Sanchez et al. [96]	Multialgorithm CAD system	Messidor	0.876
Rocha et al. [93]	Visual word dictionary used to train an SVM	Diaret Messidor	0.893
Deep Methods			
Abramoff et al. [14]	Expert mimicry ensemble of CNNs	Messidor	0.980
Quellec et al. [95]	Feature discovery ensemble of CNNs	Kaggle	0.954

SVM to classify DR in fundus images; those results are further discussed in Section 17.6.5 and are listed in Table 17.5.

iii. *Multiple-kernel learner* (MKL)

Multiple-kernel learning is a form of supersampling that combines the results from multiple image scales, using patches of varying sizes. Multiscale megapixels are then passed through a feature extractor, resulting in many sets of multiscale feature kernels. It has been shown that the combination of such kernels results in superior pixel classification, in terms of both accuracy and computational efficiency [97]. The MKL described in Table 17.4 trained an SVM classifier using the maximum response generated by a radial basis function that combined sets of a custom Gaussian filter kernel specialized in detecting MA and other lesions while ignoring vessels [94].

17.6.4.2 Deep Expert Mimicry

While the exact architectures vary, both deep methods described in this section utilize variations of *linear CNNs* for feature extraction and fully connected layers for final classification. The *selective sampling 9-layer linear CNN* (SSCNN) [91] method is very similar to that shown in Figure 17.17, albeit with a different deep network architecture, supersampling technique, and candidate validation algorithm. The defining elements of this architecture are as follows:

i. *Mimicry to classify MA*

This required several ophthalmologists to manually delineate MA and other major lesions within the Kaggle data set, yielding a very large and robust set of training and testing samples. This choice also further subsets this method into pixel-level supervision. From the quantification of MA in a fundus image, and prior knowledge of DR severity provided as data set labels, a subsequent algorithm could be developed to assess DR severity based on MA features, as discussed in Section 17.6.5.

ii. *Selective supersampling of negative megapixels*

Negative megapixels were weighted for minibatch sampling, while positive megapixels were randomly selected. The effect of selective sampling on training efficacy was investigated. When selective sampling was implemented, training concluded after roughly 60 epochs, whereas nonselective sampling terminated after 170 epochs. Even though it was trained for almost one-third the time, the network trained with selective sampling after 60 epochs greatly outperformed the 170-epoch nonselective sampling network (AUC of 0.979 vs. 0.966 on Messidor after removing low-quality images) [91].

iii. *Candidate validation by a dynamic algorithm*

Seed points for MA candidates were selected from the local maxima of RoIs smoothed with a Gaussian filter. The algorithm's cost function is the gradient magnitude of the smoothed RoI, which it uses to assign a probability equal to the resulting average probability of all pixels within the candidate.

17.6.4.3 Deep Feature Discovery

On the other hand, the feature discovery approach is the driver for the *sensitivity criterion derivation from a 25-layer linear CNN* method. The underlying architecture of the network was based on one of the top networks implemented in the 2015 Kaggle competition, replacing the prior optimization function with an Adam optimizer (see Section 17.5.3). The 25-layer linear CNN was trained to classify DR severity at the image level, and the sensitivity criterion was used to derive pixel-based features of the underlying fundus images. The same process allows for the extraction of characteristic weights and biases from different stages of training representing the strongest signal for different pathologies.

Rather than using only the weights from the image-level training, a separate CNN is configured with a different characteristic weight and bias profile all of which can be joined together into an ensemble of CNNs; we refer to this architecture as a *feature discovery ensemble* (FDE). The design is fairly atypical, as ensembles are usually made up of different network architectures to make up for the greatly increased computation requirements for training so many networks. The method described in Table 17.4 assembled an FDE of twelve 25-layer CNNs with a random forest classifier as the final decision system. This method utilizes a number of optimization strategies that dismiss the computational burden typically incurred when training an ensemble of CNNs, while enabling the system with additional pathology detection abilities:

i. *Pathology discovery with sparsity-enhanced sensitivity criterion*

The sensitivity criterion dictates how local features influence the target function of a given operation. In the case of CNNs, it is the relationship between image labels and pixel values that govern them. Specifically in this case, where the training data set is composed of fundus images and associated DR severity labels, the sensitivity criterion relates pixel-level indicators of disease to the severity level; it provides a mechanism by which DR pathologies, such as MA, lesions, and exudates, can be extracted as probability heat maps from the networks' filter banks without explicit training. For this method, the training loss function is modified to maximize the sparsity of nonzero pixels (pathological pixels) while retaining accuracy.

ii. *Training on image and pixel performance*

The network was trained on the Kaggle data set that only contains DR severity labels—image-level labels. Typically, network performance between iterations would be evaluated against solely with the classification agreement with the image-level data; if that were the case here, it would be quite challenging to generate accurate pathological heat maps. In addition to DR labels, network performance was simultaneously evaluated at the pixel level using the Diaret data set for each type of lesion label available (hard exudates, soft exudates, MA, and hemorrhages). Training completion was determined at the image level (DR severity).

iii. *Checkpoint-driven ensemble assembly*

Rather than training a series of networks for this ensemble, only one linear CNN was trained while carefully monitoring its performance for DR and associated pathologies, as described previously. Unsurprisingly, the performance at the pixel level fluctuated throughout training at the image level. Interestingly, performance between lesion types varied substantially, each reaching a peak performance at unique training iterations. The weights and biases from the iterations of optimal detections for each lesion type, the average of all lesions, and DR were exported and used to create six CNNs for the ensemble.

When patients have their eyes examined, both eyes are imaged; the Kaggle data set retained this data, providing an image of each eye for patients. To retain the relationship between the eyes of patients during classification of DR (separate for each eye), the features from two sets of the six CNNs—six from the current eye and six from the contralateral eye—were fed into the final decision system, in this case a random forest. The output is a blended heat map of all 12 networks for each eye. The output of each network may be weighted by its importance relative to the others in DR detection; this may require calibration but would be based on expert guidelines.

17.6.4.4 Comparing Performances

In this section, we present four algorithms: two traditional and two deep. The first of each uses *expert mimicry* to detect MA, while the second uses feature discovery, which outperformed its respective mimicking counterpart. The MKL did very well to identify MA, utilizing an approach quite similar to that of a CNN in its generation of many feature kernels and selection of only those with optimal responses. The performance of the SSCNN was on par with the MKL, without requiring manual calibration of filters to detect MA. The inclusion of a similar hierarchal implementation in the SSCNN would boost its performance above the MKL.

That the FDE is able to detect MA and other lesions based on selection criterion is a prime example for the power of deep learning and the feature discovery paradigm. At 0.978 AUC, it sets a wide margin comparative to all the other methods, and the network learned to detect the specific pathology as a by-product of learning to assess the severity of DR. In fact, the network was reported to work well in low-image-quality cases as well as out perform all other solutions in detecting the individual categories of lesions: hard exudates, soft exudates, MA, and hemorrhages [95]. To put this in perspective, the other deep method required several ophthalmologists each to manually label over 35,000 images from the Kaggle training set before it could be trained to detect MA with an AUC of 0.972, and that was all—it was not able to detect other pathologies or assess the severity of DR [91]. The FDE's ability to distill pathologies from a network trained on image labels makes it possible to discover previously unknown disease indicators, present in the fundus. The same network could be applied to any number of medical image diagnostics or even just image-label-based problems, assuming the training data exists.

17.6.5 Performance Comparison: Diabetic Retinopathy and Image Classification

The final performance section investigates traditional and deep methods for DR detection and grading as a real-world example of image classification and CAD systems. While there are many methods in the literature, algorithms here were selected for their use of AUC as a performance metric, superiority in AUC score, and agreement with content covered in this chapter thus far. As before, the techniques are organized chronologically with the oldest methods at the top. A common grading scheme for retinopathy screenings is presented in Table 17.6 [62]. This scheme is utilized by Messidor, which is the main data set discussed herein. The scheme has four severity levels, in which the highest grade includes neovascularization. In 2013, a new scheme was released by the International Clinical Diabetic Retinopathy (ICDR) following the Early Treatment Diabetic Retinopathy Study (ETDS) classifications [98], and is detailed in Table 17.7 [69]. Newer studies and data sets, such as the Kaggle data set, follow this new scheme, which separates neovascularization into a separate grade.

17.6.5.1 Traditional Algorithms

It was not until 2011 that a fully functional *multialgorithm CAD system* was tested end to end and shown to perform DR screenings on par with experts [96]. Since, much of research has gone into the improvement of these components, as described in earlier sections of this chapter.

TABLE 17.6

Messidor Grading Schemes for Retinopathy and Risk of Macular Edema

Grade	Description
Retinopathy Grade	
0	Normal, healthy fundus; no MA or hemorrhages
1	Less than 5 MAs and no hemorrhages
2	Less than 15 MAs and less than 5 hemorrhages
3	Greater than 15 MAs or hemorrhages or neovascularization
Risk of Macular Edema	
0	No visible hard exudates
1	Shortest distance between macula and hard exudates greater than one papilla diameter
2	Shortest distance between macula and hard exudates equal or less than one papilla diameter

Source: Adapted from E. Decencière, X. Zhang, G. Cazuguel, B. Lay, B. Cochener, C. Trone, P. Gain, R. Ordonez, P. Massin, A. Erginay, B. Charton, and J.-C. Klein, Feedback on a Publicly Distributed Database: The Messidor Database, *Image Analysis & Stereology*, 33(3):231–234, Aug 2014, ISSN 1854-5165, doi: 10.5566/ias.1155, URL http://www.ias-iss.org/ojs/IAS/article/view/1155.

The system followed a similar process as described in Figure 17.11 and Figure 17.16:

i. *Preprocessing*: Template matching was used to normalize the field of view for the data set.
ii. *Quality verification*: A Gaussian filter bank was applied to the data set and used to assign a quality to score to all the images. This process used clustering similar to the visual word dictionary method (Section 17.6.4) to assess the presence of morphologies. The score represented the probability the image was good quality. No image correction or removal was done.
iii. *Vessel segmentation*: A Gaussian filter bank fed into the supervised pixel classifier.
iv. *Optic disc detection*: The optic disc would conflict with bright lesion detection if not removed from the image. Intensity features were used to find the centre pixel and corresponding region of optic disc.
v. *Lesion candidate detectors*: A Gaussian filter bank was applied to the image and fed into a supervised pixel classifier. Lesion candidates were assigned a shape, structure, colour, and contrast. Two classifiers were trained, one for red lesions and the other for bright lesions.

TABLE 17.7

ICDR Severity Levels for Retinopathy and Risk of Macular Edema

Measure	Score	Description
DR Severity Level		
No apparent DR	0	Normal, healthy fundus; no abnormalities
Mild NPDR	1	MA only
Moderate NPDR	2	More than just MA but less than severe NPDR
Severe NPDR	3	No signs of PDR and *any* of the following:
		• >20 intraretinal hemorrhages in each of 4 quadrants
		• Definite venous beading in ≥2 quadrants
		• Prominent intraretinal microvascular abnormalities in ≥1 quadrant
PDR	4	One or more of the following: neovascularization and/or vitreous or preretinal hemorrhage
Macular Edema Severity Level		
Macular edema	0	Exudates or apparent thickening within 1 disc diameter from the fovea
No macular edema	1	No exudates and no apparent thickening within 1 disc diameter from the fovea

Source: Adapted from M. D. Abràmoff, J. C. Folk, D. P. Han, J. D. Walker, D. F. Williams, S. R. Russell, P. Massin, B. Cochener, P. Gain, L. Tang, M. Lamard, D. C. Moga, G. Quellec, and M. Niemeijer. Automated Analysis of Retinal Images for Detection of Referable Diabetic Retinopathy, *JAMA Ophthalmology*, 131(3):351–357, 2013, doi: 10.1001/jamaophthalmol.2013.1743, URL +http://dx.doi.org/10.1001/jamaophthalmol.2013.1743.

vi. *Decision system for DR grading*: The outputs of ii–v were fed into a k-nearest-neighbour classifier to determine final DR grade.

17.6.5.2 Deep Algorithms

Both of the deep methods possess a similar architectural philosophy to DR severity assessments: pathology-sensitive ensemble CNNs. Much like in Section 17.6.4, the earlier network implements the expert mimicry approach, whereas the second uses the feature discovery approach; the latter is in fact the same algorithm as was presented previously.

The *expert mimicry ensemble* (EME) followed a similar process as the CAD system discussed above; however, the authors augmented the ICDR severity levels to further scrutinize extreme cases of DR, as shown in Table 17.8 [14]. This work used a series of CNNs as feature detectors, each optimized for different morphologies and pathologies in normal anatomy and diseased cases, as well as quality. While the exact number and architecture of the networks were not published, the examples of these networks in the published study include optic disc, fovea, hemorrhages, exudates, and neovascularization [14]. The system used one set of detectors to discover where the pathological features existed within the fundus, and another set to determine whether the feature was a type of pathology. The output of all the networks was fed into two separate random forest algorithms, one for *referable DR* and the other for *vision-threatening DR*, each trained to classify the images against expert annotated training references. The quality detector resulted in 4% of the samples being rejected. The FDE is the same algorithm that was discussed in Section

TABLE 17.8

2016 Adaptation of ICDR Grading Schemes

Grade	Description
DR Severity Level	
No DR	ICDR level 0 (no DR) or 1 (mild DR) and no macular edema
Referable DR	ICDR level 2 (moderate NPDR), 3 (severe NPDR), 4 (PDR), or macular edema
Vision-threatening DR	ICDR level 3 (severe NPDR), 4 (PDR), or macular edema
Macular Edema Severity Level Unchanged	

Source: Adapted from M. D. Abràmoff, Y. Lou, A. Erginay, W. Clarida, R. Amelon, J. C. Folk, and M. Niemeijer, Improved Automated Detection of Diabetic Retinopathy on a Publicly Available Dataset through Integration of Deep Learning Detection of Diabetic Retinopathy, *Investigative Ophthalmology & Visual Science*, 57(13):5200, 2016, doi: 10.1167/iovs.16-19964, URL +http://dx.doi.org/10.1167/iovs.16-19964.

17.6.4, which used the selection criterion to distill pixel-level heat maps of pathologies while being trained on image-level labels for DR severity.

17.6.5.3 Comparing Performances

Upon reflection of the traditional methods and their deep counterparts, the main difference is the use of curated filter banks versus the use of a CNN to automatically generate them, selecting only the best for use at test time. It is worth noting that until deep learning became more prominent in this field, perhaps due to the 2015 Kaggle competition, algorithms were unable to reliably achieve an AUC score above 0.900. Since, deep learning methods have been game changing in the area of DR severity assessments, achieving significant and quantifiable improvements in SP and SN, reflected in an ever-growing gap in AUC: 0.893 for traditional methods and 0.980 for deep networks.

While the EME may appear to be superior in terms of AUC at first glance, it is in fact the FDE that is the best algorithm discussed herein. There are three key methodological factors that must be taken into consideration in order to properly compare the performances of the CNNs in this section:

i. *Data set*: It is known that the Kaggle database is far more challenging than Messidor, as can be seen in Table 17.4 for the SSCNN's AUC scores of 0.972 against Messidor and 0.894 against Kaggle. The SSCNN was trained and tested against Messidor in the same manner, but with Kaggle instead of a private data set. It is more than likely the EME would perform in the same range, if applied to the Kaggle data set, resulting in an AUC score near 0.900 for assessing DR severity; the FDE would easily surpass it with its AUC of 0.954.

ii. *Fidelity*: EME used a quality assessment network to automatically find and drop 4% of the sample set that it would have likely misclassified. While this is a good performance strategy, it greatly bolsters the final AUC of the network to something unrealistic for real-world settings. Conversely, the FDE was shown to perform well on poor-quality images and did not drop any images at any stage of training or at test time.

iii. *Generalizability*: Feature discovery approaches are able to generalize across many problem spaces and require substantially less effort by the experts when creating a training set for the network. The FDE was trained on image labels whereas the EME needed each pathological and anatomical feature to first be labelled by three experts in the field; that is a lot of time and energy!

Now, with these considerations in mind, the comparisons made between the FDE and *9-layer CNN* in Section 17.6.4 may be applied when comparing the FDE to the EME.

17.7 Outlook

Deep learning enables superior performance compared to traditional methods and humans with regard to accuracy, sensitivity, and area under the ROC. In many cases, DNNs will benefit from preprocessed images; however, this is often not a necessary step. In this way, they are much simpler to initialize than many traditional methods, which require strict preprocessing and a careful calibration of many parameters for a hand-selected set of features several orders of magnitude smaller than DNNs automatically produce. DNNs are beginning to outperform second observers, and methods for streamlining their architecture following training enable their use in mobile and remote settings, presenting a fantastic opportunity for applications in settings and communities that will truly benefit from the use of CAD systems. The continued development of unsupervised deep learning methods and artificial intelligence hints at a future with fewer misdiagnoses, reduced risk of disease complications, and better a better quality of care for all.

References

1. M. D. Abramoff, M. K. Garvin, and M. Sonka. Retinal imaging and image analysis. *IEEE Reviews in Biomedical Engineering*, 3:169–208, 2010. ISSN 1937-3333. doi: 10.1109/RBME.2010.2084567.
2. M. D. Abramoff, M. Niemeijer, and S. R. Russell. Automated detection of diabetic retinopathy: Barriers to translation into clinical practice. *Expert Review of Medical Devices*, 7(2):287–296, 2010. doi: 10.1586/erd.09.76. URL http://dx.doi.org/10.1586/erd.09.76.
3. M. D. Abràmoff, Y. Lou, A. Erginay, W. Clarida, R. Amelon, J. C. Folk, and M. Niemeijer. Improved automated detection of diabetic retinopathy on a publicly available dataset through integration of deep learning detection of diabetic retinopathy. *Investigative Ophthalmology & Visual Science*, 57(13):5200, 2016. doi: 10.1167/iovs.16-19964. URL +http://dx.doi.org/10.1167/iovs.16-19964.
4. R. Achanta, A. Shaji, K. Smith, A. Lucchi, P. Fua, and S. Süsstrunk. Slic superpixels compared to state-of-the-art superpixel methods. *IEEE Transactions on Pattern Analysis and Machine Intelligence*, 34(11):2274–2282, Nov 2012. ISSN 0162-8828. doi: 10.1109/TPAMI.2012.120.

5. S. A. Ali Shah, A. Laude, I. Faye, and T. B. Tang. Automated microaneurysm detection in diabetic retinopathy using curvelet transform. *Journal of Biomedical Optics*, 21(10):101404, 2016. doi: 10.1117/1.JBO.21.10.101404. URL http://dx.doi.org/10.1117/1.JBO.21.10.101404.

6. A. Almazroa, S. Alodhayb, E. Osman, E. Ramadan, M. Hummadi, M. Dlaim, M. Alkatee, K. Raahemifar, and V. Lakshminarayanan. Agreement among ophthalmologists in marking the optic disc and optic cup in fundus images. *International Ophthalmology*, 1–17, 2016. doi: 10.1007/s10792-016-0329-x.

7. R. Annunziata, A. Garzelli, L. Ballerini, A. Mecocci, and E. Trucco. Leveraging multiscale Hessian-based enhancement with a novel exudate inpainting technique for retinal vessel segmentation. *IEEE Journal of Biomedical and Health Informatics*, 20(4):1129–1138, July 2016. ISSN 2168-2194. doi: 10.1109/JBHI.2015.2440091.

8. G. Azzopardi, N. Strisciuglio, M. Vento, and N. Petkov. Trainable cosfire filters for vessel delineation with application to retinal images. *Medical Image Analysis*, 19(1):46–57, 4 2015. doi: 10.1016/j.media.2014.08.002. URL http://dx.doi.org/10.1016/j.media.2014.08.002.

9. Y. Bengio, A. Courville, and P. Vincent. Representation learning: A review and new perspectives. *IEEE Transactions on Pattern Analysis and Machine Intelligence*, 35(8):1798–1828, Aug 2013. ISSN 0162-8828. doi: 10.1109/TPAMI.2013.50.

10. J. Canny. A computational approach to edge detection. *IEEE Transactions on Pattern Analysis and Machine Intelligence*, PAMI-8(6):679–698, Nov 1986. ISSN 0162-8828. doi: 10.1109/TPAMI.1986.4767851.

11. L. M. Chalupa and J. S. Werner. *The Visual Neurosciences*. MIT Press, 2004.

12. D. C. Cireşan, U. Meier, J. Masci, L. M. Gambardella, and J. Schmidhuber. Flexible, high performance convolutional neural networks for image classification. In *Proceedings of the Twenty-Second International Joint Conference on Artificial Intelligence—Volume Two*, IJCAI'11, pages 1237–1242. AAAI Press, 2011. ISBN 978-1-57735-514-4. doi: 10.5591/978-1-57735-516-8/IJCAI11-210. URL http://dx.doi.org/10.5591/978-1-57735-516-8/IJCAI11-210.

13. M. J. Cree. The Waikato Microaneurysm Detector. *The University of Waikato, Tech. Rep*, 2008.

14. M. J. Cree, E. Gamble, and D. Cornforth. Colour normalisation to reduce inter-patient and intra-patient variability in microaneurysm detection in colour retinal images. *Proc. Workshop Digital Image Comput.*, pages 163–168, 2005.

15. J. Cuadros and G. Bresnick. Eyepacs: An adaptable telemedicine system for diabetic retinopathy screening. *Journal of Diabetes Science and Technology*, 3(3):509–516, 2009.

16. E. Decencière, X. Zhang, G. Cazuguel, B. Lay, B. Cochener, C. Trone, P. Gain, R. Ordonez, P. Massin, A. Erginay, B. Charton, and J.-C. Klein. Feedback on a publicly distributed database: the messidor database. *Image Analysis & Stereology*, 33(3):231–234, Aug. 2014. ISSN 1854-5165. doi: 10.5566/ias.1155. URL http://www.ias-iss.org/ojs/IAS/article/view/1155.

17. F. C. Delori and K. P. Pflibsen. Spectral reflectance of the human ocular fundus. *Applied Optics*, 28(6):1061–1077, Mar 1989. doi: 10.1364/AO.28.001061. URL http://ao.osa.org/abstract.cfm?URI=ao-28-6-1061.

18. T. G. Dietterich. *Ensemble Methods in Machine Learning*, pages 1–15. Springer, Berlin, 2000. ISBN 978-3-540-45014-6. doi: 10.1007/3-540-45014-9_1. URL http://dx.doi.org/10.1007/3-540-45014-9_1.

19. J. R. Ehrlich and N. M. Radcliffe. The role of clinical parapapillary atrophy evaluation in the diagnosis of open angle glaucoma. *Clinical Ophthalmology*, 4:971–976, 2010. ISSN 1177-5467. doi: 10.2147/opth.s12420. URL http://europepmc.org/articles/PMC2938276.

20. M. F. El-Bab, N. Shawky, A. Al-Sisi, and M. Akhtar. Retinopathy and risk factors in diabetic patients from al-Madinah al-Munawarah in the Kingdom of Saudi Arabia. *Clinical Ophthalmology (Auckland, N.Z.)*, 6:269–276, 2012. doi: 10.2147/OPTH.S27363. URL http://www.ncbi.nlm.nih.gov/pmc/articles/PMC3284208/.

21. Y. Fengshou. *Extraction of Features from Fundus Images for Glaucoma Assessment*. PhD thesis, National University of Singapore, 2011.

22. A. Foi and G. Boracchi. Anisotropically foveated non-local image denoising. In *2013 IEEE International Conference on Image Processing*, pages 464–468, Sept 2013. doi: 10.1109/ICIP.2013.6738096.

23. M. Fraz, P. Remagnino, A. Hoppe, B. Uyyanonvara, A. Rudnicka, C. Owen, and S. Barman. Blood vessel segmentation methodologies in retinal images—a survey. *Computer Methods and Programs in Biomedicine*, 108(1):407–433, 2012a. ISSN 0169-2607. doi: 10.1016/j.cmpb.2012.03.009. URL http://www.sciencedirect.com/science/article/pii/S0169260712000843.

24. M. Fraz, R. Welikala, A. Rudnicka, C. Owen, D. Strachan, and S. Barman. Quartz: Quantitative analysis of retinal vessel topology and size—an automated system for quantification of retinal vessels morphology. *Expert Systems with Applications*, 42(20):7221–7234, 2015. ISSN 0957-4174. doi: 10.1016/j.eswa.2015.05.022. URL http://www.sciencedirect.com/science/article/pii/S0957417415003504.

25. M. M. Fraz, P. Remagnino, A. Hoppe, B. Uyyanonvara, A. R. Rudnicka, C. G. Owen, and S. A. Barman. An ensemble classification-based approach applied to retinal blood vessel segmentation. *IEEE Transactions on Biomedical Engineering*, 59(9):2538–2548, Sept 2012b. ISSN 0018-9294. doi: 10.1109/TBME.2012.2205687.

26. K. A. Goatman, M. J. Cree, J. A. Olson, J. V. Forrester, and P. F. Sharp. Automated measurement of microaneurysm turnover. *Investigative Ophthalmology & Visual Science*, 44(12):5335, 2003. doi: 10.1167/iovs.02-0951. URL +http://dx.doi.org/10.1167/iovs.02-0951.

27. I. Goodfellow, Y. Bengio, and A. Courville. *Deep Learning*. MIT Press, 2016. URL http://www.deeplearningbook.org.

28. A. Graves. *Supervised Sequence Labelling*, pages 5–13. Springer, Berlin, 2012. ISBN 978-3-642-24797-2. doi: 10.1007/978-3-642-24797-2_2. URL http://dx.doi.org/10.1007/978-3-642-24797-2_2.

29. D. Green and J. Swets. *Signal Detection Theory and Psychophysics*. Wiley, 1966.
30. S. Hijazi, R. Kumar, and C. Rowen. Using convolutional neural networks for image recognition. *Cadence Design Systems—White Paper*, 2015. URL http://www.multimediadocs.com/assets/cadence_emea/documents/using_convolutional_neural_networks_for_image_recognition.pdf.
31. G. E. Hinton, S. Osindero, and Y.-W. Teh. A fast learning algorithm for deep belief nets. *Neural Computation*, 18(7):1527–1554, 2017/03/22 2006. doi: 10.1162/neco.2006.18.7.1527. URL http://dx.doi.org/10.1162/neco.2006.18.7.1527.
32. M. J. Hirsch and R. E. Wick. *The Optometric Profession*. Chilton Book Company, 1968.
33. A. Hoover, V. Kouznetsova, and M. Goldbaum. Locating blood vessels in retinal images by piecewise threshold probing of a matched filter response. *IEEE Transactions on Medical Imaging*, 19(3):203–210, March 2000. ISSN 0278-0062. doi: 10.1109/42.845178.
34. D. Huang, E. A. Swanson, C. P. Lin, J. S. Schuman, W. G. Stinson, W. Chang, M. R. Hee, T. Flotte, K. Gregory, C. A. Puliafito et al. Optical coherence tomography. *Science*, 254(5035):1178, 1991.
35. D. H. Hubel and T. N. Wiesel. Receptive fields of single neurones in the cat's striate cortex. *The Journal of Physiology*, 148(3):574–591, 1959.
36. D. H. Hubel and T. N. Wiesel. Receptive fields, binocular interaction and functional architecture in the cat's visual cortex. *The Journal of Physiology*, 160(1):106–154, 1962.
37. D. H. Hubel and T. N. Wiesel. Receptive fields and functional architecture of monkey striate cortex. *The Journal of Physiology*, 195(1):215–243, 1968.
38. H. Jelinek and M. J. Cree. *Automated Image Detection of Retinal Pathology*. CRC Press, Boca Raton, FL, 2009. ISBN 978-0-8493-7556-9. doi: 10.1201/9781420037005. URL http://dx.doi.org/10.1201/9781420037005.
39. J. B. Jonas, W. M. Budde, and S. Panda-Jonas. Ophthalmoscopic evaluation of the optic nerve head. *Survey of Ophthalmology*, 43(4):293–320, 1999.
40. J. P. Jones and L. A. Palmer. An evaluation of the two-dimensional Gabor filter model of simple receptive fields in cat striate cortex. *Journal of Neurophysiology*, 58(6):1233–1258, 1987.
41. T. Kauppi, V. Kalesnykiene, J.-K. Kamarainen, L. Lensu, I. Sorri, A. Raninen, R. Voutilainen, H. Uusitalo, H. Kälviäinen, and J. Pietilä. The diaretdb1 diabetic retinopathy database and evaluation protocol. In *Proceedings of the British Machine Vision Conference*, pages 1–10. BMVA Press, 2007. ISBN 1-901725-34-0. doi: 10.5244/C.21.15.
42. D. P. Kingma and J. Ba. Adam: A method for stochastic optimization. *CoRR*, abs/1412.6980, 2014. URL http://arxiv.org/abs/1412.6980.
43. R. Klein, B. E. Klein, S. E. Moss, M. D. Davis, and D. L. DeMets. The Wisconsin epidemiologic study of diabetic retinopathy: II. prevalence and risk of diabetic retinopathy when age at diagnosis is less than 30 years. *Archives of Ophthalmology*, 102(4):520–526, 1984.
44. E. Kohner, I. Stratton, S. Aldington, R. Turner, D. Matthews, U. P. D. S. U. Group et al. Microaneurysms in the development of diabetic retinopathy (ukpds 42). *Diabetologia*, 42(9):1107–1112, 1999.
45. G. Kovács and A. Hajdu. A self-calibrating approach for the segmentation of retinal vessels by template matching and contour reconstruction. *Medical Image Analysis*, 29(3):24–46, 2016. doi: 10.1016/j.media.2015.12.003. URL http://dx.doi.org/10.1016/j.media.2015.12.003.
46. V. Lakshminanarayanan, A. Raghuram, J. W. Myerson, and S. Varadharajan. The fractal dimension in retinal pathology. *Journal of Modern Optics*, 50(11):1701–1703, 2003. doi: 10.1080/09500340308235515.
47. V. Lakshminarayanan. Stochastic eye movements while fixating on a stationary target. *Stochastic Processes and their Applications*, pages 39–49, Springer-Narosa, New Delhi, 1999.
48. V. Lakshminarayanan. New results in biomedical image processing. In *2012 International Conference on Fiber Optics and Photonics*, pages 1–3. Optical Society of America, Dec 2012a. doi: 10.1364/PHOTONICS.2012.T1A.2.
49. V. Lakshminarayanan. The global problem of blindness and visual dysfunction. In *Photonic Innovations and Solutions for Complex Environments and Systems*, volume 8482 of SPIE Optical Engineering+ Applications, pages 84820A–84820A. International Society for Optics and Photonics, Oct. 2012b. doi: 10.1117/12.928050.
50. V. Lakshminarayanan and L. S. Varadharajan. *Special Functions for Optical Science and Engineering*, volume TT103. SPIE Press, 2015.
51. V. Lakshminarayanan, J. Zelek, and A. McBride. "Smartphone science" in eye care and medicine. *Optics & Photonics News*, 26(1):44–51, Jan 2015. doi: 10.1364/OPN.26.1.000044. URL http://www.osa-opn.org/abstract.cfm?URI=opn-26-1-44.
52. V. Lakshminarayanan, R. Rasheed, and A. Boudriva, editors. *Ibn al Haytham: Founder of Vision Science?* CRC Press, Boca Raton, FL, 2017.
53. B. S. Y. Lam, Y. Gao, and A. W. C. Liew. General retinal vessel segmentation using regularization-based multiconcavity modeling. *IEEE Transactions on Medical Imaging*, 29(7):1369–1381, July 2010. ISSN 0278-0062. doi: 10.1109/TMI.2010.2043259.
54. G. R. Lanckriet, N. Cristianini, P. Bartlett, L. E. Ghaoui, and M. I. Jordan. Learning the kernel matrix with semidefinite programming. *Journal of Machine Learning Research*, 5(1):27–72, January 2004.
55. Y. LeCun, Y. Bengio, and G. Hinton. Deep learning. *Nature*, 521(7553):436–444, 05 2015. doi: 10.1038/nature14539. URL http://dx.doi.org/10.1038/nature14539.
56. H. Leopold, J. Orchard, V. Lakshminarayanan, and J. Zelek. A deep learning network for segmenting retinal vessel morphology. In *2016 38th Annual International Conference of the IEEE Engineering in Medicine and Biology Society (EMBC)*, page 3144, Aug 2016. doi: 10.1109/EMBC.2016.17752973.
57. H. A. Leopold, J. Orchard, J. Zelek, and V. Lakshminarayanan. Use of Gabor filters and deep

networks in the segmentation of retinal vessel morphology. *Proc. SPIE Biomedical Optics*, 10068, 2017. doi: 10.1117/12.2252988. URL http://dx.doi.org/10.1117/12.2252988.
58. H. A. Leopold, J. Orchard, J. Zelek, and V. Lakshminarayanan. Segmentation and feature extraction of retinal vascular morphology. *Proc. SPIE Medical Imaging*, 10133, 2017. doi: 10.1117/12.2253744. URL http://dx.doi.org/10.1117/12.2253744.
59. Q. Li, B. Feng, L. Xie, P. Liang, H. Zhang, and T. Wang. A cross-modality learning approach for vessel segmentation in retinal images. *IEEE Transactions on Medical Imaging*, 35(1):109–118, Jan 2016. ISSN 0278-0062. doi: 10.1109/TMI.2015.2457891.
60. P. Liskowski and K. Krawiec. Segmenting retinal blood vessels with deep neural networks. *IEEE Transactions on Medical Imaging*, 35(11):2369–2380, Nov 2016. ISSN 0278-0062. doi: 10.1109/TMI.2016.2546227.
61. C. A. Lupascu, D. Tegolo, and E. Trucco. Fabc: Retinal vessel segmentation using adaboost. *IEEE Transactions on Information Technology in Biomedicine*, 14(5):1267–1274, Sept 2010. ISSN 1089-7771. doi: 10.1109/TITB.2010.2052282.
62. S. Marĉelja. Mathematical description of the responses of simple cortical cells. *JOSA*, 70(11):1297–1300, 1980.
63. D. Marr. *Vision: A Computational Investigation into the Human Representation and Processing of Visual Information*. W.H. Freeman, San Francisco, 1982. ISBN 0716712849.
64. D. Marr and E. Hildreth. Theory of edge detection. *Proceedings of the Royal Society of London B: Biological Sciences*, 207(1167):187–217, 1980.
65. D. Martin, C. Fowlkes, D. Tal, and J. Malik. A database of human segmented natural images and its application to evaluating segmentation algorithms and measuring ecological statistics. In *Proceedings Eighth IEEE International Conference on Computer Vision. ICCV 2001*, volume 2, pages 416–423, 2001. doi: 10.1109/ICCV.2001.937655.
66. M. D. Abràmoff, J. C. Folk, D. P. Han, J. D. Walker, D. F. Williams, S. R. Russell, P. Massin, B. Cochener, P. Gain, L. Tang, M. Lamard, D. C. Moga, G. Quellec, and M. Niemeijer. Automated analysis of retinal images for detection of referable diabetic retinopathy. *JAMA Ophthalmology*, 131(3):351–357, 2013. doi: 10.1001/jamaophthalmol.2013.1743. URL +http://dx.doi.org/10.1001/jamaophthalmol.2013.1743.
67. M. Melinščak, P. Prentašić, and S. Lončarić. Retinal vessel segmentation using deep neural networks. In *VISAPP 2015 (10th International Conference on Computer Vision Theory and Applications)*, 2015.
68. N. Mirza, T. Reynolds, M. Coletta, K. Suda, I. Soyiri, A. Markle, H. Leopold, L. Lenert, E. Samoff, A. Siniscalchi, and L. Streichert. Steps to a sustainable public health surveillance enterprise. *Online Journal of Public Health Informatics*, 5(2), 2013. ISSN 1947-2579. URL http://128.248.156.56/ojs/index.php/ojphi/article/view/4703.
69. T. T. Nguyen, J. J. Wang, A. R. Sharrett, F. A. Islam, R. Klein, B. E. Klein, M. F. Cotch, and T. Y. Wong. Relationship of retinal vascular caliber with diabetes and retinopathy the multi-ethnic study of atherosclerosis (mesa). *Diabetes Care*, 31(3):544–549, 2008.
70. M. Niemeijer, B. Van Ginneken, M. J. Cree, A. Mizutani, G. Quellec, C. I. Sánchez, B. Zhang, R. Hornero, M. Lamard, C. Muramatsu et al. Retinopathy online challenge: automatic detection of microaneurysms in digital color fundus photographs. *IEEE Transactions on Medical Imaging*, 29(1):185–195, 2010.
71. A. Osareh, M. Mirmehdi, B. Thomas, and R. Markham. Classification and localisation of diabetic-related eye disease. In *European Conference on Computer Vision*, pages 502–516. Springer, 2002.
72. C. W. Oyster. *The Human Eye*. Sinauer, Sunderland, MA, 1999.
73. G. Quellec, K. Charrière, Y. Boudi, B. Cochener, and M. Lamard. Deep image mining for diabetic retinopathy screening. *ArXiv e-prints*, Oct. 2016.
74. H. A. Quigley, R. M. Hohman, E. M. Addicks, and W. R. Green. Blood vessels of the glaucomatous optic disc in experimental primate and human eyes. *Investigative Ophthalmology & Visual Science*, 25(8):918–931, 1984.
75. E. Ricci and R. Perfetti. Retinal blood vessel segmentation using line operators and support vector classification. *IEEE Transactions on Medical Imaging*, 26(10):1357–1365, Oct 2007. ISSN 0278-0062. doi: 10.1109/TMI.2007.898551.
76. D. Ringach and R. Shapley. Reverse correlation in neurophysiology. *Cognitive Science*, 28(2):147–166, 2004. ISSN 0364-0213. doi: http://dx.doi.org/10.1016/j.cogsci.2003.11.003. URL http://www.sciencedirect.com/science/article/pii/S0364021303001174. Rendering the Use of Visual Information from Spiking Neurons to Recognition.
77. A. Rocha, T. Carvalho, H. F. Jelinek, S. Goldenstein, and J. Wainer. Points of interest and visual dictionaries for automatic retinal lesion detection. *IEEE Transactions on Biomedical Engineering*, 59(8):2244–2253, Aug 2012. ISSN 0018-9294. doi: 10.1109/TBME.2012.2201717.
78. M. S. Roy, R. Klein, B. J. O'Colmain, B. E. Klein, S. E. Moss, and J. H. Kempen. The prevalence of diabetic retinopathy among adult type1 diabetic persons in the United States. *Archives of Ophthalmology*, 122(4):546–551, 2004. doi: 10.1001/archopht.122.4.546. URL +http://dx.doi.org/10.1001/archopht.122.4.546.
79. D. E. Rumelhart and D. Zipser. Feature discovery by competitive learning. In *Parallel Distributed Processing: Explorations in the Microstructure of Cognition*, vol. 1, pages 151–193. MIT Press, 1986.
80. C. I. Sánchez, M. Niemeijer, A. V. Dumitrescu, M. S. A. Suttorp-Schulten, M. D. Abràmoff, and B. van Ginneken. Evaluation of a computer-aided diagnosis system for diabetic retinopathy screening on public data. *Investigative Ophthalmology and Visual Science*, 52(7):4866, 2011. doi: 10.1167/iovs.10-6633. URL +http://dx.doi.org/10.1167/iovs.10-6633.
81. P. H. Scanlon, C. Foy, R. Malhotra, and S. J. Aldington. The influence of age, duration of diabetes, cataract, and pupil size on image quality in digital photographic retinal screening. *Diabetes Care*, 28(10):2448–2453, 2005.

ISSN 0149-5992. doi: 10.2337/diacare.28.10.2448. URL http://care.diabetesjournals.org/content/28/10/2448.

82. D. A. Sim, P. A. Keane, A. Tufail, C. A. Egan, L. P. Aiello, and P. S. Silva. Automated retinal image analysis for diabetic retinopathy in telemedicine. *Current Diabetes Reports*, 15(3):14, 2015. ISSN 1539-0829. doi: 10.1007/s11892-015-0577-6. URL http://dx.doi.org/10.1007/s11892-015-0577-6.

83. J. V. Soares, J. J. Leandro, R. M. Cesar, H. F. Jelinek, and M. J. Cree. Retinal vessel segmentation using the 2-D Gabor wavelet and supervised classification. *IEEE Transactions on Medical Imaging*, 25(9):1214–1222, Sept 2006. ISSN 0278-0062. doi: 10.1109/TMI.2006.879967.

84. R. Srivastava, L. Duan, D. W. Wong, J. Liu, and T. Y. Wong. Detecting retinal microaneurysms and hemorrhages with robustness to the presence of blood vessels. *Computer Methods and Programs in Biomedicine*, 138(3):83–91, March 2017. doi: 10.1016/j.cmpb.2016.10.017. URL http://dx.doi.org/10.1016/j.cmpb.2016.10.017.

85. J. Staal, M. D. Abràmoff, M. Niemeijer, M. A. Viergever, and B. van Ginneken. Ridge-based vessel segmentation in color images of the retina. *IEEE Transactions on Medical Imaging*, 23(4):501–509, April 2004. ISSN 0278-0062. doi: 10.1109/TMI.2004.825627.

86. R. Szeliski. *Computer Vision: Algorithms and Applications*. Springer-Verlag, New York, NY, USA, 1st edition, 2010. ISBN 1848829345, 9781848829343.

87. G. Tangelder, N. Reus, and H. Lemij. Estimating the clinical usefulness of optic disc biometry for detecting glaucomatous change over time. *Eye*, 20(7):755–763, 2005. doi: 10.1038/sj.eye.6701993. URL http://dx.doi.org/10.1038/sj.eye.6701993.

88. D. Thapa, K. Raahemifar, W. R. Bobier, and V. Lakshminarayanan. Comparison of super-resolution algorithms applied to retinal images. *Journal of Biomedical Optics*, 19(5):056002, 2014. doi: 10.1117/1.JBO.19.5.056002. URL http://dx.doi.org/10.1117/1.JBO.19.5.056002.

89. D. Thapa, K. Raahemifar, W. R. Bobier, and V. Lakshminarayanan. A performance comparison among different super-resolution techniques. *Computers & Electrical Engineering*, 54:313–329, 2016. ISSN 0045-7906. doi: http://dx.doi.org/10.1016/j.compeleceng.2015.09.011. URL http://www.sciencedirect.com/science/article/pii/S0045790615003183.

90. *Federal Food, Drug, and Cosmetic Act (FD&C Act)*. United States House of Representatives, March 2013.

91. M. J. van Grinsven, B. van Ginneken, C. B. Hoyng, T. Theelen, and C. I. Sánchez. Fast convolutional neural network training using selective data sampling: Application to hemorrhage detection in color fundus images. *IEEE Transactions on Medical Imaging*, 35(5):1273–1284, May 2016. ISSN 0278-0062. doi: 10.1109/TMI.2016.2526689.

92. M. J. J. P. van Grinsven, Y. T. E. Lechanteur, J. P. H. van de Ven, B. van Ginneken, C. B. Hoyng, T. Theelen, and C. I. Sánchez. Automatic drusen quantification and risk assessment of age-related macular degeneration on color fundus imagesautomatic drusen quantification on color fundus images. *Investigative Ophthalmology & Visual Science*, 54(4):3019, 2013. doi: 10.1167/iovs.12-11449. URL +http://dx.doi.org/10.1167/iovs.12-11449.

93. T. Y. Wong, R. Klein, A. R. Sharrett, B. B. Duncan, D. J. Couper, J. M. Tielsch, B. E. Klein, and L. D. Hubbard. Retinal arteriolar narrowing and risk of coronary heart disease in men and women: The atherosclerosis risk in communities study. *JAMA*, 287(9):1153–1159, 2002. doi: 10.1001/jama.287.9.1153. URL +http://dx.doi.org/10.1001/jama.287.9.1153.

94. C.-Y. Yang, C. Ma, and M.-H. Yang. Single-image super-resolution: A benchmark. In D. Fleet, T. Pajdla, B. Schiele, and T. Tuytelaars, editors, *Computer Vision— ECCV 2014: 13th European Conference, Zurich, Switzerland, September 6–12, 2014, Proceedings, Part IV*, pages 372–386, Cham, 2014. Springer International Publishing. ISBN 978-3-319-10593-2. doi: 10.1007/978-3-319-10593-2_25. URL http://dx.doi.org/10.1007/978-3-319-10593-2_25.

95. J.-J. Yang, J. Li, R. Shen, Y. Zeng, J. He, J. Bi, Y. Li, Q. Zhang, L. Peng, and Q. Wang. Exploiting ensemble learning for automatic cataract detection and grading. *Computer Methods and Programs in Biomedicine*, 124:45–57, 2016. ISSN 0169-2607. doi: http://dx.doi.org/10.1016/j.cmpb.2015.10.007. URL http://www.sciencedirect.com/science/article/pii/S0169260715002679.

96. R. A. Young and R. M. Lesperance. The Gaussian derivative model for spatial–temporal vision: II. Cortical data. *Spatial Vision*, 14(3):321–389, 2001.

97. R. A. Young, R. M. Lesperance, and W. W. Meyer. The Gaussian derivative model for spatial–temporal vision: I. Cortical model. *Spatial Vision*, 14(3):261–319, 2001.

98. Z. Zhang, R. Srivastava, H. Liu, X. Chen, L. Duan, D. W. Kee Wong, C. K. Kwoh, T. Y. Wong, and J. Liu. A survey on computer aided diagnosis for ocular diseases. *BMC Medical Informatics and Decision Making*, 14(1):80, 2014. ISSN 1472-6947. doi: 10.1186/1472-6947-14-80. URL http://dx.doi.org/10.1186/1472-6947-14-80.

18

Dictionary Learning Applications for HEp-2 Cell Classification

Sadaf Monajemi, Shahab Ensafi, Shijian Lu, Ashraf A. Kassim, Chew Lim Tan,
Saeid Sanei, and Sim-Heng Ong

CONTENTS

18.1 Introduction 369
18.2 Feature Extraction 370
18.3 Sparse Coding and Dictionary Learning 371
    18.3.1 Compressed Sensing 371
    18.3.2 Dictionary Learning 372
18.4 Adaptive Distributed Dictionary Learning 372
    18.4.1 Distributed Dictionary Learning 372
    18.4.2 Diffusion Adaptation Method 374
    18.4.3 Selection of the Combination Weights 374
18.5 Experiments and Results 376
    18.5.1 Data Sets and Evaluation Methods 376
    18.5.2 Classification Results 376
    18.5.3 Computational Cost 378
18.6 Conclusion 379
References 379

18.1 Introduction

The goal of this chapter is to review and study the application of dictionary learning (DL) methods for diagnosis of autoimmune diseases (ADs). According to the American Autoimmune Related Diseases Association (AARDA), ADs are among the top mortality causes as the result of immune system failure to recognize the body's normal protein as *self*. In these diseases, the immune system produces another type of antibody, called *autoantibody*, directed against that protein. This response of the immune system to the individual's own tissues is called *autoimmunity*, and the related diseases are named ADs. Early diagnosis of ADs plays a crucial role in the treatment process of these diseases.

Antinuclear antibodies (ANAs), which are found in many disorders including autoimmunity, cancer, and infection, are a kind of antibody that binds to the contents of the cell nucleus. By screening the blood serum, the presence of ANA can be confirmed, which in turn leads to diagnosis of some autoimmune disorders. According to the American College of Rheumatology, the gold standard test for detecting and qualifying ANA is called indirect immunofluorescence (IIF), which uses the human epithelial type-2 (HEp-2) tissue. In the IIF test, antibodies are first stained in HEp-2 tissue and then bond to a fluorescent chemical compound. The antibodies bound to the nucleus can have different patterns, which can be observed using microscope imaging. Using these patterns, the severity and phase of the ADs can be assessed [1]. However, due to variations in image quality, the interpretation of fluorescence patterns can be quite challenging. To overcome this challenge and interpret the patterns more consistently, automated methods are essential for cell classification.

Computer aided diagnosis systems have attracted much attention for HEp-2 cell classification and AD diagnosis. With the help of these systems, the cost and time of diagnosis can be reduced, and the results would be reproducible by different physicians. One of the main steps for an advanced HEp-2 cell classification process is

DL and sparse coding. In the DL process, the input images are first divided into small patches. Due to the sparse nature of patch-based image classification, DL and sparse coding approaches have been widely used for HEp-2 cell classification [2–5]. The next step is to extract the features of these patches and combine them to generate the final feature vectors for classification. Several HEp-2 cell classification methods have been proposed in recent years. In one of the earliest studies [6], Otsu's global thresholding technique [7] has been used for cell segmentation and extraction of texture features for cell classification. Texture and statistical features have been also utilized in Ref. [8], where the cells are classified via self-organizing maps. In another study, binary classifiers and spectral textural features are aggregated, and a reliability measure is introduced for classification [9]. Intensity level and staining pattern classification techniques have been reported in Refs. [10–12].

Most of the works described above use their own data sets. This makes a fair comparison of different methods a nearly impossible task for other researchers. The first publicly available HEp-2 cell classification data set was released at the 2012 International Conference on Pattern Recognition and is referred to as the ICPR2012 data set in this chapter [3]. Subsequently, an extended data set was introduced at the 2013 International Conference on Image Processing, referred to as the ICIP2013 data set. Two classification tasks have been suggested for the ICPR2012 and ICIP2013 benchmarking data sets [3]. In the first task, each cell of the specimen image is classified independently without considering the class of its neighboring cells, while in the second task, it is assumed that most of the cells in a particular specimen image belong to one of the classes [3]. Therefore, each specimen image is associated with a class by considering all the cells in that image.

Several studies have used these two public data sets for HEp-2 cell classification. In the submission that won the ICPR2012 contest, the authors considered an extension of local binary patterns (LBPs) as features and used the linear support vector machine (SVM) for cell classification [13]. Morphological features [14–16] and different histograms [5,17,18] have been successfully used to further enhance the classification results. In another study, the scale-invariant feature transform (SIFT) and LBP features have been combined with a discriminative sparse representation to classify HEp-2 cells. In Ref. [19], the authors proposed to use Fisher tensors on a Riemannian manifold and a learned bag of words (BoW) dictionary to classify the cells using SVM.

Although all the above studies provide some success for classification of the HEp-2 cells in the recent benchmarking tests (ICPR2012 and ICIP2013), there are still several challenges that need to be overcome. One of the fundamental challenges is the large number of image patches that have different sizes and shapes, which adds to the complexity of the problem. Moreover, most of the proposed methods use artificial features (including SIFT, LBP, and histograms of different patches) with several parameters to be tuned manually. The choice of the parameter values (such as the size and number of patches, smoothing parameters, and number of histogram bins) can affect the performance of the method. Parameter tuning is a complex and tedious task that can lead to massive memory and computational requirements. This is especially troublesome for the cases with a large number of training images, which is the case for the ICIP2013 data set. One of these parameters is the optimal size of the dictionary, which is heavily dependent on the data and visual features. In general, employing biologically inspired overcomplete dictionaries, where the dimension of the dictionary is much larger than the feature dimension, results in better classification accuracy [20]. However, employing high-dimensional dictionaries together with high-dimensional features not only reduces classification speed significantly, but also is a huge burden on machine memory and CPU.

In this chapter, we study and implement an adaptive distributed dictionary learning (ADDL) method. This distributed method tackles the HEp-2 cell classification problem in a less memory intensive and computationally efficient manner compared to the other methods. Here, we partition the dictionary matrix and the coding vector into N blocks, with each block associated with a subdictionary and a subvector. A connected network of N nodes can be formed with these blocks, where each node is in charge of updating its own subdictionary. Moreover, each node can be connected to a number of neighboring nodes, where the neighbors share their information to update the subdictionaries. Essentially, we observe that the DL problem can be reformulated as a distributed learning task over networks. Afterwards, the diffusion adaptation strategy [21,22] can be used to solve this distributed problem. We combine the information of neighboring nodes in an adaptive manner, which enables the nodes to learn about the usefulness of the received information and helps them to ignore misleading information.

18.2 Feature Extraction

As the first step of the automated cell image classification, we should extract the image features. Using intensity values directly as features has some problems. First, there are two significantly different intensity levels, namely, *positive* and *intermediate* levels, in each data set. In particular, the intensity values of positive images are greater than those of intermediate images, and consequently, positive cells can be easily seen but not the intermediate cells. Second, the ICIP2013 data set comprises gray-level images, while the ICPR2012 data set comprises color images. Finally, gray-level analysis might be inaccurate due to the noisy nature of the images.

To extract the appearance features of the images containing different types of HEp-2 cells, the SIFT [23] and speeded-up robust features (SURF) [24] are utilized. SIFT features are computed by down-sampling the image at different smoothed image levels, and the method provides better features in the presence of illumination changes (in positive and intermediate intensity levels). SURF is computed using the Hessian matrix and performs better in the presence of image rotation and blur [25]. Therefore, these two types of features complement each other, and combining them generates features with better representation and discrimination capability.

The standard SIFT feature extraction method first detects the corners [26,27] to capture the points of interest and then extracts the features of these points [28]. Since the HEp-2 cell patterns within the immunofluorescence images are usually very small, this approach is not suitable to solve the cell classification problem. This is due to the fact that the number of interest points for the homogeneous cell with homogeneous visual patterns is much lower than that of the centromere class that contains several shiny points, as shown in Figure 18.6. Therefore, we extract *grid* SIFT and SURF features, where grids are applied over the entire cell region, to capture the visual features as illustrated in Figure 18.1. Here, the entire cell image is divided into overlapping patches, where the SIFT and SURF features are captured and combined in each patch. In particular, 192 features are extracted for each patch containing 64 SURF and 128 SIFT features.

18.3 Sparse Coding and Dictionary Learning

Recently, there has been an increasing interest in sparse coding and DL in computer vision and image processing research for classification tasks. With the help of DL methods, the input data can be represented using as few components as possible, reconstructed by a linear combination of a few dictionary elements (atoms). These atoms construct the dictionary and are not necessarily orthogonal. It is important to note that in the DL method, the dictionary is inferred and constructed from the input data, while in the traditional approaches (such as the Fourier or wavelet transforms), the dictionaries are predefined. Learning a dictionary based on the input data can improve the sparsity, resulting in a sparse representation of the input signal. Before formulating the DL problem, we introduce the compressed sensing (CS) theory to give a better understanding of sparse coding.

18.3.1 Compressed Sensing

An essential step in signal and image processing is sampling a signal for transmission and reconstruction. Based on the well-known Shannon–Nyquist sampling theorem, the sampling rate should be more than twice the highest frequency of the signal for perfect signal reconstruction. On the other hand, CS theory states that a signal can be recovered with fewer samples than the Shannon–Nyquist sampling rate if there is prior knowledge about its sparsity. The sparsity of a signal $s \in \mathbb{R}^M$ means that it has only very few nonzero samples.

In the CS method, where the input signal s is sparse, it can be reconstructed by a linear combination of the columns of dictionary D:

$$s = Dz$$
$$\text{subject to} \quad \|z\|_0 \ll M, \tag{18.1}$$

where M is the dimension of the input signal s and z consists of the sparse coefficients of the dictionary bases. It should be noted that $\|z\|_0$ is the ℓ_0-norm of z,

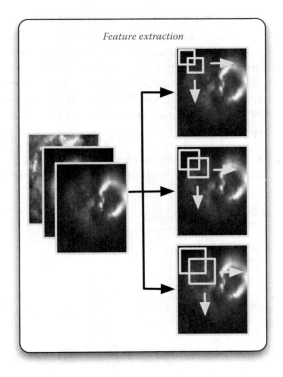

FIGURE 18.1
Grids are applied over the cell image, and the image is divided into overlapping patches, where in each patch, SIFT and SURF features are captured and combined together.

which represents the number of nonzero elements and is the sparsity measure. It has been shown that having the ℓ_0-norm constraint makes the problem NP-hard (nondeterministic polynomial time). Therefore, it can be replaced by ℓ_1-norm to relax the problem.

18.3.2 Dictionary Learning

Based on what was mentioned earlier in the overview, the input feature vector F_t for the DL algorithm is constructed by the extracted features of image patches. The DL problem can now be formulated as

$$\min_{z_t, D}(\| F_t - Dz_t \|^2 + \lambda \| z_t \|_1 + \frac{\beta}{2} \| z_t \|_2^2), \qquad (18.2)$$

where F_t is the $M \times 1$ input feature vector at time t, D is the $M \times K$ dictionary matrix, z_t is the $K \times 1$ sparse code vector, and λ and β are the adjustable penalty (regularization) terms. As mentioned earlier, the role of the L_1-norm term $\| z_t \|_1$ is to promote sparsity of the code vector, while the Euclidean norm $\| z_t \|_2$ ensures that the estimated values are small.

The main procedure of the DL problem consists of two stages:

- Sparse coding: The dictionary (D) is fixed, and the problem in Equation 18.2 is reformulated as

$$\min_{z_t}(\| F_t - Dz_t \|^2 + \lambda \| z_t \|_1 + \frac{\beta}{2} \| z_t \|_2^2) \qquad (18.3)$$

This is a linear regression problem with ℓ_1-norm and ℓ_2-norm regularization on the coefficients.

- Dictionary update: The coefficients are assumed to be fixed, and the dictionary is updated by

$$\min_{D}(\| F_t - Dz_t \|^2) \qquad (18.4)$$

Several methods have been proposed in the literature to solve the optimization problem of the DL such as K-SVD [29] and method of optimal directions (MOD) [30]. However, to solve the optimization problem in Equation 18.2, we introduce a computationally efficient distributed DL method.

18.4 Adaptive Distributed Dictionary Learning

Figure 18.2 shows an overview of the method for HEp-2 cell classification. SURF and SIFT features of the image patches are extracted and used as the inputs of the distributed DL. After learning the dictionary, the sparse codes of the image patches are combined using spatial pyramid matching (SPM) [31] and used for cell classification. As shown in Figure 18.2, there are three pyramid layers, where each input image is divided into 1, 4, and 16 regions. The final feature vector is obtained by applying max-pooling to the sparse codes of each region. Afterwards, a multi-class linear SVM can be learned for cell image classification. In the following, we introduce the distributed DL method that has been employed for this classification problem.

18.4.1 Distributed Dictionary Learning

As discussed earlier, the DL task can be quite intensive in terms of memory and computational requirements. It has been shown that learning the dictionary in a distributed manner can help overcome these challenges. We employ the recently proposed approach for distributed DL presented in Refs. [32,33]. Here, the dictionary matrix D and the coding vector z are partitioned into blocks as

$$D = [D_1 \ldots D_N], \quad z = \text{col}\{z_1, \ldots, z_N\}, \qquad (18.5)$$

where D_k is a subdictionary of size $M \times P_k$ and z_k is a subvector of size $P_k \times 1$. It should be noted that the sizes

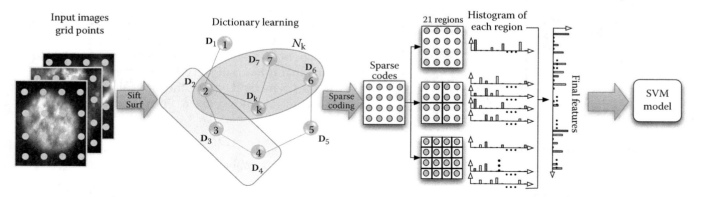

FIGURE 18.2
Proposed ADDL framework.

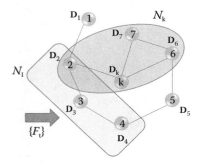

FIGURE 18.3
Sample of a connected network where each agent k is responsible for learning a subdictionary D_k. The agents connected with solid lines to agent k are its neighbors represented by N_k and can share information with each other. Moreover, it can be seen that the input data F_t is presented to a subset of agents represented by N_I.

of the subdictionaries add up to the total size of the dictionary:

$$P_1 + \ldots + P_N = K \quad (18.6)$$

Now we are able to form a connected network of N nodes where each node k is in charge of updating its own subdictionary D_k. In this way, the dictionary matrix D is distributed over the network. Figure 18.3 represents an example of such a network where each node is connected to a number of neighboring agents. Each node in the network can share information with and receive information from its neighbors. Moreover, the input features F_t might be presented only to a subset of the nodes (shown by $\mathcal{N}_I$) rather than all of them, which can further enhance the computational advantage of the method. The experiments in Section 18.5 show that the case where the input data is only presented to a subset of agents is computationally more efficient while retaining comparable performance with other methods. This is in fact due to the distributed nature of the network where the agents are allowed to interact and cooperate with their neighbors, resulting in dispersion of information over the network.

Considering Equation 18.5 in the DL problem, Equation 18.2 can be reformulated as

$$\min_{z_t, D} \left(\| F_t - \sum_{k=1}^{N} D_k z_{k,t} \|^2 + \sum_{k=1}^{N} (\lambda \| z_{k,t} \|_1 + \frac{\beta}{2} \| z_{k,t} \|_2^2) \right)$$

$$(18.7)$$

The linear combination of the subdictionaries D_k represents the input feature vector F_t. It should be noted that the first term of Equation 18.7 ensures that the reconstruction error is small, while the role of the second term is to make the code vector sparse and small.

To solve Equation 18.7 in a distributed manner, the cost function should have a *sum-of-costs* form. Specifically, distributed methods can be applied to the problem at hand if the cost function of the optimization task,

$J^{glob}(\omega)$, can be reformulated as the aggregation of individual cost functions of the agents $J_k(\omega)$:

$$J^{glob}(\omega) = \sum_{k=1}^{N} J_k(\omega) \quad (18.8)$$

Since the problem in Equation 18.7 does not follow the sum-of-costs form represented in Equation 18.8, it is not feasible to use distributed methods to solve this problem directly. However, it has been shown that its dual problem has a distributed form following Equation 18.8 [32]. After solving the dual problem, the optimal primal variables $\{D_k\}$ and z can be reconstructed. The dual problem can be formulated as [32]

$$\min_{v}(-g(v, F_t)) = \| v \|_2^2 - v^T F_t + \sum_{k=1}^{N} \mathcal{S}_{\frac{\lambda}{\beta}}\left(\frac{D_k^T v}{\beta}\right), \quad (18.9)$$

where v is an auxiliary vector in the dual problem of size $M \times 1$, λ and β are the regularization coefficients in Equation 18.2, and $\mathcal{S}_{\frac{\lambda}{\beta}}(x)$ is

$$\mathcal{S}_{\frac{\lambda}{\beta}}(x) \triangleq -\frac{\beta}{2} \cdot \| \mathcal{T}_{\frac{\lambda}{\beta}}(x) \|_2^2 - \lambda \cdot \| \mathcal{T}_{\frac{\lambda}{\beta}}(x) \|_1^2 + \beta \cdot x^T \mathcal{T}_{\frac{\lambda}{\beta}}(x),$$

$$(18.10)$$

where $\mathcal{T}_\gamma(x)$ is a soft-thresholding operator on vector x and is formulated as follows for the n_{th} element of the vector:

$$[\mathcal{T}_\gamma(x)]_n \triangleq (|[x]_n| - \gamma)_+ \mathrm{sgn}([x]_n). \quad (18.11)$$

Here $(x)_+ = \max(x,0)$ and $\mathrm{sgn}(x)$ is the signum function. The dual function in Equation 18.9 can be considered as the global cost function $J^{glob}(v; F_t)$. As a result, the individual cost function for each agent k can be formulated as (see Refs. [32,34] for details)

$$J_k(v; F_t) \triangleq \begin{cases} -\dfrac{v^T F_t}{|\mathcal{N}_I|} + \dfrac{1}{N} \| v \|_2^2 + \mathcal{S}_{\frac{\lambda}{\beta}}\left(\dfrac{D_k^T v}{\beta}\right), & k \in \mathcal{N}_I \\[6pt] \dfrac{1}{N} \| v \|_2^2 + \mathcal{S}_{\frac{\lambda}{\beta}}\left(\dfrac{D_k^T v}{\beta}\right), & k \notin \mathcal{N}_I \end{cases},$$

$$(18.12)$$

where $|\mathcal{N}_I|$ is the cardinality of $\mathcal{N}_I$.

It can be observed that the individual cost functions $J_k(v; F_t)$ add up to the cost function in Equation 18.9. Here, the dual problem for estimating v^o can be rewritten as

$$v^o = \min_{v} \sum_{k=1}^{N} J_k(v; F_t) \quad (18.13)$$

The dual problem follows the form of Equation 18.8 and can be solved using distributed learning methods.

Several distributed learning methods have been proposed in the literature, such as incremental strategies [35,36], consensus strategies [37,38], and diffusion adaptation strategies [21,39,40]. It has been shown that among these, the diffusion algorithm is robust, scalable, and capable of real-time adaptation and learning. Since diffusion strategies also have superior performance and stability compared to consensus methods [22], we use this approach to solve the distributed optimization problem in Equation 18.13. The details of this strategy are explained in the following section.

18.4.2 Diffusion Adaptation Method

A distributed network consists of N nodes where each node k is connected to a number of neighboring agents represented by N_k. As shown in Figure 18.3, each node in the network can share information with and receive information from its neighbors. Moreover, each node is associated with an individual cost function to minimize. The aggregation of all these individual cost functions forms the global cost function of the network, similar to Equation 18.8. In order to minimize the cost function, the nodes employ the diffusion adaptation strategy, which consists of two steps: the adaptation step and the combination step. During the adaptation step, each node k updates its own estimate via a gradient descent step. This estimate is an intermediate estimate represented by $\psi_{k,i}$, which is further updated in the combination step. In the combination step, the neighbors share their intermediate estimates, and each node k combines all the intermediate estimates received form its neighbors to obtain its own final estimate, $v_{k,i}$, in the ith time instant (further explanation can be found in Ref. [21]). Hence, the diffusion adaptation strategy can be formulated as

$$\psi_{k,i} = v_{k,i-1}$$
$$- \mu \nabla_v J_k(v_{k,i-1}; F_t) \quad \text{(adaptation step)} \quad (18.14)$$

$$v_{k,i} = \sum_{\ell \in \mathcal{N}_k} a_{\ell k}(i) \psi_{\ell,i} \quad \text{(combination step)} \quad (18.15)$$

where $v_{k,i}$ is node k's estimate of the optimal solution v_t^o at iteration i, $\psi_{k,i}$ is the intermediate estimate, and $\mu > 0$ is the sufficiently small fixed updating step-size. The weight $a_{\ell k}(i)$ in Equation 18.15 is the weight that node k assigns to the information received from node ℓ at time instant i. These weights are called *combination weights* and are shown in Figure 18.4. The combination weights $a_{\ell k}(i)$ must satisfy

$$\sum_{\ell \in \mathcal{N}_k} a_{\ell k}(i) = 1, \quad a_{\ell k}(i) > 0 \text{ if } \ell \in \mathcal{N}_k,$$
$$a_{\ell k}(i) = 0 \text{ if } \ell \notin \mathcal{N}_k \quad (18.16)$$

Several methods have been proposed to design the combination weights, and it has been shown that these weights can have a major impact on the performance of

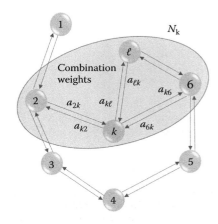

FIGURE 18.4
Example of a connected network where the neighboring nodes share information with each other. The combination weight $a_{\ell k}(i)$ is the weight that node k assigns to the information received from node ℓ at time instant i.

the diffusion algorithm [41,42]. In Section 18.4.3, we introduce an adaptive method to learn the weights over time, and we show its impact on the performance of the diffusion strategy.

After sufficient iterations of Equations 18.14 and 18.15, the optimal dual variable v_t^o is estimated. The sparse codes z_t^o and the subdictionaries $D_{k,t}$ that are the optimal primal variables for the DL problem can further be calculated as [32]

$$z_{k,t}^o = \arg\max_{z_k} [(D_k^T v_t^o)^T z_k - (\lambda \| z_k \|_1 + \frac{\beta}{2} \| z_k \|_2^2)], \quad (18.17)$$

$$D_{k,t} = \Pi_{\mathcal{D}_k}(D_{k,t-1} + \mu \cdot v_t^o z_{k,t}^o), \quad (18.18)$$

where $\Pi_{\mathcal{D}_k}(\cdot)$ projects the solution to the constraint set $\mathcal{D}_k$ as

$$\Pi_{\mathcal{D}_k}(\mathcal{D}) \triangleq \arg\min_{x \in \mathcal{D}_k} \| x - \mathcal{D} \|_2 \quad (18.19)$$

Moreover, it should be noted that the index t denotes the tth data point and the index i indicates the ith iteration of the diffusion method to derive the optimal solution for the tth data point.

In the next section, we propose an adaptive approach to design the combination weights in Equation 18.15. These weights play an important role in combining the information received from the other nodes of the network, which can affect the performance of the algorithm.

18.4.3 Selection of the Combination Weights

Designing the combination weights in Equation 18.15 is an important part of the diffusion strategy as it affects the performance of the network. One of the most common approaches for designing the combination weights is to have static weights that are constant over time. For instance, consider the case where the weights are uniform and the combination step (Equation 18.15) is simply an averaging over all the estimates:

$$a_{\ell k} = \frac{1}{|\mathcal{N}_k|} \quad \text{if} \quad \ell \in \mathcal{N}_k \quad \text{(uniform combination weights)} \quad (18.20)$$

In this way, the nodes assign the same weight to all their neighbors without considering the reliability of the received information [32,34,43]. The previously proposed distributed DL methods design the combination weights in a static manner.

However, it is important to design the combination weights in such a way that the nodes can learn about the objectives of their neighbors over time [41,42]. Hence, the combination weights must be estimated in a manner that helps each node to ignore misleading information and cooperate only with neighbors having the same objective. Here, we employ an adaptive approach to estimate these weights to tackle the DL problem. We do so by following the approach that minimizes the instantaneous mean-square deviation (MSD) of the network, defined as [44]

$$MSD(i) \triangleq \frac{1}{N} \sum_{k=1}^{N} \mathbb{E} \| \tilde{v}_{k,i} \|^2, \quad (18.21)$$

where $\tilde{v}_{k,i} \triangleq v_t^o - v_{k,i}$ represents the error vector of node k at time i. The combination weights $a_{\ell k}(i)$ can be estimated by solving the optimization problem:

$$\min_{\{a_{\ell k}(i)\}} MSD(i) = \frac{1}{N} \sum_{k=1}^{N} \mathbb{E} \| \tilde{v}_{k,i} \|^2 \quad (18.22)$$

As shown in Ref. [44], the optimal solution can be approximated by

$$a_{\ell k}(i) \approx \begin{cases} \dfrac{\| v_{k,i-1} - \psi_{\ell,i} \|^{-2}}{\sum_{n \in \mathcal{N}_k} \| v_{k,i-1} - \psi_{n,i} \|^{-2}}, & \ell \in \mathcal{N}_k \\ 0, & \text{otherwise} \end{cases} \quad (18.23)$$

It is important to observe that the combination weight $a_{\ell k}(i)$ in Equation 18.23 is inversely proportional to the distance between the estimate of node k and the intermediate estimate $\psi_{\ell,i}$ of node ℓ. This means that the combination weights are designed in such a way that the nodes assign higher weights to those of their neighbors that have the similar objectives while ignoring the misleading information of the other neighbors. Hence, with the help of this combination method, the nodes are able to benefit from cooperation by continuously learning about the objective of their neighbors, which results in distinguishing between the useful and misleading information.

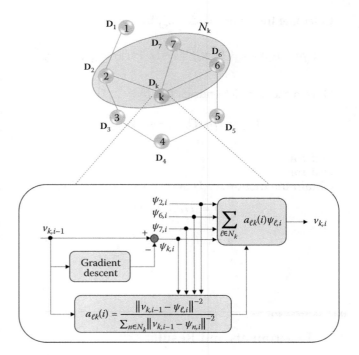

FIGURE 18.5
Diagram of the proposed adaptive diffusion method to solve the dual problem of the dictionary learning method. In the combination step, each node combines the estimates of its neighbors by the adaptive weights $a_{\ell k}(i)$ proposed in Equation 18.23. Afterwards, the optimal dual variable is used to obtain the primal variables of the dictionary learning problem according to Equations 18.17 and 18.18.

By exploiting the similarity among the nodes with similar objectives, we arrive at a more discriminative dictionary, which leads to better classification results (Section 18.5). A schematic of the diffusion adaptation method to solve the DL problem is shown in Figure 18.5. The summary of the proposed ADDL method is given in Algorithm 1.

Algorithm 1: The proposed ADDL method for HEp-2 cell classification

Require: Subdictionaries D_k are initialized randomly and projected onto the constraint set. The dual solution is initialized as $v_{k,0} = 0$ for all $k = 1,\ldots,N$.
Set the values for β, λ, and μ.
for each input feature sample F_t **do**
calculate the optimal dual variable v_t^o until convergence by

$$\psi_{k,i} = v_{k,i-1} - \mu \nabla_v J_k(v_{k,i-1}; F_t)$$

$$a_{\ell k}(i) \approx \begin{cases} \dfrac{\| v_{k,i-1} - \psi_{\ell,i} \|^{-2}}{\sum_{n \in \mathcal{N}_k} \| v_{k,i-1} - \psi_{n,i} \|^{-2}}, & \ell \in \mathcal{N}_k \\ 0, & \text{otherwise} \end{cases}$$

$$v_{k,i} = \sum_{\ell \in \mathcal{N}_k} a_{\ell k}(i) \psi_{\ell,i}$$

for each agent k **do**

Calculate the sparse codes $z_{k,t}^o$ by

$$z_{k,t}^o = \mathrm{argmax}_{\mathbf{z}_k}[(\mathbf{D}_k^T \mathbf{v}_t^o)^T z_k - (\lambda \| z_k \|_1 + \frac{\beta}{2} \| z_k \|_2^2)]$$

Obtain the subdictionaries $\mathbf{D}_{k,t}$ by:

$$\mathbf{D}_{k,t} = \Pi_{\mathcal{D}_k}(\mathbf{D}_{k,t-1} + \mu \cdot \mathbf{v}_t^o z_{k,t}^o)$$

end for
end for
Obtain the dictionary $\mathbf{D}$ and sparse codes z by:

$$\mathbf{D} = [\mathbf{D}_1 \ldots \mathbf{D}_N]$$

$$z = col\{z_1, \ldots, z_N\}$$

18.5 Experiments and Results

Here, we evaluate the performance of the ADDL method for HEp-2 cell classification on the two publicly available data sets, ICPR2012 [3] and ICIP2013 [2].

18.5.1 Data Sets and Evaluation Methods

ICPR2012: This data set consists of 28 HEp-2 specimen images where each image is of size 1388 × 1038 with 24-bit RGB pixels. Each of the 28 images contains just one of the six stained patterns including centromere (Ce), coarse-speckled (Cs), cytoplasmatic (Cy), fine-speckled (Fs), homogeneous (H), and nucleolar (N), as illustrated in Figure 18.6(a). The mask of each cell in each image and the cell labels are provided. Also, there are two levels of intensity images, which are called intermediate and positive images. In total, there are 1455 cells in the given 28 images, which are divided into 721 images for training and 734 images for testing in our experiments [3].

Two evaluation strategies have been performed in the literature and defined by the data set providers, including *test set* evaluation and *leave-one-specimen-out* (LOSO). The test set evaluation uses the provided training and test set, while the LOSO method uses all the cells in one specimen image for test and the rest of the cells for training.

ICIP2013: There are 419 sera of patients in this data set with approximately 100–200 cell images extracted from each patient serum. The total number of extracted cell images is 68,429, which consists of 13,596 cell images for training (publicly available) and 54,833 for testing (privately maintained by the organizers*).

---
* http://i3a2014.unisa.it/?page_id=126

Each annotated cell image contains the information about the cell pattern, intensity level (positive or intermediate), mask, and image ID (the category of the cell). Figure 18.6(b) shows some examples of these cell images. It can be seen that at the *cell level*, there is a six-class classification problem, where the classes are centromere, Golgi, homogeneous, nucleolar, nucleolar membrane (NuMem), and speckled, while at the *specimen level*, mitosis spindle class is added, and there are seven classes in the classification task.

As the test set is privately maintained by the organizers, two evaluation methods have been widely used in the literature. The first evaluation approach is the HSM method [4], where 600 cells (300 for Golgi class) from each class are used in the training data set and the rest of the images are used for testing. The other method is LOSO, as performed for the ICPR2012 data set.

18.5.2 Classification Results

ICPR2012
Table 18.1 shows the classification results for the proposed ADDL method and the comparison with other dictionary- and non-dictionary-based methods. We utilize the ADDL method with $\lambda = 45$, $\beta = 0.1$, and $\mu = 0.002$.

The ADDL results are reported for two forms of adaptive and fixed uniform weights according to Equations 18.23 and 18.20, respectively. It can be observed that the ADDL method with adaptive weights (72%) outperforms the uniformly weighted (69%) by 3% on average based on the test set evaluation. The best accuracy for positive images is reported by the SNPB method (82%), but for the intermediate-level images, the ADDL with adaptive weights outperforms other methods at 63%. The specimen level accuracy is also comparable with other methods at 86%.

With the LOSO evaluation method, 89% and 85% average cell level accuracies are obtained for adaptive and uniform weighted ADDL respectively, where the adaptive method outperforms other methods. At the specimen level, the accuracy of the ADDL method with adaptive weights is 93%, which is also obtained by Ref. [14].

It should be noted that due to the low number of input images in the ICPR2012 data set (28 images in total), the obtained accuracies are comparable with other methods. However, by increasing the number of input images, as in the case of ICIP2013, the advantage of the ADDL method can be seen clearly, which is described in the next subsection. Moreover, it will be shown in Section 18.5.3 that the ADDL method benefits from a lower computational and memory cost.

ICIP2013
Table 18.2 presents the results for the ICIP2013 data set. The values of the parameters λ, β, and μ are the same as

Dictionary Learning Applications for HEp-2 Cell Classification

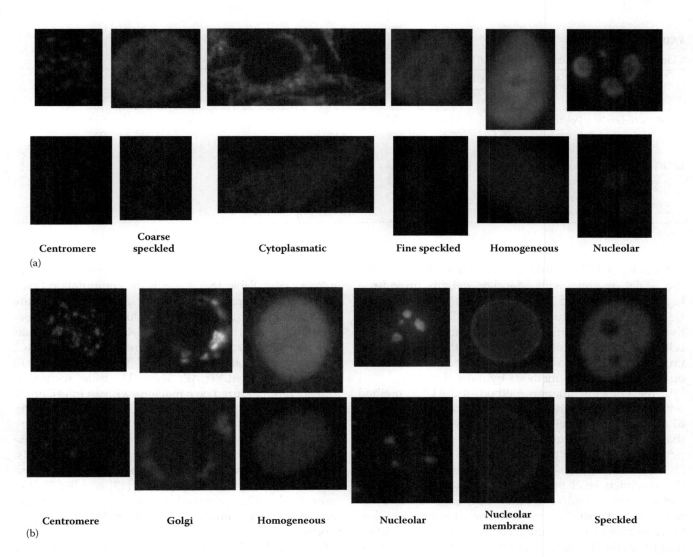

FIGURE 18.6
Samples of the *cell level* images of the (a) ICPR2012 and (b) ICIP2013 data sets in six classes with different sizes. The first rows are the positive- and the second rows are the intermediate-intensity-level images.

TABLE 18.1

Classification Accuracies for the ICPR2012 Data Set by Using Test Set and LOSO Evaluation Methods

	ICPR2012 (%)		ADDL Adaptive	ADDL Uniform	Ensafi [28]	SNPB [45]	Kastaniotis [46]	Nosaka [13]	DiCataldo [14]
Test set	Cell level	Positive	80	78	81	**82**	70	79	60
		Intermediate	**63**	59	62	59	31	58	35
		Average	**72**	69	**72**	70	51	69	48
	Specimen level		86	79	86	**93**	86	79	**93**
LOSO	Cell level	Positive	94	91	91	92	72	80	**95**
		Intermediate	**83**	78	72	70	55	60	80
		Average	**89**	85	82	81	64	70	88
	Specimen level		**93**	79	79	86	79	86	**93**

TABLE 18.2

Classification Accuracies for the ICIP2013 Data Set by Using HSM and LOSO Evaluation Methods

	ICIP2013 (%)		ADDL Adaptive	ADDL Uniform	Ensafi [47]	SNPB [45]	Gragnaniello [48]	Manivannan [49]	HSM [4]	Larsen [50]
HSM	Cell Level	Positive	**97.9**	95.4	95.8	96.8	–	–	95.5	–
		Intermediate	**89.4**	87.6	87.9	88.8	–	–	80.9	–
		Average	**93.7**	91.5	91.9	92.8	–	–	88.2	–
LOSO	Cell Level	Positive	**88.5**	84.2	83.4	83.8	–	–	–	–
		Intermediate	**74.7**	71.4	71.2	72	–	–	–	–
		Average	**81.6**	77.8	77.3	77.9	81.1	80.3	–	78.7
	Specimen Level		**90.4**	86.7	88	89.2	86.7	89.9	–	–

those in ICPR2012. It can be observed from this table that the ADDL method with adaptive weights provides a higher accuracy and outperforms the other methods. Using the HSM evaluation method, ADDL with adaptive weights gave an average accuracy of 93.7% for cell level images, which is 2% higher than ADDL with uniform weights and other DL methods. Moreover, it outperforms non-DL methods by more than 5%.

In the LOSO evaluation method, the ADDL method with adaptive weights gave an average accuracy of 81.6% in the cell level images, which is 4% higher than that achieved with uniform weights. Additionally, its performance is better than the other DL methods and 4% higher than other classification methods. The results of the specimen level images also show that the ADDL method with adaptive weights outperforms the other methods, with 90.4% accuracy. The superior classification accuracy can be explained by the use of combination weights in the diffusion adaptation method that enables the nodes to share information and solve the optimization problem in a cooperative manner. Note that the accuracies for the ICPR2012 and ICIP2013 data sets are different because the quality and amount of images within the two data sets are very different.

18.5.3 Computational Cost

DL is a computationally expensive and time consuming task, which can be especially troublesome as the size of the input data increases. Here, we evaluate the efficiency of the ADDL method in terms of computational and memory cost. Table 18.3 presents the computation times of different DL procedures. All the algorithms share the same training, testing set, and evaluation method. Specifically, the algorithms in Refs. [45,47] are implemented in-house; therefore, we can compare them in Table 18.3 for computational cost analysis. These methods are implemented in MATLAB. These DL tasks were performed and measured on a machine with an Intel Core i7 CPU and 16 GB RAM with a 64-bit operating system. As shown in Table 18.3, the proposed ADDL method, by giving the information to a single node, takes less time (56.64 seconds) than that taken by the other methods. This is more than two and five times better than those achieved by the methods of Ensafi [47] and SNPB [45], respectively. The proposed method takes 15 seconds more when the information is given to all nodes to process.

For the ICIP2013 data set, the computation time of the proposed method for calculating the dictionary is 286.21 seconds when the information is given to a single node to process. This is 47 seconds lower than the case where the information is passed to all the nodes to process. However, it can be observed that the computation time of the proposed method is lower than those of the other DL methods. For instance, the ADDL method is about 9 and 20 times faster than the Ensafi [47] and SNPB [45] methods, respectively. Hence, the proposed method can enhance the performance of the DL task in terms of both computational cost and classification accuracy.

TABLE 18.3

Computation Times of Different Dictionary Learning Methods

Dictionary Learning Computation Time (sec)	ADDL All Nodes	ADDL Single Node	Ensafi [47]	SNPB [45]
ICPR2012	71.63	56.64	126.34	354.38
ICIP2013	333.73	286.21	2751.91	5742.64

18.6 Conclusion

In this chapter, we studied the application of DL methods for HEp-2 cell classification with the aim of AD diagnosis. Since the DL problem can be computationally intensive and time consuming, we proposed an ADDL method. This method benefits from lower computational cost, which is an important advantage in solving classification problems. We applied the ADDL method to HEp-2 cell images and observed that this method can enhance the accuracy of the cell classification problem while reducing the computational time. This is due to the fact that the dictionary is learned in a distributed manner in this method. This approach is also a foundation for big data analysis where the information is available on the nodes of a computer cluster or cloud.

References

1. J. M. González-Buitrago and C. González. Present and future of the autoimmunity laboratory. *Clinica Chimica Acta*, 365(1):50–57, 2006.
2. P. Foggia, G. Percannella, A. Saggese, and M. Vento. Pattern recognition in stained HEp-2 cells: Where are we now? *Pattern Recognition*, 47(7):2305–2314, 2014.
3. P. Foggia, G. Percannella, P. Soda, and M. Vento. Benchmarking HEp-2 cells classification methods. *IEEE Transactions on Medical Imaging*, 32(10):1878–1889, 2013.
4. X.-H. Han, J. Wang, G. Xu, and Y.-W. Chen. High-order statistics of microtexton for HEp-2 staining pattern classification. *IEEE Transactions on Biomedical Engineering*, 61(8):2223–2234, 2014.
5. X. Kong, K. Li, J. Cao, Q. Yang, and L. Wenyin. HEp-2 cell pattern classification with discriminative dictionary learning. *Pattern Recognition*, 47(7):2379–2388, 2014.
6. P. Perner, H. Perner, and B. Müller. Mining knowledge for HEp-2 cell image classification. *Artificial Intelligence in Medicine*, 26(1):161–173, 2002.
7. N. Otsu. A threshold selection method from gray-level histograms. *Automatica*, 11(285-296):23–27, 1975.
8. Y.-C. Huang, T.-Y. Hsieh, C.-Y. Chang, W.-T. Cheng, Y.-C. Lin, and Y.-L. Huang. HEp-2 cell images classification based on textural and statistic features using self-organizing map. In *Intelligent Information and Database Systems*, pages 529–538. Springer, Berlin, Heidelberg, 2012.
9. P. Soda and G. Iannello. Aggregation of classifiers for staining pattern recognition in antinuclear autoantibodies analysis. *IEEE Transactions on Information Technology in Biomedicine*, 13(3):322–329, 2009.
10. R. Hiemann, N. Hilger, J. Michel, J. Nitschke, A. Boehm, U. Anderer, M. Weigert, and U. Sack. Automatic analysis of immunofluorescence patterns of HEp-2 cells. *Annals of the New York Academy of Sciences*, 1109(1):358–371, 2007.
11. U. Sack, S. Knoechner, H. Warschkau, U. Pigla, F. Emmrich, and M. Kamprad. Computer-assisted classification of HEp-2 immunofluorescence patterns in autoimmune diagnostics. *Autoimmunity Reviews*, 2(5):298–304, 2003.
12. P. Soda, G. Iannello, and M. Vento. A multiple expert system for classifying fluorescent intensity in antinuclear autoantibodies analysis. *Pattern Analysis and Applications*, 12(3):215–226, 2009.
13. R. Nosaka and K. Fukui. HEp-2 cell classification using rotation invariant co-occurrence among local binary patterns. *Pattern Recognition*, 47(7):2428–2436, 2014.
14. S. Di Cataldo, A. Bottino, I. U. Islam, T. F. Vieira, and E. Ficarra. Subclass discriminant analysis of morphological and textural features for HEp-2 staining pattern classification. *Pattern Recognition*, 47(7):2389–2399, 2014.
15. G. V. Ponomarev, V. L. Arlazarov, M. S. Gelfand, and M. D. Kazanov. ANA HEp-2 cells image classification using number, size, shape and localization of targeted cell regions. *Pattern Recognition*, 47(7):2360–2366, 2014.
16. A Sriram, S. Ensafi, S. F. Roohi, and A. A. Kassim. Classification of human epithelial type-2 cells using hierarchical segregation. In *International Conference on Control Automation Robotics Vision (ICARCV)*, pages 323–328, 2014.
17. L. Shen, J. Lin, S. Wu, and S. Yu. HEp-2 image classification using intensity order pooling based features and bag of words. *Pattern Recognition*, 47(7):2419–2427, 2014.
18. A. Wiliem, C. Sanderson, Y. Wong, P. Hobson, R. F. Minchin, and B. C. Lovell. Automatic classification of human epithelial type 2 cell indirect immunofluorescence images using cell pyramid matching. *Pattern Recognition*, 47(7):2315–2324, 2014.
19. M. Faraki, M. T. Harandi, A. Wiliem, and B. C. Lovell. Fisher tensors for classifying human epithelial cells. *Pattern Recognition*, 47(7):2348–2359, 2014.
20. M. Rehn and F. T. Sommer. A network that uses few active neurons to code visual input predicts the diverse shapes of cortical receptive fields. *Journal of Computational Neuroscience*, 22(2):135–146, 2007.
21. A. H. Sayed. Adaptation, learning, and optimization over networks. *Foundations and Trends in Machine Learning*, 7(4-5):311–801, 2014.
22. S.-Y. Tu and A. H. Sayed. Diffusion strategies outperform consensus strategies for distributed estimation over adaptive networks. *IEEE Transactions on Signal Processing*, 60(12):6217–6234, 2012.
23. D. G. Lowe. Distinctive image features from scale-invariant keypoints. *International Journal of Computer Vision*, 60(2):91–110, 2004.
24. H. Bay, A. Ess, T. Tuytelaars, and L. Van Gool. Speeded-up robust features (surf). *Computer Vision and Image Understanding*, 110(3):346–359, 2008.
25. L. Juan and O. Gwun. A comparison of SIFT, PCA-SIFT and SURF. *International Journal of Image Processing (IJIP)*, 3(4):143–152, 2009.
26. C. Harris and M. Stephens. A combined corner and edge detector. In Alvey Vision Conference, volume 15, page 50. Manchester, UK, 1988.

27. J. Shi and C. Tomasi. Good features to track. In *IEEE Conference on Computer Vision and Pattern Recognition*, pages 593–600. IEEE, 1994.
28. S. Ensafi, S. Lu, A. A. Kassim, and C. L. Tan. Automatic CAD system for HEp-2 cell image classification. In *International Conference on Pattern Recognition (ICPR)*, pages 3321–3326, 2014.
29. M. Aharon, M. Elad, and A. Bruckstein. K-SVD: An algorithm for designing overcomplete dictionaries for sparse representation. *IEEE Transactions on Signal Processing*, 54(11):4311–4322, 2006.
30. K. Engan, S. O. Aase, and J. H. Husoy. Method of optimal directions for frame design. In *IEEE International Conference on Acoustics, Speech, and Signal Processing (ICASSP)*, volume 5, pages 2443–2446. IEEE, 1999.
31. S. Lazebnik, C. Schmid, and J. Ponce. Beyond bags of features: Spatial pyramid matching for recognizing natural scene categories. In *IEEE Computer Society Conference on Computer Vision and Pattern Recognition*, volume 2, pages 2169–2178, 2006.
32. J. Chen, Z. J. Towfic, and A. H. Sayed. Dictionary learning over distributed models. *IEEE Transactions on Signal Processing*, 63(4):1001–1016, 2015.
33. S. Monajemi, S. Ensafi, S. Lu, A. A. Kassim, C. L. Tan, S. Sanei, and S-H Ong. Classification of HEp-2 cells using distributed dictionary learning. In *Proceedings of the European Signal Processing Conference (EUSIPCO)*, 2016.
34. J. Chen, Z. J. Towfic, and A. H. Sayed. Online dictionary learning over distributed models. In *IEEE International Conference on Acoustics, Speech and Signal Processing (ICASSP)*, pages 3874–3878, 2014.
35. D. P. Bertsekas. A new class of incremental gradient methods for least squares problems. *SIAM Journal on Optimization*, 7(4):913–926, 1997.
36. A. Nedic and D. P. Bertsekas. Incremental subgradient methods for nondifferentiable optimization. *SIAM Journal on Optimization*, 12(1):109–138, 2001.
37. A. Nedic and A. Ozdaglar. Distributed subgradient methods for multi-agent optimization. *IEEE Transactions on Automatic Control*, 54(1):48–61, 2009.
38. L. Xiao and S. Boyd. Fast linear iterations for distributed averaging. *Systems & Control Letters*, 53(1):65–78, 2004.
39. J. Chen and A. H. Sayed. Distributed Pareto-optimal solutions via diffusion adaptation. In *IEEE Statistical Signal Processing Workshop (SSP)*, pages 648–651, 2012.
40. S. Monajemi, S. Sanei, and S-H. Ong. Advances in bacteria motility modelling via diffusion adaptation. In *European Signal Processing Conference (EUSIPCO)*, pages 2335–2339, 2014.
41. J. Chen, C. Richard, and A. H. Sayed. Diffusion LMS over multitask networks. *IEEE Transactions on Signal Processing*, 63(11):2733–2748, 2015.
42. S. Monajemi, S. Sanei, S.-H. Ong, and A. H. Sayed. Adaptive regularized diffusion adaptation over networks. In *IEEE International Workshop on Machine Learning for Signal Processing (MLSP)*, pages 1–5, 2015.
43. Z. J. Towfic, J. Chen, and A. H. Sayed. Dictionary learning over large distributed models via dual-ADMM strategies. In *IEEE International Workshop on Machine Learning for Signal Processing (MLSP)*, pages 1–6, 2014.
44. X. Zhao and A. H. Sayed. Clustering via diffusion adaptation over networks. In *International Workshop on Cognitive Information Processing (CIP)*, pages 1–6, 2012.
45. S. Ensafi, S. Lu, A. A. Kassim, and C. L. Tan. Sparse non-parametric Bayesian model for HEp-2 cell image classification. In *IEEE International Symposium on Biomedical Imaging (ISBI)*, pages 679–682, 2015.
46. I. Theodorakopoulos, D. Kastaniotis, G. Economou, and S. Fotopoulos. HEp-2 cells classification via sparse representation of textural features fused into dissimilarity space. *Pattern Recognition*, 47(7):2367–2378, 2014.
47. S. Ensafi, S. Lu, A. A. Kassim, and C. L. Tan. A bag of words based approach for classification of HEp-2 cell images. In *Pattern Recognition Techniques for Indirect Immunofluorescence Images (I3A)*, pages 29–32, 2014.
48. D. Gragnaniello, C. Sansone, and L. Verdoliva. Biologically-inspired dense local descriptor for indirect immunofluorescence image classification. In *Pattern Recognition Techniques for Indirect Immunofluorescence Images (I3A)*, pages 1–5, 2014.
49. S. Manivannan, W. Li, S. Akbar, R. Wang, J. Zhang, and S. J. McKenna. HEp-2 cell classification using multi-resolution local patterns and ensemble SVMs. In *Pattern Recognition Techniques for Indirect Immunofluorescence Images (I3A)*, pages 37–40, 2014.
50. A. B. L. Larsen, J. S. Vestergaard, and R. Larsen. HEp-2 cell classification using shape index histograms with donut-shaped spatial pooling. *IEEE Transactions on Medical Imaging*, 33(7):1573–1580, 2014.

19

Computational Sequence- and NGS-Based MicroRNA Prediction

R.J. Peace and James R. Green

CONTENTS

19.1 Introduction: Overview of the Chapter 381
19.2 miRNA Biology 383
19.3 Experimental Determination of miRNA through Sequencing 384
19.4 Pattern Classification 385
    19.4.1 Performance Estimation 385
    19.4.2 Selecting Training and Testing Data 388
    19.4.3 Feature Selection 388
    19.4.4 Class Imbalance Correction 388
    19.4.5 Splitting of Training and Testing Data 389
    19.4.6 Model Selection 389
    19.4.7 Model Training 389
    19.4.8 Prediction Pipeline 390
19.5 The State of the Art in Computational miRNA Techniques 390
    19.5.1 The State of the Art in NGS-Based miRNA Prediction 391
        19.5.1.1 sRNA Data Set Pipelines 393
        19.5.1.2 NGS Experiments for miRNA Discovery in Species of Interest 394
        19.5.1.3 Analysis of the miRDeep2 miRNA Classification Pipeline 395
    19.5.2 The State of the Art in de novo miRNA Prediction 396
        19.5.2.1 Data Set Generation 396
        19.5.2.2 Classifier Selection and Training 397
        19.5.2.3 Feature Selection 398
        19.5.2.4 Reporting of Results 399
    19.5.3 Previous Assessments of the State of the Art 400
19.6 Discussion of the miRNA Prediction State of the Art 400
    19.6.1 Redundancy in Feature Sets 400
    19.6.2 Lack of Prevalence-Corrected Reporting 401
    19.6.3 Independent Analysis of the Effectiveness of SMOTE Class Imbalance Correction 402
    19.6.4 Failure of miRNA Predictors to Generalize to Cross-Species Negative Data 402
    19.6.5 Moving to Genome-Scanning Data Sets for Genome-Scale Experiments 403
    19.6.6 Future Research Directions in the Big Data Era 404
19.7 Concluding Remarks 405
References 405

19.1 Introduction: Overview of the Chapter

MicroRNAs (miRNAs) are short (18–23 nt), non-coding RNAs that play central roles in cellular regulation by modulating the post-transcriptional expression of messenger RNA (mRNA) transcripts [1]. MicroRNAs have been shown to be involved in several critical cellular processes including regulating expression of proteins involved in biological development [2], cell differentiation [3], apoptosis [4], cell cycle control [5], stress response [6], and disease pathogenesis [7]. Recent studies have also highlighted the role of miRNA in the cellular adaptation to severe environmental stresses (such as freezing, dehydration, and anoxia) in tolerant animals [8–10]. Due to the biological importance of miRNAs, the ability to accurately predict their sequences is of great significance. The discovery of novel miRNA sequences leads to new knowledge regarding biological pathways, through studying the target sequences and co-expression of these novel miRNAs. Discovering a greater number of novel miRNAs during miRNA prediction studies increases the amount of pathway knowledge that can be gained as a result of these studies. Methods of computational prediction of miRNAs fall into two categories: *de novo* miRNA prediction, wherein genomic sequence data sets are mined for miRNA, and next-generation sequencing (NGS-based) miRNA prediction, wherein transcriptomic data sets are mined for miRNA. This chapter examines the state of the art in both of these approaches. The prediction of target mRNA for a given miRNA is a field unto itself and has been reviewed extensively elsewhere (e.g., Ref. [11]).

Within the past few years, the field of *de novo* miRNA prediction has largely been conducted within an artificial scenario that does not accurately represent real-world applications of the field. In this scenario, all data comes from one of a small set of model organisms (species that have been widely studied due to their relevance to human life, or because of their availability in laboratory settings, and for which there is a large amount of genomic annotation available including a large set of known miRNA); the prevalence of miRNA vs. pseudo-miRNA within genomes is on the order of 1:10; and all miRNAs conform to very simple RNA secondary structure criteria. These assumptions are invalid for real-world applications of miRNA prediction, where one is often attempting to identify all miRNAs within the unannotated genome of a species that may be phylogenetically distant from any model species, and class imbalance is on the order of 1:1000.

In the field of *de novo* miRNA prediction, performance is measured by the ability of classifiers to differentiate pre-miRNA hairpin structures from pseudo-miRNA hairpin structures. However, prior to applying the classifier to a putative sequence, it is first tested by a pre-filtering stage. This often takes the form of a test for the presence of a stable hairpin within the secondary structure of the sequence. Training and test sets consisting of miRNA hairpin structures and pseudo-miRNA hairpin structures are extracted from larger genomic data sets during pre-filtering stages of a pipeline, and performance is measured only on the extracted hairpin data sets. The extraction of hairpin structures during data pre-filtering affects the pipeline's overall performance, but this effect is often not quantified, as performance is only measured on those data that pass the pre-filtering stage. In reality, the performance of *de novo* classification pipelines is dependent on pre-filtering performance as much as it is dependent on classifier performance. If a real miRNA sequence was not considered for classification because it was rejected during pre-filtering, then this represents a failure of the miRNA prediction method to identify that miRNA. Conversely, the failure of the pre-filtering stage to remove negative sequences will exacerbate the class imbalance problem and may lead to a false positive prediction. Such failures are often not included in performance assessment, and their measurement is critical to improvement of *de novo* miRNA pipelines. Following the adage that "we cannot improve what we cannot measure," it is recommended that a comprehensive evaluation framework be developed that assesses both pre-filtering and classification stages. Through application of such a comprehensive evaluation framework, opportunities for improving the overall system performance will become clear.

In the field of NGS-based miRNA prediction, state-of-the-art techniques fail to integrate all known lines of evidence that can be used to differentiate miRNA from non-miRNA and therefore limit achievable classification performance. These NGS-based methods have failed to leverage many advanced sequence-based features developed recently for *de novo* prediction methods. This represents a need and an opportunity to create a novel miRNA prediction method that integrates both expression-based and sequence-based features to improve classification performance.

The prediction of miRNAs from either genomic sequence or NGS experiments represents a "big data" challenge. For sequence-based methods, when examining the human genome for regions containing putative miRNA sequences, one must consider all sliding windows of approximately 100 bp in length. As opposed to many other bioinformatics predictions, where we can leverage simple sequence motifs to identify regions of interest, miRNAs are solely characterized by a hairpin structural motif. Therefore, the secondary structure of each window 100 bp in length must be computed, from a genome of 3 billion base pairs. Even after this pre-filtering stage, we are still left with millions of miRNA-like hairpin regions that must then be examined in detail through the computation of complex sequence-based features. Some widely used features require randomization experiments to establish the statistical significance of the observed structural free energy, resulting in order-of-magnitude increases in computational requirements. If one considers NGS-based methods, again we are confronted with significant computational requirements. Technological advancements in NGS instruments continue to lead to order-of-magnitude increases in the volume of data generated during each experiment. It is now common to see read depths (i.e., the number of times a given genomic location is sequenced by the instrument) ranging in the 1000s, where they used to be on the order of 10 a few years ago. While this leads to much greater sensitivity and data quality, the computational complexity of processing and aligning all these reads continues to grow. Therefore, computational complexity of miRNA analysis algorithms can no longer be ignored.

This chapter reviews the state of the art in both *de novo* and NGS-based miRNA prediction and provides suggestions for how the field can continue to advance. This chapter consists of seven sections. Section 19.2 provides the reader with an introduction to miRNA biology and a generic pattern classification pipeline. Section 19.3 briefly describes the process by which miRNAs are discovered via wet-lab experimental means, thereby providing the gold standard by which computational techniques are trained and evaluated. Section 19.4 reviews key concepts from the field of pattern classification, including performance metrics commonly used in miRNA prediction. Section 19.5 reviews the state of the art in computational miRNA prediction. Subsection 19.5.1 focuses on NGS-based prediction of miRNAs, while Subsection 19.5.2 focuses on *de novo* miRNA prediction techniques. This review highlights areas in which improvements should be made in order for miRNA prediction to produce higher-quality results in real-world scenarios. Section 19.6 discusses key weaknesses in the state of the art and provides suggestions for how to improve. Concluding remarks are provided in Section 19.7.

19.2 miRNA Biology

MicroRNAs are non-coding RNA (ncRNA) of length ~21 nt that are present within all animal and plant species as well as most viruses. miRNAs regulate gene expression through post-transcriptional binding with 3′ untranslated regions of mRNA, silencing the bound RNA [1]. It is estimated that between 60% and 90% of all mammalian mRNAs are regulated by miRNAs [12]. Through the regulation of critical proteins, miRNAs play a role in biological development [2], cell differentiation [3], apoptosis [4], cell cycle control [5], stress response [6], disease pathogenesis [7], and cellular adaptation to severe environmental stresses in tolerant animals [8–10]. Deregulation of miRNA is related to the onset of leukemia [13–15], and disruption of miRNA regulation is linked to heart disease and heart failure [16,17].

Most animal miRNAs share the following biogenesis: miRNA-encoding genes are transcribed within the nucleus by RNA polymerase II [18]. The resulting RNA structure, known as the primary miRNA (pri-miRNA), is an imperfect hairpin with a length of several hundred nucleotides [19]. Within this pri-miRNA structure is a shorter hairpin structure known as the precursor miRNA (pre-miRNA). Animal pre-miRNAs are typically between 70 and 100 nucleotides in length. Plant pre-miRNAs tend to be longer, reaching upward of 250 nucleotides. One or both arms of the pre-miRNA hairpin structure contain mature miRNA, sequences of approximately 18–23 nt that bind to mRNA untranslated regions to perform the regulatory function of the miRNA. Figure 19.1 demonstrates the pre-miRNA structure and mature miRNA sequences of the *H. sapiens* miRNA has-mir-1-1.

After transcription, the hairpin structure of the pri-miRNA is then cleaved by the Microprocessor complex, leaving the pre-miRNA hairpin intact. The Microprocessor complex consists of the protein DiGeorge Syndrome Critical Region 8 (DGCR8) and the enzyme Drosha. DGCR8 recognizes the pri-miRNA hairpin structure and directs Drosha toward the hairpin arms. Drosha contains an RNase III domain, which performs the cleaving action on the pri-miRNA structure [20].

Some pre-miRNAs are formed through RNA splicing and do not undergo cleaving by the Microprocessor complex [21].

After nuclear processing, the pre-miRNA structure is transported out of the nucleus by the exportin-5 protein

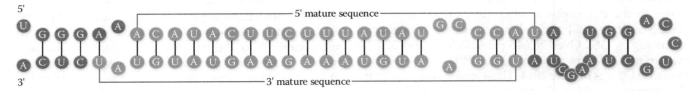

FIGURE 19.1
Hairpin structure and sequence of miRNA has-mir-1-1. Mature miRNA sequences are highlighted.

and is further processed within the cytoplasm. Dicer, a second RNase III enzyme, cleaves the pre-miRNA structure, removing the loop of the hairpin. The resulting structure is known as a miRNA:miRNA* duplex; one or both strands of this duplex are functional mature miRNAs. The two strands of the duplex are separated, and any strands that are not mature miRNAs (known as passenger strands) are discarded [22].

Mature miRNA strands are bound to Argonaute proteins in an RNA-induced silencing complex (RISC), which guide the miRNA to target mRNA regions. The binding of miRNA to mRNA inhibits ribosomal activity, preventing the expression of the protein that is regulated by the miRNA [23].

Plant miRNA biogenesis differs from animal miRNA biogenesis significantly. A Dicer homolog known as Dicer-Like 1 is responsible for all cleaving action on plant miRNA. Cleaving of pre-miRNA into miRNA:miRNA* duplexes is performed within the nucleus of plants, and the duplexes are transported into the cytoplasm [24].

Relative to most genetic sequences, miRNAs are extremely highly conserved during evolution [25,26]. In particular, the subsequences representing mature miRNA sequences are highly conserved across diverse species [27]. Because of their high level of conservation, miRNAs are considered significant phylogenetic markers. It is believed that miRNA developed independently in plants and in animals due to the differences in biogenesis between these domains [28].

The first miRNA, lin-4, was discovered in 1993 by Lee, Feinbaum, and Ambros [29]. Lin-4 was found to regulate the LIN-14 protein in *C. elegans* by binding to a repeated sequence in the 3' untranslated region of the lin-14 mRNA. Upon its discovery, it was not known that lin-4 was a member of a larger class of regulatory ncRNA, and no further miRNA discoveries were made until 2000. The second miRNA to be discovered, let-7, was found to repress lin-41 mRNA, also in *C. elegans* [30]. Following the discovery of both lin-4 and let-7, it was determined that these two ncRNAs belong to a larger family of regulatory ncRNAs [31], and the term *microRNA* was coined [32]. Within a year of the discovery of let-7, dozens of miRNAs were discovered within many organisms including *H. sapiens* [32–34]. The number of known miRNAs has grown dramatically since 2000, due partly to the development of computational tools that predict miRNA [35] and advancements in sequencing technologies [36]. Over 28,645 pre-miRNAs and 35,828 mature miRNAs are registered in miRBase, the miRNA database [37], as of June 2014. These miRNAs come from 223 species, of which *H. sapiens* is the most highly represented; 2588 of the mature miRNA sequences are from *H. sapiens*.

19.3 Experimental Determination of miRNA through Sequencing

In the past decade, NGS experiments have become the primary tool by which experimental prediction and validation of miRNA are performed. NGS experiments examine the RNA transcripts present in a sample of tissue (the sample's transcriptome), resulting in data sets that consist of a list of reads; each read corresponds to an RNA transcript that was present in the tissue sample at the time of the experiment. Expression of a given miRNA will vary between experiments, depending on the sample tissue type and the conditions under which the sample was collected. NGS data sets reflect expression levels; RNA transcripts that are highly expressed in the tissue sample will be recorded multiple times in a data set. When NGS data sets are analyzed, reads with very high similarity and overlap are grouped into read stacks; consensus sequences for these stacks and the number of reads in the stack are used to summarize large amounts of repeated sequence information.

Small RNA (sRNA) NGS experiments are used as a basis for NGS-based miRNA discovery. sRNA NGS experiments select for RNA transcripts whose lengths coincide with mature miRNA sequence lengths (18–25 nt). Within this read length range, a transcriptome will contain reads that correspond to the three Dicer products: mature miRNA, miRNA* passenger strands, and pre-miRNA hairpin loop segments. Along with these Dicer products, the transcriptome will also contain reads that derive from other ncRNAs such as piwi-interacting RNA and small nucleolar RNA, and from mRNA degradation products. If, for a given miRNA, read stacks are present in the transcriptome that align to the three Dicer products of the miRNA (mature miRNA, miRNA*, and loop), then this is considered experimental evidence that the miRNA exists within the genome and was expressed in the sample. Typically, the read stack that matches the mature miRNA will have the highest read count, as the miRNA* and loop products are degraded after Dicer processing, while the mature miRNA product is retained and incorporated into the miRISC complex.

Figure 19.2 shows a modified output of the miRDeep2 miRNA prediction algorithm. This output contains (a) the identity of the miRNA and the total read depth corresponding to the pre-miRNA region; (b) the predicted hairpin structure of the miRNA structure; and (c) the normalized read depth at each nucleotide in the sequence. Note that most reads align well to the mature miRNA sequence or the miRNA* sequence, as is expected when reads are the result of Dicer processing.

Computational Sequence- and NGS-Based MicroRNA Prediction

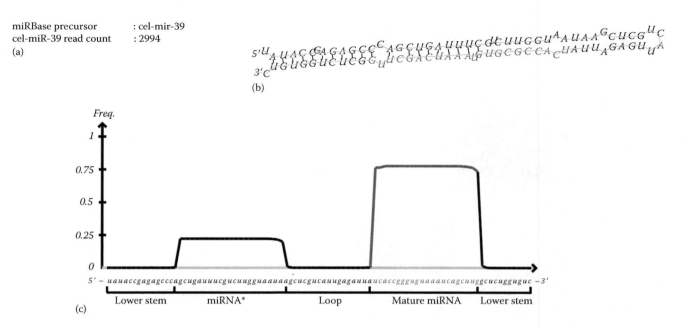

FIGURE 19.2
NGS-based miRNA prediction information, as output by the miRDeep2 pipeline. The following information is presented: (a) the identity of the miRNA and the total read depth corresponding to the pre-miRNA region; (b) the predicted hairpin structure of the miRNA structure; and (c) the normalized read depth at each nucleotide in the sequence. Most reads align to either the mature miRNA or the miRNA*Dicer products.

19.4 Pattern Classification

The goal of pattern classification is to build models that are capable of predicting the class of unlabeled data sets based on features of each datum that discriminate between classes of data. In the context of this chapter, pattern classification is used to classify an RNA sequence as either representing a pre-miRNA hairpin ("miRNA" = positive class) or not representing a pre-miRNA hairpin ("pseudo-miRNA" = negative class). Pattern classification is a form of supervised learning; classification models are trained on a set of labeled data (e.g., RNA sequences representing known miRNA and RNA sequences bearing resemblance to miRNA but that are known to perform other roles) and then use information gained from the training data in order to classify future unlabeled data (e.g., unannotated RNA sequences).

Pattern classification classifies data by means of features—numerical or categorical values that can be derived from each datum in training and testing data sets [38]. A set of d features that are derived from a datum comprise a feature vector of length d. For example, an RNA sequence may be represented by a simple feature vector of length 5 containing the percentage of A, G, C, and U within the sequence, and the total sequence length. Feature vectors can be represented as a point in d-dimensional space; classification models place decision boundaries within this d-dimensional space that attempt to discriminate between feature vectors representing different classes.

In pattern classification, training is the act of selecting appropriate labeled data representing positive and negative classes, and building a classification model from this data that is able to predict the class of future unlabeled data not found in the training data. Figure 19.3 demonstrates a typical training and testing pipeline. Careful attention must be paid to each step of the training pipeline, as small methodological errors in each step may alter classification results drastically. Furthermore, the quality of a classification model is affected by each step in the training pipeline.

19.4.1 Performance Estimation

Performance estimation is central to the success of classification experiments because all decisions made within the training pipeline are made with the goal of maximizing classifier performance with regard to performance metrics. No single metric exists that can fully describe the performance of a classifier. As with other elements of a classification experiment, selection of appropriate metrics for performance estimation of a classifier is dependent on the data and goal of the classifier.

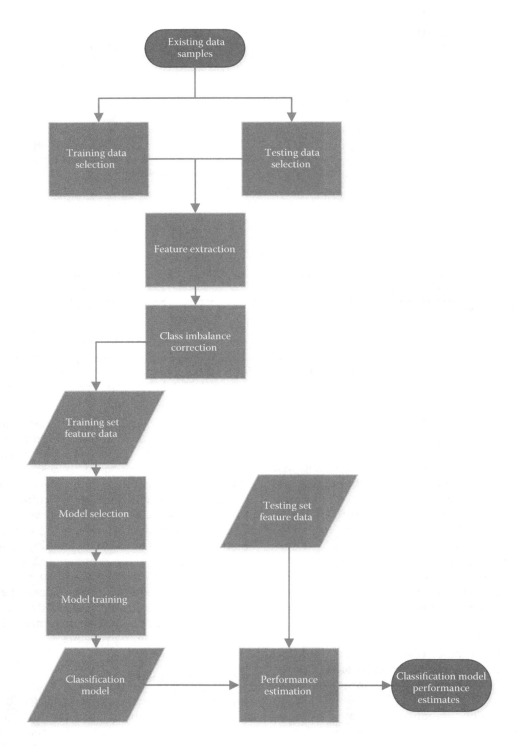

FIGURE 19.3
Training pipeline for a pattern classification model.

The confusion matrix is the basis of all classification performance reporting. This matrix describes the number of test samples of each true class that fall into each predicted class. For a binary classification problem such as miRNA prediction, the confusion matrix is a 2 × 2 matrix that describes the number of true positives, true negatives, false positives, and false negatives reported by a classifier for a given test set. Figure 19.4 shows the form of a binary confusion matrix. Here, TP, TN, FN, and FP are the counts of true positives, true negatives, false negatives, and false positives, respectively.

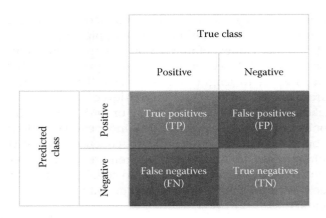

FIGURE 19.4
Confusion matrix for a binary classifier.

From the confusion matrix, several metrics can be derived that elucidate elements of the performance of a classifier. Sensitivity (also known as recall) and specificity are two of the most common of these metrics.

Sensitivity, Sn, otherwise known as recall, is the percentage of truly positive samples that the classifier correctly identifies as positive. Sensitivity describes the classifier's ability to recover positive samples, and estimates the percentage of positives that will be recovered if a complete data set is predicted. It is derived as follows:

$$Sn = \frac{TP}{TP + FN}. \quad (19.1)$$

Specificity, Sp, is the percentage of truly negative samples that the classifier correctly identifies as being from the negative class. Specificity describes the classifier's ability to filter out negative samples. It is derived as follows:

$$Sp = \frac{TN}{TN + FP}. \quad (19.2)$$

Precision (Pr), or positive predictive value, is an alternative to specificity that is especially useful when class imbalance is large. Precision is the percentage of predicted positive samples that are truly positive. Practically, precision estimates the proportion of true positives among all samples predicted to be from the positive class. It is derived as follows:

$$Pr = \frac{TP}{TP + FP}. \quad (19.3)$$

In the case of a test data set in which the class imbalance does not accurately reflect reality, prevalence-corrected precision may be used. Prevalence-corrected precision is derived from the performance over the positive samples in the test data (Sn), performance over the negative test samples (Sp), and the relative prevalence, ρ, of the negative class (e.g., for a ratio of 1000:1 negatives:positives, ρ = 1000). Prevalence-corrected precision is defined as

$$(\text{prevalence-corrected}) \, Pr = \frac{Sn}{Sn + \rho(1 - Sp)}. \quad (19.4)$$

Accuracy, Acc, is the percentage of data that are correctly classified, regardless of class. It is derived as follows:

$$Acc = \frac{TP + TN}{TP + TN + FP + FN}. \quad (19.5)$$

Accuracy estimates the performance of the classifier across all data; however, it is not always useful as a summary metric. When class imbalance is present, accuracy is biased toward the majority class; in extreme cases, the accuracy and specificity of a classifier converge, and Sn has little impact on Acc.

An alternative to accuracy that is prevalent in the field of miRNA prediction is the geometric mean (GM) of sensitivity and specificity. GM is derived as follows:

$$GM = \sqrt{Sn*Sp}. \quad (19.6)$$

GM is insensitive to class imbalance; classifier performance over the positive and negative classes are weighted equally regardless of the true size of each class.

The above metrics describe the performance of classifiers at a single parameter set. By varying parameters, such as the decision threshold in the case of a classifier producing a continuous score or confidence for each sample, a classifier can move from permissive (identifying all data as positive) to restrictive (identifying all data as negative). Receiver operator characteristics (ROC) curves and precision–recall (PR) curves illustrate the performance of a classifier across all values for such a parameter.

ROC curves plot Sn against (1 − Sp). Moving from left to right, the classifier varies from maximally restrictive to maximally permissive. A classifier that randomly selects the class of each datum would trace a straight line from the bottom left to the top right of the plot. Increases in classification performance move the classifier's ROC curve toward the upper left corner of the plot. Figure 19.5 shows the ROC plot of a classifier that performs better than random. Because both sensitivity and specificity are insensitive to class imbalance, the ROC curve shares the same property; a ROC curve describes classifier performance in a way that is agnostic to (or ignorant of) class imbalance.

PR curves plot Pr against Sn. Unlike ROC curves, PR curves are sensitive to class imbalance; a single curve describes the performance of a classifier only at a specific class imbalance. For problem domains with high class imbalances, PR curves better elucidate a classifier's real-world performance. Moving from the top to the bottom of a PR curve, the classifier varies from maximally restrictive to maximally permissive. As classification performance

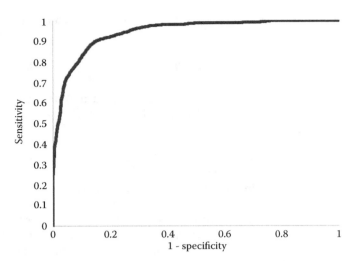

FIGURE 19.5
A typical ROC plot. Classifier sensitivity is measured against 1 − specificity. Ideal classification performance occurs in the top left corner (high sensitivity and specificity). Random classification results in a curve along the diagonal.

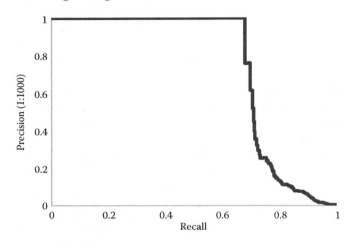

FIGURE 19.6
A typical precision–recall plot. Here, classifier recall is measured against prevalence-corrected precision at a class imbalance of 1:1000. Ideal classification performance occurs in the top right corner, where precision and recall are both high.

improves, the PR curve for the classifier moves toward the top right corner of the plot. Typically, an acceptable precision threshold is chosen from a PR curve, and classifiers are measured based on the recall that they achieve at this precision threshold. Figure 19.6 shows a typical PR curve.

19.4.2 Selecting Training and Testing Data

In order to train a classification model, labeled positive and negative data must be collected as input to the model. Data availability, reliability of data, reliability of labels, and class imbalance within the data (i.e., ratio of samples that belong to each class) are all data quality issues that may affect classification performance and must be addressed within the training pipeline.

Within the context of miRNA prediction, training data consists of RNA sequences that are typical of miRNA with respect to length and secondary (hairpin) structure. Labeled positive data is retrieved from miRNA databases such as miRBase [37] or miRTarBase [39] containing sequences proven (by independent wet-lab experiments) to be true miRNA sequences. Negative data samples are often extracted from genomic regions known to have functions other than coding for miRNA, such as protein-coding regions or other classes of ncRNA.

19.4.3 Feature Selection

Selecting appropriate features with which to train a classifier is a crucial step of the training process, as the shape of the feature vector defines the space in which data are situated and decision boundaries are formed. Features are selected based on their ability to discriminate data from different classes either in isolation or in combination with other features. An ideal feature would separate positive and negative data into two distinct groups; however, real data rarely contains such features, and no such feature exists for miRNA prediction. Increasing the number of features in the feature vector typically increases the separation between classes but does so at the cost of model complexity (the number of free parameters used to define the classification model). In general, more complex models are more likely to *overfit* training data, resulting in overly complex decision boundaries that do not generalize well to future unlabeled data [38]. Additionally, as the number of features increases, data quality is reduced as per the curse of dimensionality [40]. Therefore, a trade-off must be made during feature selection—multiple features are required in order to discriminate between classes, but model complexity should be minimized when possible.

Features typically used in miRNA prediction include RNA sequence composition, minimum free energy (MFE) of RNA secondary structure, and robustness of observed RNA secondary structure. When miRNA prediction is performed on transcriptomic data, expression of Dicer products and mature miRNA duplex stability can also be used as features.

19.4.4 Class Imbalance Correction

In many classification problems, such as miRNA prediction, classification is performed in the presence of significant class imbalance: negative samples outnumber positive samples by a large margin. For example, in the human genome, there are approximately 11 million pseudo-miRNA sequences that have a length and secondary structure typical of miRNA, and only approximately 2000 known miRNA precursors, leading to a class imbalance of >5000:1. Classifiers that are naively trained on imbalanced data sets are biased toward the dominant (negative) class, affecting overall classification performance. This bias is detrimental when it is important to correctly classify samples from both the majority and minority classes; when class imbalance is extreme, a classifier that seeks to maximize accuracy may do so by correctly classifying all majority data at a severe cost to correct prediction rates on the minority class. Several techniques have been developed that correct the learning bias that occurs when training data contains a high class imbalance. Synthetic data can be added to the minority class in order to balance the number of samples in each class, the majority class can be undersampled in order to achieve the same balance, or classification models can be built to handle the imbalanced training data set as is, using differential weighting of errors made on each class for example.

One widely used example of a class correction algorithm is the synthetic minority oversampling technique (SMOTE) [41]. SMOTE oversamples the minority class by a factor b through the following algorithm:

For each sample s_i in the minority data set S within a given feature space:

For each of the b nearest samples from S, n_j (where $i = 1$ through b):

Generate a new sample whose location in the feature space is a random point on the line between s_i and n_j.

Class imbalance must be taken into account not only during the training of classifiers but also during the evaluation of trained classifiers, as discussed below.

19.4.5 Splitting of Training and Testing Data

Because classification models are built in order to optimally separate their training data, performance on test data outside of the training data set will tend to decrease relative to performance observed on the training data. As mentioned previously, overly complex models in particular will perform very well on training data but may generalize very poorly to other data sets. For this reason, in order to estimate the applicability of a classifier, it is necessary to train using only a portion of the available labeled data. A portion of the data must be held out as testing data in order to estimate the generalizability of the model. Splitting of labeled data into training and testing can be performed several ways. A simple 70%/30% split is common. n-fold cross-validation (n-CV) is the most widely used method for generating training and testing data sets. During n-fold cross-validation, available labeled data is separated into n even partitions. n classifiers are trained, each on $n - 1$ partitions and each with a distinct partition as holdout test data. The total performance of the n classifiers on the holdout partitions is then used to estimate real-world performance of a classifier trained on the training data set.

19.4.6 Model Selection

At the core of each classifier is a model that places decision boundaries within a given feature space. Many such models exist, and these models employ vastly different algorithms for the determination of decision boundaries. The most common classification model type in miRNA prediction is the support vector machine (SVM), which draws a plane that optimally separates the classes within a feature space, maximizing the margin between the decision boundary and the nearest training points from each class (dubbed "support vectors"). Other common model types are decision trees, which perform classification based on a series of *20-questions-*style decisions ("Is feature x larger or smaller than value y?"). Each question is modeled as a node in a tree graph. Leaf nodes of the tree each contain a class prediction. Random forests are made up of collections of tens or hundreds of simple decision trees trained on subsets of the training data that ultimately vote on the class prediction for an input test feature vector. Other less commonly used classifiers within miRNA prediction include artificial neural networks, K nearest-neighbor classifiers, hidden Markov models, and Bayesian models. Multiple classifiers of the same or different model types can also be grouped into ensemble models, in which each component classifier casts a vote regarding the class of input data.

Model selection is a non-trivial decision within pattern classification because of the "no free lunch theorem," which states that no classification model is inherently superior to any other model when all possible classification problems are considered [42]. The quality of a model for a given classification problem is most often a function of the degree of agreement between the characteristics of the data and the underlying assumptions of the model.

19.4.7 Model Training

Following selection of the model type, the parameters of the classification model should be optimized in order to describe optimal decision boundaries. Many models, such as SVMs, rely on a kernel function that determines the general shape of decision boundaries. Gaussian functions, radial basis functions, and linear functions are typical kernel functions. Selection of an appropriate kernel function can affect performance greatly. Some models require the optimization of model-specific hyperparameters such as the number of trees in a random forest. Most models are trained using a cost function, which specifies the relative penalty of misclassification of positive and negative samples. Many models can be trained using different training algorithms that vary with regard to generated models and training speed. Finally, stopping criteria for training algorithms must be set in order to produce models that balance training set accuracy and generalizability. Each of these factors can affect the quality of a classification model.

19.4.8 Prediction Pipeline

Once a classification model has been built for a given training set, it is capable of predicting the classes of data in unlabeled sets.

Relative to model training, prediction of unlabeled data using an existing model is a straightforward procedure. Each pattern in the unlabeled data set is converted into a feature vector, using the features selected during the model training procedure. Feature vectors are passed through the classification model, which outputs a classification. Some classification methods also output confidence values, typically between 0 and 1, which state the confidence with which the classifier believes that a pattern belongs to a given class. In some applications, the computational runtime of a classifier is critical in order to achieve real-time predictions. This is not particularly the case for miRNA prediction.

19.5 The State of the Art in Computational miRNA Techniques

Computational prediction of pre-miRNA sequences can be broadly categorized into three major areas—homology-based techniques, NGS-based techniques, and *de novo* machine learning techniques.

Homology-based techniques represent the earliest efforts at prediction of miRNAs [43–45]. These techniques discover miRNAs based on similarity between these miRNAs and existing miRNAs. While not capable of discovering taxon- or species-specific miRNAs with novel sequences, these techniques provide a means for discovering miRNA homologs across species. The earliest homology-based miRNA prediction techniques include MiRscan, which was used to successfully predict many miRNAs in *C. elegans* and *C. briggsae* genomes [43]; srnaloop, which predicted additional miRNAs within the *C. elegans* genome [45]; and miRseeker, which played a similar important role within the *D. melanogaster* and *D. pseudoobscura* genomes [44]. The MIRcheck [46] and findMiRNA [47] algorithms were leveraged to discover 23 and 13 new miRNAs within the *A. thaliana* genome, respectively. MicroHARVESTER [48], like MIRcheck and findMiRNA, specializes in plant miRNA homologs. Most improvements in the state of the art in homology-based miRNA prediction occurred between 2003 and 2006, including all of the above studies. These studies are summarized by Berezikov et al. [35]. Improvements in homology-based miRNA prediction techniques were still being pursued as recently as 2013, when improvements on the state of the art were made by the miR-Explore algorithm [49].

NGS-based techniques focus on analysis of NGS data in order to identify miRNA at the transcriptome level. These techniques are relatively new and have been growing in popularity since their inception in 2008. Of these techniques, miRDeep [50] and the updated miRDeep2 [51] have emerged as *de facto* standards. Deep sequencing techniques provide higher confidence miRNA predictions, as the expression of RNA sequences that match the predicted miRNA constitutes experimental evidence in favor of the miRNA identification. Class imbalance within a transcriptome is less extreme than that of a genome, making identification easier since only expressed sequence regions need be explored. However, deep sequencing techniques are limited to identification of those miRNAs that are expressed under the specific experimental conditions at the time of data collection, and these techniques are biased toward identification of highly expressed miRNAs [52]. These techniques are reviewed in Section 19.5.2 of this chapter.

De novo machine learning techniques predict miRNA based only on information that can be derived from unannotated RNA sequence data. These techniques employ training data sets of known miRNAs and pseudo-miRNAs (miRNA-like hairpins that are not functional miRNAs) in order to predict RNA sequences that correspond to functional miRNAs. *De novo* miRNA prediction techniques require no annotation and can function on either genomes or transcriptomes. These techniques are reviewed in Section 19.5.3 of this chapter.

The three major branches of miRNA prediction are complementary when searching for miRNA within a given species of interest. Homology-based techniques provide a small set of high-quality miRNA identifications. NGS-based prediction methods only consider RNA sequences that are expressed, leading to potentially lower overall recall but higher precision. *De novo* machine learning techniques examine an arbitrary sequence (up to the entire genome) for one or more putative miRNAs. Therefore, they possess the highest potential recall but suffer from low precision due to high class imbalance. When discovering novel miRNAs within a species of interest, each of these three prediction techniques provides unique prediction information.

19.5.1 The State of the Art in NGS-Based miRNA Prediction

In this section, we examine the state of the art in miRNA prediction techniques for transcriptomic data sets. We will examine 15 prediction methods spanning 2008 to 2015, as well as 9 RNA sequencing pipelines that incorporate miRNA prediction methods, and 4 experiments that apply miRNA prediction methods to transcriptome data in order to predict novel miRNAs within a species of interest.

NGS-based miRNA prediction methods can be broadly categorized using the following distinctions:

- The method is specific to the animal or plant kingdoms, or is applicable to multiple kingdoms.
- The method's primary prediction is based on sequence homology, or miRNA characteristics, or both methods are used. Homology-based methods provide high-confidence predictions of typical miRNAs, while methods using miRNA characteristics may discovery additional novel miRNAs.
- The method maps ncRNA reads to a reference genome during its prediction pipeline or performs analysis without a reference genome. Mapping to a reference genome removes many false positives from data sets; however, genomic data are not available for all species.
- The method employs a machine learning algorithm during its prediction pipeline, or does not.
- The method requires that reads form a miRNA: miRNA* duplex, or does not. The discovery of a canonical miRNA:miRNA* duplex greatly increases miRNA prediction confidence; however, sensitivity is reduced as duplexes are not present for all miRNAs, especially those that are not highly expressed.

Table 19.1 provides a summary view of the 15 transcriptomic miRNA prediction methods with respect to the above criteria.

miRDeep [50], introduced by Friedlander et al. in 2008, was the first computational miRNA prediction method that made use of transcriptome data sets. miRDeep scores stacks of RNA sequence reads based on the likelihood that these reads are the result of miRNA biogenesis. Once a putative pre-miRNA region is determined surrounding a read stack, RNA secondary structure is computed, and the MFE is computed as a discriminating feature. The miRDeep study places a strong emphasis on the ability of miRDeep to estimate the quality of its results based on a statistical model of its outputs. miRDeep was initially applied to a *C. elegans* data set, discovering four novel miRNAs that were successfully validated.

No citation for MIREAP is available, and therefore knowledge of the method is limited; however, it is used

TABLE 19.1

Summary of Methods for NGS-Based miRNA Prediction

Method	Year	Kingdom(s)	Homology/Characteristics/Both	Reference Genome Required	Machine Learning Used	miRNA Duplex Required
miRDeep [50]	2008	Animal	Characteristics	Yes	No	No
MIREAP (no citation available)	2008	Animal	Characteristics	Yes	No?	No?
MiRMiner [25]	2008	Animal	Both	No	Yes	Yes
miRExpress [53]	2009	Animal + plant	Homology	No	No	No
miRTRAP [54]	2010	Animal	Characteristics	Yes	No	No
miRanalyzer [55]	2011	Animal + plant	Both	Yes	Yes	No
miRDeep-P [56]	2011	Plant	Both	Yes	No	No
miRDeep2 [51]	2012	Animal	Both	Yes	No	No
McRUM for miRNA [57]	2013	Animal + plant	Characteristics	No	Yes	No
MiRPlex [58]	2013	Animal + plant	Characteristics	No	Yes	Yes
miRDeep* [59]	2013	Animal	Characteristics	Yes	No	No
miRPlant [60]	2014	Plant	Characteristics	Yes	No	No
MIRPIPE [61]	2014	Animal + plant	Homology	No	No	No
miRdentify [62]	2014	Animal	Characteristics	No	No	Yes
miRNA and piRNA [63]	2015	Animal	Characteristics	No	Yes	No

within larger pipelines [64] and has been employed to successfully discover novel miRNA [65]. From MIREAP's description, it combines position and depth of sRNA reads with a miRNA biogenesis model.

The MiRMiner [25] miRNA prediction method first identifies known miRNAs by using BLAST [66] to search observed RNA sequence reads against known mature and star miRNAs within miRBase. Three criteria are applied to the ungapped sequence alignment: no more than 2 nt difference in matching sequence length, 100% sequence identity in positions 2–7 of the seed sequence, and no more than three mismatches in the remaining alignment. Reads corresponding to (fragments of) known non-miRNAs are removed by searching against National Center for Biotechnology Information "nr" database and excluding matches over 95% identity. Remaining reads are examined for phylogenetic conservation across multiple species. Conserved sequences are folded via Mfold [67]; novel miRNAs are predicted for those resulting hairpin structures that pass the three classic structural criteria laid out by Ambros et al. [68].

miRExpress [53] focuses on quantifying the expression levels of miRNA within data sets. As a step toward this goal, it identifies likely miRNA sequences within transcriptomic data sets. The miRExpress algorithm uses a customized implementation of the Smith–Waterman algorithm [69] in order to align sequence data with known miRNAs. The miRExpress alignment method identified 79 previously unknown *H. sapiens* miRNAs within two transcriptomic data sets.

miRTRAP [54] detects miRNA within transcriptomic data sets using the characteristic read patterns of miRNA biogenesis as well as information retrieved from the larger genomic context of the miRNA. This larger context includes the number of miRNAs in the region of the candidate miRNA, the number of non-miRNA ncRNAs within this region, and anti-sense reads that match to the miRNA reads. miRTRAP was applied to a *C. intestinalis* data set and recovered 36 miRNAs that were homologous to known miRNAs as well as 20 novel miRNA families.

miRanalyzer [55] uses an ensemble of five random forest classifiers in order to predict novel miRNA. As a pre-screening step, bowtie [70] is used to identify reads that match to known miRNAs and known mRNAs. miRanalyzer uses a set of pre-processed genome files for alignment during its prediction pipeline; 25 animal and 6 plant genomes are available. Feature sets used for classification differ for plant and animal species. No novel miRNAs have been identified by miRanalyzer.

miRDeep-P [56] is an extension of the miRDeep algorithm that targets plant sRNA data sets. This method is a straightforward extension, altering only the values of miRDeep's decision rules and adding additional prediction criteria based on Meyers's observations of plant miRNA [71]. Meyers's criteria represent a base set of criteria that all plant sequences must meet in order to be accepted for annotation as miRNAs. miRDeep-P recovered 18 novel miRNAs from *Arabidopsis* sRNA data sets that were successfully validated.

miRDeep2 [51] moves forward from the original miRDeep, implementing an updated prediction algorithm while maintaining the ability to estimate prediction quality. miRDeep2 increases the number of RNA secondary structure criteria used during pre-filtering of putative pre-miRNA regions. miRDeep2 was originally applied to seven organisms within the animal kingdom, and discovered on average 14 novel miRNAs within each data set. miRDeep2 has since been applied to transcriptomic data sets from a wide range of species, resulting in many novel miRNA predictions. Currently, miRDeep2 is the *de facto* standard for NGS-based identification of miRNAs. The miRDeep2 pipeline is analyzed in greater detail in Section 19.5.1.3 of this chapter.

Menor, Baek, and Poisson implemented a novel kernel-based learning algorithm, the "multi-class relevance units machine" (McRUM) [57], with the goal of classifying several types of ncRNA from transcriptomic data sets. Reads are classified as miRNA, piwi-interacting RNA (piRNA), or other ncRNA based on k-mer representations of the reads. This method was trained and tested on data sets of known ncRNA but not applied to real-world transcriptomic data.

MiRPlex [58] implements an SVM for classification of miRNA:miRNA* duplexes found in transcriptomic data; size, stability, and composition features distinguish miRNA duplexes from non-miRNA duplexes. The MiRPlex classifier was trained using positive data consisting of known duplexes from animals in miRBase and negative data consisting of duplexes from a *D. melanogaster* sRNA data set that do not match known miRNAs. MiRPlex was applied to four animal sRNA data sets, and a separately trained model was applied to a single plant sRNA data set. No novel miRNAs were reported from these experiments.

Like miRDeep2, miRDeep* [59] extends the original miRDeep algorithm. Also like miRDeep2, focus is placed on improvements in the precursor extraction methodology and increases in RNA secondary structure criteria. Unlike miRDeep2, miRDeep* also focuses strongly on usability; all elements of the miRDeep* pipeline are implemented within a Java graphical user interface. An analysis of the performance of miRDeep* against other popular methods was performed by the miRDeep* authors, wherein performance of miRDeep* and miRDeep2 were found to be comparable. In the initial study, four novel miRNAs were discovered from prostate cancer data sets using the miRDeep* algorithm.

The authors of miRDeep* also produced a plant-specific miRNA predictor, miRPlant [60]. miRPlant implements a very similar pipeline to miRDeep-P; however, the precursor extraction technique used differs. miRDeep-P presents the user with a graphical Java interface, much like miRDeep*. While no novel miRNAs were recorded using miRPlant, the authors of mimPlant demonstrated that, using miRPlant, performance on three plant sRNA data sets is improved relative to miRDeep-P.

MIRPIPE [61] aims to improve the viability of miRNA prediction within sRNA data sets originating from niche model organisms. Reads are aligned to each other, and then read stacks are matched to known miRNA using simple matching criteria. In independent experiments, MIRPIPE performs similarly to methods that align sRNA reads to the genome for the identification of known miRNAs, though the homology-based nature of MIRPIPE is not likely to generalize to novel predictions on niche species.

miRdentify [62] combines stringent miRNA:miRNA* duplex requirements with RNA secondary structure filters in order to generate sets of high-confidence miRNA predictions without a reference genome. RNA sequences from transcriptomic data sets that form strongly paired duplexes are considered to be candidate miRNA duplexes; no requirement is made that the two sequences that form the duplex are co-located in a genome. The miRdentify method is designed to work with large sRNA data sets encompassing multiple experiments. miRdentify compares favorably to miRDeep2, miRDeep*, and miRanalyzer when applied to the study's test data set, and identified two novel miRNAs in *H. sapiens* chromosome Y (ChrY).

Menor, Baek, and Poisson again apply a machine learning approach to miRNA prediction in sRNA data sets in their 2015 study [63]. The k-mer feature set used in the authors' previous study was augmented with additional sequence features, and systematic feature selection was employed. Classification is performed using McRUMs, as in the previous study, and a Gaussian kernel. Like the MiRPlex study, candidate miRNAs are read stacks that form duplexes, and no mapping to the genome is performed. The authors compare their method to MiRPlex, as these two methods share a common set of candidate miRNAs, and conclude that their method increases the number of miRNAs recovered from four animal data sets tenfold relative to MiRPlex.

19.5.1.1 sRNA Data Set Pipelines

Because NGS sRNA data sets contain reads pertaining to multiple types of biologically relevant ncRNAs (miRNA, piRNA, small nucleolar RNA [snoRNA], small nuclear RNA [snRNA]), several methods have been developed that incorporate multiple ncRNA prediction methods within larger pipelines. These pipelines aim to perform comprehensive analysis on sRNA data sets. With respect to miRNA, these pipelines have one or more of the following goals:

1. Detection of known miRNAs
2. Prediction of novel miRNAs
3. Detection or prediction of miRNA targets
4. Quantification of expression levels of miRNAs across multiple experiments

Additionally, these pipelines contain graphical user interfaces (GUIs), which simplify the analysis of sRNA data and present visualizations of the data. We have identified the following major axes upon which ncRNA pipelines can be measured, with respect to miRNA prediction:

- Which input file formats are supported by the pipeline?
- Does the pipeline detect known miRNAs?
- If the pipeline detects known miRNAs, what algorithm is used for the detection?
- Does the pipeline predict novel miRNAs?
- If the pipeline predicts novel miRNAs, what algorithm is used for this prediction?

Table 19.2 examines nine sRNA data set pipelines along these five axes.

In 2008, Moxon et al. introduced the UEA sRNA Toolkit, a toolkit for analyzing plant sRNA data sets [72]. This toolkit uses a custom algorithm, miRCat, for the prediction of miRNA precursors within sRNA data sets. miRCat maps reads to the genome, then searches for pairs of reads that form miRNA:miRNA* duplexes, and finally analyzes the hairpin structure of the resulting precursor candidates using criteria defined in Ref. [73].

DSAP [74], the deep-sequencing sRNA pipeline, is a web service that performs pre-processing of sRNA data sets, matches reads to known ncRNAs including miRNAs via the BLAST algorithm [66], and quantifies miRNA expression level differences across multiple experiments. Finally, for each miRNA in the data set, DSAP identifies all species that have miRNAs from the same family.

Moxon et al. advanced their toolkit in 2012, creating the UEA sRNA workbench [75]. This work extends the previous effort in the areas of visualization and the profiling of expression levels of miRNAs across multiple experiments. miRNA candidates that match known miRNAs in miRBase are also highlighted as such by the toolkit.

miREvo [76] performs three major functions for sRNA data sets: identification of novel miRNA, detection of homologs to known miRNA, and profiling of miRNA

TABLE 19.2
Summary of sRNA Pipelines for Examination of NGS Data Sets

Method	Year	Detects Known miRNA	Algorithm for miRNA Detection	Predicts Novel miRNA	Algorithm for miRNA Prediction	File Formats Accepted
UEA sRNA Toolkit [72]	2008	No	N/A	Yes	miRCat	FASTA
DSAP [74]	2010	Yes	BLAST [66]	No	N/A	TSV
UEA sRNA workbench [75]	2012	Yes	Unknown	Yes	miRCat	FASTA
miREvo [76]	2012	Yes	Novel Whole Genome Alignment algorithm	Yes	miRDeep2	Unknown
mirTools 2.0 [64]	2013	Yes	SOAP [81]	Yes	miRDeep, MIREAP	FASTA, SAM, BAM
MiRGator [77]	2013	Yes	Bowtie	Yes	miRDeep2	Unknown
miRspring [78]	2013	Yes	SAMTOOLS [82]	No	N/A	BAM
CAP-miRSeq [79]	2014	Yes	miRDeep2	Yes	miRDeep2	FASTA, FASTQ
ISRNA [80]	2014	Yes	BLAST [66]	No	N/A	FASTQ, txt

expression across multiple species or multiple experiments. Emphasis is placed on the analysis of miRNA homologs across species in order to estimate the evolutionary rate of these miRNAs. miRDeep2 is implemented within the miREvo pipeline in order to predict novel miRNA, while the identification of homologs is performed using a novel whole genome alignment algorithm.

mirTools 2.0 [64] detects known miRNAs, transfer RNA (tRNAs), snRNAs, snoRNAs, ribosomal RNA (rRNAs), and piRNAs from within sRNA data sets and profiles these ncRNAs across multiple experiments. miRNA targets are also predicted, and functional annotation of targets is performed. Finally, novel miRNAs and piRNAs are predicted from within the data sets using miRDeep and MIREAP.

MiRGator v3.0 [77] is a web portal that provides users access to sequence editing, counting, sorting, and ordering tools; miRNA and miRNA target identification tools; and miRNA–target co-expression information. Seventy-three human sRNA data sets have been curated by the MiRGator software, providing an existing library of miRNA data.

miRspring [78] implements an index-compression algorithm in order to store sRNA data sets in relatively small file sizes (approximately 3 MB per data set) and then leverages this efficient file format for miRNA analysis. miRspring provides the user with information regarding the global Dicer processing of all miRNAs within a data set and the Dicer processing details of specific miRNAs within the data set. Special attention is paid to the identification of isomiRs.

CAP-miRSeq [79], the comprehensive analysis pipeline for miRNA sequencing data, is a tool that performs pre-processing, alignment, miRNA detection, miRNA quantification, visualization, differential expression analysis, and variant detection in miRNA coding regions. Particular emphasis is placed on the user-friendliness and practicality of the program. Here, miRDeep2 is used both for its detection of known miRNAs and its prediction of novel miRNAs.

ISRNA [80], the Integrative Short Reads Navigator, is an online toolkit that provides the user with data set–wide statistics including genomic location, length distribution, and nucleotide composition bias. It also provides expression data and genomic location for known miRNAs and other known ncRNAs identified within the data. Again, emphasis is placed on the user-friendliness of the toolkit.

19.5.1.2 NGS Experiments for miRNA Discovery in Species of Interest

Several studies exist in the literature that apply NGS techniques along with NGS-based miRNA prediction methods in order to discover novel miRNAs within tissue samples of species of interest.

miRDeep2 is the most commonly applied deep sequencing analysis pipeline for miRNA prediction studies. Yin et al. analyzed miRNA within differently aged rat brain samples using miRDeep2 for miRNA prediction [83]. Differential analysis of miRNA was performed using DESeq2 [84], while MiRNA target prediction was performed using Targetscan [85]. This experiment discovered 547 known miRNAs within the tissue and predicted 171 candidate novel miRNAs, though only 3 of these novel miRNAs were experimentally validated. Cowled et al. applied the miRDeep2 prediction pipeline to the black flying fox [81]. Like the study of Yin et al., this study uncovered several hundred (222) known miRNAs and discovered 177 novel miRNAs within the tissue, though none of these miRNAs were experimentally validated.

Gu et al. applied the MIREAP algorithm to the prediction of miRNAs in a maize endosperm deep sequencing data set [65]. miRNAs were mapped to the maize genome using short oligonucleotide alignment program (SOAP) [86], while known miRNAs were identified using BLAST [66]. Target prediction was performed using the Web MicroRNA Designer (WMD3) software. This effort resulted in the recovery of 95 known miRNAs and the discovery of 18 novel miRNAs, which were validated through reverse transcription polymerase chain reaction (RT-PCR).

For the identification of miRNAs within hexaploid wheat data sets, Agharboui et al. have developed a pipeline that combines the HMMiR [87] and MiPred [88] sequence-based miRNA prediction algorithm with the miRdup* [89] mature miRNA prediction algorithm. Further analysis includes expression profile filtering using Meyers's criteria [71] and target gene prediction using the TAPIR software [90]. One hundred ninety-nine candidate miRNA were discovered in this study, demonstrating that sequence-based miRNA prediction methods are applicable to deep sequencing data sets.

19.5.1.3 Analysis of the miRDeep2 miRNA Classification Pipeline

Within the field of NGS-based miRNA prediction, miRDeep2 has emerged as the *de facto* standard prediction method. As described previously, the majority of miRNA prediction on NGS data sets is performed using the miRDeep2 pipeline. A recent independent review of the field also recommended that miRDeep2 be used for NGS-based miRNA prediction [91]. Considering its wide adoption, in this section, we briefly describe the miRDeep2 algorithm for prediction of miRNAs.

The miRDeep algorithm can be dissected into two primary steps: a pre-processing step wherein NGS read data are mapped to a genome and candidate pre-miRNA sequences are extracted from the genome at read loci; and a scoring step wherein candidate pre-miRNAs are given a numerical score based on the structural stability of the pre-miRNA sequence, and expression profile within the miRNA sequence (as described below).

The pre-processing step of the miRDeep algorithm first identifies candidate mature miRNAs by identifying local read depth maxima in the genome that match the expected sequence length of a mature miRNA (18–25 nt). By default, only the 50,000 deepest read stacks are considered during the selection of candidate mature miRNAs. The mature miRNA sequence is then extended in each direction twice: once by 70 nt in the 5' direction and 10 nt in the 3' direction, and once by 10 nt in the 5' direction and 70 nt in the 3' direction. The two resulting sequences represent candidate pre-miRNAs for which the mature miRNA rests on the 3' and the 5' arm of the hairpin, respectively.

Candidate pre-miRNA sequences are then filtered based on the following criteria:

1. RNAfold must predict a hairpin structure for the candidate pre-miRNA sequence that contains no bifurcations.
2. The miRDeep2 pre-processing algorithm attempts to identify a mature miRNA, miRNA*, and loop product within the candidate pre-miRNA sequence. If this attempt fails, the candidate pre-miRNA is rejected. The miRNA* sequence is defined as the sequence that pairs to the candidate mature miRNA sequence in the predicted RNAfold [92] structure, taking into consideration a 2 nt overhang on the 3' end of each sequence in the duplex. The loop product is defined as the subsequence of the candidate pre-miRNA that is between the mature miRNA and miRNA* products.
3. At least 60% of the bases in the stem region of the candidate pre-miRNA must be paired.
4. At least 90% of the NGS reads in the pre-miRNA sequence region must match a Dicer product. A match to a Dicer product is defined as a read that aligns to the miRNA sequence with a starting 5' position within ±2 nt of the candidate mature miRNA, miRNA*, or loop region, and a terminating 3' position within ±5 nt of the same region.
5. The length of the mature miRNA and miRNA* sequences must match to within 6 nt.

Candidate pre-miRNA sequences that do not meet these criteria are discarded prior to scoring.

One weakness of the miRDeep pre-processing algorithm is its inability to predict large or small miRNAs. The minimum length of a pre-miRNA sequence as determined by miRDeep is 98 nt, and the maximum length of a pre-miRNA sequence is 10 5 nt. Only 10.7% of miRNAs within miRBase v21.0 fall within this length range; in this respect, the miRDeep pre-processing algorithm does not accurately reflect the biogenesis of miRNA.

The numerical score given to a candidate miRNA is a simple rules-based algorithm. No mention of rigorous training or testing is presented in the miRDeep manuscript; therefore, it can be assumed that the rules were developed and tuned by hand, with the goal of optimizing performance across training data sets. Details of the miRDeep scoring algorithm are not described in the miRDeep manuscript; only the features used are described [50]. The score assigned to a candidate miRNA is the sum of the following five terms:

1. The candidate miRNA is given a starting score of −6.
2. The candidate miRNA's score is increased by 0.5 for each read that matches one of the candidate miRNA's Dicer products, using the matching rules described previously in preprocessing filter 2. If no reads match to the miRNA* region, the contribution for this score is limited to 6.
3. The candidate miRNA's score is adjusted by the log odds of the following probability distributions: P(MFE of the candidate miRNA structure is derived from a distribution of known miRNA MFEs) and P(MFE of the candidate miRNA is derived from a distribution of background training sample MFEs).
4. If any reads within the miRNA region match to the star region as per miRDeep2's Dicer processing rules, +3.9 is added to the score. If not, 1.3 is subtracted from the score.
5. If the randfold algorithm [93], using default parameters, finds the MFE of the miRNA structure to be significant ($p \leq 0.05$), +1.6 is added to the score. If not, 2.2 is subtracted from the score.

Optional scoring parameters are as follows:

6. If the seed sequence of the candidate mature miRNA (defined by miRDeep2 as the six nucleotides at the 5' end of the mature miRNA sequence) matches a known mature miRNA exactly, +3 is added to the score. If not, 0.6 is subtracted from the score.
7. The number of paired bases in the lower stem portion of the miRNA is counted, and the score is adjusted based on the number of pairs present. The lower stem portion of the miRNA is defined as the 10-nt-length duplex that is directly adjacent to the mature miRNA duplex, opposite the loop region. Figure 19.1 provides a visual representation of the lower stem portion of the miRNA.

Any candidate miRNA with a final score ≥0 is highlighted by the miRDeep2 algorithm as a true (predicted) miRNA.

The primary weakness of the miRDeep scoring algorithm, in our estimation, is the unbound contribution of term 2, reflecting the read depth of the miRNA sequence. In a modern NGS experiment, tens of thousands of reads map to single miRNAs. As a result, term 2 of this scoring algorithm solely dictates the classification of many high-abundance candidate miRNAs. Furthermore, as NGS technology improves and read depth continues to increase, the contribution of this term will increase accordingly. For a given decision threshold, the decision boundaries of the miRDeep scoring algorithm will shift relative to the average read depth of the experiment. In turn, increasing average read depth increases the false positive rate of the algorithm.

19.5.2 The State of the Art in de novo miRNA Prediction

In this section, we examine 24 published methods for *de novo* miRNA prediction. These methods were published between 2005 and 2014. The major elements of the classification pipeline—data set generation, feature set generation, classifier selection, training methodology, and reporting of results—are studied in order to represent the state of the art in miRNA prediction from the perspective of pattern classification and to highlight elements of the state of the art that require improvement.

Previous reviews of the field of *de novo* miRNA prediction [52], [94–97] have defined the major challenges of the field as follows: data set generation, specifically the generation of negative data sets; classifier selection and training, with an emphasis on class imbalance correction; and feature set selection. Improvement in these areas is defined as the path toward improved prediction performance. In this review, we examine the state of the art in miRNA prediction methods with respect to each of these areas. We also examine the reporting of results in miRNA prediction studies, as we feel that improvements in, and standardization of, performance reporting would result in miRNA prediction methods that are more appropriate for real-world applications.

19.5.2.1 Data Set Generation

In general, positive training data set generation is not a major focus of miRNA prediction methods. Positive training data sets throughout nearly all studies consist of pre-miRNA sequences that are drawn from the miRBase database. One exception is the mirnaDetect method, whose major novel contribution is an improvement to training set selection [98]. This method demonstrates a substantial improvement in performance through training set selection.

Conversely, *de novo* miRNA prediction studies employ a variety of methods and standards for generating negative training and test data. In general, this negative training and test data consists of sequences that form miRNA-like hairpin structures, as predicted by an RNA folding package such as RNAfold [92], Mfold [67], or UNAfold [99]. Negative data are extracted from annotated functional genomic regions, commonly coding regions, as these regions likely do not produce miRNAs. ncRNAs that share structural similarity to miRNA are also commonly used as negative training and test data.

Nam et al. [100] generated negative training data from chromosomes 16 through 19 of the human genome, based on RNAfold structure prediction and the following criteria: sequence length between 64 and 90 nt; stem length (number of pairs of bases in the miRNA hairpin stem, see Figure 19.1) above 22 nt; bulge size (number of unpaired bases in the miRNA hairpin stem, see Figure 19.1) under 15 nt; loop size between 3 and 20 nt, and an MFE of at most −25 kcal/mol. These strict criteria were not widely adopted.

The negative data set of Sewer et al. [101] in their 2005 study consisted of random subsequences of tRNA, rRNA, and mRNA. The size of their data set and length of the RNA subsequences were not specified.

Xue et al. [102] in their 2005 study generated negative training data from within known human coding regions. Sequences were considered to be miRNA-like if they formed single-loop hairpin structures with an MFE of at most −15.0 kcal/mol and at least 18 paired bases in the hairpin stem. These MFE and paired base criteria were chosen because all known human miRNAs at the time fell within these criteria. The resulting data set, consisting of 8494 hairpins, has become a standard data set on which miRNA prediction methods are trained and tested to this day. The MFE and paired base criteria introduced by Xue et al. have similarly become a *de facto* standard for validation of miRNA-like hairpins. The data set introduced by Xue et al., or subsets thereof, was used in 11 of 21 studies published afterward. The hairpin criteria introduced by Xue et al. were the basis of hairpin extraction in a further three studies.

In 2009, Batuwita et al. [103] extended the 8494 coding region data set with 754 ncRNA sequences. These ncRNA sequences were used by 7 of the 13 studies published after 2009.

Yousef et al., in studies from 2006 and 2008 [104,105], predicted hairpins from within 3′ untranslated regions (3′-UTR) regions of genes as annotated by UTRdb [106] as opposed to predicting hairpins from within coding regions or ncRNAs. This practice was only adopted by two studies since 2006 [107,108], and one of these studies used 3′-UTR in addition to coding region data [108].

Gudys et al. [109] in 2012 introduced an alternative to hairpin prediction for negative data set generation, wherein sequences are randomly selected from genomes and mRNA. Sequences are selected such that the length distribution matches that of known miRNA. This method produces data sets that contain some sequences that do not resemble miRNA structurally and, as a result, has not been adopted by any other miRNA prediction studies.

The majority of miRNA prediction studies use data sets that are specific to a species. Of the 24 examined studies, 21 use data sets that are specific to humans. Six of the 24 studies use data sets that are specific to other model species of interest. Multi-species positive data sets are common, appearing in 14 studies. Multi-species negative data sets are far less common, appearing in six studies [98,104,109–112].

Class imbalance is present in all data sets, favoring the negative class. This is consistent with real-world miRNA prediction, where pseudo-hairpins outnumber miRNAs. However, while actual real-world class imbalances are expected to be as high as 5000:1, typical class imbalances used in miRNA prediction studies are on the order of 1:10, with some data sets approaching even (1:1) artificially "balanced" class ratios [87,88], while the most extreme observed class imbalances are between 1:100 and 1:200 [104,109].

19.5.2.2 Classifier Selection and Training

De novo miRNA prediction has been performed using a variety of classification methods including SVMs, hidden Markov models, naïve Bayes methods, random forests, random walk rankings, one-class clustering methods, K nearest-neighbor classification, and linear dimensionality reduction. Table 19.3 lists the classifier types used by each of the studies. SVMs are the most commonly used classifier and are present in 13 of the 24 studies. Random forests and hidden Markov models are the second and third most popular classifiers, with four and three appearances respectively. K nearest-neighbor is used in two studies, once as a member of a multi-classifier ensemble method [112] and once as a proposed one-class clustering alternative to standard miRNA prediction [104]. No other classification model is used by more than a single study. Notably, no miRNA prediction study has employed artificial neural network classifiers.

Cross-validation is the dominant method of training for miRNA predictors. Sixteen of the 24 studies employ cross-validation in order to train classifiers. Fivefold and tenfold cross-validation are both common, being used by eight and six studies respectively. The second most common training methodology is a simple holdout test using one training data set and one testing data set. This method is used by five studies. Fivefold boosting and out-of-bounds estimation are each used by one study (Refs. [107] and [88], respectively). Additionally, four studies use the LIBSVM parameter grid search in order to optimize SVM hyper-parameters. This grid search uses cross-validation over training sets in order to estimate parameter performance for a range of hyper-parameter values.

As previously stated, miRNA prediction is performed on data sets that contain significant class imbalance. Several methods are employed in order to train classifiers on these imbalanced data sets, including asymmetrical misclassification penalties, undersampling of negatives,

TABLE 19.3

Classifier Selection and Training Experiments for 24 miRNA Prediction Methods

Year	Lead Author	Classifier Type(s)	Training Experiment(s)	Class Correction
2005	Nam	Paired hidden Markov model (HMM)	5-CV	None
2005	Sewer	SVM	Single training set	Asymmetrical misclassification penalties
2005	Xue	SVM	Single training set	Undersampling of negatives
2006	Yousef	Naïve Bayes	Single training set, 5-CV	Undersampling of negatives
2007	Ng	SVM	LIBSVM grid parameter search, 5-CV	None
2007	Jiang	RF	Out-of-band (OOB) estimation	Undersampling of negatives
2008	Yousef	SVM, naïve Bayes, one class (K-means, Gaussian, principle component analysis (PCA), K-nearest-neighbour (KNN))	10% holdout repeated 20 times	Undersampling of negatives
2008	Xu	Random walk ranking	One-class systems built using 1–50 positive training samples	None
2009	Kadri	hierarchical hidden Markov models (HHMM)	10-CV	None
2009	Oulas	Profile HMM	Boosting (fivefold) on positive only	Trained positive only
2009	Batuwita	SVM	Outer 5-CV	SMOTE to correct training imbalance
2010	Mathelier	Simple filters	Manual optimization	None
2010	Ding	Ensemble SVM	Outer 3-CV	Ensemble trained on partitioned negatives
2010	Zhao	SVM	Parameter testing from LIBSVM	Undersampling of negatives
2011	Han	SVM	5-CV	Undersampling of negatives
2011	Wang	SVM	5-CV	Undersampling of negatives
2011	Xiao	RF	10-CV	Undersampling of negatives
2011	Xuan	SVM	5-CV	Undersampling of negatives
2012	Liu	SVM	Single training set	Undersampling of negatives
2013	Gudys	SVM	Stratified 10-CV	ROC-select
2013	Lertampaiporn	Ensemble (4× SVM, 4× RF, 4× KNN)	10 × 5-CV	SMOTE
2013	Shakiba	Linear dimensionality reduction	Parameter grid searches, 10-CV	LDR
2013	Wei	SVM	LIBSVM grid parameter search, 10-CV	Undersampling of negatives
2014	Zou	Hierarchical RF	10-CV	Bagging through RF

SMOTE [41], bagging, and one-class prediction. Simple undersampling is the most common technique and is used by 11 of the 24 studies. SMOTE is used to balance classes in two studies, and one-class prediction is used in two studies. Outside of these, no class correction technique is used by more than one study. Six studies do not describe any kind of explicit class correction.

19.5.2.3 Feature Selection

Features for *de novo* miRNA prediction can be broadly classified as sequence features, structure and thermodynamic features, and global or intrinsic features. The sequence category of features contains sequence motifs and k-grams (words of k consecutive nucleotides). The structure category of features contains metrics related to the predicted RNA hairpin structure of a miRNA sequence, such as MFE. Global or intrinsic features are sequence-wide features such as GC-content.

Early feature sets consisted largely of sequence and structure motifs (see below), global features, and stem and loop metrics such as number of unmatched nucleotides in the stem or size of the terminal hairpin loop. A commonly used sequence–structure motif feature set is the triplet feature set, which contains 32 features, each representing the prevalence of a nucleotide along with the pairing pattern of the nucleotide alongside its two flanking nucleotides. For example, the triplet A.((represents the nucleotide A, whose 5' neighbor is unpaired, who itself is paired, and whose 3' neighbor is paired. This feature set was introduced by Xue et al. [102].

MFE was introduced as a feature by Jiang et al. in 2007 and has since gained prominence as a defining feature of miRNAs. Since 2007, the majority of studies have used MFE prominently in their miRNA prediction methods. Recently, Lertampaiporn et al. have debated the usefulness of MFE features [112], promoting alternative features based on genetic robustness. MFE still plays a role in their prediction method, however.

Early miRNA prediction studies make little use of systematic feature selection methods. Between 2005 and 2008, no systematic feature selection methods were used, outside of testing the inclusion of MFE in a feature set by Jiang et al. [88]. Over time, systematic feature selection methods became more popular. This is largely due to the inclusion of many MFE-based features and Z-features. The high number of sequence and structure motifs used in studies also necessitates systematic feature selection methods. Since 2009, 11 of 16 studies made use of systematic feature selection methods, including F-score [108], clustering [113], information gain [113], genetic algorithms [112,114,115], SVM weight measurement [116], floating forward search [117], and correlation feature selection [112]. No single feature selection method dominates in the field of miRNA prediction.

Out of the 24 studies, Batuwita et al. [103] were the first to employ a systematic feature selection method. Their initial feature set contains 29 global and intrinsic features as introduced by Xue et al. [102] and 19 new features that contain several thermodynamic metrics. After feature selection, a set of 21 features remains. Much like the data sets generated by Batuwita et al., this set of features appears commonly in future studies, with 5 studies using these 21 features in their feature sets.

Feature sets for miRNA prediction typically contain between 20 and 40 features. 14 out of 23 feature sets lie within this range. Three studies contain very large feature sets that include at least 1000 sequence or structure motifs [104,105,116]. Two studies use feature sets that contain fewer than 10 features [111,117]. One study employs a feature set containing 98 features [97]. Finally, three studies do not report feature set size [87,107,118]. Table 19.4 summarizes the feature set size and selection methods for 24 miRNA prediction methods.

19.5.2.4 Reporting of Results

As with many pattern classification problems, the most highly reported results in the field of miRNA prediction are those of sensitivity (Sn) and specificity (Sp). These two metrics are reported almost universally among miRNA prediction studies, appearing directly in 21 of 24 studies. The second most commonly used performance metric within the field is accuracy (Acc). Acc appears in 11 studies [88,98,102,107,108,110,111,116,118–120], and

TABLE 19.4

Feature Set Size and Selection Methods for 24 miRNA Prediction Methods

Year	Author	Total # Features	Final # Features	Feature Selection Method
2005	Nam	N/A	N/A	N/A (HMM)
2005	Sewer	40	40	None
2005	Xue	32	32	None
2006	Yousef	>1000	>1000	None
2007	Ng	29	29	None
2007	Jiang	32 + 2	32 + 2	Tested addition of MFEs
2008	Yousef	>1000	>1000	None
2008	Xu	36	36	None
2009	Kadri	NA	NA	Two alphabets tested
2009	Batuwita	48	21	Subset selection methods
2009	Oulas	Unknown	Unknown	Unknown
2010	Mathelier	5	5	None
2010	Ding	65	32	F-score
2010	Zhao	36	36	None
2011	Han	48	27	Clustering, info gain
2011	Wang	124	20	Genetic algorithms (GA)
2011	Xiao	24	24	None
2011	Xuan	48	22	GA
2012	Liu	29,734	1300	Highest SVM weight
2013	Gudys	28	28	None
2013	Lertampaiporn	103	20	3-CV test on many methods; correlation feature selection + GA chosen
2013	Shakiba	48	7	Floating forward search
2013	Wei	98	98	Subset testing; deciding on full set
2014	Zou	Unknown	Unknown	RF random feature selection

in two of the studies which do not report Sn and/or Sp directly [102,118].

One study makes use of the information retrieval metrics of precision and recall [121] in place of the binary classification metrics Sn and Sp.

Beginning with the microPred study by Batuwita et al. [103], a shift occurred toward reporting the GM of Sn and Sp as the primary performance result of a miRNA prediction technique. GM is insensitive to class imbalance, which, as discussed above, is substantial in miRNA prediction. Therefore, it was argued that GM is a superior performance metric to the previously used Acc metric, which is highly sensitive to class imbalance. Since its introduction into the miRNA prediction field, GM has seen common usage, appearing in 7 of 13 studies in this period [98,108,109,112,113,115,117]. In the same period, 7 out of 13 studies reported Acc [98,108,111,116,118–120]. Neither Acc nor GM represents a definitive performance metric for miRNA prediction, as both are reported equally in recent studies. Table 19.5 summarizes the performance metrics reported by 24 miRNA prediction methods.

19.5.3 Previous Assessments of the State of the Art

Since 2009, a number of review articles have captured the ongoing state of the art in computational methods for miRNA, including *de novo* miRNA prediction. One of the common themes among reviews in this period is a suggestion to improve negative set generation methodology. Mendes et al. [52] in 2009 recommend that miRNA prediction models consider more strongly "what is not miRNA." Yousef et al. [96] state that "defining the negative class is a major challenge in developing ML [machine learning] algorithms for miRNA identification." Li et al. [97] emphasize the negative class, stating that "it is usually straightforward to select positive examples (e.g., taking the known miRNAs), whereas it is harder to construct negative examples." Gomes et al. [95] emphasize data set generation in general, saying that "a careful choice of positive and negative data sets is crucial." Kleftogiannis et al. [94] state that negative set selection is one of the problems that needs solving in order to improve accuracy of miRNA prediction; however, they feel that "the selection of pseudo hairpins is also simple as they can be downloaded from RefSeq genes."

Outside of negative set generation, there is very little agreement on future directions for miRNA prediction. Standardization of data sets, and public availability of data sets, is highlighted in two studies [94,95]. Mendes et al. [52] suggest that a better understanding of miRNA biogenesis is needed for improved prediction performance. Kleftogiannis et al. [94] suggest that miRNA prediction should be integrated into larger pipelines that

TABLE 19.5

Performance Metrics Reported by 24 miRNA Prediction Methods

Year	Author	Performance Metrics Reported
2005	Nam	Sn, Sp of 5-CV; # miRNA predicted and recovered from chromosomes
2005	Sewer	Viral pre-miRNA prediction results; Sn, Sp on test set
2005	Xue	Accuracy on test sets
2006	Yousef	Sn, Sp on test sets and holdout from training sets
2007	Kwang	Sn, Sp, Acc on holdout human and pooled
2007	Jian	Acc, Sn, Sp, Matthews' correlation coefficient (MCC)
2008	Yousef	Sn, Sp, MCC
2008	Xu	Precision and recall (in ~1:2 class imbalance test data)
2009	Kadri	Sn, Sp, MCC, false discovery rate (FDR)
2009	Oulas	Sn, Sp, average of Sn/Sp (Acc @ 1:1)
2009	Batuwita	Sn, Sp, GM
2010	Mathelier	ROC curves; area under the curve (AUC), Acc, MCC
2010	Ding	Sn, Sp, GM, Acc
2010	Zhao	Acc, Sn, Sp
2011	Han	GM, Sn, Sp
2011	Wang	Sn
2011	Xiao	Sn, Sp, Acc
2011	Xuan	Sn, Sp, GM
2012	Liu	Sn, Sp, Acc, AUC
2013	Gudys	Sn, Sp, GM
2013	Lertampaiporn	Sn, Sp, GM
2013	Shakiba	Sn, Sp, GM
2013	Wei	Sn, Sp, GM, Acc
2014	Zou	Acc

have web interfaces. Several studies suggest improvements to feature selection [94,97].

19.6 Discussion of the miRNA Prediction State of the Art

In this section, we examine several outstanding issues with the state of the art in miRNA prediction, which have been reached through our independent assessment of the field. Many of these issues stem from the pattern classification approaches chosen by miRNA prediction studies, as described above.

19.6.1 Redundancy in Feature Sets

Feature sets for miRNA prediction are often built using a scattershot approach wherein many similar features are

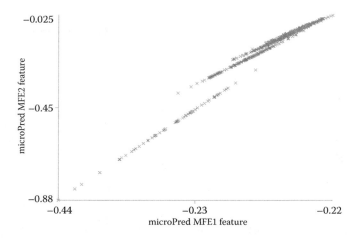

FIGURE 19.7
Correlation between features MFE1 and MFE2 in widely used microPred feature set. Data from microPred positive training set.

proposed and feature selection methods are used to reduce the length of the final feature vector. For example, feature sets often contain many features that relate to hairpin structure MFE. This technique has produced highly redundant feature sets such as the commonly used 21-feature set that formed the basis of the microPred classifier. This feature set contains many features that are highly correlated; Figure 19.7, for example, demonstrates the correlation between microPred MFE features 1 and 2.

We believe that the dominance of SVM and random forests (RF) for miRNA prediction is partly due to the presence of highly correlated features such as the widely used MFE features. At the same time, standard feature sets are likely suboptimal due to the presence of redundant features. We recommend that future miRNA prediction studies take into account feature redundancy when choosing feature sets, as this may open the door for more accurate classification using methods that underperformed previously. Explicitly handling redundant features may also lead to improved performance using existing methods, as redundant features reduce classification performance in the general sense regardless of classifier used [38].

19.6.2 Lack of Prevalence-Corrected Reporting

The estimated real-world class imbalance when predicting miRNA within eukaryote genomes is on the order of at least 1:1000. For example, 2588 *H. sapiens* miRNAs have been discovered, while the number of hairpins in the *H. sapiens* genome that meet common miRNA criteria is approximately 11 million. Because of this extremely high class imbalance, false positive rates of classifiers have a high impact on real-world performance. At 1:1000 real-world class imbalance, a classifier that operates at 90% specificity can achieve at most a 1-in -101 success rate on its predictions (i.e., Pr < 0.01). Realistically, any miRNA prediction experiment must be performed at very high (>99.9%) specificity in order for experimental validation of predictions to be feasible.

The most commonly reported performance metrics among miRNA prediction studies—sensitivity, specificity, test set accuracy, and GM—do not account for real-world class imbalance. Sensitivity and specificity ignore class imbalance. Test set accuracy reflects performance at the class imbalance of the test set and not the real-world class imbalance. Because test set class imbalance for all miRNA predictors is several orders of magnitude lower than real-world class imbalance, optimal accuracy occurs at an operating point that is not optimal for experimental validation.

With regard to real-world applicability, GM—which was introduced as an improvement on accuracy for miRNA prediction performance reporting—is much worse than test set accuracy. GM disregards class imbalance completely and assigns essentially equal importance to sensitivity and specificity even though specificity clearly has a greater impact on overall system performance, given the prevalence of the negative class. When optimizing for GM, the ideal classifier performance occurs when sensitivity and specificity are equal. For miRNA prediction, harmony between sensitivity and specificity is not a reasonable operating point. Small (1%) changes in specificity have a large impact on success rates of experimental validation techniques (i.e., precision). Conversely, high sensitivity is not a strict requirement for the successful prediction of miRNAs: recovering even 50% of all miRNAs within a genome would be a hugely successful result for a miRNA prediction study. For these reasons, current performance metrics for miRNA prediction are inadequate.

In order to encourage the production of classifiers which that tuned for the prediction of miRNAs in real-world scenarios, we believe that the field of miRNA prediction must shift toward prevalence-corrected performance reporting, i.e., the reporting of results at the expected class imbalance of real-world data sets. Prevalence-corrected reporting is not currently used to report performance in any miRNA prediction study.

In particular, prevalence-corrected precision and recall curves—common in the field of information retrieval where large class imbalances are often seen [122]—are well suited to miRNA prediction. These curves plot the recall of a classifier (synonymous with sensitivity) against the (prevalence-corrected) precision. Unlike specificity, which measures the chance of reporting a pseudo-miRNA as negative, precision measures the chance of a positive classification being truly positive. Precision therefore informs a user of miRNA prediction software of the success rate one can expect when conducting experimental validation of miRNA predictions. The combination of recall and precision allows the user to weigh the cost of validation against the number of expected miRNA predictions made. None of specificity, accuracy, or GM provides a measure of performance of classification on pseudo-miRNA that is directly useful to a user of miRNA prediction software.

As summary statistics, recall at a precision of 50% (Re@Pr50) describes the performance of a classifier at an experimental success rate that is acceptable for validation experiments. Recall at 90% precision (Re@Pr90) provides a more conservative estimate of classifier performance, which includes only miRNA that are predicted with extremely high confidence.

The use of prevalence-corrected performance metrics is necessary for the development of optimal miRNA predictors. Currently, classifiers underperform at real-world class imbalances because they are tuned for unrealistic performance metrics.

19.6.3 Independent Analysis of the Effectiveness of SMOTE Class Imbalance Correction

Recent publications have shown conflicting results regarding the effectiveness of the SMOTE class imbalance correction method for miRNA prediction. Batuwita and Palade demonstrate an increase in performance when SMOTE is used to correct class imbalance during the training of the microPred classifier [103]. However, Gudys et al. found that the optimal training pipeline for their ROC-select meta-classifier does not include SMOTE [109]. We believe that class imbalance correction increases performance of miRNA classifiers; however the metric of GM does not elucidate this increase in performance because of its narrow focus. Figures 19.8 and 19.9 detail the performance of two miRNA predictors that are trained identically except for the inclusion or exclusion of SMOTE for class imbalance correction.

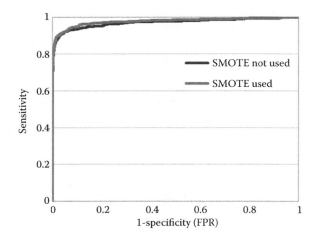

FIGURE 19.8
ROC curve demonstrating classifier performance when SMOTE is used and when SMOTE is not used for class imbalance correction.

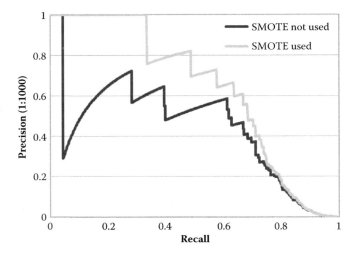

FIGURE 19.9
PR curve corrected for 1:1000 class imbalance demonstrating classifier performance when SMOTE is used and when SMOTE is not used for class imbalance correction.

These classifiers were trained using the microPred data set and feature set and the LIBSVM classification library [123] and then applied to a holdout data set consisting of 282 *A. carolinensis* miRNAs from miRBase 19 and 500 *A. carolinensis* coding region pseudo-miRNA hairpin loops.

In general, performance is increased with the addition of SMOTE. At acceptable real-world precision levels, Re@Pr50 is increased by 11.4%, and recall at high precision is increased sevenfold (see Figure 19.9). However, the peak GM for the classifier—the metric employed by Gudys' ROC-select classifier—shows almost no improvement when SMOTE is used. Peak GM is 0.938 when SMOTE is used and 0.937 when SMOTE is not used (see Figure 19.8). In spite of increasing overall performance of the classifier, SMOTE would be disregarded in this case by a study that uses peak GM as a metric, as it increases classifier complexity without improving the primary performance metric.

This experiment demonstrates the narrow applicability of GM as a miRNA prediction metric, while also providing evidence in favor of the use of SMOTE for class imbalance correction in the field.

19.6.4 Failure of miRNA Predictors to Generalize to Cross-Species Negative Data

The most commonly used metric for performance of miRNA prediction methods in recent years is classifier GM as reported on 10-CV test data. As shown in the previous section, optimizing for GM could potentially produce classifiers that perform suboptimally in the general sense. 10-CV experiments use a single data set for training and testing; therefore, iteratively optimizing one's 10-CV results is a potential source of overfitting. At the same time, emphasis is placed within miRNA

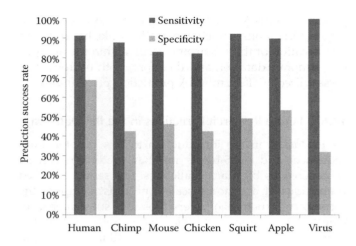

FIGURE 19.10
microPred classification results on independent holdout data sets representing multiple species.

prediction on the ability of predictors to achieve high cross-species recall—with no regard for cross-species specificity [103,109]. Because of the narrow focus of miRNA prediction performance metrics, miRNA predictors often fail to generalize to negative data sets outside of those on which they are trained.

Figure 19.10 shows the results of the microPred classifier on independent data sets that represent a range of species across various taxa. While sensitivity is maintained across all species, specificity is low, especially on non-human species. On four of the seven test sets, specificity is below random, demonstrating a complete failure of the classifier to generalize.

19.6.5 Moving to Genome-Scanning Data Sets for Genome-Scale Experiments

Within the field of *de novo* miRNA prediction, many studies have emphasized that the development of high-quality negative sets and classification performance within the field has benefitted from this increase in data set quality. Data set selection has proven to be a crucial step in the miRNA prediction pipeline. However, little attention has been paid to the methodology for selecting positive data. miRNA prediction studies use as positive training data known miRNAs from databases such as miRBase or miRTarBase, which contain experimentally validated miRNA sequences. Negative training data are miRNA-like hairpin structures that were extracted from larger genomic regions such as annotated coding regions. In other words, the pre-miRNA structures of positive data were created *in vivo* by Drosha, while the pre-miRNA structures of negative data were created *in silico* by an RNA folding algorithm. As illustrated in Figure 19.11, *in silico* RNA folding algorithms and *in vivo* RNA cleaving can produce substantially different pre-miRNA sequences, while each containing the mature miRNA.

No real-world experiments use a combination of positive data that are cleaved by Drosha and negative data that are computationally generated. For the prediction of unannotated genomic data, which we feel is the most pertinent use case of *de novo* miRNA prediction, all test data (i.e., both miRNA hairpins and pseudo-miRNA hairpins) are extracted in a single *in silico* genomic scan, which produces candidate regions corresponding to computationally predicted hairpin structures. Similarly, in the realm of NGS-based miRNA prediction, both miRNA hairpins and pseudo-miRNA hairpins (other types of transcribed RNA that form miRNA-like hairpins) are cleaved *in vivo*, and the resulting expressed reads are examined.

MiRNA prediction models have been trained to differentiate between miRNAs that are cleaved *in vivo* by Drosha and pseudo-miRNAs that are cleaved *in silico* by RNAfold. The predictive power of these models is, in part, based on the difference between miRNA biogenesis

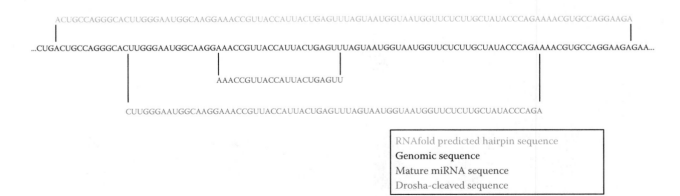

FIGURE 19.11
RNA sequences representing miRNA has-mir-451a. The sequence that was cleaved by Drosha and the sequence that was predicted by RNAfold differ in their start and end positions; however, both contain the mature miRNA.

and computational RNA secondary structure prediction. For reasons stated above, this predictive power does not apply to real-world data sets.

In combination with the previously described issues regarding class imbalance and lack of generalization, the lack of genome- or transcriptome-specific data sets should be addressed in order to provide useful, realistic *de novo* miRNA prediction pipelines. In particular, the need for *de novo* analysis of unannotated genomes is an important use case because these studies are complementary to deep sequence specific studies using methods such as miRDeep. Some recent studies have focused on the development of miRNA prediction in genomic data sets [124–126]; however, these studies do not address any of what we feel are the most important aspects of genome-wide miRNA prediction. Training data sets contain a mixture of Drosha-derived positives and computationally derived negatives; performance is measured at unrealistic class imbalances; and performance on negative sets outside of training species is not examined. An important step in the advancement of the state of the art in miRNA prediction is the development of genome-scanning miRNA prediction methods that properly address these issues. A major step toward this goal is to observe performance across the complete genomic miRNA prediction pipeline. As shown in Figure 19.12, both extraction of miRNA-like hairpins and classification of these hairpins occur within the analysis of a genomic data set, and therefore, both of these processes directly affect miRNA prediction performance.

19.6.6 Future Research Directions in the Big Data Era

As discussed in the introduction to this chapter, both sequence- and NGS-based miRNA prediction can be considered as big data challenges. For sequence-based predictions at a genome scale, one must compute the secondary structure of every possible subsequence of length 100 within a 3-billion-base-pair genome. For human, this results in 11 million miRNA-like hairpin structures, which then must each be examined in detail using computationally complex sequence-based features prior to applying the final classifier. Some of these features require extensive randomization experiments to achieve statistical comparison with structures arising from a null model. Similarly, NGS-based methods must contend with the rapidly increasing volume of sequence read data arising from NGS experiments. Clearly, parallel computing is applicable here due to the inherent data parallelism. Successful applications of parallel

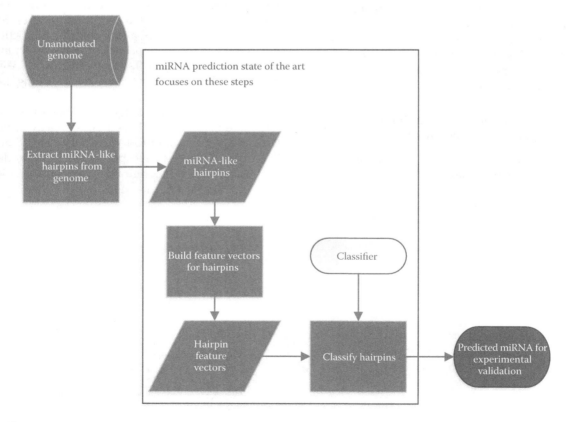

FIGURE 19.12
Prediction pipeline for miRNA within an unannotated genome.

computing in the field of miRNA prediction include multi-threading on shared-memory architectures (e.g., Ref. [127]), deployment to massively parallel architectures such as GPU (e.g., Ref. [128]), and message passing approaches for leveraging potentially cloud-based heterogeneous architectures (e.g., Refs. [129,130]). Other than data *volume*, big data challenges must also contend with data *veracity*. Here, this review has emphasized issues of data quality and pre-filtering prior to applying predictors. These issues are critical due to the high class imbalance observed in this domain. Lastly, data *velocity* is also relevant here. If we are to imagine a data-driven analysis-in-the-loop approach to directing NGS experiments, as has been done in the field of "real-time" mass spectrometry [131], then the runtime of miRNA analysis tools must be reduced by orders of magnitude to keep up with the ever-increasing measurement speed of NGS instruments.

However, this ever-increasing volume of data available for miRNA prediction also presents an opportunity. New so-called deep learning approaches are emerging in several fields and have recently been applied to miRNA prediction (e.g., Ref. [132,133]). These approaches are characterized by learning, not only the decision rule but also the manner in which data are represented within the system. Given sufficient data from which to learn, these approaches are able to optimize the feature extraction layer directly from the data, often resulting in significant gains in classification accuracy. We expect that deep learning will emerge as a highly effective approach to miRNA prediction and other related fields within bioinformatics.

19.7 Concluding Remarks

This chapter has summarized the state of the art in computational prediction of miRNAs. Both NGS-based and *de novo* methods have been examined. Critical shortcomings in modern methods were identified, largely centering on the improper treatment of class imbalance, particularly during evaluation of prediction methods. Several suggestions for how the field may continue to improve in both sensitivity and precision have been discussed. Reflecting the diversity of miRNA biogenesis and processing among various taxa, the use of species-specific classifiers are expected to be highly effective, as we have demonstrated in Ref. [134]. Ultimately, we believe that an effective convergence of both sequence-based (*de novo*) and NGS-based methods will result in the most effective approach to miRNA prediction.

References

1. D. T. Humphreys, B. J. Westman, D. I. K. Martin, and T. Preiss, "MicroRNAs control translation initiation by inhibiting eukaryotic initiation factor 4E/cap and poly(A) tail function," *Proc. Natl. Acad. Sci. U. S. A.*, vol. 102, no. 47, pp. 16961–6, Nov. 2005.
2. A. La Torre, S. Georgi, and T. A. Reh, "Conserved microRNA pathway regulates developmental timing of retinal neurogenesis," *Proc. Natl. Acad. Sci. U. S. A.*, vol. 110, no. 26, pp. E2362–70, Jun. 2013.
3. M. T. N. Le, H. Xie, B. Zhou, P. H. Chia, P. Rizk, M. Um, G. Udolph, H. Yang, B. Lim, and H. F. Lodish, "MicroRNA-125b promotes neuronal differentiation in human cells by repressing multiple targets," *Mol. Cell. Biol.*, vol. 29, no. 19, pp. 5290–305, Oct. 2009.
4. C. Körner, I. Keklikoglou, C. Bender, A. Wörner, E. Münstermann, and S. Wiemann, "MicroRNA-31 sensitizes human breast cells to apoptosis by direct targeting of protein kinase C epsilon (PKCepsilon)," *J. Biol. Chem.*, vol. 288, no. 12, pp. 8750–61, Mar. 2013.
5. Y. W. Iwasaki, K. Kiga, H. Kayo, Y. Fukuda-Yuzawa, J. Weise, T. Inada, M. Tomita, Y. Ishihama, and T. Fukao, "Global microRNA elevation by inducible Exportin 5 regulates cell cycle entry," *RNA*, vol. 19, no. 4, pp. 490–7, Apr. 2013.
6. Y. Maistrovski, K. K. Biggar, and K. B. Storey, "HIF-1α regulation in mammalian hibernators: Role of non-coding RNA in HIF-1α control during torpor in ground squirrels and bats," *J. Comp. Physiol. B.*, vol. 182, no. 6, pp. 849–59, Aug. 2012.
7. A. Kowarsch, C. Marr, D. Schmidl, A. Ruepp, and F. J. Theis, "Tissue-specific target analysis of disease-associated microRNAs in human signaling pathways," *PLoS One*, vol. 5, no. 6, p. e11154, Jan. 2010.
8. K. K. Biggar, S. F. Kornfeld, Y. Maistrovski, and K. B. Storey, "MicroRNA regulation in extreme environments: differential expression of microRNAs in the intertidal snail *Littorina littorea* during extended periods of freezing and anoxia," *Genom. Proteom. Bioinform.*, vol. 10, no. 5, pp. 302–9, Oct. 2012.
9. K. K. Biggar and K. B. Storey, "Evidence for cell cycle suppression and microRNA regulation of cyclin D1 during anoxia exposure in turtles," *Cell Cycle*, vol. 11, no. 9, pp. 1705–13, May 2012.
10. C.-W. Wu, K. K. Biggar, and K. B. Storey, "Dehydration mediated microRNA response in the African clawed frog *Xenopus laevis*," *Gene*, vol. 529, no. 2, pp. 269–75, Oct. 2013.
11. C. Barbato, I. Arisi, M. E. Frizzo, R. Brandi, L. Da Sacco, and A. Masotti, "Computational challenges in miRNA target predictions: To be or not to be a true target?," *J. Biomed. Biotechnol.*, vol. 2009, no. 803069, 2009.
12. K. C. Miranda, T. Huynh, Y. Tay, Y.-S. Ang, W.-L. Tam, A. M. Thomson, B. Lim, and I. Rigoutsos, "A pattern-based method for the identification of MicroRNA binding sites and their corresponding heteroduplexes," *Cell*, vol. 126, no. 6, pp. 1203–17, Sep. 2006.

13. M. Mraz and S. Pospisilova, "MicroRNAs in chronic lymphocytic leukemia: from causality to associations and back," *Expert Rev. Hematol.*, vol. 5, no. 6, pp. 579–81, 2012.
14. V. Balatti, Y. Pekarky, and C. M. Croce, "Role of microRNA in chronic lymphocytic leukemia onset and progression," *J. Hematol. Oncol.*, vol. 8, no. 1, pp. 8–13, 2015.
15. B. Kusenda, M. Mraz, J. Mayer, and S. Pospisilova, "MicroRNA biogenesis, functionality and cancer relevance," *Biomed. Pap. Med. Fac. Univ. Palacky, Olomouc, Czech. Repub.*, vol. 150, no. 2, pp. 205–215, 2006.
16. Y. Zhao, J. F. Ransom, A. Li, V. Vedantham, M. von Drehle, A. N. Muth, T. Tsuchihashi, M. T. McManus, R. J. Schwartz, and D. Srivastava, "Dysregulation of cardiogenesis, cardiac conduction, and cell cycle in mice lacking miRNA-1-2," *Cell*, vol. 129, no. 2, pp. 303–317, 2007.
17. J.-F. Chen, E. P. Murchison, R. Tang, T. E. Callis, M. Tatsuguchi, Z. Deng, M. Rojas, S. M. Hammond, M. D. Schneider, C. H. Selzman, G. Meissner, C. Patterson, G. J. Hannon, and D.-Z. Wang, "Targeted deletion of Dicer in the heart leads to dilated cardiomyopathy and heart failure," *Proc. Natl. Acad. Sci. U. S. A.*, vol. 105, no. 6, pp. 2111–2116, 2008.
18. Y. Lee, M. Kim, J. Han, K.-H. Yeom, S. Lee, S. H. Baek, and V. N. Kim, "MicroRNA genes are transcribed by RNA polymerase II," *EMBO J.*, vol. 23, no. 20, pp. 4051–60, Oct. 2004.
19. X. Cai, C. H. Hagedorn, and B. R. Cullen, "Human microRNAs are processed from capped, polyadenylated transcripts that can also function as mRNAs," *RNA*, vol. 10, no. 12, pp. 1957–66, Dec. 2004.
20. R. I. Gregory, T. P. Chendrimada, and R. Shiekhattar, "MicroRNA biogenesis: Isolation and characterization of the Microprocessor complex," *Methods Mol. Biol.*, vol. 342, pp. 33–47, Jan. 2006.
21. E. Berezikov, W. J. Chung, J. Willis, E. Cuppen, and E. C. Lai, "Mammalian mirtron genes," *Mol. Cell*, vol. 28, no. 2, pp. 328–336, 2007.
22. E. Lund and J. E. Dahlberg, "Substrate selectivity of exportin 5 and Dicer in the biogenesis of microRNAs," *Cold Spring Harb. Symp. Quant. Biol.*, vol. 71, no. 0, pp. 59–66, Jan. 2006.
23. T. M. Rana, "Illuminating the silence: Understanding the structure and function of small RNAs," *Nat. Rev. Mol. Cell Biol.*, vol. 8, no. 1, pp. 23–36, Jan. 2007.
24. C. Lelandais-Brière, C. Sorin, M. Declerck, A. Benslimane, M. Crespi, and C. Hartmann, "Small RNA diversity in plants and its impact in development," *Curr. Genom.*, vol. 11, no. 1, pp. 14–23, 2010.
25. B. M. Wheeler, A. M. Heimberg, V. N. Moy, E. A. Sperling, T. W. Holstein, S. Heber, and K. J. Peterson, "The deep evolution of metazoan microRNAs," *Evol. Dev.*, vol. 11, no. 1, pp. 50–68, Jan. 2009.
26. N. Fahlgren, S. Jogdeo, K. D. Kasschau, C. M. Sullivan, E. J. Chapman, S. Laubinger, L. M. Smith, M. Dasenko, S. A. Givan, D. Weigel, and J. C. Carrington, "MicroRNA gene evolution in *Arabidopsis lyrata* and *Arabidopsis thaliana*," *Plant Cell*, vol. 22, no. 4, pp. 1074–89, Apr. 2010.
27. N. Warthmann, S. Das, C. Lanz, and D. Weigel, "Comparative analysis of the MIR319a microRNA locus in *Arabidopsis* and related Brassicaceae," *Mol. Biol. Evol.*, vol. 25, no. 5, pp. 892–902, May 2008.
28. H. B. Shaffer, P. Minx, D. E. Warren, A. M. Shedlock, R. C. Thomson, N. Valenzuela, J. Abramyan, C. T. Amemiya, D. Badenhorst, K. K. Biggar, G. M. Borchert, C. W. Botka, R. M. Bowden, E. L. Braun, A. M. Bronikowski, B. G. Bruneau, L. T. Buck, B. Capel, T. a Castoe, M. Czerwinski, K. D. Delehaunty, S. V Edwards, C. C. Fronick, M. K. Fujita, L. Fulton, T. a Graves, R. E. Green, W. Haerty, R. Hariharan, O. Hernandez, L. W. Hillier, A. K. Holloway, D. Janes, F. J. Janzen, C. Kandoth, L. Kong, A. J. de Koning, Y. Li, R. Literman, S. E. McGaugh, L. Mork, M. O'Laughlin, R. T. Paitz, D. D. Pollock, C. P. Ponting, S. Radhakrishnan, B. J. Raney, J. M. Richman, J. St John, T. Schwartz, A. Sethuraman, P. Q. Spinks, K. B. Storey, N. Thane, T. Vinar, L. M. Zimmerman, W. C. Warren, E. R. Mardis, and R. K. Wilson, "The western painted turtle genome, a model for the evolution of extreme physiological adaptations in a slowly evolving lineage," *Genome Biol.*, vol. 14, no. 3, p. R28, Mar. 2013.
29. R. C. Lee, R. L. Feinbaum, and V. Ambros, "The C. elegans heterochronic gene lin-4 encodes small RNAs with antisense complementarity to lin-14," *Cell*, vol. 75, no. 5, pp. 843–54, Dec. 1993.
30. B. J. Reinhart, F. J. Slack, M. Basson, A. E. Pasquinelli, J. C. Bettinger, A. E. Rougvie, H. R. Horvitz, and G. Ruvkun, "The 21-nucleotide let-7 RNA regulates developmental timing in *Caenorhabditis elegans*," *Nature*, vol. 403, no. 6772, pp. 901–6, Feb. 2000.
31. A. E. Pasquinelli, B. J. Reinhart, F. Slack, M. Q. Martindale, M. I. Kuroda, B. Maller, D. C. Hayward, E. E. Ball, B. Degnan, P. Müller, J. Spring, A. Srinivasan, M. Fishman, J. Finnerty, J. Corbo, M. Levine, P. Leahy, E. Davidson, and G. Ruvkun, "Conservation of the sequence and temporal expression of let-7 heterochronic regulatory RNA," *Nature*, vol. 408, no. 6808, pp. 86–89, 2000.
32. A. M. Lagos-quintana, R. Rauhut, W. Lendeckel, and T. Tuschl, "Identification of novel genes Coding for RNAs of Small expressed RNAs," *Science (80)*, vol. 294, no. 5543, pp. 853–858, 2001.
33. N. C. Lau, L. P. Lim, E. G. Weinstein, and D. P. Bartel, "An abundant class of tiny RNAs with probable regulatory roles in *Caenorhabditis elegans*," *Science*, vol. 294, no. 5543, pp. 858–862, 2001.
34. J. Fleenor, S. Xu, C. Mello, A. Fire, M. Cell, R. C. Lee, and V. Ambros, "An extensive class of small RNAs in *Caenorhabditis elegans*," *Science*, vol. 294, no. 5543, pp. 862–864, 2001.
35. E. Berezikov, E. Cuppen, and R. H. a Plasterk, "Approaches to microRNA discovery," *Nat. Genet.*, vol. 38 Suppl, no. May, pp. S2–7, Jun. 2006.
36. E. Pettersson, J. Lundeberg, and A. Ahmadian, "Generations of sequencing technologies," *Genomics*, vol. 93, no. 2, pp. 105–111, 2009.
37. A. Kozomara and S. Griffiths-Jones, "miRBase: Integrating microRNA annotation and deep-sequencing data," *Nucleic Acids Res.*, vol. 39, no. October 2010, pp. 152–157, Jan. 2011.

38. R. O. Duda, P. E. Hart, and D. G. Stork, *Pattern Classification*, vol. 9. John Wiley & Sons, 2012.
39. S.-D. Hsu, Y.-T. Tseng, S. Shrestha, Y.-L. Lin, A. Khaleel, C.-H. Chou, C.-F. Chu, H.-D. H.-Y. Huang, C.-M. Lin, S.-Y. Ho, T.-Y. Jian, F.-M. Lin, T.-H. Chang, S.-L. Weng, K.-W. Liao, I.-E. Liao, C.-C. Liu, and H.-D. H.-Y. Huang, "miRTarBase update 2014: An information resource for experimentally validated miRNA–target interactions," *Nucleic Acids Res.*, vol. 42, no. Database issue, pp. D78–85, Jan. 2014.
40. G. Hughes, "On the mean accuracy of statistical pattern recognizers," *IEEE Trans. Inf. Theory*, vol. 14, no. 1, 1968.
41. N. Chawla and K. Bowyer, "SMOTE: Synthetic minority over-sampling technique," *J. Artif. Intell. Res.*, vol. 16, pp. 321–357, 2002.
42. D. H. Wolpert, "A lack of a priori distinctions between learning algorithms," *Neural Comput.*, vol. 8, no. 7, pp. 1391–1420, 1996.
43. L. P. Lim, N. C. Lau, E. G. Weinstein, A. Abdelhakim, S. Yekta, M. W. Rhoades, C. B. Burge, and D. P. Bartel, "The microRNAs of *Caenorhabditis elegans*," *Genes Dev*, pp. 991–1008, 2003.
44. E. Lai, P. Tomancak, R. Williams, and G. Rubin, "Computational identification of *Drosophila* microRNA genes," *Genome Biol*, vol. 4, no. 7, pp. 1–20, 2003.
45. Y. Grad, J. Aach, G. D. Hayes, B. J. Reinhart, G. M. Church, G. Ruvkun, and J. Kim, "Computational and experimental identification of *C. elegans* microRNAs," *Mol. Cell*, vol. 11, no. 5, pp. 1253–1263, May 2003.
46. M. W. Jones-Rhoades and D. P. Bartel, "Computational identification of plant microRNAs and their targets, including a stress-induced miRNA," *Mol. Cell*, vol. 14, no. 6, pp. 787–99, Jun. 2004.
47. A. Adai and C. Johnson, "Computational prediction of miRNAs in *Arabidopsis thaliana*," *Genome Res.*, pp. 78–91, 2005.
48. T. Dezulian, M. Remmert, J. F. Palatnik, D. Weigel, and D. H. Huson, "Identification of plant microRNA homologs," *Bioinformatics*, vol. 22, no. 3, pp. 359–60, Mar. 2006.
49. B. Sebastian and S. E. Aggrey, "miR-Explore: Predicting microRNA precursors by class grouping and secondary structure positional alignment," *Bioinform. Biol. Insights*, vol. 7, pp. 133–42, Jan. 2013.
50. M. R. Friedländer, W. Chen, C. Adamidi, J. Maaskola, R. Einspanier, S. Knespel, and N. Rajewsky, "Discovering microRNAs from deep sequencing data using miRDeep," *Nat. Biotechnol.*, vol. 26, no. 4, pp. 407–15, Apr. 2008.
51. M. R. Friedländer, S. D. Mackowiak, N. Li, W. Chen, and N. Rajewsky, "miRDeep2 accurately identifies known and hundreds of novel microRNA genes in seven animal clades," *Nucleic Acids Res.*, vol. 40, no. 1, pp. 37–52, Jan. 2012.
52. N. D. Mendes, A. T. Freitas, and M.-F. Sagot, "Current tools for the identification of miRNA genes and their targets," *Nucleic Acids Res.*, vol. 37, no. 8, pp. 2419–33, May 2009.
53. W.-C. Wang, F.-M. Lin, W.-C. Chang, K.-Y. Lin, H.-D. Huang, and N.-S. Lin, "miRExpress: Analyzing high-throughput sequencing data for profiling microRNA expression," *BMC Bioinform.*, vol. 10, no. 1, p. 328, Jan. 2009.
54. D. Hendrix, M. Levine, and W. Shi, "miRTRAP, a computational method for the systematic identification of miRNAs from high throughput sequencing data," *Genome Biol.*, vol. 11, no. 4, p. R39, Jan. 2010.
55. M. Hackenberg, N. Rodríguez-Ezpeleta, and A. M. Aransay, "miRanalyzer: An update on the detection and analysis of microRNAs in high-throughput sequencing experiments," *Nucleic Acids Res.*, vol. 39, no. Web Server issue, pp. W132-8, Jul. 2011.
56. X. Yang and L. Li, "miRDeep-P: A computational tool for analyzing the microRNA transcriptome in plants," *Bioinformatics*, vol. 27, no. 18, pp. 2614–2615, 2011.
57. M. Menor, K. Baek, and G. Poisson, "Multiclass relevance units machine: Benchmark evaluation and application to small ncRNA discovery," *BMC Genom.*, vol. 14 Suppl 2, no. Suppl 2, p. S6, Jan. 2013.
58. D. Mapleson, S. Moxon, T. Dalmay, and V. Moulton, "MiRPlex: A tool for identifying miRNAs in high-throughput sRNA datasets without a genome," *J. Exp. Zool. B. Mol. Dev. Evol.*, vol. 320, no. 1, pp. 47–56, Jan. 2013.
59. J. An, J. Lai, M. L. Lehman, and C. C. Nelson, "MiRDeep*: An integrated application tool for miRNA identification from RNA sequencing data," *Nucleic Acids Res.*, vol. 41, no. 2, pp. 727–737, 2013.
60. J. An, J. Lai, A. Sajjanhar, M. L. Lehman, and C. C. Nelson, "miRPlant: An integrated tool for identification of plant miRNA from RNA sequencing data," *BMC Bioinform.*, vol. 15, no. 1, p. 275, Jan. 2014.
61. C. Kuenne, J. Preussner, M. Herzog, T. Braun, and M. Looso, "MIRPIPE: Quantification of microRNAs in niche model organisms," *Bioinformatics*, vol. 30, no. 23, pp. 3412–3, Dec. 2014.
62. T. B. Hansen, M. T. Venø, J. Kjems, and C. K. Damgaard, "miRdentify: High stringency miRNA predictor identifies several novel animal miRNAs," *Nucleic Acids Res.*, vol. 42, no. 16, p. e124, Jan. 2014.
63. M. S. Menor, K. Baek, and G. Poisson, "Prediction of mature microRNA and piwi-interacting RNA without a genome reference or precursors," *Int. J. Mol. Sci.*, vol. 16, no. 1, pp. 1466–81, Jan. 2015.
64. J. Wu, Q. Liu, X. Wang, J. Zheng, T. Wang, M. You, Z. Sheng Sun, and Q. Shi, "mirTools 2.0 for non-coding RNA discovery, profiling, and functional annotation based on high-throughput sequencing," *RNA Biol.*, vol. 10, no. 7, pp. 1087–92, Jul. 2013.
65. Y. Gu, Y. Liu, J. Zhang, H. Liu, Y. Hu, H. Du, Y. Li, J. Chen, B. Wei, and Y. Huang, "Identification and characterization of microRNAs in the developing maize endosperm," *Genomics*, vol. 102, no. 5–6, pp. 472–478, 2013.
66. T. Madden, "The BLAST Sequence Analysis Tool," in *The NCBI Handbook*, National Center for Biotechnology Information (US), 2003.
67. M. Zuker, "Mfold web server for nucleic acid folding and hybridization prediction," *Nucleic Acids Res.*, vol. 31, no. 13, pp. 3406–15, Jul. 2003.
68. V. Ambros, B. Bartel, D. P. Bartel, C. B. Burge, J. C. Carrington, X. Chen, G. Dreyfuss, S. R. Eddy, S. Griffiths-Jones, M. Marshall, M. Matzke, G. Ruvkun, and T. Tuschl, "A uniform system for microRNA annotation," *RNA*, vol. 9, no. 3, pp. 277–9, Mar. 2003.

69. T. F. Smith and M. S. Waterman, "Identification of common molecular subsequences," *J. Mol. Biol.*, vol. 147, no. 1, pp. 195–197, Mar. 1981.
70. B. Langmead, C. Trapnell, M. Pop, and S. L. Salzberg, "Ultrafast and memory-efficient alignment of short DNA sequences to the human genome," *Genome Biol.*, vol. 10, no. 3, p. R25, Jan. 2009.
71. B. C. Meyers, M. J. Axtell, B. Bartel, D. P. Bartel, D. Baulcombe, J. L. Bowman, X. Cao, J. C. Carrington, X. Chen, P. J. Green, S. Griffiths-Jones, S. E. Jacobsen, A. C. Mallory, R. A. Martienssen, R. S. Poethig, Y. Qi, H. Vaucheret, O. Voinnet, Y. Watanabe, D. Weigel, and J.-K. Zhu, "Criteria for annotation of plant MicroRNAs," *Plant Cell*, vol. 20, no. 12, pp. 3186–90, Dec. 2008.
72. S. Moxon, F. Schwach, T. Dalmay, D. MacLean, D. J. Studholme, and V. Moulton, "A toolkit for analysing large-scale plant small RNA datasets," *Bioinformatics*, vol. 24, no. 19, pp. 2252–2253, 2008.
73. M. W. Jones-Rhoades, D. P. Bartel, and B. Bartel, "MicroRNAs and their regulatory roles in plants," *Annu. Rev. Plant Biol.*, vol. 57, pp. 19–53, 2006.
74. P.-J. Huang, Y.-C. Liu, C.-C. Lee, W.-C. Lin, R. R.-C. Gan, P.-C. Lyu, and P. Tang, "DSAP: Deep-sequencing small RNA analysis pipeline," *Nucleic Acids Res.*, vol. 38, no. Web Server issue, pp. W385-91, Jul. 2010.
75. M. B. Stocks, S. Moxon, D. Mapleson, H. C. Woolfenden, I. Mohorianu, L. Folkes, F. Schwach, T. Dalmay, and V. Moulton, "The UEA sRNA workbench: A suite of tools for analysing and visualizing next generation sequencing microRNA and small RNA datasets," *Bioinformatics*, vol. 28, no. 15, pp. 2059–61, Aug. 2012.
76. M. Wen, Y. Shen, S. Shi, and T. Tang, "miREvo: An integrative microRNA evolutionary analysis platform for next-generation sequencing experiments," *BMC Bioinform.*, vol. 13, p. 140, Jan. 2012.
77. S. Cho, I. Jang, Y. Jun, S. Yoon, M. Ko, Y. Kwon, I. Choi, H. Chang, D. Ryu, B. Lee, V. N. Kim, W. Kim, and S. Lee, "MiRGator v3.0: A microRNA portal for deep sequencing, expression profiling and mRNA targeting," *Nucleic Acids Res.*, vol. 41, no. D1, pp. 252–257, 2013.
78. D. T. Humphreys and C. M. Suter, "miRspring: A compact standalone research tool for analyzing miRNA-seq data," *Nucleic Acids Res.*, vol. 41, no. 15, p. e147, Aug. 2013.
79. Z. Sun, J. Evans, A. Bhagwate, S. Middha, M. Bockol, H. Yan, and J.-P. Kocher, "CAP-miRSeq: A comprehensive analysis pipeline for microRNA sequencing data," *BMC Genom.*, vol. 15, no. 1, p. 423, Jan. 2014.
80. G.-Z. Luo, W. Yang, Y.-K. Ma, and X.-J. Wang, "ISRNA: An integrative online toolkit for short reads from high-throughput sequencing data," *Bioinformatics*, vol. 30, no. 3, pp. 434–6, Mar. 2014.
81. C. Cowled, C. R. Stewart, V. A. Likic, M. R. Friedländer, M. Tachedjian, K. A. Jenkins, M. L. Tizard, P. Cottee, G. A. Marsh, P. Zhou, M. L. Baker, A. G. Bean, and L. Wang, "Characterisation of novel microRNAs in the Black flying fox (*Pteropus alecto*) by deep sequencing," *BMC Genom.*, vol. 15, no. 1, p. 682, Jan. 2014.
82. H. Li, B. Handsaker, A. Wysoker, T. Fennell, J. Ruan, N. Homer, G. Marth, G. Abecasis, and R. Durbin, "The sequence alignment/map format and SAMtools," *Bioinformatics*, vol. 25, no. 16, pp. 2078–9, Aug. 2009.
83. L. Yin, "Discovering novel microRNAs and age-related nonlinear changes in rat brains using deep sequencing," *Neurobiol. Aging*, vol. 36, no. 2, pp. 1037–1044.
84. S. Anders and W. Huber, "Differential expression analysis for sequence count data," *Genome Biol*, vol. 11, no. 10, p. R106, 2010.
85. B. P. Lewis, C. B. Burge, and D. P. Bartel, "Conserved seed pairing, often flanked by adenosines, indicates that thousands of human genes are microRNA targets," *Cell*, vol. 120, no. 1, pp. 15–20, 2005.
86. R. Li, Y. Li, K. Kristiansen, and J. Wang, "SOAP: Short oligonucleotide alignment program," *Bioinformatics*, vol. 24, no. 5, pp. 713–714, 2008.
87. S. Kadri, V. Hinman, and P. V Benos, "HHMMiR: Efficient de novo prediction of microRNAs using hierarchical hidden Markov models," *BMC Bioinform.*, vol. 10 Suppl 1, p. S35, Jan. 2009.
88. P. Jiang, H. Wu, W. Wang, W. Ma, X. Sun, and Z. Lu, "MiPred: Classification of real and pseudo microRNA precursors using random forest prediction model with combined features," *Nucleic Acids Res.*, vol. 35, no. Web Server issue, pp. W339-44, Jul. 2007.
89. M. Leclercq, A. B. Diallo, and M. Blanchette, "Computational prediction of the localization of microRNAs within their pre-miRNA," *Nucleic Acids Res.*, vol. 41, no. 15, pp. 7200–7211, 2013.
90. E. Bonnet, Y. He, K. Billiau, and Y. Van de Peer, "TAPIR, a web server for the prediction of plant microRNA targets, including target mimics," *Bioinformatics*, vol. 26, no. 12, pp. 1566–8, Jun. 2010.
91. V. Williamson, A. Kim, B. Xie, G. O. McMichael, Y. Gao, and V. Vladimirov, "Detecting miRNAs in deep-sequencing data: A software performance comparison and evaluation," *Brief. Bioinform.*, vol. 14, no. 1, pp. 36–45, Jan. 2013.
92. R. Lorenz, S. H. Bernhart, C. Höner zu Siederdissen, H. Tafer, C. Flamm, P. F. Stadler, and I. L. Hofacker, "ViennaRNA Package 2.0," *Algorithms Mol. Biol.*, vol. 6, p. 26, 2011.
93. E. Bonnet, J. Wuyts, P. Rouzé, and Y. Van de Peer, "Evidence that microRNA precursors, unlike other non-coding RNAs, have lower folding free energies than random sequences," *Bioinformatics*, vol. 20, no. 17, pp. 2911–7, Nov. 2004.
94. D. Kleftogiannis, A. Korfiati, K. Theofilatos, S. Likothanassis, A. Tsakalidis, and S. Mavroudi, "Where we stand, where we are moving: Surveying computational techniques for identifying miRNA genes and uncovering their regulatory role," *J. Biomed. Inform.*, vol. 46, no. 3, pp. 563–73, Jun. 2013.
95. C. P. C. Gomes, J.-H. Cho, L. Hood, O. L. Franco, R. W. Pereira, and K. Wang, "A Review of Computational Tools in microRNA Discovery.," *Front. Genet.*, vol. 4, p. 81, Jan. 2013.
96. M. Yousef, L. Showe, and M. Showe, "A study of microRNAs in silico and in vivo: Bioinformatics approaches to microRNA discovery and target identification," *FEBS J.*, vol. 276, no. 8, pp. 2150–6, Apr. 2009.

97. L. Li, J. Xu, D. Yang, X. Tan, and H. Wang, "Computational approaches for microRNA studies: A review," *Mamm. Genome*, vol. 21, nos. 1–2, pp. 1–12, Feb. 2010.
98. L. Wei, M. Liao, Y. Gao, R. Ji, Z. He, and Q. Zou, "Improved and promising identification of human microRNAs by incorporating a high-quality negative set," *IEEE/ACM Trans. Comput. Biol. Bioinform.*, pp. 1–12, Nov. 2013.
99. N. R. Markham and M. Zuker, *UNAFold: Software for Nucleic Acid Folding and Hybridization*, vol. 453. Totowa, NJ: Humana Press, 2008.
100. J.-W. Nam, K.-R. Shin, J. Han, Y. Lee, V. N. Kim, and B.-T. Zhang, "Human microRNA prediction through a probabilistic co-learning model of sequence and structure," *Nucleic Acids Res.*, vol. 33, no. 11, pp. 3570–81, Jan. 2005.
101. A. Sewer, N. Paul, P. Landgraf, A. Aravin, S. Pfeffer, M. J. Brownstein, T. Tuschl, E. van Nimwegen, and M. Zavolan, "Identification of clustered microRNAs using an ab initio prediction method," *BMC Bioinform.*, vol. 6, p. 267, Jan. 2005.
102. C. Xue, F. Li, T. He, G.-P. Liu, Y. Li, and X. Zhang, "Classification of real and pseudo microRNA precursors using local structure–sequence features and support vector machine," *BMC Bioinform.*, vol. 6, p. 310, Jan. 2005.
103. R. Batuwita and V. Palade, "MicroPred: Effective classification of pre-miRNAs for human miRNA gene prediction," *Bioinformatics*, vol. 25, no. 8, pp. 989–95, Apr. 2009.
104. M. Yousef, S. Jung, L. C. Showe, and M. K. Showe, "Learning from positive examples when the negative class is undetermined—MicroRNA gene identification," *Algorithms Mol. Biol.*, vol. 3, p. 2, Jan. 2008.
105. M. Yousef, M. Nebozhyn, H. Shatkay, S. Kanterakis, L. C. Showe, and M. K. Showe, "Combining multi-species genomic data for microRNA identification using a naive Bayes classifier," *Bioinformatics*, vol. 22, no. 11, pp. 1325–34, Jun. 2006.
106. G. Grillo, A. Turi, F. Licciulli, F. Mignone, S. Liuni, S. Banfi, V. A. Gennarino, D. S. Horner, G. Pavesi, E. Picardi, and G. Pesole, "UTRdb and UTRsite (RELEASE 2010): A collection of sequences and regulatory motifs of the untranslated regions of eukaryotic mRNAs," *Nucleic Acids Res.*, vol. 38, no. Suppl. 1, pp. 75–80, 2009.
107. A. Oulas, A. Boutla, K. Gkirtzou, M. Reczko, K. Kalantidis, and P. Poirazi, "Prediction of novel microRNA genes in cancer-associated genomic regions—A combined computational and experimental approach," *Nucleic Acids Res.*, vol. 37, no. 10, pp. 3276–87, Jun. 2009.
108. J. Ding, S. Zhou, and J. Guan, "MiRenSVM: Towards better prediction of microRNA precursors using an ensemble SVM classifier with multi-loop features," *BMC Bioinformatics*, vol. 11, no. Suppl 11, p. S11, Jan. 2010.
109. A. Gudyś, M. Szcześniak, M. Sikora, and I. Makalowska, "HuntMi: An efficient and taxon-specific approach in pre-miRNA identification," *BMC Bioinform.*, vol. 14, no. 1, p. 83, Jan. 2013.
110. K. L. Ng and S. K. Mishra, "De novo SVM classification of precursor microRNAs from genomic pseudo hairpins using global and intrinsic folding measures," *Bioinformatics*, vol. 23, no. 11, pp. 1321–1330, Jun. 2007.
111. A. Mathelier and A. Carbone, "MIReNA: Finding microRNAs with high accuracy and no learning at genome scale and from deep sequencing data," *Bioinformatics*, vol. 26, no. 18, pp. 2226–34, Sep. 2010.
112. S. Lertampaiporn, C. Thammarongtham, C. Nukoolkit, B. Kaewkamnerdpong, and M. Ruengjitchatchawalya, "Heterogeneous ensemble approach with discriminative features and modified-SMOTEbagging for pre-miRNA classification," *Nucleic Acids Res.*, vol. 41, no. 1, p. e21, Jan. 2013.
113. K. Han, "Effective sample selection for classification of pre-miRNAs," *Genet. Mol. Res.*, vol. 10, no. 1, pp. 506–18, Jan. 2011.
114. L. Wang, J. Li, R. Zhu, L. Xu, Y. He, and R. Zhang, "A novel stepwise support vector machine (SVM) method based on optimal feature combination for predicting miRNA precursors," *Afr. J. Biotechnol.*, vol. 10, no. 74, pp. 16720–16731, Nov. 2011.
115. P. Xuan, M. Z. Guo, J. Wang, C. Y. Wang, X. Y. Liu, and Y. Liu, "Genetic algorithm-based efficient feature selection for classification of pre-miRNAs," *Genet. Mol. Res.*, vol. 10, no. 2, pp. 588–603, Jan. 2011.
116. X. Liu, S. He, G. Skogerbø, F. Gong, and R. Chen, "Integrated sequence–structure motifs suffice to identify microRNA precursors," *PLoS One*, vol. 7, no. 3, p. e32797, Jan. 2012.
117. N. Shakiba and L. Rueda, "MicroRNA identification using linear dimensionality reduction with explicit feature mapping," *BMC Proc.*, vol. 7, no. Suppl 7, p. S8, Dec. 2013.
118. Q. Zou, Y. Mao, L. Hu, Y. Wu, and Z. Ji, "miRClassify: An advanced web server for miRNA family classification and annotation," *Comput. Biol. Med.*, vol. 45, pp. 157–60, Mar. 2014.
119. D. Zhao, Y. Wang, D. Luo, X. Shi, L. Wang, D. Xu, J. Yu, and Y. Liang, "PMirP: A pre-microRNA prediction method based on structure–sequence hybrid features," *Artif. Intell. Med.*, vol. 49, no. 2, pp. 127–32, Jun. 2010.
120. J. Xiao, X. Tang, Y. Li, Z. Fang, D. Ma, Y. He, and M. Li, "Identification of microRNA precursors based on random forest with network-level representation method of stem–loop structure," *BMC Bioinform.*, vol. 12, no. 1, p. 165, 2011.
121. Y. Xu, X. Zhou, and W. Zhang, "MicroRNA prediction with a novel ranking algorithm based on random walks," *Bioinformatics*, vol. 24, no. 13, pp. i50-8, Jul. 2008.
122. D. M. W. Powers, "Evaluation: From precision, recall and F-factor to ROC, informedness, markedness & correlation," *J. Mach. Learn. Technol.*, vol. 2, no. 1, pp. 37–63, Jan. 2007.
123. C. Chang and C. Lin, "LIBSVM: A library for support vector machines," *ACM Trans. Intell. Syst.*, vol. 2, no. 3, p. 27:1-27:27, 2011.
124. Y. Wu, B. Wei, H. Liu, T. Li, and S. Rayner, "MiRPara: A SVM-based software tool for prediction of most probable microRNA coding regions in genome scale sequences," *BMC Bioinform.*, vol. 12, no. 1, p. 107, Jan. 2011.
125. S. Tempel and F. Tahi, "A fast ab-initio method for predicting miRNA precursors in genomes," *Nucleic Acids Res.*, vol. 40, no. 11, p. e80, Jun. 2012.

126. J. Meng, "Prediction of plant pre-microRNAs and their microRNAs in genome-scale sequences using structure–sequence features and support vector machine," *BMC Bioinform.*, vol. 15, no. 1, p. 423, 2014.
127. E. Andrés-León, R. Núñez-Torres, and A. M. Rojas, "miARma-Seq: A comprehensive tool for miRNA, mRNA and circRNA analysis," *Sci. Rep.*, vol. 6, p. 25749, May 2016.
128. J. Kim, E. Levy, A. Ferbrache, P. Stepanowsky, C. Farcas, S. Wang, S. Brunner, T. Bath, Y. Wu, and L. Ohno-Machado, "MAGI: A Node.js web service for fast microRNA-Seq analysis in a GPU infrastructure," *Bioinformatics*, vol. 30, no. 19, pp. 2826–2827, Oct. 2014.
129. T. Vergoulis, M. Alexakis, T. Dalamagas, M. Maragkakis, A. G. Hatzigeorgiou, and T. Sellis, "TARCLOUD: A Cloud-Based Platform to Support miRNA Target Prediction," in *International Conference on Scientific and Statistical Database Management*, 2012, pp. 628–633.
130. H. Hyungro Lee, Y. Youngik Yang, H. Heejoon Chae, S. Seungyoon Nam, D. Donghoon Choi, P. Tangchaisin, C. Herath, S. Marru, K. P. Nephew, and S. Sun Kim, "BioVLAB-MMIA: A Cloud Environment for microRNA and mRNA Integrated Analysis (MMIA) on Amazon EC2," *IEEE Trans. Nanobiosci.*, vol. 11, no. 3, pp. 266–272, Sep. 2012.
131. A. Zerck, E. Nordhoff, A. Resemann, E. Mirgorodskaya, D. Suckau, K. Reinert, H. Lehrach, and J. Gobom, "An iterative strategy for precursor ion selection for LC-MS/MS based shotgun proteomics research articles," *J. Proteome Res.*, pp. 3239–3251, 2009.
132. P. Mamoshina, A. Vieira, E. Putin, A. Zhavoronkov, and R. Analytics, "Applications of deep learning in biomedicine," *Mol. Pharm.*, vol. 13, pp. 1445–1454, 2016.
133. S. Park, S. Min, H.-S. Choi, and S. Yoon, "deepMiRGene: Deep neural network based precursor microRNA prediction," *arXiv*, vol. 1605.00017, no. [cs.LG], 2016.
134. R. J. Peace, K. K. Biggar, K. B. Storey, and J. R. Green, "A framework for improving microRNA prediction in non-human genomes," *Nucleic Acids Res.*, vol. 43, no. 20, p. e138, Nov. 2015.

20
Bayesian Classification of Genomic Big Data

Ulisses M. Braga-Neto, Emre Arslan, Upamanyu Banerjee, and Arghavan Bahadorinejad

CONTENTS

20.1 Introduction 411
20.2 Classification of Expression Rank Data 413
    20.2.1 Top-Scoring Pair Classifiers 413
    20.2.2 Bradley–Terry Model 414
    20.2.3 Bayesian Inference for the Bradley–Terry Model 414
    20.2.4 Bayesian Top-Scoring Pairs 415
    20.2.5 Expression Classification Experimental Results 415
20.3 Classification of LC-MS Protein Data 416
    20.3.1 Model for LC-MS Protein Data 416
    20.3.2 Bayesian Inference for the LC-MS Model 418
        20.3.2.1 Overview of Inference Procedure 418
        20.3.2.2 Prior Calibration via ABC Rejection Sampling 418
        20.3.2.3 Posterior Sampling via an ABC-MCMC Procedure 419
        20.3.2.4 Optimal Bayesian Classifier 420
    20.3.3 LC-MS Protein Data Classification Experimental Results 421
20.4 Classification of 16S rRNA Metagenomic Data 422
    20.4.1 Model for 16S rRNA Metagenomic Data 423
    20.4.2 Bayesian Inference for the 16S rRNA Model 423
    20.4.3 16S rRNA Metagenomic Data Classification Experimental Results 424
20.5 Summary and Concluding Remarks 424
References 425

20.1 Introduction

Genomic applications in the life sciences experienced an explosive growth with the advent of high-throughput measurement technologies, which are capable of delivering fast and relatively inexpensive profiles of gene and protein activity on a genome-wide or proteome-wide scale. Gene expression microarrays [1,2] can measure the activity of tens of thousands of transcripts, while next-generation sequencing (NGS) techniques such as RNA-seq [3,4] can generate millions of transcripts reads, all in parallel and in a single assay. Likewise, "shotgun" proteomics based on mass spectrometry (MS) can detect and quantify thousands of low-abundance peptide species in complex proteomic mixtures [5,6].

The application of pattern recognition and machine learning to the voluminous amount of data produced by the aforementioned high-throughput technologies promises to revolutionize medical applications by facilitating the discovery of molecular *biomarkers*. These are measurements of gene expression or protein abundance that can be used to classify disease (e.g., "tumor" vs. "normal"), cancer subtypes (e.g., "basal-like" vs. "luminal-like" breast cancer), and clinical outcomes (e.g., "poor prognosis" vs. "good prognosis"). Biomarkers can be used to improve diagnosis, guide targeted therapy, and monitor therapeutic response [7]. However, the rate of discovery of successful biomarkers in clinical practice has been unsatisfactory [8]. An important impediment to progress is that while high-throughput data in genomics features a very large number of measurements, it typically consists of a small number of sample points, due to cost or availability issues [9]. While data matrices in other big data applications may be "wide," with the number of cases far exceeding the number of variables, the nature of big data in genomics is that data matrices are "tall," with the number of variables far exceeding the number of cases. In other words, in genomics applications, one typically has to deal with small-sample, high-dimensional data sets. See Figure 20.1 for an illustration.

While the large number of measurements presents opportunities, it also creates unique challenges in the application of classification techniques, which must be carried out with judgment to avoid pitfalls [10]. The mistaken application of complex classification methods to small-sample data will produce classifiers that are overly adjusted to the small training set and are therefore likely to perform poorly on future data, a phenomenon known as overfitting. On the other hand, though the error of the optimal classifier (i.e., the lower bound in classification performance) cannot increase, and often decreases, with higher dimensionality, the error of the classifier designed from training data will usually decrease at first, achieve a minimum, and then start to increase as more variables are added; this is known in the pattern recognition and machine learning literature as the *peaking phenomenon*, or *Hughes phenomenon* [11]. The situation is made more dangerous by the fact that methods commonly used for classifier error estimation will perform poorly with small-sample data in high-dimensional spaces [12], so the researcher may get the impression that performance is improving as more variables are added, when in fact the opposite is true.

The most effective way to mitigate the small-sample, high-dimensionality problem is to use external (prior) knowledge together with the data. In particular, this calls for a model-based approach, where the model codifies the complement of physical, logical, and empirical (historical) knowledge about the problem, up to a set of unknown parameters that must be estimated from the data. Classical (frequentist) methods in statistics treat the parameters as deterministic quantities and propose estimators that have good properties "on average" (i.e., over a large number of hypothetical data sets obtained by repetitions of the experiment). The Bayesian approach, on the other hand, codifies the uncertainty over the parameters in a prior distribution; hence, it treats the parameter as a random variable. The Bayesian approach is conditional on the available data, and therefore, the estimators it proposes need not make any reference to alternative data sets generated by a repetition of the experiment. For a comprehensive treatment of the Bayesian approach to inference, the reader is referred to Ref. [13].

In this chapter, we demonstrate the application of a Bayesian model-based approach to accurate classification of small-sample, high-dimensional data from three important genomics applications: rank expression data, liquid-chromatography mass-spectromic (LC-MS) protein expression data, and 16S rRNA sequencing metagenomic data (these applications have appeared previously in recent publications—Refs. [14–16], respectively). The methodology for rank expression data is based on the *top-scoring pair* (TSP) classifiers first introduced in Ref. [17] and employs the Bradley–Terry model [18] to the expression ranks in order to select TSPs for classification; the overall algorithm is called the Bayesian top-score pair (BTSP) classification rule [14]. The approach to LC-MS classification data is based on a model for the LC-MS pipeline proposed in Ref. [19] and on the optimal Bayesian classifier (OBC) approach proposed in Ref. [20]. Finally, the metagenomic classifier uses a model for 16S rRNA sequencing data proposed in

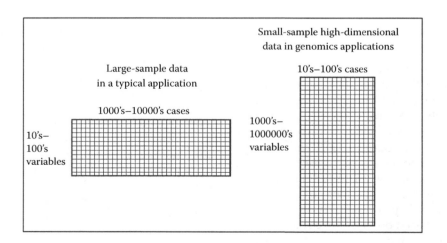

FIGURE 20.1
Genomic big data corresponds to the tall matrix on the right. It contains thousands, or even millions, of measurements, but a small number of cases.

Ref. [16] and once again the OBC method. All three approaches share the fact that a model for the data is first introduced, and then Bayesian inference of key model parameters is performed by deriving a posterior parameter distribution using the data and a prior parameter distribution. In the results presented in this chapter, the priors are *noninformative*, i.e., no prior knowledge is directly used to construct the priors. If such knowledge is available, then performance is expected to improve. A key point in all Bayesian approaches is the computation of the posterior densities, which can be difficult. The BTSP classification algorithm for expression rank data utilizes a Gibbs sampling method suggested in Ref. [21], while the Bayesian LC-MS classification rule uses an approximate Bayesian computation (ABC) Markov chain Monte Carlo (MCMC) algorithm [22]. Both previous approaches are approximate. The 16S rRNA metagenomics classification algorithm, however, computes the posterior densities analytically (in closed form) and is therefore exact.

20.2 Classification of Expression Rank Data

Our first example of application of Bayesian techniques to classification of small-sample, high-dimensional data concerns expression data presented as ranks, that is, each data point is an integer representing the relative order position in the sorted expression profile for each case. There are many advantages in using ranks instead of raw expression values, including invariance to normalization, robustness against noise, and interpretability of the results [23]. To fix ideas, we consider here that the measurements are gene expression values, though the same methodology would apply to protein abundances, metabolite concentrations, and so on.

We consider a Bayes extension of the TSP classifier, first introduced by D. Geman and collaborators in Ref. [17] and further studied in Refs. [23–25]. The TSP classifier is a simple algorithm that uses pairs of genes for which the expression ranks are inverted between one class and the other. This is illustrated in Figure 20.2. The basic TSP classifier is based on a single pair of genes, while the k-TSP classifier is an extension that uses a majority vote among a set of k pairs. We show below that a Bayesian extension of these algorithms, called BTSP and k-BTSP, respectively, can perform better than the frequentist TSP algorithms, as well as some other popular classification rules, for typical small-sample, high-dimensional genomic data sets.

20.2.1 Top-Scoring Pair Classifiers

Consider a gene expression profile $\mathbf{X} = (X_1,\ldots,X_d) \in R^d$, where X_i represents the expression value of the ith gene. We are interested in the probabilities $p_{ij}(k) = P(X_i < X_j | Y = k)$, for $k = 0,1$, where Y is the class label. A TSP classifier seeks the pair of genes that maximizes the TSP score $\hat{\Delta}_{ij} = |\hat{p}_{ij}(0) - \hat{p}_{ij}(1)|$, where $\hat{p}_{ij}(k)$ is the sample estimate of $p_{ij}(k)$. After choosing the pair (i^*, j^*) with with the highest score, and assuming $\hat{p}_{i^*j^*}(0) > \hat{p}_{i^*j^*}(1)$, the TSP classifier is simply

$$\psi_{TSP}(\mathbf{x}) = \begin{cases} 0, & \text{if } x_{i^*} < x_{j^*}, \\ 1, & \text{otherwise,} \end{cases} \quad (20.1)$$

for a fixed $x = (x_1,\ldots,x_d) \in R^d$. If $\hat{p}_{i^*j^*}(0) \leq \hat{p}_{i^*j^*}(1)$, then the labels 0 and 1 are flipped in the previous equation.

Several variations of the TSP classifier have been proposed [23,25,26]. Among these, the most well-known is the k-TSP classifier [24], which, instead of using one pair of genes as in TSP, selects the top k pairs according to $\hat{\Delta}_{ij}$ and performs majority voting to classify the given data point:

$$\psi_{kTSP}(\mathbf{x}) = \arg\max_{m \in \{0,1\}} \sum_{r=1}^{k} I(\psi_r(\mathbf{x}) = m), \quad (20.2)$$

where ψ_r is the TSP classifier based on pair r, for $r = 1,\ldots,k$. Typically, an odd k is chosen, in order to avoid ties.

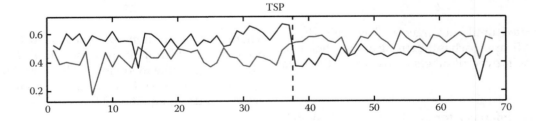

FIGURE 20.2

Expression values of a pair of genes selected by the TSP classifier. Class 0 and 1 cases are separated by the dashed vertical line. We can see that the dark gray gene expression is mostly higher than the light gray gene expression in class 0, whereas the opposite is true in class 1. Reproduced from B. Afsari, U.M. Braga-Neto, and D. Geman, Rank discriminants for predicting phenotypes from RNA expression, *Annals of Applied Statistics*, 8(3):1469–1491, 2014.

20.2.2 Bradley–Terry Model

We now briefly describe the model we use for rank expression data, which is known as the Bradley–Terry (BT) model [18]. The BT model stipulates that

$$P(X_i > X_j | \lambda_i, \lambda_j) = \pi_{ij} = \frac{\lambda_i}{\lambda_i + \lambda_j}, \quad (20.3)$$

where parameter $\lambda_i > 0$ indicates the propensity of the expression of gene i being larger than the expression of gene j. As expected, Equation 20.3 implies that $\pi_{ij} + \pi_{ji} = 1$. The BT model can also be identified as a logistic model by a nonlinear reparametrization of the parameters, $\lambda_i = e^{\beta_i}$, in which case

$$\pi_{ij} = \frac{1}{1 + e^{-(\beta_i - \beta_j)}} = \eta(\beta_i - \beta_j), \quad (20.4)$$

where $\eta(\alpha) = 1/(1+e^{-\alpha})$ is the inverse logit function. There are numerous applications of the BT model, such as ranking of chess and Go players by their respective international federations, estimation of the influence of scientific journals, and more.

Given sample data, let w_{ij} denote the number of times $X_i > X_j$ over the data. Also let $w_i = \sum_{j=1}^{n} w_{ij}$ be the number of "wins" by gene i and n_{ij} be the number of comparisons between i and j over the data. These statistics can be used to estimate the model parameters $\{\lambda_i\}$ by maximum likelihood (ML). If pairings are assumed independent, the log-likelihood function for the BT model is given by

$$\ell(\lambda) = \sum_{1 \le i \ne j \le M} w_{ij} \log \lambda_i - w_{ij} \log(\lambda_i + \lambda_j)$$

$$= \sum_{1 \le i \ne j \le M} w_i \log \lambda_i - \sum_{1 \le i < j \le M} n_{ij} \log(\lambda_i + \lambda_j). \quad (20.5)$$

The ML estimator can be found by an iterative procedure [27],

$$\lambda_i^{(k+1)} = w_i \left(\sum_{i \ne j} \frac{n_{ij}}{\lambda_i^{(k)} + \lambda_j^{(k)}} \right)^{-1}, \quad i = 1, \ldots, M, \quad (20.6)$$

which is repeated until convergence. The main drawback with the ML approach is the strong assumption that no gene may win over all others across the entire data set.

20.2.3 Bayesian Inference for the Bradley–Terry Model

To overcome the difficulties associated with ML methods, Caron and Doucet [21] introduced a Bayesian approach to the inference of the parameters $\{\lambda_i\}$ in the BT model, which we briefly describe next.

The BT model has a "Thurstonian" interpretation: for each pair $1 \le i < j \le K$ and for each pair comparison $k = 1, 2, \ldots, n_{ij}$, consider two independent random variables $U_{ki} \sim \text{Exp}(\lambda_i)$ and $U_{kj} \sim \text{Exp}(\lambda_j)$, where $\text{Exp}(\lambda)$ denotes an exponential distribution with rate parameter λ. Then a simple calculation reveals that

$$P(U_{ki} < U_{kj}) = \frac{\lambda_i}{\lambda_i + \lambda_j}. \quad (20.7)$$

In order to get a simpler complete log-likelihood, new latent variables Z_{ij} are introduced:

$$Z_{ij} = \sum_{k=1}^{n_{ij}} \min\{U_{kj}, U_{ki}\}. \quad (20.8)$$

Owing to the facts that $V_k = \min\{U_{kj}, U_{ki}\} \sim \text{Exp}(\lambda_i + \lambda_j)$ and that V_k and V_l are independent for $k \ne l$, we conclude that $Z_{ij} \sim \text{Gamma}(n_{ij}, \lambda_i + \lambda_j)$, a Gamma distribution with shape parameter n_{ij} and rate parameter $\lambda_i + \lambda_j$. The resulting density of the variables $Z = \{Z_{ij}\}$ given the data D and vector of parameters $\boldsymbol{\lambda}$ is

$$p(\mathbf{z}|\mathcal{D}, \boldsymbol{\lambda}) = \prod_{1 \le i < j \le M} \text{Gamma}(z_{ij}; n_{ij}, \lambda_i + \lambda_j), \quad (20.9)$$

$$s.t. n_{ij} > 0$$

and the resulting complete data log-likelihood is

$$\ell_c(\boldsymbol{\lambda}) = \sum_{1 \le i \ne j \le M} w_{ij} \log(\lambda_i) - \sum_{1 \le i < j \le M} (\lambda_i + \lambda_j) z_{ij}$$

$$s.t. w_{ij} > 0 \qquad s.t. n_{ij} > 0$$

$$+ (n_{ij} - 1) \log z_{ij} - \log \Gamma(n_{ij}), \quad (20.10)$$

where Γ is the Gamma function.

The prior for $\boldsymbol{\lambda}$ is assigned as in Refs. [28,29]:

$$p(\boldsymbol{\lambda}) = \prod_{i=1}^{M} \text{Gamma}(\lambda_i; a, b), \quad (20.11)$$

where a and b are hyperparameters. At this point, one needs to sample from the posterior $p(\boldsymbol{\lambda}, \mathbf{Z} | \mathcal{D})$. Canon and Doucet suggest the following Gibbs sampling scheme to accomplish that [21]. First, update $\mathbf{Z}$ from the previous value of $\boldsymbol{\lambda}$ using Equation 20.8:

$$Z_{ij}^{t+1} | \mathcal{D}, \boldsymbol{\lambda}^t \sim \text{Gamma}(n_{ij}, \lambda_i^t + \lambda_j^t); \quad (20.12)$$

Next, sample the new value of $\boldsymbol{\lambda}$ from $p(\boldsymbol{\lambda} | \mathbf{Z}, \mathcal{D}) \propto p(\boldsymbol{\lambda})\ell(\boldsymbol{\lambda}, \mathbf{z})$, which has a Gamma distribution with known parameters:

$$\lambda_i^{t+1} | \mathcal{D}, \mathbf{Z}^{t+1}$$

$$\sim \text{Gamma}\left(a + w_i, b + \sum_{\substack{i < j \\ \text{s.t. } n_{ij} > 0}} z_{ij}^{t+1} + \sum_{\substack{i > 1 \\ \text{s.t. } n_{ij} > 0}} z_{ji}^{t+1} \right). \quad (20.13)$$

20.2.4 Bayesian Top-Scoring Pairs

In this section, we describe the application of the BT model to the design of TSP classifiers. Let $\boldsymbol{\lambda}^0$ and $\boldsymbol{\lambda}^1$ be the class-specific parameters sampled from the posterior distributions for each class separately, as described in the previous section, and define

$$\pi_{ij}^0 = \frac{\lambda_i^0}{\lambda_i^0 + \lambda_j^0} \quad \text{and} \quad \pi_{ij}^1 = \frac{\lambda_i^1}{\lambda_i^1 + \lambda_j^1} \quad (20.14)$$

The main goal is to find pairs that swap frequently between classes in terms of $\boldsymbol{\lambda}^0$ and $\boldsymbol{\lambda}^1$. For this purpose, we define the *Bayesian TSP score*,

$$\Omega_{ij} = \left| \pi_{ji}^0 < \pi_{ji}^1 \right| = \left| \frac{\lambda_j^0}{\lambda_i^0 + \lambda_j^0} - \frac{\lambda_j^1}{\lambda_i^1 + \lambda_j^1} \right|, \quad (20.15)$$

and choose the best pair according to this score,

$$(i^{**}, j^{**}) = \arg\max_{(i,j) \in S} \Omega_{ij} \quad (20.16)$$

where S is whole feature space. If $\pi_{i^{**}j^{**}}^0 < \pi_{i^{**}j^{**}}^1$, the *BTSP* classifier is defined as

$$\psi_{BTSP}(\mathbf{x}) = \begin{cases} 0, & \text{if } x_{i^{**}} < x_{j^{**}}, \\ 1, & \text{otherwise} \end{cases} \quad (20.17)$$

If $\hat{p}_{i^{**}j^{**}}(0) \leq \hat{p}_{i^{**}j^{**}}(1)$, then the labels 0 and 1 are flipped in the previous equation.

As in the case of TSP, the BTSP classification rule can be extended by choosing more than one pair. We choose an odd number k of disjoint TSPs according to Ω_{ij} and define a classifier as in Equation 20.2, where ψ_r this time denotes the BTSP classifier based on pair r, for $r = 1, \ldots, k$. We call this the *k-BTSP classifier*.

20.2.5 Expression Classification Experimental Results

In this section, we report the results of the numerical experiment in Ref. [14]. We used 12 genomic data sets, summarized in Table 20.1, to compare the proposed Bayesian TSP classification rules against the conventional TSP classifiers as well as other well-known classification rules. A variance filter was employed to reduce the number of genes in all data sets to 2000. In order to consider small training sample size effects, classifiers were trained on 20% of the data and tested on the remaining 80%. The procedure was repeated 50 times, and the average accuracy was recorded. This produces an accurate estimate of the classification error, due to the large size of the testing set and little overlap between the training sets [42]. The Gibbs sampler was run for 1500 iterations with 300 burn-in iterations. We found that the choice of hyperparameters of the prior distribution in Equation 20.11 does not change results dramatically: b is a scaling parameter, and the likelihood is invariant to rescaling, so changing b does not have a large effect, while a may be fixed or can be calculated by a Metropolis–Hasting algorithm—in both cases, accuracies did not change much.

To determine the number of pairs to be used in k-TSP, we used the methods described in Refs. [23] and [24], leading to two classification rules, which we called k-TSP1 and k-TSP2, respectively. The **switchBox** R package was used for the k-TSP1 analysis [43], and the **ktspair** R package was used for the k-TSP2 analysis [44]. In each case, we limit the number of pairs to at most nine. Over all experiments, we observed that k-TSP1 uses an average of 4.02 genes, or about two pairs, while k-TSP2 uses an average of 17.8 genes, or nearly all nine pairs. By contrast, we used a fixed number $k = 9$ of pairs for the k-BTSP classifier. Linear support vector machine (SVM), naive Bayes (NB), and decision tree (DT) classification algorithms were also implemented. The SVM employed recursive feature elimination (RFE) to reduce the number of genes to 100, while the other two used all 2000 genes.

TABLE 20.1

Gene Expression Data Sets Used in the Expression Classification Experiment

Study	Number of Genes	Class 1 Size	Class 2 Size	Reference
Colon	2000	22	40	Alon et al. (1999) [30]
Leukemia$_1$	7129	25	47	Golub et al. (1999) [31]
DLBCL	7129	58	19	Shipp et al. (2002) [32]
Lung	12,533	150	31	Gordon et al. (2002) [33]
Breast	22,283	62	42	Chowdary et al. (2006) [34]
Leukemia$_2$	12,564	24	24	Armstrong et al. (2002) [35]
Monocytes	26,496	49	47	Maouche et al. (2008) [36]
Squamous	12,625	22	22	Kuriakose et al. (2004) [37]
Sarcoma	43,931	37	31	Price et al. (2007) [38]
Huntington's	22,283	14	17	Borovecki et al. (2005) [39]
CNS	7129	25	9	Pomeroy et al. (2002) [40]
Myeloma	12,625	137	36	Tian et al. (2003) [41]

TABLE 20.2

Accuracy Rates Obtained in the Expression Classification Experiment

Study	TSP	k-TSP$_1$	k-TSP$_2$	BTSP	k-BTSP	DT	NB	SVM
Colon	0.7112	0.7326	0.7392	0.7304	**0.8132**	0.6504	0.6344	0.7268
Leukemia$_1$	0.8789	0.8893	0.9072	0.8720	**0.9168**	0.8091	0.855	0.8896
DLBCL	0.779	0.8216	0.8412	0.7777	0.8761	0.7408	0.7795	**0.877**
Lung	0.9529	0.9609	0.968	0.9246	**0.9773**	0.9284	0.9678	0.9734
Breast	0.9356	0.9412	0.9465	0.9322	**0.9672**	0.8726	0.8918	0.9134
Leukemia$_2$	0.8936	0.9178	0.9331	0.8878	**0.9573**	0.8436	0.8826	0.9486
Monocytes	0.981	0.9838	0.989	0.9896	**0.9901**	0.8866	0.9884	0.9888
Squamous	0.796	0.8062	0.824	0.7668	0.8388	0.77	0.6871	**0.8708**
Sarcoma	0.8258	0.8233	0.847	0.8289	**0.8640**	0.7564	0.8133	0.852
Huntington's	0.76	0.804	0.7728	0.6248	**0.816**	0.6824	0.6008	0.7728
CNS	0.7188	0.7222	0.7274	0.6540	**0.7540**	0.6044	0.7103	0.7418
Myeloma	0.7009	0.7228	0.7349	0.6755	0.7472	0.7053	**0.7865**	0.7689
Average	0.8278	0.8438	0.8525	0.8068	**0.8765**	0.7708	0.7997	0.8603

The accuracy results are displayed in Table 20.2. We observe that k-BTSP obtained the best overall average accuracy rate and has the best accuracy rate over more individual data sets compared to all other classification rules. There does not seem to be a conclusive difference between the accuracy rates of the conventional and Bayesian TSP classifiers, but there is a significant improvement of the k-BTSP classifier over both conventional k-TSP and TSP classifiers.

20.3 Classification of LC-MS Protein Data

For another example of application of Bayesian techniques to classification of small-sample, high-dimensional data, we consider a Bayesian classifier for LC-MS proteomic profiles. MS has become the analytical tool of choice to characterize complex proteome mixtures. A mass spectrometer measures the concentration of ionized peptides at a range of mass-to-charge ratios (m/z). MS instruments consist of three modules: an ionization source, which produces ions by attaching one or more charges to each peptide, a mass analyzer, which separates the ions according to their mass-to-charge ratios, and a detector, which captures the ions and measures the intensity of each ion species [5]. Liquid chromatography (LC) is often coupled with MS to achieve additional separation of peptides and thus reduce the complexity of an individual mass spectrum. Before entering the mass spectrometer, peptide species pass through an LC column with different speeds depending on their physicochemical properties and interactions with the solvent [45]. A single LC-MS experiment usually produces hundreds to thousands of mass spectra sampled during the LC elution process. In this section, we review the Bayesian approach to classification of LC-MS proteomic profiles introduced in Ref. [15]. This classification algorithm performs Bayesian inference of the parameters of the label-free LC-MS model proposed in Ref. [19] using an ABC-MCMC algorithm [22], and then applies the OBC method introduced in Ref. [20].

20.3.1 Model for LC-MS Protein Data

As was the case with the expression rank classifier in Section 20.2, in order to apply a Bayesian approach, we first need to adopt a model for the data, which in this case is the label-free LC-MS model proposed in Ref. [19]. We briefly describe this model next. Two sample classes are considered, control (class 0) and treatment (class 1). There are n sample profiles from each class, sharing N_{pro} protein species from a specified proteome, which is typically input into the model as a FASTA file [46]. As argued in Ref. [47], protein concentration in the control sample is best described as a Gamma distribution,

$$\gamma_l = \Gamma(k, \theta), \quad l = 1, 2, \ldots, N_{pro}, \quad (20.18)$$

where the shape k and scale θ parameters are assumed to be uniform random variables, such that $k \sim \text{Unif}(k_{low}, k_{high})$ and $\theta \sim \text{Unif}(\theta_{low}, \theta_{high})$. The values for k_{low}, k_{high}, θ_{low}, and θ_{high} are chosen to adequately reflect the dynamic range of protein abundance levels.

According to whether or not there is a significant difference in abundance between control and treatment populations, proteins are divided into biomarker (differentially expressed) proteins and background (not differentially expressed) proteins. The difference in abundance for biomarker proteins is quantified by the fold change,

$$f_l = \begin{cases} a_l, & \text{if the } l\text{-th protein is overexpressed,} \\ 1/a_l, & \text{if the } l\text{-th protein is underexpressed,} \\ 1, & \text{otherwise} \end{cases} \quad (20.19)$$

The multivariate Gaussian distribution is recommended as the model for protein concentration variations in each class [48]. Accordingly, the protein expression level for the l-th protein in the j-th sample profile is modeled as

$$c_{lj}^{pro} \sim \begin{cases} \mathcal{N}\left(\left[\gamma_1, \gamma_2, \ldots, \gamma_{N_{pro}}\right], \Sigma\right), & \text{if } j \in \text{class 0,} \\ \mathcal{N}\left(\left[\gamma_1 f_1, \gamma_2 f_2, \ldots, \gamma_{N_{pro}} f_{N_{pro}}\right], \Sigma\right), & \text{if } j \in \text{class 1} \end{cases} \quad (20.20)$$

In this chapter, we assume a diagonal covariance matrix $\Sigma = [\sigma_{lk}^2]_{N_{pro} \times N_{pro}}$ such that protein concentrations are mutually independent (the results will still be approximately valid as long as the proteins are only weakly correlated):

$$\Sigma = \begin{bmatrix} \sigma_{11}^2 & 0 & \cdots & 0 \\ 0 & \sigma_{22}^2 & \cdots & 0 \\ \vdots & \vdots & \ddots & \vdots \\ 0 & 0 & \cdots & \sigma_{N_{pro}N_{pro}}^2 \end{bmatrix}, \quad (20.21)$$

where

$$\sigma_{lk}^2 = \begin{cases} \sigma_{ll}^2, & \text{if } l = k \text{ and } l, k = 1, 2, \ldots, N_{pro}, \\ 0, & \text{otherwise,} \end{cases} \quad (20.22)$$

and

$$\sigma_{ll}^2 = \varphi \times \gamma_l^2, \quad l = 1, 2, \ldots, N_{pro}. \quad (20.23)$$

The coefficient of variation φ is calibrated based on the observed data.

In order to perform *in silico* tryptic digestion of the protein samples, we use the peptide mixture model from openMS [49]. Let Ω_i be the set of all proteins that contain the i-th peptide. If there are N_{pep} peptide species, in total, across all proteins in a given sample, then their molar concentrations are given as

$$c_{ij}^{pep} = \sum_{k \in \Omega_i} c_{kj}^{pro}, \quad i = 1, 2, \ldots, N_{pep}, j = 1, 2, \ldots, 2n \quad (20.24)$$

In general, ion abundance in MS data bears the signature of the concentration of a peptide type, say i in sample j. Taking measurement uncertainty factors into consideration, one may envisage that the expected readout μ_{ij} of the abundance of said peptide can be modeled as

$$\mu_{ij} = c_{ij}^{pep} e_i \kappa, \quad i = 1, 2, \ldots, N_{pep}, \quad j = 1, 2, \ldots, 2n, \quad (20.25)$$

where e_i denotes peptide efficiency factor and κ represents the LC-MS instrument response factor [19].

The true peptide abundance differs from its readout due to noise. Accordingly, the actual abundance of a peptide v_{ij} is modeled as $v_{ij} = \mu_{ij} + \epsilon_{ij}$, where ϵ_{ij} is additive Gaussian noise and follows the distribution

$$\epsilon_{ij} \sim \mathcal{N}(0, \alpha \mu_{ij}^2 + \beta \mu_{ij}), \quad i = 1, 2, \ldots, N_{pep}, \quad j = 1, 2, \ldots, 2n, \quad (20.26)$$

where α and β specify the quadratic dependence of the noise variance on the expected abundance [19,50].

Peptide signals observed in mass spectra are in fact the result of true signals with interfering noise signals and also signals from other peptides. Therefore, the signal-to-noise ratio (SNR) affects the detectability of true positive rate (TPR) greatly. To take account of this, we describe the SNR as

$$SNR = \frac{E[v]^2}{\text{Var}(v)} = \frac{1}{\alpha + \frac{\beta}{\mu}}. \quad (20.27)$$

Taking interfering signals in consideration, the TPR of peptides is defined as

$$TPR = (t \times SNR^p + b) \times o_{ij}, \quad (20.28)$$

where o_{ij} is an overlapping factor. If algorithms like NITPICK, BPDA, and BPDA2d are used, then $o_{ij} \approx 1$ [19].

In order to reduce the number of peptides and thus the computation complexity of the algorithm, we consider three peptide filters, in order: (1) nonunique peptides present in more than one protein of the proteome in study are discarded; (2) peptides with missing value rates greater than 0.7 are discarded; (3) among the remaining peptides, those having correlation larger than 0.6 with all other peptides are kept.

The output of the first round of MS (MS1) provides information about detected peptides, their abundances, and related characteristics. The process of filtering these data and compiling the parent protein abundances from the raw peptide data is called *protein abundance roll-up*. To obtain the identities of the parent proteins from captured peptide sequence information, one will often use a second round of MS and search available MS/MS (MS2) databases. Alternatively, the *accurate mass and time* (AMT) approach matches peptides to databases using the monoisotopic mass and elution time predictors, obviating the need of a second step of MS [51]. We assume here that data is available in the form of rolled-up abundances, whereby the readout of protein l in sample j can be written as

$$x_{lj} = \frac{1}{\kappa n_l} \sum_{i \in \mathcal{N}_l} v_{ij}, \quad l = 1, 2, \ldots, N_{pro}, \quad j = 1, 2, \ldots, 2n, \quad (20.29)$$

where κ is the instrument response factor, $\mathcal{N}_l$ is the set of all peptides present in protein l that are retained after the filtering scheme described in the previous paragraph, and n_l is the number of peptides in set $\mathcal{N}_l$. The protein abundance is set to zero when less than two peptides pass the previous filters.

20.3.2 Bayesian Inference for the LC-MS Model

Bayesian analysis for complex models used in recent applications involve intractable likelihood functions, which has prompted the development of new algorithms generally called approximate Bayesian computation (ABC). In this approach, one generates candidate parameters by sampling from the prior distribution and creating a model-based simulated data set. If the data set conforms to the observed data set, the candidate can be retained as a sample from the posterior distribution. Thus, one can avoid evaluating the likelihood function, which is essential for classical Bayesian posterior simulation methods. The ABC approach can be implemented via rejection sampling, MCMC, and sequential Monte Carlo methods [22]. Utilizing the LC-MS proteomics model described in the last section, we first do prior calibration of the hyperparameters using an ABC approach via rejection sampling and then use the ABC method implemented via an MCMC procedure to obtain samples from the posterior distribution of the protein concentrations in order to derive the ABC-MCMC classifier for LC-MS data.

20.3.2.1 Overview of Inference Procedure

The sample data $S = S_0 \cup S_1$ consist of two subsamples S_0 and S_1, corresponding to the control and treatment groups, respectively, where each subsample contains n protein abundance profiles. Given the sample data, the total number of proteins N_{pro} is reduced via feature selection (e.g., ranking by the two-sample t-test statistic) to a tractable number d of selected proteins. According to the adopted LC-MS model, described in Section 20.3.1, the protein abundance profiles are a function of (1) the baseline protein concentration vector $\gamma = (\gamma_1, \ldots, \gamma_d)$; (2) the prior hyperparameters k,θ,φ,f, consisting of shape and scale parameters of the Gamma distribution in Equation 20.18, the fold change parameters in Equation 20.19, and the coefficient of variation in Equation 20.23; and (3) the LC-MS instrument-related parameters $\kappa,\alpha,\beta,\mathbf{e}, b,t,p$, which are assumed to be known for a given instrument (see Table 20.3 to see the value of these parameters in our numerical experiment). Figure 20.3 displays the relationship among these various parameters.

Our approach consists of treating γ as the "hidden parameter vector," posterior samples of which are

TABLE 20.3

LC-MS Parameters Used in LC-MS Protein Data Classification Experiment

Parameter	Symbol	Value/Range
Instrument response	κ	5
Noise severity	α,β	0.03,3.6
Peptide efficiency factor	e_i	[0.1–1]
Peptide detection algorithm	b,t,p	0,0.0016,2

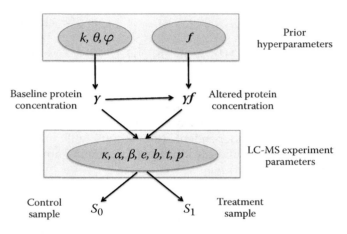

FIGURE 20.3
Relationship among all parameters of the LC-MS model (see text).

obtained using an ABC-MCMC sampling method, after a step of calibration of the hyperparameters using ABC rejection sampling. The samples from the posterior allow us to calculate the OBC for the problem. All these steps are described in detail in the sequel.

20.3.2.2 Prior Calibration via ABC Rejection Sampling

Calibration of the hyperparameters k,θ,φ,f is accomplished using the ABC rejection sampling method. Unlike Ref. [52], which proposed using discarded features to perform prior calibration for an MCMC Implementation of the OBC, here we use the selected features, as we need to calibrate the fold change as well, which is specific to each selected protein.

First, we calibrate k, θ, and φ using the control sample only, since these parameters are common across control and treatment populations and f has not been calibrated yet. The procedure used is displayed in Algorithm 1. In this algorithm, ϵ is the error tolerance. It has been proved [53] that smaller ϵ gives better approximation of posterior $p(k \mid S_n)$. However, this must be balanced against the possibility that $P(||\mathbf{T}(S_0^{(t)}), \mathbf{T}(S_0)|| < \epsilon) \approx 0$, which would prevent convergence to the posterior.

Algorithm 1 Prior calibration of k, θ and φ using ABC rejection sampling.

1. Generate M_{cal} triplets of parameters of $\{k^{(t)}, \theta^{(t)}, \varphi^{(t)}\}$ such that

    $$k^{(t)} \sim \text{Unif}(k_{low}, k_{high}),\ \theta^{(t)} \sim \text{Unif}(\theta_{low}, \theta_{high}),\ \varphi^{(t)}$$
    $$\sim \text{Unif}(\varphi_{low}, \varphi_{high})$$

    for $t = 1, \ldots, M_{cal}$.
2. Simulate a control sample set $S_0^{(t)}$ of size n for each triplet $\{k^{(t)}, \theta^{(t)}, \varphi^{(t)}\}$, for $t = 1, 2, \ldots, M_{cal}$.
3. Accept the triplet $\{k^{(t)}, \theta^{(t)}, \varphi^{(t)}\}$ if $\|\mathbf{T}(S_0^{(t)}) - \mathbf{T}(S_0)\| < \epsilon$, for $t = 1, \ldots, M_{cal}$, where $\|\cdot\|$ denotes the Euclidean norm and $\mathbf{T}$ denotes the vector sample mean.
4. Let $\mathcal{A} = \{\{k^1, \theta^1, \varphi^1\}, \ldots, \{k^{n_a}, \theta^{n_a}, \varphi^{n_a}\}\}$ be the set of all accepted triplets. The calibrated k can be approximated as follows:

    $$k_{cal} = \int_{k_{low}}^{k_{high}} k p(k|S_n) dk \approx \frac{1}{n_a} \sum_{a=1}^{n_a} k^a$$

Similar Monte Carlo integrations are performed to calculate θ_{cal} and φ_{cal}.

Next we calibrate the fold change parameter $f = (f_1, \ldots, f_d)$ for each selected protein. If sample size is large ($n > 50$), then the simple sample estimate

$$f_{l,cal} = \frac{T_l(S_1)}{T_l(S_0)}, \text{ for } l = 1, \ldots, d, \qquad (20.30)$$

where T_l denotes the sample mean for the l-th selected protein only, is fairly accurate and may be used as the prior calibration. However, for smaller sample sizes, we follow the steps enumerated in Algorithm 2.

Algorithm 2 Prior calibration of f_l, $l = 1, \ldots, d$, using ABC rejection sampling.

1. Generate M_{cal} baseline expression values $\gamma_l^{(t)} \sim \Gamma(k_{cal}, \theta_{cal})$ for $t = 1, \ldots, M_{cal}$.
2. Simulate a control sample $S_0^{(t)}$ of size n using the baseline expression mean $\gamma_l^{(t)}$, for $t = 1, \ldots, M_{cal}$ (in fact, only the abundances for the l-th protein need be generated).
3. Accept $\gamma_l^{(t)}$ if $|T_l(S_0^{(t)}) - T_l(S_0)| < \epsilon_1$ and $\rho_l(S_0^{(t)}, S_0) > 1 - \epsilon_2$, where T_l denotes the sample mean and ρ_l denotes the sample correlation for the abundances of the l-th protein only.
4. Generate M_{cal} fold change parameters $f_l^{(t)}$ such that

    if $T_l(S_1)/T_l(S_0) \geq 1$, then $f_l^{(t)}$

    $$\sim \text{Unif}\left(a_{low}, a_{high}\right)$$

    if $T_l(S_1)/T_l(S_0) < 1$, then $f_l^{(t)}$

    $$\sim \text{Unif}\left(1/a_{high}, 1/a_{low}\right)$$

    for $t = 1, \ldots, M_{cal}$.
5. Simulate a treatment sample $S_1^{(t)}$ of size n using the altered expression mean $f_l^{(t)} \gamma_l^{(t)}$, for $t = 1, 2, \ldots, M_{cal}$ (in fact, only the abundances for the l-th protein need be generated).
6. Accept $f_l^{(t)} \gamma_l^{(t)}$ if $|T_l(S_1^{(t)}) - T_l(S_1)| < \epsilon_1$ and $\rho_l(S_1^{(t)}, S_1) > 1 - \epsilon_2$.
7. Let n_a^0 be the number of accepted baseline expression means in step 3, and let n_a^1 be the number of accepted altered expression means in step 6. Define

    $$\lambda^0 = n_a^0/M_{cal}, \text{ the rate of acceptance}$$
    of control means

    $$\lambda^1 = n_a^1/M_{cal}, \text{ the rate of acceptance}$$
    of treatment means

8. If $\lambda^0 > \lambda^1$, then assign $f_{l,cal} = 1$ (i.e., background protein) and return from the algorithm.
9. Otherwise, $f_{l,cal} \neq 1$ (i.e., marker protein). For all the accepted altered expression means, we perturb each of the fold changes $f_l^* = f_l + N_l$, where N_l is zero-mean Gaussian noise with a small variance. With these perturbed fold changes, we again apply the ABC rejection algorithm, this time with error tolerances $\epsilon'_1 < \epsilon_1$ and $\epsilon'_2 < \epsilon_2$. If desired, one can apply step 9 repeatedly, until the desired performance is achieved.
10. The mean of all accepted fold change parameters in step 9 is a reasonably accurate fold-changed f_{cal} for the given protein.

20.3.2.3 Posterior Sampling via an ABC-MCMC Procedure

After prior calibration, we would like now to draw samples from the posterior distribution of the protein baseline expression vector $\gamma = (\gamma_1, \ldots, \gamma_d)$, namely, $p(\gamma|S_n) \propto p(S_n|\gamma)p(\gamma)$, in order to derive the OBC. In our case, no closed-form expressions for either the likelihood function or posterior distribution exist, so Bayesian analysis is performed using an ABC-MCMC procedure, described in Algorithm 3. After a burn-in interval of t_s time steps, the Markov chain is assumed to have become stationary, and $\gamma^{(t_s+1)}, \ldots, \gamma^{(t_s+M)}$ may be considered to be samples from the baseline expression posterior distribution $p(\gamma|y = 0, S_n)$, while $\gamma^{(t_s+1)} \mathbf{f}_{cal}, \ldots, \gamma^{(t_s+M)} \mathbf{f}_{cal}$ (where vector multiplication is defined as componentwise multiplication) may be taken to be samples from the altered expression posterior distribution $p(\gamma|y = 1, S_n)$.

Algorithm 3 Posterior sampling of γ using an ABC-MCMC procedure.

1. Generate $\gamma^{(0)} = (\gamma_0, \gamma_1, \ldots, \gamma_d)$ such that $\gamma_l \sim \Gamma(k, \theta)$ $l = 1, 2, \ldots, d$.
2. Simulate control and treatment samples $S_0^{(0)}$ and $S_1^{(0)}$ of size n using the $\gamma^{(0)}$ and $\gamma^{(0)} f_{cal}$, respectively (where vector multiplication is defined as componentwise multiplication).
3. Accept $\gamma^{(0)}$ if $||T(S_0^{(0)}) - T(S_0)|| < \epsilon_0$ and $||T(S_1^{(0)}) - T(S_1)|| < \epsilon_1$, otherwise repeat steps 1 and 2 until the condition is met.
   For $t = 0, 1, \ldots, t_s, t_{s+1}, \ldots, t_s + M$ where t_s is the burn-in period, repeat:
4. Generate $\gamma^{(t+1)} \sim g(\gamma; \gamma^{(t)})$, where the proposal density $g(\gamma; \gamma^{(t)})$ is multivariate Gaussian $\mathcal{N}_d(\gamma^{(t)}, \sigma^2 I_d)$, with a small variance σ^2.
5. Simulate control and treatment samples $S_0^{(t+1)}$ and $S_1^{(t+1)}$ of size n using $\gamma^{(t+1)}$ and $\gamma^{(t+1)} f_{cal}$, respectively.
6. Let

$$q = \begin{cases} \min\left(1, \dfrac{p(\gamma^{(t+1)})g(\gamma^{(t)}; \gamma^{(t+1)})}{p(\gamma^{(t)})g(\gamma^{(t+1)}; \gamma^{(t)})}\right), & \text{if} ||T(S_0^{(t+1)}) - T(S_0)|| < \epsilon_0 \text{ and} ||T(S_1^{(t+1)}) - T(S_1)|| < \epsilon_1 \\ 0, & \text{otherwise,} \end{cases}$$

where $p(\cdot)$ is the Gamma prior for protein baseline expression.

7. Accept $\gamma^{(t+1)}$ with probability q, or let $\gamma^{(t+1)} = \gamma^{(t)}$ with probability $1 - q$.

20.3.2.4 Optimal Bayesian Classifier

In this section we review briefly the OBC method, which is the approach used to obtain a classifier from the previous Bayesian parameter estimation step. The OBC minimizes the expected error over the space of all classifiers under assumed forms of the class-conditional densities. Ordinary Bayes classifiers minimize the misclassification probability when the underlying distributions are known. However, optimal Bayesian classification trains a classifier from data assuming the underlying distributions are not known exactly but are, rather, part of an uncertainty class of distributions, each having a weight based on the prior and the observed data.

Let $\psi: R^d \rightarrow \{0, 1\}$ be a classifier that takes a protein abundance profile $X \in R^d$ into one of the two labels 0 or 1. The *error* of the classifier is the probability of a mistake given the sample data:

$$\varepsilon[\psi] = P(\psi(X) \neq Y | S), \quad (20.31)$$

where $Y \in \{0, 1\}$ denotes the true label corresponding to X.

Now, in the Bayesian setting presently adopted, the joint distribution of (X, Y) depends on a random parameter vector θ. In this case, the classification error $\varepsilon_\theta[\psi]$ also becomes a random variable, as a function of θ. The expected value of the classification error over the posterior distribution of θ becomes the quantity of interest:

$$E_{\theta | S}[\varepsilon_\theta[\psi]] = E_{\theta | S}[P(\psi(X) \neq Y | \theta, S)] \quad (20.32)$$

The OBC [20] is the classifier that minimizes the quantity in Equation 20.32:

$$\psi_{OBC} = \arg \min_{\psi \in \mathcal{C}} E_{\theta | S}[\varepsilon_\theta[\psi]], \quad (20.33)$$

where $\mathcal{C}$ is the space of classifiers. It was shown in Ref. [20] that the OBC is given by

$$\psi_{OBC}(x) = \begin{cases} 1, & \text{if } E_{\theta|S}[c]p(x|Y=1) > (1 - E_{\theta|S}[c])p(x|Y=0), \\ 0, & \text{otherwise,} \end{cases}$$

(20.34)

where $c = P(Y = 1 | \theta)$ is the prior probability of class 1, and

$$p(x|Y = y) = \int_\Theta p(x|\theta, Y = y)p(\theta|Y = y, S)d\theta, \quad y = 0, 1, \quad (20.35)$$

are the *effective class-conditional densities*. In the previous equation, $p(x | \theta, Y = y)$ are the ordinary class-conditional densities for each fixed value of θ, and $p(\theta | Y = y, S)$ are the parameter posterior probabilities, for $y = 0, 1$.

In the present case of the LC-MS model discussed in Section 20.3.1, the random parameter vector θ corresponds to the baseline expression vector γ. We approximate the integral in Equation 20.35 using the MCMC samples $\gamma^{(t_s+1)}, \ldots, \gamma^{(t_s+M)}$ from the posterior distribution of γ, obtained with Algorithm 3:

$$p(x|Y = y) \approx \frac{1}{M} \sum_{t=t_s+1}^{t_s+M} p(x|\gamma^{(t)}, Y = y), \quad y = 0, 1 \quad (20.36)$$

Now, the densities $p(x | \gamma^{(t)}, Y = y, S)$, $y = 0, 1$, cannot be directly determined for the LC-MS model, and hence we approximate them using a kernel-based approach. For each MCMC sample $\gamma^{(t)}$, we simulate control and treatment samples $S_0^{(t)}$ and $S_1^{(t)}$ of size n based on $\gamma^{(t+1)}$ and $\gamma^{(t+1)} f_{cal}$, respectively. Let $S_0^{(t)} = \{x_1^{(t)}, \ldots, x_n^{(t)}\}$ and $S_1^{(t)} = \{x_{n+1}^{(t)}, \ldots, x_{2n}^{(t)}\}$. Then

$$p(x|\gamma^{(t)}, Y = y) \approx \frac{1}{n} \sum_{j=ny+1}^{ny+n} \frac{1}{h^d} K\left(\frac{x - x_j^{(t)}}{h}\right), \quad y = 0, 1, \quad (20.37)$$

where K is a zero-mean, unit-covariance, multivariate Gaussian density, and $h > 0$ is a suitable kernel bandwidth parameter.

In addition, we assume that c is known (e.g., from epidemiological data) and fixed, so $E[c\,|\,S] = c$. After some simplification, the resulting OBC, called the ABC-MCMC Bayesian classifier [15], is a kernel-based classifier given by

$$\psi_{\text{ABC-MCMC}}(x)$$
$$= \begin{cases} 1, & \text{if } c\sum_{t=t_s+1}^{t_s+M}\sum_{j=1}^{n}K\left(\frac{x-x_j^{(t)}}{h}\right) > (1-c)\sum_{t=t_s+1}^{t_s+M}\sum_{j=n+1}^{2n}K\left(\frac{x-x_j^{(t)}}{h}\right), \\ 0, & \text{otherwise} \end{cases}$$

(20.38)

20.3.3 LC-MS Protein Data Classification Experimental Results

We demonstrate the application of the proposed ABC-MCMC classification rule to a synthetic LC-MS data set generated from a subset of the human proteome, containing around 4000 drug targets, which was compiled as a FASTA file from DrugBank [54]—this is the same proteome that was used in the numerical experiments of [19]—and compare its performance against that of popular classification rules [55]: linear support vector machines (LSVMs), linear discriminant analysis (LDA), and 3-nearest neighbors (3NN), which are known to perform well with small-sample data and avoid overfitting.

We select randomly among these data 500 proteins to play the role of background proteins, along with 20 proteins to serve as biomarkers. Synthetic LC-MS protein abundance data was generated using realistic sample preparation, LC-MS instrument characteristics, and protein quantification parameters—see Table 20.3. These are the "LC-MS experiment parameters" of Figure 20.3, which are assumed to be known and are held constant throughout the simulation. (For the peptide efficiency factor, values uniformly distributed in the indicated range are randomly generated for each peptide and then held constant.) As argued in Ref. [19], the values and ranges adopted in Table 20.3 adequately represent the peptide mixture, peptide abundance mapping, peptide detection and identification, and protein abundance roll-up, which is typical in an LC-MS workflow.

The hyperparameter priors for k, θ, φ, f are the uniform distributions shown in Table 20.4 (except where noted below). The lower and upper bounds of each interval are chosen while keeping in consideration that, in practice, the dynamic range of protein expression level has approximately 4 orders of magnitude [19]. The synthetic sample data was generated using as parameters the middle point of each interval: $k = 2, \theta = 1000, \varphi = 0.4$, and $a_l = 1.55$ (again, except where noted below).

TABLE 20.4

Hyperparameter Priors Used in LC-MS Protein Data Classification Experiment

Parameter	Symbol	Range/Value
Shape (Gamma distribution)	k	Unif(1.6,2.4)
Scale (Gamma distribution)	θ	Unif(800,1200)
Coefficient of variation	φ	Unif(0.3,0.5)
Fold change	a_l	Unif(1.5,1.6)

We consider sample sizes from $n = 10$ through $n = 50$ per class, and select $d = 3,5,8,$ or 10 proteins from the original 520 proteins using the two-sample t-test (notice that background proteins could be erroneously selected by the t-test, especially for small sample sizes, which makes the experiment realistic). For the MCMC step, $M = 10000$ samples from the posterior distribution of γ were drawn, after a burn-in stage of $t_s = 3000$ iterations, which confers a high degree of accuracy to the approximation. A constant value $c = 0.5$ was assumed in Equation 20.38. A total of 12 runs of the experiment were run for each combination of sample size, dimensionality, and parameter settings, and the average true error rate for each classification rule was obtained using a large synthetic test set containing 1000 sample points. This is a comprehensive simulation, given the relatively large computational burden required for accurate prior calibration and ABC-MCMC computation.

Figure 20.4 displays the average error rates of the various classification rules under various experimental conditions. In Figure 20.4(a), the average error rates are plotted against sample size for a fixed number of selected proteins $d = 8$. We can see that, as expected, the errors of all classifiers tend to go down as sample size increases, but the ABC-MCMC classifier has the smallest expected error at small sample sizes, in agreement with the expectation that the Bayesian approach should be superior in small-sample scenarios. In Figure 20.4(b), the average error rates are plotted against number of selected proteins for a fixed sample size $n = 10$ per class. Here we can see that, as the number of selected proteins increases, the error rates tend to go down at first, but then appear to increase slightly, which would be in agreement with the well-known peaking phenomenon of classification [11]. We can see that the ABC-MCMC classifier error rate is smallest when d is large, which once again agrees with the expectation that Bayesian methods perform comparatively well under small-sample scenarios (here, small n/d ratio). In order to investigate the effect of very noisy background proteins in the LC-MS channel on the classification error rates, Figure 20.4(c) plots the average error rates against the value of the coefficient of variation φ used to generate the LC-MS data, with both the sample size and the

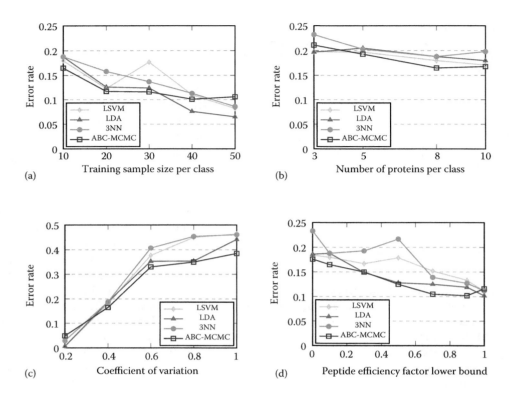

FIGURE 20.4
Average classification error rates in LC-MS protein data classification experiment. (a) Varying sample size and fixed number of selected proteins $d = 8$. (b) Varying number of selected proteins and fixed sample size $n = 10$ per class. (c) Fixed sample size $n = 10$ per class, fixed number of selected proteins $d = 8$, and varying coefficient of variation φ. (d) Fixed sample size $n = 10$ per class, fixed number of selected proteins $d = 8$, and varying lower bound a for the peptide efficiency factor $e_i \sim \text{Unif}(a,1)$.

dimensionality being kept fixed, at $n = 10$ per class and $d = 8$, respectively. To accommodate this change, the hyperparameter prior for φ is changed from the value displayed in Table 20.4 to $\text{Unif}(\varphi_0 - 0.1, \varphi_0 + 0.1)$, where φ_0 is the value used to generate the data. It can be seen that, as φ increases, all error rates approach the no-information value 0.5. However, the average error rate of the ABC-MCMC classification rule approaches this limit rather more slowly than the others, indicating superiority in classifying noisy data. Finally, in order to quantify the impact of varying the peptide efficiency factor on the classification error rates, Figure 20.4(d) plots the average error rates against the lower bound in the range for e_i displayed in Table 20.3. The value is changed from $a = 0.1$ to a varying value between 0 and 1. The peptide efficiency factor affects how many ions an instrument can detect for a given peptide. Larger values for e_i imply a smaller transmission loss for the corresponding peptide. Increasing the lower bound a will uniformly increase efficiency for all peptides, which corresponds to a better LC-MS instrument. We can see that, indeed, the average error rates tend to decrease with an increasing lower bound on the peptide efficiency factor, though somewhat modestly (all other things being equal). We can also observe that among all algorithms, the ABC-MCMC classification rule displays the smallest error rate over nearly the entire range in the plot.

20.4 Classification of 16S rRNA Metagenomic Data

In our third example of the application of Bayesian techniques to classification of small-sample, high-dimensional data, we consider the characterization of microbial communities by 16S rRNA gene amplicon sequencing. This is a problem that has received a large degree of interest recently, in part due to the emergence of high-throughput sequencing technology [56]. Microbial metagenomics provides a means to determine what organisms are present without the need for isolation and culturing. NGS, applied to microbial metagenomics, has transformed the study of microbial diversity [57]. For amplicon reads, it is possible to classify sequence reads against known taxa and determine a list of those organisms that are present and the read frequency associated with them [58]. In this case, an unsupervised strategy can be used to identify proxies to traditional taxonomic units by clustering sequences, so-called operational taxonomic units (OTUs). In this section, we first present a model for the 16S rRNA abundance data measured by amplicon sequencing proposed in Ref. [16]. This model allows the analytical (i.e., closed-form) derivation of the posterior probabilities given the data, which is described next. The actual classification rule employs the OBC algorithm discussed in the section on classification of LC-MS protein data. The performance of the proposed approach is compared to other classifiers such as a nonlinear support vector machine (SVM), random forests (RFs)—the latter is considered to be the de facto standard for metagenomics classification—and Metaphyl [59], as a function of varying sample size and classification difficulty, by means of a numerical experiment using synthetic data.

20.4.1 Model for 16S rRNA Metagenomic Data

In Ref. [16], a Dirichlet-multinomial–Poisson model for 16S rRNA data was proposed. We assume that each M-dimensional microbial abundance vector $\mathbf{x} = (x_1, \ldots, x_M)$, where x_j is the number of reads corresponding to the j-th OTU, is multinomially distributed:

$$p(\mathbf{x}|\boldsymbol{\theta}) = p_{\text{MULTI}}(\mathbf{x}|\mathbf{p},N) = N! \Pi_{j=1}^{M} \frac{p_j^{x_j}}{x_j!}, \quad (20.39)$$

where the probability vector $\mathbf{p} = (p_1, \ldots, p_M)$ follows a Dirichlet distribution with parameter $\boldsymbol{\alpha}_D = (\alpha_{D1}, \ldots, \alpha_{DM})$, and $N = \sum_{j=1}^{M} x_j$ is the total number of sequencing reads, which is assumed to have a Poisson distribution with parameter λ, which in turn follows a Gamma distribution with parameters α, β. See Figure 20.5 for a diagram of the model.

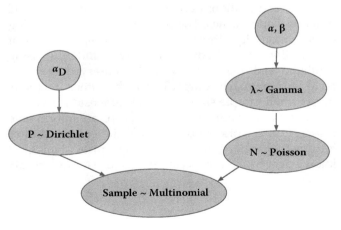

FIGURE 20.5
Dirichlet-multinomial–Poisson model for 16S rRNA metagenomic data.

20.4.2 Bayesian Inference for the 16S rRNA Model

The structure of the model proposed in the previous section allows the exact calculation of the posterior probabilities, avoiding therefore the approximations introduced by MCMC methods. Let the training data be $S = \{\mathbf{x}_1^0, \ldots, \mathbf{x}_{n_0}^0, \mathbf{x}_1^1, \ldots, \mathbf{x}_{n_1}^1\}$, where n_0 and n_1 are the numbers of training points from class 0 and 1, respectively. The posterior distributions for the parameters θ for each class are determined as follows:

$$p(\boldsymbol{\theta}|Y = y, S) \propto p(S|\boldsymbol{\theta}, Y = y) p(\boldsymbol{\theta}|Y = y)$$
$$= p(\mathbf{p}^y|\boldsymbol{\alpha}_D) p(\lambda|\alpha, \beta) \Pi_{i=1}^{n_y} p(\mathbf{x}_i^y|p_1^y, \ldots, p_M^y, N_i^y) p(N_i^y|\lambda),$$
$$= \frac{\Gamma\left(\sum_{j=1}^{M} \bar{\alpha}_j^y\right)}{\Pi_{j=1}^{M} \Gamma\left(\bar{\alpha}_j^y\right)} \frac{(\beta + n_y)^{\left(\sum_{i=1}^{n_y} N_i^y + \alpha\right)}}{\Gamma\left(\sum_{i=1}^{n_y} N_i^y + \alpha\right)}$$
$$\times e^{-\lambda(n_y + \beta)} \lambda^{\sum_{i=1}^{n_y} N_i^y + \alpha + 1} \Pi_{j=1}^{M} p_j^{\bar{\alpha}_j^y - 1}, \quad y = 0, 1,$$

(20.40)

where $N_i^y = \sum_{j=1}^{M} x_{ij}^y$ is the number of reads for each profile, $i = 1, \cdots, n_y$, $y = 0,1$, and $\bar{\alpha}_j^y = \alpha_{Dj} + \sum_{i=1}^{n_y} x_{ij}^y$, $j = 1, \cdots, M$, $y = 0,1$.

Plugging Equations 20.39 and 20.40 into Equation 20.35 leads to an integration that can be accomplished analytically, without a need for MCMC methods (please see Ref. [16] for the details), leading to the effective class-conditional densities:

$$p(\mathbf{x}|Y = y) = \frac{1}{\Pi_{j=1}^{M} \Gamma(x_j + 1)} \frac{\Gamma\left(\sum_{j=1}^{M} \bar{\alpha}_j^y\right)}{\Pi_{j=1}^{M} \Gamma(\bar{\alpha}_j^y)}$$
$$\times \frac{(\beta + n_y)^{\left(\sum_{i=1}^{n_y} N_i^y + \alpha\right)}}{\Gamma\left(\sum_{i=1}^{n_y} N_i^y + \alpha\right)} \frac{\Pi_{j=1}^{M} \Gamma(x_j + \bar{\alpha}_j^y)}{\Gamma\left(N + \sum_{j=1}^{M} \bar{\alpha}_j^y\right)}$$
$$\times \frac{\Gamma\left(\sum_{i=1}^{n_y} N_i^y + \alpha + N\right)}{(n_y + \beta + 1)^{\left(\sum_{i=1}^{n_y} N_i^y + \alpha + N\right)}}, \quad y = 0,1$$

(20.41)

In addition, we assume that the parameter c is beta distributed with hyperparameters β_0, β_1, independently of the parameters $\boldsymbol{\theta}$ (prior to observing the data). It can be shown that the posterior distribution $p(c|S)$ is also beta with hyperparameters $\beta_0 + n_0$ and $\beta_1 + n_1$, in which case $E_{\boldsymbol{\theta}|S}[c] = \frac{n_0 + \beta_0}{n + \beta_0 + \beta_1}$. This completes the specification of the OBC classifier in Equation 20.34.

20.4.3 16S rRNA Metagenomic Data Classification Experimental Results

Synthetic OTU abundance data was generated using the strategy proposed in Ref. [59], which considers a common phylogenetic tree that relates OTUs in all the 16S rRNA metagenomic profiles. To generate metagenomic profiles for class k, the phylogeny tree is traversed systematically. A decision is made for each internal node v about what fraction of species would come from each of the subtrees rooted at the child nodes of v. Two parameters are assigned to each node v for each class k. Let μ_v^k denote the average proportion of species that correspond to the subtree rooted at the left child node of v in the k-th class, and let $(\sigma_v^k)^2$ denote the variance of this proportion within the class. A new metagenomic profile is generated by sampling the proportions of species at each node v according to the normal distribution $N(\mu_v^k, (\sigma_v^k)^2)$. The parameter values μ_v^k are in turn sampled from the normal distribution $N(\tilde{\mu}_v, \tilde{\sigma}_v^2)$, where $\tilde{\sigma}_v^2$ characterizes the variance between the classes, while $\tilde{\mu}_v$ are base values that are initialized randomly.

The within- and between-class variances can be controlled by using the parameters σ_v^2 and $\tilde{\sigma}_v^2$, respectively. The exact values of $\tilde{\sigma}_v^2$ and σ_v^2 are sampled at each tree node v according to $N(0, \tilde{\lambda}d(v))$ and $N(0, \lambda d(v))$, where $d(v)$ is the distance between v and the tree root. Note that the parameters $\tilde{\lambda}$ and λ influence the difficulty of the classification problem, which is proportional to λ and inversely proportional to $\tilde{\lambda}$. We consider three sample sizes for the training data, $n = 30, 50, 70$, with class prior probability $c = 0.5$. The sample sizes n_0 and n_1 are determined according to the class prior probability as $n_0 = cn$, and $n_1 = n - n_0$. Table 20.5 displays the values of all parameters used in the simulation. Classifier accuracy is obtained by testing each designed classifier on a large synthetic test data set and averaging the results over a large number of iterations using different synthetic training data sets.

Figure 20.6 compares the performance of the proposed Bayesian classifier, using a uniform prior, against that of the nonlinear SVM, RF, and MetaPhyl classifiers. We observe that the LSVM classifier clearly performs the worst, probably due to the highly nonlinear nature of the data, while the RF and Metaphyl classifier have comparable performance. The proposed Bayesian classifier clearly displays the best performance over different values of between- and within-class variances that defines the difficulty of the classification. As the sample size increases, classification performance improves for all classification rules, as expected.

TABLE 20.5

Values of Parameters in the Numerical Experiment

Parameter	Symbol	Value
Number of OTUs	M	128
Dirichlet parameters	α_{Dj}	1
Gamma parameters	α, β	1,1
Within-class variance	λ	0.5, 1, 1.5
Between-class variance	$\tilde{\lambda}$	0.5, 1

20.5 Summary and Concluding Remarks

The Bayesian approach is a very attractive solution to the big data problem in genomics, as classical methods, including ML approaches, are overfitting in the case where sample sizes are small and the dimensionality of the feature space is large. This can be understood by considering that Bayesian methods integrate the likelihood over the parameter space using the prior density, whereas ML approaches are punctual and therefore more susceptible to noise. Classical approaches such as ML need a larger sample size in order to "average out" the effects of noise. In this chapter, we presented a model-based approach to the classification of genomic big data that employs the Bayesian paradigm. We reviewed models for three different modalities of genomic data, namely, gene expression rank data, LC-MS protein abundance data, and 16S rRNA metagenomic data, and developed classifiers based on the posterior distributions derived by either exact or MCMC numerical approaches. The performance of the resulting classification algorithms was contrasted against that of state-of-the art methods such as SVM and RFs, using synthetic and real data sets. We observed that the Bayesian approach outperforms the other methods under small sample sizes and high dimensionality. One limitation of our approach is the use of noninformative priors; future work should include the construction of informative priors using available prior biological knowledge, which is expected to produce more accurate classifiers.

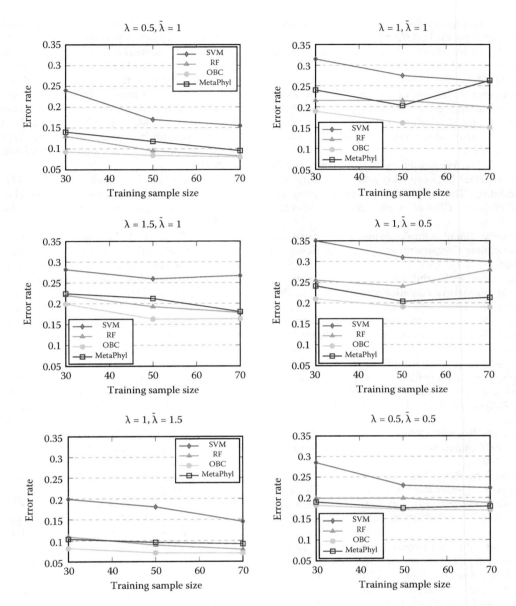

FIGURE 20.6
Comparison with SVM, RF, MetaPhyl, and OBC on simulated data sets for varying number of training samples and within- and between-class variances. Each metagenomic profile consists of 128 OTUs.

References

1. M. Schena, D. Shalon, R.W. Davis, and P.O. Brown. Quantitative monitoring of gene expression patterns via a complementary DNA microarray. *Science*, 270:467–470, 1995.
2. D.J. Lockhart, H. Dong, M.C. Byrne, M.T. Follettie, M.V. Gallo, M.S. Chee, M. Mittmann, C. Wang, M. Kobayashi, H. Horton, and E.L. Brown. Expression monitoring by hybridization to high-density oligonucleotide arrays. *Nature Biotechnology*, 14(13):1675–1680, 1996.
3. A. Mortazavi, B.A. Williams, K. McCue, L. Schaeffer, and B. Wold. Mapping and quantifying mammalian transcriptomes by RNA-Seq. *Nature Methods*, 5(7):621–628, 2008.
4. N. Ghaffari, M.R. Yousefi, C.D. Johnson, I. Ivanov, and E.R. Dougherty. Modeling the next generation sequencing

sample processing pipeline for the purpose of classification. *BMC Bioinformatics*, 14:307, 2013.
5. R. Aebersold and M. Mann. Mass spectrometry–based proteomics. *Nature*, 422(6928):198–207, 2003.
6. T. Nilsson, M. Mann, R. Aebersold, and J.R. Yates et al. Mass spectrometry in high-throughput proteomics: Ready for the big time. *Nature Methods*, 7:681–685, 2010.
7. D.C. Anderson and K. Kodukula. Biomarkers in pharmacology and drug discovery. *Biochem. Pharmacol.*, 87(1):172–188, 2014.
8. A.G. Paulovich, J.R. Whiteaker, A.N. Hoofnagle, and P. Wang. The interface between biomarker discovery and clinical validation: The tar pit of the protein biomarker pipeline. *Proteomics. Clinical Applications*, 2:1386–1402, 2008.
9. E.R. Dougherty. Small sample issues for microarray-based classification. *Comparative and Functional Genomics*, 2:28–34, 2001.
10. U.M. Braga-Neto. Fads and fallacies in the name of small-sample microarray classification. *IEEE Signal Processing Magazine*, 24(1):91–99, 2007.
11. G.F. Hughes. On the mean accuracy of statistical pattern recognizers. *IEEE Transactions on Information Theory*, IT-14(1):55–63, 1968.
12. U.M. Braga-Neto. Small-sample error estimation: Mythology versus mathematics. In J.T. Astola, I. Tabus, and J. Barrera, editors, *Proceedings of SPIE: Mathematical Methods in Pattern and Image Analysis*, volume 5916. San Diego, 2005.
13. C.P. Robert. *The Bayesian Choice: From Decision-Theoretic Foundations to Computational Implementation*. Springer, 2nd edition, 2007.
14. E. Arslan and U.M. Braga-Neto. A Bayesian approach to top-scoring pairs classification. In *Proceedings of the 42nd IEEE International Conference on Acoustics, Speech and Signal Processing (ICASSP'2017), New Orleans*, March 2017, 2017.
15. U. Banerjee and U.M. Braga-Neto. Bayesian ABC-MCMC classification of liquid-chromatography mass spectrometry data. *Cancer Informatics*, Suppl. 5:175–182, 2017.
16. A. Bahadorinejad, I. Ivanov, J. Lampe, M. Hullar, R.S. Chapkin, and U.M. Braga-Neto. Bayesian classification of microbial communities based on 16S rRNA metagenomic data. Submitted, 2018.
17. D. Geman, C. d'Avignon, D. Q Naiman, and R.L. Winslow. Classifying gene expression profiles from pairwise mRNA comparisons. *Statistical Applications in Genetics and Molecular Biology*, 3(1):1–19, 2004.
18. R.A. Bradley and M.E. Terry. Rank analysis of incomplete block designs: I. The method of paired comparisons. *Biometrika*, 39(3/4):324–345, 1952.
19. Y. Sun, U.M. Braga-Neto, and E.R. Dougherty. A systematic model of the LC-MS proteomics pipeline. *BMC Genomics*, 13:S2, 2011.
20. L. Dalton and E.R. Dougherty. Optimal classifiers with minimum expected error within a Bayesian framework—Part I: Discrete and Gaussian models. *Pattern Recognition*, 46(5):1301–1314, 2013.
21. F. Caron and A. Doucet. Efficient Bayesian inference for generalized Bradley–Terry models. *Journal of Computational and Graphical Statistics*, 21(1):174–196, 2012.
22. G.W. Peters, Y. Fan, and S.A. Sisson. On sequential Monte Carlo, partial rejection control and approximate Bayesian computation. Technical report, University of New South Wales, 2009. arxiv:0808.3466v2.
23. B. Afsari, U.M. Braga-Neto, and D. Geman. Rank discriminants for predicting phenotypes from RNA expression. *Annals of Applied Statistics*, 8(3):1469–1491, 2014.
24. A. Choon Tan, D.Q. Naiman, L. Xu, R.L. Winslow, and D. Geman. Simple decision rules for classifying human cancers from gene expression profiles. *Bioinformatics*, 21(20):3896–3904, 2005.
25. X. Lin, B. Afsari, L. Marchionni, L. Cope, G. Parmigiani, D. Naiman, and D. Geman. The ordering of expression among a few genes can provide simple cancer biomarkers and signal BRCA1 mutations. *BMC Bioinformatics*, 10(1):1, 2009.
26. H. Wang, H. Zhang, Z. Dai, M.-S. Chen, and Z. Yuan. TSG: A new algorithm for binary and multi-class cancer classification and informative genes selection. *BMC Medical Genomics*, 6(1):1, 2013.
27. D.R. Hunter. MM algorithms for generalized Bradley–Terry models. *Annals of Statistics*, pages 384–406, 2004.
28. I.C. Gormley and T.B. Murphy. Exploring voting blocs within the Irish electorate: A mixture modeling approach. *Journal of the American Statistical Association*, 103(483):1014–1027, 2008.
29. J. Guiver and E. Snelson. Bayesian inference for Plackett-Luce ranking models. In *Proceedings of the 26th Annual International Conference on Machine Learning*, pages 377–384. ACM, 2009.
30. U. Alon, N. Barkai, D.A. Notterman, K. Gish, S. Ybarra, D. Mack, and A.J. Levine. Broad patterns of gene expression revealed by clustering analysis of tumor and normal colon tissues probed by oligonucleotide arrays. *Proceedings of the National Academy of Sciences*, 96(12):6745–6750, 1999.
31. T.R. Golub, D.K. Slonim, P. Tamayo, C. Huard, M. Gaasenbeek, J.P. Mesirov, H. Coller, M.L. Loh, J.R. Downing, M.A. Caligiuri et al. Molecular classification of cancer: Class discovery and class prediction by gene expression monitoring. *Science*, 286(5439):531–537, 1999.
32. M.A Shipp, K.N. Ross, P. Tamayo, A.P. Weng, J.L. Kutok, R.C.T. Aguiar, M. Gaasenbeek, M. Angelo, M. Reich, G.S. Pinkus et al. Diffuse large B-cell lymphoma outcome prediction by gene-expression profiling and supervised machine learning. *Nature Medicine*, 8(1):68–74, 2002.
33. G.J. Gordon, R.V. Jensen, L.-L. Hsiao, S.R. Gullans, J.E. Blumenstock, S. Ramaswamy, W.G. Richards, D.J. Sugarbaker, and R. Bueno. Translation of microarray data into clinically relevant cancer diagnostic tests using gene expression ratios in lung cancer and mesothelioma. *Cancer Research*, 62(17):4963–4967, 2002.

34. D. Chowdary, J. Lathrop, J. Skelton, K. Curtin, T. Briggs, Y. Zhang, J. Yu, Y. Wang, and A. Mazumder. Prognostic gene expression signatures can be measured in tissues collected in RNA later preservative. *The Journal of Molecular Diagnostics*, 8(1):31–39, 2006.

35. S.A. Armstrong, J.E. Staunton, L.B. Silverman, R. Pieters, M.L. den Boer, M.D. Minden, S.E. Sallan, E.S. Lander, T.R. Golub, and S.J. Korsmeyer. MLL translocations specify a distinct gene expression profile that distinguishes a unique leukemia. *Nature Genetics*, 30(1):41–47, 2002.

36. S. Maouche, O. Poirier, T. Godefroy, R. Olaso, I. Gut, J.-P. Collet, G. Montalescot, and F. Cambien. Performance comparison of two microarray platforms to assess differential gene expression in human monocyte and macrophage cells. *BMC Genomics*, 9(1):1, 2008.

37. M.A. Kuriakose, W.T. Chen, Z.M. He, A.G. Sikora, P. Zhang, Z.Y. Zhang, W.L. Qiu, D.F. Hsu, C. McMunn-Coffran, S.M. Brown et al. Selection and validation of differentially expressed genes in head and neck cancer. *Cellular and Molecular Life Sciences CMLS*, 61(11):1372–1383, 2004.

38. N.D. Price, J. Trent, A.K. El-Naggar, D. Cogdell, E. Taylor, K.K. Hunt, R.E. Pollock, L. Hood, I. Shmulevich, and W. Zhang. Highly accurate two-gene classifier for differentiating gastrointestinal stromal tumors and leiomyosarcomas. *Proceedings of the National Academy of Sciences*, 104(9):3414–3419, 2007.

39. F. Borovecki, L. Lovrecic, J. Zhou, H. Jeong, F. Then, H.D. Rosas, S.M. Hersch, P. Hogarth, B. Bouzou, R.V. Jensen et al. Genome-wide expression profiling of human blood reveals biomarkers for Huntington's disease. *Proceedings of the National Academy of Sciences of the United States of America*, 102(31):11023–11028, 2005.

40. S.L. Pomeroy, P. Tamayo, M. Gaasenbeek, L.M. Sturla, M. Angelo, M.E. McLaughlin, J.Y.H. Kim, L.C. Goumnerova, P.M. Black, C. Lau et al. Prediction of central nervous system embryonal tumour outcome based on gene expression. *Nature*, 415(6870):436–442, 2002.

41. E. Tian, F. Zhan, R. Walker, E. Rasmussen, Y. Ma, B. Barlogie, and J.D. Shaughnessy Jr. The role of the Wnt-signaling antagonist DKK1 in the development of osteolytic lesions in multiple myeloma. *New England Journal of Medicine*, 349(26):2483–2494, 2003.

42. U.M. Braga-Neto and E.R. Dougherty. Is cross-validation valid for microarray classification? *Bioinformatics*, 20(3):374–380, 2004.

43. B. Afsari, E.J. Fertig, D. Geman, and L. Marchionni. switchBox: An R package for k-top scoring pairs classifier development. *Bioinformatics*, btu622, 2014.

44. J Damond. ktspair: K-top scoring pairs for microarray classification. R package version, 1, 2011.

45. Y. Sun, J. Zhang, U.M. Braga-Neto, and E.R. Dougherty. BPDA—a Bayesian peptide detection algorithm for mass spectrometry. *BMC Bioinformatics*, 11:490, 2010.

46. D.J. Lipman and W.R. Pearson. Rapid and sensitive protein similarity searches. *Science*, 227(4693):1435–1441, 1985.

47. Y. Taniguchi et. al. Quantifying *E. coli* proteome and transcriptome with single-molecule sensitivity in single cells. *Science*, 329:533, 2010.

48. P. Lu, C. Vogel, R. Wang, X. Yao, and E. M. Marcotte. Absolute protein expression profiling estimates the relative contributions of transcriptional and translational regulation. *Nature Biotechnology*, 25:117–24, 2007.

49. M. Sturm, A. Bertsch, C. Gröpl, A. Hildebrandt, R. Hussong, E. Lange, N. Pfeifer, O. Schulz-Trieglaff, A. Zerck, K. Reinert, and O. Kohlbacher. OpenMS—An open-source software framework for mass spectrometry. *BMC Bioinformatics*, 9:163, 2008.

50. M. Anderle, S. Roy, H. Lin, C. Becker, and K. Joho. Quantifying reproducibility for differential proteomics: Noise analysis for protein liquid chromatography–mass spectrometry of human serum. *Bioinformatics*, 20(18):3575–3582, 2004.

51. L. Pasa-Tolic, C. Masselon, R.C. Barry, Y. Shen, and R.D. Smith. Proteomic analyses using an accurate mass and time tag strategy. *Biotechniques*, 37(4):621–624, 626–633, 636 passim, 2004.

52. J. Knight, I. Ivanov, and E.R. Dougherty. MCMC implementation of the optimal Bayesian classifier for non-Gaussian models: Model-based RNA-Seq classification. *BMC Bioinformatics*, 15:401, 2014.

53. S.A. Sisson and Y. Fan. Likelihood-free Markov chain Monte Carlo. In S. Brooks, A. Gelman, G.L. Jones, and X.-L. Meng, editors, *Handbook of Markov Chain Monte Carlo*. Chapman & Hall/CRC Press, Boca Raton, FL, 2010.

54. C. Knox, V. Law, T. Jewison et. al. Drugbank 3.0: A comprehensive resource for 'omics' research on drugs. *Nucleic Acids Research*, 39:D1035–41, 2011.

55. A. Webb. *Statistical Pattern Recognition*. John Wiley & Sons, New York, 2nd edition, 2002.

56. S. Marguerat and J. Bahler. RNA-Seq: From technology to biology. *Cellular and Molecular Life Science*, 67(4):569–579, 2010.

57. W Streit, R Schmitz, and Tao Jiang. Metagenomics—The key to the uncultured microbes. *Current Opinion in Microbiology*, 20(7):49–498, 2004.

58. Q Wang, GM Garrity, JM Tiedje, and JR Cole. Naive Bayesian classifier for rapid assignment of rRNA sequences into the new bacterial taxonomy. *Applied and Environmental Microbiology*, 11(73):5261–5267, 2007.

59. O. Tanaseichuk, J. Borneman, and T. Jiang. Phylogeny-based classification of microbial communities. *Bioinformatics*, 20(6):924–930, 2013.

21
Neuroelectrophysiology of Sleep and Insomnia

Ramiro Chaparro-Vargas, Beena Ahmed, Thomas Penzel, and Dean Cvetkovic

CONTENTS

21.1 Big Data Breakthroughs: Genomics 429
21.2 Big Data Breakthroughs: Neuroimaging 430
21.3 Big Data Breakthroughs: Neuroelectrophysiology 431
    21.3.1 Sleep Analysis 431
    21.3.2 Insomnia Analysis and Related Disorders 433
21.4 Case Study: Insomnia Biosignal Processing and Automated Detection 434
    21.4.1 Clinical Data 434
    21.4.2 Big Data Science Model 434
        21.4.2.1 Problem Understanding 435
        21.4.2.2 Data Understanding 435
        21.4.2.3 Data Preparation 435
        21.4.2.4 Modeling 436
        21.4.2.5 Evaluation 438
        21.4.2.6 Deployment 438
21.5 Open Problems and Future Directions 439
Acknowledgements 440
References 440

21.1 Big Data Breakthroughs: Genomics

Genomics is an epitome of Big Data prowess, whereas the challenges come from the storage of petabytes till their timely processing. The advent of next-generation DNA sequencing (NGS) is smoothly turning the health and biological sciences into a data-driven realm. Undoubtedly, Big Data open-source platform Hadoop, described by White [1], is a determinant factor in the proliferation of high-profile initiatives for genomics research. For instance, Schatz's [2] project CloudBurst and its sibling Crossbow pioneered single-nucleotide polymorphism (SNP) identification by sorting genome alignments and genotyping using SoapSNP. Crossbow aligned 3 billion combinations of paired and unpaired reads and ordered 3.7 million SNPs in the genome with 99% accuracy. The genotyping of the equivalent 110 GB of compressed sequence data required 3 hours of processing on a 320-core cluster. In the same direction, CloudAligner presented a cloud-based application with higher performance in the partition and parallel processing of genome and reads, as Nguyen et al. [3] presented. CloudAligner handled reads larger than 30 base pairs, which exceeded the local-based approach of Crossbow. Finally, in Langmead et al. [4], Myrna computed the differential expression in 1.1 billion RNA sequence reads in less than 2 hours using cloud-located computation resources. It is noteworthy that Myrna was funded via a collaborative grant between Amazon Web Services and the National Institutes of Health (NIH). Despite of the potential and availability of Big Data technologies to confront state-of-the-art problems in bioinformatics, large-scale undertakings are still elusive. The barriers are related to the specialized profiles required for the development and operation of the applications built on top of the Big Data and cloud computing platforms. And, current bioinformatic computational tools are designed sequentially, as opposed to the parallel computation required by these technologies. Table 21.1 outlines some up-to-date Hadoop-based bioinformatics initiatives.

TABLE 21.1
Hadoop-Based Bioinformatics Initiatives

Function	Algorithm	Description
Genomic sequence mapping	Cloud Aligner	Mapping short reads generated by NGS
	CloudBurst	Parallel read-mapping for NGS data to the human genome
	SEAL	Align, manipulate, and analyze short DNA sequence reads
	BlastReduce	Parallel short DNA sequence read-mapping and SNP discovery
Genomic sequencing analysis	Crossbow	Whole genome resequencing analysis
	Contrail	Assembly of large genomes from short sequencing reads
	CloudBrush	A distributed genome assembler based on string graphs
RNA sequence analysis	Myrna	Calculating differential gene expression in large RNA
	FX	Estimation of gene expression levels and variant calling
	Eoulsan	RNA sequence data analysis of differential expression
Sequence file management	BAM	Library for scalable manipulation of aligned NGS data
	SeqWare	Tool set for NGS: laboratory information management system (LIMS), Pipeline, and Query Engine
	GATK	Gene analysis tool kit for NGS
Phylogenetic analysis	MrsRF	Calculate the all-to-all Robinson–Foulds distances
	Nephele	Group sequence clustering into genotypes based on a distances
Graphical processing unit (GPU) bioinformatics	GPU-BLAST	Accelerate overall algorithm processing
	SOAP3	Short sequence read and ultrafast alignment
Search engine	Hydra	Protein sequence database search engine
	CloudBlast	Basic Local Alignment Search Tool (Blast) in the cloud
Miscellaneous	BioDoop	Handling Fasta streams and wrappers for Blast
	BlueSNP	Intensive analyses for large genotype to phenotype
	Quake	DNA sequence error detection and correction
	YunBe	Gene set analysis for biomarker identification in the cloud
	PeakRanger	Detecting regions from chromatin immunoprecipitation (ChIP)

Source: O'Driscoll, A. et al., *Journal of Biomedical Informatics,* 46:774–781, 2013.

Notwithstanding, the Big Data Genomics (BDG) project and its core tool set ADAM deserve a special mention, as per Laserson [5]. This crowdsourcing initiative enables the scientific community to perform mapped reads, mark duplicates, base quality score calibration, indel realignment, and variant calling. Overcoming the sequential computing of previous bioinformatics tools, ADAM is Hadoop Distributed File System (HDFS)– and Hadoop-compliant to automatically parallelize jobs across multinode clusters without manually splitting source files or rescheduling tasks. Foremost, it has nominated the Avro schema as the standard format for reads, base observations, variants, genotypes, and sequence features, as described by White [1]. The Avro format file provides higher computational performance and data modeling add-ons over the custom ASCII format, countering its poor compression capabilities.

21.2 Big Data Breakthroughs: Neuroimaging

Neuroimaging is another exemplary domain where Big Data technologies have played a disruptive role. Brain study has fostered multiple cutting-edge advances in understanding its functioning, including electrode-based recordings of neuron activity and image stacking of the brain's anatomy. Through the capture of discharges of large sets of neurons across large regions of the cortex of active subjects, the brain's anatomy and physiology have had unseen progress in the last century. Today, the US presidential–endorsed BRAIN* initiative represents the next frontier in neuroscience, as discussed by Jorgenson et al. [6]. The state of the art allows examining individual genes, molecules, synapses, and neurons at high resolution. Alas, the main challenge refers to how to connect the dots and what lies in between thousands of millions of neuronal circuits. Thus, the BRAIN initiative moves toward the accelerated development of technologies for mapping the circuits of the brain; measuring fluctuating patterns of electrochemical activity flowing through those circuits; and understanding how their interaction shapes our unique cognitive and behavioral traits. Seven priority research areas have been identified to achieve those goals, which are described straightforwardly and summarized in Table 21.2.

- **Discovering diversity:** aims to characterize cell types of the nervous system by developing tools to record, mark, and manipulate defined neurons in the living brain. Also, new genomic and nongenomic tools should be integrated to deliver genes, proteins, and chemicals to scoped cells in human and nonhuman organisms.

---

* Brain Research through Advancing Innovative Technologies.

TABLE 21.2
BRAIN Initiative's Priority Research Areas

#	Research Area	Aim
1	Discovering diversity	Identify and provide experimental access to the different brain cell types to determine their roles in health and disease
2	Maps at multiple scales	Generate circuit diagrams that vary in resolution from synapses to the whole brain
3	The brain in action	Produce a dynamic picture of the functioning brain by developing and applying improved methods for large-scale monitoring of neural activity
4	Demonstrating causality	Link brain activity to behavior by developing and applying precise interventional tools that change neural circuit dynamics
5	Identifying fundamental principles	Produce conceptual foundations for understanding the biological basis of mental processes through development of new theoretical and data analysis tools
6	Advancing human neuroscience	Develop innovative technologies to understand the human brain and treat its disorders; create and support integrated human brain research networks
7	From BRAIN initiative to the brain	Integrate new technological and conceptual approaches produced in areas 1 to 6 to discover how dynamic patterns of neural activity are transformed into cognition, emotion, perception, and action in health and disease

Source: Jorgenson, L. A. et al., *Philosophical Transactions of the Royal Society B*, 370(20140164):1–12, 2014.

- **Maps at multiple scales:** attempts to map connected neurons in local circuits and distributed brain systems. It requires faster, less expensive, and scalable technologies for anatomic reconstruction of neural circuits from whole human brain imaging to dense synaptic reconstruction at the subcellular level.
- **The brain in action:** seizes the recording of dynamic neuronal activity from high-resolution neural networks during long time periods and across all areas of the brain. The existing and emerging technologies include bioelectromagnetism, optics, molecular genetics, and nanoscience.
- **Demonstrating causality:** pursues migrating from observation to causation by deliberated activation and inhibition of neuron populations within behavioral context. The new generation of tools should rely upon optogenetics, chemogenetics, and biochemical and electromagnetic modulation.
- **Identifying fundamental principles:** examines the complex and nonlinear brain functions, which demand advanced modeling and statistics methods to cope with new types of data growing at increasing rates.
- **Advancing human neuroscience:** undertakes diagnostic brain monitoring to decipher mechanisms of human brain disorders, the effect of therapy, and the leverage of diagnostics.
- **From BRAIN initiative to the brain:** realizes the comprehensive understanding of brain functions by the application of the new technologies and conceptual structures promoted by the initiative.

The development of the BRAIN initiative's research goals implies the generation of humongous amounts of data, which are a perfect match for Big Data technologies. Laserson's [7] Thunder project for processing of large amounts of time series data, particularly neuroimaging data, is a direct response to neuroscience's demand for sophisticated toolboxes. Thunder performs heavy matrix computations and distributed implementations of some statistical techniques upon large spatial and temporal data sets. Its Python-related interface makes Thunder accessible and malleable to a nonexpert audience about neuroscience topics. Based on Thunder, recent research works rendered bidimensional and tridimensional images of zebra fish neurons, allowing their categorization by neural behavior.

21.3 Big Data Breakthroughs: Neuroelectrophysiology

Biomedical signal processing, discussed by Mesin et al. [8] and Pereda et al. [9], and quantitative electroencephalogram (EEG), discussed by Thakor and Tong [10], have set the foundations of automated analysis of the electrical functioning of the brain. The computer-assisted methods of Merica and Fortune [11] and Ogilvie [12] have the potential to enhance the characterization of complex processes like imaginary thinking, psychiatric pathologies, or sleep and its related disorders. The remainder of this chapter focuses on breakthroughs of electrophysiological analysis of sleep and insomnia disorders based on biosignal processing and Big Data technologies.

21.3.1 Sleep Analysis

The European SIESTA project discussed by Klosh et al. [13] pioneered the collaborative analysis of sleep from eight European laboratories, including 196 polysomnogram (PSG) recordings from 98 patients affected by depression, general anxiety disorder, insomnia, Parkinson's disease, period limb movements, and sleep

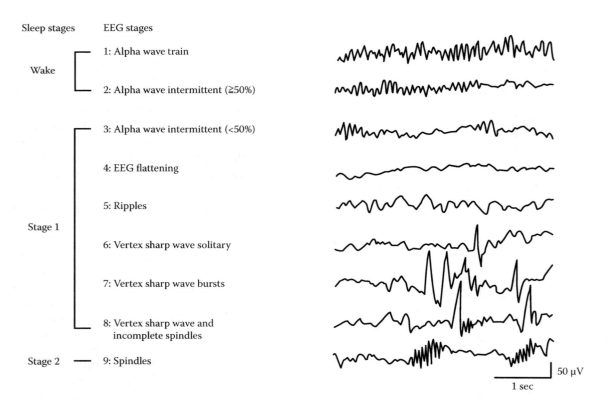

FIGURE 21.1
EEG signals and transients associated with the process of falling asleep. (From Ogilvie, R. D. *Sleep Medicine Reviews*, 5(3):247–270, 2001.)

apnea. The study aimed to assess the interrater and intersubject variability using the state of the art to characterize regular sleep patterns and related disorders. The study revealed that overall rate of agreement among scorers was 0.6816, which in terms of Cohen's κ index* indicated a substantial agreement. For related disorders, the index values ranged between 0.6138 for Parkinson's disease and 0.8176 for generalized anxiety disorder with insomnia, as per Danker-Hopfe et al. [14]. However, the scoring of sleep onset stages obtained the lowest rate of agreement, i.e., 0.349. Moreover, the transition from awake stage to light and deep sleep recurrently required a subjective estimation, due to intersubject variability.

In the work of Rodenbeck et al. [15], recommendations were introduced for more reliable expert scoring and standardized algorithms supported on computational platforms. A multidisciplinary task force provided a consorted interpretation of relevant EEG patterns and transients. Each pattern was independently analyzed to indicate onset and offset cues of amplitude, frequency, duration, scalp derivations, morphology, and surrounding events. Thus, the American Academy of Sleep Medicine (AASM) standard by Iber et al. [16] was released extending the repertoire of rules, cortical derivations, and relevant channels, moving from EEG to PSG analysis. The scoring rules led to a more consistent definition of sleep macrostructure supported on PSG signals, standardizing five sleep stages: wake (W), nonREM1 (N1), nonREM2 (N2), nonREM3 (N3), and rapid eye movement (REM). As per the nomenclature of Marzano et al. [17], each sleep stage is characterized by a single or combined EEG frequency bands—delta δ (0.5–4 Hz), theta θ (4–7 Hz), alpha α (7–12 Hz), sigma ς (12–16 Hz), beta β (16–18 Hz), and gamma γ (18–32 Hz)—accompanied by transients, such as electrooculographic (EOG) slow eye movements (SEMs), and EEG transients like vertex shape waves (VSWs), K-complexes (KCs), sleep spindles (SSs), sawtooth waves (SWs), electromyographic (EMG) body movements (BMs), and chin tension. Figure 21.1 shows the progressive changes in the EEG signals and transients associated with the process of falling asleep, as referred to in Ogilvie [12].

Clinical sleep specialists are trained to examine individual 30-second EEG epochs at a time for an overnight sleep. The stage scoring is ultimately determined using the supplementary activity from additional PSG channels, especially EOG and EMG. So, sleep specialists reconstruct an overnight sleep pattern based on the stage scoring of consecutive 30-second epochs. This outcome is known as a hypnogram. The implicit challenges

---

* One of the most popular statistics to measure the interrater agreement based on a numerical and categorical rating. The scores range from 0–0.2, light; 0.21–0.4, fair; 0.41–0.6, moderate; 0.61–0.8, substantial; to 0.81–1, perfect agreement.

inherited from expert-based scoring to computer-assisted methods are as follows: (i) a subject's recording requires the analysis of multiple channels for several hours for several nights, which turn into burdensome periods of assessment; (ii) subject-specific traits hinder the application of generalized models to characterize larger populations, a challenge known as intersubject variability; and (iii) differing backgrounds and criteria of scorers lead to interrater variability.

21.3.2 Insomnia Analysis and Related Disorders

Computer-assisted analysis not only has been applied to the characterization of regular sleep patterns but also has taken part in the characterization of other complex behavioral, psychological, and psychiatric pathophysiologies. Psychiatric disorders are associated with quantifiable physiological symptoms, such as dysregulation in the autonomic system and neuroendocrine imbalances, as Mayers and Baldwin [18] claimed. For instance, depression is distinguished by the absence of deep sleep, difficulty staying asleep, increased heart rate, hyperactivity of the sympathetic nervous system, hypothalamic pituitary overactive state, and self-regulated body temperature variations, among others. Staner et al. [19] investigated the relation between hyperarousals, sleep disturbance, and abnormal levels of arousals in depressive insomnia patients. The study extracted EEG frequency bands during sleep onset and first non rapid eye movement (NREM) from control, primary insomnia, and depressive insomnia groups. Their results showed that hyperarousals characterized the primary insomnia with stronger EEG activity during sleep onset, while depressive insomnia patients reported fewer slow waves sleep (SWS). Iverson et al. [20] firstly posited heart rate as a marker for supporting diagnosis of mental illnesses. Jurysta et al. [21] attempted to establish the relationship between cardiac activity and EEG δ power in major depression disorder. Alas, a nonsignificant bound was found between control and patient cohorts. Neither parasympathetic nor sympathetic parameters differed in both groups. Conversely, Udupa et al. [22] identified increased sympathetic and decreased parasympathetic activity in major depression patients in comparison to control subjects. The study reported the alteration of cardiac function and EEG δ band in the primary insomnia and depression patients. The symptoms of heightened physiological arousal (hyperarousal) are allegedly features to differentiate control, insomnia, and depression pathologies.

As per the American Psychiatric Association's [23] *Diagnostic and Statistical Manual (DSM-IV-TR)*, insomnia is distinguished by sleep disturbances promoting impairments of daytime functioning or clinically significant distress. It is noteworthy to set apart primary and secondary insomnia. For the former, the diagnostic criteria use as major traits inhibition to fall asleep or maintain sleep, or both. Moreover, the disturbance should not concur with another mental disorder. The *DSM-IV-TR* supports the International Classifications of Sleep Disorders diagnosis, which defines primary insomnia as the same as psychophysiological insomnia. According to it, the most common causes are high concentration of overnight arousals and behavioral factors like stressful lifestyle. For secondary insomnia, the landscape is more perplexing. Its diagnostic criteria include inhibition of sleep accompanied by psychiatric disorders or aside pathologies. However, it is hard to determine whether the insomnia disorder's root cause is either the prescribed medication or the psychiatric pathology or psychological distress, or all of the above. A patient's medical history could disclose correlations between side effects of medication, clinical distress, and sleep patterns. Even then, the APA [23] states that sleep disturbance and insomnia are common symptoms of most psychiatric disorders.

In a 2006 study, Morin et al. [24] conducted a telephone survey across Québec territory (Canada) with 2001 participants to estimate the prevalence of insomnia symptoms and syndromes in the general population. The study found that 29.9% of the interviewees attested to insomnia symptoms, although only 9.5% met medical criteria for insomnia syndrome. Later, Petrovsky et al. [25] proposed a proof of concept based on sleep deprivation to relate behavioral (e.g., chronic fatigue) to psychiatric impairments (e.g., psychosis). Their findings pointed out that sleep deprivation and quantitative biomarkers can model psychiatric disorders by mimicking their pathological origin. Those biomarkers have turned up in the mainstream of the most recent computer-assisted approaches. Unfortunately, it is unlikely that insomnia markers can be characterized exclusively using PSG recordings. Disordered sleep hardly matches the patient's experience of insomnia due to the ever-present intersubject variability. Such a condition disqualifies any objective criteria on PSGs to diagnose chronic insomnia. Notwithstanding, PSG analysis extracts leading metrics toward abnormal sleep patterns such as prolonged latency to sleep onset, frequency of arousals, and reduced amounts of total sleep. Surveys, subjective tests (e.g., Epworth Sleepiness Scale), and interviews play a supportive role for comprehensive assessment [23].

Sleep initiation is not the only facet of insomnia; sleep maintenance is a recurrent matter in clinical consultation. Computer-aided biosignal processing has also contributed to its characterization spotting on tighter EEG frequency bands, such as β_1 (16–18 Hz), β_2 (18–30 Hz), and β_3 (30–40 Hz). Such a fragmentation eased the extraction of features by digging into narrower regions of the spectrum. Also, the remanence of the events in lower bands might drag traces of activity in the neighboring bands of β. Cervena et al. [26] studied the power spectrum features of 10 subjects with sleep onset insomnia,

10 with sleep maintenance insomnia, and 10 control individuals. They concluded that there are significant differences between sleep onset and sleep maintenance insomnia in the EEG β_2 band. Consecutive attempts of sleep onset concentrated elevated power values in the EEG β_1 band, in contrast to EEG β_2 exhibited by sleep maintenance patients. Furthermore, differences in the EEG δ activity suggested different mechanisms of hyperarousal in the wake-promoting sleep-protective dilemma.

21.4 Case Study: Insomnia Biosignal Processing and Automated Detection

The case study focuses on the automated analysis of sleep onset periods to comply with two main objectives: (i) a computer-aided estimation of the sleep onset stages and (ii) the differentiation of regular and insomnia populations by computational techniques compliant with medical procedures, enacting a supportive role. The approach overcame the time-consuming and burdensome assessment of PSG data sets by the introduction of adaptive processing and Big Data Science for the classification task. Moreover, the approach introduced a characterization model for the differentiation of regular and disordered sleeping patterns dealing with subject-specific traits and deferring scoring criteria, i.e., intersubject and interrater variability.

21.4.1 Clinical Data

The sleep onset staging for the insomnia detection study used the PSGs recorded and provided by the Interdisciplinary Centre for Sleep Medicine, Charité Universitätsmedizin in Berlin (Germany). The ethics consideration and procedures were approved by the local ethics committee of the Charité Universitätsmedizin Berlin and Royal Melbourne Institute Technology (RMIT) Ethics Committee with written consent of the subjects and patients. Table 21.3 shows a summary of the recruited cohort and additional technical details of the recordings.

Insomnia diagnosis was performed as per the International Statistical Classification of Diseases and Related Health Problems (ICD-10) *Classification of Mental and Behavioural Disorders* of the World Health Organization [27], including anamnesis, i.e., clinical interviews, followed by the overnight PSG recordings to rule out comorbid sleep or psychiatric disorders. Specialists performed the sleep stage scoring, according to the guidelines of the AASM manual by Iber et al. [16], i.e., each 30-second epoch was mapped to a particular stage: wake (W), nonREM1 (N1), nonREM2 (N2), nonREM3 (N3), and REM. For all the PSGs, EEG montage was compliant to the 10–20 international system in Figure 21.2.

TABLE 21.3

Summary of Clinical Data and Recording Details of the Case Study

Feature	Description
Subjects	20 male control subjects
	20 male insomnia patients
PSG channels	EEG: C4-A1
	Electrooculographic (EOG): right, left
	Electromyographic (EMG): chin, left tibialis
PSG length	6–8 hours
Epoch staging	AASM manual (W/N1/N2/N3/REM)
Epoch length	30 seconds
Sampling frequency	200 Hz
Provided by	Interdisciplinary Centre for Sleep Medicine Charité Universitätsmedizin
Approval by	Ethics committee of the Charité Universitätsmedizin Berlin
	RMIT Ethics Committee

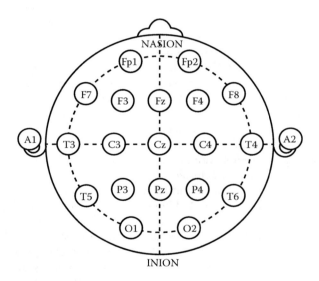

FIGURE 21.2
10–20 international system EEG montage.

21.4.2 Big Data Science Model

Big Data Science involves strong scientific and technological foundations and some crafting originality impressed on the solution model. Nonetheless, a well-defined process requires a structure to deliver Big Data Science problems with consistency, repeatability, and objectiveness. For the present case study, the chosen methodology was based on the Cross Industry Standard Process for Data Mining (CRISP-DM), shown in Figure 21.3 [28]. The CRISP-DM makes use of closed-loop iterations the rule to gain deeper knowledge of the problem's data until the best possible generalization model is found. The related stages set the grounds for the challenges expected in the upcoming

Neuroelectrophysiology of Sleep and Insomnia

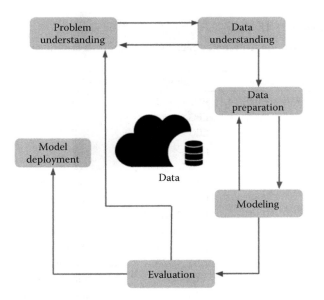

FIGURE 21.3
Cross Industry Standard Process for Data Mining. (From Shearer, C. *Journal of Data Warehousing*, 5:13–22, 2000.)

phases, expediting the refinement of constraints after the outcomes of each iteration.

21.4.2.1 Problem Understanding

Although it seems obvious, the first consideration is to understand the problem to be solved. Then, this stage is commonly subject to multiple iterations until the solution suffices for the actual question. Expert analysts' and data scientists' creativity played a crucial role in the span of insomnia detection in terms of multiple models through a bottom-up approach, i.e., statement of related subproblems to resolve the core problem. Automated insomnia detection required a multidisciplinary discussion from physics, biology, and even human sciences to appropriately grasp the implications of this undertaking. The entire Section 21.3 aimed to provide a comprehensive understanding of the sleep and insomnia characterization problem, which required a devoted literature revision and fieldwork.

21.4.2.2 Data Understanding

Once the caveats of the problem are well understood, the data is the primary source to build the solution model. It is not common practice to have structured, unambiguous, and compelling data sets. Rarely, there is an immediate match between the raw data and the problem question. Big Data's variety attribute attests to this matter. The estimation of the critical strengths and shortcomings of the available data is the critical part of this stage. The recording of PSGs was a complex process, since it involved technicians in the setup of the equipment; engineers to retrieve, process, and format the recorded data; physicians for their interpretation; and more importantly, the patients, whose collaboration was vital to the integrity of the recording. Each participant offered a different perspective of the data that needed to be taken into account. Section 21.4.1 and upcoming ones shed light on the critical aspects of PSGs.

21.4.2.3 Data Preparation

Usually, the raw format of the data needs to go through a transformation process to yield better insights. It includes cleaning and cleansing techniques, such as removal of missing values, normalization, conversion of types, scale rearrangements, categorical-to-numeric parsing, etc. At this stage, the well-known feature extraction routines take place. Unlike previous preparation techniques, the extraction routines use more sophisticated procedures to spread the raw data to different mathematical or statistical domains, where the resulting descriptors or predictors ease the separation of groups, classes, or patterns. Applied to automated sleep analysis and insomnia detection, the preparation and feature extraction stage is far from being a trivial phase.

The present approach initially aimed to represent the PSG biosignals using a collection of coefficients as high-quality features based on their frequency-domain content. The synthesis of thousands of time points into a small number of coefficients relaxed the computational demands in terms of processing load and memory capacity. The feature extraction model used state-space Time Varying Autoregressive Moving Average (TVARMA) realizations with order p,q to represent PSG signals, as in the work of Galka et al. [29]. Thus, system and observation model equations (Eqs. 21.1–21.2) ranslated each PSG epoch as follows:

$$\mathbf{x}[k] = \mathbf{x}[k-1] + \eta[k], \tag{21.1}$$

$$\mathbf{y}[k] = \mathbf{A}\mathbf{x}[k] + \mathbf{x}^H[k]\mathbf{B}v[k] \tag{21.2}$$

where $x[k]$ is the estimated vector of system coefficients understood as the synthesis vector of the PSG epoch. The system noise vector $\eta[k]$ is divided into two terms $[\eta_p[k]\ \eta_q[k]]$ to distinguish AR(p) and MA(q) noise processes, where η_p has Cauchy–Lorentz* distribution with zero translation and $\varsigma^2_{\eta_p}$ dispersion, $\eta_p \sim \mathcal{C}(0, \varsigma^2_{\eta_p}\mathbf{I})$. In turn, η_q is Gaussian distributed with zero mean and $\sigma^2_{\eta_q}$ variance, $\eta_q \sim \mathcal{N}(0, \sigma^2_{\eta_q}\mathbf{I})$. The vector $y[k]$ represents the estimated epoch, accompanied by $\mathbf{A}$ and $\mathbf{B}$ matrices in Eqs. 21.3–21.4 to perform the weighted linear combinations.

---

* Cauchy–Lorentz distribution belongs to the family of probability density functions with two-sided heavy tails, i.e., the typical Gaussian monotonically decaying values at the distant points from the distribution's mean are replaced by subtle bell-shaped curves.

The vector $v[k]$ is the Gaussian distributed observation noise with zero mean and σ_v^2 variance, $v \sim \mathcal{N}(0, \sigma_v^2 \mathbf{I})$.

$$\mathbf{A} = [\mathbf{y}[k-1], \cdots, \mathbf{y}[k-p] | 0, \cdots, 0], \quad (21.3)$$

$$\mathbf{B} = \begin{bmatrix} \mathbf{0}_{p \times p} & 0 \\ 0 & \mathbf{I}_{(q+1) \times (q+1)} \end{bmatrix} \quad (21.4)$$

The biosignal modeling allowed the tracking of sudden and slow-paced changes within PSG biosignals. State-space realizations with Cauchy–Lorentz distribution as in Eqs. 21.1–21.2 dealt with the nonlinearity, nonstationarity, and non-Gaussianity constraints of PSG signals. The Cauchy–Lorentz distribution had heavy-tailed deviations to give probabilistic relevance to abrupt fluctuations and a fastened dispersion in the central region to track smoother dynamics. A recursive Monte Carlo filter, aka particle filter, was implemented to compute the system's coefficients, whereas nonlinear and non-Gaussian models have no closed-form solutions. The particle filter built up the time-changing dynamics of biosignals with probabilistic distributions approximated by a series of samples or particles. Akaike (AIC) and Bayesian information criterion (BIC) helped to compute the optimal model orders $p = 8$ and $q = 2$ to balance out overestimations and underestimations.

The extraction of frequency-domain content implied the convolution of the non-Gaussian TVARMA(8,2) coefficients with a bank of complex Morlet wavelets. From the EEG channels, δ (0.5–4 Hz), θ (4–7 Hz), α_1 (7–9.5 Hz), α_2 (9.5–12 Hz), ς_1 (12–14 Hz), ς_2 (14–16 Hz), β_1 (16–18 Hz), β_2 (18–30 Hz), and β_3 (30–40 Hz), instantaneous power bands $\mathbf{P}[k]$ were extracted, accompanied by transients counters of VSW, KC, and SS. From the EOG channels, rapid and slow eye movements set out from the differential amplitudes of left and right derivations. And, the amplitude of EMG channels produced the features associated with the chin tension and limb movements. Table 21.4 lists the complete collection of features.

TABLE 21.4

Features for Automated Sleep Onset Analysis and Insomnia Detection

#	Feature	Channel	Event/Freq. Band (Hz)
1	$P_\delta[k]$	EEG C4-A1	0.5–4
2	$P_\theta[k]$	EEG C4-A1	4–7
3	$P_{\alpha_1}[k]$	EEG C4-A1	7–9.5
4	$P_{\alpha_2}[k]$	EEG C4-A1	9.5–12
5	$P_{\varsigma_1}[k]$	EEG C4-A1	12–14, SS
6	$P_{\varsigma_2}[k]$	EEG C4-A1	14–16, SS
7	$P_{\beta_1}[k]$	EEG C4-A1	16–18
8	$P_{\beta_2}[k]$	EEG C4-A1	18–30
9	$P_{\beta_3}[k]$	EEG C4-A1	30–40
10–11	A_{EEG}	EEG C4-A1	VSW, KC
12–13	A_{EOG}	LEOG/REOG	Slow/rapid eye movement
14	A_{EMG}	EMG chin	Chin tension
15	A_{EMG}	EMG tibialis	Limb movement

21.4.2.4 Modeling

Here, the simplest and more advanced repertoire of statistics, data mining, and machine learning techniques is crafted to make sense of the features extracted from the data, which in turn are the abstract representation of the core problem. In our case study, the model comprised two submodels. The first produced computer-assisted hypnograms using the biosignals' features to map PSG epochs to sleep onset stages: W, N1, and N2. The candidate technique consisted of an ensemble classifier, which engaged 300 tree-type weak subclassifiers, shown in Figure 21.4.

The iterative design of the CRISP methodology (Figure 21.3) led to the proposal of a second classification submodel in order to maximize the identification of regularities in the data. A fuzzy inference system (FIS), depicted in Figure 21.5, bridged the input feature space with an output space represented by the AASM

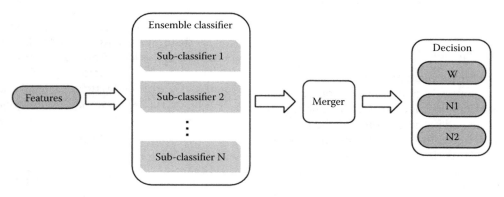

FIGURE 21.4
Diagram of the ensemble classification submodel.

Neuroelectrophysiology of Sleep and Insomnia

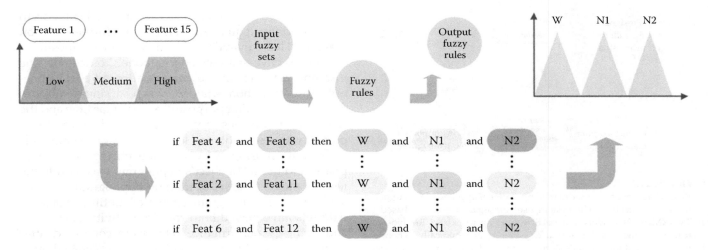

FIGURE 21.5
Diagram of fuzzy inference system consisting of input/output fuzzy sets and fuzzy rules. Each 30-second epoch is described by 15 features, whose values are mapped to three possible input fuzzy sets—low, medium, and high—using trapezoidal functions of membership. The fuzzy rules follow a Boolean logic of fuzzy variables with weighting selection. The output fuzzy sets determine the degree of membership of each 30-second epoch to a sleep onset stage.

guidelines for sleep staging, including suggestions of experienced scorers. The fuzzy logic–based classifier relied on fuzzy rules or "if–then" statements to make the classification decisions. Thus, the automatic generation of hypnograms using the ensemble and FIS submodel remains to be evaluated to opt for a definitive one to be deployed.

For the ultimate detection of insomnia, a logistic regression classifier (Eq. 21.5) was engineered using the number of transitions among W, N1, and N2 stages in the autogenerated hypnogram. Each subject's hypnogram was transformed into a sleep transition network with three vertices (one for each sleep onset stage) and weighted directed edges. Based upon the network representation, degree, adjacency, and incidence matrices, the logistic classification model performed the automated insomnia detection. Figure 21.6 depicts the graphical and matrix representation of the sleep onset transitions.

$$logit(E\{y^i|x^i\}) = logit(p^i) = ln\left(\frac{p^i}{1-p^i}\right) = \hat{\beta}x^i \quad (21.5)$$
$$= \hat{\beta}_0^i + \hat{\beta}_1^i WW^i + \ldots + \hat{\beta}_9^i N2N2^i$$

where the Bernoulli distributed variable y^i is predicted by the product of the regression coefficients $\hat{\beta}$ and i^{th}

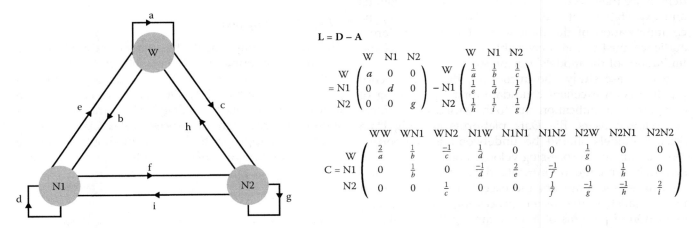

FIGURE 21.6
Sleep stage transition network and related matrices based upon an autogenerated hypnogram. A transition network consists of three vertices (i.e., one each for sleep onset stages W, N1, and N2) and nine directed edges with a weight equal to the inverse of the total number of transitions between the corresponding vertices. Three matrices are derived from the transition network: degree **D**, adjacency **A**, and incidence **C**. The adjacency **A** and incidence **C** matrices values correspond to the inverse of the total number of transitions from one stage to another. And the degree matrix **D** only includes the noninverse of the total transitions from each node to itself; therefore, it is diagonal. The Laplacian matrix **L** comes from the subtraction of the diagonal degree matrix **D** and the adjacency **A** matrix.

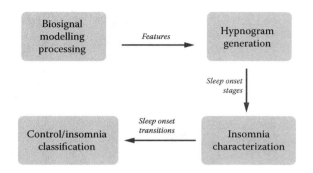

FIGURE 21.7
Block diagram of our modular computational approach for automated insomnia detection. The system uses biosignal processing based on TVARMA(8,2) models and particle filtering to extract the biosignal features, which are fed into a classifier to automatically generate a hypnogram of the subject's sleep onset period. From the hypnogram, a sleep stage transition network inputs a logistic regression classifier to differentiate control and insomnia individuals.

subject's sleep onset stage transitions $x^i = \{WW^i, WN1^i, \ldots, N2N2^i\}$.

Figure 21.7 summarizes the Big Data Science models and submodels with their by-products to prepare and model the PSG data sets.

21.4.2.5 Evaluation

A rigorous evaluation strategy warrants the validity and reliability of the proposed models to solve the original problem. The evaluation should discard large sample anomalies, model biases, false positives, and data collector subjectivities. As in most performance analyses, qualitative and quantitative criteria are accounted for. The qualitative evaluators are problem-specific dependent, since they need to satisfy the diverse backgrounds and expectations of stakeholders. For that reason, the communication of the data scientist to his/her target audience must be comprehensible about the benefits and limitations of the models' outcomes.

In this case study, the introduced Big Data Science pipeline is an excellent candidate for preprocessing of complexity simplification, due to the varied stakeholders as potential users. Big Data platforms and Big Data Science models should be understood as supportive tools, excelling in processing velocity and computational synthesis on large or intricate data sets. Health and medical sciences are not the exception. The proposed model satisfies the faster processing of PSGs and extraction of patterns of interest among the population with insomnia syndrome. However, the specialist has the final word about the usage of the models' outputs. Depending on the patient and his/her context, models' estimates could play either a conservative role or that of a risk-taking decision maker, or somewhere in between.

Quantitative performance metrics are excellent tools to guide the difficult qualitative criteria and further decision making. Their analytic and closed-form compositions enable objective analyses on the models' outcomes. The insomnia classification submodels adopted a leave-one-out cross-validation strategy, where a single subject's data set had a testing purpose at a particular time and the remaining recordings served as training data sets. More precisely, the ensemble and logistic regression submodels were built up using 39 hypnograms during the training phase, and the 40th subject's hypnogram was used to the test the model. The set of training hypnograms and the single testing hypnogram circulated until all subjects' data sets have played both roles. Sensitivity, specificity, and accuracy performance rates were computed after each round, and the overall metrics were averaged out.

Table 21.5 summarizes the comparison of performance metrics between ensemble and logistic regression classifiers. Sensitivity measures the ratio of positives correctly identified, that is, the number of insomnia patients well-classified. Specificity calculates the proportion of true negatives, i.e., control subjects, identified as such. And, accuracy accounts for the relation of true positives and true negatives over all well-classified and misclassified counts. All previous metrics are on a 0-to-1 scale, 1 being the best possible performance.

Based on the performance metrics, the classifier chosen to be deployed for automated insomnia detection is logistic regression. Both classifiers performed well identifying disordered patients (sensitivity), which is an essential requirement for this kind of medical problems. But the superior performance of the logistic regression with respect to control subjects' identification (specificity) and overall accuracy tipped the balance in its favor.

21.4.2.6 Deployment

The final stage implements the data science model over a production information system subject to automated building and testing capabilities. The deployment might go from an integrated application suite along with a robust information technology (IT) infrastructure, or simply the redesign of process rules upon existing systems. An appropriate deployment should guarantee easy and fast updating of the models, responding to new

TABLE 21.5

Performance Rates of Ensemble and Logistic Regression Submodels

Metric	Ensemble	Logistic Regression
Sensitivity	**0.89**	0.87
Specificity	0.58	**0.75**
Accuracy	0.75	**0.81**

available data and changing conditions upon the original problem statement. This is how the CRISP-DM methodology completes its iterative cycle.

The automated insomnia detection model demanded an elastic platform in terms of computation and storage power. The combination of the Amazon EC2 cloud computing platform and Apache Spark cluster-computing framework was the applied solution. Recalling the branding conditions of Big Data projects—volume, velocity, and variety—Spark was precisely the parallel computation tool delegated to the job, as Karau et al. [30] described. A single subject's PSG occupied between 400 and 500 MB in disk using European Data Format (EDF), which is common for archiving and exchanging of PSG and EEG data sets. The design, development, and testing of data models usually queue up headers and formatting sections that increase original sizes to 0.8–1 GB per data set. The case study collected 40 recordings from 20 control and 20 insomnia patients (see Section 21.4.1), thus adding 32–40 GB of raw data to prepare and model.

The Amazon EC2 service hosted the processing platform for the training and testing of the insomnia classification model. However, it could also scale out the production system for the assisted assessment of sleep patterns, as part of a standard diagnosis protocol, by processing tens of PSGs on a daily basis. On top of the Amazon EC2 virtual instances, the Spark engine orchestrated the scheduling, distributing, and monitoring of applications across one master node and three worker machines, i.e., computing cluster. Spark is equipped with higher-level packages such as SQL or machine learning; the latter implemented the logistic regression classifier for insomnia detection. The integration of specialized libraries integrated with lower-layer components benefits from optimizations as a whole unified stack. So, speed improvements in the Spark's core engine are immediately reflected in the machine learning libraries. Figure 21.8 shows the deployed master–worker configuration and the specifications of the virtual instances.

The configuration comfortably handled the volume of data, but more importantly, improved the speed of processing. In comparison, a single-node configuration required 20–22 hours to entirely process a PSG recording of 7–8 hours' length. Using the cluster configuration, the processing time was reduced to 5–6 hours for each PSG data set of the same duration. Then, the complete analysis, training, and testing of the automated insomnia detection model using 40 PSG recordings lasted 10 days, which is the average time a specialist needs to score a single PSG. A side benefit was that the computing infrastructure became disposable once the work was done, i.e., the master and working nodes were decommissioned from Spark and terminated in the

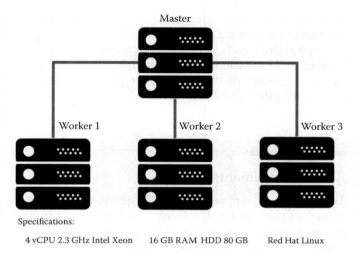

FIGURE 21.8
Spark master–worker cluster configuration. It designates one master node and three worker machines with the same processing and storage specifications.

Amazon cloud. With this on-demand Big Data Science model, the cost of analysis and classification of a single PSG was US $200 with no further depreciation of vacant infrastructure.

21.5 Open Problems and Future Directions

Academia and industry are bridging the gap around Big Data technologies in order to produce affordable solutions to health problems, as O'Driscoll et al. [31] stated. Dell Inc., the TGen,* and the NMTRC[†] have jointly developed a software as part of a personalized medicine trial for pediatric cancer, specifically neuroblastoma. Intel has partnered with NextBio to improve HDFS, Hadoop, and HBase for health sciences, turning over all contribution to the open-source community. The leader in Big Data distribution, Cloudera collaborates with the Institute for Genomics and Multiscale Biology at the Mount Sinai School of Medicine to research human and bacterial genomes, metabolic pathways of normal and disease states in the organism, and structure and function of molecules used in disease treatments, among others. Better diagnosis, understanding, and treatment of diseases supported by Big Data technologies is the motif.

Multiple fronts of improvement are still pending concerning the sophistication of Big Data Science techniques and methods. The recruitment of larger clinical cohorts

---

* Translational Genomics Research Institute.
[†] Neuroblastoma and Medulloblastoma Translational Research Consortium.

would promote the advancement of computer-based approaches in sleep clinical practice. In the same way, the critical discussion of Big Data technologies' practical benefits and limitations is expected to be an incentive for the participation of medical and engineering peers in search of further insights.

Acknowledgements

This work was supported by NPRP grant #5-1327-2-568 from the Qatar National Research Fund, which is a member of Qatar Foundation. The statements made herein are solely the responsibility of the authors.

The authors would like to thank the Interdisciplinary Centre for Sleep Medicine at the Charité Universitätsmedizin Berlin (Germany) for undertaking the recruitment and collection of polysomnographic recordings, as well as the sleep scorers dedicated to the assessment labors.

References

1. White, T. (2015). *Hadoop: The Definitive Guide: Storage and Analysis at Internet Scale*. O'Reilly Media, Inc., 4th edition.
2. Schatz, M. (2009). Cloudburst: Highly sensitive read mapping with MapReduce. *Bioinformatics*, 25:1363–1369.
3. Nguyen, T., Shi, W., and Ruden, D. (2011). CloudAligner: A fast and full-featured MapReduce based tool for sequence mapping. *BMC Res Notes*, 4:171.
4. Langmead, B., Hansen, K., and Leek, J. (2010). Cloud-scale RNA-sequencing differential expression analysis with Myrna. *Genome Biology*, 11.
5. Laserson, U. (2015). *Advanced Analytics with Spark: Patterns for Learning from Data at Scale*, chapter Analyzing Genomics Data and the BDG Project, pages 195–214. O'Reilly Media, Inc., 1st edition.
6. Jorgenson, L. A., Newsome, W. T., Anderson, D. J., Bargmann, C. I., Brown, E. N., Deisseroth, K., Donoghue, J. P., Hudson, K. L., Ling, G. S. F., MacLeish, P. R., Marder, E., Normann, R. A., Sanes, J. R., Schnitzer, M. J., Sejnowski, T. J., Tank, D. W., Tsien, R. Y., Ugurbil, K., and Wingfield, J. C. (2014). The BRAIN initiative: Developing technology to catalyse neuroscience discovery. *Philosophical Transactions of the Royal Society B*, 370 (20140164):1–12.
7. Laserson, U. (2015). *Advanced Analytics with Spark: Patterns for Learning from Data at Scale*, chapter Analyzing Neuroimaging Data with PySpark and Thunder, pages 217–236. O'Reilly Media, Inc., 1st edition.
8. Mesin, L., Holobar, A., and Merletti, R. (2011). *Advanced Methods of Biomedical Signal Processing*. IEEE Press.
9. Pereda, E., Quiroga, R. Q., and Bhattacharya, J. (2005). Nonlinear multivariate analysis of neurophysiological signals. *Progress in Neurobiology*, 77(12):1–37.
10. Thakor, N. V. and Tong, S. (2004). Advances in quantitative electroencephalogram analysis method. *Annual Review of Biomedical Engineering*, 6(1):453–495.
11. Merica, H. and Fortune, R. D. (2004). State transitions between wake and sleep, and within the ultradian cycle, with focus on the link to neuronal activity. *Sleep Medicine Reviews*, 8(6):473–485.
12. Ogilvie, R. D. (2001). The process of falling asleep. *Sleep Medicine Reviews*, 5(3):247–270.
13. Klosh, G., Kemp, B., Penzel, T., Schlogl, A., Rappelsberger, P., Trenker, E., Gruber, G., Zeithofer, J., Saletu, B., Herrmann, W., Himanen, S., Kunz, D., Barbanoj, M., Roschke, J., Varri, A., and Dorffner, G. (2001). The SIESTA project polygraphic and clinical database. *Engineering in Medicine and Biology Magazine, IEEE*, 20(3):51–57.
14. Danker-Hopfe, H., Kunz, D., Gruber, G., Klsch, G., Lorenzo, J. L., Himanen, S. L., Kemp, B., Penzel, T., Rschke, J., Dorn, H., Schlgl, A., Trenker, E., and Dorffner, G. (2004). Interrater reliability between scorers from eight European sleep laboratories in subjects with different sleep disorders. *Journal of Sleep Research*, 13(1): 63–69.
15. Rodenbeck, A., Binder, R., Geisler, P., Danker-Hopfe, H., Lund, R., Raschke, F., Wee, H.-G., and Schulz, H. (2006). A review of sleep EEG patterns. Part I: A compilation of amended rules for their visual recognition according to Rechtschaffen and Kales. *Somnologie*, 10(4):159–175.
16. Iber, C., Ancoli-Israel, S., Jr., A. L. C., and Quan, S. F. (2007). The AASM manual for the scoring of sleep and associated events. Technical report, American Academy of Sleep Medicine.
17. Marzano, C., Moroni, F., Gorgoni, M., Nobili, L., Ferrara, M., and Gennaro, L. D. (2013). How we fall asleep: Regional and temporal differences in electroencephalographic synchronization at sleep onset. *Sleep Medicine*, 14(11):1112–1122.
18. Mayers, A. and Baldwin, D. (2006). The relationship between sleep disturbance and depression. *International Journal of Psychiatry in Clinical Practice*, 10(1):2–16.
19. Staner, L., Cornette, F., Maurice, D., Viardot, G., Bon, O. L., Haba, J., Staner, C., Luthringer, R., Muzet, A., and Macher, J.-P. (2003). Sleep microstructure around sleep onset differentiates major depressive insomnia from primary insomnia. *Journal of Sleep Research*, pages 319–330.
20. Iverson, G., Gaetz, M., Rzempoluck, E., McLean, P., Linden, W., and Remick, R. (2005). A new potential marker for abnormal cardiac physiology in depression. *Journal of Behavioral Medicine*, 28(6):507–511.
21. Jurysta, F., Kempenaers, C., Lancini, J., Lanquart, J.-P., Van De Borne, P., and Linkowski, P. (2010). Altered interaction between cardiac vagal influence and delta sleep EEG suggests an altered neuroplasticity in patients suffering from major depressive disorder. *Acta Psychiatrica Scandinavica*, 121(3):236–239.

22. Udupa, K., Sathyaprabha, T., Thirthalli, J., Kishore, K., Lavekar, G., Raju, T., and Gangadhar, B. (2007). Alteration of cardiac autonomic functions in patients with major depression: A study using heart rate variability measures. *Journal of Affective Disorders*, 100(1–3):137–141.
23. APA (2000). *Diagnostic and Statistical Manual of Mental Disorders: DSM-IV-TR*. Diagnostic and Statistical Manual of Mental Disorders: DSM-IV-TR. American Psychiatric Association.
24. Morin, C., LeBlanc, M., Daley, M., Gregoire, J., and Mrette, C. (2006). Epidemiology of insomnia: Prevalence, self-help treatments, consultations, and determinants of help-seeking behaviors. *Sleep Medicine*, 7(2):123–130.
25. Petrovsky, N., Ettinger, U., Hill, A., Frenzel, L., Meyhoefer, I., Wagner, M., Backhaus, J., and Kumari, V. (2014). Sleep deprivation disrupts prepulse inhibition and induces psychosis-like symptoms in healthy humans. *The Journal of Neuroscience*, 34(27):9134–9140.
26. Cervena, K., Espa, F., Perogamvros, L., Perrig, S., Merica, H., and Ibanez, V. (2013). Spectral analysis of the sleep onset period in primary insomnia. *Clinical Neurophysiology*, 125(5):979–987.
27. World Health Organization (1993). *The ICD-10 Classification of Mental and Behavioural Disorders: Diagnostic Criteria for Research*. ICD-10 classification of mental and behavioural disorders/World Health Organization.
28. Shearer, C. (2000). The CRISP-DM model: The new blueprint for data mining. *Journal of Data Warehousing*, 5:13–22.
29. Galka, A., Wong, K. F. K., Ozaki, T., Muhle, H., Stephani, U., and Siniatchkin, M. (2011). Decomposition of neurological multivariate time series by state space modelling. *Bulletin of Mathematical Biology*, 73:285–324.
30. Karau, H., Konwinski, A., Wendell, P., and Zaharia, M. (2015). *Learning Spark: Lightning-fast Data Analysis*. O'Reilly Media, Inc., 1st edition.
31. O'Driscoll, A., Daugelaite, J., and Sleator, R. D. (2013). Big data, Hadoop and cloud computing in genomics. *Journal of Biomedical Informatics*, 46:774–781.
32. Provost, F. and Fawcett, T. (2013). *Data Science for Business*. O'Reilly Media, Inc., 1st edition.

22
Automated Processing of Big Data in Sleep Medicine

Sara Mariani, Shaun M. Purcell, and Susan Redline

CONTENTS
- 22.1 Introduction 444
  - 22.1.1 Sleep Medicine and Sleep Data 444
  - 22.1.2 Research Goals 445
  - 22.1.3 The Big Data Approach 445
  - 22.1.4 Analytical Considerations for Sleep Signal Data: Manual versus Automatic Methods 446
  - 22.1.5 Analysis Approaches 446
  - 22.1.6 The National Sleep Research Resource: An Exemplar for Sleep Big Data 447
    - 22.1.6.1 Publicly Available Sleep Data Sets 447
    - 22.1.6.2 Tools 447
- 22.2 Automated Analysis of the Sleep Electroencephalogram 448
  - 22.2.1 Automated Methods for Macroanalysis of the EEG 448
    - 22.2.1.1 Spectral Analysis 448
    - 22.2.1.2 Time-Frequency and Time-Scale Approaches 449
    - 22.2.1.3 EEG-Based Automatic Sleep Staging 449
  - 22.2.2 Automated Methods for the Detection of Transient Events 450
    - 22.2.2.1 K Complexes 450
    - 22.2.2.2 Spindles 450
    - 22.2.2.3 Arousals 450
- 22.3 Heart Rate Variability Analysis 451
  - 22.3.1 Analysis in the Time Domain 452
  - 22.3.2 Analysis in the Frequency Domain 452
  - 22.3.3 Nonlinear Methods 453
    - 22.3.3.1 Detrended Fluctuation Analysis 453
    - 22.3.3.2 Sample Entropy and Multiscale Entropy 453
    - 22.3.3.3 Lempel–Ziv Complexity 454
- 22.4 Cardiorespiratory Measures 454
  - 22.4.1 Automated Methods for the Detection of Respiratory Events 454
    - 22.4.1.1 Automatic Apnea Detection from Airflow Signals 454
    - 22.4.1.2 Automatic Apnea Detection from PPG Analysis 455
    - 22.4.1.3 Snoring Sounds 455
  - 22.4.2 Cardiopulmonary Coupling 455
- 22.5 Challenges and Needs 456
  - 22.5.1 Noise, Artifacts, and Missing Data 456
  - 22.5.2 Data Harmonization 457
- 22.6 Emerging Applications of Sleep and Big Data 458
  - 22.6.1 Genomics and Sleep 458

22.6.2 Precision Medicine and Other Applications 458
22.7 Conclusions 459
References 459

22.1 Introduction

The aim of this chapter is to introduce the reader to the characteristics and needs of sleep medicine, and the ways in which a big data approach can be beneficial in this field. We start with an overview of exemplar research questions and goals relevant to sleep medicine, and specifically focus on analysis of biomedical signals, a source for "big data analytics." In the second section, we focus on methods for automating the analysis of the electroencephalogram (EEG), the electrocardiogram (ECG), and respiratory signals given the relevance of these signals to a wide range of research, including studies of cognition and alertness [1–3], cardiovascular disease [4–6], and sleep disorders [7–10]. The last section highlights the multiple challenges of automated analysis of polysomnography (PSG) signals and proposes strategies to overcome them. Throughout the chapter, we will use the National Sleep Research Resource (NSRR, http://sleepdata.org), a National Institutes of Health (NIH)–funded sleep signal repository (R34-HL114473), as an example of application of big data methods in the field, reporting instances of both goals and methods that can be generalized to future applications.

22.1.1 Sleep Medicine and Sleep Data

Sleep is a naturally occurring, periodic decreased state of consciousness, characterized by distinct changes in brain wave frequency and amplitude, often accompanied by a decrease in heart rate and blood pressure. Sleep is a multiorgan phenomenon, which can be evaluated from a number of possible perspectives, as schematically shown in Figure 22.1. Sleep-related disorders have a major effect on quality of life and can impact daily performance and cognition, including memory, learning, concentration, and productivity. Healthy sleep is also critical for cardiometabolic health. Sleep medicine defines the diagnosis and treatment of sleep-related disorders, such as sleep-disordered breathing (SDB), narcolepsy, nocturnal frontal lobe epilepsy, periodic limb movements, rapid eye movement (REM) behavior disorder, and others, while also encompassing the research on the

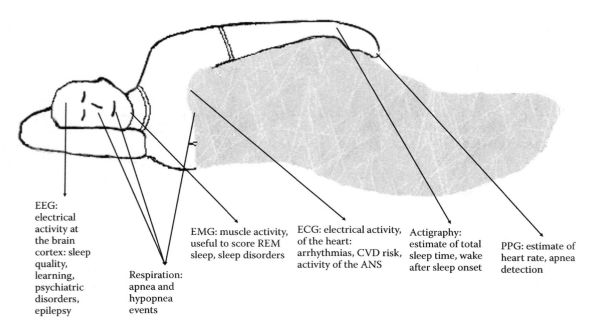

FIGURE 22.1
Sleep is a multiorgan phenomenon. A number of different signals can be monitored and evaluated to assess sleep quality and diagnose sleep-related disorders. (EMG: electromyogram, REM: rapid eye movement, CVD: cardiovascular disease, ANS: autonomic nervous system, PPG: photoplethysmogram.)

relationship of sleep-related traits with other pathologies, such as cardiovascular disease and psychiatric disorders.

The field of sleep medicine is a rich and diverse one, bringing together a variety of medical specialties and scientific disciplines, including neuroscience, pulmonology, psychiatry, epidemiology, genetics, public health, and others. The fundamental means of characterizing sleep and sleep disorders is through PSG, the simultaneous collection of multiple types of physiological data during sleep. Typically, data are collected that characterize brain activity, cardiac function, breathing and oxygen levels, and body and limb movements. PSG has traditionally been performed in specialized facilities (called "sleep laboratories"), where individuals undergo monitoring with EEG, electrooculography (EOG; measuring eye movements), chin electromyogram (EMG), ECG, respiration, and limb movement sensors. Other devices and technologies are also commonly used, such as home sleep apnea testing devices (HSATs), and actigraphy. HSATs are used for diagnosing sleep apnea. Typical HSAT signals are nasal airflow, oral airflow, respiratory effort, and oximetry. Actigraphy, typically recorded using a wrist accelerometer, is used to provide a noninvasive approach for estimating sleep time, latency, and quality over multiple days and nights of monitoring. Other data sources relevant to the practice of sleep medicine include those on patient-reported outcomes from questionnaires, adherence records from treatment devices, clinical information obtained by direct measurement or extracted from electronic medical records, output from "wearables," and genomic and biomarker data [11].

22.1.2 Research Goals

By optimally harnessing the large amounts of data typically generated during sleep assessments, there is the potential to generate high-quality evidence that can improve diagnosis, classification, and monitoring of sleep-related disorders. The large variety and volume of sleep-related data also provide opportunities to better understand pathophysiological processes and clinical manifestations of sleep disorders, helping to classify patients with specific disease profiles. Especially when used in combination with other physiological, clinical, genomic, and molecular data, quantitative sleep data can help identify phenotypes that reflect specific etiological processes and predict outcomes and likelihood of treatment responses of specific individuals or subgroups. Such data can inform the application of "precision medicine"—helping to identify which patients can benefit from alternative treatments for sleep disorders or to provide predictive information on prognosis, such as which patients are at increased risk of stroke, heart failure, or sleepiness. The ultimate goal of precision medicine is to empower clinicians with tools for improved decision making for individual patients.

22.1.3 The Big Data Approach

Strategies for the effective use of the large volumes and variety of sleep data are under development and include methods for improving data collection, integration, and analysis.

The majority of current sleep clinical studies tend to have two substantial weaknesses. First, they are often small in size, limiting statistical power (especially for "hard clinical outcomes" such as stroke or death), increasing the likelihood of erroneous inferences, and limiting generalizability (including difficulties in replicating results). Second, they lack diversity of patient characteristics, failing to encompass the necessary sex, race, and age diversity to fully understand variation in health and disease across the population, as well as limiting understanding of responses to interventions within population subgroups, potentially leading to inappropriate treatment recommendations in underrepresented groups [11,12].

As described by Budhiraja and colleagues [11], a big data approach "allows exploration of the effects of individual differences and complex interactions with the goal of facilitating identification and management of individuals according to their unique characteristics. Big data can be defined by three Vs: volume—the quantity of data, which is proportional to the statistical power for analyses; velocity—the rate at which the data are generated and analyzed; and variety—different sources, types, and formats of data." Big data must be combined with proper data analytics tools to identify new disease patterns or with predictive algorithms to identify subgroup-specific outcomes.

One approach for providing access to large and varied sources of data is to develop a platform and strategy for aggregating data from multiple studies or cohorts, supported by appropriate query and analysis tools. Integration of data originally obtained from different studies can generate large samples of individuals with wide variation in demographic and disease characteristics, allowing for generalizability and subgroup analyses. Large volumes of data also can enhance statistical power. Data variety can be achieved by integrating multidimensional data that encompass physiological signals, genetic data, sociodemographic variables, and other data types; applying multivariate analyses to optimize the informational content; and identifying temporal and other patterns of associations.

To maximize value to researchers, data should be stored on widely accessible—preferably open-access—web portals and be retrievable through user-friendly interfaces. Integrated query and search tools should be put in place to assist with report generation, data visualization, exploration, and query management within and across data sets and across data types, and allow retrieval of raw physiologic and summary statistics. In addition, secure and automated data pathways should be set up to merge data from different sources (PSG, portable devices, wearables, etc.).

Despite the many benefits of using aggregated data, there are concerns and necessary precautions. Data must be sufficiently harmonized to minimize the impact of misclassification related to the data source or collection modality. Ideally, sleep data should be collected with similar sensors and data collection procedures and processed and scored using standardized approaches. If that is not possible, metadata (defining the characteristics of the data themselves) should be sufficiently detailed to allow a thorough understanding of key aspects of how the data were collected, processed, and analyzed [11]. Provenance (origin) of data would specify the sensors used for PSG, scoring criteria, and data interpretation.

The use of controlled and structured vocabularies is an important component of large data sets, allowing the user to understand data elements and identify which terms are comparable across data sets and which terms may need to be redefined or calibrated for use in cross-study analyses. Defining the relationships among data elements, and their labels ("terms"), can facilitate this process. Highly structured systems for displaying the relationships among terms are called ontologies. Section 22.5.2 will focus on data harmonization and sleep domain ontologies.

22.1.4 Analytical Considerations for Sleep Signal Data: Manual versus Automatic Methods

Traditionally, sleep analysis has been performed mostly manually, with sleep experts visually inspecting multiple signals one study at a time, with individual scoring of respiratory events, sleep stages, arousals, and other phenomena. Recognition and diagnosis of clinical conditions often apply rule-of-thumb-type criteria, qualitative assessments, and a trained eye. Automated processes for sleep analyses (stage and event identification) have been incorporated into commercial software packages but typically call for editing by a trained scorer. Neither approach allows for the efficient analysis of large samples. The reliability of visual analysis is also influenced by the intrinsic subjectivity of manual scoring. Interscorer agreement in sleep staging has been estimated to vary between 0.6 and 0.9 in terms of Cohen's kappa (κ) [13–15]; agreement in visual "K complex" (a characteristic sleep EEG waveform) detection is only approximately 50% [16]; for sleep spindle scoring, the $\kappa = 0.52$ [17], and for arousals, is $\kappa = 0.47$ [18].

However, reliability varies according to scorer expertise and quality control procedures utilized within the given laboratory. Traditional approaches also typically only extract a limited number of features from the data, such as the number of apnea events or the amount of time spent in each sleep stage, and do not fully take advantage of the potential of quantitative data collected over time, including measures that capture the dynamic changes across the night and interrelationships among physiological signals.

Efficient multicohort, multivariate analysis of sleep studies can potentially be achieved by the use of automated analysis of the continuously collected data with extraction of informative quantitative features. Features can be extracted from a single signal, e.g., EEG power, or from interrelationships across multiple channels (characterizing "cross-talk").

A quantitative, data-driven approach to sleep data analysis can provide objectively derived information that should be able to be replicated over time and across studies. However, automated methods must be sufficiently accurate to identify clinically important features, while at the same time guaranteeing robustness to noise and artifacts, as described later in this chapter. Large data sets can provide useful training sets for developing and validating automatic methods.

22.1.5 Analysis Approaches

PSG signals can be analyzed in a variety of ways, including in the time domain and the frequency domain, and also through indices of nonlinear dynamics. These signals also present the opportunity for application of state-of-the-art and emerging computational methods to investigate the cross-talk (coupling) of these physiologic variables, which, in turn contains information about physiologic states in health and disease. For example, the coupling among respiration, EEG, and heart rate variability (HRV) varies with sleep stage and with sleep apnea. Central and obstructive apneas have distinct dynamical EEG–ECG "signatures" [10].

The automatic analysis of sleep records is often designed to detect events of interest, such as sleep stages, arousals, and respiratory events, without involvement of the human scorer. The algorithms to achieve this typically rely on three general steps: (i) the preprocessing of the signal(s), which includes noise rejection, extraction of derived time series, and segmentation (e.g., epochs); (ii) the extraction of features to characterize each segment, through analysis in the time or frequency domain; and (iii) the classification of events based on these descriptors. The classification step can be performed either in a *deductive* fashion (based on predefined rules) or following the method of *inductive* learning, or machine learning, where the algorithm "discovers" the characteristics of the

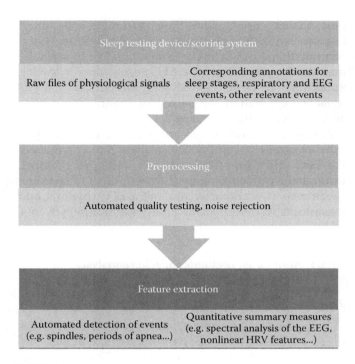

FIGURE 22.2
Schematic representation of a pipeline for the automatic analysis of sleep recordings. Files are generated from the acquisition system and sleep scoring software; signal files are linked with their respective annotation files; and, after a preprocessing step for the elimination of artifacts and interferences, features can be extracted by both detection of transient events and computation of summary measures to characterize the signals throughout the night.

various classes from the data. Inductive learning can be either supervised (e.g., when examples for each class are provided to "train" the algorithm) or unsupervised, as in clustering, when unlabeled examples are provided to the algorithm, which learns to group them based on intraclass and interclass differences in their features.

Automated PSG analyses ideally would extract meaningful features from the data in a computationally efficient fashion, and do so in a manner that is robust to artifacts that can arise from unsupervised analysis. Data extraction, integration, processing, and analysis steps can be organized in an ordered "pipeline," as illustrated in Figure 22.2.

22.1.6 The National Sleep Research Resource: An Exemplar for Sleep Big Data

22.1.6.1 Publicly Available Sleep Data Sets

An example of a multistudy resource for the analysis of sleep big data is the NSRR (http://sleepdata.org), an NIH-funded online repository to facilitate research and training in sleep medicine, bioengineering, epidemiology, and other disciplines [19]. Established through a National Health Lung and Blood Institute resource grant, the NSRR supports the goals of the broader Big Data to Knowledge (BD2K) initiative of the NIH by providing researchers and trainees with access to deidentified, annotated physiologic signals from curated sleep studies linked to data aggregated from large cohort studies and clinical trials [11]. In addition, a suite of web-based tools is available on the NSRR to explore data within and across data sets, identify subsets of data using clearly mapped terms, and facilitate data analysis and discovery of associations among physiological systems and clinical outcomes. The NSRR encourages interactions across the sleep community, and partners with other resources, such as PhysioNet [20], BioLINCC [21], and dbGAP [22].

Data sets that have been made available through this resource are derived from major sleep research studies, including the Sleep Heart Health Study (SHHS), the Cleveland Family Study, the Heart Biomarker Evaluation in Apnea Treatment, the Study of Osteoporotic Fractures, the Outcomes of Sleep Disorders in Older Men Study (MrOS Sleep Study), the Cleveland Children's Sleep and Health Study, the Honolulu Asian American Aging Sleep Study, the Childhood Adenotonsillectomy Trial, and the Multi-Ethnic Study of Atherosclerosis-Sleep Study. As of 2017, over 31,000 studies were deposited, with a goal of 50,000 by 2019. Other available data include demographic information, anthropometric parameters, blood pressure, lung function, medical history elements, and neurocognitive testing results [11]. A cross–data set query interface was developed to facilitate the collection, management, and sharing of data. A canonical data dictionary that maps study-specific data variables to a standardized set of variable definitions is used to standardize terminology across studies.

22.1.6.2 Tools

Signal processing tools play critical roles in the analysis of sleep data. In addition to providing access to raw and processed data, the NSRR encompasses a variety of open-source computational tools to analyze PSG signals. These include tools for opening and viewing PSG signals in European Data Format (EDF) and corresponding annotations; deidentifying and standardizing signals contained within EDF files; translating sleep annotations generated in different vendor formats to sleep domain standardized terminology [19,23]; translating annotations from XML to CSV files and listing sleep staging and scoring events; visualizing hidden patterns of physiological signals [24,25]; and detecting noise using entropy-based algorithms [26]. Signal processing methods can be combined and employed as "building blocks" to create an efficient pipeline for automated signal analysis. For example, NSRR tools support an EEG analysis pipeline that includes an artifact detection step, an ECG artifact decontamination step, and power spectrum and coherence computation.

22.2 Automated Analysis of the Sleep Electroencephalogram

The EEG is the primary signal used for sleep stage analysis. It is typical to record the EEG at sampling frequencies in the 128–512 Hz range, which allows the inspection of components at frequencies up to 64–256 Hz, respectively. The key EEG rhythms that are relevant to sleep analysis are delta (0.5–4 Hz), theta (4–8 Hz), alpha (8–13 Hz), sigma (11–16 Hz), and beta (16–30 Hz). These "macroscopic" rhythms are accompanied by transient events that constitute the sleep "microstructure," such as arousals, K complexes, and spindles. Arousals are transient shifts to higher-frequency and lower-amplitude waves, lasting from less than a second to several seconds. K complexes are biphasic events composed of a well-delineated negative sharp wave immediately followed by a positive component standing out from the background EEG, with a total duration ≥0.5 s, usually maximal in amplitude when recorded using frontal derivations [27]. Sleep spindles are brief distinct bursts of activity in the sigma frequency range, having a characteristic waxing and waning shape [27]. Each of these events represents distinct neurophysiological processes, with potential clinical relevance.

Sleep stages are typically assigned to each 30 s epoch of the recording period based on visual inspection of the EEG signals as well as eye movements (EOG) and chin muscle activity (EMG). Based on their frequency and amplitude characteristics, as well as the presence of spindles and K complexes, stages are classified broadly into periods of REM sleep or non-REM (NREM) sleep, and are further classified into stages N1, N2, and N3, as recommended by the 2007 Visual Scoring Task Force of the American Association for Sleep Medicine (AASM) [27]. Stage N1 is identified as slowing of the wake alpha rhythm, usually preceded by slow rolling eye movements; stage N2 shows sleep spindles and K complexes; stage N3, or slow-wave sleep, is characterized by >20% of the epoch duration consisting of delta waves that exceed a peak-to-peak amplitude of 75 μV.

While traditional sleep staging has a strong historical basis and literature to guide interpretation of normative values, it has limitations. The assignment of a single stage to arbitrarily defined epochs creates artificial discontinuities in the data, can obscure within-epoch variation in signal characteristics, and does not readily capture the dynamic properties of changes in EEG over time. Visual scoring also is time consuming and subject to scorer variability and subjectivity. Modern tools and computational power make it now possible to scale the quantitative analysis of the EEG to big data, allowing application of objective and high-throughput analysis tools and generation of a variety of quantitative metrics across different time domains.

A large-scale, automated analysis of sleep data typically involves two components: a quantitative analysis of the EEG (qEEG) on a macroscopic basis, for example, through the extraction of epoch-by-epoch, or sleep stage-specific, spectral characteristics of the EEG rhythms, and the automated detection and analysis of transient events.

22.2.1 Automated Methods for Macroanalysis of the EEG

22.2.1.1 Spectral Analysis

The application of spectral analysis to the EEG was proposed early in the study of this signal [28] and continues to be used commonly in research. The information content of the signal, originally in the time domain, is represented in the frequency domain, that is, its power spectral density or power spectrum is calculated. The approaches to power spectrum computation can be grouped into two major categories: nonparametric and parametric. Nonparametric approaches include Fourier spectral analysis, which decomposes the signal into a sum of sinusoids with different frequency and phase characteristics. The Fourier transform of a signal $x(t)$ is $X(f) = \int_{-\infty}^{\infty} x(t)e^{-j2\pi ft}dt$, translating the signal from the domain of time $x(t)$ to that of frequency $X(f)$. Parametric approaches operate under the assumption that the signal is the output of a given process [29], where the input is white noise and/or the output itself at previous instants, and the output is the transformation of the input based on parameters, namely, the model coefficients, which must be estimated. The parametric approach involves choosing a model to represent the data under observation; estimating its coefficients based on the available data; and estimating the power spectrum through substitution of the model coefficients into the model equations [30]. Autoregressive (AR) models are often used due to their ability to describe a wide range of signals [31].

Among its numerous applications, EEG spectral analysis is used to identify markers that associate with psychiatric disease, characterize risk of cognitive or mood disorders, and identify drug responses, and for genetic association testing [32].

Graphical interfaces can assist with generation, editing, and interpretation of EEG spectral analysis data. A visualization tool and analytical pipeline for spectral analysis, named SpectralTrainFig.m, can be found in the NSRR. It is implemented in MATLAB and also available in open-source format on MathWorks (http://www.mathworks.com/matlabcentral/fileexchange/49852-spectraltrainfig) and GitHub (https://github.com/nsrr/SpectralTrainFig). A graphical interface allows the user to input EDF files and annotations in XML format and returns the power spectral density of each EEG signal, computed through Fourier analysis, both on an epoch-by-epoch basis and summarized by sleep state (NREM, REM, total sleep time). Coherence analysis between EEG leads can also be performed. The program includes an artifact detection step, based on delta and beta band descriptors [33], and a template-subtraction algorithm for decontaminating the EEG signals from the ECG artifact [34]. Results are summarized in Excel spreadsheets and a Power Point file containing a series of figures that allow visual review of the results for each sleep study.

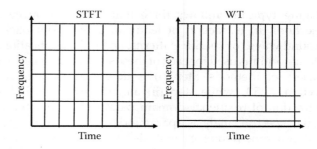

FIGURE 22.3

The time-frequency plane for the STFT (left) and WT (right). Every (t,f) point is associated with a certain resolution Δt on the horizontal axis and Δf on the vertical axis. Though the area $\Delta t \Delta f$ of each cell is constant in both cases, STFT has constant resolution in both frequency and time throughout the whole plane, while for the WT, low frequencies are sampled with large time steps, and high frequencies are sampled with small time steps. Adapted from Oropesa E, Cycon HL, Jobert M, *Sleep Stage Classification Using Wavelet Transform and Neural Network*, International Computer Science Institute, 1999.

22.2.1.2 Time-Frequency and Time-Scale Approaches

The NSRR pipeline computes power on 30 s long epochs, addressing the nonstationary nature of the signal, that is, the variability of its frequency components and amplitude throughout the night. However, some features could vary on even smaller time windows, or shorter events may be of interest, requiring application of methods that further localize the frequency content of the signal in time.

The short-time Fourier transform (STFT) is an alternative to the Fourier transform that allows for temporal dependence, by applying the Fourier transform to only a portion of the signal contained in a moving window over time. The STFT of a signal $x(\tau)$ is defined as

$$X(t,f) = \int_{-\infty}^{\infty} x(\tau) w^*(\tau - t) e^{-j2\pi f \tau} d\tau,$$

where w represents the moving window, centered around τ, and (t, f) are the time and frequency under observation. In the discrete time case, the data are segmented, and each segment is Fourier-transformed, obtaining the STFT for each point in time and frequency. The frequency resolution, that is, the smallest distance between two different components that can be discriminated in the frequency-domain representation, is related to the time resolution, or the number of samples in the observation window: wider time windows provide more refined frequency resolution than shorter windows [31,35]. For the STFT, these values are constant throughout the analysis of the whole signal, as illustrated in the left panel of Figure 22.3.

A variant of STFT is wavelet analysis, a method introduced to address the need to analyze slower frequency components of a signal with longer observation windows than faster frequency components, which can be reliably analyzed using shorter time windows.

The wavelet transform (WT) is defined as $X(t,a) = \int_{-\infty}^{\infty} x(\tau) h_{t,a}^*(\tau) d\tau$, where $h(\tau)$ is referred to as the mother wavelet and $h_{t,a}(\tau) = \frac{1}{\sqrt{|a|}} h\left(\frac{\tau - t}{a}\right)$ are stretched and compressed versions of the mother wavelet. The time-frequency resolution depends on the scaling parameter a: for smaller a, $h(\tau)$ has a narrow time window and therefore allows us to see fast frequencies. When a increases, the time window applied to $h(\tau)$ increases, allowing for the observation of slower frequencies [31,35].

The distribution of the resolution is schematically represented in the right panel of Figure 22.3.

An open-source MATLAB toolbox for time-frequency signal analysis can be found at http://tfd.sourceforge.net/ [36].

22.2.1.3 EEG-Based Automatic Sleep Staging

Automatic scoring of sleep stages can allow large volumes of sleep studies to be analyzed efficiently and with objective processes than minimize interscorer and intrascorer differences. Automated methods also may allow for application of techniques that analyze signal data continuously, without forcing classification of stages into discrete fixed 30 s epochs.

Automated sleep analysis is currently implemented, with various degrees of success, in a range of commercial sleep analysis software, and a number of approaches have been presented in the literature. EEG-based automated staging typically follows three main steps: (i) artifact rejection; (ii) extraction of features, either from background waves or specific patterns; and (iii) classification. The algorithms used for each of these steps can consist of a variety of methods. Step (i) is described in detail in Section 22.5.1, while step (ii) almost always includes extraction of frequency-domain parameters from traditional spectral analysis (parametric or nonparametric) and from time-frequency analysis. Time-domain features, such as entropy measures, have also been used [37]. Features can be extracted on a 30 s epoch basis, matching the visual scoring scheme. A training set of features and matching labels for each epoch can be employed for training the algorithm, so that it can be employed on an independent test set and validated. Alternatively, the classification can be performed on epochs of arbitrary length, obtaining a profile of sleep depth, which can subsequently be mapped to the standard sleep stages. A variety of classifiers have been used in the literature, including neural networks, support vector machines, K-means clustering, fuzzy logic, and hidden Markov models [38].

Studies have analyzed the performance of some commercially available methods. For instance, a neural-network-based algorithm named BioSleep (Oxford Instruments, UK) was shown to have a per-epoch agreement with expert annotation of $\kappa = 0.47$ [39] in a cohort of 114 individuals with suspected obstructive sleep apnea. Another algorithm, named ASEEGA (Physip, France) employs spectral features and uses an adaptive fuzzy logic iterative system to repeatedly update the sleep stage pattern definitions. It was shown to have $\kappa = 0.72$ for epoch-by-epoch agreement with visual scoring in a group of 15 healthy individuals [40].

Developing robust automated sleep staging procedures is influenced by the size of the training set and its capability to cover enough cases to be representative of the populations under study; the differential performance of the algorithms in healthy and pathological sleep; and the accuracy and reliability of the visual scoring used for algorithm development [38].

22.2.2 Automated Methods for the Detection of Transient Events

22.2.2.1 K Complexes

K complexes are thought to be associated with arousal during sleep, can occur both spontaneously and as an evoked response caused by auditory or somatosensory stimuli, and can appear concomitantly with apneic events in patients with sleep apnea. K complexes are a feature used to identify stage N2 of sleep.

One approach to the automatic identification of these waveforms is a method that employs matched filters. A template of the shape to be recognized is created, either following a mathematical description of the waveform or by averaging a number of synchronized occurrences of the event. Then, the template is correlated with the signal, or, equivalently, the signal is convolved with a mirrored, conjugate version of the template, and the time instants of occurrence of peaks in the convolution results will represent the locations of the waveform in the signal [41–43]. This approach, however, is limited by the highly stochastic nature of the EEG, resulting in variation in shape of the K complexes. Alternative approaches are based either on mathematical models of the generation of EEG phenomena, as the one introduced by Da Rosa and colleagues [44], or on the extraction of characteristic features of the K complex waveform, followed by classification through artificial neural networks, as in the work by Bankman and colleagues [45].

22.2.2.2 Spindles

Sleep spindles are discrete events observed in the scalp EEG signal that are generated as a result of interactions between several regions of the brain including thalamic and corticothalamic networks [46]. They appear as brief, powerful bursts of synchronous neuronal firing in thalamocortical networks and are a defining feature of stage N2 sleep. Sleep spindles are involved in new learning and are considered an index of brain plasticity and function, including intellectual abilities and memory consolidation during sleep [47].

Automated spindle detection has been performed on central (C3, C4, Cz), frontoparietal (Fp1, Fp2), parietal (P3, P4, Pz), and occipital (O1, O2) leads. Spindles from frontal derivations have been shown to be particularly relevant to memory consolidation [48].

The simplest approach to automated spindle detection is based on a series of basic steps summarized in Table 22.1.

One open-source toolbox for automatic spindle detection is FASST, fMRI Artefact rejection and Sleep Scoring Toolbox, developed by the University of Liege, Belgium [61]. The spindles module in FASST uses the Fz, Cz, and Pz leads. The toolbox was developed in MATLAB and is available at http://www.montefiore.ulg.ac.be/~phillips/FASST.html. Warby and colleagues [62] have reported a comprehensive comparison and evaluation of the performances of six different automated spindle detectors, including the method used by FASST [49].

22.2.2.3 Arousals

Arousals are defined as abrupt changes in frequency content of the EEG lasting at least 3 s and preceded by 10 s of sleep. The change in frequency can include theta, alpha, and beta rhythms.

Since arousals typically correspond temporally to other physiological events (obstructive sleep apnea,

TABLE 22.1

Prototypical Steps for Automated Spindle Detection

Step	Methods
Filtering of the EEG signal in the sigma band	• Finite-impulse response filter [49] with band pass (11–16 Hz) [50,51] • Separate analyses for slow (8–12 Hz) and fast (12–15 Hz) spindles [47] • Pass-band determined for each case based on peak spindle frequency from the power spectrum [52] • Wavelet decomposition [53,54]
Computation of signal amplitude	• Root mean square of the signal over a moving time window (examples are 0.25 [55] or 0.1 s [49]) • Envelope of the signal [3]
Superimposition of a threshold for spindle recognition	1. Amplitude threshold: can either be fixed [50,51] or depend on the signal properties, such as percentiles [55] or mean and standard deviation [56] 2. Minimum duration threshold (typically 0.5 s)
Detection of spindle start and end points	• First and last data point that overcomes the threshold • Methods that take into account the shape of the spindles, by using a second, lower threshold, [57] or by employing derivatives of the envelope [58]
Elimination of overlapping events	Candidate spindles that overlap with artifacts or other detected events are discarded, including alpha activity interference, muscle artifacts, saturated signals, unusual increase of EEG (e.g., EOG interferences), abrupt transitions, and movement artifacts [56].
Elimination of false detections	1. Extraction of features from the candidate spindles. 2. Training of classifiers on labeled examples for detection. Examples are fuzzy detection combined with Bayesian analysis [59], and linear discriminant analysis with Mahalanobis distance approach [60].

movements), they are normally recognized employing a combination of signals, including heart rate, EMG, and respiratory signals. Since EEG frequencies are very variable in REM sleep, a concomitant increase in EMG activity (usually chin) is also required to score a variation in EEG frequency as an arousal during REM sleep. Arousals can prove to be difficult to score reliably by manual procedures due to underlying variation in EEG frequencies.

The automatic detection of arousals is typically based on short-time frequency analysis of the EEG, potentially combined with the analysis of other signals. The suggested EEG leads for this type of analysis are a central and an occipital lead, to facilitate detection of beta and alpha activity, respectively [63]. The common steps are as follows:

1. Segmentation of the EEG into windows, that can be of fixed length (e.g., 1 s), depending on the local characteristics of the signal. Adaptive segmentation methods, like Spectral Error Measure and nonlinear energy operator [64,65], which break up the signals into quasi-stationary segments of variable length, can be applied.
2. Extraction of power spectral features from each window. Examples of features are the power in the alpha band [66], power in the extended beta band (16–40 Hz), and maximum amplitude [67]. Features extracted from multiple EEG leads can be combined by averaging over the leads.

   An alternative to points 1 and 2 is the use of time-scale methods, like the wavelet transform, to obtain the spectral power in the frequency bands of interest with variable time-frequency resolution [68].

3. Possible integration with information from other channels:

   a. Flow channel: local flow maxima indicating respiratory restoration [66]
   b. EMG: normalized power with respect to the background activity, after filtering out low-frequency components [68]
   c. Heart rate: local increases

4. A detection step that can be either rule-based (through the application of threshold criteria) [67] or based on machine-learning techniques, such as neural networks trained on features from scored arousals [69].

22.3 Heart Rate Variability Analysis

HRV analysis is the study of the fluctuations of heart rate during a certain time frame. HRV reveals information about the activity of the autonomous nervous system (ANS), which can help characterize phenomena occurring with stressors associated with a wide range of disturbances, such as sleep apnea [70,71], periodic limb movements [72], REM behavior disorder [73], narcolepsy [74], and insomnia [75].

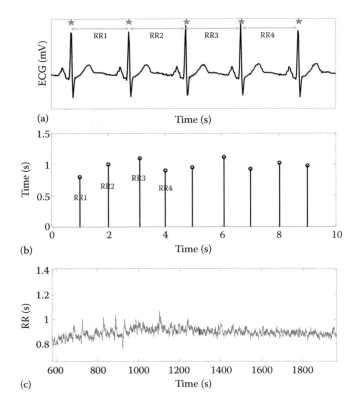

FIGURE 22.4
Extraction of RR time series from the ECG. (a) R peak detection (b) Interbeat interval (RR) time series (c) RR time series: a few minutes of trace.

The time series of the interbeat cardiac intervals can be derived from the ECG. The most common approach is the method of Pan and Tompkins [76], which employs a series of filters to recognize the R waves on the ECG. The derived RR time series can then be filtered for outliers (beats that are nonphysiological) to obtain the NN (normal sinus–to–normal sinus beat) time series. An example is shown in Figure 22.4. The approach to analysis of RR/NN time series can be summarized into linear techniques in the time domain, techniques in the frequency domain, and nonlinear methods.

22.3.1 Analysis in the Time Domain

The Task Force of the European Society of Cardiology and the North American Society for Pacing and Electrophysiology [77] has identified the following core HRV time measures:

- Mean NN interval
- Standard deviation of the NN intervals (SDNN), over the entire recording, which reflects the contribution of all the cyclic components responsible for variability
- Standard deviation of the average NN intervals (SDANN) calculated over short periods, usually 5 minutes, which is an estimate of the longer-term dynamics
- Mean of the 5-minute standard deviations of NN intervals calculated over the entire recording (SDNN index), which is an estimate of the shorter-term dynamics
- Square root of the mean squared differences of successive NN intervals (RMSSD)
- Number of interval differences of successive NN intervals greater than 50 ms (NN50)
- Proportion of NN50 divided by the total number of intervals (pNN50)

The SDNN index, RMSSD, NN50, and pNN50 are all measures of high-frequency (short-term) HRV dynamics and are highly correlated with one another. A generalization of pNN50 was introduced by Mietus and colleagues [78], who defined pNNx as the percentage of NN intervals over the whole recording that differ by more than x ms, where $x \in [4,100]$. The case where $x = 20$, pNN20, was found to have the greatest discriminative power between a healthy heart rate and a number of pathological conditions. Software to compute the pNNx measures is freely available on PhysioNet at https://www.physionet.org/physiotools/pNNx/.

Of note, these variables can be derived from full ECG recordings, or from segments, for example, individual sleep stages or REM/NREM states. Observation windows should include a large-enough number of 5-minute windows to allow for significant computation of standard deviations.

22.3.2 Analysis in the Frequency Domain

Heart rate changes during sleep occur on a variety of time scales, reflecting a range of processes such as brief, transient responses to respiratory events and arousals, effects of sleep stage changes, and circadian rhythms. An analysis in the frequency domain, which shows the contribution of each frequency component, can help elucidate the physiological mechanisms that affect HRV and the time scales over which they operate [79].

The power spectrum of HRV is typically calculated either through parametric methods, like AR models [80], or through Fourier analysis. However, the latter requires equally spaced samples; thus, some interpolation through polynomial or spline functions is required prior to power computation. To avoid this step, which may introduce error, a spectral estimate has been introduced, named the Lomb–Scargle periodogram [81,82], which only uses the observed samples in a time series to compute the power spectrum. A tutorial on this method, along with open-source C code for implementation, can be found on PhysioNet at https://www.physionet.org/physiotools/lomb/lomb.html.

The frequency bands often studied with respect to sleep are the following:

- Very low frequency (VLF): this spectral component is in general below 0.04 Hz (sometimes centered around 0 Hz); it reflects long-term regulatory mechanisms typically associated with circadian rhythms.
- Low frequency (LF): this component ranges between 0.04 and 0.15 Hz. LF power and typically decreases with respect to HF power with sleep depth and increases in the presence of respiratory events, such as obstructive and central apnea.
- High frequency (HF): this component, ranging between 0.15 and 0.4 Hz, is the expression of the respiratory arrhythmia of the heart, which is the result of the vagal activity, and thus is considered a marker of the parasympathetic activity controlling the heart rate [83].

The LF/HF ratio, considered an index of sympathovagal balance, decreases with deepening sleep as the parasympathetic branch of the ANS begins to dominate. Similar values for this ratio are found in wakefulness and REM sleep, decreasing in stage N2 and further decreasing in N3 sleep. Reported values are 0.5–1 in NREM sleep and 2–2.5 in REM sleep [38,84].

PhysioNet features a complete toolbox for the computation of time-domain and frequency-domain HRV parameters. The C version can be found at https://physionet.org/tutorials/hrv-toolkit/ and the MATLAB version at https://www.physionet.org/physiotools/matlab/wfdb-app-matlab/. An alternative, user-friendly interface for HRV analysis is Kubios, developed at the University of Eastern Finland and available in open source at http://kubios.uef.fi/ [85].

22.3.3 Nonlinear Methods

Nonlinear techniques aid in the quantification of long-range correlations in a time series, and, in particular, in characterizing the fractal properties of the NN time series. The term *fractal* refers to complex objects with properties of self-similarity, or scale-invariance, and this concept extends to describe the dynamics of complex systems that lack a characteristic time scale. It has been shown that scale invariance seems to be a characteristic signature of healthy systems, e.g., the healthy heartbeat, while this property is progressively lost with disease and ageing [86–88].

22.3.3.1 Detrended Fluctuation Analysis

The detrended fluctuation analysis (DFA) [89] quantifies the self-similarity of a time series over multiple scales. It is performed by the following steps:

1. The heartbeat time series is integrated.
2. The integrated time series is divided into boxes of equal length, n. In each box of length n, a least-squares line is fit to the data.
3. In each box, the integrated time series is detrended by subtracting the local trend in that box.
4. For a given box size n, the characteristic size of fluctuations F(n) for this integrated and detrended time series is calculated.

This computation is repeated over all box sizes, and the relationship between F(n) and the box size n is analyzed. A linear relationship on a log–log graph indicates the presence of scaling (self-similarity) in the time series of interest [83]. The slope of the line relating log[F(n)] to log[n] determines the scaling exponent α (self-similarity parameter) of the time series [89].

Software for DFA computation is available on PhysioNet, with related tutorial, at https://www.physionet.org/physiotools/dfa/.

22.3.3.2 Sample Entropy and Multiscale Entropy

The sample entropy (SampEn) [90] measures the regularity of a time series by matching it with a given template of length m. If the template and the time series match point-wise within a tolerance r, the SampEn is the negative natural logarithm of the probability that the two also match when the template length is extended by one unit.

An extension of the SampEn is multiscale entropy (MSE) [91], which computes the SampEn of a set of derived time series, called coarse-grained time series, each representing the system dynamics on a different time scale. Coarse-grained time series are obtained by averaging samples over nonoverlapping windows of increasing width (scale). An MSE graph plots SampEn as a function of scale (i.e., coarse graining factor), as shown in Figure 22.5. Both the relative and absolute amplitude of the plots provide insight into the multiscale structure of the data.

In the analysis of heart rate dynamics, the MSE method has been shown to have the highest values for the time series of healthy subjects and lower values for subjects with congestive heart failure (CHF) [91–93].

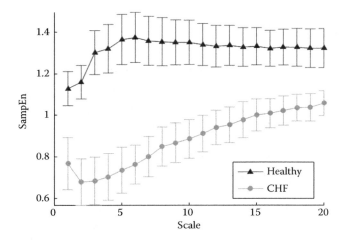

FIGURE 22.5
MSE graph, representing, for each scale, the sample entropy of the corresponding coarse-grained time series. Adapted from Costa MD, Goldberger AL, Peng C-K, Multiscale entropy analysis of complex physiologic time series, *Phys Rev Lett* 2002;89.

A tutorial and code for MSE analysis can be found on PhysioNet at https://www.physionet.org/physiotools/mse/.

22.3.3.3 Lempel–Ziv Complexity

The Lempel–Ziv (LZ) complexity is one of the several techniques used in information theory to measure algorithmic complexity, that is, the relationship between the number of operations (or the amount of time) required for the algorithm to run as a function of the length of the input series n. To compute the LZ complexity, the numeric sequence first has to be transformed into a symbolic sequence [94]; then the symbolic sequence can be parsed to obtain distinct words, and the words encoded. LZC(n) is the length of the encoded sequence from those words. The value of LZC(n) is close to n for totally random sequences [83], while it is lower for very regular time series. LZC(2) of NN time series obtained during sleep was found to be able to discriminate bipolar patients with high anxiety from healthy sleepers [95].

22.4 Cardiorespiratory Measures

22.4.1 Automated Methods for the Detection of Respiratory Events

Sleep facilitates the appearance of breathing disorders due to instability in ventilatory control, decrease in the neuromuscular tone of the upper airways, alteration in respiratory reflexes, altered carbon dioxide (CO_2) homeostasis, and decreased functional residual capacity. Respiratory events include obstructive apneas, characterized by the presence of respiratory effort while airflow completely stops; central apneas, characterized by a complete cessation of both respiratory movements and airflow for at least 10 s; and mixed apneas, a combination of the previous two. Additional events include periods of hypoventilation or airflow limitations.

Reliable automatic scoring of respiratory events during sleep could promote the generation of objectively scored events obtained on large volumes of data.

22.4.1.1 Automatic Apnea Detection from Airflow Signals

Breathing events during sleep can be detected as reductions in airflow or tidal volume. Although pneumotachography and body plethysmography are the gold standards for the assessment of these measures, less cumbersome measures are usually performed in routine PSG, typically with use of thermistors, nasal pressure sensors, and inductance abdominal and thoracic plethysmography. While thermistors and nasal pressure sensors provide relative changes in airflow, calibrated inductance plethysmography can provide quantitative assessments of changes in respiratory pattern, although calibration can be difficult to maintain across the sleep period [96].

The detection of apnea events, and their classification into central, obstructive, and mixed, has been performed following this general scheme:

1. Extraction of an amplitude measure from the raw flow signal. This can be achieved through rectification [97] (squaring or absolute value) of the signal or computation of the instantaneous power through STFT [98].

2. Apnea event detection based on thresholds on amplitude and duration. Criteria can be based on the signal being below the threshold for a required amount of time, or be linked to the speed of amplitude or power dip and its return to original value.

3. Feature extraction. Since the main difference between central and obstructive apnea lies in the absence or presence of respiratory movements, signals from inductance belts have been effectively employed for the further characterization of apnea events. One approach is the extraction of wavelet coefficients from scored apnea segments [97].

4. Automatic classification of the type of apnea. A number of classifiers can be used for this purpose, including neural networks [97],

AdaBoost, linear discriminant analysis and support vector machines [99].

Depending on the aim of the study, algorithms can focus on classifying the individual as being affected by SDB or not, rather than on detecting individual events. In this case, features can be extracted not on an event- or time-window basis but from the whole series of signals. An example is that introduced by Gutierrez-Tobal and colleagues [100], who focused on discriminating subjects with sleep apnea–hypopnea syndrome (SAHS-positive) from SASH-negative ones. Extracted features, which were derived both from the flow signal and from the derived respiratory rate variability time series, included the first four statistical moments, spectral features, and nonlinear features (LZC, central tendency measure, and SampEn). Forward stepwise logistic regression was employed for classification of each subject into one of the two categories. The best value of accuracy of this study was equal to 82%.

22.4.1.2 Automatic Apnea Detection from PPG Analysis

Photoplethysmography (PPG) is a signal from most oximeters used in HSATs as well as typically collected during PSG. The PPG waveform and its relationship with physiological systems have been studied for clinical physiological monitoring, vascular evaluation, and autonomic behavior. PPG provides a measure of the tissue blood volume, which is related to arterial vasoconstriction or vasodilation generated by the ANS and modulated by the cardiac cycle. Indeed, PPG envelope amplitude decreases as a consequence of vessel constriction generated by the activation of the sympathetic nervous system. Amplitude reduction in PPG occurs when an apnea event takes place due to sympathovagal balance changes [101]. However, not all decreases in PPG amplitude are associated with an apnea [101,102].

Some attempts at developing an automated detector of apnea events in children and adults have been made, often combining PPG with a derived measure of HRV from PPG [102]. Detection of respiratory arousals from PPG has also been performed [103]. One commercial HSAT device, Itamar's Watch-Pat™, uses PPG type data in conjunction with oxygen saturation and other data to automatically detect respiratory events.

22.4.1.3 Snoring Sounds

Respiratory sound signals that record snoring events can be used for obstructive apnea detection. The snoring phenomenon, which occurs in a majority of patients with obstructive sleep apnea, is caused by the vibration of soft tissues due to turbulent airflow through a narrow or floppy upper airway.

The complete approach to apnea detection from an audio signal of respiratory sounds can be summarized in detection of snoring episodes and classification of snoring sounds into apnea-related and non-apnea-related.

Regarding the detection of snoring episodes, one algorithm by Jané and colleagues [104] is based on segmenting the audio signal into stationary portions, with subsequent extraction of features in the time and frequency domains that are used as inputs for a neural network that classifies each segment as with or without snoring.

Identification of "pathological" snoring has used methods that take advantage of the frequency content of the signal, which differs between healthy snorers and snorers with obstructive sleep apnea. The frequency components that resonate with the upper airways chamber are named *formant frequencies*, and the signal power at those frequencies is maximized, while that at other frequencies is filtered. As the upper airways of obstructive sleep apnea (OSA) patients are typically narrower than in healthy subjects, resonating frequencies are different in the two cases [105].

Both parametric and nonparametric approaches have been used to estimate resonating frequencies. The parametric route employs linear predictive coding, which attempts to model sound samples according to an AR model, whose parameters define a filter based on the acoustic structure of the upper airways. From the equations of the model and the filter parameters, the formants can be derived [105]. The nonparametric route includes classic FFT for deriving features like spectral amplitude peaks [106], peak frequencies, soft phonation index, noise-to-harmonics ratio, power ratio [107], or wavelet analysis, which takes into account the nonstationary nature of the signal [108].

22.4.2 Cardiopulmonary Coupling

The technique of cardiopulmonary coupling (CPC), introduced by Thomas, Mietus, Peng, and Goldberger in 2005 [109–111], was devised to address the challenge of automatic detection of sleep apnea syndromes and of sleep stability solely from a continuous single-lead ECG. This spectrographic method is based on an estimation of the coupling between HRV and respiration. The method makes use of the changes in ECG waveform related to breathing. A MATLAB code for ECG-derived respiration (EDR) can be found on PhysioNet at https://physionet.org/physiotools/matlab/wfdb-app-matlab/html/edr.html. An example of the EDR output is shown in Figure 22.6.

The RR signal and the EDR signal are combined according to the following procedure [10,111]: both time series are filtered for outliers and resampled at 2 Hz. Then, the cross-power and quadratic coherence are computed on windows of 1024 samples (8.5 minutes), overlapped by 75% with one another. The product of the

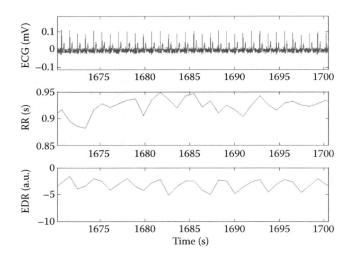

FIGURE 22.6
ECG signal with recognized R waves (top panel), interbeat interval RR time series (middle panel), and EDR signal (bottom panel) as generated by the PhysioNet function edr.m.

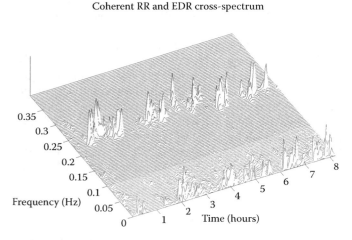

FIGURE 22.7
Example of CPC graph of one night of sleep. Peaks can be observed in the very-low-frequency range (VLF, 0–0.01 Hz), in the low-frequency range (LF, 0.01–0.1 Hz), and in the high-frequency range (HF, 0.1–0.4 Hz).

two measures is calculated, to obtain a time-frequency representation, termed a coherent cross-spectrum, as shown in Figure 22.7.

The graph shows peaks in three different frequency bands: VLF (0–0.01 Hz), LF (0.01–0.1 Hz), and HF (0.1–0.4 Hz).

The superimposition of appropriate thresholds [10] to the amplitude of these peaks allows one to automatically identify three modes during the night: very-low-frequency coupling (VLFC), indicating periods of wakefulness or REM sleep; low-frequency coupling (LFC), indicating periods of unstable NREM sleep; and high-frequency coupling, (HFC) indicating periods of stable NREM sleep [112].

A further feature of this method is its ability to score sleep apnea. Relatively pure obstructive sleep apnea results in multiple broad bands of low-frequency CPC spectral peaks, while central sleep apnea results in cyclic oscillations at a relatively fixed frequency, resulting in a narrow-band or low-frequency CPC. By superimposing appropriate thresholds [10] to the amplitude of the coherent cross-spectrum, the CPC algorithm identifies the epochs that contain obstructive apnea (elevated low-frequency coupling–broad band, or eLFC-bb) and those that contain central apnea (elevated low-frequency coupling–narrow band, or eLFC-nb).

22.5 Challenges and Needs

22.5.1 Noise, Artifacts, and Missing Data

In a completely automatic system where there is no human involvement, artifact detection is a crucial step to ensure that the derived features are informative and do not alter the results. The same is true for missing or "blank" signals, which occur in cases of sensor misplacement or interference. What would be easily identified and rejected through visual expert analysis must be similarly rejected by implementing a series of rules and criteria in the preprocessing module of any automated signal analysis. As briefly introduced in the previous sections, a number of artifact identification methods exist for different signals. Semi-deterministic signals, like the ECG, are easier to check for continuous, superimposed noise and transient artifacts than quasi-stochastic signals, like the EEG. Outlier removal in an ECG-derived interbeat interval time series can be performed comparing observed values with averages and thresholds, like in the method proposed by Mietus and colleagues and available on PhysioNet at http://physionet.incor.usp.br/tutorials/hrv-toolkit/. The detection of noise and interference in the EEG is less straightforward. Artifacts may be caused by external (e.g., electrode instability, power line interference) or internal (e.g., muscle or eye movement) factors [113]. Multiple approaches to artifact detection have been proposed, including those based on independent component analysis (ICA) [114,115], moment-based statistical methods [116], wavelet analysis [117], regression [118,119], blind source separation [120,121], averaged artifact subtraction [122], Bayesian classification [123], MSE [26], and combinations of methods [124–127]. All these methods have different strengths and limitations. For big data analysis, there is a need to consider the computational load of the algorithms, and specifically the trade-off between accuracy and efficiency.

To put in place robust criteria for artifact detection, a global, detailed knowledge of the data must be present, reflecting an understanding of the subject characteristics, setting for data collection, and instrumentation used. For example, some sensors may be more sensitive to displacement with movement or sweat, or some subjects may be more likely to have interferences between the EEG and the ECG signals. An understanding of the intrinsic data can markedly assist the user in developing procedures for minimizing residual noise. An understanding of the specific study objectives also can influence decisions regarding trade-offs in accuracy and efficiency.

An example of a completely automatic artifact detection procedure was one that was implemented in the PSYCHE project. In this project, sleep data from a wearable sensor, including the ECG, were collected at home, and then the derived RR time series were transferred to a server, where the feature extraction took place. A decision-tree system was put in place to ensure that extracted features were reliable, based on the detection of missing data and outliers in the signal. The system took into account the fact that different data lengths were necessary for the extraction of features in the time domain and in the frequency domain, and nonlinear features. The method is presented in the work of Migliorini et al. [128].

An alternative approach for automated analysis is found in the EEG processing pipeline featured on the NSRR, spectralTrainFig. This semiautomatic approach, although detailed, calls for human review and editing. The artifact detection and feature extraction module is relatively fast; however, since not all possible artifacts can be captured by the algorithm criteria, the algorithm generates a series of key figures, as that shown in Figure 22.8, which can be quickly examined by a polysomnologist, and overall noisy recordings can be detected and discarded.

22.5.2 Data Harmonization

One of the biggest challenges in the analysis of sleep data from a big data perspective is related to standardizing and understanding signals and other data elements, and the impact of data collected in different centers and with different modalities and over time. The varied sources of data relevant to sleep medicine, including PSG collected in sleep labs or at home, airflow and adherence data from continuous positive airway pressure (CPAP) machines, data from wearable devices, and questionnaires, among others, pose integration challenges.

To address differences in PSG source data, the NSRR includes applications that allow the normalization of data from different sources. For instance, EDF Editor and Translator (https://sleepdata.org/tools/edf-editor-and-translator), developed in Java, facilitates the deidentification and standardization of signals by editing EDF files and translating sleep annotations generated in different vendor formats to sleep domain standardized terminology. The NSRR Cross Cohort EDF Viewer (https://sleepdata.org/tools/edf-viewer) is an open-source MATLAB tool that enables the user to open different types of EDF files and corresponding sleep annotation files [19].

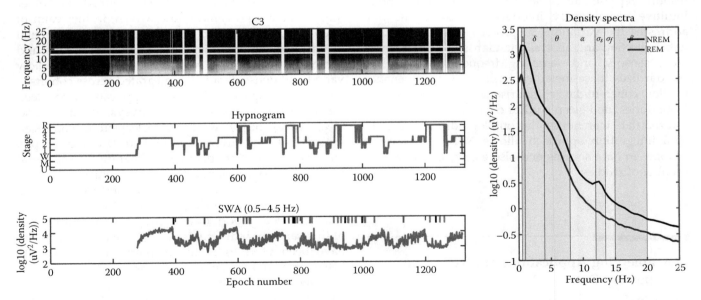

FIGURE 22.8
Example of spectral analysis adjudication panel. The figure includes a frequency vs. epoch number spectrogram (top left), a hypnogram (left middle), a slow-wave activity plot (left bottom), and signal average NREM/REM spectra (right). (From Mariani S, Tarokh L, Djonlagic I, Cade BE, Morrical M, Yaffe K, Stone KL, Loparo KA, Purcell S, Redline S, Aeschbach D. Evaluation of an automated pipeline for large scale EEG spectral analysis: The National Sleep Research Resource. *Sleep Med*, 2017 Nov 29.)

Ontologies can facilitate the use of standardized information models for harmonizing and integrating varied sleep data. Specific to sleep medicine, and as prototyped by the NSRR, a Sleep Domain Ontology (SDO) [130] contains standardized terms representing sleep disorders, medications, clinical findings, and physiological phenotypes, as well as terms representing procedures and devices used in sleep medicine, such as PSGs. SDOs can be used for data curation, federated data integration, visual query interfaces, analysis, and standardizing metadata parameters associated with sleep data collection. While the International Classification of Sleep Disorders has proposed some standards, there is not yet a nationally endorsed SDO [11]. There are several NIH-supported data warehouse systems, such as BIRN [131] and i2b2/SHRINE [132]; however, these resources do not specifically address sleep research needs including PSG data management. Adoption of SDOs by professional societies may further their use and utility across the sleep medicine community.

One of the main focuses of the NSRR is to provide an accessible, easily disseminated data integration platform and tools for sleep researchers. This includes the development of ontologies and metadata for sleep terms. The NSRR informatics infrastructure is based on two complementary and interrelated ontologies: SDO and Sleep Provenance Ontology (SPO), representing data and metadata in sleep studies. The SDO has as its aim the standardization of sleep and relevant clinical terms and the mapping of terms from the sleep studies to the SDO for orchestrating the NSRR functionalities, including resource representation, cataloging, data storage, and an intuitive visual query interface for users. The SPO addresses the variations in protocols for clinical PSG data collection and analysis standards (e.g., scoring criteria, sensors, and sampling frequencies). The SPO standardizes these metadata terms, called provenance, to allow coherent data comparisons across sleep centers, laboratories, and sleep studies over time. Unlike a centralized data warehouse, this ontology-driven federated data integration approach allows NSRR to adapt to changes in data sources and also easily scale to a large number of studies.

22.6 Emerging Applications of Sleep and Big Data

22.6.1 Genomics and Sleep

Advances in understanding the genomics of sleep and sleep disorders could shed light on fundamental mechanisms of sleep and its relationship to health. Emerging data show that many sleep traits are heritable, including sleep duration, quality and timing, and sleep apnea. Genome-wide association studies (GWAS) investigate the genetic bases of sleep phenotypes and, due to the high number of variables under exam, require very large data sets to achieve sufficient statistical power. The largest genetic analysis to date of sleep data was conducted by Lane and colleagues [133], which used self-reported chronotype data from 100,420 participants in the UK Biobank cohort. With use of fairly dense genotype data and self-reported sleep/chronotype data, these investigators identified 12 new genetic loci that implicate known components of the circadian clock machinery as well as several novel loci. The evolution of sleep big data resources may allow future studies to use more granular physiological data (such as from PSGs and actigraphy), biomarker data (such as melatonin levels), and clinical data to further expand the understanding of genetic mechanisms underlying sleep and circadian disorders.

22.6.2 Precision Medicine and Other Applications

The goal of precision medicine is to provide treatments for diseases that are tailored to an individual's unique characteristics. An initiative to achieve precision medicine in sleep is TOPMed, Trans-Omics for Precision Medicine, funded by the National Heart, Lung, and Blood Institute (NHLBI). The program enhances research on the biological bases behind heart, lung, blood, and sleep disorders by integrating whole-genome sequencing and what have been defined as "other -omics" data with clinical data, which, in the realm of sleep medicine, can include PSG data, environmental factors, and comorbidities [134]. Databases, portals, and controlled vocabularies are under development to allow varied and dense data to be integrated across cohorts.

Other examples of big data applications in sleep medicine are open-access sleep surveys, like the World Sleep Survey (http://www.worldsleepsurvey.com/), an initiative of the University of Oxford, UK, and Sleepio, an application for delivering cognitive behavioral therapy to treat insomnia. The survey, which aims to be the "largest ever survey in the world of sleep," includes about 80 questions on sleep habits and provides a personalized sleep report to each participant in return. Tens of thousands of individuals have participated, and data have been used to motivate interventions, associate behaviors with environmental factors [135], and improve the functionality of apps and self-help programs [136]. Another smartphone app, named ENTRAIN, collects data on sleep habits around the world, alongside with information about light to make inferences about global sleep trends and the effect of social pressures on sleep timing [137]. An online patient-centered portal,

MyApnea.Org, is a model for large data collection designed by a patient-researcher team comprising the Sleep Apnea Patient Centered Outcomes Network. Data consist of self-reported survey data, as well as information that can be extracted using natural language tools from forum posts, providing information on issues of importance to patients [138].

22.7 Conclusions

In this chapter, we have discussed the importance of a big data approach to sleep medicine, introduced the potential for an automated analysis approach to PSG signals, provided an overview of the tools to put this approach into practice, and reviewed some of the challenges in these activities and goals, as well as emerging opportunities to use sleep big data to support genomics, personalized medicine, and patient-centered research.

References

1. Ohayon MM, Carskadon MA, Guilleminault C, Vitiello MV. Meta-analysis of quantitative sleep parameters from childhood to old age in healthy individuals: Developing normative sleep values across the human lifespan. *Sleep* 2004;27:1255–74.
2. Petit D, Gagnon J, Fantini ML, Ferini-Strambi L, Montplaisir J. Sleep and quantitative EEG in neurodegenerative disorders. *J Psychosom Res* 2004;56:487–96.
3. Schabus M, Hödlmoser K, Gruber G et al. Sleep spindle-related activity in the human EEG and its relation to general cognitive and learning abilities. *Eur J Neurosci* 2006;23:1738–46.
4. Schafer H, Koehler U, Ploch T, Peter JH. Sleep-related myocardial ischemia and sleep structure in patients with obstructive sleep apnea and coronary heart disease. *Chest J* 1997;111:387–93.
5. Gottlieb DJ, Yenokyan G, Newman AB et al. Prospective study of obstructive sleep apnea and incident coronary heart disease and heart failure: The Sleep Heart Health Study. *Circulation* 2010;122:352–60.
6. Thayer JF, Yamamoto SS, Brosschot JF. The relationship of autonomic imbalance, heart rate variability and cardiovascular disease risk factors. *Int J Cardiol* 2010;141:122–31.
7. Zhang J, Yang XC, Luo L et al. Assessing severity of obstructive sleep apnea by fractal dimension sequence analysis of sleep EEG. *Phys A Stat Mech Appl* 2009;388:4407–14.
8. Suzuki T, Kameyama K, Inoko Y, Tamura T. Development of a sleep apnea event detection method using photoplethysmography. 2010;5258–61.
9. Palombini LO, Tufik S, Rapoport DM et al. Inspiratory flow limitation in a normal population of adults in Sao Paulo, Brazil. *Sleep* 2013;36:1663–8.
10. Thomas RJ, Mietus JE, Peng CK et al. Differentiating obstructive from central and complex sleep apnea using an automated electrocardiogram-based method. *Sleep* 2007;30:1756–69.
11. Budhiraja R, Thomas R, Kim M, Redline S. The role of big data in the management of sleep-disordered breathing. *Sleep Med Clin* 2016;11:241–55.
12. Prinz F, Schlange T, Asadullah K. Believe it or not: How much can we rely on published data on potential drug targets? *Nate Rev Drug Discov* 2011;10:712.
13. Crowell DH, Brooks LJ, Colton T et al. Infant polysomnography: Reliability. *Collaborative Home Infant Monitoring Evaluation (CHIME) Steering Committee. Sleep* 1997;20:553–60.
14. Ferri R, Bruni O, Miano S, Smerieri A, Spruyt K, Terzano MG. Inter-rater reliability of sleep cyclic alternating pattern (CAP) scoring and validation of a new computer-assisted CAP scoring method. *Clin Neurophysiol* 2005;116:696–707.
15. Rosa A, Alves GR, Brito M, Lopes MC, Tufik S. Visual and automatic cyclic alternating pattern (CAP) scoring: Inter-rater reliability study. *Arq Neuropsiquiatr* 2006;64.
16. Devuyst S, Dutoit T, Stenuit P, Kerkhofs M. Automatic K-complexes detection in sleep EEG recordings using likelihood thresholds. 2010;4658–61.
17. Wendt SL, Welinder P, Sorensen HB et al. Inter-expert and intra-expert reliability in sleep spindle scoring. *Clin Neurophysiol* 2015;126:1548–56.
18. Drinnan MJ, Murray A, Griffiths CJ, John Gibson G. Interobserver variability in recognizing arousal in respiratory sleep disorders. *Am J Respir Crit Care Med* 1998;158:358–62.
19. Dean DA, Goldberger AL, Mueller R et al. Scaling up scientific discovery in sleep medicine: The National Sleep Research Resource. *Sleep* 2016.
20. Goldberger AL, Amaral LA, Glass L et al. PhysioBank, PhysioToolkit, and PhysioNet: Components of a new research resource for complex physiologic signals. *Circulation* 2000;101:E215–20.
21. National Institutes of Health, National Heart, Lung, and Blood Institute. Biologic Specimen and Data Repository Information Coordinating Center (BioLINCC) 2015.
22. Mailman MD, Feolo M, Jin Y et al. The NCBI dbGaP database of genotypes and phenotypes. *Nat Genet* 2007;39:1181–6.
23. Mariani S, Goldberger AL, Mobley D, Redline S. The National Sleep Research Resource: A community-driven portal for the study of sleep signals 2016.
24. Burykin A, Mariani S, Henriques T et al. Remembrance of time series past: Simple chromatic method for visualizing trends in biomedical signals. *Physiol Meas* 2015;36:N95.
25. Henriques TS, Mariani S, Burykin A, Rodrigues F, Silva TF, Goldberger AL. Multiscale Poincaré plots for visualizing the structure of heartbeat time series. *BMC Med Inform Decision Making* 2016;16:1.

26. Mariani S, Borges AF, Henriques T, Goldberger AL, Costa MD. Use of multiscale entropy to facilitate artifact detection in electroencephalographic signals. 2015: 7869–72.
27. Iber C, Ancoli-Israel S, Chesson A, Quan SF. *The AASM Manual for the Scoring of Sleep and Associated Events: Rules, Terminology, and Technical Specifications.* 1st ed. Westchester, IL, 2007.
28. Dietsch G. Fourier Analysis von Electroencephalogrammen des Menschen. *Pflüger's Arch Ges Physiol* 1932; 220:106–12.
29. Bronzino, JD. *Biomedical Engineering Handbook*, vol. 2. CRC Press, Boca Raton, FL, 1999.
30. Kay SM, Marple SL. Spectrum analysis—A modern perspective. *Proc IEEE* 1981;69:1380–419.
31. Cerutti S, Marchesi C. *Advanced Methods of Biomedical Signal Processing.* John Wiley & Sons, 2011.
32. Coburn KL, Lauterbach EC, Boutros NN, Black KJ, Arciniegas DB, Coffey CE. The value of quantitative electroencephalography in clinical psychiatry: A report by the Committee on Research of the American Neuropsychiatric Association. *J Neuropsychiatry Clin Neurosci* 2006;18:460–500.
33. Buckelmüller J, Landolt H, Stassen H, Achermann P. Trait-like individual differences in the human sleep electroencephalogram. *Neuroscience* 2006;138:351–6.
34. Nakamura M, Shibasaki H. Elimination of EKG artifacts from EEG records: A new method of non-cephalic referential EEG recording. *Electroencephalogr Clin Neurophysiol* 1987;66:89–92.
35. Oropesa E, Cycon HL, Jobert M. *Sleep Stage Classification Using Wavelet Transform and Neural Network.* International Computer Science Institute, 1999.
36. Auger F, Flandrin P, Goncalves P, Lemoine O. Time-frequency toolbox for MATLAB, user's guide and reference guide. CNRS (France) and Rice University (USA), Paris 1996.
37. Liang S, Kuo C, Hu Y, Pan Y, Wang Y. Automatic stage scoring of single-channel sleep EEG by using multiscale entropy and autoregressive models. *IEEE Trans Instrum Meas* 2012;61:1649–57.
38. Roebuck A, Monasterio V, Gederi E et al. A review of signals used in sleep analysis. *Physiol Meas* 2014;35:R1.
39. Caffarel J, Gibson GJ, Harrison JP, Griffiths CJ, Drinnan MJ. Comparison of manual sleep staging with automated neural network–based analysis in clinical practice. *Med Biol Eng Comput* 2006;44:105–10.
40. Berthomier C, Drouot X, Herman Stoica M et al. Automatic analysis of single-channel sleep EEG: Validation in healthy individuals. *Sleep* 2007;30:1587–95.
41. Sherif OI, *Electrical Carleton University.* Dissertation. Engineering. Automatic detection of K-complex in sleep EEG, 1976.
42. Woertz M, Miazhynskaia T, Anderer P, Dorffner G. Automatic K-complex detection: Comparison of two different approaches. *ESRS, JSR* 2004;13.
43. Kerkeni N, Bougrain L, Bedoui MH, Alexandre F, Dogui M. Reconnaissance automatique des grapho-éléments temporels de l'électroencéphalogramme du sommeil. 2007.
44. Da Rosa A, Kemp B, Paiva T, Lopes da Silva F, Kamphuisen H. A model-based detector of vertex waves and K complexes in sleep electroencephalogram. *Electroencephalogr Clin Neurophysiol* 1991;78:71–9.
45. Bankman IN, Sigillito VG, Wise RA, Smith PL. Feature-based detection of the K-complex wave in the human electroencephalogram using neural networks. *IEEE Trans Biomed Eng* 1992;39:1305–10.
46. De Gennaro L, Ferrara M. Sleep spindles: An overview. *Sleep Med Rev* 2003;7:423–40.
47. Ayoub A, Aumann D, Horschelmann A et al. Differential effects on fast and slow spindle activity, and the sleep slow oscillation in humans with carbamazepine and flunarizine to antagonize voltage-dependent Na+ and Ca2+ channel activity. *Sleep* 2013;36:905–11.
48. Gais S, Molle M, Helms K, Born J. Learning-dependent increases in sleep spindle density. *J Neurosci* 2002;22: 6830–4.
49. Molle M, Marshall L, Gais S, Born J. Grouping of spindle activity during slow oscillations in human non-rapid eye movement sleep. *J Neurosci* 2002;22:10941–7.
50. Schimicek P, Zeitlhofer J, Anderer P, Saletu B. Automatic sleep-spindle detection procedure: Aspects of reliability and validity. *Clin Electroencephalogr* 1994;25: 26–9.
51. Crowley K, Trinder J, Kim Y, Carrington M, Colrain IM. The effects of normal aging on sleep spindle and K-complex production. *Clin Neurophysiol* 2002;113:1615–22.
52. Ruch S, Markes O, Duss SB et al. Sleep stage II contributes to the consolidation of declarative memories. *Neuropsychologia* 2012;50:2389–96.
53. Sitnikova E, Hramov AE, Koronovsky AA, van Luijtelaar G. Sleep spindles and spike–wave discharges in EEG: Their generic features, similarities and distinctions disclosed with Fourier transform and continuous wavelet analysis. *J Neurosci Methods* 2009;180:304–16.
54. Wamsley EJ, Tucker MA, Shinn AK et al. Reduced sleep spindles and spindle coherence in schizophrenia: Mechanisms of impaired memory consolidation? *Biol Psychiatr* 2012;71:154–61.
55. Martin N, Lafortune M, Godbout J et al. Topography of age-related changes in sleep spindles. *Neurobiol Aging* 2013;34:468–76.
56. Devuyst S, Dutoit T, Didier J et al. Automatic sleep spindle detection in patients with sleep disorders. 2006: 3883–6.
57. Ferrarelli F, Huber R, Peterson MJ et al. Reduced sleep spindle activity in schizophrenia patients. *Am J Psychiatr* 2007;164:483–92.
58. Wendt SL, Christensen JA, Kempfner J, Leonthin HL, Jennum P, Sorensen HB. Validation of a novel automatic sleep spindle detector with high performance during sleep in middle aged subjects. 2012:4250–3.
59. Huupponen E, Gómez-Herrero G, Saastamoinen A, Värri A, Hasan J, Himanen S. Development and comparison of four sleep spindle detection methods. *Artif Intell Med* 2007;40:157–70.
60. Anderer P, Gruber G, Parapatics S et al. An E-health solution for automatic sleep classification according to Rechtschaffen and Kales: Validation study of the

Somnolyzer 24 × 7 utilizing the Siesta database. *Neuropsychobiology* 2005;51:115–33.
61. Schrouff J, Leclercq Y, Noirhomme Q, Maquet P, Phillips C. FASST—A FMRI Artefact rejection and Sleep Scoring Toolbox. 2011.
62. Warby SC, Wendt SL, Welinder P et al. Sleep-spindle detection: Crowdsourcing and evaluating performance of experts, non-experts and automated methods. *Nat Methods* 2014;11:385–92.
63. Bonnett M, Carley D, Carskadon M. EEG arousal: Scoring rules and examples: A preliminary report from the sleep disorders atlas task force of the American Sleep Disorders Association. *Sleep* 1992;15:173–84.
64. Bodenstein G, Praetorius HM. Feature extraction from the electroencephalogram by adaptive segmentation. *Proc IEEE* 1977;65:642.
65. Agarwal R, Gotman J. Adaptive segmentation of electroencephalographic data using a nonlinear energy operator. 1999;4:199–202.
66. Sugi T, Nakamura M, Shimokawa T, Kawana F. Automatic detection of EEG arousals by use of normalized parameters for different subjects. 2003:146–7.
67. Agarwal R. Automatic detection of micro-arousals. 2006:1158–61.
68. De Carli F, Nobili L, Gelcich P, Ferrillo F. A method for the automatic detection of arousals during sleep. *Sleep* 1999;22:561–72.
69. Huupponen E, Vaerri A, Hasan J, Saarinen J, Kaski K. Sleep arousal detection with neural network. *Med Biol Eng Comput* 1996;34:219–20.
70. Guilleminault C, Pool P, Motta J, Gillis AM. Sinus arrest during REM sleep in young adults. *N Engl J Med* 1984;311:1006–10.
71. Penzel T, Kantelhardt JW, Grote L, Peter J, Bunde A. Comparison of detrended fluctuation analysis and spectral analysis for heart rate variability in sleep and sleep apnea. *IEEE Trans Biomed Eng* 2003;50:1143–51.
72. Fantini M, Michaud M, Gosselin N, Lavigne G, Montplaisir J. Periodic leg movements in REM sleep behavior disorder and related autonomic and EEG activation. *Neurology* 2002;59:1889–94.
73. Lanfranchi PA, Fradette L, Gagnon J, Colombo R, Montplaisir J. Cardiac autonomic regulation during sleep in idiopathic REM sleep behavior disorder. *Sleep* 2007;30:1019.
74. Dauvilliers Y, Pennestri M, Whittom S, Lanfranchi PA, Montplaisir JY. Autonomic response to periodic leg movements during sleep in narcolepsy-cataplexy. *Sleep* 2011;34:219.
75. Jurysta F, Lanquart JP, Sputaels V et al. The impact of chronic primary insomnia on the heart rate–EEG variability link. *Clin Neurophysiol* 2009;120:1054–60.
76. Pan J, Tompkins WJ. A real-time QRS detection algorithm. *IEEE Trans Biomed Eng* BME 32:230.
77. Camm AJ, Malik M, Bigger JT et al. Heart rate variability: Standards of measurement, physiological interpretation and clinical use. Task Force of the European Society of Cardiology and the North American Society of Pacing and Electrophysiology. *Circulation* 1996;93:1043–65.
78. Mietus JE, Peng CK, Henry I, Goldsmith RL, Goldberger AL. The pNNx files: Re-examining a widely used heart rate variability measure. *Heart* 2002;88:378–80.
79. Clifford GD, Azuaje F, McSharry P. *Advanced Methods and Tools for ECG Data Analysis*. Artech House, Inc., 2006.
80. Tacchino G, Mariani S, Migliorini M, Bianchi AM. Optimization of time-variant autoregressive models for tracking REM-non REM transitions during sleep. 2236.
81. Lomb NR. Least-squares frequency analysis of unequally spaced data. *Astrophys Space Sci* 1976;39:447–62.
82. Moody GB. Spectral analysis of heart rate without resampling. 1993:715–8.
83. Migliorini M, Mendez MO, Bianchi AM. Study of heart rate variability in bipolar disorder: Linear and non-linear parameters during Sleep. *Front Neuroeng* 2012;4.
84. Otzenberger H, Gronfier C, Simon C et al. Dynamic heart rate variability: A tool for exploring sympathovagal balance continuously during sleep in men. *Am J Physiol* 1998;275:H946–50.
85. Tarvainen MP, Niskanen J, Lipponen JA, Ranta-Aho PO, Karjalainen PA. Kubios HRV—Heart rate variability analysis software. *Comput Methods Programs Biomed* 2014;113:210–20.
86. Goldberger AL, Rigney DR, West BJ. Chaos and fractals in human physiology. *Sci Am* 1990;262:42–9.
87. Hausdorff JM, Mitchell SL, Firtion R et al. Altered fractal dynamics of gait: Reduced stride-interval correlations with aging and Huntington's disease. *J Appl Physiol* 1997;82:262–9.
88. Goldberger AL, Amaral LA, Hausdorff JM, Ivanov PC, Peng CK, Stanley HE. Fractal dynamics in physiology: Alterations with disease and aging. *Proc Natl Acad Sci U S A* 2002;99 Suppl 1:2466–72.
89. Peng C-K, Havlin S, Stanley HE, Goldberger AL. Quantification of scaling exponents and crossover phenomena in nonstationary heartbeat time series. *Chaos* 1995;5.
90. Richman JS, Moorman JR. Physiological time-series analysis using approximate entropy and sample entropy. *Am J Physiol Heart Circ Physiol*, 2000 Jun 1; 278(6):H2039–49.
91. Costa MD, Goldberger AL, Peng C-K. Multiscale entropy analysis of biological signals. *Phys Rev E Stat Nonlin Soft Matter Phys* 2005;71.
92. Costa MD, Peng C, Goldberger AL. Multiscale analysis of heart rate dynamics: Entropy and time irreversibility measures. *Cardiovasc Eng* 2008;8:88–93.
93. Costa MD, Goldberger AL, Peng C-K. Multiscale entropy analysis of complex physiologic time series. *Phys Rev Lett* 2002;89.
94. Ferrario M, Signorini MG, Magenes G. Comparison between fetal heart rate standard parameters and complexity indexes for the identification of severe intrauterine growth restriction. *Methods Inf Med* 2007;46:186.
95. Mariani S, Migliorini M, Tacchino G et al. Clinical state assessment in bipolar patients by means of HRV features obtained with a sensorized T-shirt. 2012.
96. Redline S, Budhiraja R, Kapur V et al. Reliability and validity of respiratory event measurement and scoring. *J Clin Sleep Med* 2007;3:169–200.

97. Fontenla-Romero O, Guijarro-Berdiñas B, Alonso-Betanzos A, Moret-Bonillo V. A new method for sleep apnea classification using wavelets and feedforward neural networks. *Artif Intell Med* 2005;34:65–76.
98. Nakano H, Tanigawa T, Ohnishi Y et al. Validation of a single-channel airflow monitor for screening of sleep-disordered breathing. *Eur Respir J* 2008;32:1060–7.
99. Morgenstern C, Schwaibold M, Randerath WJ, Bolz A, Jané R. An invasive and a noninvasive approach for the automatic differentiation of obstructive and central hypopneas. *IEEE Trans Biomed Eng* 2010;57:1927–36.
100. Gutiérrez-Tobal GC, Hornero R, Álvarez D, Marcos JV, del Campo F. Linear and nonlinear analysis of airflow recordings to help in sleep apnoea–hypopnoea syndrome diagnosis. *Physiol Measure*, 2012 Jun 27;33(7):1261.
101. Gil E, Mendez M, Vergara JM, Cerutti S, Bianchi AM, Laguna P. Discrimination of sleep-apnea-related decreases in the amplitude fluctuations of PPG signal in children by HRV analysis. *IEEE Trans Biomed Eng* 2009;56:1005–14.
102. Gil E, Bailón R, Vergara JM, Laguna P. PTT variability for discrimination of sleep apnea related decreases in the amplitude fluctuations of PPG signal in children. *IEEE Trans Biomed Eng* 2010;57:1079–88.
103. Karmakar C, Khandoker A, Penzel T, Schöbel C, Palaniswami M. Detection of respiratory arousals using photoplethysmography (PPG) signal in sleep apnea patients. *IEEE J Biomed Health Inform* 2014;18:1065–73.
104. Jané R, Solà-Soler J, Fiz JA, Morera J. Automatic detection of snoring signals: Validation with simple snorers and OSAS patients. 2000;4:3129–31.
105. Ng AK, San Koh T, Baey E, Lee TH, Abeyratne UR, Puvanendran K. Could formant frequencies of snore signals be an alternative means for the diagnosis of obstructive sleep apnea? *Sleep Med* 2008;9:894–8.
106. Herzog M, Schieb E, Bremert T et al. Frequency analysis of snoring sounds during simulated and nocturnal snoring. *Eur Arch Oto-Rhino-Laryngol* 2008;265:1553–62.
107. Hara H, Murakami N, Miyauchi Y, Yamashita H. Acoustic analysis of snoring sounds by a multidimensional voice program. *Laryngoscope* 2006;116:379–81.
108. Ng AK, San Koh T, Abeyratne UR, Puvanendran K. Investigation of obstructive sleep apnea using nonlinear mode interactions in nonstationary snore signals. *Ann Biomed Eng* 2009;37:1796–806.
109. Ibrahim LH, Jacono FJ, Patel SR et al. Heritability of abnormalities in cardiopulmonary coupling in sleep apnea: Use of an electrocardiogram-based technique. *Sleep* 2010;33:643–6.
110. Schramm PJ, Thomas RJ. Assessment of therapeutic options for mild obstructive sleep apnea using cardiopulmonary coupling measures. *J Clin Sleep Med* 2012;8:315–20.
111. Thomas R, Mietus J, Peng C, Goldberger A. An electrocardiogram-based technique to assess cardiopulmonary coupling during sleep. *Sleep* 2005;28:1151–61.
112. Thomas RJ, Mietus JE, Peng C et al. Relationship between delta power and the electrocardiogram-derived cardiopulmonary spectrogram: Possible implications for assessing the effectiveness of sleep. *Sleep Med* 2014;15:125–31.
113. Mammone N, Morabito FC. Independent component analysis and high-order statistics for automatic artifact rejection. 2005;4:2447–52.
114. Mantini D, Perrucci MG, Cugini S, Ferretti A, Romani GL, Del Gratta C. Complete artifact removal for EEG recorded during continuous fMRI using independent component analysis. *Neuroimage* 2007;34:598–607.
115. James CJ, Gibson OJ. Temporally constrained ICA: An application to artifact rejection in electromagnetic brain signal analysis. *IEEE Trans Biomed Eng* 2003;50:1108–16.
116. Junghöfer M, Elbert T, Tucker DM, Rockstroh B. Statistical control of artifacts in dense array EEG/MEG studies. *Psychophysiology* 2000;37:523–32.
117. Mammone N, La Foresta F, Morabito FC. Automatic artifact rejection from multichannel scalp EEG by wavelet ICA. *IEEE Sens J* 2012;12:533–42.
118. Schlögl A, Keinrath C, Zimmermann D, Scherer R, Leeb R, Pfurtscheller G. A fully automated correction method of EOG artifacts in EEG recordings. *Clin Neurophysiol* 2007;118:98–104.
119. Gasser T, Möcks J. Correction of EOG artifacts in event-related potentials of the EEG: Aspects of reliability and validity. *Psychophysiology* 1982;19:472–80.
120. Joyce CA, Gorodnitsky IF, Kutas M. Automatic removal of eye movement and blink artifacts from EEG data using blind component separation. *Psychophysiology* 2004;41:313–25.
121. Jung T, Makeig S, Humphries C et al. Removing electroencephalographic artifacts by blind source separation. *Psychophysiology* 2000;37:163–78.
122. Brookes MJ, Mullinger KJ, Stevenson CM, Morris PG, Bowtell R. Simultaneous EEG source localisation and artifact rejection during concurrent fMRI by means of spatial filtering. *Neuroimage* 2008;40:1090–104.
123. LeVan P, Urrestarazu E, Gotman J. A system for automatic artifact removal in ictal scalp EEG based on independent component analysis and Bayesian classification. *Clin Neurophysiol* 2006;117:912–27.
124. Gwin JT, Gramann K, Makeig S, Ferris DP. Removal of movement artifact from high-density EEG recorded during walking and running. *J Neurophysiol* 2010;103:3526–34.
125. Reilly R, Nolan H. FASTER: Fully Automated Statistical Thresholding for EEG artifact Rejection. 2010.
126. Mognon A, Jovicich J, Bruzzone L, Buiatti M. ADJUST: An automatic EEG artifact detector based on the joint use of spatial and temporal features. *Psychophysiology* 2011;48:229–40.
127. Delorme A, Sejnowski T, Makeig S. Enhanced detection of artifacts in EEG data using higher-order statistics and independent component analysis. *Neuroimage* 2007;34:1443–9.
128. Migliorini M, Mariani S, Bianchi AM. Decision tree for smart feature extraction from sleep HR in bipolar patients. 2013:5033–6.
129. Mariani S, Tarokh L, Djonlagic I, Cade BE, Morrical M, Yaffe K, Stone KL, Loparo KA, Purcell S, Redline S, Aeschbach D. Evaluation of an automated pipeline for large scale EEG spectral analysis: The National Sleep Research Resource. *Sleep Med*, 2017 Nov 29.

130. Ochs C, He Z, Perl Y, Arabandi S, Halper M, Geller J. Choosing the granularity of abstraction networks for orientation and quality assurance of the sleep domain ontology. 2013:84–9.
131. Keator DB, Grethe JS, Marcus D et al. A national human neuroimaging collaboratory enabled by the Biomedical Informatics Research Network (BIRN). *IEEE Trans Inform Technol Biomed* 2008;12:162–72.
132. Weber GM, Murphy SN, McMurry AJ et al. The Shared Health Research Information Network (SHRINE): A prototype federated query tool for clinical data repositories. *J Am Med Inform Assoc* 2009;16:624–30.
133. Lane JM, Vlasac I, Anderson SG et al. Genome-wide association analysis identifies novel loci for chronotype in 100,420 individuals from the UK Biobank. *Nat Commun* 2016;7.
134. National Heart, Lung, and Blood Institute. Trans-Omics for Precision Medicine (TOPMed) program. 2015. Anonymous (accessed August 8, 2016, at http://www.nhlbi.nih.gov/research/resources/nhlbi-precision-medicine-initiative/topmed).
135. Mets MA, Alford C, Verster JC. Sleep specialists' opinion on sleep disorders and fitness to drive a car: The necessity of continued education. *Ind Health* 2012;50:499–508.
136. Espie CA, Luik AI, Cape J et al. Digital cognitive behavioural therapy for insomnia versus sleep hygiene education: The impact of improved sleep on functional health, quality of life and psychological well-being. Study protocol for a randomised controlled trial. *Trials* 2016;17:1.
137. Walch OJ, Cochran A, Forger DB. A global quantification of "normal" sleep schedules using smartphone data. *Sci Adv* 2016;2:e1501705.
138. Redline S, Baker-Goodwin S, Bakker JP et al. Patient partnerships transforming sleep medicine research and clinical care: Perspectives from the sleep apnea patient-centered outcomes network. *J Clin Sleep Med* 2016;12:1053–8.

23

Integrating Clinical Physiological Knowledge at the Feature and Classifier Levels in Design of a Clinical Decision Support System for Improved Prediction of Intensive Care Unit Outcome

Ali Jalali, Vinay M. Nadkarni, Robert A. Berg, Mohamed Rehman, and C. Nataraj

CONTENTS

23.1 Introduction 465
23.2 Methods 467
    23.2.1 Collected Data 467
    23.2.2 Integrating Expert Physiological Knowledge 467
    23.2.3 Feature Extraction 468
    23.2.4 Feature Ranking 469
    23.2.5 Artificial Neural Network Classifier 470
23.3 Results 471
23.4 Conclusion 473
References 474

23.1 Introduction

Patients in the intensive care unit (ICU) are among the most critically ill patients in any hospital. For example, approximately 200,000 in-hospital cardiac arrests occur in the United States each year, and only 20% of these patients survive to discharge. In-hospital cardiac arrest is frequently preceded by early warning signs of clinical deterioration that can be recognized and treated by trained in-hospital staff. Those at higher risk (of cardiac arrest, for example) would be in immediate need for extensive monitoring and direct attention from healthcare providers [1]. The implementation of early warning scores leads to improved recognition and treatment of clinical deterioration in hospitalized patients in the general wards. Hence, evaluation of critical deterioration risk for ICU patients has drawn much interest from healthcare providers due to its importance in saving patients' lives [2–6]. Most of the clinically based studies have focused on providing simple scores that focus on the severity of disease or illness [7]. Basically, these scores add weights to the degree of abnormality of an organ or a disease based on the vital sign measurement, blood gas measurement, or visual inspection of the patients, and attempt to identify patients at high risk. Some of the currently available acuity scores are Acute Physiology and Chronic Health Evaluation (APACHE) III [8], Simplified Acute Physiology Score (SAPS) II [9], Modified Early Warning Scoring (MEWS) [10], Mortality Probability Models (MPM) [11], and Sequential Organ Failure (SOFA) score [12].

There are three fundamental problems with these acuity scoring methods. Firstly, they are population based and are not patient specific [13] since they have been developed to provide risk adjustments for population differences in evaluating the effects of medications, care guidelines, surgery, and other interventions that impact mortality in ICU patients [14]. Furthermore, the scores are not developed for frequent evaluation of the patients [13] but rather for estimating risk based on the first 24 hours after ICU admission. The currently available scoring systems such as SOFA and SAPS could not be updated regularly since they are mostly based on the indicators that can only be updated once per day, for instance, the worst measured value of the day, which can only be calculated at the end of the day. Hence, these scoring systems can only be updated once a day and fail to provide near-real-time predictive power [15]. In contrast, the developed clinical decision support system (CDSS) in this study can be updated as frequently as we get the measurements from the patients. This will allow the clinicians to track the trajectory of the patient more closely. Finally, the current scoring systems are mostly based on the raw recordings from the patients and do not take into account the quality of the data in the electronic health records (EHRs). As we know, there are many parameters such as measurement condition, patient movement, measurement device parameters, etc., which affect the quality of the EHR data.

Currently, the way that these scoring methods are used is that, as soon as a patient admitted to the ICU, the healthcare provider will calculate the severity score for the patient and then provides a care strategy based on the assigned score. This process is carried out on a daily basis, and the score does not change during the day. This strategy works fairly well for patients who either are very sick or gradually deteriorate during a period of a week or so. However, clinicians are more interested in detecting the patients who suddenly deteriorate during a very short period of time as these cases require immediate attention; however, they are very difficult to detect in advance and are clearly not captured by the standard scoring methods. Higher-frequency physiologic data collection in ICU environments provides the source for near-real-time information on patients that changes and evolves with patient state. Medical devices at the point of care in these environments are a rich source of these data.

The objective of this chapter is to address the aforementioned problem and propose the development of a machine learning–based CDSS for patient-specific prediction of in-hospital mortality of ICU patients. CDSS is a part of health information technology (HIT), which is widely considered to be an important game-changing paradigm shift in the delivery and administration of medicine leading to enormous gains in efficiency and improvements in effectiveness, resulting in savings of lives and cost. In general, CDSSs are capable of analyzing vast amounts of data and can provide patient-specific risk evaluation. Modern ICUs provide clinicians with a vast amount of physiological data including physiological waveforms, images, laboratory test results, and clinical notes with the purported aim of providing detailed information of every patient's physiological state. Furthermore, physiological data in the form of electrocardiograms, blood pressure, pulse, respiratory rate (RR), and other invasive parameters are routinely monitored in the ICU environment. Temporal telemetry on patients is used regularly for managing patient state by clinical staff, and the data available from the medical devices at the point of care are also regularly collected for operational and research use. A CDSS is intended to help the caregivers aggregate the different types of physiological data so provided and to discover the hidden knowledge or patterns in the data to help make the correct decision in an expedient manner.

CDSS typically uses a knowledge-based data mining approach with patient-specific information so as to be able to provide support for decision making in patient care [16–18]. In fact, a systematic review by Garg et al. [19] of 100 studies concluded that CDSS improved practitioner performance in 64% of the studies and improved patient outcomes in 13% of the studies. CDSS is a sophisticated HIT component. It requires computable biomedical knowledge, patient-specific data, and a reasoning or inferencing mechanism that combines knowledge and data to generate and present helpful information to clinicians as care is being delivered. This information must be filtered, organized, and presented in a way that supports the current workflow, allowing the user to make an informed decision quickly and take action.

Traditionally, CDSSs have been statistical in nature, although machine learning techniques have increasingly attracted attention from researchers in the CDSS field as a result of their superior performance in comparison with traditional stochastic approaches in prediction, modeling, and classification of biomedical signals [16,20–23]. *The main advantage of machine learning–based methods is that they are structurally sophisticated, which makes them suitable for analyzing increasingly complex medical data.* We have successfully implemented machine learning techniques in developing CDSS in our research on predicting complex physiological conditions [24–27].

CDSSs have indeed been implemented for ICU outcome prediction in some earlier studies [1]. However, these methods are tailored to a very specific data set; they also ignore available physiological knowledge, hence making them relatively unfriendly to the clinician. We believe CDSS not only should provide accurate and reliable decision support in a variety of settings but should also pave the way to eventually providing valuable insights for a better understanding of complex physiological conditions and diseases. One of the best studies done on the prediction of ICU outcomes is the study by Pirracchio et al. [31]. They used data from 24,000 patients collected from PhysioNet to develop their algorithm. They further validated their method using data from 200 patients collected from a different hospital. However, they excluded the do-not-resuscitate (DNR) or comfort-measures-only (CMO) patients from their data set. Hence, in this chapter, we describe a novel algorithm that successfully integrates expert physiological knowledge in a machine learning–based CDSS for providing accurate decision support to predict ICU outcome.

We aim to develop an easy-to-use—but algorithmically complex and dynamic—CDSS that addresses at least three unique challenges: (1) synthesis of data from multiple clinical systems and sources, (2) decision support that is predictive unlike most clinical decision support (CDS) interventions, and (3) impact to the outcomes and safety of high-risk complex patients.

23.2 Methods

23.2.1 Collected Data

The data used in this chapter are gathered from the publicly available PhysioNet database and consists of 4,000 patients [14]. The data were collected from first 48 hours of stay for an adult population who were admitted to the Massachusetts General Hospital (MGH). These patients were admitted for a wide variety of reasons to cardiac, medical, surgical, and trauma ICUs. Patients with directives of DNR or CMO were not excluded, hence making the prediction task much more complex. The exclusion criterion of ICU stays of less than 48 hours has been applied to the data set. Among the collected data, 512 patients died during their stay at the ICU. The available EHR system collects 42 physiological measurements and demographic information. Not all variables are available in all cases, however. Among the collected EHR data, six variables are general demographic descriptors (collected on admission), and the remainder are in time series format. The data were provided in the comma separated file format and are listed in Table 23.1. Be aware that the sources of these data are not coherent necessarily across organs. For instance, bicarbonate (HCO_3) is measured in the lab via blood draw, together with partial pressure of oxygen (PaO_2) and dioxide carbon ($PaCO_2$). Hence, the regularity of these measurements varies from, say, *RR* or mechanical ventilation tidal and minute volume settings or spontaneous measurements. Hence, the frequency of collection of these values differs. This is being mentioned to provide context in that not all parameters are collected at the same time, from the same source, and in the same frequency. This data set is a perfect example of practically available EHR-based clinical data with missing, sparse, or noisy observations, which we need to account for in our algorithm. In the feature extraction section 23.2.3, we will discuss how our algorithm deals with these issues.

TABLE 23.1
Collected Data for ICU Mortality Prediction

Measurement	Unit
Albumin	g/dL
Alkaline phosphatase (ALP)	IU/L
Alanine transaminase (ALT)	IU/L
Aspartate transaminase (AST)	IU/L
Blood urea nitrogen (BUN)	mg/dL
Bilirubin	mg/dL
Cholesterol	mg/dL
Creatinine	mg/dL
Diastolic arterial blood pressure (DABP)	mmHg
Systolic arterial blood pressure (SABP)	mmHg
Fractional inspired O_2 (FiO$_2$)	0–1
Glasgow Coma Score (GCS)	3–15
Serum bicarbonate (HCO$_3$)	mmol/L
Hematocrit (Hct)	%
Heart rate (HR)	bpm
Serum potassium (K)	mEq/L
Lactate	mmol/L
Serum magnesium (Mg)	mmol/L
Mean arterial pressure (MAP)	mmHg
Mechanical ventilation	0 or 1
Serum sodium (Na)	mEq/L
Noninvasive DABP (NIDABP)	mmHg
Noninvasive SABP (NISABP)	mmHg
Noninvasive MAP (NIMAP)	mmHg
Partial pressure of arterial CO_2 (PaCO$_2$)	mmHg
Partial pressure of arterial O_2 (PaO$_2$)	mmHg
pH	0–14
Platelets	cells/nL
Respiratory rate (RR)	bpm
O_2 saturation in hemoglobin (SaO$_2$)	%
Temperature	°C
Troponin-I (TropI)	μg/L
Troponin-T (TropT)	μg/L
Urine	mL
White blood cell count (WBC)	cells/nL

Note: Majority of these variables are obtained from laboratory analysis of blood draws. The heart rate, respiratory rate blood pressure, and FiO$_2$ values are obtained automatically from bedside monitoring and mechanical ventilation sources.

23.2.2 Integrating Expert Physiological Knowledge

Almost all the previously clinically developed acuity scoring systems group the collected data into different groups based on the respective organ, type of data, etc. We believe that grouping the data is essential for the success of the outcome prediction algorithm since it provides a more specific view of the data for the

TABLE 23.2

Organ Classification for ICU Mortality Prediction

Organ	Variables
Heart	MAP, DABP, SABP, NIDABP, NISABP, NIMAP, HR, K, cholesterol
Neuro	GCS, glucose, MAP, SaO$_2$
Lung	FiO$_2$, RR, SaO$_2$, PaO$_2$, PaCO$_2$, pH, mechanical ventilation, HCO$_3$
Liver	Bilirubin, albumin
Kidney	Creatinine, BUN, K, lactic, urine
Infection	WBC, temperature

Note: For definition of acronyms, please refer to Table 23.1.

algorithm and also makes the algorithm more intuitive for clinical use. Based on expert medical knowledge and opinion, we have hence divided the collected data into different groups based on their relationship to a particular organ. We also defined a new index called out-of-range index (ORI) based on clinical expert opinion that that the amount of time that a physiological variable is out of normal range or in a dangerous zone is as important as the number of times that it surpasses the normal limits. The normal limits for the patients are defined based on clinical expertise. We understand that the sources of the data are not coherent necessarily across various organs. For instance, HCO$_3$ is measured in the laboratory using blood sample analysis, together with PaO$_2$ and PaCO$_2$. Hence, the regularity of these measurements varies from variables such as heart rate (HR) or mean arterial blood pressure (MAP), which are calculated more frequently. Hence, the frequency of collection of these values differs. The grouping of the variables into these different organ groups helps the algorithm to better understand the physiological state of each organ and hence guarantees improved decision support. Table 23.2 shows organs and their respective variables based on expert opinion and is unique to our algorithm. Note that this grouping is more comprehensive than any previously developed acuity scoring system. Furthermore, we added another group called *infection* to account for the possibility of sepsis as it is a common condition in ICU patients.

23.2.3 Feature Extraction

One of the main steps of the machine learning algorithms is the feature extraction. Feature extraction can be done in a supervised or unsupervised manner. Unsupervised feature extraction techniques include autoencoders and deep Bayesian networks. The unsupervised feature extraction techniques are mostly used in image recognition or cases where we have high-resolution waveform data. Feature extraction is necessary to quantify the collected data and extract hidden information that exists within it. In the following section, we describe the features that we extract from the collected physiological data.

Statistical Features. The statistical features that we extract from the data include minimum, maximum, mean, variance, skewness, kurtosis, frequency of collection, maximum rate of change, and daily trend. Skewness and kurtosis are third- and fourth-order statistical moments of a random variable defined by Equation 23.1.

$$m_n(x) = E\{(x - \mu)^n\} \quad (23.1)$$

where n is the order, μ is the mean value of the data, and E is the expected value. Frequency of collection represents the number of times a variable is collected and is a measure of the level of care for the patient as more frequent data collection means that the clinicians suspect that the patient is critical and hence collect more data for closer monitoring of the patient's status of health. Maximum rate of change reflects the sharpest change in a variable or recording and is a strong indicator of sudden changes in the patient's health status. Daily trend is the difference between the first and last recordings of the day and is a good indicator of the patient's change of health state.

Out-of-Range Index and Number of Alarms. The ORI is a measure of both the amplitude differences of a measurement within its normal range and the time that the measurement goes out of normal range. The normal range limits of the collected data are presented in Table 23.3. We first defined ORI as a new feature in our previously published paper on prediction of neurological outcome after cardiac surgery in neonates, where the results showed that it is an excellent predictor of outcome [27]. Furthermore, the ORI is a clinically intuitive measure as clinicians believe that the amount of time that a physiological variable is out of normal range or in a dangerous zone is as important as the number of times that it surpasses the normal limits. For each variable, we also calculate the number of alarms, which is equal to the number of times that a measurement has crossed the normal range thresholds as defined in Table 23.3.

Sample Entropy. *Another feature that we extracted for more regularly collected data including HR, systolic blood pressure (SABP), diastolic blood pressure (DABP), and MAP is sample entropy.* Sample entropy (SampEn), as defined by Ref. [29], is a measure of signal complexity and irregularity and is the negative natural logarithm of the conditional probability that a signal window with length N, having repeated itself within a tolerance r for m points, will also repeat itself for $m + 1$ points, without allowing self-matches. Sample entropy has been used in the literature to evaluate the cyclic behavior of HR and blood pressure [30–32].

Demographic Features. Demographic data can represent very valuable information about a patient,

TABLE 23.3

Normal Physiological Range for the Collected Data

Measurement	Lower Limit	Upper Limit
Albumin	2	NA
Bilirubin	NA	6
BUN	NA	50
Creatinine	NA	2.5
DABP	30	120
GCS	4	NA
Glucose	60	150
Hct	30	NA
K	3	8
Lactate	NA	4
MAP	70	150
Mg	NA	5
Na	120	160
pH	7.1	7.6
PaCO$_2$	35	70
PaO$_2$	60	200
Platelets	50	800
RR	6	30
SABP	60	180
HCO$_3$	22	26
SaO$_2$	95	NA
Temperature	35.5	38.5
WBC	5	20

Note: The term NA in the table means that we did not use the respective limit for the variable. The limits of noninvasive blood pressure measurements are the same as invasive measurements.

which we quantify here for use in our algorithm. The extracted features from demographic data include age, sex, weight, and type of ICU. Also, we calculated body surface area (BSA) from height and weight. BSA is frequently used to titrate infusion and normalize physiologic parameters, including cardiac output. Each patient will be assigned a number between 1 and 7 based on his/her respective age group. The designed age groups are (1) < 30, (2) 30–40, (3) 40–50, (4) 50–60, (5) 60–70, (6) 70–80, and (7) > 80.

Design of a decision support system based on the features extracted from the measurements makes it less sensitive to noise in the data. The extracted features from the data such as statistical features, ORI, and sample entropy are more robust to the added noise and are designed to reduce the effect of the noise. To illustrate this, let us assume that a very low oxygen saturation recording is noisy. This number is used in the calculation of the SOFA, and SAPS directly, and therefore will skew the SOFA and SAPS. However, one noisy recording has very little effect on the mean of the data, negligible effect on the skewness and the kurtosis of the data, negligible effect on the ORI, and very little effect in general. Hence, the proposed decision support system is more robust to the noise in the data and the quality of the EHR data.

23.2.4 Feature Ranking

As mentioned earlier, the extracted features contain important undiscovered relationships of the inputs with the output. We could extract as many features as possible from the data and then form tens of thousands of pairs from these features. Using a large number of features leads to a well-known problem of overfitting in machine learning algorithms, leading to grossly inaccurate models. This situation hence exposes us to a very challenging problem: how do we efficiently pare the features down to a smaller subset by identifying just the most important features? This process of systematic reduction is called the feature ranking problem and attempts to sort the features based on their correlation with outcome. There have been several methods of feature ranking proposed in the literature; among them are the ensemble feature ranking method, feature ranking based on the support vector concept [33,34], and mutual information–based ranking [27,35]. In this study, we rank the extracted features from each organ based on their mutual information with the outcome. The mutual information between x_i and output w_k is then defined as

$$I(x_i; w_k) = \sum_{i \in X} \sum_{k \in \{\pm 1\}} p(x_i, w_k) \log \frac{p(x_i, w_k)}{p(x_i)p(w_k)} \quad (23.2)$$

where w_k represents the classes, survived or not survived, and $p(x_i, w_k)$ is the joint probability distribution of x_i and w_k.

We approach the feature ranking problem using mutual information as an optimization problem that seeks to find a subset S_{opt} from the whole feature set S by maximizing the information content $I(s; w_k)$ between the feature set and the output. In this we maximize the mutual information of subset S_i using the following equation:

$$I(x_i; w_k) = \frac{1}{|S|} \sum_{x_i \in S} I(x_i; w_k)$$

$$- \frac{1}{|S-1|^2} \sum_{x_i, x_j \in S} I(x_i; x_j) x$$

$$= \{x : x \in S_i \subset S\} \quad (23.3)$$

In order to maximize $I(x_i; w_k)$, we have to maximize the first part of the right-hand side of Equation 23.3, which represents the mean of the mutual information of each of the features and the class. In addition, we need to minimize the second part of the right-hand side of the equation, which represents the mutual information between the features themselves. We seek to find the features that have high mutual information with output and low mutual information with each other to ensure that the optimal feature set is robust and does not contain redundant information and features. The mutual information of a set takes into account both mutual information of features with the output and also mutual information of features with each other. If two features have high mutual information, i.e., are highly correlated, the mutual information of a set will keep the feature with higher mutual information with outcome and will remove the feature with lower mutual information with the outcome as this feature is redundant and does not contribute to the overall accuracy of the algorithm.

23.2.5 Artificial Neural Network Classifier

Much of the original excitement for the application of computational intelligence to biomedicine and biomedical systems research originated from the development of artificial neural networks (ANNs), which are designed to mimic the performance of the human nervous system. The nonlinear properties of an ANN are ideal for pattern recognition or texture classification. Successes in pattern analyses using ANN have been demonstrated in many disciplines. A neural network can be used to explore the relationships among several physiologic variables and predict outcomes of almost all types of diseases [31,36,37]. In cardiovascular system research, neural networks are applied very widely. Neural networks are used for modeling and fault-diagnosing the cardiovascular system [38,39], and heart arrhythmia detection and classification [40].

The inputs of the ANN classifier are the selected features from each patient, and the output of the classifier is the probability of death as an outcome of ICU stay for the respective patient. We use the backpropagation method with a fixed learning rate of 0.1 to train the classifier. The designed neural network is composed of an input layer with a hidden layer and an output layer. The input and hidden layers are composed of 100 and 50 neurons respectively, while the output layer is composed of 2 neurons. We selected tangent sigmoid function as the activation function for input and hidden layers while, we used log sigmoid activation function for the output layer.

We have designed the structure of the neural network in such a way as to reflect the importance of the status of the health of different organs in the eventual outcome. As shown in Figure 23.1, we have designed the neural network in such a way that the cardiac and neural features are connected to all three layers of the network while pulmonary and renal features are connected to two layers, and the rest of the features are only connected to the first layer. This is based on clinical knowledge [41] and predisposes the algorithm toward a

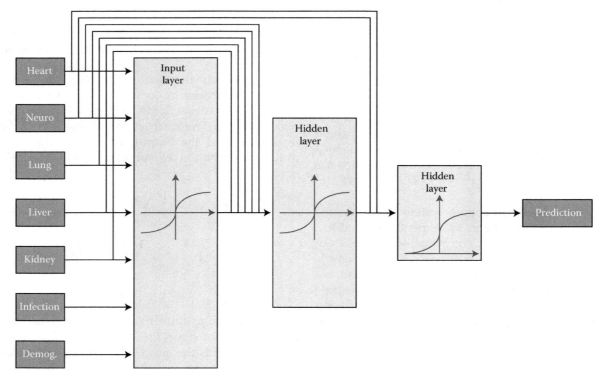

FIGURE 23.1
Schematic design of neural network for ICU outcome prediction.

model in which cardiac and neural factors are more important than those of the other organs.

23.3 Results

The evaluation metric used in this study is the *F*-score, which is a measure of a classifier's accuracy. It is the harmonic mean of the positive predictive value and sensitivity of the classification and has been widely used in the context of computer-aided diagnostics in medical decision-making systems. Since it is the mean of sensitivity and positive predictive value, the *F*-score reduces algorithmic bias and is calculated using the following equation:

$$F = 2 \times \frac{Se \times PP}{Se + PP} \quad (23.4)$$

where Se and PP are sensitivity and positive predictive values of a classifier and are defined using

$$Se = \frac{TP}{TP + FN} \quad (23.5)$$

$$PP = \frac{TP}{TP + FP} \quad (23.6)$$

where TP, FN, and FP are true positive, false negative, and false positive values respectively.

The outcome of this study is "in-hospital-mortality," so the outcome is 1 if the patient dies in the hospital even if outside ICU, and it is 0 if the patient survives during the hospitalization. The data is for a period of 48 hours after admission to the ICU.

To address the problem of missing and sparse data, we replaced the missing measurements with their respective mean for the population. This method has been implemented successfully in previous studies [1].

We use two methods of validation to measure the performance of the algorithm. These validation methods are (1) the *k*-fold cross-validation method and (2) the leave-out validation method. We use eightfold validation for testing the accuracy of the algorithm. Eightfold validation means that we divide the data into eight equally sized sections; in this case, each section comprises 500 patients. We then train the classifier with seven sections of the data and validate with the remaining section. We repeat this procedure eight times until the classifier is validated with all the available data. At each repetition, we calculate the *F*-score based on the results of the classifier. We finally average the eight calculated *F*-scores and report it as the final result of the classifier. Since the CMO and DNR patients were not excluded from the database, this will affect the performance of the classifier. Hence, by using the eightfold validation techniques and averaging the results, we can mitigate the bias induced by the inclusion of the CMO and DNR in a certain group of the data set. We also used the leave-out validation method to test the accuracy of the classifier. In this method, we use the data from the first 3,000 patients for feature selection and training the algorithm. It means that we form the best subset of the features based only on the training data. We then test the classifier on the last 1,000 patients.

Table 23.4 shows the selected features based on the proposed algorithm for each organ for the two different validation methods. The results of this table show that the proposed feature ranking method is very robust in selecting the features subsets as only there are three features different between the two validation methods. The frequency of collection has been used in the literature for ICU outcome prediction, but we can observe from Table 23.4 that it has not been picked up as an important indicator of the outcome and that it is not a useful indicator of outcome.

Furthermore, as mentioned earlier, the DNR and CMO patients were not excluded from the data set, and in fact, we are blind to these patients. Since DNR or limitation of support was not excluded, this makes it really difficult to predict the outcome, hence decreasing the accuracy of the methods because death may not have been due to

TABLE 23.4

Selected Organ Features for Two Different Validation Methods

Method	Organ	Variables
Leave-out	Heart	Minimum DABP, K ORI, HR daily trend, maximum rate of change for MAP, SABP alarms
	Neuro	MAP sample entropy, GCS alarms, glucose ORI, MAP alarms, GCS daily trend, SaO_2 variation index
	Lung	RR sample entropy, FiO_2 alarms, RR daily trend, $PaCO_2$ ORI, pH alarm
	Liver	Albumin alarms, albumin ORI, bilirubin alarms
	Kidney	BUN alarms, BUN risk, creatinine alarms, K ORI
	Infection	WBC ORI, temperature alarms
K-fold	Heart	HR sample entropy, minimum DABP, K ORI, HR daily trend, maximum rate of change for MAP, SABP alarms
	Neuro	MAP sample entropy, GCS alarms, glucose ORI, MAP alarms, GCS daily trend, SaO_2 variation index
	Lung	RR sample entropy, FiO_2 alarms, maximum rate of change for HCO_3, $PaCO_2$ ORI, pH alarm
	Liver	Albumin alarms, albumin ORI, bilirubin alarms
	Kidney	BUN alarms, BUN risk, creatinine alarms, K ORI
	Infection	WBC ORI, temperature alarms

TABLE 23.5

Comparison of Prediction Accuracies of This Chapter's Predictive Analytic Method and Traditional Acuity Scores

Method	Group No.	Died	TP	FP	FN	TN	F-Score	SOFA	SAPS II
K-fold	1	63	28	45	35	392	0.41	0.28	0.27
	2	72	38	52	34	376	0.47	0.25	0.31
	3	69	31	42	38	389	0.43	0.24	0.30
	4	56	22	43	34	401	0.37	0.26	0.25
	5	70	31	40	39	390	0.43	0.31	0.32
	6	58	20	32	38	410	0.37	0.27	0.31
	7	53	22	28	31	429	0.39	0.29	0.27
	8	71	35	35	36	394	0.49	0.29	0.28
Leave-out	1	141	66	75	71	788	0.46	0.26	0.29
Current	SOFA	512	164	443	348	3045	0.28	–	–
	SAPS	512	144	406	368	3082	0.31	–	–

poor physiological state but could have been due to poor prognosis (for example, end-stage cancer, etc.).

In order to reduce the number of design parameters for the algorithm, we designed the ANN to have two outputs. One is the probability of death, and the other one is the probability of survival. The output with higher probability is reported as outcome. The threshold for SAPS score is 29% probability, while the threshold for SOFA is 9 to predict death.

Table 23.5 presents the breakdown of the results for each of validation methods. As discussed earlier, the reported F-score for the eightfold validation method is the mean of the calculated F-scores for each group. As we can see from this table, the mortality rate for the groups in the eightfold validation varies from 10.6% to 14.4%; this rate for the leave-out validation is 14.1% and for the entire collected data is 12.8%. We can see that this is an unbalanced data set where the output of interest is significantly lower in comparison with the neutral output. An unbalanced data set makes the prediction problem even more complicated. It can be observed from Table 23.5 that our algorithm constantly outperforms the SOFA and SAPS II scoring systems.

To better describe the performance of the developed decision support system, we have performed additional analysis. The receiver operating characteristic (ROC) curves for eightfold validation, leave-out validation, and SOFA and SAPS scoring systems are shown in Figures 23.2 and 23.3. These figures further show that the designed algorithm outperforms SOFA and SAPS. Table 23.6 presents the breakdown of the sensitivity and positive predictive value results for each of the validation

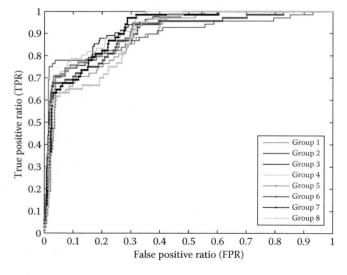

FIGURE 23.2
Receiver operating characteristic (ROC) results for eightfold validation method.

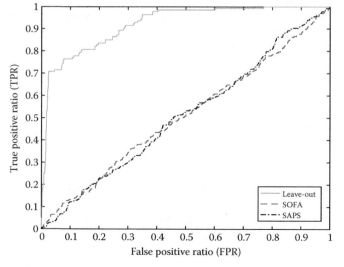

FIGURE 23.3
Receiver operating characteristic (ROC) results for leave-out validation method.

TABLE 23.6
Comparison of Prediction Accuracies of This Chapter's Predictive Analytic Method and Traditional Acuity Scores

Method	Group No.	Se	PP	AUROC	H-Statistics
K-fold	1	44.4%	38.7%	0.8906	75
	2	52.8%	42.2%	0.8940	76
	3	44.9%	42.5%	0.9268	78
	4	39.3%	33.9%	0.8840	74
	5	44.3%	43.7%	0.9140	83
	6	34.5%	38.4%	0.8891	82
	7	41.5%	44.0%	0.9208	79
	8	49.3%	50.0%	0.9278	85
Leave-out	1	48.2%	46.8%	0.9116	81
Current	SOFA	32.0%	27.0%	0.4805	205
	SAPS	28.1%	34.6%	0.5493	139

Note: AUROC is area under the receiver operating characteristic curve.

methods as well as the Hosmer–Lemeshow H-statistic results. The sensitivity and predictive values for the eightfold validation are 43.9 ± 0.06 and 41.6 ± 0.05 respectively.

Figure 23.3 results show that the area under the ROC curve (AUROC) for the SOFA and SAPS is very close to 0.5, which translates to no predictive performance for these scoring systems. In comparison, Pirracchio et al. [31] showed that the AUROC for SOFA and SAPS is around 0.7–0.8. The reason for this difference could be related to the fact that we are blind to the CMO and DNR patients and their numbers in this data set are relatively high. The results of this study show that the proposed method is robust to the presence of the DNR and CMO patients in the data set.

The final comparison of our results with SOFA and SAPS scoring systems shows approximately 55% and 45% improvement in the outcome prediction accuracy. To statistically compare our developed algorithm with SOFA and SAPS, we performed a t-test on the data. Table 23.6 represents the results obtained from performing t-test comparison analysis.

TABLE 23.7
Comparison of Developed Decision Support Algorithm and SOFA and SPAS Scoring Systems Using T-Test Comparison Analysis

Compared Methods	p-value
K-fold vs. SOFA	0.007
K-fold vs. SAPS	0.002
Leave-out vs. SOFA	0.0009
Leave-out vs. SAPS	0.01

The results represented in the Table 23.7 indeed show a significant improvement over the current status, which we believe represents a promising way forward.

23.4 Conclusion

In this work, we addressed the problem of ICU outcome prediction. One of the main objectives of this study is to increase ICU outcome prediction accuracy in order to help clinicians identify patients at high risk of mortality and morbidity and hence enable them to plan possible treatments in a timely manner to avoid losing the patient. We believe that, although the current acuity scoring systems can perform well in detecting very sick patients, they fail to detect the patients that are at high risk for critical deterioration but are not critically ill. The algorithm that we presented in this chapter is able to outperform standard acuity scoring systems in the ICU such as SOFA and SAPS by more than 45%. Although this algorithm has similarities with the acuity scoring systems, which makes it acceptable for clinical use, it is capable of analyzing the data at deeper levels to extract the hidden information that exists in the data.

We have been able to fully integrate physiological knowledge in the design of this algorithm to make it more clinically intuitive and to also improve the accuracy of the predictions. Some of the most important aspects of physiological knowledge integration are categorizing the data into different related organs, use of ORI as one of the critical features that we extract from the data, and design of the neural network classifier in a way that mimics the importance of the organs in the body. We believe that this information is fundamentally important for improving the performance of the algorithm and to make it clinically intuitive.

Additionally, in this chapter, we have demonstrated the potential of CDSSs in solving complex problems in the ICUs. The CDSSs are mostly driven by data, and their performance depends heavily on the collected data; however, integrating expert opinion and available physiological understanding is crucial to the success of these methods in real-life clinical settings as we believe that this is the only way to make clinicians relate to these systems and trust them for day-to-day clinical use. We understand that just outperforming common acuity scoring systems does not assure robustness of the proposed CDSS. However, we need to consider that these scoring systems are currently used by clinicians and no other gold standard method exists to replace these systems in order to effect an improvement. Hence, we hope that the results of this chapter show the potential of

CDSSs in solving ICU outcome prediction as well other similar complex medical decision-making processes.

References

1. A.E.W. Johnson, N. Dunkley, L. Mayaud, A. Tsanas, A.A. Kramer, and G.D. Clifford. Patient specific predictions in the intensive care unit using a Bayesian ensemble. In *Computing in Cardiology (CinC), 2012*, pages 249–252. IEEE, 2012.
2. C.B. Laramee, L. Lesperance, D. Gause, and K. Mcleod. Intelligent alarm processing into clinical knowledge. In *Proceedings of Annual Conference of the IEEE Engineering in Medicine and Biology Society*, pages 6657–6659, 2006.
3. R. Ceriani, M. Mazzoni, F. Bortone, S. Gandini, C. Solinas, G. Susini, and O. Parodi. Application of the sequential organ failure assessment score to cardiac surgical patients. *Chest*, 123(4):1229–1239, Apr 2003.
4. K.I. Halonen, V. Pettilä, A.K. Leppäniemi, E.A. Kemppainen, P.A. Puolakkainen, and R.K. Haapiainen. Multiple organ dysfunction associated with severe acute pancreatitis. *Critical Care Medicine*, 30(6):1274–1279, Jun 2002.
5. V. Lemiale, M. Resche-Rigon, E. Azoulay, and Study Group for Respiratory Intensive Care in Malignancies Groupe de Recherche en Réanimation Respiratoire du patient d'Onco-Hématologie. Early non-invasive ventilation for acute respiratory failure in immunocompromised patients (IVNIctus): Study protocol for a multicenter randomized controlled trial. *Trials*, 15:372, 2014.
6. J. Lee and R.G. Mark. An investigation of patterns in hemodynamic data indicative of impending hypotension in intensive care. *Biomedical Engineering Online*, 9:62, 2010.
7. C.W. Hug and P. Szolovits. ICU acuity: Real-time models versus daily models. *Proceedings of AMIA Annual Symposium*, 2009:260–264, 2009.
8. W.A. Knaus, D.P. Wagner, E.A. Draper, J.E. Zimmerman, M. Bergner, P.G. Bastos, C.A. Sirio, D.J. Murphy, T. Lotring, and A. Damiano. The APACHE III prognostic system. risk prediction of hospital mortality for critically ill hospitalized adults. *Chest*, 100(6):1619–1636, Dec 1991.
9. J.R. Le Gall, S. Lemeshow, and F. Saulnier. A new simplified acute physiology score (SAPS II) based on a European/North American multicenter study. *JAMA*, 270(24):2957–2963, 1993.
10. C.P. Subbe, M. Kruger, P. Rutherford, and L. Gemmel. Validation of a modified early warning score in medical admissions. *Quarterly Journal of Medicine*, 94(10):521–526, Oct 2001.
11. S. Lemeshow, J. Klar, D. Teres, J.S. Avrunin, S.H. Gehlbach, J. Rapoport, and M. Rué. Mortality probability models for patients in the intensive care unit for 48 or 72 hours: A prospective, multicenter study. *Critical Care Medicine*, 22(9):1351–1358, Sep 1994.
12. U. Janssens, C. Graf, J. Graf, P. W. Radke, B. Königs, K. C. Koch, W. Lepper, J. vom Dahl, and P. Hanrath. Evaluation of the SOFA score: A single-center experience of a medical intensive care unit in 303 consecutive patients with predominantly cardiovascular disorders. sequential organ failure assessment. *Intensive Care Medicine*, 26(8):1037–1045, Aug 2000.
13. S. Ahmad, A. Tejuja, K.D. Newman, R. Zarychanski, and A.J. Seely. Clinical review: A review and analysis of heart rate variability and the diagnosis and prognosis of infection. *Critical Care Medicine*, 13(6):232, 2009.
14. I. Silva, G. Moody, D.J. Scott, L.A. Celi, and R.G. Mark. Predicting in-hospital mortality of ICU patients: The physionet/computing in cardiology challenge 2012. In *Computing in Cardiology (CinC), 2012*, pages 245–248. IEEE, 2012.
15. C. Carle, P. Alexander, M. Columb, and J. Johal. Design and internal validation of an obstetric early warning score: Secondary analysis of the intensive care national audit and research centre case mix programme database. *Anaesthesia*, 68(4):354–367, Apr 2013.
16. K.C. Tan, Q. Yu, C.M. Heng, and T.H. Lee. Evolutionary computing for knowledge discovery in medical diagnosis. *Artificial Intelligence in Medicine*, 27(2):129–154, 2003.
17. R.B. Haynes, N.L. Wilczynski, and Computerized Clinical Decision Support System (CCDSS) Systematic Review Team. Effects of computerized clinical decision support systems on practitioner performance and patient outcomes: Methods of a decision-maker–researcher partnership systematic review. *Implementation Science*, 5:12–20, 2010.
18. J. Eberhardt, A. Bilchik, and A. Stojadinovic. Clinical decision support systems: Potential with pitfalls. *World Journal of Surgical Oncology*, 105(5):502–510, 2012.
19. A.X. Garg, N.K.J. Adhikari, H. McDonald, M.P. Rosas-Arellano, P.J. Devereaux, J. Beyene, J. Sam, and R.B. Haynes. Effects of computerized clinical decision support systems on practitioner performance and patient outcomes: A systematic review. *JAMA: The Journal of American Medical Association*, 293(10):1223–1238, 2005.
20. V. Podgorelec, P. Kokol, and M.M. Stiglic. Searching for new patterns in cardiovascular data. In *Proceedings 15th IEEE Symp. Computer-Based Medical Systems (CBMS)*, pages 111–116, 2002.
21. A.C. Stasis, E.N. Loukis, S.A. Pavlopoulos, and D. Koutsouris. A decision tree-based method, using auscultation findings, for the differential diagnosis of aortic stenosis from mitral regurgitation. In *Proceedings Computers in Cardiology (CinC)*, pages 769–772, 2003.
22. G. Dounias and D. Linkens. Adaptive systems and hybrid computational intelligence in medicine. *Artificial Intelligence in Medicine*, 32(3):151–155, 2004.
23. A. Singh and J.V. Guttag. A comparison of non-symmetric entropy-based classification trees and support vector machine for cardiovascular risk stratification. In *Proceedings IEEE International Conference Engineering in Medicine and Biology Society (EMBC)*, pages 79–82, 2011.
24. B. Samanta, G.L. Bird, M. Kuijpers, R.A. Zimmerman, G.P. Jarvik, G. Wernovsky, R.R. Clancy, D.J. Licht, J.W. Gaynor, and C. Nataraj. Prediction of periventricular

leukomalacia. Part II: Selection of hemodynamic features using computational intelligence. *Artificial Intelligence in Medicine*, 46(3):217–231, 2009.
25. A. Jalali, D.J. Licht, and C. Nataraj. Application of decision tree in the prediction of periventricular leukomalacia (pvl) occurrence in neonates after heart surgery. *Proceedings IEEE International Conference Engineering in Medicine and Biology Society*, 2012:5931–5934, 2012.
26. A. Jalali, D.J. Licht, and C. Nataraj. Discovering hidden relationships in physiological signals for prediction of periventricular leukomalacia. *Proceedings IEEE International Conference Engineering in Medicine and Biology Society*, 2013:7080–7083, 2013.
27. A. Jalali, E.M. Buckley, J.M. Lynch, P.J. Schwab, D.J. Licht, and C. Nataraj. Prediction of periventricular leukomalacia occurrence in neonates after heart surgery. *IEEE Journal of Biomedical and Health Informatics*, 18(4):1453–1460, Jul 2014.
28. R. Pirracchio, M.L. Petersen, M. Carone, M.R. Rigon, S. Chevret, and M.J. van der Laan. Mortality prediction in intensive care units with the super ICU learner algorithm (SICULA): A population-based study. *Lancet Respir Med*, 3(1):42–52, Jan 2015.
29. D.E. Lake, J.S. Richman, M.P. Griffin, and J.R. Moorman. Sample entropy analysis of neonatal heart rate variability. *American Journal of Physiology—Regulatory, Integrative and Comparative Physiology*, 283(3):R789–R797, 2002.
30. J.R. Moorman, J.B. Delos, A.A. Flower, H. Cao, B.P. Kovatchev, J.S. Richman, and D.E. Lake. Cardiovascular oscillations at the bedside: Early diagnosis of neonatal sepsis using heart rate characteristics monitoring. *Physiological Measurements*, 32(11):1821–1832, 2011.
31. G. Camps-Valls, B. Porta-Oltra, E. Soria-Olivas, J.D. Martín-Guerrero, A.J. Serrano-López, J.J. Pérez-Ruixo, and N.V. Jiménez-Torres. Prediction of cyclosporine dosage in patients after kidney transplantation using neural networks. *IEEE Transactions on Biomedical Engineering*, 50(4):442–448, Apr 2003.
32. H.M. Al-Angari and A.V. Sahakian. Use of sample entropy approach to study heart rate variability in obstructive sleep apnea syndrome. *IEEE Transactions on Biomedical Engineering*, 54(10):1900–1904, 2007.
33. Q. Zhou, W. Hong, G. Shao, and W. Cai. A new SVM-RFE approach towards ranking problem. In *Proceedings IEEE International Conference on Computing and Intelligent Systems (ICIS)*, volume 4, pages 270–273, 2009.
34. Y.-W. Chang and C.-J. Lin. Feature ranking using linear SVM. In *WCCI Workshop on Causality*, pages 53–64, 2008.
35. H. Peng, L. Fulmi, and C. Ding. Feature selection based on mutual information criteria of max-dependency, max-relevance, and min-redundancy. *IEEE Transactions on Pattern Analysis and Machine Intelligence*, 27(8):1226–1238, 2005.
36. J. Dayhoff and J.M. DeLeo. Artificial neural networks: opening the black box. *Cancer*, 91(8 Suppl):1615–1635, Apr 2001.
37. C.-L. Chi, W.N. Street, and W.H. Wolberg. Application of artificial neural network–based survival analysis on two breast cancer datasets. *AMIA Annu Symp Proc*, pages 130–134, 2007.
38. R. Das, I. Turkoglu, and A. Sengur. Effective diagnosis of heart disease through neural networks ensembles. *Expert Systems with Applications*, 36(4):7675–7680, 2009.
39. B. Sekar, M. Dong, J. Shi, and X. Hu. Fused hierarchical neural networks for cardiovascular disease diagnosis. *IEEE Sensors Journal*, (99), 2011. Early access.
40. P. Ghorbanian, A. Jalali, A. Ghaffari, and C. Nataraj. An improved procedure for detection of heart arrhythmias with novel pre-processing techniques. *Expert Systems*, 29(5):478–491, 2011.
41. G. Rocker, D. Cook, P. Sjokvist, B. Weaver, S. Finfer, E. McDonald, J. Marshall, A. Kirby, M. Levy, P. Dodek, D. Heyland, and G. Guyatt. Clinician predictions of intensive care unit mortality. *Critical Care Medicine*, 32(5):1149–1154, 2004.

24

Trauma Outcome Prediction in the Era of Big Data: From Data Collection to Analytics

Shiming Yang, Peter F. Hu, and Colin F. Mackenzie

CONTENTS

24.1 Introduction 477
24.2 Automated Medical Big Data Monitoring 478
    24.2.1 Data from a Real Trauma Center 478
    24.2.2 Redundant VS Collection System with a Diagnosis Viewer 478
24.3 Data Analytics in Trauma Patients 480
    24.3.1 Features Based on Descriptive Statistics 480
    24.3.2 Features Based on Variability 481
    24.3.3 Features Based on Correlations 483
    24.3.4 Features Based on Entropy 484
24.4 Machine Learning Framework and Its Applications 485
    24.4.1 Feature Selection 486
    24.4.2 Performance Metrics 487
24.5 Computing Issues 488
24.6 Discussion 489
References 489

24.1 Introduction

Traumatic brain injury (TBI) and hemorrhage shock are the leading causes of morbidity and mortality after injury both on the battlefield and in civilian care [1,2]. In parallel with new clinical care protocol and technique development, increasing efforts in continuous physiological monitoring and data analysis are reported [3,4]. The ultimate goal of Big Data's application in this area is to optimize the limited medical resource allocation through comprehensive analysis of a large amount of data, hence improving patients' outcomes and reducing healthcare cost.

Early recognition and mitigation of secondary injury and hemorrhage could ameliorate the effect of brain injury or prevent death caused by massive bleeding. Various statistical modeling and machine learning methods have been used to explain TBI outcome associations and to predict future conditions with massive data sets. Research and funding agencies, including National Institute of Neurological Disorders and Stroke (NINDS) and the U.S. Department of Defense (DoD), are also investing more resources in developing reliable TBI outcome and transfusion prediction models and practically usable tools based on intensive analysis of large data collections. Military medicine considers these approaches as the future way to develop combat casualty autonomous resuscitation [5,6] and enhance real-time field decision making [7].

The volume of real-time physiological patient data has proliferated with each advance in computer hardware and medical sensor technology. High-fidelity data are streamed into physiological monitors for care planning, clinical decision support, quality improvement, and remote patient monitoring. Processing and extracting useful and actionable knowledge from these data also require consideration of the techniques used to store, manage, and analyze massive data. In this chapter, we will review continuous physiological data collection, signal processing, and data analysis methods for predicting trauma patient outcomes related to actionable therapeutic interventions and translating this into autonomous resuscitation. In Section 24.2, we describe a reliable large-scale physiological data collection system, which is the basis of any Big Data study. In Section 24.3, we review some medical signal processing and feature extraction methods for signals typically used in predicting trauma patients' short- and long-term outcomes. In Section 24.4, we provide a description of a comprehensive framework for a machine learning approach to TBI outcomes and blood transfusion prediction, with example projects for the purpose of early detection and automated resuscitation.

24.2 Automated Medical Big Data Monitoring

Modern hospitals are often equipped with bedside monitors collecting various physiological data in a real-time, continuous, and automated way. Data ranging from routine intermittent observations to high-fidelity waveforms can be recorded and streamed into monitors for care planning, clinical decision support [8], quality improvement [9], and reduced hospital mortality [10]. With massive storage capability, those data can also be stored as part of the electronic health records (EHRs) for retrospective analyses such as physiological pattern discovering [11,12] and prediction modeling [13,14]. One example is the PhysioBank, a large collection of biomedical databases, which inspires studies in cardiovascular time series dynamics, modeling intracranial pressure (ICP) for noninvasive estimation, and more [15–18]. However, in a busy resuscitation or healthcare environment, collecting more complete data is not the top priority of health providers. System failure or manual data entry errors can result in missing values, causing difficulty in application of decision-support algorithms; or such failures can cause a loss of data associated with rare events. Therefore, a reliable system that simplifies and automates the hospital-wide data collection process is necessary.

24.2.1 Data from a Real Trauma Center

In the R Adams Cowley Shock Trauma Center, a level I regional trauma center located in downtown Baltimore, Maryland, 94 GE Marquette Solar 7000/8000® (General Electric, Fairfield, CT) patient vital sign (VS) monitors are networked to provide collection of real-time patient VS data streams in 13 trauma resuscitation units (TRU) and 9 operating room (OR), 12 postanesthesia care unit (PACU), and 60 intensive care unit (ICU) individual bed/monitor units. Each patient monitor collects real-time 240 Hz waveforms and 0.5 Hz trend data, which are transferred via secure intranet to a dedicated BedMaster® server (Excel Medical Electronics, Jupiter, FL) and archived [19]. This process generates approximately 20 million data points/day/bed or roughly 30 terabits/year of data. Physiological data collected through this system, when they are displayed on the GE Marquette monitor, include electrocardiographic (ECG), photoplethysmographic (PPG), carbon dioxide (CO_2), arterial blood pressure (ABP), and ICP data, among others. Trends include heart rate (HR), respiratory rate (RR), temperature, oxygen saturation (SPO_2), end-tidal CO_2 ($EtCO_2$), and ICP, among many others. They cover the categories of brain pressure, cardiac, perfusion, and respiratory.

In addition to continuous data, other heterogeneous data with different formats and temporal resolutions are also collected and organized in databases, including ordinal or categorical data [e.g., Glasgow Coma Scale (GCS), age, sex]; radiological images; text (medical records, clinical notes); bed movement (admission, discharge and transfer time, and location); and other important data (adverse events, treatments, response, etc.).

24.2.2 Redundant VS Collection System with a Diagnosis Viewer

These heterogeneous data are available and need to be collected with different formats and temporal resolutions. In the current hospital environment, these data are in a loosely organized decentralized network. One approach to manage VS waveforms and trend data, used at a level I trauma center that admits more than 8000 severely injured patients annually, is to design a triple-redundant data collection server for high fault tolerance to maximize these data collection rates to nearly 100%.

To minimize the impact of individual server collection failure, we installed three dedicated BedMaster servers in parallel to simultaneously collect physiological patient data from the network of patient monitors described above. Figure 24.1 diagrams the data streams from multiple individual bed units to three BedMaster servers arranged in parallel. This triple modular redundancy architecture permits fast switch over time and high system availability [20]. One server is selected as a principal or "backbone" server. When it fails, values from a second sever will fill in. When two servers fail, values from the third one will be used.

The triple-redundant data collection system could increase data availability. However, a tool for fast system diagnosis is still lacking. To address the need for ongoing

Trauma Outcome Prediction in the Era of Big Data

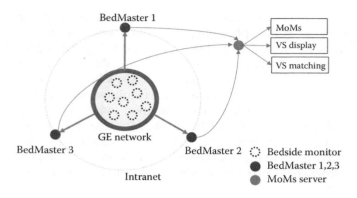

FIGURE 24.1
The MoMs system architecture with triple modular redundancy design using three BedMaster servers.

system status monitoring and real-time presentation critical clinical data, we developed the MoMs (Monitor of Monitors) information representation layer over the VS collection system. Using the current data collecting architecture and a minimum-instrument approach, we stream the most recent record from the BedMaster server from each bed to a dedicated data server, the MoMs server (Figure 24.1). A high-performance database hosts those data items labeled with data server name, bed unit, time stamp, and admission status.

The front end (MoMs viewer) is designed as a web-based application so that users can access it from any location in the hospital. IP address white list and user login modules are used for information security. Each bed collection status is summarized and pushed to the MoMs viewer through the Ajax (asynchronous Java-Script and XML) techniques every minute. All participating patient bed units are represented by individual cells in each of three spreadsheet blocks representing one of the three redundant BedMaster servers. Figure 24.2 shows a block of the web-based monitoring system corresponding to bedside collections from monitors in the TRU, ORs, and neurotrauma critical care (NTCC), and multitrauma critical care (MTCC) units. The background color of each cell represents the associated bed's data collection status. Green indicates that the data stream has been alive in the last 5 minutes. Yellow indicates that the last time stamp from data from that bed/monitor is 5 minutes to 4 hours old and that a problem may exist. Dark red indicates a time stamp gap

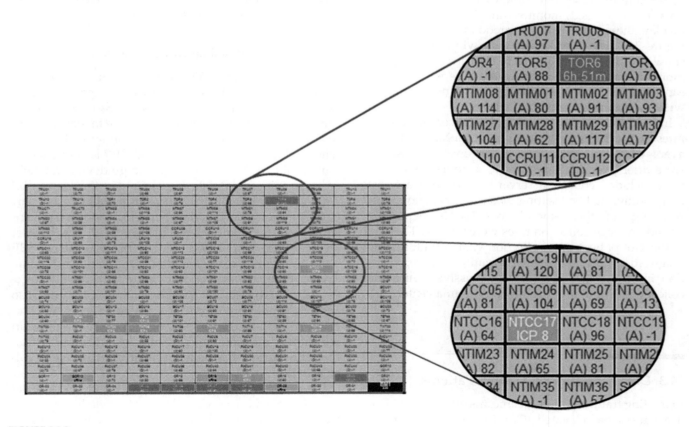

FIGURE 24.2
A portion of MoMs viewer for data collection status. Light gray (shown): collection is active (within last 5 minutes); silver (not shown): collection was active 5 minutes to 4 hours ago; dark gray (shown): no data collection has occurred in more than 4 hours; gray (shown): bedside collection is offline. In each cell, letter "A" means admitted; letter "D", discharged. The medium gray background cell indicates a patient with an intracranial monitor in place; ICP value appears in white.

greater than 4 hours and that action should be taken to remedy the problem. Report of an elapsed data collection gap includes the duration of collection failure.

There are different elements in each colored cell to indicate bed unit occupancy. Often, nurses may press a bedside button for admission (A) to or discharge (D) from this bed. This allows for a cross-check on potential causes for information gaps such as the device being temporarily inoperable or no patient being monitored. In addition, bed occupancy can be verified by real-time physiological values, such as HR. It can be used as second evidence for us to infer if a bed unit is currently occupied by a patient. If one bed unit is offline, the gap between now and its last reported time will be shown. Figure 24.2 shows one example in the unit OR-6, which is highlighted in a red cell with a time gap of 6 hours and 51 minutes.

The easy configuration of the MoMs dashboard viewer also allows it to be used for identifying and displaying clinical information of special research interest. For example, ICP monitoring is an important VS for traumatic brain-injured patients and is not often collected due to its invasive nature. To receive early notification of ICP-monitored cases, the MoMs viewer can extract ICP data from all bed/monitor units' data streams and display these data using a predefined color code. In Figure 24.2, those pink cells with white bold font text show real-time ICP values from the corresponding bed/monitor units. For example, at the time we viewed the MoMs system, the unit NTCC-17 was monitoring ICP with an instant value of 8 mmHg.

In a 12-month study period, single-server collection rates in the 3-month period before MoMs deployment averaged 81.4%. Of the 18.6% collections lost, most (18%) were brief periods, 5 minutes to 4 hours. Reasons for gaps included collection server failure, software instability, individual bed setting inconsistency, and monitor servicing. In the 6-month post-MoMs deployment period, average collection rates were 99.96%. The triple-redundant patient data collection system with real-time diagnostic information summarization and representation improved the reliability of massive clinical data collection to near 100% in a level I trauma center.

24.3 Data Analytics in Trauma Patients

In treating trauma patients, clinicians often wish to know if a patient has elevated ICP; if some lifesaving interventions are needed in the near future, such as massive transfusion; or if a patient has unstable neurologic status. Although there is a high volume of data monitored for trauma patients, they are often underused to answer those questions. Those data are often only used for bedside instant physiological status view. Due to the difficulty in storing, accessing, and analyzing, those data streams are hardly used beyond the VS readings or instant waveform morphism analysis. Clinicians often utilize those medical data in an empirical way, which may consume their extra attention while not fully unlocking the value of the data. Validated, automated medical data processing and analysis could aggregate massive amounts of data and assist clinicians to quickly recognize changes in physiological status and prioritize care.

Processing medical data/signals is an important initial step before building predictive models. In prehospital and even in hospital, data are collected in a hostile environment, which adds noise to the original signals. Numerous methods and algorithms have been studied to remove outliers and noise from the signal, to smooth or sample from high rate signals, and to explore signal characteristics in time, frequency, and joint domains [4,21–23]. In this section, we assume that data/signals have been preprocessed as "clean data" and focus on a few topics on feature design, feature selection, and model evaluation, which are less tackled in many medical data processing handbooks.

24.3.1 Features Based on Descriptive Statistics

Maintaining patients' VSs within normal ranges is a basic task for clinicians. Some treatment protocols also give guidelines on the thresholds of VSs to be watched during patient care. For example, for the management of severe TBI, it is recommended to begin initial treatment when ICP is above 20 mmHg, and CPP is suggested to remain above 70 mmHg [24]. The guidelines for field triage of injured patients recommended the use of SBP <90 mmHg or RR <10 or >29 breaths per minute as a part of the physiological criteria when considering if a patient should be transported to a facility with the highest level of care [25]. When adding the age factor, SBP <110 mmHg was recommended in the National Trauma Triage Protocol for geriatric trauma patients (age > 65 years old) [25,26]. Empirical thresholds are also used as simple predictors or decision triggers by clinicians. In field triage, shock index (SI) > 1 is used as an indicator of circulation failure and a predictor of critical bleeding, since it has high specificity and is easy to calculate [27].

Those thresholds that pass muster with the clinical experts indeed can serve as domain knowledge to design features from physiological time series. Such type of features may hold clinical meanings that are easy to interpret by humans. To quantify the cumulative effect of VS away from normal range, the "pressure-times-time dose" (PTD) is defined as the integrated area enclosed by the VS curve and the threshold line within a given time interval. Sometimes, to compare between patients, it is also calculated as averaged PTD in unit time, which is the PTD normalized by time duration. Even thought two patients may be monitored for different time durations, their PTDs in unit time are still comparable. In predicting TBI patients' outcomes, the PTDs of ICP >20 mmHg and CPP < 60 mmHg have been shown to be good predictors of in-hospital mortality and length of ICU stay [28].

Similar to the idea of PTD, some descriptive statistics of VS within a given time period are also informative for clinician use or may contribute to outcome prediction in a model. With the assumption that the observed data are approximately normally distributed, mean and variance are often used to sketch the VS value distribution. Standard deviation (SD) is used to quantify the variability of observed VS data. The coefficient of variance, which is the SD divided by the mean, is a unit-less value that is suitable to compare between data sets with widely different means. Robust statistics, such as percentiles or quartiles, are also used to quantify the shape of the VS data distribution. The median (50th percentile or 2nd quartile) is one of the most commonly used statistics in VS feature calculation.

Those descriptive statistics, plus the PTD value of VS in a given time period, have the following merits. First, those features can easily incorporate domain knowledge and hence possess straightforward meaning. Stein et al. described a scheme of designing 588 features from 9 typically used VSs, using various thresholds and time windows [14]. Second, they could aggregate large amounts of data or simplify complex data as a few summary quantities, which can be used by many statistical prediction models. For example, to identify patients with critical bleeding risk, Mackenzie et al. summarized patients' first 15 minutes 240 Hz PPG by applying the above descriptive statistics to the peak-to-valley distances of PPG. In such a way, each patient's 216,000 waveform data points are aggregated into dozens of features as inputs for logistic regression [29]. Third, those values can be calculated efficiently even if the data are of high volume and high velocity.

Despite these listed merits, we need to understand that some of these features rely on certain assumptions. For example, normality is often assumed when we use the mean and SD to sketch the shape of VS data distribution. Besides, we may assume that the thresholds suggested by guidelines can be applied to any patient group at any time period. In reality, the thresholds may vary along time, no matter how slow. Also, some hidden factors may result in a large difference in threshold values between age groups or sex groups. Therefore, we need to be careful when incorporating clinical prior knowledge into those features, by examining the applicable conditions.

24.3.2 Features Based on Variability

Continuous noninvasive ECG and PPG sensors are ubiquitous in both prehospital emergency medical service and in-hospital healthcare for TBI patients. Waveforms measured from both sensors capture rich information on cardiovascular, circulatory, and respiratory systems. Heart rate variability (HRV), derived from ECG waveform, has long been used in studies of prehospital lifesaving interventions [30] and neurologic disorder [31,32], and the association between the autonomic nervous system (ANS) and cardiovascular mortality is well established [33]. ANS is a complex life-sustaining system, which plays a role in nearly every organ and disease [34], including regulating cardiac activity, respiration, and pupillary response. ANS dysfunction is a potential complication following severe TBI [35]. While neurological diseases can lead to many changes in cardiac function, the major changes noted are arrhythmias and repolarization changes. Goldstein et al. [36] noted that either increased ICP or decreased CPP can be associated with ANS dysfunction. Baguley et al. [37] observed that in TBI, patients with or without dysautonomia showed HRV differences compared to the control group.

Another universally used sensor, pulse oximetry, generates a PPG waveform that carries rich physiological information, such as HR, SPO_2, and even RR [38]. The peaks in PPG are almost synchronized and corresponding to the R peaks from ECG in the time domain. Therefore, the peak–peak interval from PPG can be used as an alternative to the normal-to-normal (NN) interval calculated from ECG recordings. Lu et al. compared HRV and PPG variability (PPGV) in both time- and frequency-domain-based on data collected from 20 healthy subjects. They found that PPGV was highly correlated to HRV and could serve as an alternative measurement [39].

To find QRS peaks from ECG, the Pan–Tompkins method is used [40]. The ECG signal is smoothed with the Savitzky-Golay filter to increase the signal-to-noise ratio. The R peaks are then detected based on a threshold that is adaptive to the signal. To detect PPG signal peaks and valleys, the Savitzky–Golay filter is applied to smooth the signal [41]. The peaks can be found through two rounds of searching. In the first round, the peaks are roughly found through MATLAB built-in routine "findpeaks." The median distance between two consecutive peaks PD_{median} is then calculated. In the second round, any small peaks within a range of $0.6 \times PD_{median}$ from a large peak would be ignored. Figure 24.3 shows a typical PQRST segment from ECG, with identified peaks and five items that we use for calculation. The NN interval illustrated by item 1 is the time interval between two consecutive R peaks. HRV variables in the time domain and

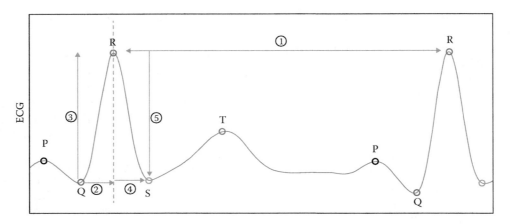

FIGURE 24.3
An exemplary ECG segment with identified P, Q, R, S, T peaks. Five items from the segment are used for ECG feature calculation. Item 1 is the NN interval. Items 2 and 4 are Q-to-R rising time and R-to-S falling time. Items 3 and 5 are Q-to-R rising amplitude and R-to-S falling amplitude.

nonlinear dynamics can be calculated based on the Task Force of the European Society of Cardiology and the North American Society of Pacing and Electrophysiology [33]. From items 2 and 4 in Figure 24.3, the rising time from Q to R and the falling time from R to S can be calculated. Similarly, from items 3 and 5, the rising and falling amplitudes from Q to R and R to S can be calculated.

Because signals may be collected when the patient goes through resuscitation, or has significant movement, artifacts exist and do not reflect the patient's real physiological status. To flag out signals with a large amount of artifacts, signal quality can be evaluated based on R peaks in ECG and the peaks in PPG. The assumption is that signals have normal distributed RR intervals. RR intervals from segments of low quality are detected as outliers using the z-test. Figure 24.4 illustrates two segments of signals. The left side shows ECG and PPG with low signal quality flagged with horizontal bars. The right side shows precisely identified peaks in both signals.

PPGV and waveform morphology features can also be designed similarly with expansion based on PPG unique characteristics. Figure 24.5 (top subplot) shows a normal PPG segment with identified peaks and valleys. Item 1 illustrates a peak-to-peak time interval, which is analogous to the NN interval in ECG. PPGV variables and morphology features were calculated from items 1–4 as they were for ECG. PPG waveform also has a unique dicrotic notch, and its shape has been studied and shown to be related to arterial stiffness and aging [42]. To measure the deceleration and acceleration near the dicrotic notch, the first and second derivatives of PPG were calculated through three-point central difference [43,44]. Item 6 in Figure 24.5 (shadowed area) depicts the region where PPG height changes its moving direction while reducing acceleration and speed until it reaches a local peak. The duration and amplitude change in this movement were derived as two features [45].

Similarly, blood pressure variability (BPV) has been derived from noninvasive blood pressure (NIBP) as a cardiovascular risk factor [46]. The ABP waveform also

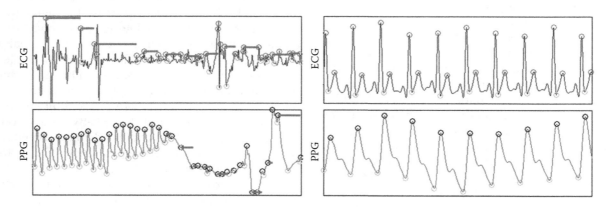

FIGURE 24.4
Illustration of ECG and PPG segments of bad and good signal quality, evaluated by z-test on NN intervals. On left side, horizontal bars flag ECG and PPG segments of bad signal quality.

Trauma Outcome Prediction in the Era of Big Data

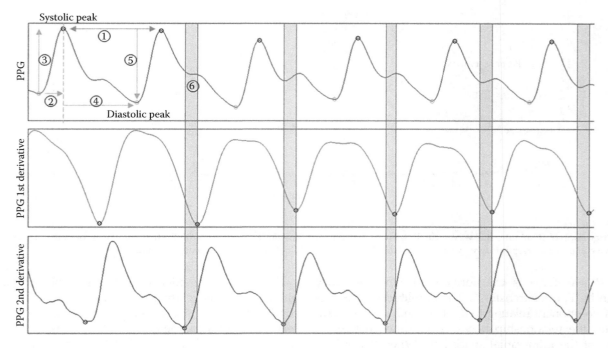

FIGURE 24.5
(Top) An exemplary PPG segment with identified peaks and valleys. (Middle) the first derivative of PPG signal. (Bottom) The second derivative of PPG signal. Item 1 is the peak–peak interval. Item 2 is the valley–peak rising time. Item 3 is the valley–peak rising amplitude. Items 4 and 5 are peak–valley falling time and amplitude. Item 6 is the notch area, where the time duration and amplitude change can be calculated.

carries a pulse wave. With identified NN intervals, NIBP or even ABP could be used to calculate BPV in the time domain and frequency domain, as well as the nonlinear dynamics features.

24.3.3 Features Based on Correlations

Components in a biological system interact with each other in a sophisticated way. Discovering patterns that sketch those interactions could help to understand and forecast the biological system's behavior. Correlation is a type of statistical quantity that can indicate predictive relationship, linear or functional, bivariate or multivariate. Hence, it is a nice tool to explore system component interactions. In this section, we review the applications of different correlations in quantifying physiological system behaviors.

Maintaining normal cerebral perfusion and oxygenation is important in managing severe TBI patients. Monitoring of the cerebral autoregulation could inform clinicians if a patient has lost the ability to maintain a constant perfusion when blood pressure changes [47]. Cerebrovascular pressure-reactivity index (PRx) was proposed as an indicator of loss of autoregulatory reserve [48,49]. PRx is calculated as a moving correlation coefficient between the mean arterial pressure (MAP) and ICP. Given a short time window, about 40 consecutive data points of MAP and ICP in 4–5 minutes are used for calculation [50]. When cerebral autoregulation is intact, cerebral blood flow (CBF) remains a normal constant speed and does not change significantly with mean blood pressure. In such a situation, PRx should be close to zero, indicating no or weak linear correlation between MAP and ICP. When cerebral autoregulation is damaged after severe head injury, CBF increases or decreases with blood pressure. The absolute value of PRx moves away from zero, indicating strong linear correlation between MAP and ICP. In this way, PRx can serve to continuously monitor the existence of cerebral autoregulation. When indicators of autoregulation are plotted against cerebral perfusion pressure (CPP), a U-shaped curve is generated, consistent with a loss of autoregulation in conditions of hypoperfusion or hyperemia. A study in 2002 by Steiner et al. took advantage of this relationship to construct curves of CPP against PRx, and hypothesized that the minima of these would indicate an ideal CPP at which pressure reactivity is maximized [51]. This group and others since have validated this model by finding a correlation between patients' deviation away from this optimal CPP and eventual neurologic outcome.

Another example of using linear correlation in brain trauma study is the calculation of pressure–volume compensatory reserve index, called RAP. It is the moving linear correlation coefficient (R) between the amplitude (A) of a frequency component corresponding to HR in the ICP waveform (Figure 24.6) and the mean ICP pressure (P). Given a short time window of length 4–5 minutes,

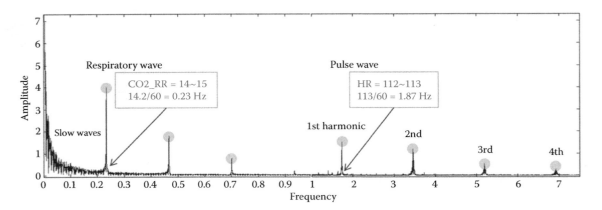

FIGURE 24.6
Fourier transform of 10 minutes ICP waveform. Pulse wave components show four harmonics, with the first one corresponding to heart rate. The other two components, respiratory wave and slow waves, are both identifiable from the frequency domain.

the ICP waveform is transformed into the frequency domain by Fourier transform. The amplitude of the frequency component relating to the HR can be found. RAP indicates the relationship between ICP and changes in volume of the intracranial space [49]. When ICP is low (e.g., ICP < 20 mmHg), near-zero RAP means that a change in cerebral blood volume has no or small impact on ICP, which indicates a good pressure–volume compensatory reserve. After severe head injury, mean pulse amplitude (AMP) and ICP may be negatively correlated, indicating the loss of cerebral autoregulation.

24.3.4 Features Based on Entropy

Physiological data are a succession of values outputted from the information source, namely, the patient. Entropy is a measure of the amount of information from the source. Given a discrete random variable X with its all-possible-element set S_X and probability function $p(x)$, the Shannon entropy $H(X)$ of the random variable X is defined as

$$H(X) = -\sum_{x \in S_X} p(x) \log p(x). \quad (24.1)$$

Given two random variables X and Y, the joint entropy is a generalization to measure the uncertainty of their joint distribution, which is defined as

$$H(X,Y) = -\sum_{x \in S_X} \sum_{y \in S_Y} p(x,y) \log p(x,y). \quad (24.2)$$

It can also be extended to the joint distribution of a set of more than two random variables. The conditional entropy of random variable X given Y measures the uncertainty of x when y is known, which is defined as

$$H(X|Y) = -\sum_{x \in S_X} \sum_{y \in S_Y} p(x,y) \log p(x|y). \quad (24.3)$$

There are more extensions derived from above definitions, such as relative entropy, mutual information, conditional mutual information, etc. [52]. They could serve as useful mathematical tools to extract features from one or jointly more medical time series to quantify the uncertainty and interactions of the systems that they represent.

Approximate entropy (ApEn) and sample entropy (SampEn) are two widely used entropies in measuring similarity in time series, especially in HRV [53,54]. For a physiological time series, even short ones, ApEn can be used to estimate the rate of new information it generates [53,55]. Given a time series $x_1, x_2, \ldots x_n$, its Takens embedding vectors with dimension m are $X_m(i) = (x_i, x_{i+1}, \ldots, x_{i+m-1})$, where $i = 1, \ldots, n - m + 1$ [56]. The distance $d(X_m(i), X_m(j))$ between any two such points $X_m(i)$ and $X_m(j)$ is the maximum absolute element-wise difference between them. With a preset threshold r, B_i is the number of $X_m(j)$ such that $d(X_m(i), X_m(j)) \le r$. Let $\Phi^m(r) = (n - m + 1)^{-1} \sum_{i=1}^{n-m+1} \log(B_i/(n-m+1))$. The ApEn is defined as $ApEn = \Phi^m(i) - \Phi^{m+1}(i)$. Larger ApEn means that the time series is more irregular or a smaller chance of repeated template sequences in the time series. However, ApEn overestimates the similarity since it counts $X_m(i)$ itself when finding B_i, which is called the self-match [53]. The SampEn was designed to reduce the bias caused by self-matching. The SampEn of HRV has been shown to be useful to detect sepsis in the neonatal ICU [54].

The interactions between physiological time series can also be sketched by the permutation entropy, which is the complexity of a symbolic sequence after embedding the time series into a high-dimensional space. First we need to define the ordinal pattern. Given a sequence of numeric elements $x_1, x_2, \ldots x_n$, its ordinal pattern is the permutation $\pi = (i_1, i_2, \ldots i_n)$ that sorts the elements in ascending order such that $x_{i_1} \le x_{i_2} \le \ldots \le x_{i_n}$. Given a time series $x_{1...N}$ and a time window of length L, the order L permutation entropy can be calculated as follows. Let π_t be the ordinal pattern (i.e., the sorting permutation) for the segment of the time series under the sliding window of length L that ends at x_t, i.e., the subsequence $x_{t-L+1}, \ldots, x_t$. Let $S_L = \{\pi_t\}$ be the set of all those unique (alphabet) ordinal patterns π_t. The time series $x_{1...N}$ corresponds to the sequence $\langle \pi_t: t = L, \ldots, N \rangle$ of $N - L + 1$ ordinal patterns from the alphabet S_L. The entropy of this sequence of ordinal patterns is the permutation entropy of the time series $x_{1...N}$. For example, the Shannon permutation entropy is defined in Equation 24.1,

$$H_L = -\sum_{\pi_k \in S_L} P(\pi_k) \log(P(\pi_k)), \quad (24.4)$$

where $P(\pi_k)$ is the frequency of π_k in the sequence $\langle \pi_t \rangle$. In the work presented here, we use the Rényi entropy with parameter α of the sequence $\langle \pi_t \rangle$ defined as

$$R_L^\alpha = \frac{1}{1-\alpha} \log\left(\sum_{\pi_k \in S_L} P(\pi_k)^\alpha\right). \quad (24.5)$$

The parameter α in the Rényi entropy acts as a selector of probabilities. It assigns almost equal weight to each possible probability when α is sufficiently close to zero. When α is larger, it puts more weight on higher probabilities. We can use this parameter to assign different weights on events of different probabilities.

Permutation-based partitions are more robust to noise and other nonlinear distortions and artifacts than value-based fixed-size partitions of the state space, since they depend on the relative order rather than the exact values of the time series. Furthermore, in order to obtain reliable entropy estimates with fixed-size partitions, one needs long time series (in the order of 2^m in order to cover all blocks of such fixed-size partitions); permutation-based entropy estimates do not require long time series. The robustness of permutation entropy makes it particularly attractive for mining VSs collected in real clinical settings, without expensive preprocessing and cleaning of such signals.

24.4 Machine Learning Framework and Its Applications

In the medical field, statistical models are traditionally used to test casual hypotheses, with the purpose focused on explanation [57]. Consider that a new surgical procedure is being tested for its effect of improving patients' long-term neurologic outcome. Doctors may firstly form a statistical null hypothesis that the new procedure causes no statistically significant improvement of patients' long-term neurologic outcome compared with the old procedure. Sample size and confounding factors will be identified, before the randomized case enrollment. With all collected data, an appropriate statistical testing method will be used to test if the null hypothesis should be rejected at a certain significance level. There are many other statistical models to explain the association between factors and outcomes, such as multivariate regression and Bayesian models. No matter what the approaches, model interpretability is always emphasized in the explanatory studies.

Another type of purpose is to predict. In daily patient care, clinicians also want to know, for example, if a patient with blunt injury needs massive transfusion after admission, or if a TBI patient will need a decompressive craniectomy procedure after 6 hours. With all available data already observed for those patients, reliable prediction can assist clinicians in making early decisions, such as calling the blood bank to prepare a sufficient amount of blood product or scheduling early the use of the OR and notifying the operation team. As the decisions based on the predictions have immediate consequences and cost, the prediction must be of high accuracy. In such situations, interpretability could be sacrificed for high prediction accuracy.

Machine learning methods are a collection of algorithms that can discover patterns from data and use them to predict future values. They can be roughly categorized as supervised, unsupervised, and semisupervised learning. Usually, there are preprocessing steps to convert raw data into a feature set that characterizes the observed object or task. In the past, an experienced clinician distilled concise and practical rules from years of observation and clinical practice. With massive data, such a process can be accelerated with automated machine learning algorithms. However, the algorithms may generate counterintuitive models, misled by outliers, missing values, biased data, and incorrect assumptions [58,59]. Therefore, *a priori* knowledge and model validation are essential in building and selecting models. Moreover, a machine learning method could learn from a training data set and attempt to minimize its error on that set. For practical use, we hope the learned model also makes small errors on previously unseen data. A prediction model that can perform well on new data has good generalization ability.

For comprehensive learning algorithms, interested readers can find useful information from Murphy, Bishop, or Hastie's books and their references [58,60,61]. In this section, we focus more on methods that increase the model generalization and the metrics of evaluating the models. With those tools, we could better validate the models and create actionable models.

24.4.1 Feature Selection

Instead of using raw data as inputs, machine learning methods often use variables that characterize the data or learning task's properties (features). Feature selection is an important part of data preprocessing before building a simple and robust model. In many applications, including the Big Data scenario, we can find many attributes to characterize the data, without knowing which ones may contribute to the improved accuracy of prediction. Features of little or no importance not only increase the model complexity but also have a negative effect on parameter estimation for those features that are of importance to the prediction [62]. With massive data, such overfitting is more likely to happen than underfitting.

One class of feature selection methods, which are called the filter-type methods, are model-independent. They evaluate features according to a certain criterion, before adopting any learning algorithm. There are various criteria that can be used. For example, zero or near-zero variance features are often excluded. This is convenient and works in most situations. However, we should be careful, if there is evidence showing that such near-zero variance features may have strong prediction of the outcome. For example, a certain type of drug, like barbiturates, is only used as a last resort when patients are in critical condition. If one binary feature indicates if a patient was administrated barbiturates in hospital, it may have almost near-zero variance since most patients would not receive this drug. But this feature could be a strong predictor of unfavorable outcome, such as mortality.

Some filter-type feature selection methods use the known outcome in the training set. For classification problems, one feature selection approach often seen from medical literature is the use of receiver operating characteristic (ROC) analysis. Besides ROC, information gain can also be used to evaluate each feature's importance to decrease the uncertainty about the outcome. For continuous value prediction, correlation could be used to find a subset of features that each has strong correlation with the outcome while this subset of features are weakly correlated internally. Those methods have low computational complexity, which grows linearly with the number of features. Since the selection is independent of the learning algorithm, the filter-type feature selection methods are good for initial fast screening of a large amount of features to filter out possibly nonimportant ones. Also, the ignorance of interaction between features may result in redundant selection.

Another class of feature selection methods, which are called the wrapper-type methods, search for the important features within the context of models. Given a learning method and a performance metric, the wrapper methods search for an optimal subset of the features that maximizes model performance. Arguably, stepwise selection is the most frequently used wrapper-type feature selection method. In the forward selection step, each feature that is not in the current model is added to a temporary model. The new model is evaluated based on a certain criterion. For example, the most debated criterion is the use of p-value as the inclusion condition. If the hypothesis test calculates that the p-value of the newly added term is less than the inclusion threshold, then keep the feature in the next round of evaluation. Other inclusion criteria include the Akaike Information Criteria (AIC) and Bayesian Information Criteria (BIC), which are considered better than p-value-based selection. The search halts when there are no features that meet the inclusion criteria remaining outside the model. In the backward selection step, the full model that contains all candidate features, or all selected features from the forward selection, has features iteratively removed based on the exclusion criteria, which could be p-value, AIC, BIC, etc.

Permutation is another way to evaluate feature importance in a model. Intuitively, if a feature is of real importance, then randomly permuting its values will greatly reduce the model performance. Given a trained model and performance metric, the trained baseline model has its performance P_b. Each feature is randomly permuted, while other features remain unchanged. The model is then trained again with the altered ith feature and evaluated for a new performance P_{ri}. The change of performance metric is calculated as $\Delta P_i = |P_b - P_{ri}|$. To have a robust estimation of the performance change, we can repeat such random permutation many times and use the averaged difference. In this way, we can rank all features according to the change of performance metric. Then the selected features are optimal in terms of that metric.

The third class of feature selection methods, which are called the embedded methods, train the model and select features simultaneously. Many machine learning algorithms have such inherent feature selection mechanism, such as random forest, relevance vector machines, and decision trees. A more general approach is to add a penalty term for the model complexity to the cost function using a Lagrangian multiplier. Take a data set $D = \{(x_i, y_i)\}$, where $x \in R^d$ and $y \in R$. A learning method finds the optimal parameter set w through minimizing a loss function f by $\min_w f(x, y, w)$. Through a Lagrangian multiplier, the optimization target can be extended to be

$$\min_w f(x, y, w) + \lambda \|w\|_1, \qquad (24.6)$$

where $\lambda > 0$ is called the regularization parameter. The penalty term $\|w\|_1$ is the ℓ_1 norm of the model parameter set w. Equation 24.6 is well known as the least absolute shrinkage and selection operator (LASSO) [63]. It has a nice property to suppress variable coefficients, which introduces sparsity while learning the parameter set. It has been shown to be efficient in creating parsimonious and robust models [60,64].

All three classes of feature selection methods have their own applicable scenarios. The filter types are computationally efficient. Because they can be independent of the model, during the initial stage of experiment design in medical studies, those methods can be used to quickly filter out possibly useless variables, even if there are no known outcome labels yet. This can help reduce the amount of expensive data collection. However, filter methods lack knowledge of the interaction between features and thus may include redundant features. The wrapper-type methods iteratively search for the optimal subset of features by testing each variable's "contribution" to a selected model performance metric. The search scheme takes variables' interactions into consideration. But each evaluation iteration requires the model to be trained again, which could be time consuming. The embedded methods have much more desired advantages. First, they include feature interaction for consideration. Second, with efficient optimization solvers, parameter learning and regularization can be done at the same time efficiently. Moreover, through special design of the regularization terms, not only feature importance but also the structure among feature groups could be selected [65].

24.4.2 Performance Metrics

An appropriate model performance metric allows defining the right learning objectives and can increase the chance of building generative models. Two levels of performance are crucial in creating a good predictive model. First, we need to evaluate the model's generalization capability, namely, can we have a high expectation on its performance on future unseen data? Second, we want to know the model's performance on the training data, namely, how well has this model learned from what it could observe?

In practice, we always have finite observations. One reason is that collecting data could be expensive. Another reason is that it is infeasible to sample the entire population in most situations. To know models' expected performance on unseen data, we have mainly two approaches, called external and internal validations. The former uses data from external sources, such as from other geologically different clinical facilities or from other historical time points. The latter reserves an internal small portion of the collected data and sets it aside only for testing.

External validation is often desired in clinical studies. Data collected from one regional hospital may still have its sampling bias, mainly influenced by its local demographics. Also, data collected from a civilian hospital may not represent the military population. Hence, it is important to validate models built from single-center studies for their generalization. There are many large-scale, high-quality clinical databases maintained for public use. PhysioNet is a large physiological database archiving data contributed from worldwide [66]. The National Trauma Data Bank (NTDB) is the largest U.S. trauma registry data set, which has been used in study trauma injury epidemiology and validation of guidelines from clinical organizations [67–69].

Despite our being in the Big Data era, collecting large amounts of clinical data is still expensive, and a tremendous amount of effort is required to store and process massive data. Hence, it is luxury to keep a large part of the data outside of training. Internal validation is necessary for most single-center studies, especially when external data are not available. The internal validation randomly partitions the collected data into training and testing. A typical validation scheme is the k-fold cross-validation. A data set is randomly partitioned into roughly k equal-size nonoverlapping subsets. In the ith validation, where $i = 1...k$, the ith subset is used as a testing set, and the remaining data are used for training. Such process is iterated over all k subsets. After one round of k-fold cross-validation, there are k model evaluations. Often, we can repeat such k-fold cross-validation N times by randomly partitioning the set again. With the $k \times N$ evaluations, we can use their averaged performance or other robust statistics (e.g., SD, median, etc.) to estimate the model's future performance on unseen data. Such random subset sampling is a simulation of possibly different distributions of new data.

Commonly, mean squared error (MSE) or root mean square error (RMSE) is used in evaluating continuous outcomes. Accuracy or the confusion matrix is typically used for discrete outcomes. When prediction problems have some special issues, such as imbalance in data labels or different preference on incorrectly predicted cases, we need to carefully select the performance metrics. In this section, we discuss the ROC curve and the precision and recall curve (PRC), the two types of model performance metrics that are widely used in medical classification problems.

Accuracy as a model performance metric has some drawbacks in medical data analysis. It could be misleading when the data labels (e.g., positive and negative outcomes in binary classification) are highly imbalanced. A classifier could cheat to achieve high accuracy by just predicting all instances to be the most frequent class label. Unfortunately, in many medical problems, the outcome of interest often is a small portion. For example, in a regional level I trauma center, massive transfusion rate could be as low as 1.3%–2.2% in adult patients [70].

A classifier could predict no massive transfusion for all patients and achieve >98% accuracy. However, mispredicting for a patient who needs massive transfusion may result in severe clinical and social outcome.

To alleviate this problem, accuracy is broken into four parts for binary classifications, which is often described as the confusion matrix. The correctly predicted positive cases are true positives (TPs). The incorrectly predicted positive cases are false positives (FPs). Similarly, the true negatives (TNs) are correctly predicted negative cases; and the false negatives (FNs) are incorrectly predicted negative cases [71,72].

Given a set of instances for prediction, a classifier gives corresponding predicted values, called prediction scores. If we sort all the scores in ascending order and use each unique value as a threshold, all cases with score values smaller than the threshold are classified as negative, whereas those that are higher will be classified as positive. Then for each threshold, we can calculate the false positive rate (FPR) and the true positive rate (TPR), which constitutes a point in a 2-D coordinate. After iterating all possible thresholds, a full curve is generated, called the ROC curve. The PRC curve is created in a similar way, with positive predicted values [PPVs = TP/(TP + FP), also known as precision] and TPR (also known as recall).

The ROC and PRC curves provide a full-spectrum evaluation of a classifier by visualizing FPR, TPR, or PPV, TPR for all possible thresholds, instead of giving a single decision point. The ROC curve is widely used to evaluate prediction models in many medical studies, while the PRC curve is much less frequently seen. Davis and Goadrich [73] proved that ROC and PRC curves have one-to-one correspondence, given a data set with finite positive and negative cases. If one ROC curve has all its points above or equal to another ROC curve, it is said that the first ROC curve dominates the second one; and this is true if and only if the first PRC curve dominates the second PRC curve. However, in many situations, ROC curves from multiple models built from the same data set are twined or close to each other. When the data set is highly imbalanced, i.e., the negative cases significantly outnumber the positive cases, the PRC curve has better ability to identify classifiers that have good performance in predicting the positive cases.

From one of our projects, we took 1191 injured patients and collected continuous physiological VSs to predict the next 3-hour blood product use. Among the 1191 patients, only 7.2% received blood product. As an illustration, Figure 24.7 compares the ROC and PRC curves of three methods we used. To focus on the use of the performance curves, we skip the details of the models here. As we can see from Figure 24.7a, the darker ROC curve (Method 1) dominates the other two, as does the blue PRC curve. The other two ROC curves (Methods 2 and 3) are close to

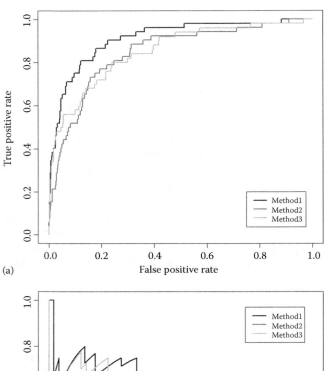

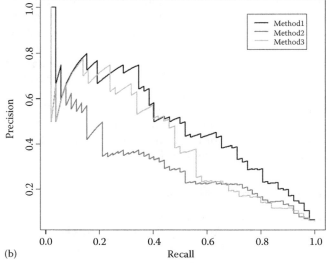

FIGURE 24.7
(a) Example ROC curves of three methods. (b) Example PRC curves of three methods.

each other and intertwined. From Figure 24.7a, the two methods have very similar performance, but from Figure 24.7b, we can observe that the light gray PRC curve (Method 3) has higher precision than the gray PRC curve (Method 2). Therefore, in comparing the methods, we may consider that Methods 1 and 3 have better performance than Method 2.

24.5 Computing Issues

Storing, processing, and learning from massive medical data require intensive calculation. First, having a high-performance file format to store and organize large amounts of data is critical for the input and output (I/O) of data processing. For a typical TBI patient staying 7 days in a trauma center, five 240 Hz waveforms are monitored and up to 700 million data points (equivalent to 1-gigabyte disk size, if data are stored in 12-bit format) would be collected. Traditional spreadsheet-based data management becomes less efficient within such a Big Data scenario. Hierarchical Data Format (HDF) is a high-performance data format that offers on-the-fly data compression and high I/O performance. It also supports reading from or writing to a subset of a data set, without loading the entire data file into memory.

Second, utilizing the independence among tasks allows for parallel data processing, thus making full use of multicore or many-core machines. There are mainly two levels of parallelism in typical medical prediction model training. One is between subjects, and the other is within subject. Often, the feature extraction from a patient's data is independent of others. At this level, we can distribute study cases evenly to all computing units. Within each subject, many tasks can also be done simultaneously. For example, features derived from single variables can be calculated on separate cores. Features from moving windows are also highly parallelizable. In the model learning steps, repeated cross-validation is commonly adopted, to test and validate these models' performance on new data and to prevent potential overfitting. A balanced training and testing model prediction is used to see if the model can be generalized to new previously unused data. For example, with multiple combinations of five outcomes, six feature groups, and 10-fold cross-validation repeated 10 times, about 1500–3000 multiples of model calculations and 100–300 model comparisons and statistical tests are required. Parallel training and testing can be used to speed up the learning process.

Many programming languages and scientific data analysis libraries support parallel computing, such as OpenMP (Open Multi-Processing), MPI (Message Passing Interface) for CPU parallelism, and Compute Unified Device Architecture (CUDA) for graphics processing unit (GPU) parallelism. Moreover, nowadays, cloud computing services provide scalable on-demand use of shared computing resources, including CPUs/GPUs, memory, storage, and security. The end users from hospital or medical research institutes can be freed from the burden of building and maintaining expensive high-performance data processing equipment and infrastructures. Giant vendors, such as Amazon Web Services (AWS), Microsoft Azure, and Google Cloud Platform (GCP), are finding more and more applications in medical data mining and machine learning.

24.6 Discussion

An unprecedented volume of data is generated daily in trauma patient care. However, one cruel fact is that healthcare resources are still very limited in both the field and the hospital. Matched blood product, operation rooms, and experienced healthcare providers such as surgeons, anesthesiologists, and nurses are always scarce. The ultimate goal of Big Data in trauma patient care is to intelligently optimize the allocation of limited healthcare resources, by reliable prediction of needs for lifesaving interventions and early decision on therapeutic plans. With automated data processing and informative data aggregation, useful knowledge from massive data can be used by clinicians in a simple way for decision making or prioritizing care in the busy hospital environment.

In this chapter, we reviewed a few critical components in large-scale medical data analysis, including reliable data collection, feature extraction, feature selection, and model evaluation. To distill reliable predictive models from massive medical data, generalizable good performance on unseen data and interpretability are the top two most important factors to consider. Clinical expert knowledge could assist in designing meaningful and useful features that may be associated with outcomes of special interest, such as transfusion, use of operation room, length of stay, and other actionable lifesaving interventions. Moreover, thorough testing with internal and external data provides some evidence of how the models may perform, before the learned models could be deployed.

References

1. Holcomb, John B. "Optimal use of blood products in severely injured trauma patients." *ASH education program book*, no. 1 (2010): 465–469.
2. Perkins, Jeremy G., Alec C. Beekley. "Damage control resuscitation." In: Savitsky E, Eastridge B, eds. *Combat Casualty Care: Lessons Learned from OEF and OIF*. Department of the Army, Office of the Surgeon General, Borden Institute, Washington, DC, 2012: 121–164.
3. Koht, Antoun, Tod B. Sloan, and J. Richard Toleikis. *Monitoring the Nervous System for Anesthesiologists and Other Health Care Professionals*. Springer, New York, NY, USA, 2012.
4. Reddy, Chandan K., and Charu C. Aggarwal, eds. *Healthcare Data Analytics*, vol. 36. CRC Press, Boca Raton, FL, 2015.
5. Palmer, Ronald W. *Integrated diagnostic and treatment devices for enroute critical care of patients within theater*. No. RTO-MP-HFM-182. Army Medical Research And Materiel Command Fort Detrick MD, 2010.

6. DuBose, Joseph J., Gallinos Barmparas, Kenji Inaba, Deborah M. Stein, Tom Scalea, Leopoldo C. Cancio, John Cole, Brian Eastridge, and Lorne Blackbourne. "Isolated severe traumatic brain injuries sustained during combat operations: demographics, mortality outcomes, and lessons to be learned from contrasts to civilian counterparts." *Journal of trauma and acute care surgery* 70, no. 1 (2011): 11–18.
7. Provost, Foster, and Tom Fawcett. "Data science and its relationship to big data and data-driven decision making." *Big data 1*, no. 1 (2013): 51–59.
8. Garg, Amit X., Neill KJ Adhikari, Heather McDonald, M. Patricia Rosas-Arellano, P. J. Devereaux, Joseph Beyene, Justina Sam, and R. Brian Haynes. "Effects of computerized clinical decision support systems on practitioner performance and patient outcomes: A systematic review." *Jama* 293, no. 10 (2005): 1223–1238.
9. Kipnis, Eric, Davinder Ramsingh, Maneesh Bhargava, Erhan Dincer, Maxime Cannesson, Alain Broccard, Benoit Vallet, Karim Bendjelid, and Ronan Thibault. "Monitoring in the intensive care." *Critical care research and practice* 2012 (2012).
10. Schmidt, Paul E., Paul Meredith, David R. Prytherch, Duncan Watson, Valerie Watson, Roger M. Killen, Peter Greengross, Mohammed A. Mohammed, and Gary B. Smith. "Impact of introducing an electronic physiological surveillance system on hospital mortality." *BMJ quality & safety* (2014): bmjqs-2014.
11. Stein, Deborah M., Megan Brenner, Peter F. Hu, Shiming Yang, Erin C. Hall, Lynn G. Stansbury, Jay Menaker, and Thomas M. Scalea. "Timing of intracranial hypertension following severe traumatic brain injury." *Neurocritical care* 18, no. 3 (2013): 332–340.
12. Kahraman, Sibel, Richard P. Dutton, Peter Hu, Lynn Stansbury, Yan Xiao, Deborah M. Stein, and Thomas M. Scalea. "Heart rate and pulse pressure variability are associated with intractable intracranial hypertension after severe traumatic brain injury." *Journal of neurosurgical anesthesiology* 22, no. 4 (2010b): 296–302.
13. Stein, Deborah M., Peter F. Hu, Megan Brenner, Kevin N. Sheth, Keng-Hao Liu, Wei Xiong, Bizhan Aarabi, and Thomas M. Scalea. "Brief episodes of intracranial hypertension and cerebral hypoperfusion are associated with poor functional outcome after severe traumatic brain injury." *Journal of trauma and acute care surgery* 71, no. 2 (2011): 364–374.
14. Stein, Deborah M., Peter F. Hu, Hegang H. Chen, Shiming Yang, Lynn G. Stansbury, and Thomas M. Scalea. "Computational gene mapping to analyze continuous automated physiological monitoring data in neuro-trauma intensive care." *Journal of trauma and acute care surgery* 73, no. 2 (2012): 419–425.
15. Goldberger, Ary L., Luis AN Amaral, Leon Glass, Jeffrey M. Hausdorff, Plamen Ch Ivanov, Roger G. Mark, Joseph E. Mietus, George B. Moody, Chung-Kang Peng, and H. Eugene Stanley. "PhysioBank, PhysioToolkit, and PhysioNet components of a new research resource for complex physiological signals." *Circulation* 101, no. 23 (2000): e215–e220.
16. Kashif, Faisal M., George C. Verghese, Vera Novak, Marek Czosnyka, and Thomas Heldt. "Model-based noninvasive estimation of intracranial pressure from cerebral blood flow velocity and arterial pressure." *Science translational medicine* 4, no. 129 (2012): 129ra44.
17. Li-wei, H. Lehman, Ryan P. Adams, Louis Mayaud, George B. Moody, Atul Malhotra, Roger G. Mark, and Shamim Nemati. "A physiological time series dynamics-based approach to patient monitoring and outcome prediction." *IEEE journal of biomedical and health informatics* 19, no. 3 (2015): 1068–1076.
18. Saeed, Mohammed, Mauricio Villarroel, Andrew T. Reisner, Gari Clifford, Li-Wei Lehman, George Moody, Thomas Heldt, Tin H. Kyaw, Benjamin Moody, and Roger G. Mark. "Multiparameter Intelligent Monitoring in Intensive Care II (MIMIC-II): A public-access intensive care unit database." *Critical care medicine* 39, no. 5 (2011): 952.
19. Excel Medical Electronics, LLC, BedMasterEx Operator's Manual, Jupiter, FL, 2013.
20. Shooman, Martin L. *Reliability of Computer Systems and Networks: Fault Tolerance, Analysis, and Design*. John Wiley & Sons, New York, 2003.
21. Khawaja, Antoun. *Automatic ECG Analysis Using Principal Component Analysis and Wavelet Transformation*. Univ.-Verlag Karlsruhe, 2006.
22. Kamath, Markad V., Mari Watanabe, and Adrian Upton, eds. *Heart Rate Variability (HRV) Signal Analysis: Clinical Applications*. CRC Press, Boca Raton, FL, 2012.
23. Van Drongelen, Wim. *Signal Processing for Neuroscientists: An Introduction to the Analysis of Physiological Signals*. Academic Press, 2006.
24. Brain Trauma Foundation, American Association of Neurological Surgeons; Congress of Neurological Surgeons. Guidelines for the management of severe head injury. *Journal of neurotrauma* 24, Suppl 1 (2007): S1–S106.
25. Sasser, Scott M., Richard C. Hunt, Mark Faul, David Sugerman, William S. Pearson, Theresa Dulski, Marlena M. Wald et al. Guidelines for field triage of injured patients: Recommendations of the National Expert Panel on Field Triage, 2011. *MMWR. Recommendations and reports: Morbidity and mortality weekly report. Recommendations and reports/Centers for Disease Control*, 61(RR-1), 1–20, 2012.
26. Brown, Joshua B., Gestring, Mark L., Forsythe, Raquel M., Stassen, Nicole A., Billiar, Timothy R., Peitzman, Andrew B., & Sperry, Jason L. "Systolic blood pressure criteria in the National Trauma Triage Protocol for geriatric trauma: 110 is the new 90." *The journal of trauma and acute care surgery* 78, no. 2 (2015): 352.
27. Olaussen, Alexander, Todd Blackburn, Biswadev Mitra, and Mark Fitzgerald. "Shock Index for prediction of critical bleeding post-trauma: A systematic review." *Emergency medicine Australasia* 26, no. 3 (June 2014): 223–228.
28. Kahraman, S., Dutton, R. P., Hu, P., Xiao, Y., Aarabi, B., Stein, D. M., & Scalea, T. M. "Automated measurement of 'pressure times time dose' of intracranial hypertension best predicts outcome after severe traumatic brain

injury." *Journal of trauma and acute care surgery* 69, no. 1 (2010a): 110–118.

29. Mackenzie, Colin F., Wang, Yulei, Hu, Peter F., Chen, Shihyu Y., Chen, Hegang H., Hagegeorge, George,... & ONPOINT Study Group. "Automated prediction of early blood transfusion and mortality in trauma patients." *Journal of trauma and acute care surgery* 76, no. 6 (2014): 1379–1385.

30. Liu, Nehemiah T., John B. Holcomb, Charles E. Wade, Mark I. Darrah, & Jose Salinas. "Utility of vital signs, heart rate variability and complexity, and machine learning for identifying the need for lifesaving interventions in trauma patients." *Shock* 42, no. 2 (2014): 108–114.

31. Ernst, Gernot. *Heart Rate Variability*. Springer, London, 2014.

32. Acharya, U. Rajendra, K. Paul Joseph, Natarajan Kannathal, Choo Min Lim, and Jasjit S. Suri. "Heart rate variability: A review." *Medical and biological engineering and computing* 44, no. 12 (2006): 1031–1051.

33. Task Force of the European Society of Cardiology. "Heart rate variability standards of measurement, physiological interpretation, and clinical use." *European heart journal* 17 (1996): 354–381.

34. Aslanidia, Theodoros. Perspectives of autonomic nervous system perioperative monitoring—Focus on selected tools. *International Archives of Medicine* 8, no. 22 (2015): 1–7.

35. Goodman, Brent, Bert Vargas, and David Dodick. "Autonomic nervous system dysfunction in concussion." *Neurology* 80, no. 7 Supplement (2013): P01–265.

36. Goldstein, Brahm, Mark H. Kempski, Donna BA DeKing, Christopher Cox, David J. DeLong, Mary M. Kelly, and Paul D. Woolf. "Autonomic control of heart rate after brain injury in children." *Critical care medicine* 24, no. 2 (1996): 234–240.

37. Baguley, Ian J., Roxana E. Heriseanu, Kim L. Felmingham, and Ian D. Cameron. "Dysautonomia and heart rate variability following severe traumatic brain injury." *Brain Injury* 20, no. 4 (2006): 437–444.

38. Allen, John. "Photoplethysmography and its application in clinical physiological measurement." *Physiological measurement* 28, no. 3 (2007): R1.

39. Lu, Sheng, He Zhao, Kihwan Ju, Kunson Shin, Myoungho Lee, Kirk Shelley, and Ki H. Chon. "Can photoplethysmography variability serve as an alternative approach to obtain heart rate variability information?" *Journal of clinical monitoring and computing* 22, no. 1 (2008): 23–29.

40. Pan, Jiapu, and Willis J. Tompkins. "A real-time QRS detection algorithm." *IEEE transactions on biomedical engineering* 3 (1985): 230–236.

41. Press, William H., Saul A. Teukolsky, William T. Vetterling, and Brian P. Flannery, Chapter 14: Statistical description of data. In: *Numerical Recipes the Art of Scientific Computing*, 3rd ed., Cambridge University Press, 2007.

42. Voss, Andreas, Rico Schroeder, Andreas Heitmann, Annette Peters, and Siegfried Perz. "Short-term heart rate variability—influence of gender and age in healthy subjects." *PloS ONE* 10, no. 3 (2015): e0118308.

43. Yousef, Q., M. B. I. Reaz, and Mohd Alauddin Mohd Ali. "The analysis of PPG morphology: Investigating the effects of aging on arterial compliance." *Measurement science review* 12, no. 6 (2012): 266–271.

44. Elgendi, Mohamed, Ian Norton, Matt Brearley, Derek Abbott, and Dale Schuurmans. "Detection of a and b waves in the acceleration photoplethysmogram." *Biomedical engineering online* 13, no. 1 (2014): 1.

45. Nitzan, Meir, Anatoly Babchenko, Boris Khanokh, and David Landau. "The variability of the photoplethysmographic signal—A potential method for the evaluation of the autonomic nervous system." *Physiological measurement* 19, no. 1 (1998): 93.

46. Voss, Andreas, Matthias Goernig, Rico Schroeder, Sandra Truebner, Alexander Schirdewan, and Hans R. Figulla. "Blood pressure variability as sign of autonomic imbalance in patients with idiopathic dilated cardiomyopathy." *Pacing and clinical electrophysiology* 35, no. 4 (2012): 471–479.

47. Aaslid, Rune, Kari-Fredrik Lindegaard, Wilhelm Sorteberg, and Helge Nornes. "Cerebral autoregulation dynamics in humans." *Stroke* 20, no. 1 (1989): 45–52.

48. Czosnyka, Marek, Piotr Smielewski, Peter Kirkpatrick, David K. Menon, and John D. Pickard. "Monitoring of cerebral autoregulation in head-injured patients." *Stroke* 27, no. 10 (1996): 1829–1834.

49. Balestreri, M., M. Czosnyka, L. A. Steiner, E. Schmidt, P. Smielewski, B. Matta, and J. D. Pickard. "Intracranial hypertension: What additional information can be derived from ICP waveform after head injury?" *Acta neurochirurgica* 146, no. 2 (2004): 131–141.

50. Czosnyka, Marek, and John D. Pickard. "Monitoring and interpretation of intracranial pressure." *Journal of neurology, neurosurgery & psychiatry* 75, no. 6 (2004): 813–821.

51. Steiner, Luzius A., Marek Czosnyka, Stefan K. Piechnik, Piotr Smielewski, Doris Chatfield, David K. Menon, and John D. Pickard. "Continuous monitoring of cerebrovascular pressure reactivity allows determination of optimal cerebral perfusion pressure in patients with traumatic brain injury." *Critical care medicine* 30, no. 4 (2002): 733–738.

52. Cover, Thomas M., and Joy A. Thomas. *Elements of information theory*. John Wiley & Sons, New York, 2012.

53. Richman, Joshua S., and J. Randall Moorman. "Physiological time-series analysis using approximate entropy and sample entropy." *American journal of physiology—Heart and circulatory physiology* 278, no. 6 (2000): H2039–H2049.

54. Lake, Douglas E., Joshua S. Richman, M. Pamela Griffin, and J. Randall Moorman. "Sample entropy analysis of neonatal heart rate variability." *American journal of physiology—regulatory, integrative and comparative physiology* 283, no. 3 (2002): R789–R797.

55. Pincus, Steven M. "Approximate entropy as a measure of system complexity." *Proceedings of the national academy of sciences* 88, no. 6 (1991): 2297–2301.

56. Takens, Floris. "Detecting strange attractors in turbulence." In *Dynamical systems and turbulence, Warwick 1980*, pp. 366–381. Springer, Berlin, 1981.

57. Shmueli, Galit. "To explain or to predict?" *Statistical science* (2010): 289–310.
58. Bishop, Christopher M. *Pattern Recognition and Machine Learning*. Springer, New York, 2006.
59. Lantz, Brett. *Machine Learning with R*. Packt Publishing Ltd, Birmingham, 2013.
60. Murphy, Kevin P. *Machine Learning: A Probabilistic Perspective*. MIT Press, Massachusetts, 2012.
61. Hastie, Trevor J.. Robert John Tibshirani, and Jerome H. Friedman. *The Elements of Statistical Learning: Data Mining, Inference, and Prediction*. Springer, New York, 2011.
62. Kuhn, Max, and Kjell Johnson. *Applied Predictive Modeling*. Springer, New York, 2013.
63. Tibshirani, Robert. "Regression shrinkage and selection via the lasso." *Journal of the royal statistical society. series B (methodological)* (1996): 267–288.
64. Hastie, Trevor, Robert Tibshirani, and Martin Wainwright. *Statistical Learning with Sparsity: The Lasso and Generalizations*. CRC Press, Boca Raton, FL, 2015.
65. Yuan, Ming, and Yi Lin. "Model selection and estimation in regression with grouped variables." *Journal of the royal statistical society: Series b (statistical methodology)* 68, no. 1 (2006): 49–67.
66. Moody, George B. "PhysioNet: Research resource for complex physiological signals." Available at http://ecg.mit.edu/george/publications/physionet-jecg-2009.pdf, retrieved Jul. 15, 2016.
67. Webman, Rachel B., Elizabeth A. Carter, Sushil Mittal, Jichaun Wang, Chethan Sathya, Avery B. Nathens, Michael L. Nance, David Madigan, and Randall S. Burd. "Association between trauma center type and mortality among injured adolescent patients." *JAMA pediatrics* (2016).
68. Tinkoff, Glen, Thomas J. Esposito, James Reed, Patrick Kilgo, John Fildes, Michael Pasquale, and J. Wayne Meredith. "American Association for the Surgery of Trauma Organ Injury Scale I: Spleen, liver, and kidney, validation based on the National Trauma Data Bank." *Journal of the American college of surgeons* 207, no. 5 (2008): 646–655.
69. Millham, Frederick H., and Wayne W. LaMorte. "Factors associated with mortality in trauma: Re-evaluation of the TRISS method using the National Trauma Data Bank." *Journal of trauma and acute care surgery* 56, no. 5 (2004): 1090–1096.
70. Parimi, Nehu, Peter F. Hu, Colin F. Mackenzie, Shiming Yang, Stephen T. Bartlett, Thomas M. Scalea, and Deborah M. Stein. "Automated continuous vital signs predict use of uncrossed matched blood and massive transfusion following trauma." *Journal of trauma and acute care surgery* 80, no. 6 (2016): 897–906.
71. Fawcett, Tom. "ROC graphs: Notes and practical considerations for researchers." *Machine learning* 31, no. 1 (2004): 1–38.
72. Krzanowski, Wojtek J., and David J. Hand. *ROC Curves for Continuous Data*. CRC Press, Boca Raton, FL, 2009.
73. Davis, Jesse, and Mark Goadrich. "The relationship between Precision-Recall and ROC curves." In *Proceedings of the 23rd international conference on Machine learning*, pp. 233–240. ACM, 2006.

25

Enchancing Medical Problem Solving through the Integration of Temporal Abstractions with Bayesian Networks in Time-Oriented Clinical Domains

Kalia Orphanou, Athena Stassopoulou, and Elpida Keravnou

CONTENTS

25.1 Introduction	493
25.2 Temporal Abstraction of Time-Oriented Clinical Data	494
25.2.1 Basic Temporal Abstractions	494
25.2.2 Complex Temporal Abstractions	495
25.2.3 Clinical Systems Using Temporal Abstractions	495
25.3 Bayesian Networks and Time	496
25.3.1 Dynamic Bayesian Networks	497
25.3.2 Networks of Probabilistic Events in Time	499
25.3.3 Continuous-Time Bayesian Networks	499
25.3.4 Irregular-Time Bayesian Networks	500
25.4 Integration of Temporal Abstractions with Bayesian Networks	501
25.5 The Clinical Domain of Coronary Heart Disease: Test Bed Data Set	502
25.6 DBN-Extended Model	503
25.6.1 Utilization of Basic Temporal Abstractions	503
25.6.2 Construction of the DBN-Extended Model	504
25.6.3 DBN without TAs	505
25.6.4 Training in the Presence of Class Imbalance	506
25.6.5 Experiments and Analysis	507
25.7 Incorporating Complex Temporal Abstractions in Naïve Bayes Classifiers	508
25.7.1 Extraction of Temporal Association Patterns	509
25.7.2 Naïve Bayesian Classifier Using Periodic Temporal Association Patterns	510
25.7.3 Extracting the Most Frequent Temporal Association Patterns	510
25.7.4 Dealing with Imbalanced Data and Overlapping Classes	511
25.7.5 Comparing the Performance of the Periodic Classifiers with a Baseline Classifier	512
25.8 Applications in Biomedical Signal Processing	513
25.9 Discussion and Conclusions	513
25.10 Open Problems and Future Directions	514
References	514

25.1 Introduction

Computer-based medical problem solving has been for decades, and continues to be, a particularly challenging field of research attracting attention from diverse scientific communities (artificial intelligence, databases, medical informatics, and others). As time is a key aspect in modeling and reasoning with dynamic processes, it could not make an exception in the case of modeling and reasoning with disease processes, irrespective of whether the task at hand is to explain a situation, as currently observed, through classification or diagnosis; to predict its future development, through prognosis or risk assessment; or to overall manage it, through therapeutic actions and the monitoring of their outcome.

Medical knowledge is usually expressed in as general a form as possible such as associations, rules, causal models, etc. Moreover, the medical history of some individual, from birth to death, is nothing other than a repository of time-stamped data/information of diverse format/content and storage medium (manual documents, computer-based records, others). Technology developments enabled the collection and storage in electronic means (e.g., electronic health records) of large volumes of such data, while the development of data mining, (or more specifically temporal data mining) techniques enabled the extraction of new medical "knowledge" improving current understanding and thus pushing the frontiers of medical knowledge in given domains. Equally important such developments can support personalized medicine. However, manually managing the medical history (or part of this history) of some individual in order to yield optimal decision making for the particular individual not only is a nonstarter, but undoubtedly is prone to errors. Such data are invariably expressed at different levels of semantic detail and sampling frequency; they could have gaps or be excessively voluminous, etc. In short, they are not amenable to direct processing and reasoning with.

To perform any medical problem solving, medical knowledge should be matched against medical data. Temporal abstraction (TA) methods, which by now represent a relatively ripe technology, by combining statistical and heuristic methods, aim to close the abstraction gap between the specific raw patient data and the general medical knowledge [1]. They glean out the useful information/patterns from multivariate, time-stamped longitudinal data, in order to facilitate specific uses including higher-level problem solving [2]. The generated more abstract information is of different types that can be roughly divided into basic and complex TAs.

Given the inherent uncertainty and incompleteness of medical knowledge and medical data, it is not surprising that probabilistic models became a popular representation for reasoning with disease processes. Bayesian networks (BNs) [3–5] belong to the family of such probabilistic models, and in fact, they have been widely used in many clinical domains as they can handle well uncertainty in medical knowledge and data. BNs were successfully applied in diverse fields such as medical diagnosis [6], forecasting [7–10], computer vision [11–13], language understanding [14,15], risk assessment and management [16–18], speech recognition [19,20], and troubleshooting [21], amongst others.

The motivation for this work is drawn from the fact that both Bayesian models and TA demonstrated their effectiveness as stand-alone engines predominantly for medical problem solving and for medical data processing respectively, but not in conjunction. Thus, the key research hypothesis that we want to investigate is whether the integration of TA with Bayesian models can yield notable performance improvements in medical problem solving, capitalizing on the complementarities of these two distinct technologies.

25.2 Temporal Abstraction of Time-Oriented Clinical Data

Temporal data can be represented using the following primitives [22]: time points, also denoted as events, and time intervals, denoted as episodes. An event is characterized by a pair (t, v), where t is the time point of occurrence and v is the value that determines the state of a parameter, while an episode is characterized by a pair (I, L), where I is the interval of occurrence of the episode and L is the label associated with the temporal pattern [22].

The TA task aims to describe a set of time-series data and external events through sequences of context-specific temporal intervals (episodes). TA techniques are divided into two main categories: basic and complex TAs. Basic TA algorithms take as input time-point events and output episodes on the basis of some predefined rules. The derived episodes can then be combined to form complex temporal patterns using temporal relation operators.

25.2.1 Basic Temporal Abstractions

The most popular types of basic TAs are state, trend, and persistence [23]. State TAs determine the state of an individual parameter based on predefined categories during the time period where the state is valid (e.g., heart rate is stable for 20 minutes). Predefined states used for state abstractions are based on domain expert knowledge, such as low, high, stable, and unstable. Trend TAs, on the other hand, represent the rate of change of the state of a parameter by identifying increase, decrease, and stationary patterns in time series (e.g., glucose is increasing for 2 consecutive days). The most frequently used method for deriving trend abstractions is a piecewise linear representation of raw temporal data. Segmentation algorithms are divided into three groups [24]: sliding window, top-down, and bottom-up algorithms. The choice of algorithms depends on the characteristics of the data set such as the length of the time series, the expected level of noise, and so on.

Persistence TA techniques merge state TAs for the whole interval period in which their value persists over time. They include the personal and family medical history of the patient. For example, when someone is diagnosed with a particular disease at time t, we assume that the patient has a history of the disease from $t + 1$ onwards.

25.2.2 Complex Temporal Abstractions

Complex TAs take as input two or more interval sets (basic or other complex TAs) and combine them into a new interval set associated with a new TA. The goal is to discover temporal relationships between discovered patterns of TAs. Each complex pattern (time interval) is defined using starting and ending points of the involved episodes. Bellazzi et al. [25] state that complex abstractions can be used in two different ways: (a) to represent the persistence of complex clinical situations and (b) to detect complex temporal patterns that can not be detected using basic abstractions.

A popular technique for discovering complex temporal patterns is using Allen's 13 temporal relationships, as shown in Figure 25.1 [26], to express any relationship held between time intervals. Some of the relations are mirror images of each other, e.g., "Y started by X" is the same as the relation: "X starts Y". A disjunctive operator PRECEDES was also defined in Refs. [23,27], which expresses the temporal relationships among patterns. Given two episodes $e1$, characterized by $[I_1 = \{Istart_1, Iend_1\}, L_1]$, and $e2$, characterized by $[I_2 = \{Istart_2, Iend_2\}, L_2]$, the PRECEDES relationship holds (i.e., "$e1$ PRECEDES $e2$") if "$Istart_1 <= Istart_2$" and "$Iend_1 <= Iend_2$." According to this definition, the PRECEDES operator disjunctively synthesizes the following six Allen's operators: OVERLAPS, FINISHED-BY, MEETS, BEFORE, EQUALS, and STARTS (see Figure 25.2).

Many automated tools were proposed in the literature to discover automatically frequent complex temporal

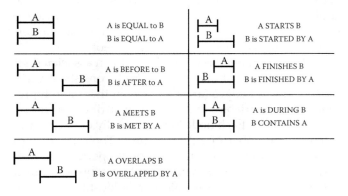

FIGURE 25.1
Allen's interval relationships (From Allen, J. Towards a general theory of action and time. *Artificial Intelligence* 23, 2 (1984), 123–154.)

Relation	Example
A PRECEDES F	aaaaaaa *Finished-by* ffff
	aaaaaa *Overlaps* fffffff
	aaaaa *Meets* fffff
	aaaaa *Before* fffff
	aaaaaa *Equals* fffff
	aaaa *Starts* fffff

FIGURE 25.2
PRECEDES temporal operator.

patterns derived from the conjunction of temporal relations between pairs of intervals. Temporal relation patterns that represent the temporal relationship among TA concepts are a category of complex TAs. These types of patterns were mainly used as features in classifiers. KarmaLego [28–30] is a mining pattern tool used for the mining and the discovery of temporal relation patterns. It implements a knowledge-based temporal abstraction (KBTA) [2] framework for deriving basic TAs, and then a miner algorithm is iteratively applied to the derived abstractions to detect temporal interval relation patterns (TIRPs). The temporal relation operators used for the detection of TIRPs include versions of Allen's 13 temporal relations. Similarly, the approach of Batal et al. [31–33] derives temporal patterns from basic or other complex TAs using some disjunctive temporal operators from Allen's algebra. They detect temporal patterns based on the sliding window method. More specifically, the width of the sliding window (w) is the maximum pattern duration, as specified by the user. The algorithm only considers temporal relationships that can be observed within this window. This algorithm is based on the assumption that events that occur far enough from each other have no temporal relationship. Another method proposed by Sacchi et al. [23] uses the disjunctive relation operator PRECEDES, presented above, for discovering frequent temporal interval patterns, which can form a basis for the discovery of a set of temporal association rules (TARs).

25.2.3 Clinical Systems Using Temporal Abstractions

A number of intelligent clinical systems employing TA techniques are reported in the literature. Such systems differ on the TA techniques applied to the data, the method that they use for acquiring the knowledge that is needed for deriving the given abstractions, and whether they are able to perform abstractions for single or multiple patients. It is noted that in machine learning, or more generally, the automatic or semi-automatic generation of new medical knowledge, multiple cases/patients need to be processed. The generated knowledge forms the model for solving problems with respect to particular patients. In the context of medical problem solving/decision making, TA mainly concerns the generation of abstractions from the records of a single patient. However, the process is of direct relevance to knowledge generation as well, since the abstractions from a representative set of individual patients can lead (be abstracted) to useful knowledge. The ability to perform abstractions from multiple patients, irrespective of purpose, entails the generalizations of abstractions from single patients or the computation of abstract concepts directly from the records of a population of patients or a combination of such techniques.

The ventilator management system (VM) [34] is one of the first temporal reasoning systems in medicine that uses TA techniques to interpret online patients' data for monitoring purposes in intensive care units (ICUs). The TOPAZ [35] and IDEFIX [36] systems are also two of the earlier clinical systems developed. The TOPAZ system summarizes patient data by utilizing an integrated interpretation model approach. The IDEFIX system is a knowledge-based system that examines low-level data such as laboratory values or symptoms from patients who have systemic lupus erythematosus (ARAMIS database) [37]. It infers high-level data such as renal failure using TA methods.

In addition, the RÈSUMÈ system [38] implements the KBTA [39] to create TAs. For the TA process, time-stamped patient data, clinical events, and the domain knowledge base are given as input. The RÈSUMÈ system was evaluated for the purpose of summarizing patient data in many clinical domains such as oncology [38], monitoring of children's growth [40], and management of insulin-dependent diabetes [41]. An extension of the RÈSUMÈ system is the RASTA system [42], which uses a distribution algorithm that allows the task of generating abstractions to be distributed over many computers. As it follows, RASTA is able to work on very large data sets and to abstract data streams from multiple patients. RASTA is used as a temporal reasoning component of clinical decision support systems. Another methodology that was developed to create abstractions on multiple patients is called probabilistic temporal abstraction (PTA) [43]. The PTA is an extension of the KBTA method that was implemented in the RÈSUMÈ system.

Two other recent frameworks applied to clinical domains are PROTEMPA [44] and Java Time Series Abstractor (JTSA) [22]. The JTSA software is used for extracting TAs following time-series data preprocessing. It is a stand-alone application for the definition and execution of a complete time-series analysis workflow to detect temporal patterns. The framework incorporates an algorithm taxonomy that includes both algorithms for time-series preprocessing and algorithms for TAs detection. The resulting TAs are episodes defined by a pair: interval of occurrence and a label for the pattern. JTSA is suitable for a large variety of applications from chronic disease management to ICU monitoring. It was applied on a diabetic data set to extract relevant patterns from data related to the long-term monitoring of diabetic patients [22]. The JTSA framework was implemented for the analysis of a single patient; however, it can be extended for the analysis of multiple patients by implementing a custom application.

PROTEMPA is a software library with a modular architecture [44]. It has four modules that provide (a) a framework for defining TA primitives and using these primitives for deriving abstractions from time-stamped data (Algorithm Source), (b) a framework connected to a knowledge base for specifying algorithm parameters for deriving low-level abstractions and interval relationships for the derivation of complex TAs (Knowledge Source), (c) a connection to an existing data store that stores the time-point data and the derived abstractions (Data Source), and (d) a data processing environment for managing the abstraction-finding routines (Abstraction Finder). It was applied for the diagnosis and management of HELLP syndrome.*

TA techniques therefore extract qualitative discrete data utilizing domain/expert knowledge. For the integration of TAs with temporal Bayesian networks (TBNs), the constructed TBN will be used as an inference engine and the temporal abstracted concepts as its input. In the next section, temporal extensions of BNs applied to clinical domains are reviewed.

25.3 Bayesian Networks and Time

Bayesian networks (BNs) [3–5], also known as belief networks, are graphs that belong to the family of probabilistic models. They were introduced as a knowledge representation and inference approach under uncertainty through the use of probability theory.

A BN is a directed acyclic graph with an associated set of probability distributions. The graph consists of vertices (nodes) and edges (arcs). Nodes on the graph represent random variables that denote an attribute, a feature or a hypothesis. Each node has a mutually exclusive and exhaustive number of values (states). These nodes represent the variables of interest for a specific problem such as a disease or a symptom. Arcs represent direct dependencies (or cause–effect relationships) among

---

* HELLP is a dangerous complication of pregnancy that appears during the latter part of the third trimester or after childbirth.

variables. An arc (arrow) from *A* to *B* indicates that the value taken by variable *B* depends on the value taken by variable *A* (or that *B* is influenced, or caused, by *A*).* The strength of these dependencies is quantified by conditional probabilities. As an example, a BN applied in a medical domain could represent cause–effect relationships between a disease, its causes, and its symptoms. Given some symptoms and some of the disease causes (risk factors), the network can be used to compute the probability of the presence of a disease.

More formally, a BN with a set of variables $A = [A^1, \ldots, A^n]$ consists of

1. A network structure, which encodes the probabilistic dependencies among the variables.
2. The network parameters, which are a set of local probability distributions $\Pr(A^i \mid parents(A^i))$ associated with each node. Each probability distribution quantifies the effect of parents on the node. These are given as tables, in the case of discrete variables, which are called conditional probability tables (CPTs). For variables that do not have any parents (i.e., the roots), their prior probability distribution is defined.

The network structure and the set of local probability distributions define the joint probability distribution (JPD) for *A* as

$$\Pr(A^1, \ldots, A^n) = \prod_{i=1}^{n} \Pr\left(A^i \mid parents\left(A^i\right)\right)$$

As an example, let us consider three discrete variables represented as nodes *A*, *B*, and *C* and the network structure as shown in Figure 25.3. The network is quantified with prior probabilities for the root nodes *A* and *B* and a CPT for node *C*, which defines $\Pr(C \mid A, B)$ over all possible combinations of values for *A*, *B*, and *C*. The JPD is therefore defined as $\Pr(A, B, C) = \Pr(C \mid A, B) \cdot \Pr(A) \cdot \Pr(B)$ considering all the possible values of the variables.

A variety of extensions to these networks introduce temporality into BNs to model temporal phenomena and reasoning through time. In temporal extensions of BNs, the initial (prior) model (at time t_0) should be constructed in cooperation with domain experts as in atemporal BNs. The knowledge acquired from domain experts will be used to define domain variables and the dependencies between variables. The transition model includes the computation of parameters, which should be either in a probability density function form or in discrete form depending on the time representation.

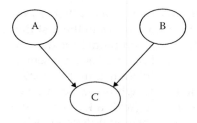

FIGURE 25.3
A BN model with three variables *A*, *B*, *C*, where *A* and *B* are the parents of *C*.

Temporal probabilistic networks were recently used in many medical problems such as diagnosis [45–47], treatment selection [46,48], therapy monitoring [49,50], and prognosis [45,47,51–53]. This is due to the fact that they can deal well with uncertainty in time-series medical data and they allow one to learn about causal relationships and dependencies of clinical features.

The most popular temporal extension of BN is a dynamic Bayesian network (DBN) [54], which uses a discrete-time representation. Extensions of DBN for decision making are dynamic influence diagrams (DIDs) [55,56] and partially observable Markov decision processes (POMDP) [48,57]. Known models that use interval-based representation of time are networks of probabilistic events in discrete time (NPEDTs) [58], modifiable temporal Bayesian networks (MTBNs) [59], probabilistic temporal networks (PTNs) [60], and temporal node Bayesian networks (TNBNs) [47]. Additionally, examples of models using continuous-time representation are the continuous-time Bayesian networks (CTBN) [61] and Berzuini's network of dates model [62]. Irregular-time Bayesian networks (ITBN) [63] are a new temporal extension of BNs, which are able to deal with processes occurring irregularly through time.

Part of the diabetes example proposed in Ref. [64] is used for illustrating the network structure of each TBN. In this example, three variables will be represented through the BN models: *MEAL*, *CHO*, and *BG*. *MEAL* represents the amount of meal intake, which can take values from 0 to 100 g, whereas *CHO* represents the rates of carbohydrate remaining in the gut after taking a meal, which can take values from 0 to 8 mmol/kg. *BG* represents the predicted blood glucose concentration at the end of the hour, which can take values from 0 to 20 mmol/l. Edges are introduced to represent interactions that occur within the model for predicting the blood glucose concentration.

25.3.1 Dynamic Bayesian Networks

A DBN is a network with the repeated structure of a BN for each time slice over a certain interval [65]. A DBN is

---
* If there is an arrow from node *A* to node *B*, *A* is said to be the parent of *B*.

a tuple (B_0, B_1), where B_0 is a BN that represents the prior distribution for the variables in the first time slice and B_1 represents the transition model for the variables in two consecutive time slices. DBNs represent the change of variable states at different time points. A node can be either a hidden node, whose values are never observed, or an observed node (with a known value). Arcs represent the local or transitional dependencies among variables. Intraslice arcs represent the dependencies within the same time slice like in an atemporal BN. Inter-slice arcs connect nodes between time slices and represent their temporal evolutions. Every node in the second slice and each node in the first slice that is not a root node has an associated CPT.

A DBN cannot represent processes that evolve at different time granularities, but it represents the whole system at the finest possible granularity. DBNs are assumed to be time invariant, which means that the network structure per time slice and across time slices does not change. Furthermore, it is assumed that DBNs use the Markovian property: conditional probability distribution of each variable at time n, for all $n > 1$, depends only on the parents from the same time slice or from the previous time slice but not from earlier time slices [18].

Let $A_n = [A_n^1, A_n^2, \ldots, A_n^m]$, $m \geq 1$, denote a set of variables at time n, with m being the number of variables; the parameters of a DBN are defined as follows [54]:

- The initial state distribution $\Pr(A_0)$ at time slice zero, such as $\Pr(BG_0)$ in the case of the diabetes example as introduced in Figure 25.4.
- The transition probability for the variables in two consecutive time slices:

  $\Pr(A_n|A_{n-1}) = \Pi_{i=1}^m \Pr(A_n^i|\text{parents}(A_n^i))$, where parents$(A_n^i)$ denotes the parent set of A_n^i from the same time slice n and the parent set of A_n^i from the previous time slice.

- The JPD for N consecutive time slices is defined by

  $\Pr(A_0, \ldots, A_N) = \Pi_{n=0}^N \Pi_{i=1}^m \Pr(A_n^i|\text{parents}(A_n^i))$.

In the diabetes example of Figure 25.4, an edge is introduced between variables MEAL and CHO within the same time slice to indicate the effect of the amount of meal intake on the rate of carbohydrate remaining in the gut, and between CHO and BG at different time slices, indicating that the rate of carbohydrate remaining in the gut at a certain time t will affect the blood glucose levels at time $t + 1$. We assume for simplicity that each time slice represents the time period of 4 hours; thus, the total number of time slices is six, and therefore, $n = [0, \ldots, 5]$. CHO_{n-1} and $MEAL_{n-1}$ for $n > 1$ are the observable variables that affect the probability distribution of their children BG_n, CHO_n, and CHO_{n-1} respectively. BG_n is the hidden node at time n represented in the corresponding time slice. Thus, the JPD is computed by

$\Pr(MEAL_0, \ldots, BG_5)$

$= \Pr(CHO_0|MEAL_0) \cdot \Pi_{n=1}^5 \Pr(MEAL_n)$

$\cdot \Pr(CHO_n|CHO_{n-1}, MEAL_n)$

$\cdot \Pr(BG_n|BG_{n-1}, CHO_{n-1})$

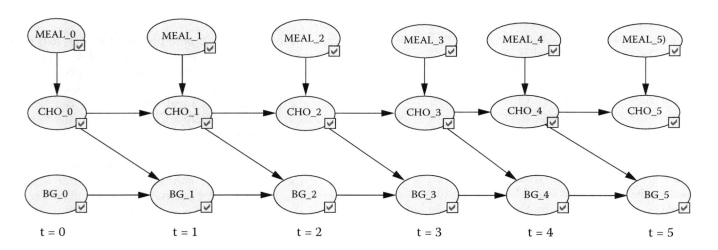

FIGURE 25.4
Dynamic Bayesian network with six time slices ($N = 5$) representing the diabetes example. The duration interval between two consecutive time slices is 4 hours.

There are several applications of DBNs in the medical domain, for performing various clinical tasks such as diagnosis and prognosis. They represent medical knowledge explicitly in terms of causes and effects as obtained from data, domain experts, and medical literature. Considerable work on dynamic models in medicine was carried out by Cao and collaborators, who successfully used a combination of graphical models with Markov chains to solve problems in different medical domains such as colorectal cancer management [66], neurosurgical ICU monitoring [46,67], and palate management [68]. Other applications of DBNs in medicine include forecasting sleep apnea [69] and diagnosis and decision making after monitoring patients suffering from renal failure and treated by hemodialysis [50]. The Pittsburgh cervical cancer screening model (PCCSM) [51] is another DBN system used for the prediction of the risk of cervical precancer and cervical cancer for patients undergoing cervical screening.

25.3.2 Networks of Probabilistic Events in Time

NPEDT is another temporal probabilistic model proposed by Galan et al. [58] that represents discrete-time events. Nodes in the model represent temporal random variables, which denote the presence or absence of an event at each time instant. For example, if variable B represents the event "abnormal glucose levels," $B[a]$ means that the patient had abnormal glucose levels at instant a. The links in the network represent the causal temporal relationships between events. The CPT of each variable represents the probability of occurrence of the child event given that its parent events occurred at any possible time point. Thus, the CPT represents the most probable delays between the occurrence of a parent event and the corresponding child event. In a family of n parents $A^1, \ldots, A^n$ and one child B, the CPT of B given its parents is defined by:

$$P(B[t_B]|A^1[t_1],\ldots,A^n[t_n]) \text{ with} \\ t_B \epsilon [0,\ldots,n_B, \text{never}], t_A \epsilon [0,\ldots,n_A, \text{never}] \quad (25.1)$$

where $[0 - n_B]$ is the temporal range of B and $[0 - n_A]$ is the temporal range of A.

In these models, each temporal random variable A can take on a set of values $v[i]$, $i \in \{a,\ldots,b,\text{never}\}$ indicating the presence or absence of an event at the particular time interval, where a and b are instants representing the limits of the temporal range of interest for A. Time is discretized, and it is divided into a discrete number of intervals of constant duration. Time representation (seconds, minutes, weeks, etc.) depends on the particular problem.

The main difference from a DBN model is that in the NPEDT model, each value of a variable represents the instant at which a certain event may occur within a certain temporal range of interest and not the state of the real-world property. Therefore, events that can take n state values are represented with n different variables.

For illustrating the network structure of NPEDT in Figure 25.5, we assume for simplicity of the model that all variables apart from MEAL can have only two states: {high, normal}. Let us assume that the discretized state values of MEAL are $a1: 0 - 30g$, $a2: 30 - 60g$, and $a3: 60 - 90g$ represented by variables $MEALa1$, $MEALa2$, and $MEALa3$ respectively. P_{ij}, in our example, represents the probability $Pr(BG[i] | CHO[j])$, i.e., the probability that glucose levels are high at time instant i ($BG[i]$) given that the rates of carbohydrate remaining in the gut at instant j ($CHO[j]$) are high (e.g., $CHO[j] > 50g$). The selected temporal range of interest for the given example is 1 day (24 hours). This period is divided into 4-hour intervals.

To the best of our knowledge, the NasoNet system [36] is the only medical system developed using the NPEDT approach. It models the evolution of a nasopharyngeal cancer to assist oncologists in the diagnosis and prognosis of this type of cancer in a patient.

25.3.3 Continuous-Time Bayesian Networks

CTBNs [61] are graphical models that represent structured stochastic processes whose states change continuously over time. Let A be a set of local variables $A^1, \ldots, A^n$. Each A^i has a finite domain of values $Val(A^i)$. A CTBN over A consists of two components: the first is an initial distribution J_1, specified as a BN over A, and the second is a continuous transition model, specified as

- A directed graph G whose nodes are $A^1, \ldots, A^n$, and $Par(A^i)$ denotes the parents of A^i in G
- A conditional intensity matrix (CIM), $Q_{A^i|Par(A^i)}$ for each variable $A^i \in A$, which represents the state changes of variables through time. If $A^i = x_i$, then it stays in state x_i for an amount of time exponentially distributed with parameter q_i^x, where $q_i^x = \sum_{j \neq i} q_{ij}^x$. Intuitively, the intensity q_i^x gives the probability of leaving state x_i, and the intensity q_{ij}^x gives the probability of transitioning from x_i to x_j.

Time is explicitly represented in a CTBN, and it is able to represent processes evolving at different granularities. Nodes in the network represent variables that evolve in continuous time, and the values of each variable depend on the state of its parents in the graph. Figure 25.6 displays the CTBN network structure for the diabetes

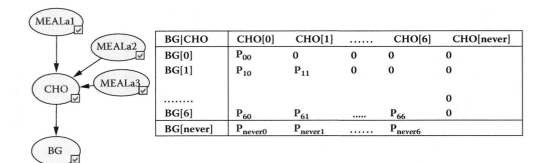

FIGURE 25.5
The NPEDT model for the diabetes example with a temporal range of 24 hours, divided into 4-hour intervals. A general CPT for $BG \mid CHO$ over all possible time instants is included.

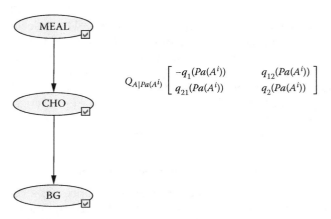

FIGURE 25.6
Graph structure of a CTBN model for the diabetes example. The $Q_{A^i \mid Par(A^i)}$ matrix describes the transient behavior of a particular variable A^i with two values, conditioned on its parents $Par(A^i)$, e.g., BG, whose domain is two values (v_1, v_2), conditioned on its parent variable CHO with two possible states.

example. The structure of the CTBN is the same as in atemporal BNs. A CIM, $(Q_{A^i \mid Par(A^i)})$, represents the transient behavior over time for a particular variable, e.g., BG.

One application of CTBNs in medicine is for the domain of colon cancer. The model was applied to extract information about recurrence of cancer from a sample of colon cancer patient records [52]. Another application of the CTBN was as a joint diagnostic and prognostic model for diagnosing cardiogenic heart failure and predicting its likely evolution [45].

25.3.4 Irregular-Time Bayesian Networks

ITBNs [70] generalize DBNs such that each time slice may span over a time interval. The goal of an ITBN is to model, learn, and reason about structured stochastic processes that produce qualitative and quantitative data irregularly along time.

Variable states are represented through a random vector $T_j : j \in \mathbb{N}$ indexed by the order of time points of interest. Any function $\vec{X}$ from T, such that $\vec{X}_t$ is a random variable for each $t \in T$, is called an irregular-time stochastic process. A vector of time offsets denoted by $(\delta_t \in \mathbb{R} : 1 \leq t \leq m)$ represents the time differences expressing a delayed effect between nodes of the same time slice. Knots (unused time points) in the network are used for changing dynamics between consecutive slices.

ITBNs have the ability to compute probabilities given evidence from the far past in one step, which expresses the long-distance effects. It is the first model that applies semi-parametric methods [71], and in particular time-varying coefficient models [72], to longitudinal data analysis. A semiparametric conditional probability distribution for each variable is computed by using vectors to parameterize the varying coefficient model of the predictor (see varying coefficient models for more details [73]).

ITBNs use a new method to represent random time points in a vector T_j where only the time points of interest need to be considered and treated as evidence. Thus, ITBNs do not require a constant time granularity for representing the changes of process states. Figure 25.7 represents the ITBN model for the diabetes example. Each patient eats five times a day; thus, five time slices are used, each one representing the process of meal intake and the estimates of blood glucose levels based on the rates of carbohydrate remaining in the gut. Time offsets (δ) represent the time taken for the carbohydrate CHO to be absorbed by the gut. The ITBN model was applied to a diabetes data set [74], monitoring the glucose levels of patients [70]. Another ITBN application concerns the investigation of the pharmacokinetics of the drug cefamandole.

A comparison of temporal extensions of BNs based on their knowledge representation and acquisition, their time representation, the computational demands, and their applications in medicine described above is displayed in Table 25.1. In the following section, we discuss

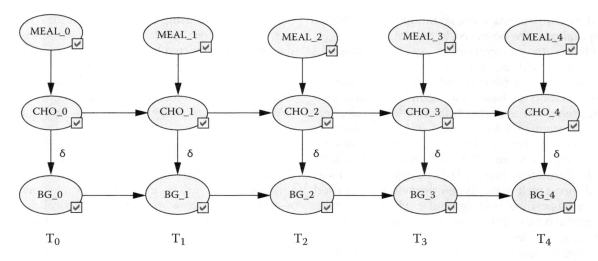

FIGURE 25.7
An ITBN model for the diabetes example represented in five time slices where δ is the time taken for carbohydrate absorption from the gut.

TABLE 25.1

Comparison of Temporal Extensions of Bayesian Networks

	DBN	NPEDT	ITBN	CTBN
Time representation	Discrete time (time slices), absolute time	Interval-based, absolute time	Irregular time (using a time vector), absolute time	Continuous time, absolute time
Computational demands	High memory consumption in parameters and inference	Learning parameters are linear to learning the number of parents	Less memory consumption than DBN	Exact inference is intractable
Applications in medicine	Head injury management, colorectal cancer management, ICU monitoring, forecasting sleep apnea	Prediction of nasopharyngeal cancer (NasoNet system)	Diabetes monitoring	Prediction of colon cancer, diagnosis of cardiogenic heart failure
Knowledge acquisition	Data, expert	Data (compound parameters), expert (structure and net parameters)	Data	Data, expert
Granularity	Constant	Constant	Multiple	Multiple

the benefits of integrating TA techniques with TBNs. We also give an overview of our approach regarding the integration of the two technologies.

25.4 Integration of Temporal Abstractions with Bayesian Networks

BNs (both atemporal and temporal) have many applications in medical expert systems in tasks such as classification, risk assessment, and prediction. The temporal abstract concepts were shown to be helpful in various clinical tasks and domains such as summarizing and managing patient data [35,38]. A claim that we investigated in our previous works [75–77] is whether techniques for these two distinct technologies could be effectively integrated in the context of medical decision support systems, capitalizing on their complementarities. A detailed survey on TA and BN applied in medical systems and the benefits of their integration in clinical domains can be found in our work in Ref. [78].

The integration of TAs (basic and complex) and BNs (both temporal and atemporal) has important benefits for various domains where time is a significant factor and especially in clinical domains where data are uncertain and voluminous. The key benefits are that the relevant reasoning can take place at higher and multiple levels of abstraction, thus controlling computational overheads associated with reasoning at a single detailed level, as well as allowing for more conceptual modeling of the given reasoning processes. Furthermore, the interpretation of high-level data can be effectively achieved by constructing a causal model able to explain the temporal patterns observed in the data.

Considering the integration of the two areas, TA methodologies will be used to extract both basic abstractions (state or trends) and complex abstractions (any combination of basic and complex abstractions, involving temporal and structural relations and periodicity). A system of TAs can deal with continuous or discrete time. Although real time is continuous, discrete time is closer to the spirit of TAs, which is to yield out the essential information (abstract) from low-level, uncertain, and incomplete data. When discrete time is used, there is a basic time unit (granularity) and/or other higher granularities giving rise to a time model with multiple granularities.

The derived concepts will then be used for BN model development and deployment. For model development, TAs involve multiple cases, since the aim is to induce knowledge from the generated TAs of a representative set of cases. For model deployment, for purposes of problem solving, TAs refer to single cases, i.e., the cases under consideration. Learning parameters and inference algorithms will be applied to the constructed model to provide clinicians with probabilistic diagnosis, prediction, or therapy classification results.

In order to validate our claim for the proposed integration, we investigated the development of two models: a TBN incorporating basic TAs as its nodes utilizing the whole monitoring history of the patients, and a naïve Bayes model incorporating TARs, a specific type of complex TAs, extracted from the most recent patient history, as its nodes.

Regarding the development of the first model (DBN-extended), DBN was considered the most appropriate TBN selection for this integration since DBNs are the most widely known temporal extensions of BNs with considerably more clinical prognostic applications compared to other TBNs. Furthermore, a DBN uses a discrete-time representation, where it may be argued that discrete time is closer to the spirit of TAs, since the aim is to yield (abstract) the essential information at discrete granularities from low-level, uncertain, and incomplete data referring to real (dense) time. Thus, a DBN is able to represent TAs capturing the evolution of processes in a sequence of time slices.

However, a DBN cannot represent processes that evolve at different time granularities, but rather, it represents the whole system at a single granularity that tends to be the finest relevant granularity. This could be a limitation since in general, temporal abstract concepts could naturally refer to different time granularities. In addition, complex TAs, where temporal patterns of any degree of complexity may be involved, entail longer time spans. As such, the restriction to fixed time slices, corresponding to the finest relevant granularity, could be a limiting factor in using DBNs with complex TAs. Hence, in the first system developed for this work, the nodes of the DBN are restricted to basic TAs derived with respect to a single fixed granularity. This restriction is lifted in the naïve Bayes classifier system, also developed in this work, whose nodes are TARs, a category of complex TAs that require longer time slices and thus higher granularities.

Regarding the application domain for both systems, we have selected the field of coronary heart disease (CHD) as a test bed and demonstrator of the attempted integration in order to have concrete performance results. An overview of the CHD medical domain and the test bed data set used is given in the following section.

The two integration models developed, namely, DBN-extended and periodic classifier, are respectively discussed in Sections 25.6 and 25.7. It is noted that a more extensive discussion of the DBN-extended model can be found in Ref. [75], while the preliminary work on the periodic classifier is reported in Ref. [79].

25.5 The Clinical Domain of Coronary Heart Disease: Test Bed Data Set

CHD is generally caused by atherosclerosis—when plaque (cholesterol substances) builds up in the arteries. Over time, plaque can cause the hardening and narrowing of the artery walls, resulting in less blood flow to the heart. As a result, the decreased blood flow may cause a CHD event such as myocardial infarction (MI), acute coronary syndrome (ACS), angina pectoris (AP), and ischemic heart disease. The identification of key risk factors will help in detecting the disease in its early phases and preserve the occurrence of future CHD events (secondary prevention of CHD).

CHD is selected as the test bed clinical domain because it is the most common cause of death in many countries [80]. In addition, a key benefit of applying our proposed models to the CHD domain is that the high degree of uncertainty inherent in the particular domain can be addressed, in conjunction with utilizing an extensive part of, or even the entire, patient history.

The benchmark data set used in the development of the two integration models is the STULONG data set, which was collected from a longitudinal study of atherosclerosis primary prevention. The data set includes men born in Prague and who were 38–53 years old between 1926 and 1937, at the first examination. The number of visits of a single patient ranges from 1 to 20, and the follow-up time spans from 1 to 24 years. The first examination included blood pressure measuring, basic anthropometric measuring (e.g., weight and height), and electrocardiography (ECG) test. Furthermore, patients were asked about their level of education and responsibility in job; their general habits, such as smoking, physical activity, and alcohol drinking; as well as family and personal medical history related to cardiovascular diseases, chest and leg pain, and breathlessness.

Below, we present the two integration models whose development is based on the STULONG data set. The target group of both models includes both patients with a past CHD history and patients without CHD history. The DBN-extended model that is presented first aims to assess the risk of developing CHD, while the periodic classifier model aims to predict whether CHD is currently present or not.

25.6 DBN-Extended Model

The first model that was developed in order to validate the benefits of the integration with TAs is the DBN-extended model. This is a DBN whose nodes represent basic TAs. The extended model was applied for both diagnosis [77] and prognosis of CHD [75] and also for the primary prevention of CHD [76]. In this section, we focus on the prognostic application of CHD, because it is this DBN-extended model that was evaluated and compared, against a standard DBN model.

Our methodology consists of five main phases:

1. Data preprocessing and feature selection
2. Derivation of basic TAs
3. Construction of the DBN-extended model
4. Application of the DBN-extended model for prognosis of a CHD event
5. Comparison of its performance against a standard DBN model, without TAs

The feature selection is based on the domain knowledge that we acquired from a CHD expert. The selected features, which are CHD risk factors, include the following: hypertension (diastolic and systolic blood pressure); cigarette smoking status (current smoker or not); dyslipidemia levels (such as total cholesterol, high-density-lipoprotein [HDL] cholesterol, low-density-lipoprotein [LDL] cholesterol, and triglycerides levels); obesity; diabetes; history features (such as past personal history and family history); age; medicines treating high cholesterol and hypertension; diet (if they follow any diet or not); and exercise (if they regularly exercise or not). The incorporation of these features into the DBN-extended model is explained in the following subsections.

However, before going into this discussion, it is noted that a key issue for model construction is the choice of the total observation period for all patients. In the STULONG data set, this ranges from 1 to 24 years. Thus, by selecting a temporal range of 21 years, we include the majority of patients with a CHD event within 19–21 years after their first examination. For patients whose total observation period is less than 21 years, all feature values are considered unknown for any years beyond their specific observation period.

25.6.1 Utilization of Basic Temporal Abstractions

Basic TA techniques such as state, trend, and persistence were applied to the test bed data set. TAs were derived on a fixed 3-year interval period. This period was selected since at least two examinations are needed in order to have any abstractions. It is noted that the finer granularity of the data set is 1 year; however, most cases do not have examinations on an annual basis.

State TAs of systolic and diastolic blood pressure values are "poorly controlled", "well controlled" hypertension, and "no hypertension." The hypertension variable is defined by the "poorly controlled" label if the patient has a history of hypertension and his/her blood pressure levels are above the standard limits, whereas the "well controlled" label is when a patient has a history of hypertension and his/her blood pressure levels are normal. Otherwise, it is defined by the "no hypertension" label.

The most frequent label for the corresponding feature for that period is selected as the label of the derived TA. When multiple values occur equally frequently, the last value in the predefined time period is selected. One of the assumptions used in deriving state abstractions is that the abstraction value of a variable with missing raw values at any time within the interval period is defined to be the same as its last known value. All the derived state TAs are displayed in Table 25.2.

Trend abstractions of a variable were generated by comparing two or more consecutive feature values during the interval period of 3 years (one to three examinations) and selecting the most frequent trend value for the corresponding feature for that period. Again, when multiple values occurred equally frequently, the last value in the predefined time period was selected. We also used a combination of trends and state abstractions in order to define the ratio of change of a particular variable based on its value. Trend abstraction values are

- "Normal" (N) when the variable state is normal and its trend ratio is decreasing or steady
- "NormalIncreasing" (NI) when the variable state is normal and its trend ratio is increasing

TABLE 25.2

State TA Variables and Their Values

Variable	Code	Value 1	Value 2	Value 3
Smoking	Smoking	Nonsmoker	Current smoker	
Hypertension medicines	medBP	Taken	Not taken	
Dyslipidemia medicines	medCH	Taken	Not taken	
Hypertension	HT	No hypertension	Well controlled	Poorly controlled
Dyslipidemia	Dyslipidemia	Absent	Present	
Obesity	Obesity	Absent	Present	
Age	AGE	Young	Old	Very old
Diet	DIET	Following a diet	Not following a diet	
Exercise	Exercise	Exercising	Not exercising	

Note: Code is the variable name in the DBN model.

- "AbnormalDecreasing" (AD) when the variable state is abnormal and its trend ratio is decreasing
- "Abnormal" (A) when the variable state is abnormal and its trend ratio is increasing or steady

It should be noted that, contrary to the rest of the variables, HDL is considered a CHD risk factor when its levels are low; thus, its trend values are: "Normal" (N), "NormalDecreasing" (ND), "AbnormalIncreasing" (AI), and "Abnormal" (A). The resulting trend abstractions are displayed in Table 25.3.

Persistence TAs include the personal and family medical history of the patient and diabetes. For example, when someone is diagnosed with a CHD event at time t, he has a history of CHD from $t + 1$ onwards; thus, the value of HistoryEvent variable is "present" from $t + 1$ until the end of the monitoring process. Similarly, when

TABLE 25.3

Trend TA Variables and Their Values in a Sorted List

Variable Name	Code	Value 1	Value 2	Value 3	Value 4
Low-density-lipoprotein cholesterol	LDL	N	NI	AD	A
Triglycerides	TRIG	N	NI	AD	A
High-density-lipoprotein cholesterol	HDL	N	ND	AI	A
Total cholesterol	TCH	N	NI	AD	A

Note: Variable code is the variable name in the DBN model.

TABLE 25.4

Persistence TA Variables and Their Values

Variable Name	Variable Code	Value = 1	Value = 2
Diabetes	Diabetes	Present	Absent
Past personal history	HistoryEvent	Present	Absent
Family history	FH	Present	Absent
History of hypertension	HHT	Present	Absent

Note: Variable code is the variable name in the DBN-extended model.

the patient is diagnosed with diabetes at time t, we assume that diabetes persists over time, so the patient is diagnosed with diabetes for time t and onwards. The family history (FH) and history of hypertension (HHT) are examples of persistence TAs for the whole representation time period, since their value does not change through time. The resulting persistence TAs are displayed in Table 25.4.

25.6.2 Construction of the DBN-Extended Model

The network structure, as displayed in Figure 25.8, was designed by incorporating prior information elicited from medical experts and medical literature [81,82]. The derived basic TAs described in the previous subsection form the nodes of DBN-extended. The DBN framework enables us to combine all the observations of a patient and predict a probability for the hypothesis that the patient will suffer with a future CHD event, given the values of all the observed nodes. Each time slice in the network represents the time interval period of 3 years since this is the selected period for deriving the TAs.

The model consists of 19 variables, which constitute the nodes of the DBN, out of which 17 are observed and 2 are hidden. Hidden variables are the class attribute Pred_Event, representing the occurrence of a CHD event in the last 3 years of the total observation period (21 years), and the Dyslipidemia node. Both of these variables take two values: "present" and "absent."

The Pred_Event variable is a terminal node, represented outside of the temporal network, and it only connects to its parents in the last time slice of the unrolled network. It represents the occurrence of a future CHD event; thus, it is not repeated in every time slice, but its value is inferred at the end of the inference process [83]. In order to simplify the parameter estimation process [84], Dyslipidemia was introduced as a common parent node of TCH, HDL, LDL, and TRIG, which are indirect risk factors to the class attribute, and CurrentEvent. Dyslipidemia is a common cause of TCH, HDL, LDL, TRIG, and CurrentEvent at each time slice. At time slice $t = 5$, Dyslipidemia is a common cause of TCH, HDL, LDL, TRIG, and the class variable. If there is no evidence or information about Dyslipidemia

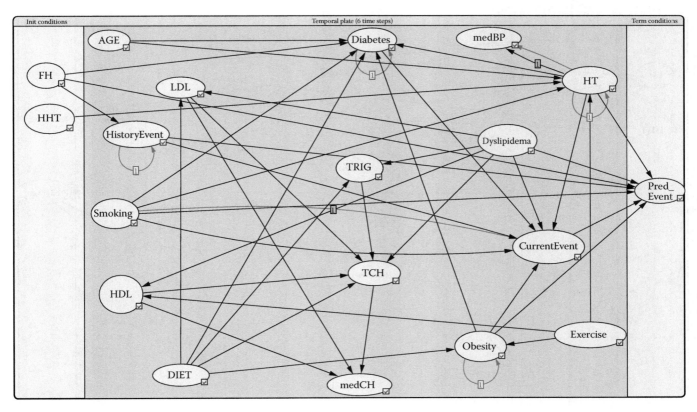

FIGURE 25.8
The structure of the DBN-extended model representing only the basic temporal abstractions. An arc labeled as 1 between the variables denotes an influence that takes one time step.

(hidden), then the presence of one or more symptoms (TCH, LDL, HDL, TRIG) will increase the chances of Dyslipidemia, which in turn will increase the probability of the effects CurrentEvent and Pred_Event.

The variable family history (FH) is not repeated since it is modeled only as an initial condition and its value does not change over time. It is therefore shown in the network of Figure 25.8 to be outside the temporal plate. The single-digit numbers on the arcs denote the temporal delay of the influence of the cause node to the effect. For example, an arc labeled as 1 between the variables history of CHD (HistoryEvent) and itself denotes that the patient's CHD history at the current time slice t is influenced by the patient's past CHD history (at time slice $t-1$). On the other hand, the arc without a label connecting the CHD risk factors (hypertension, obesity, etc.) to the CHD event denotes an instantaneous influence at the same time slice.

After defining the structure of the network, we also had to define the conditional probabilities that quantify the arcs of the network. Since the structure of a DBN is invariant for all times $t \in \{0, ..., 5\}$, the parameters of the network are fixed through all the time slices. All of the parameters were learned from data using the expectation maximization algorithm (EM) [85].

Once the network structure was defined and the network was quantified with the learned conditional probability distributions, the next step was to predict the probability of the class node: Pred_Event. Each variable in the network was instantiated with the corresponding feature value. The DBN was unrolled for six time slices $t = [0, ..., 5]$ in order to represent the observation period of 21 years. Then, it performed prognosis and derived the belief in the class variable Pred_Event, which represents CurrentEvent at $t = 6$. More specifically, the model derived by:

$$P(\text{Pred\_Event}|\text{Smoking}_5, \text{HT}_5, \text{Dyslipidemia}_5, \text{Obesity}_5, ...)$$

i.e., the probability of having a CHD event (or not) given the evidence at the last time slice.

25.6.3 DBN without TAs

In addition to the DBN-extended model, a standard DBN model without TAs was developed to represent time points rather than high-level TAs. Applying this model to the STULONG data set, each time slice of the network represents a patient examination. Although the total number of examinations of all patients is 22, most of

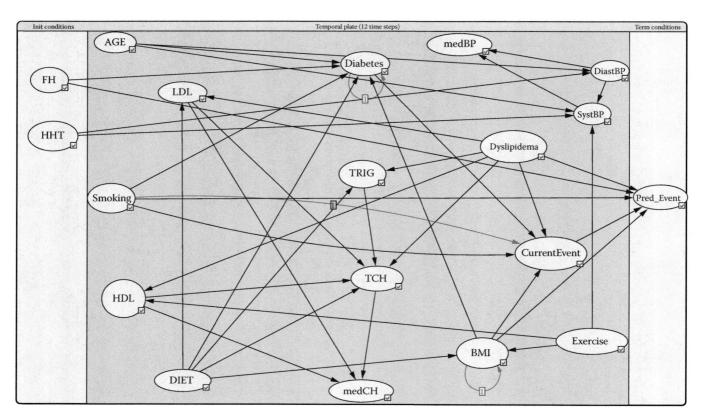

FIGURE 25.9
DBN model representing low-level data (without temporal abstractions).

the patients had a CHD event during or before their thirteenth examination. Consequently, the chosen total temporal range for this network was 13 examinations. We used EM to handle the relevant missing data.

The selected features of the DBN-without-TAs model are risk factors (RFs) of CHD, as with the DBN-extended system. The main difference is the removal of features that represent high-level concepts such as hypertension, history of CHD event, and obesity, and the addition of new features representing time-point events such as body mass index (BMI) and systolic and diastolic blood pressure, as displayed in Figure 25.9. We also discretized the values of the continuous features in the data set, using a discretization technique for clinical domains, called "domain-based discretization". Using a domain-based technique, the values of each continuous attribute are divided into a particular number of bins based on clinical domain knowledge [86]. All the selected features and the resulting discretization feature values of the continuous features are displayed in Table 25.5.

25.6.4 Training in the Presence of Class Imbalance

One important problem in the data mining field is to deal with imbalanced data sets. The data sets present a class imbalance when there are many more examples of one class (majority class) than of the other (minority class). In the current data set, individuals who did not suffer a CHD event (majority class) at a particular time period are many more than those who suffered a CHD event (minority class) [87].

In our system, we evaluated our classifier by applying two oversampling methods as well as a combination of oversampling with undersampling on the training data set. More specifically, we applied the following resampling methods: (a) SMOTE-N (Synthetic Minority Oversampling Technique for nominal features) that generates synthetic examples to be added to the minority class; (b) random oversampling where minority cases are randomly chosen for duplication; (c) random oversampling combined with clustering undersampling; and (d) SMOTE-N oversampling on the minority class was combined with clustering undersampling the majority class. Clustering undersampling [88], uses the k-means algorithm to divide the majority class samples into K (K > 1) clusters and then make K subsets of majority class samples, where each cluster is considered to be one subset of the majority class. All the subsets of the majority class are separately combined with the minority class samples to make K different training data sets. In our implementation, K was chosen to be equal

TABLE 25.5

Features Selected for the Nontemporal Abstraction Model and Discretization of Continuous Features

Variable	Variable Code	Value = 1	Value = 2	Value = 3
Smoking	Smoking	Nonsmoker	Current smoker	
Medicines for reducing cholesterol	medCH	Taken	Not taken	
Medicines for reducing blood pressure	medBP	Taken	Not taken	
Systolic blood pressure	SBP	Normal: <120	Prehypertension: [120–140]	Hypertension: >140
Diastolic blood pressure	DBP	Normal: <80	Prehypertension: [80–90]	High: >90
Dyslipidemia	Dyslipidemia	Absent	Present	
Disease	CurrentEvent	Absent	Present	
Predict coronary heart disease	Class variable	Pred_Event	Absent	Present
Diabetes	Diabetes	Absent	Present	
Family history	FH	Absent	Present	
History of coronary heart disease	HHD	Absent	Present	
Body mass index	BMI	Normal weight: <25	Overweight: 25–30	Obesity: >30
Low-density-lipoprotein cholesterol	LDL	Normal: <100 mg	High: 100–160 mg	Very high: >160 mg
Triglycerides	TRIG	Normal: <150 mg	High: 150–200 mg	Very high: >200 mg
High-density-lipoprotein cholesterol	HDL	Normal: <40 mg	High: 40–60 mg	Very high: >60 mg
Total cholesterol	TCH	Normal: <200 mg	High: 200–240 mg	Very high: >240 mg
Age	AGE	<50 years	50–60 years	>60 years
Diet	DIET	Following a diet	Not following a diet	
Exercise	Exercise	Exercising	Not exercising	

to 2. At the end, we obtained the following training data sets for each classifier:

1. Training data set 1 (*D*1): no resampling. In this experiment, the training data were not altered.
2. Training data set 2 (*D*2): random oversampling of the minority cases until we got a 1:1 ratio of the minority class to the majority class.
3. Training data set 3 (*D*3): oversampling using the SMOTE-N technique. We applied oversampling to the minority class cases until we got a 1:1 ratio of the minority class to the majority class.
4. Training data set 4 (*D*4): oversampling using the SMOTE-N technique and undersampling. We applied clustering undersampling to the majority class cases, and then we applied SMOTE-N until we got a 1:1 ratio.
5. Training data set 5 (*D*5): oversampling using random oversampling technique and clustering undersampling. We applied clustering undersampling, and then we applied random oversampling to the minority class cases until we got a 1:1 ratio.

25.6.5 Experiments and Analysis

We constructed five DBN-extended networks, one for each experiment (*C*1, *C*2, *C*3, *C*4, and *C*5, corresponding to the five data sets, *D*1, *D*2, *D*3, *D*4, and *D*5 respectively).

The networks have the same structure but differ in their parameters, i.e., prior probabilities and the CPTs according to the respective data set: each time a new training data set was introduced, new network parameters were derived using training on the new set.*

For the DBN-without-TAs model, we obtained the same training data sets as for the DBN-extended, and we constructed five networks, one for each experiment (*M*1, *M*2, ..., *M*5). In order to evaluate the prognostic performance of both developed models, we applied the 10-fold cross-validation technique to the five data sets, and then we adopted the following metrics: precision, recall, and the F_1 score [89].

To avoid or minimize skew-biased estimates of performance, we normalized the performance scores of precision and F_1 score by normalizing, to a given degree of skew (target skew ratio = 1) [90]. The values of precision, recall, and F_1 score that were obtained from the evaluation of the DBN-extended model for each of the five training data sets are given in Table 25.6. As can be seen from Table 25.6, the DBN-extended system yields promising results, with precision reaching as high as 72%, recall reaching as high as 75%, and a combined F_1 score of 74%. These are the best results that were derived by training the system with data set *D*5, which applied random oversampling combined with clustering

---

* The models presented in this chapter were created and tested using the SMILE inference engine and GeNIe available at https://dslpitt.org/genie/ [date accessed: 31 January 2017].

TABLE 25.6

The Performance for All Five DBN-Extended Models Corresponding to the Five Training Data Sets

Evaluation Metrics	C1	C2	C3	C4	C5
Precision	0.8555	0.5966	0.5057	0.5090	**0.7207**
Recall	0.1167	0.6667	0.5167	0.6000	**0.75**
F_1 score	0.2053	0.6297	0.5111	0.5508	**0.7351**

TABLE 25.7

The Performance for All Five DBN-without-TAs Models Corresponding to the Five Training Data Sets

Evaluation Metrics	M1	M2	M3	M4	M5
Precision	0.9265	**0.5115**	0.4329	0.5502	0.6350
Recall	0.0833	**0.6250**	0.3750	0.2917	0.333
F_1 score	0.1529	**0.5626**	0.4019	0.3812	0.4372

undersampling. The lowest F_1 score is obtained when the system is trained with the original, highly imbalanced data set $D1$ (no resampling). As expected, this yields a low recall due to the high false negative, as the classifier fails to recognize the minority cases (CHD present), which are instead classified as majority cases (CHD absent). In many medical domains such as CHD, recall is more important than precision since the false negative diagnosis (i.e., failure to predict a CHD event) has much more serious consequences for the patient than a false positive.

The values of precision, recall, and F_1 score that were obtained from the evaluation of the DBN model without TAs for each of the five training data sets are given in Table 25.7. As can be seen from Table 25.7, the extended DBN model outperforms the prognostic DBN model without TAs and further supports the belief that DBNs can be effectively integrated with TAs.

Next, we present the periodic classifier that incorporates complex TAs, namely, periodic temporal associations, and compare its performance against a baseline classifier that incorporates temporal associations but does not take into consideration their potential level of periodicity. Through this experiment, we therefore show that where long time periods are significant, complex TAs of a high order of complexity are called for.

25.7 Incorporating Complex Temporal Abstractions in Naïve Bayes Classifiers

The second model that we developed to validate the benefits of the potential integration is a naïve Bayes classifier whose features represent frequent TARs. As already mentioned, TA techniques are divided into two main categories: basic and complex TAs. TARs are considered as complex TAs, since they are frequent temporal patterns derived from other TAs. They are mostly used for detecting the precedence of the corresponding abstractions in a sequence of patterns. In general, a TAR is a temporal association between two episodes, the antecedent and the consequent. However, herein, a relationship is specified through a temporal operator (e.g., before), which holds between an antecedent, consisting of a single event (e.g., hypertension) or multiple co-occurring events (e.g., hypertension AND high cholesterol), and a consequent of a single event (e.g., total cholesterol increasing).

The main aim of the development of the second system is to utilize the recurrence patterns in patient records of frequent TARs in order to validate our claim that higher-level TAs such as periodic TARs can improve the classification performance and overall medical problem solving in domains where long time periods are of significance. The notion of horizontal support that was introduced in Refs. [28,30] represents the number of times that a particular temporal pattern is found in some patient's record. We use this measure to define the periodicity of a given TAR.

The development of the second system consists of the following steps:

- Data preprocessing and feature selection
- Derivation of temporal patterns using basic TA algorithms
- Use of a pattern mining algorithm to detect and mine the most frequent TARs
- Building the naïve Bayes classifiers incorporating simple and periodic TARs as their attributes
- Evaluation of the classifiers

The first phase consists of data preprocessing and feature selection. In medical data sets, usually, the most recent patterns are the most significant ones. Relying on this assumption, for each patient, we consider the last observation before the first diagnosis of the disease (for patients who were diagnosed) or the last visit (for patients who were not diagnosed with CHD). Based on the relevant medical literature [91,92], the patient history is taken to be up to 10 years prior to the last observation of the patient. In addition, the derivation of basic TAs would require a time period of at least two examinations, i.e., at least 2 years, while the derivation of complex TAs would require even longer time periods. As such, the selected data set is further reduced by removing records of patients who had less than three examinations, i.e., that spanned less than 3 years.

The resulting target group consists of 709 patients, out of which 154 were diagnosed with the disease.

The data are characterized with missing values, and in order to handle them, we use the *missforest* [93] method, a nonparametric imputation method based on the random forest algorithm [94]. For each variable, missforest fits a random forest on the observed part of the data available for that variable, and then predicts the missing part of the data. The algorithm repeats these two steps until a stopping criterion is met or the user-specified maximum of iterations is reached.

Regarding the feature selection process, we base our selection of features on the domain knowledge that we acquired from a CHD expert and from medical literature. The selected features, which are known to be CHD risk factors, are the same as for the DBN-extended model, as defined in Section 25.6. We use these features for preprocessing time series and representing them as temporal patterns through basic TAs. We then apply the TAR mining algorithm to these TAs in order to identify the most frequent temporal pattern associations.

25.7.1 Extraction of Temporal Association Patterns

After extracting basic TAs, we apply a TAR mining algorithm to detect TARs in our data [95]. TARs are the output of the basic TA algorithms using relational temporal operators, which are defined on Allen's interval relationships (see Figure 25.1). In the current study, we exclusively use the disjunctive temporal operator PRECEDES as defined in Refs. [23,27] and illustrated in Figure 25.2.

A TAR mining software tool [95] is used to discover and mine all the temporal rules extracted from basic TAs. The tool takes as input time-point events or TA episodes and discovers frequent TARs among the events/episodes on the basis of the given temporal relationship. The discovered TARs are sorted based on the support and the confidence index. The support index is defined as the proportion of cases verifying the TAR (SR) over the total number of cases involved in the study (S) (see Equation 25.2), while the confidence index is defined as the ratio between the support of a TAR ($A \rightarrow C$) and the support of the antecedent (A) as defined in Equation 25.3. More specifically, the confidence represents the conditional probability for a case, to verify the TAR given that the antecedent is detected for that case. The higher the values of support and confidence are, the more frequent is the occurrence of the particular TAR in the data set.

$$\text{Support} = SR/S \quad (25.2)$$

$$\text{Confidence} = \text{Support}(A \rightarrow C)/\text{Support}(A) \quad (25.3)$$

Besides the temporal operators that define the temporal relationships among the TAs, the software tool also uses three temporal constraints (*left shift, right shift, gap*) to properly control the mutual distance between the antecedent (A) and the consequent (C) of a TAR As already mentioned in Section 25.2, an episode is characterized by a pair (I, L), where I is an interval period consisting of a start time (Istart) and an end time (Iend) and L is the label of the pattern.

The parameter *left shift* is defined as the maximum allowed distance between the start time of the antecedent ($Istart_1$) and the start time of the consequent ($Istart_2$). The *right shift* is defined as the maximum allowed distance between the end time of the antecedent ($Iend_1$) and the end time of the consequent ($Iend_2$). The *gap* is similarly defined as the maximum allowed distance between $Iend_1$ and $Istart_2$. Using any temporal operators apart from MEETS and BEFORE, the *gap* is a negative number since $Istart_2 < Iend_1$. Figure 25.10 graphically displays the meaning of the three temporal constraints.

Such parameters are used as user-defined temporal constraints to define a temporal relationship, in order to reduce the number of extracted TARs. After the extraction of TARs, the number of selected TARs can be reduced by specifying temporal constraints based on these parameters, e.g., gap < 1 year. The assumption of using these temporal constraints is that the frequent occurrence of two or more episodes close in time is more likely to represent a potential cause-and-effect relationship than where the events are separated by a longer interval [95].

The method used for selecting the features (TARs) to be incorporated in the classifiers is described in Section 25.7.2.

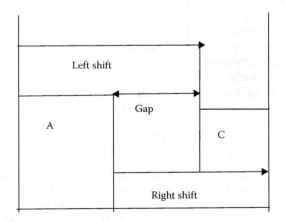

FIGURE 25.10
Representation of the three temporal constraints (left shift, gap, right shift) used in the TAR mining tool, to constrain the mutual distances between the antecedent (A) and the consequent (C).

25.7.2 Naïve Bayesian Classifier Using Periodic Temporal Association Patterns

Naïve Bayes classifiers belong to the family of probabilistic graphical models based on Bayes theorem with the "naïve" assumption that the effect of an attribute value on a given class is independent of the values of the other attributes [96,97]. This assumption is called class conditional independence. In our study, the class variable is the *Disease* (CHD event), and all the TAR attributes are connected to the class variable as shown in Figure 25.11.* The *Disease* is a binary node taking the values (a) 0: absence of CHD event and (b) 1: presence of CHD event. The goal is to classify the presence or absence of a CHD event given the pattern of occurrence of each TAR in the relevant history of the case in question.

The parameters (conditional probability distributions) of the classifier are learned from data using the maximum likelihood algorithm [98]. Once the network structure is defined and the network is quantified with the learned conditional probability distributions, the next step is to predict the probability of the class variable. Each feature in the network is instantiated with the corresponding feature value. Then the model derives the belief

$$P(\text{Disease}|tar1, tar2, tar3, \ldots, tarn)$$

25.7.3 Extracting the Most Frequent Temporal Association Patterns

As already mentioned, our experiments use exclusively the PRECEDES operator for the discovery of TARs. There are a variety of ways to effectively select the number of the frequent TARs that will be represented in the network. As a further step, we overview how the classification performance changes due to the TAR selection process. We compare the network performance in the presence or absence of inverse TARs and in the presence or absence of TARs that define the precedence of two decreasing variables. Consequently, we develop three periodic classifiers where the features are selected, after removing the generated TARs with confidence or support less than 0.5, as follows:

- Periodic classifier $C1$: We select the 10 most frequent TARs, as displayed in Table 25.8.
- Periodic classifier $C2$: We select the 10 most frequent TARs. Out of these, we then select the most predictive ones for the disease in accordance with medical knowledge. This is done by excluding those TARs that define the precedence of two decreasing variables (HDL is an exception) such as "Total cholesterol decreasing PRECEDES Triglycerides decreasing." Such associations are not considered good predictors for the disease because in general, their occurrence is less frequent with respect to patients diagnosed with the disease. The resulting number of TARs is eight, and they are displayed in Table 25.9.
- Periodic classifier $C3$: We select the 20 most frequent TARs. From these, we exclude those TARs that define the precedence of two decreasing variables (HDL is an exception), as in Classifier $C2$. Furthermore, in the case of having inverse TARs, i.e., TARs associating the same risk factors but in reverse order antecedent–consequent (e.g., one TAR says "Total cholesterol decreasing PRECEDES BMI increasing," and the other says "BMI increasing PRECEDES Total cholesterol decreasing"), we keep the TAR with the highest confidence (their support is equal). The above handling of inverse TARs is primarily heuristically driven with the objective of having as diverse risk factors in the implicated TARs as possible. After the screenings discussed above, the total number of TARs left is nine, as displayed in Table 25.10.

The next step is to determine, for each patient, the periodicity of each TAR (the number of its distinct occurrences) with respect to the relevant history of the patient. Finally, three naïve Bayes classifiers are developed whose attributes represent the selected complex periodic TARs. More specifically, discretization techniques are used for categorizing the order of periodicity of each TAR [99]. One well-known measure that characterizes the purity of the class membership of different variable states is information content or entropy [100]. The number and range of classes that result in the minimum total weighted entropy are chosen to quantize each variable. This minimum entropy principle is applied on all the variables (nodes) of the network.

For example, for classifier $C2$, for TARs $tar1 - tar6$, the discretization resulted in four orders for their potential periodicity, as follows:

- 0: no periodicity, i.e., the TAR does not occur.
- 1: low periodicity, i.e., the TAR occurs once or twice.
- 2: moderate periodicity, i.e., the TAR recurs between three and four times.
- 3: high periodicity, i.e., the TAR recurs at least five times.

---
* The models presented in this chapter were created and tested using the GeNIe Modeler available at http://www.bayesfusion.com/ [date accessed: 31 January 2017].

TABLE 25.8

Selected TARs for Classifier C1

TAR Code	Trend TA Code 1	Relation Operator	Trend TA Code 2
tar1	Systolic blood pressure = decreasing	PRECEDES	Triglycerides = increasing
tar2	Systolic blood pressure = increasing	PRECEDES	Total cholesterol = increasing
tar3	BMI = increasing	PRECEDES	Triglycerides = decreasing
tar4	BMI = increasing	PRECEDES	Systolic blood pressure = increasing
tar5	BMI = increasing	PRECEDES	Total cholesterol = decreasing
tar6	Triglycerides = increasing	PRECEDES	Systolic blood pressure = decreasing
tar7	Triglycerides = decreasing	PRECEDES	Systolic blood pressure = increasing
tar8	Systolic blood pressure = increasing	PRECEDES	Triglycerides = decreasing
tar9	Systolic blood pressure = decreasing	PRECEDES	Total cholesterol = decreasing
tar10	Total cholesterol = decreasing	PRECEDES	Systolic blood pressure = decreasing

TABLE 25.9

Selected TARs for Classifier C2

TAR Code	Trend TA Code 1	Relation Operator	Trend TA Code 2
tar1	Systolic blood pressure = decreasing	PRECEDES	Triglycerides = increasing
tar2	Systolic blood pressure = increasing	PRECEDES	Total cholesterol = increasing
tar3	BMI = increasing	PRECEDES	Triglycerides = decreasing
tar4	BMI = increasing	PRECEDES	Systolic blood pressure = increasing
tar5	BMI = increasing	PRECEDES	Total cholesterol = decreasing
tar6	Triglycerides = increasing	PRECEDES	Systolic blood pressure = decreasing
tar7	Triglycerides = decreasing	PRECEDES	Systolic blood pressure = increasing
tar8	Systolic blood pressure = increasing	PRECEDES	Triglycerides = decreasing

TABLE 25.10

Selected TARs for Classifier C3

TAR Code	Trend TA Code 1	Relation Operator	Trend TA Code 2
tar1	Triglycerides = decreasing	PRECEDES	Total cholesterol = increasing
tar2	LDL = increasing	PRECEDES	Systolic blood pressure = increasing
tar3	HDL = increasing	PRECEDES	Triglycerides = increasing
tar4	Systolic blood pressure = decreasing	PRECEDES	Triglycerides = increasing
tar5	Systolic blood pressure = increasing	PRECEDES	Total cholesterol = increasing
tar6	BMI = increasing	PRECEDES	Triglycerides = decreasing
tar7	BMI = increasing	PRECEDES	Systolic blood pressure = increasing
tar8	BMI = increasing	PRECEDES	Total cholesterol = decreasing
tar9	HDL = low	PRECEDES	Total cholesterol = increasing

For TARs *tar7* and *tar8*, whose periodicity is relatively lower, the discretization resulted in three orders for their potential periodicity, as follows:

- 0: no periodicity, i.e., the TAR does not occur.
- 1: minimal occurrence, i.e., the TAR occurs once.
- 2: periodic occurrence, i.e., the TAR recurs twice.

25.7.4 Dealing with Imbalanced Data and Overlapping Classes

In the STULONG data set, as already mentioned in Section 25.6.4, individuals who did not suffer a CHD event (majority class) are many more than those who suffered a CHD event (minority class). The situation becomes even more complicated, since the imbalance problem is combined with the class overlapping problem. Class overlapping is the case when samples from different classes have very similar characteristics [101].

The approach that we follow in this study, for tackling both issues, is to use the undersampling based on clustering (SBC) technique [102] to remove examples from the majority class of the data set in order to select a balanced sample. The ratio of the whole data set was 555/154 (around 4:1), whereas the resulting balanced sample ratio is 154/154. The SBC technique is usually used to handle class overlapping. It divides the whole data set into a specific number of clusters (k); in our case we selected $k = 4$, as with this number of clusters, we gain the best performance. Then, the method selects a suitable number of majority class samples from each cluster by considering the ratio of the number of majority class samples to the number of minority class samples in it. The technique is applied until a balanced subset of the data set is obtained.

25.7.5 Comparing the Performance of the Periodic Classifiers with a Baseline Classifier

For the evaluation of the performance of all the developed classifiers, we apply the 10-fold cross-validation technique, and we adopt metrics that are commonly used for imbalanced data sets: precision, recall, the F_1 score, and also the area under the Receiver Operating Characteristic (ROC) curve (AUC) [99].

As displayed in Table 25.11, the periodic classifier C2 has the best obtained results, in contrast to the other periodic classifiers. Although the precision of C2 is slightly lower than the precision of the other classifiers, in many medical domains such as CHD, recall is more important than precision since a false negative diagnosis (i.e., failure to predict a CHD event) has substantially more serious consequences for the patients than a false positive. In addition, this indicates that excluding those TARs that define the precedence of two decreasing variables is significant for the diagnosis of CHD.

In order to test our claim that for medical domains where long time periods are significant, higher-order, i.e., more complex, TAs can yield better performance, we compare the performance of the periodic classifier with the highest classification performance, C2, against that of another naïve Bayes classifier whose features are simple TARs, expressing the occurrence or not of the same set of TARs. This is referred to as the baseline classifier. Consequently, for the development of the baseline classifier, we omit the step of assessing the periodicity of each TAR. Thus, both classifiers have the same network structure (see Figure 25.11); however, the nodes of the baseline classifier are binary, taking the following two values:

- 0: the TAR does not occur at all in the relevant patient history.
- 1: the TAR occurs at least once in the relevant patient history (but the recurrence pattern is not categorized).

The SBC technique is also applied to the baseline classifier data set until a balanced subset of the data set is obtained. The goal of both classifiers is to predict the value of the class variable.

For the evaluation of the performance of both developed classifiers, we apply the 10-fold cross-validation technique, and we adopt the same metrics as in the comparison of the periodic classifiers: precision, recall, the F_1 score, and also the AUC [99].

As displayed in Table 25.12, the periodic classifier C2 has the best obtained results, in contrast to the baseline classifier. The higher performance of the periodic classifier in contrast to the baseline classifier further supports our belief that complex TARs can improve the classification of CHD. Thus, it is important for the classifier to detect and consider the recurrence patterns of the discovered TARs, in particular, higher-order periodicity, and not just to detect simple TARs.

For the development of the second proposed model, we implemented time-series classification by incorporating TARs and their recurrence patterns (periodic TARs) as features into a naïve Bayes classifier,

TABLE 25.11

The Performance for the Periodic Classifiers

	Periodic Classifier C1	Periodic Classifier C2	Periodic Classifier C3
Precision	0.79	**0.77**	0.78
Recall	0.81	**0.84**	0.80
F-score	0.80	**0.80**	0.79
AUC	0.84	**0.85**	0.83

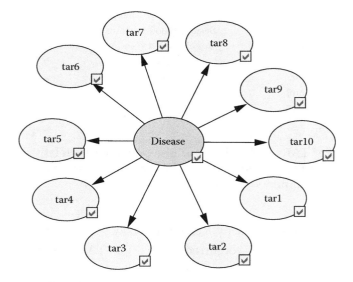

FIGURE 25.11
Naïve Bayes structure representing TARs as features.

TABLE 25.12

The Performance for the Periodic Classifiers C2 and the Baseline Classifier

	Periodic Classifier C2	Baseline Classifier
Precision	**0.79**	0.65
Recall	**0.81**	0.73
F-score	**0.80**	0.69
AUC	**0.84**	0.75

constructed for the purpose of CHD diagnosis. A notable strength of this system, in contrast to DBN-extended, is that the developed classifier can work with irregularly sampled temporal data sets.

25.8 Applications in Biomedical Signal Processing

Biomedical signal processing methods aim to extract clinically, biochemically, or pharmaceutically relevant information mainly in order to improve medical diagnosis but also to be used for patient monitoring [103,104]. Biological signals usually include time-series data, which may also be repetitive over time (periodic signals), such as the signals generated from a beating heart, respiration, and pulse rate [103]. The proposed integration can be intrinsically linked to the processing of biomedical time-series signals with the purpose of improving high-level medical problem solving.

For a given problem solving task in biomedical signal processing, the more appropriate of the two developed integration models could be selected: (a) a TBN integrated with basic TAs or (b) an atemporal BN integrated with complex TAs. In fact, both integration models could be kept in some hybrid system and deployed synergistically. Moreover, TA can be applied to a biomedical data set consisting of neural and cardiac rhythms, to detect the most significant and frequent temporal patterns from the data. Basic TAs may determine the change of the state of different parameters over time (qualitative, persistence), the rate of change (trend), or even the periodicity of their state (e.g., heart rate), while complex TAs may refer to the temporal precedence among different parameters that represent the symptoms or risk factors of a chronic disease (e.g., CHD). The derived patterns can then be incorporated into a Bayesian model (temporal or not) and used for prediction of the risk of the disease. In addition, the proposed integration can be potentially applied to the analysis of big temporal health-care data. The integration models can successfully handle missing data, uncertainty, class imbalance, and overlapping class problems, which are inherent characteristics of biomedical data. Missing data can be handled using a learning algorithm, such as EM [85], or some other statistical methods such as missforest [93]. Regarding the class imbalance and class overlapping problems, many techniques can be applied as aforementioned in Sections 25.6.4 and 25.7.4. Temporal patterns can then be extracted from the cleaned data and represented as nodes in a BN for classification or prediction.

The proposed integration could also handle data emerging from mobile health care (e.g., Fitbit) such as time-series data collected from the traces and logs of mobile health trackers (heart rate, blood pressure, sleep pattern). The detection of temporal patterns from these data can be utilized for diagnosis, prognosis, or treatment purposes. The identification of trends in patient histories over time and temporal precedence relations among potential risk factors can help medical experts to reach evidence-based decisions with the aim of reducing the risk of the disease. Such applications are especially useful for the health care of individuals suffering with chronic diseases.

Overall, the proposed integration models could have wide applications in domains where time-series biomedical and health-care data are available. Such data are invariably expressed at different levels of semantic detail and sampling frequency; they could have gaps or be excessively voluminous, etc.; and they are not amenable to direct processing and reasoning with. The integration models can utilize the abstract, and hence more useful, information extracted from the data to significantly improve medical problem solving. The integration models can be efficiently applied to both regular and irregular-sampled time-series biomedical data.

25.9 Discussion and Conclusions

The main goal of this work concerns the integration of TAs with BNs with a view to enhancing computer-based support to clinicians. The performance gains of the proposed integration have been illustrated through two integration models for the medical domain of CHD. The first model is a DBN whose nodes represent basic TAs sampled at regular time intervals. The model is applied for the prognosis of a potential CHD event. The results obtained by comparing its performance with a DBN representing time points demonstrate the effectiveness of the proposed integration.

Regarding the second model, the incorporation of TARs, which are complex temporal patterns, in a naïve Bayes classifier, shows promising results as the performance of this classifier, whose features are periodic TARs, is noticeably better than that of a naïve Bayes classifier that uses just temporal associations without taking into consideration their recurrence patterns (periodic TARs). This classifier is constructed for the purpose of CHD diagnosis. An advantage of the periodic naïve Bayes classifier is that it can work with irregularly sampled temporal data sets and multiple granularities, in contrast to the DBN-extended model that uses fixed time slices and hence a single granularity. The comparative results of the performance of the proposed two models, against more standard approaches, demonstrate the viability of our hypothesis that the integration of BNs with TAs can yield performance improvement in medical problem solving. The additional benefit of applying both systems to the CHD domain is that the interpretation of the obtained probabilistic results could give insights as to how causal dependencies and temporal relationships between abstract concepts may influence the risk of a CHD event. Moreover, a notable strength of our approach is that we show that the integration can be achieved in two different ways: either by integrating a temporal BN with basic TAs or by integrating an atemporal BN with complex TAs.

25.10 Open Problems and Future Directions

In this chapter, we have presented the development of two integration models, in order to validate the claim that the integration of TA and BNs has many benefits in medical problem solving. The evaluation results demonstrate the effectiveness of our proposed approach and provide a promising direction for future work.

As future work, we plan to discover frequent TARs on each class separately, to detect the ones that discriminate better the target classes, i.e., patterns that are more frequent in records of patients diagnosed with the disease and less frequent in records of patients not diagnosed. Furthermore, we plan to investigate the detection of even higher-order TAs over the detected periodic TARs. In addition, some other factors can be taken into consideration combined with periodicity, such as the time period of the TAR occurrence, by assigning a higher impact to the most recent TARs, the mean duration of all the occurrences of the TAR, or the duration of the most recent one. We also plan to apply a significance test to find out which of the different periodic classifiers developed in the second system are the most informative ones.

Finally, planned future work includes the detection of the risk factors that have higher impact on the disease, in contrast to others. Exploring new medical domains such as other chronic diseases, or even other application domains such as emotional detection, is also a future consideration.

References

1. Combi, C., Keravnou-Papailiou, E., and Shahar, Y. *Temporal Information Systems in Medicine*. Springer Science & Business Media, 2010.
2. Shahar, Y. Dynamic temporal interpretation contexts for temporal abstraction. *Annals of Mathematics and Artificial Intelligence* 22, 1–2 (Jan. 1998), 159–192.
3. Pearl, J. Probabilistic Reasoning in Intelligent Systems: Networks of Plausible Inference. *Morgan Kaufmann*, 1988.
4. Koller, D., and Friedman, N. *Probabilistic Graphical Models: Principles and Techniques*. MIT Press, 2009.
5. Weber, P., Medina-Oliva, G., Simon, C., and Iung, B. Overview on Bayesian networks applications for dependability, risk analysis and maintenance areas. *Engineering Applications of Artificial Intelligence* 25, 4 (2012), 671–682.
6. Spiegelhalter, D. J., Franklin, R. C. G., and Bull, K. Assessment, criticism and improvement of imprecise subjective probabilities for a medical expert system. In *Proceedings of the 5th Annual Conference on Uncertainty in Artificial Intelligence* (Amsterdam, The Netherlands, 1990), UAI '89, North-Holland Publishing Co., pp. 285–294.
7. Abramson, B., Brown, J., Edwards, W., Murphy, A., and Winkler, R. L. Hailfinder: A Bayesian system for forecasting severe weather. *International Journal of Forecasting* 12, 1 (1996), 57–71.
8. Verduijn, M., Peek, N., Rosseel, P., de Jonge, E., and de Mol, B. Prognostic Bayesian networks: I: Rationale, learning procedure, and clinical use. *Journal of Biomedical Informatics* 40, 6 (2007), 609–618.
9. Jansen, R., Yu, H., Greenbaum, D., Kluger, Y., Krogan, N. J., Chung, S., Emili, A., Snyder, M., Greenblatt, J. F., and Gerstein, M. A Bayesian networks approach for predicting protein-protein interactions from genomic data. *Science* 302, 5644 (2003), 449–453.
10. Sun, S., Zhang, C., and Yu, G. A Bayesian network approach to traffic flow forecasting. *IEEE Transactions on Intelligent Transportation Systems* 7, 1 (2006), 124–132.
11. Lehrmann, A. M., Gehler, P. V., and Nowozin, S. A nonparametric Bayesian network prior of human pose. In *Proceedings of the IEEE International Conference on Computer Vision* (2013), pp. 1281–1288.
12. Stassopoulou, A., and Caelli, T. Building detection using Bayesian networks. *International Journal of Pattern Recognition and Artificial Intelligence* 14, 6 (2000), 715–733.
13. Park, S., and Aggarwal, J. K. A hierarchical Bayesian network for event recognition of human actions and interactions. *Multimedia Systems* 10, 2 (2004), 164–179.
14. Charniak, E., and Goldman, R. P. A semantics for probabilistic quantifier-free first-order languages, with particular application to story understanding.

In *Proceedings of the 11th International Joint Conference on Artificial Intelligence (IJCA)* (1989), N. S. Sridharan, Ed., Morgan Kaufmann Publishers Inc., pp. 1074–1079.
15. Christensen, L. M., Haug, P. J., and Fiszman, M. Mplus: A probabilistic medical language understanding system. In *Proceedings of the ACL-02 Workshop on Natural Language Processing in the Biomedical Domain* (2002), vol. 3, Association for Computational Linguistics, pp. 29–36.
16. Stassopoulou, A., Petrou, M., and Kittler, J. Application of a Bayesian network in a GIS based decision making system. *International Journal of Geographical Information Science 14* (1998), 23–45.
17. Lee, E., Park, Y., and Shin, J. G. Large engineering project risk management using a Bayesian belief network. *Expert Systems with Applications 36*, 3 (2009), 5880–5887.
18. Fenton, N. E., and Neil, M. D. *Risk Assessment and Decision Analysis with Bayesian Networks*. No. 1. CRC Press, Queen Mary University of London, UK, November 2012.
19. Zweig, G., and Russell, S. Probabilistic modeling with Bayesian networks for automatic speech recognition. *Australian Journal of Intelligent Information Processing 5*, 4 (1999), 253–260.
20. Zweig, G. Bayesian network structures and inference techniques for automatic speech recognition. *Computer Speech & Language 17*, 2 (2003), 173–193.
21. Horvitz, E., Breese, J., Heckerman, D., Hovel, D., and Rommelse, K. The lumiere project: Bayesian user modeling for inferring the goals and needs of software users. In *Proceedings of the 14th Conference on Uncertainty in Artificial Intelligence* (1998), G. F. Cooper and S. Moral, Eds., Morgan Kaufmann Publishers Inc., pp. 256–265.
22. Sacchi, L., Capozzi, D., Bellazzi, R., and Larizza, C. JTSA: An open source framework for time series abstractions. *Computer Methods and Programs in Biomedicine 121*, 3 (2015), 175–188.
23. Sacchi, L., Larizza, C., Combi, C., and Bellazzi, R. Data mining with temporal abstractions: Learning rules from time series. *Data Mining and Knowledge Discovery 15*, 2 (2007), 217–247.
24. Keogh, E. J., Chu, S., Hart, D., and Pazzani, M. Segmenting Time Series: A Survey and Novel Approach. In *Data Mining In Time Series Databases*, M. Last, A. Kandel, and H. Bunke, Eds., vol. 57 of Series in Machine Perception and Artificial Intelligence. World Scientific Publishing Company, 2004, ch. 1, pp. 1–22.
25. Bellazzi, R., Larizza, C., and Riva, A. Temporal abstractions for interpreting diabetic patients monitoring data. *Intelligent Data Analysis 2*, 1–4 (1998), 97–122.
26. Allen, J. Towards a general theory of action and time. *Artificial Intelligence 23*, 2 (1984), 123–154.
27. Bellazzi, R., Larizza, C., Magni, P., and Bellazzi, R. Temporal data mining for the quality assessment of hemodialysis services. *Artificial Intelligence in Medicine 34*, 1 (May 2005), 25–39.
28. Moskovitch, R., and Shahar, Y. Medical temporal-knowledge discovery via temporal abstraction. In *AMIA Annual Symposium Proceedings* (San Francisco, USA, 2009), vol. 2009, pp. 452–456.
29. Moskovitch, R., and Shahar, Y. Fast time intervals mining using the transitivity of temporal relations. *Knowledge and Information Systems 42*, 1 (2015), 21–48.
30. Moskovitch, R., and Shahar, Y. Classification of multivariate time series via temporal abstraction and time intervals mining. *Knowledge and Information Systems 45*, 1 (2015), 35–74.
31. Batal, I., Sacchi, L., Bellazzi, R., and Hauskrecht, M. Multivariate time series classification with temporal abstractions. In *Florida Artificial Intelligence Research Society Conference* (2009).
32. Batal, I., Fradkin, D., Jr., J. H. H., Moerchen, F., and Hauskrecht, M. Mining recent temporal patterns for event detection in multivariate time series data. In *The 18th ACM SIGKDD International Conference on Knowledge Discovery and Data Mining, KDD '12, Beijing, China, August 12–16, 2012* (2012), pp. 280–288.
33. Batal, I., Valizadegan, H., Cooper, G. F., and Hauskrecht, M. A temporal pattern mining approach for classifying electronic health record data. *ACM Transactions on Intelligent Systems and Technology (TIST) 4*, 4 (2013), 63.
34. Fugan, L. VM2: Representing time-dependent relations in a medical setting. *PhD thesis*, Stanford University, 1980.
35. Kahn, M., Fagan, L., and Sheiner, L. Combining physiologic models and symbolic methods to interpret time-varying patient data. *Methods of Information in Medicine 30*, 3 (August 1991), 167–168.
36. de Zegher-Geets, I. Idefix: Intelligent summarization of a time-oriented medical database. Tech. Rep. KSL-88-34, Knowledge Systems, AI Laboratory, 1987.
37. Fries, J. F., and McShane, D. ARAMIS: A National Chronic Disease Data Bank System. *Proceedings of the Annual Symposium on Computer Application in Medical Care* (1979), 798–801.
38. Shahar, Y., Musen, M. et al. Résumé: A temporal-abstraction system for patient monitoring. *Computers and Biomedical Research 26* (1993), 255–273.
39. Shahar, Y. A framework for knowledge-based temporal abstraction. *Artificial Intelligence 90*, 1–2 (1997), 79–133.
40. Kuilboer, M. M., Shahar, Y., Wilson, D. M., and Musen, M. A. Knowledge reuse: Temporal-abstraction mechanisms for the assessment of children's growth. *Proceedings of the Annual Symposium on Computer Application in Medical Care* (1993), 449–453.
41. Shahar, Y., and Musen, M. Knowledge-based temporal abstraction in clinical domains. *Artificial Intelligence in Medicine 8*, 3 (1996), 267–298.
42. O'Connor, M. J., Grosso, W. E., Tu, S. W., and Musen, M. A. Rasta: A distributed temporal abstraction system to facilitate knowledge-driven monitoring of clinical databases. *Studies in Health Technology and Informatics*, 1 (2001), 508–512.
43. Ramati, M., and Shahar, Y. Probabilistic abstraction of multiple longitudinal electronic medical records. In *Proceedings of the 10th conference on Artificial Intelligence in Medicine* (Berlin, Heidelberg, 2005), AIME'05, Springer-Verlag, pp. 43–47.

44. Post, A. R., and Jr., J. H. H. Model formulation: PROTEMPA: A method for specifying and identifying temporal sequences in retrospective data for patient selection. *JAMIA 14*, 5 (2007), 674–683.
45. Gatti, E., Luciani, D., and Stella, F. A continuous time Bayesian network model for cardiogenic heart failure. *Flexible Services and Manufacturing Journal* (2011), 1–20.
46. Charitos, T., van der Gaag, L. C., Visscher, S., Schurink, K. A. M., and Lucas, P. J. F. A dynamic Bayesian network for diagnosing ventilator-associated pneumonia in ICU patients. *Expert Systems with Applications 36*, 2 (Mar. 2009), 1249–1258.
47. Arroyo-Figueroa, G., and Sucar, L. Temporal Bayesian network of events for diagnosis and prediction in dynamic domains. *Applied Intelligence 23*, 2 (2005), 77–86.
48. Hauskrecht, M., and Fraser, H. S. F. Planning treatment of ischemic heart disease with partially observable Markov decision processes. *Artificial Intelligence in Medicine 18*, 3 (2000), 221–244.
49. Galan, S. F., Aguado, F., Díz, F. J., and Mira, J. Nasonet, modeling the spread of nasopharyngeal cancer with networks of probabilistic events in discrete time. *Artificial Intelligence in Medicine 25*, 3 (2002), 247–264.
50. Rose, C., Smaili, C., and Charpillet, F. A dynamic Bayesian network for handling uncertainty in a decision support system adapted to the monitoring of patients treated by hemodialysis. In *Proceedings of the 17th IEEE International Conference on Tools with Artificial Intelligence* (Washington, DC, USA, 2005), ICTAI '05, IEEE Computer Society, pp. 594–598.
51. Austin, R., Onisko, A., and Druzdzel, M. The Pittsburgh cervical cancer screening model: A risk assessment tool. *Archives of Pathology & Laboratory Medicine 134*, 5 (2010), 744–750.
52. Sandilya, S., and Rao, R. B. Continuous-time Bayesian modeling of clinical data. In *Proceedings of the fourth SIAM International Conference on Data Mining (SDM)* (2004), pp. 22–24.
53. Dagum, P., and Galper, A. Forecasting sleep apnea with dynamic network models. In *Proceedings of the Ninth Conference Annual Conference on Uncertainty in Artificial Intelligence (UAI-93)* (San Francisco, CA, 1993), Morgan Kaufmann, pp. 64–71.
54. Murphy, K. P. Dynamic Bayesian Networks: Representation, Inference and Learning. *PhD thesis*, Dept. Computer Science, UC Berkeley, 2002.
55. Provan, G. Tradeoffs in constructing and evaluating temporal influence diagrams. In *Proceedings of the 9th International Conference on Uncertainty in Artificial Intelligence* (1993), Morgan Kaufmann Publishers Inc., pp. 40–47.
56. Shachter, R. D. Probabilistic Inference and Influence Diagrams. *Operations Research 36*, 4 (1988), 589–604.
57. Peek, N. Explicit temporal models for decision-theoretic planning of clinical management. *Artificial Intelligence in Medicine 15* (1999), 135–154.
58. Galan, S. F. Networks of probabilistic events in discrete time. *International Journal of Approximate Reasoning* (2002).
59. Aliferis, C., and Cooper, G. A new formalism for temporal modeling in medical decision-support systems. In *Proceedings of the Annual Symposium on Computer Application in Medical Care* (1995), American Medical Informatics Association, pp. 213–217.
60. Santos, E., and Young, J. D. Probabilistic temporal networks: A unified framework for reasoning with time and uncertainty. *International Journal of Approximate Reasoning 20* (1999), 263–291.
61. Nodelman, U., Shelton, C., and Koller, D. Continuous time Bayesian networks. In *Proceedings of the Eighteenth Conference on Uncertainty in Artificial Intelligence (UAI)* (2002), pp. 378–387.
62. Berzuini, C. Representing time in causal probabilistic networks. In *Proceedings of the Fifth Annual Conference on Uncertainty in Artificial Intelligence* (1990), North-Holland Publishing Co., pp. 15–28.
63. Ramati, M., and Shahar, Y. Irregular-time Bayesian networks. In *Proceedings of the 26th Conference Annual Conference on Uncertainty in Artificial Intelligence (UAI-10)* (Corvallis, Oregon, 2010), P. Grünwald and P. Spirtes, Eds., AUAI Press, pp. 484–491.
64. Andreassen, S., Benn, J. J., Hovorka, R., Olesen, K. G., and Carson, E. R. A probabilistic approach to glucose prediction and insulin dose adjustment: Description of metabolic model and pilot evaluation study. *Computer Methods and Programs in Biomedicine 41*, 3–4 (1994), 153–165.
65. Charitos, T. Reasoning with dynamic networks in practice. *PhD thesis*, Utrecht University, Netherlands, 2007.
66. Cao, C., Leong, T., Leong, A., and Seow, F. Dynamic decision analysis in medicine: A data-driven approach. *International Journal of Medical Informatics 51*, 1 (1998), 13–28.
67. Peelen, L., de Keizer, N., Jonge, E., Bosman, R., Abu-Hanna, A., and Peek, N. Using hierarchical dynamic Bayesian networks to investigate dynamics of organ failure in patients in the intensive care unit. *Journal of Biomedical Informatics 43*, 2 (2010), 273–286.
68. Xiang, Y., and Poh, K.-L. Time-critical dynamic decision making. In *UAI* (1999), pp. 688–695.
69. Dagum, P., Galper, A., and Horvitz, E. Dynamic network models for forecasting. In *Proceedings of the Eighth Conference on Uncertainty in Artificial Intelligence* (San Francisco, CA, USA, 1992), UAI 92, Morgan Kaufmann Publishers Inc., pp. 41–48.
70. Ramati, M. Irregular Time Markov Models. *PhD thesis*, Ben Gurion University, 2010.
71. Fahrmeir, L., and Raach, A. A Bayesian semiparametric latent variable model for mixed responses. *Psychometrika 72*, 3 (2007), 327–346.
72. Fan, J., and Zhang, W. Statistical estimation in varying coefficient models. *The Annals of Statistics 27*, 5 (1999), 1491–1518.
73. Hastie, T., and Tibshirani, R. Varying-coefficient models. *Journal of the Royal Statistical Society. Series B (Methodological)* (1993), 757–796.
74. Hand, D. J., and Crowder, M. J. *Practical Longitudinal Data Analysis (Chapman & Hall/CRC Texts in Statistical Science)*, 1st ed. Chapman & Hall, 1996.
75. Orphanou, K., Stassopoulou, A., and Keravnou, E. DBN-extended: A dynamic Bayesian network model extended

with temporal abstractions for coronary heart disease prognosis. *IEEE Journal of Biomedical and Health Informatics 20*, 3 (2016), 944–952.

76. Orphanou, K., Stassopoulou, A., and Keravnou, E. Risk assessment for primary coronary heart disease event using dynamic Bayesian networks. In *Artificial Intelligence in Medicine—15th Conference on Artificial Intelligence in Medicine, AIME 2015, Pavia, Italy, June 17–20, 2015. Proceedings* (2015), pp. 161–165.

77. Orphanou, K., Stassopoulou, A., and Keravnou, E. Integration of temporal abstraction and dynamic Bayesian networks for coronary heart diagnosis. In *STAIRS 2014—Proceedings of the 7th European Starting AI Researcher Symposium, Prague, Czech Republic, August 18–22, 2014* (2014), pp. 201–210.

78. Orphanou, K., Stassopoulou, A., and Keravnou, E. Temporal abstraction and temporal Bayesian networks in clinical domains: A survey. *Artificial Intelligence in Medicine 60*, 3 (2014), 133–149.

79. Orphanou, K., Dagliati, A., Sacchi, L., Stassopoulou, A., Keravnou, E., and Bellazzi, R. Combining naive Bayes classifiers with temporal association rules for coronary heart disease diagnosis. In *2016 IEEE International Conference on Healthcare Informatics, ICHI 2016, Chicago, IL, USA, October 4–7, 2016* (2016), pp. 81–92.

80. McCullough, P. A. Coronary artery disease. *Clinical Journal of the American Society of Nephrology 2*, 3 (2007), 611–616.

81. Wilson, P. W., D Agostino, R. B., Levy, D., Belanger, A. M., Silbershatz, H., and Kannel, W. B. Prediction of coronary heart disease using risk factor categories. *Circulation 97*, 18 (1998), 1837–1847.

82. Conroy, R., Pyörälä, K., Fitzgerald, A. e., Sans, S., Menotti, A., De Backer, G., De Bacquer, D., Ducimetiere, P., Jousilahti, P., Keil, U. et al. Estimation of ten-year risk of fatal cardiovascular disease in Europe: The score project. *European Heart Journal 24*, 11 (2003), 987–1003.

83. Hulst, J. Modeling physiological processes with Dynamic Bayesian networks. *Master's thesis*, Man-machine interaction group Delft University of Technology, Delft, Netherlands, 2006.

84. Jensen, F. V. *An introduction to Bayesian networks*, vol. 210. UCL Press, London, 1996.

85. Moon, T. The expectation-maximization algorithm. *IEEE Signal Processing Magazine 13*, 6 (1996), 47–60.

86. Gennest, J., and Libby, P. Lipoprotein disorders and cardiovascular disease. *Braunwald's Heart Disease: A Textbook of Cardiovascular Medicine*. 9th ed. Saunders Elsevier, Philadelphia, PA, 2011.

87. Stassopoulou, A., and Dikaiakos, M. D. Web robot detection: A probabilistic reasoning approach. *Computer Networks 53*, 3 (2009), 265–278.

88. Rahman, M. M., and Davis, D. Cluster based under-sampling for unbalanced cardiovascular data. In *Proceedings of the World Congress on Engineering* (2013), vol. 3.

89. Olson, D. L., and Delen, D. *Advanced Data Mining Techniques*, 1st ed. Springer Publishing Company, Incorporated, 2008.

90. Jeni, L. A., Cohn, J. F., and De La Torre, F. Facing imbalanced data—Recommendations for the use of performance metrics. In *Humane Association Conference on Affective Computing and Intelligent Interaction (ACII)* (2013), IEEE Computer Society, pp. 245–251.

91. Lloyd-Jones, D. M., Wilson, P. W., Larson, M. G., Beiser, A., Leip, E. P., D'Agostino, R. B., and Levy, D. Framingham risk score and prediction of lifetime risk for coronary heart disease. *The American Journal of Cardiology 94*, 1 (2004), 20–24.

92. Hippisley-Cox, J., Coupland, C., Robson, J., and Brindle, P. Derivation, validation, and evaluation of a new qrisk model to estimate lifetime risk of cardiovascular disease: Cohort study using qresearch database. *BMJ 341* (2010).

93. Stekhoven, D. J., and Buehlmann, P. Missforest—Non-parametric missing value imputation for mixed-type data. *Bioinformatics 28*, 1 (2012), 112–118.

94. Breiman, L. Random forests. *Machine Learning 45*, 1 (Oct. 2001), 5–32.

95. Concaro, S., Sacchi, L., Cerra, C., Fratino, P., and Bellazzi, R. Mining healthcare data with temporal association rules: Improvements and assessment for a practical use. In *Artificial Intelligence in Medicine*. Springer, 2009, pp. 16–25.

96. Bishop, C. M. *Pattern Recognition and Machine Learning*. Springer, 2006.

97. Duda, R. O., Hart, P. E., and Stork, D. G. *Pattern Classification* (2nd Edition). Wiley-Interscience, 2000.

98. Domingos, P., and Pazzani, M. On the optimality of the simple Bayesian classifier under zero-one loss. *Machine Learning 29*, 2-3 (Nov. 1997), 103–130.

99. Tan, P.-N., Steinbach, M., and Kumar, V. *Introduction to Data Mining*, (First Edition). Addison-Wesley Longman Publishing Co., Inc., Boston, MA, USA, 2005.

100. Mitchell, T. M. *Machine learning*. McGraw-Hill series in computer science. McGraw-Hill, Boston, MA, Burr Ridge, IL, Dubuque, IO, 1997.

101. Batista, G. E., Prati, R. C., and Monard, M. C. Balancing strategies and class overlapping. In *Advances in Intelligent Data Analysis VI*. Springer, 2005, pp. 24–35.

102. Yen, S.-J., and Lee, Y.-S. Cluster-based under-sampling approaches for imbalanced data distributions. *Expert Systems Applications 36*, 3 (2009), 5718–5727.

103. Ibrahimy, M. I. Biomedical signal processing and applications. In *Proceedings of the International Conference on Industrial Engineering and Operations Management, Dhaka, Bangladesh* (2010).

104. Chang, H.-H., and Moura, J. M. Biomedical signal processing. *Biomedical Engineering and Design Handbook. McGraw Hill (June 2009)* (2010), 559–579.

26
Big Data in Critical Care Using Artemis

Carolyn McGregor

CONTENTS

26.1 Introduction: Background and Driving Forces 519
26.2 Data Frequency 520
26.3 Data Availability 521
26.4 Data Quality 521
26.5 Data Acquisition 522
26.6 Data Transmission 522
26.7 Real-Time Analytics 522
26.8 Data Persistence 522
26.9 Knowledge Discovery 523
26.10 Visualisation 524
26.11 Secondary Use of Data and Consent 524
26.12 Artemis 524
26.13 Artemis Data Persistence 525
26.14 Artemis Deployments 525
26.15 Clinical Applications 527
    26.15.1 Late-Onset Neonatal Sepsis 527
    26.15.2 Apnoea of Prematurity 528
    26.15.3 Retinopathy of Prematurity 529
    26.15.4 Automated Partial Premature Infant Pain Profile (PIPP) Scoring 529
    26.15.5 Anemia of Prematurity 529
26.16 Conclusion 529
References 530

26.1 Introduction: Background and Driving Forces

Critical care units provide an acute care setting for complex interdisciplinary teams of healthcare workers to care for patients in a critical condition. Critical care units include neonatal intensive care units (NICUs) that care for premature and ill term infants, paediatric ICUs that care for critically ill children, adult intensive care units (ICUs) that care for critically ill adults, neurological ICUs that care for critical patients with brain injuries, and cardiac care units (CCUs) that care for critical patients with heart issues. Within this setting, clinicians are assisted in forming differential diagnoses and ultimately making diagnoses and prognoses by medical devices that provide continuous monitoring of physiological data. Those signals exhibit trends that reflect the underlying physiology of the patient [1].

In any country, critical care units are among the most costly aspects of the healthcare system. In Canada between 1999–2000 and 2003–2004, 15.9% of inpatient direct expenses were from critical care patients, but they accounted for only 8.1% of inpatient days [2].

Baby and Ravikumar [3] note that by 2015, it was expected that the average amount of data per hospital would have increased from 167 TB to 665 TB. This was largely as a result of imaging data and did not include the data streaming from medical devices that output data on a second-by-second basis.

Data analytics within critical care is considered a big data problem due mainly to the streaming nature of the data created by the various medical devices connected to the patient and the velocity of the data produced. Beyond the speed of sampling of the physiological data, the events in human physiology such as the frequency of the beating heart also have significant velocity.

As a source for potential new clinical knowledge discovery, high-speed physiological data are proving to be one of the most untapped resources in healthcare today based on the growing body of research studies demonstrating common physiological patterns for a range of conditions. Many medical devices produce data streams at frequencies of a reading a second or faster. In addition, from a human physiology perspective, in neonatal intensive care, for example, a premature newborn infant's heart beats approximately 7000 times an hour, and yet traditional charting on paper, or within an electronic health record (EHR), includes one number per hour of an indicative heart rate for that hour. The heuristics employed to determine the number to write are as much qualitative as quantitative, and part of the function is to express overall stability or instability hour to hour [4].

Providing healthcare workers with univariant displays of raw physiological signals over time does not provide clinical decision support that associates physiological data behaviour with pathophysiology information that is known to be associated with the onset of a given medical conditions to aid in the creation of differential diagnoses and the ultimate determination of a diagnosis and prognosis. Temporal abstraction provides a means to achieve such descriptions by translating low-level quantitative data, such as the stream of heart rate values over time, to higher-level qualitative information, such as recording the start and end of a bradycardic (low-heart-rate) or tachycardia (high-heart-rate) event [5].

In the first 10 years of the new millennium, a range of researchers were reporting on early research studies showing the potential for the discovery of new pathophysiological behaviours in physiological data streams that had the potential to be earlier-onset detectors for those conditions. Examples of this work include research for a small set of devastating conditions within the neonatal population such as late-onset neonatal sepsis (a form of hospital-acquired infection) [6], pneumothorax (collapsed lung) [7], intraventricular haemorrhage (bleeding in the brain) [8,9], and periventricular leukomalacia (death of brain tissue) [10].

Analysis of the approaches within these studies for data acquisition, data transmission, real-time data analytics, data persistence, and retrospective knowledge discovery revealed that there was a limited nature for the tools used to extract, store, and analyse the data collected [11]. The studies were usually performed autonomously and based on a research approach or clinical question that was focused around the analysis of just one physiological stream, such as using an electrocardiogram (ECG); focused on just one clinical condition, such as sepsis; or tuned for a given patient using neural networks to determine the onset of instability in a patient [12]. In addition, commercialisation of any temporal abstractions from physiological data streams within critical care had been provided within "black box" regulatory body–approved medical devices located at the patient's bedside [13].

Opportunities abound for the continuation of this exploration for many other conditions, but in order to enable this, robust big data infrastructures to support clinical research and real-time clinical decision support are required to perform this function.

This chapter provides an introduction to the application of big data analytics within the context of critical care medicine. The characteristics of the data that are created by various medical devices are introduced within the context of its frequency, availability, and quality. This is then followed by an introduction to the key components required for big data analytics in this setting relating to data acquisition, data transmission, real-time data analytics, data persistence, and knowledge discovery. A flagship big data analytics platform, Artemis, is then introduced, and details for the Artemis deployments to date are provided. The clinical research studies that have been enabled by Artemis are then presented.

26.2 Data Frequency

There are many medical devices used within critical care. Some are used to monitor the patient's physiological state; others such as ventilators provide breathing support. Infusion pumps can be used to provide medication, fluids, and/or nutrition. One common physiological stream that is captured from a medical monitor is ECG, which provides details on the functioning of the heart. Within the adult population, this is usually sampled at a speed of at least 200 readings a second, and within the neonatal space, devices recording readings of between 500 and 1000 readings a second are used. These readings are used to construct a waveform that is displayed on a monitor for the device. At 1000 readings a second, this translates to 86.4 million readings a day per patient [11]. Average heart rate values are derived from this signal by the same medical device usually at a sampling rate of one reading a second. A chest wall movement waveform can be derived from the electrodes used for a three-lead ECG. This is sampled at 62.5 readings a second within the Philips IntelliVue device. A respiratory rate is derived from this waveform each second. This results in 86,400 readings for heart rate and respiration rate being created each day.

Electroencephalogram (EEG) monitoring is at one order of magnitude higher than ECG with channels outputting 10,000 readings a second.

Similarly to ECG, a plethysmography waveform enables the derivation of an arterial blood oxygen saturation estimation that is available at a reading a second. The plethysmography wave can also be used to derive an average pulse rate each second.

Smart infusion pumps (SIPs) contribute to the big data problem as they can provide more than 60 different types of data every 10 seconds. One SIP can generate 4.4 MB of data per hour, 106 MB of data per day, and 3 GB of data monthly. Preterm infants, for example, can be connected to up to 13 SIPs, which results in 39 GB of drug infusion data for that patient per month [14].

Current practices for medical device procurement in healthcare organisations see medical devices in critical care units often having a 10- to 15-year life span as their procurement is from a capital budget within a healthcare organisation with many competing requirements for new medical equipment [11].

Many medical devices historically were designed to have a rolling memory of 72 hours. This would accommodate the Monday morning review of Friday and weekend data [11].

26.3 Data Availability

Medical device procurement within healthcare organisations with critical care units is usually through a request-for-tender process that provides a structured invitation process for suppliers to submit a bid to supply their products. The request-for-tender document details the functionality required from the medical device product, and these processes are usually coordinated by the biomedical engineering division within the healthcare organisation. As it has only been in the last decade that systems have emerged to be able to process medical device data at the speed it is generated, request-for-tender documents have not historically included required functionality to enable the extraction of data from the medical device by the healthcare organisation for use in other systems. As a result, there are many medical devices in use within healthcare organisations for which it is impossible for the data to be extracted from the medical device beyond the visual analysis of the data displayed on the device screen.

For example, the Philips IntelliVue series monitors have an Ethernet and a serial output port. These medical devices are being used in several of the partnering hospitals where Artemis, detailed later in this chapter, has been implemented. Within some of these hospitals, such as at the Hospital for Sick Children, Toronto, the Ethernet port is being utilised by the Philips central monitoring software, limiting any other data acquisition. Analytics systems such as Artemis, detailed later in this chapter, are required to use the serial port or the acquisition of data via the central monitoring software. Upon investigation, it was discovered that the data that would be available from the central monitoring software would be at a reduced sampling speed, and so in the case of Artemis, the data were acquired via the device's serial port. Acquisition of data from the serial port requires serial-to-Ethernet conversion. In the case of Artemis, to date, that has been completed at the bedside, enabling the data to be transmitted to Artemis via either a wired or wireless connection over Ethernet.

In the Artemis implementation within the Children's Hospital of Fudan University, also discussed later in this chapter, however, central monitoring was not implemented, and so either port was available for use.

26.4 Data Quality

Artefacts are extraneous signals that can be caused by a range of factors and result in randomly varying amplitudes and frequencies for varying durations. These interferences impact the quality of physiologic signals acquired in critical care units. Longitudinal studies assessing the black box, proprietary approaches for artefact detection (AD) within original equipment manufacturer (OEM) medical devices have concluded that they contain relatively simplistic built-in data preprocessing for AD that can in some cases oversimplify the data provided to the healthcare worker [1]. Artefacts increase the rate of false alarms, which results in alarm fatigue and can lead to incorrect diagnosis.

Signal artefact impacts the quality of temporal abstractions and derived analytics that can be determined from raw signals. In addition, differing AD and handling techniques between OEM patient monitors present a challenge to enabling repeatable studies where different medical devices are used by different healthcare facilities. In the past, when the goal was to display the physiological data at the bedside to support moment-to-moment patient stability analysis, such nuances were of less impact. In that context, the data that could be acquired from the medical devices were enabled for remote replication of the raw signals. In an era where analytics can be derived from these raw signals, ensuring the same baseline or the explicit declaration of AD techniques is a necessary step to ensure that comparable analytics are derived from the raw physiological data provided from the OEM post artefact processing.

Some artefacts are introduced by clinical interventions at the bedside. Within the context of neonatology, for example, such artefacts can be from events such as vascular access, suctioning, routine care, feeding, and reintubation [15]. It has been demonstrated that these clinical events can create common physiological patterns. Vascular access and reintubation were found to result in noxious stimuli causing heart rate (HR) ≥160, mean arterial blood pressure (MAP) ≥55, respiratory rate (RR) ≥40, and arterial blood oxygen saturation (SpO2) <90 [16,17].

Nizami et al. [1] concludes that a more open, white box approach to AD approaches is required not only to provide visibility to the techniques used for AD but to ensure that the most appropriate AD is chosen for any given clinical decision support system algorithm that is utilising physiological data post artefact processing.

26.5 Data Acquisition

Critical care units internationally are supported by a range of medical devices that are supplied by a range of manufacturers internationally. The initial paradigm for the delivery of information from the medical device was via a display monitor on the device. While some devices provided the ability to output data, this was not a primary function as systems to ingest and process such high-frequency data were not initially available. Medical device procurement procedures until recently did not consider assessing the functionality of device connectivity as part of the device procurement decision making. As a result, critical care units contain an array of devices that either have no mechanism to output data beyond the screen display or have a combination of serial, Ethernet (wired and wireless), and USB connectivity options. Many standards exist for data output from these devices, and as a result, a wide range of output formats exist with varying degrees of transparency and documentation available that detail the device output.

This landscape has given rise to the growing market of organisations who provide the service of medical device connectivity. Their primary purpose initially was to automate slow frequency data sampling to automate the clinical charting function. However, new opportunities exist for them to feed data to other systems such as big data–based clinical decision support systems.

26.6 Data Transmission

As a result of the array of medical devices available on the market and the different combinations of connectivity options provisioned by each device, a range of data transmission protocols have been proposed and utilised. Health Level Seven (HL7) provides "a framework and related standards for the exchange, integration, sharing, and retrieval of electronic health information" (http://www.hl7.org). Within the context of streaming data, HL7 can be utilised to package streaming data tuples for transmission to a remote clinical decision support system providing Health Analytics-as-a-Service. An example of the encoding of physiological data within an HL7 packet is presented in Figure 26.1 [18]. A series of physiological data values for a given second are packaged within the HL7 packet.

The Digital Imaging and Communication in Medicine (DICOM) standard provides a protocol for the transmission of images such as radiological images between systems. The DICOM standard has been used to create packages of time-series data streams using the 1999 DICOM Supplement 30, and the original standard, although these packages are mostly limited to short-duration data collections, and particularly to ECG data [19]. However, that method did not include continuous or long-term collection of time-series physiological data. Eklund, McGregor, and Smith [20] detail a method for continuous transmission of physiological data through a continuous stream of DICOM packets.

26.7 Real-Time Analytics

In order to be effective as a clinical decision support tool to detect pathophysiological behaviours in physiological data streams that have the potential to be earlier-onset detectors for a range of medical conditions, analytics on the physiological data streams must be performed in real time. Temporal abstractions must be performed on the raw streams of data from the medical devices, and those abstractions need to be merged and analysed together along with other clinical information for more complex temporal abstractions that relate more closely to real-world pathophysiological behaviours. For example, a bradycardic event, that is, an episode of reduced heart rate, may not have occurred in isolation but may be part of a central apnoea event.

Systemic platforms are required to enable the deployment of a range of real-time analytics–enabled clinical decision support tools for a range of medical conditions.

26.8 Data Persistence

In recent years, other research activities in other industry sections have also attempted the use of conventional

```
MSH|^~\&|GTWY|HOSP|RECV_APP|HOSP|20090826040011-04||ORU^R01|128957328110
MSH|182811|P|2.3
PID|||UV_01234567
PD1|
PV1|||NICU^BEDXX^BEDXX^^^^^^85&BEDXX||||^^^^^^^^^^^^^^~|^^^^^^^^^^^^^^~|
PV1||||||||~||^^^^^^^^^^^^^^~||72C037532909F37|||||||||||||||||||||||||
PV1|||20090823110850
ORC|||CA361400F18E1A9^GTWY||GTWYUSID
OBR|||CA361400F18E1A9^GTWY|^USID for all Monitor
OBR|||data^GTWY|||20090825235911
OBX|1|NM|1.6.10.0^MF Alarms Suspended||0|^^^^^^||||F|||20090825235904
OBX|2|NM|2.1.1.0^Heart Rate||161|beats/min||||F|||20090825235904
OBX|3|NM|2.1.2.0^Displayed Lead 1||1|2.1.2.0||||F|||20090825235904
OBX|4|NM|3.9.1.1^NIBP Mean||61|mmHg||||F|||20090825235733
OBX|5|NM|3.9.2.1^NIBP Sys||74|mmHg||||F|||20090825235733
OBX|6|NM|3.9.3.1^NIBP Dias||56|mmHg||||F|||20090825235733
OBX|7|NM|3.9.4.1^NIBP Pulse rate||151|beats/min||||F|||20090825235733
OBX|8|NM|3.9.4.2^NIBP Pulse Source||1|3.9.4.2||||F|||20090825235733
OBX|9|NM|6.1.1.1^SPO2||94|%||||F|||20090825235904
OBX|10|NM|6.1.2.0^SpO2 Pulse rate||162|beats/min||||F|||20090825235903
OBX|11|NM|6.1.3.0^ART Waveform Index||177|6.1.3.0||||F|||20090825235903
OBX|12|NM|7.1.1.0^RESP Rate||46|br/min||||F|||20090825235904
```

FIGURE 26.1
Example of physiological data within HL7. (From McGregor, C., A Cloud Computing Framework for Real-time Rural and Remote Service of Critical Care. In *Computer-Based Medical Systems*, CD–ROM, 6 pages, 2011a.)

database management systems or Hadoop/MapReduce-based approaches for the development of big data analysis systems broadly [21]. To fit streaming data within the relational model, the stream of data is segmented to a primitive element of a tuple [22]. At minimum, the tuple contains the date and time for that value within the stream, information about the type of stream, the entity generating the stream, and the stream value. Tuples are stored using the date and time, stream type, and entity as a multipart key and with the value as a nonkey attribute. While streaming functions have then been proposed within the relational database management systems to enable viewing of these data as a stream, the approach is highly cumbersome due to the relational database model approach for storage for the elemental tuple.

The challenge of efficient and effective storage and processing of streaming data exists beyond this context of health and medicine to many other industry sectors, where there is a growing volume of sensor data being acquired. In addition, the Internet of Things with direct device-to-device connectivity also presents the general case for which this research could be applied. Neither the traditional relational database model, the multidimensional database model, nor the big data platforms such as Hadoop/MapReduce proposed to date are built on the premise that data will arrive in a stream and has the potential to do so for months, years, and even decades at a time. While there currently exists a range of stream computing software techniques to process multiple interrelated real-time data streams as the data are acquired and before they are stored, the techniques for storage and retrieval of those data to support real-time and retrospective analytics are highly inefficient.

26.9 Knowledge Discovery

Research directions in knowledge discovery have focused on research relating to the creation of temporal abstractions and ontologies to support their definition [12].

Traditional data mining techniques are not effective when applied to streaming data. Whether it be the raw physiological data streams, other medical streaming data, or the slower frequency temporal abstraction streams, each element in the data stream is treated as its own discrete autonomous entity, with no ability to associate it with the values before or after it in the stream. As a result, research in the area of temporal data mining has emerged. More details in the area of knowledge discovery are provided later in the chapter within the Artemis example.

26.10 Visualisation

In 2014, Chest published a three-part series by Halpern on innovative designs for the smart ICU [23–25]. Part 3, "Advanced ICU Informatics," provided contextual information for the current and future ICU informatics landscape [25]. However, the paradigm for content display was focused around medical device displays and did not consider new approaches for integrated information display of termporal abstractions and complex temporal abstractions from multiple sources and alternative locations for display.

Kamaleswaran and McGregor [26] present a systematic review of papers published up to August 2016 to identify novel visual representations of physiologic data that address cognitive, analytic, and monitoring requirements in critical care environments. That review found that visual representations are plagued with user-preference and interaction challenges. The results spanned two decades where there was continued positive influence of graphical representations. They found that while many of the studies showed their potential to improve clinical care, with largely positive results, there are still concerns as to the efficacy relating to reproducibility as well as translatability to the unit especially since few studies used real patient data to evaluate their prototypes.

In terms of visualisation versus visual analytics, they found that only seven of the visualisations had some ability for interactive selection or filtering to provide visual analytics functionality beyond just visualisation.

Turnkey off-the-shelf tools for the visualisation and visual analytics of temporal abstractions and complex temporal abstractions from medical device data streams are still not available.

26.11 Secondary Use of Data and Consent

Secondary usage of health data is defined as the use of personal health information collected for purposes unrelated to the initial purpose of providing direct delivery of healthcare to the patient/data subject and can include a range of activities such as research, analysis, quality and safety measurement, payment, provider accreditation, and commercial activities [4]. Retrospective and some prospective clinical research studies require secondary use of personal health data, and this enables the expansion of current knowledge and understanding of healthcare and its delivery.

In the work of McGregor et al. [4], a public opinion pilot survey deployed in Australia and Canada in 2009 explored the opinion of patients regarding the possibility of analytics on physiological data, as part of a larger study on secondary use of health data broadly. Thirty attitudinal statements using a seven-point Likert scale for responses and two open-ended questions were contained in the pilot survey. Four hundred and eighty-two self-administered surveys were distributed to residential blocks in regional New South Wales and Darwin, Australia. A further 250 surveys were distributed to sample populations in Ontario, Canada [4].

The Australian and Canadian pilot surveys achieved response rates of 34.8% and 21.5% respectively. Question 26 in the survey explored the reuse of data captured through physiological devices: "If I was in hospital and a medical device was used to care for me—like a heart monitor or oxygen saturation monitor—I would agree for the information displayed on the screen to be saved in an anonymous way and used for medical research purposes." Both the Australian and Canadian responses showed strong support for this form of secondary use.

This was the first study of its kind to consider public opinion of this form of secondary use of data and demonstrates public support for this form of clinical knowledge discovery.

26.12 Artemis

Artemis is a big data analytics platform for online health analytics that supports concurrent multipatient, multistream, and multidiagnosis and temporal analysis in real time to support clinical management and retrospective knowledge discovery to support clinical research. It was conceived as a means to satisfy a need to provide such a systemic framework instantiated through a platform to support big data analytics in critical care medicine.

It was designed, created, and deployed by a multi-institutional, multidisciplinary research team from the University of Ontario Institute of Technology (UOIT) and IBM TJ Watson Research Center. The UOIT team, led by McGregor, brought research expertise in health informatics with clinicians on information technology use in NICUs; event stream processing; and acquisition of data from medical sensors, data analytics, data warehousing, temporal abstraction, and data mining. The research team from IBM TJ Watson Research Center brought industrial research experience in distributed computing, ubiquitous computing, pervasive healthcare, machine learning, and a new form of stream computing platform that had initially been created for a sensor analytics application in defence [27].

Big Data in Critical Care Using Artemis

The acquisition of real-time medical monitoring data is performed by the data acquisition component. Clinical information system data are also acquired by the data acquisition component. The online analysis component provides an environment for real-time analytics of the data forwarded by the data acquisition component in real time. Each instantiated Artemis platform to date as detailed in the Artemis deployment section that follows employs IBM's InfoSphere streams. Clinical rules are described as "graphs" in InfoSphere streams and deployed as instances of graphs for a given patient. Post processing, the initially acquired signals together with the derived analytics data are stored within the data persistency component [9]. Instances of the Artemis Platform to date have utilised IBM's DB2 software for the database management component due to the highly structured nature of the data and the requirement for real-time reliability, although some research has been performed to review the applicability of Hadoop for data management of historical data. Retrospective analysis of stored physiological data for new knowledge discovery is enabled in Artemis through the knowledge extraction component. This can be used for the discovery of previously unknown condition onset physiological behaviours (exploratory data mining) or testing of previously assessed hypotheses for further validation (explanatory data mining) [18]. For the knowledge extraction component within instances of the Artemis platform to date, McGregor's temporal data mining approach, known as Service based Multidimensional Temporal Data Mining (STDM$^r_0$) [28], has been used. Artemis is presented in Figure 26.2.

26.13 Artemis Data Persistence

Within Artemis, three approaches for the storage of medical device sensor data have been proposed for both the real-time streaming data storage as well as the knowledge discovery data storage. The first approach was to use a relational database management system, DB2, and storing the data in the format the data were extracted from the medical device. Second, we proposed using DICOM and proposed a means to join DICOM packets together as the mechanism to create continuous packages of medical device data [20]. Recently we have assessed other big data storage approaches such as Hadoop for data persistence, and this is ongoing work.

26.14 Artemis Deployments

Artemis was initially deployed in the NICU at the Hospital for Sick Children, Toronto, (SickKids) in August 2009. Physiological data are acquired from the Philips IntelliVue MP70 neonatal monitors that are fixed at each bed space in the NICU [27]. The clinical practice within the SickKids NICU dictates that all neonatal patients are monitored using a three-lead ECG. The Capsule Tech medical device connectivity solution was used in this instance of Artemis to acquire data from the Philips IntelliVue MP70 neonatal monitors. As a result, the ECG was collected along with the derived signals such as

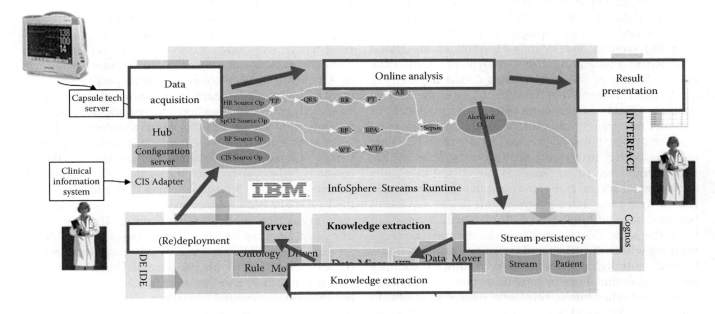

FIGURE 26.2
Artemis framework. (From McGregor, C., A Cloud Computing Framework for Real-time Rural and Remote Service of Critical Care. In *Computer-Based Medical Systems*, CD–ROM, 6 pages, 2011.)

heart rate, respiration rate, and chest impedance for breath detection together. The frequency of acquisition of data for the ECG was 500 Hz. The chest impedance wave frequency was 62 Hz. The derived heart rate and respiration rate frequency from this device is a reading every 1024 milliseconds. An arterial blood oxygen saturation sensor connected to the infant's foot provides arterial blood oxygen saturation data that are acquired from the Philips device at the frequency it is generated, which is a reading every 1024 milliseconds. Systolic, diastolic, and mean blood pressure data are also acquired only when the relevant sensors to collect this are attached to the infant as this is not compulsory.

In addition to the physiological data, clinical information is also collected. In the initial deployment, the clinical data were collected from the Clinical Information Management System (CIMS), which was the EHR software at the time. A range of clinical data were collected from CIMS. In addition, an adapter was created to connect to the ADT (admission/discharge/transfer) system that was accessed every 20 minutes to determine the current occupant of each bed space in the NICU. This was necessary as infants move locations during their stay in the SickKids NICU based on their gestation age and to balance nurse ratio loads.

The platform's data acquisition and online analysis components are located at SickKids. The real-time data persistency is also located at SickKids. The data persistency, knowledge extraction, and redeployment components reside at the UOIT [29]. Daily bulk batch data transfer is used to copy the most recent day's data from SickKids to UOIT.

The Artemis deployment in the NICU at SickKids has collected data for more than 1000 neonatal patients.

The second Artemis deployment was in partnership with the Women and Infant's Hospital Rhode Island (WIHRI) in Providence. This deployment enabled the clinical testing of the first Artemis Cloud deployment model and commenced in April 2010. One of the goals of Artemis Cloud was to demonstrate the potential for Artemis to remotely monitor neonatal infants in rural and remote locations. NICUs and in fact most ICUs are usually located in tertiary healthcare facilities within highly populated urban areas. While critical care units and neonatal special care nurseries can exist in rural and remote areas, they have a significantly lower number of bed spaces. In addition, the healthcare workers may not have the same degree of intensivist training as those working within the urban tertiary units. Transport teams located centrally or within tertiary healthcare facilities provide over the phone or sometimes video consultative support for other urban, rural, and remote locations. In some cases, remote viewing of the raw medical monitor data is available; however, patients are often transported to tertiary centres for monitoring by on-site intensivists [18]. Earlier-onset detection of changes in patient trajectory is a significant potential benefit of the deployment of systems such as Artemis in rural and remote locations.

Artemis Cloud is a framework for the provision of Artemis through the Software-as-a-Service and Health Analytics-as-a-Service models [18]. Artemis Cloud enables interfacing with Artemis through a series of web services. This framework is shown in Figure 26.3.

The function of the web services as shown in Figure 26.3 are detailed below [18]:

- *Clinical rule web service*—enables the definition of clinical rules for the deployment
- *Physiological web service*—enables the transmission of XML packets of physiological data that have been encoded in XML
- *Clinical web service*—enables the delivery of other clinical information from the clinical information system in HL7 format
- *Analyse web service*—enables the interaction with the knowledge extraction component

In the WIHRI Artemis Cloud deployment, spot readings were taken each minute from the bedside SpaceLabs devices and fed in raw form as HL7V2 data packets to the data acquisition component, implemented in Mirth, of the Artemis Cloud platform located at the UOIT. The transmission of the data was through a secure Internet tunnel. In this instantiation, all components except the data acquisition were located in the Health Informatics Research Laboratory at UOIT. Data from 203 patients were collected representing 10.6 patient-years of data [13].

A third instance of the Artemis Platform was created in partnership with SickKids containing only the data persistency, knowledge extraction, and redeployment components. This was utilised to demonstrate the efficacy of Artemis to be used solely as a retrospective environment for knowledge extraction and knowledge discovery. Previously collected 30-second spot reading data for 1151 patients were loaded into this Artemis instance [13].

A fourth instance of the Artemis platform, an Artemis Cloud instance, was created to support a clinical research collaboration with the Children's Hospital of Fudan University, Shanghai, China [30], and located at the UOIT. The beds were connected to the physiological web service provisioned by the Capsule Tech software. The purpose of the research was to assess the quality of service that could be achieved by Artemis Cloud in this context. Challenges with the Virtual Private Network (VPN) connection together with the bandwidth of the hospital intranet created limitations in delivery of data to Artemis Cloud that impacted the quality of service Artemis Cloud could provide due to the reduced data quality received. As a result, a local instance of Artemis

Big Data in Critical Care Using Artemis 527

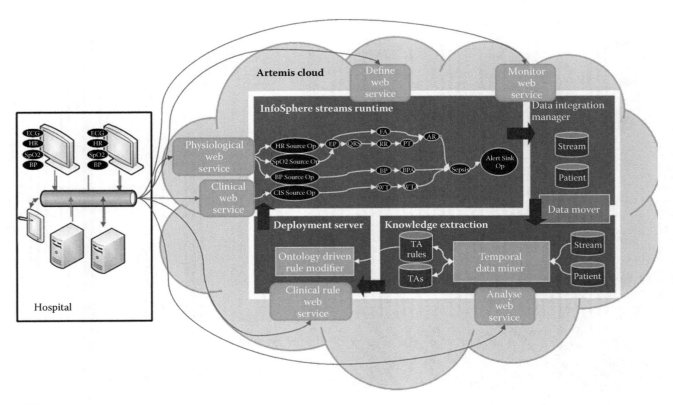

FIGURE 26.3
Artemis Cloud framework. (From McGregor, C., A Cloud Computing Framework for Real-time Rural and Remote Service of Critical Care. In *Computer-Based Medical Systems*, CD-ROM, 6 pages, 2011.)

was installed on servers within the Children's Hospital of Fudan University to support further clinical research studies.

The Expanded Artemis Cloud platform was created in partnership with the Southern Ontario Smart Computing Innovation Platform (SOSCIP) as a commercialisable deployment of Artemis. Mathematical models for deployment load estimates based on arrival patterns at two collaborating NICUs have been created [31,32]. Two hospitals are currently being brought online within this Expanded Artemis Cloud environment, and this is ongoing work.

26.15 Clinical Applications

There have been several clinical research studies that have now utilised Artemis. Some of these are detailed within this section.

26.15.1 Late-Onset Neonatal Sepsis

The first clinical research study to utilise Artemis was a study for earlier-onset detection of late-onset neonatal sepsis [33]. That work demonstrated the potential of real-time analytics using InfoSphere streams by performing second-by-second analytics on heart rate, respiration, blood oxygen, and blood pressure (when available) and constructing a risk score. A number of features were derived in real time from the physiological data streams and then analysed to determine the Late Onset Neonatal Sepsis (LONS) onset score [6,29,33,34]. The levels of the LONS onset score were defined as follows:

- *Level 0: no specified features present*
- *Level 1: presence of features reporting reduced heart rate variability (HRV), or bradycardia, or a significant downward drift of heart rate*
- *Level 2: presence of level 1 features plus the presence of a respiratory pause as determined by fall in respiratory rate below the defined threshold*
- *Level 3: presence of level 2 features plus a blood oxygen desaturation threshold breach or a significant downward drift of blood oxygen desaturation*
- *Level 4: presence of level 3 features plus a blood pressure threshold breach on any of the readings or a significant downward drift of any of the blood pressure readings*

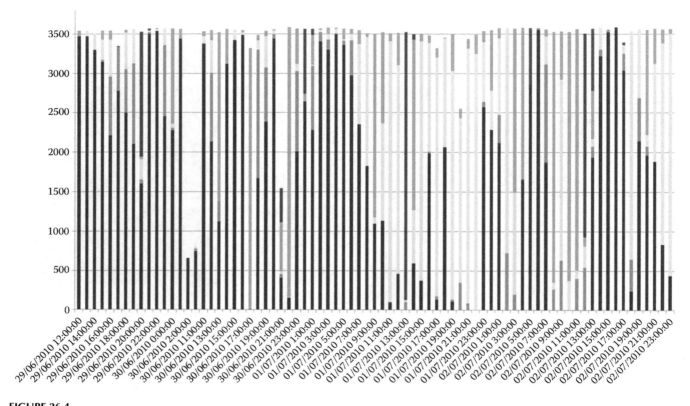

FIGURE 26.4
LONS onset score hourly summary distribution for one patient who developed LONS. (From McGregor et al., 2013 Real-time multidimensional temporal analysis of complex high volume physiological data streams in the neonatal intensive care unit data acquisition online analysis result presentation (re) deployment knowledge extraction data persistency. In *MedInfo* (pp. 362–366). Copenhagen, 2013. http://doi.org/10.3233/978-1-61499-289-9-362)

These scores were then assessed on an hourly basis. Figure 26.4 presents an example of this second-by-second scoring when aggregated within a stack bar on an hourly basis. The dark blue represents the amount of time within the hour where there was a score of 0. The light blue represents the amount of time within the hour where there was a score of 1. The yellow represents the amount of time within the hour where there was a score of 2. The pink represents the amount of time within the hour where there was a score of 3, and the red represents the amount of time within the hour where there was a score of 4.

In addition to that study, an approach for scoring HRV and respiration rate variability was discovered using the STDM$^n_0$ approach [6,34], and the potential effectiveness of that approach was demonstrated utilising data sets from the first three deployments of Artemis.

In this approach, variability was calculated by assessing the absolute difference between two consecutive heart rate and respiratory rate values and determining whether that was less than a given threshold. The hourly score was determined based on the total number of minutes where this difference was less than the threshold. Optimal threshold values of 3 for respiration rate and 4 for heart rate were determined taking into account known normal variations in these parameters for neonates. Figure 26.5 provides an example of this scoring approach for a patient who developed LONS. It shows the results of this scoring approach for a patient who underwent surgery and was subsequently diagnosed with LONS. In the postoperative period, the patient had low HRV and Respiration Rate Variability (RRV); however, while the RRV recovers to an acceptable baseline, the HRV does not.

26.15.2 Apnoea of Prematurity

This study sought to automate the classification of situations for irregular readings in heart rate, blood oxygen, and breathing, collectively known as apnoea and spells [35–37]. In these scenarios, the infant experiences pauses in breathing and reductions in heart rate and blood oxygen saturation. Single-stream temporal abstractions were created, and these were assessed together with other clinical information to construct the complex temporal abstractions that correlated to the 10 different forms of apnoea and spells. A greater than 95% accuracy was achieved when compared to the clinical inspection and annotation.

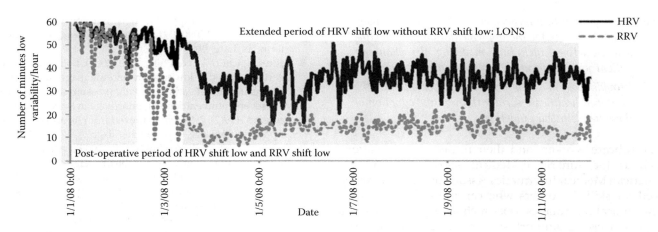

FIGURE 26.5
Example of HRV and RRV for a patient who developed LONS. (From McGregor, C. et al., Variability analysis with analytics applied to physiological data streams from the neonatal intensive care unit. In *25th IEEE International Symposium on Computer-Based Medical Systems (CBMS 2012)* (p. 5). Rome, 2012.)

26.15.3 Retinopathy of Prematurity

Retinopathy of prematurity (RoP) is a clinical condition that babies can develop when they are born before 32 weeks gestation and can lead to blindness and other lifelong retina damage [38,39]. Artemis at SickKids has been used to demonstrate the potential to provide real-time analytics for hourly oxygen target range achievement [38] and to support clinical research studies into the correlation between RoP stages and oxygenation [39].

26.15.4 Automated Partial Premature Infant Pain Profile (PIPP) Scoring

The objective of this research was to demonstrate the potential for Artemis to provide a partial pain score through the automated generation of analytics for the portion of the pain score that assesses physiological variables [40].

26.15.5 Anemia of Prematurity

The objective of this research was to perform a retrospective analysis using the knowledge extraction component of Artemis utilising $STDM^n{}_0$ to determine if there was an association between blood transfusion for anemia of prematurity and changes in the neonates' HRV where HRV was calculated utilising the approach detailed by McGregor et al. [6]. That research showed that there was, on average, an 8% improvement in HRV immediately after transfusion. This research demonstrated the potential of that HRV for deployment as a real-time algorithm to support clinical management of anemia of prematurity and blood transfusion efficacy [41].

In addition, we have demonstrated the potential of automated sleep–wake cycling through a clinical research study that analysed CFM data to detect sleep–wake cycling in neonates [42].

26.16 Conclusion

McGregor [11] notes that "the use of big data solutions on high frequency physiological data has the potential to be equally disruptive for healthcare." She notes that "significant progress within this area has been hampered by a lack of systemic computing and information technology and informatics research." Artemis and Artemis Cloud present approaches to address this need. In particular, Artemis Cloud with its Health Analytics-as-a-Service model demonstrates how this new form of analytics can be provided as a service, thus relieving the healthcare organisation from the burden of having to support such a platform in house.

She further proposes that more research is required relating to the approaches to

1. *Extract, securely transmit, process and store the data streams together with any derived features and abstractions;*
2. *Support retrospective and prospective clinical research studies using generalized systemic approaches such as Artemis*
3. *Provide effective and appropriate visualization for the various critical care healthcare roles*
4. *Support clinical evaluation of these new earlier-onset behaviours through mechanisms such as clinical trial*
5. *Provide mechanisms to translate the knowledge to the bedside without the introduction of another black box at the bedside*

6. *Support quality improvement initiatives to manage through change for the implementation of new clinical guidelines including modeling the changing patient journey together with training*
7. *Provide tools to easily measure and demonstrate the improved healthcare service through the adoption of these new clinical guidelines*

Healthcare workers and their respective associations such as the American Medical Association and the American Medical Informatics Association are seeing the need for skilled workers who can not only work with new clinical informatics tools such as big data analytics clinical decision support systems at the bedside but also contribute to create new knowledge for their use [43,44].

The prevalence of medical devices within and outside of healthcare is driving the need for more effective use of this information. Significant opportunities exist to service this need in critical care medicine and beyond to other areas of healthcare and for wellness within society broadly.

References

1. Nizami, S., Green, J. R., & McGregor, C. (2013). Implementation of artifact detection in critical care: A methodological review. *IEEE Reviews in Biomedical Engineering*, 6, 127–142. http://doi.org/10.1109/RBME.2013.2243724
2. Leeb, K., Jokovic, A., Sandhu, M., & Zinck, G. (2006). Intensive care in Canada. *Healthcare Quarterly*, 9, 32–33. Retrieved from http://www.ncbi.nlm.nih.gov/pubmed/16548431
3. Baby, K., & Ravikumar, A. (2014). *Big data: An ultimate solution in health care. International Journal of Computer Applications*, 106(10), 28–31.
4. McGregor, C., Heath, J., & Choi, Y. (2015). Streaming physiological data: General public perceptions of secondary use and application to research in neonatal intensive care. *MEDINFO 2015: EHealth-Enabled Health: Proceedings of the 15th World Congress on Health and Biomedical Informatics*, 216, 453–457. http://doi.org/10.3233/978-1-61499-564-7-453
5. Stacey, M., McGregor, C., & Tracy, M. (2007). An architecture for multi-dimensional temporal abstraction and its application to support neonatal intensive care. In *Annual International Conference of the IEEE Engineering in Medicine and Biology Society* (pp. 3752–3756). Lyon, France.
6. McGregor, C., Catley, C., & James, A. (2012). Variability analysis with analytics applied to physiological data streams from the neonatal intensive care unit. In *25th IEEE International Symposium on Computer-Based Medical Systems (CBMS 2012)* (p. 5). Rome.
7. McIntosh, N. (2000). Clinical diagnosis of pneumothorax is late: use of trend data and decision support might allow preclinical detection. *Pediatric Research*, 48(3), 408–15. Retrieved from http://journals.lww.com/pedresearch/Abstract/2000/09000/Clinical_Diagnosis_of_Pneumothorax_Is_Late_Use_of.25.aspx
8. Fabres, J., Carlo, W., & Phillips, V. (2007). Both extremes of arterial carbon dioxide pressure and the magnitude of fluctuations in arterial carbon dioxide pressure are associated with severe intraventricular hemorrhage in Preterm Infants. *Pediatrics*, 119(2), 299–305. Retrieved from http://pediatrics.aappublications.org/cgi/content/abstract/119/2/299
9. Tuzcu, V., Nas, S., & Ulusar, U. (2009). Altered heart rhythm dynamics in very low birth weight infants with impending intraventricular hemorrhage. *Pediatrics*. Retrieved from http://pediatrics.aappublications.org/cgi/content/abstract/123/3/810
10. Shankaran, S., & Langer, J. (2006). Cumulative index of exposure to hypocarbia and hyperoxia as risk factors for periventricular leukomalacia in low birth weight infants. *Pediatrics*. Retrieved from http://pediatrics.aappublications.org/cgi/content/abstract/118/4/1654
11. McGregor, C. (2013). Big data in neonatal intensive care. *Computer*, 46(6), 54–59. http://doi.org/10.1109/MC.2013.157
12. Stacey, M., & McGregor, C. (2007). Temporal abstraction in intelligent clinical data analysis: A survey. *Artificial Intelligence in Medicine*, 39, 1–24. http://doi.org/10.1016/j.artmed.2006.08.002
13. McGregor, C., Catley, C., James, A., & Padbury, J. (2011). Next generation neonatal health informatics with Artemis. In *User Centred Networked Health Care* (pp. 115–119). IOS Press.
14. Bressan, N., James, A., & McGregor, C. (2013). *Integration of drug dosing data with physiological data streams using a cloud computing paradigm*. In 35th Annual International Conference of the IEEE EMBS (pp. 4175–8). Osaka, Japan.
15. Percival, J., McGregor, C., Percival, N., Kamaleswaran, R., & Tuuha, S. (2010). A framework for nursing documentation enabling integration with EHR and real-time patient monitoring. In *23rd IEEE International Symposium on Computer-Based Medical Systems* (pp. 468–73).
16. Bressan, N., James, A., & McGregor, C. (2012). Physiological data stream analytics to evaluate noxious stimuli in the newborn infant. In *23rd Meeting of the European Society for Computing and Technology in Anaesthesia and Intensive Care (ESTAIC 2012)* (pp. 17–18). Timisoara, Romania. http://doi.org/10.1002/ana.222
17. Bressan, N., McGregor, C., Blount, M., Ebling, M., Sow, D., & James, A. (2012). Identification of noxious events for newborn infants with a neural network. In *Archives of Disease in Childhood 2012* (p. volume 97:A458, supplement 2, abstract number 1618).
18. McGregor, C. (2011). A cloud computing framework for real-time rural and remote service of critical care. In *Computer-Based Medical Systems* (CD-ROM, 6 pages).
19. Ling-Ling, W., Ni-Ni, R., Li-xin, P., & Gang, W. (2005). Developing a DICOM middleware to implement ECG conversion and viewing. In *27th Annual International Conference of the Engineering in Medicine and Biology Society* (pp. 6953–6956).
20. Eklund, J. M., McGregor, C., & Smith, K. P. (2008). *A method for physiological data transmission and archiving to*

support the service of critical care using DICOM and HL7. Conference Proceedings: Annual International Conference of the IEEE Engineering in Medicine and Biology Society. *IEEE Engineering in Medicine and Biology Society. Conference*, 2008, 1486–9. http://doi.org/10.1109/IEMBS.2008.4649449

21. Golab, L., & Johnson, T. (2014). Data stream warehousing. In *2014 IEEE 30th International Conference on Data Engineering (ICDE)* (pp. 1290–1293).

22. Jain, N., Zdonik, S., Mishra, S., Srinivasan, A., Gehrke, J., Widom, J., ... Tibbetts, R. (2008). Towards a streaming SQL standard. In *Proceedings of the VLDB Endowment* (vol. 1, pp. 1379–1390). http://doi.org/10.14778/1454159.1454179

23. N. A. Halpern, "Innovative Designs for the Smart ICU Part 1: From Initial Thoughts to Occupancy," Chest, vol. 145 no. 2, pp. 399–403, 2014.

24. N. A. Halpern, "Innovative Designs for the Smart ICU Part 2: The ICU," Chest, vol. 145 no. 4, pp. 903–912, 2014.

25. N. A. Halpern, "Innovative Designs for the Smart ICU Part 3: Advanced ICU informatics," Chest, vol. 145 no. 3, pp. 646–658, 2014.

26. Kamaleswaran, R., & Mcgregor, C. (2016). A review of visual representations of physiologic data: *JMIR Medical Informatics*, 4(4), 1–20. http://doi.org/10.2196/medinform.5186

27. Blount, M., Ebling, M., Eklund, J. M., James, A., McGregor, C., Percival, N., ... Sow, D. (2010). Real-time analysis for intensive care. *IEEE Engineering in Medicine and Biology Magazine*, 29 (2)(April), 110–118.

28. McGregor, C. (2011). System, method and computer program for multi-dimensional temporal data mining. WO Patent 2,011,009,211. US Patent Office. Retrieved from http://www.wipo.int/pctdb/en/wo.jsp?IA=CA2010001148

29. Blount, M., Mcgregor, C., James, A., Tuuha, S., Percival, J., & Percival, N. (2010). On the integration of an artifact system and a real-time healthcare analytics system. In *ACM International Health Informatics Symposium* (pp. 647–655).

30. Kamaleswaran, R., Thommandram, A., Zhou, Q., Eklund, J. M., Cao, Y., Wang, W. P., & McGregor, C. (2013). Cloud framework for real-time synchronous physiological streams to support rural and remote critical care. In *26th IEEE International Symposium on Computer-Based Medical Systems (CBMS 2013)* (pp. 473–476). Porto, Portugal.

31. Hayes, G., Khazaei, H., El-khatib, K., Mcgregor, C., & Eklund, J. M. (2015). Design and analytical model of a platform-as-a-service cloud for healthcare. *Journal of Internet Technology*, 16(1), 139–150. http://doi.org/10.6138/JIT.2014.16.1.20131203b

32. Khazaei, H., & Mench-bressan, N. (2015). Health informatics for neonatal intensive care units: An analytical modeling perspective. *IEEE Journal of Translational Engineering in Health and Medicine*, 3(August).

33. McGregor, C., James, A., Eklund, M., Sow, D., Ebling, M., & Blount, M. (2013). Real-time multidimensional temporal analysis of complex high volume physiological data streams in the neonatal intensive care unit data acquisition online analysis result presentation (re) deployment knowledge extraction data persistency. In *MedInfo* (pp. 362–366). Copenhagen. http://doi.org/10.3233/978-1-61499-289-9-362

34. McGregor, C., Catley, C., Padbury, J., & James. A. (2013). Late onset neonatal sepsis detection in newborn infants via multiple physiological streams. *Journal of Critical Care*, 28(1), e11–e12. http://doi.org/10.1016/j.jcrc.2012.10.037

35. Pugh, E., Thommandram, A., Ng, E., McGregor, C., Eklund, M., Narang, I., ... James, A., (2013). Classifying neonatal spells using real-time temporal analysis of physiological data streams—algorithm development. *Journal of Critical Care*, 28(1), e9. http://doi.org/10.1016/j.jcrc.2012.10.033

36. Thommandram, A., Eklund, J. M., McGregor, C., Pugh, J. E., & James, A. G. (2014). 2014 IEEE International Congress on Big Data A Rule-Based Temporal Analysis Method for Online Health Analytics and Its Application for Real-Time Detection of Neonatal Spells. In *IEEE Big Data*. Anchorage, Alaska. http://doi.org/10.1109/BigData.Congress.2014.74

37. Thommandram, A., Pugh, J. E., Eklund, J. M., McGregor, C., & James, A. G. (2013). Classifying neonatal spells using real-time temporal analysis of physiological data streams: Algorithm development. In *2013 IEEE Point-of-Care Healthcare Technologies (PHT)* (pp. 240–43). Bangalore, India.

38. Cirelli, J., Graydon, B., McGregor, C., & James, A. (2013). Analysis of continuous oxygen saturation data for accurate representation of retinal exposure to oxygen in the preterm infant. In K. Courtney, O. Shabestari, & A. Kuo (Eds.), *Enabling Health and Healthcare through ICT* (Technology, pp. 126–131). Vancouver: IOS Press.

39. Fernando, K. E. S., McGregor, C., & James, A. G. (2016). *Correlation of retinopathy of prematurity and blood oxygen saturation in neonates using temporal data mining: A pilot study.* In *38th Annual International Conference of the IEEE Engineering in Medicine and Biology Society* (pp. 1).

40. Naik, T., Thommandram, A., Fernando, K. E. S., Bressan, N., James, A. G., & McGregor, C. (2014). A method for a real-time novel premature infant pain profile using high rate, high volume data streams. In *27th IEEE International Symposium on Computer-Based Medical Systems (CBMS 2014)* (pp. 34–7). http://doi.org/10.1109/CBMS.2014.29

41. Pugh, J. E., Keir, A., McGregor, C., & James, A. (2013). The impact of routine blood transfusion on heart rate variability in premature infants. In *AMIA* (pp. 1–1).

42. Eklund, J. M., Fontana, N., Pugh, E., Mcgregor, C., Yielder, P., James, A., ... Mcnamara, P. (2014). Automated sleep–wake detection in neonates from cerebral function monitor signals. In *27th IEEE International Symposium on Computer-Based Medical Systems (CBMS 2014)* (pp. 22–27). http://doi.org/10.1109/CBMS.2014.36

43. Detmer, D. E., Lumpkin, J. R., & Williamson, J. J. (2009). Defining the medical subspecialty of clinical informatics. *Journal of the American Medical Informatics Association: JAMIA*, 16(2), 167–168. http://doi.org/10.1197/jamia.M3094

44. Gardner, R. M., Overhage, J. M., Steen, E. B., Munger, B. S., Holmes, J. H., Williamson, J. J., & Detmer, D. E. (2009). Core content for the subspecialty of clinical informatics. *Journal of the American Medical Informatics Association: JAMIA*, 16(2), 153–157. http://doi.org/10.1197/jamia.M3045

27

Improving Neurorehabilitation of the Upper Limb through Big Data

José Zariffa

CONTENTS

27.1 Introduction 533
27.2 Need for Large-Scale Practice-Based Evidence in Neurorehabilitation..............................534
27.3 Strategies for Large-Scale Data Gathering in Neurorehabilitation 535
    27.3.1 Robotic Rehabilitation Devices 535
    27.3.2 Wearable Sensors 535
    27.3.3 Environmental Sensors...537
27.4 Understanding the Neurorehabilitation Process through Big Data Approaches 538
    27.4.1 Insights to Date from Large Neurorehabilitation Data Sets..................................538
    27.4.2 Nonlinear Modeling to Uncover New Relationships between Different Aspects of Function in Neurorehabilitation Data Sets 541
    27.4.3 Use of Longitudinal Recovery Modeling to Inform Clinical Decision Making....542
27.5 Outlook 543
    27.5.1 Big Data Approaches Will Open Unexplored Directions in Neurorehabilitation....543
    27.5.2 Challenges 544
27.6 Conclusion 544
References 545

27.1 Introduction

The objective of the neurorehabilitation process is to aid the recovery of function after injuries to the nervous system, such as in spinal cord injury (SCI) or stroke. The focus is on achieving outcomes that enable the injured individual to participate in society and enjoy a high quality of life [1]. Neurorehabilitation is, by nature, a highly individualized process, in which a multidisciplinary clinical team attempts to address each patient's unique impairments, priorities, and recovery profile over time. The heterogeneous nature of the process makes it very challenging to extract evidence that can guide clinical best practices and maximize outcomes. In order to overcome this limitation of current neurorehabilitation practice, it will be necessary to leverage new technologies to enable large-scale data gathering and paint an accurate picture of current clinical practices and outcomes. Powerful analytic approaches must then be called upon to disentangle the complex relationships between patient characteristics, interventions, environments, and outcome metrics, and finally optimize the outcomes of the neurorehabilitation process. The science of big data, with its emphasis on large, continuously growing volumes of data from mixed sources, has much to offer for extracting new insights from the large and complex data sets that will become increasingly prevalent in neurorehabilitation.

This chapter will review the need for large-scale practice-based evidence in neurorehabilitation, discuss strategies for collecting the relevant data, and examine how big data approaches can be used to extract key relationships and help to move neurorehabilitation practice forward. The relationships between these concepts are illustrated in Figure 27.1. Studies and examples

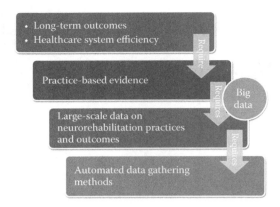

FIGURE 27.1
Role of big data in improving the upper limb neurorehabilitation process.

are focused in particular on the rehabilitation of upper limb function after neurological injury.

27.2 Need for Large-Scale Practice-Based Evidence in Neurorehabilitation

Inpatient rehabilitation after neurological injuries is a complex multidisciplinary process. Because rehabilitation emphasizes individualized plans rather than "one-size-fits-all" approaches, large randomized controlled trials (RCTs) focused on a well-defined intervention are difficult to perform, and strong evidence for the efficacy of many current practices is lacking [2–4]. Even when RCTs are conducted, the study sample is often more homogeneous than the general clinical population, and this may affect the generalizability of the results. Practice-based evidence (PBE) may therefore be a more appropriate paradigm for rehabilitation evidence than traditional RCTs [5]. PBE can be defined as a "prospective observational cohort design that seeks to categorize very systematically an exhaustive array of interventions and determine which interventions are most strongly associated with outcomes, taking into account a large number of patient characteristics that may also influence outcomes" [5]. While RCTs may still be required in certain circumstances, notably when new devices or drugs are introduced and direct evidence of their safety and effectiveness is needed for regulatory purposes, PBE has the potential to play a major role in identifying best practices for the care and interventions delivered during neurorehabilitation.

A few studies have attempted to gather large-scale PBE in order to gain a better understanding of the neurorehabilitation process. For SCI, the SCIRehab study prospectively collected a detailed breakdown of the interventions delivered by multiples disciplines to 1,376 individuals across six facilities drawn from the SCI Model Systems [6,7]. The resulting data has been used to elucidate how different components of the rehabilitation process contribute to outcomes, after controlling for individual demographic and injury characteristics. For examples, the associations were examined between outcomes such as Functional Independence Measure (FIM) scores and the total time spent with each of seven rehabilitation disciplines during the inpatient stay (occupational therapy, physical therapy, speech therapy, therapeutic recreation, social work/case management, psychology, and nursing education and care coordination) [8]. The use of time within each discipline was also studied, for example, showing that time spent on range of motion (ROM)/stretching in physical and occupational therapy sessions was associated with relatively lower function, while other interventions such as gait training, home management skills, and strengthening exercises were associated with better functional outcomes [9, 10]. A conceptually related effort in stroke was the Post-Stroke Rehabilitation Outcomes Project (PSROP), which used data from 1,291 patients across seven sites to try to understand how rehabilitation practice is related to outcomes [11]. Some examples of their findings included confirming the benefits of starting therapy earlier [12], and more rapidly progressing to higher-order activities with a focus on community integration [13–15]. Another notable outcome from the PSROP was to highlight the variability that exists in the practices of different rehabilitation centers, for example, in the use of medications during stroke rehabilitation [16].

The SCIRehab and PSROP studies highlighted a desire in the neurorehabilitation community to deepen our understanding of the complex interactions of factors that contribute to patient outcomes. At the same time, both of these studies generated large volumes of data (for instance, details on 462,455 interventions in the case of SCIRehab), and the analyses conducted to date have by no means been exhaustive. There is a clear need for a next generation of analysis techniques that can better parse the relationships that govern these large, heterogeneous data sets, and a big data perspective is highly relevant in this regard.

There are intrinsic limitations in the labour-intensive strategies used by SCIRehab and PSROP to collect large-scale PBE on the rehabilitation process. These studies were by necessity limited to a small number of sites, which may affect the generalizability of their results. More crucially, they were snapshots of clinical practices at a particular period in time and cannot capture evolutions in practices over time. In order to create a wider-reaching and more dynamic PBE base for neurorehabilitation, automated data collection approaches will need to be implemented. A number of technological innovations in recent years have brought us closer to this goal and are reviewed in the next section.

27.3 Strategies for Large-Scale Data Gathering in Neurorehabilitation

27.3.1 Robotic Rehabilitation Devices

Rehabilitation robotics is an area of research that has developed over the last 20 years, initially motivated by the concept that robots could help to provide higher intensities of therapy and thus improve functional outcomes [17–19]. The primary focus of the field has accordingly been on therapeutic effectiveness; however, an increasing number of studies are examining the correlations between movement parameters measured by robots (including smoothness, ROM, accuracy, speed, and force produced) and clinical measures. In stroke survivors, clinical assessments that have been studied in relation to robot-derived measures include the Fugl-Meyer test [20–25], the Motor Power Score [22,26], the Motor Status Score [22,23], and the Action Research Arm Test (ARAT) [24]. A number of moderate and statistically significant correlations have been identified [20,22,24]. As can be expected, the use of multiple linear regression to combine information from multiple robot metrics yields stronger correlations that regression models based on individual metrics [21,23,26]. In these studies and in this population, measures related to ease and smoothness of movement (e.g., involving velocity profiles, trajectory errors, and measures of jerk) have generally been found to be most predictive of clinical scores. More recent studies, reviewed below in Section 27.4.2, have extended this line of investigation using nonlinear modeling of the relationships between robotic measures and clinical assessments.

In individuals with SCI, manual assessments have also been related to robotic measures (ROM, grip strength, and movement smoothness) [27]. The best relationships were found in predicting the Prehension Performance component of the Graded Redefined Assessment of Strength, Sensibility and Prehension (GRASSP), and the Spinal Cord Independence Measure III (SCIM), with adjusted R^2 values of 0.78 and 0.77, respectively. In contrast to the findings in stroke, this study found that variables describing ROM and grip ability were more important than smoothness measures in achieving good prediction of clinical scores (GRASSP, SCIM, and ARAT).

While the aforementioned studies focused on analyzing measurements from therapeutic robots, several groups have also proposed custom mechatronic platforms dedicated to assessment [28–30], the best-known of which is the KINARM exoskeleton [30]. Similarly, the ReJoyce telerehabilitation robot includes a Hand Function Test software module that turns the therapeutic device into an assessment platform, the results of which have been shown to correlate with ARAT and, to a lesser extent, Fugl-Meyer scores [31].

Multiple studies have demonstrated strong psychometric properties (e.g., concurrent validity, high test–retest reliability) for robot-derived metrics in stroke survivors [25,32] and individuals with SCI [33]. In summary, automated assessment in neurological populations is a field of research that initially stemmed from robotic rehabilitation but is now attracting attention in its own right [34,35].

To date, the costs of most robotic rehabilitation platforms are sufficiently high that widespread deployment in the home is not likely. The work described in this section is therefore primarily relevant to assessing function in the clinic or laboratory environment. Applications include developing new outcome measures for clinical trials, informing the inpatient plan of care, and characterizing the profile of recovery after neurological injuries. From the point of view of gathering large-scale PBE in neurorehabilitation, robotic rehabilitation platforms can make two key contributions. First, they can provide more frequent assessments than what is possible with manual approaches, making it possible to capture the recovery process with a finer time resolution, and thus potentially providing insights into the optimal timing of different interventions. Second, robotic platforms that couple therapeutic and assessment functionalities offer an excellent opportunity to understand the optimal strategies for progressing the difficulty of the therapeutic exercises, based on quantitative data on the user's function and performance [36–38]. Nonetheless, there are limitations to robotic platforms in terms of scaling up the data gathering process. As a result of cost and relatively modest therapeutic effectiveness in many studies [39], the adoption of these platforms by rehabilitation centers is not currently widespread, limiting the number of patients with access to them outside of clinical trials. As noted above, these devices are primarily deployed to clinics or laboratories, limiting their ability to capture information about rehabilitation strategies targeting the home or community. There is a need for complementary technologies that can provide lower-cost solutions to capture information in a wider variety of environments. A variety of strategies based on wearable sensors are under investigation to fill this need and are reviewed in the next section.

27.3.2 Wearable Sensors

In the context of neurorehabilitation, we define a wearable sensor as a body-worn device that is capable of measuring some aspect of impairment or function and that does not impose substantial restrictions on the user's movements (e.g., as a result of wires, weight, etc.).

The predominant technology underlying wearable sensing in neurorehabilitation to date has been accelerometry, which in its simplest form provides a means to quantify the amount of movement. Accelerometers can be used to monitor overall levels of physical activity [40], which is of interest in the rehabilitation context as well as for nonmedical activity trackers. For instance, accelerometers have been used to investigate how activity levels in the community for several clinical populations compare to physician estimates [41]. Wrist-worn accelerometers have also been studied extensively as a strategy to measure the amount of upper limb activity [42,43]. For example, in studies with stroke survivors, a frequently used metric is the ratio of movement between the impaired and unimpaired arms [44–50]. Similarly, in SCI, accelerometer measures can reveal, for example, the amount of asymmetry in the impairment [51]. Here we emphasize that the activity measures derived from accelerometers are focused on the quantity of arm movement. While this has been found to be a valid indicator of upper limb use [42,52], it does not directly reveal what tasks were attempted or accomplished with these movements, or provide insights into the type and quality of the movements.

The range of applications of this technology can be further expanded by using inertial measurement units (IMUs), which combine accelerometers with gyroscopes, and sometimes magnetometers. By combining linear and angular motion measurements with appropriate calibration and signal processing procedures, IMU-based motion capture is possible [53–55], resulting in the ability to estimate specific postures instead of simply the amount of movement. The natural target applications for these wearable devices are activity recognition and movement analysis, which in the context of rehabilitation can be used in multiple ways. Firstly, IMU measurements have been used to predict clinical assessment scores. For example, scores have been predicted for subsets of tasks in the Functional Ability Scale of the Wolf Motor Function Test in individuals with stroke [56], and associations have been identified between IMU-derived upper limb kinematics and SCIM and FIM subscores in individuals with SCI [57]. Secondly, an alternative approach to trying to estimate scores for existing assessments is to derive novel metrics based on the kinematic data provided by the wearable sensors [58–60]. This approach is appealing because it more fully takes advantage of the capabilities of new technology, but on the other hand, it may face a steeper road to clinical adoption, since new metrics must be validated and may take time to gain acceptance. Beyond use as an assessment tool, IMU-based wearable systems have been explored as a means to provide real-time feedback on rehabilitation exercises [61]. For instance, Lam and colleagues developed a system to quantify movement quality and repetitions during rehabilitation exercises, with a focus on creating an interface both useful for providing feedback to patients and also beneficial to physical therapists for quantifying their patients' movements and tracking their progress [62]. Identifying the number and type of exercises being performed is of great interest from the point of view of gathering PBE on the neurorehabilitation process.

Activity recognition methods using IMU-based systems also have the potential to provide information about a person's daily activities in the community. For example, these devices have been used to gain insight into wheelchair usage patterns [63]. This type of information may yield insights into profiles of recovery beyond the clinic, as well as inform the timing and nature of outpatient plans of care.

While IMUs can be used to identify and analyze movement patterns for a variety of applications, one area in which they are limited is the analysis of hand function. IMUs and accelerometers for tracking upper limb activity are typically worn on the wrist, and as such are much more reflective of arm movements than of finger movements and are not able to capture information related to fine manipulation. One device by Friedman and colleagues sought to partially overcome this limitation by using magnetometers to track a magnetic ring on the index finger, complementing the information provided by a wrist-worn accelerometer [64]. This approach revealed clear but very task-dependent relationships between arm and hand usage [65]. Instrumented gloves can describe hand posture in more detail and have been the topic of a few studies in rehabilitation [66–69]. Potential limitations of these devices are that they may interfere with certain tasks and may be difficult to don for some patients with hand contractures, and consequently, it is not clear whether these devices have sufficient usability to accommodate large-scale data gathering in either the clinic or the home. Expanding on the concept of the instrumented glove, the use of electronic textiles (e-textiles) is also appealing for monitoring rehabilitation activities. The term *e-textiles* refers to garments that are able to perform sensing or actuating functions by virtue of electronic components embedded in the fabric, or of textiles that have electronic properties themselves. This technology could eventually be used to achieve wearable motion capture or monitoring of electromyographic (EMG) signals; however, a recent systematic review found that, despite a number of studies describing e-textiles for rehabilitation applications, only a very small proportion of these had tested the garments on individuals with neurological impairments [70,71]. The suitability of e-textile garments in this context therefore still needs to be established.

Recently, the use of wearable cameras that record the user's point of view (egocentric video) has also been proposed as a possible avenue to monitor hand function outside of the laboratory or clinic environment [72,73]. The advantages of video are that it offers a rich source of data and can capture information about the objects in the environment as well as about the hand. This approach therefore has the potential to capture information about functional use of the hand, which is different from the movement information collected by IMUs. The drawback of the egocentric video approach is the complexity of the algorithms required to extract meaningful information from recordings obtained with a moving camera in unconstrained environments. On the other hand, the availability of computer vision techniques dedicated to analyzing able-bodied hand use in egocentric videos has increased rapidly in recent years as a result of commercial egocentric cameras (e.g., GoPro, Google Glass) [72, 74–82], and this work may find applications in rehabilitation problems. In addition to their use in assessing hand function, wearable cameras are under active investigation for their potential in activity recognition applications [83–87]. Both of these lines of inquiry are of considerable interest from the point of view of capturing an individual's rehabilitation process in an automated manner and capturing PBE in a variety of settings.

It is clear that wearable sensors will have a significant role to play in the future of neurorehabilitation. Nonetheless, despite their relatively low degree of invasiveness, any device that requires setup time and that is expected to be used repeatedly over long periods of time is likely to face occasional problems with compliance. As a possible way around this problem, an intriguing alternative strategy for large-scale data collection about the rehabilitation process is to integrate sensors into the environment itself, as detailed in the next section.

27.3.3 Environmental Sensors

We define an environmental sensor here as one that can collect information about movements and activities, without being in direct contact with the patient. A system that meets this definition could enable data collection that does not interfere at all with either the patient's or the therapist's tasks during rehabilitation activities. Note that a rich literature exists on the topic of ambient sensors for monitoring various physiological variables in healthcare applications [88], but here we restrict ourselves to the monitoring of motor function.

Interest in environmental sensors for rehabilitation has increased in recent years as a consequence of the availability of low-cost commercial devices capable of performing markerless motion capture, most notably the Kinect (Microsoft Corp., Redmond, WA, USA). As mounting evidence describes the psychometric properties of these devices' skeletal tracking measurements and provides an improved understanding of their strengths and limitations [89–95], a number of rehabilitation applications have been envisioned. These can be broken down roughly into two categories: using the Kinect as a platform for rehabilitation exercises in virtual environments or using the Kinect as a tool for motion analysis and assessment.

Similarly to the use of robotic platforms discussed above, Kinect-based exercises could make it possible to more finely track the time course of recovery, as well as to study adaptive interventions that progress according to the user's performance. In order to make progress in this direction, the feasibility of guiding exercises based on the Kinect or similar sensors has been investigated by a number of groups [96–101]. The results of these studies suggest that markerless motion capture could be integrated into interactive, automated rehabilitation systems for use in the clinic or the home. In must be noted, however, that few reports to date have gone beyond feasibility to describe the impact of such approaches on functional outcomes. More work will therefore be required to ascertain the clinical potential of these ideas.

As a motion analysis tool, the Kinect can be used to derive clinical information from the kinematic data. Possibilities include defining novel metrics derived from the kinematic information and exploring the relationships of these metrics with clinical assessments [102–104], or using machine learning to predict the clinical assessment scores directly from the skeletal tracking information (e.g., joint angles, ROM) [105]. As with the Kinect-based rehabilitation systems, however, the development of Kinect-based upper limb assessments is still a recent field of study, and there is a need for more high-quality evidence before a better picture can be drawn of the role that this technology can play in rehabilitation.

It is worth noting that marker-based motion capture techniques have been used to investigate the function of the upper limb in neurorehabilitation and have produced evidence that kinematic descriptors may provide valid and reliable assessments in this context [106–112]. While marker-based strategies are too cumbersome to be integrated on a large scale into clinical rehabilitation processes, the evidence that they have produced could still be directly applicable if the appropriate kinematic information could be extracted using markerless motion capture devices such as the Kinect. However, translating marker-based findings in this manner will be contingent on the accuracy and reliability of the markerless skeletal tracking, and existing studies on these points suggest that further technical or methodological advances are likely to be required in order to improve the measurements [94].

As the associated technological challenges are progressively addressed, the combination of robotic or virtual environment platforms, environmental sensors, and wearable systems may well unlock in the near future an explosion of practice-based data about how neurorehabilitation is delivered and how clinical processes and patient characteristics interact to determine outcomes. As such, the way that we approach and analyze these data sets must adapt accordingly and begin to draw from a big data perspective.

27.4 Understanding the Neurorehabilitation Process through Big Data Approaches

Apart from a very small number of exceptions such as the SCIRehab and PSROP efforts, clinical studies in the stroke and SCI populations may involve a few hundred participants at the very most and are typically much smaller than that. Viewed in this light, one may argue that these data sets do not truly fall under the purview of "big data." However, the new data collection approaches described above will soon create a situation where this is no longer true. Each patient evaluation may further include multiple types of clinical assessments, possibly combined with imaging, electrophysiological, and demographic data, resulting in data sets with high dimensionality. Likewise, a record of the multidisciplinary interventions delivered will need to keep track of a large number of pieces of information (e.g., types, durations, and repetitions of different exercises). To date, no neurorehabilitation study has truly combined large-scale PBE with the power of analytics drawn from big data applications. That being said, a number of forward-thinking investigations have been conducted that give us a glimpse of how large data sets can be used to gain new insight into recovery following neurological injuries and the associated clinical processes. These efforts are summarized in Table 27.1 and discussed in more detail below.

27.4.1 Insights to Date from Large Neurorehabilitation Data Sets

A number of examples of how large data sets can lead to improvements in our understanding of neurorehabilitation can be drawn from the European Multicenter Study about Spinal Cord Injury (EMSCI). The EMSCI project is using a prospective design to characterize spontaneous recovery after SCI and understand relationships between different aspects of function. The network of participating centers further serves as a platform for the investigation of new treatment strategies. Each individual in the EMSCI database, which contains over 3,200 people to date, undergoes a set of neurological, neurophysiological, and functional assessments at fixed time points (acute and 4, 12, 24, and 48 weeks after injury). The intent of EMSCI is different from those of studies such as SCIRehab and PSROP. Whereas the latter focused on detailing the interventions delivered and improving our understanding of the clinical process of rehabilitation, EMSCI studies have focused, rather, on injury characteristics, their changes over time, and ways to measure these changes. Numerous relationships have been uncovered as a result of these efforts that can guide clinical processes.

At a first level, describing the expected course of recovery after SCI is crucial because it provides a very important baseline for the design of clinical trials. By understanding the changes that different assessments are expected to exhibit over the months after injury, under the current standard of care, we can determine which outcome measures and end points in a trial are most likely to reveal that the intervention produced a change. The EMSCI database has been used to establish such baselines for recovery [113,114]. Some of this work has focused on standard clinical assessments such as the International Standards for Neurological Classification of SCI (ISNCSCI) [113–116], while a few studies have also examined recovery profiles for electrophysiological [117,118] and functional [119,120] measures. This data can also be used to gain further insights into the influence of demographic factors on recovery, for example, age [121].

In parallel to describing changes in the injury itself, the EMSCI data have been valuable for understanding the relationships between many of the existing neurological and functional assessments for SCI. In doing so, we can gain new perspectives on the interactions between body structures and function, activity, and participation after SCI (following the model of the International Classification of Functioning, Disability and Health [122]). Studies in this direction have primarily focused on understanding at what point measured neurological recovery translates into functional benefits. For example, links have been established between the neurological or motor level of an SCI (i.e., neurological assessments) and the SCIM, which evaluates function by quantifying independence in activities of daily living [123,124]. It is also of interest to understand what types of functions are impacted by specific aspects of recovery: a study by Rudhe and colleagues established a link between the recovery of hand function and the ability to perform self-care activities (feeding, bathing, dressing, and grooming), which is of great interest from the perspective of interventions for restoring upper limb function [125].

With the availability of such a large data set, a greater range of analyses can be leveraged to answer questions with direct applications to neurorehabilitation processes and research. For example, studies by Tanadini et al. [126,127] and Velstra et al. [128,129] have examined

TABLE 27.1

Examples of Large-Scale Initiatives Contributing to Our Understanding of the Neurorehabilitation Process

Initiative	Individuals Included	Objectives	Selected Examples of Use
SCIRehab	1,376	To use observational PBE methods to relate the details of the rehabilitation process to outcomes in individuals with SCI, after controlling for individual demographic and injury characteristics [6]	• Examining associations between functional outcomes and total time spent with different rehabilitation disciplines during the inpatient stay [8] • Examining associations between functional outcomes and a range of treatment variables within individual disciplines (e.g., physical therapy, occupational therapy [9,10])
PSROP	1,291	To understand the impact of each stroke rehabilitation activity or intervention on patient outcomes at discharge, controlling for patient differences including medical and functional status on admission [11]	• Examining the impact of the timing of rehabilitation initiation on functional outcomes [12] • Understanding relationships between occupational therapy activities and upper limb functional outcomes [14]
EMSCI	>3,200	• To uncover the relationships between electrophysiological, neurophysiological, and functional measurements after SCI • To determine the prognostic value of these measurement outcomes • To examine the mechanisms of spontaneous recovery • To provide a framework to investigate the efficacy of new treatment strategies [159,160]	• Describing the expected course of recovery after cervical SCI [113] • Understanding relationships between upper limb impairment and function [123,125] • Using prognostic indicators to stratify potential participants in clinical trials [127]
RHSCIR	>5,000	To provide a resource to • Study the epidemiology of SCIs • Study the effectiveness of a variety of current and proposed treatments • Facilitate the implementation of clinical trials • Evaluate the quality of care delivery • Facilitate the implementation of evidence-based practices • Facilitate service delivery to the SCI community [130]	• Examining the relationship between the timing of surgery and outcomes after SCI [131] • Assessing whether ICD codes need to be supplemented with more information to characterize SCI [132]
VISTA	>82,000 (>10,800 for VISTA-Rehab)	To provide a resource to study questions concerning • The expected spontaneous recovery of specific population subgroups after stroke • The optimal selection of patients for clinical trials • The choice of outcome measures for clinical trials • The influence of country, comorbidities, and other factors on recovery [133,161]	• Identifying prognostic indicators for functional independence [134] • Quantifying interdependence of outcome scales [135]
COST	1,197	• To understand the effect of organized care and rehabilitation • To relate neurological and functional outcomes of stroke to initial stroke severity and functional disability • To characterize the recovery of upper limb function and walking • To characterize the time course of neurological and functional recovery relative to initial stroke severity • To study the mechanisms of stroke recovery • To understand the effects on recovery of various demographic, medical, and pathophysiological factors [148]	• Describing the expected course of upper limb recovery after stroke [149] • Determining the impact of a dedicated stroke unit on outcomes [152]
NSCID	>31,000	• To study the longitudinal course of traumatic SCI and factors that affect that course	• Statistical modeling of recovery trajectories [153,155]

(Continued)

TABLE 27.1 (CONTINUED)

Examples of Large-Scale Initiatives Contributing to Our Understanding of the Neurorehabilitation Process

Initiative	Individuals Included	Objectives	Selected Examples of Use
		• To identify and evaluate trends over time in etiology, demographic, and injury severity characteristics of persons who incur an SCI • To identify and evaluate trends over time in health services delivery and treatment outcomes for persons with SCI • To establish expected rehabilitation treatment outcomes for persons with SCI • To facilitate other research such as the identification of potential persons for enrollment in appropriate SCI clinical trials and research projects or as a springboard to population-based studies [162,163]	• Describing the expected course of recovery after SCI [142]

Abbreviations: COST, Copenhagen Stroke Study; EMSCI, European Multicenter Study about Spinal Cord Injury;NSCID, National Spinal Cord Injury Database PSROP, Post-Stroke Rehabilitation Outcomes Project; RHISCR, Rick Hansen Spinal Cord Injury Registry; VISTA, Virtual International Stroke Trials Archive.

how indicators at early time points after injury can be used to predict the expected progression of recovery and to stratify a very heterogeneous pool of individuals into more homogeneous subgroups. This subdivision is of interest for designing clinical trials that minimize sources of variability. These investigations used conditional inference trees, an unbiased recursive partitioning method, to break a large heterogeneous data set into progressively more homogeneous subgroups. Briefly, this is accomplished by first identifying the individual predictor that has the strongest association with the ultimate clinical outcome of interest, finding the dichotomous split on this predictor that results in two subgroups that are as distinct as possible, and then iteratively repeating this process on the subgroups until no more predictors have a statistically significant relationship with the target outcome. In this manner, individuals in a clinical trial can be stratified into homogeneous subgroups, which makes it easier to detect the effect of an intervention, while at the same time the investigators are not forced to limit enrollment by having very narrow inclusion criteria.

The approaches that have been developed in these studies could not have been arrived at without the large-scale data collected by the EMSCI project. An important observation is that the benefit of the large sample size here comes not from being able to detect small effects in a relatively homogeneous sample, but rather, from being able to describe the highly variable nature of neurorecovery after SCI. By teasing out relationships between the different aspects of injury and function, this variability no longer appears random, but can be understood and used to guide clinical processes in a manner that is both individualized and evidence-based.

Another illustration of how large-scale data gathering can be put to use in neurorehabilitation can be found in the Rick Hansen Spinal Cord Injury Registry (RHSCIR) [130]. Like EMSCI, RHSCIR is a prospective, multisite observational data collection effort aiming to improve our understanding of care after SCI. RHSCIR puts the emphasis on linking data from multiple stages of healthcare system utilization after SCI (acute care, rehabilitation, and long-term community follow-up), and in doing so aims to reveal how healthcare processes impact on outcomes after neurological injuries. This registry has been used, for example, to examine the relationship between the timing of surgery and outcomes after SCI [131]. Another interesting application has been to understand whether standard diagnostic information found in clinical records, such as International Classification of Diseases (ICD) codes, is adequate for characterizing neurological injuries, or if it needs to be supplemented with more specialized information [132].

While EMSCI and RHSCIR are prospective studies, large rehabilitation data sets can alternatively be created by retrospectively amalgamating data from various sources. This is the strategy employed by the Virtual International Stroke Trials Archive (VISTA), which collates anonymized data from clinical trials. This resource can be used for exploratory analyses in the design of trials or to provide insights into the expected progression of recovery after stroke under the current standard of care [133]. VISTA includes multiple topic areas related to stroke, including rehabilitation (VISTA-Rehab). The use of retrospective analysis on one hand results in less controlled and standardized data (for example, the trials in the database do not all use the same outcome measures) but, on the other hand, allows for a much larger resource than would otherwise be possible. Getting more use out of data already collected is also an appealing aspect of VISTA. Examples of how VISTA has been used to answer questions related to the rehabilitation process include the identification of prognostic indicators for upper limb function [134] and to better understand the relationships between stroke outcome scales [135].

27.4.2 Nonlinear Modeling to Uncover New Relationships between Different Aspects of Function in Neurorehabilitation Data Sets

The studies described in the previous section have so far provided compelling demonstrations of the benefits of large data sets in neurorehabilitation. Nonetheless, these approaches are recent, and there is considerable scope for bringing to bear techniques from machine learning and data science to better understand these growing data sets and transform them into clinically actionable information, as well as to integrate the contributions of automated data gathering through the technologies reviewed in the previous section.

While much insight has been gained to date from linear modeling methods, the underlying processes or relationships that are of interest in neurorehabilitation are almost certainly not linear. In order to deepen our ability to model and understand neurorehabilitation processes, it will be important to incorporate nonlinear models. On the other hand, as the complexity of our models increases, it becomes a challenge to apply them to small data sets because of the risk of overfitting. In other words, the usefulness of more complex models increases with the amount of data available. Examples of how increased sample sizes can form the basis for more complex nonlinear modeling of neurorehabilitation constructs can be found in the robotic rehabilitation literature. For instance, Krebs and colleagues examined the use of measurements from a robotic rehabilitation platform to establish biomarkers of motor function [136] in a population of 208 stroke survivors. After extracting a set of arm kinematic measures, they used them to create models that could predict clinical measures. In contrast to previous efforts using multilinear regression [21,23,26], this study used nonlinear models in the form of artificial neural networks combined with a feature selection algorithm based on artificial ant colony systems [137]. In addition to showing that the nonlinear modeling outperformed the multilinear regression results, they showed that the robot-derived scores were more sensitive for measuring recovery over time than the manual clinical assessments. This analysis included historical data from 2,937 patients extracted from VISTA. Demonstrating the good sensitivity of robot-based assessments has implications for designing clinical trials with fewer participants and suggests that using sensor-based biomarkers can improve our ability to track the underlying recovery process.

The notion of robot-derived biomarkers can be further extended to include a prognostic component. In addition to estimating an individual's current clinical status, is it possible to construct biomarkers that predict how they will recover in the future? Mostafavi et al. explored this question using the KINARM robotic assessment platform [138]. They sought to predict FIM scores at 2 weeks and at 3 months poststroke, using a set of clinical assessments and biomarkers derived from robotic measurements of upper limb function, all obtained at 2 weeks postinjury in 85 stroke survivors. The prediction of the 2-week FIM scores was therefore a concurrent prediction task, whereas the prediction of the 3-month scores was a prognostication task. Nonlinear models were used in combination with a feature selection scheme based on a Fast Orthogonal Search (FOS). Both clinical and robotic indicators were found to be able to predict the FIM scores at both time points, though accuracy for the 3-month scores was unsurprisingly lower ($r = 0.774$ for the 2-week scores vs. $r = 0.685$ for the 3-month scores, using the robotic indicators).

When seeking to identify sensor-derived biomarkers that can predict clinical measures and guide the recovery process, we need to consider their predictive ability but also the feasibility of acquiring the necessary data. In that respect, it is desirable to gain more insights into the relationships between different aspects of a person's disability, and large data sets can help us uncover quantitative relationships that are otherwise difficult to obtain. For instance, while we may be interested in an individual's ability to perform a functional task, it may be difficult to automate the evaluation of such a complex construct, which can reflect contributions from the neurological impairment but also the environment, compensatory postures, and motivation. As a result, robotic or sensor-derived biomarkers more often measure aspects of the impairment, such as muscle strength or the ability to perform certain movement primitives. The extent to which these simpler measures of impairment are informative about the ability to perform tasks is therefore an important question. Zariffa and colleagues leveraged machine learning to address this point, using a data set of clinical assessments from 129 individuals with SCI [139]. They demonstrated that classifiers could be constructed to quantitatively predict the scores that would be obtained when performing functional tasks in the GRASSP, using as inputs motor scores from 10 upper extremity muscles (also evaluated in the GRASSP). These results provide support for the idea that simple measures of impairment can form the basis for functionally relevant biomarkers. They further demonstrate that large data sets that contain multidimensional data about injury characteristics provide an opportunity to uncover relationships that can guide the efficient collection and use of information in the neurorehabilitation process.

27.4.3 Use of Longitudinal Recovery Modeling to Inform Clinical Decision Making

The desire to establish prognostic estimates of recovery is not new, and this pursuit is also in a position to benefit from the availability of large-scale data sets. To date, in addition to describing the expected course of recovery under existing standards of care [113,140–142], much effort has gone into identifying biomarkers in the early stages after injury that could predict the eventual outcomes for a given individual. The nature of these biomarkers has been varied, and they have included, for example, data from clinical examinations, functional performance, electrophysiology, imaging, and protein markers in cerebrospinal fluid (for stroke, these topics are reviewed in Refs. [143,144]; for SCI, examples include Refs. [129,145–147]). Of particular note here is the Copenhagen Stroke (COST) Study [148], a prospective, community-based study that tracked 1,197 acute stroke patients for 6 months. The outcomes of this initiative included the establishment of recovery profiles after stroke [149], the characterization of prognostic indicators such as initial stroke severity or the presence of certain comorbidities [150,151], and a study of the benefits of rehabilitation in dedicated stroke units [152]. Building on these previous efforts to describe the progress of recovery over time, the creation of statistical models of longitudinal recovery has recently found several applications in neurorehabilitation.

Traditional clinical studies most often seek to interpret evidence at the group level, for example, by comparing control participants to those receiving an intervention. As discussed earlier, a PBE approach instead emphasizes prospective observational studies designed to capture the variability in practice and to understand relationships between specific elements of the care or environment, patient characteristics, and outcomes. A natural next step of collecting PBE is to then apply this evidence on an individual basis to guide clinical decision making. In other words, when a person arrives at a rehabilitation center, how can their injury and demographic information be combined with existing evidence to project their likely course of recovery and plan the delivery of care accordingly? A quantitative approach to this question is to develop statistical models capable of encoding the impact of numerous factors on recovery profile. Pretz and colleagues proposed a solution to this task, in the form of individual growth curve (IGC) analysis [153]. For this analysis, they used data drawn from the National Spinal Cord Injury Database (NSCID), which collects information on individuals with SCI admitted to SCI Model Systems [154]. A sample of 4,504 individuals was used to construct the IGC models, which described recovery trajectories as negative exponential functions. The outcome measure used was a Rasch-transformed subset of the FIM. The models provided estimates of these scores at admission and once recovery had plateaued, as well as the time required to reach this plateau. A host of demographic and injury characteristic factors, as well as time to admission and rehabilitation length of stay, were investigated for their associations with these three features of the recovery profile. The results were encoded into an interactive tool that allowed the user to investigate how particular factors (e.g., age, injury severity, etiology, etc.) could be expected to affect recovery, as well as to create a profile of a given individual and gain insight into his/her potential for recovery. A similar study was subsequently conducted focusing instead on life satisfaction, which used analogous methods to model IGCs for Rasch-transformed Satisfaction With Life Scale scores [155]. While this study examined similar covariates as the FIM-based work described previously, the variance that the models were able to explain was smaller. The authors suggested that important predictors of life satisfaction may not have been included in the model, which was limited to information available in the NSCID. This observation underscores the importance of developing data gathering strategies that can allow us to take into consideration all relevant aspects of the neurorehabilitation process.

Beyond helping to support the planning of care, the availability of longitudinal recovery models has interesting implications for evaluating interventions. If we expect that recovery will follow a particularly trajectory, but a group receiving an intervention displays consistent deviations from this trajectory, we may gain insights about the effectiveness of this intervention. For example, Massie et al. used kinematic measurements obtained several times a week from a rehabilitation robot to characterize motor performance changes over time. They used random coefficient modeling (RCM) to capture the longitudinal variations in the data and further used these models to compare two groups of stroke survivors receiving different treatment protocols [156]. In fact, it has been suggested that modeling recovery profiles may lead to more efficient clinical trial designs requiring smaller sample sizes. Forsyth and colleagues presented an analysis of this concept using an example drawn from pediatric traumatic brain injury rehabilitation [157]. Nonlinear mixed effects (NLME) models were constructed to describe the recovery trajectories, based on frequent simple assessments. They showed that hypothesis testing based on the parameters of the NLME models could reduce the sample size needed to achieve the desired study power by up to five times, in comparison to the traditional paradigm of comparing assessments at fixed time points. Nonetheless, the variability that exists in longitudinal recovery profiles [158] must be carefully taken into account in these types of modeling approaches, and the incorporation and reliable collection of relevant covariate factors will be paramount to the progress of these methods.

As new technologies provide greater and greater amounts of data, the statistical modeling of longitudinal recovery profiles is likely to gain in importance. These methods relate to multiple aspects of improving neurorehabilitation outcomes through large-scale PBE. They can facilitate tailored care through predictions of individual recovery trajectories; they can help to identify factors that are likely to impact outcomes in a given situation, which will help to optimize resource allocation in the healthcare system; and they can help to leverage improved knowledge about the neurorehabilitation process for the design of more efficient clinical trials.

Lastly, while the statistical modeling of longitudinal recovery is essentially descriptive, a complementary approach is to seek computational models that can provide a more mechanistic perspective on this process. While some attempts have been made in this direction (reviewed in Ref. [144]), few to date have had the opportunity to draw on large data sets. The prospect of computational models of neurological recovery based on richer clinical data sets is an interesting one, which could lead to new interventions or the optimization of existing rehabilitation approaches.

27.5 Outlook

27.5.1 Big Data Approaches Will Open Unexplored Directions in Neurorehabilitation

Big data approaches are poised to have a significant impact on neurorehabilitation because they can provide the means to deal with the tremendous variability inherent in this process. For example, interventional clinical studies must attempt to detect an effect despite differences in the injury characteristics between individuals, but also despite potential differences in the care being delivered to each participant. Each individual in a study will be receiving care from multiple rehabilitation disciplines (e.g., physical therapy, occupational therapy, nursing, etc.), each of which will develop a plan of care that takes into consideration that patient's injury but also a myriad of other factors, such as comorbidities, personal goals, and environment. In contrast, conducting an effective RCT requires as much standardization as possible. As a result, neurorehabilitation RCTs are highly challenging endeavours, and many of the current practices in rehabilitation are not based on strong evidence.

A way out of this conundrum is to dramatically increase the knowledge base about the neurorehabilitation process, so that a meaningful point of reference can be available for any given individual. How did an intervention change the course of recovery compared to what was expected for this patient's injury and demographics? Can we compare the outcomes of many different patients with similar injuries and demographics but different interventions to arrive at clinical best practice guidelines? Answering these types of questions will require both the means to collect PBE data on large scales and the means to make sense of it. The strategies discussed in this chapter for achieving these goals will enable us to ask entirely new questions about the neurorehabilitation process.

The use of big data will moreover greatly improve our understanding of the relationships between the various assessments, diagnostic tools, and biomarkers available to us, as has already begun to happen in some of the studies discussed above (e.g., from the EMSCI database). These relationships can then be leveraged to streamline clinical processes, by understanding what data actually provides actionable information at what time and by reducing the redundancy in the data that we collect. Existing efforts to define standardized sets of common data elements [164] can be informed and refined as large data sets and machine learning methods improve our ability to accurately predict certain pieces of data from others (e.g., functional performance from motor scores, as described above [139]). These strategies have the potential to both reduce healthcare costs (by reducing the amount of data collected) and improve outcomes (by more effectively putting to use the information that we have).

From a technical point of view, a central theme in the development of big data analytics in other fields has been the ability to process data that is continually coming in, as opposed to off-line processing of batches of data [165]. This is a different paradigm than anything that has been done in neurorehabilitation to date, but has clear appeal. For example, while the SCIRehab or PSROP projects may provide a very detailed and informative picture of the neurorehabilitation process, that picture will always only be about the activities that occurred within the limited timeframe of the study, in the participating centers. As practice evolves over time, this picture may become increasingly dated. In contrast, a big data framework could allow for the incorporation of continuously generated PBE data (using, for example, some of the technologies discussed earlier). In this way, a dynamic model of neurorehabilitation could emerge, which both reflects evolving practices and can truly take advantage of increased data collection abilities as the technologies discussed earlier become more widespread.

Another common issue that algorithms designed for big data must contend with is missing data. Incorporating data from multiple sources, combined with possible errors in data entry, storage, and communication, inevitably results in data sets with missing and corrupted entries. The situation is no different in neurorehabilitation, where data collection spread across numerous individuals, devices, and sites is sure to result in a certain number of problematic entries. An increased reliance on automated rather than manual data collection processes can reduce but not eliminate this problem. Advances in data cleaning and imputation strategies [165] are thus quite relevant in this context. In that sense, drawing from advances in big data will facilitate the transition from an evidence base built on well-controlled RCTs to a PBE paradigm that must deal with less controlled but more ecologically valid data collection.

27.5.2 Challenges

The collection and analysis of large-scale, high-dimensionality data in a clinical context do not come without a set of challenges. One issue that must remain at the forefront of these discussions is that of privacy. Pooling data from multiple sources will of course require stringent deidentification processes, but more than that, the possibility of individuals becoming identifiable from their data increases as more data is collected about them. In other words, it may be impossible to identify individuals based only on their score on a single clinical scale, but if that score is combined with a host of other clinical assessments, demographic information, and imaging data, the risk of a privacy breach becomes much higher. The large-scale PBE approaches advocated here cannot proceed until data security and organization frameworks are in place that can provide the required safeguards [166]. Further privacy concerns exist from the point of the clinicians. With increased opportunities for data gathering embedded in clinical processes, it could conceivably become feasible to extract and review the practices of individual staff members. The possibility of such practices can legitimately be seen as problematic, and data management infrastructures will need to take into account the privacy of all parties involved, lest it create obstacles to the adoption of technology for the tracking of interventions.

The infrastructure challenges are certainly not limited to privacy considerations. In order to be able to extract meaningful relationships between different components of neurorehabilitation, these cannot be tracked in separate systems that do not communicate with each other. All of the large-scale studies discussed above and summarized in Table 27.1 were made possible by the incredible collaboration of a large number of clinicians and investigators but also by the underlying presence of custom information technology infrastructures. The design of these infrastructures themselves required close collaboration with interdisciplinary clinical teams, as exemplified by the discipline-specific entry forms that were developed for the SCIRehab and PSROP studies [167,168]. Pursuing big data strategies will expand these ambitions even further to include greater numbers of sites, more data elements, and a move of these paradigms from research studies into clinical processes. The availability of suitable infrastructure will be central to this undertaking. The integration with electronic health records is an appealing concept but one certainly not without obstacles [169].

In terms of the analysis of large-scale PBE, some of the challenges inherent in most rehabilitation research will persist. In particular, we know that outcomes are affected by a constellation of factors, and separating out the influences of injury, environment, social support, motivation, concomitant injuries, and other issues will usually not be completely possible. Even with large, multidimensional data sets, we are unlikely to ever be able to systematically capture all of the relevant factors. Nonetheless, this is exactly the challenge that big data approaches can be used to address, and there is reason for optimism that with novel data gathering strategies combined with analysis techniques capable of fully leveraging this data, it will still be possible to identify strong and clinically meaningful relationships despite the variability in the neurorehabilitation process.

Lastly, great care must be taken in the interpretation of analyses conducted on very large data sets. As the amount of data grows, even very weak relationships or effect sizes may be found to be statistically significant [170]. These findings must be contextualized by neurorehabilitation professionals so that their clinical significance can be ascertained. Even as the amount and quality of scientific evidence grow, the delivery of care in neurorehabilitation must always rely strongly on the experience of the clinician in applying this evidence to each patient's unique situation. Similarly, the availability of large data sets leads to the temptation to examine huge numbers of possible relationships, leading to multiple comparison bias and an important increase in type I errors [171]. As a result, many conclusions may be erroneous and not replicable. While large-scale PBE evidence would enable us to gain new insights into many new aspects of the neurorehabilitation process, the use of carefully formulated *a priori* questions in the analysis remains essential.

27.6 Conclusion

The complex nature of the neurorehabilitation process makes RCTs challenging to perform, and as a result, many current practices are lacking strong evidence to support their use. PBE offers an alternative route to improving the effectiveness of the care delivered after neurological injuries. While existing efforts in this direction have yielded many valuable insights, their logistical complexity has been considerable. The resulting data sets are rich in their size and the number of factors captured, but these benefits create a proportional need for appropriate means of analysis.

New developments in sensor technologies and big data analysis offer exciting opportunities to overcome many of these obstacles. Through progress in rehabilitation technologies such as robotics, wearable sensors, and markerless motion capture, our ability to capture the details of the upper limb interventions being delivered is likely to increase considerably. For this to happen, however, a number of technical challenges remain to be solved before these technologies can be seamlessly integrated into clinical environments and reliably provide the desired information. Sensor data will further need to be combined with other sources of data such as electronic health records that can provide demographics and clinical diagnostic information, but again these efforts can only truly succeed if underpinned by the development of an information infrastructure. Lastly, big data approaches must be applied to extract actionable insights from large, high-dimensional data sets.

By putting into place large-scale collection and analysis of PBE in neurorehabilitation, we can gain the knowledge needed to optimize the care delivery process for each individual patient. By delivering the right interventions, to the right person, in the right manner, and at the right time, the resources of the healthcare system can be allocated more effectively, and long-term outcomes after neurological injuries can be improved.

References

1. M. Selzer et al., *Textbook of Neural Repair and Rehabilitation: Volume 1, Neural Repair and Plasticity*. Cambridge University Press, 2014.
2. M. L. Sipski and J. S. Richards, "Spinal cord injury rehabilitation: State of the science," *Am. J. Phys. Med. Rehabil.*, vol. 85, pp. 310–342, 2006.
3. L. A. Harvey, J. V. Glinsky and J. L. Bowden, "The effectiveness of 22 commonly administered physiotherapy interventions for people with spinal cord injury: A systematic review," *Spinal Cord*, 2016.
4. J. M. Veerbeek et al., "What is the evidence for physical therapy poststroke? A systematic review and meta-analysis," *PloS ONE*, vol. 9, p. e87987, 2014.
5. S. L. Groah et al., "Beyond the evidence-based practice paradigm to achieve best practice in rehabilitation medicine: A clinical review," *PM&R*, vol. 1, pp. 941–950, 2009.
6. G. Whiteneck et al., "New approach to study the contents and outcomes of spinal cord injury rehabilitation: The SCIRehab Project," *J. Spinal Cord Med.*, vol. 32, p. 251, 2009.
7. M. P. Dijkers, G. G. Whiteneck and J. Gassaway, "CER, PBE, SCIRehab, NIDRR, and other important abbreviations," *Arch. Phys. Med. Rehabil.*, vol. 94, p. S66, 2013.
8. G. Whiteneck et al., "Relationship of patient characteristics and rehabilitation services to outcomes following spinal cord injury: The SCIRehab project," *J. Spinal Cord Med.*, vol. 35, pp. 484–502, November 1, 2012.
9. L. Teeter et al., "Relationship of physical therapy inpatient rehabilitation interventions and patient characteristics to outcomes following spinal cord injury: The SCIRehab project," *J. Spinal Cord Med.*, vol. 35, pp. 503–526, 2012.
10. R. Ozelie et al., "Relationship of occupational therapy inpatient rehabilitation interventions and patient characteristics to outcomes following spinal cord injury: The SCIRehab Project," *J. Spinal Cord Med.*, vol. 35, pp. 527–546, 2012.
11. G. DeJong et al., "Opening the black box of poststroke rehabilitation: Stroke rehabilitation patients, processes, and outcomes," *Arch. Phys. Med. Rehabil.*, vol. 86, pp. 1–7, 2005.
12. S. A. Maulden et al., "Timing of initiation of rehabilitation after stroke," *Arch. Phys. Med. Rehabil.*, vol. 86, p. S40, December 1, 2005.
13. N. K. Latham et al., "Physical therapy during stroke rehabilitation for people with different walking abilities," *Arch. Phys. Med. Rehabil.*, vol. 86, p. S50, December 1, 2005.
14. L. G. Richards et al., "Characterizing occupational therapy practice in stroke rehabilitation," *Arch. Phys. Med. Rehabil.*, vol. 86, pp. S60, December 1, 2005.
15. S. D. Horn et al., "Stroke rehabilitation patients, practice, and outcomes: Is earlier and more aggressive therapy better?" *Arch. Phys. Med. Rehabil.*, vol. 86, pp. S114, December 1, 2005.
16. B. Conroy et al., "An exploration of central nervous system medication use and outcomes in stroke rehabilitation," *Arch. Phys. Med. Rehabil.*, vol. 86, p. S81, December 1, 2005.
17. H. I. Krebs and B. T. Volpe, "Rehabilitation robotics," *Handb. Clin. Neurol.*, vol. 110, pp. 283–294, 2013.
18. P. S. Lum et al, "Robotic approaches for rehabilitation of hand function after stroke," *Am. J. Phys. Med. Rehabil.*, vol. 91, p. 242, November, 2012.
19. N. Norouzi-Gheidari, P. S. Archambault and J. Fung, "Effects of robot-assisted therapy on stroke rehabilitation in upper limbs: Systematic review and meta-analysis of the literature," *J. Rehabil. Res. Dev.*, vol. 49, pp. 479–496, 2012.
20. B. Rohrer et al., "Movement smoothness changes during stroke recovery," *J. Neurosci.*, vol. 22, pp. 8297–8304, September 15, 2002.
21. J. J. Chang et al., "The constructs of kinematic measures for reaching performance in stroke patients," *JBME*, vol. 28, pp. 65–70, 2008.
22. R. Colombo et al., "Assessing mechanisms of recovery during robot-aided neurorehabilitation of the upper limb," *Neurorehab. Neural Repair*, vol. 22, pp. 50–63, 2008.

23. C. Bosecker et al., "Kinematic robot-based evaluation scales and clinical counterparts to measure upper limb motor performance in patients with chronic stroke," *Neurorehabil. Neural Repair*, vol. 24, pp. 62–69, January, 2010.
24. O. Celik et al., "Normalized movement quality measures for therapeutic robots strongly correlate with clinical motor impairment measures," *IEEE Trans. Neural Syst. Rehabil. Eng.*, vol. 18, pp. 433–444, August, 2010.
25. M. Gilliaux et al., "Using the robotic device REAplan as a valid, reliable, and sensitive tool to quantify upper limb impairments in stroke patients," *J. Rehabil. Med.*, vol. 46, pp. 117–125, January 30, 2014.
26. H. I. Krebs et al., "Robot-aided neurorehabilitation: From evidence-based to science-based rehabilitation," *Top. Stroke Rehabil.*, vol. 8, pp. 54–70, 2002.
27. J. Zariffa et al., "Relationship between clinical assessments of function and measurements from an upper-limb robotic rehabilitation device in cervical spinal cord injury," *IEEE Trans. Neural Syst. Rehabil. Eng.*, vol. 20, pp. 341–350, May, 2012.
28. G. Van Dijck, J. Van Vaerenbergh and M. M. Van Hulle, "Posterior probability profiles for the automated assessment of the recovery of patients with stroke from activity of daily living tasks," *Artif. Intell. Med.*, vol. 46, pp. 233–249, July, 2009.
29. D. Pani et al., "A device for local or remote monitoring of hand rehabilitation sessions for rheumatic patients," *IEEE J. Transl. Eng. Health Med.*, vol. 2, pp. 1–11, 2014.
30. A. M. Coderre et al., "Assessment of upper-limb sensorimotor function of subacute stroke patients using visually guided reaching," *Neurorehabil. Neural Repair*, vol. 24, pp. 528–541, 2010.
31. A. Prochazka and J. Kowalczewski, "A fully automated, quantitative test of upper limb function," *J. Mot. Behav.*, vol. 47, pp. 19–28, 2015.
32. R. Colombo et al., "Test–retest reliability of robotic assessment measures for the evaluation of upper limb recovery," *IEEE Trans. Neural Syst. Rehabil. Eng.*, vol. 22, pp. 1020–1029, September, 2014.
33. U. Keller et al., "Robot-assisted arm assessments in spinal cord injured patients: A consideration of concept study," *PloS ONE 10(5)*, p. e0126948. 2015.
34. S. Balasubramanian et al., "Robotic assessment of upper limb motor function after stroke," *Am. J. Phys. Med. Rehabil.*, vol. 91, p. 255, November, 2012.
35. A. de los Reyes-Guzman et al., "Quantitative assessment based on kinematic measures of functional impairments during upper extremity movements: A review," *Clin. Biomech. (Bristol, Avon)*, vol. 29, pp. 719–727, August, 2014.
36. P. Kan et al., "The development of an adaptive upper-limb stroke rehabilitation robotic system," *J. Neuroeng. Rehabil.*, vol. 8, p. 1, 2011.
37. J. Metzger et al., "Assessment-driven selection and adaptation of exercise difficulty in robot-assisted therapy: A pilot study with a hand rehabilitation robot," *J. Neuroeng. Rehabil.*, vol. 11, p. 1, 2014.
38. Y. Choi et al., "Feasibility of the adaptive and automatic presentation of tasks (ADAPT) system for rehabilitation of upper extremity function post-stroke," *J. Neuroeng. Rehabil.*, vol. 8, p. 1, 2011.
39. J. M. Veerbeek et al., "Effects of robot-assisted therapy for the upper limb after stroke—A systematic review and meta-analysis," *Neurorehabil. Neural Repair*, pp. 107–121, 2016.
40. K. R. Evenson, M. M. Goto and R. D. Furberg, "Systematic review of the validity and reliability of consumer-wearable activity trackers," *Int. J. Behav. Nutr. Phys. Act.*, vol. 12, p. 1, 2015.
41. van den Berg-Emons, Rita J, J. B. Bussmann and H. J. Stam, "Accelerometry-based activity spectrum in persons with chronic physical conditions," *Arch. Phys. Med. Rehabil.*, vol. 91, pp. 1856–1861, 2010.
42. M. Noorkoiv, H. Rodgers and C. I. Price, "Accelerometer measurement of upper extremity movement after stroke: A systematic review of clinical studies," *J. Neuroeng Rehabil.*, vol. 11, p. 144, October 9, 2014.
43. K. S. Hayward et al., "Exploring the role of accelerometers in the measurement of real world upper-limb use after stroke," *Brain Impair.*, vol. 17, pp. 16–33, 2016.
44. G. Uswatte et al., "Validity of accelerometry for monitoring real-world arm activity in patients with subacute stroke: evidence from the extremity constraint-induced therapy evaluation trial," *Arch. Phys. Med. Rehabil.*, vol. 87, pp. 1340–1345, October 01, 2006.
45. G. Thrane et al., "Arm use in patients with subacute stroke monitored by accelerometry: Association with motor impairment and influence on self-dependence," *J. Rehabil. Med.*, vol. 43, pp. 299–304, March 01, 2011.
46. S. C. van der Pas et al., "Assessment of arm activity using triaxial accelerometry in patients with a stroke," *Arch. Phys. Med. Rehabil.*, vol. 92, pp. 1437–1442, September 01, 2011.
47. T. Wang et al., "Validity, responsiveness, and clinically important difference of the ABILHAND questionnaire in patients with stroke," *Arch. Phys. Med. Rehabil.*, vol. 92, pp. 1086–1091, 2011.
48. W. Liao et al., "Effects of robot-assisted upper limb rehabilitation on daily function and real-world arm activity in patients with chronic stroke: A randomized controlled trial," *Clin. Rehabil.*, vol. 26, pp. 111–120, 2012.
49. E. Taub et al., "Constraint-induced movement therapy combined with conventional neurorehabilitation techniques in chronic stroke patients with plegic hands: A case series," *Arch. Phys. Med. Rehabil.*, vol. 94, pp. 86–94, 2013.
50. R. R. Bailey and C. E. Lang, "Upper extremity activity in adults: Referent values using accelerometry," *J. Rehabil. Res. Dev.*, vol. 50, p. 1213, 2014.
51. M. Brogioli et al., "Novel sensor technology to assess independence and limb-use laterality in cervical spinal cord injury," *J. Neurotrauma*, 2016.
52. M. Brogioli et al., Multi-day recordings of wearable sensors are valid and sensitive measures of function and independence in human spinal cord injury. *J. Neurotrauma*, 2016.

53. D. Roetenberg, H. Luinge and P. Slycke, "Xsens MVN: Full 6DOF human motion tracking using miniature inertial sensors," *Xsens Motion Technologies BV, Tech.Rep*, 2009.
54. R. Zhu and Z. Zhou, "A real-time articulated human motion tracking using tri-axis inertial/magnetic sensors package," *IEEE Trans. Neural Syst. Rehabil. Eng.*, vol. 12, pp. 295–302, 2004.
55. H. J. Luinge, P. H. Veltink and C. T. Baten, "Ambulatory measurement of arm orientation," *J. Biomech.*, vol. 40, pp. 78–85, 2007.
56. V. T. Cruz et al., "A novel system for automatic classification of upper limb motor function after stroke: An exploratory study," *Med. Eng. Phys.*, vol. 36, pp. 1704–1710, 2014.
57. F. Trincado-Alonso et al., "Kinematic metrics based on the virtual reality system Toyra as an assessment of the upper limb rehabilitation in people with spinal cord injury," *Biomed. Res. Int.*, vol. 2014, p. 904985, 2014.
58. A. de los Reyes-Guzmn et al., "Novel kinematic indices for quantifying upper limb ability and dexterity after cervical spinal cord injury," *Med. Biol. Eng. Comput.*, pp. 1–12, 2016.
59. S. B. Thies et al., "Movement variability in stroke patients and controls performing two upper limb functional tasks: A new assessment methodology," *J. Neuroeng. Rehabil.*, vol. 6, p. 1, 2009.
60. Z. Zhang, Q. Fang, and X. Gu, "Objective assessment of upper limb mobility for post-stroke rehabilitation," *IEEE Trans. Biomed. Eng.*, pp. 1–10, 2015.
61. F. Wittmann et al., "Self-directed arm therapy at home after stroke with a sensor-based virtual reality training system," *J. Neuroeng. Rehabil.*, vol. 13, p. 75, 2016.
62. A. W. Lam et al., "Automated rehabilitation system: Movement measurement and feedback for patients and physiotherapists in the rehabilitation clinic," *Human–Comput. Interact.*, 2015.
63. W. L. Popp et al., "A novel algorithm for detecting active propulsion in wheelchair users following spinal cord injury," *Med. Eng. Phys.*, vol. 38, pp. 267–274, 2016.
64. N. Friedman et al., "The manumeter: A wearable device for monitoring daily use of the wrist and fingers," *IEEE J. Biomed. Health Inform.*, vol. 18, pp. 1804–1812, 2014.
65. J. B. Rowe et al., "The variable relationship between arm and hand use: A rationale for using finger magnetometry to complement wrist accelerometry when measuring daily use of the upper extremity," in *Engineering in Medicine and Biology Society (EMBC), 2014 36th Annual International Conference of the IEEE*, 2014, pp. 4087–4090.
66. M. F. de Castro and A. Cliquet Jr, "An artificial grasping evaluation system for the paralysed hand," *Med. Biol. Eng. Comput.*, vol. 38, pp. 275–280, 2000.
67. A. S. Merians et al., "Innovative approaches to the rehabilitation of upper extremity hemiparesis using virtual environments," *Eur. J. Phys. Rehabil. Med.*, vol. 45, pp. 123–133, 2009.
68. N. Friedman et al., "Retraining and assessing hand movement after stroke using the MusicGlove: Comparison with conventional hand therapy and isometric grip training," *J. Neuroeng. Rehabil.*, vol. 11, p. 1, 2014.
69. N. P. Oess, J. Wanek and A. Curt, "Design and evaluation of a low-cost instrumented glove for hand function assessment," *J. Neuroeng. Rehabil.*, vol. 9, p. 1, 2012.
70. R. McLaren et al., "A review of e-textiles in neurological rehabilitation: How close are we?" *J. Neuroeng Rehabil.*, vol. 13, June 21, 2016.
71. T. Giorgino et al., "Assessment of sensorized garments as a flexible support to self-administered post-stroke physical rehabilitation," *Eur. J. Phys. Rehabil. Med.*, vol. 45, pp. 75–84, March 1, 2009.
72. J. Zariffa and M. R. Popovic, "Hand contour detection in wearable camera video using an adaptive histogram region of interest." *J. Neuroeng. Rehabil. 10(1)*, p. 114, 2013.
73. J. Likitlersuang and J. Zariffa, "Interaction detection in egocentric video: Towards a novel outcome measure for upper extremity function," *Journal of Biodemical and Health Informatics*, vol. 22(2), p. 561–569, 2018.
74. A. Betancourt et al., "A dynamic approach and a new dataset for hand-detection in first person vision," in *Computer Analysis of Images and Patterns*, September 2, 2015, pp. 274–287.
75. A. Betancourt et al., "Left/right hand segmentation in egocentric videos," *Comput. Vision Image Understand.*, 2016.
76. Cheng Li and K. M. Kitani, "Pixel-level hand detection in ego-centric videos," in *IEEE Conference on Computer Vision and Pattern Recognition*, 2013, pp. 3570–3577.
77. D. Huang et al., "How do we use our hands? Discovering a diverse set of common grasps," in *Proceedings of the IEEE Conference on Computer Vision and Pattern Recognition*, 2015, pp. 666–675.
78. M. Cai, K. M. Kitani and Y. Sato, "A scalable approach for understanding the visual structures of hand grasps," in *International Conference on Robotics and Automation*, 2015.
79. G. Serra et al., "Hand segmentation for gesture recognition in EGO-vision," in *Proceedings of the 3rd ACM International Workshop on Interactive Multimedia on Mobile & Portable Devices*, 2013, pp. 31–36.
80. S. Bambach et al., "Lending A hand: Detecting hands and recognizing activities in complex egocentric interactions." *Presented at IEEE International Conference on Computer Vision (ICCV)*, 2015.
81. J. Likitlersuang and J. Zariffa, "Arm angle detection in egocentric video of upper extremity tasks," in World Congress on Medical Physics and Biomedical Engineering, June 7–12, 2015, Toronto, Canada, 2015, pp. 1124–1127.
82. T. Vodopivec, V. Lepetit, and P. Peer, "Fine hand segmentation using convolutional neural networks," *arXiv Preprint arXiv:1608.07454*, 2016.
83. A. Fathi, X. Ren, and J. M. Rehg, "Learning to recognize objects in egocentric activities," in *IEEE Conference on Computer Vision and Pattern Recognition (CVPR)*, 2011, pp. 3281–3288.
84. T. Ishihara et al., "Recognizing hand-object interactions in wearable camera videos." *IEEE International Conference on Image Processing (ICIP)*, 2015.

85. H. Pirsiavash and D. Ramanan, "Detecting activities of daily living in first-person camera views," in *IEEE Conference on Computer Vision and Pattern Recognition (CVPR)*, 2012, pp. 2847–2854.
86. Y. Yan et al., "Egocentric daily activity recognition via multitask clustering," *IEEE Trans. Image Process.*, vol. 24, pp. 2984–2995, 2015.
87. T. Nguyen, J. Nebel and F. Florez-Revuelta, "Recognition of activities of daily living with egocentric vision: A review," *Sensors*, vol. 16, p. 72, 2016.
88. P. Rashidi and A. Mihailidis, "A survey on ambient-assisted living tools for older adults," *IEEE J. Biomed. Health Inform.*, vol. 17, pp. 579–590, 2013.
89. G. Kurillo et al., "Evaluation of upper extremity reachable workspace using Kinect camera," *Technol. Health Care*, vol. 21, pp. 641–656, 2013.
90. A. V. Dowling et al., "An adaptive home-use robotic rehabilitation system for the upper body," *IEEE J. Transl. Eng. Health Med.*, vol. 2, p. 2100310, 2014.
91. B. Bonnechere et al., "Validity and reliability of the Kinect within functional assessment activities: Comparison with standard stereophotogrammetry," *Gait Posture*, vol. 39, pp. 593–598, 2014.
92. A. Mobini, S. Behzadipour, and M. Saadat, "Test–retest reliability of Kinect's measurements for the evaluation of upper body recovery of stroke patients," *Biomed. Eng. Online*, vol. 14, p. 1, 2015.
93. M. E. Huber et al., "Validity and reliability of Kinect skeleton for measuring shoulder joint angles: A feasibility study," *Physiotherapy*, vol. 101, pp. 389–393, 2015.
94. R. P. Kuster et al., "Accuracy of KinectOne to quantify kinematics of the upper body," *Gait Posture*, vol. 47, pp. 80–85, 2016.
95. H. Mousavi Hondori and M. Khademi, "A review on technical and clinical impact of Microsoft Kinect on physical therapy and rehabilitation," *J. Med. Eng.*, vol. 2014, 2014.
96. D. Webster and O. Celik, "Systematic review of Kinect applications in elderly care and stroke rehabilitation," *J. Neuroeng. Rehabil.* vol. 11, p. 108, 2014.
97. H. Sin and G. Lee, "Additional virtual reality training using Xbox Kinect in stroke survivors with hemiplegia," *Am. J. Phys. Med. Rehabil.*, vol. 92, pp. 871–880, 2013.
98. I. Pastor, H. A. Hayes and S. J. Bamberg, "A feasibility study of an upper limb rehabilitation system using Kinect and computer games," in *2012 Annual International Conference of the IEEE Engineering in Medicine and Biology Society*, 2012, pp. 1286–1289.
99. R. Proffitt and B. Lange, "The feasibility of a customized, in-home, game-based stroke exercise program using the Microsoft Kinect Sensor," *Int. J. Telerehabil.*, vol. 7, p. 23, 2015.
100. K. J. Bower et al., "Clinical feasibility of interactive motion-controlled games for stroke rehabilitation," *J. Neuroeng. Rehabil.*, vol. 12, p. 1, 2015.
101. M. Pirovano et al., "Intelligent Game Engine for Rehabilitation (IGER)," *IEEE Trans. Comput. Intell. AI Games*, vol. 8, pp. 43–55, 2016.
102. E. V. Olesh, S. Yakovenko, and V. Gritsenko, "Automated assessment of upper extremity movement impairment due to stroke," *PloS ONE*, vol. 9, p. e104487, 2014.
103. A. Ozturk et al., "A clinically feasible kinematic assessment method of upper extremity motor function impairment after stroke," *Measurement*, vol. 80, pp. 207–216, 2016.
104. R. J. Adams et al., "Assessing upper extremity motor function in practice of virtual activities of daily living," *IEEE Trans. Neural Syst. Rehabil. Eng.*, vol. 23, pp. 287–296, 2015.
105. W. Kim et al., "Upper extremity functional evaluation by Fugl–Meyer Assessment Scoring using depth-sensing camera in hemiplegic stroke patients," *PloS One*, vol. 11, p. e0158640, 2016.
106. S. K. Subramanian et al., "Validity of movement pattern kinematics as measures of arm motor impairment poststroke," *Stroke*, vol. 41, pp. 2303–2308, 2010.
107. M. A. Murphy, C. Willn and K. S. Sunnerhagen, "Kinematic variables quantifying upper-extremity performance after stroke during reaching and drinking from a glass," *Neurorehabil. Neural Repair*, vol. 25, pp. 71–80, 2011.
108. M. Alt Murphy, C. Willen and K. S. Sunnerhagen, "Movement kinematics during a drinking task are associated with the activity capacity level after stroke," *Neurorehabil. Neural Repair*, vol. 26, pp. 1106–1115, 2012.
109. M. Alt Murphy, C. Willen, and K. S. Sunnerhagen, "Responsiveness of upper extremity kinematic measures and clinical improvement during the first three months after stroke," *Neurorehabil. Neural Repair*, June 13, 2013.
110. J. M. Wagner, J. A. Rhodes, and C. Patten, "Reproducibility and minimal detectable change of three-dimensional kinematic analysis of reaching tasks in people with hemiparesis after stroke," *Phys. Ther.*, vol. 88, pp. 652–663, 2008.
111. T. S. Patterson et al., "Reliability of upper extremity kinematics while performing different tasks in individuals with stroke," *J. Mot. Behav.*, vol. 43, pp. 121–130, 2011.
112. M. Caimmi et al., "Using kinematic analysis to evaluate constraint-induced movement therapy in chronic stroke patients," *Neurorehabil. Neural Repair*, vol. 22, pp. 31–39, 2008.
113. J. D. Steeves et al., "Extent of spontaneous motor recovery after traumatic cervical sensorimotor complete spinal cord injury," *Spinal Cord*, vol. 49, pp. 257–265, February, 2011.
114. J. Zariffa et al., "Characterization of neurological recovery following traumatic sensorimotor complete thoracic spinal cord injury," *Spinal Cord*, vol. 49, pp. 463–471, March, 2011.
115. M. R. Spiess et al., "Conversion in ASIA impairment scale during the first year after traumatic spinal cord injury," *J. Neurotrauma*, vol. 26, pp. 2027–2036, November, 2009.
116. J. Zariffa et al., "Functional motor preservation below the level of injury in subjects with American Spinal Injury Association Impairment Scale grade A spinal cord injuries," *Arch. Phys. Med. Rehabil.*, vol. 93, pp. 905–907, May, 2012.
117. J. A. Petersen et al., "Spinal cord injury one-year evolution of motor-evoked potentials and recovery of leg motor function in 255 patients," *Neurorehabil. Neural Repair*, vol. 26, pp. 939948, 2012.
118. M. Spiess et al., "Evolution of tibial SSEP after traumatic spinal cord injury: Baseline for clinical trials," *Clin. Neurophysiol.*, vol. 119, pp. 1051–1061, 2008.

119. B. Wirth et al., "Changes in activity after a complete spinal cord injury as measured by the Spinal Cord Independence Measure II (SCIM II)," *Neurorehabil. Neural Repair*, vol. 22, pp. 279–287, 2008.
120. F. Högel, O. Mach, and D. Maier, "Functional outcome of patients 12 and 48 weeks after acute traumatic tetraplegia and paraplegia: Data analysis from 2004–2009," *Spinal Cord*, vol. 50, pp. 517–520, 2012.
121. M. Wirz and V. Dietz, "Recovery of sensorimotor function and activities of daily living after cervical spinal cord injury: The influence of age," *J. Neurotrauma*, vol. 32, pp. 194–199, 2015.
122. World Health Organization. *International Classification of Functioning, Disability and Health (ICF)*, 2001.
123. J. L. Kramer et al., "Relationship between motor recovery and independence after sensorimotor-complete cervical spinal cord injury," *Neurorehabil. Neural Repair*, vol. 26, pp. 1064–1071, 2012.
124. H. J. van Hedel and A. Curt, "Fighting for each segment: Estimating the clinical value of cervical and thoracic segments in SCI," *J. Neurotrauma*, vol. 23, pp. 1621–1631, November, 2006.
125. C. Rudhe and H. J. van Hedel, "Upper extremity function in persons with tetraplegia: Relationships between strength, capacity, and the Spinal Cord Independence Measure," *Neurorehabil. Neural Repair*, vol. 23, pp. 413–421, June, 2009.
126. L. G. Tanadini et al., "Identifying homogeneous subgroups in neurological disorders: Unbiased recursive partitioning in cervical complete spinal cord injury," *Neurorehabil. Neural Repair*, vol. 28, pp. 507–515, July 01, 2014.
127. L. G. Tanadini et al., "Toward inclusive trial protocols in heterogeneous neurological disorders: Prediction-based stratification of participants with incomplete cervical spinal cord injury," *Neurorehabil. Neural Repair*, vol. 29, pp. 867–877, October 01, 2015.
128. I. M. Velstra et al., "Prediction and stratification of upper limb function and self-care in acute cervical spinal cord injury with the Graded Redefined Assessment of Strength, Sensibility, and Prehension (GRASSP)," *Neurorehabil. Neural Repair*, vol. 28, pp. 632–642, September, 2014.
129. I. Velstra et al., "Predictive value of upper limb muscles and grasp patterns on functional outcome in cervical spinal cord injury," *Neurorehabil. Neural Repair*, vol. 30, pp. 295–306, 2016.
130. V. K. Noonan et al., "The Rick Hansen Spinal Cord Injury Registry (RHSCIR): A national patient-registry," *Spinal Cord*, vol. 50, pp. 22–27, 2012.
131. M. F. Dvorak et al., "The influence of time from injury to surgery on motor recovery and length of hospital stay in acute traumatic spinal cord injury: An observational Canadian cohort study," *J. Neurotrauma*, vol. 32, pp. 645–654, May 01, 2015.
132. V. K. Noonan et al., "The validity of administrative data to classify patients with spinal column and cord injuries," *J. Neurotrauma*, vol. 30, pp. 173–180, 2013.
133. M. Ali et al., "The virtual international stroke trials archive," *Stroke*, vol. 38, pp. 1905–1910, 2007.
134. H. Hallevi et al., "Recovery after ischemic stroke: Criteria for good outcome by level of disability at day 7," *Cerebrovasc. Dis.*, vol. 28, pp. 341–348, 2009.
135. F. C. Goldie et al., "Interdependence of stroke outcome scales: Reliable estimates from the Virtual International Stroke Trials Archive (VISTA)," *Int. J. Stroke*, vol. 9, pp. 328–332, 2014.
136. H. I. Krebs et al., "Robotic measurement of arm movements after stroke establishes biomarkers of motor recovery," *Stroke*, vol. 45, pp. 200–204, January 01, 2014.
137. S. Izrailev and D. K. Agrafiotis, "Variable selection for QSAR by artificial ant colony systems," *SAR QSAR Environ. Res.*, vol. 13, pp. 417–423, 2002.
138. S. M. Mostafavi et al., "Robot-based assessment of motor and proprioceptive function identifies biomarkers for prediction of functional independence measures," *J. Neuroeng. Rehabil.*, vol. 12, p. 1, 2015.
139. J. Zariffa et al., "Predicting task performance from upper extremity impairment measures after cervical spinal cord injury," *Spinal Cord*, May 31, 2016.
140. D. T. Wade et al., "The hemiplegic arm after stroke: Measurement and recovery," *J. Neurol. Neurosurg. Psychiatr.*, vol. 46, pp. 521–524, 1983.
141. H. S. Jrgensen et al., "Outcome and time course of recovery in stroke. Part II: Time course of recovery. The Copenhagen Stroke Study," *Arch. Phys. Med. Rehabil.*, vol. 76, pp. 406–412, 1995.
142. R. J. Marino et al., "Upper- and lower-extremity motor recovery after traumatic cervical spinal cord injury: An update from the National Spinal Cord Injury Database," *Arch. Phys. Med. Rehabil.*, vol. 92, pp. 369–375, March, 2011.
143. G. Kwakkel and B. J. Kollen, "Predicting activities after stroke: What is clinically relevant?," *Int. J. Stroke*, vol. 8, pp. 25–32, 2013.
144. D. J. Reinkensmeyer et al., "Computational neurorehabilitation: Modeling plasticity and learning to predict recovery," *J. Neuroeng. Rehabil.*, vol. 13, p. 1, 2016.
145. B. K. Kwon et al., "Cerebrospinal fluid biomarkers to stratify injury severity and predict outcome in human traumatic spinal cord injury," *J. Neurotrauma*, August 15, 2016.
146. J. R. Wilson et al., "A clinical prediction model for long-term functional outcome after traumatic spinal cord injury based on acute clinical and imaging factors," *J. Neurotrauma*, vol. 29, pp. 2263–2271, 2012.
147. F. Kuhn et al., "One-year evolution of ulnar somatosensory potentials after trauma in 365 tetraplegic patients: Early prediction of potential upper limb function," *J. Neurotrauma*, vol. 29, pp. 1829–1837, 2012.
148. H. S. Jrgensen, "The Copenhagen Stroke Study experience," *J. Stroke Cerebrovasc. Dis.*, vol. 6, pp. 5–16, 1996.
149. H. Nakayama et al., "Recovery of upper extremity function in stroke patients: The Copenhagen Stroke Study," *Arch. Phys. Med. Rehabil.*, vol. 75, p. 394, 1994.
150. H. S. Jorgensen et al., "Outcome and time course of recovery in stroke. Part I: Outcome. The Copenhagen Stroke Study," *Arch. Phys. Med. Rehabil.*, vol. 76, pp. 399–405, 1995.

151. H. S. Jorgensen et al., "Effect of blood pressure and diabetes on stroke in progression," *Lancet*, vol. 344, pp. 156–159, 1994.
152. H. S. Jorgensen et al., "The effect of a stroke unit: Reductions in mortality, discharge rate to nursing home, length of hospital stay, and cost a community-based study," *Stroke*, vol. 26, pp. 1178–1182, 1995.
153. C. R. Pretz et al., "Using Rasch motor FIM individual growth curves to inform clinical decisions for persons with paraplegia," *Spinal Cord*, vol. 52, pp. 671–676, September, 2014.
154. Y. Chen et al., "Current research outcomes from the spinal cord injury model systems," *Arch. Phys. Med. Rehabil.*, vol. 92, pp. 329–331, 2011.
155. C. R. Pretz et al., "Trajectories of life satisfaction following spinal cord injury," *Arch. Phys. Med. Rehabil.*, 2016.
156. C. L. Massie et al., "A clinically relevant method of analyzing continuous change in robotic upper extremity chronic stroke rehabilitation," *Neurorehabil. Neural Repair*, p. 1545968315620301, 2015.
157. R. Forsyth et al., "Efficient rehabilitation trial designs using disease progress modeling: A pediatric traumatic brain injury example," *Neurorehabil. Neural Repair*, vol. 24, pp. 225–234, 2010.
158. J. A. Semrau et al., "Examining differences in patterns of sensory and motor recovery after stroke with robotics," *Stroke*, vol. 46, pp. 3459–3469, 2015.
159. A. Curt, M. E. Schwab and V. Dietz, "Providing the clinical basis for new interventional therapies: Refined diagnosis and assessment of recovery after spinal cord injury," *Spinal Cord*, vol. 42, pp. 1–6, January, 2004.
160. https://www.emsci.org/.
161. http://vistacollaboration.org/.
162. S. L. Stover, J. Michael, and B. K. Go, "History, implementation, and current status of the National Spinal Cord Injury Database," *Arch. Phys. Med. Rehabil.*, vol. 80, pp. 1365–1371, 1999.
163. https://www.nscisc.uab.edu/nscisc-database.aspx.
164. F. Biering-Srensen et al., "Common data elements for spinal cord injury clinical research: A National Institute for Neurological Disorders and Stroke project," *Spinal Cord*, vol. 53, pp. 265–277, 2015.
165. K. Slavakis, G. B. Giannakis, and G. Mateos, "Modeling and optimization for big data analytics: (statistical) learning tools for our era of data deluge," *IEEE Signal Process. Mag.*, vol. 31, pp. 18–31, 2014.
166. V. K. Noonan et al., "Meeting the privacy requirements for the development of a multi-centre patient registry in Canada: The Rick Hansen Spinal Cord Injury Registry," *Healthcare Policy*, vol. 8, p. 87, 2013.
167. J. Gassaway, G. Whiteneck, and M. Dijkers, "Clinical taxonomy development and application in spinal cord injury research: The SCIRehab Project," *J. Spinal Cord Med.*, vol. 32, p. 260, 2009.
168. G. DeJong et al., "Toward a taxonomy of rehabilitation interventions: Using an inductive approach to examine the "black box" of rehabilitation," *Arch. Phys. Med. Rehabil.*, vol. 85, pp. 678–686, 2004.
169. M. R. Cowie et al., "Electronic health records to facilitate clinical research," *Clin. Res. Cardiol.*, pp. 1–9, 2016.
170. R. M. Kaplan, D. A. Chambers, and R. E. Glasgow, "Big data and large sample size: A cautionary note on the potential for bias," *Clin. Transl. Sci.*, vol. 7, pp. 342–346, 2014.
171. A. M. Jette, "The post-stroke rehabilitation outcomes project," *Arch. Phys. Med. Rehabil.*, vol. 86, pp. 124–125, 2005.

28
Multimodal Ambulatory Fall Risk Assessment in the Era of Big Data

Mina Nouredanesh and James Tung

CONTENTS

28.1 Introduction 551
28.2 Risk Factors for Falls 552
28.3 Fall Risk Assessment: What Has Been Done So Far? 553
    28.3.1 Non-Sensor-Based Methods for Supervised Fall Risk Assessment 553
    28.3.2 Sensor-Based Methods for Supervised Fall Risk Assessment 554
    28.3.3 Unsupervised Fall Risk Assessment Methods 556
28.4 A New Approach for Unsupervised FRA 560
    28.4.1 System Infrastructure 560
    28.4.2 Machine Learning–Based Detection of Compensatory Balance Responses Using Wearable Sensors 562
        28.4.2.1 Data Acquisition 562
        28.4.2.2 Data Processing and Analysis 562
        28.4.2.3 Machine Learning Techniques 565
        28.4.2.4 Results 566
    28.4.3 Wearable Vision Detection of Environmental Fall Risks Using Machine Learning Techniques 567
        28.4.3.1 Data Acquisition 567
        28.4.3.2 Preprocessing 568
        28.4.3.3 Gabor Barcodes Approach 568
        28.4.3.4 Convolutional Neural Networks 571
28.5 How Big Data Could Help Us Develop Better FRA and FP Tools 572
28.6 Conclusion 575
References 576

28.1 Introduction

One of the most important public health problems worldwide is falls, which are the leading cause of injury-related hospitalizations and a major cause of disability and death among seniors. It is estimated that one in three persons over the age of 65 falls at least once each year [1,2] and 50% of seniors over age 85 suffer one fall per year [3]. In addition to physical consequences, such as hip fracture and traumatic brain injury, falls can lead to negative mental health outcomes, such as fear of falling, loss of autonomy, and depression. Not only do these harm the injured individuals, but they also affect family and care providers and incur expenses to health care systems. For example, the direct cost associated with hip fracture alone is estimated at $1.1 billion per year (≈30,000 fractures) in Canada [4].

The adverse consequences of falls can be reduced among seniors with the help of fall prevention activities and cutting-edge technologies; therefore, many researchers have endeavored to develop fall risk assessment (FRA), fall prevention (FP), and fall detection (FD) tools. As the initial step for prevention programs and interventions, FRAs are conducted to identify individuals at highest risk of falling by determining intrinsic (e.g., muscle weakness, neurological deficits) and extrinsic (e.g., poor lighting, inappropriate footwear) risk factors (detailed in Section 28.2). By identifying individual risks, FRAs inform clinical decisions on the most appropriate prevention interventions to, ultimately, reduce fall incidence. Current FRA methods largely comprise of self-reported responses to questionnaires and performance on standard gait and balance assessments conducted in the clinic. While these methods have demonstrated utility in measuring functional capabilities under ideal conditions (i.e., gait speed in straight-line walking), these may not reflect the complex, multifactorial conditions of everyday life where falls occur. Furthermore, clinical assessments of fall risk are episodic, providing snapshots in time and leaving long gaps between measurements.

With the emergence of ubiquitous computing and sensing technologies, big data is being produced by everyday objects around us. Our bodies are no exception, generating diverse types of data ranging from mobility (e.g., walking speed) to cardiac activities that can be recorded and transmitted by sensors and mobile devices for further analysis. We anticipate that technological advances in sensors, machine learning, signal processing, computer vision, and big data will lead to discovery of new solutions for many medical problems. Importantly, the current lack of clinical information on a day-to-day basis hinders our understanding of disease trajectories on multiple time scales, including risks for falls. These emerging technologies have the potential to collect data in everyday environments during free-living activities (FLAs) over long periods to address these knowledge gaps.

In this chapter, current and emerging FRA methods are described and critically evaluated. Moreover, the role of big data and state-of-the-art sensors in transitioning from episodic, supervised assessment (i.e., under controlled conditions in the clinic or laboratory) to unsupervised conditions (i.e., everyday life over long periods) will be described (Section 28.5). As an example of the transition to unsupervised assessment, we describe a novel approach using wearable sensors, signal and image processing, and machine learning techniques (Section 28.4). Finally, the challenges with sensor-based FRA methods are discussed, looking prospectively at new opportunities for collecting big data toward better understanding of fall etiology and prevention.

28.2 Risk Factors for Falls

Unlike other ages, falls among older adults tend to occur as a result of multiple interacting factors [5], as each older person faces unique combinations of risk factors according to his/her life circumstances. As a basis for fall prediction and intervention programs, specific knowledge about individual risk factors is needed [6]. In other words, clinicians aim to understand what puts each senior at risk of falling (e.g., muscle weakness) to inform the selection and timing of FP interventions (e.g., strengthening program). Risk factors for falls are generally categorized into intrinsic or biological (e.g., acute or chronic illness), and extrinsic or environmental, (e.g., slippery floor; see Figure 28.1). Importantly, there may be interactions between multiple risk factors that can have complicated effects and need to be considered (e.g., foot sensation deficits × slippery terrain) [7].

Intrinsic, or patient-related, risk factors include advanced age, chronic diseases, muscle weakness, gait

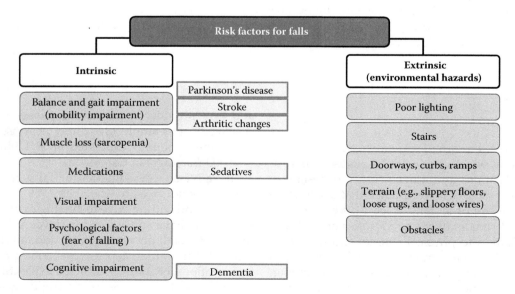

FIGURE 28.1
Common intrinsic and extrinsic fall risk factors.

disorders, mental status alterations, and medications [8]. Gait and balance deficits are prominently featured as key intrinsic risk factors. Extrinsic or environmental (e.g., terrain, obstacles, lighting) factors originate outside of the body and are responsible for approximately one-third to half of all falls [9,10]. Terrain (e.g., irregular, cracked, or slippery surfaces) and visual conditions (e.g., poor lighting) pose high risk for falls [11,12] and are often targets for intervention. Even walking on a familiar route can lead to falls as a consequence of poor building design and inadequate consideration [11]. Additionally, the risk from extrinsic factors can be aggravated by intrinsic and behavioral risk factors (e.g., cognitive deficits, being hurried or inattentive, difficulty or discomfort during a task, or moving beyond limits of stability [13]).

28.3 Fall Risk Assessment: What Has Been Done So Far?

In general, we consider two main categories of FRA methods (Figure 28.2, upper panel): 1) supervised fall risk assessment (SFRA) methods (Sections 28.3.1 and 28.3.2), which are conducted under the supervision of a trained assessor (e.g., nurse, physician) in a nursing home, laboratory, or clinical setting, and 2) unsupervised FRA (UFRA) (Section 28.3.3), which applies technologies (i.e., ambient, wearable, mobile sensing) to assess the risk of falling in an ambulatory fashion within the home environment or community setting and without supervision. While we identify these two classes of FRA approaches for the purpose of this chapter, we recognize a spectrum of approaches, as illustrated in Figure 28.2, lower panel.

28.3.1 Non-Sensor-Based Methods for Supervised Fall Risk Assessment

SFRA methods are generally performed by a geriatrician, nurse, or therapist in a clinical setting or institution encompassing a combination of three main assessments: 1) comprehensive medical assessments, 2) screening and/or questionnaire-based assessments (using screening instruments or forms), and 3) functional mobility assessments [14].

Comprehensive medical assessments: Comprehensive medical assessments typically focus on identifying intrinsic risk factors that can be treated to reduce the likelihood of falling [15]. Medical history and charts provide detailed notes of previous fall history, cognition, balance control behavior, gait, muscle strength, chronic diseases, mobility, nutrition, and medications [16]. While these assessments are resource intensive (i.e., time-consuming [17] and often involving a team of clinicians [18]), their contributions can identify common risk factors and provide historical context to adapt intervention plans (e.g., alternate medications).

Screening and questionnaire-based assessments: Screening and questionnaire-based assessments document

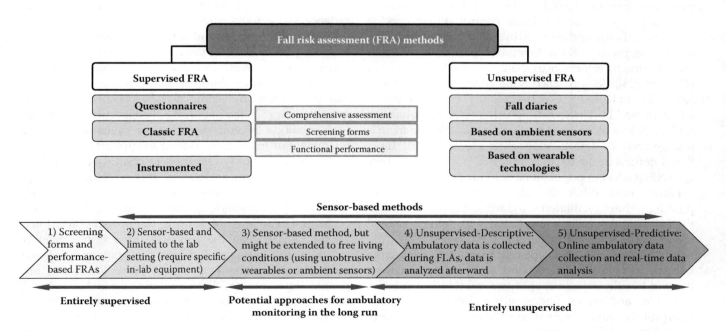

FIGURE 28.2
Upper panel: Two main FRA methods: supervised and unsupervised FRA methods. Lower panel: Instead of binary categorization of FRA methods, a spectrum of methods ranging from entirely supervised to unsupervised settings and free-living activities (FLAs) is considered. The majority of FRA methods are concerned with supervised FRA. Big data is expected to play an important role in steps 4 and 5.

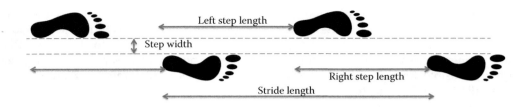

FIGURE 28.3
Example spatiotemporal gait parameters.

the presence of specific factors to rapidly generate a fall risk score. In contrast to a full comprehensive medical assessment, these tools are fast and easy to administer. Many tools have been developed, including the Morse Fall Scale [19], STRATIFY [20], Resident Assessment Instrument (RAI) [21], Hendrich Fall Risk Model [22], High Risk for Falls Assessment Form [23], and Royal Melbourne Hospital Risk Assessment Tool [24]. For instance, STRATIFY scores 2-month fall history, mental alteration, frequent toileting, visual impairment, psychotropic medication use, and mobility issues with a threshold of score 2 or above to indicate increased fall risk. These instruments are commonly used by nurses upon admission to a hospital or long-term care facility to identify high-risk patients and are periodically updated (e.g., per shift, daily, or weekly) [14].

Functional mobility assessments: Functional mobility assessments provide standardized measures of capabilities, such as mobility (dynamic balance, gait), posture, visual acuity, sensation, and vital signs (heart rate, blood pressure). Patients typically perform a series of tasks requiring postural and balance control under varying levels of challenge and assessed by a clinician using a rubric and/or stopwatch. Such functional performance scales include Tinetti Performance Oriented Mobility Assessment (POMA) [25], Berg Balance Scale (BBS) [26], Functional Reach [27], or Dynamic Gait Index [28]. For instance, BBS is a widely used 14-item tool designed to assess static balance and fall risk in older adult populations [26]. While there is a large body of data to support its use to identify functional deficits in the balance-impaired populations, there is a ceiling effect in populations without disabilities [29].

Limitations: While the aforementioned assessment paradigms have facilitated standardized measurement of the intrinsic risk factors associated with falling, several major limitations are apparent. First, assessments are typically conducted in an episodic manner, leaving large unmeasured gaps in time. Second, many assessment tools depend heavily on clinical judgement, which can introduce bias and errors. For example, different staff may interpret the same observation differently (i.e., interrater reliability) [30]. Furthermore, emerging research indicates that movement kinematics (e.g., joint angles, toe clearance, spatiotemporal gait parameters) that are difficult to observe with the naked eye are more sensitive to subtle deficits. Thus, there is a need to develop objective FRA tools capable of capturing sensitive measures in an objective manner. Sensors that are able to capture information related to gait and balance control behavior, body motion, muscle activities, and ground reaction forces can provide objective, quantitative measures for FRA [31].

28.3.2 Sensor-Based Methods for Supervised Fall Risk Assessment

Gait analysis background: In clinical and research settings, gait performance is a widely accepted measure of functional ability. Consisting of repetitive cycles (Figure 28.3), gait is a process commonly described as starting from the initial contact (IC) of the lead leg, with a short period of double support (DS) until the following leg leaves the ground. After IC of the following leg, and another DS period, a full stride is completed when the lead leg leaves the ground indicated by toe-off (TO). Measured gait parameters are generally categorized into spatial (e.g., step size, step width; see Figure 28.3) and temporal (e.g., cadence, stride time) parameters. Comparing an individual's successive strides and extraction of gait parameters including stride-to-stride variability, asymmetry, index of harmonicity, harmonic ratios, and entropy offers sensitive measures to identify abnormal patterns in gait. Measured gait parameters are significantly associated with prospective falls [32], leading to the development of specific FRA methods by employing sensor systems to record movement or force data. Aiming to support SFRA, these methods are typically restricted to clinical or laboratory conditions due to constraints such as fixed equipment (e.g., force plates), requirement to wear markers, and/or specific testing conditions (e.g., standing in tandem stance).

Ground reactions: Measuring the pressure and force interaction between the feet and the ground, ground reactions are excellent indicators of gait and balance control. Force plates (e.g., AMTI Force Platforms) are precise instruments that measure three-dimensional ground reaction forces and torques. These measures can be used to estimate the location of the center of pressure (COP) under the feet, a commonly used indicator of balance control. Some companies, such as Biodex,* have

---

* http://www.biodex.com

developed specialized fall risk screening tools (e.g., the Biodex Falls Screening and Conditioning Program 2010) based on ground reaction measures. Treadmills instrumented with force plates are another option as a repeatable and sensitive method to analyze common spatiotemporal gait parameters and ground reaction forces (e.g., BalanceTutor-MediTouch* and FDM-THM-S, Zebris Medical GmbH).

Pressure-sensitive floor mats (e.g., GaitRite M2,[†] CIR Systems Inc.) or insoles inserted into footwear (e.g., Tekscan Sway Analysis Module,[‡] NovelPedar[§]) are alternative methods capable of measuring ground reaction forces and deviations in COP. While pressure-sensitive mats can sense over a larger area and can also measure the pressure distribution under the feet, they lack the resolution sensitivity of force plates. For example, Ayena et al. [33] proposed and evaluated an instrumented version of the One-Leg Standing (OLS) score using pressure-sensitive insoles, demonstrating an inverse relationship between measured OLS and risk of falling.

Similar to most sensor systems for SFRA, the aforementioned equipment is constrained to customized environments (e.g., laboratory) and requires trained operators to administer tests and interpret results. Such systems are typically too costly for use in standard clinics, but fall clinics may refer patients for in-depth evaluation if they observe balance impairment that warrants a comprehensive assessment.

Optical motion capture: In laboratory settings, 3-D optical motion capture systems allow the derivation of spatiotemporal gait variables by tracking markers attached to body landmarks. In popular systems, such as CODA** [34] or Vicon (Vicon Ind., Oxford, UK), multiple optical sensors track body-worn markers and calculate position based on localization techniques to estimate the movement kinematics. Variables, such as joint angles, step length, and toe clearance, are calculated and can be displayed in real time. Limitations of motion capture systems are the requirement to wear markers on the body, limited working volumes, and stringent calibration requirements. The complexity and high cost of these systems limit them mainly to research purposes.

Posturography: Patients may be referred for posturography, a general term that covers a range of techniques (i.e., force plates, pressure-sensitive mats) used to quantify postural control in upright stance in either static or dynamic conditions. Computerized dynamic posturography (CDP) is a noninvasive specialized clinical assessment technique used to objectively identify abnormalities by challenging and quantifying the sensorimotor mechanisms (i.e., visual, vestibular, somatosensory) involved in the control of posture and balance. For example, the EquiTest system is a commercially available CDP system launched by NeuroCom International, Inc. (NeuroCom EquiTest 2010).* Posturography requires complex and bulky equipment and is an expensive, cumbersome, and time-consuming method; however, it provides one of the most detailed balance evaluations in a supervised manner.

Wearable sensors for SFRA: Advancements in microelectromechanical systems (MEMS) technology have accelerated manufacturing of miniaturized, inexpensive, and low-power wearable sensors, capable of quantifying body kinematics, electrophysiology [e.g., electromyography (EMG)], and environmental data. In this section, we describe wearable sensors developed and evaluated for FRA use under supervised conditions.

Wearable inertial measurement units: Inertial measurement units (IMUs) are small, easy-to-use packages typically consisting of accelerometer, gyroscope, and/or magnetometer sensors. Despite the potential of wearable IMUs to record kinematic behavior in an unsupervised manner, the vast majority of uses to date aim to instrument classic FRA functional assessments (discussed in Section 28.3.1) to provide more accurate and objective metrics. In particular, numerous studies have investigated the association between sensor-derived gait parameters and retrospective and/or prospective falls. Howcroft et al. [31] provided an exhaustive, methodical review of 40 studies employing IMUs to develop FRA methods in geriatric populations. Most of these studies proposed methods to estimate spatiotemporal gait parameters (e.g., walking speed) and stability-related parameters [e.g., variability in gait rhythm, root mean square (RMS)] extracted from acceleration signals. Activities include frequently used functional assessments such as 1) level-ground walking, 2) Timed Up and Go test (TUG), 3) sit-to-stand (STS) transitions, 4) standing postural sway, 5) left–right alternating step test (AST), and/or 6) uneven-ground walking. All of these studies were conducted in a clinical setting; however, they differ from each other in terms of the setup and sensor placement (e.g., lower lumbar spine, chest, leg) with few standards for sensor specifications, signal processing, and analysis.

Wearable surface electromyography: Muscles (particularly in the lower limbs) are involved in coordinated mechanisms to generate forces required for body movement and maintaining static and dynamic balance. EMG refers to the measurement of the electrical signal of the muscle using electrodes, typically on the skin surface. Emerging wearable surface EMG (sEMG)

---

* http://meditouch.co.il/balancetutor/
[†] http://www.gaitrite.com
[‡] http://www.tekscan.com
[§] http://www.novel.de
** Cartesian Optoelectronic Dynamic Anthropometer (CODA) optical systems from Codamotion (Charnwood Dynamics Ltd.), http://www.codamotion.com.

* http://www.resourcesonbalance.com

fabrics (e.g., instrumented garments by Athos*) may be employed to study muscle coordination and fatigue on balance control behavior.

Although there are numerous studies investigating accelerometry-based methods, fewer research has examined the use of sEMG sensors for FRA. For example, Wong et al. [35] used sEMG to record the activity patterns of bilateral tibialis anterior and gastrocnemius muscles in older adults ($n = 23$) in five static balance challenges (e.g., Romberg eyes open/closed, sharpened Romberg eyes open/closed, and single-leg standing). By measuring bilateral co-contraction about the ankle, elevated co-contraction was significantly associated with an increased risk for falls in older adults. Similarly, Bounyong et al. [36] employed wearable EMG sensors to estimate the fall risk of older adults based on co-contraction of the lower limbs during walking. The results indicated that thigh co-contraction during the stance phase can be predictive of falling experience with 65% accuracy. Although there are many commercially available wearable EMG sensors (Figure 28.4), the effectiveness of sEMG-based techniques is restricted by many factors including fatigue, sweat accumulation underneath the electrodes, and heterogeneity in body characteristics (e.g., body fat mass). Even after normalization, intertrial variability renders magnitude measurements difficult to interpret, while muscle timing measures (e.g., reaction time) are more reliable.

28.3.3 Unsupervised Fall Risk Assessment Methods

Despite advances in SFRA methods leading to current prevention best practices, falls remain a major priority in geriatric medicine and public health. In particular, the heterogeneous nature of older adults' health, lifestyle, and behaviors are major barriers to comprehensive assessments. Circumstances logged in fall diaries and examined in security video capturing footage of institutional falls reveal complex interactions between intrinsic and extrinsic risk factors at the individual level. While SFRA methods capable of quantifying intrinsic capabilities have advanced significantly, the ability to quantify detailed lifestyle and behavioral risks remains lacking. Furthermore, patients tend to perform to the best of their ability in supervised assessments conducted with a clinician, which may not be representative of patients' actual behavior. A promising approach to generate more comprehensive assessments is to examine patients' behavior in FLAs and in an unsupervised manner. This section describes the literature regarding UFRA methods.

Fall diaries: As the earliest unsupervised assessment method, fall diaries are manual logs that report the incidence and circumstances of falls in an effort to document individual risks [37]. While these methods are rapid

---

* https://www.liveathos.com/

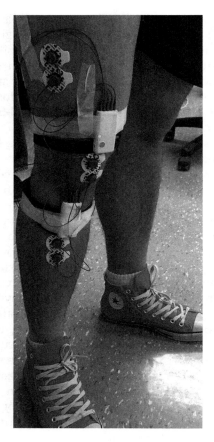

FIGURE 28.4
Using SHIMMER (Sensing Health with Intelligence, Modularity, Mobility, and Experimental Reusability) wearable sensors for detection of compensatory balance reactions (discussed in Section 28.4.2). The sensors are capable of tracking EMG of major lower-body muscle groups and wirelessly transmit to a mobile application via Bluetooth.

and easy to conduct, the quality of data is limited [38]. Reporting fall incidents among older adults relies on accurate recall to describe the event; however, individuals are likely to misreport and/or underreport fall events due to recall difficulties, sustained injuries, or fear of falling [32]. Furthermore, subtle behaviors (e.g., trips, stumbles) may be missed due to low sensitivity to minor events.

Sensor-based UFRA: Unobtrusive ambient or wearable sensor systems (WSSs) are capable of measuring and analyzing natural patterns of mobility behavior in the home and community (see Figure 28.5). Unsupervised methods afford more frequent sampling over longer periods of time under real-world conditions toward more comprehensive risk assessment. Compared to supervised methods, there are relatively few studies evaluating the effectiveness of sensor-based methods in collecting and analyzing data from FLA. These studies, evaluating gait characteristics over periods ranging from 2 days to several years [32,39–43], are discussed in the sections that follow.

Ambient sensors: Ambient sensors for FRA are systems mounted in the home or community to monitor individual

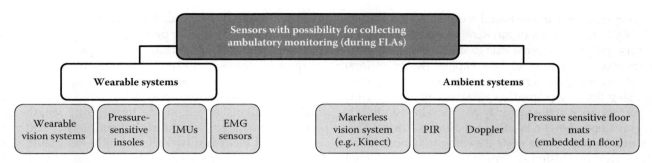

FIGURE 28.5
Sensors that can be used for collecting big data for ambulatory fall risk assessment.

behaviors. Once installed, these systems are considered passive, without the need for user interaction or the need to wear any equipment (e.g., sensors, markers). Such systems include markerless vision-based depth and motion measurement systems (e.g., Microsoft Kinect, machine vision), passive infrared (PIR) cameras, Doppler radar systems, microphones, pressure sensors, and vibration sensors embedded in the walls, ceiling, floor, or furniture. Ambient sensors allow seniors to live independently and safely in their own homes, reduce the need for expensive care facilities, and also enable caregivers to detect not only adverse events such as falls but also continuously assess and monitor the risk of such events.

The Intelligent Assistive Technologies and Systems Laboratory (IATSL) at the University of Toronto has been developing a Microsoft Kinect–based tool for balance assessment and monitoring longitudinal changes in movement patterns. In a case study (2014 [44,45]) a 64-year-old male with total hip replacement surgery on his right side performed two functional tasks: 1) walking and 2) STS. Spatiotemporal gait parameters (e.g., stance time, step length, stride length, cadence) were estimated using lower extremity 3-D ankle motions along the depth axis. Starting from 1 day before the surgery, gait characteristics improved significantly from 1 week to 6 weeks. Although only a case study, this study suggests that a markerless vision system along with an automated algorithm has the potential to be integrated into a patient's home to monitor changes in mobility.

Similarly, researchers at the University of Missouri have been designing ambient sensor systems to estimate the risk of falling from everyday living environments [46–50]. After laboratory validation, Stone et al. [46] examined the utility of a Kinect-based method to measure gait parameters in an independent-living apartment [47]. To address issues related to ambient sensing in a home with multiple residents, a resident model identification technique was first proposed. The model used outputs of the Kinect system to form a feature set in the form of x_i = {$height_i$, $walking\ speed_i$, $stride\ time_i$, $stride\ length_i$} to identify each resident. In the case that the *stride length* and *stride time* were not available due to visual occlusion (e.g.,

furniture), they were estimated based on the three nearest neighbors. These issues (i.e., multiple residents, missing data due to fixed sensors) reflect common limitations of ambient sensing systems. For evaluation, the aforementioned Kinect-based system was deployed in 12 apartments in an independent care facility for older adults with 15 residents (ages ranging from 67 to 97 years). To validate the Kinect-based method, two traditional FRA measures, habitual gait speed (HGS) and TUG tests, along with others such as the Short Physical Performance Battery (SPPB) were collected once a month over a period ranging from 4 months to 2 years. Validation results indicate that the Kinect-based mean gait speed was as accurate as the HGS measured traditionally and also emphasize that FLA speed variability was considerably high compared to explicit performance testing [51].

In another University of Missouri study, Rantz et al. [48] implemented and evaluated an ambient sensor system comprising a pulse-Doppler radar system, Microsoft Kinect, and two orthogonal web cameras to estimate gait parameters. Mounted in the apartments of 19 older adults (mean age = 87 years) at a senior living facility for 2 years, the radar and Kinect-based estimates were highly correlated with six classic FRAs: 1) HGS, 2) TUG, 3) Multidimensional Functional Reach, 4) Short Physical Performance Battery, 5) BBS, and 6) single-leg stance. Another approach to unobtrusive and ongoing gait assessment involves using PIR motion sensors. Using a method to estimate gait speed from the pattern and time intervals of the PIR sensor firings, the system captured gait speed in the homes of 76 community-dwelling older adults. Longitudinal assessment of daily gait speed was used to find the relations to function (decline in motor and cognitive functioning) [52].

While ambient sensors have provided valuable information on movement behavior in residential settings, there are several key limitations to be considered. First, testing location and frequency are limited to a specific place. While 50% of falls occur in the home, capturing mobility behavior in novel, dynamic environments may provide an important perspective on adaptive capabilities. Second, privacy concerns is an important barrier for the acceptance of camera-based ambient sensor systems [53]. While some types of ambient sensors (e.g., PIR) do not raise privacy concerns due to the lack of identifiable information, they tend to lack capabilities to estimate spatiotemporal gait parameters beyond walking speed (e.g., step time, step length, gait symmetry) [46]. Third, the performance of the majority of vision-based methods is highly dependent on lighting conditions. Additionally, dynamic objects in a senior's dwelling may obstruct the field of view of a vision system (e.g., Kinect) [48]. Finally, installation, calibration, and maintenance of the such systems in nursing home cares or dwellings of older adults is a time-consuming and expensive process.

Wearable sensors and emerging approaches for UFRA: As discussed in Section 28.3.2, WSSs can efficiently capture and analyze mobility data to advance classic FRA methods by reducing subjectivity and improving sensitivity. Emerging research suggests that measurement of gait in uncontrolled settings using WSSs is a feasible and promising approach; however, this has not been investigated in depth [54]. Major barriers to the adoption of WSS have been the size and weight of the sensors, which have limited their suitability for long-term monitoring, and the ability to interpret logged data from FLA due to task variability and noise. This section presents an overview on the most recent studies that employ WSSs for ambulatory monitoring of the elderly in unsupervised settings, including IMUs and pressure shoe insoles.

Wearable IMUs: As discussed earlier (Section 28.3.2), wearable IMUs are popular WSSs able to collect and stream kinematic signals (i.e., body's acceleration, angular velocity) to a smartphone, mobile device, or computer to measure factors related to fall risk. Previous efforts using IMU signals have examined features related to physical activity (e.g., total activity duration, number of steps taken), gait parameters (e.g., step-to-step consistency), and spectral components (e.g., amplitude and width of dominant frequency) to assess fall risk [42,55,56]. In the work of Weiss et al., 71 community-living older adults wore a 3-D accelerometer on their lower back for 3 consecutive days, and found that step-to-step consistency was lower in the fallers (i.e., two or more falls in 6 months) in the vertical axis and higher in the mediolateral direction compared to controls [42]. Follow-up studies [55] using the same data set found that local dynamic stability measures (λ) distinguished between elderly fallers and nonfallers, which resulted in a better classification performance compared to the gait-related features [56].

In a series of larger studies, Rispen, Schooten, and Dieen et al. demonstrated that daily-life accelerometry can identify the fallers with a good accuracy. These studies examined data collected from $n = 113$ (Ref. [57]), $n = 169$ (Ref. [32]), and $n = 319$ (Ref. [58]) participants, using a triaxial accelerometer (DynaPort MoveMonitor) worn on the back at L5, for 1 week (in Refs. [32,58]) and 2 weeks (in Ref. [57]). The prospective fall incidence was followed up for 6–12 months, and the predictive ability of questionnaires and classic tests as well as gait characteristics (extracted from acceleration signals in free-living duration) were assessed. Gait characteristics include walking speed, stride time, stride length, gait intensity, symmetry, smoothness, complexity, and variability. Fall incidence was significantly associated with lower number of strides per day, lower total duration of daily locomotion, and a higher power in the dominant frequency in the mediolateral direction. Falls were also associated with the inability to use public transportation, using a walking aid, lower grip strength, higher fear of falling, and a higher depression score.

Using a pendant accelerometer sensor (Senior Mobility Monitor research prototype from Philips Research Europe, Netherlands) at their sternum, Brodie et al. [59] examined 14 weeks of walking data from 18 independent-living older adults (mean age = 83 years). Using a wavelet-based decision tree algorithm to identify consecutive heel strikes, or 1 gait cycle, the authors considered the following gait domains: 1) quantity (e.g., steps/day from walks ≥3 steps, steps/walk (mean of walks ≥8 steps), walks/day (of walks ≥8 steps), and longest walk); 2) daily-life cumulative exposure from ambulation (e.g., exposure from walks less than 7 s and less than 60 s); 3) intensity (e.g., cadence, vigor); and 4) gait quality (e.g., variability or standard deviation of step times). The results demonstrated that nonfallers completed more and longer walks compared to fallers. Furthermore, IMU data suggested that incorrect weight shifting during transfer movements is a major cause of falls. They concluded that using a freely worn device and wavelet-based analysis tools allows long-term monitoring of walks, and 1 week's monitoring is sufficient to reliably assess the long-term propensity for falling.

Pressure-sensitive insoles: There has been increasing interest in extending the use of pressure-sensitive insoles [60–65] (discussed in Section 28.3.2) to real-life situations (i.e., FLAs) in an unsupervised manner. Pressure-sensitive insoles (e.g., F-Scan system, Tekscan, Boston, MA)* are advantageous due to the ability to capture dynamic and interaction forces between foot and footwear on every step [66].

Recently, Howcroft et al. [60] proposed a combined insole- and accelerometer-based classification model of faller status in older adults. One hundred older individuals (age = 75.5 ± 6.7 years, 24% fallers based on 6-month retrospective fall occurrence) wore pressure-sensing insoles and 3-D accelerometers on the head, pelvis, and left and right shanks, while walking 7.62 m under single-task and dual-task conditions. Fall risk

---

* http://www.tekscan.com

classification models were assessed for all sensor combinations after extraction of temporal- and spectral-based features. After testing various machine learning techniques to classify fall occurrence, the best performance was 84% accuracy using a multilayer perceptron neural network with input parameters from pressure-sensing insoles and head, pelvis, and left shank accelerometers. While the older adults who participated in this study walked under the researchers' supervision and only for a short period of time, the results indicate potential to be extended to unsupervised situations.

In another study, 10 healthy community-dwelling elderly (mean age ≈ 70) wore an instrumented shoe consisting of 1) an IMU and 2) a pressure-sensing insole able to measure pressure under eight regions of the foot [65]. Participants wore the shoe over a 4-hour period to collect free-living data. Data captured from the IMU was used to detect walking and classify different locomotion types, including incline walking and stair climbing, and insole data was employed to distinguish sitting from standing. For gait analysis, spatiotemporal parameters including stride velocity, stride length, cadence, interstride gait cycle time variability, and foot clearance parameters [i.e., maximal heel clearance [HC] and minimum toe clearance (TC)] were extracted from locomotion periods with at least 20 steps. The findings revealed that heel and toe clearance were moderately and weakly correlated to stride velocity, respectively. The authors emphasized the importance of foot clearance metrics in providing new insights on a subject's performance in daily-life assessment, particularly in obstacle negotiation and fall avoidance.

Although many studies have explored the usability of the WSS in gait and fall risk assessment, a major limitations is inconsistency in defining bouts of walking. For instance, in Ref. [59], eight consecutive steps were determined to be a reliable bout of walking compared to five steps [46,47], 60 s duration [42], and 10 s [57]. In Section 28.4, we discuss a novel sensor-based ambulatory FRA method that does not rely on spatiotemporal gait parameters by proposing other markers to indicate risk for falling.

Wearable vision systems: As mentioned in Section 28.2, between 25% and 75% of falls in older people involve an environmental component (e.g., obstacles, terrain, lighting). The context of the physical environment is an important consideration when analyzing ambulatory gait metrics from the unsupervised home and community setting. The exposure to environmental hazards, such as curbs, carpets, and pets, may only be investigated in real-world situations. While SFRAs rely on controlling external conditions, current UFRA tools neglect measurement of extrinsic factors, which may explain why current clinical prediction models provide only poor to fair predictive ability [32]. Achieving contextual awareness during FLAs may offer a promising avenue to improve estimates of fall risk.

Likely attributable to technical barriers, there are few studies exploring methods to extract environmental risks in an unsupervised manner. Recently, advances in machine learning, state-of-the-art mobile vision systems, and wearable egocentric cameras (e.g., GoPro, Autographer) enable a new form of capturing the physical context of human experience. First-person perspective photos and videos captured by these cameras can provide rich and objective evidence of a person's everyday activities to capture contextual information of the physical environment. To the authors' knowledge, only a few studies in the literature have developed FRA methods combining analysis of gait and balance control behavior in an ambulatory fashion, while detecting the associated environmental hazards. Wearable vision systems could play an important role specifically in development of such novel FRA strategies.

In a related study, Robinovich et al. [67] employed fixed digital video cameras in common areas (i.e., dining rooms, lounges, hallways) to examine the circumstances of falls in two long-term care facilities in British Columbia, Canada. Capturing 227 fall videos from 130 individuals (mean age = 78 years), the researchers reviewed each fall video with a validated questionnaire that probed the cause of imbalance and activity at the time of falling. The most frequent causes of falling were incorrect weight shifting ($n = 93$), trip or stumble ($n = 48$), hit or bump ($n = 25$), loss of support ($n = 25$), and collapse ($n = 24$). Slipping accounted for only six falls. The three activities associated with the highest proportion of falls were forward walking (54 of 227 falls, 24.41%), standing quietly (29 falls, 13.41%), and sitting down (28 falls, 12.41%). While this study employed vision-based sensors to examine falls, poor resolution limited information about the associated movement behavior prior to and following the event.

Aiming to capture both movement behavior and associated environmental context, Taylor et al. [68] studied 12 participants (mean age = 70.9 years) with IMUs attached above the ankles (around both shanks) and an Autographer wearable camera (Autographer, Cambridge, UK) worn around the participants' necks. Metrics for each gait event were computed, and the Autographer images corresponding to gait events were manually annotated to obtain contextual information, such as walking outdoors on pavement, indoors on carpet, and indoors on polished or hardwood flooring. The results demonstrated that fallers spent significantly less time walking under regular conditions and outdoors and were slightly slower in general. However, larger group differences were observed when the participants were regrouped according to mobility levels determined from baseline assessments using traditional methods.

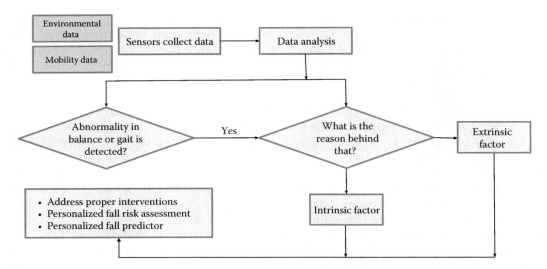

FIGURE 28.6
Proposed approach for fall risk assessment.

Although the aforementioned studies have investigated or employed new methods to measure gait characteristics in different environments, none have proposed *automated* methods of detecting environmental fall risk factors. These methods, employing either egocentric (i.e., patient-worn) or ambient (i.e., in-home) cameras, require intensive resources to manually annotate each video. While laborious, these studies form a foundation to pursue new opportunities for developing automated methods.

28.4 A New Approach for Unsupervised FRA

This chapter section describes a new FRA approach, adopting an unsupervised method using wearable IMUs and wearable egocentric cameras (see Figure 28.6). As discussed earlier, most studies exploring UFRA methods have focused on the evaluation of spatiotemporal parameters of gait to assess the risk for falls. Considering the heterogeneity in the cause and circumstances of falls, often a result of interactions between intrinsic factors (e.g., gait instability) and environmental circumstances (e.g., tripping obstacle) as described in Section 28.2, new approaches for assessing risk are warranted.

As an alternative to measuring spatiotemporal gait parameters, we consider examining compensatory balance responses (CBRs). Sometimes called near-falls, CBRs are defined as reactions to recover stability following a loss of balance, potentially resulting in a fall if sufficient recovery mechanisms are not activated [38] (Figure 28.7). Laboratory evidence supports the view that impaired ability to execute compensatory stepping reactions associated with aging and age-related pathology is a strong risk factor for falls [69], and early reviews of perturbation-based balance training to improve CBR performance in older adults suggest protective benefits. While one report estimates that CBRs are more frequent than falls [70], little is known about the prevalence and behavior of CBRs performed in everyday life activities. Our proposed approach is to explore whether the frequency and cause of CBRs in FLAs can provide new perspectives of fall risk [71]. Considering CBRs as predictors of potential falls, understanding of their prevalence and nature may provide a new lens to provide more detailed assessments of conditions leading to losses of balance for a specific individual.

However, the detection of CBRs during normal walking or FLAs is not an easy task. At the Neural and Rehabilitation Engineering (NRE) Laboratory at the University of Waterloo, we aim to develop new FRA tools to 1) detect and track the frequency of CBRs and 2) identify the extrinsic factors associated with the detected CBRs. As depicted in Figure 28.22, the long-term goal of this study is to develop new FRA and FP tools by automating detection of cause and frequency of CBRs (Figure 28.22), over a specific assessment period.*

28.4.1 System Infrastructure

A multimodal wearable/ambient sensor system platform was designed for the acquisition and analysis of wearable sensor data. This platform permits the collection of big data from a range of sensor streams (e.g., egocentric

---
* The study has received ethics clearance and was reviewed and approved by the Office of Research Ethics (ORE), University of Waterloo.

Multimodal Ambulatory Fall Risk Assessment in the Era of Big Data

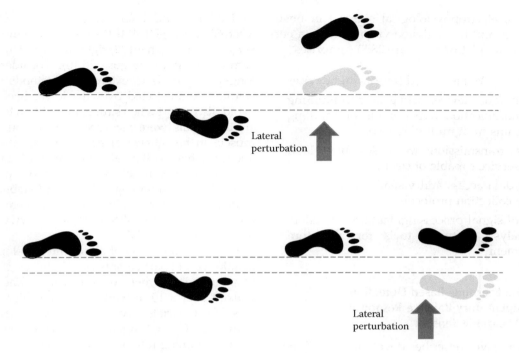

FIGURE 28.7
Types of compensatory balance responses (CBRs) to lateral perturbation: Upper panel: Right foot loaded at perturbation (arrow), eliciting a left foot side step (SS). Lower panel: Left foot loaded at perturbation (arrow), eliciting right foot crossover (CO) step.

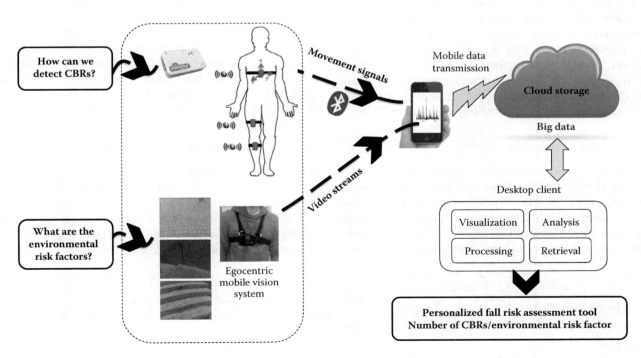

FIGURE 28.8
System infrastructure to acquire IMU + vision-based big data toward developing novel FRA methods.

cameras, IMUs, electrophysiological signals) in unsupervised settings and transmission via mobile data networks. The general platform (Figure 28.8) comprises

1. Bluetooth- or WiFi-enabled sensors and devices
2. A smartphone device, acting as a computing and communication hub capable of handling data streams from multiple sensors
3. Wireless transmission to a scalable cloud storage service capable of storing big data
4. A rapid data access and visualization client to facilitate collection protocols
5. A suite of signal processing, feature extraction, and analysis software tools for algorithm development

28.4.2 Machine Learning–Based Detection of Compensatory Balance Responses Using Wearable Sensors

In this section, we describe development of new methods to automatically identify lateral CBR episodes using wearable IMUs and sEMG sensors. While CBRs can occur in all directions, we initially focus on lateral balance recovery based on evidence indicating its importance in hip fractures [72]. Lateral CBRs include 1) the side step (SS) (or lateral CBR type 1) and 2) crossover (CO) step strategy (or lateral CBR type 2) (Figure 28.7). The proposed method is based on applying signal processing and machine learning methods on wearable IMUs and sEMG signals attached to the lower limbs to distinguish CBRs from normal walking (NW) patterns. After signal preprocessing (e.g., filtering), a pattern recognition design is described in three main phases: feature extraction, feature space dimensionality reduction (or feature selection), and applying supervised learning techniques (classification) (see Figure 28.9).

28.4.2.1 Data Acquisition

The WSS used to acquire movement data is the SHIMMER (Sensing Health with Intelligence, Modularity, Mobility, and Experimental Reusability, Shimmer, Ireland) wireless system. Each SHIMMER sensor comprises a battery, microprocessor, Bluetooth radio, 9-D IMU (3-D accelerometer, 3-D gyroscope, 3-D magnetometer), two channels of ExG [i.e., ECG (electrocardiogram), EMG], and a MicroSD card. SHIMMER sensors were mounted on the subjects' 1) right shank, 2) right thigh, and 3) sternum, and secured by straps (Figure 28.10). Considering that our long-term goal is to extend the methods to FLAs, the sternum was selected over the lower back (sacrum) to avoid discomfort when sitting. In addition to IMU signals, sEMG signals were recorded synchronously from four muscles in the right leg: 1) rectus femoris, 2) biceps femoris, and right shank—3) tibialis anterior and 4) gastrocnemius. These muscles were selected because they play an important role in maintaining balance and stabilization. Electrode placement and skin preparation were conducted in accordance with the SENIAM (http://www.seniam.org) recommendations [73].

For the initial phase of algorithm development, seven healthy young participants aged between 18 and 39 (two women and five men) were recruited. Each participant walked over a 10 m walking path at his/her preferred speed for 100 trials. Accelerometer ($n = 9$), gyroscope ($n = 9$), and sEMG ($n = 4$) signals were recorded synchronously with a sampling rate of 512 Hz and streamed wirelessly via Bluetooth to an Android mobile device (Nexus 7, Google Inc.) (Figure 28.10). In ≈60% of the trials, participants were randomly perturbed by lateral pushes to their right shoulder by a researcher walking alongside. Perturbation magnitude was calibrated to consistently elicit a CBR by slowly increasing push magnitude over a series of 5–10 training trials. Perturbation trials were labeled as CBR events in which subjects recovered their balance (and prevented a fall). Half of the perturbation trials (≈30% of total) were timed to be delivered during left leg swing to elicit an SS strategy or CBR type 1 (see Figure 28.10, right panel). The remaining perturbation trials were delivered during right swing eliciting a CO strategy or CBR type 2 (see Figure 28.7). Three acceleration and sEMG signals captured from one subject for 10 successive trials are depicted in Figures 28.11 and 28.12 respectively.

28.4.2.2 Data Processing and Analysis

Signal preprocessing: Signal preprocessing, including signal rectification, filtering, and normalization, plays a key role in detecting CBRs. In particular, corner frequency of the low-pass filtering for the sEMG-based approach is an important decision. While a corner frequency of 20 Hz is recommended for preprocessing of sEMG signals to

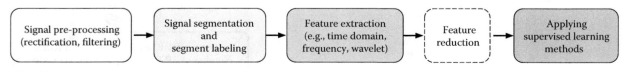

FIGURE 28.9
Machine learning approach for CBR detection. Applying feature reduction methods is not used; however, such techniques typically reduce processing time and may be beneficial for big data applications.

Multimodal Ambulatory Fall Risk Assessment in the Era of Big Data

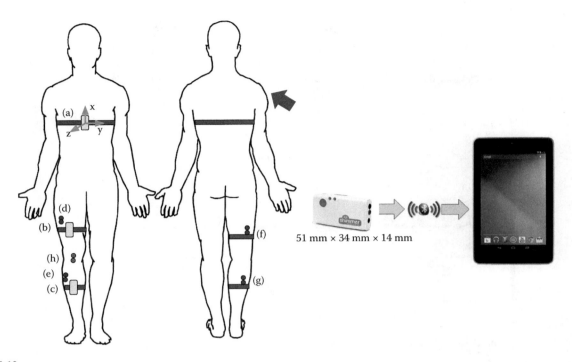

FIGURE 28.10
(a, b, and c) SHIMMER sensors on sternum, thigh, and shank. sEMG electrode placement for (d) rectus femoris, (e) tibialis anterior, (f) biceps femoris, and (g) gastrocnemius. (h) Electrodes on the patella were selected as the reference. Arrow indicates location of perturbation.

remove typical artifact sources [74], we examined alternate corner frequencies ($f_c \geq 20$) and found that a fifth-order digital Butterworth filter with high-pass frequency = 50 Hz and low-pass frequency = 200 Hz produced the most discriminative features to distinguish postural muscle activity associated with CBRs [75]. After full-wave rectification, the filter is applied on each sEMG signal and normalized to allow comparisons across subjects, and compensate for individual differences in strength, muscle tone, body fat, muscle geometry, etc. sEMG signals were normalized by dividing signal amplitudes by the RMS value of the filtered signal, recorded for each participant separately. Figure 28.12 (right) shows the sEMG signal before and after preprocessing.

Segmentation of signals: Segmentation refers to identifying a reference time to form a window, or segment, whose features are calculated. For each trial, 22 signals, including 9 acceleration signals, (3 sensors × 3 axes), 9 angular velocity signals, and 4 sEMG signals were collected. Our group examined several methods for IMU- and sEMG-based signal segmentation: 1) maximum peak sEMG, 2) maximum total acceleration of the sternum ($A_{T-Sternum}$; see Figure 28.13, Equation 28.1), and 3) maximum total sEMG (EMG_T) (Equation 28.2).

$$A_{T-Sternum} = \sqrt{A_x^2 + A_y^2 + A_z^2}, \quad (28.1)$$

where Ax, Ay, and Az are the accelerations along the x, y, and z axes, respectively (x is approximately vertical, y is approximately lateral, and z is approximately anteroposterior).

$$EMG_T = \sqrt{EMG_{TA}^2 + EMG_G^2 + EMG_{BF}^2 + EMG_{Rf}^2}, \quad (28.2)$$

where EMG_{TA}, EMG_G, EMG_{RFM}, and EMG_{BF} denote the amplitudes of the normalized, filtered sEMG signals captured from tibialis anterior, gastrocnemius, rectus femoris, and biceps femoris, respectively.

Our findings demonstrate that the maximum value of the total acceleration at the sternum $IMU_{A_{T-Sternum}}$ is reliable in representing the onset of a perturbation and compensatory reaction (Equation 28.1) [75].* To segment the data, the time index corresponding to the maximum A_T value was identified. For all acceleration (see Figure 28.13) and sEMG signals, 256 samples before and after of this point were used, corresponding to a time window of 513 samples to calculate features, and the rest of the data were ignored. For the angular velocity signals, time windows were time index corresponding to the maximum value of the angular velocity in x direction captured from the sternum IMU.

Feature extraction: Extraction of features from acquired IMU and sEMG signals is a critical step in our machine learning–based method for recognition of CBR patterns. Candidate features, described in the following sections, were drawn from previous studies investigating activity recognition and decision support problems.

---
* $A_{T-Sternum}$ outperforms others by resulting in higher detection accuracies [75,76].

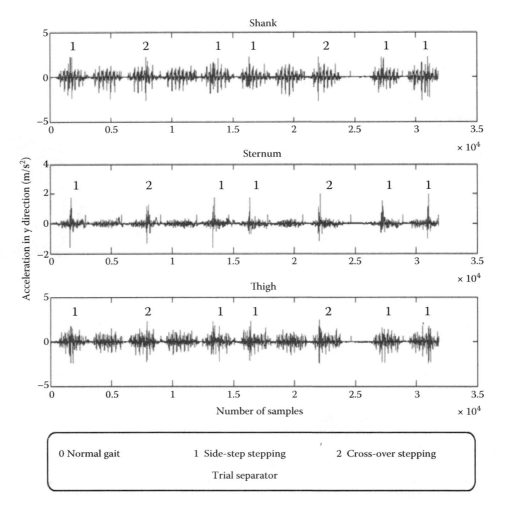

FIGURE 28.11
Sample acceleration signals in the y direction for 10 successive trials from shank, sternum, and thigh IMUs. Numbers 1 and 2 indicate elicited side step and crossover stepping strategies, respectively. The red markers separate the trials.

Acceleration-based features: The following features of acceleration signals were extracted for each segment of nine (3 sensors × 3 axes) captured signals: 1) RMS value of the acceleration signal, 2) variance of the acceleration signal, 3) range of the acceleration signal (i.e., maximum peak-to-peak signal value), and 4) the maximum absolute value of the acceleration signal, all reflecting limb movement intensity; 5) absolute value of the signal mean, an indicator of body segment orientation; 6) skewness and 7) kurtosis features to determine the shape and dynamics of the acceleration signal; 8) mean of the acceleration derivative, 9) variance of the acceleration derivative, reflecting higher order jerk measures, and 10) total acceleration (A_T) of shank, thigh, and sternum sensors over the full period of time for each trial, as an indicator of baseline activity.

Angular velocity–based features: The following features for each of the nine angular velocity signals (i.e., 3 sensors × 3 axes) were extracted from each segment: 1) mean absolute value (MAV), 2) variance, 3) maximum of absolute value, 4) maximum of the absolute value of the signal derivative, 5) mean of the absolute value of the signal derivative, and 6) variance of the signal derivative.

Surface electromyography–based features: For sEMG, the following time domain features (except entropy) are described by Phinyomark et al. [77]: 1) maximum peak, 2) RMS (signal power), 3) square integral, 4) integrated sEMG, 5) waveform length, 6) MAV, 7) modified MAV, 8) variance, 9) zero crossing (ZC), 10) slope sign changes, 11) V-order, 12) log-Detector (logDetect), and 13) (Shannon) entropy. Slope sign changes and ZC features were obtained from filtered signal, prior to rectification. Entropy was calculated using the wentropy function in MATLAB. Overall, we extracted 4 × 13 (number of muscles × number of features) to form a 52-dimensional feature space representing properties of muscle activities during regular or CBR stepping.

Multimodal Ambulatory Fall Risk Assessment in the Era of Big Data

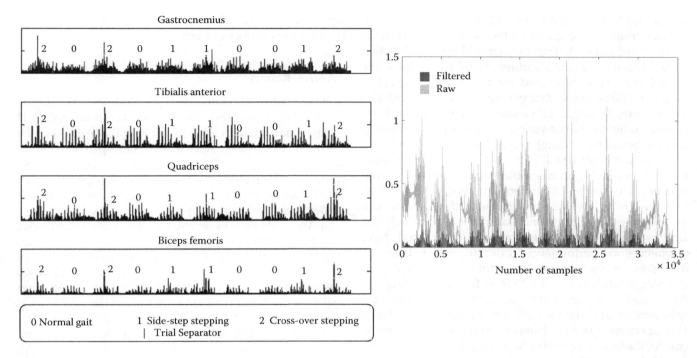

FIGURE 28.12
Left: Representative sEMG signals (rectified, filtered at 50–200 Hz) for 10 successive trials from four muscles. Numbers 0, 1, and 2 indicate normal walking (NW), elicited side step (SS), and crossover (CO) stepping trials, respectively. Right: Sample raw thigh sEMG signal for 10 trials and the resulting signal after detrending (removing DC value from signal), rectification, and filtering.

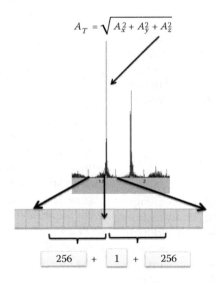

FIGURE 28.13
The procedure for signal segmentation. The sample corresponding to the maximum $A_{T-Sternum}$ is determined in each trial. The $A_{T-Sternum}$ is plotted for each data point captured from sternum accelerometer for three successive trials.

28.4.2.3 Machine Learning Techniques

Using the features extracted from IMU and sEMG data, this work investigated the application of two supervised machine learning methods, artificial neural networks (ANNs), and random forest (RF) classification approaches to classify each segment into one of three classes: NW, CBR type 1, and CBR type 2.

Artificial neural networks: In machine learning, an ANN is a model inspired by biological neural systems, which can be used to classify data sets or approximate unknown functions based on a large number of inputs. Feedforward networks consist of a series of layers: 1) the first layer, which has a connection from the network input; 2) the hidden layers, which are connected to their previous layers; and 3) the output layer, which indicates the resulting class number estimated by the network. A two-layer feedforward network with tan-sigmoid transfer function in the hidden layer, and a softmax transfer function in the output layer was developed for the current three-class problem. The acquired data was split 70%, 15%, and 15% for training, testing, and validation of the network (to examine whether the network is generalizing and also stop training before overfitting), respectively.

Random forest: RFs [78] are a powerful classification method, employing an ensemble learning method by constructing a collection of randomly trained decision trees. RFs have many advantages over standard classifiers, such as robustness against nonlinear relationships between features and higher generalization power, and have exhibited good performance in dealing with imbalanced data sets. In the training procedure, the RF starts by choosing a random subset (D') from the training data (D). At the node n, the entered training data (D_n) is iteratively split into left and right subsets (see Figure 28.14) (D_l and D_r, respectively), using a threshold (t) and split function ($SF(x_i)$), for the feature vector (x), using Equation 28.3, where $t \in (min(SF(x_i)), max(SF(x_i)))$ is randomly chosen by the $SF(x_i)$.

$$D_l = \{i \in D_n | SF(x_i) < t\}$$
$$D_r = \frac{D_n}{D_l} \quad (28.3)$$

Among several candidates that are randomly created by the $SF(v_i)$ and t at the split node, the candidate that maximizes the information gain (ΔE) about the corresponding node is selected. The ΔE is calculated by entropy estimation, according to Equation 28.4, where $E(D)$ denotes the Shannon entropy of the classes in the set of training [79].

$$\Delta E = -\frac{|D_l|}{|D_r|} E(D_l) - \frac{|D_r|}{|D_n|} E(D_r) \quad (28.4)$$

Considering that one of the most important challenges in big data analysis is to reduce processing/analysis time of such massive data, RFs include parallel and independent construction of the trees enabling parallel processing. The ability to process large training sets very rapidly is appropriate for efficient training and application of real-time detection algorithms. In RF_k (see Table 28.1), k denotes the number of trees.

28.4.2.4 Results

The training data points are (x_i^p, y_i), where $x_i \in \mathbb{R}_p$ (p is the number of features), and the class labels are $y_i \in \{0, 1\}$ for the binary classification problem (1 for all CBRs and 0 for the NW episodes). For the three-class problem, labels are defined as $y_i \in \{0, 1, 2\}$, where 0, 1, and 2 denote normal walking (NW, $n_{NW} = 218$), side step (SS, $n_{side\text{-}step} = 167$ trials), and crossover (CO, $n_{crossover} = 165$ trials), respectively.

In the reported experiments, the key performance measure is average accuracy after employing 100 times classification using RF (tenfold cross-validation: 10% of data points for testing and 90% for training) and ANN methods. Table 28.1 indicates the accuracy results after applying machine learning techniques on 1) an IMU-based data matrix (138 × N), 2) an sEMG-based data matrix (52 × N), and 3) a combination of IMU- and sEMG-based feature matrix (190 × N), where N = 550 is the number of trials. All sEMG features were extracted at 50–200 Hz, and $SVA_{Sternum}$ method was used for signal segmentation.

In comparison, using only the sEMG-based method was less accurate than IMU-only classification. For binary and three-class classification, the optimal IMU-based results (96.66% for binary and 89.72% for three-class classification) outperformed the best sEMG-based results (91.40% for binary classification and 82.63% for three-class classification). While the IMU-only results were promising, there were difficulties in distinguishing SS responses from NW, largely attributable to the lack of motion in the instrumented leg, which provided support only. The use of sEMG signals was expected to capture muscle activity associated with postural muscle responses without concomitant movement (apparent to IMU sensors). The combined IMU + sEMG approach resulted in a slightly higher accuracies compared to the IMU-based approach only. The most accurate method achieved maximum mean accuracy of 97.29% and 91.02% for binary classification and three-class classification, respectively, using RF_{70} method.

Because the initial set of features is quite large (190 total) and not all of the features are equally useful, feature

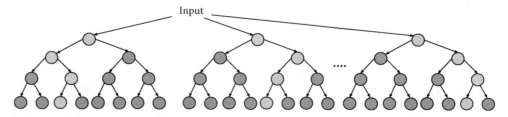

FIGURE 28.14
Algorithm of the random forest classification method.

TABLE 28.1

Results for Binary and Three-Class Classification (Including Normal Gait, CBR Type 1, and CBR Type 2) Using RF_k (Where k Denotes the Number of Trees) and ANN Methods

Features	$N_{features}$	Binary Classification		Three-Class Classification	
		ANN	RF_{30}	ANN	RF_{40}
IMU	138	93.63	96.66	89.31	89.72
EMG	52	86.63	91.40	81.39	82.63
IMU + EMG	190	89.09	96.82 (97.29RF_{70})	87.36	90.53 (91.02RF_{70})
IMU + EMG	$P = 20$	90.36	93.25	84.00	86.55
IMU + EMG	$P = 40$	91.27	94.07	80.81	88.02

Note: P refers to the number of features after employing feature selection criteria (Eq. 28.5). The point corresponding to maximum $SVA_{Sternum}$ in each trial was used for the signal segmentation.

reduction or feature selection methods may be applied. Furthermore, feature reduction can be used to reduce processing time toward a robust real-time predictor.

One method to determine the best indicators of CBR incidents is to estimate the correlation A between the features and labels from Equation 28.5.

$$A_j = |X^j \times Y|$$
$$Sort\left(A = \{A_j\}_{j=1}^{190}\right) \quad (28.5)$$

where $X = \{x_1, x_2, \cdots, x_n\}$ is the normalized data matrix (all values are between −1 and 1) where N is the number of samples, X^j is a row vector that includes the jth feature of all samples so $j = \{1, \cdots, 190\}$, and Y is the vector of labels. Based on linear correlation, the best 20 and 40 indicators (features) of CBRs were passed for the learning process. The last two rows of Table 28.1 depict the classification results after feature reduction.

28.4.3 Wearable Vision Detection of Environmental Fall Risks Using Machine Learning Techniques

While it is well recognized that hazards in the home and environment often contribute to falls and related injuries, there remains a distinct lack of sensor-based methods to identify individual risks. One of the biggest challenges for development of a fall prediction and prevention method is a lack of automated techniques to extract contextual information needed to interpret unsupervised gait and postural behavior. For example, a specific pattern in IMU signals or sEMG signals can be interpreted as gait instability or a compensatory balance reaction, or may reflect anticipatory adjustments to avoid collisions on a crowded sidewalk. Without detailed information of the mobility context, such as the presence of other pedestrians, terrain characteristics, and obstacles, the ability to interpret ambulatory gait data is constrained.

In combination with the CBR detection aim, our group is developing egocentric wearable vision systems (discussed in Section 28.3.3) to identify key environmental factors. Other techniques using egocentric cameras rely on manual identification of environmental circumstances, which are inefficient and impractical for real-time use [68,80,81]. This section examines the potential for wearable egocentric cameras, combined with image processing and machine learning techniques to automatically detect fall risk hazards. Two approaches are developed and evaluated using: 1) a Gabor filter feature extraction coupled with machine learning and 2) a convolutional neural network (CNN) approach. Specifically, we explore a classification problem to detect 17 different environmental conditions that may influence fall risk (e.g., curbs, stairs, rocks) [82,83].

28.4.3.1 Data Acquisition

A commercially available wearable egocentric camera, the GoPro Hero 4 Session (GoPro, USA), was used to collect video data. The camera can record high-resolution 1080p60 videos and 12MP photos up to 30 frames per second, is waterproof to 40 m, and features wireless streaming capabilities. The camera was attached to participants' chests and angled downward to sufficiently capture the immediate spatial environment within ≈2 steps (see Figure 28.15). Three healthy, young participants were asked to walk around the University of Waterloo campus across a wide variety of indoor and outdoor areas for 10–15 minutes. Overall, 22,001 video frames were collected (see Figure 28.16). To provide training data, each frame was manually annotated with the following labels: 1, crosswalk; 2, curbs; 3, ramp; 4, stairs (ascending); 5, stairs (descending); 6, gravel; 7, grass; 8, concrete; 9, tiles; 10, bricks; 11, carpets; 12, dirt; 13, water; 14, snow; 15, slush; 16, ice; and 17, rocks.

28.4.3.2 Preprocessing

To facilitate feature extraction and machine learning steps, each frame was 1) converted to a grayscale image, 2) resized into 128 × 128 images (Figure 28.17b), and 3) cropped by a 64 × 64 rectangular window (Algorithm 1) (Figure 28.17c). Cropping of frames was used to remove participants' torso, legs, and feet appearing in the camera's field of view. The parameters in the cropping (Algorithm 1) were achieved empirically upon visual inspection and based on effects on classification results. For initial development purposes, a subset of available data ($n = 10{,}430$ frames) was generated with a fewer number of classes (12 classes of environmental characteristics, as shown in Table 28.2). A larger data set was generated using all available video data concatenated in a single matrix with 12,382 data points (shown in Table 28.3). To alleviate the effect of imbalanced data, some frames from "concrete" and "brick" classes were randomly removed to achieve more reasonable/balanced distribution of data points. Both data sets were used to train and evaluate the two approaches explored in this work (Gabor filters and CNNs, described in the following sections). The larger data set was used to test the Gabor barcodes (GBC) approach.

FIGURE 28.15
GoPro Hero Session camera and chest harness. The camera is inclined facing down to capture terrain features.

Algorithm 1 Frame preprocessing
1: Initialize $a = b \leftarrow 128$
2: $I_{gray} \leftarrow I_{rgb}$
3: $I = \text{Normalize}(I_{gray}, a, b)$
4: Initialize $x_{min} \leftarrow 32$, $y_{min} \leftarrow 0$, $W_x = W_y \leftarrow 64$
5: $I = \text{Crop}(I, [x_{min}, y_{min}, W_x, W_y])$

28.4.3.3 Gabor Barcodes Approach

Automatic terrain type detection is a challenging problem that has attracted attention of researchers from different fields, especially robotics for precise velocity control, gait adaption of legged robots, and safe navigation. Moreover, terrain detection has applications in teleoperation during critical missions, such as urban search and rescue and bomb disposal [84–87]. A major challenge associated with terrain detection methods is the presence of unwanted/dynamic objects (i.e., noise) in real images or videos. Such objects considerably affect the accuracy of the classifier, highlighting the need for development of a classifier with high generalizability. Moreover, computational complexity plays an important role in efficiency of a real-time detection system. These methods primarily employed image processing–based methods (e.g., Hough transform, SIFT (scale-invariant feature transform), and SURF (speed up robust features) for feature extraction, which require high computational resources.

Due to the specific characteristics of our images, texture seems to be an appropriate feature for describing their contents (e.g., brick, tiles, rocks, carpet) [79]. Texture analysis has been an active research area, and numerous algorithms have been proposed based on different models, for example, gray-level co-occurrence (GLC) matrices and Markov random field (MRF) model [79,88]. In recent works, wavelets have become very popular due to their capacity to provide multiresolution analysis. In particular, the Gabor transform has mathematical and biological

FIGURE 28.16
Sample resized and cropped frames that were captured by a GoPro Session camera while walking around the University of Waterloo campus. While the gray images were used for feature extraction, for a better visualization, the color (RGB) frames are shown.

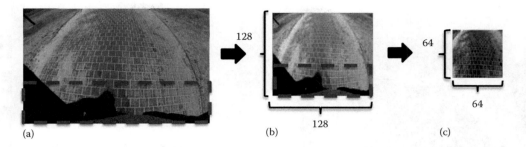

FIGURE 28.17
Preprocessing steps. From left: (a) original RGB image, (b) gray, resized image, and (c) cropped, resized image.

TABLE 28.2
Frames Extracted for 12 Classes (Video$_1$)

Crosswalk	Curb	Ramp	Stairs (asc)	Stairs (desc)	Gravel
103	78	436	81	208	253

Carpet	Water	Rock	Concrete	Tile	Brick
736	25	63	499	96	991

TABLE 28.3
Frames Extracted for 17 Classes from All Videos after Rebalancing (Video$_1$ + Video$_2$ + Video$_3$)

Crosswalk	Curb	Ramp	Stairs (asc)	Stairs (desc)	Gravel	Grass	Concrete	Tile
103	78	436	254	208	253	910	10,118 (initial)	952

Brick	Carpet	Dirt	Water	Snow	Slush	Ice	Rock
4782	2369	189	142	746	315	83	63

properties resembling the characteristics of human visual cortical cells, such as extracting texture features from images for segmentation, object detection, and biometric identification applications.

The most important property of Gabor features is their robustness against rotation, scale, and translation. Furthermore, they are robust against photometric disturbances, such as illumination changes and noise. These properties are mainly due to the fact that the parameters of Gabor filters enable us to establish invariance in this regard [89]. In the spatial domain, a two-dimensional Gabor filter is a Gaussian function, modulated by an exponential or complex sinusoidal plane wave, defined as

$$G(x,y) = \frac{f^2}{\pi \gamma \eta} \exp\left(-\frac{x'^2 + \gamma y'^2}{2\sigma^2}\right) \exp(j2\pi f x' + \phi), \quad (28.6)$$

where $x' = x \cos \theta + y \sin \theta$, $y' = x \sin \theta + y \cos \theta$, f is the frequency of the sinusoid (modulation frequency), θ represents the orientation of the normal to the parallel stripes of a Gabor function, ϕ is the phase offset, σ is the standard deviation of the Gaussian envelope, and γ is the spatial aspect ratio that specifies the ellipticity of the support of the Gabor function [90]. Given an image $I(x, y)$, the response of Gabor filter is the convolution of Gabor window with the image I given by

$$\psi_{u,v}(x,y) = \sum_s \sum_t I(x-s, y-t) * G_{u,v}(s,t), \quad (28.7)$$

where s and t are the window/mask size of the Gabor filter, u is the number of scales, and v is the number of orientations that are used in the Gabor filter bank [GFB (u, v, s, t)] (see Figure 28.18). The $\psi_{u,v}(x, y)$ forms a complex valued function including real and imaginary parts. In this study, in order to obtain Gabor features, the magnitudes of the $\psi_{u,v}$ values ($\psi_{ABS-u,v}$) are calculated. There have been several studies in the literature reporting the optimal values for the parameters of the Gabor filter bank (i.e., spatial frequencies and number of orientations) in such a way that it can effectively mimic the human visual system [91].

More recently, fast feature extraction methods from images or query search systems have used binary features (e.g., binary hashing), improving efficiency in terms of detection speed and lower storage space requirements.

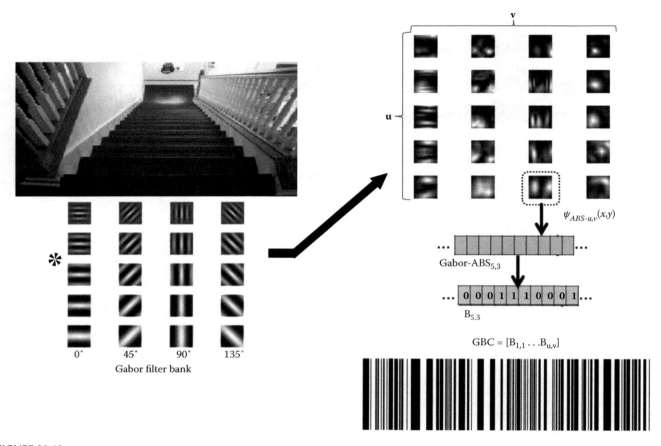

FIGURE 28.18
Generating Gabor barcodes (GBCs) from preprocessed video frame (left), including binarizing of Gabor feature vectors and appending the binary features (B's) afterward (right).

Inspired by Radon barcodes, the concept of "barcode" annotation to generate an efficient content-based binary tag has been proposed [92]. This idea is extended to generate texture-based barcodes: GBCs (Algorithm 2). The performance of GBCs was initially tested for content-based medical x-ray image classification and retrieval with 193 classes.* A total error score as low as ≈80% accuracy for the first hit was achieved [93]. Based on primary examination of the video data, a GFB comprising filters with $s = t = 11$ or 11×11 mask size, with $u = 5$ scales and $v = 8$ orientations, resulted in a good overall accuracy. The GFBs were applied to 32×32 grayscale images on both the smaller data set (Table 28.2) with 12 classes and 3669 images for initial testing, and the larger data set with 12,382 frames with 17 classes. Average processing time for generation of GBCs from an image in our database was 0.0716 s [applying GFB(5,8,11,11)], using a laptop computer with an Intel i7-3.60 GHz processor, which remarkably outperformed other image processing–based methods for feature extraction like SIFT and SURF in terms of computational complexity.

Algorithm 2 Generation of GBCs (inspired by Ref. [92])

1: Initialize Gabor barcode for image I_m: $GBC_m \leftarrow \emptyset$
2: Initialize $R_N = C_N \leftarrow 32$
3: I Normalize(I, R_N, C_N)
4: Apply Gabor filters with u scales v orientations
5: **for** all u and v **do**
6: Calculate the magnitude of $\psi_{u,v}(x, y): \psi ABS_{-u,v}(x,y) = |\psi_{u,v}(x,y)|$
7: Downsample each $\psi_{u,v}(x, y)$ with factor of 4
8: Generate row feature vectors Gabor-$ABS_{u,v,m}$
9: Typical$_{u,v,m} \leftarrow$ median(Gabor-$ABS_{u,v,m}$)
10: $B_{u,v,m}$ Gabor-$ABS_{u,v,m} \geq$ Typical$_{u,v,m}$
11: $GBC_m \leftarrow$ append(GBC_m, $B_{u,v,m}$)
12: **end for**

The resulting GBCs (see Figure 28.19 for examples) were used as features to train machine learning classifiers previously discussed in Section 28.4.2. Since GBCs are binary features, Hamming distance between the data

---

* Comprising 12,677 x-ray images for training and 1,733 x-ray images for testing.

points is readily calculated as a distance metric for a $k-$NN ($k=1$) classifier. For each of the test images, a complete search is performed to find the most similar image where similarity of an input image I_i^{query} with the corresponding barcode B_i^{query} is calculated based on Hamming distance to any other image I_j with its annotated barcode B_j:

$$\max_{j=1,2,3,\ldots,N_{test}; j \neq i} \left(1 - \frac{\left|XOR\left(B_i^{query}, B_j\right)\right|}{B_i^{query}}\right), \quad (28.8)$$

where B can be GBC.

For the $k-$NN method, 80% of the data matrix ($128{,}382 \times N_{class}$, $N_{class} \in \{12, 17\}$ of GBCs) was chosen randomly for training, and the remaining 20% was used to test classifier performance. Similarly, GBC features were used to train ANNs using 20, 30, and 40 neurons in the hidden layer, and 12 and 17 neurons in the output layer for the first and second database, respectively. In this case 70%, 15%, and 15% of data was randomly selected for training, validation, and testing of the network, respectively.

Tables 28.4 and 28.5 depict the mean accuracies after 10 times training, using RF and ANN methods (which were discussed in Section 28.4.2), and also $k-$NN, for 12 classes (only video 1) and 17 classes (three videos, V_{1+2+3}) respectively.

The $k-$NN classifier is a representative of the "lazy classifiers" group, which employs a set of training examples to choose the closest exemplars using a predefined metric, such as Euclidean distance (Equation 28.9). The output label is the mode of the labels for the k data points. While relatively simple to use, this method is heavily dependent on the structure of the training data.

$$d_{Euclidean} = \sqrt{\sum_{j=1}^{N}\left(x_i - x_j\right)^2} \quad (28.9)$$

Due to the binary nature of the GBCs, Euclidean distance (Equation 28.9) turns to the Hamming distance (Equation 28.8). In RF_n, $k-NN_n$, and ANN_n, n denotes the number of decision trees, closest neighbors, and neurons in the hidden layer, respectively. Using a combination of GBCs and Hamming distance (Equation 28.8) for identifying the class number for a test image resulted in the highest accuracy in both the 12-class (95.82%) and larger (88.51%) data sets. Considering that the random selection accuracy is 8.3% and 5.9%> for the 12- and 17-class problem, the accuracies achieved were very strong.

28.4.3.4 Convolutional Neural Networks

One of the most promising approaches for automated classification of images and videos is CNNs. Originally proposed by LeCun [94], CNN is a biologically inspired variant of ANNs with three key architectural ideas: 1) local receptive fields, 2) weight sharing, and 3) sub-sampling (i.e., max pooling). Images are inputs to the convolution layer, which extracts structure features of the input. For example, the first convolution layer extracts low-level features like edges, lines, and corners. Higher-level layers build on these lower-level features to extract more complex structures (e.g., shapes).

CNNs possess several useful properties. First, feature extraction and classification are integrated into one structure compared to traditional pattern recognition models that hand-design features. Moreover, they relatively invariant to geometric, local distortions in the image and have been shown to be invariant to pose, lighting, and surrounding clutter [95]. Finally, the network structures of CNNs lead to savings in memory requirements and computation complexity requirements while providing better performance for applications where the input has local correlation (e.g., image and speech).

CNN-based methods have been used to automatically extract information from videos (e.g., detecting human

FIGURE 28.19
Sample GBCs extracted for 1) pavement, 2) carpet, and 3) tiles.

TABLE 28.4

Mean Accuracies after 10 Times Training, Using Video$_1$ Database (3669 Data Points, 12 Classes)

$k - NN_1$	$k - NN_5$	$k - NN_15$	RF_{10}	RF_{20}	RF_{30}	ANN_{15}	ANN_{20}	ANN_{40}
95.82	92.86	81.01	88.32	92.67	93.07	91.92	92.72	93.89

Note: In RF_n, $k - NN_n$, and ANN_n, n denotes the number of decision trees, closest neighbors, and neurons in the hidden layer, respectively.

TABLE 28.5

Mean Accuracies after 10 Times Training, Using Video$_{1+2+3}$ Database (12,382 Data Points, 17 Classes)

$k - NN_1$	$k - NN_5$	RF_{10}	RF_{20}	RF_{30}	ANN_{30}	ANN_{40}	ANN_{50}
88.51	78.97	78.08	79.81	81.29	79.25	82.26	82.45

Note: In RF_n, $k - NN_n$, and ANN_n, n denotes the number of decision trees, closest neighbors, and neurons in the hidden layer, respectively.

actions [96,97], object and scene detection in autonomous driving systems). A recent CNN application aimed to detect 35 different indoor places using a GoPro wearable camera [98], primarily for navigation tasks. In comparison to the current study, the aforementioned approach was not designed to capture short-range details (e.g., stairs, crosswalks). By combining the concepts of image pyramids and local receptive fields, Phung et al. [99] proposed the pyramidal neural network (PyraNet) for classification of gender from a facial image. In PyraNet, a pyramidal layer consists of neurons arranged in a 2-D array with each neuron connected to a specific rectangular region (i.e., the receptive field) in the previous layer. The first pyramidal layer is connected to the input image, followed by one or more pyramidal layers, with the last pyramidal layer connected to 1-D layers. A 2-D neuron computes a weighted sum of inputs from its receptive field and applies a nonlinear activation function (e.g., sigmoid) to produce an output signal. The 1-D feedforward layers, which may include several layers depending on image complexity, process the features produced by the pyramidal layers. The outputs of the last 1-D layer are the final network outputs, which determine the categories/class labels of input patterns.

To train the CNN, a network with three convolutional layers and two subsampling (max-pooling) layers, followed by a fully connected multilayer perceptron (FCMLP) with 12 neurons in the output (corresponding to the number of classes [83]), was used, as shown in Figure 28.20. The filter sizes are 5 × 5, 2 × 2, 3 × 3, 2 × 2, and 6 × 6 for the C1, S2, C3, S4, and C5 layers respectively. We employed 64-, 32-, and 16-filters (resulting in similar feature maps depth) for the C1, C3, and C5 layers, respectively, with connections between each feature map in the convolution layer and its adjacent subsampling layer being one to one. There is full connection between F and C5, as shown in Figure 28.20. Activation functions for network layers are set as 1) convolution layers C1, C3, and C5: tansig; 2) subsampling layers S2 and S4: purelin; and 3) output layer F: tansig.

The resilient backpropagation (RPROP) training method for the network was chosen due to its speed among the first-order training algorithms. Weight update depends on the sign of the gradient, defined by $\Delta w_i(t) = -\text{sign}\left\{\frac{\partial E}{\partial w_i} t \times \Delta_i(t)\right\}$, where $\Delta_i(t)$ is an adaptive step specific to weight w_i defined as below ($\eta_{inc} > 1$ and $\eta_{dec} < 1$ are scalars).

$$\Delta_i(t) = \begin{cases} \eta_{inc}\Delta_i(t-1), & \text{if } \frac{\partial E}{\partial w_i}(t) \times \frac{\partial E}{\partial w_i}(t-1) > 0 \\ \eta_{dec}\Delta_i(t-1), & \text{if } \frac{\partial E}{\partial w_i}(t) \times \frac{\partial E}{\partial w_i}(t-1) < 0 \\ \Delta_i(t-1), & \text{otherwise} \end{cases}$$

The goal is to minimize the mean square error (MSE),

$$E_{\text{MSE}} = \frac{1}{K \times N_L} \sum_{K=1}^{K} \sum_{n=1}^{N_L} \left| y_n^{L,k} - d_n^K k \right|^2,$$ during the process of training. Figure 28.21 plots the resulting decreasing training error (MSE) over 100 epochs. After ≈50 epochs, the MSE error clearly converges to ≈0.08, roughly interpreted as an accuracy of 92%. In advance of an in-depth evaluation, the low MSE indicates the efficiency of the trained network in detection of objects/terrains.

28.5 How Big Data Could Help Us Develop Better FRA and FP Tools

While FP efforts remain a major challenge, the methods discussed in this chapter suggest that the capabilities to collect, store, and analyze big data from FLAs is within

Multimodal Ambulatory Fall Risk Assessment in the Era of Big Data 573

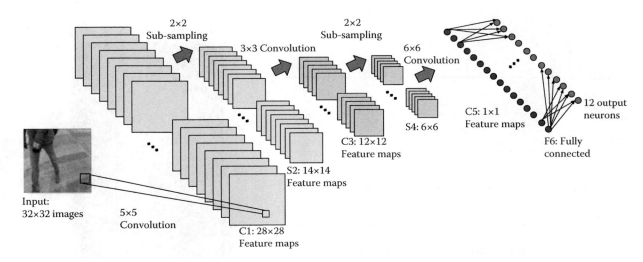

FIGURE 28.20
The developed CNN to detect 12 different fall-related environmental risks.

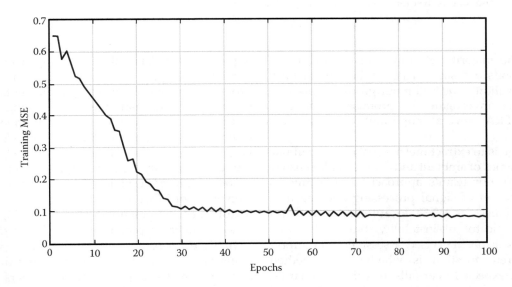

FIGURE 28.21
CNN training MSE over 100 epochs for detecting 12 different classes of environmental fall-related risk factors.

reach. For instance, the Michael J. Fox Foundation for Parkinson's Research (MJFF) and Intel Corporation initiated a collaboration to create a test group of 10,000 individuals with Parkinson's disease to track symptoms such as tremor, sleep patterns, gait, and balance using WSSs. Initial estimates considering 300 measurements per second for each person will result in a data set of 9.7 TB of data per day. Using Intel's big data analytics platform, this big data set will be used to develop new models of Parkinson's disease, particularly disruptive symptoms and events (e.g., freezing of gait). By detecting anomalies and changes in sensor and other data, the analytics platform can provide researchers with new means to measure and predict disease progression. Our team proposes a new perspective on FRA based on automated detection of frequency and environmental context of CBRs. As illustrated in Figure 28.22, unsupervised monitoring of the frequency and circumstances of CBR events may provide unique insight on individual fall risks specific to a person's own lifestyle and surroundings. For example, the imagined CBR data shown in Figure 28.22 (left panel) illustrates a potential analysis of instability requiring a recovery reaction categorized by environmental context. In this (imagined) case, observing the relatively high frequency of CBRs detected on stairs ($n = 7$), terrain transitions ($n = 9$), and thresholds ($n = 5$), may be used to prioritize intervention strategies, such as training to negotiate stairs and transitions.

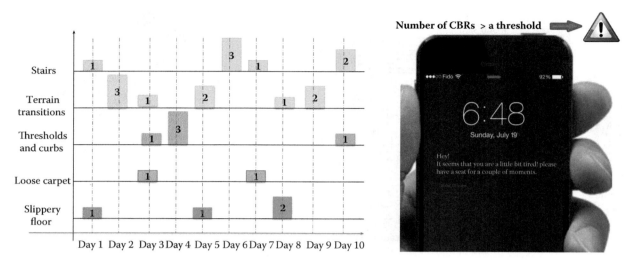

FIGURE 28.22
Potential applications of automated unsupervised FRA tools. Left: Envisioned analysis of frequency of CBRs by day and environmental factor. Right: Mobile alert to user or caregiver for circumstances of high fall risk.

However, the majority of studies examining unsupervised methods are small-scale studies to establish measurement validity, including our group's work presented here. To investigate the promise of big data methods for FRA, several important challenges are identified.

One challenge is to extract metrics sensitive to fall risk from large volumes of ambient and/or wearable sensors. As an example, our team's approach demonstrates a suitable application of signal processing and machine learning techniques for WSSs to automatically detect CBRs as an indicator of instability. Similar to other research efforts in unsupervised assessment, an important unanswered question is whether this method translates to unexpected near-falls in real-life scenarios. Our immediate efforts are focused on validation and optimizing these methods for unsupervised conditions.

Identifying contextual factors contributing to fall risk using ambient or wearable sensors is a largely unexplored area of research, likely attributable to a lack of efficient methods. In this chapter, image processing (i.e., Gabor filters) and machine learning (i.e., CNNs) techniques were explored to identify different terrain characteristics with promising initial results (>90% accuracy). Deeper examination is needed to determine whether the system is robust under more complex situations, such as unusual or unfamiliar objects that challenge generalization capabilities of the detection system, varying lighting conditions and shadows, and simultaneous presence of multiple risk factors.

To achieve large-scale studies, new methods to address limitations in computational speeds, storage capacities, and annotation methods are needed. New data processing methods (e.g., parallel processing, feature reduction) to reduce complexity and minimize storage needs are promising avenues to facilitate implementation at larger scales. Similar to many machine learning projects, correctly annotated data is critical for effective training and testing. Systematic and efficient methods to annotate large amounts of mobility- and image-based data to reduce uncertainty (i.e., veracity) remain a challenge.

To the authors' best knowledge, Philips et al. [39] at the University of Missouri are a pioneering group that has investigated the potential of analyzing big data (66 TB collected over 10 years) for FRA. Using data acquired from Kinect sensor systems installed in independent-living apartments, this study examined the association between prefall changes in gait parameters (i.e., in-home gait speed, stride length) and fall incidents with the goal of improving fall risk prediction models. Preliminary results indicate that a change in gait speed (5.1 cm/s) over time (7 days) is associated with a high probability of falling (86.3%) within the next 3 weeks (compared to 19.5%> fall probability for those with no change in gait speed). Similarly, a cumulative decrease in stride length (7.6 cm) over 7 days in in-home stride length is associated with a 50.6% probability of falling within the next 3 weeks (vs. 11.4% for those with no change). While these results indicate the potential of big data to inform the development of new models of estimating fall risk, the utility of these methods to inform interventions remains to be examined.

The aforementioned paper employed a descriptive analysis using *batch data analytics*, in which a high volume of data is collected over a relatively long period of time and then processed and analyzed afterward. While such

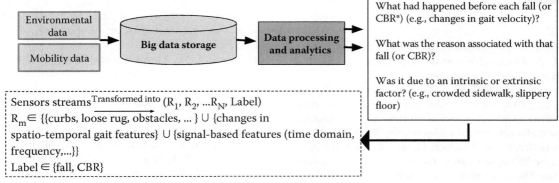

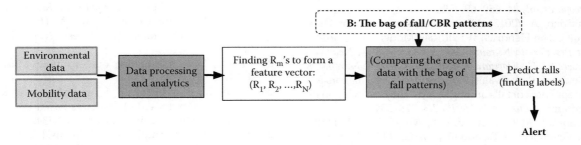

FIGURE 28.23
Upper panel: Descriptive analysis, in which batch data analytics provides correlational models of gait-related parameters and prospective/retrospective falls. Lower panel: Predictive analysis, in which the ongoing data stream is automatically assessed in real time to detect periods of high fall risk and send alerts. A variety of statistical, modeling, data mining, and machine learning techniques are used to study recent and historical data.

descriptive analysis and batch processing uncover hidden patterns and unknown correlations, real-time analytics may be preferable to identify and mitigate transient periods of high fall risk. In contrast to batch data analytics, real-time analytics processes continuous data streams as inputs with strict time constraints for analysis and decision-making outcomes (Figure 28.23). A real-time analytic platform would permit identification of periods of high fall risk for immediate interventions, such as triggering user and/or caregiver alerts (illustrated in Figure 28.22, right), assistive technologies, and/or environmental controls (e.g., lighting).

28.6 Conclusion

This chapter examines current FRA practices, highlighting key limitations, and summarizes recent research in emerging sensor systems to improve quality and frequency of quantitative assessments. With continued advancement of these sensor systems, particularly with an emphasis on unsupervised assessment, the potential to apply big data methods toward realizing the vision of comprehensive individual assessment of fall risk is apparent.

In SFRA, mounting evidence supports the utility of sensor-based systems as measurement tools to improve assessment sensitivity and reliability over traditional questionnaire-based or functional performance testing. However, a fundamental limitation of SFRA methods is the need for tightly controlled task and environmental conditions. Considering that fall risk is frequently influenced by multiple interacting factors, including the individual's own environment, lifestyle, and behavior, SFRA methods may not capture the full range of risks to inform clinical decisions.

A promising avenue to advance FRA is to monitor behavior from everyday life in an unsupervised manner and examining the emergent risks. Rapidly developing ambient and wearable sensor systems capable of collecting gait and balance behavior with high resolution over long periods of time are generating new opportunities for UFRA. Importantly, emerging big data methods will be critical to process and analyze the wave of information and advance prediction of fall risk.

References

1. World Health Organization. Ageing, & Life Course Unit. (2008). WHO global report on falls prevention in older age. *World Health Organization*.
2. Hawley-Hague, H., Boulton, E., Hall, A., Pfeiffer, K., and Todd, C. (2014). Older adults' perceptions of technologies aimed at falls prevention, detection or monitoring: A systematic review. *Int J Med Inform*, 83(6), 416–426.
3. Pfortmueller, C. A., Lindner, G., & Exadaktylos, A. K. (2014). Reducing fall risk in the elderly: risk factors and fall prevention, a systematic review. *Minerva Med*, 105 (4), 275–281.
4. M. Nikitovic, W. P. Wodchis, M. D. Krahn, and S. M. Cadarette. (2013). Direct health-care costs attributed to hip fractures among seniors: A matched cohort study. *Osteoporos Int*, 24, 659–69, Feb.
5. Hendrich, A. (2013). "Fall Risk Assessment for Older Adults: The Hendrich II Fall Risk ModelTM." *Try This: Best Practices to Nursing Care of Older Adults. New York University, The Hartford Institute for, Geriatric Nursing*, 8.
6. Stalenhoef, P. A. et al. (1997). Incidence, risk factors and consequences of falls among elderly subjects living in the community. *Eur J Publ Health* 7.3, 328–334.
7. Hornbrook, M. C., Stevens, V. J., Wingfield, D. J., Hollis, J. F., Greenlick, M. R., and Ory, M. G. (1994). Preventing falls among community-dwelling older persons: Results from a randomized trial. *The Gerontologist*, 34(1), 16–23.
8. Kronfol, N. (2012). *Biological, medical and behavioral risk factors on falls*. World Health Organisation. Available at http://www.who.int/ageing/project/falls_prevention_older_age/en/inded.html.
9. Tinetti, M. E., Speechley, M., and Ginter, S. F. (1988). Risk factors for falls among elderly persons living in the community. *N Engl J Med*, 319.26, 1701–1707.
10. Rubenstein, L. Z., Robbins, A. S., Schulman, B. L., Rosado, J., Osterweil, D., and Josephson, K. R. (1988) Falls and instability in the elderly. *J Am Geriatr Soc*, 36(3), 266–278.
11. Di Pilla, S. (2004). *Slip and Fall Prevention: A Practical Handbook*. CRC Press, Boca Raton, FL.
12. Lord, S. R., Sherrington, C., Menz, H. B., & Close, J. C. (2007). *Falls in Older People: Risk Factors and Strategies for Prevention*. Cambridge University Press.
13. Connell, B. R. (1996). Role of the environment in falls prevention. *Clin Geriatr Med*, 12, 859–880.
14. Perell, K. L. et al. (2001). Fall risk assessment measures an analytic review. *J Gerontol A Biol Sci Med Sci*, 56.12, M761–M766.
15. Rubenstein, L. V., Josephson, K. R., and Osterweil D. (1996). Falls and fall prevention in the nursing home. *Clin Geriatr Med*, 12, 881903.
16. King, M. B. and Tinetti, M. E. (1996) A multifactorial approach to reducing injurious falls. *Clin Geriatr Med*, 12, 745759.
17. Fleming, K. C., Evans, J. M., Weber, D. C., and Chutka, D. S. (1995) Practical functional assessment of elderly persons: A primary-care approach. *Mayo Clin Proc*, 70, 890910.
18. Wolf-Klein, G. P., Silverstone, F. A., Basavaraju, N., Foley, C. J., Pascaru, A., and Ma, P. H. (1988). Prevention of falls in the elderly population. *Arch Phys Med Rehabil*, 69, 689691.
19. Morse, J. M., Morse, R., and Tylko, S. (1989). Development of a scale to identify the fall-prone patient. *Can J Aging*, 8, 366377.
20. Oliver, D., Britton, M., Seed, P., Martin, F. C., and Happer, A. H. (1997) Development and evaluation of evidence based risk assessment tool (STRATIFY) to predict which elderly inpatients will fall, case-control and cohort studies. *Br Med J*, 315, 10491053.
21. Morris, J. N., Nonemaker, S., Murphy, K. et al. (1997). Commitment to change: Revision of HCFA's RAI. *J Am Geriatr Soc*, 45, 10111016.
22. Hendrich, A., Nyhuuis, A., Kippenbrock, T., and Soja, M. E. (1995) Hospital falls, development of predictive model for clinical practice. *Appl Nurs Res*, 8: 129139.
23. Fife, D. D., Solomon, P., an Stanton, M. (1984) A risk falls program: Code orange for success. *Nurs Manage*, 15, 5053.
24. Mercer, L. (1997). Falling out of favour. *Aust Nurs J*, 4, 2729.
25. Tinetti, M. E., Williams, T. F., and Mayewski, R. (1986). Fall risk index for elderly patients based on number of chronic disabilities. *Am J Med*, 80, 429434.
26. Berg, K., Wood-Dauphine, S., Williams, J. I., and Gayton, D. (1989). Measuring balance in the elderly: Preliminary development of an instrument. *Physiother Canada*, 41(6), 304–311.
27. Duncan, P. W., Wiener, D. K., Chandler, J., and Studenski, S. (1990). Functional reach: A new clinical measure of balance. *J Gerontol Med Sci*, 45A, M192M197.
28. Whitney, S. L., Hudak, M. T., and Marchetti, G. F. (2000). The dynamic gait index relates to self-reported fall history in individuals with vestibular dysfunction. *J Vestib Res*, 10, 99105.
29. Steffen, T. M., Hacker, T. A., and Mollinger, L. (2002). Age-and gender-related test performance in community-dwelling elderly people: Six-Minute Walk Test, Berg Balance Scale, Timed Up & Go Test, and gait speeds. *Phys Ther*, 82(2), 128–137.
30. Vassallo, M. et al. (2008) Fall risk-assessment tools compared with clinical judgment: An evaluation in a rehabilitation ward. *Age Ageing* 37.3, 277–281.
31. Howcroft, J., Kofman, J., and Lemaire, E. D. (2013) Review of fall risk assessment in geriatric populations using inertial sensors. *J Neuroeng Rehabil*, 10.1, 1.
32. van Schooten, K. S., Pijnappels, M., Rispens, S. M., Elders, P. J., Lips, P., and van Dien, J. H. (2015). Ambulatory fall-risk assessment: Amount and quality of daily-life gait predict falls in older adults. *J Gerontol A Biol Sci Med Sci*, 70(5), 608–615.
33. Ayena, J. C., Chapwouo, T. L. D., Otis, M. J.-D., and Menelas, B. A. J. (2015). An efficient home-based risk of falling assessment test based on Smartphone and instrumented insole. *In Proceedings of the 2015 IEEE International Symposium on Medical Measurements and Applications (MeMeA), Turin, Italy, 79, May 2015*; pp. 416421.

34. Walsh, L., Greene, B. R., McGrath, D., Burns, A., & Caulfield, B. (2011, August). Development and validation of a clinic based balance assessment technology. In *Engineering in Medicine and Biology Society*, EMBC, 2011 Annual International Conference of the IEEE; pp. 1327–1330. IEEE.
35. Nelson-Wong, E., Appell, R., McKay, M., Nawaz, H., Roth, J., Sigler, R., Third, J., and Walker, M. (2012). *Increased fall risk is associated with elevated co-contraction about the ankle during static balance challenges in older adults. Eur J Appl Physiol*, 112, no. 4, 1379–1389.
36. Bounyong, S., Adachi, S., Ozawa, J., Yamada, Y., Kimura, M., Watanabe, Y., and Yokoyama, K. (2016, January). Fall risk estimation based on co-contraction of lower limb during walking. In 2016 IEEE International Conference on Consumer Electronics (ICCE); pp. 331–332. IEEE.
37. Hill, K. et al. (1999). Falls among healthy, community-dwelling, older women: A prospective study of frequency, circumstances, consequences and prediction accuracy. *Austr N Z J Publ Health* 23.1, 41–48.
38. Maidan, I., Freedman, T., Tzemah, R., Giladi, N., Mirelman, A., and Hausdorff, J. M. (2014). *Introducing a new definition of a near fall: Intra-rater and inter-rater reliability. Gait Posture* 39, no. 1, 645–647.
39. Phillips, L. J., DeRoche, C. B., Rantz, M., Alexander, G. L., Skubic, M., Despins, L., ... and Koopman, R. J. (2016). Using embedded sensors in independent living to predict gait changes and falls. *West J Nursing Res*.
40. van Schooten, K. S., Rispens, S. M., Elders, P. J., Lips, P., van Dieen, J. H., and Pijnappels, M. (2015). Assessing physical activity in older adults: Required days of trunk accelerometer measurements for reliable estimation. *J Aging Phys Act*. 23, 917.
41. Weiss, A., Herman, T., Giladi, N., and Hausdorff, J. M. (2014). Objective assessment of fall risk in Parkinson's disease using a body-fixed sensor worn for 3 days. *PLoS One*, 9, e96675.
42. Weiss, A. et al. (2013). Does the evaluation of gait quality during daily life provide insight into fall risk? A novel approach using 3-day accelerometer recordings. *Neurorehabil Neural Repair*, 27.8, 742–752.
43. Brodie, M. A. D., Coppens, M. J. M., Lord, S. R., Lovell N. H., Gschwind, Y. J., Redmond, S. J., Del Rosario, M. B. et al. (2015). Wearable pendant device monitoring using new wavelet-based methods shows daily life and laboratory gaits are different. *Med Biol Eng Comput*, 1–12.
44. Taati, B. and Mihailidis, A. (2014). Vision-based approach for long-term mobility monitoring: Single case study following total hip replacement. *J Rehabilit Res Dev*, 51(7), 1165.
45. Dolatabadi, E., Taati, B., Parra-Dominguez, G. S., & Mihailidis, A. (2013). *A markerless motion tracking approach to understand changes in gait and balance: A case study*. In Proceedings of the Rehabilitation Engineering and Assistive Technology Society of North America Annual Conference, pp. 20–24.
46. Stone, E. E. and Skubic, M. (2011). Evaluation of an inexpensive depth camera for passive in-home fall risk assessment. *Pervasive Computing Technologies for Healthcare (PervasiveHealth)*, 2011 5th International Conference on. IEEE.
47. Stone, E. E. and Skubic, M. (2013). Unobtrusive, continuous, in-home gait measurement using the Microsoft Kinect. *IEEE Trans Biomed Eng*, 60.10, 2925–2932.
48. Rantz, M. et al. (2015). *Automated in-home fall risk assessment and detection sensor system for elders. Gerontologist*, 55, Suppl 1, S78–S87.
49. Yardibi, T., Cuddihy, P., Genc, S., Bufi, C., Skubic, M., Rantz, M., and Phillips, C. (2011, March). Gait characterization via pulse-Doppler radar. In IEEE International Conference on Pervasive Computing and Communications Workshops (PERCOM Workshops), 2011, pp. 662–667. IEEE.
50. Cuddihy, P. E., Yardibi, T., Legenzoff, Z. J., Liu, L, Phillips, C. E., Abbott, C., Galambos, C. et al. (2012). *Radar walking speed measurements of seniors in their apartments: Technology for fall prevention*. In Engineering in Medicine and Biology Society (EMBC), 2012 Annual International Conference of the IEEE, pp. 260–263. IEEE.
51. Stone, E. E. and Skubic, M. (2013, May). *Mapping Kinect-based in-home gait speed to TUG time: A methodology to facilitate clinical interpretation*. In 2013 7th International Conference on Pervasive Computing Technologies for Healthcare and Workshops, pp. 57–64. IEEE.
52. Kaye, J., Mattek, N., Dodge, H., Buracchio, T., Austin, D., Hagler, S., and Hayes, T. (2012). One walk a year to 1000 within a year: Continuous in-home unobtrusive gait assessment of older adults. *Gait Posture*, 35(2), 197–202.
53. Cardinaux, F., Bhowmik, D., Abhayaratne, C., & Hawley, M. S. (2011). Video based technology for ambient assisted living: A review of the literature. *J Ambient Intell Smart Environ*, 3(3), 253–269.
54. Del Din, S., Godfrey, A., Galna, B., Lord, S., and Rochester, L. (2016). Free-living gait characteristics in ageing and Parkinson's disease: Impact of environment and ambulatory bout length. *J Neuroeng Rehabil*, 13(1), 1.
55. Ihlen, E. A., Weiss, A., Helbostad, J. L., and Hausdorff, J. M. (2015). The discriminant value of phase-dependent local dynamic stability of daily life walking in older adult community-dwelling fallers and nonfallers. *BioMed Res Int*.
56. Ihlen, E. A., Weiss, A., Beck, Y., Helbostad, J. L., and Hausdorff, J. M. (2016). A comparison study of local dynamic stability measures of daily life walking in older adult community-dwelling fallers and non-fallers. *J Biomech*, 49(9), 1498–1503.
57. Rispens, S. M., van Schooten, K. S., Pijnappels, M., Daffertshofer, A., Beek, P. J., and van Dien, J. H. (2014). Identification of fall risk predictors in daily life measurements gait characteristics reliability and association with self-reported fall history. *Neurorehabil Neural Repair*, 1545968314532031.
58. van Schooten, K. S., Pijnappels, M., Rispens, S. M., Elders, P. J., Lips, P., Daffertshofer, A., .and van Dieen, J. H. (2016). Daily-life gait quality as predictor of falls in older people: A 1-year prospective cohort study. *PLoS ONE*, 11(7), e0158623.

59. Brodie, M. A., Lord, S. R., Coppens, M. J., Annegarn, J., & Delbaere, K. (2015). Eight-week remote monitoring using a freely worn device reveals unstable gait patterns in older fallers. *IEEE Trans Biomed Eng*, 62(11), 2588–2594.
60. Howcroft, J., Lemaire, E. D., and Kofman, J. (2016). *Wearable-sensor-based classification models of faller status in older adults*. PLoS ONE 11, no. 4, e0153240.
61. Xu, W., Huang, M.-C., Amini, N, Liu, J. J., He, L., and Sarrafzadeh, M. (2012). Smart insole: A wearable system for gait analysis." In Proceedings of the 5th International Conference on Pervasive Technologies Related to Assistive Environments, p. 18. ACM.
62. Rouhani, H., Favre, J., Crevoisier, X., and Aminian, K. (2011). Ambulatory measurement of ankle kinetics for clinical applications. *J Biomechan*, 44(15), 2712–2718.
63. Dyer, P. S. and Bamberg, S. J. M. (2011). Instrumented insole vs. force plate: A comparison of center of plantar pressure. *Engineering in Medicine and Biology Society, EMBC*, 2011 Annual International Conference of the IEEE. IEEE.
64. Redd, C. B. and Bamberg, S. J. M. (2012). A wireless sensory feedback device for real-time gait feedback and training. *IEEE/ASME Trans Mechatron*, 17.3, 425–433.
65. Moufawad El Achkar, C., Lenoble-Hoskovec, C., Paraschiv-Ionescu, A., Major, K., Bla, C., and Aminian, K. (2016). Physical behavior in older persons during daily life: Insights from instrumented shoes. *Sensors*, 16(8), 1225.
66. Bamberg, S. J. Morris, Benbasat, A. Y., Scarborough, D. M., Krebs, D. E., and Paradiso, J. A. (2008). Gait analysis using a shoe-integrated wireless sensor system. *IEEE Trans Inform Technol Biomed*, 12, no. 4, 413–423.
67. Robinovitch, S. N., Feldman, F., Yang, Y., Schonnop, R., Leung, P. M., Sarraf, T., Sims-Gould, J., and Loughin, M. (2013). *Video capture of the circumstances of falls in elderly people residing in long-term care: An observational study*. Lancet, 381, no. 9860, 47–54.
68. Taylor, K. et al. (2015). *Context focused older adult mobility and gait assessment*. 2015 37th Annual International Conference of the IEEE Engineering in Medicine and Biology Society (EMBC). IEEE.
69. Maki, B. E. and Mcilroy, W. E. (1999). Control of compensatory stepping reactions: Age-related impairment and the potential for remedial intervention. *Physiother Theor Pract*, 15.2, 69–90.
70. Mansfield, A. et al. (2010). Effect of a perturbation-based balance training program on compensatory stepping and grasping reactions in older adults: A randomized controlled trial. *Phys Ther*, 90.4, 476–491.
71. Weiss, A., Shimkin, I., Giladi, N., and Hausdorff, J. M. (2010). Automated detection of near falls: Algorithm development and preliminary results. *BMC Res Notes*, 3(1), 62.
72. Mille, M. L., Johnson, M. E., Martinez, K. M., and Rogers, M. W. (2005). Age-dependent differences in lateral balance recovery through protective stepping. *Clin Biomech*, 20(6), 607–616.
73. Hermens, H. J., Freriks, B., Merletti, R., Stegeman, D., Blok, J., Rau, G., Disselhorst-Klug, C., and Hagg, G. (1999). European recommendations for surface electromyography. *Roessingh Res Dev*, 2(8), 13–54.
74. De Luca, C. J. et al. (2010). Filtering the surface EMG signal: Movement artifact and baseline noise contamination. *J Biomechan*, 43, 1573–1579.
75. Nouredanesh, M., Kukreja, S. L., and Tung, J. (2016, October). Detection of compensatory balance responses using wearable electromyography sensors for fall-risk assessment. In *Engineering in Medicine and Biology Society (EMBC)*, 2016 IEEE 38th Annual International Conference of the, pp. 1680–1683. IEEE.
76. Nouredanesh, M. Tung, J. (2015). *Machine learning based detection of compensatory balance responses to lateral perturbation using wearable sensors*, pp. 14.
77. Phinyomark, A., Limsakul, C., and Phukpattaranont, P. (2009). A novel feature extraction for robust EMG pattern recognition. *J Comput*, 1, 71–80.
78. Breiman, L. (2001). Random forests. *Machine Learning*, 45 (1), 5–32.
79. Ko, B. C., Kim, S. H., and Nam, J.-Y. (2011). X-ray image classification using random forests with local wavelet-based CS-local binary patterns. *J Digit Imaging*, 24.6, 1141–1151.
80. Patterson, M. R. et al. (2014). Does external walking environment affect gait patterns? *Engineering in Medicine and Biology Society (EMBC)*, 2014 36th Annual International Conference of the IEEE. IEEE.
81. Reginatto, B. et al. (2015). *Context aware falls risk assessment: A case study comparison*. 2015 37th Annual International Conference of the IEEE Engineering in Medicine and Biology Society (EMBC). IEEE.
82. Nouredanesh, M., McCormick, A., Kukreja, S. L., and Tung, J. (2016, June). *Wearable vision detection of environmental fall risk using Gabor Barcodes*. In 2016 6th IEEE International Conference on Biomedical Robotics and Biomechatronics (BioRob), pp. 956–956. IEEE.
83. Nouredanesh, M., McCormick, A., Kukreja, S. L., and Tung, J. (2016). Wearable vision detection of environmental fall risks using convolutional neural networks. *arXiv* preprint arXiv:1611.00684.
84. Mou, W. and Kleiner, A. (2010). Online learning terrain classification for adaptive velocity control. *IEEE International Workshop on Safety Security and Rescue Robotics (SSRR)*, 2010. IEEE.
85. Laible, S. et al. (2012). 3d lidar- and camera-based terrain classification under different lighting conditions. *Autonomous Mobile Systems*. Springer, Berlin, 21–29.
86. Zenker, S. et al. (2013). *Visual terrain classification for selecting energy efficient gaits of a hexapod robot*. IEEE/ASME International Conference on Advanced Intelligent Mechatronics (AIM), 2013. IEEE.
87. Lu, X. and Manduchi, R. (2005). Detection and localization of curbs and stairways using stereo vision. ICRA. Vol. 5.
88. Cross, G. R. and Jain, A. K. (1983). Markov random field texture models. *IEEE Trans Pattern Anal Machine Intell*, 1, 2539.
89. Kamarainen, J.-K., Kyrki, V., and Kälviäinen, H. (2006). Invariance properties of Gabor filter-based features—Overview and applications. *IEEE Trans Image Process*, 15.5, 1088–1099.
90. Haghighat, M., Zonouz, S., and Abdel-Mottaleb, M. (2015). CloudID: Trustworthy cloud-based and cross-

enterprise biometric identification. *Expert Syst Appl*, 42.21, 7905–7916.
91. Chen, L., Lu, G., and Zhang, D. (2004). Effects of different Gabor filter parameters on image retrieval by texture. *Null*. IEEE.
92. Tizhoosh, H. R., *Barcode annotations for medical image retrieval: A preliminary investigation*, IEEE International Conference on Image Processing, pp. 818–822, doi: 10.1109/ICIP.2015.7350913, 2015.
93. Nouredanesh, M., Tizhoosh, H. R., and Banijamali, E. (2016). Gabor barcodes for medical image retrieval. *arXiv* preprint arXiv:1605.04478.
94. LeCun, Y., Bottou, L., Bengio, Y., and Haffner, P. (1998). *Gradient-based learning applied to document recognition*. Proc IEEE, 86, no. 11, 22782323.
95. Ji, S., Yang, M., and Yu, K. (2013). 3D convolutional neural networks for human action recognition. *IEEE Trans Pattern Anal Mach Intell*, 35, 1, 22131.
96. Ji, S., Xu, W., Yang, M., and Yu, K. (2013). 3D convolutional neural networks for human action recognition. *IEEE Trans Pattern Anal Mach Intell*, 35(1), 221–231.
97. Simonyan, K., and Zisserman, A. (2014). Two-stream convolutional networks for action recognition in videos. *In Advances in Neural Information Processing Systems*, pp. 568–576.
98. Zhang, F., Duarte, F., Ma, R., Milioris, D., Lin, H., and Ratti, C. (2016). Indoor space recognition using deep convolutional neural network: A case study at MIT campus. *arXiv* preprint arXiv:1610.02414.
99. Phung, S. L. and Bouzerdoum, A. (2007). A pyramidal neural network for visual pattern recognition. *IEEE Trans Neural Netw*, 18, no. 2, 329–343.

Index

Page numbers followed by f and t indicates figures and tables, respectively.

A

AARDA, see American Autoimmune Related Diseases Association (AARDA)
ABC-MCMC procedure, 419–420
ABC rejection sampling method, 418–419
Accelerometers, 536
Accountable care organizations (ACOs), 12–13
   key requirement of, 13
Accuracy, 345, 387
   vs. performance, CAD system, 345–347
Accurate mass and time (AMT) approach, 417
ACOs, see Accountable care organizations (ACOs)
Actigraphy, 445
Action Research Arm Test (ARAT), 535
Active learning (AL), 116
   for RBSE, 118–119
Active-RBSE, see Active recursive Bayesian state estimation (Active-RBSE)
Active recursive Bayesian state estimation (Active-RBSE), 116–123
   active learning for RBSE, 118–119
   BCI design example, 123–129
      balanced-tree visual presentation paradigms, 124
      decision-making component, 125–126
      EEG feature extraction and classification, 125–126
      ERP-based BCI for letter-by-letter typing, 124–125, 124f
      experimental results and discussions, 126–129
      language model, 126
      matrix presentation, 123
      rapid serial visual presentation (RSVP), 123–124
      visuospatial presentation, 123–124
   combinatorial optimization, 122–123
   hidden Markov model of order n (HMM-n), 116, 116f, 117f, 118f
   maximum a posteriori (MAP) inference method, 117
   objective functions for query optimization, 120
   proposed graphical model, 121–122
   open problems and future directions, 129–130
   probabilistic graphical model (PGM), 116–118, 117f, 118f
   submodular-monotone set functions for set optimization problems, 119–120
   type I (T-I), 116
   type II (T-II), 116
Active RSVP (ARSVP), 127
Activity monitors, 18
Acute Physiology and Chronic Health Evaluation (APACHE, APACHE-II) scores, 250, 460
ADAM, 430
Adam optimization method, 350, 351f
Adaptive distributed dictionary learning (ADDL)
   HEp-2 cell classification, 372–376, 372f
      classification results, 376–378, 377f, 377t–378t
      combination weights selection, 374–376, 375f
      computational cost, 378, 378t
      data sets and evaluation methods, 376
      diffusion adaptation method, 374, 374f
Additive treatment effect (ATE), 237
ADDL, see Adaptive distributed dictionary learning (ADDL)
ADHD, see Attention deficit hyperactivity disorder (ADHD)
ADHD-200, 36
Adjacent windowing, 212
AD Neuroimaging Initiative (ADNI), 27
   global programs, 27
ADNI, see AD Neuroimaging Initiative (ADNI)
ADNI GO study, 27
ADNI MRI Core, 27
ADNI1 study, 27
ADNI2 study, 27
ADNI3 study, 27
ADs, see Autoimmune diseases (ADs)
Affordable Care Act, the, 13, 16
Agency for Healthcare Research and Quality (AHRQ), 10
   survey data and, 10–11
AHRQ (Agency for Healthcare Research and Quality), 10
A-IPW (augmented inverse probability weighted) estimator, 240
Akaike information criterion (AIC), 201
Alzheimer's disease (AD), 25, 32
   neuroimaging-based machine learning classification of, 33t–36t
Alzheimer's disease neuroimaging initiative (ADNI), 23
Amazon, 223
Amazon Web Services, 419
Ambient sensors, for FRA, 556–558
American Academy of Sleep Medicine (AASM), 432
American Association of Retired Persons (AARP) Medigap insurance program, 16
American Autoimmune Related Diseases Association (AARDA), 369
American Clinical Neurophysiology Society (ACNS), 273
American Electroencephalographic Society, 261
American Psychiatric Association (APA), 433
   Diagnostic and Statistical Manual (*DSM-IV-TR*), 433
American Recovery and Reinvestment Act of, 2009, 13
Amplitude-based abnormalities, of newborn, 77
Analysis, complex systems, 177–178
Analytics, big data, 5
ANAs, see Antinuclear antibodies (ANAs)
Anemia of prematurity, 529–530
Annotations, manual interpretation of EEGs, 276–277, 278f
Anomalous data streams, problems of detection and identification
   biomedical signal processing applications of, 70–71
   detection of existence, 61
   nonparametric model, 63–64
   parametric model, 61–62
   semiparametric model, 62–63

identification
　nonparametric model, 67–70
　parametric model, 64–65
　semiparametric model, 65–67, 66f
nonparametric approaches
　generalized likelihood-based approach, 59
　kernel-based approach, 60
　KL divergence-based approach, 60–61
open problems and future directions, 71
overview, 57–58
types of problems, 58
　detection of existence, 58–59
　identification, 59
ANOVA, 29
Antinuclear antibodies (ANAs), 369
Apache, 4
Apache Spark, 5
Apnea, automatic detection
　from airflow signals, 454–455
　from PPG analysis, 455
Apnoea of prematurity, 529
Apparent diffusion coefficient (ADC), 22
Approximate Bayesian computation (ABC), 418
Approximations; see also DCT approximations
　defined, 154–155
Arai DCT scheme, 152
Area under the curve (AUC), CAD system, 346–347
Arousals
　automated analysis of sleep EEG, 450–451
Artefacts, in EEG, 77–78
　detection, 78
　detection and removal, 80
　features for detecting, 79
　problem of, 75–76
　removal using automated component separation, 78–79
　standard methods for handling, 78–79
　types, 78
Artemis, in critical care
　clinical applications, 527–530
　　anemia of prematurity, 529–530
　　apnoea of prematurity, 529
　　automated partial premature infant pain profile (PIPP) scoring, 529
　　late-onset neonatal sepsis, 528–529, 528f
　　retinopathy of prematurity (RoP), 529
　data persistence, 525–526
　deployments, 526–527, 527f

framework, 524–525, 525f
overview, 524–525
Artemis Cloud, 526–527, 527f
Arterial spin labeling (ASL), 25
Artifacts (ARTF), EEGs, 277
Artificial neural network (ANNs), 331
　classifier, 470–471, 470f
　wearable sensors, 565–566
ASD, see Autism spectrum disorder (ASD)
Asymmetric/asynchronous-based abnormalities, of newborn, 77
Atlas-based quantification of diseased shapes, left-ventricular analysis, 317–318, 319f
Attention deficit hyperactivity disorder (ADHD), 25
　neuroimaging-based machine learning classification of, 32, 36, 41t
Auction algorithm, for Wasserstein distance, 221
Augmented inverse probability weighted (A-IPW) estimator, 240
Autism Genetic Resource Exchange (AGRE), 26
Autism Informatics Consortium (AIC), 26
Autism Speaks, 26
Autism spectrum disorder (ASD), 25
　neuroimaging-based machine learning classification of, 32, 40t
Autism Tissue Program (ATP—a post-mortem brain donation program), 26
Autoantibody, 369
Auto-encoder (AE), 31
Autoencoders, 353
Autoimmune diseases (ADs), 369
Autoimmunity, 369
Automated anatomical labeling (AAL), 29
Automated biventricular cardiovascular modelling, from MRI, 313–324
　background, 313–315, 314t
　congenital heart disease, 322, 322f
　　biventricular finite-element model, 322–323, 323f
　　guide-point modelling, 323–324
　　guide-point modelling vs. manual contouring, 324, 324f
　left-ventricular analysis
　　atlas-based quantification of diseased shapes, 317–318, 319f
　　building an atlas of heart, 315–317, 316f
　　classification, 318–320, 320f

　　finite-element model, 315, 316f
　　regional analysis, 320–322, 321f, 321t
　　signal representation of shapes, 315
　　statistical analysis of shape, 317, 317f
Automated component separation, artefact removal using, 78–79
Automated medical big data monitoring
　TBI outcome prediction, 478
　　data from real trauma center, 478
　　redundant vs. collection system with diagnosis viewer, 478–480, 479f
Automated partial premature infant pain profile (PIPP) scoring, 529
Automated processing of big data, in sleep medicine, 443–459
　analysis approaches, 446–447, 447f
　big data approach, 445–446
　cardiorespiratory measures
　　apnea detection from airflow signals, 454–455
　　apnea detection from PPG analysis, 455
　　cardiopulmonary coupling, 455–456, 456f
　　snoring sounds, 455
　challenges and needs
　　data harmonization, 457–458
　　noise, artifacts, and missing data, 456–457, 457f
　EEG, 448
　　arousals, 450–451
　　automatic sleep staging, 449–450
　　K complexes, 450
　　macroanalysis, 448–450
　　spectral analysis, 448–449
　　spindles, 450, 451t
　　time-frequency and time-scale approaches, 449, 449f
　　transient events detection, 450–451
　emerging applications
　　genomics and sleep, 458
　　precision medicine and other applications, 458–459
　heart rate variability analysis, 451–452, 452f
　　detrended fluctuation analysis (DFA), 453
　　frequency domain analysis, 452–453
　　Lempel–Ziv (LZ) complexity, 454
　　multiscale entropy (MSE), 453–454, 454f
　　nonlinear methods, 453–454
　　sample entropy (SampEn), 453
　　time domain analysis, 452

Index

manual *vs.* automatic methods, 446
National Sleep Research Resource (NSRR)
   publicly available sleep data sets, 447
   tools, 447–448
research goals, 445
sleep data, 444–445, 444f
Automatic interpretation, of EEGs, 287
   feature extraction, 288–292, 289f, 290f, 291t
   hidden Markov models, 292–293, 292f, 293f, 293t, 294f
   machine learning approaches, 287–288
   normal/abnormal detection, 293–295, 294f, 295t
   seizure detection, 295–296
   typical static classification systems, 288
Automatic sleep staging, EEG-based, 449–450
Average amount of mutual information (AAMI) index, 260

B

Background (BCKG), EEG, 277
Backward elimination, 29, 30
Backward feature elimination, 5
Balanced-tree visual presentation paradigms, BCI system, 124
Barcode, 216
Bar-selective combination of shifted filter responses method, 354
Basis pursuit (BP) approach
   sparse signal reconstruction, 141
BAS (Bouguezel–Ahmad–Swamy) series of approximations, 153, 158, 159t
   parametric BAS, 162
Bayesian-based reconstruction, of big sparse signals, 145–148
   algorithm, 146
   in discrete sine transform (DST) domain, example, 147–148, 147f
Bayesian classification of genomic big data, 411–425
   expression rank data, 413, 413f
      Bayesian inference for Bradley–Terry model, 414–415
      Bayesian top-scoring pairs, 415
      Bradley–Terry model, 414
      expression classification experimental results, 415–416, 415t, 416t
      top-scoring pair classifiers, 413

LC-MS protein data, 416
   Bayesian inference for, 418–421, 418f, 418t
   classification experimental results, 421–422, 421t, 422f
   model for, 416–418
overview, 411–413, 412f
16S rRNA metagenomic data, 422–424
   Bayesian inference for, 423
   classification experimental results, 424t, 4234
   model for, 423, 423f
Bayesian inference, 413
   for Bradley–Terry model, 414–415
   for LC-MS protein data, 418–421, 418–421, 418f, 418f, 418t, 418t
      optimal Bayesian classifier, 420–421
      posterior sampling via an ABC-MCMC procedure, 419–420
      prior calibration via ABC rejection sampling, 418–419
      procedure overview, 418
   for 16S rRNA metagenomic data, 423
Bayesian information score (BIC), 194, 201, 227
Bayesian networks and time, TAs and, 496–501, 497f, 501t
   continuous-time Bayesian networks (CTBNs), 499–500, 500f
   dynamic Bayesian networks (DBN), 497–499, 498f
   integration, 501–502
   irregular-time Bayesian networks (ITBNs), 500–501, 501f
   networks of probabilistic events in time (NPEDT), 499, 500f
Bayesian top-score pair (BTSP) classification rule, 412
Bayesian Top-Scoring Pairs, 415
BCI system, Active-RBSE and, 123–129
   decision-making component, 125–126
   EEG feature extraction and classification, 125–126
   language model, 126
   ERP-based BCI for letter-by-letter typing, 124–125, 124f
   presentation component, 124–125
   experimental results and discussions, 126–129
   matrix-based presentation paradigm with overlapping trials, 127–128, 128f
   matrix-based presentation paradigm with single-character trials, 129, 129f
   RSVP paradigm, 127, 127f

visuospatial presentation, 123–124
   balanced-tree visual presentation paradigms, 124
   matrix presentation, 123
   rapid serial visual presentation (RSVP), 123–124
Bertsekas's auction algorithm, for Wasserstein distances, 221
Betti numbers, 216, 218
Big data, 3, 209–210
   computational challenges, 260
   defined, 3, 115
   four V's of, 14
   in FRA and FP tools, 572–572
   in health care, driving factors
      database architecture, 13–14
      the HITECH Act, Meaningful Use, and ACOs, 12–13
   information retrieval and
      DSP approach and engineering rationale, 74
      EEG, 74–75, 75f
      EEG clinical rationale and problem of artefacts, 75–76
   issues in manual interpretation of EEGs
      annotations, 276–277, 278f
      decision support systems, 280, 281f
      evaluation metrics, 279–280
      inter-rater agreement, 277–279
      locality, 276
      signal conditioning, 275–276
      unbalanced data problem, 280–282
      waveform displays, 273–275, 275f
   neurorehabilitation improvement through, *see* Neurorehabilitation
   smart medical devices and, 17–19
Big data analysis
   persistent homology for
      auction algorithm for Wasserstein distance, 221
      bottleneck matching, 221
      simplicial complex constructions and, 221–222
      stability of, 219–221, 220f
Big data analytics, 5
Big Data Genomics (BDG) project, 420
Big data projects, of human brain, 23–28
   AD Neuroimaging Initiative, 27
   Federal Interagency Traumatic Brain Injury Research, 25
   1000 Functional Connectomes Project (FCP), 24–25

Human Connectome Project (HCP), 23
   MGH/Harvard–UCLA consortium, 24
   WU–Minn–Oxford consortium, 23–24
   IMAGEN, 27–28
   National Database for Autism Research, 25–26
   SchizConnect, 26–27
Big data science, in health care
   coordination of care, 16
   emergency department revisits, 16–17
   precision medicine (PM), 14–15
   predicting disease risk, 15–16
Big Data Science model
   insomnia biosignal processing and automated detection (case study), 434–440
      data preparation, 435–436, 436t
      data understanding, 435
      deployment, 438–439, 439f
      modeling, 436–438, 437f–438f
      problem understanding, 435
Big Data to Knowledge (BD2K) initiative, 447
Big sparse signals reconstruction, compressive sensing methods for, 133–149
   Bayesian-based reconstruction, 145–148, 147f
   biomedical signal processing applications of, 148
   conditions for, 135–136
   definitions, 134–137
   measurement matrices, 137
   norm-one-based reconstruction algorithms, 141–145, 141f
      LASSO minimization, 142–143, 143f
      signal reconstruction with gradient algorithm, 143–144
      total variations, 144–145
   norm-zero-based reconstruction, 137–141
      block and one-step reconstruction, 139–141
      coherence, 139
      CoSaMP reconstruction algorithm, 140–141
      with known/estimated positions, 138
      MP reconstruction algorithm, 139
      position estimation and, 138–139
   open problems and future directions, 148–149
   overview, 133–134
binDCT, 153
Biogenesis, miRNAs, 383–384, 383f
BioLINCC, 447
Biomarkers, 411
Biomedical image segmentation, information fusion in deep CNNs for
   background, 303–305, 303f–305f
   data and implementation details
      brain tumors (BRATS), 307
      MS datasets, 307
      pre-processing and, 307–308
   experiments and results, 308–310, 308f, 309t, 310f–311f
   method
      hetero-modal image segmentation, 305–306, 306f
      interpretation as an embedding, 307
      pseudo-curriculum training procedure, 306–307
   overview, 301–303, 302f
Biomedical Informatics Research Network (BIRN), 26
Biomedical signal processing
   big sparse signals reconstruction and, 148
   DCT approximations applications, 167
      image compression, 167–169, 168f, 169f–170f
      image registration, 169–171, 171f
   functional networks modeling via graphical models and, 207
   problems of detection and identification anomalous data streams applications in, 70–71
   TAs applications, 513
   TDA applications in, 227–229
      EEG signals, 227–228
      EMG signal, 228
      human motion capture data, 228–229
Biomedical time-series data, TDA and, 210–211, 212f
Bipolar disorder (BPD), 32
   neuroimaging-based machine learning classification of, 32, 39t
Biventricular finite-element model, 322–323, 323f
Blind source separation (BSS), 78–79
   for artefact removal, 78–79, 108, 108f, 109f
Blood oxygenation level dependence (BOLD) contrast, fMRI and, 22–23
Bottleneck matchings
   Hopcroft–Karp algorithm for, 221
   persistent homology, 221
Bouguezel–Ahmad–Swamy (BAS) series of approximations, 153, 158, 159t
   parametric BAS, 162
Boundary map, 218
BPD, see Bipolar disorder (BPD)
Bradley–Terry model, 412, 414
   Bayesian inference for, 414–415
Brain; see also Human brain, big data projects of
   electrical activity, recording of, 261, 271–273, 272f, 273f
   functional connectivity measures, 259; see also Functional connectivity measures
   functional networks in, 259–260
   structural networks, 259
Brain–behavior system, 28
Brain/body computer interface (BBCI), 116
Brain imaging data, see Neuroimaging data
BRAIN initiative, 430–431, 431t
BRATS-2015 dataset, 307
Breast Cancer Family Registry (CFR), 11
BSS, see Blind source separation (BSS)
BSS algorithm, 90
BSS-SVM, 79

C

Canadian Partnership for Tomorrow Project (CPTP), 314
Canny, John, 344
Canny edge detectors, 343
   retinal analysis, 344
Canonical correlation analysis (CCA), 78–79
CAP, see Cardiac Atlas Project (CAP)
CAP-miRSeq, 394
Cardiac activity, 78
Cardiac Atlas Project (CAP), 314
Cardiac care units (CCUs), 519
Cardiac remodelling, 314
Cardiopulmonary coupling (CPC), 455–456, 456f
Cardiorespiratory measures, automated processing in sleep medicine
   apnea detection from airflow signals, 454–455
   apnea detection from PPG analysis, 455
   cardiopulmonary coupling, 455–456, 456f
   snoring sounds, 455
Cassandra, 5
CCA, see Canonical correlation analysis (CCA)
CCUs, see Cardiac care units (CCUs)

CDSS, *see* Clinical decision support system (CDSS)
Cech complex, 214–215, 215f
Center for Biomedical Research Excellence (COBRE), 26–27
Center for Information Technology (CIT), 25
Centers for Disease Control (CDC)
National Center for Health Statistics (NCHS), 10
Centers for Medicare and Medicaid Services (CMS), 13
Chain complex, 218
Change point detection, 198–199
regularization selection for, 200
Checkpoint-driven ensemble assembly, 360
Chen algorithm, 152
Childhood Adenotonsillectomy Trial, 447
Child Mind Institute, 24
CIFAR10, 345
CLAHE (contrast-limited adaptive histogram equalization), 344
Claims databases, 10
Classical multiplierless transforms, DCT approximations
Hadamard transform, 156
Walsh–Hadamard transform (WHT), 156
Classifier performance, 345
Classifiers, machine learning, 30–31
categories, 30–31
performance measures, 31–32
Class imbalance correction, miRNA prediction and, 388–389
Cleveland Children's Sleep and Health Study, 447
Cleveland Family Study, 447
Clinical and Translation Science Award (CTSA) program, NIH's, 10
Clinical decision support system (CDSS)
clinical physiological knowledge at feature and classifier levels integration in, ICU outcome, 465–467
artificial neural network classifier, 470–471, 470f
collected data, 467, 467t
feature extraction, 468–469, 469t
feature ranking, 469–470
methods, 467–471
results, 471–473, 471t–472t, 472f, 473t
Clinical information management system (CIMS), 526
Clinical registries, 11
Clinical systems, using TAs, 495–496
Clique complex, 215–216, 215f, 216f
CloudAligner, 419

CloudBurst, 419
Cloud computing, 14
Cloudwave signal format (CSF), 261
functional connectivity measures in epilepsy
computational pipeline for signal data analysis, 263–265, 264f
signal data management, 261–263, 262f
signal data processing, 263
Clustering analysis, 30
based topological network, 226
Clustering protocol, 226
CMS, *see* Centers for Medicare and Medicaid Services (CMS)
CNNs, *see* Convolutional neural networks (CNNs)
Coherence, norm-zero-based reconstruction and, 139
Coherence index, 139
Collaborative Imaging and Neuroinformatics System (COINS) Data Exchange, 26
Collaborative TMLE (C-TMLE), 244–245
Combinatorial optimization
Active-RBSE, 122–123
Compact kernel distribution (CKD), 81–82
Complex networks, dynamic processes on, 177–189
application (vulnerable communities), 182–183
biomedical signal processing application of, 189
continuous-time Markov process, 179–180
endogenous infection dominant (regime II), 183–184
equilibrium distribution and induced subgraphs, 184–185, 184f
most-probable configuration and densest subgraph, 186
most-probable configuration and induced subgraphs, 185–186
examples, 186–187, 187f, 188f
exogenous infection dominant (regime III), 187–189
healing dominant (regime I), 183
infection dominant (regime IV), 183
open problems and future direction, 189
overview, 177–178
scaled SIS process, 178–179, 179f, 180f
equilibrium distribution, 180–181
MPCP, 183
parameter regimes, 181–182, 181t

Complex systems
analysis, 177–178
synthesis, 177
Complex TAs, 495, 495f
naïve Bayes classifiers incorporation in, 508–513
extraction, 509, 509f
imbalanced data and overlapping classes, 511–512
most frequent temporal association patterns extraction, 510–511
periodic classifiers *vs.* baseline classifier, performance, 512–513, 513t
using periodic temporal association patterns, 510, 512f
Comprehensive medical assessments, falls, 553
Compressed sensing (CS), 5
HEp-2 cell classification, 371–372
Compressive sampling matching pursuit (CoSaMP) reconstruction algorithm, 140–141
Compressive sensing (CS) methods, for big sparse signals reconstruction, 133–149
Bayesian-based reconstruction, 145–148, 147f
biomedical signal processing applications of, 148
conditions for, 135–136
definitions, 134–137
measurement matrices, 137
norm-one-based reconstruction algorithms, 141–145, 141f
LASSO minimization, 142–143, 143f
signal reconstruction with gradient algorithm, 143–144
total variations, 144–145
norm-zero-based reconstruction, 137–141
block and one-step reconstruction, 139–141
coherence, 139
CoSaMP reconstruction algorithm, 140–141
with known/estimated positions, 138
MP reconstruction algorithm, 139
position estimation and, 138–139
open problems and future directions, 148–149
overview, 133–134

Computational miRNAs prediction techniques, 390–400
 de novo miRNA prediction, 396–400, 398t
 homology-based, 390
 NGS-based miRNA prediction, 391–396, 391t
 previous assessments, 400, 400t
Computational search, DCT approximations based on, 159
 improved 8-Point DCT approximation, 160–161
 for IR, 161
 for RF imaging, 160
Computed tomography (CT) imaging, 148, 301
 yearly scans in United States, 301, 302f
Computer-aided detection (CADe), 341
Computer-aided diagnostics (CAD)
 area under the curve (AUC), 346–347
 categories, 341
 HEp-2 cell classification, 369–370
 KPIs, *see* Key performance indicators (KPIs)
 receiver operating characteristic (ROC) curve, 346–347, 347f
 retinal analysis, 340–341
  Canny edge detectors, 344
  challenges, 341
  data sets, 345
  feature discovery, 342–343
  global histogram enhancement, 343–344
  image processing fundamentals, 343–345
  local histogram enhancement, 344
  mimicking experts, 341–342, 342f
  performance evaluation, 347–348
  performance *vs.* accuracy, 345–347, 346f, 346t, 347f
  superresolution techniques, 344
  traditional CAD of retinal images, 343
  watershed methods, 344–345
Computing issues, TBI outcome prediction, 489
Conditional mean outcome, longitudinal-TMLE for, 246–247, 246f
Conditions, for big sparse signals reconstruction, 135–136
Confounding, time-dependent, 245
Confusion matrix, 386–387, 386f, 387f
Congenital heart disease (CHD), 322, 322f
 biventricular finite-element model, 322–323, 323f
 guide-point modelling, 323–324
 guide-point modelling *vs.* manual contouring, 324, 324f

Congressionally Directed Medical Research Programs (CDMRP), 25
Connectome, 177
"Connectome Skyra," 23–24
Contact network, 178
Continuous outcomes, point treatment effect estimation and, 245
Continuous-time Bayesian networks (CTBNs), 499–500, 500f
Continuous-time Markov process, 179–180
Contrast-limited adaptive histogram equalization (CLAHE), 344
Convolutional layer (CL), CNNs, 335, 336f
Convolutional neural networks (CNNs), 331, 333, 335f
 background, 303–305, 303f–305f
 convolutional layer (CL), 335, 336f
 deep CNNs, *see* Deep convolutional neural networks (CNNs)
 fully connected layer (FC), 335
 human visual system *vs.*, 335–337, 336f
 linear CNNs, 353
 multichannel CNNs, 353
 pooling layer (PL), 335
 vs. RNNs, 337
 wearable sensors application of, 571–572, 573f
Coordination of care, 16
Copenhagen Stroke (COST) Study, 542
Copula GGMs
 change point detection, 198–199
 hidden variable copula GGMs, 197–198, 197f, 198f, 202–203
 regularization selection for, 200–203
 standard copula GGMs, 196–197, 201–202, 201f
Copula(s)
 defined, 195
 Gaussian, 195–196
 Sklar's theorem, 195–196
Correlation matrix, 213
Correlation measures, functional connectivity analysis, 260
Correlations, data analytics in trauma patients and, 483–484, 484f
CoSaMP (compressive sampling matching pursuit) reconstruction algorithm, 140–141
CPT, *see* Current Procedural Terminology (CPT)
Critical care/critical care units, 519
 artemis
  anemia of prematurity, 529–530
  apnoea of prematurity, 529

  automated partial premature infant pain profile (PIPP) scoring, 529
  clinical applications, 527–530
  data persistence, 525–526
  deployments, 526–527, 527f
  framework, 524–525, 525f
  late-onset neonatal sepsis, 528–529, 528f
  retinopathy of prematurity (RoP), 529
 background, 519–520
 data acquisition, 522, 523f
 data analytics, 520
 data availability, 521
 data frequency, 520–521
 data persistence, 523
 data quality, 521–522
 data transmission, 522
 driving forces, 519–520
 knowledge discovery, 524
 real-time analytics, 522–523
 secondary use of data and consent, 524
 visualisation, 524
Crossbow, 419
Cross Industry Standard Process for Data Mining (CRISP-DM), 434–435, 435f
Cross-validated area under the receiver operating curve (cv-AUC), 250
Cross-validation
 SL algorithm, 238
CSF, *see* Cloudwave Signal Format (CSF)
CTBNs, *see* Continuous-time Bayesian networks (CTBNs)
C-TMLE, *see* Collaborative TMLE (C-TMLE)
CTSA Accrual to Clinical Trials (CTSA ACT) program, 10
Current Procedural Terminology (CPT), 13

D

Data; *see also* Big data
 sources, in health care
  clinical registries, 11
  electronic health records (EHRs), 9–10
  genetic data, 12
  health insurance claims data, 10
  patient-reported outcomes measures (PROM), 11
  publically supported databases, 10–11
  sensor data, 12
  social media, 12

Index

Data acquisition
　critical care/critical care units, 522, 523f
Data analytics
　critical care/critical care units, 520
　in trauma patients, 480
　　correlations and, 483–484, 484f
　　descriptive statistics and, 480–481
　　entropy and, 484–485
　　variability and, 481–483, 482f, 483f
Data availability
　critical care/critical care units, 521
Database architecture, health care information system, 13–14
Data compression, 5
　lossless, 5
　lossy, 5
Data Dictionary, 26
Data-driven feature selection, 29–30
Data-driven nonparametric tests, 58; see also Nonparametric models
Data frequency
　critical care/critical care units, 520–521
Data persistence
　Artemis, 525–526
　critical care/critical care units, 523
Data processing
　parallel/distributed, 4–5
Data provenance, 4
Data quality, 4
　critical care/critical care units, 521–522
Data sampling, 5
Data segmentation, TDA and, 211–213, 213f
Data storage, 5
Data transmission
　critical care/critical care units, 522
Data variety, 4
Data windowing, 212–213
dbGAP, 447
DBN, see Dynamic Bayesian networks (DBN)
DBN-extended model, TAs and, 503–508
　basic techniques, 503–504, 504t
　construction of, 504–505, 505f
　experiments and analysis, 507–508
　training in presence of class imbalance, 506–507
　without TAs, 505–506
DCT approximations, 151–172
　based on integer functions
　　collection of, 157–158
　　rounded DCT, 157
　　signed DCT, 156

biomedical signal processing
　applications of, 167
　image compression, 167–169, 168f, 169f–170f
　image registration, 169–171, 171f
categories, 155
classical multiplierless transforms
　Hadamard transform, 156
　Walsh–Hadamard transform (WHT), 156
computational search-based, 159
　improved 8-Point DCT approximation, 160–161
　for IR, 161
　for RF imaging, 160
by inspection
　BAS series of approximations, 158, 159t
　modified RDCT, 158
　signed SPM transform, 159
　SPM transform, 158–159, 159t
low-complexity, 151–172
open problems and future directions, 172
overview, 152–153, 155
parametric
　FW class of, 162–163, 164t–165t
　level 1 approximation, 161–162
　parametric BAS, 162
performance comparison, 163, 165f
　discussion, 166–167, 166t, 167t
　mean square error, 166
　measures, 163, 165–166
　total error energy, 165
　transform distortion, 163
　transform efficiency, 166
　unified transform coding gain, 166
Decision-making component, BCI system, 125–126
　EEG feature extraction and classification, 125–126
　language model, 126
Decision support systems
　manual interpretation of EEGs, 280, 281f
Deep CNNs, see Deep convolutional neural networks (CNNs)
Deep convolutional neural networks (CNNs)
　background, 303–305, 303f–305f
　convolutional layer, 303
　convolution of kernels (filters), 303
　data and implementation details
　　brain tumors (BRATS), 307
　　MS datasets, 307
　　pre-processing and, 307–308
　experiments and results, 308–310, 308f, 309t, 310f–311f
　feature maps, 303

local receptive field, 303
max-pooling, 305
method
　hetero-modal image segmentation, 305–306, 306f
　interpretation as an embedding, 307
　pseudo-curriculum training procedure, 306–307
non-linear activation function, 304–305, 305f
overview, 301–303, 302f
valid mode convolution, 304
Deep expert mimicry, 359
Deep feature discovery, 360
Deep feedforward networks, 333–337, 335f–336f
　decision/classification layer, 333
　visual cortex (V1), 333, 335–337
Deep learning, 5, 30–31, 302
Deep learning, for retinal analysis, 329–363
　Adam optimization, 350, 351f
　background and context, 330–331
　computer-aided diagnostics (CAD), 340–341
　　Canny edge detectors, 344
　　data sets, 345
　　feature discovery, 342–343
　　global histogram enhancement, 343–344
　　image processing fundamentals, 343–345
　　local histogram enhancement, 344
　　mimicking experts, 341–342, 342f
　　performance evaluation, 347–348
　　performance vs. accuracy, 345–347, 346f, 346t, 347f
　　superresolution techniques, 344
　　traditional CAD of retinal images, 343
　　watershed methods, 344–345
　deep feedforward networks, 333–337, 335f–336f
　diseases of retina, 337–338
　　diabetic retinopathy (DR), 338–339, 338f, 339f
　　glaucoma, 339–340, 340f
　Gabor functions, 337
　microaneurysm detection algorithms, 354
　　deep learning approach, 356–357, 358f
　　performance comparisons, 357, 359–363, 359t, 361t, 362t
　　superpixel sampling, 356
　　traditional methods, 354–356, 355f

neurons, 331–332
performance comparison (vessel segmentation and pixel classification), 350, 352t, 353–354
  deep algorithms, 352–353
  traditional algorithms, 350, 352
pixel sampling, 348–350, 349f
recurrent neural networks, 337
retinal morphology segmentation, 350
traditional retinal segmentation, 348
visual inputs, 332, 333f
Deep neural networks (DNNs), 31, 331
Deep-sequencing sRNA pipeline (DSAP), 393
Defibrillators to Reduce Risk by Magnetic Resonance Imaging Evaluation (DETERMINE) study, 314, 317
Denoising autoencoder method, 353
De novo machine learning techniques, 390–391
  classifier selection and training, 397–398, 398t
  data set generation, 396–397
  feature selection, 398–399, 399t
  miRNA prediction, 396–400, 398t
  reporting of results, 399–400
Department of Defense(DoD), 25
Descriptive statistics, data analytics in trauma patients and, 480–481
Detection of existence, of anomalous data streams, 58–59, 61
  nonparametric model, 63–64
  parametric model, 61–62
  semiparametric model, 62–63
DETERMINE (Defibrillators to Reduce Risk by Magnetic Resonance Imaging Evaluation) study, 314, 317
Deterministic greedy algorithm, 123
Detrended fluctuation analysis (DFA), 453
DFS, *see* Distributed file system (DFS)
Diabetic macular edema (DME), 338–339
Diabetic retinopathy (DR), 338–339, 338f, 339f
  diabetic macular edema (DME), 338–339
  microaneurysms (MAs), 338
  nonproliferative DR (NPDR), 338
  proliferative DR (PDR), 338
  treatments, 338
  vs. image classification, 361–363, 361t
    comparing performances, 362–363
    deep algorithms, 362
    tradtional algorithms, 361, 362t

Diagnostic and Statistical Manual (DSM-IV-TR), 433
Diaret (Standard Diabetic Retinopathy Database), 345
Dictionary learning (DL), 369
  adaptive distributed, 372–376, 372f
    classification results, 376–378, 377f, 377t–378t
    combination weights selection, 374–376, 375f
    computational cost, 378, 378t
    data sets and evaluation methods, 376
    diffusion adaptation method, 374, 374f
  compressed sensing and, 371–372
  distributed, 372–374, 373f
  experiments and results, 372–374, 373f
  HEp-2 cell classification applications of, 369–379
  sparse coding and, 371
Diffuse optical tomography (DOT), 148
Diffuse RNFL defects, 340, 341f
Diffusion spectrum imaging (DSI), 22
Diffusion tensor imaging (DTI), 22
Digital Imaging and Communications in Medicine (DICOM), 168, 522
Digital Retinal Images for Vessel Extraction (DRIVE) database, 345, 346f
Dimensionality reduction, 5
Dirichlet-multinomial–Poisson model, 423
Discontinuity-based abnormalities, of newborn, 77
Discrete cosine transform (DCT); *see also* DCT approximations
  definition, 153–154
  domain, 137
  overview, 152
Discrete Fourier domain, 148
Discrete Fourier transform (DFT), 137
Discrete SL, 238
Discrete-time signals, 133
Disease risk
  predicting, 15–16
Distance matrix, 213
Distributed dictionary learning, 372–374, 373f
Distributed file system (DFS), 4
DL, *see* Dictionary learning (DL)
DME, *see* Diabetic macular edema (DME)
DNA sequencing, 12
DNNs, *see* Deep neural networks (DNNs)
Double robustness (DR), 240
DR, *see* Diabetic retinopathy (DR); Double robustness (DR)

DRIVE (Digital Retinal Images for Vessel Extraction) database, 345, 346f
DSI, *see* Diffusion spectrum imaging (DSI)
DTI, *see* Diffusion tensor imaging (DTI)
Dynamic Bayesian networks (DBN), 497–499, 498f; *see also* DBN-extended model
Dynamic longitudinal treatment regime, TMLE, 248–250
Dynamic processes on complex networks, 177–189
  application (vulnerable communities), 182–183
  biomedical signal processing application of, 189
  continuous-time Markov process, 179–180
  endogenous infection dominant (regime II), 183–184
  equilibrium distribution and induced subgraphs, 184–185, 184f
  most-probable configuration and densest subgraph, 186
  most-probable configuration and induced subgraphs, 185–186
  examples, 186–187, 187f, 188f
  exogenous infection dominant (regime III), 187–189
  healing dominant (regime I), 183
  infection dominant (regime IV), 183
  open problems and future direction, 189
  overview, 177–178
  scaled SIS process, 178–179, 179f, 180f
    equilibrium distribution, 180–181
    MPCP, 183
    parameter regimes, 181–182, 181t

E

Early Treatment Diabetic Retinopathy Study (ETDS), 361
ECG, *see* Electrocardiography (ECG)
Echo planar imaging (EPI), 23
EDF, *see* European Data Format (EDF)
Edge refinement, 344
Edge thinning, 344
EEG feature extraction and classification, BCI system, 125–126
EHRs, *see* Electronic health records (EHRs)
8-Point DCT approximation, improved, 160–161
8-point DCT fast algorithms, 154t
8-point KLT matrix, 153
Elastic Net, 30

Index

Electrical activity in brain, recording of, 261, 271–273, 272f, 273f
Electrocardiogram (ECG), 444
Electrocardiography (ECG), 211, 212f
Electroencephalograms (EEGs), 260, 444
　artefacts in, 77–78
　　detection, 78
　　detection and removal, 80
　　features for detecting, 79
　　problem of, 75–76
　　removal using automated component separation, 78–79
　　standard methods for handling, 78–79
　　types, 78
　automatic interpretation of, 287
　　feature extraction, 288–292, 289f, 290f, 291t
　　hidden Markov models, 292–293, 292f, 293f, 293t, 294f
　　machine learning approaches, 287–288
　　normal/abnormal detection, 293–295, 294f, 295t
　　seizure detection, 295–296
　　typical static classification systems, 288
　clinical prognostic value, 74–75
　in critical care units, 521
　future directions, 296
　manual interpretation of, big data issues
　　annotations, 276–277, 278f
　　decision support systems, 280, 281f
　　evaluation metrics, 279–280
　　inter-rater agreement, 277–279
　　locality, 276
　　signal conditioning, 275–276
　　unbalanced data problem, 280–282
　　waveform displays, 273–275, 275f
　neurodevelopmental outcomes and, 76–77
　newborn EEG abnormalities
　　amplitude-based, 77
　　asymmetric/asynchronous-based, 77
　　discontinuity-based, 77
　　EEG seizures, 77
　　features for detecting, 79
　　frequency-based, 77
　　maturation-based, 77
　　sleep state-based, 77
　　transient-based, 77
　quality measure, 98–100
　　model of artefact, 100
　　real EEG with real artefact, validation for, 100
　　real EEG with simulated artefacts, validation for, 99–100
　segment, six-way classification for, 276–277
　signal enhancement, 79
　sleep EEG, automated analysis of, 448
　　arousals, 450–451
　　automatic sleep staging, 449–450
　　EEG, 448
　　K complexes, 450
　　macroanalysis, 448–450
　　spectral analysis, 448–449
　　spindles, 450, 451t
　　time-frequency and time-scale approaches, 449, 449f
　　transient events detection, 450–451
　time-frequency analysis, see Time-frequency analysis, EEG measurement and enhancement
　TUH EEG Corpus, 282–287
　　de-identification of data, 284, 284t
　　descriptive statistics, 284–286, 285f, 286t
　　digitization and signal processing, 282–283
　　EEG report pairing, 283–284
　　released data structure, 286–287, 286f
Electroencephalography (EEG), 24, 209, 210–211
　brain electrical activity recording, 261, 271–273, 272f, 273f
　long-term monitoring (LTM), 271–272
　reconstruction of signals, 148; see also Reconstruction of big sparse signals
　of scalp, during seizures; see also Functional networks modeling, via graphical models
　　real data, 204–207, 204f–206f
　　synthetic data (non-Gaussian data), 203–204, 203t, 204t
　TDA application, 227–228
Electromyography (EMG), 210–211
Electromyography (EMG) signals, 148
　TDA application, 228
Electronic health records (EHRs), 9–10
　adoption of, 13
　categories (examples), 10
　in predicting disease risk, 15–16
　structured data, 9
　unstructured data, 9
Electronic medical records (EMRs), 71, 280
Electronic stethoscopes, 17–18
Electro-oculography (EOG), 211, 212f
Embedding, interpretation as, 307
EMD, see Empirical mode decomposition (EMD)
EMD-BSS, 79
Emergency department (ED) revisits, 16–17
　72-hour revisits, 16–17
EMG, see Electromyography (EMG)
Empirical mode decomposition (EMD), 79
EMRs, see Electronic medical records (EMRs)
Endogenous infection dominant (regime II), scaled SIS process, 182, 183–184
　equilibrium distribution and induced subgraphs, 184–185, 184f
　most-probable configuration and densest subgraph, 186
　most-probable configuration and induced subgraphs, 185–186
Endogenous infection rate, scaled SIS process, 179
Ensemble intensity gradient detection, 344
Entropy
　data analytics in trauma patients and, 484–485
　persistent, 219
Environmental sensors, 537–538
Epilepsy/epileptic disorders; see also Seizures
　defined, 260
　functional connectivity measures in functional networks in brain, 259–260
　methods, 261–265
　ontologies use, 267
　overview, 261
　performance improvement, 266–267
　related work, 260
　results, 265–266, 265f
　scalable signal data processing for, 259–267
　long-term monitoring (LTM), 271–272
　overview, 260
Epilepsy monitoring unit (EMU), 272
Epileptogenic zone, 260
Equilibrium distribution, scaled SIS process, 180–181
ERP-based BCI, for letter-by-letter typing, 124–125, 124f
　presentation component, 124–125
Estimation roadmap, TL, 236–237, 236f
ETDS, see Early treatment diabetic retinopathy study (ETDS)

Euclidean distance, 213
EuroCMR, 314
European data format (EDF), 260, 447
　functional connectivity measures in epilepsy
　　computational pipeline for signal data analysis, 263–265, 264f
　　signal data management, 262
　　signal data processing, 263
European multicenter study about spinal cord injury (EMSCI), 538, 540
European SIESTA project, 431
Evaluation metrics, manual interpretation of EEGs, 279–280
Exabytes, 3, 12
Exhaustive search, features, 86
Exogenous infection rate, scaled SIS process, 179
Expert mimicry ensemble (EME), 362
Expression rank data, 412, 413, 413f
　Bayesian top-scoring pairs, 415
　Bradley–Terry model, 414
　　Bayesian inference for, 414–415
　expression classification
　　experimental results, 415–416, 415t, 416t
　top-scoring pair classifiers, 413
eXtensible Neuroimaging Archiving Toolkit (XNAT—xnat.org) format, 25, 28
Extrinsic/environmental risk factors, for falls, 552
Eye, structure and basic function, 332, 333f; *see also* Retinal analysis
Eye movement (EYEM), EEG, 277
Eye movements and blinks, 78
EyePACS database, 345

F

Facebook, 12, 14, 116
FAIR principles, 261–262
Fall detection (FD) tools, 551
Fall prevention (FP), 551
Fall risk assessment (FRA), 551
　big data in, 572–572
　current and emerging methods, 552
　overview, 551–552
　supervised fall risk assessment (SFRA) methods, 553, 553f
　　non-sensor-based methods for, 553–554, 554f
　　sensor-based methods for, 554–556
　unsupervised FRA (UFRA), 553, 553f, 556–560, 557f
　new approach for, 560–572
Falls; *see also* Fall risk assessment (FRA)
　incidence, 551
　preventive measures, 551–552

risk factors for, 552–553, 552f
　extrinsic/environmental, 552
　intrinsic/patient-related, 552
Fast algorithms, 154, 154t; *see also* 8-point DCT fast algorithms
　defined, 154
Fast ISTA (FISTA), 142
Fast Orthogonal Search (FOS), 541
Feature discovery approach, retinal analysis, 342–343
Feature discovery ensemble (FDE), 360
Feature extraction, 468–469, 469t
　automatic interpretation of EEGs, 288–292, 289f, 290f, 291t
　BCI system, 125–126
　brain imaging data selection and, 28–30
　defined, 29
　HEp-2 cell classification, 370–371, 371f
　topological summaries based on, 218–219
Feature ranking, 469–470
Feature reduction, 29
Feature search techniques, 86–87
Feature selection, 29
　based topological network, 226–227
　data-driven, 29–30
　de novo machine learning techniques for miRNA prediction, 398–399, 399t
　knowledge-driven, 29
　miRNA prediction, 388
　model-driven, 29
　TBI outcome prediction, 486–487
　TF features for newborn EEG abnormalities detection, 85–87, 86f
　　feature search techniques, 86–87
　　filter methods, 86
　　wrapper methods, 86, 86f
Federal interagency traumatic brain injury research (FITBIR), 23, 25
Feedforward networks, 333
Feig–Winograd (FW) factorization, 152
Filter functions, 225
Filter methods, feature selection, 86
findMiRNA algorithm, 390
Finite-element modelling (FEM), 315
　of left ventricle, 315, 316f
Fisher criterion, 86
Fisher information distance, 318–319
FITBIR (Federal interagency traumatic brain injury research), 23, 25
fMRI, *see* Functional MRI (fMRI)
Forward selection, 29–30
Fourier transform matrix, 137
Four-sphere head model, 93, 93f
Fovea, 332
Foveation, 332

FP, *see* Fall prevention (FP)
FRA, *see* Fall risk assessment (FRA)
Fractal(s), 453
Fractional anisotropy (FA), 22
Framingham Offspring Study, 314
Frequency-based abnormalities, of newborn, 77
Frequency domain analysis, 452–453
Fugl- Meyer test, 535
Fully connected layer (FC), CNNs, 335
Functional connectivity measures, in epilepsy
　AAMI index, 260
　correlation measures, 260
　functional networks in brain, 259–260
　linear measures, 260
　methods
　　brain electrical activity recording, 261
　　computational pipeline for signal data analysis, 263–265, 264f
　　FAIR principles, 261–262
　　signal data management, 261–263, 262f
　　signal data processing, 263
　non-linear measures, 260
　ontologies use, 267
　overview, 261
　performance improvement, 266–267
　related work, 260
　results, 265–266, 265f
　scalable signal data processing for, 259–267
1000 Functional Connectomes Project (FCP), 24–25
Functional Independence Measure (FIM) scores, 534
Functional mobility assessments, falls, 554
Functional MRI (fMRI), 22–23, 260
　time series, 211
Functional networks, in brain, 259–260
Functional networks modeling, via graphical models, 193–207
　biomedical signal processing applications of, 207
　change point detection, 198–199
　copula GGMs
　　hidden variable copula GGMs, 197–198, 197f, 198f
　　standard copula GGMs, 196–197
　copulas and Gaussian copulas, 195–196
　numerical results
　　real data, 204–207, 204f–206f
　　synthetic data, 203–204, 203t, 204t
　open problems, and future directions, 207

Index

overview, 193–194
regularization selection
   for change point detection, 200, 200f
   for graphical model inference, 200–203
undirected graphical models and GGMs, 194–195
Function BIRN (FBIRN) Phase II, 26
Funduscope, 330
FW DCT approximations, 162–163, 164t–165t

G

Gabor barcodes approach, 568–571, 570f
Gabor function, 337
Gait analysis background, FRA and, 554
Galen of Pergamon, 330
Ganglion cells, 332
Gaussian copula(s), 195–196
Gaussian graphical models (GGMs), 194
  copula GGMs
    hidden variable copula GGMs, 197–198, 197f, 198f
    standard copula GGMs, 196–197
  Hammersley–Clifford theorem, 194–195
  undirected graphical models and, 194–195; *see also* Graphical models, functional networks modeling via
Gaussian likelihood model, 145
Gaussian radial basis function (RBF), 30
Gauss–Markov random field (GMRF), 195
Generalized likelihood-based approach anomalous data streams detection, 58, 59
Generalized likelihood ratio test (GLRT), 61–63
Generalized periodic epileptiform discharges (GPED), EEGs, 276–277
Generalized seizures, 276
Genetic data, 12
Genetic sequencing, 9
Genome-scanning data sets, *de novo* miRNA prediction, 403–404, 403f, 404f
Genome-wide association studies (GWASs), 14–15
Genomics, 3
  big data breakthroughs, 419–420, 420t
  sleep and, 458
Geometric mean (GM), 387
Geriatric Depression Scale (GDS), 19
GGMs, *see* Gaussian graphical models (GGMs)

Glaucoma, 339–340, 340f
  optic nerve head (ONH), 339, 340f
  peripapillary atrophy (PPA), 339–340, 340f
  retinal nerve fibre layer (RNFL), 340, 340f
Global cardiac magnetic resonance (CMR) registries, 314
Global histogram enhancement, retinal analysis, 343–344
Global positioning systems, 9
GLRT, *see* Generalized likelihood ratio test (GLRT)
Google, 4, 14
GoPro Hero Session camera, 567, 568f
Graded redefined assessment of strength, sensibility and prehension (GRASSP), 535
Gradient algorithm, sparse signal reconstruction with, 143–144
Graphical models, functional networks modeling via, 193–207
  biomedical signal processing applications of, 207
  change point detection, 198–199
  copula GGMs
    hidden variable copula GGMs, 197–198, 197f, 198f
    standard copula GGMs, 196–197
  copulas and Gaussian copulas, 195–196
  numerical results
    real data, 204–207, 204f–206f
    synthetic data, 203–204, 203t, 204t
  open problems, and future directions, 207
  overview, 193–194
  regularization selection
    for change point detection, 200, 200f
    for graphical model inference, 200–203
  undirected graphical models and GGMs, 194–195
Graphic processing unit (GPU) based-processing, 331
Ground reactions, sensor-based SFRA methods, 554–555
Guide-point modelling, 323–324
  vs. manual contouring, 324, 324f
GWASs, *see* Genome-wide association studies (GWASs)
Gyro sensors, 18

H

Hadamard transform, 156
Hadoop, 4, 14, 419
Hadoop-based bioinformatics initiatives, 419, 420t

Hadoop DFS, 4, 5
Hairpin structure and sequence, of miRNA, 383–384, 383f
Hammersley–Clifford theorem, 181, 194–195
HBase, 4, 5
HCP, *see* Human Connectome Project (HCP)
HCUP (Healthcare Cost and Utilization Project), 11
HDF5, *see* Hierarchical Data Format (HDF5)
Healing dominant (regime I), scaled SIS process, 182, 183
Healing rate, scaled SIS process, 179
Health/biomedical big data, 3
Health care
  big data in, driving factors
    database architecture, 13–14
    the HITECH Act, Meaningful Use, and ACOs, 12–13
  big data science in
    coordination of care, 16
    emergency department revisits, 16–17
    precision medicine (PM), 14–15
    predicting disease risk, 15–16
  sources of data in
    clinical registries, 11
    electronic health records (EHRs), 9–10
    genetic data, 12
    health insurance claims data, 10
    patient-reported outcomes measures (PROM), 11
    publically supported databases, 10–11
    sensor data, 12
    social media, 12
Healthcare Cost and Utilization Project (HCUP), 11
Health Care Cost Institute (HCCI), 10
Health care research
  TL applications in, 250–252;
    see also Targeted learning (TL)
Health data, secondary usage of, 524
Health insurance claims data, 9, 10
Health Level Seven (HL7), 522
Heart Biomarker Evaluation in Apnea Treatment, 447
Heart rate variability (HRV) analysis, 451–452, 452f
  detrended fluctuation analysis (DFA), 453
  frequency domain analysis, 452–453
  Lempel–Ziv (LZ) complexity, 454
  multiscale entropy (MSE), 453–454, 454f
  nonlinear methods, 453–454

sample entropy (SampEn), 453
time domain analysis, 452
HeMIS, *see* Heteromodal Image Segmentation (HeMIS)
Hemorrhage shock, 477
Hendrich Fall Risk Model, 553–554
HEp-2 cell, *see* Human epithelial type-2 (HEp-2) cell
Hermite transform domain, 148
Heterogeneity
in patients and disease subtype classification, 39–40
Heteromodal Image Segmentation (HeMIS), 302, 302f, 305–306, 306f; *see also* Deep convolutional neural networks (CNNs)
Hidden Markov model of order n (HMM-n), 116, 116f, 117f, 118f
Hidden Markov models
automatic interpretation of EEGs, 292–293, 292f, 293f, 293t, 294f
Hidden variable copula GGMs (HVCGGM), 197–198, 197f, 198f
copula GGMs with hidden variables, 198, 198f
hidden variable GGM, 197–198
regularization selection for, 202–203
Hidden variable GGM, 197–198
Hierarchical clustering, 30
Hierarchical data format (HDF5), 260
High-resolution TFDs
for EEG data processing, 81
multidirectional distributions, 104, 105f–106f
selection of, 81–82
High-resolution TF techniques, EEG signal representation using
high-resolution TFDs, 81–82
multidirectional distribution (MDD), 82–83
quadratic TF distributions formulation, 80, 81t
High-risk case management (HRCM) care coordination program, 16
High risk for falls assessment form, 554
Hilbert space, 223
HITECH Act (Health Information Technology for Economic and Clinical Health), 12–13
Hive, 4
HMMiR, 395
Home sleep apnea testing devices (HSATs), 445
Homogeneous mixing, 178
Homology; *see also* Persistent homology
defined, 214
Homology-based techniques
for miRNAs prediction, 390

Honolulu Asian AmericanAging Sleep Study, 447
Hopcroft–Karp algorithm, for bottleneck matchings, 221
Hou algorithm, 152
Hughes phenomenon, 412
Human brain, big data projects of, 23–28
AD Neuroimaging Initiative, 27
Federal Interagency Traumatic Brain Injury Research, 25
1000 Functional Connectomes Project (FCP), 24–25
Human Connectome Project (HCP), 23
MGH/Harvard–UCLA consortium, 24
WU–Minn–Oxford consortium, 23–24
IMAGEN, 27–28
National Database for Autism Research, 25–26
SchizConnect, 26–27
Human Connectome Project (HCP), 23
MGH/Harvard–UCLA consortium, 24
WU–Minn–Oxford consortium, 23–24
Human epithelial type-2 (HEp-2) cell classification, 369–379
adaptive distributed dictionary learning, 372–376, 372f
classification results, 376–378, 377f, 377t–378t
combination weights selection, 374–376, 375f
computational cost, 378, 378t
data sets and evaluation methods, 376
diffusion adaptation method, 374, 374f
compressed sensing, 371–372
dictionary learning, 372
distributed dictionary learning, 372–374, 373f
experiments and results, 372–374, 373f
feature extraction, 370–371, 371f
overview, 369–370
sparse coding, 371
Human motion capture data, TDA application, 228–229
Human visual pathway, 334f
Hybrid methods, 79

I

Ibn al Haytham, 330
ICA, *see* Independent component analysis (ICA)

ICD-10-CM, *see* International Classification of Diseases, Tenth Revision, Clinical Modification (ICD-10-CM)
ICDR, *see* International Clinical Diabetic Retinopathy (ICDR)
Iceland Myocardial Infarction study, 314
ICIP2013 data set, 370, 376–378
ICPR2012 data set, 370, 376–378
ICU, *see* Intensive care unit (ICU)
IF-based approach, 91–92
Image classification
vs. DR, 361–363, 361t
comparing performances, 362–363
deep algorithms, 362
tradtional algorithms, 361, 362t
Image compression
DCT approximations applications, 167–169, 168f, 169f–170f
IMAGEN, 27–28
ImageNet, 345
Image registration (IR)
DCT approximations applications, 169–171, 171f
DCT approximations for, 161
Image segmentation
biomedical, information fusion in deep CNNs for, *see* Biomedical image segmentation, information fusion in deep CNNs for
defined, 301
Image similarity metric, 170
Independent component analysis (ICA), 78
Indirect immunofluorescence (IIF), 369
Individual growth curve (IGC) analysis, 542
Inertial measurement units (IMUs), 536, 537
wearable IMUs, 558
Infection dominant (regime IV), scaled SIS process, 182, 183
Information fusion in deep CNNs for biomedical image segmentation, *see* Biomedical image segmentation, information fusion in deep CNNs for
Inherent (t, f) features, 83–85, 84f; *see also* Time-frequency (TF) features
In-home monitors, 18
In-home sensors, 12
Input data, for TDA techniques, 213
Insomnia analysis, 433–434
biosignal processing and automated detection, case study, 434–439
related disorders, 433–434

Index 593

Insomnia biosignal processing and automated detection, case study, 434–439
  Big Data Science model, 434–440
  clinical data, 434, 434f, 434t
Inspection
  DCT approximations by
    BAS series of approximations, 158, 159t
    modified RDCT, 158
    signed SPM transform, 159
    SPM transform, 158–159, 159t
Instantaneous amplitude (IA)–based features, 83
Instantaneous frequency (IF)–based features, 83
Institutional Review Board (IRB), 126
Integer functions
  DCT approximations based on collection of, 157–158
    rounded DCT, 157
    signed DCT, 156
Integrative short reads navigator (ISRNA), 394
Intelligent assistive technologies and systems laboratory (IATSL), 557
Intensity-based IR approach, 169–170
Intensive care unit (ICU), 465, 519
  outcome, clinical physiological knowledge at feature and classifier levels integration, 465–467
    artificial neural network classifier, 470–471, 470f
    collected data, 467, 467t
    feature extraction, 468–469, 469t
    feature ranking, 469–470
    methods, 467–471
    results, 471–473, 471t–472t, 472f, 473t
Interactive Autism Network (IAN—self-report database), 26
Interagency Autism Coordinating Committee (IACC), 25–26
International Classification of Diseases, Tenth Revision, Clinical Modification (ICD-10-CM), 13
International Classifications of Sleep Disorders, 433
International Clinical Diabetic Retinopathy (ICDR), 361
  grading schemes, 362t
  severity levels for retinopathy and risk of macular edema, 362t
2013 International conference on image processing, 370
2012 International conference on pattern recognition, 370

International Neuroimaging Data-sharing Initiative (INDI), 24
International Standards for Neurological Classification of SCI (ISNCSCI), 538
Internet, 3, 177
Internet of Things (IoT), 3, 4
Inter-rater agreement (IRA), manual interpretation of EEGs and, 277–279
Intrinsic/patient-related risk factors, for falls, 552
Invariants, topological, 216–218, 217f
  interpretation of (example), 216–218, 217f
Inverse DFT matrix, 137
IR, *see* Image registration (IR)
Irregular-time Bayesian networks (ITBNs), 500–501, 501f
Isometric constant, 136
ISRNA (Integrative Short Reads Navigator), 394
ISTA, *see* Iterative soft-thresholding algorithm (ISTA)
ITBNs, *see* Irregular-time Bayesian networks (ITBNs)
Iterative soft-thresholding algorithm (ISTA), 142

J

Jackson Heart Study (JHS), 314
JavaScript Object Notation (JSON) framework, 262, 264
JPEG, 168
JPEG-2000, 168
JPEG-like image compression, 168–169, 168f–170f

K

Kaggle dataset, 345
Karhunen–Loève transform (KLT)
  definition, 153–154
  overview, 152
Kavali HUMAN Project (KHP), 14, 15
K complexes, automated analysis of sleep EEG, 450
KDDM (knowledge discovery and data mining), 13
Kernel-based approach
  anomalous data streams detection, 58, 60
Kernel-based learning methods, 223–224
Kernel density estimation (KDE), 122, 224
Kernel function (K), 223
Kernel trick, 222, 223

Key performance indicators (KPIs), 345–346, 346t
  classification, 346
  false negative (FN), 346
  false positive (FP), 346
  sensitivity (SN), 346
  specificity (SP), 346
  true negative (TN), 346
  true positive (TP), 346
KINARM exoskeleton, 535
KINARM robotic assessment platform, 541
Kinect, 537
Kitab al Manazir (*The Book of Optics*), 330
KLT, *see* Karhunen–Loève transform (KLT)
k-means clustering, 30
Knowledge discovery
  critical care/critical care units, 524
Knowledge discovery and data mining (KDDM), 13
Knowledge-driven feature selection, 29
Known/estimated positions, reconstruction with, 138
Kullback–Leibler (KL) divergence-based approach
  anomalous data streams detection, 58, 60–61

L

Landmark points, 222
Landscapes, persistence, 218–219
Language model, BCI system, 126
Laplacian density function, 145
Laplacian eigenmaps, 5
Large-scale data gathering strategies, neurorehabilitation
  environmental sensors, 537–538
  robotic rehabilitation devices, 535
  wearable sensors, 535–537
Large-scale practice-based evidence, in neurorehabilitation, 534–535
LASSO, *see* Least absolute selection and shrinkage operator (LASSO)
LASSO–ISTA reconstruction algorithm, 143
LASSO minimization, sparse signal reconstruction, 142–143, 143f
Late Onset Neonatal Sepsis (LONS), 528–529, 528f
LC-MS protein data, 412, 416
  Bayesian inference for, 418–421, 418f, 418t
    optimal Bayesian classifier, 420–421
    posterior sampling via an ABC-MCMC procedure, 419–420

prior calibration via ABC
rejection sampling, 418–419
procedure overview, 418
classification experimental results,
421–422, 421t, 422f
model for, 416–418
LDA, see Linear discriminant analysis
(LDA)
Least absolute selection and shrinkage
operator (LASSO), 30, 142
algorithm, 90
Leave-one- out CV (LOOCV), 31
Lee DCT for power-of-two blocklengths,
152
Left-ventricular analysis
atlas-based quantification of diseased
shapes, 317–318, 319f
building an atlas of heart, 315–317,
316f
classification, 318–320, 320f
finite-element model, 315, 316f
regional analysis, 320–322, 321f, 321t
signal representation of shapes, 315
statistical analysis of shape, 317, 317f
Lempel–Ziv (LZ) complexity, 454
LeNet, 350
Letter-by-letter typing, ERP-based BCI
for, 124–125, 124f
AUC values, 126–127
presentation component, 124–125
probability of phrase completion
(PPC), 126
total typing duration (TTD), 126
Level 1 DCT approximation, 161–162
Linear CNNs, 353
Linear correlation measures, functional
connectivity analysis, 260
Linear discriminant analysis (LDA), 5,
30, 125, 421
Linear support vector machines
(LSVMs), 421
LinkedIn, 12, 14, 116
Liquid-chromatography mass-
spectromic (LC-MS) protein
expression data, see LC-MS
protein data
Local binary patterns (LBPs), 370
Local histogram enhancement, retinal
analysis, 344
Locality, manual interpretation of EEGs
and, 276
Localized RNFL defects, 340, 340f
Locally linear embedding, 5
Local receptive field, 303
Loeffler algorithm, 152
Logical Observation Identifiers Names
and Codes (LOINC), 13
LOINC, see Logical Observation
Identifiers Names and Codes
(LOINC)

Longitudinal recovery modeling,
neurorehabilitation and,
542–543
Longitudinal TMLE (L-TMLE), 245
Longitudinal treatment regimes, TMLE
for, 245–246
conditional mean outcome, 246–247,
246f
dynamic regimes and marginal
structural models, 248–250
two time-point example, 247–248,
248t
Long-term monitoring (LTM), 271–272
Lossless data compression, 5
Lossy data compression, 5
L-TMLE mapping, 246

M

Machine learning (ML), 5
active learning, see Active learning
(AL)
algorithms, categories, 30
applications, 115–116
automatic interpretation of EEGs,
287–288
challenges and future directions
heterogeneity in patients
and disease subtype
classification, 39–40
multimodal data fusion, 40, 41
overfitting, 36, 39
translation to real clinical utility,
41–42
classifiers, 30–31
literature, 29
for structural and functional imaging
data, 28–36
commonly used classifiers, 30–31
feature extraction and brain
imaging data selection,
28–30
neurological and psychiatric
diseases diagnostic
predictions, 32–36, 33t–36t,
37t–39t, 40t, 41t
performance evaluation and
pattern interpretation, 31–32
statistical comparison vs. machine
learning classification, 28
supervised, 30, 223
TBI outcome prediction, 485–486
feature selection, 486–487
performance metrics, 487–488,
489f
topological, 222
topological kernel and kernel-
based learning methods,
223–224
traditional algorithms, 223

unsupervised, 30, 223
wearable sensors using
ML-based detection of
compensatory balance
responses using, 562–567
wearable vision detection of
environmental fall risks,
567–572
Macroanalysis, sleep EEG, 448–450
Macula, 332
Magnetic resonance imaging (MRI),
21–23, 148, 301
advantages, 22
automated biventricular
cardiovascular modelling from,
313–324
background, 313–315, 314t
congenital heart disease, 322, 322f
left-ventricular analysis
atlas-based quantification of
diseased shapes, 317–318,
319f
building an atlas of heart,
315–317, 316f
classification, 318–320, 320f
finite-element model, 315, 316f
regional analysis, 320–322, 321f,
321t
signal representation of shapes,
315
statistical analysis of shape, 317,
317f
diffusion tensor imaging (DTI), 22
functional MRI (fMRI), 22–23
overview, 21–22
signal types, 22–23
structural MRI (sMRI), 22
yearly scans in United States, 301,
302f
Magnetoencephalography (MEG), 24
Major depressive disorder (MDD), 32
Manual contouring, guide-point
modelling vs., 324, 324f
Manual interpretation, of EEGs
annotations, 276–277, 278f
decision support systems, 280, 281f
evaluation metrics, 279–280
inter-rater agreement, 277–279
locality, 276
signal conditioning, 275–276
unbalanced data problem, 280–282
waveform displays, 273–275, 275f
Mapper algorithm pipeline, 224–225,
225f
MapReduce, 4–5, 14
Marginal mean outcome estimation
under missingness, TMLE for,
241–243, 242f
Marginal structural models (MSMs),
TMLE, 248–250

Index

Markov random field (MRF), 194
Martinos Center for Biomedical Imaging, 24
MAs, see Microaneurysms (MAs)
Master–slave architecture, 4
Matched-filtering techniques, 352
Matching pursuit (MP) approach, 138
    norm-zero-based MP reconstruction algorithm, 139
MATLAB programming, 157
Matrices, big sparse signals reconstruction, 137
Matrix decomposition-based features, 83
Matrix presentation, BCI system, 123
    experimental results and discussions
        with overlapping trials, 127–128, 128f
        with single-character trials, 129, 129f
Matrix theory, 136
Maturation-based abnormalities, of newborn, 77
Maximal marginal diversity (MMD), 86
Maximum mean discrepancy (MMD), 60
Maximum relevance, 86
Max-pooling, deep CNNs, 305
MCA, see Morphological component analysis (MCA)
MCICShare project, 27
McKinsey Global Institute in, 2011, 3
MDD, see Major depressive disorder (MDD)
Mean embedding, 60
Meaningful Use
    defined, 13
    goal of, 13
Mean square error (MSE)
    DCT approximations, 166
Measurement matrices, big sparse signals reconstruction, 137
Medical expenditure panel (MEP) survey, 11
Medicare, 10
Medicare Claims Public Use File, 10
Megapixels, 348
MEP (medical expenditure panel) survey, 11
Messidor database, 345
MGH/Harvard–UCLA consortium, 24
Microaneurysms (MAs), 338
    detection algorithms, 354
        deep expert mimicry, 359
        deep feature discovery, 360
        deep learning approach, 356–357, 358f
        multiple-kernel learner (MKL), 359
        performance comparisons, 357, 359–363, 359t, 361t, 362t

        Speeded Up Robust Features (SURF) algorithm, 357, 359
        superpixel sampling, 356
        traditional methods, 354–356, 355f
        visual word dictionary methods, 357
        Waikato Microaneurysm Detector, 354
Microelectromechanical (MEM) accelerometers, 18
MicroHARVESTER, 390
MicroRNAs (miRNAs)
    biology, 383–384, 383f
    experimental determination through sequencing, 384, 385f
    future research directions, 404–405
    genome-scanning data sets, 403–404, 403f, 404f
    lack of prevalence-corrected reporting, 401–402
    as non-coding RNA (ncRNA), 383
    overview, 381–383
    pattern classification model, 385, 386f
        class imbalance correction, 388–389
        feature selection, 388
        model selection, 389
        model training, 389–390
        performance estimation, 385–388, 388f
        prediction pipeline, 390
        training and testing data selection, 388
        training and testing data splitting, 389
    predictors failures, 402–403, 403f
    redundancy in feature sets, 400–401, 401f
    SMOTE class imbalance correction method, 402, 402f
    state of the art computational techniques, 390–400
        de novo miRNA prediction, 396–400, 398t
        NGS-based miRNA prediction, 391–396, 391t
        previous assessments, 400, 400t
Mild cognitive impairment (MCI), 27, 32
    neuroimaging-based machine learning classification of, 33t–36t
Mimicry, in retinal analysis, 341–342, 342f
    expert mimicry ensemble (EME), 362
Mind Clinical Imaging Consortium (MCICShare), 26
Mini-batch preparation, 348
Minimum redundancy, 86
Mini-Sentinel, 10

MiPred, 395
miRanalyzer, 392
miRCat algorithm, 393
MIRcheck algorithm, 390
miRDeep*, 391, 392–393
miRDeep2 miRNA prediction algorithm, 384, 392, 394–395
    limitations, 396
    pre-processing step, 395
    scoring parameters, 396
    steps, 395–396
miRDeep-P, 392
miRdentify, 393
MIREAP algorithm, 395
miREvo, 393–394
miRExpress algorithm, 392
MiRGator v3.0, 394
MiRMiner, 392
MIRPIPE, 393
MiRPlex, 392, 393
MiRscan, 390
miRspring, 394
mirTools, 2.0, 394
miRTRAP, 392
Missing outcomes, point treatment effect estimation and, 244
Mixed National Institute of Standards and Technology database (MNIST), 331, 345
ML, see Machine learning (ML)
MMD, see Maximum mean discrepancy (MMD)
Model-based approaches, 302, 352
Model-driven feature selection, 29
Modified early warning scoring (MEWS), 460
Modified RDCT (MRDCT), 158
Monte Carlo simulations, 126, 127
Morphological component analysis (MCA), 78, 79
Morphology tracking methods, 352
Morse Fall Scale, 553
Mortality Probability Models (MPM), 460
Most-probable configuration problem (MPCP), 183
    scaled SIS process, 183
Motion detectors, 12
Motor Power Score, 535
Motor Status Score, 535
MP reconstruction algorithm, norm-zero-based, 139
MRI, see Magnetic resonance imaging (MRI)
MSE, see Mean square error (MSE)
MSGC dataset, 307
Multi-atlas approaches, 301–302
Multichannel CNNs, 353
Multi-class relevance units machine (McRUM), 392

Multidirectional distribution (MDD), 82–83
Multi-ethnic study of atherosclerosis (MESA), 313–314, 315
Multi-ethnic study of atherosclerosis-sleep study, 447
Multilayer perceptrons (MLPs), 333
Multimodal data fusion
  machine learning and, 40, 41
Multi-modality imaging, 301–302
  model-based approaches, 302
  multi-atlas approaches, 301–302
Multiple-kernel learner (MKL), 359
Multiscale entropy (MSE), 453–454, 454f
Multi-voxel pattern analyses, 28
Muscle artefacts, 78
MyApnea.Org, 459
MyCarePath, 16

N

Naïve Bayes classifiers, complex TAs and, 508–513
  extraction, 509, 509f
  imbalanced data and overlapping classes, 511–512
  most frequent temporal association patterns extraction, 510–511
  periodic classifiers vs. baseline classifier, performance, 512–513, 513t
  using periodic temporal association patterns, 510, 512f
Narcolepsy, 444
National Cancer Institute, 14
National Center for Health Statistics (NCHS), CDC's, 10, 11
National Database for Autism Research (NDAR), 23, 25–26
  The Global Unique Identifier (GUID), 26
National Institute of Child Health and Human Development (NICHD), 25
National Institute of Environmental Health Sciences (NIEHS), 25
National Institute of Neurological Disorders and Stroke (NINDS), 25, 477
National Institutes of Health (NIH), 14, 23, 25, 40, 419, 444
  Clinical and Translation Science Award (CTSA) program, 10
  registries, 11
National Sleep Research Resource (NSRR), 444
  publicly available sleep data sets, 447
  tools, 447–448
National Spinal Cord Injury Database (NSCID), 542

NEO format, see Neuroscience Electrophysiology Object (NEO) format
Neonatal intensive care units (NICUs), 519, 526
Neonatal multichannel non-stationary EEG and artefacts, 90
  assumptions and functionality of model, 97–98, 98f, 99f
  EEG multichannel propagation, 93–98, 94f
  four-sphere head model, 93, 93f
  IF-based approach, 91–92
  model validation, 98, 99f
  multichannel attenuation matrix, 95
  multichannel background and seizure patterns generation, 96–97, 97f
  multichannel path lengths, solving, 96, 96f
  multichannel translation matrix modelling, 95
  newborn head model, 92–93
  newborn mono-channel non-stationary EEG modelling, 90–91
  quality measure, 98–100
Netflix, 116, 223
Networks, 177; see also Complex networks, dynamic processes on
Networks of probabilistic events in time (NPEDT), 499, 500f
Neurodevelopmental outcomes, EEG and, 76–77
Neuroelectrophysiology
  big data breakthroughs, 431–434
  insomnia analysis and related disorders, 433–434
  sleep analysis, 431–433, 432f
Neuroimaging, 3
  big data breakthroughs, 430–431, 431t
  BRAIN initiative, 430–431, 431t
Neuroimaging data
  classifiers used in, 30–31
  multimodal data fusion, 40–41
  selection of, feature extraction and, 28–30; see also Machine learning
Neuroimaging Informatics Tools and Resources Clearinghouse (NITRC), 24
Neurological ICUs, 519
Neuromorphometry by Computer Algorithm Chicago (NMorphCH), 27
Neurons, 261, 331–332
  in visual cortex process images, 332, 333f
  visual inputs, 332, 333f

Neuropsychiatric disorders
  machine learning-based diagnostic predictions of, 32–36, 33t–36t, 37t–39t, 40t, 41t
Neurorehabilitation
  big data approaches, 538
    challenges, 544
    examples, 538–541, 539t–540t
    improvements, 543–544
    longitudinal recovery modeling, 542–543
    nonlinear modeling, 541–542
  large-scale data gathering strategies
    environmental sensors, 537–538
    robotic rehabilitation devices, 535
    wearable sensors, 535–537
  large-scale practice-based evidence need, 534–535
  objective, 533
  overview, 533–534
  of upper limb through big data, 533–544, 534f
Neuroscience Electrophysiology Object (NEO) format, 260
Newborn EEG abnormalities
  amplitude-based, 77
  asymmetric/asynchronous-based, 77
  discontinuity-based, 77
  EEG seizures, 77
  features for detecting, 79
    spectral features, 79
    temporal features, 79
    TF features, 79
    TS features, 79
  frequency-based, 77
  maturation-based, 77
  sleep state-based, 77
  TF matched filter approach for detection, 88–89, 88f
  transient-based, 77
Newborn head model, 92–93
Next-generation sequencing (NGS-based) miRNA prediction, 382, 390, 391–396, 391t, 411, 419
  categories, 391, 391t
  experimental determination, 384, 385f
  experiments for miRNA discovery in species of interest, 394–395
  miRDeep2 miRNA classification pipeline analysis, 395–396
  small RNA (sRNA), 384
  sRNA data set pipelines, 393–394, 394t
Neyman–Rubin counterfactual framework, 237
n-fold cross-validation (CV), 31
n-homology, of simplicial complex, 218

Index

NICUs, *see* Neonatal intensive care units (NICUs)
Niedermeyer's Electroencephalography, 261
NIH Study of Normal Brain Development, 26
Nocturnal frontal lobe epilepsy, 444
Noise reduction, 344
Non-linear activation function, 304–305, 305f
Non-linear measures, functional connectivity analysis, 260
Nonlinear methods
    HRV analysis, 453–454
Nonlinear mixed effects (NLME) models, 542
Nonlinear modeling
    neurorehabilitation data sets, 541–542
Non-negative matrix factorization (NMF), 83
Nonparametric approaches/models
    anomalous data streams
        detection of existence of, 63–64
        generalized likelihood-based approach, 59
        identification of, 67–70
        kernel-based approach, 60
        KL divergence-based approach, 60–61
Non-parametric structural equation model (NPSEM), 237
Nonproliferative DR (NPDR), 338
Non rapid eye movement (NREM), 433
Non-sensor-based SFRA methods, 553–554, 554f
    comprehensive medical assessments, 553
    functional mobility assessments, 554
    screening and questionnaire-based assessments, 553–554
Nonstationarity, 211
Norm-one-based reconstruction algorithms, 141–145, 141f
    LASSO minimization, 142–143, 143f
    signal reconstruction with gradient algorithm, 143–144
    total variations, 144–145
Norm-zero-based reconstruction, 137–141
    block and one-step reconstruction, 139–141
    coherence, 139
    CoSaMP reconstruction algorithm, 140–141
    with known/estimated positions, 138
    MP reconstruction algorithm, 139
    position estimation and, 138–139
Northwestern University Schizophrenia Data and Software Tool (NUSDAST), 26
NPEDT, *see* Networks of probabilistic events in time (NPEDT)
NPSEM, *see* Non-parametric structural equation model (NPSEM)
NU_REDCap, 27
Nyquist rate, 5
Nyquist sampling theorem, 5
Nyquist theorem, 5

O

OASIS (Outcome and Assessment Information Set), 10
Obama, Barack, 13, 14
Objective functions, for query optimization
    Active-RBSE inference framework, 120–122
    proposed graphical model, 121–122
Obstructive sleep apnea (OSA), 455
OCT, *see* Optical coherence tomography (OCT)
One-dimensional (1-D) DCT transform, 154
ONH, *see* Optic nerve head (ONH)
Operational taxonomic units (OTUs), 423
Ophthalmology
    defined, 330
    history, 330
    retinal analysis, *see* Retinal analysis, deep learning for
Ophthalmoscope, 330
Optical coherence tomography (OCT), 330
    retina crosssection, 334f
Optic nerve head (ONH), 339, 340f
Optimal Bayesian classifier (OBC) approach, 412, 413, 420–421
OptumLabs™, 10
Outcome and assessment information set (OASIS), 10
Outcomes of Sleep Disorders in Older Men Study (MrOS Sleep Study), 447
Overfitting, 412
    machine learning and, 36, 39
    SL algorithm, 238
Overlapping trials, matrix-based presentation paradigm with, BCI system, 127–128, 128f
Overlapping windowing, 212
Oximetry, 445

P

Pandora, 116
Parallel/distributed data processing, 4–5
Parameter regimes, scaled SIS process, 181–182, 181t
    endogenous infection dominant, 182
    exogenous infection dominant, 182
    healing dominant, 182
    infection dominant, 182
Parametric BAS, 162
Parametric DCT approximations
    FW class of, 162–163, 164t–165t
    level 1 approximation, 161–162
    parametric BAS, 162
Parametric models
    anomalous data streams
        detection of existence, 61–62
        identification, 64–65
Partial (focal) seizures, 276
Pathology discovery, with sparsity-enhanced sensitivity criterion, 360
Patient-Centered Outcomes Research Institute (PCORI), 11
Patient-powered research networks (PPRNs), 11
Patient-reported outcomes measures (PROM), 9, 11
    defined, 11
    roles, 11
Pattern classification model, miRNA prediction, 385, 386f
    class imbalance correction, 388–389
    feature selection, 388
    model selection, 389
    model training, 389–390
    performance estimation, 385–388, 388f
    prediction pipeline, 390
    training and testing data selection, 388
    training and testing data splitting, 389
    training pipeline for, 386f
PBE, *see* Practice-based evidence (PBE)
PCA, *see* Principal component analysis (PCA), *see* Principal components analysis (PCA)
PCORI, *see* Patient-Centered Outcomes Research Institute (PCORI)
Peaking phenomenon, 412
Performance; *see also* Key performance indicators (KPIs)
    diabetic retinopathy *vs.* image classification, 361–363, 361t
    deep algorithms, 362
    tradtional algorithms, 361, 362t
    evaluation, 347–348
        pixel based (pixel level), 347
        region based (pixel level), 347

retinal vessel segmentation *vs.* pixel classification, 350, 352t, 353–354
deep algorithms, 352–353
traditional algorithms, 350, 352
vs. accuracy, CAD system, 345–347
Performance estimation
pattern classification model of miRNA prediction, 385–388, 388f
Performance measures
DCT approximations, 163, 165–166
mean square error, 166
total error energy, 165
transform distortion, 163
transform efficiency, 166
unified transform coding gain, 166
Performance metrics
TBI outcome prediction, 487–488, 489f
Periodic lateralized epileptiform discharges (PLED), EEGs, 276
Periodic limb movements, 444
Peripapillary atrophy (PPA), 339–340, 340f
Persistence diagram, 216
Persistence landscapes, 218–219
Persistent entropy, 219
Persistent homology
defined, 214, 218
TDA and, 214
auction algorithm for Wasserstein distance, 221
bottleneck matching, 221
complexity of simplicial complex constructions and, 221–222
feature extraction–based topological summaries, 218–219
mathematical formalism, 218
properties, 219–222
simplicial complexes, 214–216, 214f–216f
stability of, 219–221, 220f
topological invariants, 216–218, 217f
Personal health records, 10
Philips IntelliVue MP70 neonatal monitors, 526
Photoplethysmography (PPG), 455
Photoreceptors, 332
Phototransduction, 332
PhysioNet, 447
Picture archiving and communication system (PACS) database, 316
Piecewise linear frequency modulated (LFM) model, 82
Pig, 4

Pipeline (workflow)
defined, 210
time-series TDA processing pipeline, 210, 211f
biomedical time-series data, 210–211, 212f
data segmentation, 211–213, 213f
input data for, 213
Piwi-interacting RNA (piRNA), 392
Pixel based (pixel level) performance evaluation, 347
Pixel-based retinal vessel segmentation, traditional, 348, 349f
Pixel-level classification system
performance evaluation, 347–348
pixel based (pixel level), 347
region based (pixel level), 347
Pixel sampling, 348–350, 349f
mini-batch preparation, 348
random, 348
regional, 348, 350
retinal morphology segmentation, 348–354
stratified, 348
PM, *see* Precision medicine (PM)
POC (point-of-care) lab testing, 18
Point cloud, 213
Point-of-care (POC) lab testing, 18
Point treatment effect estimation, TMLE for, 243–244
continuous outcomes, 245
missing outcomes, 244
near violations of positivity assumption, 244–245
variable importance, 245
Polysomnography (PSG) signals, 444, 447
Pooling layer (PL), CNNs, 335
Position estimation, reconstruction and, 138–139
Positron emission tomography (PET), 27, 148
Posterior probability mass function (PMF), 117
Post-Stroke Rehabilitation Outcomes Project (PSROP), 534, 538
Posturography, 555
PPA, *see* Peripapillary atrophy (PPA)
Practice-based evidence (PBE)
large-scale, in neurorehabilitation, 534–535
Precision (Pr), 387
Precision medicine (PM), 14–15, 445, 458–459
goal of, 458
Precision medicine initiative (PMI), 14
Pre-clinical remodelling, 314–315
Precursor miRNA (pre-miRNA), 383
Predictive models, disease risk, 15–16
Preprocessing pipeline, 211

Pressure-sensitive insoles, 558–559
Primary miRNA (pri-miRNA), 383
Principal component analysis (PCA), 5, 223, 317
Principal components analysis (PCA), 79
Probabilistic graphical model (PGM), 116–118, 117f, 118f
Probability density functions (PDFs), 126, 195, 318
class-conditional, 122
Probability mass function (PMF), 180
Probability of phrase completion (PPC), 126, 127
Procrustes alignment, 315
Project executive committee (PEC), 28
Proliferative DR (PDR), 338
PROM, *see* Patient-reported outcomes measures (PROM)
Propensity to succeed (PTS) scores, 16
Proposed objective function, 121–122
Prospective Observational Multicenter Major Trauma Transfusion study, 250
Protein abundance roll-up, 417
Pruned exact linear time (PELT) method, 194
Pseudo-curriculum training procedure, deep CNNs, 306–307
Psychiatric diseases
machine learning-based diagnostic predictions of, 32–36, 33t–36t, 37t–39t, 40t, 41t
Publically supported databases, 10–11

Q

Quadratic discriminant analysis (QDA), 125, 126
Quadratic TF distribution (QTFD), formulation of, 80, 81t
Quantified self-movement, defined, 18
Quantitative EEG (qEEG), 279

R

Radio frequency (RF) imaging
DCT approximations for, 160
Random forest (RFs), 566
Randomized control trial (RCT), 245, 534
Random pixel sampling, 348
Random RSVP, 127
Rapid eye movement (REM), 444
Rapid serial visual presentation (RSVP), BCI system, 123–124
active RSVP (ARSVP), 127
experimental results and discussions, 127, 127f
random RSVP, 127
RDCT, *see* Rounded DCT (RDCT)

Index   599

Real clinical utility, machine learning-based neuroimaging studies, 41–42
Real-time analytics
  critical care/critical care units, 522–523
Real trauma center, data from, 478
Receiver operating characteristic (ROC) curve, 31–32, 387–388, 388f
  CAD system, 346–347, 347f
Receiver operating characteristics (ROC) analysis, 86
Reconstruction of big sparse signals
  Bayesian-based reconstruction, 145–148, 147f
  biomedical signal processing applications of, 148
  compressive sensing methods for, 133–149
  conditions for, 135–136
  definitions, 134–137
  measurement matrices, 137
  norm-one-based reconstruction algorithms, 141–145, 141f
    LASSO minimization, 142–143, 143f
    signal reconstruction with gradient algorithm, 143–144
    total variations, 144–145
  norm-zero-based reconstruction, 137–141
    block and one-step reconstruction, 139–141
    coherence, 139
    CoSaMP reconstruction algorithm, 140–141
    with known/estimated positions, 138
    MP reconstruction Algorithm, 139
    position estimation and, 138–139
  open problems and future directions, 148–149
  overview, 133–134
Recurrent neural networks (RNNs), 331, 335
  vs. CNNs, 337
Recursive feature elimination (RFE), 29–30
Reduced interference TFDs (RI-TFDs), 80
Regional analysis, left-ventricle, 320–322, 321f, 321t
Regional pixel sampling, 348, 350
Region based (pixel level) performance evaluation, 347
Registry, defined, 11; see also Clinical registries
Regularization, 125, 126

Regularization selection
  for change point detection, 200, 200f
  for graphical model inference, 200–203
Regularized discriminant analysis (RDA), 125–126
Rehabilitation robotics, 535
Reinforcement learning, 5
ReJoyce telerehabilitation robot, 535
Relevance vector machine (RVM) approach, 145
Reproducing kernel Hilbert space (RKHS), 60
Resident Assessment Instrument (RAI), 553
Resting-state fMRI (rsfMRI), 24
Restricted Boltzmann machines (RBMs), 31
Retina
  anatomy, 332, 333f
  diseases of, 337–338
    diabetic retinopathy (DR), 338–339, 338f, 339f
    glaucoma, 339–340, 340f
  OCT crosssection, 334f
Retinal analysis, deep learning for, 329–363
  Adam optimization, 350, 351f
  background and context, 330–331
  computer-aided diagnostics (CAD), 340–341
    Canny edge detectors, 344
    data sets, 345
    feature discovery, 342–343
    global histogram enhancement, 343–344
    image processing fundamentals, 343–345
    local histogram enhancement, 344
    mimicking experts, 341–342, 342f
    performance evaluation, 347–348
    performance vs. accuracy, 345–347, 346f, 346t, 347f
    superresolution techniques, 344
    traditional CAD of retinal images, 343
    watershed methods, 344–345
  deep feedforward networks, 333–337, 335f–336f
  diseases of retina, 337–338
    diabetic retinopathy (DR), 338–339, 338f, 339f
    glaucoma, 339–340, 340f
  Gabor functions, 337
  microaneurysm detection algorithms, 354
    deep learning approach, 356–357, 358f
    performance comparisons, 357, 359–363, 359t, 361t, 362t

  superpixel sampling, 356
  traditional methods, 354–356, 355f
  neurons, 331–332
  performance comparison (vessel segmentation and pixel classification), 350, 352t, 353–354
    deep algorithms, 352–353
    traditional algorithms, 350, 352
  pixel sampling, 348–350, 349f
  recurrent neural networks, 337
  retinal morphology segmentation, 350
  traditional retinal segmentation, 348
  visual inputs, 332, 333f
Retinal nerve fibre layer (RNFL), 340, 340f
Retinopathy of prematurity (RoP), 529
Reverse transcription polymerase chain reaction (RT-PCR), 395
RF imaging, see Radio frequency (RF) imaging
Rick Hansen Spinal Cord Injury Registry (RHSCIR), 540
Rips–Vietoris complex, 214–215, 215f, 222
Risk factors, for falls, 552–553, 552f
Risk function, 59
RNA-induced silencing complex (RISC), 384
RNFL, see Retinal nerve fibre layer (RNFL)
RNNs, see Recurrent neural networks (RNNs)
Robotic rehabilitation devices, 535
Rounded DCT (RDCT), 157
  and SDCT, collection of, 157–158
Royal Melbourne Hospital Risk Assessment Tool, 554

S

SAE, see Stacked auto-encoders (SAE)
Saliency, 332
Sample entropy (SampEn), 453
SBS, see Sequential backward selection (SBS)
Scalable signal data processing, for functional connectivity measurement in epilepsy, 259–267
  AAMI index, 260
  accurate measurement of signal correlation, 266
  correlation measures, 260
  functional networks in brain, 259–260
  linear measures, 260

methods
- brain electrical activity recording, 261
- computational pipeline for signal data analysis, 263–265, 264f
- FAIR principles, 261–262
- signal data management, 261–263, 262f
- signal data processing, 263
- non-linear measures, 260
- ontologies use, 267
- overview, 261
- performance improvement, 266–267
- related work, 260
- results, 265–266, 265f

Scaled susceptible-infected-suceptible (SIS) process, 178–179, 179f, 180f
- configuration, 178
- continuous-time Markov process, 179–180
- dynamics rules, 178–179
- endogenous infection dominant (regime II), 182, 183–184
  - equilibrium distribution and induced subgraphs, 184–185, 184f
  - most-probable configuration and densest subgraph, 186
  - most-probable configuration and induced subgraphs, 185–186
- endogenous infection rate, 179
- equilibrium distribution, 180–181
- exogenous infection dominant (regime III), 182, 187–189
- exogenous infection rate, 179
- healing dominant (regime I), 182, 183
- healing rate, 179
- infection dominant (regime IV), 182, 183
- MPCP, 183
- parameter regimes, 181–182, 181t
  - endogenous infection dominant, 182
  - exogenous infection dominant, 182
  - healing dominant, 182
  - infection dominant, 182
- topology-dependent process, 181
- topology-independent process, 181

Scalp EEG, during seizures; see also Functional networks modeling, via graphical models
- real data, 204–207, 204f–206f
- synthetic data (non-Gaussian data), 203–204, 203t, 204t

SchizConnect, 26–27

Schizophrenia (SZ), 25
- neuroimaging-based machine learning classification of, 37t–38t

SCIRehab study, 534, 538
Screening and questionnaire-based assessments, falls, 553–554
SDCT, see Signed DCT (SDCT)
SDNN index, 452
Searchlight, 30
Secondary usage of health data, 524
Segmentation of data, TDA and, 211–213, 213f
Seizures, 260; see also Epilepsy/epileptic disorders
- detection, automatic interpretation of EEGs and, 295–296
- generalized, 276
- newborn EEG abnormalities, 77
  - detection results, 101–102
  - multichannel patterns generation, 96–97, 97f
- partial (focal), 276
- scalp EEG during; see also Functional networks modeling, via graphical models
  - real data, 204–207, 204f–206f
  - synthetic data (non-Gaussian data), 203–204, 203t, 204t

Selective sampling 9-layer linear CNN (SSCNN) method, 359
Semiparametric models
- anomalous data streams
  - detection of existence, 62–63
  - identification, 65–67, 66f

Senapati, Pati, and Mahapatra (SPM) transform, 158–159, 159t
Sensitivity (SN), 346, 387
Sensitivity criterion, pathology discovery with, 360
Sensor-based SFRA methods, 554–556
- gait analysis background, 554
- ground reactions, 554–555
- posturography, 555
- wearable surface electromyography, 555–556, 556f

Sensor-based UFRA, 5556
Sensor data, 9, 12
Sensors, 3, 12
Sentiment analysis, 12
Sequencing, miRNAs determination, 384, 385f
Sequential backward selection (SBS), 86–87
Sequential forward floating selection (SFFS), 87
Sequential forward selection (SFS), 86
Sequential Organ Failure (SOFA) score, 460

Service based Multidimensional Temporal Data Mining (STDMn0), 525
72-hour ED revisits, 16–17
SFFS, see Sequential forward floating selection (SFFS)
SFRA methods, see Supervised fall risk assessment (SFRA) methods
SFS, see Sequential forward selection (SFS)
Shannon–Nyquist sampling theorem, 371
SHIMMER (Sensing Health with Intelligence, Modularity, Mobility, and Experimental Reusability) wearable sensors, 556f, 562, 563f
Short oligonucleotide alignment program (SOAP), 395
Short Physical Performance Battery (SPPB), 557
Short-time Fourier transform (STFT), 148, 449
Shrinkage, defined, 125
SickKids, 526
SID (State Inpatient Database), 11
SIGFRIED methodology, 228
Signal conditioning, manual interpretation of EEGs and, 275–276
Signal processing, see Biomedical signal processing
Signal types, MRI, 22–23
Signed DCT (SDCT), 153, 156, 157
- and RDCT, collection of, 157–158
Signed SPM transform, 159
Simplicial complexes, persistent homology, 214–216, 214f–216f
- Cech and Rips–Vietoris complexes, 214–215, 215f
- chain complex, 218
- clique complex, 215–216, 215f, 216f
- complexity of constructions, 221–222
- n-homology of, 218
- topological invariants of, 216–218, 217f
Simplification, topological, 224–227, 225f
- cluster analysis–based topological network, 226
- feature selection-based topological network, 226–227
Simplified Acute Physiology Score (SAP, SAP-II), 250, 460
Single-character trials, matrix-based presentation paradigm with, BCI system, 129, 129f
Single-nucleotide polymorphisms (SNPs), 15, 419
Sklar's theorem, 195–196
SL, see Super learning (SL)

Index

Sleep, defined, 444
Sleep analysis, 431–433, 432f
Sleep data, 444–445, 444f
    harmonization, 457–458
    manual *vs.* automatic methods, 446
Sleep-disordered breathing (SDB), 444
Sleep Heart Health Study (SHHS), 447
Sleep medicine, automated processing of big data in, 443–459
    analysis approaches, 446–447, 447f
    big data approach, 445–446
    cardiorespiratory measures
        apnea detection from airflow signals, 454–455
        apnea detection from PPG analysis, 455
        cardiopulmonary coupling, 455–456, 456f
        snoring sounds, 455
    challenges and needs
        data harmonization, 457–458
        noise, artifacts, and missing data, 456–457, 457f
    EEG, 448
        arousals, 450–451
        automatic sleep staging, 449–450
        K complexes, 450
        macroanalysis, 448–450
        spectral analysis, 448–449
        spindles, 450, 451t
        time-frequency and time-scale approaches, 449, 449f
        transient events detection, 450–451
    emerging applications
        genomics and sleep, 458
        precision medicine and other applications, 458–459
    heart rate variability analysis, 451–452, 452f
        detrended fluctuation analysis (DFA), 453
        frequency domain analysis, 452–453
        Lempel–Ziv (LZ) complexity, 454
        multiscale entropy (MSE), 453–454, 454f
        nonlinear methods, 453–454
        sample entropy (SampEn), 453
        time domain analysis, 452
    manual *vs.* automatic methods, 446
    National Sleep Research Resource (NSRR)
        publicly available sleep data sets, 447
        tools, 447–448
    research goals, 445
    sleep data, 444–445, 444f

Sleep state-based abnormalities, of newborn, 77
Sliding Windows and 1-Dimensional Persistence Scoring (SW1PerS), 213
Slow waves sleep (SWS), 433
Small RNA (sRNA) NGS experiments, 384
Smart infusion pumps (SIPs), 521
Smart medical devices
    activity monitors, 18
    big data and, 17–19
    computer algorithms, 18–19
    electronic stethoscopes, 17–18
    in-home monitors, 18
    point-of-care (POC) lab testing, 18
Smartphone-based ophthalmology, 330
Smart search algorithms, 86–87
S-method (SM), 81
Smith–Waterman algorithm, 392
SMOTE, *see* Synthetic minority oversampling technique (SMOTE)
SMOTE class imbalance correction, 402, 402f
SNOMED-CT, *see* Systematized Nomenclature of Medicine—Clinical Terms (SNOMED-CT)
Snoring sounds, 455
SNPs, *see* Single-nucleotide polymorphisms (SNPs)
SoapSNP, 419
Social media, 3, 4
    data sources in health care, 9, 12
Soft-thresholding rule, 142
Southern Ontario Smart Computing Innovation Platform (SOSCIP), 527
Spark, 5
Sparse coding, dictionary learning and, 371
Specificity (SP), 346, 387
Spectral analysis, automated analysis of sleep EEG, 448–449
Spectral features, newborn EEG abnormalities detection, 79
Speeded up robust features (SURF) algorithm, 357, 359
Spike and/or sharp wave (SPSW), EEGs, 276
Spinal Cord Independence Measure III (SCIM), 535
Spinal cord injury (SCI), 533
Spindles, automated analysis of sleep EEG, 450, 451t
SPM (Senapati, Pati, and Mahapatra) transform, 158–159, 159t
Spotify, 116

sRNA data set pipelines, 393–394, 394t
16S rRNA sequencing metagenomic data, 412–413, 422–424
    Bayesian inference for, 423
    classification experimental results, 424t, 4234
    Dirichlet-multinomial–Poisson model for, 423
    model for, 423, 423f
Stability, of of persistent homology, 219–221, 220f
Stability surface (SS) method, 202–203
Stacked auto-encoders (SAE), 31
Standard copula GGMs, 196–197
    regularization selection for, 201–202, 201f
Standard deviation of the NN intervals (SDNN), 452
Standard Diabetic Retinopathy Database (Diaret), 345
Standardized terminologies, 13
STARE (Structured Analysis of the Retina) database, 345
State Inpatient Database (SID), 11
Stationarity, defined, 211
Statistical analysis of shape, left-ventricle, 317, 317f
Statistical comparison *vs.* machine learning classification, 28
Steady-state free precession (SSFP) imaging, 317
Stereotactic electroencephalography (SEEG), 261
Stratified pixel sampling, 348
STRATIFY, 553, 554
Stroke, 533
Structural and functional imaging data machine learning for, 28–36
    commonly used classifiers, 30–31
    feature extraction and brain imaging data selection, 28–30
    neurological and psychiatric diseases diagnostic predictions, 32–36, 33t–36t, 37t–39t, 40t, 41t
    performance evaluation and pattern interpretation, 31–32
    statistical comparison *vs.* machine learning classification, 28
Structural MRI (sMRI), 22, 27
Structural networks, brain, 259
Structured analysis of the retina (STARE) database, 345
Structured query language (SQL) searches, 13, 14
Study of osteoporotic fractures, 447

Submodular-monotone set functions
  Active-RBSE inference framework, 119–120
  definitions, 119, 120
  discrete derivative, 119
  for set optimization problems, 119–120
  theorem, 119–120
SuperLearner, 237, 238
Super learning (SL), 235, 237–240
  algorithm, 238
  cross-validation, 238
  discrete SL, 238
  health care research applications of, 250–251
  overfitting, 238
  overview, 237–238
  in practice, 238–239
  reproducible research, 239–240
  SuperLearner, 237, 238
  and TMLE, combination of, 250
  verifiability, 240
Superpixel sampling, MAs detection, 356
  sampling with foveation, 356
  selective sampling, 356
Superresolution techniques, retinal analysis, 344
Supervised fall risk assessment (SFRA) methods, 553, 553f
  non-sensor-based methods for, 553–554, 554f
  sensor-based methods for, 554–556
Supervised learning, 30, 223
Support vector machines (SVMs), 29, 223, 370, 389
SVMs, *see* Support vector machines (SVMs)
The Swedish CArdioPulmonary BioImage Study (SCAPIS), 314
SW1PerS (Sliding Windows and 1-Dimensional Persistence Scoring), 213
Synthesis, complex systems, 177
Synthetic minority oversampling technique (SMOTE), 389
Systematized Nomenclature of Medicine—Clinical Terms (SNOMED-CT), 13
SZ, *see* Schizophrenia (SZ)

T

TAPIR software, 395
Targeted learning (TL)
  estimation roadmap, 236–237, 236f
  health care research applications of, 250–251
  identifying assumptions, 237
  open problems and future directions, 251–252
  overview, 235–236
  super learning, 237–240
    algorithm, 238
    cross-validation, 238
    discrete SL, 238
    overfitting, 238
    overview, 237–238
    in practice, 238–239
    reproducible research, 239–240
    SuperLearner, 237, 238
    verifiability, 240
  targeted minimum loss-based estimation, 240–250;
    see also Targeted minimum loss-based estimation (TMLE)
Targeted minimum loss-based estimation (TMLE), 235, 240–250
  collaborative TMLE (C-TMLE), 244–245
  double robustness (DR), 240
  health care research applications of, 250–251
  longitudinal TMLE (L-TMLE), 245
  for longitudinal treatment regimes, 245–246
    conditional mean outcome, 246–247, 246f
    dynamic regimes and marginal structural models, 248–250
    two time-point example, 247–248, 248t
  for marginal mean outcome estimation under missingness, 241–243, 242f
  overview, 240–241
  for point treatment effect estimation, 243–244
    continuous outcomes, 245
    missing outcomes, 244
    near violations of positivity assumption, 244–245
    variable importance, 245
  and SL, combination of, 250
  stages, 240–241
TAs, *see* Temporal abstractions (TAs)
Task-evoked fMRI, 24
TBI, *see* Traumatic brain injury (TBI)
TBI outcome prediction, 477–489
  automated medical big data monitoring, 478
    data from real trauma center, 478
    redundant *vs.* collection system with diagnosis viewer, 478–480, 479f
  computing issues, 489
  data analytics in trauma patients, 480
    correlations and, 483–484, 484f
    descriptive statistics and, 480–481
    entropy and, 484–485
    variability and, 481–483, 482f, 483f
  machine learning framework and applications, 485–486
    feature selection, 486–487
    performance metrics, 487–488, 489f
  overview, 477–478
TDA, *see* Topological data analysis (TDA)
TechAmerica Foundation, 115
Technological challenges, 3–5
  analytics, 5
  data quality, 4
  data storage/compression/ sampling, 5
  data variety, 4
  parallel/distributed data processing, 4–5
Temporal abstractions (TAs), 494
  Bayesian networks and time, 496–501, 497f, 501t
    continuous-time Bayesian networks (CTBNs), 499–500, 500f
    dynamic Bayesian networks (DBN), 497–499, 498f
    irregular-time Bayesian networks (ITBNs), 500–501, 501f
    networks of probabilistic events in time (NPEDT), 499, 500f
  biomedical signal processing applications of, 513
  clinical domain of CHD (test bed data set), 502–503
  complex, incorporation in naïve Bayes classifiers, 508–513
    extraction, 509, 509f
    imbalanced data and overlapping classes, 511–512
    most frequent temporal association patterns extraction, 510–511
    periodic classifiers *vs.* baseline classifier, performance, 512–513, 513t
    using periodic temporal association patterns, 510, 512f
  DBN-extended model, 503–508
    basic techniques, 503–504, 504t
    construction of, 504–505, 505f
    experiments and analysis, 507–508
    training in presence of class imbalance, 506–507
    without TAs, 505–506

Index

integration with Bayesian networks, 501–502
of time-oriented clinical data
  basic TAs, 494–495
  clinical systems using TAs, 495–496
  complex TAs, 495, 495f
Temporal features, newborn EEG abnormalities detection, 79
Terminologies, standardized, 13
3-nearest neighbors (3NN), 421
Thresholding, 344
Time-delay embedding technique, 213
Time-dependent confounding, 245
Time domain analysis, 452
Time–frequency analysis, EEG measurement and enhancement
  artefacts, 77–78
    features for detecting, 79
    standard methods for handling, 78–79
    types, 78
  big data and information retrieval
    DSP approach and engineering rationale, 74
    EEG and its clinical prognostic value, 74–75
    EEG clinical rationale and the problem of artefacts, 75–76
  BSS algorithms, 90
  classification, 89–90
  data acquisition and labeling, 79–80
  neonatal multichannel non-stationary EEG and artefacts, 90
    assumptions and functionality of model, 97–98, 98f, 99f
    EEG multichannel propagation, 93–98, 94f
    four-sphere head model, 93, 93f
    IF-based approach, 91–92
    model validation, 98, 99f
    multichannel attenuation matrix, 95
    multichannel background and seizure patterns generation, 96–97, 97f
    multichannel path lengths, solving, 96, 96f
    multichannel translation matrix modelling, 95
    newborn head model, 92–93
    newborn mono-channel non-stationary EEG modelling, 90–91
    quality measure, 98–100
  neurodevelopmental outcomes, 76–77
  newborn EEG abnormalities
    amplitude-based, 77
    asymmetric/asynchronous-based, 77
    discontinuity-based, 77
    EEG seizures, 77
    features for detecting, 79
    frequency-based, 77
    maturation-based, 77
    sleep state-based, 77
    transient-based, 77
  pre-processing (artefact detection and removal), 80
  results for stage 1 experiments
    newborn EEG abnormality detection using (t, f) matched filters, 102–104, 103t–104t
    newborn EEG abnormality detection using TF-based features, 101–102, 101t–102t, 103f
  results for stage 2 experiments
    BSS-based artefact removal, 108, 108f, 109f
    high-resolution TFDs (multi-directional distributions), 104, 105f–106f
    machine learning, 104–108, 106t–107t
  signal representation using high-resolution TF techniques
    high-resolution TFDs, 81–82
    multidirectional distribution (MDD), 82–83
    quadratic TF distributions formulation, 80, 81t
  TF features for abnormalities and artefacts detection
    feature fusion, 87, 87f–88f
    feature selection, 85–87, 86f
    inherent (t, f) features, 83–85, 84f
    translation approach-based TF features, 85
  TF matched filter approach for newborn abnormality detection, 88–89, 88f, 89t
Time–frequency analysis, EEG measurement and enhancement, 74–108
Time-frequency approach, automated analysis of sleep EEG, 449, 449f
Time-frequency (TF) features, newborn EEG abnormalities detection, 79
  feature fusion, 87, 87f–88f
  feature selection, 85–87, 86f
    feature search techniques, 86–87
    filter methods, 86
    wrapper methods, 86, 86f
  inherent (t, f) features, 83–85, 84f
  results of experiments, 101–102
  translation approach-based TF features, 85
Time-oriented clinical data, TAs of
  basic TAs, 494–495
  clinical systems using TAs, 495–496
  complex TAs, 495, 495f
Time-scale approach, automated analysis of sleep EEG, 449, 449f
Time-scale (TS) features, newborn EEG abnormalities detection, 79
Time series, defined, 211
Time-series TDA processing pipeline, 210, 211f
  biomedical time-series data, 210–211, 212f
  data segmentation, 211–213, 213f
  input data for, 213
TL, *see* Targeted learning (TL)
TMLE, *see* Targeted minimum loss-based estimation (TMLE)
TOPMed, 458
Topological data analysis (TDA), 209–230
  biomedical signal processing applications of, 227–229
    EEG signals, 227–228
    EMG signal, 228
    human motion capture data, 228–229
  overview, 209–210
  persistent homology, 214
    auction algorithm for Wasserstein distance, 221
    bottleneck matching, 221
    complexity of simplicial complex constructions and, 221–222
    feature extraction–based topological summaries, 218–219
    mathematical formalism, 218
    properties, 219–222
    simplicial complexes, 214–216, 214f–216f
    stability of, 219–221, 220f
    topological invariants, 216–218, 217f
  time-series TDA processing pipeline, 210, 211f
    biomedical time-series data, 210–211, 212f
    data segmentation, 211–213, 213f
    input data for, 213

topological machine learning, 222
  topological kernel and kernel-based learning methods, 223–224
  traditional algorithms, 223
topological simplification, 224–227, 225f
  cluster analysis–based topological network, 226
  feature selection-based topological network, 226–227
Topological invariants, 216–218, 217f
  interpretation of (example), 216–218, 217f
Topological kernel methods, 223–224
Topological machine learning (ML), 222
  topological kernel and kernel-based learning methods, 223–224
  traditional algorithms, 223
Topological network
  cluster analysis-based, 226
  feature selection-based, 226–227
Topological simplification, 224–227, 225f
  cluster analysis–based topological network, 226
  feature selection-based topological network, 226–227
Topology-dependent process, 181
Topology-independent process, 181
Top-scoring pair (TSP) classifiers, 412
  expression rank data, 413
Total error energy, for DCT approximations, 165
Total typing duration (TTD), 126, 127
Total variations, sparse signal reconstruction, 144–145
Traditional pixel-based retinal vessel segmentation, 348, 349f
Transform distortion measure, DCT approximations, 163
Transform efficiency, DCT approximations, 166
Transient-based abnormalities, of newborn, 77
Transient events, automated analysis of sleep EEG, 450–451
  arousals, 450–451
  K complexes, 450
  spindles, 450, 451t
Translation approach-based TF features, 85
Traumatic brain injury (TBI), 25, 477; see also TBI outcome prediction

TUH EEG Corpus (TUH-EEG), 273, 282–287
  de-identification of data, 284, 284t
  descriptive statistics, 284–286, 285f, 286t
  digitization and signal processing, 282–283
  EEG report pairing, 283–284
  released data structure, 286–287, 286f
Twitter, 12
Two time-point example, L-TMLE algorithm, 247–248, 248t
Type 2 diabetes mellitus (T2DM) predictive models, 15–16
Type I error probability $e_1(\delta)$, 59
Type II error probability $e_2(\delta)$, 59

U

UEA sRNA Toolkit, 393
UK Biobank, 314
Unbalanced data problem, manual interpretation of EEGs, 280–282
Undirected graphical models; see also Graphical models, functional networks modeling via
  and GGMs, 194–195
  Hammersley–Clifford theorem, 194–195
Unified transform coding gain, DCT approximations, 166
Unsupervised FRA (UFRA), 553, 553f, 556–560, 557f
  ambient sensors, 556–558
  new approach for, 560–572
    ML-based detection of compensatory balance responses, 562–567
    system infrastructure, 560–562
    wearable vision detection of environmental fall risks using ML techniques, 567–572
  pressure-sensitive insoles, 558–559
  sensor-based UFRA, 556
  wearable IMUs, 558
  wearable sensors and emerging approaches for, 558
  wearable vision systems, 559–560
Unsupervised learning, 30, 223
Upper limb, neurorehabilitation improvement through big data, 533–544, 534f
  big data approaches, 538
  challenges, 544
  examples, 538–541, 539t–540t

improvements, 543–544
longitudinal recovery modeling, 542–543
nonlinear modeling, 541–542
large-scale data gathering strategies
  environmental sensors, 537–538
  robotic rehabilitation devices, 535
  wearable sensors, 535–537
large-scale practice-based evidence need, 534–535
objective, 533
overview, 533–534
of upper limb through big data, 533–544, 534f
U.S. Department of Defense (DoD), 477
US Army Medical Research and Material Command (USAMRMC), 25
US Department of Health and Human Services, 10
US Food and Drug Administration, 10

V

Valid mode convolution, 304
van der Laan, Mark, 235
Variability, data analytics in trauma patients and, 481–483, 482f, 483f
Variable importance measures, point treatment effect estimation and, 245
Vetterli–Nussbaumer algorithm, 152
Vineyards, 219
Virtual International Stroke Trials Archive (VISTA), 540–541
Visual cortex (V1), 333
Visual inputs, neurons, 332, 333f
Visualisation, critical care/critical care units, 524
Visual system, 333
  CNNs vs., 335–337, 336f
Visual word dictionary methods, 357
Visuospatial presentation, BCI system, 123–124
  balanced-tree visual presentation paradigms, 124
  matrix presentation, 123
  rapid serial visual presentation (RSVP), 123–124
von Helmholtz, Hermann, 330

W

Waikato Microaneurysm Detector, 354
Walsh–Hadamard transform (WHT), 156

Wang factorization, 152
Wasserstein distance, auction algorithm for, 221
Watershed methods, retinal analysis, 344–345
Waveform displays, manual interpretation of EEGs, 273–275, 275f
Wavelet-BSS, 79
Wavelet domain, 148
Wearable IMUs, 558
Wearable sensors, 535–537
    ML-based detection of compensatory balance responses using, 562–567
    for UFRA, 558
    wearable vision detection of environmental fall risks using ML techniques, 567–572
Wearable surface electromyography, 555–556, 556f
Wearable vision systems, 559–560
Web MicroRNA Designer (WMD3) software, 395
Whole-gene sequencing, 12
WHT (Walsh–Hadamard transform), 156
Wigner–Ville distribution (WVD), 80, 81
Wilcoxon signed-rank sum test, 127
Windowing, data, 212–213
Wireless body area networks (WBANs), 148
Witness complex, 222
Wrapper methods, feature selection, 86, 86f
Wrist-worn accelerometers, 536
WU–Minn–Oxford consortium, 23–24

Y

Yahoo!, 14
Yottabyte (10^{24}) data system, 3

Z

Zettabyte (10^{21}) data system, 3

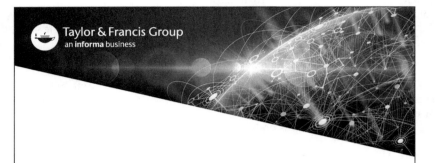